Biopolymers

for Medical and Pharmaceutical Applications

Edited by A. Steinbüchel
and R. H. Marchessault

Related Titles

A. Steinbüchel

Biopolymers. 10 Volumes + Index

2003

ISBN 3-527-30290-5

Single "Biopolymers" Volumes:

M. Hofrichter, A. Steinbüchel

Vol. 1: Lignin, Humic Substances and Coal

2001

ISBN 3-527-30220-4

T. Koyama, A. Steinbüchel

Vol. 2: Polyisoprenoids

2001

ISBN 3-527-30221-2

Y. Doi, A. Steinbüchel

Vol. 3a: Polyesters I – Biological Systems and Biotechnological Production

2002

ISBN 3-527-30224-7

Y. Doi, A. Steinbüchel

Vol. 3b: Polyesters II – Properties and Chemical Synthesis

2002

ISBN 3-527-30219-0

Y. Doi, A. Steinbüchel

Vol. 4: Polyesters III – Applications and Commercial Products

2002

ISBN 3-527-30225-5

E. J. Vandamme, S. De Baets, A. Steinbüchel

Vol. 5: Polysaccharides I – Polysaccharides from Prokaryotes

2002

ISBN 3-527-30226-3

S. De Baets, E. J. Vandamme, A. Steinbüchel

Vol. 6: Polysaccharides II – Polysaccharides from Eukaryotes

2002

ISBN 3-527-30227-1

S. R. Fahnestock, A. Steinbüchel

Vol. 7: Polyamides and Complex Proteinaceous Materials I

2003

ISBN 3-527-30222-0

S. R. Fahnestock, A. Steinbüchel

Vol. 8: Polyamides and Complex Proteinaceous Materials II

2003

ISBN 3-527-30223-9

S. Matsumura, A. Steinbüchel

Vol. 9: Miscellaneous Biopolymers and Biodegradation of Polymers

2003

ISBN 3-527-30228-X

A. Steinbüchel

Vol. 10: General Aspects and Special Applications

2003

ISBN 3-527-30229-8

A. Steinbüchel

Cumulative Index

2003

ISBN 3-527-30230-1

H. F. Mark

Encyclopedia of Polymer Science and Technology

Part 1: Volumes 1–4, **2003**, ISBN 0-471-28824-1
Part 2: Volumes 5–8, **2003**, ISBN 0-471-28781-4
Part 3: Volumes 9–12, **2004**, ISBN 0-471-28780-6

Biopolymers
for Medical and Pharmaceutical Applications

Edited by A. Steinbüchel and R. H. Marchessault

Volume 1
Humic Substances, Polyisoprenoids, Polyesters, and Polysaccharides

WILEY-VCH

WILEY-VCH Verlag GmbH & Co. KGaA

Editors:

Prof. Dr. Alexander Steinbüchel
Institut für Mikrobiologie
Westfälische Wilhelms-Universität
Corrensstrasse 3
48149 Münster
Germany

Prof. Dr. Robert H. Marchessault
Department of Chemistry
McGill University
801 Sherbrooke Street West
Montreal, Quebec, H3A 2K6
Canada

1st Edition 2005
1st Reprint 2006

Library of Congress Card No: applied for

British Library Cataloguing-in-Publication Data: A catalogue record for this book is available from the British Library

Bibliographic information published by Die Deutsche Bibliothek
Die Deutsche Bibliothek lists this publication in the Deutsche Nationalbibliografie; detailed bibliographic data is available in the Internet at <http://dnb.ddb.de>.

Printed in the Federal Republic of Germany
Printed on acid-free paper.

Composition: Konrad Triltsch Print und digitale Medien GmbH, Ochsenfurt-Hohestadt
Printing: Strauss GmbH, Mörlenbach
Bookbinding: Litges & Dopf Buchbinderei GmbH, Heppenheim

ISBN-13 978-3-527-31154-5
ISBN-10 3-527-31154-8

Introductory Preface

General Aspects

Biopolymers include broad families of biosynthesized materials. Living organisms are able to synthesize an overwhelming variety of polymers which can be classified as members of one of eight classes according to their chemical formulae: (1) polynucleotides or nucleic acids, e.g., DNA and RNA; (2) polyamides, e.g., protein and other poly(amino acids); (3) polysaccharides such as cellulose and starch; (4) polyesters, i.e., aliphatic polyoxoesters, better known as polyhydroxyalkanoates (PHA); (5) polythioesters such as poly(3-mercapto-propionate); (6) polyanhydrides, with polyphosphate as the only known example; (7) polyisoprenoids, principally rubber-like polymers; and (8) polyphenols, with lignin as the most abundant example. Within each category the chemistry of the repeating units, which are covalently linked, the type of linkage between the repeating units, and their structural arrangement (i.e., linear, cyclic, branched) distinguish the members. The study of these characteristics is ongoing, and some organic materials such a lignin do not have well-defined structures. Today's analytical tools have allowed rapid progress in identifying moieties. Furthermore, the physical organization (such as double-helical, or the complex three-dimensional arrangement of folded proteins) is readily available by x-ray diffraction. The polymorphism in a glycan such as cellulose or starch is usually a function of the material origin.

One other important aspect should be emphasized: only the polynucleotides and the proteins among the polyamides are monodisperse, and each polymer species exhibits a well-defined molecular weight, because these biopolymers are synthesized by template-dependent biosynthesis processes. All other polymers mentioned above are more or less polydisperse, because biosynthesis is template-independent. This also provides an opportunity of directing the average molecular weight to a certain extent, e.g., by varying the enzyme-to-substrate ratio in the cells and thereby modulating various properties of the biopolymer.

Among the biopolymers produced are the many used for various applications in industry. Until recently, biopolymers were the product of a biosynthetic process. Microorganisms are capable of synthesizing biopolymers belonging to classes 1–6, whereas eukaryotes synthesize mainly biopolymers belonging to classes 1–3 and 6–8. Whether derived from natural sources such as plants, animals, or algae by extraction processes, by controlled microbial fermentations in stirred tank reactors yielding extracellular or intracellular biopolymers, or by enzymatic synthesis, biosynthesis was the producing method. Today,

Biopolymers for Medical and Pharmaceutical Applications. Edited by A. Steinbüchel and R. H. Marchessault

ISBN: 3-527-31154-8

chemical synthesis can be also used on a biopolymer as starting material leading to a monomer which is then again transformed to a biopolymer by a chemical polymerization process. Polylactide (PLA) as produced by Cargill-Dow Inc. is a well-known example. In this case starch is fermented to lactic acid which is then converted to dimeric lactide and polymerized by ring-opening. Thus, agriculture and fermentation combined with polymer science lead to large-scale production of model biopolymers. Recently, Dupont Inc., a major plastics producer, and Tate & Lyle Inc. announced the fermentative production of 1,3-propanediol which will be a component of the new aromatic polyester Sorona.

In this overview, we will concentrate on the "workhorse" biopolymers prominent in biomedical applications. Thus, a polysaccharide such as heparin is one of the most widely used drugs in surgical practice. Excipients, prostheses (sutures, stents, surgical screws) and tissue culture matrices are fields where biopolymer materials with their biocompatible/biodegradable properties are major contributors. Biomedical uses of biopolymers often involve neutralization of toxic products from some physiological malfunction, e.g., bilirubin in babies suffering from jaundice. Wound dressings can be chemically modified cellulose or antimicrobial chitosan. Iron oxide-filled gutta-percha, a natural rubber with malleability, is used in dentistry to fill the void of a root canal. Because none of the buccal enzymes attack gutta-percha, it offers lifelong protection, and the iron oxide provides x-ray visibility. Important vaccines depend on biopolymers as the antigen to produce antibodies.

Drug Delivery

This field is highly innovative in terms of materials to assist delivery, excipients, and technology which allows fast or slow release of drugs. Analgesics, which often involve as much as five or six tablets a day, can be reduced to a single dose by using appropriate excipients, which are frequently based on carbohydrate polymers.

The most widely used excipient in the pharmaceutical industry is an HCl-hydrolyzed high-purity wood pulp (2.5 *N* HCl, 15 min reflux at the boiling point). This excipient, invented half a century ago, was originally intended as a low-calorie food additive. However, its properties as a tablet and the high-speed tableting process it enabled led to general adoption. The actual tablets are about 40% Avicel and 40% lactose, while the remaining components are adjuvants to aid the tableting and release processes. When compression-molded, the fine white powder AVICEL forms a marble-like tablet which can resist an oxyacetylene torch. In water, the tablet disperses in a few minutes, thereby releasing the drug. Large pharmaceutical companies dedicate a major development effort to understanding and controlling this valuable material.

Polysaccharide excipients are also based on starch, frequently the high-amylose hybrid with 70% amylose and 30% amylopectin, which consist of linear or branched chains of glucose, respectively. Unlike AVICEL, the starch-based excipient maintains its tablet shape in water but swells to an elastomeric hydrogel over a period of about 20–25 h at 37 °C. During this period the gel slowly releases its drug dose.

Many other polysaccharides are used in drug delivery. The water-soluble cellulose ether derivatives hydroxypropyl and carboxymethyl cellulose are adjuvants in this health-related technology. Polyelectrolyte biopolymers often form complexes with each other, i.e., they form coacervates by the interaction of their cationic and anionic functionalities, leading to molecular encapsulation. Ink, for example was locked in the familiar gelatin–gum Arabic capsular coating for paper, thereby providing a pressure-sensitive writing process.

Prostheses

A surgical suture is a classic example of a bioabsorbable/biocompatible prosthesis. The development of polyglycolide (PGA) sutures some 25 years ago by Davis and Geck Inc. in the United States was a major breakthrough. For centuries the familiar "catgut," a collagenous suture, was the only compatible/bioabsorbable material available to surgeons. Made from animal intestines, it was not uniform in diameter and strength. Surgeons yearned for a continuous filament suitable for the fine threads needed for eye surgery. PGA is a synthetic bioabsorbable polymerization product which is melt-spun to uniform strength-oriented fibers, ideal for surgery. Sold under the trade name Dexon, it was soon in competition with Ethicon Inc.'s Vicryl. The copolymeric composition of the latter (70% glycolide/30% lactide) has a 30–40 °C lower melting temperature, which eliminates thermodegradation during melt/spinning. Other companies produce similar polymers. Ironically, the claimed biodegradation *in vivo* is in fact an aqueous hydrolysis. In similar trial applications, the bacterial polyesters poly(3-hydroxybutyrate), PHB, and the copolymer poly(3-hydroxybutyrate-*co*-3-hydroxyvalerate) proved to be much slower to bioabsorb, requiring several months compared to a few weeks for surgical use. It was soon appreciated that the polyhydroxyalkanoate (PHA) backbone compared to PGA or PLA slowed the hydrolysis due to an inductive effect. PHAs are therefore recommended as surgical pins where long life is desirable. Several other biocompatible/bioabsorbable synthetic sutures are now marketed.

Both synthetic and biosynthetic aliphatic polyesters have niches in orthopedic usage. The ability to use conventional polymer synthesis, plastic molding, and machining methods for thermoplastic biopolymers are the key properties. The hydrophobic character of these polyesters and the nontoxic byproducts of degradation *in vivo* favor polyesters over polysaccharides in these applications. When greater permanence is required, aromatic polyesters such as Dacron are used, e.g., in artery repair.

Tissue culture is the growth of cells (tissue) separated from the organism. Cells in living tissue provide a platform for motility, differentiation, and proliferation. Tissue replacement efforts require a biopolymer matrix that mimics the developing living tissue, especially biodegradability. Culture studies using osteoblastic and epithelial cell lines have revealed the excellent biocompatibility of the PHB matrix. Similar studies with mouse fibroblast cells revealed that irradiation (UV or plasma) post-treatments on PHB films and fibers could reduce cell adherence problems. The obvious use of the aliphatic biopolyesters (PHAs) in numerous applications such as organ replacement, e.g., heart valves, biocompatible bone and bone cement implants for osteoporosis, etc., has encouraged biomedical replacement studies. The ongoing advances in fundamental microbiology, chemistry, and mechanics of biopolymer materials, both synthetic and biosynthetic, have set the stage for progress and understanding of regeneration of functional living tissue.

Supramolecular Polymers

A fast developing field of chemistry involving synthetic and natural polymers goes under the name of supramolecular. Prime examples are the "host–guest" complexes of small molecules with the glucan called cyclodextrin, α-CD. The latter is the product of starch dextrins acted on by glucosyl transferase. These complexes are major players in drug delivery and enzyme mimicry. Recently, a supramolecular version of α-CD, called polyrotaxane, has been synthesized. It is the simple threading of α-CD rings onto a linear polymer such as polyethylene glycol. Multiple α-DCs on the polymeric template lead to a molecular shish kebab, and when the α-CDs are cross-linked, a molecular tube is obtained. This kind of manipulation where noncovalent forces lead to self-assembly, as in the well-known surfactant "micelles," is the "stuff" of supramolecular chemistry.

Not since Staudinger and Carothers convincingly established macromolecular chemistry has there been such a concerted effort toward a new giant molecule concept. It is the chemistry of the intermolecular bond based on noncovalent forces, as distinct from molecular chemistry. At first it was small molecules in host–guest complexes with descriptive family names such as kryptands, calixarenes, sarcophagines, etc. However, supramolecular polymers very soon were recognized by their numerous examples in existing fields such as liquid crystals or biological helicoids, all deserving the supramolecular label. Supramolecular hydrogen-bonded polymers occur when repeating units are joined into linear chains by hydrogen bonds. Columnar assemblies stabilized by supramolecular bonds perpendicular to the monomer surface could be shown to form nematic organizations similar to rod-like colloids. However, conformational variations add flexibility features at the repeating unit level, thereby allowing a wide variety of supramolecular shapes.

Double-helical DNA is supramolecular with hydrogen bonding and π stacking as the intermolecular bonds. Besides the double helix with its nanoscale distance for the intermolecular bonds, DNA teaches that the value of the hydrogen bond pairing between bases is essential for self-assembly. Thus, building blocks chosen at will offer an inexhaustible range of supramolecular design choices.

Scope of This Book

Taking the importance of biopolymers for biomedical and pharmaceutical applications into account, we carefully selected 32 chapters from the recently published ten-volume *Biopolymers*. These chapters have been collected in two volumes. This two-volume spin-off product contains in Volume 1: three chapters focusing on polyphenols and natural rubber, four chapters dealing with various polyesters including the polyanhydride polyphosphate, and 14 chapters focusing on the various polysaccharides; Volume 2 contains eight chapters focusing on polyamides and complex proteinaceous materials and three chapters dealing with special aspects of biopolymers. The polymers from these selected chapters are either already being successfully used for biomedical and pharmaceutical applications or have the potential to be used in the future for such applications, or else the contents of the chapters are in other respects relevant to the subject.

In compiling this handbook, it has been our intention to provide the scientific and industrial community with a comprehensive view of the current state of knowledge on biopolymers and their derivatives in medicine and pharmacy. This handbook attempts to review what is currently known about these fascinating biopolymers in terms of their discovery, occurrence, chemical and physical properties, analysis, biosynthesis, molecular genetics, physiological role, fermentative production, isolation, purification, and applications. With the title more focused and at a price more affordable than that of the complete *Biopolymers* series, this two-volume handbook will be of interest in particular to medium-sized laboratories that are interested or active in this area, and to libraries. We are convinced that each chapter is in some way appropriate to the overall topic. We will feel it has been successful if some of these chapters stimulate readers to become interested in and solve specific problems, or make the field more accessible to newcomers.

Acknowledgments

Whatever is accomplished with these books is of course the achievement of the authors. We are most grateful to all of them for devoting so much of their valuable time to this endeavor and for sharing their knowledge and insights so generously. We are also particularly grateful to the authors of the selected chapters for allowing the contents of their *Biopolymers* contributions to be included in this new title.

Last but not least, we would like to thank Wiley-VCH for publishing this new handbook with their customary professionalism and excellence, and for their outstanding help throughout the gestation and birth of this handbook. Special thanks are due to Mrs. Karin Dembowsky, who initiated the *Biopolymers* series, to Dr. Andreas Sendtko, who continued and finalized it, and to many others at Wiley-VCH for their initiative, constant efforts, helpful suggestions, constructive criticism, and wonderful ideas.

Alexander Steinbuechel — Münster and Montreal

Robert H. Marchessault — May 2005

Contents

Volume 1

Biopolymers for Medical and Pharmaceutical Applications. Edited by A. Steinbüchel and R. H. Marchessault

ISBN: 3-527-31154-8

I
Humic Substances and Polyisoprenoids

Biopolymers for Medical and Pharmaceutical Applications. Edited by A. Steinbüchel and R. H. Marchessault

ISBN: 3-527-31154-8

1 Medical Aspects and Applications of Humic Substances

Prof. Dr. Renate Klöcking[1]
Dr. rer. nat. Björn Helbig[2]
[1] Institute for Antiviral Chemotherapy, Friedrich Schiller University Jena, Winzerlaer Str. 10, 07745 Jena, Germany, Tel: +49 (0) 361 – 7411482, Fax: +49 (0) 361 – 7411166, E-mail: "rkloeck@zmkh.ef.uni-jena.de" and Nordhäuser Str. 78, 99089 Erfurt, Germany
[2] Institute for Antiviral Chemotherapy, Friedrich Schiller University Jena, Winzerlaer Str. 10, 07745 Jena, Germany, Tel: +49 (0) 3641 – 657307, Fax: +49 (0) 3641 – 657301

AA	Arachidonic acid
BFD	Blackfoot disease
BOP	Catechol oxidation product
CC_{50}	Half-maximal cytotoxic concentration

Biopolymers for Medical and Pharmaceutical Applications. Edited by A. Steinbüchel and R. H. Marchessault

ISBN: 3-527-31154-8

CMV	Cytomegalovirus
2,5-DHBQOP	2,5-Dihydroxybenzoquinone oxidation product
2,5-DHPOP	2,5-Dihydroxyphenylacetic acid oxidation product
3,4-DHPOP	3,4-Dihydroxyphenylacetic acid oxidation product
2,5-DHTOP	2,5-Dihydroxytoluene oxidation product
3,4-DHTOP	3,4-Dihydroxytoluene oxidation product
CHOP	Chlorogenic acid oxidation product
DNA	Deoxyribonucleic acid
GENOP	Gentisinic acid oxidation product
HA	Humic acids
HS	Humic substances
HSV-1	Herpes simplex virus type 1
HSV-2	Herpes simplex virus type 2
HYDROP	Hydroquinone oxidation product
HYKOP	Hydrocaffeic acid oxidation product
IC_{50}	Half-maximal inhibition concentration
IL-1	Interleukin-1
KOP	Caffeic acid oxidation product
M.W.	Molecular weight
MX	Mutagen 'X'
POP	Protocatechuic acid oxidation product
RSV	Respiratory syncytial virus
TNF-α	Tumor necrosis factor-alpha
UV	Ultraviolet

1 Introduction

Humic substances (HS) comprising one of the largest reservoirs of carbon in nature may originate from different sources. For example, they can be formed as final products of biosynthetic pathways in micro-organisms, degradation and transformation products in plants, synthetic oxidation products of phenolic compounds and as polymers resulting from roasting processes, e.g. coffee roasting. Surprisingly, besides their brown color, which is responsible for the high UV absorption of HS, polymers of the HS type have several features in common that enable them to interact with other biopolymers as well as with low-molecular weight organic and inorganic compounds and, in particular, with metals, thus forming chelate complexes. Fractal structures, neighboring carboxyl and hydroxyl groups, reduction-oxidation and association-dissociation potentials are some of the most important features of HS as they cause HS to be important biogeochemical components of the Earth's surface. Besides their traditional use as fuel and organic fertilizers, HS are substrates for medical preparations, and also starting materials in the synthesis of specialized industrial products.

In this chapter, we will focus on medical and veterinary-medical applications of HS, and follow this with a discussion of several aspects of environmental health. Finally, we will look ahead to the possibilities of preparing novel biopolymers of the HS type, and to their potential use and application.

2
Historical Outline

The balneotherapeutic use of peat represents the most significant medical application of HS with regard to volume, therapeutic spectrum and tradition. Heavily degraded high moor peat, which is abundant in HS, has been used therapeutically long ago in Babylonia and the Roman Empire, where the inhabitants already recognized the healing effects of mud (Priegnitz, 1986). As health clinics' specialties, mud baths were offered in Europe in the early 19th century. Traditional indications for mud therapy are gynecological and rheumatic diseases (Baatz, 1988; Kleinschmidt, 1988; Kovarik, 1988; Lent, 1988). Beside mud baths which consist of peat pulp, baths with suspended peat material as well as drinking cures were also applied, the latter especially in case of gastric, intestinal or hepatic diseases (Kallus, 1964). The most frequent indications of peat therapy currently offered by health clinics in Germany are summarized in Table 1.

These conditions comprise various disorders of the musculoskeletal and gynecological systems, as well as skin diseases. Acute inflammatory and infectious diseases as well as malign tumors are usually regarded as contraindications. The primary effect of high-temperature peat therapy is the unique depth hyperthermia, which improves blood circulation and regeneration processes in the patient being treated. Depth warming is caused by the special physical consistency of the mud bath. As therapeutic effects have also been reported for low-temperature peat therapy (Balasheva and Gadzhi, 1971), HS (and possibly other chemical compounds) are also strongly suggested to participate in the healing effect by both chemical and biochemical effects. In a recent study, Bellometti et al. (1997) were able to show a favorable influence of mud bath therapy on osteoarthritis, a rheumatic condition characterized by the progressive destruction of cartilage. It was shown that peat therapy influenced the state of chondrocytes as well as the level of inflammation markers interleukin-1 (IL-1)

Tab. 1 Selected indications of peat therapy in the effect of which humic substances are probably involved

Diseases	***Indications***	***Major therapeutic effects***
Musculoskeletal diseases	Degenerative and deforming arthroses Gout Spondylopathies, e.g. Morbus Bechterew, Osteoporosis Muscular rheumatism Rheumatoid arthritis (polyarthritis) Rehabilitation after operations and accidents	Depth hyperthermia improves blood circulation and regeneration processes
Gynecological diseases	Chronic inflammatory diseases Hormonal imbalances Low back pain Adhesions Sterility Climacteric complaints	Depth hyperthermia Estrogenic effect and/or support of endogenous estrogen production Prophylaxis of thrombosis through release of tissue-type plasminogen activator (profibrinolytic effect)
Skin diseases	Chronic eczema Neurodermatitis Psoriasis	Activation of skin metabolism and regeneration processes, improvement of blood circulation

and tumor necrosis factor alpha (TNF-α). Iubitskaia and Ivanov (1999) demonstrated clearly a substantial contribution of HA to the balneotherapy of osteoarthritis patients. Using sodium humate (instead of peat) they observed analgesic, anti-inflammatory and lipid modulatory effects. Moreover, due to the low concentration of the HA preparation and the lack of effects of other mud factors, sodium humate procedures proved to be well tolerated by the patients.

The question of whether, and to what extent, HS are transferred to the patient during the mud bath has still to be answered. Application conditions such as temperature, ionic strength and pH value are thought to influence the balneotherapeutic effect.

3 Pharmacological Effects of Humic Substances with Potential Use in Medicine

In spite of the predominantly positive experience with balneological peat therapy, only limited knowledge is available of the physiologic and pharmacological effects of peat components. Nevertheless, our understanding of the biologic effects of HS with regard to their antiviral activity, interactions with isolated enzymes, effects on blood coagulation and fibrinolysis, estrogenic activity as well as toxicologically significant interactions with environmentally harmful substances has been considerably extended during the last decades. In this section, we will discuss some of these effects with regard to their potential therapeutic uses.

3.1 Antiviral Activity

Studies on the antiviral effect of HS were initiated after the successful combat against foot-and-mouth disease by means of peat dust-containing litter (Schultz, 1962, 1965). Preliminary in-vitro studies with Coxsackie A9 virus, influenza A virus and herpes simplex virus type 1 (HSV-1) have already shown that HS are effective against both naked and enveloped DNA viruses (Klöcking and Sprößig, 1972, 1975; Thiel et al., 1977). The same is true for synthetic HA derived from polyphenolic compounds, which in part are superior to natural HA in their effect (Thiel et al., 1976, 1981; Klöcking et al., 1983; Eichhorn et al., 1984; Hils et al., 1986). One of the antivirally most active synthetic polymers is the oxidation product of caffeic acid, KOP, the effect of which on HSV-1 *in vitro* compared with that of naturally occurring peat HA is shown in Figure 1.

Further investigations corroborate the ability of HA-like polymers to inhibit selectively viruses for human immunodeficiency virus type 1 (HIV-1) and type 2 (HIV-2), cytomegalovirus (CMV) and vaccinia virus (Schols et al., 1991; Neyts et al., 1992). No inhibition was found against poliovirus type 1, Semliki forest virus, parainfluenza virus type 3, reovirus type 1 and Sindbis virus. Adenovirus

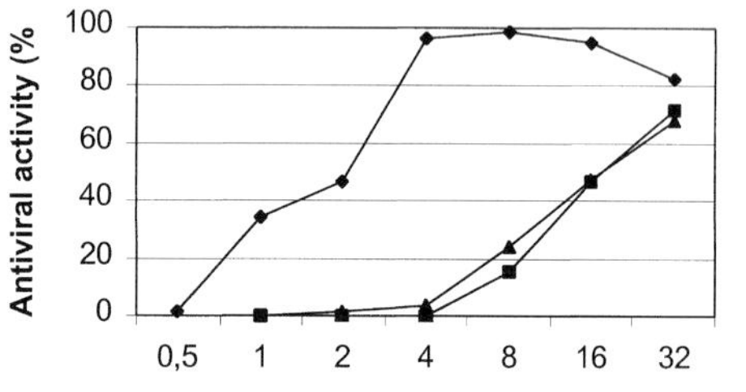

Fig. 1 Antiviral activity to HSV-1 of the synthetic HA-like polymer KOP and naturally occurring peat water HA (as sodium and ammonium salts, respectively). Test substances were added to Vero cells immediately before virus infection. After incubation at 37 °C for 120 h, cell viability was detected using the XTT-based tetrazolium reduction assay EZ4U according to Klöcking et al. (1995).

type 2 and ECHO virus type 6 showed little or no response to natural HA. Half-maximum Anti-HSV-1 inhibition concentrations (IC_{50}) and half-maximum cytotoxic concentrations (CC_{50}) of HA and HA-like polymers are summarized in Table 2, indicating the selective antiviral effect of the polymers tested.

With most viruses, the inhibitory effect of HA and HA-like polymers is directed specifically against an early stage of virus replication, namely virus attachment to cells (Klöcking and Sprößig, 1975; Schols et al., 1991; Neyts et al., 1992). As for CMV, it appears likely that the polyanionic HA occupy positively charged domains of the viral envelope glycoproteins, which are necessary for virus attachment to the cell surface (Neyts et al., 1992).

The effect of HA and HA-like polymers on an early stage of herpesvirus replication has been confirmed by the results of animal experiments. The number of lesions in the cornea of HSV-1-infected rabbits was strongly reduced when a solution of the synthetic HA-like polymer KOP (1%) was applied into the conjunctival sac of the eye along with or immediately after the infectious agent. However, KOP had no effect on the developing lesions when applied 1 and 24 hours later, respectively (Klöcking, 1994). Current interest is directed to the prophylactic effect of HA-type substances on recurrent HSV infection. It is known that topical application of KOP may significantly reduce or even completely suppress experimentally induced herpes in the mouse ear (Dürre and Schindler, 1992), though the mechanistic basis of this effect remains to be elucidated.

A low-molecular weight HA-like polymer (HS 1500, M.W. = 1500 Daltons), synthesized from hydroquinone was found to strongly inhibit HIV-1 *in vitro* (Schneider et al., 1996). Studies on the mechanism of action revealed virus penetration into host cells as the target of the anti-HIV-1 activity. HS 1500 has passed a panel of preclinical tests including eye irritation according to Draize, as well as pregnancy risk in rats. Neither sensitizing nor irritating effects were detectable in concentrations of up to 10% HA (Wiegleb et al., 1993; Lange et al., 1996a,b). SP-303 is another phenolic polymer with antiviral activity, which was isolated from a

Tab. 2 Half-maximal cytotoxic concentrations (CC_{50}) and half-maximal antiviral inhibitory concentrations (IC_{50}) of humic acids and humic acid-like polymers

Test substance	*M.W. (Da)*	*Starting compound*	*CC_{50} (μg/mL)*	*IC_{50} (μg/mL)*
BOP	5300	Catechol	69	26
3,4-DHTOP	3800	3,4-Dihydroxytoluene	>128	42
POP	8000	Protocatechuic acid	70	9.6
3,4-DHPOP	6000	3,4-Dihydroxyphenylacetic acid	>128	9.6
HYKOP	6000	Hydrocaffeic acid	>128	8.0
KOP	6000	Caffeic acid	>64	2.3
CHOP	14 000	Chlorogenic acid	>128	6.4
HYDROP	5200	Hydroquinone	>128	3.7
2,5-DHTOP	4700	2,5-Dihydroxytoluene	>128	1.6
GENOP	5500	Gentisinic acid	>128	2.2
2,5-DHPOP	5500	2,5-Dihydroxyphenylacetic acid	>256	0.7
2,5-DHBQOP	3400	2,5-Dihydroxybezoquinone	>512	322
Sodium humate	7500	not known	>128	18.2
Ammonium humate	7900	not known	108	17.8

Euphorbiaceae shrub. The polymer inhibits a panel of respiratory viruses, such as parainfluenza virus type 1, respiratory syncytial virus, influenza A viruses, and influenza B viruses (Gilbert et al., 1993; Wyde et al., 1993). Hemagglutination and other studies suggested that SP-303 at least partially inactivates viruses by direct interaction with virus or host cell lipid membranes. SP-303 at antiviral concentrations did not induce interferon or inhibit virus attachment; however, it abolished RSV penetration into host cells (Barnard et al., 1993). Administered as a small-particle aerosol to influenza A/HK virus-infected mice and RSV-infected cotton rats, SP-303 at 0.5–9.4 mg/kg/day for 3–4 days increased both the percentage and duration of survival of mice. Taken together, results so far show that HS are promising candidates for prophylactic rather than therapeutic use in the treatment of viral diseases.

3.2
Anti-inflammatory Effect and Pro-inflammatory Properties

Various healing effects of peat therapy were attributed to the anti-inflammatory activity of HS. Taugner (1963) showed in the rat paw edema model that sodium humate significantly inhibits the development of various edemas compared with untreated controls. As found by Klöcking et al. (1968) in the same model, ammonium humate isolated from peat water exceeds the anti-inflammatory effect of sodium humate and is twice as effective as acetylsalicylic acid and aminophenazone, respectively. Amosova et al. (1990), in evaluating the biologic activity of HA from Tambukan therapeutic mud in animals, found HA (10 mg/kg) to suppress both phases of the inflammatory process: the exudation (by 44 %) and the proliferation process (by 50–55%).

The anti-inflammatory effect of HS has been supported by a plausible biochemical explanation. As demonstrated by Schewe et al. (1991), naturally occurring HA, and even more synthetic HA-like polymers, inhibit the lipoxygenase pathway of the arachidonic acid (AA) cascade. AA is an integral part of the cell membrane, and the substrate for the synthesis of eicosanoid-based inflammation mediators such as leukotrienes, thromboxane and prostacyclin. Recently, HA (sodium humate) as well as various synthetic HA-like polymers were also found to suppress the heat-induced (42 °C, 6 h) AA release of human promonocytic U937 cells (Dunkelberg et al., 1997; Klöcking et al., 1997). The inhibition of AA release was most pronounced in cells treated with nontoxic concentrations (20 μg/mL) of 3,4-DHPOP (96%) and 3,4-DHTOP (92%), respectively (Table 3). CHOP, sodium humate, KOP and BOP protected cells from membrane damage at 65–90%. These findings may be indicative for membrane-protective activities of HA type substances.

Unlike 5-lipoxygenase, phospholipase A_2 (porcine pancreas), the rate-limiting key enzyme of the AA cascade, is strongly activated at low HA concentrations (0.1–

Tab. 3 Influence of naturally occurring humic acids (sodium humate) and of synthetic humic acid-like polymers on the heat-induced (42 °C, 6 h) [^{3}H]arachidonic acid (AA) release of U937 cells. AA release of untreated control cells = 100%. MEC, Maximum effect concentration; SD, Standard deviation; *Significant, $p \leq 0.05$.

Polymer	*MEC μg/mL*	*[^{3}H]AA release % of controls ± SD*
Sodium humate	80	26.5 ± 8.9*
BOP	160	10.1 ± 14.2*
3,4-DHTOP	20	8.1 ± 10.1*
3,4-DHPOP	20	4.3 ± 2.4*
KOP	40	18.1 ± 7.1*
CHOP	80	34.9 ± 12.6*

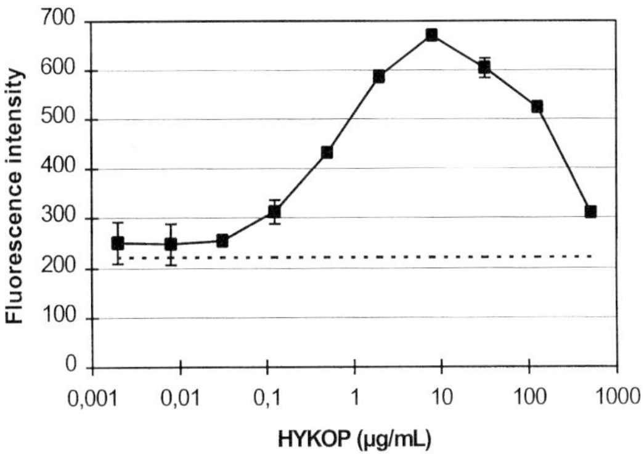

Fig. 2 Influence of the synthetic HA-like polymer HYKOP on the PLA_2-catalyzed hydrolysis of NBD-C_6-HPC using the method of Bennett et al. (1991). Dotted line = Reference value (without HYKOP).

1 µg/mL) and normalized or slightly inhibited at high HA concentrations (Klöcking et al., 1999). Figure 2 shows the typical concentration-dependent course of the dose–response curve for HYKOP, the oxidation product of hydrocaffeic acid. The shape of the dose–response curve suggests HS to have a regulatory function on PLA_2 activity.

Little is known about the influence of different molecular weight fractions of HS on inflammation. Pro-inflammatory activity has found to be associated with the synthetic low-molecular weight HS 1500, which activates human neutrophiles similar to TNF-α (Zeck-Kapp et al., 1991). Liang et al. (1998), while investigating rabbit articular chondrocytes, revealed an inhibitory effect of the ethyl acetate fraction of the commercial Aldrich HA (100–500 µg/mL) on the survival of chondrocytes. Cell injuries were attributed primarily to $O_2^{\bullet -}$ production, which is converted into H_2O_2, thus initiating lipid peroxidation followed by cell necrosis.

In referring to the function of HS as electron donor–acceptor system, Jurcsik (1994) discussed the behavior of HS as the consequence of a 'buffering effect'; this means that HA are able to produce as well as to bind activated oxygen species. This regulatory system is assumed to be important for the favorable influence of HS on wound healing and killing of cancer cells (Jurcsik, 1994).

3.3 Influence on Blood Coagulation and Fibrinolysis

Prophylaxis and therapy of fusions after tubal or ovarian inflammations as well as the postoperative treatment of sterility operations in order to prevent secondary adhesions and repeated occlusions of the ovarian duct are important indications of peat therapy. Fusions are caused by postoperatively reduced degradation of fibrin to soluble fibrin degradation products. As shown by Mesrogli et al. (1988) in laparotomied rats, postoperative baths in peat extract, peat pulp or HA have a clear adhesion-inhibiting effect. A possible explanation of this effect could be the activated fibrin degradation due to the HA-induced release of tissue-type plasminogen activator (t-PA). t-PA is regarded as the regulator of the antithrombotic defense mechanism. It activates plasminogen to plasmin, which splits insoluble fibrin to soluble fibrinogen degradation products (Klöcking, 1991). In addition, HA inhibit the coagulation enzyme thrombin, thereby suppressing the formation of fibrin monomers from fibrinogen (Klöcking, 1994; Klöcking et al., 1999).

Compared with other polyanionic compounds (heparin, pentosanpolysulfate), the anticoagulant effect of HA was found to be less pronounced.

3.4 Estrogenic Activity

Since the first detection of estrogenic substances in the bitumen fraction of peat by Aschheim and Hohlweg (1933), attempts were made to identify the nature of these substances. The assumption that steroid hormones are responsible for the colpotropic effect of peat could hardly be confirmed by chemical analysis. Therefore, the question arose whether, in addition to the lipid-soluble hormones, other peat components might contribute to the estrogenic activity of peat. Studies in castrated ICR (Institute of Cancer Research, USA) mice showed both naturally occurring peat humic acids and synthetic HA-like polymers to be positive in the Allen-Doisy test (Klöcking et al., 1992). The estrogenic activity of sodium humate was found to be 1/3000 of the estriol standard preparation. Referring to the HA content of dry peat, the estrogenic activity of the high-moor peat studied was 5000 times as high as has been supposed to date. Although the components responsible for the estrogenic activity of peat remain under discussion, the results suggest that HA – provided that they can penetrate the skin – may contribute substantially to the estrogenic effect of peat. These findings may have also implications for the use of HS in dermatology and cosmetics.

4 Veterinary-Medical Applications of Humic Substances

In veterinary medicine, HS are successfully applied as drugs for prophylaxis and therapy of gastrointestinal diseases in small animals. Furthermore, HS are utilized as antidotes to prevent intoxication (Kühnert et al., 1989). In order to bind and possibly metabolize as-yet resorbed poisons in the stomach-gastrointestinal tract, HA are given orally as a 20–30% watery solution or suspension in a dosage of 0.5–1.0 g/kg twice daily for 3–5 days (Kühnert, 1996). As observed by Ridwan (1977), a HA concentration of 0.1% is sufficient to reduce significantly the incorporation of lead and cadmium in rats, thus minimizing the risk of heavy metal intoxication. Experiments in mice have shown that orally administered lead humate is less toxic than lead acetate (Klöcking, R., 1980). Opposing results have been obtained following parenteral application of the same compounds. Clearly, the application route is an important factor in deciding whether a metal bound to HS is toxic or detoxified.

5 Humic Substances and Environmental Health

5.1 Mutagenicity

As HS are naturally widespread in the environment and present in surface water, studies on their genotoxic potential are justified, particularly as by-products of chlorination and ozonization in HS containing drinking water are known to be extremely active in bacterial genotoxicity tests (Meier et al., 1987; Meier, 1988). Mutagenesis studies carried out on fractions of drinking water samples have shown that 3-chloro-4-(dichloromethyl)-5-hydroxy-(5H)-furanone (MX) is one of the main chlorination intermediate products, responsible for more than 20% of observed mutagenicity (Holmbom, 1984; Kronberg et al., 1985, 1991). However, in-vivo mutagenicity tests have provided conflicting results, possibly due to the great reactivity and instability of the furanones formed (Dayan, 1993). Furanones occur also in foods, where they appear mainly as a result of Maillard reactions between sugars and amino acids during heating. Furthermore, they play an important role in the flavor of fruits and as an essential antioxidant food component (ascorbic acid, vitamin C) for humans (Colin-Slaughter, 1999). Although furanones are mutagenic to bacteria and cause DNA damage in laboratory animals, these compounds are, in practice, very effective anti-carcinogenic agents in the diets of animals which are being treated with known cancer-inducing compounds such as benzo[a]pyrene or azoxymethane. Evidence for the desmutagenic activity of HA has also been reported by Cozzi et al. (1993), De Simone et al. (1993), and Ferrara et al. (2000).

5.2 Protection against Ionizing Irradiation

Oris et al. (1990) were able to show that dissolved humic materials at concentrations in the range of 1 to 7 μg/mL significantly reduced acute photo-induced toxicity in fish (*Pimephales promelas*) and daphnia (*Daphnia magna*). The phenomenon is explained by selective attenuation of the active wavelengths of solar UV radiation by dissolved HS.

A protective effect of HA to injuries caused by an external whole-body ^{60}Co gamma irradiation in female Wistar rats has been reported in a World Patent Application (WO 9858655). HA were extracted from a 3000- to 7000-years-old fen peat standardized by topographic and paleobotanical characterization. The HA preparation (240 mg/animal/day) was applied by gastric intubation to female Wistar rats of 190–220 g bodyweight for 7 days before irradiation (7 Gy), followed by an additional 4-week treatment with the same dose after irradiation. No injury of the hemopoietic system occurred in the HA-treated animals. Lower dosages of HA (90 mg/animal/day) were also effective, albeit to a lesser extent. The HA-containing preparation is intended to improve the regeneration of the hemopoietic system in case of accidental radiation effects, and possibly to mitigate against injuries due to chemotherapy.

A therapeutic effect of sodium humate given as a single dose to experimental mongrel rats 5–10 min following irradiation with lethal doses of ^{60}Co led to 43% survival of animals after 60 days (Pukhova et al., 1987).

In addition to the protection against radiation-induced injuries and the supporting effect on tissue regeneration, HA show indirect detoxifying effects, e.g. by preventing the photoactivation of polycyclic aromatic hydrocarbons. As demonstrated by Nikkila et al. (1999), HA reduced the toxicity of UV-B-irradiated pyrene to *Daphnia magna* in a dose-dependent fashion. The effect was assumed to be due to the decrease in the photomodification of the dissolved pyrene by diminishing the light penetration into the water, and possible interaction with the intact parent compound.

5.3 Blackfoot Disease

Artesian drinking water containing a high concentration of greenish-blue fluorescent HS and/or arsenic has been implicated as one of the etiological factors of Blackfoot disease (BFD), which occurs endemically in the southwest coast of Taiwan (Lu, 1990). Clinically, BFD is a peripheral vascular disorder with symptoms similar to those of arteriosclerosis obliterans and thrombotic vasculopathy. The disease can be induced experimentally in mice receiving fluorescent HS at a daily dose of 5 mg per 20 g body mass for at least 22 days. The pathogenesis of the disease has not yet fully established. In-vitro studies with purified commercial HA revealed a destruction of human erythrocytes at (relatively high) HA concentrations of 50 – 100 μg/mL, probably due to the generation of oxidative stress (Cheng et al., 1999). Recently, Gau et al. (2000) demonstrated an inhibition of lipopolysaccharide-induced expression of NF-κB in HA-pretreated cultured human umbilical vein endothelial cells.

6
Outlook and Perspectives

The impact of HS on the quality of human health is increasingly recognized as an important subject of future research work. Investigations of HS aimed at the molecular structure and mechanism of action encompass specialized investigations within such diverse fields as physical, analytical, environmental and food chemistry, cell biology, molecular genetics, pharmacology and toxicology.

As outlined in this chapter, some of the naturally occurring or synthetically prepared biopolymers of the HA type have the potential of highly effective drugs. Therefore, in addition to the classic use of peat in balneotherapy and veterinary medicine, the application of isolated HS as well as synthetic HA-like polymers may play a considerable role in future. There are a large number of phenolic compounds which can be transformed into HA-like substances targeted for special functions such as antivirally active agents, heavy metal-chelating compounds, toxic chemical-binding polymers and substances protecting against ionizing radiation. However, the use of HS as therapeutic drugs make high demands on pharmacologically evidenced efficacy, toxicological safety standards and a clearly defined chemical composition of the preparation used.

To elucidate the chemical structure of synthetic HA that originate from comparatively simple individual starting compounds remains an important goal for the near future. The results will definitely stimulate and facilitate the much more complicated exploration of natural HS.

7 References

Amosova, Ya. M., Kosyanova, Z. F., Orlov, D. S., Tikhomirova, K. S., Shinkarenko, A. L. (1990) Humic acids in the therapeutic muds with a special reference to their physiological activity, *Kurortol. Fizioter.* **27** (4), 1–6.

Aschheim, S., Hohlweg, W. (1933) Über das Vorkommen östrogener Wirkstoffe in Bitumen, *Dtsch. Med. Wochenschr.* **59**, 12–14.

Baatz, H. (1988) Moortherapie in der Frauenheilkunde, in: *Moortherapie – Grundlagen und Anwendungen* (Flaig, W., Goecke, C., Kauffels, W., Eds.), pp. 161–168. Wien, Berlin: Ueberreuter.

Balasheva, L. I. and Gadzhi, F. I. (1971) Experience in the treatment of patients with rheumatism in inactive phase using low-temperature diluted mud baths, *Vopr. Kurortol. Fizioter. Lech. Fiz. Kult.* **36**, 240–242.

Barnard, D. L., Huffman, J. H., Meyerson, L. R., Sidwell, R. W. (1993) Mode of inhibition of respiratory syncytial virus by a plant flavonoid, SP-303, *Chemotherapy (Basel)* **39**, 212–217.

Bellometti, S., Giannini, S., Sartori, L., Crepaldi, G. (1997) Cytokine levels in osteoarthrosis patients undergoing mud bath therapy, *Int. J. Clin. Pharmacol. Res.* **17**, 149–153.

Bennett, E. R., Yedgar, S., Lerer, B., Ebstein, R. P. (1991) Phospholipase A_2 activity in Epstein-Barr virus-transformed lymphoblast cells from schizophrenic patients, *Biol. Psychiatry* **29**, 1058–1062.

Cheng, M. L., Ho, H. Y., Chiu, D. T. Y., Lu, F. J. (1999) Humic acid-mediated oxidative damages to human erythrocytes: a possible mechanism leading to anemia in Blackfoot disease, *Free Radic. Biol. Med.* **27**, 470–477.

Colin-Slaughter, J. (1999) The naturally occurring furanones: formation and function from pheromone to food, *Biol. Rev. Camb. Philos. Soc.* **74**, 259–276.

Cozzi, R., Nicolai, M., Perticone, P., De Salvia, R., Spuntarelli, F. (1993) Desmutagenic activity of natural humic acids: inhibition of mitomycin C and maleic hydrazide mutagenicity, *Mutat. Res.* **299**, 37–44.

Dayan, A. D. (1993) Carcinogenicity and drinking water, *Pharmacol. Toxicol.* **72**, 108–115.

De Simone, C., Piccolo, A., De Marco, A. (1993) Effects of humic acids on the genotoxic activity of maleic hydrazide, *Fresenius Environ. Bull.* **2**, 157–161.

Dunkelberg, H., Klöcking, H.-P., Klöcking, R. (1997) Suppression of heat-induced [^{3}H]arachidonic acid release in U937 cells by humic acid-like polymers, *Pharmacol. Toxicol.* **80** (Suppl. III), 175–176.

Dürre, K., Schindler, S. (1992) Austestung antiviraler Substanzen an der rezidivierenden kutanen Herpes-simplex-Virus-Infektion der Maus unter Berücksichtigung des Einflusses von Penetrationsvermittlern. Dissertation, Medizinische Akademie Erfurt.

Eichhorn, U., Klöcking, R., Helbig, B. (1984) Anwendung von ^{51}Cr-markierten FL-Zellen zur Testung der antiviralen Aktivität von Phenolkörperpolymerisaten gegen Coxsackieviren *in vitro*, *Dtsch. Gesundh. wes.* **39**, 1514–1519.

Ferrara, G., Loffredo, E., Simeone, R., Senesi, N. (2000) Evaluation of antimutagenic and desmutagenic effects of humic and fulvic acids on root tips of *Vicia faba*, *Environ. Toxicol.* **15**, 513–517.

Gau, R.-J., Yang, H.-L., Chow, S.-N., Suen, J.-L., Lu, F.-J. (2000) Humic acid suppresses the LPS-induced expression of cell surface adhesion proteins through the inhibition of NF-κB activation, *Toxicol. Appl. Pharmacol.* **166**, 59–67.

Gilbert, B. E., Wyde, P. R., Wilson, S. Z., Meyerson, L. R. (1993) SP-303 small-particle aerosol treatment of influenza A virus infection in mice and respiratory syncytial virus infection in cotton rats, *Antiviral Res.* **21**, 37–45.

Hils, J., May, A., Sperber, M., Klöcking, R., Helbig, B., Sprößig, M. (1986) Hemmung ausgewählter Influenzavirusstämme der Typen A und B durch Phenolkörperpolymerisate, *Biomed. Biochim. Acta* **45**, 1173–1179.

Holmbom, B., Voss, R. H., Mortimer, R. D., Wong, A. (1984) Fraction, isolation and characterization of Ames mutagenic compounds in Kraft chlorination effluents, *Environ. Sci. Technol.* **18**, 333–337.

Iubitskaia, N. S., Ivanov, E. M. (1999) Sodium humate in the treatment of osteoarthrosis patients, *Vopr. Kurortol. Fizioter. Lech. Fiz. Kult.* **1999**(5), 22–24.

Jurcsik, I. (1994) Possibilities of applying humic acids in medicine (wound healing and cancer therapy), in: Humic Substances in the Global Environment (Senesi, N., Miano, T.M., Eds), pp. 1331–1336. Amsterdam, London, New York, Tokyo: Elsevier.

Kallus, J. (1964) Gastritis, das gastroduodenale Ulkus und Neydhartinger Moor, in: *Bericht über den 8. Internationalen Kongreß für universelle Moor- und Torfforschung*, pp.111–114. Linz: Länderverlag.

Kleinschmidt, J. (1988) Moortherapie bei rheumatischen Erkrankungen, in: *Moortherapie – Grundlagen und Anwendungen* (Flaig, W., Goecke, C., Kauffels, W., Eds.), pp. 216–224. Wien, Berlin: Ueberreuter.

Klöcking, H.-P. (1991) Influence of natural humic acids and synthetic phenolic polymers on fibrinolysis, in: *Humic Substances in the Aquatic and Terrestrial Environment* (Allard, B., Boren, H: Grimvall, A., Eds.), Vol. 33 of Lecture Notes in Earth Sciences (Bhattacharji, S., Friedmann, G. M., Neugebauer, H. J., Seilacher, A., Eds.), pp. 423–428. Berlin, Heidelberg, New York, London, Paris, Tokyo, Hong Kong, Barcelona: Springer-Verlag.

Klöcking, H.-P. (1996) Anti-factor IIa–activity of humic acid-like polymers derived from p-diphenolic compounds. In: *Humic Substances and Organic Matter in Soil and Water Environments: Characterization, Transformations and Interactions*: Proceedings of the 7th International Conference of the International Humic Substances Society at the University of the West Indies, St. Augustine, Trinidad and Tobago, 3–8 July 1994 (Clapp, C. E., Hayes, M. H. B., Senesi, N., Griffith, S. M., Eds), pp. 411–415. Birmingham: Dawn Printers.

Klöcking, H.-P., Dunkelberg, H., Klöcking, R. (1997) Substances of the humic acid type prevent U937 cells from heat-induced arachidonic acid release, in: *Modern Aspects in Monitoring of Environmental Pollution in the Sea* (Müller, W. E. G., Ed.), pp. 156–158. Erfurt: Akademie gemeinnütziger Wissenschaften.

Klöcking, H.-P., Helbig, B., Klöcking, R. (1999) Antithrombin activity of synthetic humic acid-like polymers derived from o-diphenolic starting compounds, *Thromb. Haemost.* Suppl. 299–300.

Klöcking, R. (1980) Giftung und Entgiftung von Schwermetallen durch Huminsäuren, *Arch. Exper. Veterinärmedizin* **34**, 389–393.

Klöcking, R. (1994) Humic substances as potential therapeutics, in: *Humic Substances in The Global Environment* (Senesi, N., Miano, T. M., Eds.) pp. 1245–1257. Amsterdam, London, New York, Tokyo: Elsevier.

Klöcking, R., Helbig, B. (1999) Report on the Workshop of DGMT, Section IV, October 7–8, 1999, Bad Elster, Germany, *Telma*, **29**, 239–243.

Klöcking, R., Sprößig, M. (1972) Antiviral properties of humic acids, *Experientia* **28**, 607–608.

Klöcking, R., Sprößig, M. (1975) Wirkung von Ammoniumhumat auf einige Virus-Zell-Systeme, *Z. Allg. Mikrobiol.* **15**, 25–30.

Klöcking, R., Hofmann, R., Mücke, D. (1968) Tierexperimentelle Untersuchungen zur entzündungshemmenden Wirkung von Humaten, *Arzneim. Forsch.* **18**, 941–942.

Klöcking, R., Sprößig, M., Wutzler, P., Thiel, K.-D., Helbig, B. (1983) Antiviral wirksame Huminsäuren und huminsäureähnliche Polymere, *Z. Physiother.* **33**, 95–101.

Klöcking, R., Fernekorn, A., Stölzner, W. (1992) Nachweis einer östrogenen Aktivität von Huminsäuren und huminsäureähnlichen Polymeren, *Telma* **22**, 187–197.

Klöcking, R., Schacke, M., Wutzler, P. (1995) Primärscreening antiherpetischer Verbindungen mit EZ4U, *Chemotherapie J.* **4**, 141–147.

Kovarik, R. (1988) Über die Anwendung von Präparaten aus Torf, bzw. Huminstoffen bei gynäkologischen Erkrankungen, in: *Moortherapie – Grundlagen und Anwendungen* (Flaig, W., Goecke, C., Kauffels, W., Eds.), pp. 177–197. Wien, Berlin: Ueberreuter.

Kronberg, L., Holmbom, B., Tikkanen, L. (1985) Mutagenic activity in drinking water and humic water after chlorine treatment, *Vatten* **41**, 106–109.

Kronberg, L., Christman, R. F., Singh, R., Ball, L. M. (1991) Identification of oxidized and reduced forms of the strong bacterial mutagen (Z)-2-chloro-3-(dichloromethyl)-4-oxobutenoic acid (MX) in extracts of chlorine-treated water, *Environ. Sci. Technol.* **25**, 99–104.

Kühnert, M. (1996) Vergiftungen, in: *Lehrbuch der Pharmakologie und Toxikologie für die Veterinärmedizin* (Frey, H. H., Löscher, W., Eds.), p. 675. Stuttgart: Ferdinand Enke.

Kühnert, M., Fuchs, V., Golbs, S. (1989) Pharmakologisch-toxikologische Eigenschaften von Huminsäuren und ihre Wirkungsprofile für eine veterinärmedizinische Therapie, *DTW Dtsch. Tierärztl. Wochenschr.* **96**, 3–10.

Lange, N., Faqi, A. S., Kühnert, M., Haase, A., Hoke, H., Seubert, B. (1996a) Influences of a low molecular humic substance on pre- and postnatal development of rats, *DTW Dtsch. Tierärztl. Wochenschr.* **103**, 6–9.

Lange, N., Kühnert, M., Haase, A., Hoke, H., Seubert, B. (1996b) The resorptive behavior of a low-molecular weight humic substance after a single oral administration to the rat, *DTW Dtsch. Tierärztl. Wochenschr.* **103**, 134–135.

Lent, W. (1988) Bericht über die Moorforschung und -anwendung in der DDR, Polen, Tschechoslowakei und UdSSR, in: *Moortherapie – Grundlagen und Anwendungen* (Flaig, W., Goecke, C., Kauffels, W., Eds.), pp. 169–176. Wien, Berlin: Ueberreuter.

Liang, H. J., Tsai, C. L., Lu, F. J. (1998) Oxidative stress induced by humic acid solvent extraction fraction in cultured rabbit articular chondrocytes, *J. Toxicol. Environ. Health* **54**, 477–489.

Lu, F. J. (1990) Fluorescent humic substances and blackfoot disease in Taiwan, *Appl. Organomet. Chem.* **4**, 191–195.

Mesrogli, M., Maas, A., Schneider, J. (1988) Stellenwert der Moortherapie in der Sterilitätsbehandlung, in: *Moortherapie: Grundlagen und Anwendungen* (Flaig, W., Goecke, C., Kauffels, W., Eds.), pp. 225–235. Wien, Berlin: Ueberreuter.

Meyer, J. R. (1988) Genotoxic activity of organic chemicals in drinking water, *Mutat. Res.* **196**, 211–245.

Meyer, J. R., Knohl, R. B., Coleman, W. E., Ringhand, H. P., Munch, J. W., Kaylor, W. H., Streicher, R. P., Kopfler, F. C. (1987) Studies on the potent bacterial mutagen, 3-chloro-4-(dichloromethyl)-5-hydroxy-2(5H)-furanone, *Mutat. Res.* **189**, 363–373.

Neyts, J., Snoeck, R., Wutzler, P., Cushman, M., Klöcking, R. Helbig, B., Wang, P., De Clercq, E. (1992) *Antiviral Chem. Chemother.* **3**, 215–222.

Nikkila, A., Penttinen, S., Kukkonen, J. V. (1999) UV-B-Induced acute toxicity of pyrene to the waterflea *Daphnia magna* in natural freshwaters, *Ecotoxicol. Environ. Safety* **44**, 271–279

Oris, J. T., Hall, A. T., Tylka, J. D. (1990) Humic acids reduce the photo-induced toxicity of anthracene to fish and *Daphnia*, *Environ. Toxicol. Chem.* **9**, 575–583.

Priegnitz, H. (1986) *Wasserkur und Badelust.* Leipzig: Koehler & Amelang.

Pukhova, G. G., Druzhina, N. A., Stepchenko, L. M., Chebotarev, E. E. (1987) Effect of sodium humate on animals irradiated with lethal doses, *Radiobiologiia* **27**, 650–653.

Ridwan, F. N. J. (1977) Untersuchungen zum Einfluß von Huminsäuren auf die Blei- und Cadmium-Absorption bei Ratten. Dissertation, Universität Göttingen.

Schewe, C., Klöcking, R., Helbig, B., Schewe, T. (1991) Lipoxygenase-inhibitory action of antiviral polymeric oxidation products of polyphenols, *Biomed. Biochim. Acta* **50**, 299–305.

Schneider, J., Weis, R., Manner, C., Kary, B., Werner, A., Seubert, B. J., Riede, U. N. (1996) Inhibition of HIV-1 in cell culture by synthetic humate analogues derived from hydroquinone: mechanism of inhibition, *Virology* **218**, 389–395.

Schols, D., Wutzler, P., Klöcking, R., Helbig, B., De Clercq, E. (1991) *J. Acq. Immun. Defic. Synd.* **4**, 677–685.

Schultz, H. (1962) Die viricide Wirkung der Huminsäuren im Torfmull auf das Virus der Maul- und Klauenseuche, *Dtsch. tierärztl. Wochenschr.* **69**, 613–616.

Schultz, H. (1965) Untersuchungen über die viricide Wirkungsweise der Huminsäuren im Torfmull, *Dtsch. tierärztl. Wochenschr.* **72**, 294–297.

Taugner, B. (1963) Tierexperimentelle Untersuchungen über ein Natriumhuminat-Salicylsäure-Bad, *Arzneimittelforschung* **13**, 329–333.

Thiel, K.-D., Klöcking, R., Helbig, B. (1976) In-vitro-Untersuchungen zur antiviralen Aktivität enzymatisch oxidierter o-Diphenolverbindungen gegenüber Herpes simplex-Virus Typ 1 und Typ 2, *Zbl. Bakt. Hyg., I. Abt. Orig,* **A 234**, 159–169.

Thiel, K.-D., Klöcking, R., Schweizer, H., Sprößig, M. (1977) Untersuchungen *in vitro* zur antiviralen Aktivität von Ammoniumhumat gegenüber Herpes simples Virus Typ 1 und Typ 2, *Zbl. Bakt. Hyg., I. Abt. Orig,* **A 239**, 304–332.

Thiel, K.-D., Helbig, B., Klöcking, R., Wutzler, P., Sprößig, M., Schweizer, H. (1981) Vergleich der In-vitro-Wirkung von Ammoniumhumat und enzymatisch oxidierter Chlorogen- und Kaffeesäure gegenüber Herpesvirus hominis Typ 1 und Typ 2, *Pharmazie* **36**, 50–53.

Wiegleb, K., Lange, N., Kühnert, M. (1993) The use of the HET-CAM test for the determination of the irritating effects of humic acids, *DTW Dtsch. Tierärztl. Wochenschr.* **100**, 412–416.

Wyde, P. R., Meyerson, L. R., Gilbert, B. E. (1993) In vitro evaluation of the antiviral activity of SP-303, an Euphorbiaceae shrub extract, against a panel of respiratory viruses, *Drug Dev. Res.* **28**, 467–472.

Zeck-Kapp, G., Nauck, M., Riede, U. N., Block, L., Freudenberg, N., Seubert, B. (1991) Low-molecular humic substances as pro-inflammatory cell signals, *Verh. Dtsch. Ges. Path.*75, 504.

2
Melanin

Prof. Dr. Joan M. Henson
Department of Microbiology, Montana State University, Bozeman, MT 59717, USA,
Tel: +1-4069944690; Fax: +1-4069944926; E-mail:

cAMP	cyclic adenosine monophosphate
DHI	5,6 dihyroxyindole
DHICA	5,6 dihydroxy-2-carboxylic acid
DHN	dihydroxynaphthalene
DOPA	L-dihydroxyphenylalanine

Biopolymers for Medical and Pharmaceutical Applications. Edited by A. Steinbüchel and R. H. Marchessault

ISBN: 3-527-31154-8

GHB	γ–glutaminyl-4-hydroxybenzene
HJ	hydroxyjuglone
MSH	melanin stimulating hormone
nm	nanometer
NMR	nuclear magnetic resonance
SD	scytalone dehydratase
THN	tetra- or trihydroxynaphthalene
TRP	tyrosinase-related protein
UV	ultraviolet light

1 Introduction

Most of the dark material in nature is either lignin or melanin. While lignin biosynthesis is restricted to plants, many organisms, including plants, animals, insects, and microorganisms, produce melanin (reviewed in Bell and Wheeler, 1986; Hill, 1992; Riley, 1997; Butler and Day, 1999; Henson et al., 1999; Cooper and Szaniszlo, 1997; Bagnara, 1998; Quevedo and Holstein, 1998). Melanins are darkly pigmented phenolic or indolic polymers that provide structural strength or protect organisms from environmental stress. In higher organisms melanins may also serve as an antibiotic, provide ornamentation for sexual display and/or camouflage, and improve optical efficiency. In addition, melanin probably plays an as yet unidentified role in mammalian hearing.

Chemical criteria defining a pigment as melanin include dark color, general insolubility, bleaching by oxidizing agents, and resistance to degradation by concentrated acids. Another criterion used to define a polymer as melanin is its free radical character that is detectable by electron spin resonance. However, melanin extraction involves the use of hot solvents or acids that can alter melanin polymers or destroy their associations with other biological material. Thus, the chemical structures of melanins in their natural state are still unknown. In addition, melanins are usually heterogeneous and their composition depends not only on their monomeric units, which can differ, but also on environmental conditions during polymerization. Nevertheless, because of their ubiquity throughout nature and their physiological and even sociological importance, melanins continue to be intensively examined.

2 Historical Outline

The term melanin (from μελανοσ, melanos, Greek for dark) was in use by early nineteenth century anatomists who were searching for the chemical basis of racial differences among humans (reviewed by Riley, 1997; Prota et al., 1998; Ray and Eakin, 1975). In 1895 Bourquelot and Bertrand discovered that the mushroom *Russula nigricans* produced a colorless substrate, later identified as tyrosine, that darkened when oxidized by a fungal enzyme that Bertrand named tyrosinase. Soon thereafter tyrosinase was discovered in human skin and melanomas, and the tyrosine-derived pathway for dihydroxyphenylalanine (DOPA) melanin synthesis, also known as the Raper-Mason pathway, was proposed by Raper in 1927. Early studies of animal melanins demonstrated their chemical composition, free radical nature, and cation exchange properties (White, 1958; Kikkawa et al., 1955). Aspergillin, the black fungal spore pigment of *Aspergillus niger*, was first described by Linossier in 1891 although its biosynthetic pathway was not proposed until 1975 (Ray and Eakin). In 1960 Allport and Bu'Lock found that dihydroxynaphthalene (DHN) was the main component of *Daldinia concentrica* melanin and Wheeler and associates elucidated the DHN melanin pathway of many ascomycetous fungi (reviewed in Bell and Wheeler, 1986). In recent years, several other fungal, bacterial, and animal melanin precursors and pathways have been described (Coon et al., 1994; Smythies and Galzigna, 1998; reviewed in Bell and Wheeler, 1986 and Prota, 1992).

3 Occurrence and Structure

3.1 Fungi

Many fungal species produce melanin during one or more stages of their life cycles (reviewed in Bell and Wheeler, 1986; Cooper and Szaniszlo, 1997; Butler and Day, 1999; Henson et al., 1999). For example, 9% of the 256 known marine ascomycetes produce melanin (McKeown et al., 1996). Besides hyphae, fungal structures that are often melanized include conidia (asexual spores), sexual spores, appressoria (and related hyphopodia), fruiting structures such as perithecia and basidiocarps, rhizomorphs (highly differentiated cord-like structures that form interconnective networks between mycelia of some basidiomycetes), and sclerotia (hard, differentiated masses of hyphae, often with a rind-like outer layer). Hyphae of some species are constitutively melanized, whereas in others, hyphal melanization is induced by environmental factors such as heavy metals or UV light. Most, if not all, melanizing ascomycetous fungi synthesize DHN melanin in hyphae (Figure 1), where it is deposited as granules between the outer mucilage and chitin cell wall layers, or alternatively it

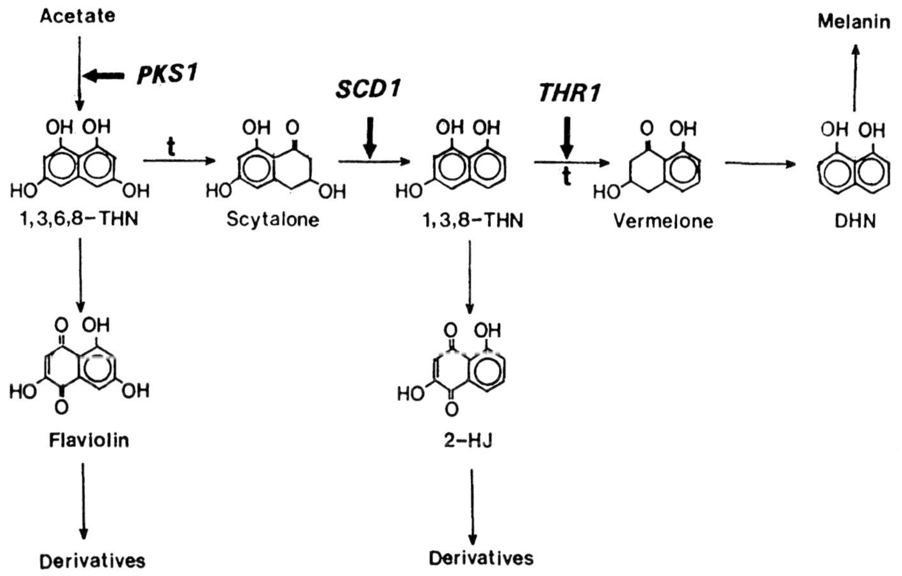

Fig. 1 The dihydroxynaphthalene (DHN) melanin biosynthetic pathway.Gene designations are indicated in bold capitals (see Section 5.3), THN = tri- and tetrahydroxynaphthalene, HJ = hydroxyjuglone, and **t** = sites of where tricyclazole inhibition occurs. Figure adapted from Bell and Wheeler (1986) with permission.

becomes the outermost, granular cell wall layer. Melanization may occur to a greater extent in aging hyphal cell walls, with little melanin deposited at hyphal tips (Frederick et al., 1999; Henson et al., 1999). Whether melanin precursors or polymers are transported to the cell wall via membrane-bound vesicles is not known. Although cytoplasmic, "melanosome-like" structures are apparent in some melanizing fungi (Franzen et al., 1999; Sänchez-Mirt et al., 1997), the contents of these electron-dense organelles have not been established, and it is possible they are osmolyte or storage inclusions instead of organelles involved in melanogenesis.

Some species produce melanized sexual or asexual spores and in most cases melanin is localized in the outer layer of the conidial spore coat. Spore ornamentation, or spore striations may also be melanized; what processes occur to distribute cytoplasmic melanin granules into specific regions of the coat are completely unknown (McKeown et al., 1996; Bunning and Griffiths, 1984).

In some fungal species, melanin is also deposited in the cell walls of appressoria and hyphopodia, which are adhesive cells from which an infection hypha emerges from the adhesive side of the cell to penetrate the substratum or tissue underneath (reviewed in Howard, 1997; Money et al., 1998; Bechinger, 1999). It is interesting that in those species that accumulate melanin in both hyphal and appressorial cell walls, the orientation of the chitin and melanin is often reversed; that is, in the hyphae chitin is closest to the cytoplasmic membrane, whereas in melanized appressoria, the melanin layer is adjacent to the cytoplasmic membrane. What determines this order of deposition is unknown.

Melanin is not required for growth or development of hyphae or other fungal structures where it occurs, and is therefore often referred to as a secondary metabolite. One notable exception, however, is that viable ascospores fail to develop in non-melanized *Ophiostoma piliferum* perithecia (Zimmerman, 1995).

Little is known of the structure of fungal melanins. Most fungal cell wall melanin is composed of 30–200 nm granules, with the notable exception of appressorial and hyphopodial melanin, which are smooth in appearance (Bell and Wheeler, 1986; Henson et al., 1999; Howard, 1997). Melanin polymers of γ-glutaminyl-4-hydroxybenzene (GHB) extracted from spores of the mushroom *Agaricus bisporus* and GHB melanin synthesized with tyrosinase in vitro had 50–100 nm granules that often fused into larger granules; in addition, amorphous, plate-like structures were observed (Hegnauer et al., 1985).

DHN melanin synthesized in vitro with purified laccase was >60 kD in molecular mass (Edens et al., 1999). Subsequent studies demonstrated that this melanin contains some nitrogen, apparently from laccase incorporation during synthesis; DHN melanin also cross-reacted with monoclonal antibodies made against L-DOPA melanin (Rosas et al., 2000).

3.2
Animals

Melanin in vertebrates is synthesized in membrane-bound organelles called melanosomes within melanocytes, which mature in different tissues after migrating from the neural crest as melanoblasts during embryogenesis (reviewed in Prota, 1992; Reedy et al., 1998). Mammalian melanin occurs in various tissues, including skin, hair, eye, inner ear and mucous membranes (reviewed in Boissy, 1998). In addition, neuromelanin occurs in neurons of the substantia nigra of the mammalian brain, but unlike melanin of other tissues, neuromelanin is not present at birth, but accumulates with age. Melanin is also a major component of melanomas, melanocyte cancers of humans and other melanin-producing animals. Melanin is distributed to a lesser extent in birds, amphibians and reptiles, where it is located in skin and/or feathers (reviewed in Bagnara, 1998). Amphibians and reptiles also have melanin-synthesizing liver cells. Among invertebrates, most of the dark pigments are still chemically uncharacterized. However, melanin is observed in the cuticle, wing, and serum of some insect species, and in the ink glands of cephalopods.

Melanocytes produce at least two types of melanin; pheomelanin is reddish-yellow due to cysteine residues that react with dopaquinone to produce cysteinyldopa that is polymerized (Figure 2). In contrast, eumelanin is derived from dopachrome, lacks cysteine, and has more carbonyl groups that absorb more in the red part of the spectrum, causing it to be darker than pheomelanin (Figure 2; Riley et al., 1997; Sarna and Swartz, 1998; Nofsinger et al., 1999). Light-skinned people generally have more pheomelanin than dark-skinned people. Extracted eumelanin typically has a protein "coat" associated with it that is not removed by the purification process. While 8–50% of extracted melanin is protein, autooxidized melanin is protein-free. Eumelanin extracted from the cuttlefish, *Sepia officinalis,* is particulate, and granules range in size from 70–460 nm, but even these granules are composed of smaller particles of ~60 nm. The size and shape of eumelanin particles depends, in part, on the size and shape of their melanosomes, which can differ in different tissues and different skin types. For example, darker skin has melanosomes and melanin particles that are larger than those of light skin. Neuromelanin is composed of 5-*S*-cysteinyldopa and dopaminochrome in amorphous granules of varying

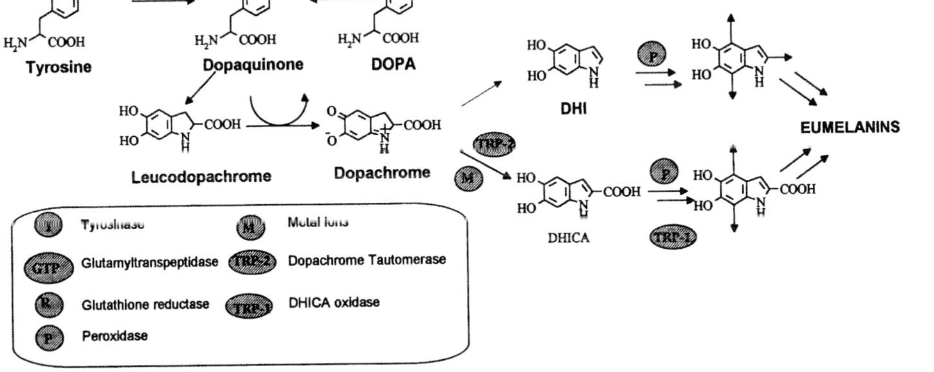

Fig. 2 Biosynthetic pathways for eumelanin (DOPA), pheomelanin, and related metabolites Intermediates in the pathways are also directly oxidized into melanin as well. Regulatory factors are highlighted. Adapted from Prota et al. (1998) with permission.

sizes and shapes that are not membrane-bound in melanosomes.

Most chemical and structural analyses of melanin are conducted with synthetic or extracted L-DOPA or tyrosine-derived melanins (reviewed in Sarna and Swartz, 1998; Prota et al., 1998; Ito, 1998). Nuclear magnetic resonance (NMR) and infrared spectroscopy are useful to determine chemical composition and covalent linkages between melanin and other macromolecules (Nilges, 1998; Bridelli et al., 1999, and references within). The smallest molecular units of DOPA melanin, as shown by several techniques, including x-ray scattering, light scattering, scanning tunneling microscopy (reviewed in Nilges, 1998; Gallas et al., 1999), and matrix-assisted laser desorption/ionization mass spectroscopy (Bertazzo et al., 1999, and references within), are oligomers of 5–8 covalently-linked indolequinone units arranged in planes that are stacked and interact by van der Waals forces (Figures 3 and 4). The structure(s) of melanins derived from precursors other than tyrosine derivatives have not been elucidated.

3.3
Other Organisms

Bacteria that produce melanin include species of *Aeromonas, Azotobacter, Mycobacterium, Micrococcus, Bacillus, Legionella, Streptomyces, Vibrio, Rhizobium, Proteus, Azospirillum,* and *Shewanella* (Coon et al., 1994, and references within; Coyne and Al-Harth, 1992, and references within; Mercado-Blanco et al., 1993; Agoti et al., 1996, and references therein). In each case, melanin production is enhanced by tyrosine, and the tyrosine

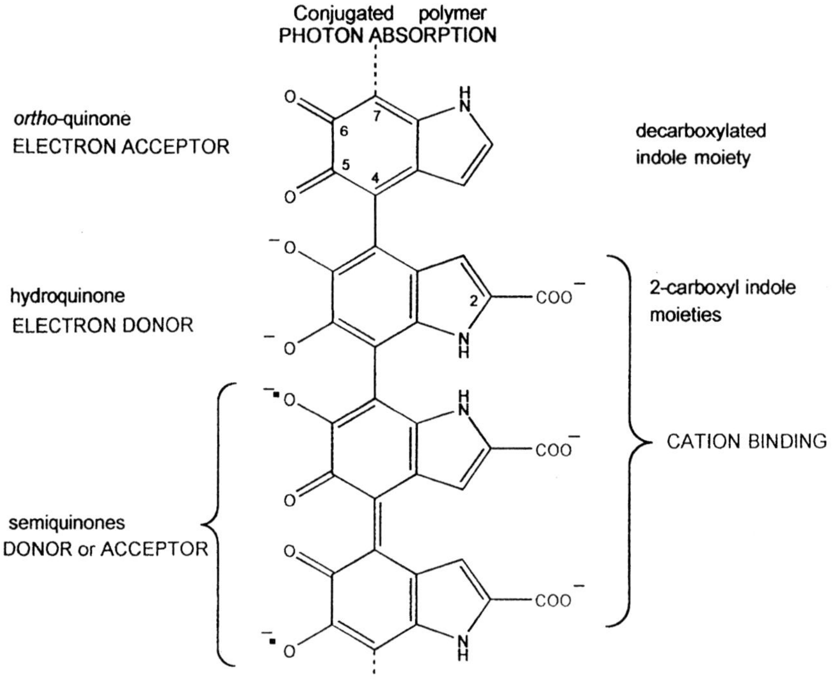

Fig. 3 Structure of DOPA melanin oligomers. The functions of different moieties of melanin are indicated by capitol letters. Reproduced with permission from Riley (1997).

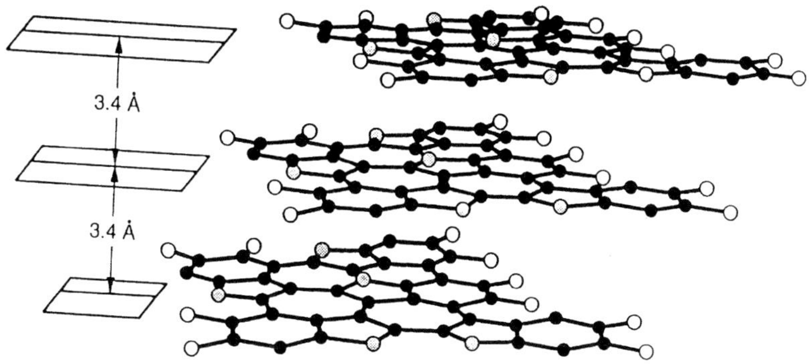

Fig. 4 Structure of stacked DOPA melanin oligomers. Hydroxyl groups (clear circles) are extended outward. Adapted from Zajac et al. (1994) with permission.

derivatives, homogentisic acid and L-DOPA, were identified as components of *Shewanella colwelliana* and *Bacillus thuringiensis* melanin, respectively. The cellular location of bacterial melanin is usually unreported; however, one study utilizing the Fontana-Masson assay for melanin demonstrated silver precipitation throughout the bacterial cytoplasm, suggesting melanin synthesis occurs within the cytoplasm (Agodi et al., 1996).

In plants, melanin synthesis is associated with bruising or browning of wounded tissue. It is proposed that mechanical impact releases polyphenol oxidase from its compartmentalization to cause oxidation of tyrosine and cysteine derivatives to produce melanin (Stevens et al., 1998).

4 Function

Because of its remarkable photochemistry, most studies of melanin function focus on its role in photoprotection (reviewed in Sarna and Swartz, 1998; Gilchrest et al., 1998; Prota et al., 1998). Melanin is dark because its aromatic rings absorb most of the visible spectrum of light. Melanin also absorbs in the UV range of the spectrum and this property can protect underlying tissue from the damaging effects of UV. For example, melanosomes in human epidermal cells form "caps" positioned over cell nuclei, which consequently undergo less UV-induced DNA damage than cells without melanin caps (Kobayashi et al., 1998). People with dark skin experience less photodamage and melanoma than people with light skin. Melanin also confers photoprotection to collagen in vitro (Menter et al., 1998). In addition, fungal melanin protects melanized hyphae and spores from UV (Romero-Martinez et al., 2000; Frederick et al., 1999; reviewed in Henson et al., 1999). In fungal-algal symbioses, the melanized fungal partner protects the photosynthetic pigments of the algal partner from photobleaching and other photodamage (Kappen, 1973).

Melanins may also act as antioxidants by scavenging free radicals (Smythies and Galzigna, 1998; Rözanowska et al., 1999; Blarzino et al., 1999); for example, melanized conidia of the human pathogen *Sporothrix schenckii* were less sensitive to oxygen and nitrogen radicals than unmelanized mutant conidia (Romero-Martinez et al., 2000). In addition, melanin protects animal fungal pathogens from being killed by the oxidative burst inside the phagolysosome of the host neutrophil (Schnitzler et al., 1999; Tsai et al., 1999; Jacobson et al., 1995). Melanin acts as a redox buffer and can reduce superoxide anion to H_2O_2 or oxidize it to O_2, depending on the redox state of the melanin (Jacobson and Hong, 1998; reviewed in Sarna and Swartz, 1998; Prota et al., 1998). In the presence of Fe^{2+} or Fe^{3+} ions, melanin precursors interact with superoxide anions to produce hydroxyl radicals that lead to semiquinones that polymerize into melanin, thus diverting the highly toxic superoxide anion (Novellino et al., 1999). Melanization is often observed in amphibian liver cells that have high superoxide dismutase activity, a finding that suggests liver melanin or its precursors may also trap free radicals.

The energy of light absorbed by melanin is released almost entirely as heat and thus, can be thermoprotective (reviewed in Hill, 1992; Riley, 1997; Sarna and Swartz, 1998; Prota et al., 1998). Melanin protects some fungi from both high and low temperature (Rosas and Casadevall, 1997; Henson et al., 1999) and is associated with thermoregulation in mammals and insects (Quevedo and Holstein, 1998; Bagnara, 1998). Melanin of lichens may help warm them during spring thaws and allow them to grow in lower than ambient temperatures (Kershaw, 1974). Melanin may also simply scatter light and thereby act as a passive screen to protect underlying tissue from photodamage; this is a proposed function for ocular melanin in vertebrates (Bridelli, 1998; reviewed in Boissy, 1998).

Melanin can protect organisms from antibiotics or other drugs. However, it is this affinity for drugs that causes skin sensitivity to sunlight during treatment with certain antibiotic substances, and binding of drugs by melanin can reduce the dose delivered to the intended target (Zemel et al., 1999; Knörle et al., 1998; reviewed in Larsson, 1998). Protection from plant phenolics, that are toxic for colonizing or invading fungi or bacteria, is another proposed function for melanin (Mercado-Blanco, 1993; Henson et al., 1999). Because of its many hydroxyl groups, melanin binds cationic heavy metals (reviewed in Henson et al., 1999). For example, lichens can grow directly on uranium ore, and uranium is associated with melanin of the fungal partner of the lichen (McLean et al., 1998). Evidence that melanin protects against heavy metal toxicity is lacking, and it may serve instead to sequester essential metal ions, such as Fe and Cu ions (reviewed in Henson et al., 1999; Niecke et al., 1999). Regardless of its purpose, melanin binding of metals from rock and ore causes metal mobilization and has an impact of geochemical cycling of metals (Jongmans et al., 1997; de Leo, et al., 1999, and references within). Metal binding also affects the color and function of melanin (Kikkawa et al., 1955; Krol and Liebler, 1998).

Lichens are resistant to desiccation, and fungal melanin is believed to be responsible for their survival during dry conditions (Kappen, 1973, and references within). Mycorrhizal fungal melanin may also be associated with drought tolerance of host plants (Worley and Hacskaylo, 1959). Melanin is required for survival of fungal sclerotia and protects them from drought, UV, temperature extremes, and predation (Hawke and Lazarovits, 1994).

Although melanin is generally protective, it has been described as "a double-edged sword" because it can release damaging free radicals (reviewed in Sarna and Swartz, 1998; Prota et al., 1998). Melanin is unique as a biopolymer in its free radical character. It is estimated that in metal-free, fully hydrated, synthetic DOPA melanin at pH 7, there is one free radical per 1500 melanin subunits (monomers). However, the concentration of free radicals is influenced by metal binding, redox reactions, and polymer hydration. Both eumelanin and pheomelanin lose their antioxidant character when complexed with Fe^{+3}, and pheomelanin becomes a pro-oxidant, or a photosensitizer (Krol and Liebler, 1998). It is postulated that iron binding by neuromelanin causes greater free radical production and contributes to the pathogenesis of Parkinson's disease, where heavily pigmented neurons in the substantia nigra are selectively degraded (Shima et al., 1997). The size and shape of melanin granules influences their photochemistry; small eumelanin granules produce radicals after absorbing UV light, whereas larger granules do not (Nofsinger et al., 1999). Inappropriate melanin dispersion and deposition also causes ocular pathology, perhaps by altering the size of the melanin granules (Küchle et al., 1998). In addition, melanin or its precursors can inhibit enzyme reactions, and even melanin precursors may be toxic or mutagenic to cells (Hiraku et al., 1998; Giambernardi et al., 1998; Larsson, 1998; Kipp and Young, 1999). Some organisms use the toxicity of melanin to their advantage; for example, quinones of the eumelanin pathway are involved in immobilizing bacteria in some insects (Charalambidis et al., 1995), and cuttlefish ink melanin, sprayed when the animal is frightened, may serve as a predator antagonist (Prota, 1992).

Nearly all lower vertebrates can change their color physiologically; that is, they can redistribute pigment granules, usually to match their background for camouflage to become less visible to prey or predators (reviewed in Bagnara, 1998). Cuttlefish ink may allow it to evade predators by limiting visibility. Birds also use melanin for camouflage, as well as for ornamentation in sexual display, and in some species melanin patterns signal sexual maturity (Niecke et al., 1999). In addition, insects use tyrosine-derived melanin for wing and body coloration (Koch and Kaufmann, 1995; Weirnasz, 1995), which can serve as camouflage (Grant, 1995, and references within), affect their choice of mates (Weirnasz, 1995), or signal colony behavioral stages (Tawfik et al., 1999; Wilson, 2000; Sword, 1999). Bacterial melanin may also serve as an attractant to recruit oysters to an environment where bacteria can colonize them (Coon et al., 1994).

Melanin confers structural strength and examples of rigidity provided by melanin include melanized insect cuticles and injured banana peels that toughen when they blacken. Many fungal structures, such as spore walls, hyphopodia, perithecia, and sclerotia also increase their tensile strength upon melanization (reviewed in Henson et al., 1999 and Butler et al., 1999). In fungal phytopathogens that must penetrate their host via melanized appressoria, melanin is a required virulence factor. Melanin of appressoria and hyphopodia limits cell wall permeability allowing osmolytes to accumulate and turgor pressure to build such that mechanical penetration of the host by the infection peg can occur (Howard, 1997; Bechinger et al., 1999). The pressure exerted on the substratum by a melanized appressorium is approximately 17 $\mu N\ \mu m^{-2}$ or the equivalent force needed for a human to lift a killer whale in her palm (Bechinger et al., 1999; Money, 1999).

Melanin is also a required virulence factor for several pigmented fungal pathogens of animals (reviewed in Cooper and Szaniszlo, 1997; Tsai et al., 1999, and references within). In addition to protecting cells from the oxidative burst in phagocytes, melanin helps them evade phagocytosis by an as yet unknown mechanism (Romero-Martinez et al., 2000; Tsai et al., 1999). Hyphal melanin permits invasive growth through high percentage agar, and may be necessary for invasive growth in animal tissue (Brush and Money, 1999).

Finally, the function of some melanin, such as that found in the inner ear of higher vertebrates, is unknown; however, the occurrence of hearing defects in people with pigmentation disorders suggests that this pigment is important for hearing (Boissy, 1998; Benedito et al., 1997).

5
Synthesis

The chemical composition of growth media influences the type and amount of melanin synthesized by either fungal or animal cells in culture (Eagen et al., 1997; Smit et al., 1997; Franzen et al., 1999). Some fungi, such as *Cryptococcus neoformans*, only produce melanin in culture when their growth medium is amended with a suitable substrate, such as L-DOPA. These organisms do not synthesize melanin precursors, but produce phenol oxidase(s) that polymerize exogenous phenolic compounds into fungal cell walls. Whether fungi produce melanin in their hosts is still in question because both animal and plant hosts produce phenolic compounds that oxidize around invading fungal cells, giving the appearance of melanin (Nosanchuk et al., 1999; Liu et al., 1999; Sticher et al., 1997). For example, lignin, which is similar in chemical structure to melanin, is deposited around plant-invading hyphae, and this deposition may impart a "melanized" appearance to intracellular, invasive hyphae.

5.1
Precursors

Precursors for natural melanins are shown in Figures 1 and 2; however, other tyrosine derivatives or phenolics can form melanin-like polymers (Mosca et al., 1998, 1999; Schraermeyer, 1994). Due to autooxidation and/or broad substrate specificity of polymerizing oxidases, many other molecules, including other melanin pathway intermediates, macromolecules, or metals can become incorporated into natural melanins (Napolitano et al., 1996; Nakano et al., 2000; Riley, 1997; Mosca et al., 1999), and most melanins are considered heterogeneous in composition and structure.

5.2
Pathways

Hyphal melanin is usually, if not always, DHN (Figure 1; reviewed in Butler and Day, 1999; Henson et al., 1999; Bell and Wheeler, 1986). The DHN pathway, or pentaketide pathway begins with head-to-tail conjoining of five ketide subunits obtained from five acetate precursors. After cyclization, tetrahydroxynaphthalene undergoes a series of reduction/oxidation steps to produce the precursor, 1,8-dihydroxynaphthalene. Studies with inhibitors of the DHN pathway suggest that hyphal melanin in many fungi is synthesized by a similar, if not identical pathway. Non-hyphal melanins in other fungal structures are derived from different pathways. The black conidial pigment of *Aspergillus niger* is water-soluble due to its conjugation to protein and/or sugars and has several undefined pigment components, none of which are products of DHN biosynthesis (Ray and Eakin, 1975; Bell and Wheeler, 1986). The blue-green conidial pigment of *A. fumigatus*, in contrast, is synthesized by a combination of the DHN pathway with other oxidases (Tsai et al., 1999). *Ustilago* spp. produce melanin derived from catechol (Lichter and Mills, 1998), whereas other basidiomycetes produce GHB melanin (Pierce and Rast, 1995; reviewed in Bell and Wheeler, 1986). Except for the DHN pathway, however, the enzymatic steps involved in synthesis of fungal melanin precursors are unknown. Moreover, the terminal oxidases that convert monomers into polymers in each pathway are still undetermined.

In 1928 Raper first demonstrated that in the presence of oxygen, tyrosine is oxidized to dopa and then dopaquinone by tyrosinase (reviewed in Prota et al., 1998). He proposed that dopachrome was an autooxidation product of dopaquinone, and by stopping oxidation with sulfuric acid was able to purify 5,6-dihydroxyindole (DHI) and 5,6-dihydroxyindole-2-carboxylic acid (DHICA). In the 1940s Mason (1948) proposed that melanin was a homopolymer of DHI. Nicolaus (1962) demonstrated in the 1960s that melanin is a heteropolymer that included pathway intermediates directly oxidized into the heterogeneous polymer. During the same decade Nicolaus and Prota (1967) purified pheomelanin from the red feathers of roosters and demonstrated that it can be synthesized from cysteine coupled to dopaquinone. A current scheme for animal melanin biosynthesis is shown in Figure 2. It includes a recently proposed alternate route for production of cysteinyldopa via glutathionyldopa, but the in vivo derivation of cysteinyldopa is still unknown. It should be noted that other oxidases, including peroxidases, lipoxygenases, catalases, if present, catalyze oxidation and polymerization reactions of melanin metabolites (Blarzino et al., 1999; Rosei et al., 1998).

5.3 Enzymes, Genes and their Regulation

Genes required for melanin biosynthesis in bacteria reside either on plasmids or the bacterial chromosome (Mercado-Blanco, 1993). Little is known about the regulation of bacterial melanin genes. Fungal melanin pathways are controlled by various environmental stimuli, and are often differentially expressed in different tissues. For example, some fungi undergo melanogenesis in response to heavy metals or low pH (reviewed in Kubo et al., 1999; Butler and Day, 1999; Henson et al., 1999; Lichter and Mills, 1998). Genes responsible for melanin biosynthesis via the DHN pathway have been cloned from several melanized fungal species, and their sequences, as well as cross-species complementation suggest that the pathway is similar in different species. However, the timing and distribution of melanin biosynthesis are regulated differently in various species. For example, some species produce both melanized hyphae and spores, whereas others express the DHN pathway only during specific developmental stages. Progress toward identifying regulatory transcription factors that control pathway gene expression is underway (Kubo et al., 1999). The only enzyme of the DHN pathway that has been crystallized is scytalone dehydratase (Wawrzak et al., 1999; Zheng and Bruice, 1998). It is of particular interest because it is the target of several fungicides that are used to control *Magnaporthe grisea*, the causative agent of rice blast disease. In addition, this enzyme is unusual because it catalyzes the dehydration of melanin intermediates without using metal ions or other cofactors, and a water molecule stabilizes the transition state of the reaction.

Genes, enzymes, and substrates involved in melanin biosynthesis continue to be identified, and we are just beginning to understand regulatory mechanisms of melanogenesis (Benedito et al., 1997; Haavik, 1997; Tief et al., 1998; reviewed in Pawelek and Chakraborty, 1998, Shibahara et al., 1998, and Solano and García-Borrón, 1998). Melanin biosynthesis in insects and higher animals is hormonally controlled (Tawfik et al., 1999; Lu et al., 1998; Hearing, 1998). Melanocyte stimulating hormone (MSH) is a small peptide produced by the pituitary gland in mammals. Melanocytes bind MSH with receptors and transcription of tyrosinase and eumelanin biosynthesis is stimulated. DNA damage due to UV irradiation or even the dinucleotide pTpT, a signal of DNA damage, increases the response of melanocytes to MSH (Gilchrest et al., 1998). What regulates the amount of pheomelanin synthesis is not known, although it is correlated with the amount of melanocyte cAMP (Lu et al., 1998).

At least 50 genes related to animal pigmentation have been cloned and sequenced (reviewed in Shibahara et al., 1998; and Hearing, 1998). Many of these genes encode regulatory proteins that regulate melanogenesis or pigmentation distribution at the tissue level. Others are proteins involved in melanosome biosynthesis and/or regulation. Others are involved in the switch from eumelanin to pheomelanin synthesis (Furumura et al., 1998). Enzymes of the DOPA pathway include tyrosinase, and tyrosinase-related proteins 1 and 2 (TRP-1 and TRP-2), all members of a tyrosinase gene family (Figure 2). Tyrosinase, which catalyzes the oxidation of tyrosine to 3,4-dihydroxyphenylalanine (DOPA), is rate limiting for melanin biosynthesis. Both tyrosinase and TRP-1 oxidize dopa, and TRP-2 tautomerizes dopachrome. Tyrosinase is regulated both at the transcriptional and post-transcriptional level, and by substrate and inhibitor concentrations. It is becoming increasingly clear that melanin synthesis in animals is not just a matter of regulating tyrosinase activity, however, and understanding the regulatory mechanisms of melanogenesis and its modulation is a field for innovative melanin research.

6
Biodegradation

Melanins are known for their invincibility, and because of their resistance to microbial degradation, melanins contribute to the humic fraction of soils (Henson et al. 1999, and references within). Yet several studies demonstrate the biodegradability of melanins. A melanolytic fungus could degrade DOPA melanin and utilize it as a sole nitrogen source (Liu et al., 1995); however, the enzymes involved in degradation were not identified. Fungal gnats and another fungus were able to degrade melanized sclerotia of *Sclerotinia sclerotiorum*, but again, specific enzymes were not implicated in this process and it is not clear if the melanin layer of sclerotial rinds was really degraded (Gracia-Garza et al., 1997). The lignin-degrading enzymes of *Phanerochaete chrysosporium*, however, completely oxidized (bleached) either synthetic DOPA or fungal DHN melanins (Butler and Day, 1998). Recombinant manganese peroxidase alone degraded both lignin and melanin, which probably reflects their similarities in chemical structure and genesis (radical polymerization).

7
Production and Applications

Most melanin research is directed toward inhibiting melanization, as several phytopathogens require melanin for virulence. Especially targeted is causative agent of rice blast disease, *Magnaporthe grisea*, whose appressoria must be melanized for host penetration. Hence, the agrochemical industry is continually searching for new compounds that inhibit the fungal DHN melanin biosynthetic pathway without inhibiting plant growth. Most of the world market for these chemicals occurs in areas with high rice production. Among the melanin biosynthesis inhibitors, tricyclazole (Eli Lily Co.; Greenfield, IN), pyroquilon (Novartis; Bagsvaerd, Denmark), phthalide (Kureha Chemical Industry; Tokyo, Japan) and carpropamid (Bayer and Nihon Bayer Agrochem; Yuki, Japan) are currently applied to agricultural fields. All inhibit THN reductase, except for carpropamid, which inhibits scytalone dehydratase. Dupont Agricultural Products is developing an additional SD inhibitor (Jennings et al., 1999, and references within), and compounds that inhibit earlier steps in the pathway are in development as well (Kim et al., 1998). Commercial DOPA melanin extracted from *Sepia officinalis* is available from Sigma (St. Louis, MO). Phenol oxidases such as tyrosinase or laccase also are commercially available from Sigma, as is L-DOPA, and DOPA melanin is easily synthesized in vitro by combining L-DOPA with either oxidase. Dihydroxynaphthalenes other than 1,8 DHN are commercially available from Fluka Chemika (Ronkonkoma, NY), but 1,8 DHN is not currently commercially available, and must be synthesized. Laccase was used to synthesize DHN melanin (Edens et al., 1999), and tyrosinase would probably oxidize DHN into melanin as well. Applications for synthetic melanins include their use as protective agents for other organisms, such as biocontrol agents that are UV sensitive. In addition, melanins and the enzymes that catalyze melanin polymerization are also being developed to remove phenolic pollutants and heavy metals from contaminated water or soil (Payne and Sun, 1994; Majcherczyk et al., 1998). Synthetic melanins are also useful in imaging procedures used to locate melanized cells, including the neurons of the substantia nigra and metastatic melanomas (Sadzot et al., 1999; Williams et al., 1994).

8 Outlook and Perspectives

Recent studies have increased our understanding of the chemical and physical properties of melanin. In addition, genes involved in melanin biosynthesis and its regulation are now identified and we are learning more of the ecology and function of various melanins. Melanin may be exploited for its protective properties to prolong the lives of some organisms, such as biocontrol agents. Other compounds that interfere with its synthesis or promote its degradation may help prevent plant disease. An understanding of animal melanogenesis and its regulation will aid the development of better sunscreens and will improve our detection and successful treatment of melanomas and other pigmentation disorders.

9 Patents

Although patents for melanin synthesis inhibitors are held by their manufacturers, other melanin applications and/or processes are currently unpatented or have patents pending.

10
References

Agodi, A., Stefani, S., Corsaro, C., Campanile, F., Gribaldo, S., Sichel, G. (1996) Study of a melanic pigment of *Proteus mirabilis, Res. Microbiol.* **147**, 167–174.

Allport, D. C., Bu'Lock, J. D. (1960) Biosynthetic pathways in *Daldinia concertrica, J. Chem. Soc.* 654–662.

Bagnara, J. T. (1998) Comparative anatomy and physiology of pigment cells in nonmammalian tissues, in: *The Pigmentary System* (Nordlund, J. J., Boissy, R. E., Hearing, V. J., King, R. A., Ortonne, J.- P., Eds.), pp. 9–40. Oxford: Oxford University Press.

Bell, A. A., Wheeler, M. H. (1986) Biosynthesis and functions of fungal melanins, *Annu. Rev. Phytopath.* **24**, 411–451.

Bechinger, C., Giebel, K.-F., Schnell, M., Leiderer, P., Deising, H. B., Bastmeyer, M. (1999) Optical measurements of invasive forces exerted by appressoria of a plant pathogenic fungus, *Science* **285**, 1896–1899.

Benedito, E., Jiménez-Cervantes, C., Pérez, D., Cubillana, J. D., Solano, F., Jiménez-Cervantes, J., Meyer zum Gottesberge, A. M., Lozano, J. A., García-Borrón, J. C. (1997) Melanin formation in the inner ear is catalyzed by a new tyrosine hydroxylase kinetically and structurally different from tyrosinase, *Biochim. Biophys. Acta* **1336**, 59–72.

Bertazzo, A., Costa, C. V. L., Graziella, A., Favretto, D., Traldi, P. (1999) Application of matrix-assisted laser desorption/ionization mass spectroscopy to the detection of melanins formed from dopa and dopamine, *J. Mass Spec.* **34**, 922–929.

Blarzino, C., Mosca, L., Foppoli, C., Coccia, R., de Marco, C., Rosei, M. (1999) Lipoxygenase/H_2O_2-catalyzed oxidation of dihydroxyindoles: synthesis of melanin pigments and study of their antioxidant properties, *Free Radic. Biol. Med.* **26**, 446–453.

Boissy, R. E. (1998) Extracutaneous melanocytes, in: *The Pigmentary System* (Nordlund, J. J., Boissy, R. E., Hearing, V. J., King, R. A., Ortonne, J.- P., Eds.), pp. 59–73. Oxford: Oxford University Press.

Bourquelot, E., Bertrand, A. (1895) Le bleuissement et le noircissement des champignons. *C. R. Soc.* Biol (Paris) **2**, 582–584.

Bridelli, M. G. (1998) Self-assembly of melanin studied by laser light scattering, *Biophys. Chem.* **73**, 227–239.

Bridelli, M. G., Tampellini, D., Zecca, L. (1999) The structure of neuromelanin and its iron binding site studied by infrared spectroscopy, *FEBS Lett.* **457**, 18–22.

Brush, L., Money, N. P. (1999) Invasive hyphal growth in *Wangiella dermatitidis* is induced by stab inoculation and shows dependence upon melanin synthesis, *Fung. Gen. Biol.* **28**, 190–200.

Bunning, S. E., Griffiths, D. A. (1984) Spore development in species of *Periconia* II. *P. byssoides* and *P. igniaria, Trans. Br. Mycol. Soc.* **82**, 397–404.

Butler, M. J., Day, A. W. (1999) Fungal melanins: a review, *Can. J. Microbiol.* **44**, 1115–1136.

Butler, M. J., Day, A. W. (1998) Destruction of fungal melanins by ligninases of *Phanerochaete chrysosporium* and other white rot fungi, *Int. J. Plant Sci.* **159**, 989–995.

Charalambidis, N. D., Bournazos, S. N., Lambroupousou, M., Marmaras, V. J. (1994) Defense and melanization depend on the eumelanin pathway, occur independently and are controlled differentially in developing *Ceratitis capitata, Insect. Biochem. Molec. Biol.* **24**, 655–662.

Coon, S. L., Kotob, S., Jarvis, B. B., Wang, S., Fuqua, W. C., Weiner, R. M. (1994) Homogentisic acid is the product of *melA*, which mediates melanogenesis in the marine bacterium *Shewanella colwelliana* D, *Appl. Environ. Microbiol.* **60**, 3006–3010.

Cooper, C. R., Szaniszlo, P. J. (1997) Melanin as a virulence factor in dematiaceous pathogenic fungi, in: *Host–Fungus Interplay* (Vanden Bossche, H., Stevens, D. A., Odds, F. C., Eds.), pp. 81–93. Bethesda, MD: National Foundation for Infectious Diseases.

Coyne, V. E., Al-Harth, L. (1992) Induction of melanin biosynthesis in *Vibrio cholerae, Appl. Environ. Microbiol.* **58**, 2861–2865.

deLeo, F., Urzi, C., de Hoog, G. S. (1999) Two *Conidiosporium* species from rock surface, *Studies in Mycology* **43**, 70–79.

Eagen, R., Brisson, A., Breuil, C. (1997) The sap-staining fungus *Ophiostoma piceae* synthesizes different types of melanin in different growth media, *Can. J. Microbiol.* **43**, 592–595.

Edens, W. A., Goins, T. Q., Dooley, D., Henson, J. M. (1999) Purification and characterization of a secreted laccase of *Gaeumannomyces graminis* var. *tritici, Appl. Environ. Microbiol.* **65**, 3071–3074.

Franzen, A. J., de Souza, W., Farina, M., Alviano, C. S., Rozental, S. (1999) Morphometric and densitometric study of the biogenesis of electron-dense granules in *Fonsecaea pedrosoi, FEMS Microbiol. Let.* **173**, 395–402.

Frederick, B. A., Caesar-Tonthat, T.-C., Wheeler, M. H., Sheehan, K. B., Edens, W. A., Henson, J. M. (1999) Isolation and characterization of *Gaeumannomyces gramins* var. *graminis* melanin mutants, *Mycol. Res.* **103**, 99–110.

Furumura, M., Sakai, C., Potterf, B., Vieira, W. D., Barsh, G. S., Hearing, V. J. (1998) Characterization of genes modulated during pheomelanogenesis using differential display, *Proc. Nat. Acad. Sci. USA* **95**, 7374–7378.

Gallas, J. M., Littrell, K. C., Seifert, S., Zajac, G. W., Thiyagarajan, P. (1999) Solution structure of copper ion-induced molecular aggregates of tyrosine melanin, *Biophys. J,* **77**, 1135–1142.

Giambernardi, T. A., Rodeck, U., Klebe, R. J. (1998) Bovine serum albumin reverses inhibition of RT-PCR by melanin, *Biotechniques* **25**, 564–565.

Gilchrest, B. A., Park, H.- Y., Eller, M. S., Yaar, M. (1998) The photobiology of the tanning response, in: *The Pigmentary System* (Nordlund, J. J., Boissy, R. E., Hearing, V. J., King, R. A., Ortonne, J.-P., Eds.), pp. 359–389. Oxford: Oxford University Press.

Gracia-Garza, J. A., Reeleder, R. D., Paulitz, T. C. (1997) Degradation of sclerotia of *Sclerotinia sclerotiorum* by fungus gnats (*Bradysia coprophila*) and the biocontrol fungi *Trichoderma spp, Soil Biol. Biochem.* **29**, 123–129.

Grant, B., Owen, D. F., Clarke, C. A. (1995) Decline of melanic moths, *Nature* **373**, 565.

Haavik, J. (1997) L-DOPA is a substrate for tyrosine hydroxylase, *J. Neurochem.* **69**, 1720–1728.

Hawke, M. A., Lazarovits, G. (1994) Production and manipulation of individual microsclerotia of *Verticillium dahliae* for use in studies of survival, *Phytopathology* **84**, 883–890.

Hearing, V. J. (1998) Regulation of melanin formation, in: *The Pigmentary System* (Nordlund, J. J., Boissy, R. E., Hearing, V. J., King, R. A., Ortonne, J.- P., Eds.), pp. 423–438. Oxford: Oxford University Press.

Hegnauer, H., Nyhlén, L. E., Rast, D. M. (1985) Ultrastructure of native and synthetic *Agaricus bisporus* melanins—implications as to the compartmentalization of melanogenesis in fungi, *Exp. Mycol.* **9**, 221–229.

Henson, J. M., Butler, M. J., Day, A. W. (1999) The dark side of the mycelium: melanins of phytopathogenic fungi, *Annu. Rev. Phytopath.* **37**, 447–471.

Hill, H. Z. (1992) The function of melanin or 6 people examine an elephant, *BioEssays* **14**, 46–49.

Hiraku, Y., Yamasaki, M., Kawanishi, S. (1998) Oxidative DNA damage induced by homogentisic acid, a tyrosine metabolite, *FEBS Lett.* **432**, 13–16.

Howard, R. J. (1997) Breaching the outer barriers – cuticle and cell wall penetration, in: *The Mycota* Vol.5A *Plant Relationships* (Carroll, G., Tudzynski, P., Eds.), pp. 43–60. New York: Springer Verlag.

Ito, S. (1998) Advances in chemical analysis of melanins, in: *The Pigmentary System* (Nordlund, J. J., Boissy, R. E., Hearing, V. J., King, R. A., Ortonne, J.- P., Eds.), pp. 439–450. Oxford: Oxford University Press.

Jacobson, E. S., Hove, E., Emery, H. S. (1995) Antioxidant function of melanin in black fungi, *Infect. Immun.* **63**, 4944–4945.

Jacobson, E. S., Hong, J. D. (1997) Redox buffering by melanin and Fe(II) in *Cryptococcus neoformans, J. Bacteriol.*, **179**, 5340–5346.

Jennings, L. D., Wawrzak, Z., Amorose, D., Schwartz, R. S., Jordan, D. B. (1999) A new potent inhibitor of fungal melanin biosynthesis identified through combinatorial chemistry, *Bioorg. Med. Chem. Let.* **9**, 2509–2514.

Jongmans, A. G., van Breemen, N., Lundström, U., van Hees, P. A. W., Finlay, R. D., Srinivasan, M., Unestam, T., Giesler, R., Melkerud, P.-A., Olsson, M. (1997) Rock–eating fungi, *Nature* **389**, 682–683.

Kappen, L. (1973) Response to extreme environments, in: *The Lichens* (Ahmadjian, V., Hale, M. E., Eds.), pp. 310–380. New York: Academic Press.

Kershaw, K. A. (1974). Studies on lichen-dominated systems. XII. The ecological significance of thallus color, *Can. J. Bot.*, **53**, 660–667.
Kikkawa, H., Ogita, Z., Fujito, S. (1955) Nature of pigments derived from tyrosine and tryptophan in animals, *Science* **121**, 43–47.
Kim, J. -C., Min, J.-Y., Kim, H. T., Kim, B. S., Kim, Y. S., Kim, B. T., Yu, S. H., Yamaguchi, I., Cho, K. Y. (1998). Target site of a new antifungal compound KC 10017 in the melanin biosynthesis of *Magnaporthe grisea*, *Pest. Biochem. Physiol.* **62**, 102–112.
Kipp, C., Young, A. R. (1999) The soluble eumelanin precursor 5,6-dihydroxyindole-2-carboxylic acid enhances oxidative damage in human keratinocyte DNA after UVA, *Photochem. Photobiol.* **70**, 191–198.
Knörle, R., Schniz, E., Feuerstein, T. J. (1998) Drug accumulation in melanin: an affinity chromatographic study, *J. Chrom.* **714**, 171–179.
Kobayashi, N., Nakagawa, A., Muramatsu, T., Yamashina, Y., Shirai, T., Hashimoto, M. W., Ishigaki, Y., Ohnishi, T., Mori, T. (1998) Supranuclear melanin caps reduce ultraviolet induced DNA photoproducts in human epidermis, *J. Invest. Dermatol.* **110**, 806–810.
Koch, P. B., Kaufmann, N. (1995) Pattern specific melanin synthesis and DOPA decarboxylase activity in a butterfly wing of *Precis coenia* Hubner, *Insect Biochem. Mol. Biol.* **25**, 73–82.
Krol, E. S., Liebler, D. C. (1998) Photoprotective actions of natural and synthetic melanins, *Chem. Res. Toxicol.* **11**, 1434–1440.
Kubo, Y., Takano, Y., Tsuji, G., Horino, O., Furusawa, I (2000) Regulation of melanin biosynthesis genes during appressorium formation of *Colletotrichum lagenarium*, in: *Host Specificity, Pathogenicity, and Host Pathogen Interactions of Colletotrichum* (Freeman, S., Prusky, D., Dickman, M., Eds.) pp. 99–113. St. Paul, MN: American Phytopathology Society.
Küchle, M., Mardin, C. Y., Nguyen, N. X., Martus, P., Naumann, G. O. H. (1998) Quantification of aqueous melanin granules in primary pigment dispersion syndrome, *Am. J. Ophthalmol.* **126**, 425–431.
Larsson, B. (1998) The toxicology and pharmacology of melanins in: *The Pigmentary System* (Nordlund, J. J., Boissy, R. E., Hearing, V. J., King, R. A., Ortonne, J.- P., Eds.), pp. 373–390. Oxford: Oxford University Press.
Lichter, A., Mills, D. (1998) Control of pigmentation of *Ustilago hordei*: the effect of pH, thiamine, and involvement of the cAMP cascade, *Fung. Gen. Biol.* **25**, 63–74.
Linossier, G. (1891) L'aspergilline, pigment des spores de l'*Aspergillus niger*, *C. R. Acad. Sci.* **112**, 489–492.
Liu, L., Wakamatsu, K., Ito, S., Williamson, P. R. (1999) Catecholamine oxidative products, but not melanin, are produced by *Crytpococcus neoformans* during neuropathogenesis in mice, *Infect. Immun.* **67**, 108–112.
Liu, Y.-T., Lee, S.-H., Liao, Y.-Y. (1995) Isolation of a melanolytic fungus and its hydrolytic activity on melanin, *Mycologia* **87**, 651–654.
Lu, D., Chen, W., Cone, R. D. (1998) Regulation of melanogenesis by the MSH receptor, in: *The Pigmentary System* (Nordlund, J. J., Boissy, R. E., Hearing, V. J., King, R. A., Ortonne, J.- P., Eds.), pp. 183–198. Oxford: Oxford University Press.
Majcherczyk, A., Johannes, C., Huttermann, A. (1998) Oxidation of polycyclic aromatic hydrocarbons (PAH) by laccase of *Trametes versicolor*, *Appl. Environ. Microbiol.* **62**, 4563–4567.
Mason, H. S. (1948) The chemistry of melanin III. Mechanism of the oxidation of dihydroxyphenylalanine by tyrosinase, *J. Biol. Chem.* **172**, 83–92.
McKeown, T. A., Moss, S. T., Jones, E. B. G. (1996) Ultrastructure of the melanized ascospores of *Corollospora fusca*, *Can. J. Bot.* **74**, 60–66.
McLean, J., Purvis, O. W., Williamson, B. J. (1998) Role for lichen melanins in uranium remediation, *Nature* **391**, 649–650.
Menter, J. M., Patta, A. M., Hollins, T. D., Moore, C. L., Willis, I. (1998) Photoprotection of mammalian acid-soluble collagen by cuttlefish sepia melanin in vitro, *Phytochem. Photobiol.* **68**, 532–537.
Mercado-Blanco, J., García, F., Fernandez-López, M., Olivares, J. (1993) Melanin production by *Rhizobium meliloti* GR4 is linked to nonsymbiotic plasmid pRmeGR4b: cloning, sequencing, and expression of the tyrosinase gene *mepA*, *J. Bacteriol.* **175**, 5403–5410.
Money, N. P. (1999) To perforate a leaf of grass, *Fung. Gen. Biol.* **28**, 146–147.
Money, N. P., Caesar-TonThat, T.-C., Frederick, B. A., Henson, J. M. (1998) Melanin synthesis is associated with changes in hyphopodial turgor, permeability, and wall rigidity in *Gaeumannomyces graminis* var. *graminis*, *Fung. Gen. Biol.* **24**, 240–251.
Mosca, L., Blarzino, C., Coccia, R., Foppoli, C., Rosei, M. A. (1998) Melanins from tetrahydroisoquinolines: spectroscopic characteristics, scavenging activity and redox transfer properties, *Free Radic. Biol. Med.* **24**, 161–167.
Mosca, L., De Marco, C., Fontana, M., Rosei, M. A. (1999) Fluorescence properties of melanins from

opioid peptides, *Arch. Biochem. Biophys.* **371**, 63–69.

Nakano, T., Miyake, K., Ikeda, M., Mizukami, T., Katsumata, R. (2000) Mechanism of the incidental production of a melanin-like pigment during 6-demethylchlortetracycline production in *Streptomyces aureofaciens, Appl. Environ. Microbiol.* **66**, 1400–1404.

Napolitano, A., Palumbo, A., d'Ischia, M., Prota, G. (1996) Mechanism of selective incorporation of the melanin seeker 2-thiouracil into growing melanin, *J. Med. Chem.* **39**, 5192–5201.

Nicolaus, R. A. (1962) Biogenesi delle melanine, *Rass. Med. Sperim.* **9**, 1–32.

Niecke, M., Heid, M., Krüger, A. (1999) Correlations between melanin pigmentation and element concentration in feathers of white-tailed eagles *Haliaeetus albicilla, J. Ornithol.* **140**, 355–362.

Nilges, M. J. (1998) Advances in physical analysis of melanins, in: *The Pigmentary System* (Nordlund, J. J., Boissy, R. E., Hearing, V. J., King, R. A., Ortonne, J.- P., Eds.), pp. 451–460. Oxford: Oxford University Press.

Nofsinger, J. B., Forest, S. E., Simon, J. D. (1999) Explanation for the disparity among absorption and action spectra of eumelanin, *J. Phys. Chem. B* **103**, 11428–11432.

Nosanchuk, J. D., Valadon, P., Feldmesser, M., Casadevall, A. (1999) Melanization of *Cryptococcus neoformans* in murine infection, *Mol. Cell. Biol.* **19**, 745–750.

Novellino, L., Napolitano, A., Prota, G. (1999) 5,6-dihydroxindoles in the Fention reaction: a model study of the role of melanin precursors in oxidative stress and hyperpigmentary processes, *Chem. Res. Toxicol.* **12**, 985–992.

Pawelek, J. M., Chakraborty, A. K. (1998) The enzymology of melanogenesis, in: *The Pigmentary System* (Nordlund, J. J., Boissy, R. E., Hearing, V. J., King, R. A., Ortonne, J.- P., Eds.), pp. 391–400. Oxford: Oxford University Press.

Payne, G. G., Sun, W.-Q. (1994) Tyrosinase reaction and subsequent chitosan adsorption for selective removal of a contaminant from a fermentation recycle stream, *Appl. Environ. Microbiol.* **60**, 397–401.

Pierce, J. A., Rast, D. M. (1995) A comparison of native and synthetic mushroom melanins by fourier-transform infrared spectroscopy, *Phytochemistry* **39**, 49–55.

Prota, G. (1992) *Melanins and Melanogenesis*. Academic Press: San Diego, CA.

Prota, G., Nicoluas, R. A. (1967) Struttura e biogenesi delle creomelanine. Nota I. Isolamento e propieta de; pigment delle plume, *Gazz. Chim. Ital.* **97**, 666–684.

Prota, G., d'Ischia, M., Napolitano, A. (1998) The chemistry of melanins and related metabolites, in: *The Pigmentary System* (Nordlund, J. J., Boissy, R. E., Hearing, V. J., King, R. A., Ortonne, J.- P., Eds.), pp. 307–332. Oxford: Oxford University Press.

Quevedo, W. C., Holstein, T. J. (1998) General biology of mammalian pigmentation, in: *The Pigmentary System* (Nordlund, J. J., Boissy, R. E., Hearing, V. J., King, R. A., Ortonne, J.-P., Eds.), pp. 43–58. Oxford: Oxford University Press.

Raper, H. S. (1927 The tyrosinase-tyrosine reaction, *Biochem. J.* **21**, 89–96.

Raper, H. S. (1928). The aerobic oxidases, *Physiol. Rev.* **8**, 245–282.

Ray, A. C., Eakin, R. E. (1975) Studies on the biosynthesis of aspergillin by *Aspergillus niger, Appl. Microbiol.* **30**, 909–915.

Reedy, M. V., Parichy, D. V., Erickson, C. A., Mason, K. A., Frost-Mason, S. K. (1998) Regulation of melanoblast migration and differentiation in: *The Pigmentary System* (Nordlund, J. J., Boissy, R. E., Hearing, V. J., King, R. A., Ortonne, J.- P., Eds.), pp. 75–96. Oxford: Oxford University Press.

Riley, P. A. (1997) Melanin, *Int. J. Biochem. Cell Biol.* **11**, 1235–1239.

Romero-Martinez, R., Wheeler, M., Guerrero-Plata, A., Rico, G., Torres-Guerrero, H. (2000) Biosynthesis and functions of melanin in *Sporothrix schenckii*, Inf. immun. 68: 3696–3706.

Rosas, Á. L., Nosanchuk, J. D., Gomez, B. L., Edens, W. A., Henson, J. M., Casadevall, A. (2000) Serological methods for the analysis of melanins, *J. Immunol. Meth.*, 244: 69–80.

Rosei, M. A., Blarzino, C., Coccia, R., Foppoli, C., Mosca, L., Cini, C. (1998) Production of melanin pigments by cytochrome c/H_2O_2 system, *Int. J. Biochem. Cell Biol.* **30**, 457–463.

Rözanowska, M., Sarna, T., Land, E. J., Truscott, T. G. (1999) Free radical scavenging properties of melanin interaction of eu- and pheo-melanin models with reducing and oxidising radicals, *Free Radic. Biol. Med.* **26**, 518–525.

Sadzot, B., Sheldon, J., Flesher, J., Dannals, R. F., Schachat, A. P., Frost, J. J (1999) Tracers for imaging melanin with positron emission tomography, *Synapse* **31**, 5–12.

Sänchez-Mirt, A., Romero, H., Fernandez-Zeppendelft, G. (1997) Growth and morphology of *Cladophialophora (Cladosporium) carrionii, J. Mycol. Med.* **7**, 1–4.

Sarna, T., Swartz, H. M. (1998) The physical properties of melanins, in: *The Pigmentary System*

(Nordlund, J. J., Boissy, R. E., Hearing, V. J., King, R. A., Ortonne, J.- P., Eds.), pp. 333–357. Oxford: Oxford University Press.
Schnitzler, N., Peltroche-Llacsahuanga, H., Bestier, N., Zündorf, J., Lütticken, R., Haase, G. (1999) Effect of melanin and carotenoids of *Exophiala (Wangiella) dermatitidis* on phagocytosis, oxidative burst, and killing by human neutrophils, *Infect. Immun.* **67**, 94–101.
Schraermeyer, U. (1994) Transformation of albumin into melanin by hydroxyl radicals, *Comp. Biochem. Physiol.* **108C**, 281–288.
Shibahara, S., Yasumoto, K.-I., Takahashi, K. (1998) Genetic regulation of the pigment cell in: *The Pigmentary System* (Nordlund, J. J., Boissy, R. E., Hearing, V. J., King, R. A., Ortonne, J.-P., Eds.), pp. 251–274. Oxford: Oxford University Press.
Shima, T., Sarna, T., Swartz, H. M., Stroppolo, A., Gerbasi, R., Zecca, L. (1997) Binding of iron to neuromelanin of human substantia nigra and synthetic melanin: an electron paramagnetic resonance spectroscopy study, *Free Radic. Biol. Med.* **23**, 110–119.
Smit, N. P. M., Van der Meulen, H., Koerten, H. K., Kolb, R. M., Mommaas, A. M., Lentjes, E. G. W. M., Pavel, S. (1997) Melanogenesis in cultured melanocytes can be substantially influenced by L-tyrosine and L-cysteine, *J. Invest. Dermatol.* **109**, 796–800.
Solano, F., García-Borrón, J. C. (1998) Advances in enzymatic analysis of melanogenesis, in: *The Pigmentary System* (Nordlund, J. J., Boissy, R. E., Hearing, V. J., King, R. A., Ortonne, J.- P., Eds.), pp. 461–471. Oxford: Oxford University Press.
Sticher, L., Mauch-Mani, B., Metraux, J. P. (1997) Systemic acquired resistance, *Annu. Rev. Phytopath.* **35**, 235–270.
Smythies, J., Galzigna, L. (1998) The oxidative metabolism of catecholamines in the brain: a review, *Biochim. Biophy. Acta* **1380**, 159–162.
Stevens, L. H., Davelaar, E., Kolb, R. M., Pennings, E. J. M., Smit, N. P. M. (1998) Tyrosine and cysteine are substrates for blackspot synthesis in potato, *Phytochemistry* **49**, 703–707.
Sword, G. A. (1999) Density-dependent warning coloration, *Nature* **397**, 217.
Tawfik, A. I., Tanaka, S., DeLoof, A., Schoofs, L., Baggerman, G., Waelkens, E., Derua, R., Milner, Y., Yerushalmi, Y., Pener, M. P. (1999) Identification of the gregarization-associated dark-pigmentotropin in locusts through an albino mutant, *Proc. Natl. Acad. Sci.* **96**, 7083–7087.
Tief, K., Schmidt, A., Beermann, F. (1998) New evidence for presence of tyrosinase in substantia nigra, forebrain and midbrain, *Mol. Brain Res.* **53**, 307–310.
Tsai, H.-F., Wheeler, M. H., Chang, Y. C., Kwon-Chung, K. J. (1999) A developmentally regulated gene cluster involved in conidial pigment biosynthesis in *Aspergillus fumigatus*, *J. Bacteriol.* **181**, 6469–6477.
Wawrzak, Z., Sandalova, T., Steffens, J. J., Basarab, G. S., Lundqvist, T., Lindqvist, Y., Jordan, D. B. (1999) High-resolution structures of scytalone dehydratase-inhibitor complexes crystallized at physiological pH, *Proteins: Struct. Funct. Gen.* **35**, 425–439.
White, L. P. (1958) Melanin: a naturally occurring cation exchange material, *Nature* **182**,1427–1428.
Wiernasz, D. C. (1995) Male choice on the basis of female melanin pattern in *Pieris* butterflies, *Anim. Behav.* **49**, 45–51.
Williams, R. F., Siegle, R. L., Pierce, B. L., Floyd, L. J. (1994) Analogs of synthetic melanin polymers for specific imaging applications, *Invest. Rad.* **29S2**, 5116–5119.
Wilson, K. (2000) How the locust got its stripes: the evolution of density-dependent aposematism, *Tree* **15**, 88–89.
Worley, J. F., Hacskaylo, E. (1959) The effect of available soil moisture on the mycorrhizal association of virginia pine, *For. Sci.* **5**, 267–268.
Zajac, G. W., Gallas, J. M., Cheng, J., Eisner, M., Moss, S. C., Alvarado-Swaisgood, A. E. (1994) The fundamental unit of synthetic melanin: a verification by tunneling microscopy of X-ray scattering results, *Biochim. Biophys. Acta* **1199**, 271–278.
Zemel, E., Loewenstein, A., Lei, B., Lazar, M., Perlman, I. (1995) Ocular pigmentation protects the rabbit retina from gentamicin-induced toxicity, *Invest. Ophthal. & Vis. Sci.* **36**, 1875–1884.
Zheng, Y.-J., Bruice, T. C. (1998) Role of a critical water in scytalone dehydratase-catalyzed reaction, *Proc. Nat. Acad. Sci. USA* **95**, 4158–4163.
Zimmerman, W. C., Blanchette, R. A., Burnes, T. A., Farrell, R. L. (1995) Melanin and perithecial development in *Ophiostoma piliferum*, *Mycologia* **87**, 857–863.

3
Biochemistry of Natural Rubber and Structure of Latex

Dr. Dhirayos Wititsuwannakul[1], Dr. Rapepun Wititsuwannakul[2]
[1] Department of Biochemistry, Faculty of Science, Mahidol University, Rama 6 Road, Bangkok 10400, Thailand; Tel: +66-022455195; Fax: +66-022480375; E-mail: scdwt@mahidol.ac.th
[2] Department of Biochemistry, Faculty of Science, Prince of Songkla University, Hat-Yai, Songkla 90110, Thailand; Tel: +66-074211030; Fax: +66-074446656; E-mail: wrapepun@ratree.psu.ac.th

Biopolymers for Medical and Pharmaceutical Applications. Edited by A. Steinbüchel and R. H. Marchessault

ISBN: 3-527-31154-8

AOS	active oxygen species
ATPase	adenosine 5′-triphosphatase
BI	bursting index
cDNA	complementary DNA
DCPTA	2(3,4-dichlorophenoxy) triethylamine
DMADP	dimethylallyl diphosphate
DOPA	3,4-dihydroxyphenylalanine
FDP	farnesyl diphosphate
GDP	geranyl diphosphate
GGDP	geranylgeranyl diphosphate
GPC	gel permeation chromatography
HMGCoA	3-hydroxy-3-methyl-glutaryl coenzyme A

HMGR	HMGCoA reductase
HMGS	HMGCoA synthase
IDP	isopentenyl diphosphate
kD	kilodalton
mRNA	messenger RNA
MVA	mevalonic acid
MWD	molecular weight distribution
NAD	nicotinamide adenine dinucleotide
NADH	nicotinamide adenine dinucleotide, reduced form
NADP	nicotinamide adenine dinucleotide phosphate
NADPH	nicotinamide adenine dinucleotide phosphate, reduced form
NRL	natural rubber latex
pI	isoelectric point
PIP-DP	polyisoprenyl diphosphate
PR proteins	pathogenesis-related proteins
REF	rubber elongation factor
SALB	South American leaf blight
SRPP	small rubber particle protein
TLC	thin layer chromatography
TSC	total solids content
UDP	uridine diphosphate

1 Introduction

Natural rubber and other polyisoprenoids obtained from plants are high-molecular weight hydrocarbon polymers consisting almost entirely of five-carbon isoprene units, C_5H_8. These polyisoprenoids are major components of the latex synthesized by specially differentiated cells of the plants and other living organisms. Detailed studies on structure of these polymers have shown that double bonds in the rubbers from *Hevea brasiliensis* (*Hevea* rubber tree) and *Parthenium argentatum* (Guayule) are in the *cis* configuration, *cis*-1,4-polyisoprene, and those from gutta and chicle are in the *trans* configuration, *trans*-1,4-polyisoprene. These natural polyisoprenes are synthesized by enzyme-catalyzed polymerization of isoprene units to various different degrees, resulting in a wide range of molecular weights that are dependent upon the sources from which they are derived. *Hevea* rubber is a typical high-molecular weight *cis*-polyisoprene with a very wide range of molecular weight distribution. Guayule rubber is also of high-molecular weight *cis*-polyisoprene, but having a narrow molecular weights distribution and physical properties slightly different from that of *Hevea* rubber. Gutta and chicle are *trans*-polyisoprenes with much lower molecular weights as compared to the other two rubbers (Tanaka, 1991). In general, the molecular weight of these different polyisoprenes can range from a few thousand up to several million Daltons.

Numerous plants belonging to several different families can form rubber latex. The rubber with high-molecular weight polyisoprenes are produced in the latex of about 300 genera of Angiosperms. The milky latex fluid will flow from these plants after a slight incision of the tissues. In *Parthenium argentatum* (Guayule), the latex is produced and stored in the parenchyma cells (Backhaus and Walsh, 1983). However, more commonly in most plants, the latex is produced and stored in the tubular structure known as laticifers. The most prominent and well studied among these plants is *Hevea brasiliensis*. The laticiferous system of the rubber tree has recently been described and extensively reviewed (de Fay et al., 1989).

The latex composition may vary to a large extent among the different plant species. The polymers in the suspension of which it is formed may contain varying proportions of rubber and of different compounds in addition. Of some 12,500 species of laticiferous plants, about 7000 are found to produce polyisoprenes. In most case the polyisoprene is mixed with resin, making it difficult to use when the content of the resin is high. A limited number of rubber-producing plants can be suitably utilized, and only a few species are cultivated and have economic importance. Among them, *H. brasiliensis* is proved to be the best rubber producer. A few hundred milliliters of latex can be obtained from each tree by simply incising the bark, the common practice of tapping the rubber trees.

Hevea brasiliensis is used commercially for the production of natural rubber that is used industrially for various finished products. High-yielding clones of *Hevea* have been selected and developed, leading to the high productivity of the cultivars in rubber plantations, especially in Southeast Asia. To a large extent, plant breeders have bred out the regulatory mechanism that controls the supply of photosynthetically derived carbon sources that are subsequently channeled into rubber formation. This has resulted in plants that produce far more rubber than the native clones of the *H. brasiliensis*. The discovery that latex production in *Hevea* can be stimulated by the plant hormone, ethylene, is clearly of advantage to rubber planters, and has led to a great deal of research into understanding the hormonal stimulation mechanism. Today, an ethylene generator (Ethephon) is commonly used to stimulate *Hevea* latex production. The effect of high rubber production on growth is quite considerable in rubber trees of high-yielding clones, as evidenced in the reduction in girth increment of the trees.

The latex, which contains the rubber particles, accumulates in specialized cells or vessels known as laticifers. In *Hevea*, the rubber is formed and stored in the rings of laticifers in the bark. Anatomoses between adjacent vessels (Figure 1) in the rings allow the latex from a large area of the cortex to drain upon tapping. The opening of the latex vessels by tapping cuts causes the latex to flow out due to the high turgor pressure inside. The latex flow will continue for a certain period of time and subsequently stop due to coagulation of the rubber particles and

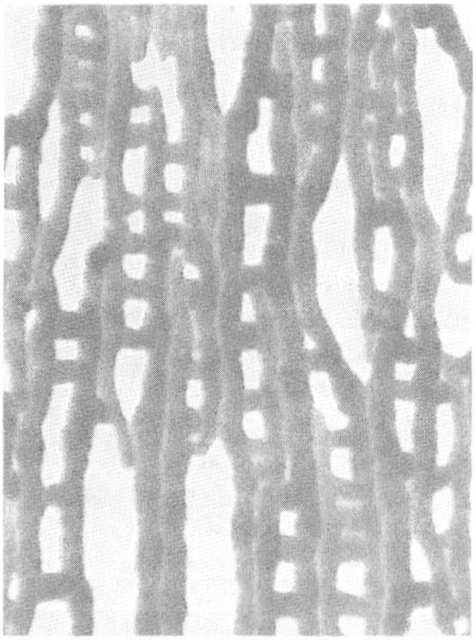

Fig. 1 The latex vessels of *Hevea brasiliensis* laticifers showing anastomoses between adjacent vessels which allow the latex from a large area of the cortex to drain upon tapping of the rubber tree. (Adapted from Zhao, X-Q. (1987) *J. Natl. Rubb. Res.* **2**, 94–98, with permission.).

formation of flocs that leads to plugging at the end of the vessels. The latex is specialized cytoplasm containing several different organelles in addition to rubber particles. These organelles include nuclei, mitochondria, fragments of the endoplasmic reticulum and ribosomes. In addition to these minor components, there are two other major specialized particles which are uniquely characteristic of *Hevea* latex, namely the lutoids and Frey-Wyssling particles.

Lutoids – a major component of latex – are osmotically active spheres which are 1 – 3 μm in diameter. They are gray in color, and surrounded by a single layer membrane. Biochemical analyses revealed that the lipids of lutoids membrane are rich in phosphatidic acids and saturated fatty acyl residues. The name lutoid (yellow) turned out to be a misnomer, because as originally isolated they were contaminated with the yellow Frey-Wyssling particles which contain β-carotene. Frey-Wyssling particles represent a minor component of latex as compared to lutoids. These Frey-Wyssling particles are surrounded by a double-layer membrane and contain many membrane or tubular structures (Figure 2) as well as β-carotene, which is responsible for the characteristic orange to yellow color layer on fractionation by ultracentrifugation of the fresh latex. However, the detailed functions of these particles in latex are still very little known at the present.

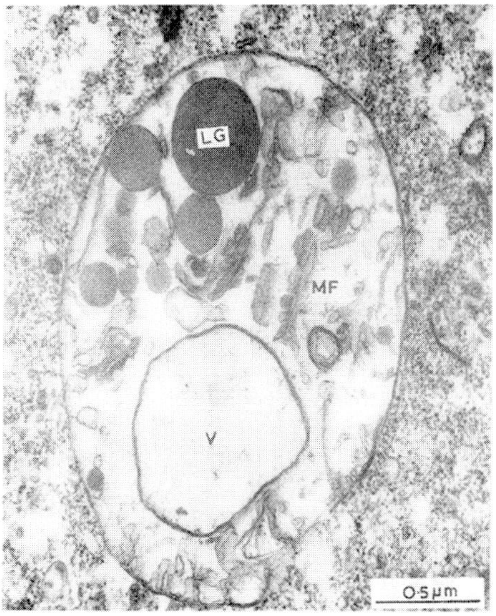

Fig. 2 Frey-Wyssling particles prepared from *Hevea* latex showing the bounded membrane with lipid globules (LG), vesicles (V) and membrane fragments (MF) inside the particles. (Adapted from Gomez, J. B., Hamzah, S. (1989) *J. Nat. Rubb. Res.* **4**, 75–85, with permission.).

This chapter will describe in detail the natural rubber and the latex of *Hevea brasiliensis*. Although the rubbers and latex of other plants and organisms have also been included, they are only covered briefly and less extensively as they will be covered in detail by other authors in this series. This chapter is aimed at providing only background information for comparison with *Hevea* latex. As most of the studies have been performed extensively with the *Hevea* latex over a long period, the focus will be on the plentiful information and details available on this material. Further emphasis will be placed on health problems related to the latex finished products, and to latex allergy caused by latex proteins. Reports of the natural rubber latex allergy have increased steadily since the first reported case in 1979. This is a problem of major health concern and deserves serious attention.

2 Historical Outline

The history of natural rubber and its industrial utilization can be described in two phases. The first phase is the early discovery of natural rubber and its uses in making simple waterproof products. This is followed by improvements in the properties of rubber to make it more suitable and versatile for better-quality products. The second phase is the transfer of the rubber tree from its original site and the establishment of the rubber plantations around the world, especially the predominant rubber plantation industry in Southeast Asia.

2.1 Discovery of Rubber and Improvement of Rubber Properties

Objects made from rubber were first noticed and recorded when Christopher Columbus and his entourage discovered America. The first botanical description of *Hevea* was made by Fresneau in 1747. Subsequently, the names of Hancock, Macintosh, Goodyear, Dunlop and several others have been associated with the history of rubber. Despite the long period since the first discovery of rubber usage by native Indians, rubber is still being regarded as being a strategic material attributed to our civilization. Industrial revolution doubtless provided a major input in contributing to the versatility of natural rubber with regard to the invention and manufacture of rubber products for our daily needs and well-being.

The historical perspective on rubber and rubber trees, as related to the development of the rubber industry, has recently been recounted (Truscott, 1995). Reference to the 'elastic material' was first recorded in the 1500s, when Spanish explorers discovered native inhabitants of the New World using a milky substance to create bouncing toys and also to render their foot coverings and food-handling objects impermeable to water. The milky latex sap was collected from rubber trees, which oozed the rubber sap when their bark was slashed. Unfortunately, samples of these waterproof items did not fare well during the long return journey to Europe. The products degraded, and the resultant gummy substance was of little interest. Indeed, the material was virtually forgotten in recorded history until 1777, when Joseph Priestley pushed a piece of this elastic material across a lead pencil mark on a piece of paper and the mark was rubbed out – thus, the term 'rubber' was first coined. In 1823, Charles Macintosh dissolved lumps of rubber in benzene and poured rubber solution between two pieces of fabric. When the rubber solidified, the resulting material was waterproof and fairly flexible. This simple procedure thus marked the inauguration of the rainwear industry. Although of great use to people living in a temperate, wet climate, the rubberized material was seen to become stiff and brittle in the cold, and sticky and soft in the heat. Both these extremes of conditions led to material breakdown and were tolerated only because an alternative was not yet available.

In 1839, while attempting to create more thermally compatible rubber products, Charles Goodyear inadvertently dropped elemental sulfur into a pan of liquid rubber heating on a stove. The resultant product maintained its elasticity remarkably well when exposed to temperature extremes (Mernagh, 1986), and thus the process of 'vulcanization' was born. The introduction of a practical method of vulcanization, by both Thomas Hancock and Charles Goodyear, changed the nature of rubber and created a new demand for rubber with properties that were satisfactory for industrial use. The incorporation of sulfur, plus admixture of lead oxide and heat, produced a rubber which was proof against hot air and cold air temperatures, and also resistant to melting. Hence, the explosive growth of the rubber industry was started with vulcanized rubber (Brydson, 1978), and this has subsequently resulted in innumerable applications of rubber products, all of which have been supported by patient research not only into the properties of rubber but also improvements in existing manufacturing processes. Vulcanization was much improved by the use of organic accelerators which reduced both the time taken and the amount of sulfur required for the process.

The next hurdle was to move rubber and latex manufacturing into the industrial age. This feat was accomplished by John Dunlop in 1886 while developing a method to produce pneumatic tires. The birth of the tire industry intensified the demand for rubber, the progress of research, and launched the development in manufacturing of latex. Initially, latex was coagulated and processed into compressed blocks and large sheets to await solvent dissolution before pouring it into molds or making it into dipped products. Johnson introduced the method of preservation of liquid latex with ammonia in 1953. This method, together with development of a 50% reduction in the water content of bulk latex by centrifugation, has made feasible the commercial export of concentrated liquid latex (Allen and Jones, 1988).

2.2 History of Rubber Trees and Rubber Plantations

Initially, *Hevea* was not the only candidate for domestication among the many rubber-producing trees. Between 1870 and 1914 four main species underwent extensive trials, including *Ficus elastica* (Asia), *Castilla elastica* (Mexico and West Indies), Ceara rubber *Manihot glaziovii* (Tanganyika and Asia) and *Funtumia elastica* (Africa). None of these plants, however, was sufficiently productive to compete with *Hevea*. The era for rubber production from cultivated *Hevea* and the organized rubber plantation was thus begun.

Although more than 2000 species of plants produce latex, almost 99% of the latex used commercially is harvested from *H. brasiliensis* trees. In 1876, the British East India Company collected 70,000 *Hevea* seeds from their native habitat in the Amazon basin of South America. The seeds were transported to greenhouse facilities in England, where only 2500 germinated. The majority of seedlings were sent to Ceylon (now Sri Lanka), but 22 plants being were taken to Singapore where they formed the foundation for the plantations in Southeast Asia. In retrospect, this endeavor was fortunate because in the early 1900s, the fungus *Microcyclus ulei* virtually destroyed the indigenous-source wild rubber from the jungle trees of the Amazonian rain forest. The South American leaf blight (SALB) caused by this fungus has virtually wiped out commercial plantations there. The blight remain endemic today (de Camargo et al., 1976), and the region has ceased to be a major economic contributor to the world supply of rubber, with most rubber raw material now being produced in Southeast Asia. Thailand has become the number one rubber producer in the world since the early 1990s, though parts of West Africa, South China and Vietnam are increasingly contributing to the world rubber supply.

The organized production of natural rubber to meet industrial needs is now 150 years old. This was preceded by many centuries of localized rubber use in the tropics. Over these years the once simple activity of gathering wild rubber to make a few useful objects has developed into the extensive and complex plantation industry of today. Natural rubber production history (Baulkwill, 1989) can be described in four phases:

1) Pre-industrial but often sophisticated use of rubber from wild plants in tropical America.
2) The organized gathering and export of wild rubber in response to demand from nineteenth-century manufacturers.
3) The establishment of cultivated *Hevea* plantations as the era of the motor vehicle began, and the cultivated crop's domina-

tion of the market in the first half of the twentieth century.
4) The emergence of synthetic rubber in large quantities in the Second World War and its subsequent capture of two-thirds of the market.

An early rubber-based industry, namely the rubber clothing business, boomed and subsequently collapsed due to the unsatisfactory properties of raw rubber, and only the discovery of vulcanization was able to solve this problem. The most startling changes in the demand for rubber in industry came with the adaptation of Dunlop's pneumatic rubber tires for bicycles and motor-powered vehicles. In fact, after 1900 the automobile tire industry began to create a demand for the rubber raw material which 10 years later the wild-rubber industry would be unable to satisfy. This led to the new finding of raw material from wild to cultivated rubber. Early production of rubber was from the lower Amazon (Brazil's Para State), but the domestication of *Hevea brasiliensis* in the East (Asia) has been the most spectacular event in the rubber industry. During the space of only 40 years (1870–1914), a novel plantation industry of some 21 million acres was created to meet the new industrial demand. The successful transfer of *Hevea brasiliensis* to Asia, and the subsequent establishment of commercial rubber plantations there, was the result of many favorable circumstances, the ecological suitability for *Hevea* of several Asian countries in the tropical rain forest belt being the most important contributory factor.

3
Latex of *Hevea brasiliensis*

The *Hevea* latex collected by regular tapping consists of the cytoplasm expelled from the latex vessels, and is similar to the latex in situ. The cytoplasmic nature of tapped latex was firmly established by electron microscopy studies (Dickenson, 1969). Latex is the cytoplasm of an anastomosed cell system which is specialized in the synthesis of *cis*-polyisoprenes. The latex usually contains 25–50% dry matter, 90% of which is made up of rubber. Tapping severs a number of latex vessel rings, and the latex which flows out comprises the contents of vessels at different stages of development. All the organelles as occurring in the latex vessels can be found in the tapped latex. The major particles most common in latex are the rubber particles, the lutoids, and Frey-Wyssling particles which are less numerous than the other two. The composition of latex is about 30–40% rubber, 10–20% lutoids, and 2–3% other substances. The structure and composition of fresh latex has been elucidated by high-speed centrifugation (Moir, 1959). Generally, the latex can be fractionated into three distinct zones (Figure 3) by ultracentrifugation. The top fraction consists almost entirely of rubber, the middle fraction is the metabolic active aqueous phase of latex called C-serum, and the relatively heavy bottom fraction consists mainly of lutoids. The yellow, lipid-containing Frey-Wyssling complexes are normally found at the upper border of the sedimented bottom fraction.

3.1
Composition of *Hevea* Latex

Separation of the latex into three major zones is by ultracentrifugation, after which their chemical compositions can be determined. The top rubber fraction contains, in addition to the rubber hydrocarbon, the proteins and lipids associated with the rubber particles. The serum phase contains most of the soluble substances normally found in the cytosol of plant cells. The bottom fraction can be studied by repeated freezing and thawing of the

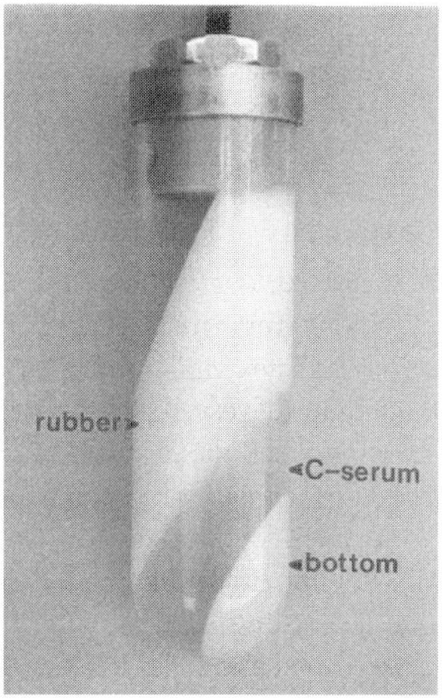

Fig. 3 Fractionation of fresh *Hevea* latex by ultracentrifugation into three major separated zones with the upper rubber layer, the aqueous phase C-serum and the bottom fraction containing lutoid particles.

Tab. 1 Composition of fresh natural rubber latex

Component	*Content* (%, wt/vol)*
Rubber hydrocarbon	25–45
Protein	1–2
Carbohydrate	1–2
Lipids	0.9–1.7
Organic solutes	0.4–0.5
Inorganic substances	0.4–0.6

* The % content of the components varies according to clonal variations of the rubber clones.

lutoids. In this manner the membranes of the lutoids are ruptured and their liquid content, referred to as B-serum, can be analyzed. B-serum has been found to contain proteins and other nitrogen compounds as well as metal ions. It can thus be visualized that the latex is a cytoplasm system consisting of particles of rubber hydrocarbon dispersed in an aqueous serum phase. There are also numerous non-rubber particles called lutoids. The rubber particles are made up of rubber hydrocarbon surrounded by a protective membrane layer consisting of proteins and lipids. The total composition of fresh latex, apart from water, can be summarized as shown in Table 1. Besides the rubber hydrocarbon which is the major component of the latex, various other components (proteins, lipids, carbohydrates and inorganic substances) are also present which play important roles in the latex metabolism and functions.

The details on the compositions of the three major latex fractions (rubber particles, C-serum, and lutoids) and their functions will be extensively presented here. In general, the latex composition can be collectively categorized into two major components, the rubber and non-rubber constituents. The non-rubber constituents of latex will be briefly outlined first.

3.2
Non-Rubber Constituents of *Hevea* Latex

As mentioned, in addition to the rubber, latex contains numerous non-rubber constituents (Figure 4), all of which are present and distributed in all three latex fractions. Proteins and lipids are found associated with the rubber particles, while C-serum contains substances normally found in the cytosol (carbohydrates, proteins, amino acids, inositols, enzymes and intermediates of various biochemical processes, including rubber biosynthesis). Lutoids contain specific substances unique to its functions. Details of the non-rubber constituents have been reviewed recently (Subramaniam, 1995). The following brief discussion will refer to proteins, carbohydrates, and lipids, as well as the inorganic substances.

3.2.1
Proteins

Apart from rubber hydrocarbon and water, proteins and carbohydrates are present in

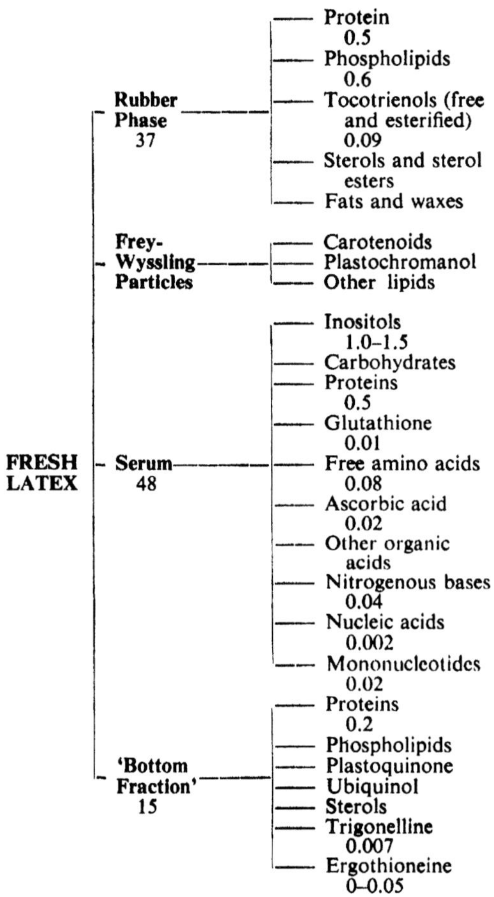

Fig. 4 Organic non-rubber consituents of fresh latex. Values positioned beneath the components indicate their approximate concentration in g per 100 g latex. (Adapted from Archer, B. L. et al. (1969) *J. Rubb. Res. Inst. Malaya* **21**, 560–569, with permission.).

highest proportion in latex. The proteins content in latex shows clonal variations and can range from 1% to more than 1.8% in different samples of latex. In a typical latex sample about 25–30% of the proteins are found in the rubber phase, with 45–50% in the serum phase and about 25% in the bottom fraction. The amount of proteins in the rubber phase is less variable than the total amount of proteins in different samples of latex. The serum proteins consists of around 19 anionic and five cationic proteins (Tata and Moir, 1964). The major protein is α-globulin with an isoelectric point (pI) of 4.8. There are seven anionic and six cationic proteins in the bottom fraction, the major proteins being hevein (>50%, pI 4) and hevamine (~30%, pI 9). The amino acid sequences of hevein (Walujono et al., 1975) and rubber elongation factor (REF), a 14 kDa protein in the rubber phase (Dennis et al., 1989), have been established. Recent report showed more than 200 different proteins are present in the latex (Alenius et al., 1994b), suggesting that the protein composition in the latex is highly complex.

3.2.2 Carbohydrates

Sucrose supply and utilization for latex production plays a very important role in the metabolism of latex, and has been reviewed by Tupy (1989). Of important note and quite unique to the latex is the presence of quebrachitol, an inositol derivative. Quebrachitol (1-methyl inositol) is the most abundant, and makes up about 75–95% of the total carbohydrates present in latex, being found mainly in the serum phase. The large amount and ubiquitous presence of this compound is a unique characteristic of the *Hevea* latex. The reason for its accumulation and its physiological function in latex is not known, but it has been postulated to serve an active role in rubber biosynthesis (Bealing, 1976). About six to seven other carbohydrates components are found in small amounts; these are mostly common sugars for various metabolic processes in the latex.

3.2.3 Lipids

Lipids in the latex play an important role in the stability of rubber particles. They are found not only associated with the rubber particles, but also throughout the latex fractions. The lipids content of latex also shows clonal variations (Ho et al., 1976). The neutral lipids constitute more than 59% of the total lipids, while the other components are mainly phospholipids and glycolipids. The presence of proteolipids have also been reported. The lipids occur mainly in the rubber particles and the bottom fraction. Several neutral lipids found are triglycerides (also mono- and diglycerides), free fatty acids and esters, sterols, and lipid-soluble vitamins (carotenoids and tocotrienols). Phosphatidylcholine is the major phospholipid, while phosphatidylethanolamine and phosphatidylinositol are found in smaller amounts. Phosphatidic acids are also reported as important components of the lutoid membrane (DuPont et al., 1976). Fatty acids in the latex occur mostly in the esterified form, the major fatty acids being C_{16}, C_{18} and C_{20} (palmitic, palmitoleic, stearic, oleic, linoleic, linolenic, and arachidonic acids). Also present is a rare furanoic acid (10,13-epoxy-11-methyloctadeca-10,12-dienoic acid) which is found mainly in the triglycerides fraction. Rubber latex is only the second known plant source of this furanoic acid (Hasma and Subramaniam, 1978). It is somewhat unusual to find this fatty acid in the latex, and its physiological function is not yet known. This situation is similar to that seen with quebrachitol, which is present in large quantities, though its function has yet to be determined.

3.2.4
Inorganic Substances

In addition to the major non-rubber constituents described above, several inorganic components are also present. Of these, potassium is the most abundant element in the latex, its concentration being of the order of a few thousand part per million (ppm). The next most common element (a few hundred ppm) is magnesium, this being mainly contained in the lutoids. Magnesium was found to reduce the mechanical stability of the latex (Philpott and Wesgarth, 1953). Magnesium and potassium also show clonal variations. Other elements occurring in much smaller concentrations are the more common sodium, calcium, iron, copper, manganese, and zinc. Also present is rubidium, though the function of this element in latex is not yet known.

3.3
Rubber Particles in *Hevea* Latex

Rubber hydrocarbon is the major component of *Hevea* latex, the rubber content varying from 25% to 45% as dry content of latex. The average molecular weight ranges from 200 to 600 kDa. The rubber molecules are found as particles in the latex (Figure 5), these consisting mainly of rubber (90%) in association with lipophilic molecules (mainly lipids and proteins), forming the film that encloses the rubber particles (Ho et al., 1976). This film carries negative charges and is responsible for the stability of rubber particles when suspended in aqueous serum. The size of the particles ranges from 5 nm to 3 μm, and they are spherical in shape. They also show plasticity as they have polygonal shapes in mature laticifers, where the particles are numerous. The size distribution, as determined by electron microscopy, showed a maximum distribution of 0.1 μm particles (Gomez and Moir, 1979), each of which may contain several hundred rubber molecules. Molecular weight analyses using gel permeation chromatography showed a bimodal distribution of rubber of low and high molecular weights, with average values of 100–200 kDa and 1000–2500 kDa, respectively (Subramaniam, 1976). The other main component of rubber particles is the enclosing membrane, which consists of lipids, proteins, and enzymes. These components contribute colloidal charge to the rubber

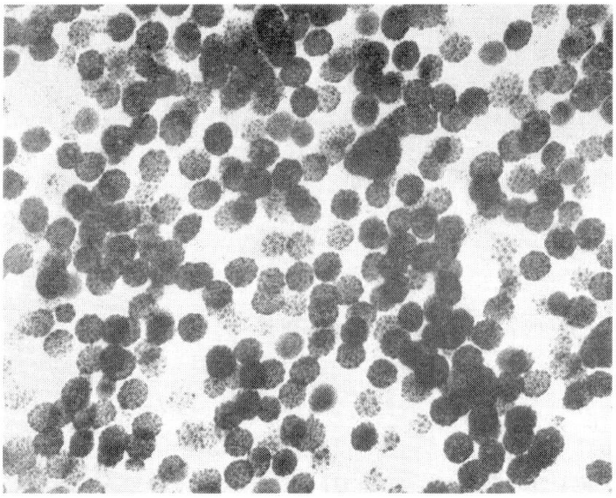

Fig. 5 Electron micrograph of rubber particles showing the spherical shape and uniform structure of the separated small rubber particles. (Adapted from Dickenson, P. B. (1969) *J. Rubb. Res. Inst. Malaya* **21**, 543–559, with permission.).

particle and their stability in the latex. Each component will be described in greater detail later in the chapter.

3.4 Rubber Particle Membrane

The rubber particles are commonly found in association with lipids, which is thought to be of membrane nature. Microscopically, the particles appear to have a uniform structure, with the rubber molecules enclosed by a thin film (Southorn, 1961). When examined by electron microscopy, the rubber particles appear homogeneous, and have a uniform internal structure, but are surrounded by a film that is more opaque than the polyisoprenes contained inside (Lau et al., 1986). Analyses of the nature of the film enclosing the rubber particles show the presence of phospholipids and proteins, together with neutral lipids similar to the membrane structure. The detailed description of rubber particles and other components has been reviewed (d'Auzac and Jacob, 1989), and the composition of the membrane of rubber particles separated and purified by ultracentrifugation analyzed. The membrane components comprise lipids, proteins, enzymes, and charges (see below).

3.4.1 Rubber Particle Lipids

Analyses of rubber particles purified by ultracentrifugation showed them to contain up to 3.2% total lipids, of which ~2.1% are neutral lipids expressed as rubber weight (Ho et al., 1976). Separation of neutral lipid showed it to be composed of at least 14 different substances. Triglycerides were the most abundant, accounting for almost 45% of the neutral lipids, while sterols, sterol esters and fatty acid esters constituted about 40%. Other neutral lipids present in trace amount were diglycerides, monoglycerides, and free fatty acids. In addition, tocotrienols and some phenolic substances were also found to be associated with rubber particles (Ho et al., 1976).

Phospholipids are important components of the rubber particles. Marked differences between clones in neutral lipids content of the rubber phase were noted. In contrast, the phospholipid content did not vary much among the different clones (Ho et al., 1976). Analyses of the phospholipids consistently showed three spots on thin-layer chromatography (TLC) separation; these were identified as a considerable quantity of phosphatidylcholine and smaller quantities of phosphatidylethanolamine and phosphatidylglycerol. Phosphatidic acid was found to be predominant in the membrane of lutoids, but was not detected on the rubber particles, even though the precaution was taken of inhibiting phospholipase D activity (Dupont et al., 1976). In addition, the presence of sphingolipids and glycolipids has also been reported. The stability of rubber particles suspension in latex is dependent on the negative charges film of proteins and phospholipids (Philpott and Wesgarth, 1953).

3.4.2
Rubber Particle Proteins

Proteins are found as indigenous components of the film enclosing the rubber particle. Together with lipids, these proteins form the membrane of particles which contribute to their stability. The pI of these proteins ranges from 3.0 to 5.0 which is characteristic for surface proteins. In an electric field, particles will move toward the anode, indicating that they have net negative charge on the surface (Verhaar, 1959). Anionic soaps do not affect the particles' colloidal stability, but cationic soaps cause flocculation, probably due to neutralization of the surface charge. These proteins can be considered as intrinsic or peripheral, depending on their binding and affinity with rubber particles. One of the most plentiful proteins in latex is α-globulin with pI of 4.5. It was found both in the cytosol and adsorbed onto the particles' surface, and might contribute to their colloidal stability in latex (Archer et al., 1963a). A protein group of hydrophobic nature was also found in rubber particles, and proteolipids have been isolated and characterized (Hasma, 1987). This protein was suggested to be a component of the polar lipid backbone that forms part of the membrane of rubber particles. The protein content of rubber particles was recently refined for a more accurate quantitative analysis (Yeang et al., 1995) in light of the concern over rubber protein allergens.

Other proteins with enzyme activity have also been described. One interesting and well-characterized enzyme is rubber transferase, which was first detected on the washed rubber particle surface (Audley and Archer, 1988). This enzyme is involved in the synthesis and formation of rubber molecules on the particles' surface, and was found to be distributed between the cytosol and rubber particles, similar to the case of α-globulin. It has been isolated and purified from latex C-serum, but was active only when adsorbed onto particles for the chain elongation process of rubber molecules (Light and Dennis, 1989). Another important protein found to be actively involved in rubber synthesis was a 14 kDa protein referred to as rubber elongation factor (REF). The amino acid sequence of REF has been determined, and the molecular cloning of the *REF* gene also carried out (Attanyaka et al., 1991). More recently, molecular cloning of a protein that is tightly bound onto the small rubber particles has been reported (Oh et al., 1999). The cloned cDNA encodes a 24 kDa protein which is tightly bound onto the rubber particles. This protein was suggested to be active in the synthesis of rubber, together with the REF. The 24 kDa protein has been reported previously as a potent latex allergen, and is always found together with REF, bound so tightly to particles that they cannot be removed even by extensive washing. In addition to these tightly bound and well-characterized proteins, some peripheral proteins have also been described, though their function is unknown. The implication of these particle-bound proteins causing allergic reactions towards rubber products, especially latex gloves, will be discussed later.

3.4.3
Rubber Particle Enzymes

Some of the rubber particle proteins with enzyme activity were described briefly in the preceding section. Washed rubber particles and the bound rubber synthesis enzymes have been studied and reviewed (Audley and Archer, 1988). The formation of rubber molecules, at least in terms of the elongation steps, occurs at the particles' surface (Archer et al., 1982), and rubber transferase is the enzyme responsible for this process. This enzyme was also found in the C-serum, and is most likely distributed between the two fractions (McMullen and McSweeney, 1966). It has been isolated from the C-serum and purified for enzyme characterization (Archer and Audley, 1967). The enzyme was without activity in the absence of washed rubber particles, remaining inactive while not adsorbed onto the particles, even when the latter have been purified by gel filtration and washed repeatedly. The reaction catalyzed by rubber transferase appears to be essentially chain extension or elongation of the pre-existing rubber molecule, although a role in the formation of new rubber molecules has also been suggested (Lynen, 1969). However, the mechanism by which new rubber particles are formed is still unknown at present.

To date, it has not been possible to demonstrate the polymerization of isopentenyl diphosphate (IDP) in vitro except on the pre-existing rubber particles. The reaction that occur in vivo must do so at some other sites as a prerequisite on initiation that preludes the formation of new rubber molecules. Some reports have suggested vehemently that membrane phospholipids may have a key role in enabling the combined functions of different transferase enzymes to operate in vivo for new rubber formation (Keenan and Allen, 1974; Baba and Allen, 1980). It appears that some phospholipids, membranes, or other amphipathic micelles are essential for the initiation and formation of new rubber molecules. The rubber particles' surface was also suggested as being the site of IDP isomerase (Lynen, 1969), the enzyme catalyzing the conversion of IDP into dimethylallyl diphosphate (DMADP). The IDP isomerase is essential for the formation of DMADP and the chain initiation of new rubber molecules. It has long been suggested to be present in latex, but only indirect evidence has been provided for its detection. More recently, the direct detection and characterization of this enzyme was reported (Koyama et al., 1996), the isomerase being located in the cytoplasm or C-serum of latex, and not with the rubber particles as was previously suspected. The enzyme was activated in the presence of reducing agents and detergents (Koyama et al., 1996). These findings might provide partial support for the possible initiation of rubber chain elongation at a site other than on the rubber particles.

3.4.4
Rubber Particle Charges

The colloidal stability of latex is attributed to the presence of surface charges on the rubber particles. The film or membrane surrounding the particles provide them with a negative charge, as shown by surface potential or zeta potential (Southorn and Yip, 1968). The particle membrane composition of proteins, phospholipids, and other substances has been determined as described above, and shown to be of a negatively charged nature (Ho et al., 1976). The reduction in phospholipid content of a cloned latex known for its instability has also been noted, and this observation was subsequently extended to show that the lipid content of rubber particles correlated positively with colloidal stability of the latex (Sherief and Sethuraj, 1978). The colloidal stability was reduced by magnesium released from the damaged or ruptured lutoids (Philpott and Wesgarth, 1953) as the surface charges were neutralized. The effect of inorganic cations was investigated in relation to flocculation of the rubber particles and to plugging of the latex vessels, with a negative effect on latex flow (Yip and Gomez, 1984).

3.5 C Serum of *Hevea* Latex

The C-serum fraction of the centrifuged latex (see Figure 3) is the aqueous phase of laticiferous cytoplasm, and this can be considered as the latex cytosol. This cytosol is not fundamentally different from the normal cytosol of plant cells. Analyses of the latex cytosol fraction obtained by ultracentrifugation show that it contains various different organelles and particles. All glycolytic enzymes (d'Auzac and Jacob, 1969) and other common cytosolic enzymes including those of the isoprenoid pathway (Suvachittanont and Wititsuwannakul, 1995; Koyama et al., 1996) have been detected, indicating that the cytosol is active in a number of metabolic processes.

The involvement of C-serum in rubber biosynthesis was also noted (Tangpakdee et al., 1997a), as was the importance of calcium binding protein (calmodulin) in controlling many different metabolic processes. Calmodulin was found to activate HMG-CoA reductase in the bottom fraction, and purification and characterization of calmodulin from C-serum (Wititsuwannakul et al., 1990b) showed it to have an important role in the regulation of latex metabolism. Moreover, a highly positive correlation between calmodulin level and latex yield was also demonstrated (Wititsuwannakul et al., 1990b). The composition of latex cytosol has been reviewed (d'Auzac and Jacob, 1989), and the presence of many high-molecular weight compounds, low-molecular weight organic solutes and mineral elements are also well documented (see below).

3.5.1 High-Molecular Weight Compounds in C Serum

The high-molecular weight compounds in C-serum (see Figure 4) are mainly proteins and specific nucleic acids. The distribution of proteins in whole latex is approximately 20% in the rubber phase, 20% in the bottom fraction, and 60% in the C-serum (Archer et al., 1963a). A large number of proteins are present, as revealed by gel electrophoresis (Tata and Edwin, 1970). One C-serum protein that is present at the highest level is α-globulin, the pI of this (4.55) being similar to that of α-globulin from other sources. α-Globulin has high binding affinity for adsorption onto rubber particles, and was suggested to be one of the particle proteins contributing to the particles' colloidal stability (Archer et al., 1963a). Several enzymes present in C-serum were reported, as described above; thus, those proteins with enzymatic activities are also the major component. These enzymes and their metabolic activities in latex will be discussed in details below.

Nucleic acids are another group of another high-molecular weight compounds present in the latex C-serum, while the presence of in-vitro active ribosomes in latex showed that latex cytosol also contains many nucleic acids (McMullen, 1962). These nucleic acids (soluble RNA and DNA) were found mainly in the serum fraction of centrifuged fresh latex (Tupy, 1969). Analyses of the nucleic acids show both clonal variation and treatment of rubber trees as expressed in the latex. Wound-induced accumulation of mRNA for hevein protein has also been reported (Broekaert et al., 1990), this being the defense response that results from tapping of the rubber trees. The latex cytosol was also found to show differential expression of laticifer-specific genes that have been extensively characterized (Kush et al., 1990). These genes were also found to be influenced by latex production and hormone treatments. Specific mRNAs separated as poly(A) RNA was shown to be induced and elevated by the effect of ethylene, with an ethylene-induced increase and accumulation of glutamine synthetase and mRNA level having been noted (Pujade-Renaud et al., 1994).

3.5.2
Low-Molecular Weight Compounds in C Serum

The metabolic importance of C-serum components is well noted (Jacob et al., 1989). Latex cytosol contain numerous groups of low-molecular weight organic solutes (see Figure 4) that are important in latex metabolic activities, either as precursors or metabolic intermediates. They comprise sugars, amino acids, nucleotides, and nitrogenous bases among many others. Sucrose is the main sugar in latex, together with smaller amounts of glucose, fructose and raffinose. These sugars are important for glycolysis and have been shown to be very active in latex (d'Auzac and Jacob, 1969; Tupy, 1973). In addition, inositols are also prominently present in latex cytosol. Quebrachitol (1-methyl inositol) is found in very high content, comprising >75% of the total carbohydrates in the latex. The presence of quebrachitol is closely correlated with the *cis*-polyisoprene level in the latex, and it was suggested that quebrachitol might undergo catabolism for rubber biosynthesis as an alternative to sucrose (Bealing, 1969, 1976). The pentose phosphate pathway was also found to be very active (Tupy and Resing, 1969) in supplying NADPH for synthesis of rubber in the latex. Channeling of sucrose for different metabolic processes in latex was supported from these findings (Bealing, 1976). C-serum contains all the common amino acids, but in different proportion (Yong and Singh, 1976). The concentration in cytosol (~30 mM) is similar to that of amino acids in lutoid B-serum. However, the distribution of neutral, acidic and basic amino acids in the cytosol is quite different from that in lutoid serum, the acidic amino acids being predominant in the C-serum and the basic amino acids in B-serum. The major amino acids in the cytosol are constituted mainly (up to 80%) by alanine, aspartic, and glutamic acid and its amide (Yong and Singh, 1976).

The latex cytosol contains all the known nucleotides for metabolic activities of latex. Adenine nucleotides attribute to an average energy charge of around 0.6 in the C-serum, while average values for ATP, ADP, and AMP were around 125, 133, and 55 nmol g^{-1} dry rubber, respectively (d'Auzac and Jacob, 1989). Concentrations of coenzymes in the latex (coenzyme A, NAD, and NADP) have been measured. Coenzyme A is indispensable in the biosynthesis of lipids and isoprenoids, while concentrations of cytosolic NAD and NADPH are 70 μM and 2 μM, respectively (Archer et al., 1969). The low content of NADPH in latex might serve as regulatory cofactor in latex. NADPH is a specific electron donor for the reduction of HMG-CoA to mevalonic acid (MVA) by the enzyme HMG-CoA reductase (HMGR), a possible rate-limiting step in the biosynthesis of rubber (Benedict, 1983). The important role of HMGR in *Hevea* rubber biosynthesis has been extensively investigated. A correlation between diurnal variations of HMGR activity and rubber content has been verified (Wititsuwannakul, 1986), as has the purification of *Hevea* latex HMGR (Wititsuwannakul et al., 1990a) and activation by cytosolic calmodulin (Wititsuwannakul et al., 1990b). Likewise, a correlation between levels of HMGR and rubber yield has also been documented (Wititsuwannakul et al., 1988).

In addition, the presence of UDP-glucosamine and UDP-galactosamine in latex have been reported (Archer et al., 1969). These nucleotides may involved in the glycosylation of glycoproteins in latex, though this was also implied for the lectin function of latex. The *Hevea* lectin was recently isolated, purified and characterized (Wititsuwannakul et al., 1998), and implicated in rubber coagulation and vessels plugging (Gidrol et al., 1994). In addition to these key molecules, the latex cytosol also contains several other small molecules, including many organic acids of which malic and citric acids comprise almost 90% of the total organic acids in latex. Several reducing agents are also present in latex cytosol, the major participants being glutathione, cysteine, and ascorbic acid (0.72, 0.44, and 1.9–3.9 mM, respectively) (d'Auzac and Jacob, 1989). These reducing compounds are essential for maintaining the redox potential of latex.

3.6 Lutoids of *Hevea* Latex

Lutoids are major membrane-bound particles that are sedimented in the bottom fraction (see Figure 3) of the centrifuged fresh latex (Moir, 1959). They are considered as polydispersed lysosomal vacuoles and have been extensively discussed (d'Auzac and Jacob, 1989). The lutoid content of latex is quite considerable, and has been assigned a number of important functions in the latex. Lutoids constitute ~20% by volume of fresh latex, whereas the rubber phase forms an average of ~30–40%. Lutoids are spherical in shape (Figure 6), and their diameter is larger than that of rubber particles. Lutoids are enclosed by single membrane, the lipids composition of which has been determined. They play an important role in the colloidal stability of latex due to the negatively charged membrane which is very rich in phosphatidic acids content (Dupont et al., 1976). Lutoids are considered to be similar to lysosomes, with a high acid hydrolase content (Jacob et al., 1976), and play a major role in the coagulation of rubber particles in colloidal suspensions of latex (Southorn and Edwin, 1968). The mechanism of this coagulant effect is due to the release of cations and proteins from ruptured lutoids (Pakianathan and Milford, 1973). The storage function of lutoids was shown in the accumulation of several proteins, enzymes and solutes with active membrane transport activity (Chrestin and Gidrol, 1986). Lutoids are an important latex component, both with regard to volume and the functions of their

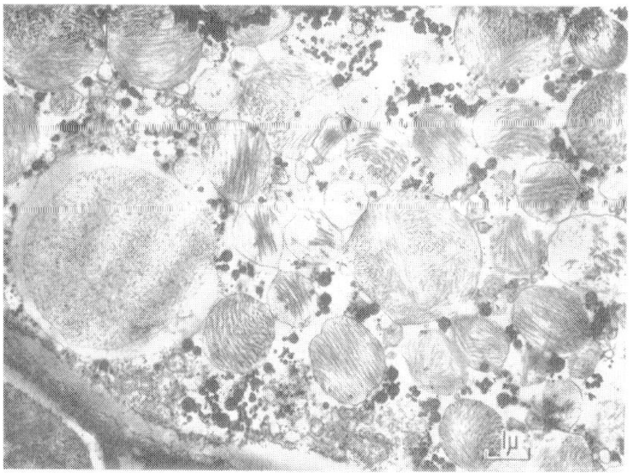

Fig. 6 Lutoid particles of the fresh *Hevea* latex showing the spherical shape of a larger diameter than the rubber particles, with full development of the enclosed proteinaceous microfibrils. The unit membrane surrounding the particles is contorted because of slight plasmolysis as the lutoids are quite osmosensitive. (Adapted from Dickenson, P. B. (1969) *J. Rubb. Res. Inst. Malaya* **21**, 543–559, with permission.).

chemicals and enzymes (d'Auzac and Jacob, 1989). Their composition can be considered as having two distinct components, namely the lutoids membrane and the internal contents, referred to as B-serum.

3.6.1
Lutoids Membrane

Cytological observation, together with biochemical and physiological research on lutoids membrane have provided data concerning the structure, composition, and role of these organelles in latex. Electron microscopy showed the micellar nature of the lutoids membrane structure (Gomez and Southorn, 1969), this being a highly osmosensitive, single-layer membrane of 8–10 nm thickness. Analysis of the membrane chemical composition showed the phospholipid content to constitute 37.5% of the weight of proteins. The lutoid membrane has been shown to be very rich in phosphatidic acids, these accounting for 82% of the total phospholipids fraction. The fatty acids composition (analyzed by methanolysis) of phospholipids showed a predominance of saturated fatty acids, mainly palmitic and stearic, with unsaturated fatty acids (oleic, linoleic acid) also present (Dupont et al., 1976). There were almost equivalent quantities of saturated and unsaturated fatty acids in the membrane. The exceptionally high phosphatidic acid content may explain the high negative surface charges of the lutoid (Southorn and Yip, 1968). The relative abundance of saturated fatty acids in lutoids stands out clearly from that in the membranes of other plant organelles. The relative rigidity and fragility of the lutoids membrane to osmotic shocks, and its low resistance to mechanical stress (Pakianathan et al., 1966), can be partly explained by the membrane's fatty acids composition. The analysis of rubber particles membrane carried out under the same conditions shows that the membrane is totally free of phosphatidic acids, but does contain mainly phosphatidylcholine and phosphatidylethanolamine (Ho et al., 1976). The other major lutoid membrane components are proteins, several of which are enzymes important in the lutoids' function. Thus, the lutoid membrane has been shown to play an essential and highly complex role in latex (d'Auzac and Jacob, 1989; Paardekooper, 1989).

3.6.2
Lutoids Membrane Proteins and Enzymes

Several proteins are present in the lutoid membrane, and many of these are active enzymes. One well-characterized membrane enzyme is ATPase (Moreau et al., 1975). The electron transport activity of lutoids has been linked to ATPase, which is activated by several anions, this in turn leading to an accumulation of anions within the lutoid compartment. ATPase also operates as a proton pump to maintain proton gradients between the lutoids and latex cytosol – a function which was demonstrated by Chrestin and Gidrol (1986) and extensively reviewed (Chrestin et al., 1986). Other membrane enzymes included NADH-cytochrome c reductase; this functions in an outward proton-pumping redox system that tends to reduce the concentration of protons in lutoids and hence acidify the cytosol (Moreau et al., 1975). NADH-quinone reductase (d'Auzac et al., 1986) has also been described as being responsible for the production of superoxide ions. The generation of free radicals and the consequent effect on lutoid integrity and latex colloid stability has been the subject of recent interest and speculation.

Recently, the enzyme HMG-CoA reductase has been purified from lutoids membrane by solubilization with mild detergent (Wititsuwannakul et al., 1990a). Characterization of the purified enzyme was carried out by determining its molecular structure and properties. The native enzyme was found to be a tetramer of four 44 kDa subunits – as found for other plant specimens – and membrane-bound (Bach, 1986). HMG-CoA was found to be activated by reducing agents for maximum activity. In C-serum, regulation of this enzyme by calmodulin was also studied in detail (Wititsuwannakul et al., 1990b), the results suggesting important interactions of C-serum contents in order to maintain optimal lutoids membrane activities. Rubber formation by the bottom fraction was also recently reported (Tangpakdee et al., 1997b), and this help to explain the participation of lutoids in the rubber biosynthesis process. A number of other different lutoid functions have been reviewed, and various lutoid membrane enzymes characterized, by d'Auzac and Jacob (1989), and therefore these aspects will be described here only briefly. Details of the structural proteins of the lutoid membrane are poorly understood, and still await further characterization.

3.6.3 B-Serum of Lutoids and Its Composition

The inside content of lutoids is called B-serum, and this comprises of several different components, including microfibrils, proteins and enzymes, as well as various small-molecule compounds. Microfibrils (Figure 7) are cluster of proteins with helical structure when seen under the electron microscope (Archer et al., 1963b; Dickenson, 1965). Characterization of their structure showed the presence of carbohydrate up to 4% and the presence of acidic proteins with a pI of 4 (Audley, 1965; Gomez, 1976). These fibril proteins may completely fill the space inside the lutoids, and it has been speculated that such proteins may function as a nitrogen reserve (Dickenson, 1969) that may be degraded by lutoid proteases. Microhelical, spring-shaped proteins have also been identified; these consist of basic proteins of 22 kDa and an acidic assembling protein of ~160 kDa in microhelical cluster formations (Gomez and Tata, 1977). Clonal variation of microhelical content was also observed. In addition to the clustered insoluble proteins, other soluble proteins have also been characterized, with electrophoretic analyses showing the presence of at least seven to eight different proteins. Among these, hevein was the most abundant, comprising 70% of the total proteins (Archer, 1960; Audley, 1966).

Hevein is quantitatively the most important anionic proteins, its special feature being that it is a very small protein (5 kDa). Hevein

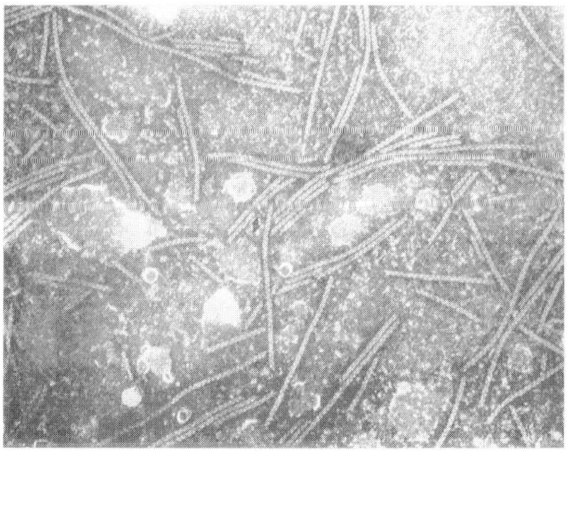

Fig. 7 The extracted microfibrils in B serum of the lutoids, showing the long proteinaceous helical microfibrils. The microfibrils are clusters of proteins with double-helical structure as observed by electron microscopy. (Adapted from Dickenson, P. B. (1969) *J. Rubb. Res. Inst. Malaya* **21**, 543–559, with permission.).

has been purified, crystallized, and characterized (Tata, 1976), and its amino acids composition and sequence determined (Walujono et al., 1975). Its high sulfur content is contributed by cystine alone. In light of its low molecular weight and high cysteine content, hevein has been considered possibly to function as a protease inhibitor similar to those detected in the vacuoles of potato and other *Solanaceae*. However, no protease inhibitor activity of hevein could be detected by a variety of tests (Walujono et al., 1975). The role of these fibril proteins and hevein, and the significance of their presence in lutoid, is not understood, although recently the involvement of hevein in the coagulation of rubber particles was suggested (Gidrol et al., 1994). Besides hevein, two other basic proteins that occur in high concentration in lutoids have been detected and characterized as hevamine A and hevamine B (Archer et al., 1969). The lutoids have considerable influence on the metabolism of latex and its regulation by means of the exchanges which take place between their contents and the latex C-serum. Several enzymes are present in B-serum that are important in metabolic processes and homeostasis of the latex.

3.6.4
B-Serum Enzymes of the Lutoids

Several typical vacuole enzymes can be detected in B-serum of the lutoids. Lysozyme, which is commonly found in egg white and hydrolyzes mucopolysaccharides and bacterial cell walls, was detected in the latex bottom fraction (Archer et al., 1969). It was located in B-serum and considered to be the same as the hevamines, which were shown to be abundant basic protein in lutoids (Tata et al., 1976). B-serum lysozyme has been purified and characterized, and the enzyme's kinetics investigated and partial amino acid composition and sequence determined (Tata et al., 1983). Lysozyme is generally implicated in the defense against antimicrobials, and several acid hydrolases (phosphodiesterase, acid phosphatase, ribonuclease, cathepsin, β-glucosidase and β-galactosidase) were reported to be present in B-serum (Archer et al., 1969). The presence of these enzymes shows similarity between lysosomes in animals, and lutoids. The role of the lutoids' content in stability and the flow of latex was demonstrated (Southorn and Edwin, 1968), and shown to cause rubber particles to coagulate, leading to the plugging of latex vessels and stopping of latex flow (see below). Acid phosphatase was found to be a lutoid-specific enzyme and to appear in the latex cytosol only after the lutoids had been damaged or ruptured. The ratio of free acid phosphatase and the total enzyme activities reflects the stability of lutoids; this ratio is referred to as the bursting index (BI), and influences the flow of latex. A high correlation between BI and vessel plugging, as measured by latex flow duration, has been documented and reported for seasonal and clonal variations (Yeang and Paranajothy, 1982).

Details of oxidation-reduction enzymes in B-serum have been reviewed (d'Auzac and Jacob, 1989). These enzymes include catalase, phenol oxidase and tyrosinase, while peroxidase enzymes such as monophenol, polyphenol and DOPA oxidases have been detected in B-serum as cationic proteins. These enzymes are always found together with their inhibitors, and it was suggested that compartmentalization in the lutoids might occur in order for these enzymes to be regulated. In addition, the enzymes of pathogenesis-related proteins such as chitinase and β-1,3-glucanase were also found to be accumulated in lutoids (Churngchow et al., 1995). These enzymes appeared to be induced as a wounding response following repeated tappings, and served as a defense against subsequent attack by pathogenic microbes. Normally, these enzymes are either absent or at very low level in healthy plants, and this can also be applied to the high level of lysozyme (Tata et al., 1983). Other B-serum enzymes have been investigated and their details reviewed (d'Auzac and Jacob, 1989).

3.7
Lutoids and Colloidal Stability of Latex

The contribution of lutoids to latex colloidal stability, which results from negative charges carried by the lutoid structural components, was referred to earlier. The more important of these structural components include rubber particles and lutoids, and integrity of the latter is recognized as being essential to maintain latex stability. Their content of acidic serum, divalent cations, and positively charged proteins might suggest that lutoids have a destabilizing role in latex, and this is certainly true while the lutoids remain undamaged or unruptured. However, the lutoids are highly osmosensitive, and consequently, after tapping, a certain amount of their content is released from the ruptured lutoids into latex cytosol. The role of lutoids as coagulants has been demonstrated by the formation of microflocs consisting of lutoids fragment and rubber particles (Southorn, 1969). The lutoids B-serum is effectively capable of provoking the microflocs formation of a dilute suspension of rubber (Southorn and Yip, 1968). It was shown that C-serum proteins are mainly anionic, while those of B-serum proteins are mainly cationic (Moir and Tata, 1960). It is clear that placing the B-serum proteins in contact with negatively charged particles of rubber results in neutralization of the surface charges, and destabilization of the latex colloidal solution (Southorn and Yip, 1968; Sherief and Sethuraj, 1978). These early studies form the basis for further analysis in rubber particle coagulation caused by purified hevein (Gidrol et al., 1994). The importance of lutoids with regard to latex flow duration, vessel plugging and the effect on rubber yield will be described in greater detail below.

3.7.1
Lutoids and Latex Vessels Plugging

Rubber yield can be ascribed to two parameters: latex flow duration; and latex synthesis or regeneration capacity. Lutoids are implicated to influence the time of latex flow and latex vessels plugging (Paardekooper, 1989), and thus integrity of the lutoids is essential for latex colloidal stability. Damage to, or rupture of, the lutoids causes coagulation of rubber particles, and this leads to stopping of latex flow. Several studies have provided evidence that the major cause of vessel plugging during latex flow is damage to the lutoids. This was caused by osmotic shock as a result of different turgor pressures between the inside and outside of vessels during flow after tapping. During the initial period of fast flow, the damaged lutoids are swept out of the vessels before they suffering irreversible damage. During the subsequent slower flow, the lutoids suffer greater damage within the vessels and aggregate with rubber particles to form flocs which accumulate near the cut ends, thus initiating the plugging process (Pakianathan, 1966; Pakianathan and Milford, 1973). In whole latex, breakage of lutoids by ultrasonic treatment resulted in the formation of flocs of rubber and a damaged lutoids fragment (Southorn and Edwin, 1968). Therefore, plugging within the vessels is caused by the release of B-serum from ruptured lutoids. Fresh latex always contains microflocs, the formation of which is confined to the area of damaged lutoids where the content of B-serum is momentarily high, and is limited by a stabilizing effect of the C-serum. The latex is envisaged as a dual colloidal system in which negatively charged particles, rubber particles and lutoids, are dispersed in the neutral C-serum containing anionic proteins (Southorn and Yip, 1968). The two antagonistic systems can only exist as long as they are separated by the intact lutoids membrane. Lutoids damage results in interaction between its cationic contents and anionic surfaces of the rubber particles, causing the formation of flocs.

It is observed that the enzyme acid phosphatase is released in the B-serum when lutoids are damaged. The ratio of free and total acid phosphatase – the bursting index – indicates the percentage of lutoids in a sample that had ruptured. The BI was inversely related to the osmolarity of latex, and the first fraction of latex collected after tapping was higher than subsequent fractions (Pakianathan et al., 1966). It was found that damage to the bottom fraction was greatest in the first flow fraction after tapping. A high correlation between plugging index, intensity of plugging, and the lutoid BI was found when studying seasonal variations in the flow pattern (Yeang and Paranjothy, 1982). A positive correlation between total cyclitols in C-serum and plugging index with increased lutoids damage was also noted (Low, 1978). Lutoids could be disrupted by the mechanical shearing forces to which the latex is subjected when flowing through the vessels under high-pressure gradients after tapping (Yip and Southorn, 1968). Difference between clones in the composition of protective film on rubber particles may be partly responsible for the flocs formation. Rubber particles are strongly protected by a complex film of protein and lipid materials (see above). Marked differences between clones in their neutral lipid content of the rubber phase in which phospholipids content did not differ much showed a negative correlation with the plugging index. Long-flowing or slow-plugging clones have a high neutral lipid content in the rubber phase, as much as five times that of the fast-plugging clones (Ho et al., 1976). A negative correlation between the phospholipid content of lutoid membrane and BI, leading to rapid plugging, was documented (Sherief and Sethuraj, 1978). These findings suggested that, in addition to the effect of lutoids behavior, latex vessel plugging is also influenced by the lipids of the rubber particles. Thus, it can be seen that the plugging of vessels and any subsequent effect on the latex yield is quite a complex process. Nonetheless, other factors in the process have also been speculated upon and are awaiting elucidation.

3.8 *Hevea* Latex Metabolism

Hevea latex is a unique system consisting of a specialized cytoplasm (Dickenson, 1965; Gomez and Moir, 1979), the organization and composition of which reflects the biological functions of *Hevea* specialized laticiferous tissue. It has long been shown that fresh latex can synthesize rubber in vitro from labeled precursors such as acetate or MVA (Park and Bonner, 1958; Kekwick et al., 1959). The enzymes' activity and their location with substrates and affectors make it possible to study metabolism and the biochemical function of latex as a 'rubber factory'. Latex as obtained by tapping is not so destructive compared with the general method for preparing cell cytoplasm, and this makes it very suitable to study latex metabolic functions. Latex metabolism has been studied in various different aspects, especially the rubber biosynthesis and other related metabolic processes (Jacob et al., 1989). Tapping of rubber trees for latex makes it necessary for laticiferous tissue to make up for the lost materials between successive tappings, and the regeneration of cell materials increases along with latex production. If regeneration is not sufficiently effective, this can be a limiting factor for latex production (Jacob et al., 1986). Intense metabolic activity and numerous enzymes are required to be involved in at least four important processes for latex regeneration:

- the catabolism which provide energy and reducing capacity for the anabolic processes;
- the activity of anabolic pathways for various syntheses including isoprenoids;
- the mechanisms associated with regulatory systems and homeostasis; and
- the supply of nutrients to the zones or subcellular components in which cell materials are regenerated.

The specificity of laticiferous tissue is so organized that its particular major function is directed to the production of rubber in the latex.

The formation of rubber in *Hevea* laticifers seems to be a very complex control system, and several questions regarding the regulation of rubber biosynthesis have been investigated. The continuing requirement of carbon, NADPH and the need of ATP for rubber biosynthesis must place a high degree of demands on the metabolic economy of the tissues. It has been calculated that the required rate of regeneration of rubber in alternate daily tapping is of the order of 1 μmol isoprene unit per mL latex per minute (Bealing, 1976). The capacity of latex to incorporate acetate and MVA has been found to fluctuate markedly with the season (Bealing, 1976). The control mechanism in the formation of rubber appears to be a complex and intricate process. The interesting feature of *Hevea* metabolism is not only that for rubber formation, but also that it can be stimulated to produce rubber and other many components of latex by repeated tapping. This replenishment is not called for in the untapped tree, and although a few terpenoids have been shown to suffer catabolism in plants, there is no evidence for the breakdown of rubber in vivo. The enhanced rubber yield by ethylene – a hormone associated with response to wounding in plants – has been extensively studied and reviewed (d'Auzac, 1989a,b). The activation of specific genes may shed some light on the mode of action of the tapping stimulus and ethylene activation, and this topic will be discussed in the following text.

3.8.1
Rubber and Isoprenoids Biosynthesis in *Hevea brasiliensis*

Hevea rubber biosynthesis and its control has been often reviewed (Archer and Audley, 1987; Kekwick, 1989; Kush, 1994; Tanaka et al., 1996), and so it is only briefly described here as related to the function of specialized laticifers, together with a discussion on the key regulatory enzymes. An interesting aspect of *Hevea* laticifers is the fine-tuning in terms of compartmentalization of function or the division of labor for rubber biosynthesis pathway. It was shown that the laticifers have a differential gene expression profile (Kush et al., 1990; Kush, 1994). The genes involved in rubber synthesis are highly expressed in the latex as compared to those in the leaves. The specialized differential expression serves a two-fold function. First, that the desired enzymes for rubber synthesis are expressed in the very tissues where formation is taking place. Localizing the rubber synthesis activity in laticifers allows other, different, metabolic processes in other tissues to operate at their optimum and to be well-balanced with the whole of the plant's functions. Second, specialized functions and well-coordinated divisions of labor thus appear to be well organized for specific channeling of precursors and metabolites for different metabolic pathways. The tissue and cell differentiations destined to perform certain functions to best fit the metabolic distribution can thus clearly be seen in *Hevea brasiliensis*. This is somewhat different for other rubber-producing plants such as guayule.

Hevea brasiliensis is unlike some plants, for example *Parthenium argentatum* (guayule), where rubber synthesis takes place in the cytosol of the parenchymal tissues along with other metabolic processes required for orderly parenchymal cell functions. The extraction of rubber from guayule plants is difficult because of the relatively low abundance of rubber particles in the cell, and the limits imposed by the cell volume (Backhaus and Walsh, 1983). The procedure of obtaining rubber from guayule is very destructive because of the subcellular localization of rubber. Guayule rubber can only be obtained by crushing the stems and extracted the rubber, together with all other cellular materials and impurities of all cell types. This is one of the prime reasons that the *Hevea* rubber tree is the only commercially viable source of natural rubber in the world. Even though some 2000 species of plants that produce rubber of varying types and quantities are known (Backhaus, 1985; Mahlberg, 1993), none is used or found comparable to *Hevea* with regard to the superior quality of the rubber produced. This is in contrast to the *Hevea* laticifers, which contain a high content of latex that gushes out after the latex vessels are opened, either by a small excision in the bark or by tapping (which is less destructive to the plant tissues). The flow of latex is due to the high turgor inside the laticifers compared with the outside. After latex has flown out of the cut vessels for some time, the flow will stop due to coagulation of rubber particles forming latex plugs either at the vessel ends or at the wounding site when tapped. A lectin-like small protein, hevein, which is localized in lutoids (Gidrol et al., 1994) has been shown to play an important role in the plugging of latex vessels. Lutoids have been shown to possess very fragile membranes that burst in response to tapping because of the difference in turgor pressure on the lutoid, and its properties. The tapped latex contains a vast number of intact organelles, which in turn makes it an excellent specimen for the study of differential and specialized metabolic functions in plants. As *Hevea* laticifers are an anastomosed system, the latex in essence represents the cytoplasm of a single cell type. The rubber yields are the results of two contributing factors, the latex flow properties and regenerating synthetic capacity.

3.8.2
Isoprenoids Biosynthesis

Besides rubber, *Hevea* laticifers also synthesize a number of diverse isoprenoids (Kush, 1994). Different plants have the capacity to synthesize certain isoprenoid compounds for specific functions, from simple isoprenoids to the more complicated versions such as natural rubber. The plants produce this wide range of isoprenoids in different amounts in specific organelles, and at different stages of growth and development. Since the diverse isoprenoid compounds are produced by a more or less conserved biosynthetic pathway (Randall et al., 1993), plants must execute a control mechanism to ensure that synthesis of the necessary isoprenoids occurs in the right place and at the right time. Such control is very likely mediated through regulatory enzymes, and attempts have been made to understand the regulatory mechanism of the isoprenoids and polyisoprenoids biosynthesis as well as their interrelationship (Kekwick, 1989; Mahlberg, 1993). The pathway used for the formation of isoprenoids in plants is similar to the sterol biosynthetic pathway that was elucidated in animals and yeast (Cornforth et al., 1966, 1972; Clausen et al., 1974; Taylor and Parks, 1978). The isoprenoid biosynthesis may be viewed as the pathway from acetyl-CoA via MVA and IDP to long-chain prenyl diphosphate. A large number of branch points can lead to a variety of diverse isoprenoids in plants (Kleinig, 1989; Randall et al., 1993).

Recently, the oligoprenoid and polyprenoid in *Hevea* latex were examined (Koyama et al., 1995; Tangpakdee et al., 1997a). Although the chain length of B-serum isoprenoids showed several components of C_{15}-C_{60} which were more or less of equal proportion, the C-serum isoprenoids were quite different. The major chain length in the C-serum isoprenoids was the C_{20}-GGDP as analyzed by autoradiography. Moreover, only a few isoprenoids were detected in the C-serum as compared with several in the B-serum. The C_{15}-farnesyl diphosphate (FDP) in C-serum was present in much less quantity than the C_{20}, though the difference between the two remains unclear. It can be assumed that the components in B-serum may be the intermediates preluding the formation of rubber, as was recently reported on rubber formation in the fresh bottom fraction of centrifuged latex (Tangpakdee et al., 1997b). The major C-serum of C_{20}-GGDP, and to a lesser extent C_{15}-FDP, might be the substrates for the prenylation of proteins, as the presence of prenylated proteins has been increasingly reported to occur in plant cells. However, the exact role of these isoprenoids in the two latex sera remains to be elucidated. More recently, the polyprenoids of the dolichols group and other group in *Hevea* latex were analyzed by two-dimensional TLC (Tateyama et al., 1999) It was found that the chain length of dolichols in *Hevea* ranges from C_{65} to C_{105}. The analysis of dolichols of the *Hevea* seeds, root, shoots, and leaves of different ages was also carried out, with comparison of the differences being examined to understand the changes associated with growth and development. The function of dolichols is commonly known to associate with the glycosylation process, and it is assumed that the presence of dolichols in *Hevea* is no exception. The role of glycoproteins has received much attention in the light of reports on the presence of lectin in *Hevea* latex (Wititsuwannakul, D. et al., 1997; Wititsuwannakul, R. et al., 1997), the glycoproteins having been found to play important roles in latex metabolism and colloidal stability.

The enzymes HMG-CoA synthase (HMGS) and HMG-CoA reductase (HMGR) have been implicated as essential regulatory enzymes in the biosynthesis of IDP (Brown and Goldstein, 1980; Bach, 1986; Goldstein and Brown, 1990). Similar roles have also been implicated in plants, although conclusive evidence for the regulatory role of HMGS is not yet available. Downstream of IDP in the formation of specific isoprenoids varies, depending on the end products and subcellular compartmentalization. The role of HMGR in the regulation of isoprenoids and rubber biosynthesis in *Hevea brasiliensis* has been well documented (Wititsuwannakul, 1986). Diurnal variations of HMGR levels and the dry rubber contents of latex were conclusively shown with high corresponding and positive correlations (Wititsuwannakul, 1986). The enzyme was located as membrane-bound and purified from the membrane of lutoids (Wititsuwannakul et al., 1990a). The purified enzyme was then analyzed and characterized, and found to be similar to the HMGR from other plant specimens. *Hevea* HMGR was activated by calmodulin, the calcium-binding protein involved in myriad regulatory processes, located in the latex C-serum (Wititsuwannakul et al., 1990b). A positive correlation between HMGR activity and calmodulin levels was demonstrated, corresponding with the levels of dry rubber content. Comparison between the high-yielding and low-yielding clones also showed clonal variations of calmodulin level in the same direction, and correspondingly. In the rubber tree, this compartmentalization is highly specialized for the syntheses of the rubber and isoprenoids, as demonstrated by the regulatory mechanism of lutoid HMGR by C-serum calmodulin in the control of rubber biosynthesis (Wititsuwannakul,

1986; Wititsuwannakul, 1990a,b) and the difference in isoprenoids distribution in B-serum and C-serum.

3.9
Factors Affecting Rubber and Latex Yields

Rubber yield is the result of two contributory factors, namely latex flow properties and regenerating synthetic capacity. It is generally observed that the latex yield after tapping depends first on the duration of latex flow. Subsequently, regeneration of the laticifer's content between the two tappings limits in turn the quantity of latex being collected. Therefore, flow and regeneration constitute the two most important limiting factors of latex production in *Hevea*. Put another way, we can examine this aspect as the synthesis capacity and the latex flow characters that influence the latex yield, and this has been shown to be a phenomenon of clonal variations.

3.9.1
Rubber Latex Regeneration Capacity

The utilization of sucrose as a carbon source, together with the enzymes of related metabolic pathways and the effect of pH changes, have been shown repeatedly to correlate with latex production. This subject has been extensively reviewed (Tupy, 1989), and will be outlined briefly to provide a cohesive view of the results referred to earlier in this chapter. Rubber biosynthesis, and the role of the key enzymes in various different pathways involved, has been outlined elsewhere. Here, we focus mainly on the regeneration of latex between successive tappings. Incision of the bark, causing the latex to exude from the latex vessels, results in a set of processes to be initiated and activated. Migration then occurs of reserves or their products from their production or accumulation zones to the demanding zones where regeneration of the rubber and latex constituents. These phenomena take place progressively, and the reconstitution of the latex to be collected at tapping requires a certain amount of time which depends on the amount being exuded. The essential function of laticiferous cells is the synthesis of *cis*-polyisoprene, which forms over 90% of cell contents. It is the metabolism of this highly specialized cellular environment and of the regeneration of latex between two tappings of *Hevea* which can be a limiting factor for the latex and rubber production.

Several apparent parameters were found to influence the latex regeneration metabolism. These controlling factors were commonly analyzed and compared for their effects in the tapped latex. The importance of sufficient availability of the sucrose in latex was extensively studied as precursor and having essential role for polyisoprenoids synthesis (Tupy, 1989). Sucrose metabolism and its utilization is controlled by invertase, and the levels of enzyme activity were positively correlated with latex production (Yeang et al., 1984; Low and Yeang, 1985). Some key enzymes affected by ions (Mg^{2+} and phosphate) and thiol (SH) groups for reducing conditions were documented in relation to the rubber biosynthesis. These included invertase, puruvate kinase and phosphoenol pyruvate (PEP) carboxylase. The activity levels of these enzymes influence the latex production capacity, as has been detailed and summarized earlier (Jacob et al., 1986; Tupy, 1989). The intracellular pH is the essential factor in isoprenoid metabolic regulation, while increased activities of invertase and PEP carboxylase were found under the alkaline pH of the laticifer contents, and this correlated with the latex and rubber yield (Yeang et al., 1986; Tupy, 1989).

Significant positive correlation between these parameters and latex production were documented. The relationships between the total solids content, reduced thiols content, pH of the latex, and the latex yields among different clones and the seasons have been analyzed and extensively characterized. Highly significant positive correlations were found between these three key parameters and the latex yield (Jacob et al., 1986). All correlations analyzed were of high statistical significance ($P < 0.001$), as reported in extensive studies (Jacob et al., 1986), and hence these reflect the latex regeneration capacity of the laticifers which requires the coordination and interaction of several parameters in intricate manner. In addition, the induction and activation of protein synthesis was also noted for certain enzymes (Broekaert et al., 1990; Kush et al., 1990; Goyvaert et al., 1991; Kush, 1994; Pujade-Renaud et al., 1994) as the key factor on enzyme levels for latex regeneration. This is also the important key parameter in addition to the availability of sugars and the alkaline cytosol conditions of the laticifers and the latex cytosol (Tupy, 1989). Thus, it can be clearly seen that latex regeneration is a highly complex process that requires the participation of a myriad of parameters in a coordinated and synchronized manner for the orderly function of the laticifers and latex cytosol. Moreover, there are as yet some unidentified parameters to be added to the list as research progresses in this area.

3.9.2
Ethylene and Latex Yield

The mechanism of ethylene activation on rubber biosynthesis and enhanced latex yield has been the subject of many studies and has been recently reviewed (d'Auzac, 1989a,b; Kush, 1994). Ethylene has been implicated in mediating the signal for wounding response in the plant defense (Ryan, 1984), as well as in the stimulation of latex production (d'Auzac, 1989a,b). A protective role against various fungal and bacterial invasions has been proposed, this view on defense function having been developed in connection with the presence of high concentrations of antifungal and antibacterial substances in latex. The presence of lytic enzymes (chitinase, glucanase, lysozyme and others) has been demonstrated in the latex (John, 1993; Churngchow et al., 1995), and a protease inhibitor has also been found in latex, probably for defense purposes (Archer, 1983). The accumulation of these protective substances in latex is not constitutive, but rather an induction as a consequence of wounding responses to the tapping, for defense functions. Ethylene has been shown to play the key role in defense response (Koiwa et al., 1997), although its effect on the enhancement of latex yield remains poorly understood.

It is well documented that rubber trees, like any other plant, will respond to phytohormone treatment. The effect of ethylene on latex yield has been extensively investigated in *Hevea brasiliensis,* using a molecular biological approach. Although the biochemistry of *Hevea* polyisoprene synthesis is, in general, relatively well understood, it was only recently that studies on the molecular biology of the system were initiated (Broekaert et al., 1990; Kush et al., 1990). Cloning of various genes upstream of IDP was possible due to the availability of the heterologous probes (Dennis et al., 1989; Broekaert et al., 1990; Attanyaka et al., 1991; Chye et al., 1991) and to the highly conserved nature of this pathway. However, in the biosynthesis of natural rubber, molecular analysis downstream of IDP was more difficult because little was known or previously characterized. The cloning of prenyltransferase and REF (Light and Dennis, 1989; Attanyaka et al., 1991; Goyvaerts et al., 1991) that are involved in the final step of rubber biosynthesis are examples of studies of this type. Nonetheless, a number of questions remain unanswered about the process, some of which concern the exact mechanism of polyisoprene elongation of IDP when added in the *cis* configuration (Light et al., 1989; Cornish, 1993). In addition to these is the effect of hormone treatment on gene expression, as well as specific genes activation related to rubber biosynthesis and the increase in latex production stimulated by ethylene.

The stimulatory effect of ethylene on latex leads to various induced metabolic changes. It has been shown that a wide variety of substances and treatments can stimulate latex production. Each substance might have a specific role on the laticifers, and the end result is a combined, cumulative effect. It has been shown in vitro that the acid phosphatases released from lutoids hydrolyzed the key substances needed for the biosynthesis of rubber (Archer et al., 1963b). It would also be interesting to determine which molecules or enzymes are involved in the improvement of lutoids stability. It was noted that a direct correlation existed between the stability of lutoids and in vitro rubber biosynthesis, as well as a prolongation of latex flow (Archer et al., 1963b; Southorn and Edwin, 1968). Ethylene stimulates latex production, increasing duration of latex flow after tapping, by activating the metabolism involved in latex regeneration (Coupe and Chrestin, 1989). It is clear, therefore, that stimulation resulted in an increased stability of the lutoids. After tapping, regeneration in situ leads to the reconstitution of exuded latex before the next tapping. In quantitative terms, 100 mL of latex that is exuded is completely regenerated within ~60 h (Pujade-Renaud et al., 1994). This corresponds to a net synthesis of ~50 g of dry rubber and 1–2 g of protein during this period. Thus, highly intense rates of metabolic activity such as energy-generating catabolic pathways as well as anabolic processes are required. These include a large increase in glycolysis (Tupe, 1973), as well as increases in the adenylate pool and polysome and RNA contents (Amalon et al., 1992) – all of which are indicators of metabolic activation leading to increased latex production. The rise in the level of transcripts of several enzymes (chitinase, glutamine synthetase, and hevein) was shown to be increased by ethylene (Pujade-Renaud et al., 1994).

The response to ethylene can provide an ideal system to study the signal transduction in *Hevea*. An intriguing question is how the signal of ethylene becomes transduced. A number of speculative suggestions have been made, including the involvement of calmodulin or other calcium-binding proteins (Wititsuwannakul et al., 1990b). Another possibility is the role of protein prenylation. The rapidly expanding family of prenylated proteins include those that are involved in signal transduction (Cox and Der, 1992), as well as proteins involved with intracellular vesicular transport, cytoskeleton organization, and cell growth control and polarity (Cox and Der, 1992). It might be possible that a specific chaperone may need to be prenylated in order to act as a messenger, as has been documented in other plants (Zhu et al., 1993). It is therefore logical to consider that an important role for the prenylation in latex might exist. As elaborate metabolic machinery has been adapted for the biosynthesis of isoprenoids in specialized laticifers of *Hevea*, it is possible that the non-rubber isoprenoids may serve a regulatory function for metabolism. Intensive tapping or over-stimulation by ethylene leads to increased consumption of sucrose, leading in turn to an imbalance of carbon sources relationships. The exhaustion of carbohydrate reserves is one of principal reasons for physiological fatigue of stress and reduced capacity to hormone response. This results in the increased superoxide anion production, in turn causing free radical-based damage of various membranes, and especially those of the lutoids. Over-stimulation by ethylene can result in a phenomenon of coagulation of latex in situ called 'bark dryness', after which no more latex is obtained (Chrestin, 1989).

3.9.3
Latex Flow and Affecting Parameters

Latex flow has been outlined above briefly as being related to the roles of lutoids in latex. Latex regeneration and latex flow characters are interrelated, and the importance of flow with regard to latex yield or productivity is stressed by the fact that part of the effect on ethylene stimulation is prolongation of latex flow time (Abraham and Taylor, 1967; Abraham et al., 1968). Stimulation causes more intensive drainage from latex vessels due to the high turgor pressure inside the laticifers (Pakianathan et al., 1976). Ethylene stimulation was also found to increase sucrose utilization by activating the invertase enzyme and its level, leading to enhanced latex regeneration capacity (Low and Yeang, 1985). The enhanced capacity would certainly lead to the increased latex volume being stored in the laticifers, and the higher turgor pressure with the enhanced flow character (Pakianathan et al., 1976). Several mechanisms or causes have been suggested to explain the increased flow elongation as a consequence of enhanced latex regeneration capacity are (d'Auzac, 1989b), with a number of points identified as contributory factors:

1) An increase in osmotic and turgor pressures in the laticifers as a result of increased latex volume is due to an effect of plant growth regulators and ethylene stimulation.
2) The lowering of latex viscosity by a slight reduction in rubber concentration may explain the effect of ethylene on enhanced flow.
3) An increase in the elasticity of the laticifer's cell wall leads to a reduction in the collapse of vessels.
4) Modification of laticifer's permeability by ethylene will allow higher influx of water and solute, and hence better drainage of the latex vessels.
5) Plugging of the vessel ends is delayed by increased latex stability, hence leading to longer flow.
6) Ethylene stimulation increases the bark drainage area of the latex flow.
7) The increased lutoids stability by ethylene leads to a delay in coagulation of the rubber particles and hence a slowing effect on vessel plugging.

There is a limit to the enhancement of latex flow following ethylene stimulation. After a certain duration of flow time, the flow will stop as a result of the vessels becoming plugged; this is an attempt by the plant to contain the damage that might occur to laticifer metabolism due to exhaustion of its contents, and is analogous to the wounding response as a protective mechanism. Upon neutralization of their negative surface charges by various factors, rubber particles clump together to form microflocs. This leads to plugging of the vessels ends, and hence stops the latex flow. Floc formation caused by release of the contents of the lutoids that were damaged at the tapping cut, as well as during the flow, has been well documented (Southorn and Yip, 1968; Southorn, 1969, d'Auzac, 1989a). This was shown by the inverse relationship between the BI of the lutoids and the duration of latex flow (Yeang and Paranajothy, 1982; Yeang and David, 1984), as described above. Any delay in obstruction of the latex vessels due to slow plugging is associated with stability of luoids. Although the effects of ethylene stimulation on enhanced latex flow has been reviewed (Coupe and Chrestin, 1989), several other parameters may also have such a positive effect on latex flow and hence increase its yield. A high total solids content (TSC) in the latex can limit flow due to high viscosity, while increased water exchange in laticifers may cause a reduction in latex TSC, in turn lowering latex viscosity and promoting flow. Latex also contains thiols in the form of cysteine and glutathione which play an important role in the metabolism of laticifers, the thiols providing stability to the lutoids when they are damaged by free radicals generated in latex (Coupe and Chrestin, 1989). Thus, it is clear that latex flow is very much influenced by the lutoids, whether in terms of reduced latex flow time after lutoids rupture, or prolonged flow time by improved lutoid stability and integrity. Although much emphasis has been placed on ethylene however, several other factors are known to be involved in latex flow, notably the rubber particles and C-serum. In addition, other factors have yet to be identified that might provide a better understanding of this complex and complicate process.

4 Latex of Other Plants

Numerous plants can synthesize rubber in the latex, even though *Hevea* latex is the only suitable and viable source of natural rubber produced commercially. The formation of rubber in plants has been reviewed (Backhaus, 1985) and will be described only briefly here. It is commonly known that rubber latex is not only formed in the *Hevea brasiliensis*, but it is also synthesized and accumulated in over 2000 species of plants representing about 300 genera from seven families (Archer and Audley, 1973). The rubber-producing plant families are *Apocynaceae, Asteraceae, Asclepiadaceae, Euphorbiaceae, Loranthaceae, Moraceae*, and *Sapotaceae*. The latex and rubber composition varies among all these species. Rubber is composed of plant secondary metabolites that are produced in large quantities by certain plant cells, but has no known function to the plants. A metabolic function has been speculated upon, but specific details have yet to be indicated. The presence of rubber in various plants does not appear to deter herbivore feeding, insect attack or diseases caused by microbes. Rubber does not serve as an energy source, despite tremendous resources being allocated to its formation. There is no evidence that plants possess the enzymes capable of degrading rubber (Archer and Audley, 1973), and it appears that once rubber is formed, it remains in the cells. The enzyme responsible for the turnover of carbon in the rubber has yet to be identified and characterized. The ability of these plants to synthesize rubber is based on the presence of the key enzyme, rubber transferase, which causes *cis*-polymerization of isoprene units (IDP) into the long hydrocarbon chain of rubber. These are assembled and accumulated as rubber particles ranging in size from 5 nm to 3 μm. These particles are enclosed or surrounded by a lipophilic film. The presence of enzymes and the processes for rubber formation are well documented, and in some of these plants, for example *Parthenium argentatum* (guayule) and *Hevea brasiliensis* (rubber tree), detailed studies have been performed on the rubber synthesis process (Benedict, 1983). *P. argentatum* and *H. brasiliensis* represent plant species which accumulate large quantities of high-molecular weight rubber, albeit by different intercellular routes. Rubber synthesis takes place in parenchyma cells in guayule (Backhaus and Walsh, 1983), but in specialized latex vessel cells (laticifers) in *Hevea* (Archer et al., 1982). Other plants produce far less rubber than *Hevea* or guayule; they usually produce rubber of inferior quality and of a much lower molecular weight than *Hevea*, and this limits their commercial potential.

The molecular weights of rubber from various species were examined and compared. The average molecular weight of rubber molecules can vary considerably with the species. The reason for *Hevea* and guayule being unique among rubber-producing plants is that they produce rubber with a predominantly high molecular weight average of $3-7\times10^5$, which is critical for commercial utilization. Other rubber-producing plants produce inferior quality rubber with a molecular weight of $\sim5\times10^4$ or less. The molecular weight distribution (MWD) of rubber extracted from *Lactuca, Carissa, Candelilla*, and *Asclepias* compared with guayule, as demonstrated by gel-permeation chromatography (GPC), showed a value at least 100 times lower than that for guayule ($\leq10^4$ as compared with 10^6 for guayule). An interesting feature of molecular weight is that *Hevea* and guayule exhibit bimodal MWD, with a major peak at $\sim7\times10^5$ and a minor peak at $\sim0.5-1.0\times10^5$. This suggests either that polymerization in these two plants either undergoes a two-step process, or that two forms of rubber transferase may exist. In contrast, rubber in other plants uniformly shows a single peak of low molecular weight of $\sim10^4$ or less, indicating a lower degree of polymerization in these plants and different properties of the rubber transferase and activity in *Hevea* and guayule.

4.1 Latex of Guayule

Rubber biosynthesis has been extensively studied in *Hevea brasiliensis* (rubber tree) and *Parthenium argentatum* (guayule). Guayule rubber is a typical high-molecular weight *cis*-polyisoprene having a narrow MWD and physical properties that are slightly different from those of *Hevea* rubber. However, little is known of the structural differences between the rubbers from these two major sources. Recently, sunflower rubber was characterized and reported as a prominent rubber-producing member of higher plants (Tanaka, 1986). Low-molecular weight rubber from sunflower could serve as appropriate model for characterization of the chemical structure of *Hevea* and guayule rubbers. The findings showed that the fundamental structure of guayule rubber is similar to that of *Hevea* rubber, as deduced from the sunflower rubber structure elucidation (Tanaka, 1986). The guayule shrub synthesizes two types of secondary compounds – rubber and resin – in relatively large amounts. The rubber accumulates in parenchymal cells of the stem (Backhaus and Walsh, 1983) as opposed to the latex vessels in *Hevea*. Clearly, the amount of rubber that a guayule plant is capable of accumulating is limited by the total volume of its parenchymal tissues (Backhaus and Walsh, 1983). Cells which store rubber are found in the secondary cortex, phloem and xylem rays (Mehta, 1982; Mehta and Hanson, 1983). Treatment with a bioregulator, 2(3,4-dichlorophenoxy) triethylamine (DCPTA), was reported to increase the rubber content in guayule by two- to six-fold (Yokoyama et al., 1977), the increase of rubber in the phloem being more pronounced in winter samples in cells of the cortex.

An outstanding feature of rubber biosynthesis in guayule is that it is stimulated by low night-time temperatures (Bonner, 1967). Exposure of plants to 27°C during day-time, followed by 7°C at night for four months induced a three- to four-fold increase in rubber compared with control plants grown at a constant 27°C day and night. The low night temperature specifically induced the synthesis of *cis*-polyisoprene, with no effect on the rate of *trans*-isoprenoid resins formation (Goss et al., 1984). Exposure of guayule plants to low night-time temperatures stimulated the formation of rubber particles in cortical parenchymal cells. This suggested that the enzymatic potential for synthesizing rubber is greater in the stems from cold-exposed plants (Goss et al., 1984). The mechanism of the cold induction of guayule rubber is not known, but has been suggested as induced expression of genes coding for enzymes involved in rubber synthesis (Bonner, 1975). Regulation of guayule rubber content and biomass by water stress has also been noted (Bucks et al., 1984, 1985a, b). Several studies on guayule indicated that water stress will result in increased rubber content in the shrub (Miyamoto et al., 1984; Bucks, et al., 1985b). It is commonly known that rubber particles isolated from *Hevea* latex are enclosed with a film of phospholipids and proteins, as described above. It is clear that rubber particles contain protein factors necessary for the polymerization reactions in rubber formation. The same can also be observed for rubber particles of guayule (Backhaus and Walsh, 1983; Backhaus, 1985).

4.2
Possible Role or Functions of the Latex in Plants

Latex is produced in about 2000 plant species, and in varying degrees of quality and quantity. Rubber, which is the major and most important isoprenoid polymer in the latex, has no clear physiological function within the plants, and the specific and exact function of latex is the subject of much speculation. Details of the nature and possible role of latex rubber polyisoprene in plants have been presented only briefly (Archer, 1980), and solid evidence and strong support are still required to substantiate the roles as discussed. To date, several hypotheses and speculations have been proposed to explain the role or function played by rubber and latex in plants. The initial suggestion that the latex protected the plants from animal attack, but this has been discarded due to a lack of corroborative evidence. The proposition that rubber in latex acted as a reserve food supply or carbon storage reservoir was without strong support. This is not in accord with the fact that rubber in the latex, once it has been formed, is not further metabolized (Bonner and Galston, 1947). There is no evidence for the breakdown of rubber in vivo to substantiate this proposition. On the other hand, the suggestion that the formation of rubber in latex is a fermentative process analogous to the anaerobic production of ethanol has also been proposed (Ritter, 1954; Bealing, 1965). However, the absence of stoichiometric coupling of the oxidative and reductive phases of the biosynthesis processes has been extensively analyzed and discussed in detail (Archer and Audley, 1973), which precluded this explanation.

The main question needing to be addressed is why the plants, especially *H. brasiliensis*, expend so much energy and resources to synthesize natural rubber at all. This should lead to a search to determine whether latex has an important role in plant metabolism per se. The control mechanism in the formation of rubber appears to be a complex and intricate process. The interesting feature of *Hevea* metabolism is not only that for rubber formation, but also that it can be stimulated to produce rubber and other many components of latex by repeated tapping (Kush, 1994). This replenishment is not called for in the untapped tree, and although a few terpenoids have been shown to suffer catabolism in plants, there is no evidence for the breakdown of rubber in vivo. These are intriguing questions to most plant physiologists and biochemists, but we still do not have a clear answer to them. It was recently proposed that the latex may play a protective role against various fungal and bacterial attacks and the invasion of plant tissues by such pathogenic microbials (Kush, 1994). This view has been proposed because of the very high concentrations of antifungal and antibacterial substances such as chitinase, lysozyme and other pathogenesis-related proteins (PR proteins) being present in the latex (John, 1993). The latex of *H. brasiliensis,* which contains high levels of both chitinases and chitinases/lysozymes substantiates this suggestion (Martin, 1991). More recently, the role of *cis*-1,4-polyisoprene in the *Hevea* latex as having antioxidant function has been proposed (Tangpakdee et al., 1999). In latex, this compound is presumed to act as a free radicals acceptor or scavenger in laticifers to protect against damage caused by active oxygen species (AOS). However, the evidence to support this proposal remains superficial and is not sufficiently well documented to substantiate this claim.

Ethylene has been implicated in mediating signal for wounding response in plant defense (Ryan, 1984) and in the stimulation of latex production (d'Auzac, 1989b). A protective role for latex against various fungal and bacterial invasions has been proposed, this having been developed in connection with the presence of high concentrations of antimicrobial substances in the latex. Lytic enzymes (chitinase, glucanase, lysozyme and some other PR proteins) have been shown to be present in the latex (John, 1993; Churngchow et al., 1995), while the presence of a protease inhibitor in *Hevea* latex, probably for defense function, has also been reported (Archer, 1983). The accumulation of these protective substances in latex is not constitutive, but is induced as a consequence of the wounding response after tapping of the rubber trees. Ethylene has been shown to play a key role in defense response (Koiwa et al., 1997), although the effect of ethylene on enhanced latex yield is still not well understood. However, the role of latex as a result of induction by ethylene in defense function appears to be the most acceptable hypothesis at present. Thus, it seems likely that the end product of polyisoprene synthesis or rubber formation in latex has no known function per se, and the exact role of rubber and latex in plants remains an open question. In the case of laticiferous plants such as *H. brasiliensis,* it is possible that other processes occurring in laticifers may be of greater importance in terms of plant physiology than are polyisoprene synthesis and rubber formation.

5
Latex of Fungi

As mentioned earlier, the formation of polyisoprene rubber is confined to higher plants of dicotyledon angiosperms. Over 2000 species, representing about 300 genera of higher plants, are capable of the synthesis and accumulation of polyisoprene in the form of rubber latex. These natural polyisoprenes have been characterized as the *cis*- or *trans*-configuration by NMR analytical techniques. In addition, at last two fungal genera, *Lactarius* and *Peziza,* were also found capable of rubber biosynthesis (Stewart et al., 1955). Unknown species of fungal genera *Lactarius* and *Peziza* were reported to produce low-molecular weight polyisoprenes. Rubber extracted from four species of *Lactarius* was characterized by infrared analysis as *cis*-polyisoprene with an intrinsic viscosity of 0.43 – 1.02. Very few studies were carried out on the rubber from fungi, and hence the characterization of fungal rubber is little known. Here, discussion will be limited to the minimal information and reports available. Studies on the structure and biosynthetic mechanism of rubber from mushroom (Tanaka et al., 1990) have provided some details on the nature and characteristics of the rubber produced by the lower living organisms.

Fungal rubber was reviewed with regard to elucidating its structure and characterizing the initiation and terminal units of low molecular weight rubber. These simple studies with the fungal small molecule rubber may serve as a model for more advanced investigations into the rubber of higher plants, which has a much greater molecular weight. The most extensive studies were made with *Lactarius* rubber, using NMR analysis, to characterize the rubber biosynthetic process in fungi (Tanaka et al., 1990). A few species of *Hygrophorus, Russula,* as well as *Lactarius,* were found to produce rubber. Among different *Lactarius* species, *Lactarius subplinthogalus* was found to produce *cis*-polyisoprene rubber as high as 7% of the dry weight of sporocarps. The biosynthesis of *Hevea* natural rubber has been assumed to proceed from DMADP by the successive addition of IPP in the *cis* configuration. However, there was no direct evidence either to prove or verify the initiation step. Nor has any biochemical investigation been carried out on the termination mechanism. A ^{13}C-NMR method was used to analyze the structure of both terminal units, and the alignment of *cis* and *trans* isoprene units along the polymer chain. This has provided important information on *cis*- polyisoprene synthesized in the nonlaticiferous cells. The rubbers from goldenrod and sunflower leaves consist of a dimethylallyl group and about two or three isoprene units in the *trans* configuration, followed by a long sequence of isoprene units in the *cis* configuration, aligned in that order. This is the evidence that biosynthesis of rubber in these plants starts with *trans* GGDP or *trans* FDP as the initiating species. This initiation step is similar to that observed with other *cis*-polyisoprenes. Using ^{13}C-NMR analyses, the presence of terminal *trans* units were shown for *Hevea* rubber and other wild rubbers (Tanaka, 1989).

Rubbers from fungi were characterized both for the structure of terminal units and mechanism of rubber biosynthesis in the fungal latex. Comparison was then made with other rubber-producing species of higher plants, which contain mostly *cis*-polyisoprene as rubber in the latex. The rubbers from *Lactarius volemus* and *Lactarius chrysorrheus* (*Lactarius* rubber) were characterized (Tanaka et al., 1990) and found to be similar to *cis*-polyisoprenes in latex of other higher plants, as previously reported. The structure of terminal units and alignment of the isoprene units provide information on initiation and termination mechanisms of rubber formation in fungi. Polymerization of *Lactarius* rubber starts from *trans* FDP and proceeds by successive condensation of IDP to form isoprene units in the *cis* configuration. Termination is presumed to occur by direct esterification of polyisoprenyl diphosphate (PIP-DP), or after dephosphorylation followed by esterification with fatty acids. Chemical or biochemical modification of both terminal groups and the main chain also occur on the fungus rubber (Tanaka et al., 1990). Information on the termination mechanism may provide a clue as to the process of controlling the molecular weight of rubber. The study on low-molecular weight rubber from fungi may also serve as a model for elucidation of the molecular weight controlling mechanism in *Hevea* rubber, as well as in the rubber from other higher plants. It is expected that greater attention will be paid to studying fungal rubber as a possible novel source of natural products other than the rubber that might be present in the fungal latex. So far, information on the structure and composition of fungal latex has not been obtained in great quantity, except for the demonstration of the presence of rubber in fungal latex (Tanaka et al., 1990). The plentiful information from *Hevea* latex study would be very helpful to investigate the structure and composition of fungal latex, especially the nonrubber constituents. Whether the nonrubber constituents of fungal latex are similar or different from those of *Hevea* latex will be interesting, and this subject clearly requires further study.

6
Diseases Related to Natural Rubber Latex (Latex Allergy)

Natural rubber latex (NRL) from *Hevea brasiliensis* has been used for a very long time in the industrial production of items that are used on a daily basis and, being an inert material should pose no danger to health. However, the latex in which rubber forms a major component contains a myriad of substances that are foreign to our immune system. These non-rubber substances are also found tightly bound or associated with the rubber particles and hence will always be present with rubber in the finished products. As mentioned above, rubber particles are enclosed or surrounded by a protective film that is composed mainly of lipids and proteins. These lipids are similar to the common lipids in our body, and so would not be expected to be recognized as foreign substances by our immune system. However, the rubber particle proteins are quite different from proteins in our bodies, and hence will be recognized as 'foreign' by our immune system, and this is where health problems begin. Frequent exposure to these foreign proteins will cause sensitization in the individuals or users, and this in turn leads to an allergic response. Allergy to natural rubber has been identified as being caused by the proteins present in the rubber products, especially in the case of rubber gloves and other rubber products than are used frequently. Indeed, allergy to natural rubber has recently become a major health concern and will be discussed in detail below.

6.1
Latex Allergy

Reports of NRL immediate hypersensitivity have increased steadily since the first reported case in 1979 (Nutter, 1979; Granady and Slater, 1995), and NRL protein allergy has been the subject of recent review (Hamann, 1993). Specific immunoglobulin IgE and IgG4 antibodies to a series of NRL proteins have been detected (Alenius et al., 1991, 1992) in the sera of NRL-sensitized individuals. Different sources of NRL proteins, and purification methods coupled with nonstandardized pooled, sensitized sera have contributed to the detection of a wide range of antigens. Antigens in the regions of 14 kDa and 30 kDa seem the most reproducible. The initial presentation after cutaneous exposure is typically contact urticaria. Aerosolized protein bound to cornstarch powder may lead to rhinitis, conjunctivitis and bronchospasm. Mucosal and intraoperative parenteral exposure seem to lead to the greatest risk of massive histamine release and anaphylaxis. Although a standardized reagent is not available, the most sensitive diagnostic test remains the prick test. Treatment for the NRL protein-sensitized individual is NRL product avoidance. Non-NRL alternatives should be considered routine among risk group, such as spina bifida and atopic patients with hand eczema, in order to prevent unnecessary sensitization.

Each year, NRL-containing medical and consumer products are safely and efficaciously used on billions of occasions. The challenge remains to identify specifically the NRL allergens, and to reduce their levels below a threshold which either induces sensitivity or elicits a reaction. The increase in prevalence of latex hypersensitivity since 1979 remains an issue of debate, and it is most likely that the situation has arisen due to a complex combination of factors. The most significant contributor to the rise in prevalence has been the vast increase (by several hundred) in the number of (inexperienced) NRL glove manufacturers during the period 1986–7. This was due to public awareness of AIDS, coupled with an increased demand for protection against the condition. Allergen levels rapidly reached unprecedented levels and exceeded sensitizing thresholds. Vast numbers of people, who previously had been using NRL without problems for decades, were needlessly sensitized. Hypersensitivity reactions to rubber products are common. Most early references to latex as initiator of immune responses describe cases of contact dermatitis, a delayed hypersensitivity (type IV). This type of reaction, in cases involving latex exposure, is typically localized to the hands after the use of rubber gloves. The allergic lesions are localized to the areas of skin in direct contact with the rubber product. Between 1979 and 1988, numerous reports appeared in the literature of IgE-mediated (type I) immediate reaction to latex (Nutter, 1979; Dooms-Goossens, 1988), the studies pointing to the fact that the allergenic reactions were due to specific IgE antibodies directed to allergenic proteins in the *Hevea* latex (Wrangsjo et al., 1986).

6.2 Latex Protein Allergens

Characterization of NRL allergens has been recently reviewed (Kurup et al., 1992, 1995). A number of proteins have been detected in NRL and in latex finished products (Dalrymple and Audley, 1992; Makinen-Kiljunen et al., 1992). Several of these proteins have specific reactivity with patient serum antibodies. A large number of peptides has been reported in non-ammoniated *Hevea* latex. Although more than 240 different polypeptides can be separated by two-dimensional gel electrophoresis (2D-PAGE) of the latex serum, less than 25% of these peptides showed reactivity with IgE from patients with latex allergy (Alenius et al., 1994a). A protein with molecular weight of 58 kDa was isolated from latex gloves with the same monomers of 14.6 kDa, and was shown to have complete homology with REF (Czuppon et al., 1993a). REF was implicated as the major allergen in latex, and a 14-amino acid synthetic peptide representing the N-terminal domain of REF reacted with IgE antibodies in a majority of latex-sensitive patients (Czuppon et al., 1993b). Although REF was considered the major allergen in NRL, several other studies showed a large number of polypeptides associated with development of the latex allergy (Kurup et al., 1993a; Alenius et al., 1994a). The major polypeptide among these is hevein, which has chitin-binding properties. Significant variation in the number and quantity of proteins in various extracts of latex gloves has been reported (Chambeyron et al., 1992). SDS-PAGE demonstrated a wide range of peptides, from 5 kDa to 200 kDa in various extracts. IgE antibody in the sera of latex-allergic patients showed a wide range of reactivity in immunoblot studies. Predominant protein allergens showed molecular sizes ranging from 5 kDa to 100 kDa, and the 14 kDa peptide (REF) showed more frequent reactivity with patient sera (Yeang et al., 1996). Significant IgE binding to four latex proteins of 14, 20, 23, and 28 kDa was reported for sera from allergic patients (Yeang et al., 1996).

6.3 Proteins and Allergens in Latex Gloves

The question of whether the amount of eluting protein correlates with the allergenic reactions of a given NRL product is an issue for debate (Turjanmaa et al., 1995). It is generally agreed that both the protein and the allergen activity in latex gloves may vary considerably (Turjanmaa et al., 1988; Alenius et al., 1994b). The determination of total protein extractable from NRL products does not necessarily characterize NRL allergen activity. Besides, cross-reactivity of NRL and food allergens has been documented (Lavaud et al., 1992), with NRL-allergic patients showing IgE antibodies that bind to a number of banana and avocado proteins (Ahlroth et al., 1994). The characteristics of cross-reacting NRL and fruit allergens are as yet largely unknown. Allergens in the glove extract may be denatured or modified during the rubber aging and processing (Makinen-Kiljunen et al., 1992). Antigen cross-reaction with latex proteins is, therefore, relatively common.

Several studies have shown that there is cross-reactivity between latex and other allergens, these including a variety of fruits, vegetables, and grains. Such studies have shown variable levels of cross-reactivity between food allergens and gloves or latex (Kurup et al., 1993b). The cross-reactivity among these antigens may be explained in term of the process that is seen uniquely in plants. It was reported that the peptide sequence deduced from the mRNA accumulating in response to wounding or hormonal treatment shows homology to several chitin-binding proteins, such as the latex hevein (Broekaert et al., 1990). A chitin-binding domain has been suggested to be the building block of many proteins with diverse specificity (Chrispeels and Raikhel, 1991). Whether such a common molecule exists and can be implied in the cross-reactivity between these other allergens and latex is not known, or is not clear at present. It is possible (or likely) that a common molecule can be implicated in the cross-reactivity between these fruit or vegetable allergens and hevein, as the latter has been purified from latex and shown to have chitin-binding properties (Gidrol et al., 1994). Hevein was also found to be one of the major latex allergens. An association between latex hypersensitivity and fruit allergy would act as a precautionary measure, but this would not necessarily mean that all people allergic to fruit antigens would become allergic to latex, and vice versa.

6.4 Cross-Reactivity of Allergens

Cross-reactivity between latex and other plant source allergens has more recently been assessed in greater detail (Truscott, 1995). It was commonly observed that several proteins in NRL are similar to those found in other plant species. Hevamines, the cationic proteins in *Hevea* latex, are very similar in molecular weight, amino acid composition and cationic features to lysozymes found in fig (*Ficus*) and papaya (*Carica*) latex (Tata et al., 1976). The *Hevea* lutoid proteins, which increase during ethylene or ethylene oxide treatment as described above on ethylene stimulation of latex yield, are very similar to the lutoids found in latex of banana (*Musaceae*), avocado (*Lauraceae*), and several other fruits. These fruits have been reported to cross-react with *Hevea* latex in several clinical tests on allergenicity evaluations (Czuppens et al., 1992; Lavaud et al., 1992; Ross et al., 1992; DeCorres et al., 1993; Fisher, 1986, 1987, 1993). Interestingly, chestnuts are also found to be cross-reactive with allergenic proteins from *Hevea* (Anabarro et al., 1993; Lavaud et al., 1992). Chestnuts are often sterilized with ethylene oxide to prevent spoilage by microbial and insect infestation (Foegeding and Busta, 1991). It is conceivable that these somewhat similar lutoid proteins conjugate with the extremely reactive ethylene molecule, creating a mutually recognized hapten or carrier responsible for cross-sensitization or cross-reactivity. By taking these several observations to the next step, it is possible that ethylene becomes the common bridge linking 'transferred recognition' and subsequent cross-reactivity.

The nature of the similar (but not identical) proteins fractions in ethylene-treated fruits (avocado, banana, chestnuts) and the *Hevea* latex, which is exposed to ethylene released in the wounding response of tapping the rubber trees, might lead to the allergens' cross-reactivity. It then stands to reason that the allergenicity potential of latex is increased with further exposure to ethylene in the form of ethylene oxide, which conjugates with the non-haptenated latex proteins. The increased exposure to these hapten or carrier allergens as a whole increases the dose challenge that would be experienced from a single allergen, and this is a plausible contributing factor for reducing the amount of time required until a threshold sensitization level is reached. It may be that the sensitized healthcare worker would experience his/her first type I reaction at an earlier age when he/she would finally have had a sufficiently high exposure if the sensitization mechanism recognized each of these proteins as separate and unique allergens.

It will be interesting to note that this logical hypothesis can hold by examining other fruits and vegetables (potato, melon, carrot, radish and orange) treated with ethylene derivatives (Goren and Huberman, 1976; Christofferson and Laties, 1982; Apelbaum et al., 1984; Vreugdenhil et al., 1984). It is most probable that individuals will react to the raw vegetable rather than the cooked preparations. Receptors for ethylene must be present in these fruits and vegetables, as they are in tomato and avocado. Since they recognize the same hormone chemical switch (ethylene), the specific receptors must be very similar, potentiating the possibility of mutual immunological recognition.

6.5
Allergens and Antigens of Latex Products

The number and concentration of allergenic proteins remaining in the finished product are subjects currently undergoing serious and significant scrutiny. Processing variables that might conceivably contribute to this variability are manifold. Antigen isolation is difficult to harmonize because of the variability of antigen source, isolation techniques and the serum pools. Detection by immunological techniques usually involves incubation of the NRL-sensitive sera with various sources of antigens. These antigen sources include non-ammoniated, low-ammoniated, and high-ammoniated NRL. In addition, the antigen sources are either dry rubber latex, latex-finished products, or fresh rubber latex. These have potentially a distinct hydrolytic impact on the NRL antigen and contribute to the broad range of protein size showing serum antibodies with binding activity (Makinen-Kiljunen et al., 1992, 1993; Alenius et al., 1994a). Despite the variability of techniques used in currently published reports, proteins with molecular weight of ~14.5 kDa and 30 kDa appear with significant frequency to merit closer consideration (Yeang et al., 1996).

6.6
Remarks on *Hevea* Latex Usage and Allergy

Development of the medical glove as a natural rubber finished product has revolutionized the protection of both healthcare providers and patients. The manufacture of latex gloves involves the combination of a complex raw material, a variety of chemical additives, and multiparameter process to meet customer requirements. Dermal compatibility issues related to these chemicals have been known since the 1930s, when the first case of irritant and allergic contact dermatitis due to a variety of rubber products were first documented. With occupational exposure to hepatitis and HIV, there has been a re-emphasis on the importance of hand protection. As would be expected, the increased glove usage has resulted in an increase in the frequency of irritant and allergic contact dermatitis (delayed type IV). However, it is the occurrence and rising incidence of immediate type I hypersensitivity to allergenic proteins from NRL that have caused the greatest concern.

Research into this issue has been slow because sensitized individuals recognize different peptide allergens, and not all have routine (consistent) latex contact. This seems to indicate that the cause of latex allergy is varied and multifaceted. NRL allergy has proved to be an important allergy among both healthcare and nonhealthcare professionals. The issues to focus on at present include the development of more accurate in vivo and in vitro diagnostic methods, and the standardization of techniques for the allergenicity testing of various latex gloves and other NRL products. To reach these goals, the molecular structure of important NRL allergens should be identified. Future studies should elucidate whether previously characterized rubber proteins, such as hevein, hevamine and REF are significant allergens.

6.7
Latex Protein Allergens from cDNA Clones

Attempts have been made to isolate and purify NRL proteins through cDNA cloning (Broekaert et al., 1990; Attanyaka et al., 1991; Lee et al., 1991). The two major rubber latex proteins obtained by this method, hevein and REF, are being evaluated for their immunochemical characteristics. Hevein, the major latex lutoids protein, with molecular weight ~5 kDa, was shown to have functions involving rubber particle coagulation and microfloc formation (Gidrol et al., 1994), leading to latex vessel plugging. Hevein has been shown to be the major component of the lutoids and is present at the highest concentration, representing ~70% of the total lutoids soluble B-serum proteins (Audley, 1966).

Hevein is quantitatively the most important latex B-serum protein which has been purified and characterized in detail (Tata, 1976), its amino acid composition and sequence having long been determined (Walujono et al., 1975). Hevein is a chitin-binding latex protein that has been purified and crystallized (Archer, 1960; Audley, 1966), and has also been shown to accumulate as one of the highly induced proteins in the wounding response (Broekaert et al., 1990) of rubber trees following tapping for latex collection. The protein of hevein cDNA clones has been shown to have high reactivity to IgE antibody of patients with latex allergy (Broekaert et al., 1990; Makinen-Kiljunen et al., 1992). Hevein has been implicated as a common molecule in allergen cross-reactivity. The REF, a tightly bound rubber particle protein that is involved in the rubber biosynthesis process, has been characterized and its amino acid sequence determined (Dennis et al., 1989; Light and Dennis, 1989). The nucleotide sequence of REF has been reported for a 14.6 kDa protein (Attanyaka et al., 1991). This protein was shown to be the major allergen in latex, with very high reactivity to the serum IgE antibody of latex-allergic patients (Czuppon et al., 1993b) in several tests and allergenic studies.

Most recently, the cDNA clone for small rubber particle protein (SRPP) was reported in an extensive and detailed study (Oh et al., 1999). The isolation, characterization and functional analysis of the cDNA clone encoding a small rubber particle protein in *Hevea* latex was reported and assessed in great detail. It is generally found that more proteins of SRPP are tightly bound with greater affinity towards small rubber particles than larger particles. These proteins are still found to be tightly bound to rubber particles even after extensive washing of the particles, and centrifugation (Oh et al., 1999). The characterization and functional analysis of the cDNA clone encoding for this 24 kDa protein was elucidated and found to be involved in rubber biosynthesis, its function being similar to that of REF. The SRPP might work in conjunction, or in combination, with the REF which is also present on the surface of rubber particles, together with this 24 kDa protein.

The 24 kDa protein is one of the most tightly bound on the rubber particle surface, similar to REF (Dennis et al., 1989; Light and Dennis, 1989; Oh et al., 1999). Both the REF and 24 kDa protein have been reported to be major rubber latex allergens, with very strong allergenicity and high antigenicity (Makinen-Kiljunen et al., 1992; Yeang et al., 1996, 1998; Oh et al., 1999). These two proteins are the most frequently encountered major latex allergens and pose a serious threat to sensitized latex-allergic patients (Makinen-Kiljunen et al., 1992; Yeang et al., 1998). Developments in the molecular biology and cDNA cloning of these latex allergen proteins will clearly provide important and major input to the progress, and improve our understanding of this increasingly serious health problem. The outcome of these studies will help to ease our concern for effective protection and prevention from contamination or infection, and of allergies associated with NRL finished products encountered in daily life.

7
Outlook and Perspectives

Natural rubber, an important isoprenoid polymer with no known physiological function to the plant, is produced in about 2000 plant species, and in varying degrees of both quality and quantity. Rubber is the raw material of choice for heavy-duty tires and other industrial uses that require elasticity, flexibility and resilience. *Hevea brasiliensis* has, until now, been the only commercial source of natural rubber, mainly because of its abundance in the tree, its quality, and the ease of harvesting. However, the diminishing acreage of rubber plantations and life-threatening latex allergy to *Hevea* rubber, coupled with an increasing demand, have prompted research interest in the study of rubber biosynthesis and the development of alternative rubber sources. Despite this, no better alternative source of natural rubber has been obtained; neither can any alternative match the superior quality, economic advantage, and developed expertise in the rubber plantations of *Hevea brasiliensis*. The possibility of synthetic rubber has been raised, but no satisfactory synthetics can match the superior biological and physical properties of *Hevea* rubber, even though the economic consideration in terms of higher cost can be put aside. Natural rubber is an extraordinary product of polyisoprene synthesis, and *Hevea brasiliensis*, especially when compared with other plant, is the only commercially viable source. A minor exception is guayule rubber, though commercial production of this has yet to be developed. Rubber appears to be a complex of secondary metabolites for which the producing plant has no obvious use, yet is in itself of major commercial importance. For rubber planters, as well as for rubber plant breeders, the early molecular markers of latex production can be of vital use and have immense practical implication. This can only be achieved through an understanding of the biochemistry of rubber biosynthesis and the physiology of rubber trees during their production of latex and their capacity for regeneration upon tapping. The selection of superior rubber clones may be at the seedling stage, based on findings made in multidisciplinary research. However, improvements in rubber yield and rubber quality, as well as the response to ethylene stimulation, are priority criteria for the mutual interest of both the agricultural and industrial exploitation of rubber trees.

In *Hevea brasiliensis*, rubber biosynthesis takes place on the surface of rubber particles suspended in the latex, which is the cytoplasm of laticifers. The laticifers are specialized vessels located adjacent to the phloem of rubber tree. When severed during tapping, the high turgor pressure inside the laticifers expels latex containing 30–50% (w/w) *cis*-1,4-polyisoprene. The differential expression of several rubber biosynthesis-related genes in latex has been documented. REF, a protein (enzyme?) involved in rubber synthesis, is highly expressed in laticifers. These transcribed genes are actively translated into proteins, with some 200 distinct polypeptides being present in *Hevea* latex. Genes expressed in *Hevea* latex can be divided into three groups based on the proteins that they encode: (1) defense-related proteins such as hevein, chitinase, and β-1,3-glucanase; (2) rubber biosynthesis (RB)-related proteins such as REF, HMGR, and FDP synthase; and (3) latex allergen proteins such as hevein and REF, among several other known allergens (Hev b3, Hev b4, Hev b5, and Hev b7). The biological functions of the allergenic proteins are largely unknown, but continuing molecular biology developments and cDNA cloning in the study of these latex allergen proteins will clearly provide major input to progress and a better understanding of this increasingly serious health problem. In addition, benefit will be obtained with regard to effective protection and prevention against contamination or infection, and of allergies associated with the NRL finished products encountered during daily living. Moreover, in the long term it is possible to foresee the transformed rubber tree as a mini-industry in the production of biologically and pharmaceutically important molecules, along with natural rubber. To achieve this, the development of a transformation system for rubber tree is a requisite, in order to identify the laticifer-specific *cis* elements, probably in the promoter of genes which show laticifer specific expression. Laticifer-specific gene expression has been well documented, an example being the production of rubber particle proteins expressed exclusively in rubber plants. One might consider also the use of a model system such as dandelion (Russian rubber), a rubber-producing weed, having anastomosed laticifers like the rubber tree, for transformation and promoter analysis of latex and rubber formation. This situation is similar to that with *Arabidopsis* in our understanding of plant molecular biology. To comprehend the molecular and biological aspects of rubber biosynthesis, it is important to investigate the gene expression profile in latex. Among the genes most abundantly expressed in latex were found cDNA clones encoding major rubber particle proteins of 14.6 kDa and 24 kDa, both of which cause allergenic responses in sensitized patients. The amino acid sequence of the 24 kDa protein is highly homologous to that of REF, suggesting its potential involvement in rubber synthesis.

Because rubber formation takes place in laticifers, genes highly expressed in such tissues may code for enzymes involved in rubber biosynthesis. A number of genes that are highly expressed in latex (as compared with leaves) have been identified. The latex RNAs are highly enriched in transcripts encoding rubber biosynthesis-related enzymes (20- to 100-fold) as well as defense-related proteins (10- to 50-fold). These genes, along with defense genes, are among the most abundant transcripts. The latex has been suggested to play a protective role because of the high content of defense-related proteins that it contains. However, it remains to be answered why *Hevea* allocate excessive energy and resources to rubber biosynthesis and the formation of rubber particles that are dispersed in the latex as a colloidal suspension. Each of the rubber particles contains hundreds to thousands of rubber molecules within this enclosing interface. Analysis using GPC has revealed that small rubber particles contain rubber of higher molecular weight than do large particles. Rubber particles are not simply an inert ball of rubber; rather, analyses of the particles of four rubber-producing plants (including *Hevea brasiliensis*) show that the particle surface is a mosaic of proteins, conventional membrane lipids, and other components. Among many latex proteins, two main proteins remain associated with rubber particles after repeated washing, namely REF (14 kDa) with large particles, and the 24 kDa protein, with small particles. Rubber transferase (*cis*-prenyltransferase) activity has been reported for the rubber particles-bound proteins of three rubber-producing plants, *Hevea brasiliensis*, *Parthenium argentatum*, and *Ficus elastica*. Although at least two rubber particles proteins, REF and rubber transferase, have been suggested to be involved in rubber biosynthesis in *Hevea*, the detailed mechanism is still little known. Although the gene encoding for rubber transferase has not been cloned, a full-length cDNA encoding REF has been cloned. It has been established that REF plays a functional role in rubber polymerization; however, the actual role of REF and the nature of rubber transferase in rubber elongation has not been fully assessed.

The present major concern is the health problem related to latex allergy. Some latex proteins have been shown to be allergens which create a serious biomedical problem and deserve the due attention of molecular biologists and biochemists. In the rubber industry today, an enormous amount of protein in latex (~5 g L^{-1}) is wasted after the coagulation and separation of natural rubber. This rich source of nitrogen can be used as an important by-product after suitable treatment. Not only the proteins, but also numerous nonrubber constituents in the latex, are also overlooked and discarded. Indeed, close attention should be paid to the potential use of these precious chemicals, which not only contribute to the colloidal stability of latex but also to the maintenance of important metabolic functions in the cytoplasm. In addition to *cis*-polyisoprenoids, many nonrubber isoprenoids produced in the latex may offer commercial potential. The above outline forms only part of the vast interest in *Hevea* latex, and clearly some areas of investigation have been omitted here. Nonetheless, in our resource-scarce world, attention should be perhaps be paid in the short term to a more comprehensive utilization of this type of valuable industrial by-product. The discussion presented has attempted to highlight the importance of a highly specialized branch of isoprenoids biosynthesis that occurs naturally in rubber trees. The vast potential for improving the rubber industry, together with a potential for establishing spin-off industries, is clear, with outlooks and perspectives not confined to rubber but extending to the other, non-rubber, constituents of *Hevea* latex.

Two recent developments of interest in natural rubber research have been: (1) the reduction of protein in latex finished products to minimize allergic effects; and (2) vulcanization of rubber latex by radiation, rather than with sulfur. Both biological and physical approaches have been used to remove the proteins from rubber particles in latex before its use in manufactured products. Protease treatment of the latex to hydrolyze or digest the proteins away from the rubber particles has been attempted, but this appears ineffective as the proteins tend to localize in hydrophobic regions inaccessible to the proteases. Several enzymes have been used, but none has provided any satisfactory outcome, and problems concerning steric hindrance must be addressed if optimum protease treatment is to be realized. It appears paradoxical that exogenous protein (as proteases) be added to rid the latex of endogenous proteins; indeed, the suggestion has been made that the proteases might themselves act as additional allergens. A double centrifugation process to concentrate latex offers one physical means of reducing latex proteins, but whether this technique will remove tightly bound proteins from the rubber particles remains unclear. A second approach, which is 'environmentally friendly', is to use radiation for vulcanization, rather than sulfur. This would avoiding the production of sulfur-containing waste and be beneficial to the environment, though the performance of both vulcanized rubber types must be comparable.

8 Relevant Patents

Backhaus, R.A. and Pan, Z. (1997) Rubber particle protein gene from guayule. U.S. Patent. 5 633 433.

Beezhold, D.H. (1996) Methods to remove proteins from natural rubber latex. U.S. Patent. 5 563 241.

Cornish, K. (1998) Hypoallergic natural rubber products from parthenium argentatum (gray) and other non-*Hevea brasiliensis* species. U.S. Patent. 5 717 050.

Dove, J.S. (1997) Methods for reducing allergenicity of natural rubber latex articles and articles so produced. U.S. Patent. 5 691 446.

Dove, J.S. (1998) Methods for reducing allergenicity of the natural rubber latex articles. U.S. Patent. 5 741 885.

Ji, W. (1994) Methods for extracting polyisoprenes from plants. U.S. Patent. 5 321 111.

Lui, J.H. and Shreve, D.S. (1987) Rubber polymerase and methods for their production and use. U.S. Patent. 4 638 028.

Ong, C.O. (1998) Preservation and enhanced stabilization of latex. U.S. Patent. 5 840 790.

Raikhel, N. V. et al. (1999) cDNA encoding a polypeptide including a hevein sequence. U.S. Patent. 5 900 480.

Schloman, W. W. (1999) Reduced lipid natural rubber latex. U.S. Patent. 5 998 512.

Sikora, L.A. (1991) DNA fragment encoding a rubber polymerase and its use. U.S. Patent. 4 983 729.

Tanaka, Y. et al. (1999) Process for preparing deproteinized natural rubber latex molding and deproteinizing agent for natural rubber latex. U.S. Patent. 5 910 567.

Tanaka, Y. et al (1997) Means for mechanically stabilizing deproteinized natural rubber latex. U.S. Patent. 5 610 212.

Trautman, J.C. (1998) Method of neutralizing protein allergens in natural rubber latex product formed thereby. U.S. Patent. 5 777 004.

Umland, H. and Petri, C. (1998) Preserved and stabilized natural latex, with water soluble carboxylic acid salts. U.S. Patent. 5 773 499.

9 References

Abraham, P. D., Taylor, R. S. (1967) Stimulation of latex flow in *Hevea brasiliensis, Exp. Agric.* **3**, 1–12.

Abraham, P. A., Wycherley, P. R., Pakianathan, S. W. (1968) Stimulation of latex flow in *Hevea brasiliensis* by 4-amino-3,5,6-trichloropicolinic acid and 2-chloroethane phosphonic acid, *J. Rubb. Res. Inst. Malaya* **20**, 291–305.

Ahlroth, M., Alenius, H., Makinen-Kiljunen, S., Turjanmaa, K., Reunala, Palosuo, T. (1994) Cross-reacting allergens in natural rubber latex and avocado [abstract.], *J. Allergy Clin. Immunol.* **93**, 299.

Alenius, H., Turjanmaa, K., Palosuo, T., Makinen-Kiljumem, S., Reunala, T. (1991) Surgical latex glove allergy: Characterization of rubber protein allergens by immunoblotting, *Int. Arch. Allergy Appl. Immunol.* **96**, 376–380.

Alenius, H., Reunala, T., Turjanmaa, K. (1992) Detection of IgG_4 and IgE antibodies to rubber proteins by immunoblotting in latex allergy, *Allergy Proc.* **13**, 75–79.

Alenius, H., Kurup, V., Kelly, K., Palosuo, T., Turjanmaa, K., Fink, J. (1994a) Latex allergy: Frequent occurrence of IgE antibodies to a cluster of 11 latex proteins in patients with spina bifida and histories of anaphylaxis, *J. Lab. Clin. Med.* **123**, 712–720.

Alenius, H., Makinen-Kiljunen, S., Turjanmaa, K., Palosuo, T., Reunala, Reunala, T. (1994b) Allergen and protein content of latex gloves, *Ann. Allergy* **75**, 315–320.

Allen, P. W., Jones, K. P. (1988) A historical perspective of the rubber industry, in: *Natural Rubber Science and Technology* (Roberts, A.D., Ed), pp. 1–34. New York: Oxford University Press.

Amalon, Z., Bangratz, J., Chrestin, H. (1992) Ethrel (ethylene releaser) induced increase in the adenylated pool and transtonoplast pH within latex cells, *Plant Physiol.* **98**, 1270–1276.

Anabarro, B., Garcia-Ara, M. C., Pascual, C. (1993) Associated sensitization to latex and chestnut, *Allergy* **48**, 130–131.

Apelbaum, A., Winkler, C., Seakiotakia, E. (1984) Increased mitochondrial DNA and RNA polymerase activity in ethylene-treated potato tubers, *Plant Physiol.* **76**, 461–465.

Archer, B. L. (1960) The protein of *Hevea brasiliensis* latex. 4. Isolation and characterization of crystalline hevein, *Biochem. J.* **75**, 236–240.

Archer, B. L. (1980) Polyisoprene, in: *Encyclopedia of Plant Physiology New Series, Vol. 8, Plant Products* (Bell, E.A., Charlwood, B.W., Eds.), pp. 316–327. Basel: Springer-Verlag.

Archer, B. L. (1983) An alkaline protease inhibitor from *Hevea brasiliensis* latex, *Phytochemistry* **22**, 633–639.

Archer, B. L., Audley, B. G. (1967) Biosynthesis of Rubber, in: *Advances in Enzymology. Vol. 29* (Nord, F.F., Ed.), pp. 221–257. New York: Interscience.

Archer, B. L., Audley, B. G. (1973) Rubber, gutta percha and chicle, in: *Phytochemistry Vol. 2* (Miller, L.P., Ed), pp. 310–343. New York: Van Nostrand Reinhold.

Archer, B. L., Audley, B. G. (1987) New aspects of rubber biosynthesis, *Bot. J. Linnean Soc.* **94**, 309–332.

Archer, B. L., Barnard, G., Cockbain, E. G. C., Dickenson, P. B., McMullen, A. I. (1963a) Structure, composition and biochemistry of *Hevea* latex, in: *Chemistry and Physics of Rubber-Like Substances* (Bateman, L., Ed.), pp. 41–72. London: McLaren and Sons.

Archer, B. L., Audley, B. G., Cockbain, E. G., McSweeney, G. P.(1963b) Biosynthesis of rubber, *Biochem. J.* **89**, 565–574.

Archer, B. L., Cockbain, E. G., McSweeney, G. P., Hong, T. C. (1969) Studies on the composition of latex serum and bottom fraction, *J. Rubb. Res. Inst. Malaya* **21**, 560–569.

Archer, B. L., Audley, B. G., Bealing, F. L. (1982) Biosynthesis of rubber in *Hevea brasiliensis, Plast. Rubb. Intl* **7**, 109–111.

Attanyaka, D. P. S. T. G., Kekwick, R. G. O., Franklin, F. H. C. (1991) Molecular cloning and nucleotide sequencing of rubber elongation factor from *Hevea brasiliensis, Plant Mol. Biol.* **16**, 1079–1081.

Audley, B. G. (1965) Studies of organelles in *Hevea* latex containing helical protein microfibrils, in: *Proc. Natl: Rubber Prod. Res. Assoc., Jubilee Conf.* (Mullins, L., Ed.), pp. 67–72. London: McLaren & Sons.

Audley, B. G. (1966) The isolation and chemical composition of helical protein microfibrils from *Hevea brasiliensis* latex, *Biochem. J.* **98**, 335–341.

Audley, B. G., Archer, B. L. (1988) Biosynthesis of rubber, in: *Natural Rubber Science and Technology* (Roberts, A. D., Ed.), pp. 35–62. London: Oxford University Press.

Baba, T., Allen, C. M. (1980) Prenyl transferase from *Micrococcus luteus, Arch. Biochem. Biophys.* **200**, 474–484.

Bach, T. J. (1986) Hydroxylmethylglutaryl CoA reductase, a key enzyme in phytosterol synthesis, *Lipids* **21**, 82–88.

Backhaus, R. A. (1985) Rubber formation in plants: a mini-review, *Israel J. Bot.* **34**, 283–293.

Backhaus, R. A., Walsh, S. (1983) The ontogeny of rubber formation in guayule, *Parthenium argentatum* Gray, *Bot. Gaz.* **144**, 391–400.

Baulkwill, W. J. (1989) The history of natural rubber production, in: *Rubber* (Webster, C. C., Baulkwill, W. J., Eds), pp. 1–56. Essex: Longman.

Bealing, F. J. (1965) Role of rubber and other terpenoids in plant metabolism, in: *Proc. Nat. Rubber Prod. Res. Assoc., Jubilee Conf.* (Mullins, L., Ed), pp. 113–122. London: McLaren & Sons.

Bealing, F. J. (1969) Carbohydrate metabolism in *Hevea* latex, availability and utilization of substrates, *J. Rubb. Res. Inst. Malaya* **21**, 445–455.

Bealing, F. J. (1976) Quantitative aspects of latex metabolism: possible involvement of precursors other than sucrose in the biosynthesis of *Hevea* rubber, in: *Proc. Intl. Rubb. Conf. Vol. 2*, pp. 543–565. Kuala Lumpur: RRIM.

Benedict, C. R. (1983) The biosynthesis of rubber, in: *Biosynthesis of Isoprenoid Compounds* (Porter, J.W., Spurgeon, S.L., Eds), pp. 355–369. New York: John Wiley & Sons.

Bonner, J. (1967) Rubber biosynthesis, in: *Biogenesis of Natural Compounds* (Bernfield, P., Ed.), pp. 491–552. Oxford: Pergamon.

Bonner, J. (1975) Physiology and chemistry of guayule, in: *An International Conference on the Utilization of Guayule* (McGinnies, N.G., Haas, E.F., Eds.), pp. 78–83. Tucson: University of Arizona.

Bonner, J., Galston, A. (1947) Rubber formation in plants - Review, *Bot. Rev.* **13**, 543–596.

Broekaert, W., Lee, H., Kush, A., Chua, N-.H., Raikhel, N. (1990) Wound-induced accumulation of mRNA containing a hevein sequence in laticifers of rubber tree (*Hevea brasiliensis*), *Proc. Natl: Acad. Sci. USA* **87**, 7633–7637.

Brown, M. S. Goldstein, J. L. (1980) Multivalent feedback regulation of HMG CoA reductase, a control mechanism co-ordinating isoprenoid synthesis and cell growth, *J. Lipid. Res.* **21**, 505–517.

Brydson, J. A. (1978) The historical development of rubber chemistry, in: *Rubber Chemistry*, pp. 2–9. London: Applied Science.

Bucks, D. A., Nakayama, F. S., French, O. F. (1984) Water management for guayule rubber production, *Trans. ASAE* **27**, 1763–1770.

Bucks, D. A., Nakayama, F. S., French, O. F., Legard, W. W., Alexander, W. L. (1985a) Irrigated guyule-production and water use relationships, *Agric. Water Manag.* **10**, 95–102.

Bucks, D. A., Roth, R. L., Nakayama, F. S., Gardner, B. R. (1985b) Irrigation water, nitrogen, and bioregulation for guayule production, *Trans. ASAE* **28**, 1196–1205.

Chambeyron, C., Dry, J., Leynadier, F. (1992) Study of allergenic fractions of latex allergy, *Allergy* **47**, 92–97.

Chrestin, H. (1989) Biochemical aspects of bark dryness induced by over-stimulation of rubber trees with Ethrel, in: *Physiology of rubber tree latex* (d'Auzac J., Jacob, J.-L., Chrestin, H., Eds.), pp. 431–4421. Florida: CRC Press.

Chrestin, H., Gidrol, X. (1986) Contribution of lutoidic tonoplast in regulation of cytosolic pH of latex from *Hevea brasiliensis*, in: *Proc. Int. Rubber Conf. Vol. 3*, pp. 66–87. Kuala Lumpur: RRIM.

Chrestin, H., Jacob, J.-L., d'Auzac, J. (1986) Biochemical basis for the cessation of latex flow and occurrence of physiological bark dryness, in: *Proc. Int. Rubber Conf. Vol. 3*, pp. 20–42. Kuala Lumpur: RRIM.

Chrispeels, M. J., Raikhel, N. V. (1991) Lectins, Lectin genes and their role in plant disease, *Plant Cell* **3**, 1–9.

Christofferson, R. E., Laties, G. G. (1982) Ethylene regulation of gene expression in carrots, *Proc. Natl: Acad. Sci. USA* **79**, 4060–4064.

Churngchow, N., Suntaro, A., Wititsuwannakul, R. (1995) β-1,3-Glucanase isozymes from latex of *Hevea brasiliensis, Phytochemistry* **39**, 505–509.
Chye, M. L., Kush, A., Tan, T. C., Chua, N. H. (1991) Characterization of cDNA and genomic clones encoding HMG CoA reductase from *Hevea brasiliensis, Plant Mol. Biol.* **16**, 567–577.
Clausen, M. K., Christansen K., Jensen, P. K., Behnke, O. (1974) Isolation of lipid particles from baker's yeast, *FEBS Lett.* **43**, 176–179.
Cornforth, J. W., Cornforth, R. H., Donninger, C., Popjak, G. (1966) Biosynthesis of cholesterol, steric course of hydrogen eliminations and C-C bond formation in squalene biosynthesis, *Proc. R. Soc. B London.* **163**, 429–432.
Cornforth, J. W., Clifford, K., Mallaby, R., Phillips, G. T. (1972) Stereochemistry of isopentenyl pyrophosphate isomerase, *Proc. R. Soc. B London.* **182**, 277–281.
Cornish, K. (1993) The separate role of plant *cis* and *trans* prenyl transferases in *cis*-1,4-polyisoprene biosynthesis, *Eur. J. Biochem.* **218**, 267–271.
Coupe, M., Chrestin, H. (1989) Physico-chemical and biochemical mechanisms of hormonal (ethylene) stimulation, in: *Physiology of rubber tree latex* (d'Auzac J., Jacob, J.-L., Chrestin, H., Eds.), pp. 295–319. Florida: CRC Press.
Cox, A. D., Der, C. J. (1992) Protein prenylation: more than just glue? *Curr. Opin. Cell Biol.* **4**, 1008–1016.
Czuppens, J. L., Van Durme, P., Dooms-Goosens, A. (1992) Latex allergy in patient with allergy to fruit [letter], *Lancet* **339**, 493.
Czuppon, A., Chen, Z., Baur, X. (1993a) Chemical synthesis of a peptide representing a major latex allergen, *Chest* **104**, 159S.
Czuppon, A. B., Chen, Z., Rennert, S., Engelke, T., Meyer, H. E., Heber, M., Baur, X. (1993b) The rubber elongation factor of rubber trees (*Hevea brasiliensis*) is the major allergen in latex, *J. Allergy Clin. Immunol.* **92**, 690–697.
d'Auzac, J. (1989a) Factors involved in the stopping of flow after tapping, in: *Physiology of rubber tree latex* (d'Auzac J., Jacob, J.-L., Chrestin, H., Eds.), pp. 257–285. Florida: CRC Press.
d'Auzac, J. (1989b) Historical account, in: *Physiology of rubber tree latex* (d'Auzac J., Jacob, J.-L., Chrestin, H., Eds.), pp. 289–293. Florida: CRC Press.
d'Auzac, J., Jacob, J.-L. (1969) Regulation of glycolysis in latex of *Hevea brasiliensis, J. Rubb. Res. Inst. Malaya* **21**, 417–444.
d'Auzac, J., Jacob, J.-L. (1989) The composition of latex from *Hevea brasiliensis* as a laticiferous cytoplasm, in: *Physiology of rubber tree latex* (d'Auzac J., Jacob, J.-L., Chrestin, H., Eds.), pp. 59–96. Florida: CRC Press.
d'Auzac, J., Sanier, C., Chrestin, H. (1986) Study of a NADH-quinone-reductase producing toxic oxygen from *Hevea* latex, in: *Proc. Int. Rubb. Conf., Vol. 3*, pp. 102–126. Kuala Lumpur: RRIM.
Dalrymple, S. J., Audley, B. G. (1992) Allergenic proteins in dipped products: Factors influencing extractable protein levels. *Rubb. Dev.*. **45**, 51–60.
de Camargo, A. P., Schmidt, N. C., Cardoso, R. M. G. (1976) South American leaf blight epidemics and rubber phenology in Sao Paulo, in: *Proc. Intl. Rubb. Conf. Vol. 3*, pp. 251–265. Kuala Lumpur: RRIM.
De Corres, L., Moneo, I., Minoz, D. Bernada, G., Fernandez, E., Auaicana, M., Urrutia, T. (1993) Sensitization from chestnuts and bananas in patients with urticaria and anaphylaxis from contact with latex, *Ann. Allergy* **70**, 35–39.
de Fay, E., Hebant, Ch., Jacob, J. L. (1989) Cytology and cytochemistry of the laticiferous system, in: *Physiology of rubber tree latex* (d'Auzac J., Jacob, J.-L., Chrestin, H., Eds.), pp.15–30. Florida: CRC Press.
Dennis, M. S., Henzel, W. J., Bell, J., Kohr, W., Light, D. R. (1989) Amino acid sequence of rubber elongation factor protein associated with rubber particles in *Hevea* latex, *J. Biol. Chem.* **264**, 18618–18626.
Dickenson, P. B. (1965) The ultrastructure of the latex vessel of *Hevea brasiliensis*, in: *Proc. Natl. Rubber Prod. Res. Assoc., Jubilee Conf.* (Mullins, L., Ed.), pp. 52–66. London: McLaren & Sons.
Dickenson, P. B. (1969) Electron microscopical studies of latex vessel system of *Hevea brasiliensis, J. Rubb. Res. Inst. Malaya* **21**, 543–559.
Dooms-Goossens, A. (1988) Contact urticaria caused by rubber gloves [letter], *J. Am. Acad. Dermatol.* **18**, 1360–1361.
Dupont, J., Moreau, F., Lance, C., Jacob, J. L. (1976) Phospholipid composition of the membrane of lutoids of *Hevea brasiliensis* latex, *Phytochemistry* **15**, 1215–1217.
Fisher, A.A. (1986) Contact dermatitis from foods and food additives, In: *Contact Dermatitis*, pp. 582–586. Philadelphia: Lea & Febiger.
Fisher, A. A. (1987) Contact urticaria and anaphylactoid reaction due to cornstarch surgical glove power, *Contact Dermatitis* **16**, 224–225.
Fisher, A. A. (1993) Association of latex and food allergy, *Cutis* **52**, 70–71.
Foegeding, P. M., Busta F. F. (1991) Disinfection, sterilization and preservation, in: *Chemical Food Preservatives* (Block, S. S., Ed.), pp. 816–820. Philadelphia: Lea & Febiger.

Gidrol, X., Chrestin, H., Tan, H. L., Kush, A. (1994) Hevein, a lectin like protein from *Hevea brasiliensis* (rubber tree) is involved in the coagulation of latex, *J. Biol. Chem.* **269**, 9278–9283.

Goldstein J. L., Brown, M. S. (1990) Regulation of mevalonate pathway, *Nature* **343**, 425–430.

Gomez, J. B. (1976) Comparative ultracytology of young and native vessel in *Hevea brasiliensis*, in: *Proc. Int. Rubber Conf. Vol. 2*, pp. 143–164. Kuala Lumpur: RRIM.

Gomez, J. B., Hamzah, S. (1989) Frey-Wyssling complex in *Hevea latex* – Uniqueness of the organelle, *J. Natl: Rubb. Res.* **4**, 75–85.

Gomez, J. B., Moir, G. F. J. (1979) The ultracytology of latex vessels in *Hevea brasiliensis, Malaysian Rubber Research and Development Board*. Monograph No. **8**.

Gomez, J. B., Southorn, W. A. (1969) Studies in lutoid membrane ultrastructure, *J. Rubb. Res. Inst. Malaya* **21**, 513–523.

Gomez, J. B., Tata, S. J. (1977) Further studies on the occurrence and distribution of microhelices in clones of *Hevea*, *J. Rubb. Res. Inst. Malaya* **25**, 120–124.

Goren, R., Huberman, M. (1976) Effects of ethylene and 2,4-D on the activity of cellulose isozymes in abscission zones of the developing orange fruit, *Plant Physiol.* **37**, 123–127.

Goss, R. A., Benedict, C. R., Keithly, J. H., Nessler, C. L., Stipanovic, R. D. (1984) cis-Polyisoprene synthesis in guayule plants (*Parthenium argentatum* Gray) exposed to low, non-freezing temperatures, *Plant Physiol.* **74**, 534–537.

Goyvaerts, E., Dennis, M., Light, D., Chua, N. H. (1991) Cloning and sequencing of the cDNA encoding the rubber elongation factor of *Hevea brasiliensis*, *Plant Physiol.* **97**, 317–321.

Granady, L. C., Slater, J. E. (1995) The history and diagnosis of latex allergy, in: *Immunology and Allergy Clinics of North America; Latex Allergy, Vol. 15* (Fink, J. N., Ed), pp. 21–29. Philadelphia: W. B. Saunders.

Hamann, C. P. (1993) Natural rubber latex protein sensitivity in review, *Am. J. Contact Dermatitis* **4**, 4–29.

Hasma, H. (1987) Proteolipids of the natural rubber particles, *J. Natl: Rubb. Res.* **2**, 129–133.

Hasma, H., Subramaniam, A. (1978) The occurrence of furanoid fatty acid in *Hevea brasiliensis* latex, *Lipids* **13**, 905–908.

Ho, C. C., Subramaniam, A., Yong, W. M. (1976) Lipids associated with the particles in *Hevea* latex, in: *Proc. Int. Rubber Conf.* pp. 441–456. Kuala Lumpur: RRIM.

Jacob, J. L., Moreau, F., Dupont, J., Lance, C. (1976) Some characteristics of the lutoids in *Hevea brasiliensis* latex, in: *Proc. Int. Rubber Conf.* pp. 470–483. Kuala Lumpur: RRIM.

Jacob, J. L., Eschbach, J. M., Prevot, J. C., Roussel, D., Lacrotte, R., Chrestin, H., d'Auzac, J. (1986) Physiological basis for latex diagnosis of the functioning of the laticiferous system in rubber-tree, in: *Proc. Int. Rubber Conf. Vol. 3*. pp. 43–65. Kualar Lumpur: RRIM.

Jacob, J. L., Prevot, J. C., Kekwick, R. G. O. (1989) General metabolism of *Hevea brasiliensis* latex (with the exception of isoprenoid anabolism), in: *Physiology of rubber tree latex* (d'Auzac J., Jacob, J.-L., Chrestin, H., Eds.), pp.101–144, Florida: CRC Press.

John, P. (1993) Rubber, in: *Biosynthesis of the Major Crop Products*, pp. 114–126. New York: John Wiley & Sons.

Keenan, M. V., Allen, C. M. (1974) Phospholipid activation of *Lactobacillus plantarum* undecaprenyl pyrophosphate synthetase. *Biochem. Biophys. Res. Commun.* **61**, 338–342.

Kekwick, R. G. O. (1989) The formation of polyisoprenoids in *Hevea* latex, in: *Physiology of rubber tree latex* (d'Auzac J., Jacob, J. L., Chrestin, H., Eds.), Florida: CRC Press.

Kekwick, R. G. O., Archer, B. L., Barnard, D., Higgins, G. M. C., McSweeney, G. P., Moore, C. G. (1959) Incorporation of DL-(2-^{14}C)-mevalonic acid lactone into polyisoprene, *Nature* **184**, 268–270.

Kleinig, H. (1989) The role of plastids in isoprenoid biosynthesis, *Annu. Rev. Plant Physiol. Plant Mol. Biol.* **40**, 39–59.

Koiwa, H., Bressan, R. A., Hasegawa, P. M. (1997) Regulation of protease inhibitors and plant defense, *Trends Plant Sci.* **2**, 379–384.

Koyama, T., Wititsuwannakul, D., Wititsuwannakul, R., Ogura, K. (1995) Analysis of prenyltransferase products from C-serum of *Hevea brasiliensis*, in: *Biopolymers and Bioproducts* (Svasti, J., Ed.), pp. 608–612. Bangkok: IUBMB.

Koyama, T., Wititsuwannakul, D., Asawatheratanakul, K., Wititsuwannakul, R., Ohya, N., Tanaka, Y., Ogura, K. (1996) Isopentenyl diphosphate isomerase in rubber latex, *Phytochemistry* **43**, 769–772.

Kurup, V.P., Kely, K.J., Turjanmaa, K. (1992) Characterization of latex antigen and demonstration of latex-specific antibodies by enzyme-linked immunosorbent assay in patients with latex hypersensitivity, *Allergy Proc.* **6**, 329–333.

Kurup, V. P., Kelly, K. J., Fink, J. N. (1993a) Characterization of monoclonal antibody against

latex protein associated with the latex allergy, *J. Allergy Clin. Immunol.* **92**, 638–643.
Kurup, V. P., Kelly, K. J., Turjanmaa, K. (1993b) Immunoglobulin E reactivity to latex antigens in the sera of patients from Finland and the United States, *J. Allergy Clin. Immunol.* **91**, 1128–1134.
Kurup, V. P., Murali, P. S., Kelly, K. J. (1995) Latex antigens, in: *Immunology and Allergy Clinics of North America; Latex Allergy, Vol. 15* (Fink, J. N., Ed.), pp. 45–59. Philadelphia: W. B. Saunders.
Kush, A. (1994) Isoprenoid biosynthesis: the *Hevea* factory, *Plant Physiol. Biochem.* **32**, 761–767.
Kush, A., Goyvaerts, E., Chye, M. L., Chua, N. H. (1990) Laticifer specific gene expression in *Hevea brasiliensis* (rubber tree), *Proc. Natl. Acad. Sci. USA* **87**, 1787–1790.
Lau, C. M., Gomez, J. B., Subramaniam, A. (1986) An electron microscopy study of the epoxidation of natural rubber particles, in: *Proc. Int. Rubb. Conf. Vol. 2.* pp. 525–529. Kuala Lumpur: RRIM.
Lavaud, F., Cossart, C., Reiter, V., Bernard, J., Deltour, G., Holmquist, I. (1992) Latex allergy in patient with allergy to fruit, *Lancet* **339**, 492–493.
Lee, H., Broekaert, W. F., Raikhel, N. V. (1991) Co- and post-translational processing of the hevein preproprotein of latex of the rubber tree (*Hevea brasiliensis*), *J. Biol. Chem.* **266**, 15944–15948.
Light, D. R., Dennis, M. S. (1989) Purification of prenyltransferase that elongates cis-polyisoprene rubber from the latex of *Hevea brasiliensis*, *J. Biol. Chem.* **264**, 18589–18597.
Light, D. R., Lazarus, R. A., Dennis, M. S. (1989) Rubber elongation by farnesyl pyrophosphate synthases involves a novel switch in enzyme stereospecificity, *J. Biol. Chem.* **264**, 8598–8607.
Low, F. C. (1978) Induction and control of flowing in *Hevea*, *J. Rubb. Res. Inst. Malaya* **26**, 21–32.
Low, F. C., Yeang, H. Y. (1985) Effect of ethephon stimulation on latex invertase in *Hevea*, *J. Rubb. Res. Inst. Malaya* **33**, 37–47.
Lynen, F. (1969) Biochemical problems of rubber synthesis, *J. Rubb. Res. Inst. Malaya* **21**, 389–406.
Mahlberg, P. G. (1993) Laticifers: An historical prospective, *Bot. Rev.* **59**, 1–23.
Makinen-Kilijunen, S., Turjanmaa, K., Palosuo, T., Reunala, T. (1992) Characterization of latex antigens and allergens in surgical gloves and natural rubber by immunoelectrophoretic methods, *J. Allergy Clin. Immunol.* **90**, 230–235.
Makinen-Kilijunen, S., Alenius, H., Palosuo, T., Reunala, T. (1993) Immunoblot inhibition detects several common allergens in rubber latex and banana [Abstract], *J. Allergy Clin. Immunol.* **91**, 242–242.
Martin, M. N. (1991) The latex of *Hevea brasiliensis* contains high levels of both chitinases and chitinases/lysozymes, *Plant Physiol.* **95**, 469–476.
McMullen, A. I. (1962) Particulate ribonucleoprotein components of *Hevea brasiliensis* latex, *Biochem. J.* **85**, 491–495.
McMullen, A. I., McSweeney, G. P. (1966) Biosynthesis of rubber, *Biochem. J.* **101**, 42–47.
Mehta, I. J. (1982) Stem anatomy of *Parthenium argentatum, P. incanum* and their natural hybrids, *Am. J. Bot.* **69**, 503–512.
Mehta, I. J., Hanson, G. P. (1983) Distribution of rubber and comparative stem anatomy of high and low rubber-bearing guayule (*Parthenium argentatum*) from Mexico, in: *Proceedings of the Third International Guayule Conference*, pp. 181–197. Tucson: University of Arizona.
Mernagh, L. R. (1986) Rubber, in: *The New Encyclopedia Britannica, Vol. 21* (15th edn), pp. 282–285. London.
Miyamoto, S., Piela, K., Davis, J. (1984) Water use, growth and rubber yields of guayule selections as related to irrigation regimes. *Irrig. Sci.* **5**, 95–103.
Moir, G. F. J. (1959) Ultracentrifugation and staining of *Hevea* latex, *Nature* **184**, 1626–1628.
Moir, G. F. J., Tata, S. J. (1960) The proteins of *Hevea brasiliensis* latex. III. The soluble proteins of bottom fraction, *J. Rubb. Res. Inst. Malaya.* **16**, 155–159.
Moreau, F., Jacob, J. L., Dupont, J., Lance, C. (1975) Electron transport in the membrane of lutoids from the latex of *Hevea brasiliensis*, *Biochim. Biophys. Acta* **396**, 116–124.
Nutter, A. F. (1979) Contact urticaria to rubber, *Br. J. Dermatol.* **101**, 597–598.
Oh, S. K., Kang, H., Shin, D. H., Yang, J., Chow, K-S., Yeang, H.Y., Wagner, B., Breiteneder, H., Han, K-H. (1999) Isolation, characterization, and functional analysis of a novel cDNA clone encoding a small rubber particle protein from *Hevea brasiliensis*, *J. Biol. Chem.* **274**, 17132–17138.
Paardekooper, E. C. (1989) Exploitation of rubber tree, in: *Rubber* (Webster, C. C., Baulkwill, W. J., Eds), pp. 319–414. Essex: Longman.
Pakianathan, S. W., Milford, G. F. J. (1973) Changes in the bottom fraction contents of the latex during flow in *Hevea brasiliensis*, *J. Rubb. Res. Inst. Malaya* **23**, 391–400.
Pakianathan, S. W., Boatman, S. G., Taysum, D. H. (1966) Particle aggregation following dilution of *Hevea* latex: a possible mechanism for the closure of latex vessels after tapping, *J. Rubb. Res. Inst. Malaya* **19**, 259–271.

Pakianathan, S. W., Wain, R. L., Ng, E. K. (1976) Studies on displacement area on tapping in mature *Hevea* trees, in: *Proc. Int. Rubber Conf. Vol. 3*. pp. 225–246. Kuala Lumpur: RRIM.

Park, R. B., Bonner, J. (1958) The enzymatic synthesis of rubber from mevalonic acid, *J. Biol. Chem.* **233**, 340–342.

Philpott, M. W., Wesgarth, D. R. (1953) Stability and mineral composition of *Hevea* latex, *J. Rubb. Res. Inst. Malaya* **14**, 133–148.

Pujade-Renaud, V., Clement, A., Perrot, C., Prevot, J. C., Chrestin, H., Jacob, J. L., Guern, J. (1994) Ethylene induced increase in glutamine synthetase activity and mRNA levels in *Hevea brasiliensis* latex cells, *Plant Physiol.* **105**, 127–132.

Randall, S. K., Marshall, M. S., Crowell, D. N. (1993) Protein isoprenylation in suspension-cultured tobacco cells, *Plant Cell* **5**, 433–442.

Ritter, F. J. (1954) Biosynthesis of rubber and other isoprenoid compounds, *Rubber J.* **126**, 55–71.

Ross, B. D., McCullough, J., Owenby, D. R. (1992) Partial cross-reactivity between latex and banana allergens, *J. Allergy Clin. Immunol.* **90**, 409–410.

Ryan, C. A. (1984) Defense response in plants, in: *Genes Involved in Microbe Plant Interaction* (Verma, D., Hihn, T., Eds), pp. 377–386. Berlin: Springer-Verlag.

Sherief, P. M., Sethuraj, M. R. (1978) The role of lipids and proteins in the mechanism of latex vessel plugging in *Hevea brasiliensis*, *Physiologia Plantarum* **42**, 351–355.

Southorn, W. A. (1961) Microscopy of *Hevea* latex, in: *Proc. Nat. Rubb. Conf. 1960*. pp. 766–776. Kuala Lumpur: RRIM.

Southorn, W. A. (1969) Physiology of *Hevea* latex flow, *J. Rubb. Res. Inst. Malaya* **21**, 494–512.

Southorn, W. A., Edwin, E. E. (1968) Latex flow studies, II. Influence of lutoids on the stability and flow of *Hevea* latex, *J. Rubb. Res. Inst. Malaya* **20**, 187–200.

Southorn, W. A., Yip, E. (1968) Latex flow studies. III. Electrostatic considerations in the colloidal stability of fresh *Hevea* latex, *J. Rubb. Res. Inst. Malaya* **20**, 201–215.

Stewart, W. D., Wachtel, J. J., Shipman, J. J., Yanks, J. A. (1955) Synthesis of rubber by fungi, *Science* **122**, 1271–1272.

Subramaniam, A. (1976) Molecular weight and other properties of natural rubber: a study of clonal variation, in: *Proc. Int. Rubber Conf. 1975, Vol. 4*. pp. 3–11. Kuala Lumpur: RRIM.

Subramaniam, A. (1995) The chemistry of natural rubber latex, in: *Immunology and Allergy Clinics of North America; Latex Allergy, Vol. 15* (Fink, J. N., Ed.), pp. 1–20. Philadelphia: W. B. Saunders.

Suvachittanont, W., Wititsuwannakul, R. (1995) 3-Hydroxy-methylglutaryl-Coenzyme A synthase in *Hevea brasiliensis*, *Phytochemistry* **40**, 757–761.

Tanaka, Y. (1986) Structural characterization of *cis*-polyisoprene from sunflower, *Hevea* and guayule, in: *Proc. Int. Rubb. Conf. Vol. 2*, pp. 1–9. Kuala Lumpur: RRIM.

Tanaka, Y. (1989) Structure and biosynthesis mechanism of natural polyisoprene, *Prog. Polym. Sci.* **14**, 339–371.

Tanaka, Y. (1991) Recent advances in structural characterization of elastomers, *Rubber Chem. Technol.* **64**, 325–385.

Tanaka, Y., Mori, M., Ute, K., Hatada, K. (1990) Structure and biosynthesis mechanism of rubber from fungi, *Rubber Chem. Technol.* **63**, 39–45.

Tanaka, Y., Eng, A. H., Ohya, N., Nishiyama, N., Tangpakdee, J., Kawahara, S., Wititsuwannakul, R. (1996) Initiation of rubber biosynthesis in *Hevea brasiliensis*: Characterization of initiating species by structural analysis, *Phytochemistry* **41**, 1501–1505.

Tangpakdee, J., Tanaka, Y., Ogura, K., Koyama, T., Wititsuwannakul, R., Wititsuwannakul, D. (1997a) Isopentenyl diphosphate isomerase and prenyl transferase activities in bottom fraction and C-serum from *Hevea* latex, *Phytochemistry* **45**, 261–267.

Tangpakdee, J., Tanaka, Y., Ogura, K., Koyama, T., Wititsuwannakul, R., Wititsuwannakul, D. (1997b) Rubber formation by fresh bottom fraction of *Hevea* latex, *Phytochemistry* **45**, 267–274.

Tangpakdee, J., Tanaka, Y., Jacob, J. L., d'Auzac, J. (1999) Characterization of *Hevea brasiliensis* rubber from virgin trees: A possible role of *cis*-polyisoprene in unexploited tree, *Rubber Chem. Technol.* **72**, 299–307.

Tata, S. J. (1976) Hevein: its isolation, purification and some structural aspects in: *Proc. Int. Rubber Conf. 1975, Vol. 2*. pp. 449–517. Kuala Lumpur: RRIM.

Tata, S. J., Edwin, E. E. (1970) *Hevea* latex enzymes detected by zymogram technique after starch gel electrophoresis, *J. Rubb. Res. Inst. Malaya* **23**, 1–12.

Tata, S. J., Moir, G. F. J. (1964) The proteins of *Hevea brasiliensis* latex, V. Starch gel electrophoresis of C-serum proteins, *J. Rubb. Res. Inst. Malaya* **18**, 97–101.

Tata, S. J., Boyce, A. N., Archer, B. L., Audley, B. G. (1976) Lysozymes: major component of the sedimentable phase of *Hevea brasiliensis* latex, *J. Rubb. Res. Inst. Malaya* **24**, 233–240.

Tata, S. J., Beintema, J. J., Balabaskaran, S. (1983) The lysozyme of *Hevea brasiliensis* latex, isolation purification: enzyme kinetics and a partial amino acid sequence, *J. Rubb. Res. Inst. Malaya* **31**, 35–48.

Tateyama, S., Wititsuwannakul, R., Wititsuwannakul, D., Sagami, H., Ogura, K. (1999) Dolicols of rubber plant, ginkgo and pine, *Phytochemistry* **51**, 11–15.

Taylor, F. R., Parks, L. W. (1978) Metabolic conversion of free sterols and steryl esters in *Saccharomyces cerevisiae*, *J. Bacteriol.* **126**, 531–537.

Truscott, W. (1995) The industry perspective on latex, in: *Immunology and Allergy Clinics of North America; Latex Allergy, Vol. 15* (Fink, J. N., Ed.), pp. 89–121. Philadelphia: W. B. Saunders.

Tupy, J. (1969) Stimulatory effects of 2,4-dichlorophenoxyacetic acid and 1-naphthyacetic acid on sucrose level invertase activity and sucrose utilization in the latex of *Hevea brasiliensis, Planta* **88**, 144–148.

Tupy, J. (1973) The regulation of invertase activity in latex of *Hevea brasiliensis, Exp. Bot.* **24**, 515–524.

Tupy, J. (1989) Sucrose supply and utilization for latex production, in: *Physiology of rubber tree latex* (d'Auzac J., Jacob, J.-L., Chrestin, H., Eds) pp. 179–199, Florida: CRC Press.

Tupy, J., Resing, W. L. (1969) Substrate and metabolism of carbon dioxide formation in *Hevea* latex *in vitro, J. Rubb. Res. Inst. Malaya* **21**, 456–460.

Turjanmaa, K., Laurila, K., Makinen-Kiljunen, S., Reunala, T. (1988) Rubber contact urticaria; Allergenic properties of 19 brands of latex gloves, *Contact Dermatitis* **19**, 362–367.

Turjanmaa, K., Makinen-Kiljunen, S., Reunala, T, Alenius, H., Palosuo, T. (1995) Natural rubber latex allergy: The European experience, in: *Immunology and Allergy Clinics of North America; Latex Allergy, Vol. 15* (Fink, J. N., Ed.), pp. 71–88. Philadelphia: W. B. Saunders.

Vehaar, G. (1959) Natural latex as a colloidal system, *Rubber Chem. Technol.* **32**, 1622–1627.

Vreugdenhil, D., Oerlemans, A. P. C., Steeghs, M. H. G. (1984) Hormonal regulation of tuber induction in radishes (*Raphanus sativus*): The role of ethylene, *Plant Physiol.* **62**, 175–179.

Walujono, K., Schiolma, R. A., Beintema, J. J., Mariono, A., Hahv, A. M. (1975) Amino acid sequence of Hevein, in: *Proc. Int. Rubber Conf. 1975, Vol. 2*. pp. 518–531. Kuala Lumpur: RRIM.

Wititsuwannakul, D., Wititsuwannakul, R., Pasitkul, P. (1997) Lectin binding protein in C-serum of *Hevea brasiliensis, Plant Physiol. (Suppl.)* **114**, 72.

Wititsuwannakul, R. (1986) Diurnal variation of 3-hydroxy-3-methylglutaryl coenzyme A reductase in latex of *Hevea brasiliensis* and its relationship to rubber content, *Experientia* **42**, 44–45.

Wititsuwannakul, R., Wititsuwannakul, D., Sothibanhhu, R., Suvachithanont, W., Sukonrat, W. (1988) Correlation studies on 3-hydroxy-3-methylglutaryl Coenzyme A reductase activity and dry rubber yield in *Hevea brasiliensis*, in: *C.R. Coll. Expl. Physiol. Amel. Hevea, IRCA-CIRAD, France.* 2–7 Nov. pp. 161–172. Paris: IRCA-CIRAD.

Wititsuwannakul, R., Wititsuwannakul, D., Suwanmanee, P. (1990a) 3-Hydroxy-3-methyglutaryl Coenzyme A reductase from the latex of *Hevea brasiliensis, Phytochemistry* **29**, 1401–1403.

Wititsuwannakul, R., Wititsuwannakul, D., Dumkong, S. (1990b) *Hevea* calmodulin: Regulation of the activity of latex 3-hydroxy-3-methylglutaryl coenzyme A reductase, *Phytochemistry* **29**, 1755–1758.

Wititsuwannakul, R., Wititsuwannakul, D., Pasitkul, P. (1997) Rubber latex coagulation by lutoidic lectin in *Hevea brasiliensis, Plant Physiol. (Suppl.)* **114**, 71.

Wititsuwannakul, R., Wititsuwannakul, D., Sakulborirug, C. (1998) A lectin from the bark of the rubber tree (*Hevea brasiliensis*), *Phytochemistry* **47**, 183–187.

Wrangsjo, K., Mellstrom, G., Axelsson, G. (1986) Discomfort from rubber gloves indicating contact urticaria, *Contact Dermatitis* **15**, 70–84.

Yeang, H. Y., David, M. N. (1984) Quantitation of latex vessels plugging by the intensity of plugging, *J. Rubb. Res. Inst. Malaya* **32**, 164–169.

Yeang, H. Y., Paranjothy, K. (1982) Initial physiological changes in *Hevea* latex flow characteristics associated with intensive tapping, *J. Rubb. Res. Inst. Malaya* **30**, 31–36.

Yeang, H. Y., Low, F. C., Gomez, J. B., Paranjothy, K., Sivakumaran, S. (1984) A preliminary investigation into the relationship between latex invertase and latex vessel plugging in *Hevea brasiliensis, J. Rubb. Res. Inst. Malaya* **32**, 50–54.

Yeang, H. Y., Jacob, J. L., Prevot, J. C., Vidal, A. (1986) Invertase activity in *Hevea* latex serum: Interaction between pH and serum concentration, *J. Natl. Rubb. Res.* **1**, 16–22.

Yeang, H. Y., Yusof, F., Abdullah, L. (1995) Precipitation of *Hevea brasiliensis* latex proteins with trichloroacetic acid and phosphotungstic acid in preparation for the Lowry protein assay, *Anal. Biochem.* **226**, 35–43.

Yeang, H. Y., Cheong, K. F., Sunderasan, E., Hamzah, S., Chew, N. P., Hamid, S., Hamilton, R.

G., Cardosa, M. J. (1996) The 14.6 kD rubber elongation factor (Hev b1) and 24 kD (Hev b3) rubber particle proteins are recognized by IgE from patients with spina bifida and latex allergy, *J. Allergy Clin. Immunol.* **98**, 628–639.

Yeang, H. Y., Ward, M. A., Zamri, A. S. M., Dennis, M. S., Light, D. R. (1998) Amino-acid sequence similarity of Hev. b3 to 2 previously reported 27 kDa and 23 kDa latex proteins allergenic to spina bifida patients, *Allergy* **53**, 513–519.

Yip, E., Gomez, J. B. (1984) Characterization of cell sap of *Hevea* and its influence on cessation of latex flow, *J. Rubb. Res. Inst. Malaya* **32**, 1–19.

Yip, E., Southern, W. A. (1968) Latex flow studies, VI. Effects of high pressure gradients on flow of fresh latex in narrow bore capillaries, *J. Rubb. Res. Inst. Malaya* **20**, 248–256.

Yokoyama, H., Hayman., W. J., Hsu, W. J., Poling, S. M., Bauman, A. J. (1977) Chemical bioinduction of rubber in guayule, *Science* **197**, 1076–1077.

Yong, W. M., Singh, M. M. (1976) Thin-layer chromatographic resolution of free amino acids in clonal latices of natural rubber, in: *Proc. Int. Rubber Conf. Vol. 2.* pp. 484–498. Kuala Lumpur: RRIM.

Zhao, X-Q. (1987) The significance of the structure of laticifer with relation to the exudation of latex in *Hevea brasiliensis, J. Natl. Rubb. Res.* **2**, 94–98.

Zhu, J.-K., Bressan, R. A., Hasegawa, P. M. (1993) Isoprenylation of the plant molecular chaperone ANJI facilitates membrane association and function at high temperature, *Proc. Natl. Acad Sci. USA* **90**, 8557–8561.

II
Polyesters

Biopolymers for Medical and Pharmaceutical Applications. Edited by A. Steinbüchel and R. H. Marchessault

ISBN: 3-527-31154-8

4 Applications of PHAs in Medicine and Pharmacy

Dr. Simon F. Williams[1], **Dr. David P. Martin**[2]
[1] Tepha, Inc. 303 Third Street, Cambridge, MA 02142, USA;
Tel.: +01-617-492-0505; Fax: +01-617-492-1996; E-mail: williams@metabolix.com
[2] Tepha, Inc. 303 Third Street, Cambridge, MA 02142, USA;
Tel.: +01-617-492-0505; Fax: +01-617-492-1996; E-mail: martin@metabolix.com

Biopolymers for Medical and Pharmaceutical Applications. Edited by A. Steinbüchel and R. H. Marchessault

ISBN: 3-527-31154-8

HA	hydroxyapatite
mcl-PHA	medium chain-length PHA
M_w	molecular weight
PCL	polycaprolactone
PGA	polyglycolic acid
PHA	polyhydroxyalkanoate
PLA	polylactic acid
poly(3HB)	poly-*R*-3-hydroxybutyrate
poly(3HB-*co*-3HV)	poly-*R*-3-hydroxybutyrate-*co*-*R*-3-hydroxyvalerate
poly(3HB-*co*-4HB)	poly-*R*-3-hydroxybutyrate-*co*-4-hydroxybutyrate
poly(3HO-*co*-3HH)	poly-*R*-3-hydroxyoctanoate-*co*-*R*-3-hydroxyhexanoate
poly(3HP)	poly-3-hydroxypropionate
poly(4HB)	poly-4-hydroxybutyrate
Poly(5HV)	poly-5-hydroxyvalerate
poly(6HH)	poly-6-hydroxyhexanoate
T_g	glass transition temperature
T_m	melting temperature

1 Introduction

Polyhydroxyalkanoates (PHAs) are a class of naturally occurring polyesters that are produced by a wide variety of different microorganisms (Steinbüchel, 1991). Although they are derived biologically, the structures of these polymers bear a fairly close resemblance to some of the synthetic absorbable polymers currently used in medical applications. Owing to their limited availability, the PHAs have remained largely unexplored, yet these polymers offer an extensive range of properties that extend far beyond those currently offered by their synthetic counterparts.

At the last count there were well over 100 different types of hydroxy acid monomers that had been incorporated into PHA polymers, and the list is continuing to grow (Steinbüchel and Valentin, 1995). These monomers include hydroxyalkanoate units ranging from 2- to 6-hydroxy acids substituted with a wide range of groups including alkyl, aryl, alkenyl, halogen, cyano, epoxy, ether, acyl, ester, and acid groups (see Figure 1). By no means will all of these monomers be useful or suitable for medical use; however, they provide a set of materials with properties that range from rigid and stiff to flexible and elastomeric, including polymers that degrade relatively quickly *in vivo* and others that are slow to degrade. In addition, the PHA polymers are thermoplastic in nature, with a wide range of thermal properties, and can be processed using conventional techniques (Holmes, 1988).

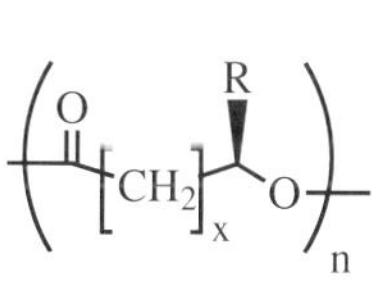

Typical values of x, n and R
x = 1 to 4
n = 1,000 to 10,000
R = alkyl group (C_mH_{2m+1})
or functionalized alkyl group

Fig. 1 General chemical structure of the PHAs.

2 Historical Outline

As a class of polymers, the PHAs are relative newcomers, with many of the different types having been discovered during only the past 20 years. One of the simplest members of the class, poly-*R*-3-hydroxybutyrate, poly(3HB), is an exception as it was first identified in 1925 and is the most well-known PHA polymer. It should be noted however, that the properties of poly(3HB) are not representative of the polymer class as a whole.

During the 1980s, the British company, Imperial Chemical Industries (ICI), developed a commercial process to produce poly(3HB), and a related copolymer known as poly-*R*-3-hydroxybutyrate-*co*-*R*-3-hydroxyvalerate, poly(3HB-*co*-3HV). These polymers were sold under the tradename of Biopol®, and were developed primarily as renewable and biodegradable replacements for petroleum-derived plastics. As a result of these activities and others (Lafferty et al., 1988), both polymers became widely available, which in turn provided opportunities for their evaluation as medical biomaterials. While these efforts have resulted in several promising clinical trials, and development efforts continue, products containing these materials have yet to be approved for *in vivo* medical use.

In 1993, ICI transferred its biological division to Zeneca, which continued to develop PHAs for commodity applications under the tradename Biopol. Zeneca, however, sold its Biopol assets to Monsanto in the mid-1990s. In 2001, an American company,

Metabolix, Inc. acquired the Biopol assets from Monsanto, and is developing transgenic approaches to the large-scale manufacture of PHAs through fermentation and agricultural biotechnology.

More recent interest in the use of PHA polymers for medical applications has arisen primarily in response to the needs of the emerging field of tissue engineering, where a much wider range of absorbable polymers are being sought for use as tissue scaffolds. In fact, in the past two years PHAs have become one of the leading classes of biomaterials under investigation for the development of tissue-engineered cardiovascular products because they can offer properties not available in existing synthetic absorbable polymers.

An American company, Tepha, Inc., is currently engaged in the development of a range of tissue-engineered products based on PHA polymers, and is expanding the number available for medical research to meet both the needs of tissue engineering and the development of more traditional medical devices. As a result of these efforts during the past two years, the number of materials currently under evaluation has expanded and now includes three additional PHA polymers, namely, poly-*R*-3-hydroxyoctanoate-*co*-*R*-3-hydroxyhexanoate (poly(3-HO-*co*-3HH)), poly-4-hydroxybutyrate (poly(4HB)), and poly-*R*-3-hydroxybutyrate-*co*-4-hydroxybutyrate (poly(3HB-*co*-4HB)). This brings the total number of PHA polymers currently under investigation for medical application to five (Figure 2).

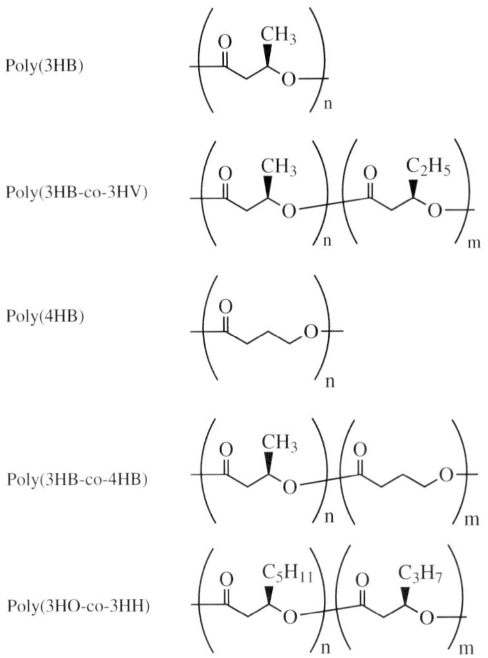

Fig. 2 Chemical structures of PHAs currently under medical investigation.

3
PHA Preparation and Properties: A Primer

3.1
Production

The PHA polymers are accumulated as discrete granules within certain microorganisms at levels reaching 90% of the dry cell mass, and can be isolated fairly readily by breaking open the cells and using either an aqueous-based or solvent-based extraction process to remove cell debris, lipid, nucleic acids, and proteins. Traditionally, these polymers have been produced by fermentation from sugars or oils, often with co-feeds, and the majority of medical studies on poly(3HB), poly(3HB-*co*-3HV), and poly(3HO-*co*-3HH), have been based on polymers derived via this route.

During the late 1980s, the genes responsible for PHA production were isolated, and this has led more recently to the development of transgenic methods for PHA production (see Williams and Peoples, 1996, and references therein). This breakthrough has provided a new means of tailoring the properties of PHA polymers to particular applications, and represents a potentially important advance in the development of technologies to produce designer biomaterials for medical use. Poly-4-hydroxybutyrate (poly(4HB)), for example, is produced using this technology. Transgenic PHA production may also prove to be important in the medical field from a regulatory standpoint, since this technology allows the production host to be selected. For example, PHAs may now be produced by fermentation in *Escherichia coli* K12, a well-characterized host used extensively by the biotechnology industry.

In general, PHA polymers are produced with relatively high molecular weights (M_w) *in vivo*. Commercial grades of poly(3HB) copolymers typically have M_w that are at least 500,000, although PHAs with much longer pendant groups (known as medium chain-length PHAs, mcl-PHAs), such as poly(3HO-*co*-3HH), typically have M_w that are closer to 100,000. Polydispersity is typically around 2.0. By isolating the enzymes responsible for PHA production, namely PHA synthases, researchers have also been able to produce PHA polymer *in vitro* with ultra-high M_w exceeding several million (Gerngross and Martin, 1995), and also *in vivo* in transgenic organisms (Kusaka et al., 1997; Sim et al., 1997).

3.2
Mechanical and Thermal Properties

As a class of polymers, the PHAs offer an extensive design space with properties spanning a large range, and usefully extending the relatively narrow property range offered by existing absorbable synthetics (Engelberg and Kohn, 1991). The mechanical properties of the five PHAs currently being investigated for medical use are shown in Table 1. The homopolymer, poly(3HB), is a relatively stiff, rigid material that has a tensile strength comparable with that of polypropylene. The introduction of a comonomer into this polymer backbone, however, significantly increases the flexibility and toughness of the polymer (extension to break and impact strength), and this is accompanied by a reduction in polymer stiffness (Young's modulus). This is evident in the poly(3HB) copolymers, poly(3HB-*co*-3HV) and poly(3HB-*co*-4HB) (Doi, 1990; Sudesh et al., 2000).

A progressive and substantial change in the mechanical properties of poly(3HB) also occurs when the pendant groups are extended from the polymer backbone. The mcl-PHA, poly(3HO-*co*-3HH), for example, shares the same backbone as poly(3HB), but in contrast is a highly flexible thermoplastic elastomer with properties comparable with those of commercially produced materials (Gagnon et al., 1992).

Extending the distance between the ester groups in the PHA backbone can also have a dramatic impact on mechanical properties. The homopolymer poly(4HB), for example, is a highly ductile, flexible polymer with an extension to break of around 1000%, compared with poly(3HB), which has an extension to break of less than 10%. Combining these different monomers to form copolymers, as in poly(3HB-*co*-4HB), produces one series of materials with a wide range of useful mechanical properties that can be tailored to specific needs. Interestingly, at levels of around 20–40% 4HB, the poly(3HB-*co*-4HB) copolymers actually behave like elastic rubbers.

The thermal properties of PHAs also span wide ranges (see Table 1). Typical melting temperatures (T_m) range from around 55 °C for poly(3HO-*co*-3HH) to about 180 °C for the poly(3HB) homopolymer. Glass transition temperatures (T_g) span the range from about –55 °C to about 5 °C. In general, T_m values decrease as the pendant groups become longer. This is particularly important in the melt processing of poly(3HB), which is unstable at temperatures just above its melting point. Incorporation of other monomers into the poly(3HB) polymer backbone yields lower-melting poly(3HB) copolymers that can be more readily processed. T_g values are also depressed by the incorporation of monomers with longer pendant groups, the depression being relatively modest in the poly(3HB) copolymers, but pronounced for the mcl-PHAs.

3.3
Sterilization of PHA Polymers

For medical use, most PHAs have been sterilized using ethylene oxide, without causing any significant changes to the physico-chemical properties of the polymers. However, low-melting PHAs such as poly(4HB) and poly(3HO-*co*-3HH) are generally sterilized using a cold cycle, particularly if the polymer has been fabricated ready for use. Residual ethylene oxide levels in poly(3HO-*co*-3HH) after cold sterilization with ethylene oxide for 8 h at 38 °C with 65% humidity have been reported to be < 1 ppm after one week (Marois et al., 1999a).

Several studies have described the effects of γ-irradiation on PHA polymers derived from 3-hydroxy acids, such as poly(3HB), poly(3HB-*co*-3HV), and poly(3HO-*co*-3HH). It has been reported that poly(3HB), unlike polyglycolic acid (PGA), can be sterilized by γ-irradiation doses on the order of 2.5 Mrad (Holmes, 1985), although it is likely that some reduction in molecular weight results from this treatment. At higher doses (10–20 Mrad) the mechanical integrity of both

Tab. 1 Mechanical and thermal properties of some representative PHAs

PHA	Poly-(3HB)	Poly(3HB-co-20%3HV)	Poly(4HB)	Poly(3HB-co-16%4HB)	Poly(3HO-co-12%3HH)
Melting temperature [T_m, °C]	177	145	60	152	61
Glass transition temperature [T_g, °C]	4	–1	–50	–8	–35
Tensile strength [MPa]	40	32	104	26	9
Tensile modulus [GPa]	3.5	1.2	0.149	n.d.	0.008
Elongation at break [%]	6	50	1000	444	380

poly(3HB) and poly(3HB-*co*-3HV) are significantly compromised (Miller and Williams, 1987). Luo and Netravali (1999) have also reported significant changes in the mechanical properties and M_w of poly(3HB-*co*-3HV) after exposure to γ-irradiation at doses of 10–25 Mrad.

Exposure of poly(3HO-*co*-3HH) to γ-irradiation at a dose of 2.5 Mrad at room temperature has also been reported to result in a loss of molecular weight on the order of 17%; this is caused by random chain scission, accompanied by some degree of physical cross-linking (Marois et al., 1999a). Thus, while γ-irradiation is generally recognized as a desirable alternative to ethylene oxide for sterilization, care must be exercised in its use on PHA polymers, and the procedures carefully validated.

A few PHA polymers may also be sterilized by steam (Baptist and Ziegler, 1965), particularly if they have T_m over 140 °C, and are thermally stable at this temperature. Holmes (1985) has reported that poly(3HB) powders can be sterilized in this manner.

4 Biocompatibility

Without doubt, the biological response to PHA polymers *in vivo* represents the most important property of these biomaterials if a medical application is being contemplated. Most of the information currently available relates to poly(3HB) and poly(3HB-*co*-3HV), and has been recently reviewed (Hasirci, 2000). A small amount of information on poly(3HO-*co*-3HH), poly(3HB-*co*-4HB), and poly(4HB) has also been published. Care should be exercised in interpreting these data however, since most studies have been based on the use of industrial rather than medical grades of PHA polymers. Notably, Garrido (1999) has described the presence of cellular debris in industrial samples of poly(3HB-*co*-3HV), Rouxhet et al. (1998) detected a number of contaminants on the surface of these samples by X-ray photoelectron spectroscopy, and Williams et al. (1999) reported that an industrial sample of poly(3HB) contained more than 120 endotoxin units per gram. Two methods to remove endotoxin have been reported recently, one being based primarily on the use of peroxide (Williams et al., 1999) and the other by use of sodium hydroxide (Lee et al., 1999).

4.1 Natural Occurrence

Some of the monomers incorporated into PHA polymers are known to be present *in vivo*, and both their metabolism and excretion are well understood. The monomeric component of poly(3HB), *R*-3-hydroxybutanoic acid, for example, is a normal metabolite found in human blood. This hydroxy acid is a ketone body, and is present at concentrations of 3–10 mg per 100 mL blood in healthy adults (Hocking and Marchessault, 1994). This monomer has been administered to obese patients undergoing therapeutic starvation to reduce protein loss (Pawan and Semple, 1983), and also evaluated as an intravenously administered energy source in both humans (Hiraide and Katayama, 1990) and piglets (Tetrick et al., 1995). There is also interest in the use of this monomer in ocular surgery as an irrigation solution to maintain the tissues (Chen and Chen, 1992).

The monomeric component of poly(4HB), 4-hydroxybutanoic acid, is also a naturally occurring substance that is widely distributed in the mammalian body, being present in the brain, kidney, heart, liver, lung, and muscle (Nelson et al., 1981). This 4-hydroxy acid has been used for over 35 years as an intravenous agent for the induction of anesthesia and for long-term sedation (Entholzner et al., 1995). It is also one of the most promising treatments for narcolepsy (Scharf et al., 1998), although unfortunately as with many hypnotics there has been some illegitimate use of this compound. However, since the half-life of the acid is short (35 min), and relatively high doses (several grams) are required to obtain any hypnotic effect, small implants of poly(4HB) could not induce general sedation, for example.

In addition to the known presence of certain PHA monomers in humans, low molecular-weight forms of poly(3HB) have also been detected in human tissues. Reusch and colleagues first identified poly(3HB) in blood serum (0.6–18.2 mg L^{-1}) complexed with low-density lipoproteins, and with the carrier protein albumin (Reusch et al., 1992). The oligomers have also been detected in human aorta (Seebach et al., 1994), and are known to form ion channels *in vivo* when complexed with polyphosphate (Reusch et al., 1997).

4.2 *In vitro* Cell Culture Testing

Relatively few studies have attempted to characterize the tissue response of PHAs caused by leachables such as impurities, additives, monomers, and degradation products. Chaput et al. (1995) evaluated the cytotoxic responses of three poly(3HB-*co*-3HV) compositions (7, 14, and 22% hydroxyvalerate) using direct contact and agar diffusion cell culture tests, and reported that the solid polymers elicited mild to moderate cellular reactions *in vitro*. However, the cytotoxicity of extracts from these polymers varied with the medium, surface-to-volume ratio, time and temperature. Dang et al. (1996) also evaluated an extract from an industrial sample of poly(3HB-*co*-3HV) in an *in vitro* cell culture test method with mouse fibroblasts, and reported that the extract appeared slightly to suppress cellular activity.

In other *in vitro* testing, Rivard et al. (1995) showed that porous poly(3HB-*co*-9%3HV) substrates (Selmani et al., 1995), when seeded with canine anterior cruciate ligament (ACL) fibroblasts, sustained a cell proliferation rate similar to that observed in collagen sponges for around 35 days, with maximal cell density occurring after 28 days. Interestingly, the poly(3HB-*co*-9%3HV) substrates maintained their structural integrity during the culturing, whereas the collagen foams contracted substantially and produced significantly less protein. In evaluating poly(3HB) as a potential drug delivery matrix, Korsatko et al. (1983a) also reported no significant differences in cellular growth with mice fibroblasts.

Saito et al. (1991) evaluated poly(3HB) sheets in an inflammatory test using the chorioallantoic membrane of the developing egg, and reported that the polymer did not cause any inflammation.

Several reports have described the effects of small, low molecular-weight, crystalline particles of poly(3HB) on the viability of cultured macrophages, fibroblasts, co-cultures of Kupffer cells and hepatocytes, and osteoblasts (Ciardelli et al., 1995; Saad et al., 1996a,b,c). These particles represent one of the degradation products expected to arise *in vivo* from the absorption of poly(3HB) and DegraPol®, a phase-segregated multiblock polyesterurethane copolymer. At low concentrations, the small poly(3HB) particles were found to be well tolerated by macrophages, fibroblasts, Kupffer cells and hepatocytes. Macrophages, Kupffer cells, and to a lesser extent fibroblasts and osteoblasts, were all found to take up (phagocytose) the small particles of poly(3HB) (1–20 μm), and evidence of biodegradation by macrophages was also found (Ciardelli et al., 1995). Hepatocytes, in contrast, demonstrated no signs of poly(3HB) phagocytosis. At high concentrations ($>10\ \mu g\ mL^{-1}$), phagocytosis of poly(3HB) particles was found to cause cell damage and cell activation in macrophages and to a lesser degree in osteoblasts, but not in fibroblasts (Saad et al., 1996a,b,c). Separately, the chondrocyte compatibility of a DegraPol foam was also evaluated *in vitro*. Rat chondrocytes were found to attach to about 60% of the foam compared with a polystyrene control, and proliferated at comparable rates (Saad et al., 1999), leading to the conclusion that the DegraPol foam had acceptable chondrocyte compatibility.

Of particular interest in the evolving field of tissue engineering was a report by Rouxhet et al. (1998) on the effect of adhesion and proliferation of monocytes-macrophages to a poly(3HB-*co*-8%3HV) film when modified by hydrolysis or coated with different proteins. As anticipated, the cells were found to have a greater affinity for the polymer surface after it had been hydrolyzed to liberate additional carboxylate and hydroxyl functions. However, it was also found that adhesion of this cell type increased significantly when fibronectin was adsorbed to the polymer surface, but not when collagen or albumin were pre-absorbed.

Cellular attachment to porous tubes made from poly(3HO-*co*-3HH) under different seeding conditions has been evaluated (Stock et al., 1998). Although dynamic cell seeding techniques were found initially to result in a higher rate of ovine smooth muscle cellular attachment compared with static seeding, higher attachment was not sustained under simulated blood flow conditions. Cell attachment to a composite material of PGA and poly(3HO-*co*-3HH) has also been reported (Sodian et al., 1999). After seeding with myofibroblasts and endothelial cells, these composites were incubated in a bioreactor under pulsatile flow. After eight days, near-confluent layers of cells were observed with the formation of extracellular matrix. Sodian et al. (2000a) also studied cellular attachment to porous samples of poly(4HB), and compared the results to those obtained with a porous poly(3HO-*co*-3HH) material and a PGA mesh. After seeding and incubating these materials with ovine vascular cells for eight days, there were significantly more cells on the PGA, although after exposure to flow no significant differences were found. A considerable amount of collagen development was noted for each sample, with the highest amounts present in the PGA meshes. Cellular attachment to a composite of poly(4HB) with a PGA mesh has also been evaluated *in vitro* recently, and compared with the mesh alone and a poly(4HB) foam (Nasseri et al., 2000). Better cell migration into the composite, and better shape retention were observed.

4.3 *In vivo* Tissue Responses

Some of the earliest investigations of the *in vivo* tissue responses to PHA polymers were made by W. R. Grace and Co. in the mid-1960s (Baptist and Ziegler, 1965). In these early studies, film strips of poly(3HB) were implanted subcutaneously and intramuscularly in rabbits, and removed after eight weeks. Examination of the implant sites revealed granulomatous foreign body reactions, but these did not affect the underlying area.

Since these early investigations, many reports have been made describing the *in vivo* tissue responses of poly(3HB) and poly(3HB-*co*-3HV) in both biocompatibility and application-directed studies. Chaput et al. (1995) described one of the longest *in vivo* studies, in which poly(3HB-*co*-3HV) films (containing 7, 14, and 22% valerate) were sterilized by ethylene oxide and implanted intramuscularly in sheep for up to 90 weeks. No abscess formation or tissue necrosis was seen in the vicinity of the implants. However, after 1 week *in vivo*, acute inflammatory reactions with numerous macrophages, neutrophils, lymphocytes and fibrocytes were observed in a capsule at the interface between the polymers and the muscular tissues. After 11 weeks, the observed reaction was less intense with a lower density of inflammatory cells present, though lymphocytes were still observed in the capsule and muscular tissues. At this stage, the capsules were reported to consist primarily of connective tissue cells, and were dense and well-vascularized with highly organized oriented fibers and fibroblastic cells aligned in parallel with the polymer surfaces. A large number of fatty cells were also observed in the capsule, as well as at the interface and in adjacent muscles after long-term implantation (at 70 and 90 weeks). Interestingly, few differences were observed between the capsules, tissue characteristics or cellular activity in terms of the compositions of the three poly(3HB-*co*-3HV) polymers.

Similar results were also observed by Gogolewski et al. (1993) when poly(3HB) and poly(3HB-*co*-3HV) samples were implanted subcutaneously in mice. Fibrous capsules of around 100 μm thickness developed after one month, and these increased to 200 μm by three months, but then thinned to 100 μm at six months. However, the number of inflammatory cells was found to increase with valerate content, and a few granulocytes were still present around blood vessels near encapsulated implants containing 22% valerate at six months. Separately, Tang et al. (1999) suggested that leachable impurities and low molecular-weight poly(3HB) are at least partly responsible for increased collagen deposition following an *in vivo* study of subcutaneous poly(3HB) implants in rats.

Williams et al. (1999) reported a 40-week subcutaneous implant study of poly(3HO-*co*-3HH) in mice. At two weeks, there was minimal reaction to the implants which had been encapsulated by a thin layer of fibroblasts, four to six cell layers thick, surrounded by collagen. There was no evidence of macrophages, and the tissue response continued to be very mild at 4, 8, 12 and 40 weeks, with the amount of connective tissue surrounding the implants remaining fairly constant. The polymer proved to be particularly inert, and could be readily removed with little tissue adherent to the implants. An extract from poly(3HO-*co*-3HH) was also tested in a standard skin sensitization test (ASTM F270), but no discernable erythema/eschar formation was observed.

Subcutaneous implants of poly(4HB) have also been reported to be well tolerated *in vivo* during the course of their degradation (Martin et al., 1999), with minimal inflammatory responses occurring.

It is worth noting that, *in vivo*, as most PHA polymers break down they release hydroxy acids that are significantly less acidic and less inflammatory than many currently used synthetic absorbable polymers (Taylor et al., 1994). For example, poly(3HB) and poly(4HB), are derived from 3- and 4-hydroxybutanoic acids (pK_a 4.70 and 4.72, respectively), that are significantly less acidic than the 2-hydroxy acids (glycolic acid, pK_a 3.83; lactic acid, pK_a 3.08) found in PGA and poly-lactic acid (PLA). Furthermore, significant differences in the mechanism of degradation of these synthetic polymers, which can degrade autocatalytically from the inside outward, can result in substantial amounts of acidic degradation products being released. In one clinical study, for example, around 5% of the patients receiving PGA screws had an inflammatory reaction to the implants that was sufficient to warrant operative drainage (Böstman, 1991).

Finally, Holmes (1988) has reported that poly(3HB) shows negligible oral toxicity, the LD_{50} being greater than 5 g kg^{-1}.

5 Biodistribution

The biodistribution of poly(3HB) microspheres in mice (Bissery et al., 1984a), and poly(3HB) granules in rats (Saito et al., 1991) has been investigated using ^{14}C-labeling and, as anticipated, results have been found to depend upon particle size. In the first study, microspheres of 1–12 μm diameter were injected intravenously into mice, and traced at 0.5, 1, and 24 h, and every seven days thereafter. After 30 min, 47% of the radioactivity was found in the lungs, 14% in the liver, and 2.1% in the spleen. After 1 h, concentrations in the lungs and liver had increased to 62 and 16%, respectively, and by 24 h there was still 60% in the lungs and 24% in the liver. Thereafter, the amounts remained fairly constant, but fell somewhat in the lungs. In the rat study, granules of 500–800 nm diameter were injected through a tail vein into rats and traced at intervals of 2.5 h, 1 day, 13 days, and 2 months. After 2.5 h, approximately 86% of the radioactivity had accumulated in the liver, with 2.5% and 2.4% of the total distributed in the spleen and lungs, respectively. During the following two months, radioactivity levels in most of the tissues decreased slowly, but steadily.

6 Bioabsorption

The rates of bioabsorption of PHA polymers *in vivo* vary considerably, and depend primarily upon their chemical compositions. Other factors such as their location, surface area, physical shape and form, crystallinity, species, and molecular weight can also be very important. While useful information can be derived from *in vitro* studies, results of *in vitro* studies with PHA polymers are not always good indicators of *in vivo* behavior.

6.1 *In vitro* Degradation

In order to investigate the mechanism of degradation of poly(3HB) and poly(3HB-*co*-3HV) *in vivo*, a number of studies have been conducted to determine their rates of hydrolysis *in vitro* (see Holland et al., 1987, 1990; Yasin et al., 1990; Knowles and Hastings, 1992; Chaput et al., 1995). These studies have used complementary techniques such as gravimetric and molecular weight analysis, as well as measurements of surface and tensile properties to monitor different aspects of degradation and develop a concept of the overall degradation process. This has led to the following general scheme of *in vitro* behavior for poly(3HB) and poly(3HB-*co*-3HV). Initially, some surface modification is observed, with water diffusing into the polymer and porosity increasing. Crystallinity also increases, but there is relatively little change in molecular weight in the first few months, and tensile properties remain fairly constant. As the porosity increases, hydrolysis of the polymer chains releases degradation products that can diffuse away more easily. The molecular weight decreases, erosion increases, and both weight and tensile strength begin to decrease more rapidly. At about one 1 year, the initial resistance to degradation is followed by an accelerated degradation with the material becoming more brittle, but not losing its physical integrity. After one year, the most apparent change in the physical appearance of the polymer is loss of surface gloss and the development of surface rugosity.

Other *in vitro* studies have examined the action of additives such as polysaccharides (Yasin et al., 1989), polycaprolactone (PCL) (Yasin and Tighe, 1992), as well as lipases, PHA depolymerases, and several extracts on PHA degradation. Although PHA depolymerases are abundant in the environment and are responsible for PHA biodegradation in soil, there is currently no evidence that these enzymes are present *in vivo*. Mukai et al. (1993) investigated the action of 16 lipases on five different PHA polymers prepared either by fermentation or synthetically, and found that none of these enzymes catalyzed the hydrolysis of poly(3HB). However, the other four PHA polymers were hydrolyzed by lipases, with the number of lipases capable of hydrolyzing the PHA polymer chains decreasing in the following order: poly-3-hydroxypropionate (poly(3HP)) > poly(4HB) > poly-5-hydroxyvalerate (poly(5HV)) > poly-6-hydroxyhexanoate (poly(6HH)). Interestingly, two lipases have been detected recently in tissue adjacent to poly(3HB) implants in rats, raising the possibility of their involvement in poly(3HB) bioabsorption (Löbler et al., 1999). The copolymer, poly(3HB-*co*-3HV), formulated as microspheres with PCL, and loaded with bovine serum albumin, has also been incubated with four different extracts *in vitro* (Atkins and Peacock, 1996a). The percentage weight loss decreased in the order newborn calf serum > pancreatin > synthetic gastric juice > Hanks' buffer, and it was speculated that the enhanced biodegradation in newborn calf serum, and surface erosion in pancreatin, must be due to enzymatic activity in these extracts.

The *in vitro* degradation of poly(3HO-*co*-3HH) has also been examined for up to 60 days (Marois et al., 1999b). When exposed to acid phosphatase and β-glucuronidase for this time period, no significant surface or chemical modifications were observed, and no significant weight loss was detected. It was concluded that this polymer, which shares a common backbone with poly(3HB), degrades slowly by chemical hydrolysis.

Degradation of poly(4HB) *in vitro* has recently been reported (Martin et al., 1999). The homopolymer is fairly resistant to hydrolysis at pH 7.4, and over a 10-week period very little degradation was observed, although a 20–40% reduction in average molecular mass did occur during this time period.

6.2
In vivo Bioabsorption

In early studies, some confusion arose around the stability of poly(3HB) and poly(3HB-*co*-3HV) *in vivo*. Korsatko et al. (1983a, 1984) and Wabnegg and Korsatko (1983) evaluated poly(3HB) for use as matrix retard tablets and reported that the polymer was degraded *in vivo* at a rate directly proportional to the elapsed time (a zero-order reaction). However, it was reported later that monofilaments derived from poly(3HB) and poly(3HB-*co*-3HV) (8 and 17% valerate) showed little, if any, loss of strength when implanted subcutaneously in rats for up to six months (Miller and Williams, 1987), except after γ-irradiation. Many subsequent studies have confirmed that poly(3HB) and poly(3HB-*co*-3HV) do degrade *in vivo*, albeit slowly (Hasirci, 2000). Typically, poly(3HB) is completely absorbed *in vivo* in 24–30 months (Malm et al., 1992b; Hazari et al., 1999a). During the first four weeks *in vivo*, the degree of crystallinity of a sample of poly(3HB) implanted in the peritoneal cavity was reported to have increased, presumably as a result of the amorphous regions of the polymer degrading more rapidly than the crystalline domains (Behrend et al., 2000a). After four weeks, crystallinity, Young's modulus, and microhardness were each shown to have decreased fairly steadily, this being consistent with a surface process.

Kishida et al. (1989) attempted to develop a method to accelerate the bioabsorption of poly(3HB) and poly(3HB-*co*-3HV) *in vivo* by adding basic compounds to the polymers. Although *in vitro* the rate of hydrolysis was found to increase, the effect *in vivo* was minimal, presumably because the basic accelerators had leached out.

The mcl-PHA, poly(3HO-*co*-3HH), also degrades slowly *in vivo*. Williams et al. (1999) reported that the molecular weight (M_w) of subcutaneous implants of poly(3HO-*co*-3HH) in mice decreased from 137,000 at implantation to around 65,000 over 40 weeks, and that there were no significant differences between the molecular weights of samples taken from the surfaces and interiors of the implants. The latter finding suggests that slow, homogeneous hydrolytic breakdown of the polymer occurs.

While poly(3HB), poly(3HB-*co*-3HV), and poly(3HO-*co*-3HH) are generally degraded slowly *in vivo*, consistent with *in vitro* observations, the homopolymer, poly(4HB) is an exception. Martin et al. (1999) found the *in vivo* degradation of this polymer to be relatively rapid, and to vary with porosity. Over a 10-week period it was reported that film, 50%, and 80% porous samples implanted subcutaneously in rats, lost 20%, 50%, and nearly 100% of their mass, respectively. The average molecular mass of the polymer also decreased significantly, but independently of sample configuration. These data suggest that the degradation of poly(4HB) *in vivo* depends in part on surface area, and that the mechanical properties of poly(4HB) implants are likely to undergo a gradual change rather than the more abrupt changes seen with other synthetic absorbables, such as PGA. This might be advantageous, for example, in tissue regeneration applications where a sudden loss of a mechanical property is undesirable, or more gradual loss of implant mass and steady in growth of new tissue are beneficial.

7 Applications

Until recently, only poly(3HB) and poly(3HB-*co*-3HV) were available commercially, and consequently the majority of investigations into applications have focused on a relatively narrow set of polymer properties within the PHA design space. This situation is beginning to change however, with more recent studies involving poly(3HO-*co*-3HH), poly(4HB) and poly(3HB-*co*-4HB).

7.1 Cardiovascular

Without doubt, the major medical use of PHAs has been in the development of cardiovascular products.

7.1.1 Pericardial Patch

One of the most advanced applications of PHA polymers in cardiovascular products has been the development of a regenerative poly(3HB) patch that can be used to close the pericardium after heart surgery, without formation of adhesions between the heart and sternum (Bowald and Johansson, 1990; Malm et al., 1992a,b; Bowald and Johansson-Ruden, 1997). These adhesions represent a significant complication if a second operation is necessary, thereby increasing the risk of rupturing the heart or a major vessel, and prolonging the overall duration of the operation. In an initial study, native pericardium was excised from 18 sheep, and replaced with a nonwoven poly(3HB) patch (Malm et al., 1992a). Patches were then harvested between 2 and 30 months after the operation, examined for adhesions, infection, and inflammatory response, and compared with controls where native pericardium had been removed and left open. Moderate adhesions were present in all the controls, whereas no adhesions developed in 14 of the animals receiving poly(3HB) patches. Interestingly, the pericardium was regenerated in all animals receiving a patch, with the surface of the patches being completely covered with mesothelium-like cells after two months, and a dense underlying collagen layer developing over 12 months. A pronounced tissue response to the patch was observed, with the polymer being slowly phagocytosed by polynuclear macrophages – a finding which has led others to question the biocompatibility of the patch (Tomizawa et al., 1994). Polymer remnants were still present after 24 months, and some macrophages were still found at 30 months, but no platelet aggregates were detected.

Following animal studies with the poly(3HB) patch, a randomized clinical study of 50 human patients admitted for bypass surgery and/or valvular replacement was undertaken (Duvernoy et al., 1995). Using computed tomography (CT), 39 of these patients (19 with the patch and 20 without) were examined for the presence of adhesions at 6 and 24 months, and a lower incidence of postoperative adhesions was reported for the group receiving PHA patches, based on the presence of fat located between the patch and the cardiac surface.

In contrast to these studies, Nkere et al (1998) found no significant difference in a short-term study of adhesion formation among calves undergoing bypass surgery with and without the poly(3HB) patch, as well as calves not undergoing bypass surgery but with their pericardium left open. It was noted however that there might be a species variation, and also that the duration of the studies was different. Also, in comparison with Malm's sheep study, the calves in this study had been subjected to bypass, which was considered more clinically relevant.

7.1.2
Artery Augmentation

Non-woven patches of poly(3HB) have been evaluated in the augmentation of the pulmonary artery as scaffolds for the regeneration of arterial tissue in low-pressure systems (Malm et al., 1994). A total of 19 lambs was used for the trial, with 13 receiving poly(3HB) patches, and six receiving Dacron patches as a control group. No aneurysms were observed in either group, and the pores of the non-woven poly(3HB) were small enough to prevent bleeding. All the patches were harvested between 3 and 24 months, and endothelial layers were found on both patch materials. Beneath the endothelium-like surface, the configuration closely resembled native artery, with smooth muscle cells, collagen and elastic fibers in the poly(3HB) explants. By contrast, a thin collagenous layer had formed under the endothelium lining of the Dacron implants due to the well-known inflammatory reaction to Dacron fiber, and a dense infiltration of lymphocytes was present. As in the case of the pericardial studies, the nonwoven poly(3HB) patch was phagocytosed by polynucleated macrophages, and macrophages persisted, even at 24 months. However, no platelet aggregates were found on the luminal surface.

Highly porous foam patches of poly(4HB) seeded with endothelial, smooth muscle cells, and fibroblasts have also been evaluated in artery augmentation, and with good results (Stock et al., 2000a). A total of six cell-seeded patches, and one unseeded control patch were implanted into the pulmonary artery of sheep. Echocardiography and examination of the cell-seeded explants at 4, 7 and 20 weeks indicated that progressive tissue regeneration had occurred, but with no evidence of thrombus, dilation, or stenosis. Examination of the control patch at 20 weeks revealed slight bulging at the site of implantation, and less tissue regeneration.

7.1.3
Atrial Septal Defect Repair

Malm et al. (1992c) also tested the efficacy of the nonwoven poly(3HB) patch in the repair of atrial septal defects created in six calves. Implants were evaluated between 3 and 12 months, and complete endothelial layers facing the right and left atrium were observed, with a subendothelial layer of collagen and some smooth-muscle cells. As before, the patch was degraded by polynuclear macrophages, with small particles of polymer still present at 12 months, and a foreign body reaction persisting. Nonetheless, the patches prompted the formation of regenerated tissue that resembled native atrial septal wall, and had sufficient strength to prevent the development of shunts in the atrial septal position.

7.1.4
Cardiovascular Stents

One of the main problems with the use of metallic stents in cardiovascular applications is the subsequent restenosis that can result from excessive growth of the blood vessel wall. This is believed to be due (at least in part) to irritation caused by the metallic stent on the vessel wall. A potential solution to this problem may lie in the development of an absorbable stent that can prevent reocclusion of the vessel in the short term, but then be absorbed so that it does not cause any persistent irritation of the vessel wall.

Attention is beginning to focus on the use of PHAs in absorbable stents as well as coatings, often in combination with drug delivery systems. Van der Giessen et al. (1996) evaluated several bioabsorbable polymers, including poly(3HB-*co*-3HV), as candidate biomaterials for cardiovascular stents. Strips of polymer were deployed on the surface of coil wire stents, and implanted in porcine coronary arteries of 2.5–3.0 mm diameter. After four weeks, most of the materials tested, including poly(3HB-*co*-3HV), had provoked extensive inflammatory responses and fibrocellular proliferation. However, these results were shown to be inconsistent with *in vitro* results, and that other factors such as implant geometry, implant design, and degradation products may have been responsible for some of the observed response. It was also noted that the polymers were not sterilized prior to implantation.

The homopolymer, poly(3HB), has also been fabricated into a cardiovascular stent (Schmitz and Behrend, 1997), and tested in a rabbit model (Unverdorben et al., 1998). It was also reported that poly(3HB) stents plasticized with triethyl citrate (Behrend et al., 2000b) and fabricated by laser cutting of a molded construct had an average elastic recoil of about 20–24% immediately after dilation, and of 27–29% after 120 h *in vitro* (Behrend et al., 1998). After implantation into the arteries of rabbits, the poly(3HB) stents instigated a temporary intimal proliferation, and were observed to degrade fairly rapidly *in vivo*.

7.1.5
Vascular Grafts

Vascular grafts are currently inserted to repair or replace compromised blood vessels in arterial or venous systems that have been subjected to damage or disease, for example atherosclerosis, aneurysmal disease or traumatic injury. For the larger-diameter vessels, synthetic grafts are frequently employed, and these can be impregnated with protein to make them completely impervious to blood, and thereby ready for anastomotic procedures. Several studies have shown, however, that when protein substrates are impregnated into grafts they may promote undesirable immunological reactions. In order to try and develop an improved sealant for a synthetic graft, Noisshiki and Komatsuzaki (1995) investigated the possible use of poly(3HB-*co*-4HB) as a graft coating, the coated grafts being implanted into dogs and examined at 2 and 10 weeks. Subsequently, it was noted that degradation of the polymer had already started after two weeks.

Marois et al. (1999c, 2000) investigated the use of poly(3HO-*co*-3HH) as an impregnation substrate. Polyester grafts impregnated with poly(3HO-*co*-3HH) were implanted in rats, and compared with both protein- and fluoropolymer-impregnated grafts for periods ranging from 2 to 180 days. No infiltration of tissue into the poly(3HO-*co*-3HH)-impregnated graft occurred because of the presence of the polymer; moreover, polymer degradation was found to be very slow with just a 30% reduction in molecular weight after six months. Tissue response, after an initial acute phase seen in all grafts, was reported as generally mild, and additional investigations with this biomaterial were recommended.

The elastomeric polymer, poly(3HO-*co*-3HH), has also been evaluated as a component of an autologous cell-seeded tissue-engineered vascular graft in lambs (Shum-Tim et al., 1999). Tubular conduits (7 mm diameter) comprising a nonwoven PGA mesh on the inside and layers of poly(3HO-*co*-3HH) outside were prepared, and seeded with a mixed cell population of endothelial cells, smooth muscle cells, and fibroblasts, obtained by the expansion of sections of carotid artery harvested from lambs. After seven days, the seeded conduits were used to replace 3–4-cm abdominal aortic segments in lambs. The lambs were sacrificed at 10 days and 5 months after surgery, and the conduits compared with unseeded controls. All control conduits became occluded during the study, but the cell-seeded tissue-engineered grafts remained patent (open to blood flow), except for one stricture, and no aneurysms were observed. This contrasts sharply with results obtained using a polyglactin-PGA composite that was limited by high porosity, stiffness, and a relatively short degradation time, and developed aneurysms within a few weeks. Histologic analysis of the poly(3HO-*co*-3HH)-PGA grafts was reported to reveal an insignificant inflammatory response, with increased cell density, collagen formation and mechanical properties that were approaching those of native aorta.

It has also been proposed that poly(3HB) might be used to repair severed blood vessels by the insertion of a tube of this material (Baptist and Ziegler, 1965).

7.1.6 Heart Valves

Perhaps the most remarkable results with PHA polymers have been obtained in the development of cell-seeded tissue-engineered heart valves. These valves promise to provide unique solutions to the deficiencies of mechanical and animal valves currently in clinical use, such as the need for anticoagulant therapy and repeat surgery to replace defective or even outgrown valves in young children. In initial studies in this area, scaffolds seeded with autologous vascular cells and based on porous PGA and PLA had been used to replace a single pulmonary valve leaflet in lambs, but attempts to replace all three pulmonary valve leaflets had failed due to the relatively high stiffness and rigidity of the PGA-PLA biomaterials. However, when the leaflets were replaced with porous poly(3HO-*co*-3HH), and sutured to a conduit composed of a poly(3HO-*co*-3HH) film sandwiched between layers of PGA mesh, these difficulties were overcome (Stock et al., 2000b). Echocardiography of the seeded constructs implanted in lambs indicated no thrombus formation with only mild, nonprogressive, valvular regurgitation up to 24 weeks after implantation. Histologic examination revealed organized and viable tissue, without thrombus formation. Notably, thrombus developed on all leaflets in the unseeded control scaffolds after four weeks. After six weeks *in vivo*, no PGA remained, but poly(3HO-*co*-3HH) was still substantially present with a slight reduction in molecular weight (26%). A variant of this scaffold, with leaflets derived from poly(3HO-*co*-3HH) sandwiched between PGA mesh, has also been evaluated under pulsatile flow *in vitro* (Sodian et al., 1999). Under these conditions, it was shown that vascular cells attached to the scaffold, proliferated, and oriented in the direction of flow after four days.

The design of the heart valve scaffolds have been further refined in subsequent studies, and other PHA polymers have also been evaluated (Sodian et al., 2000a). Sodian et al. (2000b,c,d) also described the fabrication of a functional tissue-engineered heart valve entirely from porous poly(3HO-*co*-3HH), and seeding of the scaffold with vascular cells from ovine carotid artery which resulted in cellular in-growth into the pores and formation of a confluent layer under pulsatile flow conditions.

Recently, in one of the most astonishing results of tissue engineering described to date, Hoerstrup et al. (2000) succeeded in developing a PHA-based heart valve scaffold in lambs that was completely replaced at eight weeks by a functional trileaflet heart valve. Two components contributed to this success. First, the use of poly(4HB), which was coated on a PGA mesh to provide a rapidly degrading yet flexible scaffold; and second, the use of an *in vitro* pulse duplicator system. After 20 weeks *in vivo*, the mechanical properties of the valve were reported to resemble those of the native valve, and histologic analysis showed uniform, layered cuspal tissue with endothelium (see Figure 8 of Schoen and Levy, 1999). Echocardiography demonstrated mobile, functioning leaflets without stenosis, thrombus, or aneurysm to 20 weeks, and importantly, the inner diameter of the valve construct was found to have increased from 19 mm at implant to 23 mm at 20 weeks. The latter finding is particularly exciting for the development of a tissue-engineered heart valve that can be used in young children, and will grow with the child.

7.2 Dental and Maxillofacial

7.2.1 Guided Tissue Regeneration

In guided tissue regeneration, barrier membranes are used to encourage regeneration of new periodontal ligament by creating a space or pocket that excludes gingival connective tissue from the healing periodontal wound, and also by preventing the downgrowth of epithelial tissue into the wound. Galgut et al. (1991) evaluated the histologic response of rats to poly(3HB-*co*-3HV) membranes that could be used in this application, and found that the membranes were well tolerated. Compared with Gore-Tex™ (polytetrafluoroethylene, PTFE) membranes, little downgrowth of epithelial tissue was observed with poly(3HB-*co*-3HV).

During the closure of palatal defects, mucoperiosteal flaps are frequently moved to the midline of the palate, leaving two areas of denuded bone adjacent to the dentition. These wounds heal by migration of keratinocytes and fibroblasts, as well as by wound contraction. Later, the formation of scar tissue occurs, and it is believed that attachment of this tissue to areas of the denuded bone can disturb subsequent maxillary growth, or might affect the development of dentition. Using beagle dogs, Leenstra et al. (1995) investigated the use of nonporous poly(3HB-*co*-3HV) films to keep the mucoperiosteum and bone separated until the transition of teeth was completed (at about 24 weeks). At two weeks, it was reported that one of the poly(3HB-*co*-3HV) films had deformed; however, this was attributed to the method of insertion. At 8 and 12 weeks, the films were unimpaired and surrounded by fibrous capsules, and it was concluded that poly(3HB-*co*-3HV) films were more suitable for this procedure in terms of mechanical properties and tissue response than was PLA or PCL.

7.2.2 Guided Bone Regeneration

In addition to using barrier membranes to create new periodontal ligament, membranes can also be used to generate new bone in jaw bone defects, as well as to increase the width and height of the alveolar ridge. Kostopoulos and Karring (1994a) reported successful bone regeneration in jaw bone defects in rats using poly(3HB-*co*-3HV) membranes to create spaces for bone fill. Mandibular defects were produced in rats and either covered with poly(3HB-*co*-3HV) membranes, or left uncovered. During the following 15 to 180 days, increasing bone fill was achieved with poly(3HB-*co*-3HV) membranes, whereas in the uncovered control group ingrowth of other tissues occurred, and only 35–40% of the defect area was filled with bone after 3–6 months. Kostopoulos and Karring (1994b) also used poly(3HB-*co*-3HV) membranes successfully to increase the height of the rat mandible. In two-thirds of the rats, the space created by the poly(3HB-*co*-3HV) membrane was completely filled with bone by six months, although in some cases soft tissue migrated through ruptures in the membranes, thereby inhibiting bone formation. In contrast, bone formation was negligible when membranes were not used. It was noted that while this biomaterial might form the basis of a barrier membrane, some modifications to the physical properties would be required.

Barrier membranes of poly(3HB-*co*-3HV) have been reinforced with polyglactin fibers and used to cover dental implants placed in fresh extraction sockets in dogs (Gotfredsen et al., 1994). However, very poor results were obtained. After 12 weeks, inflammatory infiltrates were seen adjacent to the poly(3HB-*co*-3HV)-polyglactin membrane that interfered with bone healing, and less bone fill was observed compared with control sites with no membrane. Given the results of Kostopoulos and Karring (1994a,b), and the known inflammatory response of tissues to PGA degradation products, the observed result might be attributed to the polyglactin component of the membrane.

7.3 Drug Delivery

The potential use of poly(3HB) and poly(3HB-*co*-3HV) in drug delivery has been evaluated in a number of studies, and the field has been reviewed several times (Holland et al., 1986; Juni and Nakano, 1987; Koosha et al., 1989; Pouton and Akhtar, 1996; Nobes et al., 1998, Scholz, 2000). Studies have included investigations of these polymers as subcutaneous implants, compressed tablets for oral administration, and microparticulate carriers for intravenous use.

7.3.1
Implants and Tablets

In the early 1980s, Korsatko et al. (1983a,b) studied the release of a model drug, 7-hydroxyethyltheophylline (HET), from poly(3HB) tablets prepared by homogeneously compounding and compressing the polymer and the drug. In both *in vitro* and *in vivo* experiments, the drug was released in a linear manner, but the rate of release from subcutaneous implants in mice was found to be about two- to three-fold slower than *in vitro*. At a drug loading of 10%, sustained release was observed for approximately 10 weeks *in vitro*, and up to 20 weeks *in vivo*. At loadings up to 30%, release could be sustained for up to 50 days; however, at substantially higher levels (60–80%) all the drug was released within 24 h. Changes in tablet compaction pressures did not alter the release rates. In subsequent studies, Korsatko et al. (1987) evaluated the influence of poly(3HB) molecular weight on drug release, and found the release of the antihypertensive drug midodrin-HCl from compressed tablets to be increased as the polymer's molecular weight increased from 3000 to 600,000. In comparison, when poly(3HB)-coated granules of the beta-blocker, Celiprolol-HCl were prepared using a fluid bed dryer system, it was found that smaller amounts of the higher molecular-weight polymers were required to slow the release of Celiprolol-HCl, presumably because of improved film formation at higher molecular weight.

Gould et al. (1987) investigated the release of fluorescein, and dextrans labeled with fluorescein, from tablets of poly(3HB-*co*-3HV) prepared by direct compression. Consistent with the observations of Korsatko et al. (1983a,b), the rate of release was found to increase significantly at higher loadings. Faster rates of release were also observed as the percentage of valerate in the copolymer was decreased, leading to reduced compressibility, and more rapid influx of fluid into the tablet. The higher molecular-weight dextrans also released more rapidly than fluorescein, but this was attributed to the creation of more porous and hydrophilic matrices and led to the testing of porosigen (pore-forming) additives as a means of controlling the rate of drug release. Two additives, microcrystalline cellulose and lactose, were tested, and both were found to enhance the rate of drug release from the modified tablets.

Release profiles of 5-fluorouracil (5-FU) from melt-pressed poly(3HB) disks have been examined *in vitro* over a range of drug concentrations (10–50%) by Juni and Nakano (1987). The rate of release was found to increase with drug loading, and complete release was observed in five days at a 50% loading. Gangrade and Price (1992) also used melt compression to prepare PHA drug delivery systems with lower porosity, but chose the lower-melting poly(3HB-*co*-3HV) copolymer, and incorporated progesterone at loadings of 5 to 50%. Consistent with the findings of others, increased loadings resulted in faster rates of release, with 85% of the drug being released in three days at a 50% loading compared with 50% at a 5% loading.

The homopolymer, poly(3HB), has been evaluated as a potential candidate for the development of a gastric retention drug delivery device that could extend the absorption period of a drug from the stomach (Cargill et al., 1989). However, further development was not pursued because no degradation of the device was observed in a 12-h period during an *in vitro* dissolution assay. Jones et al. (1994) developed an auricular poly(3HB-*co*-3HV) implant for cattle containing metoclopramide at a loading of 50% as a prophylactic treatment for fescue toxicosis (which is caused by cattle grazing on endophyte-infected fescue). The implant was prepared by melt compression, and released metoclopramide at an effective rate of 12 mg per day *in vivo*.

Compression-molded compacts of poly(3HB) loaded with tetracycline have been evaluated in the treatment of periodontal disease (Collins et al., 1989; Deasy et al., 1989). Using *in vitro* studies, poly(3HB) compacts loaded with 50% tetracycline were developed that could deliver therapeutic levels of the antibiotic for eight to nine days. Six patients with gingivitis were treated with these compacts, and their saliva was monitored for the release of tetracycline. Therapeutic levels of tetracycline were detected over the 10-day study period, and an improvement in the gingival condition from moderate to mild inflammation was reported. However, when the treatment was stopped the improvement was not maintained.

Subcutaneous implants of poly(3HB) containing gonadotropin-releasing hormone (GnRH) have been tested for their ability to release this hormone, and stimulate luteinizing hormone (LH) secretion, promote preovulatory follicle growth, and induce ovulation in acylic sheep (McLeod et al., 1988). In comparison with oil-based formulations, the poly(3HB) implants containing 40–50 μg of GnRH consistently produced elevated plasma levels of LH for periods of two to four days, and a high incidence of ovulation was obtained, particularly when two implants per animal were used.

Akhtar et al. (1989, 1991, 1992) investigated the release of a model drug, methyl red, from both solution-cast and melt-processed films of poly(3HB) and poly(3HB-*co*-3HV), and examined the effect of varying the temperature during polymer crystallization. It was found that the faster-crystallizing homopolymer, poly(3HB), was better able to trap methyl red than the slower-crystallizing poly(3HB-*co*-3HV) copolymers, contributing to a slower release of the drug from the homopolymer *in vitro*.

Hasirci et al. (1998) described the development of poly(3HB-*co*-3HV) rods containing the antibiotic Sulperazone for the treatment of osteomyelitis. Rods loaded with 20 and 50% of the drug were prepared by adding granules of the antibiotic to poly(3HB-*co*-3HV) solvent solutions, and molding the resulting pastes into rods. These rods were then introduced into rabbit tibias containing metal implants infected with *Staphylococcus aureus* (obtained from chronic osteomyelitis patients). After 15 days the infection had been eradicated. A similar approach to the treatment of osteomyelitis using poly(3HB-*co*-3HV) rods containing sulbactam-cefoperazone has also been reported (Yagmurlu et al., 1999). Recently, Korkusuz et al. (2001) evaluated the use of poly(3HB-*co*-4HB) and poly(3HB-*co*-3HV) rods as antibiotic carriers for the treatment of osteomyelitis. A bone infection, experimentally induced with *S. aureus*, was effectively treated with solvent-blended rods containing Sulperazone or Duocid. It was noted that the poly(3HB-*co*-4HB) rods were preferred as they were less rigid and easier to handle than the poly(3HB-*co*-3HV) rods.

Kharenko and Iordanskii (1999) prepared poly(3HB) tablets containing the vasodilator, diltiazem, with drug loadings up to about 45%, and monitored release rates of these delivery systems *in vitro*. Near-complete release of the drug was observed at the highest loading concentrations, with slower release being observed at lower loadings.

7.3.2
Microparticulate Carriers

A number of groups have examined the potential use of poly(3HB) and poly(3HB-*co*-3HV) polymers as microparticulate carriers for drug delivery. In general, these systems have been produced using solvent evaporation techniques, and variables such as drug loading, polymer composition, molecular weight, crystallization rate, particle size, and the use of additives have been investigated. Although there are exceptions, the following observations are fairly typical: (1) Increased valerate content in copolymers of poly(3HB-*co*-3HV) usually slows the rate of drug release, presumably because the copolymers are less crystalline than poly(3HB). Incorporation of valerate into poly(3HB) also tends to yield microparticles that are less susceptible to physical damage than poly(3HB); (2) Smaller particle sizes decrease loading capacity, but increase the rate of drug release; (3) Small changes in polymer molecular weight have little impact on release rates; however, large changes can increase crystallinity and lead to enhanced release rates; (4) Drug release from poly(3HB) and poly(3HB-*co*-3HV) is completed well before any significant degradation of the polymers has begun, so drug release is entirely diffusion controlled; and (5) Lower drug loadings reduce the release rate.

One of the earliest studies of poly(3HB) microspheres in drug delivery was carried out by Bissery et al. (1983, 1984a). Using a solvent evaporation technique, ^{14}C-labeled poly(3HB) microspheres (1 – 12 μm) were prepared and found, as expected, to concentrate primarily in the lungs of mice upon intravenous administration. However, when the microspheres were loaded with an anticancer agent, lomustine [*N*-(2-chloroethyl)-*N*′-cyclohexyl-*N*-nitrosourea; CCNU], and administered to Lewis lung carcinoma-bearing mice, little effect was observed (Bissery et al., 1985). From *in vitro* studies it was found that the drug was completely released from the microspheres in 24 h at a loading of 7.4%, compared with a release time of over 90 h when PLA microspheres were used (Bissery et al., 1984b). Notably, the poly(3HB) microspheres were found to be somewhat irregular in shape, which was attributed to the highly crystalline nature of the homopolymer.

Brophy and Deasy (1986) examined the release profiles of sulphamethizole from poly(3HB) and poly(3HB-*co*-3HV) microparticles (53 – 2000 μm) formed by grinding a solvent-evaporated matrix of these components, and reported that increased rates of release were observed when the molecular weight of poly(3HB) was increased. This observation was attributed to poor distribution of the drug in the highly crystalline poly(3HB) polymer. As anticipated, faster rates of release were also observed as the poly(3HB) particle size was decreased, and as drug loading was increased. Incorporation of valerate into the poly(3HB) polymer chain decreased the release rate, presumably on account of the improved distribution of the drug. Overcoating of the poly(3HB) microparticles with PLA reduced the burst effect (the initial, rapid release of the drug from the delivery device surface), as well as significantly reducing the overall release rate. A sustained release of sulphamethizole was demonstrated *in vivo* when poly(3HB) microparticles (425 – 600 μm) loaded at 50% were administered intravenously to dogs. Release was complete in about 24 h, and correlated well with *in vitro* results.

Slower release of several anticancer antibiotics, including doxorubicin, aclarubicin, 4′-*O*-tetrahydropyranyl doxorubicin, bleomycin, and prodrugs of 5-fluoro-2′-deoxyuridine from poly(3HB) microspheres has been reported (Juni et al., 1985, 1986, Juni and Nakano, 1987; Kawaguchi et al., 1992). For example, at a 13% loading of aclarubicin-HCl, poly(3HB) microspheres with a mean diameter of 170 μm were reported to release only 10% of the drug over five days *in vitro*. The observed release rates could, however, be increased by the incorporating fatty acid esters into the poly(3HB) microspheres. These esters were believed to facilitate drug release by forming channels in the poly(3HB) matrix (Kubota et al., 1988). Abe et al. (1992a,b) also reported faster release of an anticancer agent, lastet, from poly(3HB) microspheres when acylglycerols were incorporated into the microspheres.

In vivo, poly(3HB) microspheres containing two prodrugs of 5-fluoro-2′-deoxyuridine were reported to induce higher antitumor effects against P388 leukemia in mice when compared with administration of the free prodrugs over five consecutive days (Kawaguchi et al., 1992).

Koosha et al. (1987, 1988, 1989) reported the use of high-pressure homogenization to produce poly(3HB) nanoparticles containing prednisolone. At drug loadings up to 50%, a biphasic release pattern was observed *in vitro*, with an initial burst effect followed by a slow release of the drug that was complete in one to two days. Similar results were observed with tetracaine. Akhtar et al. (1989) has also prepared smaller poly(3HB) particles (20–40 μm) by spray-drying, and reported fairly rapid release of a model drug (methyl red) from this matrix.

Microspheres of poly(3HB) have been evaluated *in vivo* as controlled release systems for the oral delivery of vaccines that could potentially protect vaccine antigens from digestion in the gut, and target delivery to Peyer's patches (Eldridge et al., 1990). When a single oral dose of poly(3HB) microspheres containing coumarin was administered to mice, very good absorption of microspheres (of diameter < 10 μm) was observed in the Peyer's patches 48 h later.

Conway et al. (1996, 1997) have evaluated the adjuvant properties of poly(3HB) microspheres *in vivo*, and observed a potent antibody response to encapsulated bovine serum albumin (BSA), with the highest titer being generated to preparations containing particles with the smallest sizes. Linhardt et al. (1990) also considered using poly(3HB) and poly(3HB-*co*-3HV) as vaccine delivery vehicles.

The effect of several different parameters on the release of progesterone (a low molecular-weight model drug) from poly(3HB) and poly(3HB-*co*-3HV) microspheres prepared with an emulsion solvent evaporation technique has been evaluated by Gangrade and Price (1991). Using scanning electron microscopy (SEM), the use of gelatin as an emulsifier was shown to provide more spherical microspheres than when using polyvinyl alcohol, sodium lauryl sulfate or methyl cellulose, and that smoother surfaces were obtained when the solvent was switched from chloroform to methylene chloride. Generally, poly(3HB) microspheres had very rough surfaces which became smoother as valerate was incorporated. Interestingly, release of the drug from poly(3HB-*co*-9%3HV) was slower than from either poly(3HB) or poly(3HB-*co*-24%3HV). Examination of the internal surfaces of these microspheres by SEM revealed fewer cavities and less porosity for the poly(3HB-*co*-9%3HV) microspheres, and this was consistent with the slower release observed. Porosity was also found to decrease when the microspheres were prepared at increasing temperatures between 25 and 40 °C. Typically, 60–100% of the drug was released from these microspheres in 12 h at drug loadings up to 12%.

Embleton and Tighe (1992a,b) also investigated the effects of increasing the valerate content of poly(3HB-*co*-3HV), temperature, and molecular weight on microsphere formation, and obtained results consistent with those of Gangrade and Price (1991). When 10–50% PCL was incorporated into these poly(3HB-*co*-3HV) microspheres, a systematic increase in porosity was observed with increasing PCL content, that was attributed to the elution of PCL from the hardened microcapsule wall once the poly(3HB-*co*-3HV) had precipitated during formation (Embleton and Tighe, 1993). Porosity was also found to increase significantly when polyphosphate-Ca^{2+} complexes were introduced into poly(3HB-*co*-3HV) microspheres (Gürsel and Hasirci, 1995).

Atkins and Peacock (1996b) also used PCL to prepare microcapsules (21 – 200 μm) from a blend of poly(3HB-*co*-3HV)/PCL(20%) with an inner reservoir of BSA loaded in an agarose core. The protein encapsulation was, however, low (< 12 %) and only slightly influenced by loading. Release from the microcapsules loaded with up to 50% protein was observed *in vivo* for about 24 days.

Microspheres of poly(3HB) have been evaluated as potential embolization agents in renal arteries (Kassab et al., 1999), and with rifampicin as a chemoembolization agent (Kassab et al., 1997). Release of rifampicin during *in vitro* studies was consistent with other controlled release studies, with near-complete release of the drug being observed at high loadings (40%) in 24 h, and a somewhat prolonged release at lower loadings. Renal angiograms obtained before and after embolization with poly(3HB) microspheres (120 – 200 μm) using a contrast agent showed that 10 mg of the microspheres was sufficient to slow renal arterial blood flow, with subsequent partial occlusion of the pre-capillaries in two adult dogs. When a second injection was given, complete embolization was achieved. Histopathologic examination of the kidneys revealed changes consistent with renal artery obstruction and blockade of the blood supply to the kidneys.

In addition to using poly(3HB) compacts, microspheres of poly(3HB-*co*-3HV) have been used to deliver tetracycline and its hydrochloride salt for the treatment of periodontitis (Sendil et al., 1998, 1999). Depending upon valerate content, between 42 and 90% of the tetracycline was released at 100 h, with loadings up to about 11%. Interestingly, it was found that molecular weight variation in the poly(3HB-*co*-3HV) compositions tested might have influenced the observed release of tetracycline. Encapsulation efficiency of the tetracycline hydrochloride salt was significantly less than the neutral form of the antibiotic.

A novel approach for preparing PHA drug delivery systems (Nobes et al., 1998) involves encapsulating a given drug during the *in vitro* enzymatic formation of poly(3HB) granules (Gerngross and Martin, 1995). Using this technique, it was possible to encapsulate approximately 4.5% of a model drug compound, Netilmicin, compared with between 4.1 and 17% using solvent evaporation.

Wang and Lehmann (1999) recently prepared poly(3HB) microspheres containing levonorgestrel with an average particle size of 64 μm, and found that this system prolongs the release of the drug by 1.8-fold compared with administration of the drug alone. The microspheres were also found to be effective *in vivo*, inducing a contraceptive effect in mice. Chen et al. (2000) recently described microspheres (30 – 40 μm) loaded with diazepam, and reported the characteristic biphasic release pattern with an initial burst effect.

Andersson et al. (1999) disclosed the use of supercritical fluid technology to incorporate the water-insoluble *Helicobacter pylori* adhesion protein A (HpaA) into poly(3HB) particles. This was achieved by preparing an emulsion with the protein, polymer, and methylene chloride, and then extracting the organic solvent from the emulsion with supercritical carbon dioxide to induce particle formation. Particles containing 0.6% protein were produced, with sizes of 1 – 3 μm.

7.4
Prodrugs

In 1999, it was discovered that poly(4HB) could be used as a prodrug of 4-hydroxybutyrate (Williams and Martin, 2001). In this study, rats were dosed (by gavage) with low molecular-weight polymers of poly(4HB) (138 mg kg^{-1}), and the serum was assayed for the presence of monomer. The serum concentration of 4-HB increased to ~86 μM within 30 min, and remained elevated at approximately three- to five-fold the baseline value (9 μM) for about 8 h. In contrast, administration of the monomer resulted in a characteristic rapid increase to 182 μM within 30 min, followed by a rapid decrease to baseline within 2 h. The prolonged release of the monomer from poly(4HB) might potentially be beneficial in the treatment of narcolepsy, alcohol withdrawal, and several other indications. The therapeutic potential of poly(4HB) was also alluded to recently (Sudesh et al., 2000).

7.5
Nerve Repair

Partly on account of the piezoelectric properties of poly(3HB), interest has arisen in the use of this polymer for nerve repair (Aebischer et al., 1988). The basic approach involves aligning severed nerve ends within a small tube of poly(3HB), thus avoiding the need for sutures. Hazari et al. (1999a) and Ljungberg et al. (1999) evaluated the use of a nonwoven poly(3HB) sheet as a wrap to repair transected superficial radial nerves in cats for up to 12 months. Axonal regeneration was shown to be comparable with closure with an epineural suture for a nerve gap of 2–3 mm, and that the inflammatory response created by poly(3HB) was also similar to that found in primary epineural repair. In a subsequent study, the same material was used to bridge an irreducible gap of 10 mm in rat sciatic nerve, and the results were compared to an autologous nerve graft (Hazari et al., 1999b). Good axonal regeneration in the poly(3HB) conduits with a low level of inflammatory infiltration was observed over 30 days, although the rate and amount of regeneration in the poly(3HB) conduit did not fully match that of the nerve graft.

7.6
Nutritional Uses

7.6.1
Human Nutrition

Several groups have evaluated oligomeric forms of the ketone body, *R*-3-hydroxybutanoic acid, as an alternative to the sodium salt of the monomer, for potential nutritional and therapeutic uses. Use of these polymeric forms might provide controlled release systems for the monomer and, importantly, overcome the problems associated with administering large amounts of sodium ion *in vivo*. Tasaki et al. (1998) reported the results of infusing dimers and trimers of *R*-3-hydroxybutyrate into rats, as well as exposure of these compounds to human serum samples and liver homogenate. Although mixtures of these compounds were not hydrolyzed by human serum, the monomer was liberated upon exposure to the liver homogenate as well as after infusion in rats. *In vitro*, the monomer was also liberated after incubation with the enzyme carboxylesterase.

Veech (1998, 2000) and Martin et al. (2000) have evaluated oligomers and oligolides of *R*-3-hydroxybutyrate *in vivo*, and observed release of the ketone body over prolonged periods. Potential uses of these delivery systems might include seizure control, reduction of protein catabolism, appetite suppression, control of metabolic disease, use in parenteral nutrition, increased cardiac efficiency, treatment of diabetes and insulin-resistant states, control of damage to brain cells in conditions such as Alzheimer's, and treatment of neurodegenerative disorders and epilepsy.

7.6.2
Animal Nutrition

The potential use of poly(3HB) and poly(3HB-*co*-3HV) polymers as a source of animal nutrition has been evaluated *in vivo*. Brune and Niemann (1977a) initially reported studies in rats, and subsequently described the digestion of poly(3HB) in pigs (Brune and Niemann, 1977b). In the latter work, when whole cells containing poly(3HB) were used as a nutrient source, approximately 65% of the poly(3HB) was excreted. Forni et al. (1999a,b) found the digestibility of poly(3HB-*co*-3HV) treated with sodium hydroxide to be increased when compared with the untreated polymer, in both sheep and pigs. Holmes (1988) also reported that poly(3HB) is degraded in the bovine rumen, while Peoples et al. (1999) studied the digestion of poly(3HB) and poly(3HO-*co*-3HH) in broiler chicks, concluding that the available energy from these polymers lies between that provided by carbohydrates and oils.

7.7
Orthopedic

Several studies of the use of PHA polymers for internal fixation have been undertaken. Vainionpää et al. (1986) described the use of compression-molded T-plates, prepared from poly(3HB) reinforced with carbon fiber (7%), to fix osteotomies of the tibial diaphysis in rabbits. The implants were fixed to the tibia with absorbable PGA sutures, and compared with implants prepared from reinforced Vicryl®. After 12 weeks, better results were obtained with the reinforced poly(3HB) plates relative to the Vicryl plates, with the latter frequently leading to nonunion of the osteotomies, breakage, and angulation.

Doyle et al. (1991) reported that poly(3HB) can be reinforced with hydroxyapatite (HA) to increase its stiffness to a level approaching that of cortical bone (7–25 GPa). At a HA loading of 40% wt., a poly(3HB)/HA composite had a Young's modulus value of 11 GPa, although strength decreased from ~40 MPa for poly(3HB) to ~20 MPa for the composite. Under *in vitro* conditions in buffered saline at 37 °C it was noted that the modulus of the filled poly(3HB) sample decreased more rapidly, falling from 9 GPa to 4 GPa over four months at a loading of 20% wt. HA. Bending strength of the same filled sample also fell by about 50% over the same time period, from 55 MPa to around 25 MPa. The behavior of poly(3HB) filled with HA was also studied *in vivo*. No significant differences between implants derived from poly(3HB) or poly(3HB) filled with HA were observed when rivets were prepared from these materials and inserted into predrilled holes in rabbit femurs. Over time, increased amounts of new bone were found on the implant surfaces, and by six months the implants were closely encased in new cortical bone. The overall tissue responses were considered favorable, and some indication of osteogenic activity for poly(3HB) was noted. Boeree et al. (1993) and Galego et al. (2000) have also studied the mechanical properties of poly(3HB) and poly(3HB-*co*-3HV)/HA composites, and concluded that they could serve as alternatives to corticocancellous bone grafts.

Since the homopolymer poly(3HB) is piezoelectric, it might help to induce new local bone formation if used as an implant. The addition of other additives might further enhance this property, for example bioactive glasses that upon dissolution and deposition at an implant surface may encourage new bone formation. In this light, Knowles et al. (1991) studied the piezoelectric characteristics of poly(3HB-*co*-3HV) composites with glass fiber at 20, 30 and 40% wt., and found the piezoelectric potential output of these composites to be fairly close to that of bone. In subsequent studies, these composites were evaluated *in vitro* and *in vivo* (Knowles and Hastings, 1993a,b), and have also been studied with the inclusion of HA (Knowles et al., 1992). During *in vitro* studies of the poly(3HB-*co*-3HV)-glass composites, the glass component was found to be highly soluble in the polymer, and the observed weight loss was attributed to dissolution of the glass from the composites. These studies correlated well with observations *in vivo*. The poly(3HB-*co*-3HV)-glass composites were implanted subcutaneously, and as nonload-bearing femoral implants in rats. Initially, relatively high cellular activity was observed that was attributed to ions being released from the glass (and causing a soft tissue reaction), though this activity decreased with time. At four weeks in the femoral implants, cells could be seen entering the surface porosity formed by the solubilizing glass, and over time new bone was seen developing on the implant surface (Knowles and Hastings, 1993a,b). However, when composites of poly(3HB-*co*-3HV)/HA and poly(3HB-*co*-3HV)/HA-glass were implanted in rabbit femurs, and evaluated using a mechanical push-out test during the first eight weeks *in vivo*, it was found that the former bonded better than the latter. One explanation for this observation was attributed to the release of ions from the poly(3HB-*co*-3HV)/HA-glass composite inducing a soft tissue reaction that inhibited the formation of hard tissue at the implant surface (Knowles et al., 1992; Knowles, 1993).

In addition to composites with hydroxyapatite and glass, Jones et al. (2000) tested composites of poly(3HB-*co*-3HV) with tri-calcium phosphate filler in subcutaneous and femoral implants. These implants were compared with composites derived from PLA and tri-calcium phosphate, and found to degrade about four times more slowly *in vivo*.

In an *in vitro* study, Rivard et al. (1996) prepared highly porous foams of poly(3HB-*co*-3HV) and seeded these scaffolds with chondrocytes and osteoblasts. Maximal cell densities were achieved after 21 days, with cellular diffusion taking place throughout the porous foams.

7.8 Urology

In the mid-1960s it was proposed that poly(3HB) could be used to repair a ureter by inserting a short tube of this material (Baptist and Ziegler, 1965). More recently, Bowald and Johansson (1990) described the use of poly(3HB-*co*-3HV) in the development of a tube for urethral reconstruction. A solution of the poly(3HB-*co*-3HV) copolymer was used to coat thin, knitted tubes of Vicryl, and the tubes were then implanted into four dogs to replace the urethra. After six to nine months, it was claimed that a fully functional urethra tissue had been reconstructed in all animals.

7.9 Wound Management

7.9.1 Sutures

As early as the mid-1960s it was suggested that poly(3HB) could be used as an absorbable suture (Baptist and Ziegler, 1965). Others have also proposed that poly(3HB-*co*-3HV) could be used as a suture coating, and have coated braided sutures of PGA with solutions of this polymer (Wang and Lehmann, 1991).

7.9.2
Dusting Powders

Holmes (1985) has produced powders of poly(3HB) with small particle sizes, and proposed that these can be used as medical dusting powders, particularly with surgical gloves.

7.9.3
Dressings

Webb and Adsetts (1986) described wound dressings based on volatile solutions of poly(3HB) and poly(3HB-*co*-3HV) that could form thin films over wounds; these would be especially useful for emergency treatments. These films could potentially prevent airborne bacterial contamination of the wound, but would still be permeable to water vapor. As the preferred solvents are chlorinated hydrocarbons, however, these solutions might present other health hazards both to the patient and administrator. Steel and Norton-Berry (1986) also described wound dressings based on poly(3HB), their method involving the preparation of nonwoven fibrous materials of poly(3HB) and PHBV that could be used like swabs, gauze, lint or fleece.

Davies and Tighe (1995) evaluated the use of poly(3HB) fibers as a potential wound scaffold that could provide a framework for the laying down of a permanent dermal architecture. Using *in vitro* cell attachment assays with human epithelial cells, it was found that pre-treatment of the poly(3HB) fibers with either base or strong acid improved cell attachment and spreading.

Ishikawa (1996) described the implantation of poly(3HB-*co*-4HB) films in the abdominal cavity of rats between incisions in the skin and intestine to prevent coalescence. After one month, the incisions had substantially healed, and no coalescence had occurred. However, no degradation of the film *in vivo* was observed at one year.

7.9.4
Soft Tissue Repair

In early studies of the potential uses of poly(3HB) it was proposed that this polymer could be used in the surgical repair of hernias (Baptist and Ziegler, 1965).

Patches of poly(3HB) with a smooth surface on one side and a porous surface on the other have been evaluated as resorbable scaffolds for the repair of soft-tissue defects, and specifically for the closure of lesions in the gastrointestinal tract (Behrend et al., 1999). Using *in vitro* cell culture, moderate adhesion of mouse and rat intestine fibroblasts to the patch material was observed. When the patches were sutured over incisions in the stomachs of rats that had been closed with two surgical knots, adhesion of the patch to the gastric wall was found to be better than that seen with Vicryl patches, and good ingrowth and tissue regeneration was evident on the porous side of the implanted poly(3HB) patch.

8
Future Directions

With technology now in place to allow the properties of PHA polymers to be tailored to specific applications, coupled with a significant increase in the need for new absorbable biomaterials, this class of polymers currently appears to have a bright future in medicine and pharmacy. Indeed, a wide new range of applications for PHA polymers has recently been described which includes their use as suture anchors, meniscus repair devices, interference screws, bone plates and bone plating systems, meniscus regeneration devices, ligament and tendon grafts, spinal fusion cages and bone dowels, bone graft substitutes, surgical mesh and repair patches, slings, adhesion prevention barriers, skin substitutes, dural substitutes, bulking and filling agents, ureteric and urethral stents, vein valves, ocular cell implants, and hemostats (Williams, 2000; Williams et al., 2000).

9
Patents

Patents cited in studies described in this chapter are listed in Table 2.

Tab. 2 Patents referenced in this work

Patent Number	*Assignee*	*Inventor(s)*	*Title*	*Date of Publication*
WO 88/06866	Brown University Research Found.	Aebischer, P., Valentini, R.F., Galletti, P.M.	Piezoelectric nerve guidance channels	September 22, 1988
WO 99/52507	Astra Aktiebolag	Andersson, M.-L., Boissier, C., Juppo, A.M., Larsson, A.	Incorporation of active substances in carrier matrixes	October 21, 1999
US 3,225,766	W.R. Grace and Co.	Baptist, J.N., Ziegler, J.B.	Method of making absorbable surgical sutures from poly beta hydroxy acids	December 28, 1965
EP 0 349 505 A2	Astra Meditec AB	Bowald, S.F., Johansson, E.G.	A novel surgical material	January 3, 1990
EP 0 754 467 A1	Astra Aktiebolag	Bowald S.F., Johansson-Ruden, G.	A novel surgical material.	January 22, 1997
US 5,116,868	John Hopkins University	Chen, C.-H., Chen, S.C.	Effective ophthalmic irrigation solution	May 26, 1992
EP 355,453 A2.	Kanegafuchi Kagaku Kogyo Kabushiki Kaisha	Hiraide, A., Katayama, M.	Use of 3-hydroxybutyric acid as an energy source	February 28, 1990
GB 2 160 208A.	Imperial Chemical Industries, Plc.	Holmes, P.A.	Sterilised powders of poly(3-hydroxybutyrate)	December 18, 1985
US 5,480,394 WO93/05824	Terumo Kabushiki Kaisha	Ishikawa, K.	Flexible member for use as a medical bag	January 2, 1996 Apr. 1, 1993
WO 99/32536	Metabolix, Inc.	Martin, D.P., Skraly, F.A., Williams, S.F.	Polyhydroxyalkanoate compositions having controlled degradation rates	July 1, 1999
WO 00/04895	Metabolix, Inc.	Martin, D.P., Peoples, O.P., Williams, S.F.	Nutritional and therapeutic uses of 3-hydroxyalkanoate oligomers	February 3, 2000

Tab. 2 (cont.)

Patent Number	***Assignee***	***Inventor(s)***	***Title***	***Date of Publication***
JP7275344A2	Nippon Zeon Co. Ltd.	Noisshiki, Y., Komatsuzaki, S.	Medical materials for soft tissue use	October 24, 1995
WO 99/34687	Metabolix, Inc.	Peoples, O.P., Saunders, C., Nichols, S., Beach, L	Animal nutrition compositions	July 15, 1999
EP 0 770 401 A2.	Biotronik Mess- und Therapiegeräte GmbH & Co.	Schmitz, K.-P., Behrend, D.	Method of manufacturing intraluminal stents made of polymer material	February 5, 1997
US 4,603,070	Imperial Chemical Industries, Plc.	Steel, M.L., Norton-Berry, P.	Non-woven fibrous material	July 29, 1986
WO 98/41200 WO 98/41201	British Tehnology Group Ltd.	Veech, R.L.	Therapeutic compositions	September 24, 1998
WO 00/15216	BTG Int. Ltd.	Veech, R.L.	Therapeutic compositions	March 23, 2000
US 5,032,638	American Cyanamid Co.	Wang, D.W., Lehmann, L.T.	Bioabsorbable coating for a surgical device	July 16, 1991
GB 2,166,354	Imperial Chemical Industries, Plc.	Webb, A., Adsetts, J.R.	Wound dressings	May 8, 1986
WO 00/51662	Tepha, Inc.	Williams, S.F.	Bioabsorbable, biocompatible polymers for tissue engineering	September 8, 2000
WO 00/56376	Metabolix, Inc.	Williams, S.F., Martin, D.P., Skraly, F.	Medical devices and applications of polyhydroxyalkanoate polymers,	September 28, 2000
WO 01/19361A2	Tepha, Inc.	Williams, S.F., Martin, D.P.	Therapeutic uses of polymers and oligomers comprising gamma-hydroxybutyrate	March 22, 2001

10
References

Abe, H., Doi, Y., Yamamoto, Y. (1992a) Controlled release of lastet, an anticancer drug, from poly(3-hydroxybutyrate) microspheres containing acylglycerols, *Macromol. Rep.* **A29**, 229–235.

Abe, H., Yamamoto, Y., Doi, Y. (1992b) Preparation of poly(3-hydroxybutyrate) microspheres containing lastet an anticancer drug and its application to drug delivery system, *Kobunshi Ronbunshu* **49**, 61–67.

Aebischer, P., Valentini, R. F., Galletti, P. M. (1988) Piezoelectric nerve guidance channels, PCT Patent Application No. WO 88/06866.

Akhtar, S., Pouton, C. W., Notarianni, L. J., Gould, P. L. (1989) A study of the mechanism of drug release from solvent evaporated carrier systems of biodegradable P(HB-HV) polyesters, *J. Pharm. Pharmacol.* **41**, 5.

Akhtar, S., Pouton, C. W., Notarianni, L. J. (1991) The influence of crystalline morphology and copolymer composition of drug release from solution cast and melt-processed P(HB-HV) copolymer matrices, *J. Controlled Release* **17**, 225–234.

Akhtar, S., Pouton, C. W., Notarianni, L. J. (1992) Crystallization behaviour and drug release from bacterial polyhydroxyalkanoates, *Polymer* **33**, 117–126.

Andersson, M.-L., Boissier, C., Juppo, A. M., Larsson, A. (1999) Incorporation of active substances in carrier matrixes, PCT Patent Application No. WO 99/52507.

Atkins, T. W., Peacock, S. J. (1996a) *In vitro* biodegradation of poly(β-hydroxybutyrate-hydroxyvalerate) microspheres exposed to Hanks buffer, newborn calf serum, pancreatin and synthetic gastric juice, *J. Biomater. Sci. Polymer Edn.* **7**, 1075–1084.

Atkins, T. W., Peacock, S. J. (1996b) The incorporation and release of bovine serum albumin from poly-hydroxybutyrate-hydroxyvalerate microcapsules, *J. Microencapsulation* **13**, 709–717.

Baptist, J. N., Ziegler, J. B. (1965) Method of making absorbable surgical sutures from poly beta hydroxy acids, US Patent No. 3,225,766.

Behrend, D., Lootz, D., Schmitz, K. P., Schywalsky, M., Labahn, D., Hartwig, S., Schaldach, M., Unverdorben, M., Vallbracht, C., Laenger, F. (1998) PHB as a bioresorbable material for intravascular stents, *Am. J. Cardiol., Tenth Annual Symposium Transcatheter Cardiovascular Therapeutics, Abstract TCT-8*, 4S.

Behrend, D., Nischan, C., Kunze, C., Sass, M., Schmitz, K.-P. (1999) Resorbable scaffolds for tissue engineering, Proc. European Medical and Biological Engineering Conference, *Med. Biol. Eng. Comput.* **37** (Suppl.), 1510–1511.

Behrend, D., Schmitz, K.-P., Haubold, A. (2000a) Bioresorbable polymer materials for implant technology, *Adv. Eng. Mater.* **2**, 123–125.

Behrend, D., Kramer, S., Schmitz, K.-P. (2000b) Biodegradation and biocompatibility of resorbable polyester, *Zell. Interakt. Biomater.* 28–32.

Bissery, M.-C., Puisieux, F., Thies, C. (1983) A study of process parameters in the making of microspheres by the solvent evaporation procedure, *Third Expo. Congr. Int. Technol. Pharm.* **3**, 233–239.

Bissery, M.-C., Valeriote, F., Thies, C. (1984a) *In vitro* and *in vivo* evaluation of CCNU-loaded microspheres prepared from poly(L)-lactide and poly(β-hydroxybutyrate), in: *Microspheres and Drug Therapy, Pharmaceutical, Immunological and Medical Aspects* (Davis, S.S., Illum, L., McVie, J.G., Tomlinson, E., Eds.), Amsterdam: Elsevier, 217–227.

Bissery, M.-C., Valeriote, F., Thies, C. (1984b) *In vitro* lomustine release from small poly(β-hydroxybutyrate) and poly(D,L-lactide) micro-

spheres, *Proc. Int. Symp. Controlled Release Bioact. Mater.* **11**, 25–26.
Bissery, M.-C., Valeriote, F., Thies, C. (1985) Fate and effect of CCNU-loaded microspheres made of poly(D,L)lactide (PLA) or poly-β-hydroxybutyrate (PHB) in mice. *Proc. Int. Symp. Controlled Release Bioact. Mater.* **12**, 181–182.
Boeree, N. R., Dove, J., Cooper, J. J., Knowles, J., Hastings, G. W. (1993) Development of a degradable composite for orthopedic use: mechanical evaluation of an hydroxyapatite-polyhydroxybutyrate composite material, *Biomaterials* **14**, 793–796.
Böstman, O. M. (1991) Absorbable implants for the fixation of fractures, *J. Bone Joint Surg. Am.* **73**, 148–153.
Bowald, S. F., Johansson, E. G. (1990) A novel surgical material, European Patent Application No. 0 349 505 A2.
Bowald, S. F., Johansson-Ruden, G. (1997). A novel surgical material. European Patent Application No. 0 754 467 A1.
Brophy, M. R., Deasy, P. B. (1986) *In vitro* and *in vivo* studies on biodegradable polyester microparticles containing sulphamethizole, *Int. J. Pharm.* **29**, 223–231.
Brune, H., Niemann, E. (1977a) Use and compatibility of bacterial protein (*Hydrogenomonas*) with various contents of poly-beta-hydroxybutyric acid in animal nutrition. 1. Weight development and N balance in growing rats, *Z. Tierphysiol. Tierernähr. Futtermittelkd* **38**, 13–22.
Brune, H., Niemann, E. (1977b) Value and compatibility in animal nutrition of bacterial protein (*Hydrogenomonas*) containing various amounts of poly-beta-hydroxybutyric acid. 2. Body weight, nitrogen balance and fatty acid pattern of liver, muscle and kidney depot fat of growing swine, *Z. Tierphysiol. Tierernähr. Futtermittelkd* **38**, 81–93.
Cargill, R., Engle, K., Gardner, C. R., Porter, P., Sparer, R. V., Fix, J. A. (1989) Controlled gastric emptying. II. *In vitro* erosion and gastric residence times of an erodible device in Beagle dogs, *Pharm. Res.* **6**, 506–509.
Chaput, C., Yahia, L'H., Selmani, A., Rivard, C.-H. (1995) Natural poly(hydroxybutyrate-hydroxyvalerate) polymers as degradable biomaterials, *Mater. Res. Soc. Symp. Proc.* **385**, 49–54.
Chen, C.-H., Chen, S. C. (1992) Effective ophthalmic irrigation solution, US Patent No. 5,116,868.
Chen, J.-H., Chen, Z.-L., Hou, L.-B., Liu, S.-T. (2000) Preparation and characterization of diazepam-polyhydroxybutyrate microspheres, *Gongneng Gaofenzi Xuebao* **13**, 61–64.
Ciardelli, G., Saad, B., Hirt, T., Keiser, O., Neuenschwander, P., Suter, U. W., Uhlschmid, G. K. (1995) Phagocytosis and biodegradation of short-chain poly[β-3-hydroxybutyric acid] particles in macrophage cell line, *J. Mater. Sci. Mater. Med.* **6**, 725–730.
Collins, A. E. M., Deasy, P. B., MacCarthy, D., Shanley, D. B. (1989) Evaluation of a controlled-release compact containing tetracycline hydrochloride bonded to tooth for the treatment of periodontal disease, *Int. J. Pharm.* **51**, 103–114.
Conway, B. R., Eyles, J. E., Alpar, H. O. (1996) Immune response to antigen in microspheres of different polymers, *Proc. Int. Symp. Controlled Release Bioact. Mater.* **23**, 335–336.
Conway, B. R., Eyles, J. E., Alpar, H. O. (1997) A comparative study on the immune responses to antigens in PLA and PHB microspheres, *J. Controlled Release* **49**, 1–9.
Dang, M.-H., Birchler, F., Ruffieux, K., Wintermantel, E. (1996) Toxicity screening of biodegradable polymers. I. Selection and evaluation of cell culture test methods, *J. Environ. Poly. Degrad.* **4**, 197–203.
Davies, S., Tighe, B. (1995) Cell attachment to gel-spun polyhydroxybutyrate fibers, *Poly. Prepr. (Am. Chem. Soc., Div. Polym. Chem.)* **36**, 103–104.
Deasy, P. B., Collins, A. E. M., MacCarthy, D. J., Russell, R. J. (1989) Use of strips containing tetracycline hydrochloride or metronidazole for the treatment of advanced periodontal disease, *J. Pharm. Pharmacol.* **41**, 694–699.
Doi, Y. (1990) *Microbial Polyesters*, New York: VCH.
Doyle, C., Tanner, E. T., Bonfield, W. (1991) *In vitro* and *in vivo* evaluation of polyhydroxybutyrate and of polyhydroxybutyrate reinforced with hydroxyapatite, *Biomaterials* **12**, 841–847.
Duvernoy, O., Malm, T., Ramström, J., Bowald, S. (1995) A biodegradable patch used as a pericardial substitute after cardiac surgery: 6- and 24-month evaluation with CT, *Thorac. Cardiovasc. Surg.* **43**, 271–274.
Eldridge, J. H., Hammond, C. J., Meulbroek, J. A., Staas, J. K., Gilley, R. M., Tice, T. R. (1990) Controlled vaccine release in the gut-associated lymphoid tissues. I. Orally administered biodegradable microspheres target the Peyer's patches, *J. Controlled Release* **11**, 205–214.
Embleton, J. K., Tighe, B. J. (1992a) 5. Regulation of polyester microcapsule morphology, *Drug Tar-*

geting Delivery, 1 (Microencapsulation of Drugs) 45–54.

Embleton, J. K, Tighe, B. J. (1992b) Polymers for biodegradable medical devices. IX: Microencapsulation studies; effects of polymer composition and process parameters on poly-hydroxybutyrate-hydroxyvalerate microcapsule morphology, *J. Microencapsulation* **9**, 73–87.

Embleton, J. K., Tighe, B. J. (1993) Polymers for biodegradable medical devices. X. Microencapsulation studies: control of poly-hydroxybutyrate-hydroxyvalerate microcapsules porosity via polycaprolactone blending, *J. Microencapsulation* **10**, 341–352.

Engelberg, I., Kohn, J. (1991) Physico-mechanical properties of degradable polymers used in medical applications: a comparative study, *Biomaterials* **12**, 292–304.

Entholzner, E., Mielke, L., Pichlmeier, R., Weber, F., Schneck, H. (1995) EEG changes during sedation with gamma-hydroxybutyric acid, *Anesthetist* **44**, 345–350.

Forni, D., Bee, G., Kreuzer, M., Wenk, C. (1999a) Novel biodegradable plastics in sheep nutrition. 2. Effects of NaOH pretreatment of poly(3-hydroxybutyrate-*co*-3-hydroxyvalerate) on *in vivo* digestibility and on *in vitro* disappearance, *J. Anim. Physiol. Anim. Nutr.* **81**, 41–50.

Forni, D., Wenk, C., Bee, G. (1999b) Digestive utilization of novel biodegradable plastic in growing pigs, *Ann. Zootech.* **48**, 163–171.

Gagnon, K. D., Fuller, R. C., Lenz, R. W., Farris, R. J. (1992) A thermoplastic elastomer produced by the bacterium *Pseudomonas oleovorans, Rubber World* **207**, 32–38.

Galego, N., Rozsa, C., Sánchez, R., Fung, J., Vázquez, A., Tomás, J. S. (2000) Characterization and application of poly(β-hydroxyalkanoates) family as composite biomaterials, *Polym. Test.* **19**, 485–492.

Galgut, P., Pitrola, R., Waite, I., Doyle, C., Smith, R. (1991) Histological evaluation of biodegradable and non-degradable membranes placed transcutaneously in rats, *J. Clin. Periodontol.* **18**, 581–586.

Gangrade, N., Price, J. C. (1991) Poly(hydroxybutyrate-hydroxyvalerate) microspheres containing progesterone: preparation, morphology and release properties, *J. Microencapsulation* **8**, 185–202.

Gangrade, N., Price, J. C. (1992) Properties of implantable pellets prepared from a biodegradable polyester, *Drug Dev. Ind. Pharm.* **18**, 1633–1648.

Garrido, L. (1999) Nondestructive evaluation of biodegradable porous matrices for tissue engineering, in *Methods in Molecular Medicine, Vol. 18: Tissue Engineering Methods and Protocols* (Morgan, J. R., Yarmush, M. L., Eds.), Totowa, NJ: Humana Press, 35–45.

Gerngross, T. U., Martin, D. P. (1995) Enzyme-catalyzed synthesis of poly[β-(-)-3-hydroxybutyrate]: formation of macroscopic granules *in vitro, Proc. Natl. Acad. Sci. USA* **92**, 6279–6283.

Gogolewski, S., Jovanovic, M., Perren, S. M., Dillon, J. G., Hughes, M. K. (1993) Tissue response and *in vivo* degradation of selected polyhydroxyacids: Polylactides (PLA), poly(3-hydroxybutyrate) (PHB), and poly(3-hydroxybutyrate-*co*-3-hydroxyvalerate) (PHB/VA), *J. Biomed. Mater. Res.* **27**, 1135–1148.

Gotfredsen, K., Nimb, L., Hjørting-Hansen, E. (1994) Immediate implant placement using a biodegradable barrier, polyhydroxybutyrate-hydroxyvalerate reinforced with polyglactin 910. An experimental study in dogs, *Clin. Oral Impl. Res.* **5**, 83–91.

Gould, P. L., Holland, S. J., Tighe, B. J. (1987) Polymers for biodegradable medical devices IV. Hydroxybutyrate-valerate copolymers as non-disintegrating matrices for controlled-release oral dosage forms, *Int. J. Pharm.* **38**, 231–237.

Gürsel, I., Hasirci, V. (1995) Properties and drug release behaviour of poly(3-hydroxybutyric acid) and various poly(3-hydroxybutyrate-hydroxyvalerate) copolymer microcapsules, *J. Microencapsulation* **12**, 185–193.

Hasirci, V. (2000) Biodegradable biomedical polymers, in: *Biomaterials and Bioengineering Handbook* (Wise, D.L., Ed.), New York: Marcel Dekker, 141–155.

Hasirci, V., Gürsel, I., Türesin, F., Yigitel, G., Korkusuz, F., Alaeddinoglu, G. (1998) Microbial polyhydroxyalkanoates as biodegradable drug release materials, in: *Biomedical Science and Technology* (Hincal, A. A. and Kas, H. S., Eds.), New York: Plenum Press, 183–187.

Hazari, A., Johansson-Rudén, G., Junemo-Bostrom, K., Ljungberg, C., Terenghi, G., Green, C., Wiberg, M. (1999a) A new resorbable wrap-around implant as an alternative nerve repair technique, *J. Hand Surg.* **24B**, 291–295.

Hazari, A., Wiberg, M., Johansson-Rudén, G., Green, C., Terenghi, G. (1999b) A resorbable nerve conduit as an alternative to nerve autograft in nerve gap repair, *Br. J. Plast. Surg.* **52**, 653–657.

Hiraide, A., Katayama, M. (1990) Use of 3-hydroxybutyric acid as an energy source, European Patent Application No. 355,453 A2.

Hocking, P. J., Marchessault, R. H. (1994) Biopolyesters, in: *Chemistry and Technology of Biodegradable Polymers* (Griffin, G.J.L., Ed.), Glasgow: Blackie, 48–96.

Hoerstrup, S. P., Sodian, R., Daebritz, S., Wang, J., Bacha, E. A., Martin, D. P., Moran, A. M., Guleserian, K. J., Sperling, J. S., Kaushal, S., Vacanti, J. P., Schoen, F. J., Mayer, J. E. (2000) Functional trileaflet heart valves grown *in vitro*, *Circulation* **102** (suppl. III), III-44–III-49.

Holland, S. J., Tighe, B. J., Gould, P. L. (1986) Polymers for biodegradable medical devices. 1. The potential of polyesters as controlled macromolecular release systems, *J. Controlled Release* **4**, 155–180.

Holland, S. J., Jolly, A. M., Yasin, M., Tighe, B. J. (1987) Polymers for biodegradable medical devices II. Hydroxybutyrate-hydroxyvalerate copolymers: hydrolytic degradation studies, *Biomaterials* **8**, 289–295.

Holland, S. J., Yasin, M., Tighe, B. J. (1990) Polymers for biodegradable medical devices VII. Hydroxybutyrate-hydroxyvalerate copolymers: degradation of copolymers and their blends with polysaccharides under *in vitro* physiological conditions, *Biomaterials* **11**, 206–215.

Holmes, P. A. (1985) Sterilised powders of poly(3-hydroxybutyrate), UK Patent Application No. 2 160 208A.

Holmes, P. A. (1988) Biologically produced β-3-hydroxyalkanoate polymers and copolymers, in: *Developments in Crystalline Polymers* (Bassett, D.C., Ed.), London: Elsevier, 1–65, Vol. 2.

Ishikawa, K. (1996) Flexible member for use as a medical bag, US Patent No. 5,480,394.

Jones, R. D., Price, J. C., Stuedemann, J. A., Bowen, J. M. (1994) *In vitro* and *in vivo* release of metoclopramide from a subdermal diffusion matrix with potential in preventing fescue toxicosis in cattle, *J. Controlled Release* **30**, 35–44.

Jones, N. L., Cooper, J. J., Waters, R. D., Williams, D. F. (2000) Resorption profile and biological response of calcium phosphate filled PLLA and PHB7V, in: *Synthetic Bioabsorbable Polymers for Implants*, (Agrawal, C.M., Parr, J.E., Lin, S.T. Eds.), Scranton: ASTM, 69–82.

Juni, K., Nakano, M. (1987) Poly(hydroxy acids) in drug delivery, *CRC Crit. Rev. Ther. Drug Carrier Syst.* **3**, 209–232.

Juni, K., Nakano, M., Kubota, M., Matsui, N., Ichihara, T., Beppu, T., Mori, K., Akagi, M. (1985) Controlled release of aclarubicin from poly(beta-hydroxybutyric acid) microspheres, *Igaku no Ayumi* **132**, 735–736.

Juni, K., Nakano, M., Kubota, M. (1986) Controlled release of aclarubicin, an anticancer antibiotic, from poly-β-hydroxybutyric acid microspheres, *J. Controlled Release* **4**, 25–32.

Kassab, A. Ch., Xu, K., Denkbas, E. B., Dou, Y., Zhao, S., Piskin, E. (1997) Rifampicin carrying polyhydroxybutyrate microspheres as a potential chemoembolization agent, *J. Biomater. Sci., Polym. Edn.* **8**, 947–961.

Kassab, A. Ch., Piskin, E., Bilgic, S., Denkbas, E. B., Xu, K. (1999) Embolization with polyhydroxybutyrate (PHB) microspheres: *In-vivo* studies, *J. Bioact. Compat. Polym.* **14**, 291–303.

Kawaguchi, T., Tsugane, A., Higashide, K., Endoh, H., Hasegawa, T., Kanno, H., Seki, T., Juni, K., Fukushima, S., Nakano, M. (1992) Control of drug release with a combination of prodrug and polymer matrix: Antitumor activity and release profiles of 2′,3′-diacyl-5-fluoro-2′-deoxyuridine from poly(3-hydroxybutyrate) microspheres, *J. Pharm. Sci.* **81**, 508–512.

Kharenko, A. V., Iordanskii, A. L. (1999) Diltiazem release from matrices based on polyhydroxybutyrate, *Proc. Int. Symp. Controlled Release Bioact. Mater.* **26**, 919–920.

Kishida, A., Yoshioka, S., Takeda, Y., Uchiyama, M. (1989) Formulation-assisted biodegradable polymer matrices, *Chem. Pharm. Bull*, **37**, 1954–1956.

Knowles, J. C. (1993) Development of a natural degradable polymer for orthopaedic use, *J. Med. Eng. Tech.* **17**, 129–137.

Knowles, J. C., Hastings, G. W. (1992) *In vitro* degradation of a polyhydroxybutyrate/polyhydroxyvalerate copolymer, *J. Mater. Sci. Mater. Med.* **3**, 352–358.

Knowles, J. C., Hastings, G. W. (1993a) *In vitro* and *in vivo* investigation of a range of phosphate glass-reinforced polyhydroxybutyrate-based degradable composites, *J. Mater. Sci. Mater. Med.* **4**, 102–106.

Knowles, J. C., Hastings, G. W. (1993b) Physical properties of a degradable composite for orthopaedic use which closely matches bone, *Proceedings, First International Conference on Intelligent Materials* (Takagi, T., Ed.), Lancaster: Technomic, 495–504.

Knowles, J. C., Mahmud, F. A., Hastings, G. W (1991) Piezoelectric characteristics of a polyhydroxybutyrate-based composite, *Clin. Mater.* **8**, 155–158.

Knowles, J. C., Hastings, G. W., Ohta, H., Niwa, S., Boeree, N. (1992) Development of a degradable composite for orthopedic use: *in vivo* biomechanical and histological evaluation of two bioactive degradable composites based on the polyhydroxybutyrate polymer, *Biomaterials* **13**, 491–496.

Koosha, F. Muller, R. H., Washington, C. (1987) Production of polyhydroxybutyrate (PHB) nanoparticles for drug targeting, *J. Pharm. Pharmacol.* **39**, 136.

Koosha, F., Muller, R. H., Davis, S. S. (1988) A continuous flow system for *in-vitro* evaluation of drug-loaded biodegradable colloidal carriers, *J. Pharm. Pharmacol.* **40**, 131.

Koosha, F., Muller, R. H., Davis, S. S. (1989) Polyhydroxybutyrate as a drug carrier, *CRC Crit. Rev. Ther. Drug Carrier Syst.* **6**, 117–130.

Korkusuz, F., Korkusuz, P., Eksioglu, F., Gursel, I., Hasirci, V. (2001) *J. Biomed. Mater. Res.* **55**, 217–228.

Korsatko, W., Wabnegg, B., Tillian, H. M., Braunegg, G., Lafferty, R. M. (1983a) Poly-D(-)-3-hydroxybutyric acid – a biodegradable carrier for long term medication dosage. 2. The biodegradation in animals and *in vitro – in vivo* correlation of the liberation of pharmaceuticals from parenteral matrix retard tablets, *Pharm. Ind.* **45**, 1004–1007.

Korsatko, W., Wabnegg, B., Braunegg, G., Lafferty, R. M., Strempfl, F. (1983b) Poly-D(-)-3-hydroxybutyric acid (PHBA) – a biodegradable carrier for long term medication dosage. 1. Development of parenteral matrix tablets for long term application of pharmaceuticals, *Pharm. Ind.* **45**, 525–527.

Korsatko, W., Wabnegg, B., Tillian, H. M., Egger, G., Pfragner, R., Walser, V. (1984) Poly-D(-)-3-hydroxybutyric acid (poly-HBA) – a biodegradable former for long term medication dosage. 3. Studies on compatibility of poly-HBA – implantation tablets in tissue culture and animals, *Pharm. Ind.* **46**, 952–954.

Korsatko, W., Korsatko, B., Lafferty, R. M., Weidmann, V. (1987) The influence of the molecular weight of poly-D-(-)-3-hydroxybutyric acid on its use as a retard matrix for sustained drug release, *Proceedings, Third European Congress of Biopharmacology and Pharmokinetics* 234–242, Vol. 1.

Kostopoulos, L., Karring, T. (1994a) Guided bone regeneration in mandibular defects in rats using a bioresorbable polymer, *Clin. Oral Impl. Res.* **5**, 66–74.

Kostopoulos, L., Karring, T. (1994b) Augmentation of the rat mandible using guided tissue regeneration, *Clin. Oral Impl. Res.* **5**, 75–82.

Kubota, M., Nakano, M., Juni, K. (1988) Mechanism of enhancement of the release rate of aclarubicin from poly-β-hydroxybutyric acid microspheres by fatty acid esters, *Chem. Pharm. Bull.* **36**, 333–337.

Kusaka, S., Abe, H., Lee, S. Y., Doi, Y. (1997) Molecular mass of poly[β-3-hydroxybutyric acid] produced in a recombinant *Escherichia coli*, *Appl. Microb. Biotechnol.* **47**, 140–143.

Lafferty, R. M., Korsatko, B., Korsatko, W. (1988) Microbial production of poly-β-hydroxybutyric acid, in: *Biotechnology* Vol. 6b (Rehm, H.-J., Reed, G., Eds.), Weinheim: VCH, 135–176.

Lee, S. Y., Choi, J.-I., Han, K., Song, J. Y. (1999) Removal of endotoxin during purification of poly(3-hydroxybutyrate) from gram-negative bacteria, *Appl. Environ. Microbiol.* **65**, 2762–2764.

Leenstra, T. S., Maltha, J. C., Kuijpers-Jagtman, A. M. (1995) Biodegradation of non-porous films after submucoperiosteal implantation on the palate of Beagle dogs, *J. Mater. Sci. Mater. Med.* **6**, 445–450.

Linhardt, R. J., Flanagan, D. R., Schmitt, E., Wang, H. T. (1990) Biodegradable poly(esters) and the delivery of bioactive agents, *Poly. Prepr. (Am. Chem. Soc., Div. Polym. Chem.)* **31**, 249–250.

Ljungberg, C., Johansson-Ruden, G., Boström, K. J., Novikov, L., Wiberg, M. (1999) Neuronal survival using a resorbable synthetic conduit as an alternative to primary nerve repair, *Microsurgery* **19**, 259–264.

Löbler, M., Sass, M., Michel, P., Hopt, U. T., Kunze, C., Schmitz, K.-P. (1999) Differential gene expression after implantation of biomaterials into rat gastrointestine, *J. Mater. Sci. Mater. Med.* **10**, 797–799.

Luo, S., Netravali, A. N. (1999) Effect of ^{60}Co γ-radiation on the properties of poly(hydroxybutyrate-*co*-hydroxyvalerate), *Appl. Polym. Sci.* **73**, 1059–1067.

Malm, T., Bowald, S., Bylock, A., Saldeen, T., Busch, C. (1992a) Regeneration of pericardial tissue on absorbable polymer patches implanted into the pericardial sac. An immunohistochemical, ultrasound and biochemical study in sheep. *Scand. J. Thorac. Cardiovasc. Surg.* **26**, 15–21.

Malm, T., Bowald, S., Bylock, A., Busch, C. (1992b) Prevention of postoperative pericardial adhesions by closure of the pericardium with absorbable polymer patches. An experimental study. *J. Thorac. Cardiovasc. Surg.* **104**, 600–607.

Malm, T., Bowald, S., Karacagil, S., Bylock, A., Busch, C. (1992c) A new biodegradable patch for closure of atrial septal defect, *Scand. J. Thor. Cardiovasc. Surg.* **26**, 9–14.

Malm, T., Bowald, S., Bylock, A., Busch, C., Saldeen, T. (1994) Enlargement of the right ventricular outflow tract and the pulmonary

artery with a new biodegradable patch in transannular position, *Eur. Surg. Res.* **26**, 298–308.
Marois, Y., Zhang, Z., Vert. M., Deng, X., Lenz, R., Guidoin, R. (1999a) Effect of sterilization on the physical and structural characteristics of polyhydroxyoctanoate (PHO), *J. Biomater. Sci. Polymer Edn.* **10**, 469–482.
Marois, Y., Zhang, Z., Vert. M., Deng, X., Lenz, R., Guidoin, R. (1999b) Hydrolytic and enzymatic incubation of polyhydroxyoctanoate (PHO): a short-term *in vitro* study of a degradable bacterial polyester, *J. Biomater. Sci. Polymer Edn.* **10**, 483–499.
Marois, Y., Zhang, Z., Vert, M., Beaulieu, L., Lenz, R. W., Guidoin, R. (1999c) *In vivo* biocompatibility and degradation studies of polyhydroxyoctanoate in the rat: A new sealant for the polyester arterial prosthesis, *Tissue Eng.* **5**, 369–386.
Marois, Y., Zhang, Z., Vert, M., Deng, X., Lenz, R. W., Guidoin, R. (2000) Bacterial polyesters for biomedical applications: *In vitro* and *in vivo* assessments of sterilization, degradation rate and biocompatibility of poly(β-hydroxyoctanoate) (PHO), in: *Synthetic Bioabsorbable Polymers for Implants* (Agrawal, C.M., Parr, J.E., Lin, S.T., Eds.), Scranton: ASTM, 12–38.
Martin, D. P., Skraly, F. A., Williams, S. F. (1999) Polyhydroxyalkanoate compositions having controlled degradation rates, PCT Patent Application No. WO 99/32536.
Martin, D. P., Peoples, O. P., Williams, S. F. (2000) Nutritional and therapeutic uses of 3-hydroxyalkanoate oligomers, PCT Patent Application No. WO 00/04895.
McLeod, B. J., Haresign, W., Peters, A. R., Humke, R., Lamming, G. E. (1988) The development of subcutaneous-delivery preparations of GnRH for the induction of ovulation in acyclic sheep and cattle, *Anim. Reprod. Sci.* **17**, 33–50.
Miller, N. D., Williams, D. F. (1987) On the biodegradation of poly-β-hydroxybutyrate (PHB) homopolymer and poly-β-hydroxybutyrate-hydroxyvalerate copolymers, *Biomaterials* **8**, 129–137.
Mukai, K., Doi, Y., Sema, Y., Tomita, K. (1993) Substrate specificities in hydrolysis of polyhydroxyalkanoates by microbial esterases, *Biotechnol. Lett,* **15**, 601–604.
Nasseri, B. A., Pomerantseva, I., Ochoa, E. R., Martin, D. P., Oesterle, S. N., Vacanti, J. P. (2000) Comparison of three polymer scaffolds for vascular tube formation, *Tissue Eng.* **6**, 667.
Nelson, T., Kaufman, E., Kline, J., Sokoloff, L. (1981) The extraneural distribution of γ-hydroxybutyrate, *J. Neurochem.* **37**, 1345–1348.
Nkere, U. U., Whawell, S. A., Sarraf, C. E., Schofield, J. B., O'Keefe, P. A. (1998) Pericardial substitution after cardiopulmonary bypass surgery: A trial of an absorbable patch, *Thorac. Cardiovasc. Surg.* **46**, 77–83.
Nobes, G. A. R., Marchessault, R. H., Maysinger, D. (1998) Polyhydroxyalkanoates: materials for delivery systems, *Drug Delivery,* **5**, 167–177.
Noisshiki, Y., Komatsuzaki, S. (1995) Medical materials for soft tissue use, Japanese Patent Application No. JP7275344A2.
Pawan, G. L. S., Semple, S. J. G. (1983) Effect of 3-hydroxybutyrate in obese subjects on very-low-energy diets and during therapeutic starvation, *Lancet* **1**, 15–17.
Peoples, O. P., Saunders, C., Nichols, S., Beach, L. (1999) Animal nutrition compositions, PCT Patent Applications No. WO 99/34687.
Pouton, C. W., Akhtar, S. (1996) Biosynthetic polyhydroxyalkanoates and their potential in drug delivery, *Adv. Drug Del. Rev.* **18**, 133–162.
Reusch, R. N., Sparrow, A. W., Gardiner, J. (1992) Transport of poly-β-hydroxybutyrate in human plasma, *Biochim. Biophys. Acta* **1123**, 33–40.
Reusch, R. N., Huang, R., Kosk-Kosicka, D. (1997) Novel components and enzymatic activities of the human erythrocyte plasma membrane calcium pump, *FEBS Lett.* **412**, 592–596.
Rivard, C. H., Chaput, C. J., DesRosiers, E. A., Yahia, L'H., Selmani, A. (1995) Fibroblast seeding and culture in biodegradable porous substrates, *J. Appl. Biomater.* **6**, 65–68.
Rivard, C. H., Chaput, C., Rhalmi, S., Selmani, A. (1996) Bio-absorbable synthetic polyesters and tissue regeneration: a study of three-dimensional proliferation of ovine chondrocytes and osteoblasts, *Ann. Chir.* **50**, 651–658.
Rouxhet, L., Legras, R., Schneider, Y.-J. (1998) Interactions between biodegradable polymer, poly(hydroxybutyrate-hydroxyvalerate), proteins and macrophages, *Macromol. Symp.* **130**, 347–366.
Saad, B., Ciardelli, G., Matter, S., Welti, M., Uhlschmid, G. K., Neuenschwander, P., Suter, U. W. (1996a) Characterization of the cell response of cultured macrophages and fibroblasts to particles of short-chain poly[β-3-hydroxybutyric acid], *J. Biomed. Mater. Res.* **30**, 429–439.
Saad, B., Ciardelli, G., Matter, S., Welti, M., Uhlschmid, G. K., Neuenschwander, P., Suter, U. W. (1996b) Cell response of cultured macrophages, fibroblasts, and co-cultures of Kupffer cells and hepatocytes to particles of short-chain poly[β-3-hydroxybutyric acid], *J. Mater. Sci. Mater. Med.* **7**, 56–61.

Saad, B., Matter, S., Ciardelli, G., Uhlschmid, G. K., Welti, M., Neuenschwander, P., Suter, U. W. (1996c) Interactions of osteoblasts and macrophages with biodegradable and highly porous polyesterurethane foam and its biodegradation products, *J. Biomed. Mater. Res.* **32**, 355–366.

Saad, B. Neuenschwander, P., Uhlschmid, G. K., Suter, U. W. (1999) New versatile, elastomeric, degradable polymeric materials for medicine, *Int. J. Biol. Macromol.* **25**, 293–301.

Saito, T., Tomita, K., Juni, K., Ooba, K. (1991) *In vivo* and *in vitro* degradation of poly(3-hydroxybutyrate) in rat, *Biomaterials* **12**, 309–312.

Scharf, M. B., Lai, A. A., Branigan, B., Stover, R., Berkowitz, D. B. (1998) Pharmacokinetics of gammahydroxybutyrate (GHB) in narcoleptic patients, *SLEEP* **21**, 507–514.

Schmitz, K.-P., Behrend, D. (1997) Method of manufacturing intraluminal stents made of polymer material, European Patent Application No. 0 770 401 A2.

Schoen, F. J. Levy, R. J. (1999) Tissue heart valves: current challenges and future research perspectives, *J. Biomed. Mater. Res.* **47**, 439–465.

Scholz, C. (2000) Poly(beta-hydroxyalkanoates) as potential biomedical materials: an overview, *ACS Symp. Ser. (Polymers from renewable resources: biopolyesters and biocatalysis)*, **764**, 328–334.

Seebach, D., Brunner, A., Bürger, H. M., Schneider, J., Reusch, R. N. (1994) Isolation and ^{1}H-NMR spectroscopic identification of poly(3-hydroxybutanoate) from prokaryotic and eukaryotic organisms, *Eur. J. Biochem.* **224**, 317–328.

Selmani, A., Chaput, C., Paradis, V., Yahia, L'H., Rivard, C.-H. (1995) Microscopical analysis of 2D/3D porous poly(hydroxybutyrate-*co*-hydroxyvalerate) for tissue engineering, *Adv. Sci. Technol.* **12**, 165–172.

Sendil, D., Gürsel, I., Wise, D. L., Hasirci, V. (1998) Antibiotic release from biodegradable PHBV microparticles, in: *Biomed. Sci. Technol. [Proceedings, Fourth International Symposium (1998), Meeting Date 1997]*, (Hincal, A. A., Kas, H. S., Eds.), New York: Plenum Press, 89–96.

Sendil, D., Gürsel, I., Wise, D. L., Hasirci, V. (1999) Antibiotic release from biodegradable PHBV microparticles, *J. Controlled Release* **59**, 207–217.

Shum-Tim, D., Stock, U., Hrkach, J., Shinoka, T., Lien, J., Moses, M. A., Stamp, A., Taylor, G., Moran, A. M., Landis, W., Langer, R., Vacanti, J. P., Mayer, J. E. (1999) Tissue engineering of autologous aorta using a new biodegradable polymer, *Ann. Thorac. Surg.* **68**, 2298–2305.

Sim, S. J., Snell, K. D., Hogan, S. A., Stubbe, J., Rha, C., Sinskey, A. J. (1997) PHA synthase activity controls the molecular weight and polydispersity of polyhydroxybutyrate *in vivo*, *Nature Biotechnol.* **15**, 63–67.

Sodian, R., Sperling, J. S., Martin, D. P., Stock, U., Mayer, J. E., Vacanti, J. P. (1999) Tissue engineering of a trileaflet heart valve – Early *in vitro* experiences with a combined polymer, *Tissue Eng.* **5**, 489–493.

Sodian, R., Hoerstrup, S. P., Sperling, J. S., Martin, D. P., Daebritz, S., Mayer, J. E., Vacanti, J. P. (2000a) Evaluation of biodegradable, three-dimensional matrices for tissue engineering of heart valves, *ASAIO J.* **46**, 107–110.

Sodian, R., Sperling, J. S., Martin, D. P., Egozy, A., Stock, U., Mayer, J. E., Vacanti, J. P. (2000b) Fabrication of a trileaflet heart valve scaffold from a polyhydroxyalkanoate biopolyester for use in tissue engineering, *Tissue Eng.* **6**, 183–188.

Sodian, R., Hoerstrup, S. P., Sperling, J. S., Daebritz, S. H., Martin, D. P., Schoen, F. J., Vacanti, J. P., Mayer, J. E. (2000c) Tissue engineering of heart valves: *in vitro* experiences, *Ann. Thorac. Surg.* **70**, 140–144.

Sodian, R., Hoerstrup, S. P., Sperling, J. S., Daebritz, S., Martin, D. P., Moran, A. M., Kim, B. S., Schoen, F. J., Vacanti, J. P., Mayer, J. E. (2000d) Early *in vivo* experience with tissue-engineered trileaflet heart valves, *Circulation* **102** (suppl. III), III-22–III-29.

Steel, M. L., Norton-Berry, P. (1986) Non-woven fibrous material, US Patent 4,603,070.

Steinbüchel, A. (1991) Polyhydroxyalkanoic acids, in: *Biomaterials* (Byrom, D., Ed.), New York: Stockton Press, 123–213.

Steinbüchel, A., Valentin, H. E. (1995) Diversity of bacterial polyhydroxyalkanoic acids, *FEMS Microbiol. Lett.* **128**, 219–228.

Stock, U. A., Sodian, R., Herden, T., Martin, D. P., Egozy, A., Vacanti, J. P., Mayer, J. E. (1998) Influence of seeding techniques on cell-polymer attachment of biodegradable polymers – polyhydroxyalkanoates (PHA), *Abstract from the 2nd Bi-Annual Meeting of the Tissue Engineering Society*, December 4–6, Orlando, Florida.

Stock, U. A, Sakamoto, T., Hatsuoka, S., Martin, D. P., Nagashima, M., Moran, A. M., Moses, M. A., Khalil, P. N., Schoen, F. J., Vacanti, J. P., Mayer, J. E. (2000a) Patch augmentation of the pulmonary artery with bioabsorbable polymers and autologous cell seeding, *J. Thorac. Cardiovasc. Surg.* **120**, 1158–1168.

Stock, U. A., Nagashima, M., Khalil, P. N., Nollert, G. D., Herden, T., Sperling, J. S., Moran, A., Lien, J., Martin, D. P., Schoen, F. J., Vacanti, J. P., Mayer, J. E. (2000b) Tissue-engineered valved conduits in the pulmonary circulation, *J. Thorac. Cardiovasc. Surg.* **119**, 732–740.

Sudesh, K., Abe, H., Doi, Y. (2000) Synthesis, structure and properties of polyhydroxyalkanoates: biological polyesters, *Prog. Polym. Sci.* **25**, 1503–1555.

Tang, S., Ai, Y., Dong, Z., Yang, Q. (1999) Tissue response to subcutaneous implanting poly-3-hydroxybutyrate in rats, *Disi Junyi Daxue Xuebao* **20**, 87–89.

Tasaki, O., Hiraide, A., Shiozaki, T., Yamamura, H., Ninomiya, N., Sugimoto, H. (1998) The dimer and trimer of 3-hydroxybutyrate oligomer as a precursor of ketone bodies for nutritional care, *JPEN J Parenter. Enteral Nutr.* **23**, 321–325.

Taylor, M. S., Daniels, A. U., Andriano, K. P., Heller, J. (1994) Six bioabsorbable polymers: *In vitro* acute toxicity of accumulated degradation products, *J. Appl. Biomater.* **5**, 151–157.

Tetrick, M. A., Adams, S. H., Odle, J., Benevenga, N. J. (1995) Contribution of D-(-)-3-hydroxybutyrate to the energy expenditure of neonatal pigs, *J. Nutr.* **125**, 264–272.

Tomizawa, Y., Moon, M. R., Noishiki, Y. (1994) Antiadhesive membranes for cardiac operations, *J. Thorac. Cardiovasc. Surg.* **107**, 627–629.

Unverdorben, M., Schywalsky, M., Labahn, D., Hartwig, S., Laenger, F., Lootz, D., Behrend, D., Schmitz, K., Schaldach, M., Vallbracht, C. (1998) Polyhydroxybutyrate (PHB) stent-experience in the rabbit, *Am. J. Card. Transcatheter Cardiovascular Therapeutics, Abstract TCT-11*, 5S.

Vainionpää, S., Vihtonen, K., Mero, M., Pätiälä, H., Rokkanen, P., Kilpikari, J., Törmälä, P. (1986) Biodegradable fixation of rabbit osteotomies, *Acta Orthopaed. Scand.* **57**, 237–239.

Van der Giessen, W. J., Lincoff, A. M., Schwartz, R. S., van Beusekom, H. M. M., Serruys, P. W., Holmes, D. R., Ellis, S. G., Topol, E. J. (1996) Marked inflammatory sequelae to implantation of biodegradable and nonbiodegradable polymers in porcine coronary arteries, *Circulation* **94**, 1690–1697.

Veech, R. L. (1998) Therapeutic compositions, PCT Patent Application Nos. WO 98/41200 and WO 98/41201.

Veech, R. L. (2000) Therapeutic compositions, PCT Patent Application No. WO 00/15216.

Wabnegg, B., Korsatko, W. (1983) About the compatibility of retard tablets consisting of poly-D-(-)-3-hydroxybutyric acid as a carrier of active agents delivered applied parenterally, *Sci. Pharm.* **51**, 372.

Wang, D. W., Lehmann, L. T. (1991) Bioabsorbable coating for a surgical device, US Patent No. 5,032,638.

Webb, A., Adsetts, J. R. (1986) Wound dressings, UK Patent Application No. 2,166,354.

Williams, S. F. (2000) Bioabsorbable, biocompatible polymers for tissue engineering, PCT Patent Application No. WO 00/51662.

Williams, S. F., Martin, D. P. (2001) Therapeutic uses of polymers and oligomers comprising gamma-hydroxybutyrate. PCT Patent Application No. WO 01/19361A2.

Williams, S. F., Peoples, O. P. (1996) Biodegradable plastics from plants, *CHEMTECH* **26**, 38–44.

Williams, S. F., Martin, D. P., Horowitz, D. H. Peoples, O. P. (1999) PHA applications: addressing the price performance issue I. Tissue engineering, *Int. J. Biol. Macromol.* **25**, 111–121.

Williams, S. F., Martin, D. P., Skraly, F. (2000) Medical devices and applications of polyhydroxyalkanoate polymers, PCT Patent Application No. WO 00/56376.

Yagmurlu, M. F., Korkusuz, F., Gursel, I., Korkusuz, P., Ors, U., Hasirci, V. (1999) Sulbactam-cefoperazone polyhydroxybutyrate-*co*-hydroxyvalerate (PHBV) local antibiotic delivery system: *in vivo* effectiveness and biocompatibility in the treatment of implant-related experimental osteomyelitis, *J. Biomed. Mater. Res.* **46**, 494–503.

Yasin, M., Tighe, B. J. (1992) Polymers for biodegradable medical devices VIII. Hydroxybutyrate-hydroxyvalerate copolymers: physical and degradative properties of blends with polycaprolactone, *Biomaterials* **13**, 9–16.

Yasin, M., Holland, S. J., Jolly, A. M., Tighe, B. J. (1989) Polymers for biodegradable medical devices VI. Hydroxybutyrate-hydroxyvalerate copolymers: accelerated degradation of blends with polysaccharides, *Biomaterials* **10**, 400–412.

Yasin, M., Holland, S. J., Tighe, B. J. (1990) Polymers for biodegradable medical devices V. Hydroxybutyrate-hydroxyvalerate copolymers: effects of polymer processing on hydrolytic degradation, *Biomaterials* **11**, 451–454.

5
Polyanhydrides

Dr. Neeraj Kumar[1], Prof. Ann-Christine Albertsson[2], Dr. Ulrica Edlund[3], Dr. Doron Teomim[4], Aliza Rasiel[5], Prof. Abraham J. Domb[6]

[1] Department of Medicinal Chemistry and Natural Products, School of Pharmacy, The Hebrew University of Jerusalem, Jerusalem-91120, Israel; Tel.: +972-2-6757573; Fax: +972-2-6757629; E-mail: nk31@mailcity.com

[2] Department of Polymer Technology, Royal Institute of Technology, S-100 44 Stockholm, Sweden; Tel.: +46-8-7908274; Fax: +46-8-100775; E-mail: aila@polymer.kth.se

[3] Department of Polymer Technology, Royal Institute of Technology, S-100 44 Stockholm, Sweden; Tel.: +46-8-7908274; Fax: +46-8-100775; E-mail: aila@polymer.kth.se

[4] Department of Medicinal Chemistry and Natural Products, School of Pharmacy, The Hebrew University of Jerusalem, Jerusalem-91120, Israel; Tel.: +972-2-6757573; Fax: +972-2-6757629; E-mail: doront@shellcase.com

[5] Police Headquarters, Forensic Department, Jerusalem; E-mail: magoraia@012.net.il

[6] Department of Medicinal Chemistry and Natural Products, School of Pharmacy, The Hebrew University of Jerusalem, Jerusalem-91120, Israel; Tel.: +972-2-6757573; Fax: +972-2-6758959; E-mail: adomb@cc.huji.ac.il

Biopolymers for Medical and Pharmaceutical Applications. Edited by A. Steinbüchel and R. H. Marchessault

ISBN: 3-527-31154-8

ACDA	acetylenedicarboxylic acid
a_N	Hyperfine splitting constant
BTC	1,3,5-benzenetricarboxylic acid
Co	drug loading in g mL^{-1}
CPH	1,3-bis(*p*-carboxyphenoxy)hexane
CPP	1,3-bis(*p*-carboxyphenoxy)propane
PEG	polyethylene glycol
DMF	*N,N*- dimethylformamide
DSC	differential scanning calorimetry
EPR	electron paramagnetic resonance
EPRI	electron paramagnetic resonance imaging
EAD	euracic acid dimer
FA	fumaric acid
FAD	fatty acid dimer
FDA	Food and Drug Administration
g-value	Zeeman splitting constant for free electrons
Gelfoam®	absorbable gelatin sponge
Gliadel™	polyanhydride brain tumor implant containing BCNU
GPC	gel permeation chromatography
IPA	isophthalic acid
L_n	average length of sequence

MIT	Massachusetts Institute of Technology
M_w	weight average molecular weight
Mn	number average molecular weight
MRI	magnetic resonance imaging
MTX	methotrexate
NMR	nuclear magnetic resonance
PLA	poly(lactic acid)
PCL	poly(caprolactone)
PHB	poly(hydroxybutyrate)
PSA	poly(sebacic acid)
PTMC	poly(trimethylene carbonate)
PAA	poly(adipic acid)
P(RAS)	poly(ricinoleic acid succinate)
SA	sebacic acid
SEM	scanning electron microscopy
Septicin™	polyanhydride antibacterial bone implant
STDA	4,4′-stilbendicarboxylic acid
Surgicel®	oxidized cellulose absorbable hemostat
TA	terephthalic acid
T_g	glass transition temperature
TMA-gly	trimellitimide-glycine
ToF-SIMS	time-of-flight secondary ion mass spectroscopy
Vicryl®	synthetic absorbable suture
Xc	degree of crystallinity
XPS	X-ray photoelectron spectroscopy

1 Introduction

Polyanhydrides are emerging as a new class of biodegradable polymers for drug delivery, and have been successfully utilized clinically as a carrier vehicle for a number of drugs (Domb et al., 1999; Teomim et al., 1999; Stephens et al., 2000). Polyanhydrides have a hydrophobic backbone with a hydrolytically labile anhydride linkage, such that hydrolytic degradation can be controlled by manipulation of the polymer composition. They are of major interest because they show no evidence of inflammatory reactions, and are hydrolytically unstable and degraded both *in vitro* and *in vivo*, with release of nonmutagenic and noncytotoxic products (Leong et al., 1986a,b). Polyanhydrides are biocompatible, and have excellent controlled release characteristics (Tamargo et al., 1989; Chasin et al., 1990).

Pharmaceutical research has been focused on polyanhydrides derived from sebacic acid (SA), 1,3-bis(*p*-carboxyphenoxy) propane (CPP) and fatty acid dimer (FAD). Recently, the Food and Drug Administration (FDA) has approved the use of the polyanhydride poly[sebacic acid-co-1,3-bis(*p*-carboxyphenoxy) propane] to deliver the chemotherapeutic agent for treatment of brain cancer (Dang et al., 1996). The introduction of an imide group into polyanhydrides enhances the mechanical properties of polymers, thus enabling them to be used in orthopedics applications (Uhrich et al., 1995, 1997; Muggli et al., 1999; Young et al., 2000), while the presence of PEG groups in polyanhydrides increases their hydrophilicity and sets up the polymer for rapid drug release applications (Jiang and Zhu, 2000).

The main limitation of polyanhydrides is their storage stability, as they must be stored under refrigeration. Several reviews have been published on polyanhydrides for controlled drug delivery applications (Leong et al., 1989; Domb et al., 1993, 1997; Laurencin et al., 1995). This chapter reviews the chemistry, degradation behavior, biocompatibility, and medical uses of polyanhydrides.

2
Historical Outline

Bucher and Slade (1909) were the first to report on the synthesis of polyanhydrides, and some years later, Hill and Carothers (1930, 1932) synthesized aliphatic polyanhydrides and studied the behavior of diacids towards anhydride formation. These authors prepared super-anhydrides from homologous aliphatic dicarboxylic acid and used them to spin fibers with good mechanical strength. Conix (1958) reported polyanhydrides derived from aromatic acids, and found that these compounds were hydrolytically stable and had excellent film- and fiber-forming properties. Conix proposed a general method for the preparation of aromatic polyanhydrides as shown below:

Conix subsequently studied a large number of aromatic polyanhydrides and found that they had glass transition temperatures (T_g) in the range of 50 to 100 °C, were transformed into opaque porcelain-like solids, and had a resistance to hydrolysis even on exposure to alkaline solution. Yoda (1962, 1963) introduced a new class of heterocyclic crystalline compounds into the polyanhydride family. He synthesized various types of five-membered heterocyclic dibasic acids, and polymerized these compounds with acetic anhydride at 200–300 °C under vacuum and nitrogen atmosphere. The heterocyclic polymers thus obtained had melting points in the range 70–190 °C, and good fiber- and film-forming properties. Consequently, extensive research was undertaken to enhance the stability against hydrolysis and to increase the fiber- and film-forming properties. Aliphatic polyanhydrides were considered as the least important compounds due to their unstable nature against hydrolysis. In 1980, Langer was the first to exploit the hydrolytically unstable nature of the polyanhydrides for the sustained release of drugs in controlled drug delivery applications (Rosen et al., 1983), and used these compounds as biodegradable carriers in various medical devices.

3
Synthesis

Polyanhydrides have been synthesized by melt condensation of activated diacids (Domb and Langer, 1987), ring-opening polymerization (ROP), dehydrochlorina-

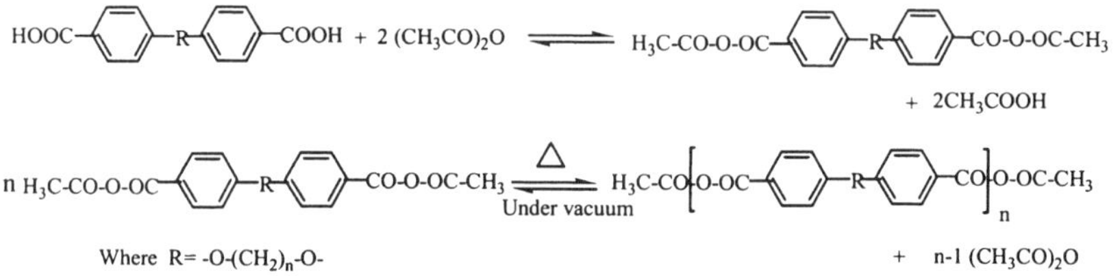

tion, and dehydrative coupling agents (Leong et al., 1987; Domb et al., 1988). In general, solution polymerization yielded low-molecular weight polymers. The most widely used method is the melt condensation of dicarboxylic acids treated with acetic anhydride:

$$HOOC\text{-}R\text{-}COOH + (CH_3\text{-}CO)_2O \xrightarrow{\text{Reflux}} \underset{(I)}{CH_3\text{-}CO\text{-}(O\text{-}CO\text{-}R\text{-}CO\text{-})_mO\text{-}CO\text{-}CH_3}$$

$$(I) \xrightarrow{180^\circ\ C/<1\ mmHg} CH_3\text{-}CO\text{-}(O\text{-}CO\text{-}R\text{-}CO\text{-})_nO\text{-}CO\text{-}CH_3$$

$m = 1 - 20$; $n = 100 - 1000$

The polycondensation takes place in two steps. In the first step, dicarboxylic acid monomers are reacted with excess acetic anhydride to form acetyl-terminated anhydride prepolymers with a degree of polymerization (Dp) ranging from 1 to 20; these are then polymerized at elevated temperature under vacuum to yield polymers with Dp ranging from 100 to over 1000. Acetic acid mixed anhydride prepolymers were also prepared from the reaction of diacid monomers with ketene or acetyl chloride.

The condensation reaction of diacetyl mixed anhydrides of aromatic or aliphatic diacids is carried out in the temperature range of 150 to 200 °C. Recently, we reported the synthesis of polyanhydride from ricinoleic acid half-esters with maleic and succinic anhydride (Teomim et al., 1999). We prepared ricinoleic acid maleate or succinate diacid half-esters which were polymerized by melt condensation (Figure 1).

A variety of catalysts has been used in the synthesis of a range of polyanhydrides. Significantly higher molecular weights, with shorter reaction time, were achieved by utilizing cadmium acetate, earth metal oxides, and $ZnEt_2$-H_2O. Except for calcium carbonate, which is a safe natural material, the use of these catalysts for the production of medical-grade polymers is limited because of their potential toxicity.

Polyanhydrides can be synthesized by melt condensation of trimethylsilyl dicarboxylates and diacid chlorides to yield polymers with an intrinsic viscosity 0.43 dl g^{-1} (Gupta, 1988). Direct polycondensation of adipic acid at a high temperature under vacuum resulted in the low-molecular weight oligomers (Knobloch and Ramirez, 1975).

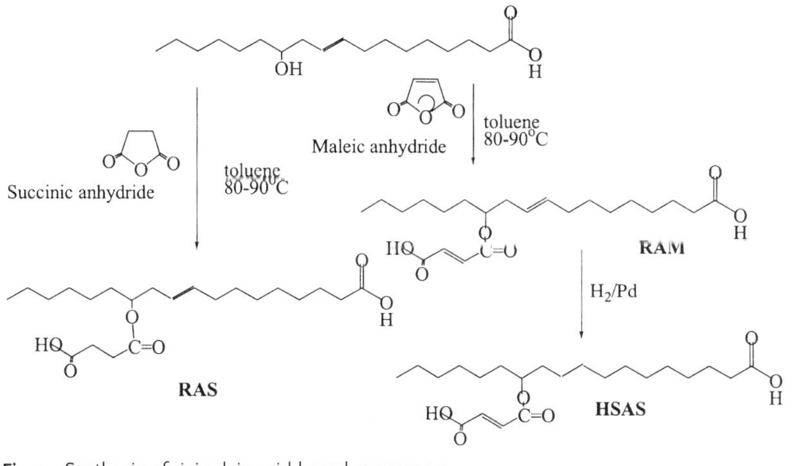

Fig. 1 Synthesis of ricinoleic acid-based monomers.

ROP offers an alternate approach to the synthesis of polyanhydrides used for medical applications. Albertsson and coworkers prepared adipic acid polyanhydride from cyclic adipic anhydride (oxepane-2,7-dione) using cationic (e.g.$AlCl_3$ and $BF_3.(C_2H_5)_2O$), anionic (e.g. $CH_3COO^-K^+$ and NaH), and coordination-type inhibitors such as stannous-2-ethylhexanoate and dibutyltin oxide (Albertsson and Lundmark, 1988, 1990; Lundmark et al., 1991). ROP takes place in two steps: (1) preparation of the cyclic monomer; and (2) polymerization of the cyclic monomers. These authors synthesized oxipane-2,7-dione (Figure 2), a cyclic monomer of adipic acid by heating the reaction mixture containing 15 g (0.1026 mol) adipic acid in 150 mL acetic anhydride for 4 h in a nitrogen atmosphere, the excess of acetic anhydride and acetic acid formed during the course of the reaction being removed by vacuum distillation. The residue was transferred to a Claisen flask, followed by addition of depolymerization catalyst [$(CH_3COO)^2$-Zn. $2H_2O$] and heated under vacuum. After removal of residual acetic anhydride at 0.9 mbar and 250 °C, $ZnCl_2$ (1 wt.%) was added at room temperature to a solution of 6 g (0.05 mol) adipic anhydride in 50 mL methylene chloride. After 6 h, the reaction was terminated by pouring the reaction mixture into dry ether to precipitate the polymer. The white precipitate thus obtained was filtered, washed with dry ether, and dried.

A variety of solution polymerizations at ambient temperature has been reported (Leong et al., 1987; Domb et al., 1988). Partial hydrolysis of terephthalic acid chloride in the presence of pyridine as an acid acceptor, yielded a polymer of *molecular weight* 2100. The use of *N*,*N*-bis(2-oxo-3-oxazolidinyl)phospharamido chloride, dicyclohexylcarbodiimide, chlorosulfonyl isocyanate, and phosgene as coupling agents produced low-molecular weight polymers. Furthermore, homo- and copolyanhydrides have also been synthesized via aqueous and nonaqueous interfacial reaction conditions, albeit with limited success. Various aromatic polymers were prepared by the phase transfer reaction having equimolar amounts of acid dissolved in aqueous base, and corresponding diacid chlorides dissolved in organic solvent (Subramanyam and Pinkus, 1985; Leong et al., 1987; Domb et al., 1988). The copolymers thus obtained present a regularly alternating structure. In the reaction between sebacoyl chloride in chloroform and sodium salt of isophthalic acid in water, a copolymer with large number of SA units was obtained. This result can be explained on the basis of a side reaction that occurs between SA and the product formed in the immiscible organic phase; the result was a polymer having a high SA content.

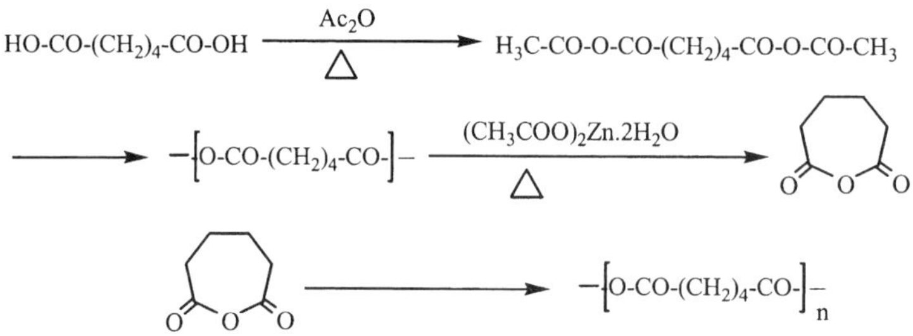

Fig. 2 Ring-opening polymerization mechanism of poly(adipic acid)anhydride.

4 Polyanhydride Structures

Since the introduction of polyanhydrides into the regime of polymers, hundreds of polyanhydride structures have been reported (Mark et al., 1969). A representative list of polymers is shown in Table 1. Polyanhydrides developed for intended use in medicine are described below.

Tab. 1 Representative polyanhydrides synthesized after 1909

Polymer structure	*Melting point (°C)*
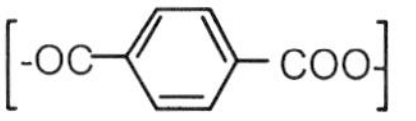	400
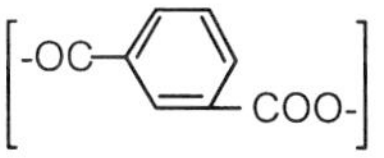	256
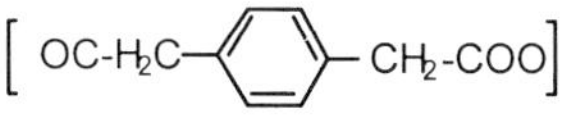	91
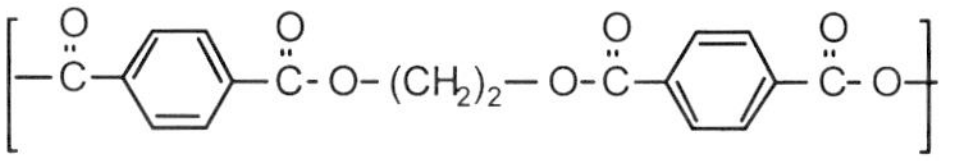	250
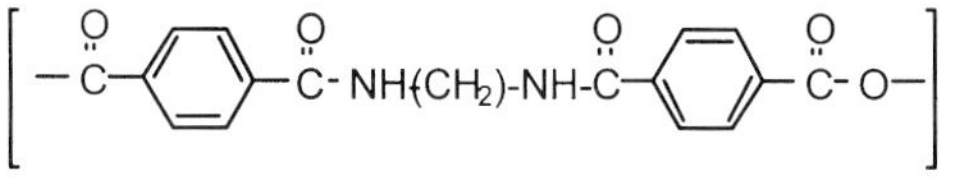	330
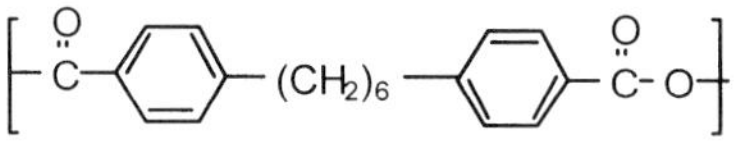	151
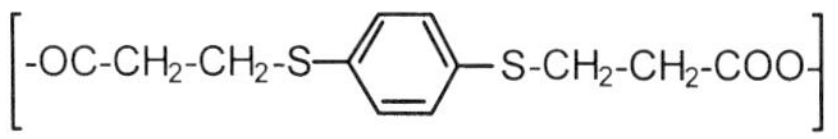	91
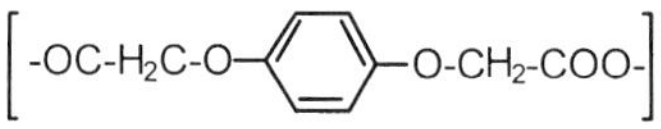	160
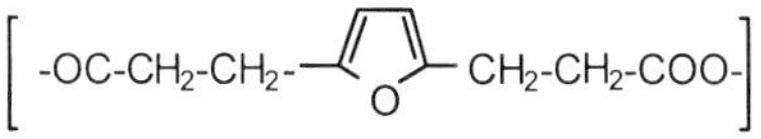	67
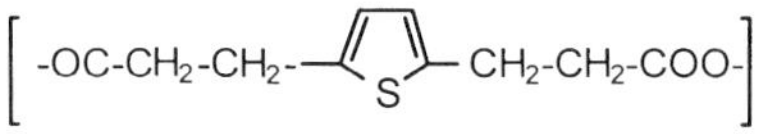	78
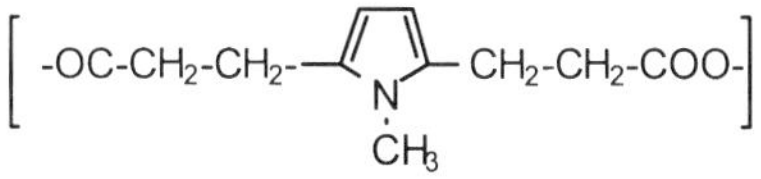	188
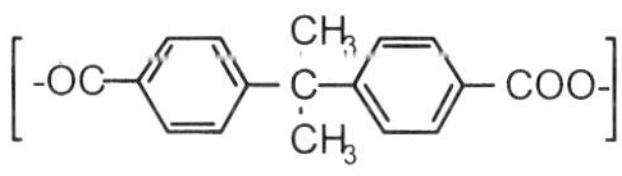	230
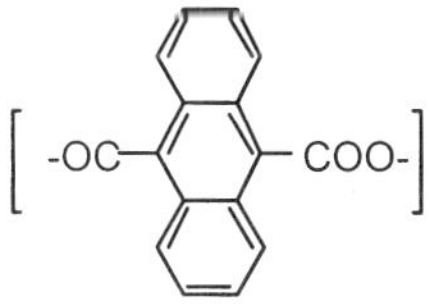	>300
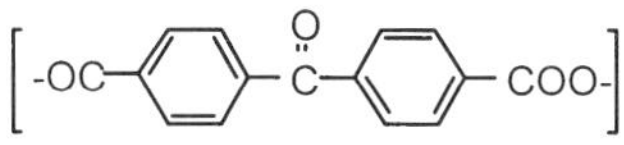	338

Tab. 1 (cont.)

Polymer structure	***Melting point (°C)***
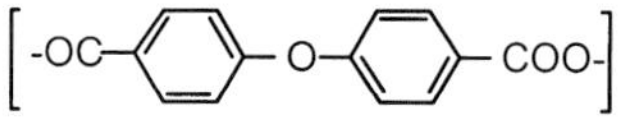	295
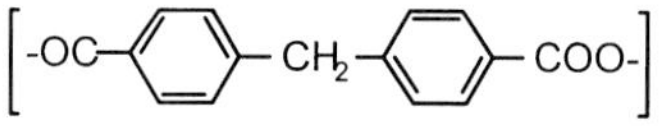	332
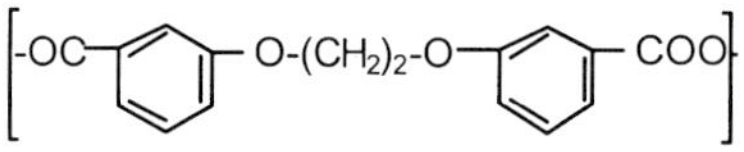	237
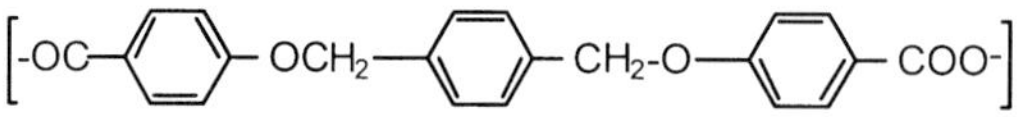	140
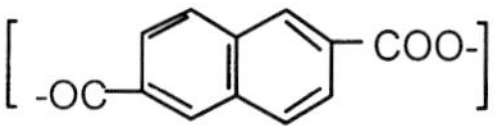	450
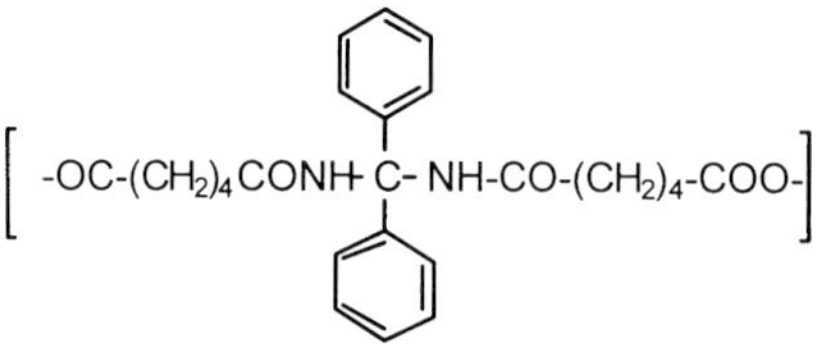	> 300
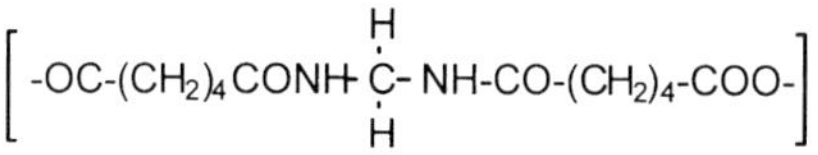	285
[OC-$(CH_2)_2$-SO_2-$(CH_2)_2$-SO_2-$(CH_2)_2$-COO]	185
[OC-$(CH_2)_2$-S-$(CH_2)_2$-S-$(CH_2)_2$-COO]	81
[OC-P-COO]	
[OC-$(CH_2)_x$-COO] x = 4–16	60–100
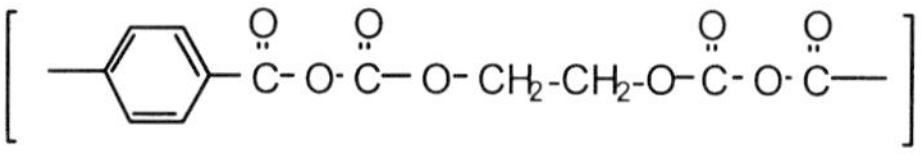	155
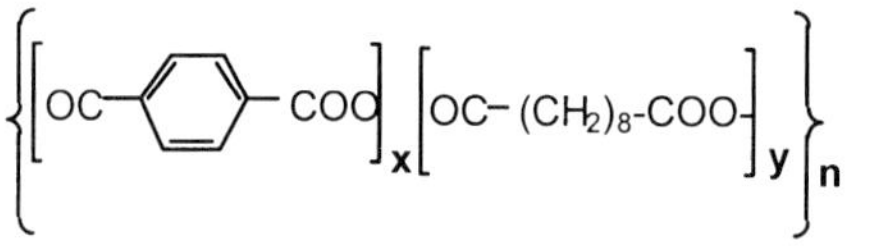	220
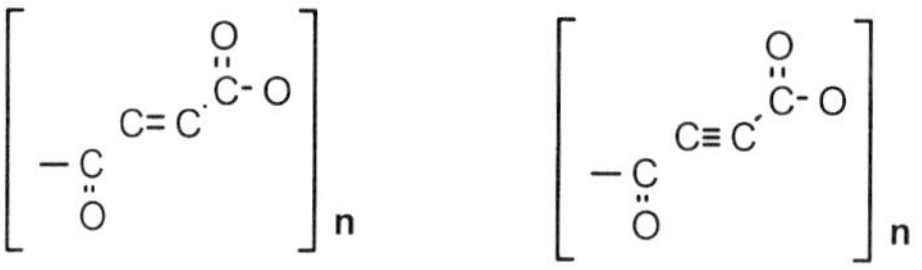	–
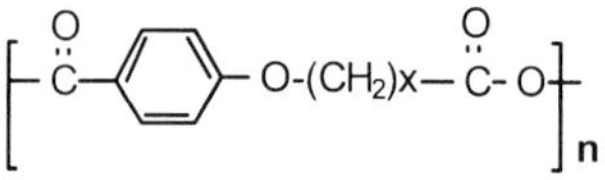 x = 1–10	< 100
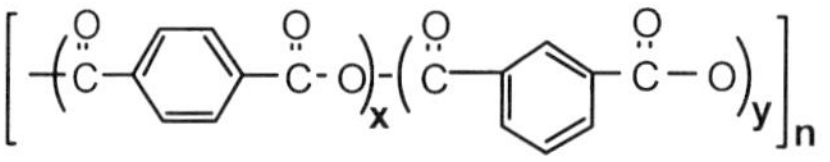	> 200

Tab. 1 (cont.)

Polymer structure	*Melting point (°C)*
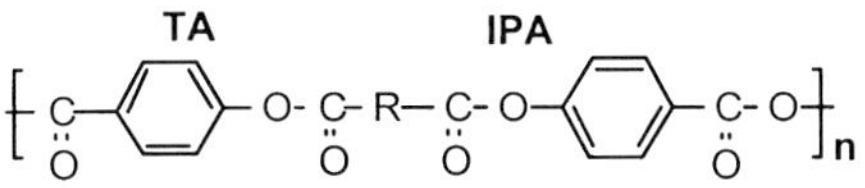	98–176
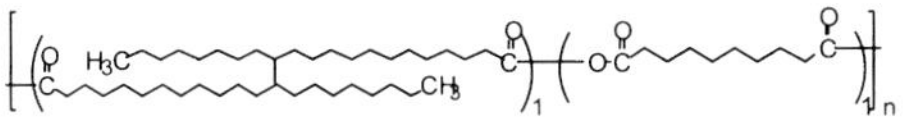	–
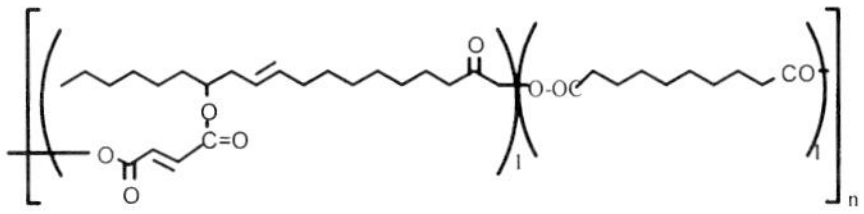	59.3
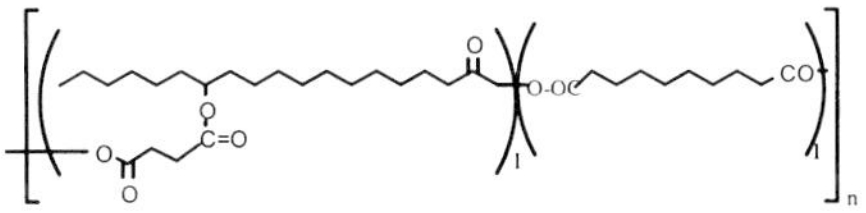	70.4
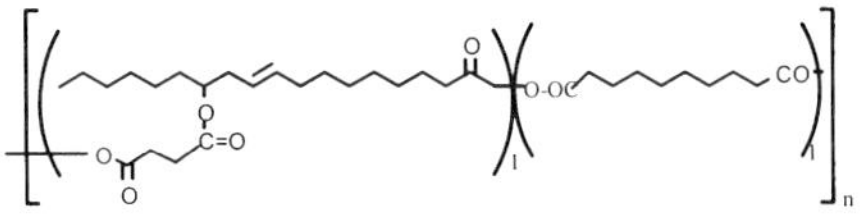	61.1
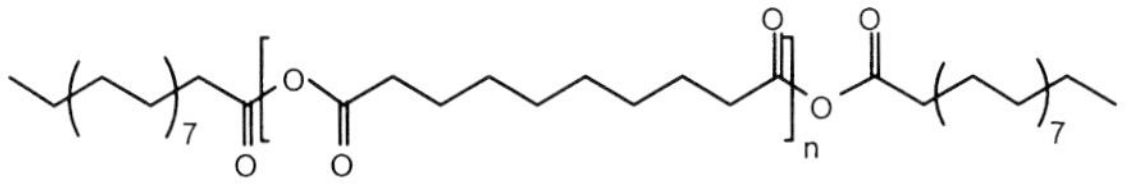	71.2–77.8
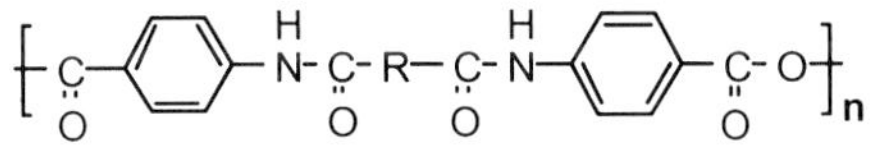 {R = $(CH_2)_3$; $(CH_2)_2$-O-$(CH_2)_8$-O-$(CH_2)_2$}G	58–177
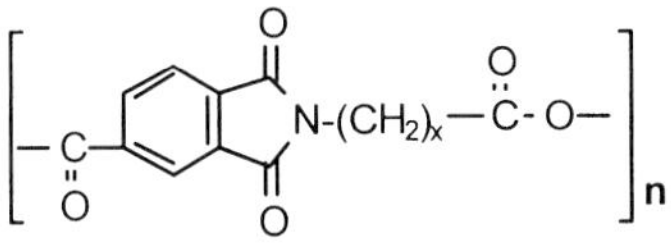	
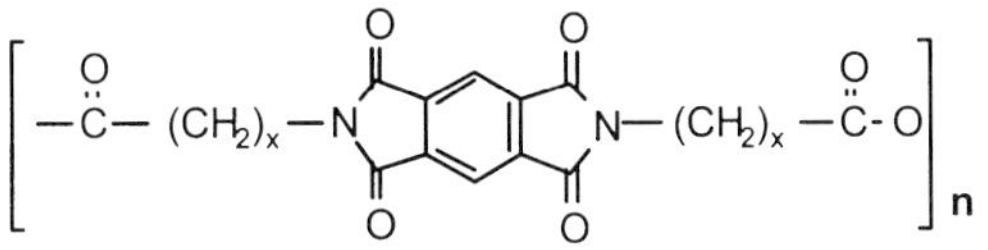 x = 1–5	
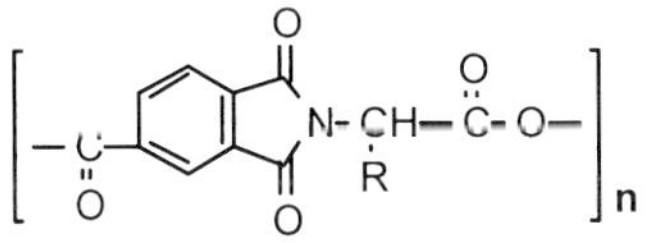 R = -CH_2-CH-$(CH_3)_2$ R = -CH2-C_6H_5-OH R = -CH_2-S-CH_2-C_6H_5	
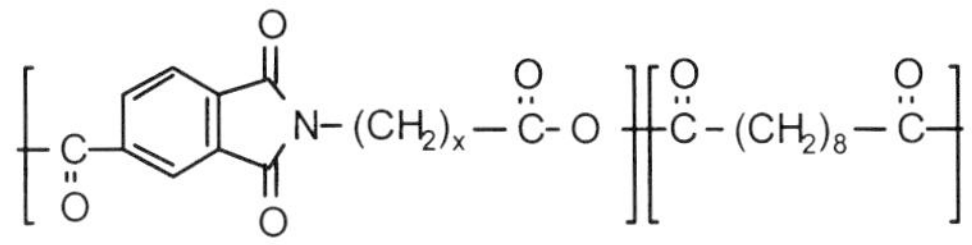	45 (1:1, x = 1)

4.1
Unsaturated Polymers

A series of unsaturated polyanhydrides were prepared by melt or solution polymerization of fumaric acid (FA), acetylenedicarboxylic acid (ACDA), and 4,4′-stilbendicarboxylic acid (STDA) (Domb et al., 1991). The double bonds remained intact throughout the polymerization process, and were available for a secondary reaction to form a cross-linked matrix. The unsaturated homopolymers were crystalline and insoluble in common organic solvents, whereas copolymers with aliphatic diacids were less crystalline and soluble in chlorinated hydrocarbons (Domb et al., 1991):

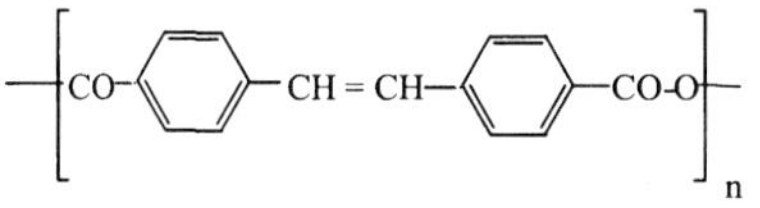

4.2
Aliphatic-aromatic Homopolymers

Polyanhydrides of diacid monomers containing aliphatic and aromatic moieties, poly-[(*p*-carboxyphenoxy)alkanoic anhydride], were synthesized by either melt or solution polymerization with *molecular weight* of up to 44,600 (Domb and Langer, 1989; Domb et al., 1989):

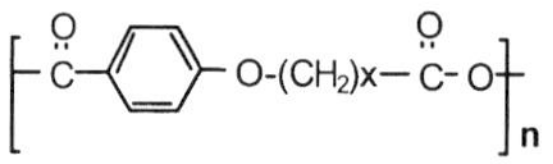

x = 1–10

The polymers of carboxyphenoxy alkanoic acid having methylene groups (n = 3, 5, and 7) were soluble in chlorinated hydrocarbons, and melted at temperatures below 100 °C. These polymers displayed a zero-order hydrolytic degradation profile for 2 to 10 weeks. The length of the alkanoic chain dictated the degradation time wherein an increasing degradation time was observed with an increasing chain length.

4.3
Soluble Aromatic Copolymers

Aromatic homopolyanhydrides are insoluble in common organic solvents, and melt at temperatures above 200 °C (Domb, 1992). These properties limit the use of purely aromatic polyanhydrides, since they cannot be fabricated into films or microspheres using solvent or melt techniques. Fully aromatic polymers that are soluble in chlorinated hydrocarbons and melt at temperatures below 100 °C were obtained by copolymerization of aromatic diacids such as isophthalic acid (IPA), terephthalic acid (TA), 1,3-bis(carboxyphenoxy)-propane (CPP), or 1,3-bis(carboxyphenoxy)-hexane (CPH).

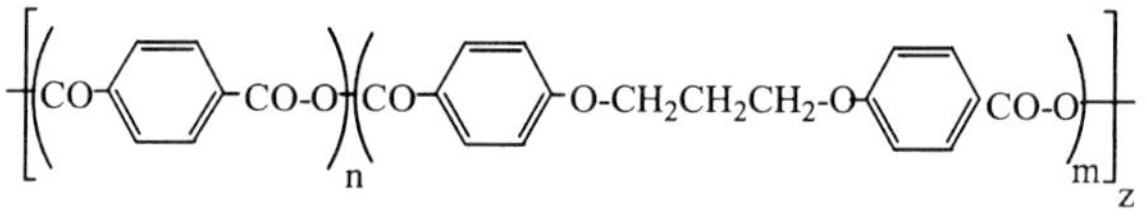

4.4
Poly(ester-anhydrides)

4,4′-Alkane- and oxa-alkanedioxydibenzoic acids were used for the synthesis of polyanhydrides (McIntyre, 1994). Polymers melted at a temperature range of 98 to 176 °C, and had a *molecular weight* of up to 12,900. Di- and tri-block copolymers of poly(caprolactone), poly(lactic acid), and poly(hydroxybutyrate) have been prepared from carboxylic acid-terminated low-molecular weight polymers copolymerized with SA prepolymers by melt condensation (Abuganima, 1996). Similarly, di-, tri- and brush copolymers of poly(ethylene glycol) with poly(sebacic anhydride) have been prepared by melt copolymerization of carboxylic acid-terminated PEG (Gref et al., 1995).

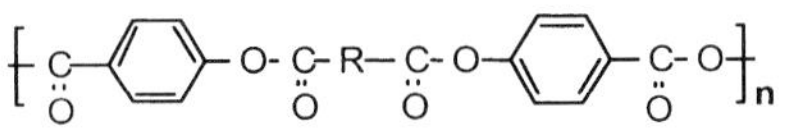

$R = (CH_2)_{2\text{–}8}$; $(CH_2)_2\text{-}O\text{-}(CH_2\text{-}CH_2\text{-}O)_2\text{-}(CH_2)_2$

4.5
Fatty Acid-based Polyanhydrides

Polyanhydrides were synthesized from dimer and trimer of unsaturated fatty acids; the details of this are described in Section 3 (Domb, 1993; Domb and Maniar, 1993a,b; Teomim and Domb, 1999). The dimers of oleic acid and eurecic acid are liquid oils containing two carboxylic acids available for anhydride polymerization. The homopolymers are viscous liquids, and copolymerization with increasing amounts of SA forms solid polymers with increasing melting points as a function of the SA content. The polymers are soluble in chlorinated hydrocarbons, tetrahydrofuran, 2-butanone, and acetone. Polyanhydrides synthesized from nonlinear hydrophobic fatty acid esters, based on ricinoleic, maleic acid, and SA, possessed desired physicochemical properties such as low melting point and hydrophobicity and flexibility to the polymer formed, in addition to biocompatibility and biodegradability. The polymers were synthesized by melt condensation to yield film-forming polymers with *molecular weight* > 100,000 (Domb and Nudelman, 1995a).

The properties of polyanhydrides were modified by the incorporation of long-chain fatty acid terminals such as stearic acid in the polymer composition; this alters the polyanhydride's hydrophobicity and decreases its degradation rate (Teomim and Domb, 1999; Teomim et al., 1999b; Domb and Maniar, 1993b).

Since natural fatty acids are monofunctional, they would act as chain terminators in the polymerization process to control the molecular weight of a polymer. A detailed analysis of the polymerization reaction shows that for up to about 10 mol.% content of stearic acid, the final product is essentially a stearic acid-terminated polymer; this also applies in the case of octanoic acid and lauric acid. In contrast, the higher concentration of acetyl stearate in the reaction mixture resulted in the formation of an increasing amount of stearic anhydride byproduct, with minimal effect on the polymer molecular weight – which remains in the range of 5000. Physical mixtures of polyanhydrides with triglycerides and fatty acids or alcohols did not form uniform blends.

4.6
Amino Acid-based Polymers

General methods for the synthesis of poly(amide-anhydrides) and poly(amide-esters) based on naturally occurring amino acids have been described (Domb, 1990). The polymers were synthesized by amidation of the amino group of an amino acid with a cyclic anhydride, or by the amide coupling of two amino acids with a diacid chloride. Low-molecular weight polymers from methylene bis(*p*-carboxybenzamide) were synthesized by melt condensation (Hartmann and Schultz, 1989). A series of

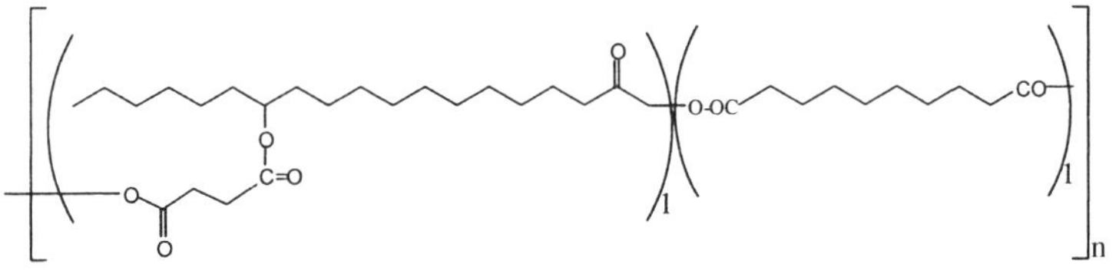

amido-containing polyanhydrides based on *p*-aminobenzoic acid were synthesized by melt condensation; these polymers melted at 58 to 177 °C and had *molecular weight* in the range 2500 to 12,400.

Poly(anhydride-*co*-imides) were synthesized by melt condensation polymerization (Uhrich et al., 1995; Staubli et al., 1991a,b). The trimellitic-amino acid polymers and its copolymers were extensively studied for use as drug carriers (Uhrich et al., 1995; Staubli et al., 1990, 1991a,b). The following amino acids were incorporated in a cyclic imide structure to form a diacid monomer: glycine, β-alanine, γ-aminobutyric acid, L-leucine, L-tyrosine, 11-aminododecanoic acid, and 12-aminododecanoic acid. During the course of the reaction, trimellitic anhydride and pyromellitic anhydride were acetylated by treating imide acid with excess of acetic anhydride, and heating at a reflux temperature until the starting material was dissolved (1 h); this was followed by cooling and drying. The reaction mixture was precipitated in cold, dry diethyl ether. The white precipitate thus obtained was re-washed with diethyl ether, dried and forwarded for melt polymerization to obtain the polymer (Domb and Langer, 1987). The homopolymers of all *N*-trimellitylimido acids containing amino acids were rigid and brittle, with *molecular weight* below 10,000 (Staubli et al., 1991a,b). Higher-molecular weight polymers were obtained by the incorporation of flexible segments, i.e., copolymers with aliphatic diacids, in the polymer backbone. Copolymers of *N*-trimellitylimido-glycine or aminodecanoic acid with either SA or 1,6-bis(*p*-carboxyphenoxy)hexane (CPH) were prepared in defined ratios. High-molecular weight copolymers ($M_w > 100{,}000$) were generally obtained with an increasing content of the SA or CPH comonomer.

4.7 Modified Polyanhydrides and Blends

The physical and mechanical properties of polyanhydrides can be altered by modification of the polymer structure, with a minor change in the polymer composition. Several such modifications include the formation of polymer blends, branched and cross-linked polymers, partial hydrogenation and reaction with epoxides.

Biodegradable polymer blends of polyanhydrides and polyesters have been investigated as drug carriers (Domb, 1993; Abuganima, 1996). Albertsson and coworkers blended poly(trimethylene carbonate) (PTMC) with poly(adipic anhydride) (PAA), and the matrix of PTMC-PAA blend was found biocompatible in *in vitro* and *in vivo* experiments, as well as being a promising candidate for controlled drug delivery (Edlund and Albertsson, 1999; Edlund et al., 2000). These authors suggested that the erosion rate of the blend-matrix could be controlled by varying the proportions of PTMC and PAA. In general, polyanhydrides of different structures form uniform blends with a single melting temperature. Low-molecular weight poly(lactic acid) (PLA), poly(hydroxybutyrate) (PHB), and poly(caprolactone) (PCL) are miscible with polyanhydrides, while high-molecular weight polyesters ($M_w > 10{,}000$) are not compatible with polyanhydrides. Uniform blends of PCL with 10 to 90% by weight of poly(dodecanedioic anhydride) (PDD) were prepared by melt mixing, and the product exhibited good mechanical strength. Differential scanning calorimetry (DSC) thermograms showed two separate peaks (at 55 °C and 75–90 °C for PCL and PDD, respectively) for all compositions. Infra-red (IR) and weight-loss measurements of the blends during hydrolysis indicated a rapid degradation of the anhydride component. After 20 days, the blends contained only PCL with some diacid degradation products, and no anhydride polymer. These studies indicated that the anhydride component degraded and released from the blend composition, without affecting the PCL degradation.

Branched and cross-linked polyanhydrides were synthesized in the reaction of diacid monomers with tri- or polycarboxylic acid branching monomers (Maniar et al., 1990). Muggli *et al.* (1999) synthesized highly cross-linked, surface-eroding polymers by the process of photopolymerization. They used anhydride monomers (e.g., SA, CPP, CPH) end-capped with methacrylate functionalities, and observed that the degradation time scale could be controlled by variation in the network composition, from about 2 days for poly(MSA) to 1 year for poly(MCPH). SA was polymerized with 1,3,5 benzenetricarboxylic acid (BTC) and poly(acrylic acid) to yield random and graft-type branched polyanhydrides. The molecular weights of the branched polymers were significantly higher (M_w 250,000) than the molecular weight of the respective linear polymer (M_w 80,000). The specific viscosities of the branched polymers were lower than those of the linear polyanhydrides, albeit with similar molecular weights. Except for the differences in molecular weight, there were no noticeable changes in the physicochemical or thermal properties of the branched polymers and the linear polymers. Drug release was faster from the branched polymers as compared with the respective linear polymer of a comparable molecular weight.

5 Characterization

The characterization of polyanhydrides, together with data obtained for their chemical composition, structure, crystallinity and thermal properties, mechanical properties, and thermodynamic and hydrolytic stability, are summarized in this section.

5.1 Composition by ^{1}H-NMR

The following copolymer characteristics have been studied by ^{1}H-NMR (Ron et al., 1991): the degree of randomness that suggests whether the polyanhydride is either a random or block copolymer; the average length of sequence (L_n); and the frequency of occurrence of specific comonomer sequences. Copolymers of CPP and SA were used as model polymer. The protons on the aromatic ring close to the anhydride groups experience a low shielding electron density, and absorb at low frequency; in contrast, the protons next to aliphatic comonomers absorb at higher frequency. Accordingly, the CPP-CPP and CPP-SA diads were represented by peaks at 8.1 and 8.0 respectively, and the triplets at 2.6 and 2.4 represent the SA-CPP and SA-SA diads, respectively. The degree of randomness, average block length, and the probability of finding the diad SA-SA or SA-CPP were calculated by integrating the ^{1}H-NMR spectral peaks of poly(CPP-SA) of various compositions. Similar data analysis was also applied to aromatic-aliphatic homopolymers (Domb et al., 1989) and to copolymers of SA with FA (Domb et al., 1991), 1,3-bis(*p*-carboxyphenoxy)hexane (CPH) (Uhrich et al., 1995), and trimellitimide derivatives (Staubli et al., 1991b).

Kim et al. (2000) used a ^{1}H-NMR spectral technique for the estimation of the percent conversion of PEG-dicarboxylate to the methacrylated derivative. This conversion was calculated by using integral ^{1}H-NMR intensities of the protons adjacent to anhydride (A), and those adjacent to carboxylic acid (B), in the following manner:

$$\text{Conversion } \% = A/(A+B) \times 100$$

These authors compared the ^{1}H-NMR spectra of PEG-succinate with dimethacrylated macromer, and observed that the chemical shifts for ethylene protons adjacent to the anhydride bond in succinate were shifted from 2.62–2.69 ppm to 2.69–2.84 ppm, corresponding to the protons in unconverted succinate. The ratio of the peak integral of two methacrylate protons to methylene protons adjacent to the ester bond in the PEG backbone was approximately 1, which was in accordance with the theoretical value.

5.2 Molecular Weight

Molecular weights of the polyanhydrides were determined by viscosity measurements and gel permeation chromatography (GPC) (Ron et al., 1991) Weight average molecular weight (M_w) of polyanhydrides were in the range of 5000 to 300,000, and a polydispersity of 2 to 15 which increased as M_w increased. The intrinsic viscosity $[\eta]$ increases with the increase in M_w. The Mark-Houwink relationship for poly(CPP-SA) was calculated from the viscosity data and the M_w values, which were determined by universal calibration of the GPC data using polystyrene standards.

$$[\eta]^{23\,°C}_{CHCl_3} = 3.88 \times 10^{-7}\ M_w^{0.658}$$

The acetic acid end group determination for molecular weight estimation was not used because the polymer may contain cyclic macromolecules with no acetate end groups (Domb et al., 1991).

5.3 Crystallinity

Since crystallinity is an important factor to control the polymer erosion, an analysis was made of the effect of polymer composition on crystallinity (Ron et al., 1991; Staubli et al., 1991b; Uhrich et al., 1995). Polymers based on SA, CPP, CPH, and FA were investigated. Crystallinity was determined by X-ray diffraction, a combination of X-ray and DSC, and data generated from ^{1}H-NMR spectroscopy and Flory's equilibrium theory. Homopolyanhydrides of aromatic and aliphatic diacids were crystalline ($>50\%$ crystallinity). The copolymers possess a high degree of crystallinity at high molar ratios of either aliphatic or aromatic diacids. The heat of fusion values for the polymers demonstrated a sharp decrease as CPP was added to SA, or vice versa. After adding one monomer, a decreasing trend in crystallinity appeared in the results of X-ray and DSC analysis, which was a direct result of the random presence of other units in the polymer chain. A detailed analysis of the copolymers of SA with the aromatic and unsaturated monomers, CPP, CPH, FA, and trimellitic-amino acid derivative was reported (Staubli et al., 1991). Copolymers with high ratios of SA and CPP, TMA-gly, or CPH were crystalline, while copolymers of equal ratios of SA and CPP or CPH were amorphous. The poly(FA-SA) series displayed high crystallinity, regardless of comonomer ratio.

5.4 Infra-red and Raman Analysis

Anhydrides present characteristic peaks in the IR and Raman spectra. In general, the carbonyl of aliphatic polymers absorb at 1740 and 1810 cm^{-1}, and aromatic polymers at 1720 and 1780 cm^{-1}. Typical IR spectra of aliphatic and aromatic polymers that contain aliphatic and aromatic anhydride bonds may present three distinct peaks, in which the aromatic peak appears at 1780 cm^{-1} and the peaks at 1720–1740 cm^{-1} show a general overlap. The presence of carboxylic acid groups in the polymer can be determined from the presence of a peak at 1700 cm^{-1}. The degradation of polyanhydrides can be estimated by the ratio of the anhydride peaks at 1810 cm^{-1} and 1700 cm^{-1}. The significance of this analysis is that one can measure the degradation of the anhydride bonds without any dissolution of the degradation products, this being dependent on the solubility of the latter.

Fourier transform (FT-IR) spectra have also been used for calculating the relative concentration of free acid to anhydride released during degradation of the polyanhydride matrix (Santos et al., 1999). In this study, the ratio of the peak area of anhydride (1860–1775 cm^{-1}) and carboxylic acid (1775–1675 cm^{-1}) bonds was plotted with time, and the concentrations of free acid were then calculated.

FT-Raman spectroscopy (FTRS) was used to characterize an homologous series of aliphatic polyanhydrides, poly(carboxyphenoxy)alkanes, and copolymers of carboxyphenoxy propane (CPP) and SA. All anhydrides show two diagnostic carbonyl bands: the aliphatic polymers have the carbonyl pairing at 1803/1739 cm^{-1}, and the aromatic polymers have the band pair at 1764 and 1712 cm^{-1} (Davies et al., 1991; Tudor et al., 1991). All the homopolymers and copolymers show different methylene bands due to deformation, stretching, rocking and twisting; the spectra for the aromatic polyanhydrides such as PCPP also have diagnostic benzene para-substitution bands. Thus, it is possible to differentiate between aromatic and aliphatic anhydrides bonding and, in conjunction with other diagnostic bands, to monitor the change in individual monomer composition within a copolymer mixture.

FTRS was also used to study the hydrolytic degradation of polyanhydrides (Davies et al., 1991). PSA rods exposed to water for 15 days were analyzed daily by FTRS. The intensity of the carbonyl anhydride band pair (1803/1739 cm^{-1}) disappeared from day 0 to 15, with the emergence of the complementary acid carbonyl band (1640 cm^{-1}) which increased in intensity over the same period. Similarly, an increase in the intensity of the C–C deformation at 907 cm^{-1} with hydrolysis reflects the gain in freedom of the methylene chain of low-molecular weight oligomers.

5.5
Surface and Bulk Analysis

The morphology of polyanhydrides was studied by scanning electron microscopy (SEM) to elucidate the mechanism of polymer degradation and drug release from polyanhydrides (Mathiowitz et al., 1990). The surface chemical structures of aliphatic polyanhydride films have been examined using time-of-flight secondary ion mass spectroscopy (ToF-SIMS) and X-ray photoelectron spectroscopy (XPS) (Davies et al., 1996). The main peak observed at 285 eV was due to the C–H bond, and the peak at 289.5 eV appeared from O–C=O. The XPS data confirmed the purity of the surface, and the elemental ratios of the experimental surface were in good agreement with the known stoichiometry of the examined polyanhydrides. ToF-SIMS spectra of the polyanhydrides show the reflection of polymer structure. Both negative- and positive-ion SIMS spectra, occurring throughout the entire series of the polyanhydrides, were examined as a result of confirmation on systematic fragmentation where radical cations were observed in the positive-ion spectra. The ion at m/z 71 may arise from fragmentation of the anhydride unit; thus, the appearance of $CH_2{=}CHCOO^-$ is observed for all the polyanhydrides. The combined use of ToF-SIMS and XPS provides a detailed insight into the interfacial chemical structure of polyanhydrides. Atomic force microscopy (AFM) was used to follow the degradation of poly(sebacic acid) and its blends with poly(lactide) (Shakesheff et al., 1994, 1995). These AFM studies reveal the surface polymer morphology to a resolution comparable with that achieved with vacuum-based SEM, and demonstrate the influence of a variety of factors, including polymer crystallinity and the pH of the aqueous environment on the kinetics of the degradation of biodegradable polymers and their blends.

5.6 Stability

The stability of polyanhydrides in solid state and in dry chloroform solution was reported (Domb and Langer, 1989). Aromatic polymers such as poly(CPP) and poly(CPM) maintained their original molecular weight for at least one year in the solid state. In contrast, aliphatic polyanhydrides such as PSA decreased in molecular weight over time. The decrease in molecular weight shows first-order kinetics, with activation energies of 7.5 Kcal mol^{-1} °K. The decrease in molecular weight was explained by an internal anhydride interchange mechanism, as revealed from elemental and spectral analysis. This mechanism was supported by the fact that the decrease in molecular weight was reversible, and heating of the depolymerized polymer at 180 °C for 20 min yielded the original high-molecular weight polymers. However, under similar conditions the hydrolyzed polymer did not increase in molecular weight. It was also found that, in many cases, the stability of polymers in the solid state or in organic solutions did not correlate with their hydrolytic stability. A similar decrease in molecular weight as a function of time was also observed among the aliphatic-aromatic-*co*-polyanhydrides and imide-*co*-polyanhydrides (Domb et al., 1989, 1991; Staubli, 1991a,b).

Mader et al. (1996) have utilized the γ-radiation technique for sterilization of polyanhydrides. In this technique, aliphatic and aromatic homo- and copolymers were irradiated at 2.5 Mrad dose under dry ice, with changes in properties being monitored before and after irradiation. The mechanical and physical properties of the polymer were found to be the same before and after irradiation, as were those monitored by ^{1}H-NMR and FTIR spectra. Using the same concept, these studies were extended for the saturated and unsaturated polyanhydrides (Teomim et al., 2001). Ricinoleic acid-based copolymers with SA and P(CPP-SA) were irradiated under dry ice and at room temperature, while P(FA-SA) was irradiated only at room temperature, under the same conditions as described earlier. It was concluded that saturated polyanhydrides are sufficiently stable during γ-irradiation, while the presence of the double bond conjugated to an anhydride bond makes it unstable and leads to the formation of free radicals. These free radical polyanhydrides transform via an anhydride interchange process into less conjugated polyanhydrides, as shown in Scheme 1.

The outcome of this process was self-depolymerization via inter- and/or intramolecular anhydride interchange to form lower-molecular weight polymers, as shown in Figure 3.

In general, polymers with high melting points and high crystallinity provided a high yield of observable radicals at room temperature. These endogenous free radicals were used to study the processes of water penetration and polymer degradation *in vivo* (Mader et al., 1996). The detection of γ-irradiation sterilization-induced free radicals *in vivo* (using EPR) might be of significance in that the changes in the mobility of the

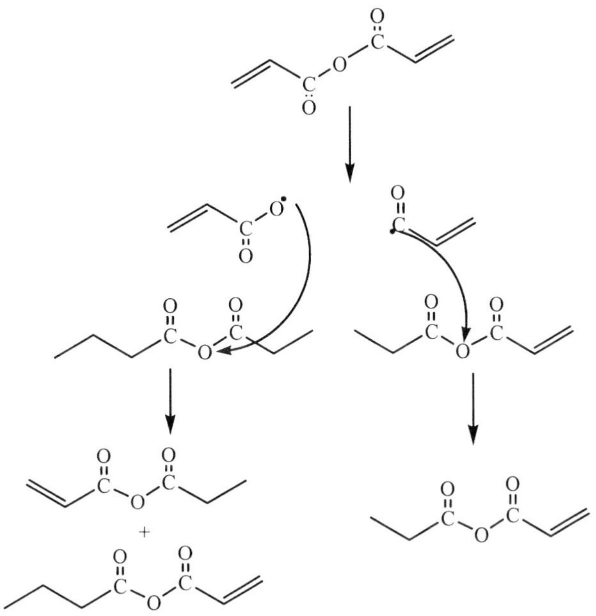

Scheme 1 Proposed mechanism of free radical formation and anhydride transformation into more stable anhydride bonds.

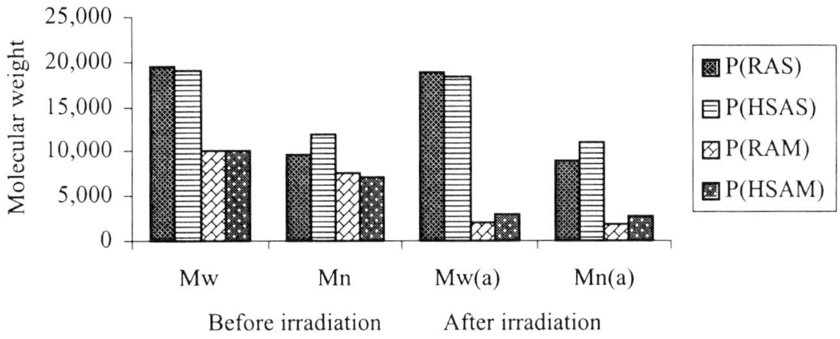

Fig. 3 Molecular weight changes of irradiated ricinoleic acid-based polymers.

radicals could be used as a tool to study drug-release kinetics in a noninvasive and continuous fashion, without any need to introduce paramagnetic species.

6
Fabrication of Delivery Systems

Polymers with low melting points and good solubility in common organic solvents (e.g., methylene chloride) allow easy dispersion of a drug into their matrix. Drugs can also be incorporated using either compression or melt-molding processes. For example, drugs can be incorporated into a slab either by melt mixing the drug into the melted polymer, or by solvent casting. Polymer slabs loaded with drug can also be prepared by compression-molding a powder that contains the drug. Similarly, the drug–polymer formulation can be molded into beads or rods. Polymer films can be prepared by casting the polymer solution containing the drug onto a Teflon-coated dish, followed by evaporation of solvent. Microsphere-based delivery systems can be formulated using a variety of common techniques, including solvent removal, hot-melt encapsulation, and spray drying (Mathiowitz and Langer, 1987; Bindschaedler et al., 1988; Pekarek et al., 1994). However, it is essential that all processes be performed under anhydrous conditions in order to avoid hydrolysis of the polymer.

7 *In vitro* Degradation and Drug Release

The degradation of polyanhydride matrices depends on many factors, including the chemical nature and hydrophobicity of the monomers used in formation of the polymer, the shape and geometry of the implants (as erosion rate depends on the surface area of the polymer), accessibility of the implants to water (as porous materials degrade rapidly in comparison with non-porous materials), the level of drug loading in the polymer matrix, and the pH of the surroundings (polymers degrade rapidly in alkaline media). However, the porosity of the implant can be changed by altering the method of fabrication. For example, a compression-molded device will degrade faster than an injection-molded device due to the former having a higher porosity in the polymer mass. There are many examples of the degradation rate of various types of polyanhydrides. Among all the polyanhydrides, the clinically tested members – poly(FAD-SA), poly(CPP-SA), poly(FA-SA) and poly(EAD-SA) – have been the focus of these studies, although other examples are also available. In general, during the initial 10 to 24 h of water incubation in aqueous medium, the molecular weight fell rapidly, with no loss in wafer mass. However, this was followed by a rapid decrease in wafer mass, accompanied by a small change in polymer molecular weight. The period of extensive mass loss starts when the number average molecular weight (M_n) of polymer approaches 2000, regardless of the initial molecular weight of the polymer. During this period (which lasts for about one week), SA – a relatively water-soluble comonomer – is released from the wafer, leaving the less soluble comonomer, CPP or FAD, which is slow to solubilize (Dang et al., 1996). Increasing the SA content in the copolymer increases the hydrophilicity of the copolymer, which in turn results in a higher erosion rate and hence a higher drug release rate. In a similar mechanism, poly(adipic anhydride) (PAA) in PMTC/PAA blends, when incubated in an aqueous medium, degrades rapidly into its diacid monomers and diffuses out of the matrix. Typical scanning electron micrographs of PTMC/PAA (Edlund and Albertsson, 1999) blends which contain different ratios of monomers and different exposure times in water are shown in Figure 4. This situation could be explained by the fact that the anhydride linkages in the polymer are hydrolyzed subsequent to penetration of water into the polymer. The water uptake depends on the hydrophobicity of the polymer, and therefore the hydrophobic polymers that prevent water uptake have slower erosion rates and lower drug release rates. This is valuable information, as the hydrophobicity of the polymer can be altered by changing the structure and/or the content of the copolymer, thereby being able to alter the drug release rate. An increasing release of methotrexate (MTX) by increasing the SA content in the fatty acid-terminated polyanhydrides (Figure 5) was recently reported (Teomim and Domb, 1999). In the P(CPP-SA) and P(FAD-SA) series of copolymers, a 10-fold increase in

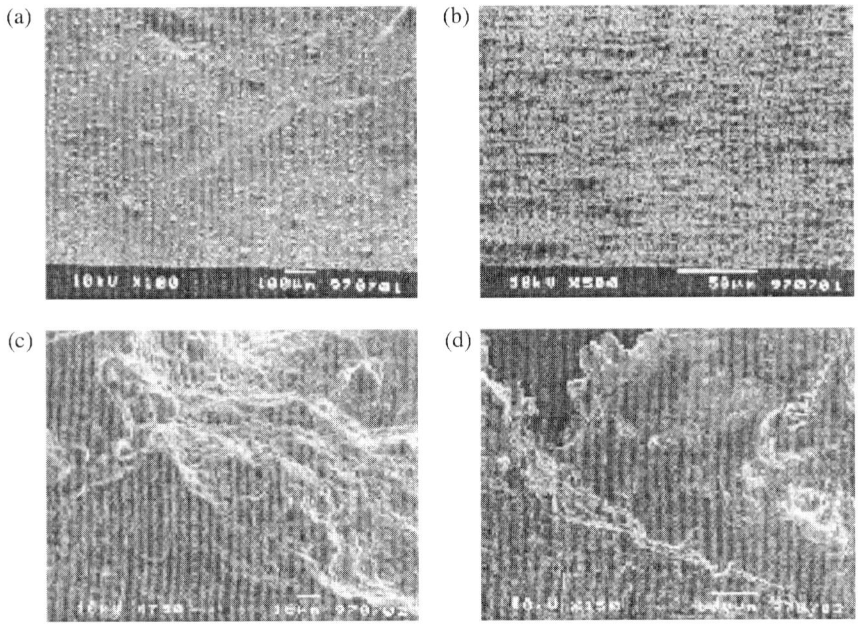

Fig. 4 Scanning electron microscopy micrographs showing the topology of PTMC-PAA films at different stages of erosion: (a) blend of 50% PTMC (medium molecular weight, MMW) and 50% PAA before incubation; (b) the same material after 2 days of water exposure; (c) the same material after 3 weeks; (d) blend of 20% PTMC (low molecular weight, LMW) and 80% PAA after 3 weeks (taken from Edlund and Albertsson, 1999).

drug release rate was achieved by alteration of the ratio of the monomers; thus, both polymers can be used to deliver drugs over a wide range of release rates.

As mentioned earlier, there is no correlation between the rate of drug release and polymer degradation expressed as % decrease in the molecular weight. At first glance, this might appear to be contradictory (D'Emanuele et al., 1994), but on closer examination it appears that a drug dispersed in the polymer matrix is released when the eroding polymer brings the drug with it into solution. Thus, the release rate would depend on the rate of erosion expressed as volume of the matrix dissolved per unit time, multiplied by the drug load, rather than the rate of polymer degradation. The implication is that drug release should correlate with weight loss, which is a more appropriate indicator of the erosion rate than the decrease in molecular weight. Another feature of surface erosion is that the molecular weight of the polymer at the surface may decrease, while the interior of the device may retain the same molecular weight. Then, the lower-molecular weight fragments so formed may not diffuse out or dissolve into the release medium. Therefore, it is not the decrease in molecular weight but the subsequent weight loss to the diffusion and erosion of molecular weight fragments, which should correlate with drug release. This also explains why drug release from these polymer devices was independent of the initial molecular weight of the copolymer (D'Emanuele et al., 1994).

Thus, the factors that affect drug release from the polyanhydride matrix include polymer composition, fabrication method,

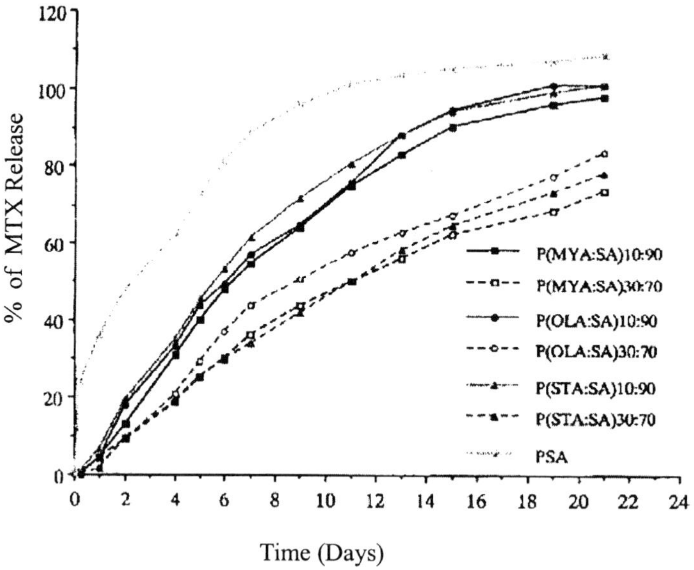

Fig. 5 *In vitro* release of methotrexate (MTX) from C_{14}-C_{18} fatty acid-terminated polymers. MTX release was conducted in phosphate buffer, pH 7.4, at 37 °C and determined by UV absorbance at 303 nm (from Teomim and Domb, 1999).

size and geometry, particle size of incorporated drug, drug solubility, and drug loading. Recently, heparin release from thin flexible sheets composed of PLA laminated polyanhydride P(EAD-SA) was reported (Teomim et al., 1999a). Drug release was observed to follow first-order kinetics, and ~50% of the drug was released within 24 h from incubated poly (EAD-SA) film. The polymer matrix coated with PLA and PLA-PEG sheets controlled the release rate, as heparin release was observed continuously for 2–3 weeks. Polyanhydride matrix coated with PLA-PEG block copolymer was more hydrophilic than the coated with PLA, and showed a faster release (Figure 6). It is observed that hydrophilic drugs are released much more rapidly than the hydrophobic drugs and suffer from a significant initial burst effect, which is a function of drug particle size. Moreover, the correlation between drug release and polymer degradation is better for injection or melt-molded devices than with compression-molded devices (Domb et al., 1994a). In a recent report, Stephens et al. (2000) investigated the *in vitro* release behavior of gentamicin from a poly(EAD-SA) matrix, and found that release rate of the drug was faster in water than in phosphate buffer, pH 7.4. The beads of polyanhydride matrix were cracked in water, and drug release was more rapid due to hydrophilicity of the drug and its diffusion through the cracked matrix. In contrast, drug release in buffer at pH 7.4 was diffusion-controlled, as no crack was observed in the bead dipped in phosphate buffer. The release of indomethacin from poly(CPP-SA) and poly(FAD-SA) was studied and found to be independent of drug loading (Gopferich and Langer, 1993). A simple model that takes into account the following kinetic steps, namely the spontaneous degradation of polymer to crystallized monomer, the creation of pores, the dissolution of monomers inside the pores, and the final release of monomer via

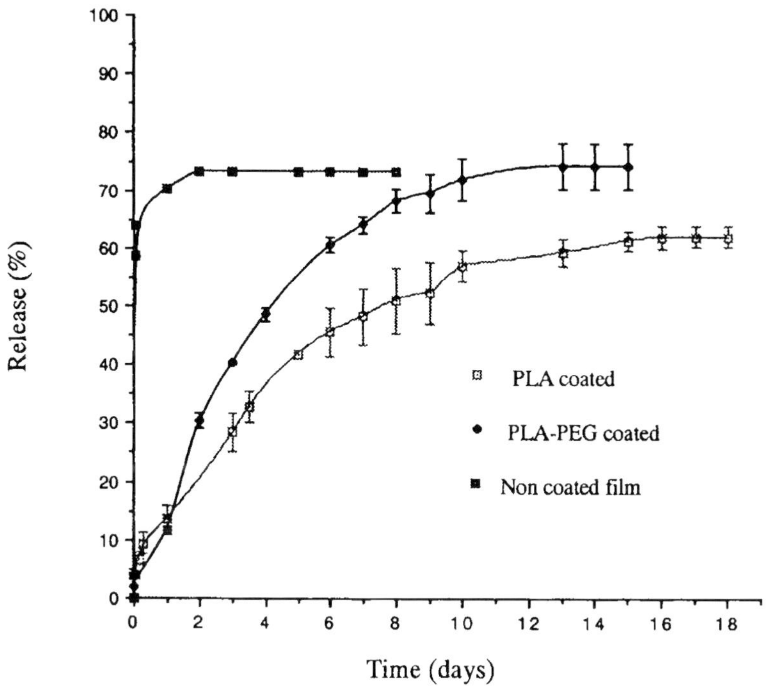

Fig. 6 *In vitro* release of heparin from laminated and coated polyanhydride films [2% (w/v) 10 x 5 x 0.1 mm³]. The release was conducted in phosphate buffer, pH 7.4, at 37 °C (from Teomim et al., 1999a).

diffusion through the pore network was proposed by Gopferich and Langer to explain this unusual release behavior (Gopferich and Langer, 1993, 1995a,b; Gopferich et al., 1995). Erosion was then simulated using a Monte Carlo method that describes these morphological changes during erosion.

Noninvasive *in vivo* and *in vitro* monitoring of drug release and polymer degradation using EPR spectroscopy was carried out by Mader et al. (1997a,b). By incorporating a nitro-oxide radical probe such as 2,2,5,5-tetramethyl 3 carboxyl pyrrollidine-1-oxyl (PCA) at 5 mmol kg^{-1}, the microviscosity and drug mobility, as well as the pH within a device, can be monitored both *in vivo* and *in vitro* using EPR methods. Low-frequency EPR spectrometry was used to study the *in vivo* and *in vitro* degradation of polyanhydrides. Tablets loaded with PCA were exposed to 0.1 M phosphate buffer pH 7.4 at 37 °C, or implanted subcutaneously into the back of rats or mice. Measurements were carried out using a standard 9.4-Ghz EPR spectrometer or a 1.1-GHz spectrometer equipped with a surface coil. EPR measurements of pH are based on the effect of protonation/deprotonation of groups located in close proximity to the radical moiety, which induces changes in the hyperfine splitting constant, a_N and the g value (the Zeeman splitting constant for free electrons). The measurements of pH values in

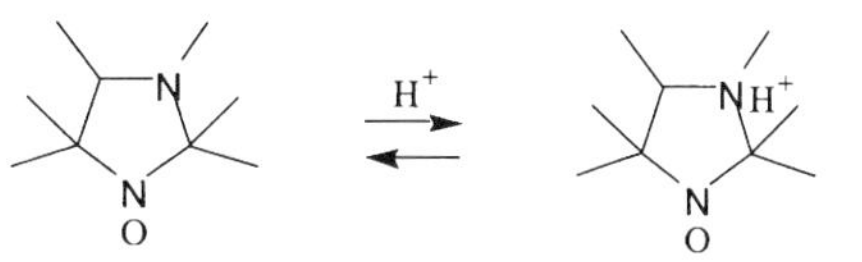

g = 2.0054; a_N = 1.557 mT   g = 2.0057; a_N = 1.43 mT

the range between 0 and 9 are possible by using probes with different pK_a values.

In vitro and *in vivo* studies have demonstrated the environment within the polyanhydride tablets as acidic (pH ≤ 4) when the degradation is carried out in 0.1 M pH 7.4 phosphate-buffered solution. One approach that has been taken to counteract the acidity is the incorporation of buffering substances into the polymer matrix during the device fabrication. This has been shown to cause an increase in pH inside the delivery system. The microviscosity and drug mobility, as well as the formation of radicals, within a device were monitored both *in vivo* and *in vitro* using EPR methods. EPR imaging (EPRI) introduces a special dimension by means of additional gradients, and is applied to characterize the degradation front and the microenvironment of polyanhydride disks loaded with a pH-sensitive nitrooxide (Mader et al., 1997a). Exposure to buffer (pH 7.4) resulted in the formation of a front of degraded polymer from outside to inside. A pH gradient was found to exist within the polymer matrix, and the pH rises with time from 4.7 to 7.4. The issue of a possible chemical reaction between amine and hydroxyl group-containing drug moieties with the anhydride bonds in the polymer during drug incorporation and release has also been investigated (Mader et al., 1997a).

8 Biocompatibility and Elimination

The Food and Drug Administration (FDA), in testing and evaluating new biomaterials, established the biocompatibility and safety of polyanhydrides following the 1986 guidelines. Several accepted criteria and tests to evaluate new biomedical materials were used to assess the safety of polyanhydrides (Braun et al., 1982; Leong et al., 1986a; Laurencin et al., 1990). In these studies, poly[bis(*p*-carboxy-phenoxy)propane anhydride (PCPP) and its copolymers with SA were tested. Neither mutagenicity nor cytotoxicity or teratogenicity was associated with the polymers or their degradation products, as evaluated by mutation assays. The tissue response of these polyanhydrides was studied by subcutaneous implantation in rats, and in the cornea of rabbits. The polymers did not provoke inflammatory responses in the tissues over a 6-week implantation period, and histologic evaluation indicated relatively minimal tissue irritation, with no evidence of local or systemic toxicity (Laurencin et al., 1990). Systemic response to the polymer was evaluated by monitoring of blood chemistry and hematologic values, and by comprehensive examination of organ tissues. Neither method revealed any significant response to the polymer. Recently, a report on the biocompatability of ricinoleic acid-based polymer (published by our laboratory; Teomim and Domb, 1999) in which biocompatability of poly(RAS-SA) was tested in rats and compared with Vicryl™ surgical suture and sham surgery. It was observed that no any abnormal gross pathological finding occurred at the implantation site after 3, 7 and 21 days of implantation. Blood chemistry and cell counts were similar for treated and untreated rats. The polymer was degraded into ricinoleic acid with constant release of incorporated drug, while the inflammatory response after subcutaneous implantation was minimal to mild, and comparable with that seen clinically with Vicryl™ absorbable sutures.

Since the CPP-SA copolymer was designed to be used clinically to deliver an anticancer agent directly into the brain for the treatment of brain neoplasm, *in vivo* safety evaluations and brain biocompatibility were assessed in rats (Tamargo et al. 1989), rabbits (Brem et al., 1989) and monkeys (Brem et al., 1988). In the rat brain study, the tissue reaction of the polymer (PCPP-SA 20:80) was compared with the reaction observed with two standard materials used in surgery, which have been extensively studied, namely Gelfoam® (an absorbable gelatin sponge) and Surgicel® (an oxidized cellulose absorbable hemostat commonly used in brain surgery). Histologic evaluation of the tissue demonstrated a small rim of necrosis around the implant, and a mild to marked cellular inflammatory reaction limited to the area immediately adjacent to the implantation site. The pathologic response associated with poly(CPP-SA) copolymer was slightly more pronounced than with Surgicel® at the earlier time points, but noticeably less marked than with Surgicel® at the later times. The reaction to Gelfoam® was essentially equivalent to that observed in control rats. In a similar brain biocompatibility study carried out in monkeys, no tissue abnormalities were noted either in computed tomography or magnetic resonance imaging (MRI) scans. Furthermore, no abnormalities were observed either in the blood chemistry or hematologic evaluations (Brem et al., 1988). There appeared to be no adverse systemic effects due to the implants as assessed by histologic evaluation of the tissue tested. Overall, no unexpected or untoward reaction to treatment was observed. Copolymers of SA with several aliphatic comonomers such as dimer of erucic acid (FAD), FA acid and isophthalic acid were also tested both subcutaneously and in the rat brain, and were also found to be biocompatible (Rock et al., 1991). The hydrolysis and elimination processes of polyanhydrides have been studied using a series of polyanhydrides derived from different linear aliphatic diacids (Domb and Nudelman, 1995b). These polymers degrade into their monomer or oligomer units at about the same rate, but differ in the water solubility of their degradation products. Polymers based on natural diacids of the general structure -[OOC-$(CH_2)_x$-CO]- (where x is between 4 and 12) were implanted subcutaneously in rats, and the elimination of polymers from the implantation site was studied. The *in vitro* hydrolysis of this polymer series was studied by monitoring the weight loss, release of monomer degradation products, and changes in content of anhydride bonds of polymer as a function of time. It was observed that, both *in vitro* and *in vivo*, the rate of polymer elimination was a function of monomer solubility. The elimination time for polymers based on soluble monomers ($x = 4-8$) was 7 – 14 days, while the polymers based on monomers with lower solubility ($x = 10-12$) were eliminated only after 8 weeks. All polymers were found to be biocompatible and useful as carriers of drugs.

The elimination of poly(CPP-SA)-based implant (Gliadel™), which is currently in clinical use for the treatment of brain cancer, was studied in rabbit and rat brains using radioactive polymer and drug (Domb et al., 1994b, 1995a). The implant is composed of *N*,*N*-bis(2-chloroethyl)-*N*-nitrosourea (BCNU) dispersed in a co-polyanhydride matrix of CPP and SA. Four groups of rabbits were implanted with wafers loaded with BCNU, one in a ^{14}C-SA-labeled polymer, another in a ^{14}C-CPP-labeled polymer, and two groups with ^{14}C-BCNU in a non-labeled polymer in which one was for a BCNU disposition study and one for a residual drug study. In the rabbits implanted with the ^{14}C-SA-labeled polymer, approximately 10% of the radioactivity was found in the urine and 2% in the feces, and about 10% remained in the device after 7 days of implantation. In contrast, only 4% of the radioactivity associated with the ^{14}C-CPP labeled polymer was found in urine and feces during this period. However, a drastic increase in CPP excretion was found after 9 days; moreover, after 21 days 64% of the implanted ^{14}C-CPP was recovered in the urine and feces, and 29% was still in the recovered wafers. Studies with radiolabeled BCNU in rabbit brain revealed that approximately 50% of the BCNU in the wafers was released in 3 days, and over 95% was released after 6 days in the rabbit brain. Excretion of this polymer after implantation in the rat brain using radiolabeled polymers showed that over 70% of the SA comonomer was excreted in 7 days, with about 40% of the SA metabolized to CO_2 (Domb et al., 1995). The elimination of poly(FAD-SA) rods loaded with 0, 10, and 20 wt.% of gentamicin sulfate after implantation in the femoral muscle and bone of dogs was studied as part of the preclinical studies for Septacin® bone implant (Domb and Amselem, 1994). Most of the polymer implant was gradually eliminated from bone and muscle within 4 to 8 weeks post implantation, with the elimination from bone being faster and leading to new bone formation in the implant site, without any polymer entrapment. The elimination rate was dependent mainly on the amount of polymer implanted. Gentamicin was released for a period of about 3 weeks, with no residual drug being detected in the polymer remnants 8 weeks post implantation. In all experiments, no local or systemic toxicity was observed.

Mader et al. (1997b) utilized the noninvasive technique of MRI to visualize both polymer erosion (*in vivo* and *in vitro*) and the physiologic response (edema, encapsulation) to the implant. MRI enables monitoring, *in vivo* and in a noninvasive manner, of the water content, implant shape, and response of the biological system to an implant in real time, without stopping the experiment. MRI images were taken during the course of degradation of slabs of PSA – a rapidly degrading polyanhydride – placed in physiologic buffer solution and implanted subcutaneously in rats. However, a water penetration front was clearly observed at day 21 *in vitro* and at day 32; thus, the bright image inside the MRI images indicated that the entire polymer matrix was filled with water. This brightness was never noticed *in vivo*. Instead, a deformation of the implant was observed, starting with the rounding of the corners (day 9), which progressively increased (days 16 and 20) and was completed at day 28.

9 Applications

Polyanhydrides have been investigated as a candidate for controlled release devices of drugs for treating eye disorders (Albertsson et al., 1996), chemotherapeutic agents (Wu et al., 1994), local anesthetics (Masters et al., 1993a,b; Maniar et al., 1994), anticoagulants (Chickering et al., 1996), neuroactive drugs (Kubek et al., 1998), and anticancer agents (Domb and Ringel, 1994). Polyanhydride poly(CPP-SA) were loaded with carmustine and implanted into the brains of rats, rabbits, and monkeys, and were found to be biocompatible and efficacious against brain tumors (Tamargo et al., 1993; Fung et al., 1998). Based on the findings of these laboratory studies, a phase I–II clinical trial was completed which showed treatment to be well tolerated in all patients, apparently without the production of any systemic side effects (Brem et al., 1991). Phase III human clinical trials have demonstrated that the site-specific delivery of BCNU (carmustine) from a poly(CPP-SA)20:80 wafer (Gliadel®) in patients with recurring brain cancer (glioblastoma multiforme) significantly prolongs patient survival (Brem et al., 1995). On the basis of the results of clinical trials, the FDA has given its first approval during the past 23 years for a new treatment for brain tumors (Brem and Lawson, 1999). Gliadel®

was subsequently approved for the treatment of brain tumors in Canada, South America, Israel, South Korea and Europe. Masters et al. (1993a,b) used a polyanhydride cylinder for the delivery of local anesthetics in close proximity to the sciatic nerve to produce a neuronal block for several days. The use of polyanhydrides in the oral delivery of insulin and plasmid DNA has also been investigated (Mathiowitz et al., 1997). During the past five years, investigations have expanded to newer polymers and other drugs such as 4-hydroperoxy cyclophosphamide (4HC), cisplatin, carboplatin, paclitaxel, as well as several alkaloid drugs, in an effort to develop a better system for treating brain tumors (Laurencin et al., 1993; Brem et al., 1994; Judy et al., 1995; Olivi et al., 1996). Carboplatin incorporated into poly(FAD-SA), prepared by mixing the drug in the melted polymer, has been evaluated for the treatment of brain tumors in laboratory animals, and has shown promising results (Olivi et al., 1996). Poly(CPP-SA) has also been used to develop a delivery system for gentamicin sulfate for the treatment of osteomyelitis (Laurencin et al., 1993; Domb and Amselem, 1994). A sustained release of gentamicin sulfate over a period of few weeks was obtained both *in vivo* and *in vitro* using this system. This delivery device, which is in the form of a chain of beads, has been undergoing human clinical trials in the USA. The effect of long-term glutamic acid stimulation of trigeminal motorneurons, using poly(FAD-SA) microspheres, has also been explored. This study was undertaken to determine the role of glutamate in possible growth disorders of the craniofacial skeleton. Pronounced skeletal changes in the snout region were observed in growing rats receiving glutamate, indicating that a sustained release of glutamic acid *in vivo* can affect the development of skeletal tissue (Hamilton-Byrd et al., 1992). Recently, the use of polyanhydrides for the delivery of heparin and the treatment of osteomyelitis have also been reported (Teomim et al., 1999a; Stephens et al., 2000).

10 Outlook and Perspectives

Polyanhydrides have now been investigated as a 'smart' biomaterial used for the short-term release of drugs for more than two decades. Over these years, intensive research has been conducted in both academia and industry which has yielded hundreds of publications and patents describing new polymer structures, studies on chemical and physical characterization of these polymers, degradation and stability properties, toxicity studies, and applications of these polymers mainly for the controlled delivery of bioactive agents. Such research also yielded a device (Gliadel®) which can be used clinically in the treatment of brain cancer. Due to the rapid degradation and limited mechanical properties, the main application for this class of polymers is for the short-term controlled delivery of bioactive agents.

The advantages and disadvantages of these polymers relate to the hydrolytic instability of the anhydride bond that degrades rapidly, thereby changing the polyanhydride into its nontoxic monomers. The main advantages of this class of polymers are:

- They may be prepared from readily available, low-cost resources that are generally considered as safe dicarboxylic acid building blocks, many of which are either body constituents or metabolites.
- They are prepared in a one-step synthesis, with no need for purification steps.
- They have a well-defined polymer structure, with controlled molecular weight;

moreover, they degrade hydrolytically into their building blocks at a predictable rate.

- They can be manipulated accordingly to release bioactive agents in a predictable rate for periods of weeks.
- They are processable, either by low melting injection molding or extrusion for mass production, and have versatile properties which can be obtained by monomer selection, composition, surface area, and additives.
- They degrade to their respective diacids, and are completely eliminated from the body within periods of weeks to months.
- They can be sterilized by terminal γ-irradiation, without there being any adverse effect on polymer properties.

It should be noted that, although these polymers have so many advantages, they also have certain disadvantages, the main ones being their short-term degradation and release periods (which may extend from weeks to a few months), and the need for specialized storage conditions (e.g., refrigeration, dry). Ironically, however, the high rate at which these polymers degrade renders them most suitable for the short-term, i.e., days to weeks, controlled delivery of bioactive agents in the form of solid implants or injectable microspheres.

On this basis, it is expected that these polymers will be developed in the coming years into implantable devices for the local and systemic delivery of drugs.

11 Patents

Patents relating to the application of polyanhydrides are listed in Table 2.

Tab. 2 Patents for the application of polyanhydrides

S. No.	*Patent No./ Issue Date*	*Title*	*Author(s)*	*Applicant(s)*
1	US04886870 12/12/1989	Bioerodible articles useful as implants and prostheses having predictable degradation rates	D'Amore, P.; Leong, K. W.; Langer, R.S.	Massachusetts Institute of Technology (USA)
2	US04898734 02/06/1990	Polymer composite for controlled release or membrane formation	Mathiowitz, E.; Langer, R. S.; Warshawsky, A.; Edelman, E.	Massachusetts Institute of Technology(USA)
3	US04906473 03/06/1990	Biodegradable poly(hydroxyalkyl)amino dicarboxylic acid) derivatives, a process for their preparation, and the use thereof for depot formulations with controlled delivery of active ingredient	Ruppel, D.; Walch, A.	Hoechst Aktiengesellschaft (Germany)
4	US04999417 03/12/1991	Biodegradable polymer compositions	Domb, A. J.	Nova Pharmaceutical Corporation (USA)
5	US05019379 05/28/1991	Unsaturated polyanhydrides	Domb, A. J.; Langer, R. S.	Massachusetts Institute of Technology (USA)
6	US05175235 12/29/1992	Branched polyanhydrides	Domb, A. J.; Maniar, M.	Nova Pharmaceutical Corporation (USA)
7	US05179189 01/12/1993	Fatty acid-terminated polyanhydrides	Domb, A. J.; Maniar, M.	Nova Pharmaceutical Corporation (USA)

Tab. 2 (cont.)

S. No.	Patent No./ Issue Date	Title	Author(s)	Applicant(s)
8	US05197466 03/30/1993	Method and apparatus for volumetric interstitial conductive hyperthermia	Marchosky, J. A.; Moran, C. J.; Fearnot,N. E.	MED Institute Inc. West Lafayette, IN
9	US05395916 03/07/1995	Biodegradable copolymer from hydroxy proline	Mochizuki, S.; Nawata, K.; Suzuki, Y.	Teijin Limited (Japan)
10	US05459258 10/17/1995	Polysaccharide-based biodegradable thermoplastic materials	Merrill, E. W.; Sagar, A.	Massachusetts Institute of Technology (USA)
11	US05473103 12/05/1995	Biopolymers derived from hydrolyzable diacid fats	Domb, A. J.; Nudelman, R.	Yissum Research Development Co. of the Hebrew University of Jerusalem (Israel)
12	US05522895 06/04/1996	Biodegradable bone templates	Mikos, A. G.	Rice University (USA)
13	US05545409 08/13/1996	Delivery system for controlled release of bioactive factors	Laurencin, C. T.; Lucas, P. A.; Syftestad, G.T. ; Domb, A. J.; Glowacki, J.; Langer, R. S.	Massachusetts Institute of Technology (USA)
14	US05618563 04/08/1997	Biodegradable polymer matrices for sustained delivery of local anesthetic agents	Berde, C. B.; Langer, R. S.	Children's Medical Center Corporation (USA)
15	US05626862 05/06/1997	Controlled local delivery of chemotherapeutic agents for treating solid tumors	Brem, H.; Langer, R. S.; Domb, A. J.	Massachusetts Institute of Technology (USA)
16	US05660851 08/26/1997	Ocular inserts	Domb, A. J.	Yissum Research Development Company of the Hebrew University of Jerusalem (ISRAEL)
17	US05716404 02/10/1998	Breast tissue engineering	Vacanti, J. P.; Atala, A.; Mooney, D. J.; Langer, R. S.	Massachusetts Institute of Technology (USA)
18	US05756652 05/26/1998	Poly (ester-anhydrides) and intermediates therefore	Storey, R. F.; Deng, Z. D.; Peterson, D. R.; Clancy, T. P.	DePuy Orthopedics, Inc. (USA)
19	US05855913 01/05/1999	Particles incorporating surfactants for pulmonary drug delivery	Hanes, J.; Edwards, D. A.; Evora, C.; Langer, R. S.	Massachusetts Institute of Technology (USA)
20	US06046187 04/04/2000	Formulations and methods for providing prolonged local anesthesia	Berde, C. B.; Langer, R. S.; Curley, J.; Castillo, J	Children's Medical Center Corporation (USA)

12
References

Abuganima, E. (1996) *Synthesis and characterization of copolymers and blends of polyanhydrides and polyesters*, M.Sc. Thesis, The Hebrew University of Jerusalem.

Albertsson, A.-C., Lundmark, S. (1988) Synthesis of poly(adipic anhydride) by use of ketenes, *J. Macromol. Sci.- Chem. A* **25**, 247–258.

Albertsson, A.-C., Lundmark, S. (1990) Synthesis of poly(adipic) anhydride by use of ketene, *J. Macromol. Sci.- Chem. A* **27**, 397–412.

Albertsson, A.-C., Carlfors, J., Sturesson, C. (1996) Preparation and characterization of poly(adipic anhydride) microspheres for ocular drug delivery, *J. Appl. Polym. Sci.* **62**, 695–705.

Bindschaedler, C., Leong, K., Mathiowitz, E., Langer, R. (1988) Poly(anhydride) microspheres formulation by solvent extraction, *J. Pharm. Sci.* **77**, 696–698.

Braun, A. G., Buckner, C. A., Emerson, D. J., Nichinson, B. B. (1982) Quantitative correspondence between the in vivo and *in vitro* activity of teratogenic agents, *Proc. Natl. Acad. Sci. USA* **79**, 2056.

Brem, H., Lawson, H. C. (1999) The development of new brain tumor therapy utilizing the local and sustained delivery of chemotherapeutic agents from biodegradable polymers, *Cancer* **86**, 197–199.

Brem, H., Tamargo, R. J., Pinn, M., Chasin, M. (1988) Biocompatibility of a BCNU-loaded biodegradable polymer: a toxicity study in primates. *Am. Assoc. Neurol. Surg.* **24**, 381.

Brem, H., Kader, A., Epstein, J. I., Tamargo, R. J., Domb, A. J., Langer, R., Leong, K. W. (1989) Biocompatibility of bioerodible controlled release polymer in the rabbit brain, *Sel. Cancer Ther.* **5**, 55–65.

Brem, H., Mahley, M. S., Vick, N. A. et al. (1991) Interstitial chemotherapy with drug polymer implants for the treatment of recurrent gliomas, *J. Neurosurg.* **74**, 441–446.

Brem, H., Walter, K. A., Tamargo, R. J., Olivi, A., Langer, R. (1994) Drug delivery to the brain, in: *Polymeric Site-Specific Pharmacotherapy* (Domb, A., Ed.), John Wiley & Sons: Chichester, 117–140.

Brem, H., Piantadosi, S., Burger, P. C. et al. (1995) Placebo-controlled trial of safety and efficacy of intraoperative controlled delivery by biodegradable polymers of chemotherapy for recurrent gliomas. The Polymer-brain Tumor Treatment Group, *Lancet* **345**, 1008–1012.

Bucher, J. E., Slade, W. C. (1909) The anhydrides of isophthalic and terephthalic acids, *J. Am. Chem. Soc.* **31**, 1319–1321.

Chasin, M., Domb, A., Ron, E., Mathiowitz, E., Leong, K., Laurencin, C., Brem, H., Grossman, S., Langer, R. (1990) Polyanhydrides as drug delivery systems, in: *Biodegradable Polymers as Drug Delivery Systems* (Chsin, M., Langer, R., Eds.), New York: Marcel Dekker, 43–70.

Chickering, D., Jacob, J., Mathiowitz, E.(1996) Poly(fumaric-co-sebacic) microspheres as oral drug delivery systems, *Biotechnol. Bioenerg.* **52**, 96–101.

Conix, A. (1958) Aromatic poly(anhydrides): a new class of high melting fiber-forming polymers, *J. Polym. Sci.* **29**, 343–353.

Dang, W. B., Daviau, T., Nowotnik, D. (1996) *In vitro* erosion kinetics of implantable polyanhydride Gliadel™, *Proc. Int. Symp. Control. Rel. Bioact. Mater.* **23**, 731–732.

Davies, M. C., Khan, M. A., Domb, A. J., Langer, R., Watts, J. F., Paul, A. (1991) The analysis of the surface chemical structure of biomedical aliphatic poly(anhydrides) using XPS and ToF-SIMS, *J. Appl. Polym. Sci.* **42**, 1597–1605.

Davies, M. C., Shakesheff, K. M., Shard, K. M., Domb, A. J., Roberts, C. J., Tendler, S. J. B.,

Williams, P. M., Tendler, S. J. B. and Williams, P. M. (1996) Surface analysis of biodegradable polymer blends of poly(sebacic anhydride) and poly-(DL-lactic acid), *Macromolecules* **29**, 2205–2212.

D'Emanuele, A., Hill, J., Tamada, J. A., Domb, A. J., Langer, R. (1994) Molecular weight changes in polymer erosion, *Pharmaceut. Res.* **9**, 1279–1283.

Domb, A. J. (1990) Biodegradable polymers derived from amino acids, *Biomaterials* **11**, 686–689.

Domb, A. J. (1992) Synthesis and characterization of bioerodible aromatic anhydride copolymers, *Macromolecules* **25**, 12–17.

Domb, A. J. (1993) Biodegradable polymer blends: screening for miscible polymers, *J. Polym. Sci.: Polym. Chem.* **31**, 1973–1981.

Domb, A. J., Amselem, S. (1994) Antibiotic delivery systems for the treatment of bone infections, in: *Polymeric Site-Specific Pharmacotherapy* (Domb, A.J., Ed.), Chichester: John Wiley & Sons, 242–265.

Domb, A. J., Langer, R. (1987) Poly(anhydrides). I. Preparation of high molecular weight polyanhydrides, *J. Polym. Sci. Polym. Chem.* **25**, 3373–3386.

Domb, A. J., Langer, R. (1989) Solid-state and solution stability of poly(anhydrides) and poly(esters), *Macromolecules* **22**, 2117–2122.

Domb, A. J., Maniar, M. (1993a) Absorbable biopolymers derived from dimers fatty acids, *J. Poly. Sci.: Polym. Chem.* **31**, 1275–1285.

Domb, A. J., Maniar, M. (1993b) Fatty acid terminated polyanhydrides, US Patent 5,179,189.

Domb, A. J., Nudelman, R. (1995a) Biodegradable polymers derived from natural fatty acids, *J. Polym. Sci.* **33**, 717–725.

Domb, A. J., Nudelman, R. (1995b) *In vivo* and *in vitro* elimination of aliphatic polyanhydrides, *Biomaterials* **16**, 319–323.

Domb, A. J., Ringel, I. (1994) Polymeric Drug Carrier Systems in the Brain, in: *Providing Pharmaceutical Access to the Brain, Methods in Neuroscience* (Flanaga, T. R., Emerich, D. F., Winn, S. R. Eds.) CRC Press: Boca Raton, FL, 169–183, Vol. 21.

Domb, A. J., Ron, E., Langer, R. (1988) Poly(anhydrides). II. One step polymerization using phosgene or diphosgene as coupling agents, *Macromolecules* **21**, 1925–1929.

Domb, A. J., Gallardo, C. F., Langer, R. (1989) Poly(anhydrides). 3. Poly(anhydrides) based on aliphatic-aromatic diacids, *Macromolecules* **22**, 3200–3204.

Domb, A. J., Mathiowitz, E., Ron, E., Giannos, S., Langer, R. (1991) Polyanhydrides IV. Unsaturated and cross-linked poly(anhydrides), *J. Polym. Sci. Part A: Polym. Chem.* **29**, 571–579.

Domb, A. J., Amselem, S., Shah, J., Maniar, M. (1993) Polyanhydrides: Synthesis and characterization, in: *Advances in Polymer Sciences* (Peppas, N. A., Langer, R., Eds.), Heidelberg: Springer-Verlag, 93–141.

Domb, A. J., Amselem, S., Langer, R., Maniar, M. (1994a) Polyanhydrides as carriers of drugs, in: *Designed to Degrade Biomedical Polymers* (Shalaby, S., Ed.), Munich: Carl Hanser Verlag, 69–96.

Domb, A. J., Rock, M., Perkin, C., Proxap, B., Villemure, J. G. (1994b) Metabolic disposition and elimination studies of a radiolabelled biodegradable polymeric implant in the rat brain, *Biomaterials* **15**, 681–688.

Domb, A. J., Rock, M., Perkin, C., Proxap, B., Villemure, J. G. (1995) Excretion of a radiolabelled biodegradable polymeric implant in the rabbit brain, *Biomaterials* **16**, 1069–1072.

Domb, A. J., Elmalak, O., Shastri, V. R., Ta-Shma, Z., Masters, D. M., Ringel, I., Teomim, D., Langer, R. (1997) Polyanhydrides, in: *Handbook of Biodegradable Polymers* (Domb, A. J., Kost, J., Weiseman, D. M., Eds.), Amsterdam: Harwood Academic Publishers, 135–159.

Domb, A. J., Israel, J. H., Elmalak, O., Teomim, D., Bentolila, A. (1999) Preparation and characterization of Carmustine loaded biodegradable disc for treating brain tumors, *Pharm. Res.* **16**, 762–765.

Edlund, U., Albertsson, A.-C. (1999) Copolymerization and polymer blending of trimethylene carbonate and adipic anhydride for tailored drug delivery, *J. Appl. Polym. Sci.* **72**, 227–239.

Edlund, U., Albertsson, A.-C., Singh, S. K., Fogelberg, I. (2000) Sterilization, storage, stability and *in vivo* biocompatibility of poly(trimethylene carbonate)/poly(adipic anhydride) blends, *Biomaterials* **21**, 945–955.

Fung, L. K., Ewend, M., Sills, A. et al. (1998) Pharmacokinetics of interstitial delivery of carmustine, 4-hydroperoxycyclophosphamide and paclitaxel from a biodegradable polymer implant in the monkey brain, *Cancer Res.* **58**, 672–684.

Gopferich, A., Langer, R. (1993) The influence of microstructure and monomer properties on the erosion mechanism of a class of polyanhydrides, *J. Polym. Sci.* **31**, 1445–1458.

Gopferich, A., Langer, R. (1995a) Modeling of polymer erosion in three dimensions: rotationally symmetric devices, *AIChE J.* **41**, 2292–2299.

Gopferich, A., Langer, R. J (1995b) Modeling monomer release from bioerodible polymers, *J. Control. Rel.* **33**, 55–69.

Gopferich, A., Karydas, D., Langer, R. (1995) Predicting drug release from cylindrical polyan-

hydride matrix discs, *Eur. J. Pharm. Biopharm.* **41**, 81–87.
Gref, R., Minamitake, Y., Peracchia, M. T., Domb, A. J., Trubetskoy, V., Torchilin, V., Langer, R. (1995) Poly(ethylene glycol) coated nanospheres, *Adv. Drug Deliv. Rev.* **16**, 215–233.
Gupta, B. (1988) Polyanhydride process from bis(trimethylsilyl) ester of dicarboxylic acid, US Patent 4,868,265.
Hamilton-Byrd, E. L., Sokoloff, A. J., Domb, A. J., Terr, L., Byrd, K. E. (1992) L-Glutamate microsphere stimulation of the trigeminal motor nucleus in growing rats, *Polym. Adv. Technol.* **3**, 337–344.
Hartmann, M., Schultz, V. (1989) Synthesis of poly(anhydride) containing amido groups, *Macromol. Chem.* **190**, 2133.
Hill, J. W. (1930) Studies on polymerization and ring formation. VI. Adipic anhydride. *J. Am. Chem. Soc.* **52**, 4110–4114.
Hill, J. W., Carothers, H. W. (1932) Studies on polymerization and ring formation. XIV. A linear superpolyanhydride and a cyclic dimeric anhydride from sebacic acid, *J. Am. Chem. Soc.* **54**, 5169.
Jiang, H. L., Zhu, K. J. (2000) Pulsatile protein release from a laminated device comprising of polyanhydrides and pH-sensitive complexes, *Int. J. Pharm.* **194**, 51–60.
Judy, K. D., Olivi, A., Buahin, K. G., Domb, A. J., Epstein, J. I., Colvin, O. M., Brem, H. (1995) Effectiveness of controlled release of a cyclophosphamide derivative with polymers against rat gliomas, *J. Neurosurg.* **82** 481–486.
Kim, B. S., Hrkach, J. S., Langer, R. (2000) Synthesis and characterization of novel degradable photocrosslinked poly(ester-anhydride) networks, *J. Polym. Sci. A: Polym. Chem.* **38**, 1277–1282.
Knobloch, J. O., Ramirez, F. (1975) *J. Org. Chem.* **40**, 1101–1106.
Kubek, M. J., Liang, D., Byrd, K. E., Domb, A. J. (1998) Prolonged seizure suppression by a single implantable polymeric TRH microdisk preparation, *Brain. Res.* **809**, 189–197.
Laurencin, C. T., Domb, A. J., Morris, C., Brown, V., Chasin, M., McConnell, R., Lange, N., Langer, R. (1990) Poly(anhydrides) administration in high doses in vivo: studies of biocompatibility and toxicology, *J. Biomed. Mater. Res.* **24**, 1463–1481.
Laurencin, C., Gerhart, T., Witschger, P., Satcher, R., Domb, A. J., Hanff, P., Edsberg, L., Hayes, W., Langer, R. (1993) Biodegradable polyanhydrides for antibiotic drug delivery, *J. Orthop. Res.* **11**, 256–262.
Laurencin, C. T., Ibim, S. E. M., Langer, R. (1995) Poly(anhydrides) in: *Biomedical Applications of Synthetic Biodegradable Polymers* (Hollinger, J. O., Ed.), Boca Raton, FL: CRC Press, 59–102.
Leong, K. W., D'Amore, P., Langer, R. (1986a) Bioerodible poly(anhydrides) as drug carrier matrices. II. Biocompatibility and chemical reactivity, *J. Biomed. Mater. Res.* **20**, 51–64.
Leong, K.W., Kost, J., Mathiowitz, E., Langer, R. (1986b) Poly(anhydrides) for controlled release of bioactive agents, *Biomaterials* **7**, 364.
Leong, K. W., Simonte, V., Langer, R. (1987) Synthesis of poly(anhydrides): melt-polycondensation, dehydrochlorination, and dehydrative coupling, *Macromolecules* **20**, 705–712.
Leong, K. W., Domb, A., Langer, R. (1989) Poly(-anhydrides), in: *Encyclopedia of Polymer Science and Engineering*, 2nd Edition, New York: John Wiley & Sons.
Lundmark, S., Sjoling, M., Albertsson, A.-C. (1991) Polymerization of oxipane-2,7-dione in solution and synthesis of block copolymer of oxipane-2,7-dione and 2-oxipanone, *J. Macromol. Sci.- Chem. A* **28**, 15–29.
Mader, K., Domb, A. J., Swartz, H. M. (1996) Gamma sterilization induced radicals in biodegradable drug delivery systems, *J. Appl. Rad. Isotop.* **47**, 1669–1674.
Mader, K., Bacic, G., Domb, A. J., Elmalak, O., Langer, R., Swartz, H. M. (1997a) Noninvasive in vivo monitoring of drug release and polymer erosion from biodegradable polymers by EPR spectroscopy and NMR imaging, *J. Pharm. Sci.* *86*, 126–134.
Mader, K., Cremmilleleux, Y., Domb, A. J., Dunn, J. F., Swartz, H. M. (1997b) *In vitro/in vivo* comparison of drug release and polymer erosion from biodegradable P(FAD-SA) polyanhydrides – a noninvasive approach by the combined use of Electron Paramagnetic Resonance Spectroscopy and Nuclear Magnetic Resonance Imaging, *Pharm. Res.* **14**, 820–826.
Maniar, M., Xie, X., Domb, A. J. (1990) Poly(anhydrides). V. Branched poly(anhydrides), *Biomaterials* **11**, 690–694.
Maniar, M., Domb, A., Haffer, A., Shah, J. (1994) Controlled release of local anesthetics from fatty acid dimer based polyanhydrides, *J. Control. Rel.* **30**, 233–239.
Mark, H.E., et al. (Eds.) (1969) Poly(anhydrides), in: *Encyclopedia of Polymer Science and Technology* New York: John Wiley & Sons, 630, Vol. 10.
Masters, D. B., Berde, C. B., Dutta, S. K., Griggs, C. T., Hu, D., Kupsky, W., Langer, R. (1993a)

Prolonged regional nerve blockade by controlled release of local anesthetic from a biodegradable polymer matrix, *Anesthesiology* **79**, 340–346.

Masters, D. B., Berde, C. B., Dutta, S., Turek, T., Langer, R. (1993b) Sustained local anesthetic release from bioerodible polymer matrices: a potential method for prolonged regional anesthesia, *Pharm. Res.* **10**, 1527–1532.

Mathiowitz, E., Langer, R. (1987) Poly(anhydride) microspheres as drug carriers. I. Hot-melt microencapsulation, *J. Control. Rel.* **5**, 13–22.

Mathiowitz, E., Kline, D., Langer, R. (1990) Morphology of poly(anhydride) microspheres delivery systems, *J. Scan. Microsc.* **4**, 329.

Mathiowitz, E., Jacob, J. S., Jong, Y. S., Carino, G. P., Chickering, D. E., Chaturvedi, P., Santos, C. A., Vijyaraghavan, K., Montgomery, S., Basset, M., Morrell, C. (1997) Biologically erodible microspheres as potential oral drug delivery systems, *Nature* **386**, 410–414.

McIntyre, J. E. (1994) British Patent 978,669.

Muggli, D. S., Burkoth, A. K., Anseth, K. S. (1999) Crosslinked polyanhydrides for use in orthopedic applications: degradation behavior and mechanics, *J. Biomed. Mater. Res.* **46**, 271–278.

Olivi, A., Awend, M.G., Utsuki, T., Tyler, B., Domb, A. J., Brat, D. J., Brem, H. (1996) Interstitial delivery of carboplatin via biodegradable polymers is effective against experimental glioma in the rat. *Cancer, Chemother. Pharmacol.* **39**, 90–96.

Pekarek, K. J., Jacob, J. S., Mathiowitz, E. (1994) One-step preparation of double-walled microspheres *Nature* **357**, 258–260.

Rock, M., Green, M., Fait, C., Gell, R., Myer, J., Maniar, M., Domb, A. (1991) Evaluation and comparison of biocompatibility of various classes of polyanhydrides, *Polym. Preprints* **32**, 221.

Ron, E., Mathiowitz, E., Mathiowitz, G., Domb, A., Langer, R. (1991) NMR characterization of erodible copolymers, *Macromolecules* **24**, 2278–2282.

Rosen, H. B., Chang, J., Wnek, G. E., Lindhardt, R. J., Langer, R. (1983) Biodegradable poly(anhydrides) for controlled drug delivery, *Biomaterials* **4**, 131–133.

Santos, S. A., Freedman, B. D., Leach, K. J., Press, D. L., Scarpulla, M., Mathiowitz, E. (1999) Poly(fumaric-co-sebasis anhydride) – A degradation study as evaluated by FTIR, DSC, GPC and X-ray diffraction, *J. Control. Rel.* **60**, 11–22.

Shakesheff, K. M., Davies, M. C., Roberts, C. J., Tendler, S. J. B., Shard, K. M., Domb, A. J. (1994) In situ AFM imaging of polymer degradation in an aqueous environment, *Langmuir* **10**, 4417–4419.

Shakesheff, K. M., Chen, X., Davies, M. C., Domb, A. J., Roberts, M. C., Tendler, S. J. B., Williams, P. M. (1995) Relating the phase morphology of a biodegradable polymer blend to erosion kinetics using simultaneous *in situ* atomic force microscopy and surface plasmon resonance analysis, *Langmuir* **11**, 3921–3927.

Staubli, A., Ron, E., Langer, R. (1990) Hydrolytically degradable amino acid containing polymers, *J. Am. Chem. Soc.* **112**, 4419–4424.

Staubli, A., Mathiowitz, E., Lucarelli, M., Langer, R. (1991a) Characterization of hydrolytically degradable amino acid containing poly(anhydride-co-imides), *Macromolecules* **24**, 2283–2290.

Staubli, A., Mathiowitz, E., Langer, R. (1991b) Sequence distribution and its effects on glass transition temperatures of poly(anhydrides-co-imides) containing asymmetric monomers, *Macromolecules* **24**, 2291–2298.

Stephens, D., Li, L., Robinson, D., Chen, S., Chang, H. C., Liu, R. M., Tian, Y. Q., Ginsberg, E. J., Gao, X. Y., Stultz, T. (2000) Investigation of the *in vitro* release of gentamicin from a polyanhydride matrix, *J. Control. Rel.* **63**, 305–317.

Subramanyam, R., Pinkus, A. G. (1985) Synthesis of poly(terephthalic anhydride) by hydrolysis of terephthaloyl chloride triethylamine intermediate adduct: characterization of intermediate adduct, *J. Macromol. Sci. Chem.* **A22**, 23.

Tabata, Y., Domb, A. J., Langer, R. (1994) Polyanhydride granules provide controlled release of water-soluble drugs with reduced initial burst, *J. Pharm. Sci.* **83**, 5–11.

Tamargo, R. J., Epstein, J. I., Reinhard, C. S., Chasin, M., Brem, H. (1989) Brain biocompatibility of a biodegradable controlled-release polymer in rats, *J. Biomed. Mater. Res.* **23**, 253–266.

Tamargo, R. J., Myseros, J. S., Epstein, J. I., Yang, M. B., Chasin, M., Brem, H. (1993) Interstitial chemotherapy of the 9L-glyosarcoma – Controlled release polymers for drug delivery in the brain, *Cancer Res.* **53**, 329–333.

Teomim, D., Domb, A. J. (1999) Fatty acid terminated polyanhydrides, *J. Polym. Sci. A: Polym. Chem.* **37**, 3337–3344.

Teomim, D., Fishbien, I., Golomb, G., Orloff, L., Mayberg, M., Domb, A. J. (1999a) Perivascular delivery of heparin for the reduction of smooth muscle cell proliferation after endothelial injury, *J. Control. Rel.* **60**, 129–142.

Teomim, D., Nyska, A., Domb, A. J. (1999b) Ricinoleic acid based biopolymers, *J. Biomed. Mater. Res.* **45**, 258–267.

Teomim, D., Mader, K., Bentolila, A., Magora, A., Domb, A. J. (2001) *Macromolecules* (in press).

Tudor, A. M., Melia, C. D., Davies, M. C., Hendra, P. J., Church, S., Domb, A. J., Langer, R. (1991) The application of the Fourier-Transform Raman spectroscopy to the analysis of poly(anhydride) home and copolymers, *Spectrochim. Acta* **47A**, 1335–1343.

Uhrich, K. E., Gupta, A., Thomas, T. T., Laurencin, C. T., Langer, R. (1995) Synthesis and characterization of degradable poly(anhydride-co-imides), *Macromolecules* **28**, 2184–2193.

Uhrich, K. E., Thomas, T. T., Laurencin, C. T., Langer, R. (1997) *In vitro* degradation characteristics of poly(anhydride-co-imides) containing trimellityimidoglycine, *J. Appl. Polym. Sci.* **63**, 1401–1411.

Wu, M. P., Tamada, J. A., Brem, H., Langer, R. (1994) *In vivo* versus *in vitro* degradation of controlled release polymers for intercranial surgery therapy, *J. Biomed. Mater. Res.* **28**, 387–395.

Yoda, N. (1962) Synthesis of poly(anhydrides). Poly(anhydrides) of five-membered heterocyclic dibasic acids, *Makromol. Chem.* **55**, 174–190.

Yoda, N. (1963) Synthesis of poly(anhydrides). Crystalline and high melting poly(amide) poly(anhydrides) of methylene bis(*p*-carboxyphenyl) amide, *J. Polym. Sci., Part A* **1**, 1323–1338.

Young, J. S., Gonzales, K. D., Anseth, K. S. (2000) Photopolymers in orthopedics: Characterization of novel crosslinked polyanhydrides, *Biomaterials* **110**, 1181–1188.

6
Polyglycolide and Copolyesters with Lactide

Prof. Dr. Michel Vert
CRBA-UMR CNRS 5473, Faculty of Pharmacy, University Montpellier 1, 15 Avenue Charles Flahault, 34060 Montpellier Cedex 2, France; Tel: +33-467-418260; Fax: +33-467-520998; E-mail: vertm@pharma.univ.montpl.fr

Biopolymers for Medical and Pharmaceutical Applications. Edited by A. Steinbüchel and R. H. Marchessault

ISBN: 3-527-31154-8

NMR nuclear magnetic resonance
PGA poly(glycolic acid) (= polyglycolide)
PLGA or PLAGA or PLA_xGA_y poly(glycolic acid-*co*-lactic acid) copolymers (= poly(glycolide-*co*-lactide) copolymers)

1 Introduction

Polyglycolide, also named poly(glycolic acid), and poly(glycolide-*co*-lactide) copolymers are polymers of the poly(α-hydroxy acid) type that are composed of building block or repeating units derived from glycolic and lactic acids. These polymers and copolymers that are sensitive to moisture became of interest as polymeric materials rather late as compared with other polymers like polyethylene or poly(vinyl chloride). When people realized that advantage could be taken of the degrading activity of water, devices were fabricated that can be degraded in an animal's body. Since then lactic acid and glycolic acid-based copolymers have attracted the attention of both scientists and industrialists in the biomedical and the pharmaceutical fields. In contrast to their homologs based only on lactic acid enantiomers that are being commercialized as therapeutic materials and environment-friendly materials as well, glycolic acid-containing copolymers have not found other fields of applications other surgery and pharmacology. In polymer science, trying to enlarge the range of properties of homopolymers by copolymerization is a basic strategy. In the case of glycolide, a number of copolymers with lactide enantiomers have been investigated.

2 Historical Outline

Glycolic acid-containing polymer chains were first synthesized by step-growth polymerization (polycondensation) of glycolic and lactic acid at the time other polycondensates like Nylon®-type polyamide were investigated (Higgins, 1950). The resulting poor polymers were rather sensitive to humidity and heat, and they were considered as useless. Fortunately, advances in polymer chemistry led chemists to discover another synthetic route, i.e. the ring-opening polymerization of the cyclic dimers of glycolic and lactic acids (Lowe, 1954). Soon after, poly(glycolic acid) was proposed by American Cyanamid as suitable polymers to make sutures (Schmitt and Polistina, 1967). Commercial sutures were developed by Davis and Geek under the registered name Dexon® in the USA and Ercedex® in Europe (Frazza and Schmitt, 1971). Soon after, Ethicon came up with a concurrent product that was a copolymer of glycolide with 8% lactide known as Glactine910® or Vicryl®, and processed to sutures and other fiber-based devices. In parallel, lactic acid-based polymers were studied from a scientific viewpoint (Kulkarni et al., 1966) until they became of industrial interest for bone surgery in the 1990s (Barber, 1998) after their known properties had been considerably improved (Kulkarni et al., 1971; Vert et al., 1984). In the meantime pharmacists found poly(glycolic acid-*co*-lactic acid) copolymers of interest to deliver drugs in a more or less controlled manner (Holland et al., 1986; Brannon-Peppas and Vert, 2000). Among the poly(glycolic acid-*co*-lactic acid) copolymer family, lactoyl-rich semi-crystalline polymers were preferred for the temporary internal fixation of bone fractures, whereas amorphous polymers appeared more convenient for tissue reconstruction and drug delivery.

3 Chemical Structures

Poly(glycolic acid-*co*-lactic acid) copolymers belong to the poly(α-hydroxy acid) family that is characterized by the following general formula:

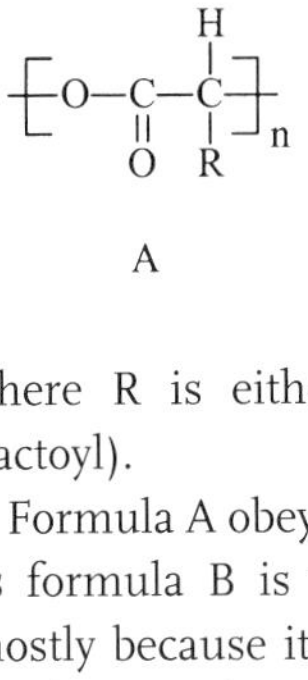

or

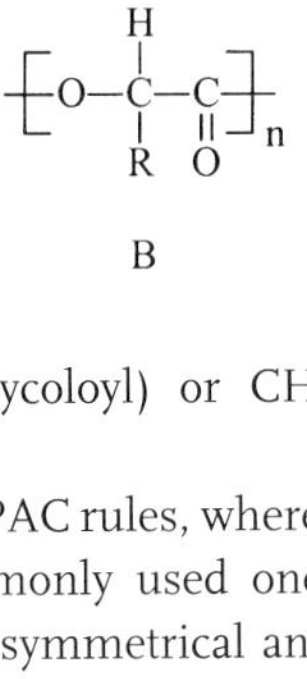

A B

where R is either H (glycoloyl) or CH_3 (lactoyl).

Formula A obeys the IUPAC rules, whereas formula B is the commonly used one, mostly because it is more symmetrical and emphasizes the ester links.

In literature, different acronyms are currently used to name these different basic chains, the most common being PLGA and PLAGA. Our preference for PLA_xGA_y was mentioned as early as 1981 (Vert et al., 1981) when we realized that the properties of a given member of the poly(α-hydroxy acid) family depend very much on many factors and that a convenient acronym has to reflect at least the gross composition in repeating units. Since then, we have identified more than 20 factors that can affect the behavior of a PLA_xGA_y polymer, i.e. much more than in the case of a regular polymer primarily because of the sensitivity to water and heat, the particular pair-addition mechanism of the ring-opening polymerization due to the dimeric structure of the cyclic monomers, and the presence of chiral centers in lactic acid-based polymers and stereocopolymers and glycolic acid-copolymers (Li and Vert, 1995; Vert et al., 1995). Of course, it is not possible to reflect the contribution of all these factors within a simple acronym, but we strongly believe that everyone ought to use the PLA_xGA_y acronyms because they reflect both the chemical composition and the enantiomeric composition.

The basic chain structures of PLA_xGA_y polymers are shown in Table 1 (Vert et al., 1984).

Lactic acid-containing stereocopolymers and copolymers are outstanding compounds because of the presence of chiral centers along the chain that generate another source of differentiation between PLA_xGA_y polymers of similar gross compositions, i.e. the distribution of chiral centers. This distribution depends on many more or less interrelated factors, including the gross composition, the nature of the monomer, the mechanism of polymerization, the differences between the reactivities of glycolic acid

Tab. 1 The PLA_xGA_y family composed of homopolymers, stereocopolymers and copolymers

Polyglycolic acid (PGA)	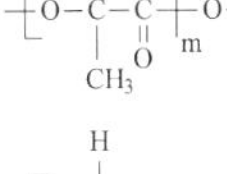
Poly(L-lactic acid) (PLA100)	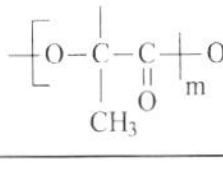
Lactic acid stereocopolymers (PLAX)	
Lactic-glycolic acid terpolymers (PLAXGAY)	

and lactic acid-type monomers, and the occurrence of configurational rearrangements caused by intrachain and interchain transesterification reactions (Lillie and Schulz, 1975; Chabot et al., 1983).

There are two main routes to make PLA_xGA_y polymer chains, i.e. step-growth polymerization or copolycondensation of glycolic and lactic acids, and the chain-growth polymerization or ring-opening polymerization of the glycolide and lactide cyclic dimers.

3.1
Step-growth Polymerization of Glycolic Acid

The step-growth polymerization of glycolic acid (GlcAc) corresponds to the following equilibrium reaction (see below)

The resulting PGA homopolymers have low molecular weight although they already present the characteristics of high molecular weight PGA because of their high crystallinity and insolubility in common solvents. Although nobody seems to have investigated the relative reactivities of glycolic acid and of lactic acid enantiomers, it is likely that the distribution of the glycoloyl and lactoyl repeating units is not random. Anyhow, the distribution resulting from the polycondensation of the hydroxy acids will be always different from that resulting from the ring-opening polymerization of glycolide with L- and/or D-lactide for reason underlined below. Under normal conditions PLA_xGA_y polymer chains obtained by polycondensation of the hydroxy acids are terminated by OH and COOH end groups in equivalent amounts.

3.2
Chain-growth Polymerization of Glycolide, the Cyclic Dimer of Glycolic Acid

The structure of the cyclic dimer of glycolic acid is symmetrical and includes two ester bonds that are equivalent provided the ring is closed. As soon as one of the ester bond is broken due to the action of a reagent, the second ester bond is stabilized and can no longer be involved in the polymerization process.

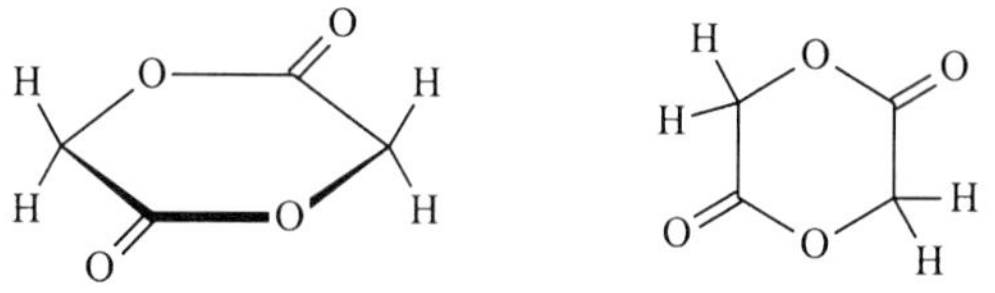

Spatial   Planar

diglycolide (glycolide or GA)

The official IUPAC nomenclature is 1,4-dioxane-2,5-dione or diglycolide. In literature people generally use glycolide for the sake of simplicity.

The polymer resulting from ring-opening polymerization is actually a polydimer according to chain growth by the addition of pairs of repeating units (Lillie and Schulz, 1975; Vert et al., 1984).

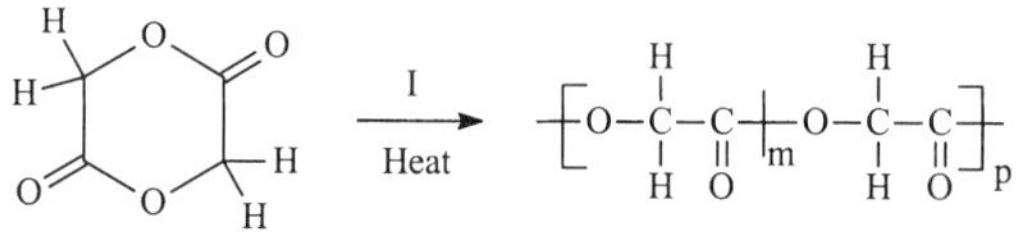

GA $\xrightarrow{I, Heat}$ HMW polydimeric PGA

The ring-opening polymerization of glycolide can be initiated by many compounds and leads to rather high molecular weight

$$n\ HO-\underset{H}{\overset{H}{C}}-COOH \quad \underset{Heat + vacuum}{\overset{Catalyst}{\rightleftharpoons}} \quad H\text{–}[O\text{–}\underset{H}{\overset{H}{C}}\text{–}\underset{O}{\overset{\|}{C}}]_n\text{–}OH\ +\ (n-1)\,H_2O$$

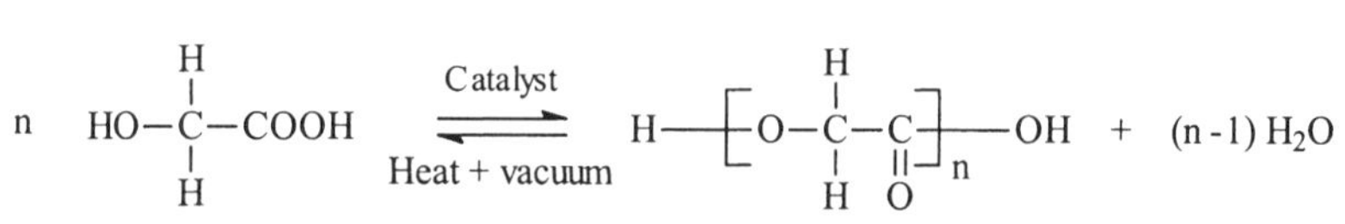

GlcAc   LMW PGA

PGA polymers whose molecular weights are difficult to measure because of the lack of practical solvents. In the case of PGA, there is no difference in the distribution of repeating units resulting from step-growth and chain-growth polymerizations. Only differences at chain ends can occur, the polycondensation leaving OH and COOH chain ends, whereas ring-opening polymerization leads to substitution by an initiator or a catalyst residue at one terminal unit, a behavior usually deduced from the features collected from lactide polymerizations (Dubois et al., 1991; Kowalski et al., 2000; Kricheldorf et al., 2000). Chain end differences, that are generally neglected in common polymers, turn out to be critical in the case of PLA_xGA_y polymers, as it will be discussed in the section on degradation. Insofar as the chain structure is concerned, the situation is much more complex when one deals with lactic acid-containing glycolic acid copolymers.

3.3 Step-growth Copolymerization of Glycolic Acid with L- and D-Lactic Acid

Low molecular weight PLA_xGA_y copolymers can be synthesized by polycondensation starting from mixtures of glycolic acid and D- and L-lactic acids. Except for the glycolic acid-rich PLA_xGA_y copolymers that are semicrystalline, most of the PLA_xGA_y copolymers are amorphous (Miller 1977). The reaction proceeds as in the case of glycolic acid alone. However, whereas L- and D-lactic acids react at similar rates basically, glycolic acid is likely to react differently thus leading to non random unit distributions. The unit distributions in low molecular weight PLA_xGA_y chains obtained by polycondensation has not been investigated so far. The chain structures of PLA_xGA_y copolymers obtained through chain-growth ring-opening polymerization are dramatically more complex than those of the chains obtained by polycondensation.

3.4 Chain-growth Copolymerization of Glycolide with L- and D-Lactides

As the lactide molecule bears two asymmetric carbon atoms, lactide exists under the form of three diastereoisomers and a racemate whose structures are shown below, see p. 184 (Holten, 1971).

As in the case of the cyclic dimer of glycolic acid, the lactic acid cyclic dimers should be named after 2,6-dimethyl-1,4-dioxane-2,5-dione or referred to as dilactides. Instead people use lactides as indicated in the formula above. Accordingly, D-lactide, L-lactide and *meso*-lactide are composed of two D-lactoyl, two L-lactoyl, and one D- and one L-lactoyl units, respectively, whereas *rac*-lactide is a 50/50 mixture of L- and D-lactides.

The diasastereosisomeric structure of dilactides and the pair addition mechanism, and the difference of reactivity between glycolic acid and lactic acid cyclic monomers lead to irregular unit distributions in a given chain. The use of high-field nuclear magnetic resonance (NMR) is a good means to distinguish between PLA_xGA_y copolymers having different chain compositions and chiral and achiral distributions, although no one has tried to fully assign the fine structures so far. Nevertheless, Figure 1 shows that NMR is very sensitive to the gross composition and to the chirality, as shown by the spectra of various PLA_xGA_y copolymers (left hand side), the comparison of copolymers of glycolide with D- and L-lactide (first top line), and the spectrum of a $PLA_{50}GA_{50}$ obtained by ring-opening polymerization of 3-methyl-1,4-dioxane-2,5-dione, a cyclic monomer composed of one lactic acid and one glycolic acid residues. The

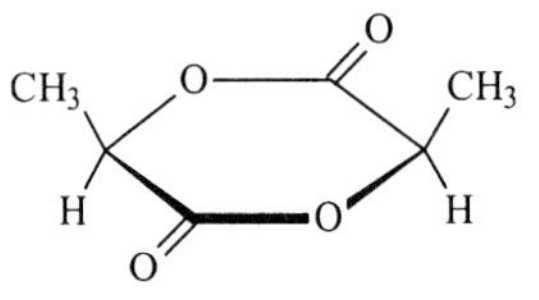

D-dilactide (D-lactide)

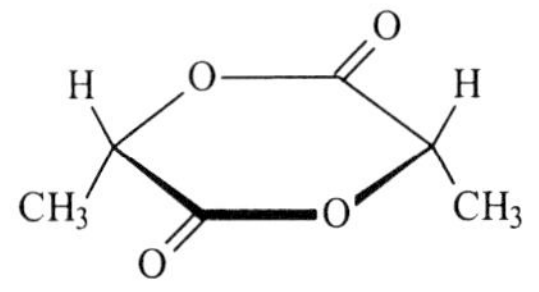

L-dilactide (L-lactide)

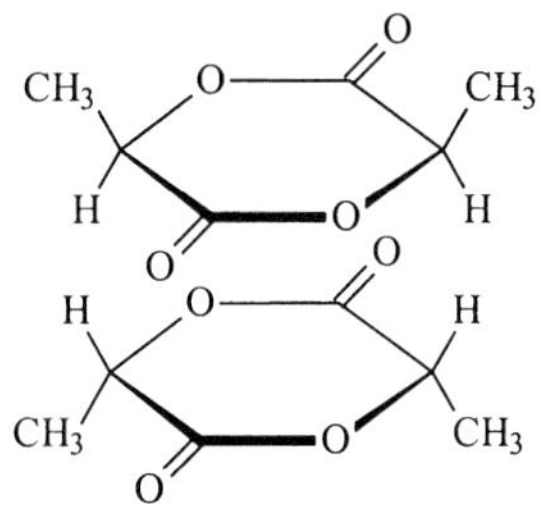

rac-dilactide (*rac*-lactide)

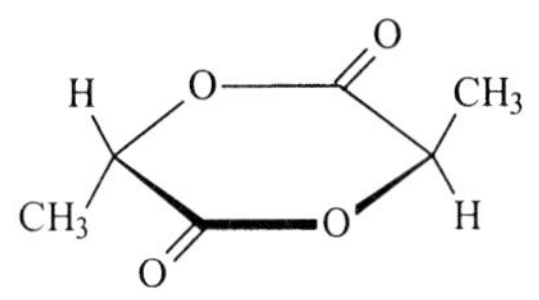

meso-dilactide (*meso*-lactide)

combination of mass spectrometry and molecule fragmentation might be good complementary tools to determine the chain fine structures (Chen et al., 2000; Adamus et al., 2000).

The characterization of PLA_xGA_y copolymer chain structures in relation to the gross composition and repeating unit distribution is very important because these factors can have dramatic consequences on the hydrolytic degradation of corresponding polyester chains as we will see later on. The ring-opening polymerization can also affect the hydrolytic degradation because of the chain end modification by initiator or catalyst residues. This was shown during the past years in the case of lactides initiated by zinc metal, zinc lactate, and stannous octoate comparatively under conditions allowing chain end detection by NMR (Schwach et al., 1997, 1998a). Last, but not least, the characterization of PLA_xGA_y copolymer fine structure prior to any other characterization is also important because of the possible occurrence of transesterification affecting repeating unit distributions or racemization at lactoyl sites affecting the chiral unit distribution, two factors of dramatic significance with respect to hydrolytic degradation.

3.5 Transesterification Rearrangements

As we have seen before, the fine chain structure of PLA_xGA_y copolymers is rather complex. In contrast to the cases of vinylics or acrylics copolymerization, where one can basically calculate unit distributions using simple kinetics equations, the copolymerization of glycolic acid with L- and D-lactic acid does not allow simple calculation, partly because the achiral and chiral unit distributions can be rearranged during the polymerization reaction, or even later on during any processing, especially in the presence of compounds that can catalyze transesterification rearrangements. The risk of transesterification reactions is very high when polymerization is conducted at high temperature and in the bulk or in solution in the

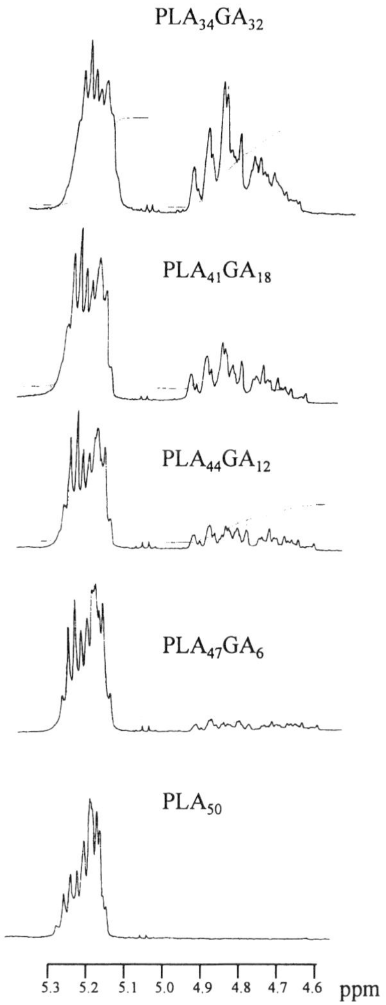

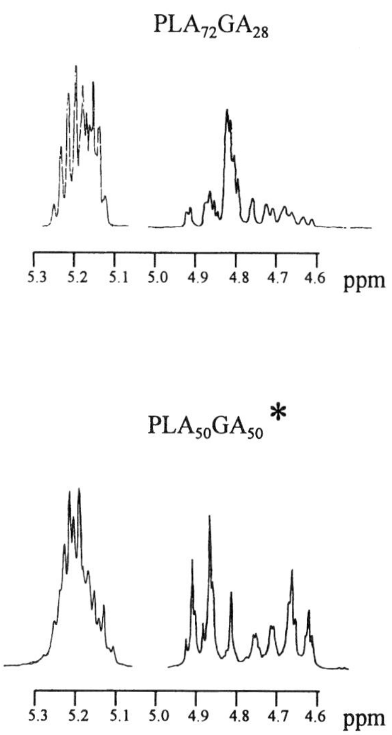

Fig. 1 The 400 MHz ^{1}H-spectra of different PLA_x-GA_y copolymers in $CDCl_3$ (Granger et al., unpublished data). The $PLA_{50}GA_{50}$* copolymer was kindly supplied by Y. Kimura from the Tokyo Institute of Technology. (Lactic acid fine structures between 5.1 and 5.3 ppm; glycolic acid fine structures between 4.6 and 4.9 ppm, resonances of the residual lactide serving as internal standard.)

presence of an initiator that can catalyze transesterification. Such a situation is well identified in the lactic acid stereocopolymer literature (Lillie and Schultz, 1975; Chabot et al., 1983; Kricheldorf and Kreiser, 1987a). Transesterification rearrangements tend to randomize the achiral and chiral repeating units. However, randomization cannot be total because of the closest neighbor dependence of the reactivity of a given ester bond taken at random within a aliphatic copolyester chain (Chabot et al., 1983). In the case of PLA_x stereocopolymers polymerized in the bulk in the presence of zinc metal, transesterification always leads to the same NMR spectrum resulting from the particularities of both polymerization and transesterification processes occurring almost simultaneously (Chabot et al., 1983). Transesterification rearrangements appeared different in the cases of stannous octoate and zinc metal initiated poly(DL-lactides) (Schwach et al., 1984). Polymer chains are also sensitive to the radiation generally used for sterilization (Chu and Williams, 1983; Vert et al., 1992)

4 Occurrence

PGA and PLA_xGA_y copolymers are synthetic polymers that do not have any equivalent in nature. As previously shown they can be produced from glycolic acid, and from mixtures of glycolic acid and D- and or L-lactic acids by step-growth polymerization or from mixtures of glycolide and lactides by chain-growth polymerization. The literature is sometimes misleading because authors do not say whether the composition of the monomer feed is defined in mol/mol or in w/w. Another source of confusion exists at the level of the gross chain composition that is defined either from the composition of the monomer feed or from the analysis of the chain composition by NMR. Any scientific contribution dealing with PLA_xGA_y copolymers should mentioned as accurately as possible the identity of the investigated compounds – a requirement underlined 20 years ago and still not respected (Vert et al., 1981)

4.1 Monomers

Glycolic, and L-, D- and *rac*-lactic acids are commercially available either as solids or as concentrated aqueous solutions. The acid aqueous solutions usually contain the dimer and trimer oligomers at equilibrium with the hydroxy acid itself (Braud et al., 1996).

Glycolide (melting temperature M_p = 86 °C) is synthesized from synthetic glycolic acid. There are different routes but the most common consists in thermal degradation of low molecular weight PGA or PLA_xGA_y copolymers obtained by polycondensation. These oligomers are heated under vacuum in the presence of a catalyst to yield a crude monomer. This process is usually referred to as ring closure. Several purification steps are necessary to obtain the highly pure monomers required to polymerize glycolide up to high molecular weight PGA (Neuwenjhuis, 1992).

Similarly, L- and D-lactide are generated during the thermal degradation of lactic acid oligomers obtained by polycondensation of L- and D-lactic acid, respectively. They melt at 97–98 °C and are soluble in many organic solvents (e.g. acetone, ethyl acetate, and benzene) that can be used for purification by crystallization from hot concentrated solutions.

rac-lactide (M_p = 126 °C) is obtained by the same method starting from *rac*-lactic acid. This lactide can be readily purified by recrystallization from hot saturated solutions in organic solvents.

meso-lactide (M_p = 46 °C) is usually extracted from the cold filtrate solution remaining after recrystallization of *rac*-lactide after evaporation of most of the solvent to end up with a hot concentrated solution. It is rather difficult to obtain pure *meso*-lactide (Chabot et al., 1983).

4.2 Polymers

PGA and PLA_xGA_y copolymers can be obtained by atmospheric or vacuum distillation of water starting from a defined monomer feed composed of glycolic and lactic acids. This route leads to compounds having very low molecular weights that are brittle and glassy or waxy and sticky depending on the composition and on the molecular weight of the copolymers.

High molecular weight PGA and PLA_xGA_y copolymers are synthesized by ring-opening polymerization of glycolide and lactides in convenient proportions either under an inert atmosphere or under vacuum. Under vacuum people ought to be careful not to distillate glycolide through the vacuum line otherwise the composition of the resulting polymer can be largely different from that of the feed. The polymerization reaction can be conducted either in the melt, i.e. at temperatures higher than the melting temperature of the resulting polymer (melt polymerization), below the polymer melting temperature (bulk polymerization), in suspension or in solution (Nieuwenhuis, 1992).

Only low molecular weight poly(α-hydroxy acid)s had been synthesized by melt polymerization at 250–270 °C when, in 1954, a route to high molecular weight PGA using $ZnCl_2$ as initiator was patented. The procedure included purification of the monomer by recrystallization and degassing of the reaction mixture prior to bulk polymerization under vacuum (Lowe, 1954).

Since then many initiator systems have been reported in literature, primarily for lactide polymerization (Kleine and Kleine, 1959). In comparison, little attention has been paid to glycolide polymerization (Kohn et al., 1983). Except for our group which used zinc metal and more recently zinc lactate to copolymerize glycolic acid and lactic acid monomers, most of the polymers reported in the literature were made using stannous octoate ($Sn(Oct)_2$) either alone or with a co-initiator of the alcohol type. Both zinc lactate and stannous octoate are approved by the American Food and Drug Administration for surgical and pharmacological applications. We selected zinc many years ago because stannous octoate is unstable and usually contains impurities, especially when the industrial product is used as received (Schwach et al., 1997). Zinc metal was also retained because it is an oligoelement with daily allowance for the metabolism of mammalian bodies (Leray et al., 1976). Stannous octoate is generally preferred because it gives faster polymerization, higher conversion ratios and higher molecular weights than zinc metal or zinc lactate (Schwach et al., 1994, 1996a). It has been also shown that stannous octoate is slightly cytotoxic (Tanzi et al., 1994). In solution polymerization and in melt and bulk polymerizations, analyses of the reaction media showed that polymers are at equilibrium with the monomers because of a ceiling temperature evaluated at 270 °C (Nieuwenhuis, 1992; Leenslag and Pennings, 1987). Thermal degradation showed that PGA- and PLA-type polymers degrade to several degradation products including carbon dioxide, monomers, and aldehydes, i.e. acetaldehyde in the case of PLA and formaldehyde in the case of PGA (McNeill and Leiper, 1985). Last, but not least, the presence of residues issued from stannous octoate was recently shown for the bulk polymerization of *rac*-lactide (Schwach et al., 1998b) and L-lactic acid-rich PLA stereocopolymer. Among these residues hydroxy tin octanoate that can initiate lactic acid polymerization and ethyl-2 hexanoic acid that hydrophobizes the polymer mass have been identified in the case of polymerizations conducted at high initiator/monomer ratios. It seems that hydrophobic residues issued from stannous octoate are difficult to remove by precipitation from organic solution either by methanol, ethanol, or water (Schwach et al., 1998b).

According to literature, the polymerization of glycolide and lactides can proceed through anionic, cationic, or coordination mechanisms. A great deal of work has been devoted to the understanding of polymerization mechanisms because they determine the chain structures and to some extent, the degradability. However, many unknowns still remain in the literature due to the complexity of lactone polymerization chemistry (Li and Vert, 1995; Li, 1999). Recently, the polymerization of lactides in the presence of stannous octoate was revisited (Kowalski et al., 2000; Kricheldorf et al.,

2000). However, the authors did not pay attention to the difference between polymerizations carried out in solution and in the bulk, a point that would deserve deeper investigation in the future because industrial polymers are synthesized through bulk polymerization. It is of interest to note that most of the kinetics investigations reported in literature dealt with lactide polymerization.

5 Biochemistry

PLA_xGA_y polymers are not biopolymers. So far no one has found any poly(α-hydroxy acid)-type polymers in living systems. Therefore, these compounds have no recognized biological or physiological function. However, because they degrade in aqueous media and because living systems function in aqueous environments, PLA_xGA_y polymers undergo hydrolytic degradation in the presence of living organisms or microorganisms. At the minimum they then generate glycolic and lactic acids as degradation by-products. These metabolites are finally bioassimilated. Whether enzymes are involved in the degradation of PLA_xGA_y polymers is still a matter of debates in literature.

6 Degradation in Abiotic and Biotic Aqueous Media

Basically the degradation of polymeric materials in a living environment can result from enzymatically mediated and/or chemically mediated cleavage. Although animal or human *in vivo* environments are different from environmental outdoor ones, there is no fundamental difference between the biodegradation of a polymer by animal cells or by microorganisms. Both involve water, enzymes, metabolites, ions, etc., which interact with the material (Vert et al., 1994a). From a general viewpoint, degradation in the presence of enzymes with or without the cells that produce these enzymes is regarded as biodegradation and can proceed through unzipping from chain ends (exoenzymes) or cleavages within main chains (endoenzymes). Anyhow, the term "degradation" should be reserved to degradations in the absence of any living cells or isolated enzymes or when the mechanisms of degradation is unknown (Li and Vert, 1995).

6.1 Degradation in Abiotic Media

For almost 20 years after the early work on high molecular weight PLA_xGA_y aliphatic polyesters, hydrolytic degradation in the absence of enzyme was regarded as homogeneous (bulk erosion) (Pitt et al., 1981), although surface erosion was claimed in a few cases (Kronenthal, 1974). This was primarily due to the fact that molecular weight changes were monitored by viscometry. During the last 12 years, the understanding of the hydrolytic degradation of PLA_xGA_y polymers advanced significantly due to the use of size exclusion chromatography that revealed the presence of two populations of partially degraded PLA_xGA_y devices of rather large sizes. This discovery led us to introducing the concept of heterogeneous degradation related to diffusion–reaction–dissolution phenomena (Li et al., 1990a; Vert, 1990)

To summarize, once a device made of a PLA_xGA_y polymer is placed in contact with an aqueous medium, water penetrates into the specimen and the hydrolytic cleavage of ester bonds starts (Figure 2).

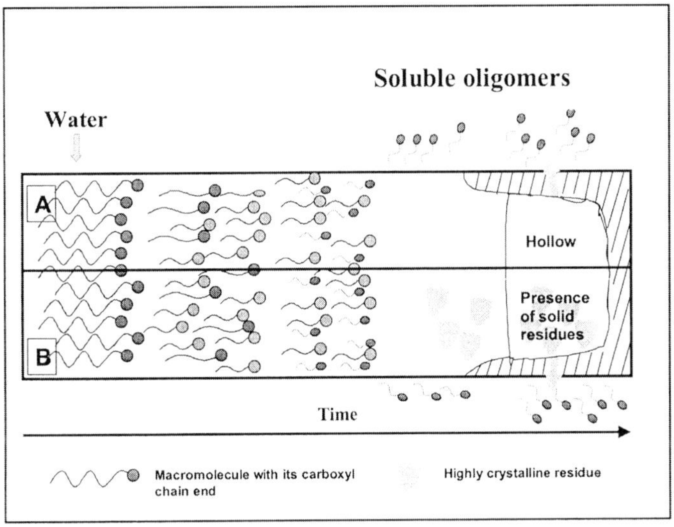

Fig. 2 Schematic representation of the fate of macromolecules forming a parallel-sided device made of a PLA_xGA_y polymer as a function of time (case A: initially amorphous device composed of a stereo-irregular PLA_xGA_y polymer that does not lead to crystalline residues; case B: either an initially semi-crystalline device or a device that is initially amorphous but composed of PLA_xGA_y polymer stereoregular enough to generate semi-crystalline low molecular weight residues because of faster degradation in the most disordered or glycolic acid-rich part of copolymer chains

Water absorption is thus a critical factor (Schmitt et al., 1994). Each cleaved ester bond generates a new carboxyl end group that, in principle, can catalyze the hydrolytic reaction of other ester bonds as proposed in the case of the homogeneous degradation mechanism (Pitt et al., 1981). For a time the partially degraded macromolecules remain insoluble in the surrounding aqueous medium, regardless of its nature and the degradation proceeds homogeneously. However, as soon as the molecular weight of some of the partially degraded macromolecules becomes low enough to allow dissolution in the aqueous medium, diffusion starts within the whole bulk, with the soluble compounds moving slowly to and off the surface while they continue to degrade. This process combining diffusion, chemical reaction and dissolution phenomena results in a differentiation between the rates of the degradations at surface and interior of the matrix. Indeed, oligomers can escape from the surface before total degradation, in contrast to those oligomers that are far inside. The result is a smaller autocatalytic effect at the carboxyl-depressed surface with respect to the bulk. In buffered media, neutralization of terminal carboxyl groups might also contribute to discriminate the surface degradation rate. This mechanism well accounted for the formation of two populations of partially degraded macromolecules with different molecular weight and thus the observation of bimodal size exclusion chromatography traces. At the end, hollow structures were observed occasionally, especially in the case A in Figure 2 representing the fate of amorphous matrices that cannot generate semi-crystalline residues upon degradation (Li et al., 1990b,c). Indeed it has been shown that degradation of PLA_xGA_y is faster within glycolic acid-rich intrachain segments (Li et al., 1990b) and in highly disordered parts of the copolymer chains. Therefore, PLA_xGA_y polymer chains lacking intrachain order cannot generate crystallizable residues. In contrast, in the case of initially crystalline or of initially amorphous but rather stereoregular PLA_x-GA_y polymer chains, no hollow structures is observable despite the faster inner degradation, the crystallization of inside degradation products leading to a more degraded but solid interior (case B in Figure 2) (Li et al., 1990b,c). The formation or not of crystalline residues was mapped and discussed in details, at least for zinc-derived PLA_xGA_y polymer (Vert et al., 1994b; Li and Vert, 1999).

Basically there are four main factors which condition the diffusion-reaction-dissolution phenomena: (1) the hydrolysis rate constant of the ester bond, (2) the diffusion coefficient of water within the matrix, (3) the diffusion coefficient of chain fragments within the polymeric matrix, and (4) the solubility of degradation products, generally oligomers, within the surrounding liquid medium from which penetrating water is issued. Any additional factors such as temperature, additives in the polymeric matrix, additives in the surrounding medium, pH, buffering capacity, size and processing history, quenching or annealing, steric hindrance, porosity, and other variables affect the general balance through their effects on the main factors listed above (Vert, 1998).

The discovery of the heterogeneous degradation mechanism helped understanding the effect of many factors, i.e. matrix morphology, chemical composition and configurational structure, molar mass, size and shape, distribution of chemically reactive compounds within the matrix, and nature of the degradation media (Vert, 1990).

The effects of chemical composition, molecular weight, molecular weight and repeating units distributions, and crystallinity are well documented in the literature (Holland et al., 1986). Recently, the differences between the degradation rates observed when comparing the degradation characteristics of films, powder, and microspheres issued from the same batch of PLA_{50} polymer were accounted for (Grizzi et al., 1995). This work shows conclusively that the smaller the size of a polymer device, the slower the degradation rate. From the same logic, one can conclude that porous systems degrade at a slower rate than plain ones, especially if dimensions of the considered device are millimetric. Glycolic acid-rich PLA_xGA_y copolymers and PGA being more hydrophilic and degrading faster in aqueous media than PLA and lactic acid-rich PLA_xGA_y copolymers, the leaching of the water-soluble oligomers from the surface of large-size devices made of a glycolic acid-rich PLA_xGA_y occurs relatively earlier than for lactic acid-rich analogs and thus the skin is thinner.

The heterogeneous degradation mechanism provides also fruitful information on the behavior one can expect from a PLA_xGA_y polymer chain in which additives or foreign molecules have been introduced. Indeed any electrostatic interaction between a PLAGA matrix and acidic, basic or amphoteric molecules can drastically affect the degradation characteristics by changing the natural acid–base equilibrium of the matrix due to the presence of chain end carboxylic groups. For acidic compounds, one can expect faster hydrolysis of ester bonds. In contrast, in the case of basic compounds, two effects can be observed: base catalysis of ester bond cleavage when the basic compound is in excess with respect to acid chain ends and decrease of the degradation rate in the opposite situation (Li et al., 1996). Examples were reported that deal with oligomers (Mauduit et al., 1996), residual monomers (Cordewener et al., 1995), coral (Li and Vert, 1996), caffeine (Li et al., 1996), and gentamycin (Mauduit et al., 1993a,b). Reviews are available (Li, 1999; Li and Vert, 1995, 1999; Brannon-Peppas, 2000)

Recently, attention was paid to the effects of the initiator on the degradation characteristics of some PLA_xGA_y polymers. It has been shown that dramatic differences in repeating unit distribution can be observed in the case of stereocopolymers such as PLA_{50} polymerized in the presence of stannous octoate or zinc metal (Schwach et al., 1994). Later on it was suggested that stannous octoate causes more or less chain end modification by esterification of alcoholic end groups by octanoic acid and generates hydrophobic residues (Zhang et al., 1994; Schwach et al., 1997) which resist purification by precipitation from an organic solution with ethanol or water (Schwach et al., 1997). Such modifications do not occur in the case of zinc metal or zinc lactate initiations. Whether similar effects can be observed in the case of polymers obtained using other initiating systems, or through solution polymerization, or in the presence of alcohols is still unknown. Indeed, scientists rarely investigated the effect of resulting structures on degradation characteristics and comparison of literature data is seldom feasible significantly. In the absence of accurate information on structure and history, it is rather difficult to imagine how much the various factors mentioned above contribute to the degradation and the release characteristics. Nevertheless, the influence of chain end hydrophobization on degradation characteristics of PLA_xGA_y polymers is well exemplified by the difference between the so-called "H" and "non-H" polymers appearing in the catalog of one of the greatest supplier of PLA_xGA_y polymers, i.e. Boeringher Ingelheim (Germany). At this point people ought to keep in mind that the degradation and release characteristics of PLA_xGA_y depend on experimental conditions and on the device's history much more than for any other type of polymeric matrix. It is then strongly recommended again that careful identification of the investigated compounds be included in any future paper in the field.

Another critical recommendation is to be careful in modeling the conditions found in real applications. To model body fluids, it is recommended to take into account physiological or outdoor environmental pH, ionic strength, and temperature as much as possible. Using 0.13 M pH 7.4 isosmolar phosphate buffer at 37 °C is a convenient choice to model human body fluids. If phosphate ions can perturb the system, the use of any other pH 7.4 isosmolar buffered medium is possible. However, one must keep in mind that, in such media, species can be present that can diffuse and modify the matrix physical characteristics and the chain end chemistry, thus affecting the general behavior through the various interrelated factors mentioned above. As the glass temperature of PGA and of PLA_xGA_y polymers is a critical factor for water uptake and for diffusion of soluble oligomers, raising the temperature to shorten degradation time can be the source of unexpected phenomena that change the regime of degradation in a glassy state to a regime of degradation in a rubbery state.

The hydrolytic degradation of PGA fibers was extensively investigated by several authors, primarily by Chu and co-workers. It was shown that degradation occurs faster in amorphous domains, thus leading to crystallinity increase. The effects of other parameters like pH, temperature, etc., were also studied (Chu et al., 1981; Chu and Browning, 1988). It is worth noting that the main characteristics of the above mechanism of heterogeneous degradation well agree with the features reported for small-size PGA sutures. Indeed, the small size and the high crystallinity of fibers and suture threads prevent the occurrence of any surface-center differentiation and any formation of hollow structures. Attempts have been made to predict molecular weight changes on random hydrolysis (Nishida et al., 2000). Successful predictions are reported for fims, i.e. for small-size devices. In contrast no theoretical derivation has been found to reflect the heterogeneous degradation mechanism, mostly because all parameters change with time and swelling (Vert, 1998).

6.2 Degradation in Biotic Media

One can define different cases where degradation occurs in the presence of living systems. The problem is then showing the involvement of enzymes in the presence of the complex abiotic hydrolysis described previously.

6.2.1 Enzyme-containing Media

Whether enzymes in aqueous solutions can degrade PLA_xGA_y polymers *in vitro* or *in vivo* has been a matter of debates in literature. Some authors claimed occurrence of enzymatic degradation (Herrman et al., 1970; Williams, 1977; Reed, 1978), while most people excluded biodegradation or relegated it to a second role (Salthouse and Matlaga, 1976; Gilbert et al., 1982; Zaikov, 1985; Schakenraad et al., 1990).

In order to make the situation clearer, one must consider different cases. Some enzymes have been found that can attack PLA_xGA_y polymer chains *in vitro* and in the absence of the cells that normally produce the enzymes. Among these enzymes, bacterial proteinase K was shown to be able to strongly accelerate the degradation rate of PLA stereocopolymers (Williams, 1981; Fukuzaki et al., 1989; Ashley and McGinity, 1989; Reeve et al., 1994; MacDonald et al., 1996; Li, 1999). L-Lactic acid units are preferentially degraded as compared to D-lactic acid ones (Reeve et al., 1994; MacDonald et al., 1996). Enzymes such as tissue esterases, pronase, and bromelain are also able to affect PLA degradation (Herrman et al., 1970; Williams, 1981). It is important to note that there is usually competition between hydrolytic and enzymatic degradations. On the other hand, enzymes can be inactive on high molecular weight compounds and become active at the later stages of degradation when the chain fragments become small and soluble in surrounding fluids.

Because glycolic acid-containing, and especially glycolic acid-rich PLA_xGA_y polymers, are very sensitive to water and heat, showing a contribution specific to enzymes is usually difficult. Enzymatic degradation should not be claimed unless mass loss and dimensional change without molar mass decrease are conclusively shown since enzymes can hardly penetrate deep in a dense polymer matrix. However, one must keep in mind that if, for any reason, the rate of nonenzymatic hydrolysis at the surface of a hydrolytically degradable matrix is greater than the rate of diffusion of water into the polymeric mass, surface erosion can be observed, and thus confusion is possible between enzymatic surface degradation and surface degradation by water in the absence of enzymes. This is unlikely in the case of PLA_xGA_y polymers for which water uptake is always rather fast as compared with degradation rates.

6.2.2 Degradation by Microorganisms

In the past decades increasing attention has been paid to the degradation of aliphatic polyesters of lactic acid-rich PLA_xGA_y polymers in relation with the problems of the accumulation of solid plastic waste in the environment. Whether these polymers can biodegrade in a wild environment or whether it is the by-products generated by abiotic hydrolysis that can be bioassimilated by fungi or bacteria are two questions currently being investigated, in particular with regard to composting.

The ability of some microorganisms to use lactic acids, PLA oligomers and polymers, and PLA_xGA_y copolymers as sole carbon and energy sources under controlled or natural conditions was investigated in details (Vert et al., 1994c; Torres et al., 1996a–c). Among 14 filamentous fungal strains tested, two strains of *Fusarium moniliforme* and one strain of *Penicillium roqueforti* were identified as able to totally assimilate L- and DL-lactic acids as well as PLA_{50} oligomers (Torres et al., 1996a,b). In contrast, PLA_{100} oligomers appeared biostable because of crystallinity. A synergistic effect was observed when both microorganisms were present in the same culture medium containing PLA_{50} oligomers (Torres et al., 1996b).

High molar mass polymers were also considered via degradation tests in soil and in selected culture media. *F. moniliforme* filaments were found to grow on the surface of a $PLA_{37.5}GA_{25}$ copolymer and through the bulk. On the other hand, five strains of different filamentous fungi were isolated from the soil as capable of bioassimilating PLA_{50} soluble oligomers in mixed cultures. Based on the results obtained by scanning electron microscopy, size exclusion chromatography, pH, and water absorption measurements, the authors proposed a mechanism including abiotic hydrolysis of macromolecules followed by bioassimilation of soluble oligomers (Torres et al., 1996c). To summarize, when a PLA_xGA_y material is placed in a degradation medium in the presence of microorganisms, only abiotic hydrolytic degradation occurs unless the few enzymes mentioned above as capable of degrading high molar mass PLA are present. Similar investigations were reported recently that came to similar conclusions (Hakkarainen et al., 2000).

Whether solid fermentation of PLA_xGA_y polymers in compost or in soil proceeds through homogeneous or heterogeneous degradation is of interest. It should depend on the size of the material as discussed before (Grizzi et al., 1985). However, one must consider also the physical state of the medium (liquid or moisture). In a liquid medium, degradation of large-size devices will be heterogeneous due to greater autocatalysis inside, whereas, in the case of a gaseous water environment, degradation of similar devices will be homogeneous since the soluble oligomers cannot diffuse off the matrix. Anyhow, sooner or later, the whole compound will be degraded to low molecular weight assimilatable oligomers. The bioassimilation process will then take place. At this time, fungal filaments will penetrate the partially degraded mass to take advantage of the internal oligomers, thus turning the abiotic degradation to a biotic one. Under wild conditions, this should happen after a long lag period. However, PLA_xGA_y polymers will be definitely degraded and assimilated since the ultimate degradation products (L- and D-lactic acids) have been shown to be metabolized by some common microorganisms (Torres et al., 1996b; Hakkarainen, 2000). In particular, it is now well known that industrial lactic acid-based polymers like Dow-Cargill ECOPLA® degrade completely and rather rapidly in a compost provided the temperature of the compost raises up to 70 °C. At lower temperature composting is much slower. Information on the biodegradation of glycolic acid-containing polymers under composting conditions is not yet available, probably because such polymers have found no potential use in the environment so far.

6.2.3 Degradation in an Animal Body or in Human

The most efficient enzyme that is capable of degrading PLA_xGA_y polymers *in vitro*, i.e. proteinase K, is a protease of bacterial origin, and thus is not present in animal and human bodies. During the pre-industrial development of PGA sutures, information on the *in vivo* behavior and on the biocompatibility of these sutures was proprietary. It is in the late 1960s that people started releasing information. In 1966 appeared a remarkable contribution dealing with poly(L-lactide) and PLA_{50} stereocopolymers (Kulkarni et al., 1966). Most of the basic information regarding biocompatibility, configuration-dependence of degradability and crystallinity, and routes of elimination after ^{14}C labeling were identified. Until 1981, when it was shown that PLA_{100} and lactic acid-rich stereocopolymers prepared in the presence of zinc metal and carefully purified can last for a year *in vivo* with less than 10% molecular weight decrease (Chabot et al., 1981), the literature had reported much shorter half-lifetime systematically (Cutright et al., 1974; Miller et al., 1977; Guilding and Reed, 1979, 1981). Despite this difference of extreme lifetimes, it had been shown that the *in vivo* degradability of PLA_xGA_y polymers, copolymers, and stereocopolymers depends on the gross composition, and on the content in chiral units (Sedel et al., 1978; Reed, 1979). Since then, most of the contributions concluded to excellent biocompatibility of PLA_xGA_y polymers (Ignatius and Claes, 1996). Year after year the literature was enriched in information on devices made of PLA_xGA_y, such as sutures (Chu and Browning, 1988), implants for bone surgery (Hollinger, 1989) or microspheres and nanospheres for drug delivery (Brannon-Peppas, 1995). The literature is now rich of thousands of articles concerning PLA_xGA_y polymers. From a general viewpoint a majority of authors agree to say that PLA_xGA_y polymers do not degrade enzymatically in an animal body. In contrast, a minority claim enzymatic contribution (Li and Vert, 1995, 1999). The demonstration is usually based on the comparison of *in vivo* and *in vitro* degradation characteristics – differences being assigned to effects due to enzymes. However, the correlation cannot be straightforward. Indeed the heterogeneous degradation mechanism tells us that one of the driving phenomena is the dissolution of oligomers in the surrounding liquid medium. One can find many reasons to see different oligomer diffusion and solubilization between a device implanted *in vivo* and the same device allowed to age in a phosphate buffer medium modeling body fluids. Phenomena like adsorption of proteins, absorption of lipids, and greater solubility of PLA_xGA_y oligomers in blood are examples of sources of differences (Sharma and Williams, 1981). Superoxide ions also contribute to *in vivo* degradation of synthetic sutures (Lee and Chu, 2000). Anyhow, it has been found that *in vitro* and *in vivo* degradation characteristics are comparable in the case of small PLA_xGA_y cylinders implanted subcutaneously in rats (Kenley et al., 1987). In agreement with these findings, PLA_xGA_y parallel-sided plates implanted intramuscularly in rabbits and similar plates allowed to age in isoosmolar pH 7.4 phosphate buffer showed similar degradation characteristics, except shape modifications that were greater *in vivo* than *in vitro* according to the effect of the mechanical stresses generated by the muscles (Therin et al., 1993). Several authors found differences and the subject will be the source of endless debate (Gogolewski et al., 1993; Mainil-Varlet et al., 1997) unless people characterize carefully the investigated compounds including processing and sterilization protocols and consequences. In the previous sections it was mentioned for PLA_{50}

matrices that stannous octoate lead to more hydrophobic polymers than zinc metal and zinc lactate (Schwach et al., 1997). Similar investigations performed on PLA_{100} tin- and PLA_{98} zinc-derived interference screws currently commercialized to fix anterior cruciate ligament autografts in human and implanted in sheep knees showed dramatic differences of behavior insofar as water absorption, molecular weight decreases, decrease of mechanical properties, and rate of degradation were concerned (Schwach and Vert, 2000). Similar features should be found for glycolic acid-containing polymers.

Nowadays everybody agrees that PLA_xGA_y polymers are remarkably biocompatible. However, it appears that glycolic acid-containing copolymers are more inflammatory than PLA_x stereocopolymers, the most critical factor being the presence of crystalline tiny particles that cause well-known foreign body reactions to particles. Accordingly, taking advantage of polymer science to keep the polymer mass amorphous during the *in vivo* degradation process is always profitable in terms of biocompatibility since it keeps the inflammatory response low. Anyhow, once formed during degradation, any tiny PLA_x-GA_y crystalline particles can be phagocytosed to undergo intracellular degradation (Woodward et al., 1985; Tabata and Ikada, 1988).

In the future, one of the key points in degradable and biodegradable polymer science is going to be the demonstration of the fate of degradation by-products in complex living media, especially mineralization to carbon dioxide and water, bioassimilation, and/or subsistence as biostable residues. The best method to monitor these fates within the complex living media (animal body, microorganism, outdoor environment) is radiolabeling. Recently, a method to tritiate lactide and glycolide by catalytic isotopic exchange in the presence of palladium on calcium carbonate was reported that paves the route to the synthesis of radioactive PLA_xGA_y polymers (Dos Santos et al., 1998; Dos Santos et al., 1999).

In conclusion, it is recommended that the demonstration of the mechanism of degradation of PLA_xGA_y polymers under any condition be based on the concordance of data issued from different analytical methods, and from a careful monitoring of the generation and the fate of degradation by-products.

7 Production

7.1 Producers

The main manufacturers of PGA and glycolic acid-rich PLA_xGA_y polymers in the world are Purac (Purasorb®) in The Netherlands, Boeringher Ingelheim (Resomer®) in Germany, and Birmingham Polymers Inc. and Alkermes in the USA. Information can be found through the Internet.

7.2 World Market

The world market of PLA_xGA_y polymers other that PLA_{100} and PLA_x stereocopolymers offered for environmental applications is rather small – the present applications being limited to small market/high-added value domains such as drug delivery systems.

7.3 Applications

Whereas PLA_{100} and PLA_x stereocopolymers and lactic acid-rich PLA_xGA_y copolymers are preferred to make implants and devices for bone surgery (bone plates, screws, pins, etc.) because they last for a long time in the human body, glycolic acid-rich PLA_xGA_y

copolymers are generally selected for drug-delivery systems, especially the amorphous compounds, because they lead to longer sustained release than semi-crystalline compounds (Mauduit et al., 1993a,b).

The interest of PGA and PLA_xGA_y copolymers for bone surgery was recognized very early, although clinical and industrial developments came much later when controlled polymerization, processing, sterilization, and storage conditions became under good control. Nowadays, several devices are on the market for the temporary internal fixation of bone fractures (screws, plates, pins) and anterior cruciate ligament autografts (interference screws) from several companies, i.e. Biomet Inc. ($PLA_{82}GA_{18}$), Bionx Implants, Inc. (self-reinforced PGA), Instrument Makar, Inc. ($PLA_{42.5}GA_{15}$), Smith & Nephew Endoscopy (PGA; $PLA_{82}GA_{18}$), and Instrument Makar, Inc. ($PLA_{42.5}GA_{15}$) (Barber, 1998; Bostman et al., 1998). As early as 1978 a totally bioresorbale composite system composed of PGA fibers reinforcing PLA_{100} matrices, the long-lasting matrix protecting the fastly degraded fibers, was patented (Vert et al., 1978; Christel et al., 1982, 1985). Later on other composite systems were proposed such as self-reinforced PGA plates where PGA fibers were embedded in a PGA matrix (Törmälä et al., 1991), and went up to clinical and industrial applications. PLA_xGA_y are also worthwhile compounds for tissue reconstruction (Sedel et al., 1978; Hollinger, 1983)

In drug delivery, the literature is full of examples of PLA_xGA_y-based drug formulations, and devices such as implants, microparticles (Anderson and Shive, 1997), and nanoparticles (Brannon-Peppas, 1995) aimed at controlling the delivery of drugs of any kind. Several reviews have been issued that cover well a very rich field (Holland et al., 1985; Li and Vert, 1999; Brannon-Peppas and Vert, 2000). A few systems are commercialized that are used to treat prostate cancer, i.e. Decapeptyl® by Ipsen-Beaufour in France and Enantone® by Takeda in Japan.

Nowadays, one of the most stimulating areas of potential applications is tissue engineering, a technique that consists of culturing cells on a degradable or biodegradable scaffold for the sake of implantation for tissue reconstruction. Tissues like skin, bone, and cartilage are the favorite targets of scientists (Agrawal et al., 1995). Among the various degradable and biodegradable scaffolds presently under investigation, those made of PLA_xGA_y three-dimensional porous devices seem to be the most attractive (Ishaug et al., 1994; Agrawal et al., 1995; Holder et al., 1997; Athanasiou et al., 1998).

7.4 Patents

Many patents involving PLA_xGA_y polymers have been delivered during the last 30 years.

The early patents dealt with polymer and copolymer synthesis: Condensation polymers of hydroxy acetic acid, US2676945; Preparation of high molecular weight polyhydroxy acetic ester, US2668162.

Later on patents dealing with PGA and glycolic acid-rich PLA_xGA_y crystalline sutures and devices were issued: Absorbale polyglycolic sutures, US3297033; Polyglycolic acid prosthetic devices, US3739773; Method for heat setting of strecht oriented polyglycolic acid filament, US3422181; Process for polymerizing a glycolide, US3442871; Absorbable poly(glycolic acid) suture of enhanced *in vivo* strength retention and a method and apparatus for preparing same, US3626942; Preparation of glycolide polymerizable into polyglycolic acid of consistently high molecular weight, US3597450; Reducing capillarity of polyglycolic acid sutures, US3867190; Coating composition for sutures, US4185637.

More recent patents are: Yttrium and rare earth compounds catalyzed lactone polymerization, US5028667; Verfahren zur Herstellung von Polyestern auf der Basis von Hydroxycarbosäuren, EP0443542A2; Biodegradable chewing gum bases including plasticized poly(DL-lactic acid) and copolymers thereof, WO0019837; Polymerization process and product, US4273920; Method for producing poly(glycolic acid) with excellent heat resistance, KR9502607.

Besides these patents dealing with polymeric matter and surgical devices, a number of patents were delivered for drug delivery systems of different kinds, namely implants, microspheres and nanospheres. The list of recent patents can be easily found using free data banks like http://ep.espacenet.com/

8 Outlook and Perspectives

PLA_xGA_y polymers constitute a large family of aliphatic polyesters that can be synthesized by different routes leading to compounds with different molecular weights, unit distributions, and terminal groups. All the members of this family are hydrolytically degradable. They are not biodegradable under normal conditions either in the animal body or in the environment, although some enzymes are able to biodegrade some members under *in vitro* model conditions. Among the whole family, PGA and glycolic acid-rich polymers are finding applications in the biomedical and pharmacological fields. They are being currently used to make commercial sutures. However, it is in the field of controlled drug delivery via implants, microspheres, and nanospheres that amorphous PLA_xGA_y polymers are showing the greatest potential for applications, and a few formulations being already on the market. Some lactic acid-rich PLA_x-GA_y polymers are being used in orthopedic surgery, especially for fixing anterior cruciate ligament autografts. In contrast to PLA_x stereocopolymers, none of the glycolic acid-containing compounds have been considered as candidates or applications as environmentally degradable polymer, so far.

9 References

Adamus, G., Sikorska, W., Kowalczuk, M., Montaudo, M., Scandola, M. (2000) Sequence distribution and fragmentation studies of bacterial copolyester macromolecules: characterization of PHBV macroinitiator by electrospray ion-trap multistage mass spectrometry, *Macromolecules* **33**, 5797–5802.

Agrawal, C. M., Niedersuer, C. C., Micallef, D. M., Athanasiou, K. A. (1995) The use of PLA–PGA polymers in orthopedics, in: *Encyclopedic Handbook of Biomaterials and Bioengineering*, New York: Marcel Dekker, 2081–2115.

Anderson, J. M., Shive, M. S. (1997) Biodegradation and biocompatibility of PLA and PLGA microspheres, *Adv. Drug Del. Rev.* **28**, 5–24.

Ashley, S. L., McGinity, J. W. (1989) Enzyme-mediated drug release from poly(D,L-lactide) matrices, *Congr. Int. Technol. Pharm.* **5**, 195–204.

Athanasiou, K. A., Schmitz, J. P., Agrawal, C. M. (1998) The effects of porosity on *in vitro* degradation of polylactic acid–polyglycolic acid implants used in repair of articular cartilage. *Tissue Eng.* **4**, 53–63.

Barber, A. F. (1998) Resorbable fixation devices: a product guide, *Orthopaedic Special Edition* **4**, 11–17.

Bostman, O., Hirvensalo, E., Vainionpaa, S., Vihtonen, K., Tormala, P., Rokkanen, P. (1990) Degradable polyglycolide rods for the internal fixation of displaced bimalleolar fractures, *Int. Orthop.* **14**, 1–8.

Brannon-Peppas, L. (1995) Recent advances on the use of biodegradable microparticles and nanoparticles in controlled drug delivery, *Int. J. Pharm.* **116**, 1–9.

Brannon-Peppas, L., Vert, M. (2000) *Polylactic and Polyglycolic Acids as Drug Delivery Carriers, in Handbook of Pharmaceutical Controlled Release Technology* (Wise, L. R., Klibanov, D. L., Mikos, A., Brannon-Peppas, L., Peppas, N. A., Trantalo, D. J., Wnek, G. E., Michael, J., Yaszemski, M. J., Eds.), New York: Marcel Dekker, pp. 90–130.

Braud, C., Devarieux; R., Garreau, H., Vert, M. (1996) Capillary electrophoresis to analyze water-soluble oligo(hydroxy acids) issued from degraded or biodegraded aliphatic polyesters, *J. Environ. Polym. Degrad.* **4**, 135–148.

Chabot, F., Vert, M., Chapelle, S., Granger, P. (1983) Configurational structires of lactic acid stereocopolymers as determined by ^{13}C–^{1}H-NMR, *Polymer* **24**, 53–59.

Chen, J., Lee, J. W., Hernandez de Gatica, N. L., Burkhart, C. A., Hercules, D. M., Gardella, J. A. (2000) Time-of-flight secondary ion mass spectrometry studies of hydrolytic degradation kinetics at the surface of poly(glycolic acid), *Macromolecules* **33**, 4726–4732.

Christel, P., Chabot, F., Leray, J. C., Morin, C., Vert, M. (1982) Biodegradable composites for internal fixation, in: *Biomaterials 1980: Advances in Biomaterials* (Winter, G. D., Gibbons, D. F., Plenk, H., Eds.), New York: John Wiley & Sons, Vol. 3, 271–280.

Christel, P., Vert, M., Chabot, F., Garreau, H., Audion, M. (1985) PGA (polyglycolic acid) fiber-reinforced PLA(polylactic acid) as an implant material for bone surgery, Proceedings, Plastic and Rubber Institute, London, p. 11.

Chu, C. C. (1981) The *in-vitro* degradation of poly(glycolic acid) sutures—effect of pH, *J. Biomed. Mater. Res.* **15**, 795–804.

Chu, C. C., Browning, A. (1988) The study of thermal and gross morphologic properties of polyglycolic acid upon annealing and degradation treatments, *J. Biomed. Mater. Res.* **22**, 699–712.

Chu, C. C., Williams, D. F. (1983) The effect of gamma irradiation on the enzymatic degradation of polyglycolic acid absorbable sutures, *J. Biomed. Mater. Res.* **17**, 1029–40.

Cordwener, F. W., Rozema, F. R., Bos, R. R. R., Boering, G. (1995) Material properties and tissue reaction during degradation of poly(96L/4D-lactide)—a study *in vitro* and in rats, *J. Mater. Sci.: Mater. Med.* **6**, 211–218.

Cutright, D. E., Perez, B., Beasley, J. D., Larson, W. J., Rosey, W. R. (1974) Degradation rates of polymers and copolymer of polylactic and polyglycolic acids, *Oral Surg.* **37**, 142–152.

Dos Santos, I., Morgat, J. L., Vert, M. (1998) *J. Labelled Cpd Radiopharm.*, **41** 1005–1012.

Dos Santos, I., Morgat, J. L., Vert, M. (1999) Glycolide deuteration by hydrogen isotope exchange using the HSCIE method, *J. Labelled Cpd Radiopharm.* **42**, 1093–1101.

Dubois, P., Jacobs, C., Jérôme, R., Teyssié, P. (1991) Macromolecular engineering of polylactones and polylactides. IV, Mechanism and kinetics of lactide homopolymerization by aluminum isopropoxide, *Macromolecules* **24**, 2266–2270.

Dunsing, R., Kricheldorf, H. R. (1985) Polylactones. 5. Polymerization of LL-lactide by means of magnesium salts, *Polym. Bull.* **14**, 491–496.

Frazza, E. J., Schmitt, E. E. (1971) A new absorbable suture, *J. Biomed. Mater. Sci.* **1**, 43–58.

Fukuzaki, H., Yoshida, M., Asano, M., Aiba, Y., Kumakura, M. (1990) Direct copolymerization of glycolic acid with lactones in the absence of catalysts, *Eur. Polym. J.* **26**, 457–461.

Gilbert, R. D., Stannett, V., Pitt, C. G., Schindler, A. (1992) The design of biodegradable polymers, in: *Development in Polymer Degradation* (Grassie, N., Ed.), London: Applied Science, Vol. IV, 259–293

Granger, P., Chabot, F., Chapelle, S., Vert, M., Unpublished data.

Gogolewski, S., Jovanovic, M., Perren, S. M., Dillon, J. G., Hughes, M. K. (1993) Tissue response and *in vivo* degradation of selected polyhydroxyacids: Polylactides (PLA), poly(3-hydroxybutyrate-*co*-3-hydroxyvalerate) (PHB/VA), *J. Biomed. Mater. Res.*, **27**, 1135–1148.

Grizzi, I., Garreau, H., Li, S., Vert, M. (1995) Biodegradation of devices based on poly(DL-lactic acid): size dependence, *Biomaterials* **16**, 305–311.

Guilding, D. K., Reed, A. M. (1979) Biodegradable polymers for use in surgery – Polyglycolic/poly(lactic acid) homo- and copolymers: 1, *Polymer* **20**, 1459–1464.

Hakkarainen, M., Karlsson, S., Albertsson, A.-C. (2000) Rapid (bio)degradation of polylactide by mixed culture of compost microorganisms—low molecular weight products and matrix changes, *Polymer* **41**, 2331–2338.

Herrman, J. B., Kelly, R. J., Higgins, G. A. (1970) Polyglycolic acid sutures, *Arch. Surg.* **100**, 486–490.

Higgins N. A. (1950) Condensation polymers of hydroxy-acetic acid, US patent 2676945, Appl. October 18.

Holder, Jr., W. D., Gruber, H. E., Moore, A. L., Culberson, C. R., Anderson, W., Burg, K. J. L., Mooney, D. J. (1998) Cellular in growth and thickness changes in poly-L-lactide and polyglycolide matrices implanted subcutaneously in the rat, *J. Biomed. Mater. Res.* **41**, 412–421.

Holland, S. J., Tighe, B. J., Gould, P. L. (1986) Polymers for biodegradable medical devices. 1. The potential of polyesters as controlled macromolecular release systems, *J. Control. Rel.* **4**, 155–180.

Hollinger, J. O. (1983) Preliminary report on the osteogenic potential of a biodegradable copolymer of polylactide (PLA) and polyglycolide (PGA), *J. Biomed. Mater. Res.* **17**, 71–82.

Holten, C. H. (1971) Intermolecular esters, in: *Lactic acid* (Holton, C. H., Ed.), Weinheim: Verlag Chemie, 221–224.

Ignatius, A. A., Claes, L. E. (1996) In vitro biocompatibility of bioresorbable polymers: poly(L,DL-lactide) and poly(L-lactide-*co*-glycolide), *Biomaterials* **17**, 831–839.

Ishaug, S. L., Yashemski, M. J., Bizios, R., Mikos, A. G. (1994) Osteoblast function on synthetic biodegradable polymers, *J. Biomed. Mater. Res.* **28**, 1445–1453.

Kenley, R. A., Ott Lee, M., Mahoney, T. R., Sanders, L. (1987) Poly(lactide-*co*-glycolide) decomposition kinetics *in vivo* and *in vitro*, *Macromolecules* **20**, 2398–2403.

Kleine, J. and Kleine, H. (1959) Über hochmolekulare, insbesondere optisch aktive Polyester der Milchsäure, ein Beitrag zur Stereochemie makromolekularer Verbindungen, *Makromol. Chem.* **30**, 23–38.

Kohn, F. E., van Ommen, J. G., Feijen, J. (1983) The mechanism of the ring-opening polymerization of lactide and glycolide, *Eur. Polym. J.* **12**, 1081–1088.

Kowalski, A., Duda, A, Penczek, S. (2000) Kinetics and mechanism of cyclic esters polymerization initiated with tin(II) octoate. 3. Polymerization of L,L-dilactide, *Macromolecules* **33**, 7359–7370.

Kricheldorf, H. R., Kreiser, I. (1987) Polylactones. 13. Transesterification of poly(L-lactide with poly(glycolide), poly(β-propiolactione) and poly(ε-caprolactone) *J. Macromol. Sci. Chem.* **A24**, 1345–1358.

Kricheldorf, H. R., Kreiser-Saunders, I., Stricker, A. (2000) Polylactones 48. Sn-Oct_2-initiated polymerizations of lactide: a mechanistic study, *Macromolecules* **33**, 702–709.

Kronenthal, R. L. (1974) Biodegradable polymers in medicine and surgery, in: *Polymers in Medicine and Surgery* (Kronenthal, R. L., User, Z., Martin, E., Eds.), New York: Plenum Press, 119–137.

Kulkarni, R. K., Moore, E. G., Hegyeli, A. F., Leonard, F. (1971) Biodegradable poly(lactic acid) polymers. *J. Biomed. Mater. Res.* **5**, 169–181.

Kulkarni, R. K., Pani, K. C., Neuman, C., Leonard, F. (1966) Polylactic acid for surgical implants, *Arch. Surg.* **93**, 839–843.

Lee, K. H., Chu, C. C. (2000) The role of superoxide ions in the degradation of synthetic absorbable sutures, *J. Biomed. Mater. Res.* **49**, 25–35.

Leenslag, J. W., Pennings, A. J. (1987) Synthesis of high molecular weight poly(L-lactide) initiated by tin-2-ethyl hexanoate, *Makromol. Chem.* **188**, 1809–1814.

Leray, J., Vert, M., Blanquaert, D. (1976) Nouveau matériau de prothèse osseuse et son application, FR7628629.

Li, S. (1999) Hydrolytic degradation characteristics of aliphatic polyesters derived from lactic and glycolic acids, *J. Biomed. Mater. Res. (Appl. Biomater.)* **48**, 342–353.

Li, S., Vert, M. (1995) Biodegradation of aliphatic polyesters, in: *Biodegradable Polymers, Principles and Applications* (Scott, G. and Gilead, D., Eds.), London: Chapman & Hall, 43–87.

Li, S. M., Vert, M. (1996) Hydrolytic degradation of coral/poly(DL-lactic acid) bioresorbable materials, *J. Biomater. Sci.: Polym. Ed.* **7**, 817–827.

Li., S., Vert., M. (1999) Biodegradable polymers: polyesters, in: *The Encyclopedia of Controlled Drug Delivery* (Mathiowitz, E., Ed.), New York: John Wiley & Sons, 71–93.

Li, S., Garreau, H., Vert, M. (1990a) Structure–property relationships in the case of degradation of solid aliphatic poly(α-hydroxy acids) in aqueous media: 3. Amorphous and semi-crystalline PLA 100, *J. Mater. Sci. Mater. Med.* **1**, 198–206.

Li, S., Garreau, H., Vert, M. (1990b) Structure–property relationships in the case of degradation of solid aliphatic poly(α-hydroxy acids) in aqueous media: 2. PLA37. 5GA25 and PLA75GA25 copolymers, *J. Mater. Sci. Mater. Med.* **1**, 131–139.

Li, S. M., Garreau, H., Vert, M. (1990c) Structure–property relationships in the case of the degradation of massive aliphatic poly-(alphahydroxy acids) in aqueous media: part 1: poly(DL-lactic acid), *J. Mater. Sci. Mater. Med.* **1**, 123–130.

Li, S. M., Girod-Holland, S., Vert, M. (1996) Hydrolytic degradation of poly(DL-lactic acid) in the presence of caffeine base, *J. Control. Rel.* **40**, 41–53.

Lillie, E., Schulz, R. C. (1975) ^1H and ^{13}C-{^1H} NMR spectra of stereocopolymers of lactide, *Makromol. Chem.* **176**, 1901–1906.

Lowe C. H. (1954) Preparation of high molecular weight polyhydroxyacetic ester, US patent 2668162, Appl. March 24, 1952.

Mainil-Varlet, P., Curtis, R., Gogolewski, S. (1997) Effect of *in vivo* and *in vitro* degradation on molecular and mechanical properties of various low-molecular weight polylactides, *J. Biomed. Mater. Res.*, **36**, 360–380.

Mauduit, J., Bukh, N., Vert, M. (1993a) Gentamycin/poly(lactic acid) mixtures aimed at sustained release local antibiotic therapy administered peroperatively: II. The case of gentamycin sulfate in high molecular weight poly(DL-lactic acid) and poly(L-lactic acid), *J. Control. Rel.* **23**, 221–230.

Mauduit, J., Bukh, N., Vert, M. (1993b) Gentamycin/poly(lactic acid) blends aimed at sustained release local antibiotic therapy administered peroperatively. III. The case of gentamycin sulfate in films prepared from high and low molecular weight poly(DL-lactic acids), *J. Control. Rel.* **25**, 43–49.

Mauduit, J., Perouse, E., Vert, M. (1996) Hydrolytic degradation of films prepared from blends of high and low molecular weight poly(DL-lactic acid)s. *J. Biomed. Mater. Res.* **30**, 201–207.

McDonald, R. T., McCarthy, S. P., Gross, R. A. (1996) Enzymatic degradability of poly(lactide): effects of chain stereochemistry and material crystallinity, *Macromolecules* **29**, 7356–7361.

McNeil, I. C., Leiper, H. A. (1985) Degradation studies of some polyesters and polycarbonates—1. Polylactide: general features of the degradation under programmed heating conditions, *Polym. Degrad. Stab.* **11**, 267–285.

Miller, R. A., Brady, J. M., Cutright, D. E. (1977) Degradation rates of oral resorbable polyesters for orthopaedic surgery, *J. Biomed. Mater. Res.* **11**, 711–719.

Nieuwenhuis, J. (1992) Synthesis of polylactides, polyglycolides and their copolymers, *Clin. Mater.* **10**, 59–67.

Nishida, H., Yamashita, M., Nagashima, M., Hattori, N., Endo, T., Tokiwa, Y. (2000) Theoretical prediction of molecular weight on autocatalytic random hydrolysis of aliphatic polyesters, *Macromolecules* **33**, 6595–6601.

Pitt, C. G., Gratzel, M. M., Kimmel, G. L., Surles, J., Schindler, A. (1981) Aliphatic polyesters. 2. The degradation of poly(DL-lactide), poly (ε-caprolactone) and their complexes *in vivo*. *Biomaterials* **2**, 215–220.

Reed, A. M. (1978) *In vivo* and *in vitro* studies of biodegradable polymers for use in medicine, PhD Thesis, Liverpool University.

Reed, A. M., Guilding, D. K. (1981) Biodegradable polymers for use in surgery—poly(glycolic)/poly(lactic acid) homo and copolymers, *Polymer* **22**, 494–498.

Reeve, M. S., McCarthy, S. P., Downey, M. J., Gross, R. A. (1994) Polylactide stereochemistry: effect on enzymatic degradability, *Macromolecules* **27**, 825–831.

Salthouse, T. N., Matlaga, B. F. (1976) Polyglactin 910 suture absorption and the role of cellular enzymes, *Surg. Gynecol. Obstet.* **142**, 544–550.

Schakenraad, J. M., Hardink, M. J., Feijen, J., Molenaar, I., Nienwenhuis, P. (1990) Enzymatic activity toward poly(L-lactic acid) implants, *J. Biomater. Med. Res.* **24**, 529–545.

Schmitt, E. A., Flanagan, D. R., Lindhart, R. J. (1994) Importance of distinct water environments in the hydrolysis of poly(DL-lactide-*co*-glycolide), *Macromolecules* **27**, 743–748.

Schmitt, E. E., Polistina R. A. (1967) Surgical sutures, US3297033.

Schmitt, E. E., Polistina R. A. (1973) Polyglycolic acid prosthetic devices, US patent 3297033.

Schwach, G., Vert, M. (2000) *In vitro* and *in vivo* degradation of lactic acid-based interference screws used in cruciate ligament reconstruction, *Int. J. Biol. Macromol.* **25**, 283–291.

Schwach, G., Coudane, J., Engel, R., Vert, M. (1997) More about the initiation mechanic of lactide polymerization in the presence of stannous octoate, *J. Polym. Sci.: Polym. Chem.* **35**, 3431–3440.

Schwach, G., Coudane, J., Engel, R., Vert, M. (1998a) Ring opening polymerization of DL-lactide in the presence of zinc-metal and zinc-lactate. *Polym. Int.* **46**, 177–182.

Schwach, G., Coudane, J., Engel, R., Vert, M. (1998b) Something new in the field of PGA/LA bioresorbable polymers, *J. Control. Rel.* **53**, 85–92.

Schwach, G., Coudane, J., Engel, R., Vert, M. (1996a) Zn lactate as initiator of DL-lactide ring opening polymerization and comparison with Sn octoate, *Polym. Bull.* **37**, 771–776.

Schwach, G., Coudane, J., Vert, M., Huet-Olivier, J. (1996b) Catalyseur et composition catalytique pour la fabrication d'un polymère biocompatible résorbable, et procédés les mettant en œuvre, FR9602140.

Schwach, G., Coudane, J., Vert, M., *Biomaterials*, in press.

Schwach, G., Engel, R., Coudane, J., Vert, M. (1994) Stannous octoate versus zinc-initiated polymerization of racemic lactide: effect of configurational structures, *Polym. Bull.* **32**, 617–623.

Sedel, L., Chabot, F., Christel, P., de Charantenay, F. X., Leray, J., Vert, M. (1978) Les implants biodégradables en chirurgie orthopédique, *Rev. Chir. Orthop.* **64**, 92–101.

Sharma, C. P., Williams, D. F. (1981) The effects of lipids on the mechanical properties of polyglycolic acid structures, *Eng. Med.* **10**, 8–10.

Tabata, Y., Ikada, Y. (1988) Macrophage phagocytosis of biodegradable microspheres composed of L-lactic acid/glycolic acid homo and copolymers, *J. Biomed. Mater. Res.* **22**, 837–858.

Tanzi, M. C., Verderio, P., Lampugnani, M. G., Resnati, M., Dejana, E., Sturanie E. (1994) Cytotoxicity of some catalysts commonly used in the synthesis of copolymers for medical use, *J. Mater. Sci.: Mater. Med.* **5**, 393–401.

Therin, M., Christel, P., Li, S., Garreau, H., Vert, M. (1992) Degradation of massive poly(α-hydroxy acids) in aqueous living medium: *in vivo* validation of *in vitro* findings, *Biomaterials* **13**, 594–600.

Törmälä, P., Vasenius, J., Vainionpää, S., Laiho, J., Pohjonen, T., Rokkanen, P. (1991) Ultra high strength absorbable self-reinforced polyglycolide (SR-PGA) composite rods for internal fixation of bone fractures: *in vitro* and *in vivo* study, *J. Biomed. Mater. Res.* **25**, 1–22.

Torres, A., Li, S., M., Roussos, S., Vert, M. (1996a) Degradation of L- and DL-lactic acid oligomers in the presence of *Fusarum moniliforme*, *J. Environ. Polym. Degrad.* **4**, 213–223.

Torres, A., Li, S., M., Roussos, S., Vert, M. (1996b) Poly(lactic acid) degradation in soil or under controlled conditions, *J. Appl. Polym. Sci.* **62**, 2295–2302.

Torres, A., Roussos, S., Li, S., M., Vert, M. (1996c) Screening of microorganisms for biodegradation of poly(lactic acid) and lactic acid-containing polymers, *Appl. Environ. Microbiol.* **62**, 2393–2397.

Vert, M. (1990) Degradation of polymeric biomaterials with respect to temporary applications, in: *Degradable Materials*, Boca Raton, FL: CRC Press, 11–37.

Vert, M. (1998) Bioresorbable synthetic polymers and their operation field, in: *Biomaterials in Surgery* (Walenkamp, G., Ed.), Stuttgart: Thieme, 97–101.

Vert., M., Chabot, F., Leray, J., Christel, P. (1978) Nouvelles pièces d'ostéosynthèse, leur préparation et leur application, FR7829878.

Vert, M., Chabot, F., Leray, J., Chrsitel, P. (1981) Bioresorbable polyesters for bone surgery. *Makromol. Chem. Suppl.* **5**, 30–41.

Vert, M., Christel, P., Chabot, F., Leray, J. (1984) Bioresorbable plastic materials for bone surgery, in: *Macromolecular Biomaterials* (Hastings G. W., Ducheyne, P., Eds.), Boca Raton, FL: CRC Press, 119–141.

Vert, M., Li, S. M., Spenlehauer, G., Guerin, P. (1992) Bioresorbability and biocompatibility of aliphatic polyesters. *J. Mat. Sci. Mat. Med.* **3**, 432–446.

Vert, M., Mauduit, J., Li, S. M. (1994a) Biodegradation of PLA/GA polymers: increasing complexity, *Biomaterials* 15, 1209–1213.

Vert, M., Li, S., Garreau, H. (1994b) Attempts to map structure and degradation characteristics of aliphatic polyesters derived from lactic and glycolic acids, *J. Biomater. Sci.: Polym. Ed.* **6**, 639–649.

Vert, M., Torres, A., Li, S. M., Roussos, S., Garreau, H. (1994c) The complexity of the biodegradation of poly(2-hydroxy acid)-type aliphatic polyesters, in: *Biodegradable Plastics and Polymers* (Doi, Y. and Fukuda, K., Eds.), Amsterdam: Elsevier, pp. 11–23

Vert, M., Li. S., Garreau, H. (1995) Recent advances in the field of lactic-acid/glycolic acid polymer-based therapeutic systems *Macromol. Symp.* **98**, 633–642.

Vert, M., Schwach, G., Engel, R., Coudane, J. (1998) Something new in the field of PLA/GA bioresorbable polymers. *J. Control. Rel.* **53**, 85–92.

Williams, D. F. (1981) Enzyme hydrolysis of polylactic acid, *Eng. Med.* **10**, 5–7.

Williams, D. F., Mort, E. (1977) Enzyme accelerated hydrolysis of polyglycolic acid, *J. Bioeng.* **1**, 231–238.

Zaikov, G. E. (1985) Quantitative aspects of polymer degradation in the living body, *J. Macromol. Sci., Rev. Macromol. Chem. Phys.* **C25**, 551–597.

Zhang, X., MacDonald, D. A., Goosen, M. F. A., Mcauley, K. B. (1994) Mechanism of lactide polymerization in the presence of stannous octoate: the effect of hydroxy and carboxylic acid substances, *J. Polym. Sci.: Polym. Chem.* **32**, 2965–2970.

7
Polylactides

Prof. Dr. Hideto Tsuji
Department of Ecological Engineering, Faculty of Engineering, Toyohashi University of Technology, Tempaku-cho, Toyohashi, Aichi 441–8580, Japan; Tel.: +81-532-44-6922; Fax: +81-532-44-6929; E-mail: tsuji@eco.tut.ac.jp

Biopolymers for Medical and Pharmaceutical Applications. Edited by A. Steinbüchel and R. H. Marchessault

ISBN: 3-527-31154-8

CLL	ε-caprolactone
[COOH]	concentration of terminal carboxyl group
DDS	drug delivery system
DLA	D-lactide
DLLA	DL-lactide
DR	draw ratio
E	Young's modulus
[ester]	ester group concentration
GPC	gel permeation chromatography
HMB	hexamethylbenzene
HMW	high-molecular weight
$[H_2O]$	water concentration
J^0_e	equilibrium compliance
k	hydrolysis rate constant
LA	lactide
Lc	crystalline thickness
LLA	L-lactide
LMW	low-molecular weight
MLA	meso-lactide
Mn	number-average molecular weight
$M_{n,t}$	M_n at hydrolysis time t
$M_{n,0}$	M_0 at hydrolysis time zero
$M_{n,s}$	peak molecular weight of the lowest molecular weight specific peak
M_v	viscosity-average molecular weight
M_w	weight-average molecular weight
Ny6	poly(ε-amino caproic acid)
P	physical property
PBS	poly(butylene succinate)
PCL	poly(ε-caprolactone)
PCL-b-PEO	poly(ε-caprolactone)-b-poly(ethylene oxide)
PDLA	poly(D-lactide), poly(D-lactic acid)
poly(DLA-GA)	poly(D-lactide-*co*-glycolide)
PDLLA	poly(DL-lactide)
PDLLA-b-PEO	poly(DL-lactide)-b-poly(ethylene oxide)
poly(DLLA-CL)	poly(DL-lactide-co-ε-caprolactone)
PEBS	poly(ethylene/butylene succinate)
PEO	poly(ethylene oxide)
PGA	polyglycolide
poly(HB-HV)	poly[(*R*)-3-hydroxybutyrate-*co*-3-hydroxyvalerate]
PHMS	poly(hexamethylene succinate)
PHO	poly(3-hydroxyoctanoate)
PLA	polylactide, poly(lactic acid)
poly(LA-GA)	poly(lactide-*co*-glycolide)
PLLA	poly(L-lactide), poly(L-lactic acid)
poly(LLA-DLA)	poly(L-lactide-*co*-D-lactide)

poly(LLA-CL)	poly(L-lactide-*co*-ε-caprolactone)
poly(LLA-GA)	poly(L-lactide-*co*-glycolide)
poly(LLA-Ly)	poly(L-lactide-*co*-lysine)
Pluronic	ABA triblockcopolymers of poly(ethylene oxide) (A) and poly(propylene oxide) (B)
PML	poly(meso-lactide)
PPHOS	polyphosphazene
PSA	poly(sebacic anhydride)
PVA	poly(vinyl alcohol)
PVAc	poly(vinyl acetate)
PDXO	poly(1,5-dioxepan-2-one)
P_0	physical property of the polymer with infinite M_n
ROP	ring-opening polymerization
R-PHB	poly[(*R*)-3-hydroxybutyrate]
R,S-PHB	poly[(*R*,*S*)-3-hydroxybutyrate]
SEM	scanning electron microscopy
SR	self-reinforced
syn-PLA	syndiotactic poly(lactide), syndiotactic poly(lactic acid)
T_a	annealing temperature
T_g	glass transition temperature
T_m	melting temperature
T_m^0	equilibrium melting temperature
T_{1C}	spin lattice relaxation time
x_c	crystallinity
$[a]^{25}_{589}$	specific optical rotation at 25 °C and a wavelength of 589 nm
ΔH_m	enthalpy of melting
ΔH_c	enthalpy of cold crystallization
Δh^0	heat of fusion (per unit mass)
Δn	birefringence
δ_p	polymer solubility parameter
δ_s	solvent solubility parameter
ε_B	elongation-at break
$[\eta]$	intrinsic viscosity
η_0	terminal viscosity
ρ_c	crystal density
σ	specific fold surface free energy
σ_B	tensile strength
σ_Y	yield stress
3D	three-dimensional

1 Introduction

During the past few decades, polylactides, poly(lactic acid)s (PLAs) and their copolymers have attracted much attention in terms of their ecological, biomedical, and pharmaceutical applications, for the following reasons.

- They can be produced from renewable resources such as starch.
- They have mechanical properties comparable with those of commercial polymers such as polyethylene, polypropylene, and polystyrene.
- They are degradable in the human body, as well as in natural environments.
- The toxicity of their degradation products (lactic acid and its oligomers) in the human body and in natural environments is very low.

PLAs can be synthesized either by the polycondensation of lactic acid, or by ring-opening polymerization (ROP) of lactide (LA) (IUPAC name: 3,6-dimethyl-1,4-dioxane-2,5-dione). Lactic acid (2-hydroxy propionic acid) is optically active, and has two enantiomeric L- and D- (*S*- and *R*-) forms. In accordance with the two enantiomeric forms, their homopolymers have respective stereoisomers, and their crystallizablility changes depending on their tacticity and optical purity. On the other hand, Ikada et al. (1987) found that stereocomplexation occurs between the enantiomeric poly(L-lactide), poly(L-lactic acid) (PLLA) and poly(D-lactide), poly(D-lactic acid) (PDLA) due to their peculiar strong interaction. The physical properties, hydrolysis and biodegradation behavior of PLAs can be controlled by altering their molecular characteristics such as molecular weight and monomer sequence distribution, and their highly ordered structures such as crystallinity and crystalline thickness, by polymer blending, additives, material shapes, etc. The recent developments to produce PLAs at low costs (US $1–2 per kg) will accelerate their use as the commodity plastics. This chapter outlines the basic aspects of synthesis, processing, structures, physical properties, hydrolysis, and biodegradation of PLAs. The commercial aspects of PLAs are described in detail in Chapters 8 and 9 of this volume.

2 Historical Outline

Pelouze (1845) first discovered the formation of a linear dimer of lactic acid, i.e., lactoyllactic acid, through the esterification reaction of lactic acid by removal of water at high temperature (130 °C), whilst Nef (1914) later confirmed the presence of lactic acid oligomers from trimer to heptamer when lactic acid was dehydrolyzed at an elevated temperature (90 °C) at reduced pressure (15 mmHg). A two-step polymerization procedure using the cyclic dimer of lactic acid, i.e. lactide (LA), for the synthesis of high-molecular weight (HMW) poly(lactide)s, i.e., poly(lactic acid)s (PLAs), was suggested by Carothers et al. (1932) and developed by Lowe (1954). Since the late 1960s, PLAs and their copolymers have been investigated for their biomedical applications such as sutures and prostheses (Wise et al., 1979), and in the 1970s the commercial biodegradable sutures Vicryl and Glactin 910 were produced from the copolymers of lactide and glycolide (Shneider, 1972). On the other hand, De Santis and Kovacs (1968) analyzed the crystal structure (α-form) of the isotactic poly(L-lactide) (PLLA). Since the late 1970s, PLAs and their copolymers have been studied for the pharmaceutical applications as the matrices for drug delivery systems (DDSs) (Wise et al., 1979).

Additional basic information concerning the synthesis, physical properties, crystallization behavior, and medical and pharmaceutical applications of PLAs and their copolymers was obtained throughout the 1980s and 1990s. Kricheldorf and colleagues (Kricheldorf et al., 1990; Kricheldorf and Kreiser-Saunders, 1996) and Dubois et al. (1997) investigated the synthesis and molecular characterization of LA homo- and copolymers, whilst Pennings and coworkers studied the crystallization (Kalb and Pennings, 1980; Vasanthakumari and Pennings, 1983) and spinning of PLLA and the physical properties of its spun and drawn fibers (Leenslag and Pennings, 1987a). Ikada and coworkers (Idada, 1998; Ikada and Tsuji, 2000) investigated the medical and pharmaceutical applications of PLAs and their copolymers, whilst Tsuji and colleagues (Tsuji and Ikada, 1999a; Tsuji, 2000a,b) studied the crystallization, physical properties, hydrolysis, and biodegradation of PLA-based materials, including a PLA stereocomplex. This stereocomplex is formed by the peculiarly strong interaction between enantiomeric PLLA and poly(D-lactic acid), poly(D-lactide) (PDLA) (Ikada et al., 1987), and it was Okihara et al. (1991) who first proposed the crystal structure of the PLA stereocomplex, where PLLA and PDLA are packed in side-by-side fashion. Li and Vert (1995) investigated the hydrolysis behavior of massive PLA-based materials, and demonstrated that the core-accelerated hydrolysis occurs by the catalytic oligomers formed by hydrolysis. During the latter 1990s, in addition to medical applications as scaffolds for tissue regeneration, PLAs have been attracting much attention in terms of ecological applications, and have been recognized as being substitutes of the commercial "petro" polymers because they can be produced from renewable resources such as starch, and at low cost (ca. US$ 1–2 per kg).

3
Synthesis and Purification of Monomers

Lactic acid is a well-known, simple hydroxy acid with an asymmetric carbon atom, and is synthesized by the bacterial fermentation of carbohydrates such as sugar from starch. As mentioned earlier, lactic acids are optically active compounds, including L- and D-forms. The melting temperature (T_m) of a 1:1 racemic mixture (compound) of L- and D-lactic acids is 52.8 °C, which is higher than that of the respective enantiomers (16.8 °C). The cyclic dimers of lactic acids, LAs, have three different forms; L-lactide (LLA) with two L-lactyl units, D-lactide (DLA) with two D-lactyl units, and meso-lactide (MLA) with one L-lactyl unit and one D-lactyl unit. Racemic lactide or DL-lactide (DLLA) is a 1:1 physical mixture or a 1:1 racemic compound (stereocomplex) of LLA and DLA, and has a T_m (124 °C) higher than that of LLA (95–99 °C), DLA (95–99 °C), and MLA (53–54 °C). The structures and T_m values of lactic acids and LAs are shown in Figure 1 (Tsuji and Ikada, 1999a). The laboratory method for LA synthesis is thermal depolymerization of low-molecular weight (LMW) PLAs, followed by distillation of LAs under reduced pressure. The crude LAs thus obtained are normally purified by recrystallization using ethyl acetate as a solvent. Purification of LAs is indispensable to increase the molecular weight of the resulting polymers. The industrial procedures for synthesis and purification of lactic acids and LAs have been described in detail (Kharas et al., 1994; Hartmann, 1998).

4
Synthesis of Polymers

A variety of LA homo- and copolymers having different molecular structures and, therefore, different physical properties, hy-

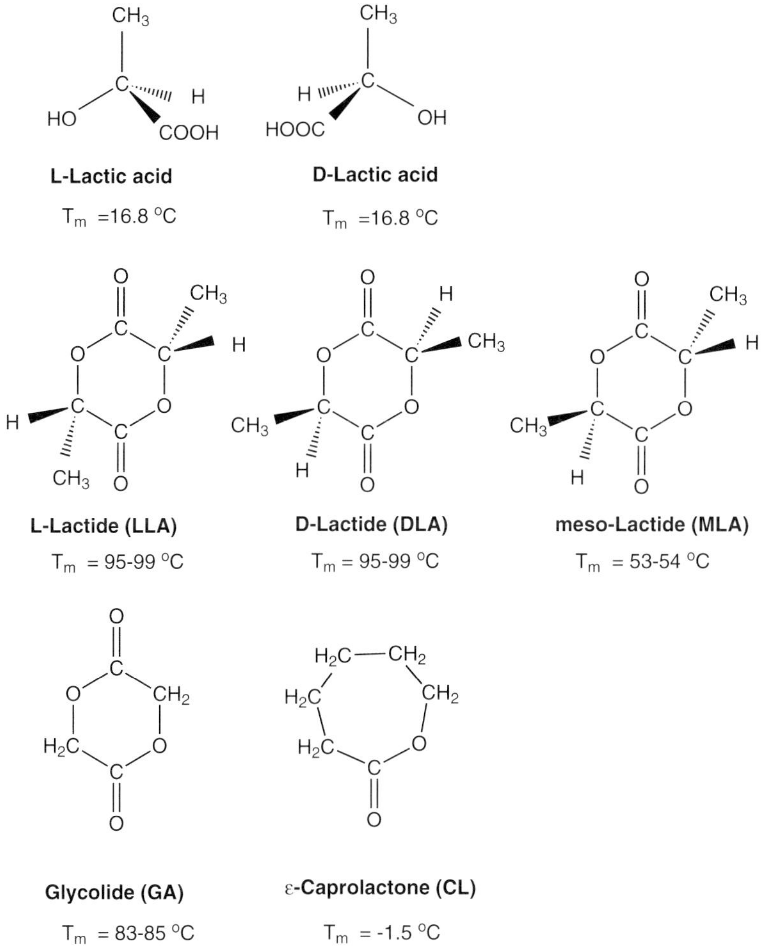

Fig. 1 Structure and melting point (T_m) of L- and D-lactic acids and lactides (LAs), glycolide (GA), and ε-caprolactone (CL).

drolyzability, and biodegradability have been synthesized using LAs, lactic acids, and comonomers.

4.1 Homopolymers

PLAs can be synthesized by two methods, as illustrated in Figure 2: (1) Polycondensation of lactic acids, and (2) ROP of LAs (Tsuji and Ikada, 1999a). The catalysts reported to be effective in increasing the molecular weight of the resulting PLAs up to the order of 10^5 and 10^6 for polycondensation (Ajioka et al., 1995; Moon et al., 2000) and for ROP (Lillie and Schulz, 1975; Kohn et al., 1984; Leenslag and Pennings, 1987b; Kricheldorf and Sumbél, 1989), respectively, are tin-based catalysts such as tin (II or IV) chloride and tin (II) bis-2-ethylhexanoic acid (stannous octoate or tin octoate). These tin-based catalysts are generally used because of their high solubility in LAs and PLA oligomers and low toxicity; moreover, they have re-

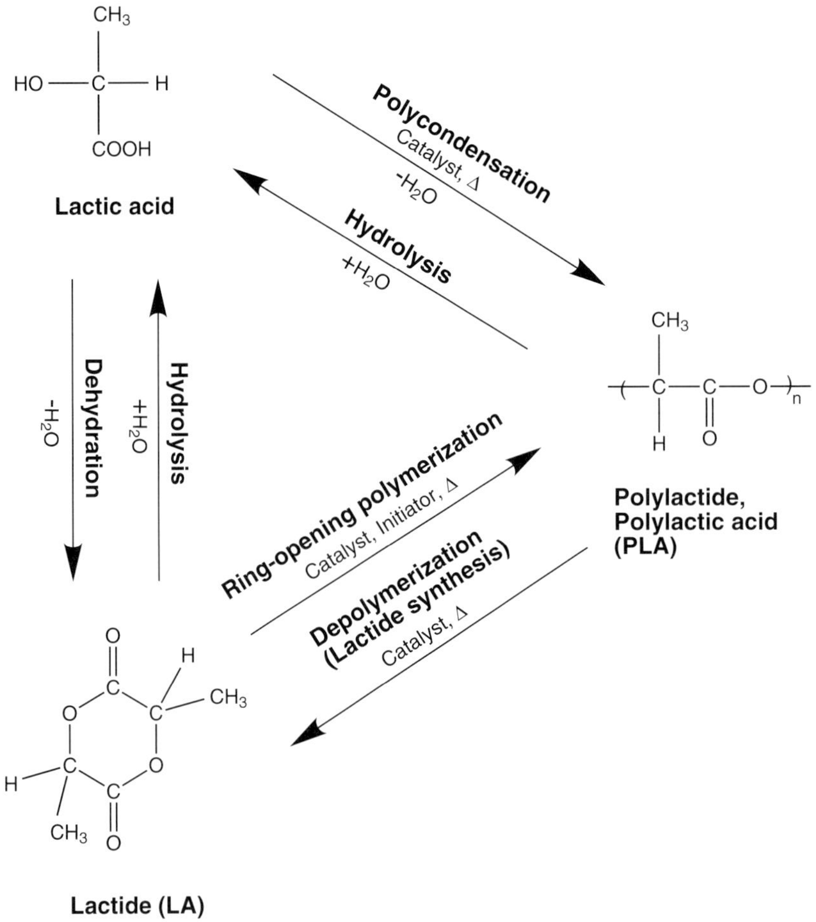

Fig. 2 Schematic representation of polycondensation and dehydration of lactic acid, synthesis and ring-opening polymerization of lactide (LA), and hydrolysis of polylactide, polylactic acid (PLA) and LA.

ceived approval from the FDA. Bulk polymerization in the presence of these catalysts is favorable to avoid the problem of racemization and transesterification during the ROP of lactide in solution. In contrast, aluminum alkoxides are known to be effective catalysts to synthesize mono-dispersed PLAs (Trofimoff et al., 1987) and their block-copolymers, utilizing the "living" nature of polymerized chains (Song and Feng, 1984). Recently, PLAs with weight-average molecular weight (M_w) as high as 2.7×10^5 and oligomeric lactic acids have been synthesized by lipase-catalyzed ROP of LAs (Matsumura et al., 1998) and polycondensation of lactic acids (Ohya et al., 1995), respectively.

Hyon et al. (1997) synthesized poly(DL-lactic acid) with a molecular weight range of 2×10^3 to 2×10^4 by polycondensation varying the reaction temperature, time, and pressure. PLAs with relatively high molecular weights are obtained by polycondensation only when water is efficiently removed from the polymerization mixture of lactic acid, PLA, and water. Such conditions involve: (1) a temperature range of 180–200 °C; (2) low pressure (< 5 mmHg); (3) long reaction time; and (4) the addition of an appropriate catalyst and azeotropic solvent of water. Ajioka et al. (1995) and Moon et al. (2000) synthesized poly(L-lactic acid) having a M_w of more than 1×10^5 using tin-based catalysts and with diphenylether as an azeotropic solvent, and using those activated by proton acids, respectively. Too high a reaction temperature results in the reduction of optical purity of the PLAs, even when optically pure L- or D-lactic acid is used as a monomer. In contrast, PLAs with very high molecular weights ($M_w > 1\times10^6$) can be synthesized using the ROP method under the following conditions (Leenslag and Pennings, 1987b): (1) low concentration of initiator; (2) relatively low temperature (< 130 °C); and (3) long reaction time. The molecular weight of PLAs synthesized by ROP of LAs can be varied by altering the concentration of initiator having a hydroxyl group, such as alcohols (Schindler et al., 1982; Tsuji and Ikada, 1999a).

PLA composed of equimolar L- and D-lactyl (a half of lactide) units, can be synthesized by polycondensation of DL-lactic acid and ROP either of DLLA or MLA (Schindler and Gaetano, 1988; Kricheldorf and Boettcher, 1993). The monomer distribution and tacticity of this stereocopolymer varies depending on the monomer, the nature of the polymerization catalyst, and the polymerization conditions such as temperature and time (Kasperczyk, 1995). Recently, Ovitt and Coates (1999) successfully synthesized syndiotactic PLA (syn-PLA) from MLA using an aluminum alkoxide, whilst Chisholm et al. (1997) proposed new assignments of resonance lines to the monomer sequences in ^{13}C- and ^{1}H-NMR spectra of poly(DL-lactide) (PDLLA) and poly(meso-lactide) (PML) in the region of the methine group (Table 1).

PLAs with different L-lactyl unit contents can be prepared by altering the enantiomeric excess of the monomers: L- and D-lactic acids (Fukuzaki et al., 1989), LLA and DLA (Tsuji and Ikada, 1992), and LLA and MLA (MacDonald et al., 1996). The stereoblock-copolymers were synthesized by two-step living ROP of LLA with DLA or DLLA (Yui et al., 1990; Spinu et al., 1996), and also by single-step ROP of DLLA using Shiff's base/aluminum methoxide as initiator (Sarasua et al., 1998; Spassky et al., 1998).

4.2 Linear Copolymers

Since the copolymers from LAs and glycolide (GA) are described in detail in Chapter 6 of this volume, the synthesis of LA copolymers is briefly outlined here. Typical comonomers of LAs often used for ROP are GA and CL (see Figure 1). The copolymers of LA with GA or CL, as well as the LA homopolymers and stereocopolymers, have been intensively studied as matrices for drug delivery systems (DDS) (Lewis, 1990; Pitt, 1990) and biodegradable scaffolds for tissue regeneration (Kharas et al., 1994; Agrawal et al., 1997). The other comonomers utilized for copolymer preparation have been summarized in reviews (for example, Kharas et al., 1994; Hartmann, 1998).

Similarly to PLA stereocopolymers, the monomer distribution in the copolymers depends strongly on the monomer pairs, the nature of catalysts, and the polymerization conditions. The assignments of resonance lines to monomer sequences in ^{13}C- and ^{1}H-NMR spectra of poly(L-lactide-*co*-glycolide) [poly(LLA-GA)] (Kasperczyk, 1996) or poly(L-lactide-*co*-ε-caprolactone) [poly(LLA-CL)] (Kasperczyk and Bero, 1991) are summarized in Table 1. The proton resonance lines that are useful in determining the respective monomer contents of the copolymers are those at around 5.2 ppm for methine group of the lactyl unit, at 4.8 ppm for the meth-

Tab. 1 Assignments of resonance lines to monomer sequences in ^{13}C- and ^{1}H-NMR spectra of stereocopolymers from DLLA or MLA (Chisholm et al., 1997), poly(LLA-GA) (Kasperczyk, 1996), and poly(LLA-CL) (Kasperczyk and Bero, 1991)

Polymer	*Nuclei*	*Monomer sequence*[a–d]	*Chemical shift [ppm]*
Stereocopolymers from DLLA or MLA	Methine carbon	iss/ssi	69.22
		sss	69.17
		sii/iis, sis	69.10
		ssi/iss	68.99
		iii,isi,iis/sii	68.91
	Methine proton	sis,iss/ssi	5.100
		sii,iis	5.089
		iis/sii	5.044
		iii	5.035
		sss, isi, ssi, iss	5.027
Poly(LLA-GA)	Lactyl carbon	LLGG	168.89
		LLLL	168.82
		GLG	168.76
	Glycolyl carbon	GGGG	166.35
		GGLL	166.21
	Methylene protone	GLGGG, GGGLG	4.96
		LGGLG, GLGGL	4.94
		GGGG	4.90
		LLGGL, LGGLL	4.88
		GGGGL, LGGGG	4.87
		LLGGG, GGGLL	4.86
		LLGLL, GLGLL, LLGLG, GLGLG	4.83
		GGGLG, GLGGG	4.80
		LGGGL, GLGGL, LGGLG	4.78
Poly(LLA-CL)	Lactyl carbon	C$\underline{L}$C	170.83
		LLL$\underline{L}$C, CLL$\underline{L}$C	170.32
		CL$\underline{L}$C	170.26
		C$\underline{L}$LC	170.22
		C$\underline{L}$LLC	170.11
		C$\underline{L}$LLL	170.08
		LL$\underline{L}$LC	169.72
		CL$\underline{L}$LC	169.67
		CL$\underline{L}$LL	169.57
		LL$\underline{L}$LLL	169.57
	Caprolactone carbon	C$\underline{C}$C	173.49
		CL$\underline{C}$C	173.44
		LL$\underline{C}$C	173.42
		C$\underline{C}$LL	172.84
		LL$\underline{C}$LL	172.77
		CL$\underline{C}$LC, LL$\underline{C}$LC	172.72
		C$\underline{C}$LC	172.53

[a]Sequences are given along the chain direction of CO to O; [b]s and i represent syndiotactic and isotactic sequences, respectively; [c]G and L represent glycolyl (a half of GA) and L-lactyl (a half of LLA) units, respectively; [d]C and L represent CL and L-lactyl (a half of LLA) units, respectively.

ylene group of the glycolyl unit, and at 4.1 ppm for the γ-methylene group of the CL unit (Tsuji and Ikada, 1994, 2000a). The monomer sequences of poly(LLA-GA) (Gilding and Reed, 1979; Kricheldorf and Kreiser, 1987a) and poly(LLA-CL) (Kricheldorf and Kreiser, 1987b; Choi et al., 1994) are known to be blocky and difficult to be randomized (Shen et al., 1996). Two-step block copolymerization was performed for LLA or DLLA with GA (Rafler and Jobman, 1994) or with CL (Song and Feng, 1984) to obtain their block copolymers. Recently, (*R*)-β-butyrolactone(B) was used as comonomer of LLA(A), and their relatively random (Hori et al., 1993) and A-B-A block (Hiki et al., 2000) copolymers were synthesized, while A-B-A block copolymers was prepared from LLA(A) and 1,5-dioxepan-2-one(B) by Stridsberg and Albertsson (2000).

Hydrophobic LA sequences have been copolymerized with hydrophilic sequences such as poly(ethylene oxide) (PEO) (Younes and Cohn, 1987), poly(propylene oxide) (Kimura et al., 1989), and their block copolymer (Pluronic™) (Yamaoka et al., 1999) to increase the hydrophilicity, flexibility, and biodegradability of PLA-based materials.

4.3 Graft Copolymers

Graft copolymers can be divided into two types: (1) PLAs or their copolymers grafted polymers; and (2) other polymers grafted PLAs, or their copolymers. The former examples involve PLLA, PDLLA, poly(DLLGA), or poly(LLA-CL) grafted hydrophilic dextrans (Li et al., 1997; Youxin et al., 1998), PVA (Onyari and Huang, 1996; Breitenbach and Kissel, 1998), pullulan (Ohya et al., 1998a), amylose (Ohya et al., 1998b), dextran (de Jong, 2000), hydrophobic vinyl polymers (Eguiburu, 1996), polyurethane (Hsu and Chen, 2000) and poly(*p*-xylylene)s (Agarwal et al., 1999), whilst the latter examples include hydrophilic poly(acrylamide) (Tsuchiya et al., 1993), poly(acrylic acid) (Södergård, 1998), and PEO (Cho et al., 1999) grafted PLLA. The proposed methods of preparing the graft copolymers containing PLAs and their copolymers chains were: (1) copolymerization of macromonomers of PLAs or their copolymers having a vinyl group with vinyl monomers; (2) ROP of LAs initiated by macromolecular polyols such as PVA and polysaccharides; and (3) irradiation-initiated polymerization of LAs from a polymer, and that of vinyl monomers or alkene oxide from PLAs or their copolymers. In some of the above-mentioned studies, graft polymerizations were performed on the surfaces of bulk polymeric materials.

4.4 Branched Polymers

Branched PLAs can be prepared by the ROP of LAs initiated with branched polyols, which was proposed by Schindler et al. (1982). The polyols utilized are glycerin (Arvanitoyannis et al., 1995), pentaerythritol (Kim and Kim, 1994), and sorbitol (Arvanitoyannis et al., 1996). Atthoff et al. (1999) synthesized well-defined 2-, 4-, 6-, and 12-arm PLLAs and PDLLAs from L- or DL-lactic acid using 2,2′-bis(hydroxymethyl)propanoic acid derivatives (hydroxy functional dendrimers) as initiators.

4.5 Cross-linked Polymers

The cross-linked PLAs were prepared by copolymerization of LAs with bicyclolactones or carbonates (Grijpma et al., 1993; Nijenhuis et al., 1996a), whilst Storey and coworkers (Storey et al., 1993; Storey and Hickey, 1994) synthesized a network structure containing PLA chains by copolymerization of methyl methacrylate and styrene with LMW three-arm methacrylate-endcapped PDLLA and by cross-linking of LMW three-arm PDLLA using tolylene-2,6-diisocyanate. Nijenhuis et al. (1996a) also investigated two methods to introduce cross-links into PLA molecules, and found that dicumyl peroxide and electron beam irradiation were effective and non-effective, respectively, for cross-linking.

5 Polymer Blending

The blending of PLAs with other polymers is commercially advantageous in order to obtain biodegradable materials with a wide variety of physical properties and hydrolysis, biodegradation, and drug-release behaviors. For this purpose, numerous polymer blends have been prepared, and these PLAs – or their copolymer-based polymer blends – can be divided into six groups, as shown in Table 2. Among these polymer blends, stereocomplexationable polymer blends have been intensively studied because of their specific properties in solution as well as in bulk (see below; Tsuji, 2000a). The miscibility of PLA-based blends has attracted much attention, and some miscible polymer blends of PLAs and their copolymers with other polymers may even become immiscible as the polymer molecular weights are increased because of decreased polarity of polymer chains and the entropy of mixing. The reported miscible polymer blends are those of PLLA with either poly[(*R*)-3-hydroxybutyrate] (R-PHB) (Koyama and Doi, 1997) or poly[(*R*,*S*)-3-hydroxybutyrate] (R,S-PHB) (Ohkoshi et al., 2000) and those from PDLLA and poly(1,5-dioxepan-2-one) (PDXO) (Edlund and Albertsson, 2000) where the number average molecular weight (M_n) of PLLA or R,S-PHB is as low as 9×10^3, and that of PDXO as low as 2.1×10^4.

6 Additives

Of the reported additives, hydroxyapatite – which is a component of mammalian bones – has been the most frequently and widely utilized for PLAs and copolymers to increase the biocompatibility and bending modulus (Hyon et al., 1985) and to accelerate bone regeneration (Higashi et al., 1986) when utilized as a scaffold for bone regeneration.

Other additives are plasticizers such as glycerin to decrease Young's modulus and increase elongation-at-break and drug-release rate (Pitt et al., 1979), cyclic monomer LAs to enhance hydrolysis (Nakamura et al., 1989), catalyst deactivators to reduce lactide formation during thermal processing (Hartmann, 1999), and nucleation reagents such as talc to accelerate crystallization (Kolstad, 1996). Renstad et al. (1998) found that the addition of SiO_2 had no significant effects on thermal degradation during processing and biodegradation, while the addition of $CaCO_3$ reduced the rates of thermal degradation during processing and biodegradation.

7 Fiber-reinforced Plastics

PGA, PLLA, and PDLLA fibers have been utilized to reinforce PLLA matrices (Vainionpää et al., 1989; Törmälä, 1992). Among these reinforced PLLA materials, those reinforced by PLLA, i.e., self-reinforced (SR) PLLA, can avoid interfacial problems occurring at the interface between the two different materials. SR-PLLA has a bending modulus of about 250–270 MPa, which is much higher than the 57 to 145 MPa of non-reinforced PLLA and this decreases to 10 MPa after 1 year of *in vivo* degradation (Majola et al., 1992).

Tab. 2 Polymer blends based on PLAs and their copolymers

Polymer pairs Biodegradable polymers		PLAs and copolymers	References
PLAs (Self blends)	HMW PDLLA	LMW PDLLA	Bodmeier et al. (1989); Asano et al. (1991); Mauduit et al. (1996)
	HMW PLLA	LMW PLLA	Von Recum et al. (1995)
PLAs (Stereocomplexationable enantiomeric blends)	PDLA (or copolymers with DLA monomer sequences)	PLLA (or copolymers with LLA monomers sequences)	Ikada et al. (1987); Murdoch and Loomis (1988a,b, 1989; Yui et al. (1990); Loomis et al. (1990); Loomis and Murdoch (1990); Dijkstra et al. (1991); Tsuji et al. (1991a–c, 1992a,b, 1994, 2000a); Tsuji and Ikada (1992, 1993, 1994, 1996a, 1999a,b); Tsuji (2000a,c); Tsuji and Miyauchi (in press); Tsuji and Suzuki (2000); Okihara et al. (1991); Brochu et al. (1995); Stevels et al. (1995); Brizzolara et al. (1996a,b); Spinu et al. (1996); Cartier et al. (1997) Sarasua et al. (1998); de Jong et al. (1998, 2000)
PLAs (Non-stereocomplexationable diastereoisomeric blends)	PDLLA	PLLA	Jorda and Wilkes (1988) Tsuji and Ikada (1992, 1995a, 1996b, 1997); Tsuji and Miyauchi (2000c); Serizawa et al. (2001)
		PDLA	Tsuji and Ikada (1992, 1997); Tsuji and Miyauchi (2000c)
		PLAs with different optical purities	Tsuji and Ikada (1992)
	LLA-rich PLAs	PLLA	Tsuji and Ikada (1992)
	DLA-rich PLAs	PDLA	Tsuji and Ikada (1992)
LA or lactic acid Copolymers	poly(LLA-GA)	PLLA	Cha and Pitt (1990); Tsuji and Ikada (1994)
	poly(DLA-GA)	PDLA	Tsuji and Ikada (1994); Matsumoto et al. (1997)
	poly(LLA-CL)	PLAs with different optical purities	Grijpma et al. (1994)
		PLLA	Hiljanen-Vainio et al. (1996)
	poly(LLA-L[illegible])	PLLA	Cook et al. (1997)
Aliphatic polyesters	PCL	PLLA	Hiljanen-Vainio et al. (1996); Yang et al. (1997); Lostocco et al. (1998); Tsuji and Ikada (1998a); Tsuji et al. (1998); Wang et al. (1998); Kim et al. (2000)
		PDLLA	Domb (1993); Zhang et al. (1995); Tsuji and Ikada (1996c) Gan et al. (1999); Meredith and Amis (2000); Aslan et al. (2000)
	R-PHB	PLLA	Blümm and Owen (1995); Koyama and Doi (1997); Yoon et al. (2000)
		PDLLA	Domb (1993); Zhang et al. (1996); Koyama and Doi (1995)
		poly(DLLA-CL)	Koyama and Doi (1996)
		PDLLA-b-PEO	Zhang et al. (1995, 1997)
	R,S-PHB	PLLA	Ohkoshi et al. (2000)
	poly(HB-HV)	PLLA	Iannace et al. (1994, 1995)
	PHO	PDLLA	Mallardé et al. (1998)
	PHMS	PLLA	Lostocco and Huang (1997)
	PBS	PLLA	Inoue et al. (1998)
	PEBS	PLLA	Liu et al. (1997)
	PCL-b-PEO	PDLLA-b-PEO	Zhang et al. (1995)
Miscellaneous Biodegradable polymers	Cellulose	PLLA	Nagata et al. (1998)
	Ny6	PLLA	Hu et al. (1994, 1995)
	PDXO	PDLLA	Edlund and Albertsson (2000)
	PEO	PLLA	Younes and Cohn (1988); Domb (1993) Nakafuku and Sakoda (1993); Nakafuku (1994, 1996); Nijenhuis et al. (1996b) Sheth et al. (1997); Yang et al. (1997); Tsuji et al. (2000a)
		PDLLA	Domb (1993)
		poly(LA-GA)	Domb (1993)
	Pluronic	PLLA	Park et al. (1992)
	PPHOS	poly(LA-GA)	Ibim et al. (1997)

Tab. 2 (cont.)

Polymer pairs *Biodegradable polymers*	*PLAs and copolymers*	*References*
PSA	PDLLA	Domb (1993); Shakesheff et al. (1994); Davies et al. (1996)
		Chen et al. (1998)
	PLA-GA	Domb (1993)
PVA	PLLA	Tsuji and Muramatsu (2001a,b)
	poly(DLLA-GA)	Pitt et al. (1992, 1993)
PVAc	PLLA	Gajria et al. (1996)

HMW: high-molecular weight; LMW: low-molecular weight; Ny6: poly(ε-aminocaproic acid); PBS: poly(butylene succinate); PCL: poly(ε-caprolactone); PCL-b-PEO: poly(ε-caprolactone)-b-poly(ethylene oxide); poly(DLA-GA): poly(D-lactide-co-glycolide); PDLLA: poly(DL-lactide); poly(DLLA-GA): poly(DL-lactide-co-glycolide); PDLLA-b-PEO: poly(DL-lactide)-b-poly(ethylene oxide); PDXO: poly(1,5-dioxepan-2-one); PEBS: poly(ethylene/butylene succinate); PEO: poly(ethylene oxide); poly(HB-HV): poly[(R)-3-hydroxybutyrate-co-3-hydroxyvalerate]; PHMS: poly(hexamethylene succinate); PHO: poly(3-hydroxyoctanoate); PLA: poly(lactide), poly(lactic acid); poly(LA-GA): poly(lactide-co-glycolide); PLLA: poly(L-lactide), poly(L-lactic acid); poly(LLA-CL): poly(L-lactide-co-ε-caprolactone); poly(LLA-GA): poly(L-lactide-co-glycolide); poly(LLA-Ly): poly(L-lactide-co-lysine); Pluronic: ABA triblockcopolymers of polyethylene oxide (A) and poly(propylene oxide) (B); PPHOS: poly(phosphazene); PSA: poly(sebacic anhydride); PVA: poly(vinyl alcohol); PVAc: poly(vinyl acetate); R-PHB: poly[(*R*)-3-hydroxybutyrate]; R,S-PHB: poly[(*R*,*S*)-3-hydroxybutyrate]

8 Molding

Solidification methods of polymers include: (1) melt-molding; (2) solution-casting or solvent evaporation; and (3) non-solvent precipitation. Of these methods, the melt molding approach is most favored for the preparation of biodegradable materials for ecological, medical, and pharmaceutical applications, as it does not require any solvents such as chloroform and methylene chloride, which are toxic and harmful to the environment and to human beings.

8.1 Thermal Treatments

The parameters for thermal treatment or crystallization from the melt include: (1) pretreatments prior to crystallization, such as melting and quenching from the melt; (2) melting temperature and time; (3) crystallization or annealing temperature and time (T_a and t_a, respectively); and (4) drawing. However, crystallization and orientation at a high temperature for a long time will cause a large decrease in molecular weight and optical purity due to thermal degradation and racemization of optically active LA units, respectively.

8.2 Pore Formation

Porous biodegradable materials are reported to enhance tissue regeneration, if they have appropriate pore size and porosity (Coombes and Meikle, 1994; Thomson et al., 1995a,b). The reported methods of preparing porous PLAs and their copolymers involve the removal of inorganic salts or organic low-molecular weight compounds from the melt-molded, solution-cast, gelled (Coombes and Heckman, 1992a,b; Coombes and Meikle, 1994), or frozen mixtures (Whang et al., 1995) of PLAs with additives or solvents and from the phase-separated mixtures of PLA/solvent/non-solvent systems (van de Witte et al., 1996a–f, 1997; Zoppi et al., 1999). Hence, a new method was proposed to prepare porous biodegradable polyesters by water-extraction of a water-soluble polymer such as PEO from their blends. It became evident that the pore size and porosity of the PLLA and poly(ε-caprolactone) (PCL) films are controllable by changing their mixing ratio and the molecular weight of the water-soluble polymer (Tsuji et al., 2000a; Tsuji and Ishizaka, 2001a). Another proposed novel method is to utilize selective enzymatic removal of one component from the aliphatic polyester blends (Tsuji and Ishizaka, 2001b).

9 Highly Ordered Structures

The highly ordered structures of PLAs include: (1) crystallinity (x_c); (2) crystalline thickness (L_c); (3) crystal structure; (4) spherulitic size and morphology; and (5) molecular orientation. As detailed in the literature, the highly ordered structures (1)–(5) of PLLA can be altered to some extent by varying the above-mentioned parameters for thermal treatment (Tsuji and Ikada, 1995b). Of these highly ordered structures, x_c and L_c are known to have crucial effects on the mechanical properties as well as the hydrolysis and biodegradation behavior of PLAs (Tsuji, 2000b).

9.1 Crystallization

PLAs and their copolymers crystallize into homo- and stereocomplex (racemic) crystallites in bulk as well as in solutions, in so far as their tacticities are sufficiently high. The homocrystallites consist of either LLA or DLA unit sequences, whereas the stereocomplex crystallites comprise both LLA and DLA unit sequences. PLAs and their copolymers can form eutectic and epitaxial crystallites with low-molecular weight organic compounds.

9.1.1 Homo-crystallization

The general crystallization behavior of PLLA in bulk and in solution has been studied by Fischer et al. (1973), Kalb and Pennings (1980), and Vasanthakumari and Pennings (1983). When crystallized in bulk directly from the melt, the PLLA spherulite growth rate becomes higher with decreasing PLLA molecular weight, and reaches a maximum at crystallization temperature around 130 °C (ca. 5 and 2.5 μm min^{-1} for PLLA with viscosity-average molecular weight (M_v) of 1.5 and 6.9×10^5, respectively) (Vasanthakumari and Pennings, 1983). However, the total crystallization rate monitored by total x_c increases with decreasing T_a, because the spherulite density increases with a decrease in T_a for a T_a range of 100–160 °C (Tsuji and Ikada, 1995b), and its maximum is observed at about 105 °C (Iannace and Nicolais, 1996). Vasanthakumari and Pennings (1983) found that the transition from regime II crystallization to regime I crystallization for PLLA occurs above 163 °C when monitored by spherulite growth rate, while Mazzullo et al. (1992) showed that the transition from regime III crystallization to regime II crystallization occurs above 140 °C when traced by total x_c and T_m of PLLA. Spherulite growth rate decreases and crystallization half-time becomes longer in the presence of comonomer unit in a polymer chain, as revealed for poly(L-lactide-*co*-meso-lactide) (Kolstad, 1996; Huang et al., 1998). The crystallization half-time and the time required for the completion of crystallization is reported to be dramatically decreased by addition of talc as a nucleating agent (Kolstad, 1996), and by quenching from the melt before crystallization (Tsuji and Ikada, 1995b), due to increased nuclear density of the spherulites.

Crystallization does not occur when the sequence length of L- or D-lactyl units becomes smaller than a critical value, which depends on the crystallization conditions and comonomer unit. The reported values are ca. 14 (Sarasua et al., 1998) and 15 (Tsuji and Ikada, 1996a) lactyl units for LA stereocopolymers from LLA and DLA, assuming random addition of LA units during polymerization without transesterification, whereas poly(LLA-GA) or poly(DLA-GA) has a rather small critical sequence length of nine lactyl units when calculated under the same assumption (Tsuji and Ikada, 1994). When LLA polymer contains flexible CL units, the critical L-lactyl unit sequence length is lowered to 7.2, and the copolymers are crystallizable even at room temperature (25 °C) (Tsuji and Ikada, 2000a). Fischer et al. (1973) showed that DLA units of LLA-rich PLAs are partially included in the crystalline region during formation of their single crystals, while Zell et al. (1998) revealed that LLA content in the crystalline region is almost the same as that in the amorphous region for a crystallized DLA-rich PLA.

The crystallization of homopolymer PLLA or PDLA will proceed in the polymer blends as if it were in the nonblended specimens as far as the constituent polymers are phase-separated (Tsuji et al., 1998), while PLLA will not crystallize in the partially miscible polymer blends with amorphous PDLLA at low PLLA contents (Tsuji and Ikada, 1995a). The induction period until the start of crystallization, spherulite growth rate, and spherulite density of PLLA in the presence of PDLLA depends on the molecular weight of PDLLA (Tsuji and Ikada, 1996b) and the polymer mixing ratio (Tsuji and Ikada, 1995a).

9.1.2 Stereocomplexation (Racemic Crystallization)

Ikada et al. (1987) found that stereocomplexation or racemic crystallization takes place between LLA unit and DLA sequences. Numerous studies have been performed concerning this stereocomplex ever since, as shown in Table 2 (Tsuji, 2000a). Stereocomplex crystallites (racemic crystallites) of PDLA and PLLA have a melting temperature 50 °C higher than those of homocrystallites either of PLLA or PDLA. The stereocomplexation between PDLA and PLLA is known to occur in solution as well as in bulk from the melt. Stereocomplex crystallization from the melt completes in the blend specimen in a period as short as 2 min at 140 °C, probably due to the high-density nuclear formation of the stereocomplex spherulites, while it takes a much longer time (1 h) to complete homocrystallization in the nonblended PLLA or PDLA specimens at the same temperature (Tsuji and Ikada, 1993).

Because of the stoichiometric ratio of 1:1 (L-lactyl unit:D-lactyl unit) of the stereocomplex crystallites (see below), the excessive L- or D-lactyl unit sequences crystallize in homo-crystallites which consist of either L- or D-lactyl unit sequences, when the PDLA content [$X_D = PDLA/(PLLA + PDLA)$] deviates from 0.5. Other important parameters affecting the stereocomplexation and homocrystallization are molecular weight and sequence lengths of L-lactyl and D-lactyl units of the polymers, and the solidification method and temperature (Tsuji, 2000a). Spinu and Gardner (1994) proposed a new approach of stereocomplexation during polymerization of monomer/polymer blends, while Radano et al. (2000) prepared 1:1 mixture containing stereocomplex phase directly by stereoselective polymerization of DLLA with a racemic catalyst.

9.1.3
Eutectic Crystallization

PLLA crystallizes into the eutectic crystals with pentaerythrityl tetrabromide (Vasanthakumari, 1981) and hexamethylbenzene (HMB) (Zwiers et al., 1983). Tonelli (1992) and Howe et al. (1994) showed that PLLA forms an inclusion compound with urea, where PLLA chains are trapped in the channels of the urea.

9.1.4
Epitaxial Crystallization

Cartier et al. (2000) revealed that PLA stereocomplex crystallizes epitaxially on HMB, while Brochu et al. (1995) reported that the stereocomplex crystallites are formed on those of homocrystallites composed either of L-lactyl or D-lactyl unit sequences in the blends from PLLA and D-lactide-rich PLA (D-lactide content = 80%) when the stereocomplex crystallization is slow and homocrystallization is rapid.

9.2
Crystal Structure

The cell parameters for non-blended PLLA and PLA stereocomplex are summarized in Table 3. Non-blended PLLA has been reported to crystallize in three forms: α (De Santis and Kovacs, 1968; Hoogsteen et al., 1990; Kobayashi et al., 1995), β (Hoogsteen et al., 1990; Puiggali et al., 2000), and γ (Cartier et al., 2000). The β-form of non-blended PLLA is reported to have a rather frustrated structure (Puiggali et al., 2000). The stereocomplex crystal has a triclinic unit cell, where the PLLA and PDLA chains taking a 3_1 helical conformation are packed side-by-side in parallel fashion.

Tab. 3 Unit cell parameters reported for nonblended PLLA and stereocomplex crystals

	Space group	*Chain orientation*	*Number of helices per unit cell*	*Helical conformation*	*a* [nm]	*b* [nm]	*c* [nm]	*α* [deg.]	*β* [deg.]	*γ* [deg.]
PLLAα-form (De Santis and Kovacs, 1968)	Pseudo-orthorhombic	–	2	10_3	1.07	0.645	2.78	90	90	90
PLLAα-form (Hoogsteen et al., 1990)	Pseudo-orthorhombic	–	2	10_3	1.06	0.61	2.88	90	90	90
PLLAα-form (Kobayashi et al., 1995)	Orthorhombic	–	2	10_3	1.05	0.61	2.88	90	90	90
PLLAβ-form (Hoogsteen et al., 1990)	Orthorhombic	–	6	3_1	1.031	1.821	0.90	90	90	90
PLLAβ-form (Puiggali et al., 2000)	Trigonal	Random up-down	3	3_1	1.052	1.052	0.88	90	90	120
PLLAγ-form (Cartier et al., 2000)	Orthorhombic	Antiparallel	2	3_1	0.995	0.625	0.88	90	90	90
Stereocomplex (Okihara et al., 1991)	Triclinic	Parallel	2	3_1	0.916	0.916	0.870	109.2	109.2	109.8

9.3 Morphology

The highly organized structures such as spherulites, single crystals, and gels are formed as result of the crystallization of PLAs and their copolymers.

9.3.1 Spherulites

The size and morphology of PLLA spherulites depend on the parameters such as crystallization temperature and time, copolymerization, and blending. PLLA spherulites become smaller with decreasing temperature and time (Marega et al., 1992; Tsuji and Ikada, 1995b). The spherulites of LA stereocopolymers retain well-defined structure, even when the monomer unit sequence length approaches the critical value for crystallization (Tsuji and Ikada, 1996a), whereas rather disturbed spherulites or assemblies of the crystallites were noted in the crystallizable PLLA/amorphous PDLLA blends having low PLLA contents (Tsuji and Ikada, 1995a). The morphology of the stereocomplex spherulites is similar to that of normal spherulites of the nonblended PDLA or PLLA when solely stereocomplexation takes place. However, the spherulites with complicated morphology are noted when stereocomplexation and homocrystallization occur simultaneously (Tsuji and Ikada, 1993, 1996a). The formation of isolated or aggregated stereocomplex spherulites occurs in suspended state in an acetonitrile mixed solution of PLLA and PDLA (Tsuji et al., 1992a).

9.3.2 Single Crystals

PLLA is reported to crystallize into lozenge-like (Fischer et al., 1973; Kalb and Pennings, 1980; Miyata and Masuko, 1997; Iwata and Doi, 1998) and hexagonal (Kalb and Pennings, 1980; Iwata and Doi, 1998) single crystals in dilute solutions, while single crystals of the stereocomplex of PLLA and PDLA having a peculiar triangular shape are formed in *p*-xylene when the solution concentration is as low as 0.04% (Okihara et al., 1991; Tsuji et al., 1992a). The mechanism of formation of the stereocomplex crystal has been proposed by Brizzolara et al. (1996a).

9.3.3 Gels

When the stereocomplexation of PDLA and PLLA occurs in concentrated chloroform and aqueous solution, the solution viscosity increases with time, and finally three-dimensional (3D) gelation or formation of microgels occurs as a result of the formation of stereocomplex microcrystallites which act as cross-links (Tsuji et al., 1991a; Tsuji, 2000c; de Jong et al., 2000).

9.3.4 Phase Structure

The microstructure analysis using high-resolution, solid-state ^{13}C-NMR spectroscopy was performed for nonblended PLLA and PLA stereocomplex specimens. Thakur et al. (1996) found five distinct resonance lines in the carbonyl region of the nonblended PLLA, which indicates the existence of five or more crystallographically inequivalent sites in the crystalline region of PLLA. These authors showed that solid-state NMR can be utilized to evaluate the x_c of PLLA and LA stereocopolymers. By contrast, we demonstrated from an evaluation of the spin lattice relaxation times (T_{1C}s) that the stereocomplexed PLA is composed of four regions: the racemic crystalline regions in both rigid and disordered states, the homocrystalline region, and the amorphous region (Tsuji et al., 1992b).

Tab. 4 Physical properties of some biodegradable aliphatic polyesters

	PLLA	*PDLLA*	*Syn-PLA*	*PLA stereocomplex*	*PCL*	*R-PHB*	*PGA*
T_m (°C)	170–190	–	151 Ovitt and Coates (1999)	220–230 Ikada et al. (1987)	60	5	225–230
T_m^0 (°C)	205 Tsuji and Ikada (1995b) 215 Kalb and Pennings (1980)	–	–	279 Tsuji and Ikada (1996a)	71, 79	180	–
T_g (°C)	50–65	50–60	34 Ovitt and Coates (1999)	65–72 Tsuji and Ikada (1999b)	–60	188, 197	40
$\Delta H_m(x_c = 100\%)$ (J g^{-1})	93 Fischer et al. (1973) 135 Miyata and Masuko 1998) 142 Loomis et al. (1990) 203 Jamshidi et al. (1988)	–	–	142 Loomis et al. (1990)	142	146	180–207
Density (g cm^{-3})	1.25–1.29	1.27	–	–	1.06–1.13	1.177–1.260	1.50–1.69
Solubility parameter (δ_p) (25 °C) (J cm^{-3})$^{0.5}$]	19–20.5, 22.7	21.1	–	–	20.8	20.6	–
$[a]_{589}^{25}$ in chloroform (deg dm^{-1} g^{-1} cm^3)	–155 ± 1	0	–	–	0	+44[a]	–
WVTR[b] (g m^{-2} per day)	32–172	–	–	–	177	13[c]	–
σ_B[d] (kg mm^{-2})	12–230	4–5[f]	–	90 Tsuji and Ikada (1999b)[e]	10–80[e]	18–20[e]	8–100[e]
E[g] (kg mm^{-2})	700–1000[e]	150–190[f]	–	880 Tsuji and Ikada (1999b)[e]	–	500–600[e]	400–1400[e]
ε_B[h] (%)	12–26[e]	5–10[f]	–	30 Tsuji and Ikada (1999b)[e]	20–120[e]	50–70[e]	30–40[e]

[a] 300 nm, 23 °C; [b] Water vapor transmission rate at 25 °C; [c] PHB-HV 94/6; [d] Tensile strength; [e] Oriented fiber; [f] Nonoriented film; [g] Young's modulus; [h] Elongation-at-break.

10
Physical Properties

The physical properties of PLLA, PDLLA, and PLA stereocomplex are listed in Table 4, together with those of R-PHB, poly(ε-caprolactone) (PCL), and poly(glycolide) (PGA). As mentioned earlier, the physical properties of polymeric materials depend on their molecular characteristics, highly ordered structures, and material morphology. The changes in thermal and mechanical properties of PLAs upon hydrolysis are connected with the structural changes in the crystalline and amorphous regions.

10.1
Thermal Properties

The T_m, equilibrium melting temperature ($T_m{}^0$), glass transition temperature (T_g), and melting enthalpy of the crystal having infinite thickness [ΔH_m ($x_c = 100\%$)] of PLAs are listed in Table 4. The thermal properties such as T_m and melting enthalpy (ΔH_m) or x_c are very important parameters reflecting the highly ordered structures of L_c, fraction of crystalline region, polymer chain packing in the amorphous region of the materials, respectively, which influence the initial mechanical properties before hydrolysis, and their change during hydrolysis. The Thompson–Gibbs expression between T_m and L_c is given by the following equation (e.g. Gedde, 1995):

$$T_m = T_m{}^0\ (1 - 2\sigma/\Delta h^0\ \rho_c L_c) \qquad (1)$$

where σ, Δh^0, and ρ_c are the specific fold surface free energy, heat of fusion (per unit mass), and crystal density, respectively. Equation (1) infers that the T_m of the polymeric materials, which increases with L_c, can be an index of L_c. Using the ΔH_m ($x_c = 100\%$) values, x_cs of the stereocomplex and homocrystals in specimens can be evaluated using the following equation:

$$x_c\ (\%) = 100 \times (\Delta H_c + \Delta H_m)/[\Delta H_m\ (x_c = 100\%)] \qquad (2)$$

where ΔH_c is the enthalpy of cold crystallization. By definition, ΔH_c and ΔH_m are negative and positive, respectively.

T_m (L_c) and x_c, become lower with decreasing crystallizable L- or D-lactyl unit sequence length in the LA homopolymers and copolymers including stereocopolymers, which can be caused by lowering the polymer molecular weight (Tsuji and Ikada, 1999b) and increasing the comonomer content (Tsuji et al., 1994). The T_g of LA stereocopolymers is almost constant at about 60 °C, irrespective of the LLA content (Tsuji and Ikada, 1996a), while the T_g of LA copolymers excluding stereocopolymers varies depending on the comonomer nature and its fraction, and the length of respective monomer sequences (Wada et al., 1991).

T_m and x_c increase with annealing (Migliaresi et al., 1991a,b; Tsuji and Ikada, 1995b) and orientation of PLLA (Jamshidi et al., 1988; Hyon et al., 1984; Fambri et al., 1997). The T_m and x_c of PLLA film can be altered in the range of 177–193 °C and 0–63%, respectively, by varying the pretreatment procedure, T_a, and t_a for thermal treatments (Migliaresi et al., 1991a,b; Tsuji and Ikada, 1995b). Similar to the effects of thermal treatment on T_m and x_c, Celli and Scandola (1992) and Cai et al. (1996) showed that T_g and the enthalpy of glass transition (ΔH_g) of PLLA and poly(L-lactide-*co*-D-lactide) [poly(LLA-DLA)] (96/4), respectively, increase with the increasing t_a at T_a below T_g. This means that PLLA and poly(LLA-DLA) (96/4) chains in the amorphous region become more densely packed by annealing at T_a below T_g.

In phase-separated PLA-based polymer blends, the thermal properties of PLA are insignificantly influenced by blending with other polymers in as much as the PLA content in the blends is high (Tsuji and Ikada, 1998a), while in partially miscible or miscible blends x_c and T_m of PLLA and T_g of the blends vary depending on the polymer mixing ratio (Koyama and Doi, 1997; Edlund and Albertsson, 2000; Ohkoshi et al., 2000). On the other hand, PLA stereocomplex crystallites have T_m and T_m^0 values much higher than those homocrystallites either of PDLA or PLLA (see Table 4). The T_g of the 1:1 blend films from PLLA and PDLA is ca. 5 °C higher than that of the nonblended PLLA and PDLA films in the M_w range from 5×10^4 to 1×10^5, where predominant stereocomplexation between PLLA and PDLA occurs in the blend film (Tsuji and Ikada, 1999b). The increased T_g is assignable to the strong interaction between L-lactyl and D-lactyl unit sequences in the amorphous region of the blend films, resulting in the dense chain packing that region.

The thermal stability of PLAs depends on the residual initiators and the molecular modification. Degée et al. (1997) demonstrated that the thermal stability is higher for PLLA synthesized using aluminum tri(isoproxide) as initiator than for that synthesized using tin(II) bis(2-ethylhexanoate). Jamshidi et al. (1988) found that thermal stability of PLLA synthesized using tin(II) bis(2-ethylhexanoate) becomes higher when its terminal carboxyl group is acetylated, and that residual monomers enhance the PLLA thermal degradation.

10.2 Mechanical Properties

The mechanical properties are crucial when PLAs and their copolymers are utilized as bulk materials; however, such properties may be controlled by varying the material parameters such as molecular characteristics and highly ordered structures.

10.2.1 Effect of Molecular Characteristics

Molecular weight is an important parameter to determine mechanical properties as given by the following equation:

$$P = P_0 - K/M_n \qquad (3)$$

where P is the physical property of a polymeric material, P_0 is P of the polymer with the infinite M_n, and K is a constant. We revealed that as-cast PLLA films have non-zero tensile strength (σ_B) below $1/M_n$ of 2.5×10^{-5} or above M_n of 4.0×10^4, and σ_B increases with $1/M_n$ (Tsuji and Ikada, 1999a) according to Eq. (3). Eling et al. (1982) showed the similar σ_B dependence of PLLA fibers on $1/M_v$, whilst Perego et al. (1996) reported that flexural strength becomes a plateau at M_v above 35,000 and 55,000 for PDLLA and amorphous-made PLLA, respectively. Ultimately, the σ_B and Young's modulus (E) of fibers from LA copolymers, poly(LLA-CL) and poly(LLA-DLA), are smaller than that from a homopolymer, PLLA, while elongation-at-break (ε_B) of poly(LLA-DLA) fiber becomes higher than that of PLLA fiber when they are melt-spun and thermally drawn (Penning et al., 1993). On the other hand, Grijpma et al. (1993) showed that the impact strength of PLLA can be increased by the cross-linking.

10.2.2 Effect of Highly Ordered Structures

The mechanical properties of PLLA vary depending on their highly ordered structure such as x_c and L_c (T_m) (Tsuji and Ikada, 1995b). An increase in x_c increases the σ_B and E of PLLA, but decreases the ε_B. The decrease in σ_B of PLLA films prepared at high T_a may be ascribed to the formation of large-sized spherulites and crystallites, in spite of their high x_c. These results indicate that the mechanical properties of PLLA can be controlled to some extent by altering the highly ordered structures. Similar to other polymers, the values of σ_B and E of PLLA fibers increase, but the ε_B value decreases with increasing degree of molecular orientation (Eling et al., 1982; Hyon et al., 1984; Horacek and Kalísek, 1994a,b; Fambri et al., 1997). Leenslag and Pennings (1987b) produced the PLLA fiber having $\sigma_B = 2.1$ GPa and E = 16 GPa by hot-drawing of a dry-spun fiber. Okuzaki et al. (1999) prepared the PLLA fiber with $\sigma_B = 275$ MPa and E = 9.1 GPa using a zone-drawing method from a LMW PLLA ($M_v = 13{,}100$). These results prove that molecular orientation, as well as molecular weight, has a major influence on the mechanical properties of PLLA.

10.2.3 Effects of Material Shapes

Pore formation is effective in lowering the Young's modulus of biodegradable polymers; reported examples relate to the compression modulus of poly(lactide-co-glycolide) [poly(LA-GA)] (50/50) (Thomson et al., 1995) and poly(HB-HV) (Chaput et al., 1995), and the tensile modulus of PCL (Tsuji and Ishizaka, 2001a).

10.2.4 Effect of Polymer Blending

In contrast to the mechanical properties of miscible polymer blends, those of the phase-separated polymer blends are discontinuous at a polymer mixing ratio where the inversion of the continuous and dispersed phases takes place. Even when phase separation occurs in the blends, σ_B, yield stress (σ_Y), and E of the blends of glassy PLLA (or PDLLA) with rubbery PEO (Sheth et al., 1997) or PCL (Tsuji and Ikada, 1996c, 1998a) can be widely varied by altering the polymer mixing ratio. The impact strength of PLLA is reported to become higher by the addition of rubbery biodegradable polymers such as PCL (Grijpma et al., 1994).

The stereocomplexation between PLLA and PDLA is reported to enhance the tensile properties of the blend films compared with those of the non-blended PLLA or PDLA film (Tsuji and Ikada, 1999b). Most likely, the microstructure formed by gelation and the inhibited growth of the spherulites may enhance the tensile properties of the blend film. The increased tensile properties of the blend films may be also caused by dense chain packing in the amorphous region due to a strong interaction between L- and D-unit sequences, as evidenced by the increased T_g of the blend films.

10.3 Electric Properties

The piezoelectric constants of PLLA, $-d_{14}$ and $-e_{14}$, increase with the draw ratio (DR) and become a maximum at a DR of about 4–5 (Ikada et al., 1996). Interestingly, the healing of fractured bone was found to be promoted under increased callus formation when drawn PLLA rods were implanted intramedullary in the cut tibiae of cats for its internal fixation (Ikada et al., 1996). This promotion is attributed to the piezoelectric current generated by the strains caused by leg movement of the cats. On the other hand, Pan et al. (1996) reported that the remnant polarization and coercive electric field of PLLA increased and decreased with temperature, respectively, their values at 130 °C being 96 mC m^{-2} and 20 MV m^{-1}, respectively. The pyroelectric constant of the PLLA specimen corona poled at an electric field of 50 MV m^{-1} and 130 °C for 5 min was 10 and 20 μC m^{-2} · K at temperatures lower and higher than T_g, respectively, but became almost zero at T_g (Pan et al., 1996).

10.4 Optical Properties

The specific optical rotations ($[a]^{25}_{589}$s) of PLLA and PDLA in chloroform are about –156 and 156 deg dm^{-1} g^{-1} cm^3, respectively (Tsuji and Ikada, 1992). The birefringence (Δn) of PLLA fibers increases from 0.015 to 0.038 with degree of molecular orientation (Hyon et al., 1984; Kobayashi et al., 1995). The intrinsic birefringence value estimated for highly drawn and annealed PLLA fibers containing solely α-form crystal using Stein's formula is 0.030–0.033 (Ohkoshi et al., 1999). Kobayashi et al. (1995) found that the gyration tensor component g_{33} of PLLA along the helical axis is extremely large $[(3.85 \pm 0.69) \times 10^{-2}]$, which corresponds to a rotary power of $(9.2 \pm 1.7) \times 10^3$ degree mm^{-1} and about two orders of magnitude larger than those of ordinary crystals. Similar rotary power of 7.2×10^3 degree mm^{-1} has also been reported by Tajitsu et al. (1999).

10.5 Surface Properties

The reported surface modification of PLAs to increase hydrophilicity are grafting hydrophilic chains such as acrylamide (AAm) (Tsuchiya et al., 1993) and peptide chains (Kimura and Yamaoka, 1996), and alkaline and enzymatic surface hydrolysis (Ishida and Tsuji, 2000). Alkaline and enzymatic treatment gives rise to scission of PLA chains through surface erosion, resulting in the increased densities of the hydrophilic terminal groups such as hydroxyl and carboxyl groups (Tsuji and Ikada, 1998b), resulting in the decrease in advancing and receding contact angles from 102 and 59° to 74 and 43°, respectively (Ishida and Tsuji, 2000). Another attempt to increase the hydrophilicity of PLLA can be made by coating PLA-b-PEO containing a hydrophilic chain (Otsuka et al., 1998).

The hydrophilicity at the PLLA film surface is increased by simple physical blending with hydrophilic poly(vinyl alcohol). The advancing and receding contact angles can be controlled in the ranges of 61–95° and 28–59°, respectively, by altering the mixing ratio of the polymers (Tsuji and Muramatsu, 2001b).

10.6 Permeability

Shogren (1997) studied the effects of the x_c of PLLA on water vapor permeability, which is an important factor when PLLA is used as food packages, containers, and bottles. The water vapor transmission rate of an amorphous PLLA film (172 g mm^{-2} per day at 25 °C) is higher than that of a crystallized PLLA film (82 g mm^{-2} per day at 25 °C) due to the low permeability of the crystalline region. Pitt et al. (1992) studied the effects of polymer mixing ratio and molecular weight of solutes on the permeability coefficients of the solutes in blends from poly(DLLA-GA) and PVA.

By contrast, when PLAs are used as the matrices of DDSs, the permeability of the matrices in the water-swollen state is an important parameter to determine the rate of drug release. The overall rate will vary according to the changes in molecular characteristics, highly ordered structures, and morphology of the matrices caused by hydrolysis.

10.7 Swelling and Solubility

It was found that a PLLA film is durable to swelling solvents having solubility parameter (δ_s) values much lower or higher than the value range of 19–20.5 $J^{0.5}$ $cm^{-1.5}$ (e.g., cyclohexane, $\delta_s = 16.8\ J^{0.5}\ cm^{-1.5}$; ethanol, $\delta_s = 26.0\ J^{0.5}\ cm^{-1.5}$), and that the polymer solubility parameter (δ_p) for PLLA is in the range of 19–20.5 $J^{0.5}\ cm^{-1.5}$ (Tsuji and Sumida, 2001). The solubility parameter value of PDLLA estimated by high-precision density measurements is 20.5 $J^{0.5}\ cm^{-1.5}$ (Siemann, 1992). At room temperature, HMW PLLA, PDLA, and PDLLA are soluble in the solvents having δ_s values in the range of 19–20.5 $J^{0.5}\ cm^{-1.5}$, such as chloroform, methylene dichloride, dioxane, and benzene, whereas these solvents become swelling-solvents when stereocomplexation occurs between PLLA and PDLA (Tsuji, 2000c). 1,1,1,3,3,3-Hexafluoro-2-propanol is a solvent for stereocomplexed PLAs, but these compounds become insoluble in it when the L_c increases. In addition to the above-mentioned solvents of PLLA, PDLLA is soluble in acetone. Well-known nonsolvents of PLAs are alcohols such as methanol and ethanol, and hence these are often used as precipitants.

10.8 Viscosity

The reported parameters affecting the intrinsic viscosity ($[\eta]$) of PLAs are M_w and M_n, L- and D-lactyl unit sequence of PLAs, and branching. The relationship between $[\eta]$ and M_n of PLLA and PDLLA is given by the following equations at 30 °C (Schindler and Harper, 1979):

For PLLA (4)

$$[\eta] = 5.45 \times 10^{-4}\ M_n^{0.73}\ \text{(chloroform)}$$

$$[\eta] = 5.72 \times 10^{-4}\ M_n^{0.72}\ \text{(benzene)} \quad (5)$$

For PDLLA (6)

$$[\eta] = 2.21 \times 10^{-4}\ M_n^{0.77}\ \text{(chloroform)}$$

$$[\eta] = 2.27 \times 10^{-4}\ M_n^{0.75}\ \text{(benzene)} \quad (7)$$

For branched PLLA (five arms) synthesized using pentaerythritol as initiator, the relationship between $[\eta]$ and M_n is given by the following equation (Kim and Kim, 1994):

$$[\eta] = 2.04 \times 10^{-4}\ M_n^{0.77}\ \text{(chloroform)} \quad (8)$$

Cooper-White and Mackay (1999) reported that PLLA melts have a critical molecular weight for entanglement of 16,000 g mol^{-1} and an entanglement density of 0.16 mmol cm^{-3} at 25 °C, and showed a dependence of terminal viscosity (η_0) on chain length to the fourth power, while equilibrium compliance (J^0_e) is independent of the molecular weight in the terminal region. By contrast, for branched PLLAs it was found that a critical molecular weight for entanglement is about four-fold that of linear PLLA, and a dependence of η_0 increase to the power of 4.6 for molecular weight (Dorgan et al., 1999).

11 Hydrolysis

PLAs, which are water-insoluble when their molecular weights are sufficiently high, are hydrolyzed at the ester groups to form LMW, water-soluble oligomers and monomers of lactic acids. The hydrolysis mechanisms and behaviors of PLAs are affected by numerous factors, including the materials and the hydrolysis media.

11.1 Hydrolysis Mechanisms

The hydrolysis mechanisms of PLAs may be separated into those of molecular chains and bulk materials.

11.1.1 Hydrolysis Mechanisms of Molecular Chains

The hydrolysis mechanisms of PLA chains are classified into two groups: catalytic and noncatalytic hydrolysis, or enzymatic and nonenzymatic hydrolysis (Figure 3) (Tsuji,

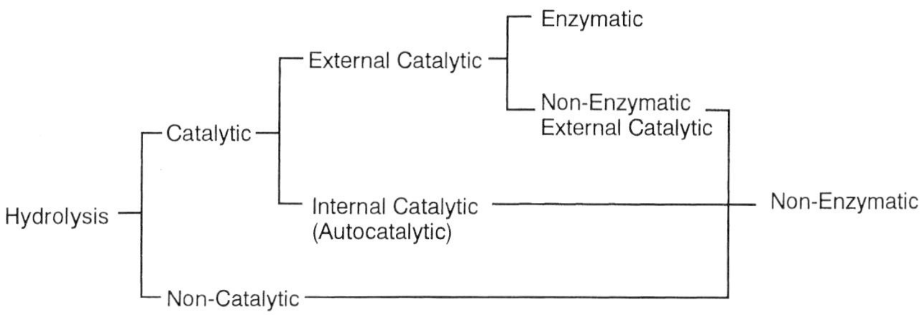

Fig. 3 Hydrolysis mechanisms of PLA chains.

2000b). Catalytic hydrolysis involves catalysis by external and internal substances (abbreviated as the external and internal catalytic hydrolysis, respectively). The representative internal catalytic hydrolysis mediated by the terminal carboxy groups is generally called autocatalytic or autocatalyzed hydrolysis.

External Catalytic Hydrolysis

The representative external catalysts for PLA hydrolysis are enzymes and alkalis. The typical hydrolysis enzymes of ester group are esterases, such as lipases. Endo- and exo-esterases catalyze the hydrolytic scission of an ester group randomly, irrespective of its position in a polymer chain (endo-chain scission) and that of ester group around a polymer chain end (exo-chain scission), respectively. However, the lipases have no significant catalytic effect on PLA hydrolysis. Williams (1981) found that proteinase K, which is a well-known hydrolytic enzyme that catalyzes endo-chain scission of poly(amino acids), can catalyze the hydrolysis of PLLA, while Makino et al. (1985) and Ivanova et al. (1997) studied carboxylic esterase (EC. 3.1.1.1) and cutinase-accelerated hydrolysis of PDLLA, respectively.

In contrast, alkali- (or base-) catalyzed hydrolysis of PDLLA (Shih, 1995) and PLLA (Tsuji and Ikada, 1998b) proceeds via an endo-chain scission mechanism, whereas acid-catalyzed hydrolysis of PDLLA demonstrated exo-chain scission which was more rapid than endo-chain scission (Shih, 1995).

Autocatalytic Hydrolysis

The hydrolytic scission of aliphatic polyester chains without any external catalysts may proceed via the combination of autocatalyzed and noncatalyzed mechanisms. For PLAs, it has been reported that the former mechanism prevails the latter mechanism. In the former mechanism, the hydrolysis is catalyzed by the terminal carboxy groups of PLAs, and its rate is proportional to the concentrations of carboxy and ester groups and water. The kinetic equation expressing the scission of molecular chains by autocatalytic hydrolysis can be derived under the above-mentioned assumption (Pitt et al., 1979):

$$d\,[COOH]/dt = k'\,[COOH]\,[H_2O]\,[ester] \quad (9)$$

where [COOH], $[H_2O]$, and [ester] are the concentrations of terminal carboxy group, water, and whole ester group in PLAs or their copolymers, respectively. If $k'\,[H_2O]\,[ester]$ is assumed to be constant, integration of Eq. (9) will give Eq. (10), coupled with the relationship of $[COOH] \propto M_n^{-1}$:

$$\ln M_{n,t} = \ln M_{n,0} - k_1\,t \quad (10)$$

where $M_{n,t}$ and $M_{n,0}$ are M_n of the polymer at hydrolysis times = t and 0, respectively, and the hydrolysis rate constant (k_1) is equal to $k'\,[H_2O]\,[ester]$. The k_1 value estimated for

PLLA is approximately $2-7\times10^{-3}$ per day, depending on the initial highly ordered structure (Tsuji et al., 2000b).

Noncatalytic Hydrolysis

When there is no catalytic effect, the kinetic equation for scission of molecular chains during hydrolysis can be expressed by the following equation:

$$d\,[COOH]/dt = k'\,[H_2O]\,[ester] \qquad (11)$$

Since the concentration terms in Eq. (11) can be assumed to be constant during initial stage of hydrolysis when the polymer molecular weight is sufficiently high, integration of Eq. (11) will give Eq. (12), coupled with the relationship of $[COOH] \propto M_n^{-1}$:

$$M_{n,t}^{-1} = M_{n,0}^{-1} + k_2\,t \qquad (12)$$

where the hydrolysis rate constant (k_2) is equal to $k'\,[H_2O]\,[ester]$.

11.1.2 Hydrolysis Mechanisms of Bulk Materials

The hydrolysis of water-insoluble bulky PLAs proceeds through surface and/or bulk erosion mechanisms, which are illustrated schematically in Figure 4 (Tsuji, 2000b). More detailed erosion mechanisms of hydrolyzable polymers are described in other monographs (e.g., Göpferich, 1997).

Surface Erosion

Surface erosion is the main route of hydrolysis when the hydrolysis media contain external catalysts, or the hydrolysis rate of the materials is extremely high compared with the diffusion rate of the hydrolysis medium. On the other hand, bulk erosion occurs when no such external catalysts are present in the hydrolysis media, or the hydrolysis rate is low. The relative contribution of these two mechanisms depends on the nature of polymer and the hydrolysis

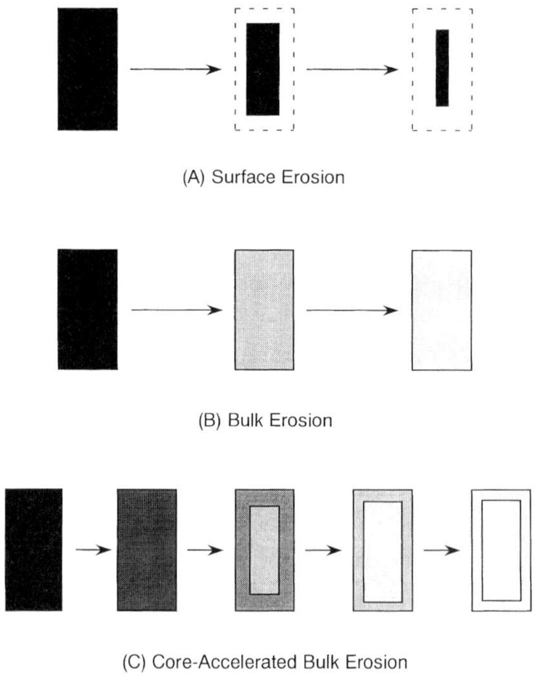

Fig. 4 Hydrolysis mechanisms of PLA bulk materials.

medium and conditions. In the case of surface erosion, hydrolytic chain scission occurs solely at the material surface and forms LMW, water-soluble oligomers, which will be removed from their surface or diffuse into the medium, while the core of the materials will remain unhydrolyzed (Figure 4A).

Bulk Erosion

The autocatalytic hydrolysis of PLLA as in phosphate-buffered solution occurs homogeneously along the material cross-section via a bulk erosion mechanism, as long as the material thickness is <2 mm (Figure 4B) (Tsuji and Ikada, 2000b; Tsuji et al., 2000b), even when the temperature is raised to 97 °C, which is higher than the T_g (60 °C) (Tsuji and Nakahara, 2001). Li et al. (1990) and Grizzi et al. (1995) demonstrated that catalytic LMW oligomers and monomers formed by hydrolysis remain at the core of PLA materials, and consequently they accelerate the hydrolysis in the core region of the PLA materials when they have a thickness >2 mm (Figure 4C). Due to this effect, the hydrolysis proceeds via two modes – slow and rapid hydrolysis at the outer layer and at the core, respectively.

Hydrolysis of Crystallized Specimens

The PLA chains in the crystalline region are more hydrolysis-resistant than those in the amorphous region. This causes predominant hydrolysis and removal of the chains in the amorphous region of the crystallized PLA, leaving the chains in the crystalline region. In addition, crystallization of PLA chains will occur during hydrolysis under increased molecular mobility in the presence of water molecules at a raised temperature, resulting in the increased x_c during hydrolysis (Li et al., 1990; Migliaresi et al., 1994; Pistner et al., 1994; Duek et al., 1999; Tsuji and Ikada, 2000b). Spherulitic structure was noticed in scanning electron microscopy (SEM) observation of the PLLA films crystallized and hydrolyzed via surface erosion in alkaline and enzyme solutions (Tsuji and Ikada, 1998b; Tsuji and Miyauchi, 2001a), which resulted from preferred hydrolysis and removal of the chains in the amorphous region inside and outside the PLLA spherulites.

The hydrolysis of the chains in the amorphous region of crystallized PLLA is accelerated in phosphate-buffered solution compared with that of the chains in the free amorphous region, as in completely amorphous specimens (Tsuji et al., 2000b). This is attributed to the increased density of catalytic terminal carboxy group in the amorphous region of the crystallized PLLA specimens during crystallization compared with that of the completely amorphous specimens.

Figure 5 shows the gel permeation chromatography (GPC) curves of crystallized PLLA films hydrolyzed enzymatically in proteinase K/Tris-buffered solution (Figure 5A) (Tsuji and Miyauchi, 2001a,b), together with those hydrolyzed in alkaline solution (Figure 5B) (Tsuji and Ikada, 1998b) and phosphate-buffered solution (Figure 5C) (Tsuji et al., 2000b; Tsuji and Ikada, 2000b). For enzymatic hydrolysis (Figure 5A), the peak ascribed to the unhydrolyzed core part remains at the same position, while LMW specific peaks appear at the molecular weights of 1, 2, and 3×10^4, which are attributed to the one, two, and three folds of PLLA chain in the crystalline region, respectively. The fact that peak ascribed to the two or three folds of PLLA chain in the crystalline region remains for as long as 70 h, even when the height of initial main peak decreased to 20% of its initial value, strongly suggests that the folding chains in the restricted amorphous region between the crystalline regions are hydrolysis-resistant in the presence of proteinase K

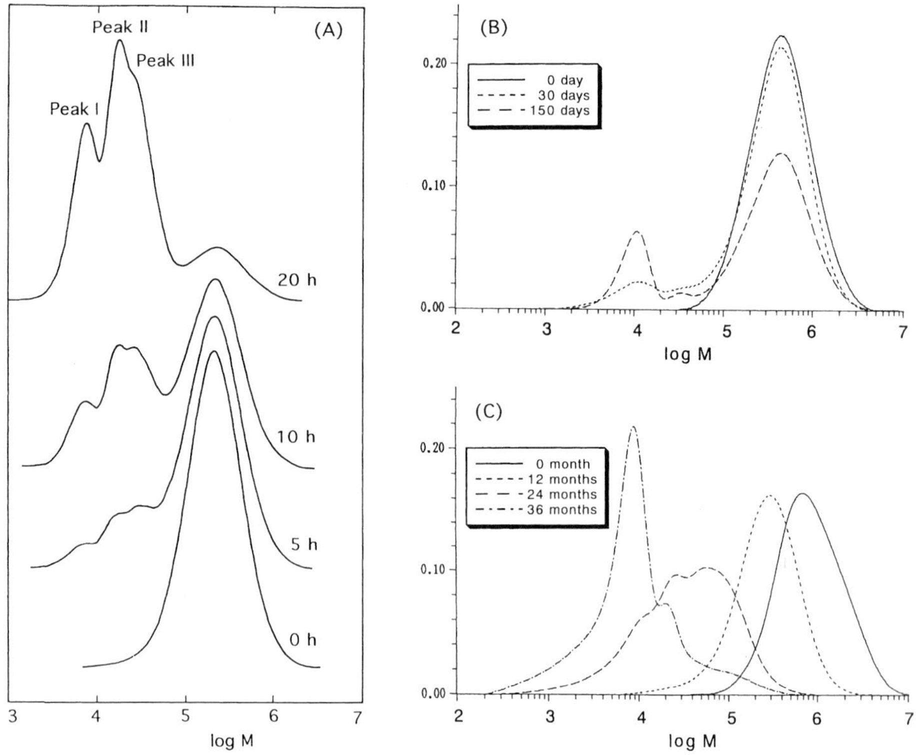

Fig. 5 Gel permeation chromatography curves of crystallized PLLA hydrolyzed in enzymatic (A), alkaline (B), and phosphate-buffered (C) solutions.

compared with tie chains and the chains with free ends. On the other hand, only the lowest molecular weight specific peak remained after long-term hydrolysis in alkaline and phosphate-buffered solutions (Figure 5B and C), representing non-selective or random scission of the chains (endo-chain scission) in the amorphous region. The dependence of x_c on the proteinase K-catalyzed hydrolysis rate of PLLA revealed that the PLLA chains in the restricted amorphous region between the crystalline regions in crystallized specimens are much more hydrolysis-resistant than those in the free amorphous region, as in a completely amorphous specimen (Tsuji and Miyauchi, 2001a,b). Iwata and Doi (1998) demonstrated that the enzymatic hydrolysis of PLLA single crystals in the presence of proteinase K occurs at their crystal edges rather than the chain fold lamellar surfaces.

Hydrolysis of amorphous specimens

Similar to the crystallized PLLA, crystallization during hydrolysis is also reported for initially amorphous PLLA specimens, resulting in the increased x_c with hydrolysis time (Li et al., 1990; Migliaresi et al., 1994; Pistner et al., 1994; Duek et al., 1999; Gonzalez et al., 1999; Tsuji et al., 2000b). Li and McCarthy (1999b) found that stereocomplexation or racemic crystallization occurs when amorphous PDLLA is hydrolyzed for a long period, or at a high temperature, to have a LMW. In contrast, due to the stabilized chain packing in the amorphous region, the increase in T_g occurs at the initial stage of hydrolysis, followed by the rapid decrease in T_g due to decreased molecular weight or enhanced molecular mobility at the late stage of hydrolysis (Gonzalez et al., 1999; Tsuji et al., 2000b).

11.2
Effects of Surrounding Media

Mason et al. (1981) studied a wide variety effects of the surrounding media on the hydrolysis of PDLLA. It is well known that the hydrolysis of PLAs is accelerated in aqueous media at high temperature (Jamshidi et al., 1986; Tsuji and Nakahara, 2001) and deviation from pH 7 (Makino et al., 1985; Tsuji and Ikada, 1998b) and in a temperature/humidity-controlled chamber at high temperature and humidity (Ho et al., 1999). Pitt and Gu (1987) studied the degradation behavior of PLLA, poly(LLA-GA), and PCL films at 37 °C in water, alcohols, and acidic and basic reagents. These authors found significant effects of the nature of degradation mediums on the degradation rates of poly(LLA-GA) and PCL, but no such effects for the PLLA degradation when their degradation was estimated by molecular weight. However, crystallization of an initially amorphous PLLA was observed during degradation in ethanol. The rapid crystallization at high temperature reduces the difference in the hydrolytic behavior between the PLLA specimens having different initial x_cs and Lc prepared by different annealing or crystallization conditions when PLLA is hydrolyzed autocatalytically in a phosphate-buffered solution (Tsuji and Nakahara, 2001).

Water-soluble oligomers and monomers of PLAs formed by hydrolysis will diffuse into surrounding media, resulting in weight loss of the materials. The water solubility of the lactic acid oligomers depend on the pH and temperature of surrounding media, which will alter the weight loss behavior of PLAs during hydrolysis. Kamei et al. (1992), using high-performance liquid chromatography, showed that 1- to 7-mers of L-lactic acid are soluble in distilled water, while Braud et al. (1996) used capillary electrophoresis to show that 1- to 9-mers of DL-lactic acid are soluble in phosphate-buffered solution at pH 6.8; however, lactic acid was the only component released from the mother PDLLA materials into aqueous medium during hydrolysis. The latter finding means that the oligomers having a degree of polymerization >2 are trapped in the materials during hydrolysis. Karlsson and Albertsson (1998) reported that high pH and high temperature are favorable for the water-soluble oligomers and monomers to diffuse out from the materials.

11.3
Effects of Material Factors

The hydrolysis behaviors of PLA materials can be controlled by altering their material parameters such as molecular characteristics and highly ordered structures.

11.3.1
Molecular Characteristics

Reducing the molecular weight of PLAs increases the density of hydrophilic terminal carboxy and hydroxy groups and molecular mobility; this will increase the diffusion rate and concentration of water in the materials and the probability of formation of water-soluble oligomers and monomers upon hydrolysis, resulting in their accelerated hydrolysis. The autocatalytic (Figure 6) and proteinase-K catalyzed (Figure 7) hydrolysis rates of initially amorphous PLLA become lower with the initial M_n (Tsuji et al., 2000b; Tsuji and Muramatsu, 2001a; Tsuji and Miyauchi, 2001c). On the other hand,

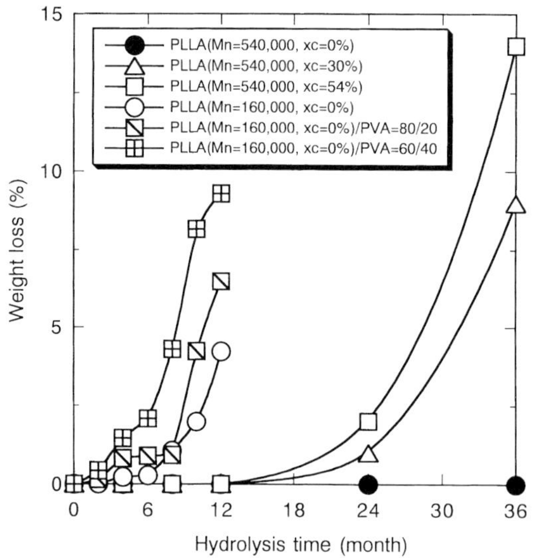

Fig. 6 Weight loss of PLAs by autocatalytic hydrolysis in phosphate-buffered solution.

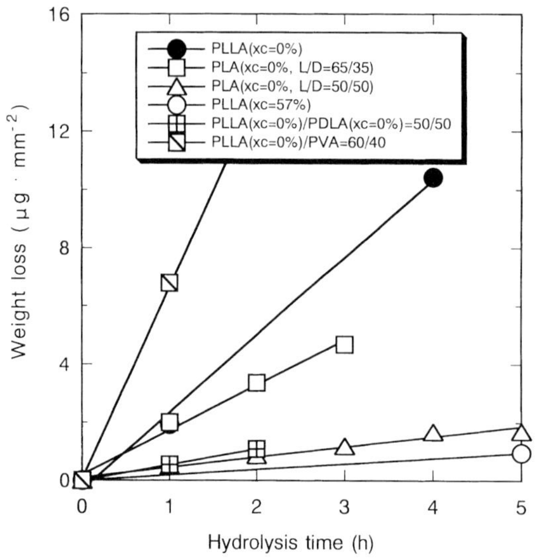

Fig. 7 Weight loss of PLAs by enzymatic hydrolysis in the presence of proteinase K.

LAs remaining unreacted during polymerization and/or formed during thermal processes accelerate the autocatalytic hydrolysis of PLA materials (e.g., Tsuji and Suzuki, 2001). Nakamura et al. (1989) and Zhang et al. (1994) confirmed the accelerated autocatalytic hydrolysis of PLAs upon addition of LAs.

The tacticity and the ratio and sequence lengths of chiral monomer units of PLAs have no significant effect on their autocatalytic hydrolysis, while proteinase K-catalyzed hydrolysis rate decreases significantly with decreasing L-lactyl unit sequence length (Reeve et al., 1994; MacDonald et al., 1996; Tsuji and Miyauchi, 2001c). Li et al. (2000)

found that the secondary effect of water-absorption of amorphous PLAs originated from the tacticity difference influences the enzymatic hydrolysis. In contrast, the autocatalytic hydrolysis rate of amorphous poly(DLLA-GA)s becomes higher with a rise in hydrophilic GA content (Hyon et al., 1998).

11.3.2
Highly Ordered Structures

Among the highly ordered structures, crystallinity is the most important parameter to determine the hydrolysis rate of PLA materials at the initial stage, whereas crystalline thickness is the dominant parameter to affect their hydrolysis rate at the late stage.

Crystallinity

When PLLA is hydrolyzed autocatalytically *in vivo* (Nakamura et al., 1989; Pistner et al., 1993, 1994) and in phosphate-buffered solution (Li et al., 1990; Duek et al., 1999; Tsuji and Ikada, 2000b; Tsuji et al., 2000b), the induction period until the start of decrease in mechanical properties and remaining weight decreases with the initial x_c. The decreases in mechanical properties, molecular weight, and remaining weight of PLLA after hydrolysis become higher with the initial x_c, when compared on the basis of the same hydrolysis time longer than 2 months (Duek et al., 1999) and 24 months (Tsuji and Ikada, 2000b; Tsuji et al., 2000b). On the other hand, when PLLA films are hydrolyzed in enzyme solution (Cai et al., 1996; Li and McCarthy, 1999a; Tsuji and Miyauchi, 2001a,b) or alkaline solution (pH 12) (Cam et al., 1995; Tsuji and Ikada, 1998b), the weight loss rate is lower for PLLA films having a higher initial x_c. Due to the high hydrolysis-resistance of the chains in the restricted amorphous region compared with that in the free amorphous region, the effect of x_c on enzymatic hydrolysis is strong and weak for the PLLA specimens having low and high x_cs, respectively (Tsuji and Miyauchi, 2001b). The hydrolysis rates of crystallized PLAs and their copolymers are determined by the combination of the relatively strong x_c effect and the weak effects of molecular characteristics such as molecular weight (PLLA) (Cam et al., 1995), L-lactyl unit sequence length (LA stereocopolymers) (Reeve et al., 1994), and copolymer composition [poly(LLA-GA)s] (Reed and Gilding, 1981).

Crystalline Thickness

Fischer et al. (1973) prepared the single crystals of L-lactide-rich PLAs in a xylene dilute solution at different temperatures to have different L_cs, and their external catalytic hydrolysis was studied in a dilute alkaline solution. In contrast, we prepared PLLA films having different L_cs by annealing at different T_as from the melt (Tsuji and Ikada, 1995b) and their external catalytic and autocatalytic hydrolysis was investigated in alkaline (Tsuji and Ikada, 1998b) and enzyme (Tsuji and Miyauchi, 2001a) solutions and a phosphate-buffered solution (Tsuji and Ikada, 2000b), respectively. The size of crystalline residue at the late stage of hydrolysis, when a large weight loss occurs, depends on the initial L_c before hydrolysis, as evidenced by the fact that the lowest molecular weight specific peak ascribed to one fold of the PLLA chain in the crystalline residue increases with the initial T_m (L_c) before hydrolysis, irrespective of the hydrolysis media. The peak molecular weight of the lowest molecular weight specific peak ($M_{n,s}$) can be converted to the L_c of PLLA lamella after hydrolysis using the following equation, assuming that the PLLA chains take 10_3 helix in the α-form unit cell having a dimension $c = 2.78$ (or 2.88) nm (fiber axis) (Hoogsteen et al., 1990; Okihara et al., 1991):

$$L_c\ (\text{nm}) = 0.278\ (\text{or}\ 0.288) \times M_{n,s}/72.1 \quad (13)$$

where 72.1 is the mass per mole of lactyl unit (a half of lactide unit). The L_c values calculated from Eq. (13) are 29 (30), 36 (37), and 54 (56) nm for the PLLA specimens having T_m values of 172, 177, and 184 °C, respectively, after hydrolysis in phosphate-buffered solution, and 23 (24), 32 (33), and 36 (37) nm for those having T_m values of 170, 177, and 182 °C, respectively, after enzymatic hydrolysis (Tsuji and Ikada, 2000b; Tsuji and Miyauchi, 2001a). Here, the values before and in parentheses were calculated using the c values of 2.78 and 2.88 nm, respectively. The obtained L_c values are comparable with 10 and 11 nm for poly-(LLA-GA) (95/5) crystallized at 120 and 140 °C, respectively, from the melt (Wang et al., 2000), and also with 11 and 16 nm for PLLA single crystals formed at 82 and 102 °C, respectively, in a dilute solution (Fischer et al., 1973). The L_c values for $c = 2.78$ nm give the following relationships:

$$T_m\,(K) = 471\,[1 - 1.59/L_c\,(nm)] \qquad (14)$$
(hydrolyzed in phosphate-buffered solution)

$$T_m\,(K) = 472\,[1 - 1.46/L_c\,(nm)] \qquad (15)$$
(hydrolyzed in the presence of proteinase K)

Comparisons between Eqs. (1) and (14) or (15) give 198 and 199 °C as T_m^0 values, which are slightly lower but almost in agreement with those obtained from the Hoffman–Weeks procedure for the melt-crystallized PLLA specimens by Kalb and Pennings (1980) (215 °C) and by ourselves (Tsuji and Ikada, 1995b) (205 °C). This agreement on T_m^0 estimated with different methods also confirms our assumption that $M_{n,s}$ is the molecular weight of one fold of the PLLA chain in the crystalline region.

Spherulite Size

It has been concluded that there is no significant effect of the spherulite size on the external catalytic and autocatalytic hydrolysis of PLLA in alkaline and phosphate-buffered solutions, respectively (Tsuji and Ikada, 1998b, 2000b), which is comparable with the result reported for the enzymatic hydrolysis of R-PHB (Kumagai et al., 1992).

Orientation

Jamshidi et al. (1986) reported that the decrease rate of residual tensile strength and remaining weight of PLLA fibers during hydrolysis in phosphate-buffered solution decreases with DR or molecular orientation, and that the disk- and column-shaped crystalline residues were formed by the hydrolysis. The formation of the disk- and column-shaped crystalline residues is attributed to the selective hydrolysis and removal of the chains in the amorphous region between the crystalline regions which are located at a common cross-sectional plane normal to the fiber axis, as proposed for PGA (Chu and Campbell, 1982).

11.3.3 Material Shapes

The porous structure of PLAs has been reported to retard their bulk autocatalytic hydrolysis in a phosphate-buffered solution, which was ascribed to the enhanced diffusion of catalytic oligomers into a surrounding medium resulting from a reduced average distance of the porous material for the elution of catalytic oligomers from the material surface compared with that of the non-porous one (Lam et al., 1994). On the other hand, in the case of the surface hydrolysis in enzymatic and alkaline solutions, the hydrolysis rate is expected to increase with increasing porosity and decreasing pore size due to the increased surface area per unit weight on the basis of the finding for porous PCL specimens (Tsuji and Ishizaka, 2001a).

11.3.4 Polymer Blending

With regard to PLAs, no polyesters are reported to be highly miscible with them in the blends, excluding the self blends from HMW and LMW PLAs and the enantiomeric and diastereoisomeric PLA blends, as long as the molecular weights of PLAs and the other polyesters to be blended are sufficiently high ($M_n > 10^5$). Therefore, most of the PLA-based polymer blends are at least partially phase-separated. The effects of polymer blending depend on the hydrolysis medium, and the effects on the hydrolysis rate of PLAs are detailed below for two different media, namely phosphate-buffered and enzyme solutions.

The hydrolysis of PLAs in a phosphate-buffered solution will be accelerated by the presence of the polymer having either a high hydrophilicity, or/and a high density of carboxyl groups. Polymers such as HMW water-insoluble PVA and PEO, which satisfy the first requirement, will increase the diffusion rate and concentration of water in the materials, resulting in accelerated hydrolysis of PLAs. This acceleration was recognized for both the partially miscible (Pitt et al., 1992; Nijenhuis et al., 1996b) and phase-separated (Tsuji and Muramatsu, 2001a) blends. The polymers such as LMW aliphatic polyesters having a high density of catalytic carboxyl groups, which satisfy the second as well as the first requirements, will enhance the autocatalytic hydrolysis. The high density of hydrophilic terminal groups will increase the water concentration and diffusion rate into the materials, resulting in accelerated hydrolysis of PLAs. Examples include the blends of HMW PLLA with LMW PLLA (von Recum et al., 1995) or LMW PCL (Tsuji and Ikada, 1998a), and those from HMW and LMW PDLLAs (Bodmeier et al., 1989; Asano et al., 1991; Mauduit et al., 1996).

By contrast, the hydrolysis of PLAs in a phosphate-buffered solution will be retarded when a specimen contains the PLA stereocomplex (Tsuji, 2000c). The retarded hydrolysis of the well-stereocomplexed PLA blend compared with that of the nonblended PLLA and PDLA is ascribed mainly to the peculiarly strong interaction between L-lactyl and D-lactyl unit sequences in the amorphous region and/or the three-dimensional micro-network structure in the blend formed by stereocomplexation in the course of solvent evaporation. They will reduce the diffusion rate of water molecules into the amorphous region and/or the concentration of catalytic oligomers formed by hydrolysis, respectively.

The proteinase K-catalyzed hydrolysis of PLLA is reported to be accelerated in the phase-separated blends by the presence of hydrophilic PEO (Sheth et al., 1997), cellulose (Nagata et al., 1998), PVA (Tsuji and Muramatsu, 2001a) and hydrophobic PCL (Tsuji and Ishizaka, 2001b), whereas decelerated in the miscible blends from L-rich PLA and hydrophobic poly(vinyl acetate) (PVAc) (Gajria et al., 1996). This means that the phase-structure of the blend rather than the hydrophilicity of the second polymers has a crucial effect on the enzymatic hydrolysis rate of PLLA, and strongly suggests that the enhanced enzymatic hydrolysis in the phase-separated blends is due to the occurrence of enzymatic hydrolysis at the interfaces of the two polymer phases as well as at the specimen surface.

12 Biodegradation

In natural environments, nonenzymatic hydrolysis is recognized as the main route of PLA degradation. However, it was confirmed that biodegradation also occurs in the presence of some microbes in natural environments (Torres et al., 1996a–c, 1999; Pranamunda et al., 1997; Karjomaa et al., 1998; Tsuji et al., 1998; Pranamunda and Tokiwa, 1999; Tokiwa et al., 1999), and that the mineralization of PLLA to carbon dioxide, rather than to lactic acid, occurs in soils (Ho and Pometto, 1999) and composts (Meinander et al., 1997).

13 Applications

PLAs have been regarded as biomedical materials and matrices in the forms of rods, plates, films, meshes, and microspheres for tissue engineering and drug delivery systems. Due to the lowered prices of PLAs, they are recently recognized as substitutes for the general-purpose "petro" polymers such as polyethylene, polypropylene, and polystyrene. The medical and ecological applications of PLAs are summarized in Tables 5 and 6 (Ikada and Tsuji, 2000), respectively. More detailed applications in medical and pharmaceutical areas are described in several monographs (e.g., Kharas et al., 1994; Hartmann, 1998; Coombes and Meikle, 1994), and the end-products of PLAs and the companies producing lactic acids, lactides, PLAs, and PLA end-products are described in Chapters 8 and 9 of this volume, and also in review articles (e.g., Datta et al., 1995; Bogaert and Coszach, 2000; Middleton and Tipton, 2000).

14 Outlook and Perspectives

The physical properties and hydrolysis and biodegradation behavior of PLAs are controllable by varying molecular characteristics, highly ordered structures, and material shapes, and also by polymer blending. It seems that PLA-based polymeric materials having excellent physical properties will be widely used as commodity plastics in the near future when PLLA is produced probably under a \$1.00/lb cost by major US companies such as Cargill-Dow Polymers. The most appropriate biodegradable polymer for the respective end uses can be selected from the PLA based polymeric materials having different physical properties, enzymatic and non-enzymatic hydrolytic behavior, and cost/performance.

Tab. 5 Hitherto medical applications of PLAs and their copolymers (Ikada and Tsuji, 2000)

Function	***Purpose***	***Examples***
Bonding	Suturing	Vascular and intestinal anastomosis
	Fixation	Fractured bone fixation
	Adhesion	Surgical adhesion
Closure	Covering	Wound cover, local hemostasis
	Occlusion	Vascular embolization
Separation	Isolation	Organ protection
	Contact inhibition	Adhesion prevention
Scaffold	Cellular proliferation	Skin reconstruction, blood vessel reconstruction
	Tissue guide	Nerve reunion
Capsulation	Controlled drug delivery	Sustained drug release

Tab. 6 Hitherto ecological applications of PLAs and their copolymers (Ikada and Tsuji, 2000)

Application	*Fields*	*Examples*
Industrial applications	Agriculture, forestry	Mulch films, temporary replanting pots, Delivery system for fertilizers and pesticides,
	Fisheries	Fishing lines and nets, fishhooks, fishing Gears
	Civil engineering and construction industry	Forms, vegetation nets and sheets, water-retention sheets
	Outdoor sports	Golf tees, disposable plates, cups, bags, and Cutlery
Composting	Food package	Package, containers, wrappings, bottles, bags, films, retail bags, six-pack rings
	Toiletry	Diapers, feminine hygiene products
	Daily necessities	Refuge bags, cups

15 Patents

The important patent articles concerning the synthesis and applications of PLAs and their copolymers are summarized in Table 7. A large number of patent articles have been issued for lactic acids, LAs formation, and synthesis of PLAs. Lowe (1954) disclosed the improved LAs purification techniques, which led to synthesis of HMW PLAs, while Selman (1967) later proposed the distillation process to obtain optically pure LAs from the mixture of LMW PLAs and LAs. Bellis (1988) claimed the recrystallization method using toluene or ethyl acetate for the purification of LAs. Datta (1989) proposed an efficient and economical process for lactic acid production and purification. Gruber and Hall (1992) claimed the continuous closed-loop system to synthesize LAs from lactic acids without waste products other than water, which leads to low-cost production of PLAs. Ohara and Okamoto (1997) proposed the recovery method of LA from high-molecular weight PLA, which may be intended to obtain LA from PLA wastes.

Enomoto et al. (1994) proposed a method of synthesizing HMW PLAs and their copolymers by direct condensation of lactic acids and other hydroxycarboxylic acids using azeotropic solvents such as diphenylether. Seppälä and Selin (1995) claimed the method to produce lactic acid-based polyurethane, which enables the chain length of the LMW PLAs and their copolymers to be extended. Vitalis (1959) and Casey and Huffman (1984) suggested the basic method to synthesize the block copolymers of poly(α-hydroxylic acid)s with flexible hydrophilic polyethers such as PEO. These block copolymers are utilized as DDS matrices containing hydrophilic drugs.

In terms of processes and applications of the materials from PLAs and their copolymers, numerous patents have been issued. Shmitt and Polistina (1967) and Shneider (1972) proposed the applications of aliphatic polyesters and poly(LA-GA), respectively, to surgical sutures. Michaels (1976) suggested the applications of PLAs to DDS matrices, whereas Pitt and Schindler (1979) proposed the application of poly(LA-CL) to the matrices for sustained subdermal delivery system. Törmälä et al. (1988) claimed the preparation method of self-reinforced high-strength PLAs, which are useful as osteosynthesis devices. Murdoch and Loomis (1988)

Tab. 7 Patent articles of PLAs and copolymers

Number of patent	*Patent holder*	*Inventor*	*Title of Patent*	*Date of publication*
USP 2 668 162	Du Pont	Lowe, C.E.	Preparation of high molecular weight polyhydroxyacetic ester	Feb. 3, 1954
USP 4 438 253	American Cyanamid	Vitalis, E.A.	Polyglycol-polyacid ester treatment of textiles	Dec. 15, 1959
USP 3 297 033	American Cyanamid	Shmitt, E.E., Polistina, R.A.	Surgical sutures	Jan. 10, 1967
USP 3 322 791	Ethicon	Selman, S.	Preparation of optically active lactide	May 30, 1967
USP 3 636 956	American Cyanamid	Shneider, A.K.	Polylactide sutures	Jan. 10, 1972
USP 3 962 414	Alza	Michaels, A.S.	Structured bioerodible drug delivery device	June 8, 1976
USP 4 148 871	–	Pitt, C.G., Schindler, A.E.	Sustained subdermal delivery of drugs using poly(ε-caprolactone) and its copolymers	Apr. 10, 1979
USP 4 438 253	American Cyanamid	Casey, D.J., Huffman, K.R.	Polyglycolic acid) / polyalkylene glycol) block copolymers and method of manufacturing the same	Mar. 20, 1984
USP 4 719 246	Du Pont	Murdoch, J.R., Loomis, G.L.	Polylactide compositions	Jan. 12, 1988
USP 4 727 163	Du Pont	Bellis, H.E.	Process for preparing highly pure cyclic esters	Feb. 23, 1988
USP 4 743 257	Materials Consultant	Törmälä, P., Rokkanen, P., Laiho, J., Tamminmäki, M., Vainionpää, S.	Materials for osteosynthesis devices	May 10, 1988
USP 4 863 506	Union Oil Company of California	Young, D.C.	Methods for regulating of the growth of plants and growth regulant composition	Sep. 5, 1989
USP 4 885 247	Michigan Biotechnology Institute	Datta R.	Recovery and purification of lactate salts from whole fermentation broth by electrodialysis	Dec. 5, 1989
Jpn. Kokai Tokkyo Koho, JP 03 183 428	Gunze	Saito, Y., Ikada, Y., Gen, J., Suzuki, M.	Hydrolyzable polylactide fishing lines and their manufacture	Aug. 9, 1991
Fr. Demande, FR 2 657 255 A1	Sederma, S.A.	Greff, D.	Description de préparations cosmétiques originales dont les principes actifs sont piégés dans un réseau polymérique greffé à la surface de particles de silice	Dec. 27, 1991
PCT Int. Appl. WO 9 201 737	Du Pont	Hammel, H.S., York, R.O.	Degradable foam materials	Feb. 6, 1992
PCT Int. Appl. WO 9 204 412 A1	Du Pont	Ostapchenko, G.J.	Films containing polyhydroxy acid	Mar. 19, 1992
EP 467 478 A2	DSM	Van den Berg, H.J.	Method for the manufacture of polymer products from cyclic esters	Jan. 22, 1992
USP 5 142 023	Cargil	Gruber, P.R., Hall, E.S.	Continuous Process for manufacture of lactide polymers with controlled optical purity	Aug. 25, 1992
USP 5 298 602	Takiron	Shikinami, Y., Hata, K.	Polymeric piezoelectric material	Mar. 29, 1994
USP 5 310 865	Mitsui Toatsu Chemicals	Enomoto, K, Ajioka, M., Yamaguchi, A.	Polyhydroxycarboxylic acid and preparation process thereof	May 10, 1994
USP 5 380 813	-	Seppälä, J., Selin, J.F.	Method for producing lactic acid based polyurethane	Jan. 10, 1995
DE 19 637 404	Shimadzu	Ohara, H., Okamoto, T.	Recovery of lactide from high-molecular-weight polylactic acid)	Mar. 20, 1997
Jpn. Kokai Tokkyo Koho, JP 09 286 909 A2	Mitsui Toatsu Chemicals	Kobayashi, N., Imon, S., Kono, A., Kuroki, T., Wanibe, H.,	Lactic acid-based polymer films for agriculture	Nov., 4, 1997
Jpn. Kokai Tokkyo Koho, JP 09 110 968 A2	Shimadzu	Tasaka, S., Kosei, E.	Polymeric electret material and its manufacture	Apr. 28, 1998

proposed the utilization of stereocomplexation to enhance the mechanical properties of PLA materials. Young (1989) suggested the utilization of L-lactic acid solution containing its dimer and oligomer to enhance growth of plants. Hammel and York (1992) disclosed the preparation method of food-grade cellular materials. Van den Berg (1992) claimed the *in-situ* polymerization method of LAs to synthesize PLA materials for reconstructive orthopedics. Shikinami and Hata (1994) proposed that piezoelectric materials could be prepared by uniaxial drawing, which is also useful for treating bone fractures, in ultrasonographic devices, and in ultrasonic flaw detectors. The other patent articles issued relating to applications of PLAs have included those for fishing lines (Saito et al., 1991), cosmetic preparations (Greff, 1991), films useful in disposable shrink-wrap packing (Ostapchenko, 1992), films for agriculture (Kobayashi et al., 1997), and electret materials (Tasaka and Kosei, 1998).

Acknowledgments

The author wishes to thank Dr. Carlos Adriel Del Carpio, Department of Ecological Engineering, Faculty of Engineering, Toyohashi University of Technology, for his helpful and kind comments on the manuscript.

16
References

Agarwal, S., Brandukova-Szmikowski, N. E., Greiner, A. (1999) Samarium (III)-mediated graft polymerization of ε-caprolactone and L-lactide in functionalized poly(p-xylylene)s: model studies and polymerization, *Polym. Adv. Technol.* **10**, 528–534.

Agrawal, C. M., Athanasiou, K. A., Heckman, J. D. (1997) Biodegradable PLA-PGA polymers for tissue engineering in orthopedics, *Mater. Sci. Forum* **250**, 115–128.

Ajioka, M., Enomoto, E., Suzuki, K., Yamaguchi, A. (1995) Basic properties of polylactic acid produced by the direct condensation polymerization of lactic acid, *Bull. Chem. Soc. Jpn.* **68**, 2125–2131.

Arvanitoyannis, I., Nakayama, A., Kawasaki, N., Yamamoto, N. (1995) Novel star-shaped polylactide with glycerol using stannous octoate or tetraphenyl tin as catalyst: 1. Synthesis, characterization and study of their biodegradability, *Polymer* **36**, 2947–2956.

Arvanitoyannis, I., Nakayama, A., Psomiadou, E., Kawasaki, N., Yamamoto, N. (1996) Synthesis and degradability of a novel aliphatic polyester based on L-lactide and sorbitol:3, *Polymer* **37**, 651–660.

Asano, M., Fukuzaki, H., Yoshida, M., Kumakura, M., Mashimo, T., Yuasa, H., Imai, K., Yamanaka, H., Kawahara, U., Suzuki, K. (1991) *In vivo* controlled release of a luteinizing hormone-releasing hormone agonist from poly(DL-lactic acid) formulations of varying degradation pattern, *Int. J. Pharm.* **67**, 67–77.

Aslan, S., Calandrelli, L., Lauriezo, P., Malinconico, M., Migliaresi, C. (2000) Poly(D,L-lactide)/poly(ε-caprolactone) blend membranes: preparation and morphological characterisation, *J. Mater. Sci.* **35**, 1615–1622.

Atthoff, B., Trollsås, M., Claesson, H., Hedrick, J. L. (1999) Poly(lactides) with controlled molecular architecture initiated from hydroxy functional dendrimers and the effects on the hydrodynamic volume, *Macromol. Chem. Phys.* **200**, 1333–1339.

Blümm, E., Owen, A. J. (1995) Miscibility, crystallization and melting of poly(3-hydroxybutyrate)/poly(L-lactide) blends, *Polymer* **36**, 4077–4082.

Bodmeier, R., Oh, K. H., Chen, H. (1989) The effect of the addition of low molecular weight poly(DL-lactide) on drug release from biodegradable poly(DL-lactide) drug delivery systems, *Int. J. Pharm.* **51**, 1–8.

Bogaert, J.-C., Coszach, P. (2000) Poly(lactic acids): a potential solution to plastic waste dilemma, *Macromol. Symp.* **153**, 287–303.

Braud, C., Devarieux, R., Garreau, H., Vert, M. (1996) Capillary electrophoresis to analyze water-soluble oligo(hydroxyacids) issued from degraded or biodegraded aliphatic polyesters, *J. Environ. Polym. Degrad.* **4**, 135–148.

Breitenbach, A., Kissel, T. (1998) Biodegradable comb polyesters: Part 1 Synthesis, characterization and structural analysis of poly(lactide) and poly(lactide-co-glycolide) grafted onto water-soluble poly(vinyl alcohol) as backbone, *Polymer* **39**, 3261–3271.

Brizzolara, D., Cantow, H.-J., Diederichs, K., Keller, E., Domb, A. J. (1996a) Mechanism of stereocomplex formation between enantiomeric poly(lactide)s, *Macromolecules*, **29** 191–197.

Brizzolara, D., Cantow, H.-J., Mülhaupt, R., Domb, A. J. (1996b) Novel materials through stereocomplexation, *J. Computer-Aided Mater. Design* **3**, 341–350.

Brochu, S. Prud'homme, R. E., Barakat, I., Jérome, R. (1995) Stereocomplexation and morphology of polylactides, *Macromolecules* **28**, 5230–5239.

Cai, H. Dave, V. Gross, R. A., McCarthy, S. P. (1996) Effects of physical aging, crystallinity, and orientation on the enzymatic degradation of poly-

(lactic acid), *J. Polym. Sci.: Part B: Polym. Phys.* **34**, 2701–2708.

Cam, D., Hyon, S.-H., Ikada, Y. (1995) Degradation of high molecular weight poly(L-lactide) in alkaline medium, *Biomaterials* **16**, 833–843.

Carothers, W. H., Dorough, G. L., Van Natta, F. J. (1932) Studies of polymerization and ring formation. X. The reversible polymerization of six-membered cyclic esters, *J. Am. Chem. Soc.* **54**, 761–772.

Cartier, L. Okihara, T., Lotz, B. (1997) Triangular polymer single crystals: stereocomplexes, twins, and frustrated structures, *Macromolecules* **30**, 6313–6322.

Cartier, L., Okihara, T., Ikada, Y., Tsuji, H., Puiggali, J., Lotz, B. (2000) Epitaxial crystallization and crystalline polymorphism of polylactides, *Polymer* **41**, 8909–8919.

Celli, A., Scandola, M. (1992) Thermal properties and physical aging of poly(L-lactic acid), *Polymer* **33**, 2699–2703.

Cha, Y. Pitt, C. G. (1990) The biodegradability of polyester blend, *Biomaterials* **11**, 108–112.

Chaput, C., DesRosiers, E. A., Assad, M., Brochu, M., Yahia, L'H., Selmani, A., Rivard, C.-H. (1995) Processing biodegradable natural polyesters for porous soft-materials, in: *Advances in Materials Science and Implant Orthopedic Surgery* (Kossowski, R., Kossovsky, N., Eds.), Dordrecht: Kluwer Academic Publishers, 229–245.

Chen, X., McGurk, S. L., Davies, M. C., Roberts, C. J., Shakesheff, K. M., Tendler, S. J. B., Williams, P. M., Davies, J., Dawkes, A. C., Domb, A. (1998) Chemical and morphological analysis of surface enrichment in a biodegradable polymer blend by phase-detection imaging atomic force microscopy, *Macromolecules* **31**, 2278–2283.

Chisholm, M. H., Iyer, S. S., Matison, M. E., McCollum, D. G., Pagel, M. (1997) Concerning the stereochemistry of poly(lactide), PLA. Previous assignments are shown to be incorrect and a new assignment is proposed, *Chem. Commun.* 1999–2000.

Cho, K. Y., Kim. C.-H., Lee, J.-W., Park, J.-K. (1999) Synthesis and characterization of poly(ethylene-glycol) grafted poly(L lactide), *Macromol. Rapid Commun.* **20**, 598–601.

Choi, E.-J, Park, J.-K., Chang, H.-N. (1994) Effect of polymerization catalysts on the microstructure of P(LLA-co-εCL), *J. Polym. Sci.: Part B: Polym. Phys.* **32**, 2481–2489.

Chu, C. C., Campbell, N. D. (1982) Scanning electron microscopic study of the hydrolytic degradation of poly(glycolic acid), *J. Biomed. Mater. Res.* **16**, 417–430.

Cook, A. D., Pajvani, U. B., Hrkach, J. S., Cannizzaro, S. M., Langer, R. (1997) Calorimetric analysis of surface reactive amino group on poly(lactic acid-co-lysin): poly(lactic acid) blends, *Biomaterials* **18**, 1417–1424.

Coombes, A. G. A, Heckman, J. D. (1992a) Gel casting of resorbable polymers. 1. Processing and applications, *Biomaterials* **13**, 217–224.

Coombes, A. G. A, Heckman, J. D. (1992b) Gel casting of resorbable polymers. 2. In-vitro degradation of bone graft substitutes, *Biomaterials* **13**, 297–307.

Coombes, A. G. A., Meikle, M. C. (1994) Bioabsorbable synthetic polymers as replacements for bone graft, *Clin. Mater.* **17**, 35–67.

Cooper-White, J. J., Mackay, M. E. (1999) Rheological properties of poly(lactides). Effect of molecular weight and temperature on the viscoelasticity of poly(L-lactic acid), *J. Polym. Sci.: Part B: Polym. Phys.* **37**, 1803–1814.

Datta, R., Tsai, S.-P., Bonsignore, P., Moon, S.-H., Frank, J. R. (1995) Technological and economic potential of poly(lactic acid) and lactic acid derivatives, *FEMS Microbiol. Rev.* **16**, 221–231.

Davies, M. C., Shakesheff, K. M., Shard, A. G., Domb, A., Roberts, C. J., Tendler, S. J. B., Williams, P. M. (1996) Surface analysis of biodegradable polymer blends of poly(sebacic anhydride) and poly(DL-lactic acid), *Macromolecules* **29**, 2205–2212.

Degée, P., Dubois, P., Jérome, R. (1997), Bulk polymerization of lactides initiated by aluminium isopropoxide, 3. Thermal stability and viscoelastic properties, *Macromol. Chem. Phys.* **198**, 1985–1995.

de Jong, S. J., van Dijk-Wolthuis, W. N. E., Kettenes-van den Bosch, J. J., Schuyl, P. J. W., Hennink, W. E. (1998) Monodisperse enantiomeric lactic acid oligomers: preparation, characterization, and stereocomplex formation, *Macromolecules* **31**, 6397–6402.

de Jong, S. J., Smedt, S. C. De, Wahls, M. W. C., Demeester, J., Kettenes-van den Bosch, J. J., Hennink, W. E. (2000) Novel self-assembled hydrogels by stereocomplex formation in aqueous solution of enantiomeric lactic acid oligomers grafted to dextran, *Macromolecules* **33**, 3680–3686.

De Santis, P., Kovacs, A. J. (1968) Molecular conformation of poly(*S*-lactic acid), *Biopolymer* **6**, 299–306.

Dijkstra, P. J. Bulte, A., Feijen, J. (1991) Block copolymers of L-lactide, D-lactide, and ε-caprolactone, *The 17th Annual Meeting of the Society for*

Biomaterials, Scottsdale, Arizona. Minneapolis, Minnesota: Society for Biomaterials.

Domb, A. J. (1993) Degradable polymer blends. I. Screening of miscible polymers, *J. Polym. Sci.: Part A: Polym. Chem. Ed.* **31**, 1973–1981.

Dorgan, J. R., Williams, J. S., Lewise, D. N. (1999) Melt rheology of poly(lactic acid): entanglement and chain architecture effects, *J. Rheol.* **43**, 1141–1155.

Dubois, P., Degée, P., Ropson, N., Jérome, R. (1997) Macromolecular engineering of polylactones and polylactides by ring-opening polymerization, in: *Macromolecular Design of Polymeric Materials* (Hatada, K., Kitayama, T., Vogl, O., Eds.), New York: Marcel Dekker, Inc., 247–272.

Duek, E. A. R., Zavaglia, C. A. C, Belangero, W. D. (1999) In vitro study of poly(lactic acid) pin degradation, *Polymer* **40**, 6465–6473.

Edlund, U., Albertsson, A.-C. (2000) Microspheres from poly(D,L-lactide)/poly(1,5-dioxepan-2-one) miscible blends for controlled drug delivery, *J. Bioact. Compatible Polym.* **15**, 214–229.

Eguiburu, J. L., Fernandez-Berridi, M. J., Roman, J. S. (1996) Graft copolymers for biomedical applications prepared by free radical polymerization of poly(L-lactide) macromonomers with vinyl and acrylic monomers, *Polymer* **37**, 3615–3622.

Eling, B., Gogolewski, S., Pennings, A.J. (1982) Biodegradable materials of poly(L-lactic acid): 1. Melt-spun and solution spun fibres, *Polymer* **23**, 1587–1593.

Fambri, L., Pegoretti, A., Fenner, R., Incardona, S. D., Migliaresi, C. (1997) Biodegradable fibres of poly(L-lactic acid) produced by melt-spinning, *Polymer* **38**, 79–85.

Fischer, E. W., Sterzel, H. J., Wegner, G. (1973) Investigation of the structure of solution grown crystals of lactide copolymers by means of chemical reaction, *Kolloid-Z. u. Z. Polym.* **251**, 980–990.

Fukuzaki, H., Yoshida, M., Asano, M., Kumakura, M. (1989) Synthesis of copoly(D,L-lactic acid) with relatively low molecular weight and *in vitro* degradation, *Eur. Polym. J.* **25**, 1019–1026.

Gajria, A. M., Davé, V., Gross, R. A., McCarthy, S. P. (1996) Miscibility and biodegradability of blends of poly(lactic acid) and poly(vinyl acetate), *Polymer* **37**, 437–444.

Gan, Z., Yu, D., Zhong, Z., Liang, Q., Jing, X. (1999) Enzymatic degradation of poly(ε-caprolactone)/poly(DL-lactide) blends in phosphate-buffer solution, *Polymer* **40**, 2859–2862.

Gedde, U. W. (1995) *Polymer Physics*, London: Chapman & Hall, Chapters 7 and 8, 131–198.

Gilding, D. K., Reed, A. M. (1979) Biodegradable polymers for use in surgery-polyglycolic/poly(lactic acid) homo- and copolymers: 1, *Polymer* **20**, 1459–1464.

Göpferich, A (1997) Mechanisms of polymer degradation and elimination, in: *Handbook of Biodegradable Polymers* (Domb, A. J., Kost, A., Wiseman, D. M., Eds.), Amsterdam: Harwood Academic Publishers, 451–471.

Gonzalez, M. F., Ruseckaite, R. A., Cuadrado, T. R. (1999) Structural changes of polylactic-acid (PLA) microspheres under hydrolytic degradation, *J. Appl. Polym. Sci.* **71**, 1223–1230.

Grijpma, D. W., Kroeze, E., Nijenhuis, A. J., Pennings, A. J. (1993) Poly(L-lactide) crosslinked with spiro-bis-dimethylene-carbonate, *Polymer* **34**, 1496–1503.

Grijpma, D. W., van Hofslot, R. D. A., Supèr, H., Nijenhuis, A. J., Pennings, A. J. (1994) Rubber toughing of poly(lactide) by blending and block copolymerization, *Polym. Eng. Sci.* **34**, 1674–1684.

Grizzi, I., Garreau, H., Li, S., Vert, M. (1995) Hydrolytic degradation of devices based on poly(DL-lactic acid) size-dependence, *Biomaterials* **16**, 305–311.

Hartmann, M. H. (1998) High molecular weight polylactic acid polymers, in: *Biopolymers from Renewable Resources* (Kaplan, D.L., Ed.), Berlin: Springer-Verlag, 367–411.

Hartmann, M. (1999) Advances in the commercialization of poly(lactic acid), *Polym. Prepr. (Am. Chem. Soc., Div. Polym. Chem.)* **40**, 570–571.

Higashi, S., Yamamuro, T., Nakamura, T., Ikada, Y., Hyon, S.-H., Jamshidi, K. (1986) Polymer-hydroxyapatite composites for biodegradable bone fillers, *Biomaterials* **7**, 183–187.

Hiki, S., Miyamoto, M., Kimura, Y. (2000) Synthesis and characterization of hydroxy-terminated (*RS*)-poly(3-hydroxybutyrate) and its utilization to block copolymerization with l-lactide to obtain a biodegradable thermoplastic elastomer, *Polymer* **41**, 7369–7379.

Hiljanen-Vainio, M., Varpomaa, P., Seppälä, J., Törmälä, P. (1996) Modification of poly(L-lactides) by blending: mechanical and hydrolytic behavior, *Macromol. Chem. Phys.* **197**, 1503–1523.

Ho, K.-L. G., Pometto, A. L., III (1999) Temperature effects on soil mineralization of polylactic acid plastic in laboratory respirometers, *J. Environ. Polym. Degrad.* **7**, 101–108.

Ho, K.-L. G., Pometto, A.L., III, Hinz, P. N. (1999) Effects of temperature and relative humidity on polylactic acid plastic degradation, *J. Environ. Polym. Degrad.* **7**, 83–92.

Hoogsteen, W., Postema, A. R., Pennings, A. J., ten Brinke, G., Zugenmaier, P. (1990) Crystal structure, conformation, and morphology of solution-spun poly(L-lactide) fibers, *Macromolecules* **23**, 634–642.

Horacek, I., Kalísek, V. (1994a) Polylactide. I. Continuous dry spinning-hot drawing preparation of fibers, *J. Appl. Polym. Sci.* **54**, 1751–1757.

Horacek, I., Kalísek, V. (1994b) Polylactide. II. Discontinuous dry spinning-hot drawing preparation of fibers, *J. Appl. Polym. Sci.* **54**, 1759–1765.

Horacek, I., Kalísek, V. (1994c) Polylactide. III. Fiber preparation by spinning in precipitant vapor, *J. Appl. Polym. Sci.* **54**, 1767–1771.

Hori, Y., Takahashi, Y. Yamaguchi, A., Nishishita, T. (1993) Ring-opening copolymerization of optically active β-butyrolactone with several lactones catalyzed by distannoxane complexes: synthesis of new biodegradable polyesters, *Macromolecules* **26**, 4388–4390.

Howe, C., Vasanthan, N., MacClamrock, C., Sankar, S., Shin, I. D., Simonsen, I. K., Tonelli, A. E. (1994) Inclusion compound formed between poly(L-lactic acid) and urea, *Macromolecules* **27**, 7433–7436.

Hsu, S.-H., Chen, W.-C. (2000) Improved adhesion by plasma-induced grafting of l-lactide onto polyurethane surface, *Biomaterials* **21**, 359–368.

Hu, L.-C., Nakata, H., Yamane, H., Kitada, T. (1994) The role of poly(L-lactide) in the degradation of poly(ε-amino caproic acid), *Kobunshi Ronbunshu* **51**, 486–492.

Hu, L.-C., Shinoda, H., Yoshida, E., Kitada, T. (1995) Hydrolysis of polyblends comprising poly(L-lactide) and poly(ε-amino-caproic acid), *Kobunshi Ronbunshu* **52**, 114–120.

Huang, J., Lisowski, M. S., Runt, J., Hall, E. S., Kean, R. T., Buehler, N., Lin, J. S. (1998) Crystallization and microstructure of poly(L-lactide-co-meso-lactide) copolymers, *Macromolecules* **31**, 2593–2599.

Hyon, S.-H., Jamshidi, K., Ikada, Y. (1984) Melt spinning of poly-L-lactide and hydrolysis of the fiber *in vitro*, in: *Polymers as Biomaterials* (Shalaby, S. W., Hoffman, A. S., Ratner, B. D., Horbett, T.A., Eds.), New York: Plenum Press, 51–65.

Hyon, S.-H., Jamshidi, K., Ikada, Y., Higashi, S., Kakutani, Y., Yamamuro, T. (1985) Bone filler from poly(lactic acid)-hydroxyapatite composites, *Kobunshi Ronbunshu* **42**, 771–776.

Hyon, S.-H., Jamshidi, K., Ikada, Y. (1997) Synthesis of polylactides with different molecular weights, *Biomaterials* **18**, 1503–1508.

Hyon, S.-H., Jamshidi, K., Ikada, Y. (1998) Effects of the residual monomer on the degradation of DL-lactide copolymer, *Polym. Int.* **46**, 196–202.

Iannace, S., Nicolais, L. (1996) Isothermal crystallization and chain mobility of poly(L-lactide), *J. Appl. Polym. Sci.* **64**, 911–919.

Iannace, S., Ambrosio, L., Huang, S. J., Nicolais, L. (1994) Poly(3-hydroxybutyrate)-co-(3-hydroxyvalerate)/poly-L-lactide blends: thermal and mechanical properties, *J. Appl. Polym. Sci.* **54**, 1525–1536.

Iannace, S., Ambrosio, L., Huang, S. J., Nicolais, L. (1995) Effect of degradation on the mechanical properties of multiphase polymer blends: PHBV/PLLA, *J. Macromol. Sci.-Pure Appl. Chem.* **A32**, 881–888.

Ibim, S. E. M., Ambrosio, A. M. A., Kwon, M. S., El-Amin, S. F., Allcock, H. R., Laurencin, C. T. (1997) Novel polyphosphazene/poly(lactide-co-glycolide) blends: miscibility and degradation studies, *Biomaterials* **18**, 1565–1569.

Ikada, Y. (1998) Tissue Engineering research trends at Kyoto University, in: *Tissue Engineering for Therapeutic Use 1* (Ikada, Y., Yamaoka, Y., Eds.), Washington, DC, Am. Chem. Soc., 1–14.

Ikada, Y., Tsuji. H. (2000) Biodegradable polyesters form medical and ecological applications, *Macromol. Rapid Commun.* **21**, 117–132.

Ikada, Y., Jamshidi, K., Tsuji, H., Hyon, S.-H. (1987) Stereocomplex formation between enantiomeric poly(lactides), *Macromolecules* **20**, 904–906.

Ikada, Y., Shikinami, Y., Hara, Y., Tagawa, M., Fukada, E. (1996) Enhancement of bone formation by drawn poly(L-lactide), *J. Biomed. Mater. Res.* **30**, 553–558.

Inoue, K., O-oya, S., Mun-soo, L., Akasaka, S., Asai, S., Sumita, M. (1998) Fractal and degradation process of biodegradable polyester blends, *Sen'i Gakkaishi* **54**, 277–284.

Ishida, T., Tsuji, H. (2000) Hydrolysis mechanisms of PLLA in enzyme and alkaline solutions, *Polym. Prepr. Jpn.* **49**, 1008.

Ivanova, T. Z., Panaiotov, I., Boury, F., Proust, J. E., Verger, R. (1997) Enzymatic hydrolysis of poly(D,L-lactide) spread monolayers by cutinase, *Colloid Polym. Sci.* **275**, 449–457.

Iwata, T., Doi, Y. (1998) Morphology and enzymatic degradation of poly(L-lactic acid) single crystals, *Macromolecules* **31**, 2461–2467.

Jamshidi, K., Hyon, S.-H., Nakamura, T., Ikada, Y., Shimidu, Y., Teramatsu, T. (1986) In vitro and in vivo degradation of poly-L-lactide fibers, in: *Biological and Biomechanical Performance of Biomaterials* (Christel, P., Meunier, A., Lee, A. J. C., Eds.), Amsterdam, The Netherlands: Elsevier Science Publisher B.V., 227–232.

Jamshidi, K., Hyon, S.-H., Ikada, Y. (1988) Thermal characterization of polylactides, *Polymer* **29**, 2229–2234.

Jorda, R., Wilkes, G. L. (1988) A novel use of physical aging to distinguish immiscibility in polymer blends, *Polym. Bull.* **20**, 479–485.

Kalb, B., Pennings, A. J. (1980) General crystallization behaviour of poly(L-lactic acid), *Polymer* **21**, 607–612.

Kamei, S., Inoue, Y., Okada, H., Yamada, M., Ogawa, Y., Toguchi, H. (1992) New method for analysis of biodegradable polyesters by high-performance liquid chromatography after alkaline hydrolysis, *Biomaterials* **13**, 953–958.

Karjomaa, S., Suortti, T., Lempiäinen, R., Selin, J.-F., Itävaara, M. (1998) Microbial degradation of poly-(L-lactic acid) oligomers, *Polym. Degrad. Stabil.* **59**, 333–336.

Karlsson, S., Albertsson, A.-C. (1998) Abiotic and biotic degradation of aliphatic polyesters from "petro" versus "green" resources, *Macromol. Symp.* **127**, 219–225.

Kasperczyk, J. E. (1995) Microstructure analysis of poly(lactic acid) obtained by lithium tert-butoxide as initiator, *Macromolecules* **28**, 3937–3939.

Kasperczyk, J. (1996) Microstructural analysis of poly[(L,L-lactide)-co-(glycolide)] by ^{1}H and ^{13}C n.m.r. spectroscopy, *Polymer* **37**, 201–203.

Kasperczyk, J., Bero, M. (1991) Coordination polymerization of lactides, 2. Micro structure determination of poly[(L,L-lactide)-co-(ε-caprolactone)] with ^{13}C nuclear magnetic resonance spectroscopy, *Makromol. Chem.* **192**, 1777–1787.

Kharas, G. B., Sanchez-Riera, F., Severson, D. K. (1994) Polymer of lactic acids, in: *Plastics from Microbes* (Mobley, D. P., Ed.), New York: Hanser Publishers, 93–137.

Kim, C.-H., Cho, K. Y., Choi, E.-J., Park, J.-K. (2000) Effect of P(lLA-co-εCL) on the compatibility and crystallization behavior of PCL/PLLA blends, *J. Appl. Polym. Sci.* **77**, 226–231.

Kim, S. H., Kim, Y. H. (1994) Biodegradable star-shaped poly-L-lactide, in: *Biodegradable Plastics and Polymers* (Doi, Y, Fukuda, K., Eds.), Amsterdam: Elsevier Science B.V., 464–469.

Kimura, Y., Yamaoka, T. (1996) Surface modification and properties of biodegradable polymers based on poly-L-lactides, in: *Surface Science of Crystalline Polymers* (Yui, N., Terano, M., Eds.), Japan: Kodansha Scientific Ltd., 163–172

Kimura, Y., Matsuzaki, Y., Yamane, H., Kitao, T. (1989) Preparation of block copoly(ester-ether) comprising poly(L-lactide) and poly(oxypropylene) and degradation of its fibre *in vitro* and in vivo, *Polymer* **30**, 1342–1349.

Kobayashi, J., Asahi, T., Ichiki, M., Okikawa, A., Suzuki, H., Watanabe, T. Fukada, E., Shikinami, Y. (1995) Structural and optical properties of poly lactic acids, *J. Appl. Phys.* **77**, 2957–2973.

Kohn, F. E., Van Den Berg, J. W. A., Van De Ridder, G., Feijen, J. (1984) The ring-opening polymerization of D,L-lactide in the melt initiated with tetraphenyltin, *J. Appl. Polym. Sci.* **29**, 4265–4277.

Kolstad, J. J. (1996) Crystallization kinetics of poly(L-lactide-co-meso-lactide), *J. Appl. Polym. Sci.* **62**, 1079–1091.

Koyama, N., Doi, Y. (1995) Morphology and biodegradability of a binary blend of poly((*R*)-3-hydroxybutyric acid) and poly((*R*,*S*)-lactic acid), *Can. J. Microbiol.* **41**(Suppl.1), 316–322.

Koyama, N., Doi, Y. (1996) Miscibility, thermal properties, and enzymatic degradability of binary blends of poly((*R*)-3-hydroxybutyric acid) with poly(ε-caprolactone-co-lactide), *Macromolecules* **29**, 5843–5851.

Koyama, N., Doi, Y. (1997) Miscibility of binary blends of poly((*R*)-3-hydroxybutyric acid) and poly((*S*)-lactic acid), *Polymer* **38**, 1589–1594.

Kricheldorf, H. R., Kreiser, I. (1987a) Polylactones. 11. Cationic copolymerization of glycolide with L,L-dilactide, *Makromol. Chem.* **188**, 1861–1873.

Kricheldorf, H. R., Boettcher, C. (1993) Polylactones. XXV. Polymerizations of racemic- and meso-D,L-lactide with Zn, Pb, Sb, and Bi salts – stereochemical aspects, *J. Macromol. Sci., Pure Appl. Chem.* **A30**, 441–448.

Kricheldorf, H. R., Kreiser, I. (1987b) Polylactones. 13. Transesterification of poly(L-lactide) with poly(glycolide), poly(β-propiolactone), and poly(ε-caprolactone), *J. Macromol. Sci.-Chem.* **A24**, 1345–1356.

Kricheldorf, H. R., Kreiser-Saunders, I., Scharnagl, N. (1990), Anionic and pseudoanionic polymerization of lactones – A comparison, *Makromol. Chem. Macromol. Symp.* **32**, 285–298.

Kricheldorf, H. R., Kreiser-Saunders, I. (1996) Polylactides – synthesis, characterization and medical application, *Macromol. Symp.* **103**, 85–102.

Kricheldorf, H. R., Sumbél, M. (1989) Polylactones-18. Polymerization of L,L-lactide with Sn(II) and Sn(IV) halogenides, *Eur. Polym. J.* **25**, 585–591.

Kumagai, Y., Kanesawa, Y., Doi, Y. (1992) Enzymatic degradation of microbial poly(3-hydroxybutyrate) films, *Makromol. Chem.* **193**, 53–57.

Lam, K. H., Nieuwenhuis, P., Molenaar, I., Esselbrugge, H., Feijen, J., Dijkstra, P. J., Shakenraad, J. M. (1994) Biodegradation of porous versus non-porous poly(L-lactic acid) films, *J. Mater. Sci. Mater. Med.* **5**, 181–189.

Leenslag., J. W., Pennings, A. J. (1987a) Synthesis of high-molecular-weight poly(L-lactide) initiated with tin 2-ethylhexanoate, *Makromol. Chem.* **188**, 1809–1814.

Leenslag., J. W., Pennings, A. J. (1987b) High-strength poly(L-lactide) fibres by a dry-spinning/hot-drawing process, *Polymer* **28**, 1695–1702.

Lewis, D. H. (1990) Controlled release of bioactive agents from lactide/glycolide polymers, in: *Biodegradable Polymers as Drug Delivery Systems* (Chasin, M., Langer, R., Eds.), New York: Marcel Dekker, 1–42.

Li, S., Vert, M. (1995) Biodegradation of aliphatic polyesters, in: *Degradable Polymers – Principles and Applications* (Scott, G., Gilead, D., Eds.), London: Chapman & Hall, 43–87.

Li, S., McCarthy, S. (1999a) Influence of crystallinity and stereochemistry on the enzymatic degradation of poly(lactide)s, *Macromolecules* **32**, 4454–4456.

Li, S., McCarthy, S. (1999b) Further investigations on the hydrolytic degradation of poly(DL-lactide), *Biomaterials* **20**, 35–44.

Li, S., Garreau, H., Vert, M. (1990) Structure-property relationships in the case of the degradation of massive poly(α-hydroxy acids) in aqueous media, *J. Mater. Sci., Mater. Med.* **1**, 198–206.

Li, S., Tenon, M., Garreau, H., Braud, C., Vert, M. (2000) Enzymatic degradation of stereocopolymers derived from L-, DL- and meso-lactides, *Polym. Degrad. Stabil.* **67**, 85–90.

Li, Y., Nothnagel, N., Kissel, T. (1997) Biodegradable brush-like graft polymers from poly(D,L-lactide) or poly(D,L-lactide-co-glycolide) and charge-modified, hydrophilic dextrans as backbone – synthesis, characterization and *in vitro* degradation properties, *Polymer* **38**, 6197–6206.

Lillie, E. Schulz, R. C. (1975) ^{1}H- and ^{13}C-{^{1}H}-NMR spectra of stereocopolymers of lactide, *Makromol. Chem.* **176**, 1901–1906.

Liu, X., Dever, M., Fair, N., Benson, R. S. (1997) Thermal and mechanical properties of poly(lactic acid) and poly(ethylene/butylene succinate) blends, *J. Environ. Polym. Degrad.* **5**, 225–235.

Loomis, G. L. Murdoch, J. R. (1990) Polylactide compositions, US Patent 4 902 515.

Loomis, G. L., Murdoch, J. R., Gardner, K. H. (1990) Polylactide stereocomplexes, *Polym. Prepr.* **31**, 55.

Lostocco, M. R., Huang, S. J. (1997) Aliphatic polyester blends based upon poly(lactic acid) and oligomeric poly(hexamethylene succinate), *J. Macromol. Sci. -Pure Appl. Chem.* **A34**, 2165–2175.

Lostocco, M. R., Borzacchiello, A., Huang, S. J. (1998) Binary and ternary poly(lactic acid)/poly(ε-caprolactone) blends: the effects of oligo-ε-caprolactones upon mechanical properties, *Macromol. Symp.* **130**, 151–160.

Lowe, C. E. (1954) Preparation of high molecular weight polyhydroxyacetic ester, US Patent 2 668 162.

MacDonald, R. T., McCarthy, S. P., Gross, R. A. (1996) Enzymatic degradability of poly(lactide): effects of chain stereochemistry and material crystallinity, *Macromolecules* **29**, 7356–7361.

Majola, A., Vainionpää, S., Rokkanen, P., Mikkola, H.-M., Törmälä, P. (1992) Absorbable self-reinforced polylactide (SR-PLA) composite rods for fracture fixation: strength and strength retention in the bone and subcutaneous tissue of rabbits, *J. Mater. Sci. Mater. Med.* **3**, 43–47.

Makino, K., Arakawa, M., Kondo, T. (1985) Preparation and *in vitro* degradation properties of polylactide microcapsules, *Chem. Pharm. Bull.* **33**, 1195–1201.

Mallardé, D., Valière, M., David, C., Menet, M., Guérin, Ph. (1998) Hydrolytic degradability of poly(3-hydroxyoctanoate) and of a poly(3-hydroxyoctanoate)/poly(*R,S*-lactic acid) blend, *Polymer* **39**, 3387–3392.

Marega, C., Marigo, A., Di Noto, V., Zannetti, R., Martorana, A., Paganetto, G. (1992) Structure and crystallization kinetics of poly(L-lactic acid), *Makromol. Chem.* **193**, 1599–1606.

Mason, N. S., Miles, C. S., Sparks, R. E. (1981) Hydrolytic degradation of poly DL-(lactide), *Polym. Sci. Technol.* **14**, 279–291.

Matsumura, S., Mabuchi, K., Toshima, K. (1998) Novel ring-opening polymerization of lactide by lipase, *Macromol. Symp.* **130**, 285–304.

Matsumoto, A., Matsukawa, Y., Suzuki, T., Yoshino, H., Kobayashi, M. (1997) The polymer-alloys method as a new preparation method of biodegradable microspheres: principle and application to cisplatin loaded microspheres, *J. Controlled Release* **48**, 19–27.

Mauduit, J., Pérouse, E., Vert, M. (1996) Hydrolytic degradation of films prepared from blends of high and low molecular weight poly(DL-lactic acid)s, *J. Biomed. Mater. Res.* **30**, 201–207.

Mazzullo, S., Paganetto, G., Celli, A. (1992) Regime III crystallization in poly-(L-lactic) acid, *Progr. Colloid Polym. Sci.* **87**, 32–34.

Meredith, J. C., Amis, E. J. (2000) LCST phase separation in biodegradable polymer blends: poly(D,L-lactide) and poly(ε-caprolactone), *Macromol. Chem. Phys.* **201**, 733–739.

Meinander, K., Niemi, M., Hakola, J. S., Selin, J.-F. (1997), Polylactides – degradable polymers for fibers and films, *Macromol. Symp.* **123**, 147–153.

Middleton, J. C., Tipton, A. J. (2000) Synthetic biodegradable polymers as orthopedic devices, *Biomaterials* **21**, 2335–2346.

Migliaresi, C., Cohn, D., De Lollis, A., Fambri, L. (1991a) Dynamic mechanical and calorimetric analysis of compression-molded PLLA of different molecular weights: effect of thermal treatments, *J. Appl. Polym. Sci.* **43**, 83–95.

Migliaresi, C., De Lollis, A., Fambri, L., Cohn, D. (1991b) The effect of thermal history on the crystallinity of different molecular weight PLLA biodegradable polymers, *Clin. Mater.* **8**, 111–118.

Migliaresi, C., Fambri, L., Cohn, D. (1994) A study on the *in vitro* degradation of poly(lactic acid), *J. Biomater. Sci., Polym. Ed.* **5**, 591–606.

Miyata, T., Masuko, T. (1997) Morphology of poly(L-lactide) solution-grown crystals, *Polymer* **38**, 4003–4009.

Miyata, T., Masuko, T. (1998) Crystallization behavior of poly(L-lactide), *Polymer* **39**, 5515–5521.

Moon, S. I., Lee, C. W., Miyamoto, M., Kimura, Y. (2000) Melt polycondensation of L-lactic acid with Sn(II) catalysts activated by various proton acids: a direct manufacturing route to high molecular weight poly(L-lactic acid), *J. Polym. Sci., Part A: Polym. Chem.* **38**, 1673–1679.

Murdoch, J. R., Loomis, G. L. (1988a) Polylactide compositions, US Patent 4 719 246.

Murdoch, J. R., Loomis, G. L. (1988b) Polylactide compositions, US Patent 4 766 182.

Murdoch, J. R., Loomis, G. L. (1989) Polylactide compositions, US Patent 4 800 219.

Nagata, M., Okano, F., Sakai, W., Tsutsumi, N. (1998) Separation and enzymatic degradation of blend films of poly(L-lactic acid) and cellulose, *J. Polym. Sci.: Part A: Polym. Chem.* **36**, 1861–1864.

Nakafuku, C. (1994) High pressure crystallization of poly(L-lactic acid) in a binary mixture with poly(ethylene oxide), *Polym. J.* **26**, 680–687.

Nakafuku, C. (1996) Effects of molecular weight on the melting and crystallization of poly(L-lactic acid) in a mixture with poly(ethylene oxide), *Polym. J.* **28**, 568–575.

Nakafuku, C., Sakoda, M. (1993) Melting and crystallization of poly(L-lactic acid) and poly(ethylene oxide) binary mixture, *Polym. J.* **25**, 909–917.

Nakamura, T., Hitomi, S., Watanabe, S., Shimizu, Y., Jamshidi, K., Hyon, S.-H., Ikada, Y. (1989) Bioabsorption of polylactides with different molecular properties, *J. Biomed. Mater. Res.* **23**, 1115–1130.

Nef, J. U. (1914) Dissoziationsvorgänge in der zuckergruppe, *Liebigs. Ann. Chem.* **403**, 204–383.

Nijenhuis, A. J. Grijpma, D. W., Pennings, A. J. (1996a) Crosslinked poly(L-lactide) and poly(ε-caprolactone), *Polymer* **37**, 2783–2791.

Nijenhuis, A. J., Colstee, E., Grijpma, D. W., Pennings, A. J. (1996b) High molecular weight poly(L-lactide) and poly(ethylene oxide) blends: thermal characterization and physical properties, *Polymer* **37**, 5849–5857.

Ohkoshi, Y., Shirai, H., Gotoh, Y., Nagura, M. (1999) Instrinsic birefringence of poly(L-lactide), *Sen'i Gakkaishi* **55**, 62–68.

Ohkoshi, I., Abe, H., Doi, Y. (2000) Miscibility and solid-state structures for blends of poly[(*S*)-lactide] with atactic poly[(*R,S*)-3-hydroxybutyrate], *Polymer* **41**, 5985–5992.

Ohya, Y., Sugitou, T., Ouchi, T. (1995) Polycondensation of α-hydroxy acids by enzymes or PEG-modified enzymes in organic media, *J. Macromol. Sci.-Pure Appl. Chem.* **A32**, 179–190.

Ohya, Y., Maruhashi, S., Ouchi, T. (1998a) Graft polymerization of L-lactide on pullulan through the trimethylsilyl protection method and degradation of the graft copolymers, *Macromolecules* **31**, 4662–4665.

Ohya, Y., Maruhashi, S., Ouchi, T. (1998b) Preparation of poly(L-lactide)-grafted amylose through the trimethylsilyl protection method and degradation of the graft copolymers, *Macromol. Chem. Phys.* **199**, 2017–2022.

Okihara, T., Tsuji, M., Kawaguchi, A., Katayama, K., Tsuji, H., Hyon, S.-H., Ikada, Y. (1991) Crystal structure of stereocomplex of poly(L-lactide) and poly(D-lactide), *J. Macromol. Sci.-Phys.* **B30**, 119–140.

Okuzaki, H., Kubota, I., Kunugi, T. (1999) Mechanical properties and structure of the zone-drawn poly(L-lactic acid) fibers, *J. Polym. Sci., Part B: Polym. Phys.* **37**, 991–996.

Onyari, J. M., Huang, S. J. (1996) Graft copolymers of poly(vinyl alcohol) and poly(lactic acid), *Polym. Prepr. (Am. Chem. Soc., Div. Polym. Chem.)* **37**, 145–146.

Otsuka, H., Nagasaki, Y., Kataoka, K., Okano, T., Sakurai, Y. (1998) Reactive-PEG-polylactide block copolymer for tissue engineering, *Polym. Prepr. (Am. Chem. Soc., Div. Polym. Chem.)* **39**, 128–129.

Ovitt, T. M., Coates, G. W. (1999) Stereoselective ring-opening polymerization of *meso*-lactide: synthesis of syndiotactic poly(lactic acid), *J. Am. Chem. Soc.* **121**, 4072–4073.

Pan, Q. Y., Tasaka, S., Inagaki, N. (1996) Ferroelectric behavior in poly-L-lactic acid. *Jpn. J. Appl. Phys.* **35**, L1442–L1445.

Park, T. G., Cohen, S., Langer, R. (1992) Poly(L-lactide)/Pluronic blends: characterization of phase separation behavior, degradation, and morphology and use as protein-releasing matrices, *Macromolecules* **25**, 116–122.

Pelouze, P. M. J. (1845) Mémoire sur l'acide lactique, *Ann. Chim. Phys., Ser.3* **13**, 257–268.

Penning, J. P., Dijkstra, H., Pennings, A. J. (1993) Preparation and properties of absorbable fibers from L-lactide copolymers, *Polymer* **34**, 942–951.

Perego, G., Cella, G. D., Bastioli, C. (1996) Effects of molecular weight and crystallinity on poly(lactic acid) mechanical properties, *J. Appl. Polym. Sci.* **59**, 37–43.

Pistner, H., Bendix, D. R., Mühling, J., Reuther, J. F. (1993) Poly(L-lactide): a long-term degradation study in vivo. Part III: Analytical characterization, *Biomaterials* **14**, 291–298.

Pistner, H., Stallforth, H., Gutwald, R., Mühling, J., Reuther, J., Michel, C. (1994) Poly(L-lactide): a long-term degradation study in vivo, Part II: Physico-mechanical behaviour of implants, *Biomaterials* **15**, 439–450.

Pitt, C. G. (1990) Poly(ε-caprolactone) and its copolymers, in: *Biodegradable Polymers as Drug Delivery Systems* (Chasin, M., Langer, R., Eds.), New York: Marcel Dekker, 71–120.

Pitt, C. G., Gu, Z.-W. (1987) Modification of the rates of chain cleavage of poly(ε-caprolactone) and related polyesters in the solid state, *J. Controlled Release* **4**, 283–292.

Pitt, C. G., Jeffcoat, A. R., Zweidinger, R. A., Schindler, A. (1979) Sustained drug delivery system. I. The permeability of poly(ε-caprolactone), poly(DL-lactic acid), and their copolymers, *J. Biomed. Mater. Res.* **13**, 497–507.

Pitt, C. G., Cha, Y., Shah, S. S., Zhu, K. J. (1992) Blends of PVA and PGLA: control of the permeability and degradability of hydrogels by blending, *J. Controlled Release* **19**, 189–200.

Pitt, C. G., Wang, J., Shah, S. S., Sik, R., Chignell, C. F. (1993) ESR spectroscopy as a probe of the morphology of hydrogels and polymer-polymer blends, *Macromolecules* **26**, 2159–2164.

Pranamuda, H., Tokiwa, Y. (1999) Degradation of poly(L-lactide) by strains belonging to genus *Amycolatopsis*, *Biotechnol. Lett.* **21**, 901–905.

Pranamuda, H., Tokiwa, Y., Tanaka, H. (1997) Polylactide degradation by an *Amycolatopsis* sp., *Appl. Environ. Microbiol.* **63**, 1637–1640.

Puiggali, J., Ikada, Y., Tsuji, H., Cartier, L., Okihara, T., Lotz, B. (2000) The frustrated structure of poly(L-lactide), *Polymer* **41**, 8921–8930.

Radano, C. P., Baker, G. L., Smith, M. R., III (2000) Stereoselective polymerization of a racemic monomer with a racemic catalyst: direct preparation of the polylactide stereocomplex from racemic lactide, *J. Am. Chem. Soc.* **122**, 1552–1553.

Rafler, G., Jobmann, M. (1994) Controlled release systems of biodegradable polymers. 3rd communication: Degradation and release properties of poly(glycolide(50)-co-lactide(50))s with random and non-random monomer distribution, *Drug Made in Germany* **38**, 20–22.

Reed, A. M., Gilding, D. K. (1981) Biodegradable polymers for use in surgery – Poly(glycolic)/poly(lactic acid) homo and copolymers: 2. *In vitro* degradation, *Polymer* **22**, 494–498.

Reeve, M. S., McCarthy, S. P., Downey, M. J., Gross, R. A. (1994) Polylactide stereochemistry: effect on enzymatic degradability, *Macromolecules* **27**, 825–831.

Renstad, R., Karlsson, S., Sandgren, Å., Albertsson, A.-C. (1998) Influence of processing additives on the degradation of melt-pressed films of poly(ε-caprolactone) and poly(lactic acid), *J. Environ. Polym. Degrad.* **6**, 209–221.

Sarasua, J.-R., Prud'homme, R. E., Wisniewski, M., Le Borgne, A., Spassky, N. (1998) Crystallization and melting behavior of polylactides, *Macromolecules* **31**, 3895–3905.

Schindler, A., Gaetano, K. D. (1988) Poly(lactate) III. Stereoselective polymerization of meso dilactide, *J. Polym. Sci.: Part C: Polym. Lett.* **26**, 47–48.

Schindler, A., Harper, D. (1979) Polylactide. II. Viscosity-molecular weight relationships and unperturbed chain dimensions, *J. Polym. Sci., Polym. Chem. Ed.* **17**, 2593–2599.

Schindler, A., Hibionada, Y. M., Pitt, C. G. (1982) Aliphatic polyesters. III. Molecular weight and molecular weight distribution in alcohol-initiated polymerizations of ε-caprolactone, *J. Polym. Sci. Polym. Chem.* **20**, 319–326.

Serizawa, T., Yamashita, H., Fujiwara, T., Kimura, Y., Akashi, M. (2001). Stepwise assembly of enantiomeric poly(lactide)s on surfaces, *Macromolecules* **34**, 1996–2001.

Shneider, A. K. (1972) Polylactide sutures, US Patent 3 636 956.

Shakesheff, K. M., Davies, M. C., Roberts, C. J., Tendler, S. J. B., Shard, A. G., Domb, A. (1994) In situ atomic force microscopy imaging of polymer degradation in an aqueous environment, *Langmuir* **10**, 4417–4419.

Shen, Y., Zhu, K. J., Shen, Z., Yao, K.-M. (1996) Synthesis and characterization of highly random copolymer of ε-caprolactone and D,L-lactide using rare earth catalyst, *J. Polym. Sci.: Part A: Polym. Chem.* **34**, 1799–1805.

Sheth, M., Kumar, R. A., Davé, V., Gross, R. A., McCarthy, S. P. (1997) Biodegradable polymer blends from poly(lactic acid) and poly(ethylene glycol), *J. Appl. Polym. Sci.* **66**, 1495–1505.

Shih, C. (1995) A graphical method for the determination of the mode of hydrolysis of biodegradable polymers, *Pharm. Res.* **12**, 2036–2040.

Shogren, R. (1997) Water vapor permeability of biodegradable polymers, *J. Environ. Polym. Degrad.* **5**, 91–95.

Siemann, U. (1992) The solubility parameter of poly(DL-lactic acid), *Eur. Polym. J.* **28**, 293–297.

Södergård, A. (1998) Preparation of poly(L-lactide-graft-acrylic acid), *Polym. Prepr. (Am. Chem. Soc. Div., Polym. Chem.)* **39**, 214–215.

Song, C. X., Feng, X. D. (1984) Synthesis of ABA triblock copolymers of ε-caprolactone and DL-lactide, *Macromolecules* **17**, 2764–2767.

Spassky, N., Pluta, C., Simic, V., Thiam, M., Wisniewski, M. (1998) Stereochemical aspects of the controlled ring-opening polymerization of chiral cyclic esters, *Macromol. Symp.* **128**, 39–51.

Spinu, M., Gardner, K. H. (1994) A new approach to stereocomplex formation between L- and D-poly(lactic acid)s, *Abstracts of Papers of the American Chemical Society,* **208**, No. Pt. 2, pp. 12 (PMSE).

Spinu, M., Jackson, C., Keating, M. Y., Gardner, K. H. (1996) Material design in poly(lactic acid) system: block copolymers, star homo- and copolymers, and stereocomplexes, *J. Macromol. Sci.-Pure Appl. Chem.* **A33**, 1497–1530.

Stevels, W. M., Ankoné, M. J. K., Dijkstra, P. J., Feijen, J. (1995) Stereocomplex formation in ABA triblock copolymers of poly(lactide) (A) and poly(ethylene glycol) (B), *Macromol. Chem. Phys.* **196**, 3687–3694.

Storey, R. F., Hickey, T. P. (1994) Degradable polyurethane networks based on D,L-lactide, glycolide, ε-caprolactone, and trimethylene carbonate homopolyester and copolyester triols, *Polymer* **35**, 830–838.

Storey, R. F., Warren, S. C., Allison, C. J., Wiggins, J. S., Puckett, A. D. (1993) Synthesis of bioabsorbable networks from methacrylate-endcapped polyesters, *Polymer* **34**, 4365–4372.

Stridsberg, K, Albertsson, A.-C. (2000) Changes in chemical and thermal properties of the tri-block copolymer poly(L-lactide-b-1,5-dioxepan-2-one-b-L-lactide) during hydrolytic degradation, *Polymer* **41**, 7321–7330.

Tajitsu, Y., Hosoya, R., Murayama, T., Aoki, M., Shikinami, Y., Date, M., Fukada, E. (1999) Huge optical rotatory power of uniaxially oriented film of poly-L-lactic acid, *J. Mater. Sci. Lett.* **18**, 1785–1787.

Thakur, K. A. M., Kean, R. T., Zupfer, J. M., Buehler, N. U., Doscotch, M. A., Munson, E. J. (1996) Solid state ^{13}C NMR studies of the crystallinity and morphology of poly(L-lactide), *Macromolecules* **29**, 8844–8851.

Thomson, R. C., Wake, M. C., Yaszemski, M. J., Mikos, A. G. (1995a) Biodegradable polymer scaffolds to regenerate organs, in: *Advances in Polymer Science: Biopolymer II* (Peppas, N. A., Langer, R. S., Eds.), Berlin: Springer-Verlag, 245–274, vol. 122.

Thomson, R. C, Yaszemski, M. J., Powers, J. M., Mikos, A. G. (1995b) Fabrication of biodegradable polymer scaffolds to engineer trabecular bone, *J. Biomater. Sci: Polym. Ed.* **7**, 23–38.

Tokiwa, Y., Konno, M., Nishida, H. (1999) Isolation of silk degrading microorganisms and its poly(L-lactide) degradability, *Chem. Lett.* 355–356.

Tonelli, A. E. (1992) Polylactides in channels, *Macromolecules* **25**, 3581–3584.

Törmälä, P. (1992) Biodegradable self-reinforced composite materials; manufacturing structure and mechanical properties, *Clin. Mater.* **10**, 29–34.

Torres, A., Li, S., Roussos, S., Vert, M. (1996a) Poly(lactic acid) degradation in soil or under controlled conditions, *J. Appl. Polym. Sci.* **62**, 2295–2302.

Torres, A., Li, S., Roussos, S., Vert, M. (1996b) Degradation of L- and DL-lactic aid oligomers in the presence of *Fusarium moniliforme* and *Pseudomonas putida*, *J. Environ. Polym. Degrad.* **4**, 213–223.

Torres, A., Li, S., Roussos, S., Vert, M. (1996c) Screening of microorganisms of biodegradation of poly(lactic acid) and lactic acid-containing polymers, *Appl. Environ. Microbiol.* **62**, 2393–2397.

Torres, A., Li, S., Roussos, S., Vert, M. (1999) Microbial degradation of a poly(lactic acid) as a model of synthetic polymer degradation mechanisms in outdoor conditions, *ACS Symp. Ser.*, **723**, 218–226.

Trofimoff, L., Aida, T., Inoue, S. (1987) Formation of poly(lactide) with controlled molecular weight. Polymerization of lactide by aluminium porphyrin, *Chem. Lett.* 991–994.

Tsuchiya, F., Tomida, Y., Fujimoto, K., Kawaguchi, H. (1993) Surface modification of poly(L-lactic

acid) by plasma discharge, *Polym. Prepr. Jpn.* **42**, 4916–4918.
Tsuji, H. (2000a) Stereocomplex from enantiomeric polylactides, in: *Research Advances in Macromolecules* (Mohan, R. M., Ed.), Trivandrum, India: Grobal Research Network, 25–48, Vol. 1.
Tsuji, H. (2000b) Hydrolysis of biodegradable aliphatic polyesters, in: *Recent Research Developments in Polymer Science* (Salamone, A. B., Brandrup, J., Ottenbrite, R.M., Editorial advisors), Trivandrum, India: Transworld Research Network, 13–37, Vol. 4.
Tsuji, H. (2000c) *In vitro* hydrolysis of blends from enantiomeric poly(lactide)s. Part 1. Well-stereocomplexed blend and non-blended films, *Polymer* **41**, 3621–3630.
Tsuji, H., Ikada, Y. (1992) Stereocomplex formation between enantiomeric poly(lactic acid)s. 6. Binary blends from copolymers, *Macromolecules* **25**, 5719–5723.
Tsuji, H., Ikada, Y. (1993) Stereocomplex formation between enantiomeric poly(lactic acid)s. 9. Stereocomplexation from the melt, *Macromolecules* **26**, 6918–6926.
Tsuji, H., Ikada, Y. (1994) Stereocomplex formation between enantiomeric poly(lactic acid)s. X. Binary blends from poly(D-lactide-co-glycolide) and poly(L-lactide-co-glycolide), *J. Appl. Polym. Sci.* **53**, 1061–1071.
Tsuji, H., Ikada, Y. (1995a) Blends of isotactic and atactic poly(lactide). I. Effects of mixing ratio of isomers on crystallization of blends from melt, *J. Appl. Polym. Sci.* **58**, 1793–1802.
Tsuji, H., Ikada, Y. (1995b) Properties and morphologies of poly(L-lactide): 1. Annealing condition effects on properties and morphologies of poly(L-lactide), *Polymer* **36**, 2709–2716.
Tsuji, H., Ikada, Y. (1996a) Crystallization from the melt of poly(lactide)s with different optical purities and their blends, *Macromol. Chem. Phys.* **197**, 3483–3499.
Tsuji, H., Ikada, Y. (1996b) Blends of isotactic and atactic poly(lactide): 2. Molecular weight effects of atactic component on crystallization and morphology of equimolar blends from the melt, *Polymer* **37**, 595–602.
Tsuji, H., Ikada, Y. (1996c) Blends of aliphatic polyesters. I. Physical properties and morphologies of solution-cast blends from poly(DL-lactide) and poly(ε-caprolactone), *J. Appl. Polym. Sci.* **60**, 2367–2375.
Tsuji, H., Ikada, Y. (1997) Blends of crystallizable and amorphous poly(lactide). III. Hydrolysis of solution-cast blend films, *J. Appl. Polym, Sci.* **63**, 855–863.
Tsuji, H., Ikada, Y. (1998a) Blends of aliphatic polyesters. II. Hydrolysis of solution-cast blends from poly(L-lactide) and poly(ε-caprolactone) in phosphate-buffered solution, *J. Appl. Polym. Sci.* **67**, 405–415.
Tsuji, H., Ikada, Y. (1998b) Properties and morphology of poly(L-lactide). II. Hydrolysis in alkaline solution, *J. Polym. Sci.: Part A: Polym. Chem.* **36**, 59–66.
Tsuji, H., Ikada, Y. (1999a) Physical properties of polylactides, in: *Current Trends in Polymer Science* (DeVries, K. L. et al., Editorial Advisory Board), Trivandrum, India: Research Trends, 27–46, Vol. 4.
Tsuji, H., Ikada, H. (1999b) Stereocomplex formation between enantiomeric poly(lactic acid)s. XI. Mechanical properties and morphology, *Polymer* **40**, 6699–6708.
Tsuji, H., Ikada, Y. (2000a) Enhanced crystallization of poly(L-lactide-co-ε-caprolactone) during storage at room temperature, *J. Appl. Polym. Sci.* **76**, 947–953.
Tsuji, H., Ikada, Y. (2000b) Properties and morphology of poly(L-lactide). 4. Effects of structural parameters on long-term hydrolysis of poly(L-lactide) in phosphate-buffered solution, *Polym. Degrad. Stabil.* **67**, 179–189.
Tsuji, H., Ishizaka, T. (2001a) Porous biodegradable polyesters. 2. Physical properties, morphology, and enzymatic and alkaline hydrolysis of porous poly(ε-caprolactone) films, *J. Appl. Polym. Sci.* **80**, 2281–2291.
Tsuji, H., Ishizaka, T. (2001b) Porous biodegradable polyesters, 3. Preparation of porous poly(ε-caprolactone) films from blends by selective enzymatic removal of poly(L-lactide), *Macromol. Biosci.* **1**, 59–65.
Tsuji, H., Miyauchi, S. (2001a) Poly(L-lactide). 6. Effects of crystallinity on enzymatic hydrolysis of poly(L-lactide) without free amorphous region, *Polym. Degrad. Stabil.* **71**, 415–424.
Tsuji, H., Miyauchi, S. (2001b) Poly(L-lactide). 7. Enzymatic hydrolysis of free and restricted amorphous regions in poly(L-lactide) films with different crystallinities and a fixed crystalline thickness, *Polymer* **42**, 4465–4469.
Tsuji, H., Miyauchi, S. (2001c) Enzymatic hydrolysis of polylactides. Effects of molecular weight, L-lactide content, and enantiomeric and diastereoisomeric polymer blending, *Biomacromolecules* **2**, 597–604.
Tsuji, H., Muramatsu, H. (2001a) Blends of aliphatic polyesters. 5. Non-enzymatic and enzymatic hydrolysis of blends from hydrophobic

poly(L-lactide) and hydrophilic poly(vinyl alcohol), *Polym. Degrad. Stabil.* **71**, 403–413.
Tsuji, H., Muramatsu, H. (2001b) Blends of aliphatic polyesters. 4. Morphology, swelling behavior, and surface and bulk properties of blends from hydrophobic poly(L-lactide) and hydrophilic poly(vinyl alcohol), *J. Appl. Polym. Sci.* **81**, 2151–2160.
Tsuji, H., Nakahara, K. (2001) Poly(L-lactide). 8. High-temperature hydrolysis of poly(L-lactide) films with different crystallinities and crystalline thicknesses in phosphate-buffered solution, *Macromol. Mater. Eng.* **286**, 398–406.
Tsuji, H., Sumida, K. (2001) Poly(L-lactide). V. Effect of storage in swelling solvents on physical properties and structure of poly(L-lactide), *J. Appl. Polym. Sci.* **79**, 1582–1589.
Tsuji, H., Suzuki, M. (2001) In vitro hydrolysis of blends from enantiomeric poly(lactide)s. 2. Well-stereocomplexed fibers and films, *Sen'i Gakkaishi* **57**, 198–202.
Tsuji, H., Hyon, S.-H., Ikada, Y. (1991a) Stereocomplex formation between enantiomeric poly(lactic acid)s. 2. Stereocomplex formation in concentrated solutions, *Macromolecules* **24**, 2719–2724.
Tsuji, H., Hyon, S.-H., Ikada, Y. (1991b) Stereocomplex formation between enantiomeric poly(lactic acid)s. 3. Calorimetric studies on blend films cast from dilute solution, *Macromolecules* **24**, 5651–5656.
Tsuji, H., Hyon, S.-H., Ikada, Y. (1991c) Stereocomplex formation between enantiomeric poly(lactic acid)s. 4. Differential scanning calorimetric studies on precipitates from mixed solutions of poly(D-lactic acid) and poly(L-lactic acid), *Macromolecules* **24**, 5657–5662.
Tsuji, H., Hyon, S.-H., Ikada, Y. (1992a) Stereocomplex formation between enantiomeric poly(lactic acid)s. 5. Calorimetric and morphological studies on the stereocomplex formed in acetonitrile solution, *Macromolecules* **25**, 2940–2946.
Tsuji, H., Horii, F., Nakagawa, M., Ikada, Y., Odani, H., Kitamaru, R. (1992b) Stereocomplex formation between enantiomeric poly(lactic acid)s. 7. Phase structure of the stereocomplex crystallized from a dilute acetonitrile solution as studied by high-resolution solid-state ^{13}C NMR spectroscopy, *Macromolecules* **25**, 4114–4118.
Tsuji, H., Ikada, Y., Hyon, S.-H., Kimura, Y., Kitada, T. (1994) Stereocomplex formation between enantiomeric poly(lactic acid)s. VIII. Complex fiber spun from mixed solution of poly(D-lactic acid) and poly(L-lactic acid), *J. Appl. Polym. Sci.* **51**, 337–344.
Tsuji, H., Mizuno, A., Ikada, Y. (1998) Blends of aliphatic polyesters. III. Biodegradation of solution-cast blends from poly(L-lactide) and poly(ε-caprolactone), *J. Appl, Polym, Sci.* **70**, 2259–2268.
Tsuji, H., Smith, R., Bonfield, W., Ikada, Y. (2000a) Porous biodegradable polyesters. I. Preparation of porous poly(L-lactide) by extraction of poly(ethylene oxide) from their blends, *J. Appl. Polym. Sci.* **75**, 629–637.
Tsuji, H., Mizuno, A., Ikada, Y. (2000b) Properties and morphologies of poly(L-lactide). III. Effects of initial crystallinity on long-term *in vitro* hydrolysis of high molecular weight poly(L-lactide), *J. Appl. Polym. Sci.* **77**, 1452–1464.
Vainionpää, S., Rokkanen, P., Törmälä, P. (1989) Surgical applications of biodegradable polymers in human tissues, *Prog. Polym. Sci.* **14**, 679–716.
van de Witte, P., Esselbrugge, H., Dijkstra, P.J., van den Berg, J. W. A., Feijen, J. (1996a) Phase transitions during membrane formation of polylactides, I. A morphological study of membranes obtained from the system polylactide-chloroform-methanol, *J. Membrane Sci.* **113**, 223–236.
van de Witte, P., Boorsma, A., Esselbrugge, H., Dijkstra, P. J., van den Berg, J. W. A., Feijen, J. (1996b) Differential scanning calorimetry study of phase transitions in poly(lactide)-chloroform-methanol system, *Macromolecules* **29**, 212–219.
van de Witte, P., Dijkstra, P. J., van den Berg, J. W. A., Feijen, J. (1996c) Phase separation processes in polymer solutions in relation to membrane formation, *J. Membrane Sci.* **117**, 1–31.
van de Witte, Dijkstra, P. J., van den Berg, J. W. A., Feijen, J. (1996d) Phase behavior of polylactides in solvent-nonsolvent mixtures, *J. Polym. Sci., Part B: Polym. Phys.* **34**, 2553–2568.
van de Witte, P., Esselbrugge, H., Dijkstra, P. J., van den Berg, J. W. A., Feijen, J. (1996e) A morphological study of membranes obtained from the systems polylactide-dioxane-methanol, polylactide-dioxane-water, polylactide-*N*-methyl pyrrolidone-water, *J. Polym. Sci., Part B: Polym. Phys.* **34**, 2569–2578.
van de Witte, P., van den Berg, J. W. A., Feijen, J., Reeve, J. L., Mchugh, A. J. (1996f) In situ analysis of solvent/nonsolvent exchange and phase separation processes during membrane formation of polylactides, *J. Appl. Polym. Sci.* **61**, 685–695.
van de Witte, P., Dijkstra, P. J., van den Berg, J. W. A., Feijen, J. (1997) Metastable liquid-liquid and

solid-liquid phase boundaries in polymer-solvent-nonsolvent system, *J. Polym. Sci., Part B: Polym. Phys.* **35**, 763–770.

Vasanthakumari, R. (1981) Eutectic crystallization of poly(L-lactic acid) and pentaerythrityl tetrabromide, *Polymer* **22**, 862–865.

Vasanthakumari, R., Pennings, A. J. (1983) Crystallization kinetics of poly(L-lactic acid), *Polymer* **24**, 175–178.

von Recum, H. A., Cleek, R. L. Eskin, S. G., Mikos, A. G. (1995) Degradation of polydispersed poly(L-lactic acid) to modulate lactic acid release, *Biomaterials* **16**, 441–447.

Wada, R., Hyon, S.-H., Nakamura, T., Ikada, Y. (1991) In vitro evaluation of sustained drug release from biodegradable elastomer, *Pharm. Res.* **8**, 1292–1296.

Wang, L., Ma, W. Gross, R. A., McCarthy, S. P. (1998) Reactive compatibilization of biodegradable blends of poly(lactic acid) and poly(ε-caprolactone), *Polym. Degrad. Stabil.* **59**, 161–168.

Wang, Z.-G., Hsiao, B. S., Zong, X.-H., Yeh, F., Zouh, J. J., Dormier, E., Jamiolkowski, D. D. (2000) Morphological development in absorbable poly(glycolide), poly(glycolide-co-lactide) and poly(glycolide-*co*-caprolactone) copolymers during isothermal crystallization, *Polymer* **41**, 621–628.

Whang, K., Thomas, C. H., Healy, K. E., Nuber, G. (1995) A novel method to fabricate bioabsorbable scaffolds, *Polymer* **36**, 837–842.

Williams, D. F. (1981) Enzymatic hydrolysis of polylactic acid, *Eng. Med.* **10**, 5–7.

Wise, D. L., Fellmann, T. D., Sanderson, J. E., Wentworth, R. L. (1979) Lactic/glycolic acid polymers, in: *Drug Carriers in Biology and Medicine* (Gregoriadis, G., Ed.), London, New York: Academic Press, 237–270.

Wisniewski, M., Le Borgne, A., Spassky, N. (1997) Synthesis and properties of (D)- and (L)-lactide stereocopolymers using the system achiral Shiff's base/aluminium methoxide as initiator, *Macromol. Chem. Phys.* **198**, 1227–1238.

Yamaoka, T., Takahashi, Y., Ohta, T., Miyamoto, M., Murakami, A., Kimura, Y. (1999) Synthesis and properties of multiblock copolymers consisting of poly(L-lactic acid) and poly(oxypropylene-*co*-oxyethylene) prepared by direct polycondensation, *J. Polym. Sci.: Part A: Polym. Chem.* **37**, 1513–1522.

Yang, J.-M., Chen, H.-L., You, J.-W., Hwang, J. C. (1997) Miscibility and crystallization of poly(L-lactide)/poly(ethylene glycol) and poly(L-lactide)/poly(ε-caprolactone), *Polym. J.* **29**, 657–662.

Yoon, J.-S., Lee, W.-S., Kim, K.-S., Chin, I.-J., Kim, M.-N., Kim, C. (2000) Effect of poly(ethylene glycol)-block-poly(L-lactide) on the poly((*R*)-3-hydroxybutyrate)/poly(L-lactide) blends, *Eur. Polym. J.* **36**, 435–442.

Younes, H., Cohn, D. (1987) Morphological study of biodegradable PEO/PLA block copolymers, *J. Biomed. Mater. Res.* **21**, 1301–1316.

Younes, H., Cohn, D. (1988) Phase separation in poly(ethylene glycol)/poly(lactic acid), *Eur. Polym. J.* **24**, 765–773.

Youxin, L., Volland, C., Kissel, T. (1998) Biodegradable brush-like graft polymers from poly(D,L-lactide) or poly(D,L-lactide-co-glycolide) and charge-modified, hydrophilic dextrans as backbone – *in vitro* degradation and controlled release of hydrophilic macromolecules, *Polymer* **39**, 3087–3097.

Yui, N., Dijkstra, P. J., Feijen, J. (1990) Stereo block copolymers of L- and D-lactides, *Makromol. Chem.* **191**, 481–488.

Zell, M. T., Padden, B. E., Paterick, A. J., Hillmyer, M. A., Kean, R. T., Thakur, K. A. M., Munson, E. J. (1998) Direct observation of stereodefect sites in semicrystalline poly(lactide) using ^{13}C solid-state NMR, *J. Am. Chem. Soc.* **120**, 12672–12673.

Zhang, L., Xiong, C., Deng, X. (1995) Biodegradable polyester blends for biomedical application, *J. Appl. Polym. Sci.* **56**, 103–112.

Zhang, L., Xiong, C., Deng, X. (1996) Miscibility, crystallization and morphology of poly(β-hydroxybutyrate)/poly(d,l-lactide) blends, *Polymer* **37**, 235–241.

Zhang, L., Deng, X., Zhao, S. Huang, Z. (1997) Biodegradable polymer blends of poly(3-hydroxybutyrate) and poly(DL-lactide)-co-poly(ethylene glycol), *J. Appl. Polym. Sci.* **65**, 1849–1856.

Zhang, X., Wyss, U. P., Pichora, D., Goosen, M. F. A. (1994) An investigation of poly(lactic acid) degradation, *J. Bioact. Compat. Polym.* **9**, 80–100.

Zoppi, R. A., Contant, S., Duek, E. A. R., Marques, F. R., Wada, M. L. F., Nunes, S. P. (1999) Porous poly(L lactide) films obtained by immersion precipitation process: morphology, phase separation and culture of VERO cells, *Polymer* **40**, 3275–3289.

Zwiers, R. J. M., Gogolewski, S., Pennings, A. J. (1983) General crystallization behaviour of poly(L-lactic acid) PLLA: 2 Eutectic crystallization of PLLA, *Polymer* **24**, 167–174.

III
Polysaccharides

Biopolymers for Medical and Pharmaceutical Applications. Edited by A. Steinbüchel and R. H. Marchessault

ISBN: 3-527-31154-8

8
Alginates from Algae

Dr. Kurt Ingar Draget[1], Prof. Dr. Olav Smidsrød[2], Prof. Dr. Gudmund Skjåk-Bræk[3]
[1] Norwegian Biopolymer Laboratory, Department of Biotechnology, Norwegian University of Science and Technology, Sem Saelands vei 6-8, N-7491 Trondheim, Norway; Tel.: +47-73598260; Fax: +47-73591283;
E-mail: Kurt.I.Draget@chembio.ntnu.no
[2] Norwegian Biopolymer Laboratory, Department of Biotechnology, Norwegian University of Science and Technology, Sem Saelands vei 6-8, N-7491 Trondheim, Norway; Tel.: +47-735-98260; Fax: +47-735-93337;
E-mail: Olav.Smidsroed@chembio.ntnu.no
[3] Norwegian Biopolymer Laboratory, Department of Biotechnology, Norwegian University of Science and Technology, Sem Saelands vei 6-8, N-7491 Trondheim, Norway. Tel.: +47-735-98260; Fax: +47-735-93340;
E-mail: Gudmund.Skjaak-Braek@chembio.ntnu.no

Biopolymers for Medical and Pharmaceutical Applications. Edited by A. Steinbüchel and R. H. Marchessault

ISBN: 3-527-31154-8

DP	degree of polymerization
EDTA	etylenediamine tetraacetic acid
G	α-L-guluronic acid
GDL	D-glucono-δ-lactone
M	β-D-mannuronic acid (M)
$N_{G>1}$	average G-block length larger than 1
NMR	nuclear magnetic resonance spectroscopy
PGA	propylene glycol alginate
pK_a	dissociation constants for the uronic acid monomers

1 Introduction

Alginates are quite abundant in nature since they occur both as a structural component in marine brown algae (*Phaeophyceae*), comprising up to 40% of the dry matter, and as capsular polysaccharides in soil bacteria (see Chapter 8 on bacterial alginates in Volume 5 of this series). Although present research and results point toward a possible production by microbial fermentation and also by post-polymerization modification of the alginate molecule, all commercial alginates are at present still extracted from algal sources. The industrial applications of alginates are linked to its ability to retain water, and its gelling, viscosifying, and stabilizing properties. Upcoming biotechnological applications, on the other hand, are based either on specific biological effects of the alginate molecule itself or on its unique, gentle, and almost temperature-independent sol/gel transition in the presence of multivalent cations (e.g., Ca^{2+}), which makes alginate highly suitable as an immobilization matrix for living cells.

Traditional exploitation of alginates in technical applications has been based to a large extent on empirical knowledge. However, since alginates now enter into more knowledge-demanding areas such as pharmacy and biotechnology, new research functions as a locomotive for a detailed further investigation of structure–function relationships. New scientific breakthroughs are made, which in turn may benefit the traditional technical applications.

2 Historical Outline

The British chemist E. C. C. Stanford first described alginate (the preparation of "algic acid" from brown algae) with a patent dated 12 January 1881 (Stanford, 1881). After the patent, his discovery was further discussed in papers from 1883 (Standford, 1883a,b). Stanford believed that alginic acid contained nitrogen and contributed much to the elucidation of its chemical structure.

In 1926, some groups working independently (Atsuki and Tomoda, 1926; Schmidt and Vocke, 1926) discovered that uronic acid was a constituent of alginic acid. The nature of the uronic acids present was investigated by three different groups shortly afterwards (Nelson and Cretcher, 1929, 1930, 1932; Bird and Haas, 1931; Miwa, 1930), which all found D-mannuronic acid in the hydrolysate of alginate. The nature of the bonds between the uronic acid residues in the alginate molecule was determined to be β1,4, as in cellulose (Hirst et al., 1939)

This very simple and satisfactory picture of the constitution of alginic acid was, however, destroyed by the work of Fischer and Dörfel (1955). In a paper chromatographic study of uronic acids and polyuronides, they discovered the presence of a uronic acid different from mannuronic acid in the hydrolysates of alginic acid. This new uronic acid was identified as L-guluronic acid. The quantity of L-guluronic acid was considerable, and a method for quantitative determination of mannuronic and guluronic acid was developed.

Alginate then had to be regarded as a binary copolymer composed of α-L-guluronic and β-D-mannuronic residues. As long as alginic acid was regarded as a polymer containing only D-mannuronic acid linked together with β-1,4 links, it was reasonable to assume that alginates from different raw materials were chemically identical and that any given sample of alginic acid was chemically homogeneous. From a practical and a scientific point of view, the uronic acid composition of alginate from different sources had to be examined, and methods for chemical fractionation of alginates had to be developed. These tasks were undertaken mainly by Haug and coworkers (Haug, 1964), as described in Section 3 below. The discovery of alginate as a block-copolymer, the correlation between physical properties and block structure, and the discovery of a set of epimerases converting mannuronic to guluronic acid in a sequence-dependent manner also are discussed further in later sections.

3
Chemical Structure

Being a family of unbranched binary copolymers, alginates consist of (1 → 4) linked β-D-mannuronic acid (M) and α-L-guluronic acid (G) residues (see Figure 1a and b) of widely varying composition and sequence. By partial acid hydrolysis (Haug, 1964; Haug et al., 1966; Haug and Larsen, 1966; Haug et al., 1967a; Haug and Smidsrød, 1965), alginate was separated into three fractions. Two of these contained almost homopolymeric molecules of G and M, respectively, while a third fraction consisted of nearly equal proportions of both monomers and was shown to contain a large number of MG dimer residues. It was concluded that alginate could be regarded as a true block copolymer composed of homopolymeric regions of M and G, termed M- and G-blocks, respectively, interspersed with regions of alternating structure (MG-blocks; see Figure 1c). It was further shown (Painter et al., 1968; Larsen et al., 1970; Smidsrød and Whittington, 1969) that alginates have no regular repeating unit and that the distribution of the monomers along the polymer chain could not be described by Bernoullian statistics. Knowledge of the monomeric composition is hence not sufficient to determine the sequential structure of alginates. It was suggested (Larsen et al., 1970) that a second-order Markov model would be required for a general approximate description of the monomer sequence in alginates. The main difference at the molecular level between algal and bacterial alginates is the presence of *O*-acetyl groups at C2 and/or C3 in the bacterial alginates (Skjåk-Bræk et al., 1986).

4
Conformation

Knowledge of the monomer ring conformations is necessary to understand the polymer properties of alginates. X-ray diffraction studies of mannuronate-rich and guluronate-rich alginates showed that the guluronate residues in homopoly-meric blocks were in the $^{1}C_{4}$ conformation (Atkins et al., 1970), while the mannuronate residues had

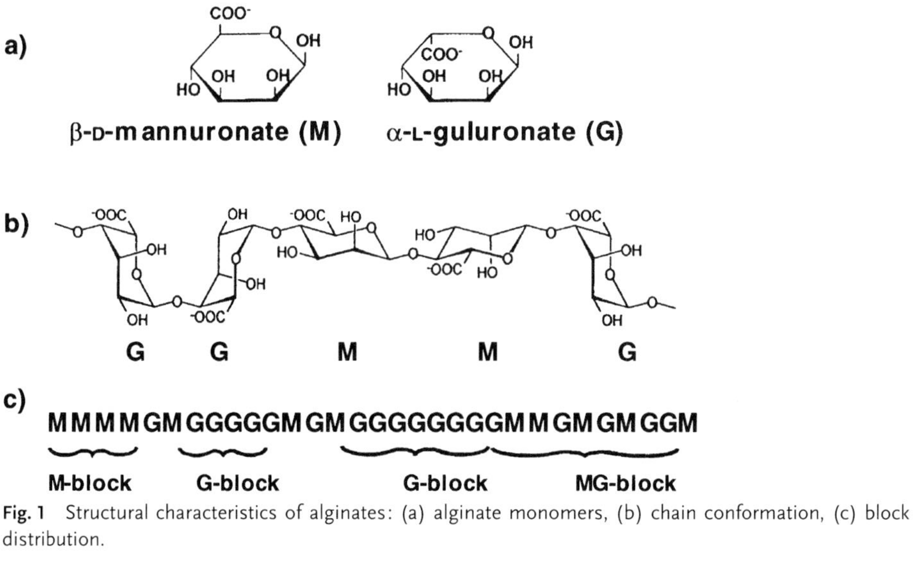

Fig. 1 Structural characteristics of alginates: (a) alginate monomers, (b) chain conformation, (c) block distribution.

the $^4C_1$ conformation (see Figure 1a). Viscosity data of alginate solutions indicated that the stiffness of the chain blocks increased in the order MG < MM < GG. This series could be reproduced only by statistical mechanical calculations when the guluronate residues were set in the $^1C_4$ conformation (Smidsrød et al., 1973) and was later confirmed by ^{13}C-NMR (Grasdalen et al., 1977). Hence, alginate contains all four possible glycosidic linkages: diequatorial (MM), diaxial (GG), equatorial-axial (MG), and axial-equatorial (GM) (see Figure 1b).

The diaxial linkage in G-blocks results in a large, hindered rotation around the glycosidic linkage, which may account for the stiff and extended nature of the alginate chain (Smidsrød et al., 1973). Additionally, taking the polyelectrolyte nature of alginate into consideration, the electrostatic repulsion between the charged groups on the polymer chain also will increase the chain extension and hence the intrinsic viscosity. Extrapolation of dimensions both to infinite ionic strength and to θ-conditions (Smidsrød, 1970) yielded relative dimensions for the neutral, unperturbed alginate chain being much higher than for amylose derivatives and even slightly higher than for some cellulose derivatives.

Another parameter reflecting chain stiffness and extension is the exponent in the Mark-Houwink-Sakurada equation,

$$[\eta] = K \cdot M^a$$

where M is the molecular weight of the polymer, $[\eta]$ is the intrinsic viscosity, and the exponent a generally increases with increasing chain extension. Some measurements on alginates (Martinsen et al., 1991; Smidsrød and Haug, 1968a, Mackie et al., 1980), yielded a-values ranging from 0.73 to 1.31, depending on ionic strength and alginate composition. Low and high a-values are related to large fractions of the flexible MG-blocks and the stiff and extended G-blocks, respectively (Moe et al., 1995).

5
Occurrence and Source Dependence

Commercial alginates are produced mainly from *Laminaria hyperborea, Macrocystis pyrifera, Laminaria digitata, Ascophyllum nodosum, Laminaria japonica, Eclonia maxima, Lessonia nigrescens, Durvillea antarctica*, and *Sargassum* spp. Table 1 gives some sequential parameters (determined by high-field NMR-spectroscopy) for samples of these

Tab. 1 Composition and sequence parameters of algal alginates (Smidsrød and Draget 1996)

Source	F_G	F_M	F_{GG}	F_{MM}	$F_{GM,MG}$
Laminaria japonica	0.35	0.65	0.18	0.48	0.17
Laminaria digitata	0.41	0.59	0.25	0.43	0.16
Laminaria hyperborea, blade	0.55	0.45	0.38	0.28	0.17
Laminaria hyperborea, stipe	0.68	0.32	0.56	0.20	0.12
Laminaria hyperborea, outer cortex	0.75	0.25	0.66	0.16	0.09
Lessonia nigrescens[a]	0.38	0.62	0.19	0.43	0.19
Ecklonia maxima	0.45	0.55	0.22	0.32	0.32
Macrocystis pyrifera	0.39	0.61	0.16	0.38	0.23
Durvillea antarctica	0.29	0.71	0.15	0.57	0.14
Ascophyllum nodosum, fruiting body	0.10	0.90	0.04	0.84	0.06
Ascophyllum nodosum, old tissue	0.36	0.64	0.16	0.44	0.20

[a] Data provided by Bjørn Larsen

alginates. The composition and sequential structure may, however, vary according to seasonal and growth conditions (Haug, 1964; Indergaard and Skjåk-Bræk, 1987). High contents of G generally are found in alginates prepared from stipes of old *Laminaria hyperborea* plants, whereas alginates from *A. nodosum*, *L. japonica*, and *Macrocystis pyrifera* are characterized by low content of G-blocks and low gel strength.

Alginates with more extreme compositions containing up to 100% mannuronate can be isolated from bacteria (Valla et al., 1996). Alginates with a very high content of guluronic acid can be prepared from special algal tissues such as the outer cortex of old stipes of *L. hyperborea* (see Table 1), by chemical fractionation (Haug and Smidsrød, 1965; Rivera-Carro, 1984) or by enzymatic modification *in vitro* using mannuronan C-5 epimerases from *A. vinelandii* (Valla et al., 1996; see Section 9.2). This family of enzymes is able to epimerize M-units into G-units in different patterns from almost strictly alternating to very long G-blocks. The epimerases from *A. vinelandii* have been cloned and expressed, and they represent at present a powerful new tool for the tailoring of alginates. It is also obvious that commercial alginates with less molecular heterogeneity, with respect to chemical composition and sequence, can be obtained by a treatment with one of the C-5 epimerases (Valla et al., 1996).

6
Physiological Function

The biological function of alginate in brown algae generally is believed to be as a structure-forming component. The intercellular alginate gel matrix gives the plants both mechanical strength and flexibility (Andresen et al., 1977). Simply speaking, alginates in marine brown algae may be regarded as having physiological properties similar to those of cellulose in terrestrial plants. This relation between structure and function is reflected in the compositional difference of alginates in different algae or even between different tissues from the same plant (see Table 1). In *L. hyperborea*, an alga that grows in very exposed coastal areas, the stipe and holdfast have a very high content of guluronic acid, giving high mechanical rigidity. The leaves of the same algae, which float in the streaming water, have an alginate characterized by a lower G-content, giving it a more flexible texture. The physiological function of alginates in bacteria will be covered elsewhere in this series.

7
Chemical Analysis and Detection

Since alginates are block copolymers, and because of the fact that their physical properties rely heavily on the sequence of these blocks, it obvious that the development of techniques enabling a sequence quantification is of the utmost importance. Additionally, molecular mass and its distribution (polydispersity) is a significant parameter in some applications.

7.1
Chemical Composition and Sequence

Detailed information about the structure of alginates became available by introduction of high-resolution ^{1}H and ^{13}C NMR-spectroscopy (Grasdalen et al., 1977, 1979; Penman and Sanderson, 1972; Grasdalen, 1983) in the sequential analysis of alginate. These powerful techniques make it possible to determine the monad frequencies F_M and F_G; the four nearest neighboring (diad) frequencies F_{GG}, F_{MG}, F_{GM}, and F_{MM}; and the eight next nearest neighboring (triad) frequencies. Knowledge of these frequencies enables, for example, the calculation of the average G-block length larger than 1:

$$N_{G>1} = (F_G - F_{MGM}) / F_{GGM}.$$

This value has been shown to correlate well with gelling properties. It is important to realize that in an alginate chain population, neither the composition nor the sequence of each chain will be alike. This results in a composition distribution of a certain width.

7.2 Molecular Mass

Alginates, like polysaccharides in general, are polydisperse with respect to molecular weight. In this aspect they resemble synthetic polymers rather than other biopolymers such as proteins and nucleic acids. Because of this polydispersity, the "molecular weight" of an alginate is an average over the whole distribution of molecular weights.

In a population of molecules where N_i is the number of molecules and w_i is the weight of molecules having a specific molecular weight M_i, the number and the weight average are defined respectively as:

$$\overline{M_n} = \frac{\Sigma_i N_i M_i}{\Sigma_i N_i}$$

$$\overline{M_w} = \frac{\Sigma_i w_i M_i}{\Sigma_i w_i} = \frac{\Sigma_i N w_i M_i^2}{\Sigma_i N_i M_i}$$

For a randomly degraded polymer, we have $\overline{M_w} \approx 2\overline{M_n}$ (Tanford, 1961). The fraction $\overline{M_w}/\overline{M_n}$ is called the polydispersity index. Polydispersity index values between 1.4 and 6.0 have been reported for alginates and have been related to different types of preparation and purification processes (Martinsen et al., 1991; Smidsrød and Haug, 1968a; Mackie et al., 1980; Moe et al., 1995).

The molecular-weight distribution can have implications for the uses of alginates, as low-molecular-weight fragments containing only short G-blocks may not take part in gel-network formation and consequently do not contribute to the gel strength. Furthermore, in some high-tech applications, the leakage of mannuronate-rich fragments from alginate gels may cause problems (Stokke et al., 1991; Otterlei et al., 1991) and a narrow molecular-weight distribution therefore is recommended.

7.3 Detection and Quantification

Detection and quantification of alginates in the presence of other biopolymers, such as proteins, are not straightforward mainly because of interference. Once isolated, a number of colorimetric methods can be applied to quantify alginate. The oldest and most common is the general procedure for carbohydrates, the phenol/sulfuric acid method (Dubois et al., 1956), but there are also two slightly refined formulas specially designed for uronic acids (Blumenkrantz and Asboe-Hansen, 1973; Filisetti-Cozzi and Carpita, 1991).

8 Biosynthesis and Biodegradation

Our knowledge of the alginate biosynthesis mainly comes from studying alginate-producing bacteria. Figure 2 shows the principal enzymes involved in alginate biosynthesis, and the activity of all enzymes (1–7) has been identified in brown algae. During the last decade, the genes responsible for alginate synthesis in *Pseudomonas* and *Azotobacter* have been identified, sequenced, and cloned. For further information on alginate biosynthesis, please see Chapter 8 on bacte-

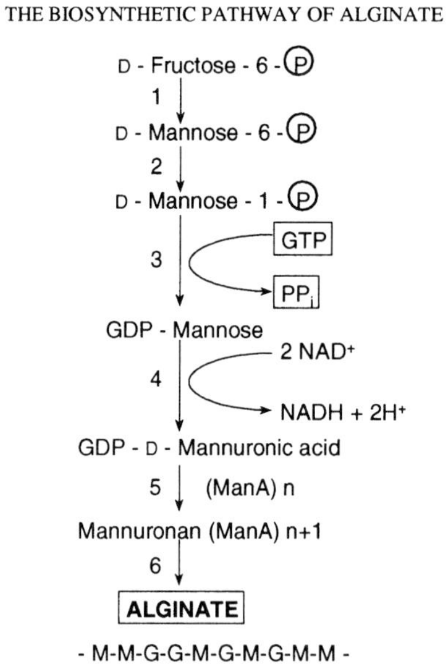

Fig. 2 Biosynthetic pathway of alginates.

rial alginates in Volume 5 of this series. Because of their potential use in alginate modification, the only enzymes we will comment on here are the alginate lyases and the mannuronan C-5 epimerases.

Alginates are not degraded in the human gastric-intestinal tract, and hence do not give metabolic energy. Some lower organisms have, however, developed lyases that degrade alginates down to single components, resulting in alginates that function as a carbon source. Alginate lyases catalyze the depolymerization of alginate by splitting the 1–4 glycosidic linkage in a β-elimination reaction, leaving an unsaturated uronic acid on the non-reducing end of the molecules. Alginate lyases are widely distributed in nature, including in organisms growing on alginate as a carbon source such as marine gastropods, prokaryotic and eukaryotic microorganisms, and bacteriophages. They also are found in the bacterial species producing alginate such as *Azotobacter vinelandii* and *Pseudomonas aeruginosa*. All of them are endolyases and may exhibit specificity to either M or G. Since the aglycon residue will be identical for both M and G, the use of lyases for structural work is limited. Table 2 lists a range of lyases and their specificities.

9
Production: Biotechnological and Traditional

There has been significant progress in the understanding of alginate biosynthesis over the last 10 years. The fact that the alginate molecule enzymatically undergoes a post-polymerization modification with respect to chemical composition and sequence opens up the possibility for *in vitro* modification and tailoring of commercially available alginates.

9.1
Isolation from Natural Sources / Fermentative Production

As already described, all commercial alginates today are produced from marine brown algae (Table 1). Alginates with more extreme compositions can be isolated from the bacterium *Azotobacter vinelandii*, which, in contrast to *Pseudomonas* species, produces polymers containing G-blocks. Production by fermentation therefore is technically possible but is not economically feasible at the moment.

9.2
Molecular Genetics and *in vitro* Modification

Alginate with a high content of guluronic acid can be prepared from special algal tissues by chemical fractionation or by *in vitro* enzymatic modification of the alginate *in vitro* using mannuronan C-5 epimerases from *A. vinelandii* (Ertesvåg et al., 1994, 1995, 1998b; Høydal et al., 1999). These epimerases, which convert M to G in the

Tab. 2 Substrate specificity and biochemical properties of some alginate lyases (Gacesa, 1992; Wong et al., 2000)

Source	*Localization*	*Sequence specifty*	*Major end-product*	*pHopt*	*Mw (kDa)*	*pI*	*Reference*
K.aerogenes	Extracellular	G↓X	Trimer	7.0	31.4	8.9	Boyd and Turvey, 1978
Enterobacter cloacae	Extracellular	G	–	7.8	32–38	8.9	Nibu et al., 1995
	Intracellular	G↓G	Dimer/pentamer	7.5	31–39	8.9	Shimokawa et al., 1997
P. aeruginosa (AlgL)	Periplasmic	M-X	Trimer	7.0	39	9.0	Boyd and Turvey, 1977
A. vinelandii	Periplasmic (AlgL)	M↓X M_{Ac}↓X	Trimer/tetramer	8.1–8.4	39	5.1	Ertesvåg et al., 1998a
	Extracellular (AlgE7)	G↓X	Tetramer- septamer	6.3–7.3	90.4	–	Ertesvåg et al., 1998a
P. alginovora	Extracellular	G↓G	–	7.5	28	5.5	Boyen et al., 1990a
	Intracellular	M↓M	–	–	24	5.8	Boyen et al., 1990a
Haliotis tuberculata	Hepato-pancreas	M↓X, G↓M	Trimer/ dimer	8	34	–	Boyen et al., 1990b
Sphingomonas sp. ALYI-III	Cytoplasmic	M↓X M_{Ac}↓X	–	5.6–7.8	38	10.16	Murata et al., 1993
Littorina sp.	Hepato-pancreas	M↓M	Trimer	5.6	~40	–	Elyakova and Favorov, 1974
A. vinelandii phage	Extracellular	M↓X M_{Ac}↓X	Trimer	7.7	30–35	–	Davidson et al., 1977

polymer chain, recently have allowed for the production of highly programmed alginates with respect to chemical composition and sequence. *A. vinelandii* encodes a family of 7 exocellular isoenzymes with the capacity to epimerize all sorts of alginates and other mannuronate-containing polymers, as shown in Figure 3, where the mode of action of AlgE4 (giving alternating introduction of G) is presented. Although the genes have a high degree of homology, the enzymes they encode exhibit different specificities. Different epimerases may give alginates with different distribution of M and G, and thus alginates with tailored physical and chemical properties can be made as illustrated in Figure 4. None of the enzymatically modified polymers, however, are commercially available at present. Table 3 lists the modular structure of the mannuronan C-5 epimerase family and its specific action.

9.3
Current and Expected World Market and Costs

Industrial production of alginate is roughly 30,000 metric tons annually, which is probably less than 10% of the annually biosynthesized material in the standing macroalgae crops. Because macroalgae also may be cultivated (e.g., in mainland China where 5 to 7 million metric tons of wet *Laminaria japonica* are produced annually) and because production by fermentation is technically possible, the sources for industrial production of alginate may be regarded as unlimited even for a steadily growing industry.

It is expected that future growth in the alginate market most likely will be of a qualitative rather than a quantitative nature. Predictions suggest that manufacturers will move away from commodity alginate production toward more refined products, e.g., for the pharmaceutical industry.

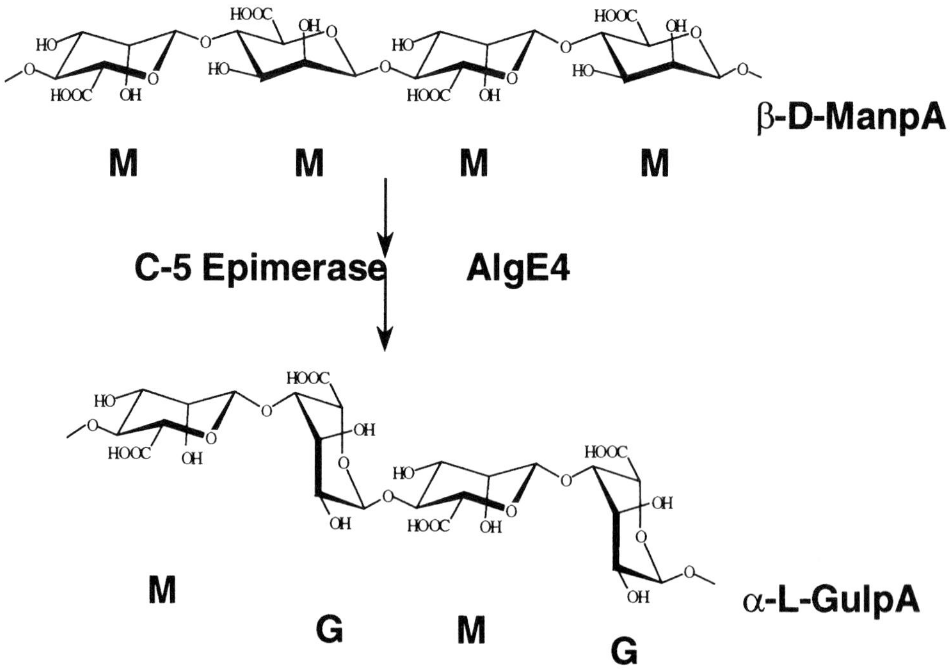

Fig. 3 Mode of action for the mannuronan C5-epimarase AlgE4.

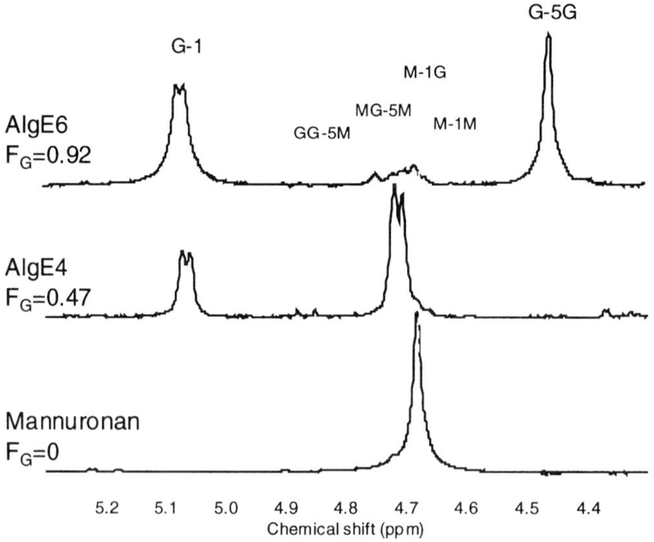

Fig. 4 Resulting chemical composition and sequence after treating mannuronan with different C5-epimerases.

Tab. 3 The seven AlgE epimerases from *A. vinelandi*[a]

Type	*[kDa]*	*Modular structure*	*Products*
AlgE1	147.2	A1 R1 R2 R3 A2 R4	Bi-functional G-blocks + MG-blocks
AlgE2	103.1	A1 R1 R2 R3 R4	G-blocks (short)
AlgE3	191	A1 R1 R2 R3 A2 R4 R5 R6 R7	Bi-functional G-block + MG-blocks
AlgE4	57.7	A1 R1	MG-blocks
AlgE5	103.7	A1 R1 R2 R3 R4	G-blocks (medium)
AlgE6	90.2	A1 R1 R2 R3	G-blocks (long)
AlgE7	90.4	A1 R1 R2 R3	Lyase activity + G-blocks + MG-blocks

[a] A – 385 amino acids; R – 155 amino acids

The cost of alginates can differ extremely depending on the degree of purity. Technical grade, low-purity alginate (containing a substantial amount of algae debris) can be obtained from around 1 USD per kilogram, and ordinary purified-grade alginate can be obtained from approximately 10 USD per kilo, whereas ultra-pure (low in endotoxins) alginate specially designed for immobilization purposes typically costs around 5 USD per gram.

9.4 Alginate Manufacturers

The alginate producers members list of the Marinalg hydrocolloid association includes six different companies. These are China Seaweed Industrial Association, Danisco Cultor (Denmark), Degussa Texturant Systems (Germany), FMC BioPolymer (USA), ISP Alginates Ltd. (UK), and Kimitsu Chemical Industries Co., Ltd. (Japan). In addition to these, Pronova Biomedical A/S (Norway) now commercially manufactures ultra-pure alginates that are highly compatible with mammalian biological systems following the increased popularity of alginate as an immobilization matrix. These qualities are low in pyrogens and facilitate sterilization of the alginate solution by filtration due to low content of aggregates.

10 Properties

The physical properties of the alginate molecule were revealed mainly in the 1960s and 1970s. The last couple of decades have exposed some new knowledge on alginate gel formation

10.1 Physical Properties

Compared with other gelling polysaccharides, the most striking features of alginate's physical properties are the selective binding of multivalent cations, being the basis for gel formation, and the fact that the sol/gel transition of alginates is not particularly influenced by temperature.

10.1.1 Solubility

There are three essential parameters determining and limiting the solubility of alginates in water. The pH of the solvent is important because it will determine the presence of electrostatic charges on the uronic acid residues. Total ionic strength of the solute also plays an important role (salting-out effects of non-gelling cations), and, obviously, the content of gelling ions in the solvent limits the solubility. In the latter case, the "hardness" of the water (i.e., the content of Ca^{2+} ions) is most likely to be the main problem.

Potentiometric titration (Haug, 1964) revealed that the dissociation constants for mannuronic and guluronic acid monomers were 3.38 and 3.65, respectively. The pK_a value of the alginate polymer differs only slightly from those of the monomeric residues. An abrupt decrease in pH below the pK_a value causes a precipitation of alginic acid molecules, whereas a slow and controlled release of protons may result in the formation of an "alginic acid gel". Precipitation of alginic acid has been studied extensively (Haug, 1964; Haug and Larsen, 1963; Myklestad and Haug, 1966; Haug et al., 1967c), and addition of acid to an alginate solution leads to a precipitation within a relatively narrow pH range. This range depends not only on the molecular weight of the alginate but also on the chemical composition and sequence. Alginates containing more of the "alternating" structure (MG-blocks) will precipitate at lower pH values compared with the alginates containing more homogeneous block structures (poly-M and poly-G). The presence of homopolymeric blocks seems to favor precipitation by the formation of crystalline regions stabilized by hydrogen bonds. By increasing the degree of alternating "disorder" in the alginate chain, as in alginates isolated from *Ascophyllum nodosum* (see Table 1), the formation of these crystalline regions is not formed as easily. A certain alginate fraction from *A. nodosum* is soluble at a pH as low as 1.4 (Myklestad and Haug, 1966). Because of this relatively limited solubility of alginates at low pH, the esterified propylene glycol alginate (PGA) is applied as a food stabilizer under acidic conditions (see Section 10.3).

Any change of ionic strength in an alginate solution generally will have a profound effect, especially on polymer chain extension and solution viscosity. At high ionic strengths, the solubility also will be affected. Alginate may be precipitated and fractionated to give a precipitate enriched with mannuronate residues by high concentrations of inorganic salts like potassium chloride (Haug and Smidsrød, 1967; Haug, 1959a). Salting-out effects like this exhibit large hysteresis in the sense that less than 0.1 M salt is necessary to slow down the kinetics of the dissolution process and limit the solubility (Haug, 1959b). The gradient in the chemical potential of water between the bulk solvent and the solvent in the alginate particle, resulting from a very high counterion concentration in the particle, is most probably the drive of the dissolution process of alginate in water. This drive becomes severely reduced when attempts are made to dissolve alginate in an aqueous solvent already containing ions. If alginates are to be applied at high salt concentrations, the polymer should first be fully hydrated in pure water followed by addition of salt under shear.

For the swelling behavior of dry alginate powder in aqueous media with different concentrations of Ca^{2+}, there seems to be a limit at approximately 3 mM free calcium ions (unpublished results). Alginate can be solubilized at $[Ca^{2+}]$ above 3 mM by the addition of complexing agents, such as polyphosphates or citrate, before addition of the alginate powder.

10.1.2
Selective Ion Binding

The basis for the gelling properties of alginates is their specific ion-binding characteristics (Haug, 1964; Smidsrød and Haug, 1968b; Haug and Smidsrød, 1970; Smidsrød, 1973, 1974). Experiments involving equilibrium dialysis of alginate have shown that the selective binding of certain alkaline earth metals ions (e.g., strong and cooperative binding of Ca^{2+} relative to Mg^{2+}) increased markedly with increasing content of α-L-guluronate residues in the chains. Poly-mannuronate blocks and alternating blocks were almost without selectivity. This is illustrated in Figures 5 and 6, where a marked hysteresis in the binding of Ca^{2+} ions to G-blocks also is seen.

The high selectivity between similar ions such as those from the alkaline earth metals indicates that some chelation caused by structural features in the G-blocks takes place. Attempts were made to explain this phenomenon by the so-called "egg-box" model (Grant et al., 1973), based upon the linkage conformations of the guluronate residues (see Figure 1b). NMR studies (Kvam et al., 1986) of lanthanide complexes of related compounds suggested a possible binding site for Ca^{2+} ions in a single alginate chain, as given in Figure 7 (Kvam 1987).

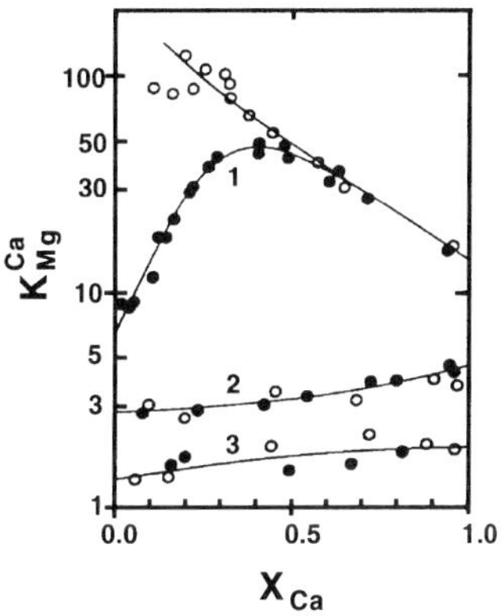

Fig. 6 Selectivity coefficients, K^{Ca}_{mg} as a function of ionic composition (X_{Ca}) for different alginate fragments. Curve 1: Fragments with 90% guluronate residues. Curve 2: Alternating fragment with 38% guluronate residues. Curve 3: Fragment with 90% mannuronate residues. •: Dialysis of the fragments in their Na^+ form. ○: Dialysis first against 0.2 M $CaCl_2$, then against mixtures of $CaCl_2$ and $MgCl_2$.

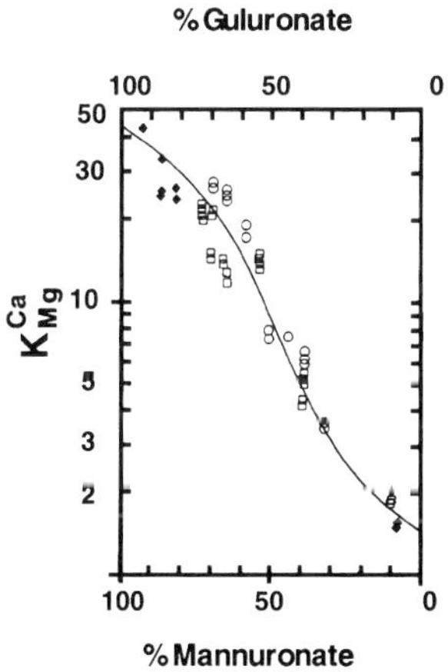

Fig. 5 Selectivity coefficients, K^{Ca}_{mg}, for alginates and alginate fragments as a function of monomer composition. The experimental points are obtained at $X_{Ca} = X_{Mg} = 0.5$. The curve is calculated using $K^{Ca}_{mg\ guluronate} = 40$ and $K^{Ca}_{mg\ mannuronate} = 1.8$

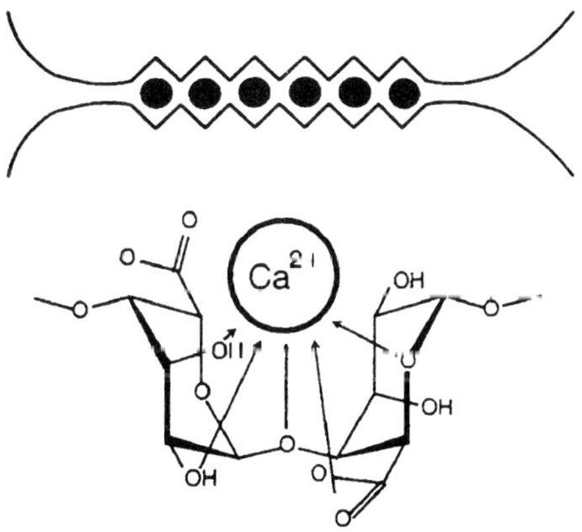

Fig. 7 The egg-box model for binding of divalent cations to homopolymeric blocks of α-L-guluronate residues, and a probably binding site in a GG-sequence.

Although more accurate steric arrangements have been suggested, as supported by x-ray diffraction (Mackie et al., 1983) and NMR spectroscopy (Steginsky et al., 1992), the simple "egg-box" model still persists, as it is principally correct and gives an intuitive understanding of the characteristic chelate-type of ion-binding properties of alginates. The simple dimerization in the "egg-box" model is at present questionable, as data from small-angle x-ray scattering on alginate gels suggest lateral association far beyond a pure dimerization with increasing [Ca^{2+}] and G-content of the alginate (Stokke et al., 2000). In addition, the fact that isolated and purified G-blocks (totally lacking elastic segments; typically DP = 20) are able to act as gelling modulators when mixed with a gelling alginate suggests higher-order junction zones (Draget et al., 1997).

The selectivity of alginates for multivalent cations is also dependent on the ionic composition of the alginate gel, as the affinity toward a specific ion increases with increasing content of the ion in the gel (Skjåk-Bræk et al., 1989b) (see Figure 6). Thus, a Ca-alginate gel has a markedly higher affinity toward Ca^{2+} ions than has the Na-alginate solution. This has been explained theoretically (Smidsrød, 1970; Skjåk-Bræk et al., 1989b) by a near-neighbor, auto-cooperative process (Ising model) and can be explained physically by the entropically unfavorable binding of the first divalent ion between two G-blocks and the more favorable binding of the next ions in the one-dimensional "egg-box" (zipper mechanism).

10.1.3
Gel Formation and Ionic Cross-linking

A very rapid and irreversible binding reaction of multivalent cations is typical for alginates; a direct mixing of these two components therefore rarely produces homogeneous gels. The result of such mixing is likely to be a dispersion of gel lumps ("fish-eyes"). The only possible exception is the mixing of a low-molecular-weight alginate with low amounts of cross-linking ion at high shear. The ability to control the introduction of the cross-linking ions hence becomes essential.

A controlled introduction of cross-linking ions is made possible by the two fundamental methods for preparing an alginate gel: the diffusion method and the internal setting method. The diffusion method is characterized by allowing a cross-linking ion (e.g., Ca^{2+}) to diffuse from a large outer reservoir into an alginate solution (Figure 8a). Diffusion setting is characterized by rapid gelling kinetics and is utilized for immobilization purposes where each droplet of alginate solution makes one single gel bead with entrapped (bio-) active agent (Smidsrød and Skjåk-Bræk, 1990). High-speed setting is also beneficial, e.g., in restructuring of foods when a given size and shape of the final product is desirable. The molecular-weight dependence in this system is negligible as long as the weight average molecular weight of the alginate is above 100 kDa (Smidsrød, 1974).

The internal setting method differs from the diffusion method in that the Ca^{2+} ions are released in a controlled fashion from an inert calcium source within the alginate solution (Figure 8b). Controlled release usually is obtained by a change in pH, by a limited solubility of the calcium salt source, and/or the by presence of chelating agents. The main difference between internal and diffusion setting is the gelling kinetics, which is not diffusion-controlled in the former case. With internal setting, the tailor-making of an alginate gelling system toward a given manufacturing process is possible because of the controlled, internal release of cross-linking ions (Draget et al., 1991). Internally set gels generally show a

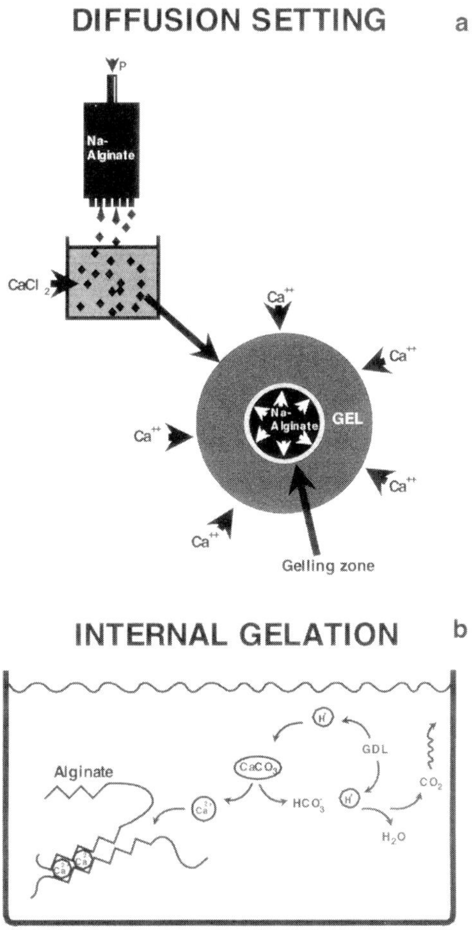

Fig. 8 Principal differences between the diffusion method exemplified by the immobilization technique and the internal setting method exemplified by the $CaCO_3$/GDL technique.

more pronounced molecular weight dependence compared with diffusion set gels. It has been reported that the internally set gels depend on molecular weight even at 300 kDa (Draget et al., 1993). This could be due to the fact that internally set gels are more calcium-limited compared with the gels made by diffusion, implying that the non-elastic fractions (sol and loose ends) at a given molecular weight will be higher in the internally set gels.

10.1.4
Gel Formation and Alginic Acid Gels

It is well known that alginates may form acid gels at pH values below the pK_a values of the uronic residues, but these alginic acid gels traditionally have not been as extensively studied as their ionically cross-linked counterparts. With the exception of some pharmaceutical uses, the number of applications so far is also rather limited. The preparation of an alginic acid gel has to be performed with care. Direct addition of acid to, e.g., a Na-alginate solution leads to an instantaneous precipitation rather than a gel. The pH must therefore be lowered in a controlled fashion, and this is most conveniently carried out by the addition of slowly hydrolyzing lactones like D-glucono-δ-lactone (GDL).

10.2
Material Properties

Since alginates are traditionally used for their gelling, viscosifying, and stabilizing properties, the features of alginate based materials are of utmost importance for a given application. Recently some quite unique biological effects of the alginate molecule itself have been revealed.

10.2.1
Stability

Alginate, being a single-stranded polymer, is susceptible to a variety of depolymerization processes. The glycosidic linkages are cleaved by both acid and alkaline degradation mechanisms and by oxidation with free radicals. As a function of pH, degradation of alginate is at its minimum nearly neutral and increases in both directions (Haug and Larsen, 1963) (Figure 9). The increased instability at pH values less than 5 is explained by a proton-catalyzed hydrolysis, whereas the reaction responsible for the

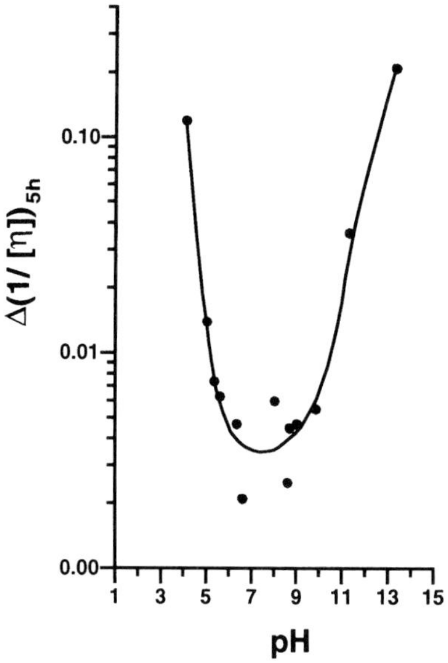

Fig. 9 Degradation of alginate isolated from *Laminaria digitata* measured as the change (Δ) in intrinsic viscosity ([η]) after 5 h at different pH and at 68°C.

degradation at pH 10 and above is the β-alkoxy elimination (Haug et al., 1963, 1967b). Free radicals degrade alginate mainly by oxidative-reductive depolymerization reactions (Smidsrød et al., 1967; Smidsrød et al., 1963a,b) caused by contamination of reducing agents like polyphenols from the brown algae. Since all of these depolymerization reactions increase with temperature, autoclaving generally is not recommended for the sterilization of an alginate solution. Since alginate is soluble in water at room temperature, sterile filtering rather than autoclaving has been recommended as a sterilization method for immobilization purposes in order to reduce polymer breakdown and to maintain the mechanical properties of the final gel (Draget et al., 1988).

Sterilization of dry alginate powder is also troublesome. The effect of γ-irradiation is often disastrous and leads to irreversible damage. It is generally believed that, under these conditions, O_2 is depleted rapidly with formation of the very reactive $OH^{\bullet}$ free radical. A short-term exposure in an electron accelerator could be an alternative to long-term exposure from a traditional ^{60}Co source. It has been shown that sterilization doses applied by ^{60}Co irradiation reduce the molecular weight to the extent that the gelling capacity is almost completely lost (Leo et al., 1990).

10.2.2
Ionically Cross-linked Gels

In contrast to most gelling polysaccharides, alginate gels are cold-setting, implying that alginate gels set more or less independent of temperature. The kinetics of the gelling process, however, can be strongly modified by a change in temperature, but a sol/gel transition will always occur if gelling is favored (e.g., by the presence of cross-linking ions). It is also important to realize that the properties of the final gel most likely will change if gelling occurs at different temperatures. This is due to alginates being non-equilibrium gels and thus being dependent upon the history of formation (Smidsrød, 1973).

Alginate gels can be heated without melting. This is the reason that alginates are used in baking creams. It should be kept in mind that alginates, as described earlier, are subjected to chemical degrading processes. A prolonged heat treatment at low or high pH might thus destabilize the gel because of an increased reaction rate of depolymerizing processes such as proton-catalyzed hydrolysis and the β-elimination reaction (Moe et al., 1995).

Because the selective binding of ions is a prerequisite for alginate gel formation, the alginate monomer composition and sequence also have a profound impact on the final properties of calcium alginate gels. Figure 10 shows gel strength as a function of

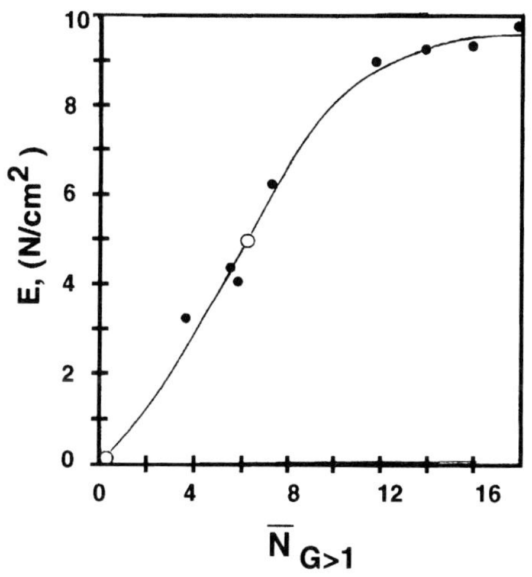

Fig. 10 Elastic properties of alginate gels as function of average G-block length.

the average length of G-blocks larger than one unit ($N_{G>1}$). This empirical correlation shows that there is a profound effect on gel strength when $N_{G>1}$ changes from 5 to 15. This coincides with the range of G-block lengths found in commercial alginates.

The polyelectrolyte nature of the alginate molecule is also important for its function, especially in mixed systems where, under favorable conditions, alginates may interact electrostatically with other charged polymers (e.g., proteins), resulting in a phase transition and altering the rheological behavior. Generally, it can be stated that if the purpose is to avoid such electrostatic interactions, the mixing of alginate and protein should take place at a relatively high pH, where most proteins have a net negative charge (Figure 11). These types of interactions also can be utilized to stabilize mixtures and to increase the gel strength of some restructured foods. In studies involving gelling of bovine serum albumin and alginate in both the sodium and the calcium form, a consid-

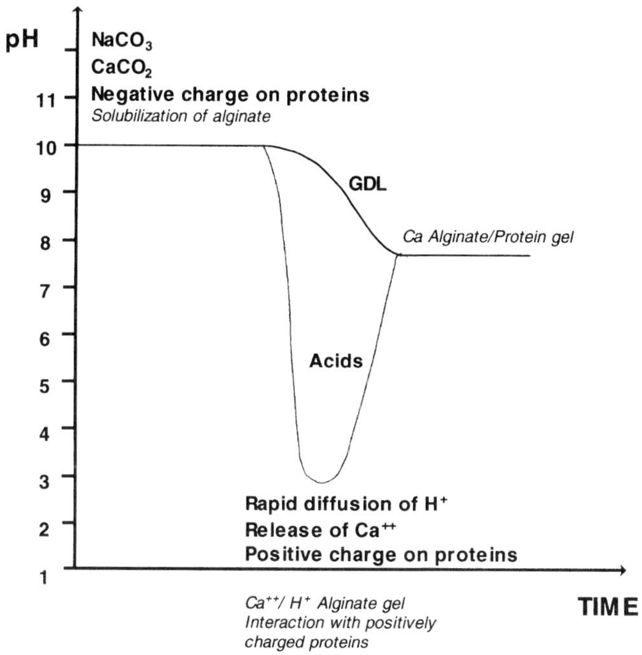

Fig. 11 Alginate/protein mixed gels exemplified by the internal gelation with $CaCO_3$. Release of Ca^{2+} is achieved by either a slow pH-lowering agent (GDL) or by a fast lowering with acids.

erable increase in Young's modulus was found within some range of pH and ionic strength (Neiser et al., 1998, 1999). These results suggest that electrostatic interactions are the main driving force for the observed strengthening effects.

An important feature of gels made by the diffusion-setting method is that the final gel often exhibits an inhomogeneous distribution of alginate, the highest concentration being at the surface and gradually decreasing towards the center of the gel. Extreme alginate distributions have been reported (Skjåk-Bræk et al., 1989a), with a five-fold increase at the surface (as calculated from the concentration in the original alginate solution) and virtually zero concentration in the center (Figure 12). This result has been explained by the fact that the diffusion of gelling ions will create a sharp gelling zone that moves from the surface toward the center of the gel. The activity of alginate (and of the gelling ion) will equal zero in this zone, and alginate molecules will diffuse from the internal, non-gelled part of the gelling body toward the zero-activity region

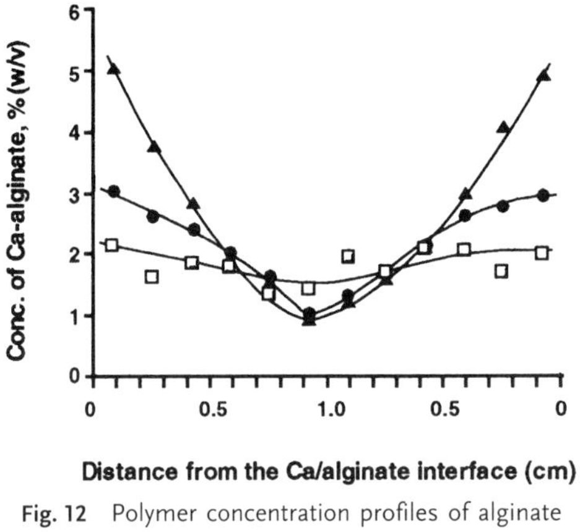

Fig. 12 Polymer concentration profiles of alginate gel cylinders formed by dialyzing a 2% (w/v) solution of Na-alginate from *Laminaria hyperborea* against 0.05 M $CaCl_2$ in the presence of NaCl. ❑: 0.2 M; ●: 0.05 M; ▲: no NaCl.

(Skjåk-Bræk et al., 1989a, Mikkelsen and Elgsæter, 1995). Inhomogeneous alginate distribution may or may not be beneficial in the final product. It is therefore important to know that the degree of homogeneity can be controlled and to know which parameters govern the final alginate distribution. Maximum inhomogeneity is reached by placing a high-G, low-molecular-weight alginate gel in a solution containing a low concentration of the gelling ion and an absence of non-gelling ions. Maximum homogeneity is reached by gelling a high-molecular-weight alginate with high concentrations of both gelling and non-gelling ions (Skjåk-Bræk et al., 1989b).

The presence of non-gelling ions in alginate-gelling systems also affects the stability of the gels. It has been shown that alginate gels start to swell markedly when the ratio between non-gelling and gelling ions becomes too high and that the observed destabilization increases with decreasing F_G (Martinsen et al., 1989).

Swelling of alginate gels can be increased dramatically by a covalent cross-linking of preformed Ca-alginate gels with epichlorohydrin, followed by subsequent removal of Ca^{2+} ions by etylenediamine tetraacetic acid (EDTA) (Skjåk-Bræk and Moe, 1992). These Na-alginate gels can be dried, and they exhibit unique swelling properties when re-hydrated. The forces affecting the swelling of a polymer network can be split into three terms. Two of these terms favor swelling and can be said to constitute what might be called "swelling pressure": (1) the mixing term (Π_{mix} = the osmotic pressure generated by polymer/solvent mixing) and (2) the ionic term (Π_{ion} = the osmotic effect of an unequal distribution of the polymer counter-ions between the inside and the outside of the gel; the Donnan equilibrium). Of these two terms, the ionic part has been shown to contribute approximately 90% of the swelling pressure, even at 1 molar ionic strength, for highly ionic gels like Na-alginate (Moe et al., 1993). The third term (Π_{el} = the reduction in osmotic pressure due to the elastic response of the polymer network) balances the "swelling pressure" so that the total of these three terms equals zero at equilibrium.

These Na-alginate gels would function well as water absorbents in hygiene and pharmaceutical applications. However, Π_{ion} depends upon the ionic strength of the solute; with increasing ionic strength, the difference in chemical potential is reduced because of a more even distribution of the mobile ions between the inside and the outside of the gel. Therefore, reduced swelling will be observed at physiological ionic conditions compared with deionized water, but this reduction will be less pronounced than that for other water-absorbing materials, such as cross-linked acrylates, as a result of the inherent stiffness of the alginate molecule itself (Skjåk-Bræk and Moe, 1992) (Figure 13).

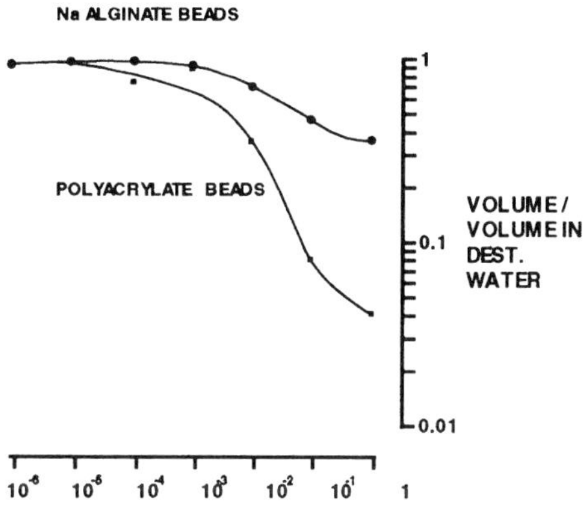

Fig. 13 Salt tolerance of covalently cross-linked Na-alginate and polyacrylate gel beads measured as swelling at different ionic strengths.

10.2.3
Alginic Acid Gels

It has been shown (Draget et al., 1994) that the gel strength of acid gels becomes independent of pH below a pH of 2.5, which equals 0.8 M GDL in a 1.0% (w/v) solution. Table 4 shows the Young's moduli of acid gels prepared (1) by a direct addition of GDL and (2) by converting an ionic cross-linked gel to the acid form by mineral acid. The modulus seems to be rather independent of the history of formation. Therefore, a most important feature of the acid gels compared with the ionic cross-linked gels seems to be that the former reaches equilibrium in the gel state.

Figure 14 shows the observed elastic moduli of acid gels made from alginates with different chemical composition, together with expected values for ionically cross-linked gels. From these data, it can be concluded that acid gels resemble ionic gels in the sense that high contents of guluronate (long stretches of G-blocks) give the strongest gels. However, it is also seen that polymannuronate sequences support alginic acid gel formation, whereas poly-alternating sequences seem to perturb this transition. The obvious demand for homopolymeric sequences in acid gel formation suggests

Tab. 4 E_{app} (kPa) for gels made from three different high-G alginates at 2% (w/v) concentration

Ca-alginate gel	*Ca-gel to alginic acid gel*	*Syneresis correction*	*Direct addition of GDL*
105 ± 4.6	52 ± 4.3	15.5 ± 0.3	15 ± 1.1
116 ± 11	64 ± 8.1	17.1 ± 1.8	17.8 ± 1.4
127 ± 6.4	79 ± 5.8	19.8 ± 1.3	20.4 ± 0.7

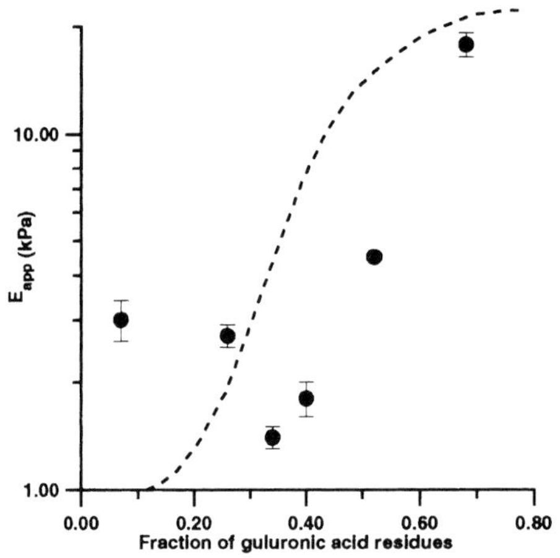

Fig. 14 Young's modulus E_{app} of alginic acid gels at apparent equilibrium as function of guluronic acid content. Dashed line refers to expected results for Ca^{2+} cross-linked alginate gels.

that cooperative processes are involved just as in the case of ionic gels (Draget et al., 1994). High molecular weight dependence has been observed, and this dependence becomes more pronounced with increasing content of guluronic acid residues.

A study of the swelling and partial solubilization of alginic acid gels at pH 4 has confirmed the equilibrium properties of the gels (Draget et al., 1996). By comparing the chemical composition and molecular weight of the alginate material leaching out from the acid gels with the same data for the whole alginate, an enrichment in mannuronic acid residues and a reduction in the average length of G-blocks were found together with a lowering of the molecular weight.

10.3 "Biological" Properties

Through a series of papers, it has been established that the alginate molecule itself has different effects on biological systems. This is more or less to be expected because of the large variety of possible chemical compositions and molecular weights of alginate preparations. A biological effect of alginate initially was hinted at in the first animal transplantation trials of encapsulated Langerhans islets for diabetes control. Overgrowth of alginate capsules by phagocytes and fibroblasts, resembling a foreign body/ inflammatory reaction, was reported (Soon-Shiong et al., 1991). In bioassays, induction of tumor necrosis factor and interleukin 1 showed that the inducibility depended upon the content of mannuronate in the alginate sample (Soon-Shiong et al., 1993). This result directly explains the observed capsule overgrowth; mannuronate-rich fragments, which do not take part in the gel network, will leach out of the capsules and directly trigger an immune response (Stokke et al., 1993). This observed immunologic response can be linked in part to $(1 \rightarrow 4)$ glycosidic linkages, as other homopolymeric di-equatorial polyuronates, like D-glucuronic acid (C6-oxidised cellulose), also exhibit this feature (Espevik and Skjåk-Bræk, 1996). The immunologic potential of polymannuronates have now been observed in *in vivo* animal models in such diverse areas as for protection against lethal bacterial infections and irradiation and for increasing non-specific immunity (Espevik and Skjåk-Bræk, 1996).

11 Applications

Given the large number of different applications, alginates must be regarded as one of the most versatile polysaccharides. These applications span from traditional technical utilization, to foods, to biomedicine.

11.1 Technical Utilization

The quantitatively most important technical application of alginates is as a shear-thinning viscosifyer in textile printing, in which alginate has gained a high popularity because of the resulting color yield, brightness, and print levelness. Alginates also are used for paper coating to obtain surface uniformity and as binding agents in the production of welding rods. In the latter case, alginate gives stability in the wet stage and functions as a plasticizer during the extrusion process. As a last example of technical applications, ammonium alginate frequently is used for can sealing. The ammonium form is used because of its very low ash content (Onsøyen, 1996).

11.2 Medicine and Pharmacy

Alginates have been used for decades as helping agents in various human-health applications. Some examples include use in traditional wound dressings, in dental impression material, and in some formulations preventing gastric reflux. Alginate's increasing popularity as an immobilization matrix in various biotechnological processes, however, demonstrates that alginate will move into other and more advanced technical domains in addition to its traditional applications. Entrapment of cells within Ca-alginate spheres has become the most widely used technique for the immobilization of living cells (Smidsrød and Skjåk-Bræk, 1990). This immobilization procedure can be carried out in a single-step process under very mild conditions and is therefore compatible with most cells. The cell suspension is mixed with a sodium alginate solution, and the mixture is dripped into a solution containing multivalent cations (usually Ca^{2+}). The droplets then instantaneously form gel-spheres entrapping the cells in a three-dimensional lattice of ionically cross-linked alginate. The possible uses for such systems in industry, medicine, and agriculture are numerous, ranging from production of ethanol by yeast, to production of monoclonal antibodies by hybridoma cells, to mass production of artificial seed by entrapment of plant embryos (Smidsrød and Skjåk-Bræk, 1990).

Perhaps the most exciting prospect for alginate gel immobilized cells is their potential use in cell transplantation. Here, the main purpose of the gel is to act as a barrier between the transplant and the immune system of the host. Different cells have been suggested for gel immobilization, including parathyroid cells for treatment of hypocalcemia and dopamine-producing adrenal chromaffin cells for treatment of Parkinson's disease (Aebisher et al., 1993). However, major interest has been focused on insulin-producing cells for the treatment of Type I diabetes. Alginate/poly-L-lysine capsules containing pancreatic Langerhans islets have been shown to reverse diabetes in large animals and currently are being clinically tested in humans (Soon-Shiong et al., 1993, 1994). Table 5 lists some biomedical applications of alginate-encapsulated cells.

Tab. 5 Some potential biomedical application of alginate-encapsulated cells

Cell type	*Treatment of*
Adrenal chromaffin cells	Parkinson's disease[a]
Hepatocytes	Liver failure[a]
Paratyroid cells	Hypocalcemia[a]
Langerhans islets (β-cells)	Diabetes[b]
Genetically altered cells	Cancer[c]

[a] Aebischer et al., 1993 [b] Soon-Shiong et al., 1993, 1994 [c] Read et al., 2000

11.3
Foods

Alginates are used as food additives to improve, modify, and stabilize the texture of foods. This is valid for such properties as viscosity enhancement; gel-forming ability; and stabilization of aqueous mixtures, dispersions, and emulsions. Some of these properties stem from the inherent physical properties of alginates themselves, as outlined above, but they also may result from interactions with other components of the food product, e.g., proteins, fat, or fibers. As an example, alginates interact readily with positively charged amino acid residues of denatured proteins, which are utilized in pet foods and reformed meat. Cottrell and Kovacs (1980), Sime (1990), and Littlecott (1982) have given numerous descriptions and formulations on alginates in food applications. A general review on this topic is given by McHugh (1987).

Special focus perhaps should be placed on restructured food based on Ca-alginate gels because of its simplicity (gelling being independent upon temperature) and because it is a steadily growing alginate application. Restructuring of foods is based on binding together a flaked, sectioned, chunked, or milled foodstuff to make it resemble the original. Many alginate-based restructured products are already on the market (see Figure 15), as is exemplified by meat products (both for human consumption and as pet food), onion rings, pimento olive fillings, crabsticks, and cocktail berries.

For applications in jams, jellies, fruit fillings, etc., the synergetic gelling between alginates high in guluronate and highly esterified pectins may be utilized (Toft et al., 1986). The alginate/pectin system can give thermoreversible gels in contrast to the purely ionically cross-linked alginate gels. This gel structure is almost independent of sugar content, in contrast to pectin gels, and therefore may be used in low calorie products.

Fig. 15 An example of alginate used in the restructuring of foods: the pimiento fillings of olives. (picture kindly provided by FMC BioPolymer).

The only alginate derivative used in food is propylene glycol alginate (PGA). Steiner (1947) first prepared PGA, and Steiner and McNeely improved the process (1950). PGA is produced by a partial esterification of the carboxylic groups on the uronic acid residues by reaction with propylene oxide. The main product gives stable solutions under acidic conditions where the unmodified alginate would precipitate. It is now used to stabilize acid emulsions (such as in French dressings), acid fruit drinks, and juices. PGA also is used to stabilize beer foam.

As for the regulatory status, the safety of alginic acid and its ammonium, calcium, potassium, and sodium salts was last evaluated by the Joint FAO/WHO Expert Committee on Food Additives (JECFA) at its 39th meeting in 1992. An ADI "not specified" was allocated. JECFA allocated an ADI of 0 to 25 mg/kg bw to propylene glycol alginate at its 17th meeting.

In the U.S, ammonium, calcium, potassium, and sodium alginate are included in a list of stabilizers that are generally recognized as safe (GRAS). Propylene glycol alginate is approved as a food additive (used as an emulsifier, stabilizer, or thickener) and in several industrial applications (used as a coating for fresh citrus fruit, as an inert pesticide adjuvant, and as a component of paper and paperboard in contact with aqueous and fatty foods). In Europe, alginic acid and its salts and propylene glycol are all listed as EC-approved additives other than colors and sweeteners.

Alginates are inscribed in Annex I of the Directive 95/2 of 1995, and as such can be used in all foodstuffs (except those cited in Annex II and those described in Part II of the Directive) under the Quantum Satis principle of the EU.

12 Relevant Patents

A search in one of the international databases for patents and patent applications yielded well above 2000 hits on alginate. It is outside the scope of this chapter, not to mention beyond the capabilities of its authors, to systematically discuss all this literature, which covers inventions for improved utilization of alginates in the technical, pharmaceutical, food, and agricultural areas. We have therefore limited this section to only a handful of *prior art* inventions (Table 6), with the present authors as co-inventors, that point toward a production of alginate and alginate fractions with novel structures and some new biomedical applications based on the physical and biological properties of certain types of alginates with specified chemical structures.

US Patent 5,459,054 may represent a large number of patents dealing with the immobilization of living cells in immuno-protective alginate capsules for implantation purposes as discussed in Section 10.2. US Patents 5,169,840 and 6,087,342 cover the use of alginates enriched with mannuronate for the stimulation of cytokine production in monocytes, which could be of future importance in the treatment of microbial infections, cancer, and immune deficiency and autoimmune diseases. This stimulating effect has been discussed and connected to the use of alginate fibers in wound-healing dressings. A closer look at these specific effects is presented in Section 9.3.

When the calcium ions in alginate gels are exchanged by covalent cross links, the resulting gel with monovalent cations as counter-ions has the ability to swell several hundred times its own weight in water or salt solutions at low ionic strength, as shown in US Patent 5,144,016. This super-absorbent system, further elaborated in Section 9.2.2, still has not found any commercial uses, mainly because of competition from similar materials based on starch and cellulose derivatives, but certain biomedical applications may be foreseen.

A patent on the genes encoding the different C5-epimerases (US Patent 5,939,289) points to the possibility of producing alginates with a large number of different predetermined compositions and sequences and opens up the possibility for the tailor-making of different alginates, as discussed in Sections 8.2 and 12.

An alternative way of manufacturing alginate fractions with extreme compositions is by using selective extraction techniques, as disclosed in Patent WO 98/51710. One possible use of such fractions as gelling modifiers is revealed in Patent WO 98/02488, where it is suggested that these purified low-molecular-weight guluronate blocks give a gel enforcement at high concentrations of calcium ions by connecting and shortening elastic segments that otherwise would be topologically restricted. In conclusion, it may be argued that this relative high rate of patent filing suggests

Tab. 6 Summary of prior art patents on alginates pointing toward specialty applications

Patent number	*Holder*	*Inventors*	*Patent title*	*Public date*
U.S. 5,459,054	Neocrine	G. Skjåk-Bræk O. Smidsrød T. Espevik M. Otterlei P. Soon-Shiong	Cells encapsulated in alginate containing a high content of α-L-guluropnic acid	2 July 1993
U.S. 5,169,840	Nobipol, Protan Biopolymer	M. Otterlei T. Espevik O. Smidsrød G. Skjåk-Bræk	Diequatorially bound β-1,4-polyuronates and use of same for cytokine stimulation	27 March 1991
U.S. 6,087,342	FMC BioPolymer	T. Espevik G. Skjåk-Bræk	Substrates having bound polysaccharides and bacterial nucleic acids	15 May 1998
U.S. 5,144,016	Protan Biopolymer	G. Skjåk-Bræk S. Moe	Alginate gels	29 May 1991
U.S. 5,939,289	Pronova Biopolymer, Nobipol	H. Ertesvåg S. Valla G. Skjåk-Bræk B. Larsen	DNA compounds comprising sequences encoding mannuroran C-5-epimerase	9 May 1995
WO 98/51710	FMC BioPolymer	M. K. Simensen O. Smidsrød K. I. Draget F. Hjelland	Procedure for producing uronic acid blocks from alginate	11 November 1998
WO 98/02488	FMC BioPolymer	M. K. Simensen K. I. Draget E. Onsøyen O. Smidsrød T. Fjæreide	Use of G-block polysaccharides	22 January 1998

that new alginate-based products are being developed and that there is continuous stable demand for alginates and their products.

13
Outlook and Perspectives

From a chemical point of view, the alginate molecule may look very simple, as it contains only the two monomer units M and G linked by the same 1,4 linkages. This simplification of its chemical structure may lead potential commercial users of alginate to treat it as a commodity like many of the cellulose derivatives. In this chapter, we have shown that alginate represents a very high diversity with respect to chemical composition and monomer sequence, giving the alginate family of molecules a large variety of physical and biological properties. This may represent a challenge to the unskilled users of alginate, but it may be an advantage for the producers and new users of alginate who are interested in developing research-based, high-value applications. When microbial alginate and epimerase-modified alginate enter into the marked in the future, the possibility of alginate being tailor-made to diverse applications will be increased even further.

We therefore see a future trend, which has already started, in which the exploitation of alginate gradually shifts from low-tech applications with increasing competition from cheap alternatives to more advanced, knowledge-based applications in the food, pharmaceutical, and biomedical areas. We then foresee continuous, high research activity in industry and academia to describe, understand, and utilize alginate-containing products to the benefit of society.

Acknowledgements

The authors would like to thank Anne Bremnes and Hanne Devle for most skillful assistance in preparing graphic illustrations and Nadra J. Nilsen for collecting the data on alginate lyases.

14
References

Aebischer, P., Goddard, M., Tresco, P. A. (1993) Cell encapsulation for the nervous system, in: *Fundamentals of Animal Cell Encapsulation and Immobilization* (Goosen, M. F. A., Ed.), Boca Raton, FL: CRC Press, 197–224.

Andresen, I.-L., Skipnes, O., Smidsrød, O., Østgaard, K., Hemmer, P. C. (1977) Some biological functions of matrix components in benthic algae in relation to their chemistry and the composition of seawater, *ACS Symp. Ser.* **48**, 361–381.

Atkins, E. D. T., Mackie, W., Smolko, E. E. (1970) Crystalline structures of alginic acids, *Nature* **225**, 626–628.

Atsuki, K., Tomoda, Y. (1926) Studies on seaweeds of Japan I. The chemical constituents of Laminaria, *J. Soc. Chem. Ind. Japan* **29**, 509–517.

Bird, G. M., Haas, P. (1931) XLVII. On the constituent nature of the cell wall constituents of *Laminaria* spp. Mannuronic acid, *Biochem. J.* **25**, 26–30.

Boyd. J., Turvey, J. R. (1977) Isolation of a poly-"-L-guluronate lyase from *Klebsiella aerogenes, Carbohydr. Res.* **57**, 163–171.

Boyd, J., Turvey, J. R. (1978) Structural studies of alginic acid, using a bacterial poly-L-guluronate lyase, *Carbohydr. Res.* **66**, 187–194.

Boyen, C., Bertheau, Y., Barbeyron, T., Kloareg, B. (1990a) Preparation of guluronate lyase from *Pseudomonas alginovora* for protoplast isolation in Laminaria, *Enzyme Microb. Technol.* **12**, 885–890.

Boyen, C., Kloareg, B., Polne-Fuller, M., Gibor, A. (1990b) Preparation of alginate lyases from marine molluscs for protoplast isolation in brown algae, *Phycologia* **29**, 173–181.

Blumenkrantz, N., Asboe-Hansen, G. (1973) New method for quantitative determination of uronic acids, *Anal. Biochem.* **54**, 484–489.

Cottrell, I.W., Kovacs, P. (1980). Alginates, in: *Handbook of Water-Soluble Gums and Resins* (Crawford, H.B., Williams, J., Eds.), Auckland, New Zealand: McGraw-Hill, 21–43.

Davidson I.W., Lawson, C.J., Sutherland, I.W. (1977) An alginate lyase from *Azotobacter vinelandii* phage, *J. Gen. Microbiol.* **98**, 223–229.

Draget, K. I., Myhre, S., Skjåk-Bræk, G., Østgaard, K. (1988) Regeneration, cultivation and differentiation of plant protoplasts immobilized in Ca-alginate beads, *J. Plant Physiol.* **132**, 552–556.

Draget, K.I., Østgaard, K., Smidsrød, O. (1991) Homogeneous alginate gels; a technical approach, *Carbohydr. Polym.* **14**, 159–178.

Draget, K.I., Simensen, M.K., Onsøyen, E., Smidsrød, O. (1993) Gel strength of Ca-limited gels made *in situ, Hydrobiologia,* **260/261**, 563–565.

Draget, K.I., Skjåk-Bræk, G., Smidsrød, O. (1994) Alginic acid gels: the effect of alginate chemical composition and molecular weight, *Carbohydr. Polym.* **25**, 31–38.

Draget, K. I., Skjåk-Bræk, G., Christensen, B. E., Gåserød, O., Smidsrød, O. (1996) Swelling and partial solubilization of alginic acid gel beads in acidic buffer, *Carbohydr. Polym.* **29**, 209–215.

Draget, K.I., Onsøyen, E., Fjæreide, T., Simensen M. K., Smidsrød O. (1997) Use of G-block polysaccharides', *Intl. Pat. Appl.#* PCT/NO97/00176.

Dubois, M., Gilles, K.A., Hamilton, J.K., Rebers, P.A., Smith, F. (1956) Colorimetric method for determination of sugars and related substances, *Anal. Chem.* **28**, 350–356.

Elyakova, L. A., Favorov, V. V. (1974) Isolation and certain properties of alginate lyase VI from the mollusk *Littorina* sp., *Biochim. Biophys. Acta* **358**, 341–354.

Ertesvåg, H., Larsen, B., Skjåk-Bræk, G., Valla, S. (1994) Cloning and expression of an *Azotobacter vinelandii* mannuronan-C-5 epimerase gene, *J. Bacteriol.* **176**, 2846–2853.

Ertesvåg, H., Høidal, H.K., Hals, I.K., Rian, A., Doseth, B., Valla, S. (1995) A family of moddular type mannuronana C-5 epimerase genes controls the alginate structure in *Azotobacter vinelandii*, *Mol. Microbiol* **16**, 719–731.

Ertesvåg, H., Erlien, F., Skjåk-Bræk, G., Rehm, B.H.A., Valla, S. (1998a) Biochemical properties and substrate specificities of a recombinantly produced *Azotobacter vinelandii* alginate lyase, *J. Bacteriol.* **180**, 3779–3784.

ErtesvågH., Høydal, H., Skjåk-Bræk, G., Valla, S. (1998b) The *Azotobacter vinelandii* mannuronan C-5 epimerase AlgE1 consists of two separate catalytic domains, *J. Biol. Chem.* **273**, 30927–30938.

Espevik, T., Skjåk-Bræk, G. (1996) Application of alginate gels in biotechnology and biomedicine, *Carbohydr. Eur.* **14**, 19–25.

Filisetti-Cozzi, T. M. C. C., Carpita, N. C. (1991) Measurement of uronic acids without interference from neutral sugars, *Anal. Biochem.* **197**, 157–162.

Fisher, F. G., Dörfel, H. (1955) Die Polyuronsäuren der Braunalgen (Kohlenhydrate der Algen), *Z. Physiol. Che.*, **302**, 186–203.

Gacesa P., (1992) Enzymic degradation of alginates, *Int. J. Biochem.* **24**, 545–552.

Grant, G. T., Morris, E. R., Rees, D. A., Smith, P. J. C., Thom, D. (1973) Biological interactions between polysaccharides and divalent cations: The egg-box model, *FEBS Lett.* **32**, 195–198.

Grasdalen, H., Larsen, B., Smidsrød O. (1977) ^{13}C-NMR studies of alginate, *Carbohydr. Res.* **56**, C11–C15.

Grasdalen, H., Larsen, B., Smidsrød, O. (1979) A PMR study of the composition and sequence of uronate residues in alginate, *Carbohydr. Res.* **68**, 23–31.

Grasdalen, H. (1983) High-field ^{1}H-nmr spectroscopy of alginate: Sequential structure and linkage conformations, *Carbohydr. Res.* **118**, 255–260.

Haug, A. (1959a) Fractionation of alginic acid, *Acta Chem. Scand.* **13**, 601–603.

Haug, A. (1959b) Ion exchange properties of alginate fractions, *Acta Chem. Scand.* **13**, 1250 1251.

Haug, A., Larsen B. (1963) The solubility of alginate at low pH, *Acta Chem. Scand.* **17**, 1653–1662.

Haug, A., Larsen, B., Smidsrød, O. (1963) The degradation of alginates at different pH values, *Acta Chem. Scand.* **17**, 1466–1468.

Haug, A. (1964) Composition and properties of alginates, Thesis, Norwegian Institute of Technology, Trondheim.

Haug, A., Smidsrød O. (1965) Fractionation of alginates by precipitation with calcium and magnesium ions, *Acta Chem. Scand.* **19**, 1221–1226.

Haug, A., Larsen B. (1966) A study on the constitution of alginic acid by partial acid hydrolysis, *Proc. Int. Seaweed Symp.* **5**, 271–277.

Haug, A., Larsen B., Smidsrød O. (1966) A study of the constitution of alginic acid by partial hydrolysis, *Acta Chem. Scand.* **20**, 183–190.

Haug, A., Smidsrød O. (1967) Precipitation of acidic polysaccharides by salts in ethanol-water mixtures, *J. Polym. Sci.* **16**, 1587–1598.

Haug, A., Larsen B., Smidsrød O. (1967a) Studies on the sequence of uronic acid residues in alginic acid, *Acta Chem. Scand.* **21**, 691–704.

Haug, A., Larsen, B., Smidsrød, O. (1967b) Alkaline degradation of alginate, *Acta Chem. Scand.* **21**, 2859–2870.

Haug, A., Myklestad, S., Larsen, B., Smidsrød, O. (1967c) Correlation between chemical structure and physical properties of alginate, *Acta Chem. Scand.* **21**, 768–778.

Haug, A., Smidsrød, O. (1970) Selectivity of some anionic polymers for divalent metal ions, *Acta Chem. Scand.* **24**, 843–854.

Hirst, E. L., Jones, J. K. N., Jones, W. O., (1939) The structure of alginic acid. Part I, *J. Chem. Soc.* 1880–1885.

Høydal, H., Ertesvåg, H., Stokke, B. T., Skjåk-Bræk, G., Valla, S. (1999) Biochemical properties and mechanism of action of the recombinant *Azotobacter vinelandii* mannuronan C-5 epimerase, *J. Biol. Chem.* **274**, 12316–12322.

Indergaard, M., Skjåk-Bræk, G. (1987) Characteristics of alginate from *Laminaria digitata* cultivated in a high phosphate environment, *Hydrobiologia* **151/152**, 541–549.

Kvam, B. J., Grasdalen, H., Smidsrød, O., Anthonsen, T. (1986) NMR studies of the interaction of metal ions with poly-(1,4-hexuronates). VI. Lanthanide(III) complexes of sodium (methyl ∀-D-galactopyranosid)uronate and sodium (phenylmethyl ∀-D-galactopyranosid)uronate, *Acta Chem. Scand.* **B40**, 735–739.

Kvam, B. (1987) Conformational conditions and ionic interactions of charged polysaccharides. Application of NMR techniques and the Poisson-Boltzmann equation, Thesis, Norwegian Institute of Technology, Trondheim.

Larsen, B., Smidsrød O., Painter T. J., Haug A. (1970) Calculation of the nearest-neighbour frequencies in fragments of alginate from the yields of free monomers after partial hydrolysis, *Acta Chem. Scand.* **24**, 726–728.

Leo, W. J., McLoughlin, A. J., Malone, D. M. (1990) Effects of terilization treatments on some properties of alginate solutions and gels, *Biotechnol. Prog.* **6**, 51–53.
Littlecott, G. W. (1982). Food gels–the role of alginates, *Food Technol. Aust.* **34**, 412–418.
Mackie, W., Noy, R., Sellen, D. B. (1980) Solution properties of sodium alginate, *Biopolymers* **19**, 1839–1860.
Mackie, W., Perez, S., Rizzo, R., Taravel, F., Vignon, M. (1983) Aspects of the conformation of polyguluronate in the solid state and in solution, *Int. J. Biol. Macromol.* **5**, 329–341.
Martinsen, A., Skjåk-Bræk, G., Smidsrød, O. (1989) Alginate as immobilization material: I. Correlation between chemical and physical properties of alginate gel beads, *Biotechnol. Bioeng.* **33**, 79–89.
Martinsen, A., Skjåk-Bræk, G., Smidsrød, O., Zanetti, F., Paoletti, S. (1991) Comparison of different methods for determination of molecular weight and molecular weight distribution of alginates, *Carbohydr. Polym.* **15**, 171–193.
McHugh, D.J. (1987). Production, properties and uses of alginates, in: Production and Utilization of Products from Commercial Seaweeds, FAO Fisheries Technical Paper No. 288 (McHugh, D. J., Ed.), Rome: FAO, 58–115.
Miawa, T. (1930) Alginic acid, *J. Chem. Soc. Japan* **51**, 738–745.
Mikkelsen, A., Elgsæter, A. (1995) Density distribution of calcium-induced alginate gels, *Biopolymers* **36**, 17–41.
Moe, S. T., Skjåk-Bræk, G., Elgsæter, A., Smidsrød, O. (1993) Swelling of covalently cross-linked ionic polysaccharide gels: Influence of ionic solutes and nonpolar solvents, *Macromolecules* **26**, 3589–3597.
Moe, S., Draget, K., Skjåk-Bræk, G., Smidsrød, O. (1995) Alginates, in: *Food Polysaccharides and Their Applications* (Stephen, A. M., Ed.), New York: Marcel Dekker, 245–286.
Murata, K., Inose, T., Hisano, T., Abe, S., Yonemoto, Y. (1993) Bacterial alginate lyase: enzymology, genetics and application, *J. Ferment. Bioeng.* **76**, 427–437.
Myklestad. S., Haug A. (1966) Studies on the solubility of alginic acid *from Ascophyllum nodosum* at low pH, *Proc. Int. Seaweed Symp.* **5**, 297–303.
Neiser, S., Draget, K. I., Smidsrød O. (1998) Gel formation in heat-treated bovine serum albumin–sodium alginate systems, *Food Hydrocolloids* **12**, 127–132.
Neiser, S., Draget, K. I., Smidsrød O. (1999) Interactions in bovine serum albumin–calcium alginate gel systems, *Food Hydrocolloids* **13**, 445–458.
Nelson, W. L., Cretcher, L. H. (1929) The alginic acid from *Macrocystis pyrifera, J. Am. Chem. Soc.* **51**, 1914–1918.
Nelson, W. L., Cretcher, L. H. (1930) The isolation and identification of *d*-mannuronic acid lactone from the *Macrocystis pyrifera, J. Am. Chem. Soc.* **52**, 2130–2134.
Nelson, W. L., Cretcher, L. H. (1932) The properties of *d*-mannuronic acid lactone, *J. Am. Chem. Soc.* **54**, 3409–3406.
Nibu, Y., Satoh, T., Nishi, Y., Takeuchi, T., Murata, K., Kusakabe, I. (1995) Purification and characterization of extracellular alginate lyase from Enterobacter cloacae M-1, *Biosci. Biotechnol. Biochem.* **59**, 632–637.
Onsøyen, E. (1996) Commercial applications of alginates, *Carbohydr. Eur.* **14**, 26–31.
Otterlei, M., Østgaard, K., Skjåk-Bræk, G., Smidsrød, O., Soon-Shiong, P., Espevik, T. (1991). Induction of cytokine production from human monocytes stimulated with alginate, *J. Immunother.* **10**, 286–291.
Painter, T. J., Smidsrød O., Haug A. (1968) A computer study of the changes in composition-distribution occurring during random depolymerisation of a binary linear heteropolysaccharide, *Acta Chem. Scand.* **22**, 1637–1648.
Penman, A., Sanderson G. R. (1972) A method for the determination of uronic acid sequence in alginates, *Carbohydr. Res.* **25**, 273–282.
Read, T.-A., Sorensen, D.R., Mahesparan, R., Enger, P.Ø., Timpl, R., Olsen, B.R., Hjelstuen, M.H.B., Haraldseth, O., Bjerkevig, R. (2000) Local endostatin treatment of gliomas administered by microencapsulated producer cells, *Nature Biotechnol.* **19**, 29–34.
Rivera-Carro H. D. (1984) Block structure and uronic acid sequence in alginates, Thesis, Norwegian Institute of Technology, Trondheim.
Schmidt, E., Vocke, F. (1926) Zur Kenntnis der Polyglykuronsäuren, *Chem. Ber.* **59**, 1585–1588.
Shimokawa, T., Yoshida, S., Kusakabe, I., Takeuchi, T., Murata, K., Kobayashi, H. (1997) Some properties and action mode of "-L-guluronan lyase from *Enterobacter cloacae* M-1. *Carbohydr. Res.* **304**, 125–132.
Sime, W. (1990). Alginates, in: *Food Gels* (Harris, P., Ed.), London: Elsevier, 53–78.
Skjåk-Bræk, G., Larsen B., Grasdalen H. (1986) Monomer sequence and acetylation pattern in

some bacterial alginates, *Carbohydr. Res.* **154**, 239–250.

Skjåk-Bræk, G., Grasdalen, H., Smidsrød, O. (1989a) Inhomogeneous polysaccharide ionic gels, *Carbohydr. Polym.* **10**, 31–54.

Skjåk-Bræk, G., Grasdalen, H., Draget, K. I., Smidsrød, O. (1989b). Inhomogeneous calcium alginate beads, in: Biomedical and Biotechnological Advances in Industrial Polysaccharides (Crescenzi, V., Dea, I. C. M., Paoletti, S., Stivala, S. S., Sutherland, I. W., Eds.), New York: Gordon and Breach, 345–363.

Skjåk-Bræk, G., Moe, S. T. (1992) Alginate gels, US Patent 5,144,016.

Smidsrød, O., Haug, A., Larsen, B. (1963a) The influence of reducing substances on the rate of degradation of alginates, *Acta Chem. Scand.* **17**, 1473–1474.

Smidsrød, O., Haug, A., Larsen, B. (1963b) Degradation of alginate in the presence of reducing compounds, *Acta Chem. Scand.* **17**, 2628–2637.

Smidsrød, O., Haug, A., Larsen, B. (1967) Oxidative-reductive depolymerization: a note on the comparison of degradation rates of different polymers by viscosity measurements, *Carbohydr. Res.* **5**, 482–485.

Smidsrød, O., Haug, A. (1968a) A light scattering study of alginate, *Acta Chem. Scand.* **22**, 797–810.

Smidsrød, O., Haug, A. (1968b) Dependence upon uronic acid composition of some ion-exchange properties of alginates, *Acta Chem. Scand.* **22**, 1989–1997.

Smidsrød, O., Whittington S. G. (1969) Monte Carlo investigation of chemical inhomogeneity in copolymers, *Macromolecules* **2**, 42–44.

Smidsrød, O. (1970) Solution properties of alginate, *Carbohydr. Res.* **13**, 359–372.

Smidsrød, O. (1973). Some physical properties of alginates in solution and in the gel state, Thesis, Norwegian Institute of Technology, Trondheim.

Smidsrød, O., Glover, R. M., Whittington, S. G. (1973) The relative extension of alginates having different chemical composition, *Carbohydr. Res.* **27**, 107–118.

Smidsrød, O. (1974) Molecular basis for some physical properties of alginates in the gel state, *J. Chem. Soc. Farad. Trans* **57**, 263–274.

Smidsrød, O., Skjåk-Bræk, G. (1990) Alginate as immobilization matrix for cells, *Trends Biotechnol.* **8**, 71–78.

Smidsrød, O., Draget, K. I. (1996) Alginates: chemistry and physical properties, *Carbohydr. Eur.* **14**, 6–13.

Soon-Shiong, P., Otterlei, M., Skjåk-Bræk, G., Smidsrød, O., Heintz, R., Lanza, R. P., Espevik, T. (1991) An immunologic basis for the fibrotic reaction to implanted microcapsules, *Transplant Proc.* **23**, 758–759.

Soon-Shiong, P., Feldman, E., Nelson, R., Heints, R., Yao, Q., Yao, T., Zheng, N., Merideth, G., Skjåk-Bræk, G., Espevik, T., Smidsrød, O., Sandford P. (1993) Long-term reversal of diabetes by the injection of immunoprotected islets, *Proc. Natl. Acad. Sci. USA* **90**, 5843–5847.

Soon-Shiong, P., Heintz, R. E., Merideth, N., Yao, Q. X., Yao, Z. W., Zheng, T. L., Murphy, M., Moloney, M. K., Schmehl, M., Harris, M., Mendez, R., Mendez, R., Sandford, P. A. (1994) Insulin independence in a type 1 diabetic patient after encapsulated islet transplantation, *Lancet* **343**, 950–951.

Stanford, E. C. C. (1881). British Patent 142.

Stanford, E. C. C. (1883a) New substance obtained from some of the commoner species of marine algæ; Algin, *Chem. News* **47**, 254–257.

Stanford, E. C. C. (1883b) New substance obtained from some of the commoner species of marine algæ; Algin, *Chem. News* **47**, 267–269.

Steginsky, C. A., Beale, J. M., Floss, H. G., Mayer, R. M. (1992) Structural determination of alginic acid and the effects of calcium binding as determined by high-field n.m.r., *Carbohydr. Res.* **225**, 11–26.

Steiner, A. B. (1947). Manufacture of glycol alginates, US Patent 2,426,215.

Steiner, A. B., McNeilly, W. H. (1950). High-stability glycol alginates and their manufacture, US Patent 2,494,911.

Stokke, B. T., Smidsrød, O., Bruheim, P., Skjåk-Bræk, G. (1991). Distribution of uronate residues in alginate chains in relation to alginate gelling properties, *Macromolecules* **24**, 4637–4645.

Stokke, B. T., Smidsrød, O., Zanetti, F., Strand, W., Skjåk-Bræk G. (1993) Distribution of uronate residues in alginate chains in relation to gelling properties 2:Enrichment of -D-mannuronic acid and depletion of "-L-guluronic acid in the sol fraction, *Carbohydr. Polym.* **21**, 39–46.

Stokke B. T., Draget K. I., Yuguchi Y., Urakawa H., Kajiwara K. (2000) Small angle X-ray scattering and rheological characterization of alginate gels. 1 Ca-alginate gels, *Macromolecules* **33**, 1853–1863.

Tanford, C. (1961). *Physical Chemistry of Macromolecules.* New York: John Wiley & Sons, Inc.

Toft, K., Grasdalen, H., Smidsrød, O. (1986). Synergistic gelation of alginates and pectins, *ACS Symp. Ser.* **310**, 117–132.

Valla, S., Ertesvåg, H., Skjåk-Bræk, G. (1996) Genetics and biosynthesis of alginates, *Carbohydr. Eur.* **14**, 14–18.
Wong, T. Y., Preston, L. A., Schiller, N. L. (2000) Alginate lyase: review of major sources and enzyme characteristics, structure-function analysis, biological roles, and applications, *Ann. Rev. Microbiol.* **54**, 289–340.

9
Alginates from Bacteria

Dr. Bernd H. A. Rehm
Institut für Mikrobiologie, Westfälische Wilhelms-Universität Münster, Corrensstraße 3, 48149 Münster, Germany; Tel.: +49-251-8339848; Fax: +49-251-8338388; E-mail: rehm@uni-muenster.de

Biopolymers for Medical and Pharmaceutical Applications. Edited by A. Steinbüchel and R. H. Marchessault

ISBN: 3-527-31154-8

CF	cystic fibrosis
EPS	extracellular polysaccharide (exopolysaccharide)
HSL	homoserine lactone
LPS	lipopolysaccharide
PMI-GMP	phosphomannose isomerase/guanosine diphosphomannose pyrophosphorylase
PMI-GMP	phosphomannose isomerase-GDP-mannose pyrophosphorylase
PMM	phosphomannomutase
PMM	phosphomannomutase
SCLM	scanning laser confocal microscopy
TCA cycle	tricarboxylic acid cycle

1 Introduction

Alginates are a family of unbranched, non-repeating copolymers consisting of variable amounts of (1-4)-linked β-D-mannuronic acid and its epimer α-L-guluronic acid. The monomers are distributed in blocks of continuous mannuronate residues (M-blocks), guluronate residues (G-blocks), or alternating residues (MG-blocks) (Figure 1). Alginates isolated from different natural sources vary in the length and distribution of the different block types. Alginates are produced by bacteria and brown seaweeds, and the mannuronate residues of the bacterial–but not those of the seaweed polymers–are acetylated to a variable extent at positions O-2 and/or O-3 (Skjåk-Bræk et al., 1986). The variability in monomer block structures and acetylation strongly affects the physico-chemical and rheological properties of the polymer, and the biological basis for the variability is therefore of both scientific and

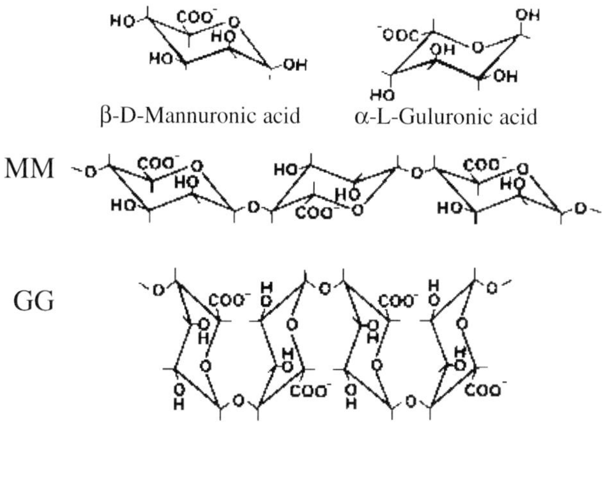

Fig. 1 The structure of alginate. MM: polymannuronate; GG: polyguluronate.

applied importance (Smidsrød and Draget, 1996). Alginate is used for a variety of industrial purposes, for example as a stabilizing, thickening and gelling agent in food production, or to immobilize cells in pharmaceutical and biotechnology industries (see below) (Onsøyen, 1996). The production is currently based exclusively on the harvesting of brown seaweeds. Several bacteria belonging to the genera *Pseudomonas* and *Azotobacter* also produce alginate (Gorin and Spencer, 1966; Linker and Jones, 1966; Govan et al., 1981; Cote and Krull, 1988), and the structures of the blocks of monomer residues are similar in alginates produced by seaweeds and those synthesized by *Azotobacter vinelandii*. In contrast, all *Pseudomonas* alginates lack G-blocks (Skjåk-Bræk et al., 1986). Most of our knowledge of the genetics of alginate biosynthesis originates from studies of *Pseudomonas aeruginosa*, mainly because of the medical relevance of this organism as an opportunistic human pathogen, particularly for patients suffering from cystic fibrosis (CF) (Govan and Harris, 1986; May and Chakrabarty, 1994). Alginate plays an important role as a virulence factor during the infectious process (Gacesa and Russell, 1990). The reason for this appears to be related to the alginate-mediated mode of biofilm growth, which causes resistant colonization of the lung. *A. vinelandii* and *P. aeruginosa* produce alginate as an extracellular polysaccharide (EPS) in vegetatively growing cells, whereas in *A. vinelandii* alginate is also involved in the differentiation process leading to a so-called cyst (Sadoff, 1975).

2 Historical Outline

The polysaccharide alginate was firstly isolated from marine macroalgae in the 19th century, but it was approximately 80 years later that a bacterial source (from *P. aeruginosa*) of the polysaccharide was identified (Linker and Jones, 1966). This alginate was later found to be similar to the commercially useful polymer obtained from marine algae (Lin and Hassid, 1966a; Linker and Jones, 1966) and also to the polysaccharide produced as a capsule by *Azotobacter vinelandii* (Gorin and Spencer, 1966). The association of mucoid *P. aeruginosa*, i.e., alginate-overproducing strains, with chronic lung infections in patients who suffer from the inherited disease CF has been well established, and is recognized as a major cause of pathogenesis in these individuals. Mucoid *P. aeruginosa* have also been isolated, albeit less frequently, from other patients, for example bronchiotatics and those with urinary tract or middle-ear infections (McAvoy et al., 1989), although not normally from individuals with infected burn sites. Although most of the mucoid isolates of *P. aeruginosa* have been obtained from clinical samples, it is clear that alginate production is important in a much wider context. Ten of 81 *P. aeruginosa* isolates from technical water systems showed a mucoid phenotype, which implies a more widespread occurrence of ecological niches for mucoid forms (Grobe et al., 1995). Alginate biosynthesis is a key factor in the establishment of stable mature biofilms of *P. aeruginosa* in a wide range of environmental situations (Nivens et al., 2001). Alginate production is fairly widespread amongst rRNA homology group I pseudomonads, as indicated by Southern hybridization experiments using various alginate biosynthesis genes as probes (Fialho et al., 1990; Fett et al., 1992; Rehm, 1996). Genomic DNA from representatives of groups II–IV gave very weak or no hybridization with the probes, except for *algC*, indicating that the ability to produce alginate is restricted to members of rRNA homology group I. This substantiates earlier physiological studies in which alginate was isolated from *Pseudomonas fluorescens, Pseudomonas putida, Pseudomonas mendocina* (Govan et al., 1981) and *P. syringae* (Fett et al., 1986; Gross and Rudolph, 1987). Alginate is also synthesized by *A. vinelandii* as part of the encystment process (Gorin and Spencer, 1966). The mature cysts are surrounded by two discrete alginate-rich layers (exine and intine) which enable the dormant cells to survive long periods of desiccation. Strains of *Azotobacter chroococcum* also produce alginate (Cote and Krull, 1988). In *Azotobacter*, abnormalities in alginate production resulted in impaired encystment, which indicated that alginate biosynthesis is required to survive and adapt to famine conditions, as was supposed for the establishment of biofilms by pseudomonads. Enzymological evidence for the biosynthesis pathway of alginate was obtained from the brown algae *Fucus gardneri* by Lin and Hassid (1966a) and from *A. vinelandii* about ten years later by Pindar and Bucke (1975). These studies indicated that fructose-6-phosphate is the first alginate precursor, which was derived from the Entner–Doudoroff pathway and from the fructose-1,6-bisphosphate aldolase reaction. The presence of the first alginate biosynthetic enzymes, phosphomannose isomerase, GDP-mannose pyrophosphorylase and GDP-mannose dehydrogenase, was demonstrated by Piggot et al. (1981) in *P. aeruginosa*. Later, Padgett and Phibbs (1986) detected another alginate biosynthesis enzyme, phosphomannomutase. The medical relevance of mucoid *P. aeruginosa* stimulated research on the genetics of alginate biosynthesis. A stable mucoid mutant of *P. aeruginosa* (8830) had to be generated by chemical mutagenesis in order to obtain nonmucoid mutants (Darzins and Chakrabarty, 1984). Complementation studies of the various mutants allowed the identification and characterization of the alginate biosynthesis genes, as well as the respective regulatory genes (May and Chakrabarty, 1994). The identification and biochemical characterization of regulatory alginate biosynthesis proteins enabled a deeper understanding of the rather complex regulatory network (Govan and Deretic, 1996; Rehm and Valla, 1997). The extracellular Ca^{2+}-dependent C-5-epimerases were the first alginate-related enzymes from *A. vinelandii*, which were biochemically and genetically characterized (Ertesvåg et al., 1994, 1999). Meanwhile, hybridization of a lambda gene library of *A. vinelandii* with the outer membrane encoding the *algE* gene from *P. aeruginosa* as a probe, led to the identification of the second alginate biosynthesis gene cluster and the first gene cluster of a biotechnologically relevant microorganism (Rehm, 1996; Rehm et al., 1996). Comparative analysis of the genetics, biochemistry and regulation of alginate biosynthesis revealed strong similarities (Rehm and Valla, 1997; Gacesa, 1998). Moreover, the plant-pathogenic *P. syringae* pv. *syringae* alginate biosynthesis gene cluster was identified and characterized, exhibiting a virtually identical alginate gene arrangement (Penaloza-Vazquez et al., 1997).

3
Chemical Structures

Alginate is composed of the uronic acid β-D-mannuronate and its C-5 epimer α-L-guluronate. These monomers may be arranged in homopolymeric (poly-mannuronate or poly-guluronate) or heteropolymeric block structures (see Figure 1). In addition, bacterial alginates are normally O-acetylated on the 2 and/or 3 position(s) of the β-D-mannuronate residues. Consequently, bacteria produce a range of alginates with different block structures and degrees of O-acetylation.

The high molecular mass of bacterial alginate and the negative charge ensure that the polysaccharide is highly hydrated and viscous. It is well established that alginates from *P. aeruginosa* do not contain polyguluronate blocks (Sherbrock-Cox et al., 1984) but those from *A. vinelandii* may do so. The block structure and degree of O-acetylation, as well as the molecular weight, determine the physico-chemical properties of alginate. Alginates which contain polyguluronate form rigid gels in the presence of Ca^{2+}, and are therefore important in structural roles, for example the outer cyst wall (exine) of *A. vinelandii*. Conversely, an absence of polyguluronate, as in *P. aeruginosa*, produces relatively flexible gels in the presence of Ca^{2+}. Extensive O-acetylation of alginate increases the water-binding capacity of the polysaccharide, which may be significant in enhancing survival under desiccating conditions.

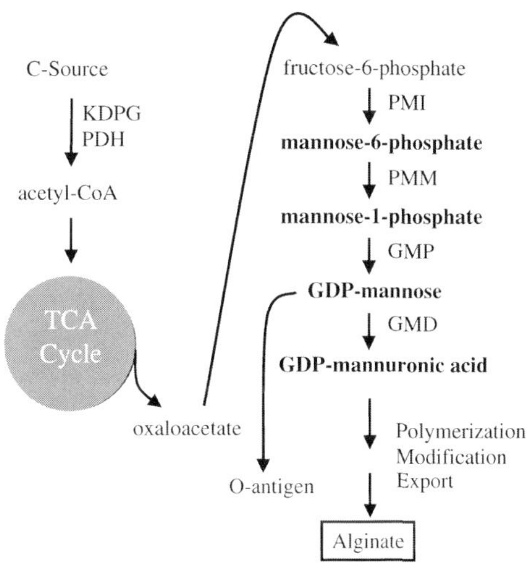

Fig. 2 Biosynthesis pathway of alginate. Oxaloacetate is converted to fructose-6-phosphate via gluconeogenesis. KDPG: ketodeoxyphosphogluconate pathway (Entner–Doudoroff pathway); PDH: pyruvate dehydrogenase; PMI-GMP: phosphomannose isomerase-GDP-mannose pyrophosphorylase; PMM: phosphomannomutase; GMD: GDP-mannose dehydrogenase. Intermediates shown in bold type are precursors of alginate.

4
Biosynthetic Pathway of the Alginate Precursor, GDP-Mannuronic Acid

A convincing pathway for alginate biosynthesis in *A. vinelandii* was first proposed by Pindar and Bucke (1975) based on the assay of individual enzyme activities. However, the corresponding enzymes in *P. aeruginosa* have proved more difficult to assay directly, and the pathway has been elucidated using a combination of complementation analyses and gene overexpression studies. Although the initial steps in the pathway are indisputable (Figure 2) (and the same applies to *A. vinelandii* and pseudomonads), there is still considerable debate about the final stages of biosynthesis and the export of alginate.

The alginate biosynthesis starts from fructose-6-phosphate in the cytosol. Radiolabeling studies have established that six-carbon growth substrates are oxidized via the Entner–Doudoroff pathway, and that the resultant pyruvate [1 mol (mol hexose)$^{-1}$] is ultimately channeled into alginate biosynthesis (Lynn and Sokatch, 1984). More detailed analyses of the labeling patterns in alginate indicate that the pyruvate derived from the oxidation of sugars is fed into the tricarboxylic acid (TCA) cycle prior to synthesis of fructose-6-phosphate and alginate (Narbad et al., 1988). Recent experiments using ^{13}C-NMR have clearly established the key role of the Entner–Doudoroff pathway in the oxidation of hexoses, and also the obligatory requirement of triose intermediates in alginate biosynthesis (Beale and Foster, 1996). The latter authors concluded that the pyruvate derived from the oxidation of hexoses feeds into alginate biosynthesis via the formation of oxaloacetate and subsequent gluconeogenesis (see Figure 2). This supports earlier data employing ^{13}C-labeled precursors, which suggested an obligatory role for the TCA cycle in the conversion of glucose to alginate (Narbad et al., 1988). This important role of the TCA cycle in alginate biosynthesis is supported by more recent genetic studies using nonmucoid mutants of *P. aeruginosa*. Levels of only the phosphorylated (active) form of the key TCA cycle enzyme succinyl-CoA synthetase are reduced in *algQ* mutants (Schlictman et al., 1994). The normally rare occurrence of mucoid forms of *P. aeruginosa* in culture can be significantly increased by growth on energy-poor media (Terry et al., 1991). However, there is a recent report that glucose can stimulate *algD* transcription and alginate production (Ma et al., 1997), though this contradicts earlier findings which proposed that glucose repression of *algD* occurs (Devault et al., 1991). Interestingly, alginate biosynthesis occurs in response to energy deprivation, despite biosynthesis of the polysaccharide being an energy-consuming process. It has been suggested that the AlgQ-controlled expression of succinyl-CoA synthetase and nucleoside-diphosphate kinase may result in a decreased pool size of GTP and hence less available GDP-mannose for alginate biosynthesis (Schlictman et al., 1994; Kim et al., 1998; Kapatral et al., 2000). However, quantification of the nucleotide sugars indicates that GDP-mannose is present in great excess (Tatnell et al., 1993), even when the GDP-mannose dehydrogenase gene was overexpressed (Tatnell et al., 1994). The initial steps in the alginate biosynthesis are related to general carbohydrate metabolism, and the intermediates are widely utilized. In particular, the intermediate GDP-mannose serves not only as a precursor for alginate biosynthesis but also for lipopolysaccharide (LPS) biosynthesis (Goldberg et al., 1993). Accordingly, the GDP-mannose dehydrogenase (AlgD) exhibits a key role in the biosynthesis of alginate. The *algD* gene is proximal to the promoter on the alginate biosynthesis gene cluster and expression is tightly controlled (Schurr et al., 1993). Analyses of nucleotide sugar pools and exopolysaccharide production clearly indicate that GDP-mannose dehydrogenase is the kinetic control point in the alginate pathway (Tatnell et al., 1994). However, the alginate biosynthesis enzyme phosphomannose isomerase/ guanosine-diphosphomannose pyrophosphorylase (PMI-GMP (AlgA)) is a bifunctional protein catalyzing the initial and third steps of alginate synthesis (see Figure 2). The PMI reaction pulls the fructose-6-phosphate out of the metabolic pool, leading to the first intermediate, mannose-6-phosphate. Phosphomannomutase (AlgC) then catalyzes the second step, resulting in the formation of mannose-1-phosphate. This is not the only reaction catalyzed by AlgC, which also exhibits phosphoglucomutase activity and which is evidently involved in rhamnolipid, LPS and alginate biosynthesis (Olvera et al., 1999). The GMP activity of PMI-GMP (AlgA) then, with concomitant GTP hydrolysis, converts mannose-1-phosphate to GDP-mannose. The enzyme favors the reverse reaction, but because of the efficient removal of the GDP-mannose in the next step, the entire pathway proceeds efficiently in the direction of alginate synthesis. The almost irreversible oxidation of GDP-mannose to GDP-mannuronic acid involves the enzyme guanosine-diphosphomannose dehydrogenase, and the reaction product is the immediate precursor for polymerization (see Figure 2). For further details on this pathway, readers are referred to the review on *P. aeruginosa* alginate synthesis by May and Chakrabarty (1994). Interestingly, the intracellular activities of key biosynthesis enzymes appear to be very low, even in extracts prepared from highly mucoid *P. aeruginosa* cells. It has been speculated that the enzymes phosphomannose isomerase-GDP-mannose pyrophosphorylase (PMI-GMP) and phosphomannomutase (PMM) may exist as an enzyme complex ("metabolon"), which allows the coupling of enzymatic reactions, as described for many metabolic pathways (Mathews, 1993).

5 Genetics of Alginate Biosynthesis

At least 24 genes have been directly implicated in alginate biosynthesis in *P. aeruginosa* (Figure 3; Table 1), and there is good evidence that others may also be involved, e.g., *glpM* (Schweizer et al., 1995). It is not possible to assign all *alg* genes identified so far as solely functioning in alginate biosynthesis, as it is now evident for example that some of the regulator genes act globally and encode proteins such as alternative sigma factors (Yu et al., 1995). Other "*alg*" genes, e.g., *algA, C,* are also required for LPS biosynthesis (Goldberg et al., 1993). The genes involved in the synthesis of the precursor GDP-mannuronic acid have all been identified and characterized, and they have been assigned the designations *algA* (encoding PMI-GMP), *algC* (encoding phosphomannomutase), and *algD* (encoding guanosine-diphosphomannose dehydrogenase). With the exception of *algC* (located at 10 min on the chromosome map; between 5,992,000 and 5,994,000 bp of the *P. aeruginosa* genome sequence), the other two genes and all other known structural genes involved in alginate biosynthesis in *P. aeruginosa* are clustered at 34 min (between positions 3,962,000 bp and 3,980,000 bp of the *P. aeruginosa* genome sequence) (Figure 3; Table 1). The biological functions of many of the gene products putatively involved in the polymerization process are poorly understood, mainly because the polymerase has

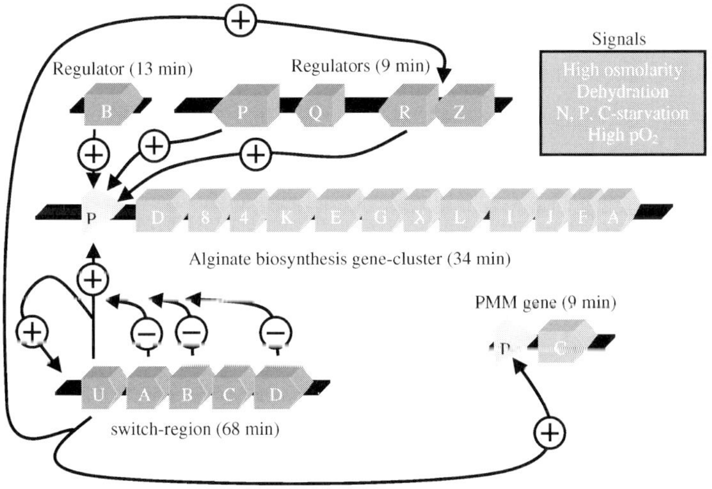

Fig. 3 Genetic organization of alginate genes and their regulation. Signals which induce alginate biosynthesis gene expression are given in the gray box. +: induction (positive effector); –: inhibition (negative effector); P: promoter; switch-region: muc-gene region (genotypic switch to alginate overproduction by mutation in *mucA*, *mucB* or *mucD*, respectively).

Tab. 1a Alginate genes from *Pseudomonas aeruginosa*

Gene	*Location*	*Gene product*
algD	34	GDP mannose dehydrogenase
alg8	34	Polymerase/export function?
alg44	34	Polymerase/export function?
*algK**	34	Polymerase/export function?
algE	34	Outer-membrane porin?
algG	34	Mannuronan C-5-epimerase
algX	34	Unidentified function but high sequence similarity to *algJ*
algL	34	Alginate lyase
algI	34	*O*-Acetylation
algJ	34	*O*-Acetylation
algF	34	*O*-Acetylation
algA	34	Phosphomannose isomerase/GDP mannose pyrophosphorylase
algB	13	Member of *ntrC* subclass of two-component regulators
algC	10	Phosphomannomutase
algH	?	Unknown function
algR1	9	Regulatory component of two-component sensory transduction system
algR2 (algQ)	9	Protein kinase or kinase regulator
algR3(algP)	9	Histone-like transcription regulator
algZ	9	AlgR cognate sensor
algU (algT)	68	Homologue of *E. coli* σ^E global stress response factor
algS (mucA)	68	Anti σ factor
algN (mucB)	68	Anti σ factor?
algM (mucC)	68	Regulator?
algW (mucD)	68	Homologue of serine protease (HtrA)

Modified according to Rehm and Valla (1997)

not been purified and no *in vitro* alginate synthesis assay has been established. However, the gene products of *alg8, alg44, algX* (formerly *alg60*) and *algK* are candidates for being subunits of the alginate polymerase. The deduced amino acid sequences of these proteins contain hydrophobic regions, suggesting a localization in the cytoplasmic membrane (Wang et al., 1987; Maharaj et al., 1993; Rehm and Valla, 1997). The *algK* gene has been identified located directly downstream of *alg44* in *P. aeruginosa* (Aarons et al., 1997). Evidence was obtained that AlgK is entirely periplasmic and is probably anchored in the cytoplasmic membrane. AlgK might be also involved in polymerization and/or export of alginate. Although most of the genes are essential for alginate biosynthesis, those encoding the epimerase and O-acetyltransferase(s) can be inactivated, provided that essential genes downstream of the mutation are expressed in trans, without abolishing alginate biosynthesis. The role of a gene (*algL*) encoding an alginate lyase in the biosynthesis operon (Boyd et al., 1993; Schiller et al., 1993; Monday and Schiller, 1996) is unknown. Contradictory data were published regarding the role of AlgL in alginate biosynthesis. Boyd and coworkers (1993) showed that a mutation in the *algL* gene had no effect on alginate biosynthesis. However, Monday and Schiller (1996) demonstrated that an *algL* mutation strongly impaired alginate biosynthesis, as

Tab. 1b Alginate genes from *Azotobacter vinelandii*

Gene	*Gene product*
alg8	Polymerase?
alg44	Polymerase/export function?
algA	Phosphomannose isomerase
	GDP mannose pyrophosphorylase
algD	GDP mannose dehydrogenase
algE1-7	Mannuronan C-5-epimerases
algG	Mannuronan C-5-epimerase
algJ	Export of alginate?
algL	Alginate lyase
algU	Homologue of *E. coli* σ^E global stress response factor
mucA	Anti σ factor
mucB	Anti σ factor
mucC	Regulator?
mucD	Homologue of serine protease (HtrA)

Modified according to Rehm and Valla (1997).

was recently confirmed by an *algL* mutant of *P. syringae* pv. *syringae* (Penaloza-Vazquez et al., 1997). The *P. aeruginosa* gene cluster at 34 min also contains the *algE* gene, encoding an outer membrane protein (Chu et al., 1991; Rehm et al., 1994a). Production of this protein is strictly correlated with the mucoid phenotype of *P. aeruginosa* (Grabert et al., 1990; Rehm et al., 1994a).

In *A. vinelandii*, our understanding of the genetics of alginate synthesis has improved greatly over the past few years. The first genes involved in alginate synthesis to be cloned and characterized encode a set of seven strongly related Ca^{2+}-dependent mannuronan C-5-epimerases, designated AlgE1 to AlgE7 (Ertesvåg et al., 1999). These proteins are structurally unusual, since they can all be described as repeats of two types of protein modules, designated A (385 amino acids) and R (153 amino acids). Each protein contains a short motif designated S at the carboxy-terminal end. The A modules are present once or twice in each protein, while the R modules are represented one to seven times. Each R module contains four to seven repeats of a nine-amino-acid sequence repeat putatively involved in the binding of Ca^{2+} ions (Ertesvåg et al., 1999). These epimerase genes (*algE1–algE7*) are clustered in the *A. vinelandii* chromosome, and they share no significant sequence homology to the *P. aeruginosa algG* gene.

It was shown recently that the *A. vinelandii* genome encodes a gene (*algD*) corresponding to the *P. aeruginosa algD* gene, and these two genes share 73% identity with each other at the protein level (Campos et al., 1996). An *A. vinelandii* gene (*algJ*) apparently corresponding to the *P. aeruginosa algE* gene was also recently identified (Rehm, 1996). AlgJ shares about 52% sequence identity with AlgE from *P. aeruginosa*, and topological modeling suggests that this protein is structurally similar to AlgE. It is also believed to be functionally equivalent, forming a pore which is involved in alginate export (Rehm, 1996). Surprisingly, the *A. vinelandii* genome also encodes a mannuronan C-5-epimerase (AlgG), which belongs to a different class from AlgE1–E7, and this epimerase seems to represent the equivalent of AlgG in *P. aeruginosa* (Rehm et al., 1996). In addition *A. vinelandii* encodes a protein

(AlgY) containing one A and one R module, but this protein appears to display no epimerase activity after expression in *Escherichia coli* (Svanem et al., 1999). Recently, the entire alginate biosynthesis gene cluster of *A. vinelandii* was identified, which revealed a similar physical organization of the *alg* genes as has been found in *P. aeruginosa* (Lloret et al., 1996; Rehm et al., 1996). A similar arrangement of *alg* genes has been also described in *Pseudomonas syringae* pv. *syringae* (Penaloza-Vazquez et al., 1997), and is also evident for *P. fluorescens* based on genome sequence analysis (B. H. A. Rehm, unpublished results). These data suggested that the physical arrangement of alginate genes in bacteria is conserved.

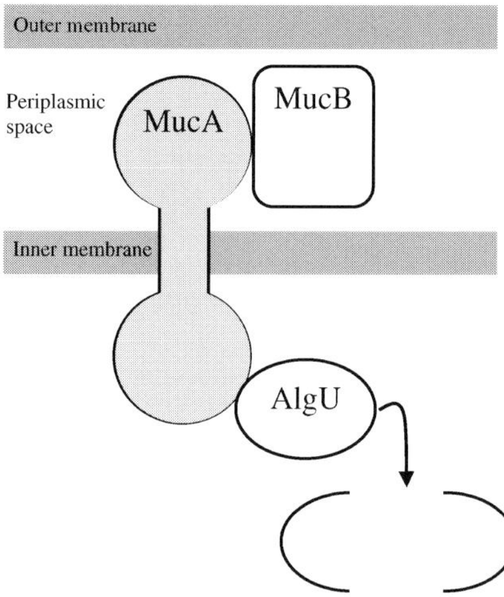

Fig. 4 Model of the action of anti-sigma factors MucA and MucB. AlgU is the alternative sigma-factor, which is required for *alg* gene expression. Binding of AlgU to MucA might expose AlgU to proteolytic digestion by MucD (Mathee et al., 1997).

6 Regulation of Alginate Biosynthesis

The regulation of alginate biosynthesis is complex, and involves specific gene products and those that act more globally (Figure 3 and Figure 4). Expression of the entire alginate biosynthesis gene cluster of *P. aeruginosa* is under the control of the *algD* promoter, and in essence this region (*algD-algA*) acts as an operon (Chitnis and Ohman, 1993), although there is sequence-based evidence for weak promoters within the gene cluster. One of the alginate biosynthesis genes, *algC*, is located at 10 min on the PAO1 map and is transcribed independently of the 34-min region, but is coordinately induced with the biosynthesis gene cluster. The regulatory genes in *P. aeruginosa* map at 9 min and 13 min, and genes responsible for a genotypic switch to alginate overproduction are found at 68 min (see Figure 3). The alginate structural genes are controlled via the positively regulated *algD* promoter (Deretic et al., 1989). In *A. vinelandii*, on the other hand, there seems to exist two additional *algD*-independent promoters, from which *alg8-algJ* and *algG-algA* are transcribed, respectively. These promoters were found to be independently regulated (Lloret et al., 1996; Vazquez et al., 1999). The genotypic switch region comprising *algU, mucA, mucB, mucC* and *mucD* (831,000–835,000, positions relative to the *P. aeruginosa* genome sequence) is found in both bacteria, and the corresponding gene sequences are highly homologous (see Figure 3). The genes have also been found to be biologically active and play a similar role in the two species (Martinez-Salazar et al., 1996). All the known *alg* genes and their corresponding proteins are listed in Table 1.

6.1 Environmentally Induced Activation of *alg* Genes

Conditions of high osmolarity, N, P or carbon starvation, dehydration, and the presence of phosphorylcholine activate a cascade of regulatory proteins in *P. aeruginosa* involved in the activation of the *algD*-promoter (Gacesa and Russell, 1990; Terry et al., 1991). Genes located at a region spanning 9 and 13 min on the *P. aeruginosa* chromosome, *algR(algR1)*, *algQ(algR2)* and *algP(algR3)*, and *algB* modulate the production of alginate and have been described as auxiliary regulators of mucoidy (Govan and Deretic, 1996). A two-component signal-transduction pathway comprising the putative sensor proteins AlgQ (kinase) and AlgZ, interacting with the cognate response regulator proteins such as AlgR and AlgB, were identified (see Figure 3) (May and Chakrabarty, 1994; Yu et al., 1997). Analysis of sequence data indicates that *algZ* encodes a sensory component of a signal transducer system, but that it lacks several expected motifs typical of histidine protein kinases (Yu et al., 1997). The best characterized of these regulators is AlgR, which gene is transcribed in response to the protein AlgU. AlgR binds to three sites upstream of the *algD* promoter and, in conjunction with AlgU, up-regulates transcription of *algD* and the downstream genes. AlgR also promotes expression of *algC*. The efficiency of AlgR is increased by phosphorylation by the cognate kinase AlgQ. AlgB also modulates *algD* expression and, based on sequence analysis, is a member of the NtrC subclass of two-component regulators (Wozniak and Ohman, 1991); however, the *algB* and *algR* regulatory systems appear to operate independently of each other (Wozniak and Ohman, 1994). Interestingly, AlgB and AlgR showed phosphorylation-independent activity on the induction of alginate biosynthesis (Ma et al., 1998). Binding of these positive regulators upstream of the *algD* promoter, presumably leads to the formation of a suprahelical structure with the aid of the histone-like AlgP protein, causing activation of transcription (Deretic and Konyecsni, 1989; Konyecsni and Deretic, 1990; Deretic et al., 1994). A comprehensive account of the inter-relationships of the regulators has been reviewed (Govan and Deretic, 1996). Furthermore, the recently identified sigma-like factor AlgU (AlgT) is responsible for the initiation of *algD* transcription (se Figure 3) (Hershberger et al., 1995). On the basis of sequence analysis (Martin et al., 1994), AlgU is a member of the σ^E class of sigma factors, i.e., analogous to RpoE of *Escherichia coli*, and is essential for alginate production. Subsequent studies have established that AlgU and RpoE are functionally equivalent (Yu et al., 1995), and that AlgU forms complexes with RNA polymerase (Schurr et al., 1995). AlgU causes an increase in alginate biosynthesis by a direct action on the *algD* promoter (see below), and indirectly by up-regulating transcription of another regulatory gene, *algR* (Martin et al., 1994). This environmentally induced transcription of the *alg* cluster and the resulting production of alginate occur only at a rather low level.

6.2 Genotypic Switch

Copious amounts of alginate are only produced in combination with inactivation (mutations) of the negative regulators of the AlgU activity (anti-sigma factors) MucA (AlgS) or MucB (AlgN) (Martin et al., 1993a,b). Mutational inactivation (genotypic switch) of MucA, MucB (Schurr et al., 1996) or MucD (Boucher et al., 1996) leads to full activity of AlgU, which allows strong transcription of the *alg* operon (Figure 4). MucA is supposed to be located in the cytoplasmic membrane interacting with MucB in the periplasm upon perception of an unknown stimulus, and transducing a signal to the cytoplasm which mediates degradation of

AlgU presumably due to the action of MucD (Mathee et al., 1997) (see Figure 4). MucD is orthologous to the *E. coli* periplasmic protease and chaperone DegP. DegP homologues are known virulence factors that play a protective role in stress responses in various species. Recently, negative regulation of AlgU by anti-sigma factors MucA and MucB and the transcriptional regulation of the *algD* gene have been also described for *A. vinelandii* (Campos et al., 1996; Martinez-Salazar et al., 1996).

7
Polymerization and Export of the Alginate Chain

Since no undecaprenol-linked intermediate has been identified in either *P. aeruginosa* or *A. vinelandii*, the polymerase–which presumably is localized as a protein complex in the cytoplasmic membrane–might synthesize alginate by an undecaprenol-independent mechanism (Sutherland, 1982). Alginate synthesis might occur similarly to bacterial cellulose synthesis. The cellulose synthase of *Acetobacter xylinum* is localized in the cytoplasmic membrane, and appears to be a protein complex of 420 kDa. This enzyme catalyzes the processive polymerization of glucose residues (from UDP-glucose), and the nascent β-(1,4)-linked glucosan chains appear to remain attached to the synthase during polymerization (Ross et al., 1991). Correspondingly, the alginate polymerase in the cytoplasmic membrane might accept the GDP-mannuronic acid at the cytosolic site while simultaneously translocating the nascent alginate chain through the cytoplasmic membrane (Rehm and Winkler, 1996). Preliminary studies using ^{14}C-GDP-mannuronic acid as substrate and defined oligomannuronates as primer have revealed that the envelope fraction of mucoid *P. aeruginosa* exhibited *in vitro* alginate polymerase activity (B. H. A. Rehm, unpublished results). Biochemical and electrophysiological characterization of AlgE revealed that it forms an anion-selective pore in the outer membrane (Rehm et al., 1994b). This pore could be partially blocked by GDP-mannuronic acid in lipid bilayer experiments. In addition, a topological model of AlgE has been developed and, according to this model, the protein is a β-barrel consisting of 18 β-strands (Rehm et al., 1994b) (Figure 5). A three-dimensional model of AlgE was developed by homology modeling, indicating pore diameters eligible for alginate export (see Figure 5). These data are consistent with the hypothesis that AlgE forms an alginate-specific pore that enables export of the nascent alginate chain through the outer membrane. Figure 6 summarizes all the findings regarding polymerization, modification and export in a model.

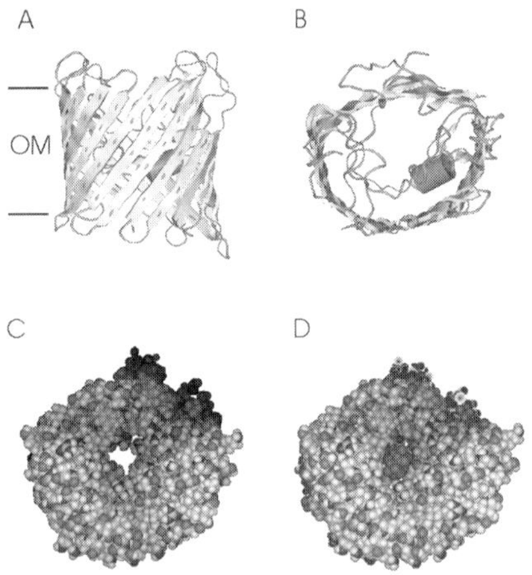

Fig. 5 Topological model of the outer membrane protein AlgE from *P. aeruginosa*. Experimental evidence was obtained that AlgE is involved in export of alginate through the outer membrane. (A) Side-view of the AlgE model (the bottom is exposed to periplasm, whereas the top is cell-surface-exposed. OM: outer membrane. (B) Top view of the AlgE model from outside the cell. (C) Top view of the AlgE model (in CPK format) from outside the cell. D, C with inserted alginate chain (dark gray).

8 Alginate-Modifying Enzymes

The alginate-modifying enzymes (the transacetylase and the C-5-mannuronan epimerase) carry N-terminal signal sequences and are mainly found in the periplasm in *P. aeruginosa* (Franklin et al., 1994). The corresponding alginate-modification reactions occur at the polymer level, presumably in the periplasm (see Figure 6). The genes encoding the transacetylase and other proteins involved in transacetylation (*algI, algJ, algF*), the epimerase (*algG*), and the lyase (*algL*) have been cloned, and the gene products have been characterized (Franklin and Ohman, 1993, 1996; Shinabarger et al., 1993; Franklin et al., 1994; Monday and Schiller, 1996). Transacetylation occurs at position(s) O-2 and/or O-3 of the mannuronic acid residue, preventing these residues from being epimerized to guluronic residues by AlgG and from degradation by AlgL (Franklin and Ohman, 1993; Franklin et al., 1994; Wong et al., 2000).

Thus, the periplasmic acetylase indirectly controls the periplasmic epimerase and lyase activity on the alginate polymer. The increasing degree of acetylation also causes the alginate polymer to have an enhanced water-binding capacity. This might be particularly important under dehydrating conditions, for example during infection and colonization of the lungs of patients with CF. The alginate lyase presumably functions as an editing protein to control the length of the polymer, or it might also serve the polymerase with alginate oligomers to prime synthesis. The lyase is not involved in providing a carbon source (Boyd et al., 1993).

8.1 Mannuronan C-5-epimerases

The G residues in alginates originate from a post-polymerization reaction catalyzed by mannuronan C-5-epimerases. In *P. aeruginosa,* and presumably also in other species belonging to this genus, there appears to be only one such enzyme encoded by the gene *algG* (Franklin et al., 1994). Like the proteins necessary for acetylation (Franklin and Ohman, 1996), the Ca^{2+}-independent AlgG is also probably located in the periplasm. The *A. vinelandii* genome also contains an *algG* homologue (Rehm et al., 1996), but this species in addition modifies its alginates by a family of extracellular Ca^{2+}-dependent epimerases that according to sequence-alignment studies are unrelated to AlgG. Seven

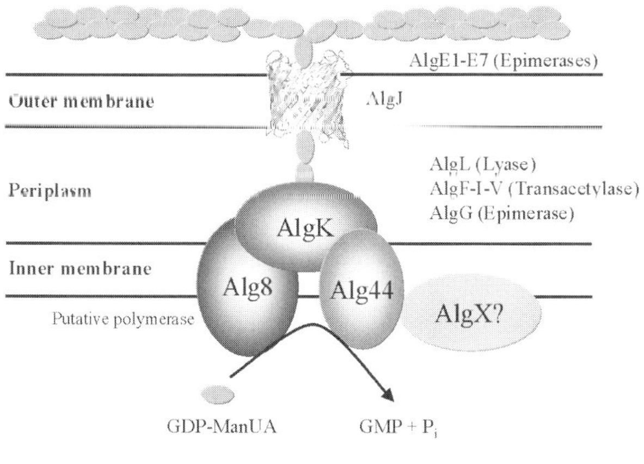

Fig. 6 Model of alginate polymerization, modification and export in *A. vinelandii* (Modified according to Rehm and Valla, 1997).

such enzymes (AlgE1–7) are now known (Ertesvåg et al., 1994, 1995; Svanem et al., 1999), and they can all be seen as composites of two structurally distinct modules, designated A and R. The A modules (about 385 amino acids) are present in one or two copies in each enzyme, while the R modules (about 155 amino acids) are present in one (AlgE4) and up to seven (AlgE3) copies. The N-terminal ends of each R module contain four to seven copies of a nine-amino-acids motif putatively involved in the binding of Ca^{2+}. In addition, *A. vinelandii* encodes a protein (AlgY) containing one A and one R module, but this protein appears to display no epimerase activity after expression in *E. coli* (Svanem et al., 1999).

8.1.1
Functional Differences

The A modules alone appeared to be sufficient both for catalyzing the epimerization reaction and for determining the epimerization pattern (Ertesvåg and Valla, 1999). It is, therefore, particularly important to understand the structure–function relationships in these modules. The epimerization patterns generated by all seven AlgE epimerases can be divided into two main groups: those which almost exclusively generate MG-blocks; and those which form G- blocks. From NMR spectroscopy analyses of the reaction products of all the enzymes, it is immediately obvious that only AlgE4 belongs to the first group. In addition, the C-terminal parts of AlgE1 and AlgE3 (AlgE1-2 and AlgE3-2) display this property, but it is not reasonable that these truncated forms are made *in vivo* in *A. vinelandii*. It is not known why the *A. vinelandii* genome encodes a specialized enzyme like AlgE4, but the physical properties of the alternating structure of its reaction product can be predicted to be quite different from the gel-forming G-block alginates (Smidsrød and Draget, 1996). All Ca^{2+}-dependent epimerases except AlgE4 are involved in the formation of G-blocks. AlgE3 and AlgE1 are composite enzymes, and due to the properties of each part they can both make long G-blocks and presumably put alternating structures between them. Comparison of the epimerization patterns of these two enzymes indicates that they share similar properties, but it was observed that less AlgE3 (measured as initial activity) compared to AlgE1, is needed to obtain high degrees of epimerization. For AlgE1 (Ertesvåg et al., 1998b) and AlgE3 the relative amount of G-blocks also increases with decreasing concentration of Ca^{2+}. This property, which was not observed for AlgE2 (Ramstad et al., 1999), probably reflects that the C-terminal part of AlgE3 displays less activity at low concentrations of Ca^{2+} than the N-terminal part, similar to what has been observed earlier for AlgE1 (Ertesvag et al., 1998b). It is also known that whole AlgE1 needs only 0.8 mM $CaCl_2$ for full activity (Ertesvåg et al., 1998b), while AlgE3 requires 3 mM $CaCl_2$, and displays less than 40% of full activity at 1 mM concentration of this cation. At low or moderate levels of epimerization, AlgE6 introduces more alternating structures than AlgE2 and AlgE5, although less than the composite enzymes AlgE1 and AlgE3. The average lengths of the G-blocks at about 40% epimerization, on the other hand, are similar to those made by AlgE2 and AlgE5. Alginate may, however, be epimerized to 90% G by AlgE6, and such highly epimerized alginate contain very long G-blocks and can be predicted to form very strong gels. This degree of epimerization has so far not been achieved using AlgE2 or AlgE5. These different epimerase specificities may indicate the requirement of the organism to generate alginate structures of importance for formation of the metabolically dormant and alginate-containing cysts which are generated under conditions of environmental stress in *A. vinelandii* (Sadoff, 1975). Furthermore, the significant alginate lyase activity of AlgE7 might be needed for the germination of the cysts (Wyss et al., 1961).

8.1.2
The Biotechnological Potential

Commercially, the alginates are harvested from different species of brown algae (Smidsrød and Draget, 1996), and it seems very likely that the composition of these products will not always meet industrial and biotechnological needs. In addition, alginate structures that are not available from seaweeds might have properties that would open the possibilities for totally new applications. By epimerizing algal alginates with the recombinantly produced epimerases, one might therefore be able either to upgrade their value or to generate completely new products. Since it is already clear that this is feasible, the questions are rather what prices are acceptable for each particular application, what the potential uses are, and how the market will react to products made using recombinant enzymes. A further extension of alginate modifications *in vitro* would be to design cells that directly synthesize the alginates of interest *in vivo*. This might lead to products of lower price, but it may prove to be more difficult to obtain the same level of control as achieved by in-vitro epimerization. In any case, the mannuronan C-5-epimerization system raises many interesting questions for basic science and applied biotechnology, and consequently the enzymes are likely to be the subject of active studies for many years to come.

8.2
O-Transacetylases

The products of *algI*, *algJ*, and *algF* from *P. aeruginosa* are required for the addition of O-acetyl groups to the alginate polymer, and mutations in *algI*, *algJ*, or *algF* resulted in production of an alginate polymer that was not O-acetylated (Shinabarger et al., 1993; Franklin and Ohman, 1996; Nivens et al., 2001). The mannuronate residues undergo modification by C-5 epimerization to form the L-guluronates and by the addition of acetyl groups at the O-2 and O-3 positions. By using genetic analysis, *algF* was identified and located upstream of *algA* in the 18-kb alginate biosynthesis operon, as a gene required for alginate acetylation (Franklin and Ohman, 1993; Shinabarger et al., 1993). An *algI*::*Tn501* mutant, which was defective in *algIJFA* because of the polar nature of the transposon insertion, produced alginate when *algA* was provided in *trans*. This indicated that the *algIJF* gene products were not required for polymer biosynthesis. To examine the potential role of these genes in alginate modification, mutants were constructed by gene replacement in which each gene (*algI*, *algJ*, or *algF*) was replaced by a polar gentamicin resistance cassette (Franklin and Ohman, 1996). Proton nuclear magnetic resonance (NMR) spectroscopy showed that polymers produced by strains deficient in *algIJF* still contained a mixture of D-mannuronate and L-guluronate, indicating that C-5 epimerization was not affected. Alginate acetylation was evaluated by a colorimetric assay and Fourier transform-infrared (FTIR) spectroscopy, and this analysis showed that strains deficient in *algIJF* produced nonacetylated alginate. Plasmids that supplied the downstream gene products affected by the polar mutations were introduced into each mutant. The strain defective only in *algF* expression produced an alginate that was not acetylated, confirming previous results. Strains missing only *algJ* or *algI* also produced nonacetylated alginates. Providing the respective missing gene (*algI*, *algJ*, or *algF*) *in trans* restored alginate acetylation. Mutants defective in *algI* or *algJ*, obtained by chemical and transposon mutagenesis, were also defective in their ability to acetylate alginate. Therefore, *algI* and *algJ* represent newly identified genes that, in addition to *algF*, are required for alginate acetylation. Once in the periplasmic space, alginate is O-acetylated by the combined action of the *algF*, *algI* and *algJ* gene products. Mutants deficient in any one of these three genes are unable to produce O-acetylated alginate, but the epimerization process and overall yields of alginate appear to be unaffected (Shinabarger et al., 1993; Franklin and Ohman, 1996). Acetyl-CoA is almost certainly the primary donor of O-acetyl groups for alginate modification; however, this metabolite is localized in the cytoplasm, whereas the O-acetylation process occurs in the periplasm. Therefore, at least one of the *algF*, *algI* or *algJ* gene products is likely to be involved in transport of O-acetyl groups across the cell membrane into the periplasmic space. Sequence data indicate that AlgI is probably a membrane-bound protein and may fulfil this role (Franklin and Ohman, 1996). AlgF has a signal peptide, which is processed by *E. coli*, indicating that it is a periplasmic protein (Shinabarger et al., 1993) and therefore could be the O-acetyltransferase. AlgJ shows remarkable similarity (30% identity, 69% similarity) to another gene product, AlgX (Monday and Schiller, 1996), which is of unknown function but is essential for alginate biosynthesis. However, neither AlgJ nor AlgX shows any significant similarity to other proteins in the databases and therefore their function in the O-acetylation process remains unresolved at this stage. The observation that a cell suspension of *P. syringae* is able to O-acetylate seaweed alginate suggests that this event is either periplasmic or extracellular (Lee and Day, 1995). It is likely that the O-acetylation of mannuronate is catalyzed by at least two enzymes each specific for either the 2- or 3-hydroxyl on the sugar ring. However, at this stage it is not known which gene products might be involved in determining these specific modifications.

8.3 Alginate Lyases

It is not clear why a degradative enzyme should be expressed concurrently with the enzymes involved in biosynthesis. One suggestion has been that the lyase may be involved in excising polysaccharide fragments from the biosynthesis complex, although there is no real evidence to support this contention. It is clear though that overexpression of the *algL* gene product (May and Chakrabarty, 1994) or of other alginate lyases (Gacesa and Goldberg, 1992) results in the release of planktonic bacteria from biofilms.

8.3.1 Reaction Mechanism

Alginate lyase catalyzes the degradation of alginate by a β-elimination mechanism, targeting the glycosidic 1,4 O-linkage between monomers. A double bond is formed between the C4 and C5 carbons of the six-membered ring from which the 4-O-glycosidic bond is eliminated, depolymerizing alginate and simultaneously yielding a product containing 4-deoxy-L-*erythro*-hex-4-enopyranosyluronic acid as the nonreducing terminal moiety (Haug et al., 1967). Gacesa (1987, 1992) proposed a catalytic mechanism for alginate lyase that described a three-step reaction to depolymerize alginate. This mechanism may also be shared with epimerase, another enzyme that acts on the alginate polymer. The two reactions differ only in the last step of the three-stage transformation of alginate. The three steps include: (1) removal of the negative charge on the carboxyl anion – essentially neutralizing the charge by a salt bridge (lysine or arginine may be the candidate residue); (2) a general base-catalyzed abstraction of the proton on C5 (aspartic acid, glutamic acid, histidine, lysine, and cysteine have been suggested for this role), where one residue may be required as the proton abstractor and another as the proton donor, although the proton may be derived from the solvent environment; and (3) a transfer of electrons from the carboxyl group to form a double bond between C4 and C5, resulting in the β-elimination of the 4-O-glycosidic bond. In the proposed mechanism for epimerase, the replacement of the proton at C5 (epimerization) takes place in step 3. The principle of the catalytic mechanism was confirmed by the identification of putative catalytic residues by structural analysis of the *Sphingomonas* lyase ALY1-III complexed with a trisaccharide (Yoon et al., 2001).

8.3.2
Function in Alginate-Producing Bacteria

Very few bacteria synthesize polysaccharides as well as the specific degrading enzymes (Kennedy et al., 1992; Sutherland, 1995). Periplasmically localized alginate lyases have been found in various species of *Pseudomonas* and *Azotobacter* that can synthesize an extracellular alginate but are unable to use alginate as a carbon or energy source. The extracellular poly(M)-rich alginate produced by these bacteria is O-acetylated at the C2 and/or C3 positions on the M residues to various degrees, which makes this polymer more resistant to degradation by the endogenously produced lyase (Nguyen and Schiller, 1989; Kennedy et al., 1992). Interestingly, alginate lyase genes in *Pseudomonas* and *Azotobacter* species are localized within their respective alginate biosynthesis gene clusters. This genetic organization has been reported in *P. aeruginosa* (Chitnis and Ohman, 1993), *P. syringae* pv. *syringae* (Penaloza-Vazquez et al., 1997), *A. vinelandii* (Rehm et al., 1996), and *A. chroococcum* (Pecina et al., 1999). This localization of the alginate lyase gene (*algL*) raises questions about the function of AlgL in alginate biosynthesis. It is possible that the lyase works as part of a polymerization complex within the periplasm, assembling the alginate exopolysaccharide for transport to the cell surface, or it may provide oligomeric alginate, which might serve as primer. Accordingly, the lyase-negative *P. aeruginosa* produced only small amounts of alginate (Monday and Schiller, 1996). However, Boyd et al. (1993) reported that lyase was not required for alginate production by *P. aeruginosa*. A similar study demonstrated that the absence of lyase activity reduced alginate production by *P. syringae* pv. *syringae* by ~50% (Penaloza-Vazquez et al., 1997). Hence, alginate lyase seems not to be essential for alginate biosynthesis, but for maximum production. May and Chakrabarty (1994) proposed that the lyase may function either as an editing protein to control the length of the alginate polymer or to provide short pieces of alginate to prime the polymerization reaction. The nascent polymannuronate in the periplasmic space would be most sensitive to endogenous lyase before acetylation and epimerization, whereas after modification, the polymer would be ready for export to the cell surface. The levels of Ca^{2+} and Mn^{2+} ions have been shown selectively to activate or inhibit alginate lyase and epimerization activity in algae (Madgwick et al., 1978). Since various cations have a strong influence on bacterial alginate lyase activity (Wong et al., 2000), this observation suggests that periplasmic ionic conditions could also regulate alginate modification in *P. aeruginosa*.

The production of lyase by *P. aeruginosa* may also be important in facilitating dissemination of the bacteria (Boyd and Chakrabarty, 1994). Alginate synthesis is increased upon attachment of the bacteria to a cell surface, resulting in stronger bacterial adhesion to the surface and colonization. However, overexpression of the lyase gene within a mucoid strain of *P. aeruginosa* led to a decrease in the length of alginate polymers and increased bacterial detachment from the surface (Boyd and Chakrabarty, 1994). Thus, cleavage of the alginate polymer within *P. aeruginosa* biofilms could enhance detachment of the bacteria, allowing them to spread and colonize new sites. In *Azotobacter* sp., alginate is produced by vegetatively growing cells as capsule, and by cells in the metabolically dormant-state in the cyst coat (Page and Sadoff, 1975, Sadoff, 1975). Both *A. vinelandii* and *A. chroococcum* strains produce M-specific endolytic lyases that are localized in the periplasmic space. These lyases may be biologically important in the differentiation of *Azotobacter* cells during encystment, when they most likely play a role in concert with epimerases to form the desiccation-resistant cyst capsule. It has now been reported that *A. vinelandii* has multicopy epimerase genes, making a gene family that is highly likely to be responsible for the synthesis of complex alginates of various polymeric composition in the cyst capsule (Svanem et al., 1999). One of these epimerases, AlgE7, also exhibited lyase activity. Although alginate-negative strains of *A. vinelandii* fail to encyst, it is not yet clear whether alginate lyase-negative strains can differentiate. An increase in alginate lyase activity just before cyst germination (Kennedy et al., 1992) suggests that lyase expression might support this process.

From the collective information on alginate and alginate lyases, it is concluded that alginate lyases are important enzymes in a broad spectrum of biological roles and applications. The lyases maintain a balance in the cell physiology of alginate-producers that efficiently use alginate as functional biopolymers and also in the natural environment, where the recycling of alginate is achieved through metabolic breakdown (for review, see Wong et al., 2000). In addition, the ability of lyases selectively to depolymerize alginate–which has become a very useful industrial polysaccharide–makes them important tools with great potential for advanced biotechnological uses.

8.3.3 Structure–Function Analysis

The cloning and sequencing of many alginate lyase genes have now enabled investigators to focus on structure–function analysis of this enzyme. DNA sequences and primary amino acid sequences are available for 23 alginate lyases (Wong et al., 2000). Various alginate lyases have been characterized with respect to enzyme properties (Wong et al., 2000; Table 2). Meanwhile, the coordinates of the three-dimensional structure of ALY1-III *Sphingomonas* sp. can be found in the Protein Data Bank (Brookhaven, NY) (Yoon et al., 1999). Based on sequence information, most alginate lyases appear to fall into three major classes according to their molecular mass: 20–35 kDa; ~40 kDa; and ~60 kDa. Analysis of amino acid sequences of alginate lyases revealed that they contain a hydrophilic central region and a hydrophobic sequence in the C terminus, with a slightly charged end. All alginate lyases share amino acid sequence similarities from 18 to 95%. Although several regions of similarity exist between these lyases, the core region residues of the lyases within the 40-kDa class

Tab. 2 Alginate lyases from Gram-negative bacteria: localization, characteristics, and sequence accession numbers (modified according to Wong et al., 2000)

Source	*Localization*	*Substrate specificity*	*Endo/exolytic*	*Cleavage site*	*Molecular mass (kDa)*	*pI*	*Opt. pH*	*Opt. T*	*Cations needed*	*GenBank accession No.*
A. chroococcum	Periplasmic	AlgL: M	Endolytic	–	43	–	–	30°C, pH 7.5	Na^+, K^+ Mg^{2+}	AJ223605
	Periplasmic	M			–	–	6.8	30°C		N/A
A. chroococcum 4A1M	Extracellular	PolyM	Endolytic	–	23–24	5.6	6	60°C	Ca^{2+}	N/A
A. vinelandii	Intracellular	M	–	–	~50		7.5	–	–	N/A
	Periplasmic	M			–	–	7.2	30°C		N/A
	Periplasmic	AlgL: M, acetyl'd (nonconsecutive M)	Endolytic not G-M	M-M and M-G,	39	5.1	8.1–8.4		Na^+, divalents not needed	AF037600
	Periplasmic	AlgE7 (epimerase and lyase): M and G	Endolytic	G-GM and G-MM	(384aa)	–	–	–	–	AF099800
A. vinelandii	N/A	M	N/A	N/A	(375aa)	N/A	N/A	N/A	N/A	AF027499
Enterobacte cloacae M-1	Extracellular	G	Endolytic	–	32–38	8.9	7.8	30°C	Ca^{2+}, Al^{3+}, Mn^{2+}	N/A
	Intracellular	G	Endolytic	7 subsites, cleaves G-G between subsites 2 and 3	31–39	8.9	7.5	40°C	–	N/A
K. aerogenes type 25	Intracellular	PolyG	Endolytic	–	28–31.6	–	7	37°C	Na^+ and K^+ (0.1–0.3 M)	N/A
	Extracellular	G	Endolytic	G-G, G-M	–	–	7	–	Mg^{2+} (0.05–0.1 M)	N/A
K. pneumoniae subsp. aerogenes	Extra/intra-celluar (R)	AlyA: polyG	–	–	28 8.9	–	–	–	N/A	25
	Extracellular (R)	AlyA: polyG	Endolytic		31.4 9.39 (calc.)	7.0 –7.6 h		50°C	Na^+	L19657
Pseudomonas sp. W7	N/A (R)	N/A	N/A	N/A	(345aa)	N/A	N/A	N/A	N/A	AF050114
P. aeruginosa	Intracellular	M, nonacetyl'd	–	–	–	–	6.2	20–40°C	–	N/A
P. aeruginosa	Intracellular	AlgL: M nonacetyl'd	–	–	53	–	8	–	Mg^{2+}, K^+, Na^+	N/A
P. aeruginosa CFl/Ml	Periplasmic	AlgL: M,	Endolytic	6 subsites, nonacetyl'd cleaves M-M between subsites 3 and 4	–	–	–	–	–	N/A
P. aeruginosa	Intracellular (R)	AlgL: M, nonacetyl'c	Endolytic	–	43.5	–	–	–	–	L14597
P. aeruginosa FRD1	Peripl.(N)/intracell.(R)	AlgL: M, nonacetyl'c	Endolytic	M-M	39	9 (calc.)	7.0	–	Mg^{2+}, Na^+	U27829/L09724
P. aeruginosa	N/A (R)	AlgY: M	N/A	N/A	(685aa)	N/A	N/A	N/A	N/A	Z54213

Tab. 2 (cont.)

Source	*Localization*	*Substrate specificity*	*Endo/exolytic*	*Cleavage site*	*Molecular mass (kDa)*	*pI*	*Opt. pH*	*Opt. T*	*Cations needed*	*GenBank accession No.*	
P. maltophilia and *P. putida*	Intracellular	M, acetyl'd	Endolytic	–	–	–	7.7–7.8	28–30°C	–	N/A	
P. syringae pv. syringae	Periplasmic	PolyM (prefers deacetylated)	Endolytic		40	8.2	7	42°C	Cations not needed	AF22020	
Pseudomonas	Intracellular	Multiple		–	–	90, 72, 60, 54	–	–	–	–	N/A
sp. OS-ALG-9	Intra/ extracellular (R)	ALY or AlyP.M	Endolytic (preferred), G		46.3	–	–	–	AlyP:—		D10336
	Intracellular	(ALY or AlyP:—)	Endolytic	–	45	–	7.5	45°C	AlyP:—		N/A
	Intracellular (R)	ALYII : M	Endolytic	–	79.8 (calc.)	8.3	7	30°C	ALYII: None required; EDTA stimulates activity	AB003330	
Sphingomonas sp.	Cytoplasmic	ALY1-I :M, nonacetyl'd, acetyl'd		Endolytic	–	60	9.03	7.5–8.5	70°C	ALY1-I, -II, -III:	2009330A
	Cytoplasmic	ALY1-II :G	Endolytic	–	25	6.82	7.5–8.5	70°C	None required;	sequenced	
	Cytoplasmic	ALY1-III : M, acetyl'd (highly active) and nonacetyl'd	Endolytic	M-M, hetero MG	38	10.16	7.5–8.5	70°C	EDTA: no effect	1QAZ	

N/A, not available

that share significant alignment are very well conserved; especially notable is the highly conserved six-amino-acid hydrophilic sequence "NNHSYW" in the center of the protein sequences. This region is also conserved in the alginate lyase ALY1-III from a *Sphingomonas* sp., which otherwise differs significantly in amino acid sequence from the other lyases in this grouping. Yoon et al. (1999) recently solved the three-dimensional structure for ALY1-III, revealing a structure with 12 α-helices, organized in a twisted α/α helix barrel composed of six inner and five outer helices. Analysis of this conformation identified a deep tunnel-like cleft, which was proposed as the catalytic site. The highly conserved NNHSYW sequence is located in the center of this cleft. As described below, studies suggest that alteration of the histidine residue in this site inactivates the lyase, providing evidence that this region is critical for enzyme activity. There is also a well-conserved nine-amino-acid hydrophobic sequence, "WLEPYCALY," in the C terminus of the lyases from *P. aeruginosa* and *Azotobacter* sp.. This nine-amino-acid block is weakly conserved in ALY1-III from *Sphingomonas* sp. Sequence and structural information indicates that this nine-amino-acid region is in H11, an inner α-helix of ALY1-III (Yoon et al., 1999).

The three-dimensional structure for alginate lyase from *Sphingomonas* sp. revealed an interesting feature of the enzyme, a disulfide bridge between Cys188 and Cys189, giving the structure a tight turn at the edge of the proposed active-site cleft (Yoon et al., 1999). This disulfide bond immediately precedes the conserved sequence of "NNHSYW" and is suggested to be very important in the maintenance of the active site conformation. This cysteine-cysteine pair in the alginate lyase of *Sphingomonas* sp. is a structural feature that is unique to this lyase, which is not surprising because the primary protein sequence of ALY1-III departs significantly from the general consensus of the majority of alginate lyases in the 40-kDa group. M-specific ALY1-III has an α-helix-rich structure and has no β-strand or sheet conformation (Yoon et al., 1999). The ALY1-III structure was used as a template to develop a threading model of the alginate lyase of *P. aeruginosa*, which exhibited 22% similarity to ALY1-III (B. H. A. Rehm, unpublished results). This model showed the tunnel-like cleft with the potential catalytic residues N171, H172, R219, Y226 closely arranged inside the cleft. Residues N171 and H172 are located in the conserved motif "NNHSYW" (Figure 7). Considering the many differences observed for the alginate lyases (primary sequence, M_W, substrate specificities, hydrophobicity profiles, etc.), it is important that additional enzymes will be biochemically analyzed as well as by X-ray crystallography. This will allow complete comparisons of structure and function between these enzymes, as well as related enzymes, and define their relationship with the structure of ALY1-III from *Sphingomonas* sp. (Yoon et al., 1999).

8.3.4
Substrate-Binding and Catalytic Sites

Experiments with defined alginate oligomers have indicated the specific substrate recognition sequences for alginate lyase action, highlighting the optimal capacity of the catalytic site for substrate units. Using a series of oligomannuronates (n D 2–9), Rehm (1998) reported that the poly(M) lyase from *P. aeruginosa* CF1/M1 would not accept oligomers that were smaller than five mannuronates for β-elimination, and that the highest enzyme activity was noted with the hexameric mannuronate oligomer, whereas the trimer was the most abundantly accumulating product. This study suggested that the catalytic site of alginate lyases probably

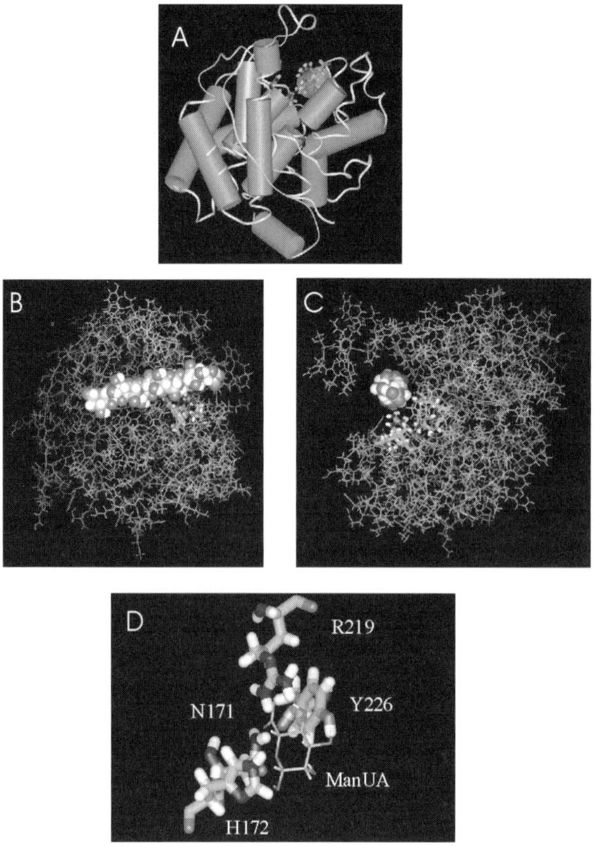

Fig. 7 Threading model of the alginate lyase from *P. aeruginosa* based on the structure of the *Sphingomonas* sp. ALY1-III alginate lyase. (A) Alginate lyase threading model indicating the secondary structure composition (cylinders represent α-helical segments) and amino acid side chain of the putative catalytic residues were demonstrated (see D). (B) The alginate lyase model in stick format complexed (docked) with the hexameric oligomannuronate and emphasis of the putative catalytic residues by thicker sticks in the tunnel-like cleft. (C) Model (B) turned around by 90°. (D) Spatial arrangement of the putative catalytic residues complexed (docked) with a mannuronic acid monomer (ManUA).

accommodates five to six residues. In the crystal structure of ALY1-III (Yoon et al., 1999), His192 is located in the highly conserved region NNHSYW, which is in the center of the proposed active-site cleft. Chemical modification of histidine residues in ALY1-III inactivated the enzyme (Yoon et al., 1999), which suggests strongly the role of histidine in the catalytic activity of ALY1-III. In ALY1-III, four conserved tryptophan residues, together with other aromatic residues, are located along the cleft of the active site, and two conserved arginine residues flank the entrance to the cleft. The charged arginine and lysine residues on both sides of the cleft and the aromatic side chains lining the active site have been suggested to be substrate-binding molecules. This active cleft can accommodate at least five residues of poly(M) along the curved surface (Yoon et al., 1999). The highly conserved sequence "INNHSY" located in the central region of the 40-kDa lyases is also highly conserved as "E/FNNVSY" in mannuronan C5-epimerases (Ertesvag et al., 1998b). This motif, found in alginate lyases and epimerases from *P. aeruginosa* (Boyd et al., 1993; Schiller et al., 1993; Franklin et al., 1994), *A. chroococcum* (Pecina et al., 1999), and *A. vinelandii* (Rehm et al., 1996; Ertesvag et al., 1998a; Svanem et al., 1999), appears to be mainly located within an average of 200 residues from the encoded N terminus of the proteins. However, this pattern of residues is not found in the G-lyases or the 30-kDa alginate lyases. Because both the alginate lyases and epimerases from *P. aeruginosa* and *A. vinelandii* are active on mannuronate units of alginate, it is highly possible that this pattern could be a binding motif for the poly(M) or the mannuronate and its glycosidic bond. The AlgG epimerases of the alginate biosynthesis operon in *A. vinelandii* (Rehm et al., 1996) and *P. aeruginosa* (Franklin et al., 1994) differ slightly from the epimerases found in the *A. vinelandii* epimerase gene family cluster (Rehm et al., 1996; Ertesvag et al., 1998a; Svanem et al., 1999). The AlgG proteins are smaller than most of the other epimerases (AlgE or AlgY), and the conserved "INNHSY" sequences from both AlgGs have greater homology between themselves than with the other epimerases. Also, the conserved sequences are located ~350 residues (instead of 200) from the encoded N termini of the respective AlgG. Through comparisons of the common "INNHSY" and "ENNVSY" motifs, it is clear that asparagine (N), serine (S), and tyrosine (Y) residues are the essential residues and are conserved throughout. A valine (V) or arginine (R) residue in the epimerases substitutes for the histidine (H) residue position in the M lyases. In the M lyases, the histidine (H) residue is highly important for catalytic activity, and this may be the residue that contributes to the main difference in catalytic mechanism between a lyase and an epimerase. However, only recently the crystal structure of ALY1-III complexed with trimeric mannuronate was obtained (Yoon et al., 2001). The binding of this substrate in the tunnel-like cleft, strongly suggested that the four residues – N191, H192, R239 and Y246 – are directly involved in substrate binding and catalysis. The following catalytic mechanism has been proposed: (1) The C5 carboxylate group is neutralized by R239 and N191; (2) subsequently, the C5 proton can more easily removed by Y246, the catalytic nucleophile, resulting in the formation of the carboxylate dianion intermediate; (3) Y246 then donates the proton to the oxygen of the glycosidic bond, which results in bond cleavage and formation of a C4–C5 double bond. H192 presumably stabilizes the carboxylate dianion intermediate during catalysis (Figure 8). A similar arrangement of the catalytic residues was obtained in the threading model of the alginate lyase from *P. aeruginosa*, supporting their role in catalysis (see Figure 7).

8.3.5
Future Applications

The therapeutic use of alginate lyase for the treatment of alginate in biofilms of *P. aeruginosa* colonizing the lungs of CF patients remains one of the most important goals of studying alginate lyase. Mrsny et al. (1996) described the complex distribution of DNA and alginate within the mucin matrix of CF sputa. The combination of alginate lyase and deoxyribonuclease demonstrated an additive reduction of the sputum viscoelasticity, which suggests that this approach deserves further study (Mrsny et al., 1994). Alginate lyases may also be used to generate defined products with potential applications in various fields. Alginates with low molecular weights act like oligosaccharides in their ability to regulate physiological processes in

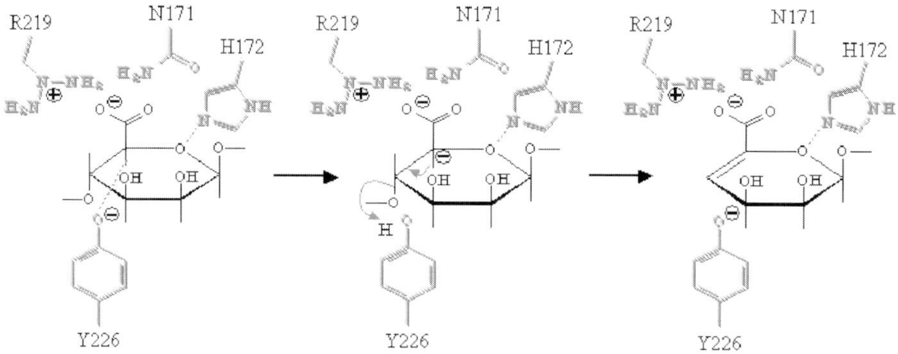

Fig. 8 The postulated catalytic mechanism of alginate lyases (see text for detailed description).

plants (Albersheim and Darvill, 1985). Oligomeric alginate (average M_W 2000 Daltons) obtained from lyase degradation of high-molecular weight alginate, can promote growth of *Bifidobacteria* sp. and thus has been proposed for use as a physiological food source (Murata et al., 1993; Akiyama et al., 1992). Furthermore, alginate lyase-degraded products (average M_W 1800 Daltons) greatly enhanced germination and shoot elongation in plants, despite repressing the growth of *Chlamydomonas* sp. and HeLa cells (Yonemoto et al., 1993). Trisaccharides (Natsume et al., 1994) or alginate lyase-lysate (Tomoda et al., 1994) has also been found to promote root growth in barley. In the presence of epidermal growth factor, dimers, trimers, and tetramers that possessed guluronic acid at the reduced ends highly induced the proliferation of keratinocytes (Kawada et al., 1997). Other alginate polymers (average M_W 230,000 Daltons) have antitumor effects and enhance the phagocytic activity of macrophages (Fujihara and Nagumo, 1993). Poly(M) block-rich alginate exhibited high antitumor activity (Fujihara and Nagumo, 1992) and could stimulate production of cytokines by human monocytes (Otterlei et al., 1991). The conformational properties of the macromolecules and the anionic character and molecular weight of the polysaccharide were suggested to be important in the effectiveness of their antitumor activity (Fujihara et al., 1984). Therefore, alginate lyases are crucial in the generation of such useful oligomeric products. The promising use of calcium alginate beads as a biomaterial in wide-ranging applications (Skjåk-Bræk and Martinsen, 1991) extends further the potential for alginates and alginate lyases. Calcium alginate, with or without a polylysine, polyarginine, or chitosan protective coating, has been used for the encapsulation of a variety of materials including drugs for controlled delivery; DNA and oligonucleotides for tumor development studies, gene delivery, gene therapy, and antisense oligonucleotide therapeutic agents; yeast cells coentrapped with lipase for the production of flavor esters; plant tissue, with the aim of developing artificial seed technology; *P. fluorescens* as a biocatalyst for phenolic-compound removal from wastewater; and entomopathogenic nematodes for agricultural biocontrol (for review, see Wong et al., 2000). Alginate in combination with other biomaterials such as hyaluronate and chitosan can be highly effective in many medical applications, including use in wound dressings impregnated with antibiotics and encapsulation of chondrocytes to engineer cartilage tissues *in vitro* for cartilage transplant and repair. Alginate with a G content >70% and average G-block length of >15, and with low polyphenol contamination, provides the most suitable characteristics for immobilization beads. Apart from generating alginate with a predictable monomeric sequence from native sources, it is possible to use a combination of D-mannuronan C5-epimerase and alginate lyase on a particular alginate substrate to engineer novel alginate polymers of defined composition (Skjåk-Bræk et al., 1986). The composition and properties of alginate gels are still being extensively studied, and the collective information will provide a guideline for the design of these novel polymers.

9
The Role of Alginate in Biofilm Formation

It has been proposed that contact with a surface may induce changes in gene expression, and there is evidence to support this idea in *P. aeruginosa*. Studies by Davies and coworkers showed that one of the genes required for the synthesis of the exopolysaccharide (EPS) alginate (*algC*) is up-regulated three- to five-fold in recently attached cells compared to their planktonic counterparts (Davies et al., 1993; Davies and Geesey, 1995). This result is not surprising because alginate – the regulation of which has been studied in depth (Govan and Fyfe, 1978; May et al., 1991) – has long been implicated as the extracellular matrix in biofilms of *P. aeruginosa*. These experiments were among the first to show surface contact induced gene expression in *P. aeruginosa*. Recent studies in the laboratory of Wozniak have taken this observation a step further. Wozniak and colleagues noted that isolates of *P. aeruginosa* from the CF lung that produced large quantities of alginate (mucoid strains) were also nonmotile, and these authors suspected a link between the two phenotypes. In a series of genetic experiments, they showed that expression of a sigma factor (AlgU or σ^{22}) required for alginate synthesis resulted in down-regulation of a key flagellar biosynthetic gene (Garrett et al., 1999). These data suggest that, on contacting the surface, flagellar synthesis is down-regulated and alginate synthesis is up-regulated. Moreover, a recent study by Whiteley and coworkers (2001) employing DNA microarray analysis demonstrated that only about 1% of the *P. aeruginosa* genes were differentially regulated in mature biofilms as compared to planktonic cells. Interestingly, none of the *alg* genes appeared to be differentially regulated, whereas biosynthesis genes for pili IV and flagella were repressed. Scanning laser confocal microscopy (SCLM) analysis confirmed this result, demonstrating that nonmucoid strains formed densely packed biofilms that were generally less than 6 μm in depth. In contrast, *P. aeruginosa* FRD1 produced microcolonies that were approximately 40 μm in depth. An *algJ* mutant strain that produced alginate lacking O-acetyl groups produced only small microcolonies. After 44 h, the *algJ* mutant switched to the nonmucoid phenotype and formed uniform biofilms, similar to biofilms produced by the nonmucoid strains (Nivens et al., 2001). Results of both the ATR/FT-IR (attenuated total reflexion/Fourier transform infrared) and SCLM analyses of nonmucoid strains demonstrated that alginate was not required for *P. aeruginosa* biofilm formation, and therefore alginate did not act as a primary adhesin for the *P. aeruginosa* cells to these surfaces. Biofilm formation by *P. aeruginosa* FRD1131 that had a *Tn501* insertion in the alginate biosynthesis gene *algD* was still possible. Strain FRD1131 showed a delay in biofilm formation but ultimately formed biofilms, indicating that alginate biosynthesis is not essential for biofilm formation.

These results demonstrate that alginate, although not required for *P. aeruginosa* biofilm development, plays a role in the biofilm structure and may act as intercellular material, required for formation of thicker three-dimensional biofilms. The results also demonstrate the importance of alginate O-acetylation in *P. aeruginosa* biofilm architecture.

The portion of the biofilm developmental pathway that concerns detachment represents an important area of future research. One possible signal for detaching may be starvation, although this has not been investigated in detail. However, Boyd and Chakrabarty (1994) reported that the enzyme alginate lyase may play a role in the detachment phase in *P. aeruginosa*. These authors showed that overexpression of alginate lyase could speed detachment and cell sloughing from biofilms (Boyd and Chakrabarty, 1994). A recent study by Allison et al. (1998) showed that a *P. fluorescens* biofilm decreased after extended incubation, which was attributed – at least in part – to the loss of EPS. Furthermore, these authors presented evidence showing that acyl-HSLs and/or another factor present in stationary-phase culture supernatants mediated this effect (Allison et al., 1998). Little else is known about the functions or regulatory pathways involved in the release of bacteria from the biofilm.

10 The Applied Potential of Bacterial Alginates

At present, all alginates used for commercial purposes are produced from harvested brown seaweeds. The prices of such alginates are generally low, and it seems to be a difficult task to establish a competitive bacterial production process in this price range. However, there are at least two factors that now make it more probable that bacterial alginates may become commercial products. The first is related to the environmental concerns associated with seaweed harvesting and processing, and the second is related to the quality of the final polymer product. The environmental impact is not the topic of this review, but the possibility of producing bacterial alginates with improved qualities will be discussed on the basis of recent scientific discoveries. Alginates with unique qualitative properties have the obvious advantage that they may potentially be sold at higher prices than the bulk materials, and such new polymer products may at least theoretically also open new kinds of markets for this polymer. The properties of pure alginates are determined by their degree of polymerization and acetylation, and by their monomer composition and sequence (Moe et al., 1995). It appears likely that it will become possible to control these three parameters in bacterial alginates. The corresponding possibility does not seem realistic for alginates obtained by the harvesting of oceanic seaweeds, and producers of such products will therefore be limited by the need to fractionate the polymer mixtures produced by these organisms in their natural environments. The degree of polymerization affects the viscosity of alginates, and it will most likely be possible – at least to some extent – to control this parameter in bacteria by strict control of the fermentation conditions and/or genetic manipulations. One possible target for genetic modifications might be the alginate lyase found in both *Pseudomonas* and *Azotobacter* species (Kennedy et al., 1992; Lloret et al., 1996; Monday and Schiller, 1996; Rehm et al., 1996). Hence, bacterial alginates might in principle provide a particular viscosity if reduced quantities of material (on a weight basis) are used. Extreme viscosities would, on the other hand, also create problems with oxygen transfer in fermentors. From a commercial point of view it is not clear if this is a fruitful approach. It now seems obvious that the most interesting possibilities offered by bacterial alginates relate to the control of their monomer composition and sequential structures. Both seaweed and bacterial alginates are heterogeneous mixtures of molecules, and their sequential monomer composition can be described by only statistical models (Stokke et al., 1991), in contrast to accurately defined protein and nucleic acid sequences. If alginate structures are to be controlled, one must therefore first understand the origin of the structural variability observed in nature. On the basis of current knowledge it appears that *Pseudomonas* species have only one mannuronan C-5-epimerase, encoded by the *algG* gene, and this enzyme is able to introduce only single guluronic acid residues into the mannuronan chain. Consequently, these alginates cannot form the divalent-cation-dependent (typically Ca^{2+}) gels formed by alginates which contain G-blocks. The potential for in-vivo manipulations of monomer structures in *Pseudomonas* alginates therefore appears to be limited. However, by knocking out the *algG* gene, it may be possible to produce pure mannuronan, and evidence for this has been presented (Franklin et al., 1994). Mannuronan is known to be a strong immunostimulant, and might have a commercial potential in applications where this property is of interest (Skjåk-Bræk and Espevik, 1996). In *A. vinelandii*, the epimerization system (AlgE1–E7) is complicated, as described above. The known secreted epimerases and their different substrate specificities might allow the *in vitro* design of alginates (see above). Specialized alginates might be made *in vitro* by first producing deacetylated poly(mannuronic acid), possibly from a *Pseudomonas* species. Deacetylated alginates are needed because early studies indicate that acetylation seems to protect them from epimerization (Skjåk-Bræk et al., 1985). The acetyl groups might be prevented from being introduced by the use of an acetylation-deficient mutant of the production strain (Franklin and Ohman, 1993, 1996; Shinabarger et al., 1993), or they might be removed later using standard chemical deacetylation procedures. Mannuronan produced in this way could then be epimerized by one particular recombinantly produced epimerase, or by combinations of such enzymes. Alternatively, one might homogenize commercially produced alginates from brown seaweeds using recombinant epimerases. It is now very probable that almost any alginate structure of applied interest can be produced by such procedures. The problems of marketing these products are therefore most likely more related to parameters such as price, and to the technical advantages of the products compared with those already available from brown seaweeds. For food and pharmaceutical applications, approval by legal authorities may also be a major obstacle. In order to evaluate the technological properties of the alginates modified *in vitro*, the enzymes must be produced in sufficient quantities to allow physical and functional studies of the modified polymers. Alginates produced as described above would probably be quite expensive and, if the price were to become a hindrance to certain applications, then the production of similar products *in vivo* using metabolic engineering techniques might well be considered.

The first applications of bacterial alginates are likely to involve their use either as immunostimulants (e.g., mannuronan, see above) or as gel-forming agents for the immobilization of cells (e.g., in tissue engineering) (Gutowska et al., 2001). Immobilized cells might be used for a variety of biotechnological production processes (see

Skjåk-Bræk and Espevik, 1996 for mini review), or in medical transplantation technologies. A variety of proposals have been suggested by different groups (Skjåk-Bræk and Espevik, 1996), and one of the most interesting examples involves the reversal of type I diabetes by immobilizing insulin-producing cells in alginate capsules. These capsules have been implanted into the body of whole animals and even humans, and the biological effects of this kind of approach are currently being evaluated (Soon-Shiong, 1995).

Acknowledgments

These studies were supported by the Deutsche Forschungsgemeinschaft, the Deutsche Fördergesellschaft der Mukoviszidose-Forschung e.V. and the Minister für Wissenschaft und Forschung des Landes Nordrhein-Westfalen.

11 References

Aarons, S. J., Sutherland, I. W., Chakrabarty, A. M., Gallagher, M. P. (1997) A novel gene, *algK*, from the alginate biosynthesis cluster of *Pseudomonas aeruginosa*, *Microbiology* **143**, 641–652.

Akiyama, H., Endo, T., Nakakita, R., Murata, K., Yonemoto, Y., Okayama, K. (1992) Effect of depolymerized alginates on the growth of bifidobacteria, *Biosci. Biotechnol. Biochem.* **56**, 355–356.

Albersheim, P., Darvill, A. G. (1985) Oligosaccharins, *Sci. Am.* **253**, 58–64.

Allison, D. G., Ruiz, B., SanJose, C., Jaspe, A., Gilbert, P. (1998) Extracellular products as mediators of the formation and detachment of *Pseudomonas fluorescens* biofilms, *FEMS Microbiol. Lett.* **167**, 179–184.

Beale, J. M., Foster, J. L. (1996) Carbohydrate fluxes into alginate biosynthesis in *Azotobacter vinelandii* NCIB 8789 – nmr investigations of the triose pools, *Biochemistry* **35**, 4492–4501.

Boucher, J. C., Martinez-Salazar, J., Schurr, M. J., Mudd, M. H., Yu, H., Deretic, V. (1996) Two distinct loci affecting conversion to mucoidy in *Pseudomonas aeruginosa* in cystic fibrosis encode homologs of the serine-protease HtrA, *J. Bacteriol.* **178**, 511–523.

Boyd, A., Chakrabarty, A. M. (1994) Role of alginate lyase in cell detachment of *Pseudomonas aeruginosa*, *Appl. Environ. Microbiol.* **60**, 2355–2359.

Boyd, A., Gosh, M., May, T. B., Shinabarger, D., Keogh, R., Chakrabarty, A. M. (1993) Sequence of the *algL* gene from *Pseudomonas aeruginosa* and purification of its alginate lyase product, *Gene* **131**, 1–8.

Campos, M.-E., Martinez-Salazar, J. M., Lloret, L., Moreno, S., Nunez, C., Espin, G., Soberon-Chavez, G. (1996) Characterization of the gene coding for GDP-mannose dehydrogenase (*algD*) from *Azotobacter vinelandii*, *J. Bacteriol.* **178**, 1793–1799.

Chitnis, C. E., Ohman, D. E. (1993) Genetic analysis of the alginate biosynthetic gene cluster of *Pseudomonas aeruginosa* shows evidence of an operonic structure, *Mol. Microbiol.* **8**, 583–590.

Chu, L., May, T. B., Chakrabarty, A. M., Misra, T. K. (1991) Nucleotide sequence and expression of the *algE* gene involved in alginate biosynthesis by *Pseudomonas aeruginosa*, *Gene* **107**, 1–10.

Cote, G. L., Krull, L. H. (1988) Characterization of the exocellular polysaccharides from *Azotobacter chroococcum*, *Carbohydr. Res.* **181**, 143–152.

Darzins, A., Chakrabarty, A. M. (1984) Cloning of genes controlling alginate biosynthesis from a mucoid cystic fibrosis isolate of *Pseudomonas aeruginosa*, *J. Bacteriol.* **159**, 9–18.

Davies, D. G., Geesey, G. G. (1995) Regulation of the alginate biosynthesis gene *algC* in *Pseudomonas aeruginosa* during biofilm development in continuous culture, *Appl. Environ. Microbiol.* **61**, 860–867.

Davies, D. G., Chakrabarty, A. M., Geesey, G. G. (1993) Exopolysaccharide production in biofilms – substratum activation of alginate gene-expression by *Pseudomonas aeruginosa*, *Appl. Environ. Microbiol.* **59**, 1181–1186.

Deretic, V., Konyecsni, W. M. (1989) Control of mucoidy in *Pseudomonas aeruginosa*: transcriptional regulation of *algR* and identification of the second regulatory gene *algQ*, *J. Bacteriol.* **171**, 3680–3688.

Deretic, V., Dikshit, R., Konyecsni, W. M., Chakrabarty, A. M., Misra, T. K. (1989) The *algR* gene, which regulates mucoidy in *Pseudomonas aeruginosa*, belongs to a class of environmentally response genes, *J. Bacteriol.* **171**, 1278–1283.

Deretic, V., Schurr, M. J., Boucher, J. C., Martin, D. W. (1994) Conversion of *Pseudomonas aeruginosa* to mucoidy in cystic fibrosis: environmental stress and regulation of bacterial virulence by alternative sigma factors, *J. Bacteriol.* **176**, 2773–2780.

Devault, J. D., Hendrickson, W., Kato, J., Chakrabarty, A. M. (1991) Environmentally regulated *algD* promoter is responsive to the cAMP receptor protein in *Escherichia coli, Mol. Microbiol.* **5**, 2503–2509.

Ertesvåg, H., Valla, S. (1999) The A modules of the *Azotobacter vinelandii* mannuronan-C-5-epimerase AlgE1 are sufficient for both epimerization and binding of Ca^{2+}, *J. Bacteriol.* **181**, 3033–3038.

Ertesvåg, H., Doseth, B., Larson, B., Skjåk-Bræk, G., Valla, S. (1994) Cloning and expression of an *Azotobacter vinelandii* mannonuran C-5-epimerase gene, *J. Bacteriol.* **176**, 2846–2853.

Ertesvåg, H., Hoidal, H. K., Hals, I. K., Rian, A., Doseth, B., Valla, S. (1995) A family of modular type mannuronan C-5-epimerase genes controls alginate structure in *Azotobacter vinelandii, Mol. Microbiol.* **9**, 719–731.

Ertesvåg, H., Frode, E., Skjåk-Bræk, G., Rehm, B. H. A., Valla, S. (1998a) Biochemical properties and substrate specificities of a recombinantly produced *Azotobacter vinelandii* alginate lyase, *J. Bacteriol.* **180**, 3779–3784.

Ertesvåg, H., Hoidal, H. K., Skjåk-Bræk, G., Valla, S. (1998b) The *Azotobacter vinelandii* mannuronan C-5-epimerase AlgE1 consists of two separate catalytic domains, *J. Biol. Chem.* **273**, 30927–30932.

Ertesvåg, H., Hoidal, H. K., Schjerven, H., Svanem, B. I., Valla, S. (1999) Mannuronan C-5-epimerases and their application for *in vitro* and *in vivo* design of new alginates useful in biotechnology, *Metab. Eng.* **1**, 262–269.

Fett, W. F., Osman, S. F., Fishman, M. L., Siebles, T. S. (1986) Alginate production by plant-pathogenic Pseudomonads, *Appl. Environ. Microbiol.* **52**, 466–473.

Fett, W. F., Wijey, C., Lifson, E. R. (1992) Occurrence of alginate gene-sequences among members of the Pseudomonad ribosomal-RNA homology groups I-IV, *FEMS Microbiol. Lett.* **99**, 151–157.

Fialho, A. M., Zielinski, N. A., Fett, W. F., Chakrabarty, A. M., Berry, A. (1990) Distribution of alginate gene-sequences in the *Pseudomonas* ribosomal-RNA homology group I-*Azomonas-Azotobacter* lineage of superfamily-B procaryotes, *Appl. Environ. Microbiol.* **56**, 436–443.

Franklin, M. J., Ohman, D. E. (1993) Identification of *alg*F in the alginate biosynthetic gene cluster of *Pseudomonas aeruginosa* which is required for alginate acetylation, *J. Bacteriol.* **175**, 5057–5065.

Franklin, M. J., Ohman, D. E. (1996) Identification of *algI* and *algJ* in the *Pseudomonas aeruginosa* alginate biosynthetic gene cluster which are required for alginate acetylation, *J. Bacteriol* **178**, 2186–2195.

Franklin, M. J., Chitnis, C. E., Gacesa, P., Sonesson, A., White, D. C., Ohman, D. E. (1994) *Pseudomonas aeruginosa* AlgG is a polymer level alginate C5-mannuronan epimerase, *J. Bacteriol.* **176**, 1821–1830.

Fujihara, M., Nagumo, T. (1992) The effect of the content of D-mannuronic acid and L-guluronic acid blocks in alginates on antitumor activity, *Carbohydr. Res.* **224**, 343–347.

Fujihara, M., Nagumo, T. (1993) An influence of the structure of alginate on the chemotactic activity of macrophages and the antitumor activity, *Carbohydr. Res.* **243**, 211–216.

Fujihara, M., Iizima, N., Yamamoto, I., Nagumo, T. (1984) Purification and chemical and physical characterization of an antitumor polysaccharide from the brown seaweed *Sargassum fulvellum*, *Carbohydr. Res.* **125**, 97–106.

Gacesa, P. (1987) Alginate-modifying enzymes: a proposed unified mechanism of action for the lyases and epimerases, *FEBS Lett.* **212**, 199–202.

Gacesa, P. (1992) Enzymic degradation of alginates, *Int. J. Biochem.* **24**, 545–552.

Gacesa, P. (1998) Bacterial alginate biosynthesis – recent progress and future prospects, *Microbiology* **144**, 1133–1143.

Gacesa, P., Goldberg, J. B. (1992) Heterologous expression of an alginate lyase gene in mucoid and non-mucoid strains of *Pseudomonas aeruginosa, J. Gen. Microbiol.* **138**, 1665–1670.

Gacesa, P., Russell, N. J. (1990) The structure and property of alginate, in: *Pseudomonas* Infection and Alginates (Gacesa, P., Russell, N. J., Eds.), London: Chapman & Hall, 29–49.

Garrett, E. S., Perlegas, D., Wozniak, D. J. (1999) Negative control of flagellum synthesis in *Pseudomonas aeruginosa* is modulated by the alternative sigma factor AlgT (AlgU), *J. Bacteriol.* **181**, 7401–7404.

Goldberg, J. B., Hatano, K., Pier, G. B. (1993). Synthesis of lipopolysaccharide-O side-chains by *Pseudomonas aeruginosa* PAO1 requires the enzyme phosphomannomutase, *J. Bacteriol.* **175**, 1605–1611.

Gorin, P. A., Spencer, J. F. T. (1966) Exocellular alginic acid from *Azotobacter vinelandii, Can. J. Chem.* **44**, 993–998.

Govan, J. R. W., Deretic, V. (1996) Microbial pathogenesis in cystic fibrosis: mucoid *Pseudo-*

monas aeruginosa and *Burkholderia cepacia*, *Microbiol. Rev.* **60**, 539–574.

Govan, J. R. W., Fyfe, J. F. M. (1978) Mucoid *Pseudomonas aeruginosa* and cystic fibrosis: resistance of the mucoid form to carbenicillin, flucloxacillin and tobramycin and the isolation of mucoid variants *in vitro*, *J. Antimicrob. Chemother.* **4**, 233–240.

Govan, J. R. W., Harris, G. S. (1986) *Pseudomonas aeruginosa* and cystic fibrosis: unusual bacterial adaptation and pathogenesis, *Microbiol. Sci.* **3**, 302–308.

Govan, J. R. W., Fyfe, J. F. M., Jarman, T. R. (1981) Isolation of alginate-producing mutants of *Pseudomonas fluorescens, Pseudomonas putida*, and *Pseudomonas mendocina*, *J. Gen. Microbiol.* **125**, 217–220.

Grabert, E., Wingender, J., Winkler, U. K. (1990) An outer membrane protein characteristic of mucoid strains of *Pseudomonas aeruginosa*, *FEMS Microbiol. Lett.* **68**, 83–88.

Grobe, S., Wingender, J., Trüper, H. G. (1995) Characterization of mucoid *Pseudomonas aeruginosa* strains isolated from technical water systems, *J. Appl. Bacteriol.* **79**, 94–102.

Gross, M., Rudolph, K. (1987) Demonstration of levan and alginate in bean-plants (*Phaseolus vulgaris*) infected by *Pseudomonas syringae* pv. *phaseolicola*, *J. Phytopathol.* **120**, 9–19.

Gutowska, A., Jeong, B., Jasionowski, M. (2001) Injectable gels for tissue engineering, *Anat. Rec.* **263**, 342–349.

Haug, A., Larsen, B., Smidsrød, O. (1967) Studies on the sequence of uronic acid residues in alginic acid, *Acta Chem. Scand.* **21**, 691–704.

Hershberger, C. D., Ye, R. W., Parsek, M. R., Xie, Z. D., Chakrabarty, A. M. (1995) The *algT* (*algU*) gene of *Pseudomonas aeruginosa*, a key regulator involved in alginate biosynthesis, encodes an alternative sigma factor (sigma^{E}), *Proc. Natl. Acad. Sci. USA* **92**, 7941–7945.

Kapatral, V., Bina, X., Chakrabarty, A. M. (2000) Succinyl coenzyme A synthetase of *Pseudomonas aeruginosa* with a broad specificity for nucleoside triphosphate (NTP) synthesis modulates specificity for NTP synthesis by the 12-kilodalton form of nucleoside diphosphate kinase, *J. Bacteriol.* **182**, 1333–1339.

Kawada, A., Hiura, N., Shiraiwa, M., Tajima, S., Hiruma, M. (1997) Stimulation of human keratinocyte growth by alginate oligosaccharides, a possible co-factor for epidermal growth factor in cell culture, *FEBS Lett.* **408**, 43–46.

Kennedy, L., McDowell, K., Sutherland, I. W. (1992) Alginases from *Azotobacter* species, *J. Gen. Microbiol.* **138**, 2465–2471.

Kim, H. Y., Schlictman, D., Shankar, S., Xie, Z., Chakrabarty, A. M., Kornberg, A. (1998) Alginate, inorganic polyphosphate, GTP and ppGpp synthesis co-regulated in *Pseudomonas aeruginosa*: implications for stationary phase survival and synthesis of RNA/DNA precursors, *Mol. Microbiol.* **27**, 717–725.

Konyecsni, W. M., Deretic, V. (1990) DNA sequence and expression analysis of *alg*P and *alg*Q, components of the multigene system transcriptionally regulating mucoidy in *Pseudomonas aeruginosa*: *algP* contains multiple direct repeats, *J. Bacteriol.* **172**, 2511–2520.

Lee, J. W., Day, D. F. (1995) Bioacetylation of seaweed alginate, *Appl. Environ. Microbiol.* **61**, 650–655.

Lin, T.-Y., Hassid, W. Z. (1966a) Pathway of alginic acid synthesis in the marine brown alga, *Fucus gardneri* Silva, *J. Biol. Chem.* **241**, 5284–5297.

Lin, T.-Y., Hassid, W. Z. (1966b) Isolation of guanosine diphosphate uronic acids from a marine brown alga, *Fucus gardneri* Silva, *J. Biol. Chem.* **241**, 3283–3293.

Linker, A., Jones, R. S. (1966) A new polysaccharide resembling alginic acid isolated from pseudomonads, *J. Biol. Chem.* **241**, 3845–3851.

Lloret, L., Barreto, R., Leon, R., Moreno, S., Martínez-Salazar, J., Espín, G., Soberón-Chávez, G. (1996) Genetic analysis of the transcriptional arrangement of *Azotobacter vinelandii* alginate biosynthetic genes: identification of two independent promoters, *Mol. Microbiol.* **21**, 449–457.

Lynn, A. R., Sokatch, J. R. (1984) Incorporation of isotope from specifically labelled glucose into alginates of *Pseudomonas aeruginosa* and *Azotobacter vinelandii*, *J. Bacteriol.* **158**, 1161–1162.

Ma, J. F., Phibbs, P. V., Hassett, D. J. (1997) Glucose stimulates alginate production and *algD* transcription in *Pseudomonas aeruginosa*, *FEMS Microbiol. Lett.* **148**, 217–221.

Ma, S., Selvaraj, U., Ohmann, D. E. Quarless, R., Hassett, D. J., Wozniak, D. J. (1998) Phsophorylation-independent activity of the response regulators AlgB and AlgR in promoting alginate biosynthesis in mucoid *Pseudomonas aeruginosa*, *J. Bacteriol.* **180**, 956–968.

Madgwick, J., Haug, A., Larsen, B. (1978) Ionic requirements of alginate-modifying enzymes in the marine alga *Pelvetia canaliculata* (L.) Dcne. et Thur, *Bot. Mar.* **21**, 1–3.

Maharaj, R., May, T. B., Wang, S. K., Chakrabarty, A. M. (1993) Sequence of the *alg8* and alg44 genes involved in the synthesis of alginate by *Pseudomonas aeruginosa*, *Gene* **136**, 267–269.

Martin, D. W., Schurr, M. J., Mudd, M. H., Deretic, V. (1993a) Differentiation of *Pseudomonas aeruginosa* into the alginate-producing form: inactivation of *mucB* causes conversion to mucoidy, *Mol. Microbiol.* **9**, 495–506.

Martin, D. W., Schurr, M. J., Mudd, M. H., Govan, J. R. W., Holloway, B. W., Deretic, V. (1993b) Mechanism of conversion to mucoidy in *Pseudomonas aeruginosa* infecting cystic fibrosis patients, *Proc. Natl. Acad. Sci. USA* **90**, 8377–8381.

Martin, D. W., Schurr, M. J., Yu, H., Deretic, V. (1994) Analysis of promoters controlled by the putative sigma-factor AlgU regulating conversion to mucoidy in *Pseudomonas aeruginosa* relationship to sigma(e) and stress-response, *J. Bacteriol.* **176**, 6688–6696.

Martínez-Salazar, J. M., Moreno, S., Najera, R., Boucher, J. C., Espín, G., Soberón-Chávez, G., Deretic, V. (1996) Characterization of genes coding for the putative sigma factor AlgU and its regulators MucA, MucB, MucC, and MucD in *Azotobacter vinelandii* and evaluation of their roles in alginate biosynthesis, *J. Bacteriol.* **178**, 1800–1808.

Mathee, K., McPherson, C. J., Ohmann, D. E. (1977) Posttranslational control of the algT (algU)-encoded sigma22 for expression of the alginate regulon in *Pseudomonas aeruginosa* and localization of its antagonist proteins MucA and MucB (AlgN), *J. Bacteriol.* **179**, 3711–3720.

Mathews, C. K. (1993) Enzyme organization in DNA precursor biosynthesis, *Prog. Nucleic Acid Res. Mol. Biol.* **44**, 167–203

May, T. B., Chakrabarty, A. M. (1994) *Pseudomonas aeruginosa*: genes and enzymes of alginate biosynthesis, *Trends Microbiol.* **2**, 151–157.

May, T. B., Shinabarger, D., Maharaj, R., Kato, J., Chu, L., Devault, J. D., Roychoudhury, S., Zielinski, N. A., Berry, A., Rothmel, R. K., Misra, T. K., Chakrabarty, A. M. (1991) Alginate synthesis by *Pseudomonas aeruginosa*: a key pathogenic factor in chronic pulmonary infections of cystic fibrosis patients, *Clin. Microbiol. Rev.* **4**, 191–206.

McAvoy, M. J., Newton, V., Paull, A., Morgan, J., Gacesa, P., Russell, N. J. (1989) Isolation of mucoid strains of *Pseudomonas aeruginosa* from non-cystic-fibrosis patients and characterization of the structure of their secreted alginate, *J. Med. Microbiol.* **28**, 183–189.

Moe, S. T., Draget, K. I., Skjåk-Bræk, G., Smidsrød, O. (1995) Alginates, in: *Food Polysaccharides and Applications* (Stephen, A. M., Ed.), New York: Marcel Dekker, Inc., 245–286.

Monday, S. R., Schiller, N. L. (1996) Alginate synthesis in *Pseudomonas aeruginosa*: the role of AlgL (alginate lyase) and AlgX, *J. Bacteriol.* **178**, 625–632.

Mrsny, R. J., Lazazzera, B. A., Daugherty, A. L., Schiller, N. L., Patapoff, T. W. (1994) Addition of a bacterial alginate lyase to purulent CF sputum *in vitro* can result in the disruption of alginate and modification of sputum viscoelasticity, *Pulm. Pharmacol.* **7**, 357–366.

Mrsny, R. J., Daugherty, A. L., Short, S. M., Widmer, R., Siegel, M. W., Keller, G.-A. (1996) Distribution of DNA and alginate in purulent cystic fibrosis sputum: implications to pulmonary targeting strategies, *J. Drug Target.* **4**, 233–243.

Murata, K., Inose, T., Hisano, T., Abe, S., Yonemoto, Y. (1993) Bacterial alginate lyase: enzymology, genetics and application, *J. Ferment. Bioeng.* **76**, 427–437.

Narbad, A., Russell, N. J., Gacesa, P. (1988) Radiolabelling patterns in alginate of *Pseudomonas aeruginosa* synthesized from specifically-labelled ^{13}C monosaccharide precursors, *Microbios* **54**, 171–179.

Natsume, M., Kamo, Y., Hirayama, M., Adachi, T. (1994) Isolation and characterization of alginate-derived oligosaccharides with root growth-promoting activities, *Carbohydr. Res.* **258**, 187–197.

Nivens, D. E., Ohman, D. E., Williams, J., Franklin, M. J. (2001) Role of alginate and its O acetylation in formation of *Pseudomonas aeruginosa* microcolonies and biofilms, *J. Bacteriol.* **183**, 1047–1057.

Nguyen L. K., Schiller, N. L. (1989) Identification of a slime exopolysaccharide de-polymerase in mucoid strains of *Pseudomonas aeruginosa*, *Curr. Microbiol.* **18**, 323–329.

Olvera, C., Goldberg, J. B., Sanchez, R., Soberon-Chavez, G. (1999) The *Pseudomonas aeruginosa algC* gene product participates in rhamnolipid biosynthesis, *FEMS Microbiol. Lett.* **179**, 85–90.

Onsøyen, E. (1996) Commercial applications of alginates. *Carbohydr. Eur.* **14**, 26–31.

Otterlei, M., Østgaard, K., Skjåk-Bræk, G., Smidsrød, O., Soon-Shoing, Espevik, T. (1991) Induction of cytokine production from human monocytes stimulated with alginate, *J. Immunother.* **10**, 286–291.

Padgett, P. J., Phibbs, P. V., Jr. (1986) Phosphomannomutase activity in wild-type and alginate-producing strains of *Pseudomonas aeruginosa*, *Curr. Microbiol.* **14**, 187–192.

Page, W. J., Sadoff, H. L. (1975) Relationship between calcium and uronic acids in the encystment of *Azotobacter vinelandii*, *J. Bacteriol.* **122**, 145–151.

Pecina, A., Pascual, A., Paneque, A. (1999) Cloning and expression of the *algL* gene, encoding the

Azotobacter chroococcum alginate lyase: purification and characterization of the enzyme, *J. Bacteriol.* **181**, 1409–1414.

Penaloza-Vazquez, A., Kidambi, S. P., Chakrabarty, A. M., Bender, C. L. (1997). Characterisation of the alginate biosynthetic gene cluster in *Pseudomonas syringae* pv. *syringae*, *J. Bacteriol.* **179**, 4464–4472.

Piggott, N. H., Sutherland, I. W., Jarman, T. R. (1981) Enzymes involved in the biosynthesis of alginate by *Pseudomonas aeruginosa*, *Eur. J. Appl. Microbiol. Biotechnol.* **13**, 179–183.

Pindar, D. F., Bucke, C. (1975) The biosynthesis of alginic acid by *Azotobacter vinelandii*, *Biochem. J.* **152**, 617–622.

Ramstad, M. V., Ellingsen, T. E., Josefsen, K. D., Hoidal, H. K., Valla, S., Skjåk-Bræk, G., Levine, D. W. (1999) Properties and action pattern of the recombinant mannuronan C-5-epimerase AlgE2, *Enzyme Microbiol. Technol.* **24**, 636–646.

Rehm, B. H. A. (1996) The *Azotobacter vinelandii* gene *algJ* encodes an outer membrane protein presumably involved in export of alginate, *Microbiology* **142**, 873–880.

Rehm, B. H. A. (1998) The alginate lyase from *Pseudomonas aeruginosa* CF1/M1 prefers the hexameric oligomannuronate as substrate, *FEMS Microbiol. Lett.* **165**, 175–180.

Rehm, B. H. A., Valla, S. (1997) Bacterial alginates: biosynthesis and applications, *Appl. Microbiol. Biotechnol.* **48**, 281–288.

Rehm, B. H. A., Winkler, U. K. (1996) Alginatbiosynthese bei *Pseudomonas aeruginosa* und *Azotobacter vinelandii*: Molekularbiologie und Bedeutung, *BIOspektrum* **4**, 31–36.

Rehm, B. H. A., Grabert, G., Hein, J., Winkler, U. K. (1994a) Antibody response of rabbits and cystic fibrosis patients to alginate-specific outer membrane protein of a mucoid strain of *Pseudomonas aeruginosa*, *Microb. Pathog.* **16**, 43–51.

Rehm, B. H. A., Boheim, G., Tommassen, J., Winkler, U. K. (1994b) Overexpression of *algE* in *Escherichia coli*: subcellular localization, purification, and ion channel properties, *J. Bacteriol.* **176**, 5639–5647.

Rehm, B. H. A., Ertesvåg, H., Valla, S. (1996) A new *Azotobacter vinelandii* mannuronan C-5-epimerase gene (*algG*) is part of an *alg* gene cluster physically organized in a manner similar to that in *Pseudomonas aeruginosa*, *J. Bacteriol.* **178**, 5884–5889.

Ross, P., Meyer, R., Benziman, M. (1991) Cellulose biosynthesis and function in bacteria, *Microbiol. Rev.* **55**, 35–58.

Sadoff, H. L. (1975) Encystment and germination in *Azotobacter vinelandii*, *Bacteriol. Rev.* **39**, 516–539.

Schiller, N. L., Monday, S. R., Boyd, C. M., Keen, N. T., Ohman, D. E. (1993) Characterization of the *Pseudomonas aeruginosa* alginate lyase gene (*algL*) – cloning, sequencing, and expression in *Escherichia coli*, *J. Bacteriol.* **175**, 4780–4789.

Schlictman, D., Kavanaughblack, A., Shankar, S., Chakrabarty, A. M. (1994) Energy metabolism and alginate biosynthesis in *Pseudomonas aeruginosa* – role of the tricarboxylic acid cycle, *J. Bacteriol.* **176**, 6023–6029.

Schurr, M. J., Martin, D. W., Mudd, M. H., Hibler, N. S., Boucher, J. C., Deretic, V. (1993) The *algD* promoter – regulation of alginate production by *Pseudomonas aeruginosa* in cystic fibrosis, *Cell. Mol. Biol. Res.* **39**, 371–376.

Schurr, M. J., Yu, H., Martinez-Salazar, J. M., Hibler, N. S., Deretic, V. (1995) Biochemical characterisation and posttranslational modification of AlgU, a regulator of stress response in *Pseudomonas aeruginosa*, *Biochem. Biophys. Res. Commun.* **216**, 874–880.

Schurr, M. J., Yu, H., Martinez-Salazar, J. M., Boucher, J. C., Deretic, V. (1996) Control of *algU*, a member of the sigma(e)-like family of stress sigma factors, by the negative regulators MucA and MucB and *Pseudomonas aeruginosa* conversion to mucoidy in cystic fibrosis, *J. Bacteriol.* **178**, 4997–5004.

Schweizer, H. P., Po, C., Bacic, M. K. (1995) Identification of *Pseudomonas aeruginosa glpM*, whose gene-product is required for efficient alginate biosynthesis from various carbon sources, *J. Bacteriol.* **177**, 4801–4804.

Sherbrock-Cox, V., Russell, N. J., Gacesa, P. (1984) The purification and chemical characterization of the alginate present in extracellular material produced by mucoid strains of *Pseudomonas aeruginosa*, *Carbohydr. Res.* **135**, 147–154.

Shinabarger, D., May, T. B., Boyd, A., Ghosh, M., Chakrabarty, A. M. (1993) Nucleotide sequence and expression of the *Pseudomonas aeruginosa algF* gene controlling acetylation of alginate, *Mol. Microbiol.* **9**, 1027–1035.

Skjåk-Bræk, G., Espevik, T. (1996) Application of alginate gels in biotechnology and medicine, *Carbohydr. Eur.* **14**, 19–25.

Skjåk-Bræk, G., Martinsen, A. (1991) Applications of some algal polysaccharides in biotechnology, in: *Seaweed Resources in Europe: Uses and Potential* (Guiry, M.D., Blunden, G., Eds.), New York: John Wiley & Sons, 219–257.

Skjåk-Bræk G., Larsen, B. Grasdalen, H., (1985) The role of O-acetyl groups in the biosynthesis of alginate by *Azotobacter vinelandii*, *Carbohydr. Res.* **145**, 169–174.

Skjåk-Bræk G., Grasdalen, H., Larsen, B. (1986) Monomer sequence and acetylation pattern in some bacterial alginates, *Carbohydr. Res.* **154**, 239–250.

Smidsrød, O., Draget, K. I. (1996) Chemistry and physical properties of alginates, *Carbohydr. Eur.* **14**, 6–13.

Svanem, B. J., Skjåk-Bræk G., Ertesvåg, H., Valla, S. (1999) Cloning and expression of three new *Azotobacter vinelandii* genes closely related to a previously described gene family encoding mannuronan C-5-epimerases, *J. Bacteriol.* **181**, 68–77

Soon-Shiong, P. (1995) Encapsulated islet cell therapy for the treatment of diabetes: intraperitoneal injection of islets, *J. Controlled Release* **39**, 399–409.

Stokke, B. T., Smidsrød, O., Bruheim, P., Skjåk-Bræk, B. (1991) Distribution of uronate residues in alginate chains in relation to alginate gelling properties, *Macromolecules* **24**, 4637–4645.

Sutherland, J. W. (1982) Biosynthesis of microbial exopolysaccharides, *Adv. Microb. Physiol.* **23**, 79–150.

Sutherland, I. W. (1995) Polysaccharide lyases, *FEMS Microbiol. Rev.* **16**, 323–347.

Tatnell, P. J., Russell, N. J., Gacesa, P. (1993) A metabolic study of the activity of GDP-mannose dehydrogenase and concentrations of activated intermediates of alginate biosynthesis in *Pseudomonas aeruginosa*, *J. Gen. Microbiol* **139**, 119–127.

Tatnell, P. J., Russell, N. J., Gacesa, P. (1994). GDP-mannose dehydrogenase is the key regulatory enzyme in alginate biosynthesis in *Pseudomonas aeruginosa*: evidence from metabolite studies, *Microbiology* **140**, 1745–1754.

Terry, J. M., Pina, S. E., Mattingly, S. J. (1991) Environmental conditions which influence mucoid conversion in *Pseudomonas aeruginosa* PAO1, *Infect. Immun.* **59**, 471–477.

Tomoda, Y., Umemura, K., Adachi, T.(1994) Promotion of barley root elongation under hypoxic conditions by alginate lyase-lysate, *Biosci. Biotechnol. Biochem.* **58**, 202–203.

Vazquez, A., Soledad, M., Guzman, J., Alvarado, A., Espin, G. (1999) Transcriptional organization of the *Azotobacter vinelandii algGXLVIFA* genes: characterization of *algF* mutants, *Gene* **232**, 217–222.

Wang, S. K., Sa-Correia, I., Darzins, A., Chakarabarty, A. M. (1987) Characterization of *Pseudomonas aeruginosa* alginate (*alg*) gene region II, *J. Gen. Microbiol.* **133**, 2303–2314.

Whiteley, M., Bangera, M. G., Bumgarner, R. E., Parsek, M. R., Teitzel, G. M., Lory, S., Greenberg, E. P. (2001) Gene expression in *Pseudomonas aeruginosa* biofilms, *Nature* **413**, 860–864.

Wong, T. Y., Preston, L. A., Schiller, N. L. (2000) Alginate lyase: review of major sources and enzyme characteristics, structure–function analysis, biological roles, and applications, *Annu. Rev. Microbiol.* **54**, 289–340.

Wozniak, D. J., Ohman, D. E. (1991) *Pseudomonas aeruginosa* AlgB, a 2-component response regulator of the NtrC family, is required for *algD* transcription, *J. Bacteriol.* **173**, 1406–1413.

Wozniak, D. J., Ohman, D. E. (1994) Transcriptional analysis of the *Pseudomonas aeruginosa* genes *algR*, *algB*, and *algD* reveals a hierarchy of alginate gene expression which is modulated by *algT*, *J. Bacteriol.* **176**, 6007–6014.

Wyss, O., Neumann, M. G., Socolofsky, M. D. (1961). Development and germination of the *Azotobacter* cyst, *J. Biophys. Biochem. Cytol.* **10**, 555–565.

Yonemoto, Y., Tanaka, H., Yamashita, T., Kitabatake, N., Ishida, Y. (1993) Promotion of germination and shoot elongation of some plants by alginate oligomers prepared with bacterial alginate lyase, *J. Ferment. Bioeng.* **75**, 68–70.

Yoon, H.-J., Mikami, B., Hashimoto, W., Murata, K. (1999) Crystal structure of alginate lyase A1-III from *Sphingomonas* species A1 at 1.78 Å resolution, *J. Mol. Biol.* **290**, 505–514.

Yoon, H.-J., Hashimoto, W., Miyake, O., Murata, K., Mikami, B. (2001) Crystal structure of alginate lyase A1-III complexed with trisaccharide product at 2.0 Å resolution, *J. Mol. Biol.* **307**, 9–16.

Yu, H., Schurr, M. J., Deretic, V. (1995) Functional equivalence of *Escherichia coli* sigma(e) and *Pseudomonas aeruginosa* AlgU. *Escherichia coli* RpoE restores mucoidy and reduces sensitivity to reactive oxygen intermediates in AlgU mutants of *Pseudomonas aeruginosa*, *J. Bacteriol.* **177**, 3259–3268.

Yu, H., Mudd, M., Boucher, J. C., Schurr, M. J., Deretic, V. (1997) Identification of the *algZ* gene upstream of the response regulator *algR* and its participation in control of alginate production in *Pseudomonas aeruginosa*, *J. Bacteriol.* **179**, 187–193.

10 Carrageenan

Dr. Ir. Fred van de Velde[1], Dr. Gerhard A. De Ruiter[2]

[1] Wageningen Centre for Food Sciences and TNO Nutrition and Food Research Institute, Carbohydrate Technology Department; present address NIZO food research, PO Box 20, 6710 BA Ede, The Netherlands; Tel.: +31-318-659-582; Fax: +31-318-650-400; E-mail: Fvelde@nizo.nl

[2] NIZO food research, Product Technology Department, PO Box 20, 6710 BA Ede, The Netherlands; Tel.: +31-318-659-636; Fax: +31-318-650-400; E-mail: DeRuiter@nizo.nl

Biopolymers for Medical and Pharmaceutical Applications. Edited by A. Steinbüchel and R. H. Marchessault

ISBN: 3-527-31154-8

AMF alkali-modified flour
ARC alternatively refined carrageenan
ERF enzyme-resistant fractions

GC	gas chromatography
GPC	gel permeation chromatography
GRAS	generally recognized as safe
HPAEC	high-performance anion-exchange chromatography
HPLC	high-performance liquid chromatography
IR	infrared
MALLS	multi-angle laser-light scattering
MMB	methylmorpholine-borane
NMR	nuclear magnetic resonance
PES	processed *Eucheuma* seaweed
PNG	Philippines natural grade
SEC	size exclusion chromatography
SRC	semi-refined carrageenan

1 Introduction

Carrageenan is the generic name for a family of gel-forming, viscosifying polysaccharides that are obtained commercially by extraction of certain species of red seaweeds (*Rhodophyceae*). The main species responsible for most of today's carrageenan production belong to the following genera:

- *Gigartina* (Argentina/Chile, France, Morocco),
- *Chondrus* (France, North Atlantic),
- *Iridaea* (Chile), and
- *Eucheuma* (Philippines/Indonesia).

Carrageenans are composed of a linear galactose backbone with a varying degree of sulfatation (between 15% and 40%). Different carrageenan types differ in composition and conformation, resulting in a wide range of rheological and functional properties. Carrageenans are used in a variety of commercial applications as gelling, thickening, and stabilizing agents, especially in food products such as frozen desserts, chocolate milk, cottage cheese, whipped cream, instant products, yogurt, jellies, pet foods, and sauces. Aside from these functions, carrageenans are used in pharmaceutical formulations, cosmetics, and industrial applications such as mining.

2 Historical Outline

For several hundred years, carrageenan has been used as a thickening and stabilizing agent in food in Europe and the Far East. In Europe the use of carrageenan started more than 600 years ago in Ireland. In the village of Carraghen on the south Irish coast, flans were made by cooking the so-called Irish moss (red seaweed species *Chondrus crispus*) in milk. The name carrageenin, the old name for carrageenan, was first used in 1862 for the extract from *C. crispus* and was dedicated to this village (Tseng, 1945). Schmidt described the extraction procedure in 1844.

Since the 19th century, Irish moss also has been used for industrial beer clarification and textile sizing. The commercial production began in the 1930s in the U.S. During that time, the trading shifted from dried seaweed meal to refined carrageenan (Therkelsen, 1993). After the Second World War, a general increase in the standard of living forced an increase in carrageenan production.

Fractionation of crude carrageenan extracts started in the early 1950s (Smith et al., 1955), resulting in the characterization of the different carrageenan types. A Greek prefix was introduced to identify the different carrageenans. In the same period, the molecular structure of carrageenans was determined (O'Neill 1955a,b). The structure of 3,6-anhydro-D-galactose in κ-carrageenan, as well as the type of linkages between galactose and anhydrogalactose rings, was determined.

Today, the industrial manufacture of carrageenan is no longer limited to extraction from Irish moss, and numerous red seaweed species are used. Traditionally, these seaweeds have been harvested from naturally occurring populations. Seaweed farming to increase the production started almost 200 years ago in Japan. Scientific information about the seaweed life cycles allowed artificial seeding in the 1950s. Today, nearly a dozen seaweed taxa are cultivated commercially, lowering the pressure on naturally occurring populations.

During the past few years, the total carrageenan market has shown a growth rate of 3% per year, reaching estimated worldwide sales of 310 million US$ in 2000. At the end of the 20th century, a few large corporations that account for over 80% of the supply dominate the carrageenan market, including:

- FMC Corporation (USA),
- CP Kelco (USA),
- Degussa (Germany),
- Danisco (Denmark),
- Ceamsa (Spain), and
- Quest International (The Netherlands).

3 Chemical Structure

3.1 General Description

Carrageenan is a high molecular mass material with a high degree of polydispersity. The molecular mass distribution varies from sample to sample, depending upon the sample history, e.g., age of the harvested seaweed, season of harvesting, way of extracting, and duration of heat treatment. Commercial (food-grade) carrageenans have a weight average molecular mass (M_w) ranging from 400–600 kDa with a minimum of 100 kDa. This minimum is set in response to reports of cecal and colonic ulceration induced by highly degraded carrageenan. In 1976 the U.S. Food and Drug Administration defined food-grade carrageenan as having a water viscosity of no less than 5 mPa·s (5 cP) at 1.5% concentration and 75°C, which corresponds to the above-mentioned 100 kDa.

Besides the traditionally extracted carrageenan, called refined carrageenan in trade, a new type of carrageenan product is promoted by a group of Philippine producers (Seaweed Industry Association of the Philippines). This product is marketed under the name Philippines natural grade (PNG). Other synonyms for this type of carrageenan product are processed *Eucheuma* seaweed (PES is the regulatory name), semi-refined carrageenan (SRC), alternatively refined carrageenan (ARC), and alkali-modified flour (AMF). These *Eucheuma* seaweeds (*E. cottonii* and *E. spinosum*) are harvested around Indonesia and the Philippines and treated with a more cost-effective process that avoids extraction of carrageenan in dilute solutions (see Section 9.3). The above-mentioned carrageenan differs from the traditionally refined carrageenan in that it contains 8% to 15% acid-insoluble matter compared with 2% in extracted carrageenan. The acid-insoluble matter consists mainly of cellulose, which is normally present in algae cell walls. Also, the heavy metal content of processed *Eucheuma* seaweed is higher than that of traditionally refined carrageenan (Imeson, 2000). The water-soluble component in PES is κ-carrageenan and is almost indistinguishable from the refined carrageenan. The molecular mass of κ-carrageenan present in PES can be slightly higher than that of refined carrageenan (Hoffmann et al., 1996).

3.2
Molecular Structure

Carrageenan is not a single biopolymer but a mixture of water-soluble, linear, sulfated galactans. They are composed of alternating 3-linked β-D-galactopyranose (G-units) and 4-linked α-D-galactopyranose (D-units) or 4-linked 3,6-anhydrogalactose (DA-units), forming the "ideal" disaccharide-repeating unit of carrageenans (see Figure 1). The sulfated galactans are classified according to the presence of the 3,6-anhydrogalactose on the 4-linked residue and the position and number of sulfate groups. For commercial carrageenan, the sulfate content falls within the range of 22% to 38% (w/w). Besides galactose and sulfate, other carbohydrate residues (e.g., xylose, glucose, and uronic acids) and substituents (e.g., methyl ethers and pyruvate groups) are present in carrageenans. Since natural carrageenan is a mixture of nonhomologous polysaccharides, the term disaccharide-repeating unit refers to the idealized structure. To describe more complex structures, a letter-code-based nomenclature for red algae galactans has been developed (Knutsen et al., 1994).

The most common types of carrageenan are traditionally identified by a Greek prefix. The three commercially most important carrageenans are called ι-, κ-, and λ-carrageenan. The corresponding IUPAC (International Union of Pure and Applied Chemistry) names and letter codes are carrageenose 2,4′-sulfate (G4S-DA2S), carrageenose 4′-sulfate (G4S-DA), and carrageenan 2,6,2′-trisulfate (G2S-D2S,6S). ι- and κ-carrageenan are gel-forming carrageenans, whereas λ-carrageenan is a thickener/viscosity builder.

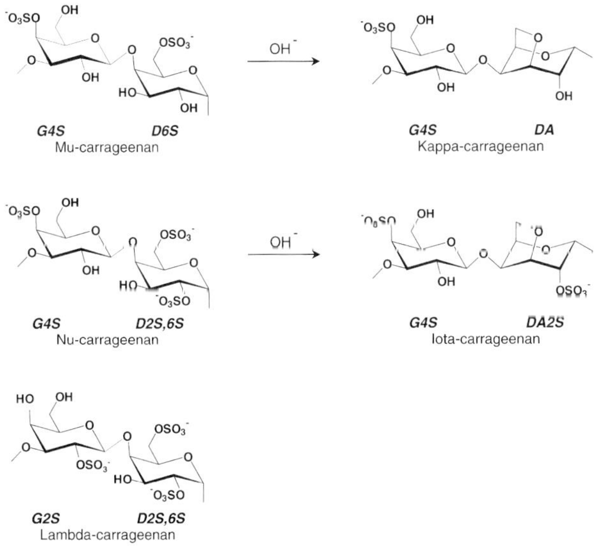

Fig. 1 Schematic representation of the different idealized repeating units of carrageenans. The letter codes refer to the alternative nomenclature (Knutsen et al., 1994).

The difference in rheological behavior between ι- and κ-carrageenan on the one hand and λ-carrageenan on the other results from the fact that the DA-units of the gelling ones have the $^{1}C_4$-conformation that results from the 3,6-anhydro bridges and λ-carrageenan does not. The natural precursors of ι- and κ-carrageenan are called ν- and μ-carrageenan (letter code G4S-D2S,6S and G4S-D6S, respectively) and are also non-gelling carrageenans with the D-units in the $^{4}C_1$-conformation as a consequence of the absence of the 3,6-anhydro bridge.

The 3,6-anhydro bridges are formed by the elimination of the sulfate from the C-6 sulfate ester of the precursors and the concomitant formation of the 3,6-anhydro bridge. *In vivo*, ι- and κ-carrageenan are formed enzymatically from their precursors, by a sulfohydrolase (see also Section 7). In industrial processing, the cyclization reaction is carried out with OH^- as a catalyst. The $^{1}C_4$-conformation of the 3,6-anhydro-D-galactopyranosyl units in ι- and κ-carrageenan allows for a helical tertiary structure, which is essential for the gel-forming properties. Occurrence of disaccharide units without the 3,6-anhydro ring and, as a consequence, with a $^{4}C_1$-conformation causes "kinks" in the regular chain and prevents the formation of helical strands, thus, preventing the gelation of the carrageenan.

4
Occurrence

All of the seaweeds that produce carrageenan as their main cell-wall material belong to the class of the red algae, or Rhodophyceae (Figure 2). Different seaweed species produce different types of carrageenans. The

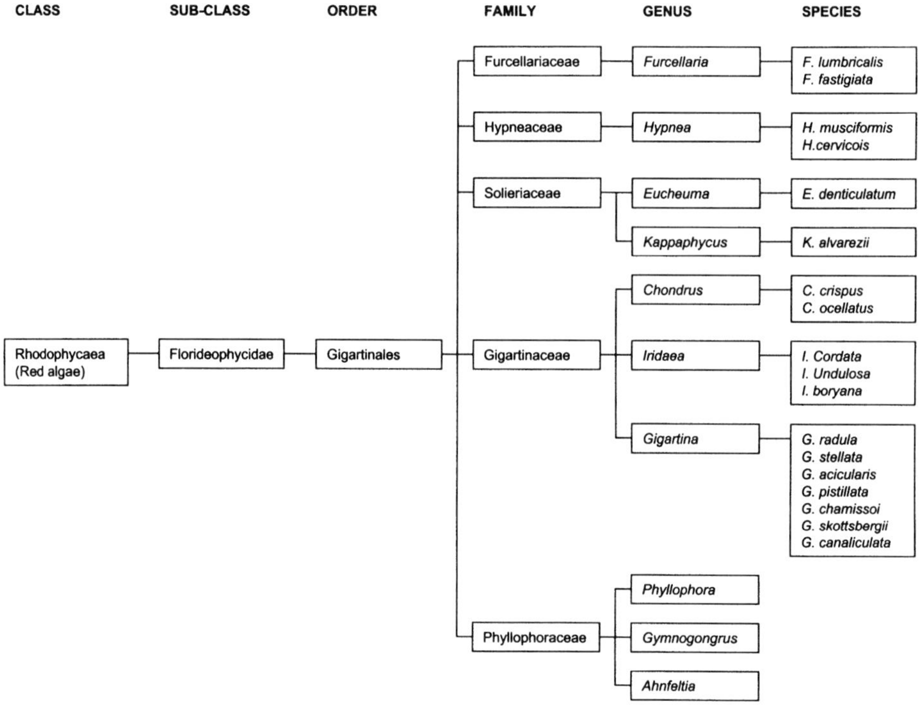

Fig. 2 Taxonomical tree of carrageenan-bearing seaweeds.

tropical seaweed *Kappaphycus alvarezii*, known in trade as *Eucheuma cottonii* (or simply cottonii), yields a relatively homogeneous κ-carrageenan after alkali treatment. This seaweed accounts for the largest production worldwide (De Ruiter and Rudolph, 1997; Rudolph, 2000). Another important species is *Eucheuma denticulatum* (trade name Spinosum), which yields ι-carrageenan upon alkali treatment. λ-Carrageenan is obtained from different species from the *Gigartina* and *Chondrus* genera. These seaweeds have a life cycle with alternating gametophytic (male and female; 1n: one set of chromosomes each) and sporophytic (2n: two sets of chromosomes) phases. The sporophytic plants produce λ-carrageenan, whereas the gametophytic plants produce a κ/ι-hybrid type of carrageenan. These κ/ι-hybrid carrageenans consist of a mixed chain containing both κ- and ι-units (VandeVelde et al., 2001) and range from almost pure ι-carrageenan to almost pure κ-carrageenan (Bixler, 1996).

Kappaphycus alvarezii (Figure 3a) has a natural distribution in the Philippines, Indonesia, and East Africa. It grows on the inner side of coral reefs in the upper subtidal zone. It has a bushy thallus consisting of numerous round branches. The surface can be both rough and smooth with a green to brown color. Its normal size is around 20–30 cm in diameter, but large ones can grow to a size of 1 m.

Eucheuma denticulatum (Figure 3b) has the same natural distribution as *K. alvarezii*. Their morphologies are also the same, except that the branches of *E. denticulatum* have spines of 3–4 mm all over the surface of the thallus and are smaller and more slender. The average size is around 75 cm in diameter.

Chondrus crispus (Figure 3c) is distributed from Norway to Morocco and from Newfoundland (Canada) to Cape Cod (Massachusetts). It grows from a basal disc that is adhered to rocks or stones. *Chondrus* is found from the upper subtidal zone down to a depth of 5 to 6 m. It has a dark violet thallus, with numerous dichotomous branches and has a size of around 25 cm (Rudolph, 2000).

Besides the above-mentioned three important species, several other species are used commercially. *Gracilaria* is harvested in many regions, especially in Chile, Indonesia, Namibia, Japan, Thailand, Taiwan, and Vietnam. *Gelidium* species are found all over the world, but the commercial harvesting is restricted to Morocco, Senegal, New Zealand, Japan, Chile, Venezuela, Spain, and Portugal. *Gigartina radula* is harvested in Chile and Argentina. *Hypnea* is found in the Caribbean Sea, Indian Ocean, and the tropical Pacific Ocean, but commercial harvesting is focused in Brazil.

Fig. 3 Photographs of the red seaweeds *Kappaphycus alvarezii* (Philippines), *Eucheuma denticulatum* (Philippines), and *Chondrus crispus* (Canada); copyright Dr. B. Rudolph (CP Kelco, Lille Skensved, Denmark).

5 Physiological Function

The majority of the red seaweeds have a cell-wall structure that is analogous to that of the higher plants: cellulosic microfibrils embedded in matrix polysaccharides. Because red algae inhabit aquatic environments, the composition and organization of their intercellular matrices differ from those of land plants. Land plants require a rigid structure to support them against the pull of gravity, making cellulose an important part of the cell wall of these plants. The crystalline structure of cellulose offers a rigid structure. In red algal cell walls, cellulose is only a minor component, varying from 1% to 8% of the dry weight of the seaweed. Marine plants require a more flexible structure to accommodate the varying stresses of tidal and wave motion. An extracellular matrix composed of mainly carrageenan gives the seaweeds the required flexible and gelatinous structure. However, the physiological significance of the composition of the intercellular matrix of red seaweeds in mechanical, hydration, and osmotic regulation in marine environments is still a matter of debate (Kloareg and Quantrano, 1988).

Attempts have been made to visualize the distribution of carrageenans over the cells, cell walls, and intercellular matrix of different red algae using different techniques, such as gold-labeled monoclonal antibodies against either κ- or ι-carrageenan or gold-labeled κ-carrageenases. The different types of carrageenan are distributed in different ways, and distributions differ from species to species. For example, in *K. alvarezii*, κ-carrageenan is located in the intercellular matrix, but not in the cell walls. In *Agardhiella subulata*, ι-carrageenan is found in both the cell walls and the intercellular matrix (Gretz et al., 1997).

6 Chemical Analysis

Chemical analysis of carrageenans is done mostly on isolated and purified carrageenan samples to reveal their (detailed) molecular structure. In the beginning, chemical modification and degradation methods were time-consuming and tedious analytical techniques. In the mid-1970s, a real boost was given by the introduction of nuclear magnetic resonance (NMR) spectroscopy, which was followed by introduction of reductive hydrolysis to preserve the 3,6-anhydro bridge. The most important analytical techniques for carrageenan determination will be discussed below. At present, there is still lack of adequate analytical techniques to determine the amounts, the polydispersity, and the purity of carrageenans in raw materials and food products. The different techniques and approaches used for this type of analysis, e.g., colorimetry, immunoassays, combined with high-performance chromatography (HPLC), and electrophoresis (Roberts and Quemener, 1999), will not be discussed in this section.

6.1 Isolation and Fractionation

6.1.1 Isolation

Commercial extraction of carrageenans is normally performed under strong alkaline conditions at elevated temperatures (see Section 9.3). However, in order to obtain specific carrageenan fractions or carrageenan with a high content of precursor units for analytical and research purposes, several mild extraction methods have been developed. Hot water extraction is probably the oldest and most applied method for the extraction of carrageenan from seaweed. Prior to carrageenan extraction, seaweeds can be extracted with organic solvents, such as acetone, alcohol, or diethyl ether, to remove undesired compounds. By choosing the appropriate extraction time and temperature, different carrageenan fractions are obtained. Salts, such as sodium chloride or sodium bicarbonate, are added to the extraction medium to fine-tune the extraction procedure. Amyloglucosidase can be added to a cooled extraction mixture to digest any floridean starch present. Dialysis may be applied to remove low-molecular-mass impurities, such as salts and/or sugars.

Centrifugation or filtration is applied to remove the undissolved cell-wall material and to obtain a clear solution. As in industrial processing, carrageenans are obtained by precipitation with ethanol or isopropanol followed by a drying step. Lyophilization is an elegant method to obtain carrageenans on a small scale. Dialysis or ion-exchange resins are applied to obtain carrageenans with specific counter ions.

6.1.2 Fractionation

A widely used separation technique is the KCl fractionation for the separation of κ-carrageenan from other types of carrageenans (Smith and Cook, 1953). The method is based on the unique property of κ-carrageenan helices, which cluster together in the presence of potassium ions forming gel structures. Both ι- carrageenan and λ-carrageenan do not have this gel-forming mechanism. The KCl fractionation was developed to separate κ- and λ-carrageenan from a hot water extract of *Chondrus crispus*. The addition of potassium ions to a sufficiently diluted carrageenan solution permits a sharp separation into a κ-carrageenan containing gel phase and a solution containing λ-carrageenan. This technique also has been applied to distinguish mixtures of κ- and ι-carrageenan from κ/ι-hybrid carrageenans (van deVelde et al., 2001).

The leaching method is a modified KCl-fractionation method for the extraction of λ-carrageenan (Stancioff and Stanley, 1969). Dried, powdered crude carrageenan extracts are stirred in a KCl solution, in which the κ-carrageenan fraction remains in swollen gel particles but the λ-carrageenan fraction leaches out. The use of stepwise, increasing KCl concentrations in both methods leads to a series of carrageenan fractions, precipitating at different salt concentrations.

6.1.3 Separation of Low-molecular-mass Fractions

Low-molecular-mass fractions of carrageenans are obtained either by unspecific acid hydrolysis or by specific enzymatic hydrolysis. Enzyme hydrolysates are important for the determination of carrageenan fine structures (Knutsen and Grasdalen, 1992). The first step is the ethanol precipitation of the higher molecular mass or enzyme-resistant fractions (ERF). In a second step, different fractions are obtained by gel filtration and reversed-phase HPLC. Improvements in the performance of chromatographic materials resulted in the development of a rapid analysis method that allowed the separation of oligosaccharides with a degree of polymerization (DP) of 2 to approximately 12 in about 20 min (Knutsen et al. 2001).

6.2 Infrared Spectroscopy

Infrared (IR) spectroscopy has been one of the most frequently used technologies. Infrared absorption results from the stretching and bending vibrations of molecular bonds and thus reflects the molecular structure of the material under study. An overview of the absorption bands of the carrageenan structural elements is given in Table 1. The broad

Tab. 1 Identification of carrageenan types by infrared spectroscopy

Vibrations	*Bond(s)/Group(s)*	*Carrageenan dimeric units*				
Wavenumbers **(cm^{-1})[a]**		ι	κ	λ	ν	μ
1240	S=O of sulfate esters	+	+	+	+	+
930	C-O of 3,6-anhydrogalacote (DA)	+	+	−	−	−
845	C-O-S on C-4 of G4S	+	+	−	+	+
830	C-O-S on C-2 of D2S,6S	−	−	+	+	−
820	C-O-S on C-6 of D2S, or D6s	−	−	+	+	+
805	C-O-S on C-2 of DA2S	+	−	−	−	

[a] Wavenumbers from Chopin and Whalen (1993) and Stancioff and Stanley (1969) and letter codes from Knutsen et al. (1994).

band at 1240 cm^{-1} is common for all sulfated polysaccharides, as it results for the S=O vibration of the sulfate groups. This band correlates with the sulfate content of the sample. Another important absorption band is the 903 cm^{-1} band, which corresponds with the presence of 3,6-anhydrogalactose bridges in the sample. The 805 cm^{-1} band is restricted to ι-carrageenan and is used to determine the ι-to-κ ratio in κ/ι-hybrid carrageenans.

Infrared spectroscopy has been used to study the composition of carrageenan samples from different origins as well as their fractions obtained by precipitation and fractionation (Correa-Diaz et al., 1990; Stancioff and Stanley 1969). This technique has been limited by the fact that only soluble carrageenans could be measured. The technique is broadened to dried, ground algal material by the introduction of FT IR diffuse reflectance spectroscopy (Chopin and Whalen, 1993). Because its accuracy is low, IR spectroscopy normally is used to quantify the different carrageenan types present in a sample without quantifying their amounts.

6.3
Nuclear Magnetic Resonance Spectroscopy

Today, NMR spectroscopy (both ^{1}H- and ^{13}C-NMR) is one of the standard tools for the determination of the chemical structure of carrageenan samples (Usov, 1998). Prior to recording the spectra, the carrageenan samples are sonicated to reduce the viscosity of the solution (high viscosity results in line broadening). Because of the low natural abundance of the ^{13}C isotope, samples for ^{13}C-NMR are prepared at relatively high concentrations (5% to 10% w/w) compared with ^{1}H-NMR samples (0.5% to 1.0% w/w).

^{13}C-NMR spectroscopy is used mostly to reveal the molecular composition of carrageenan samples, e.g., the proportion of κ-, ι-, λ-, μ-, and ν-repeating units. Quantification of ^{13}C-NMR spectra, based on the intensities of the resonances of the anomeric carbons, is feasible, provided that sufficiently long interpulse delays are used (above 1.5 s is appropriate for the most important carrageenans) (van deVelde et al., 2001). The major drawback of ^{13}C-NMR spectroscopy is its low sensitivity, which results in a lack of information about minor components (components present in amounts below 5%).

The ^{13}C-NMR spectra of κ-, ι-, μ-, and ν-carrageenan have been fully elucidated for many years and used for the determination of carrageenan samples and fractions (Ciancia et al., 1993; Usov et al., 1980). The ^{13}C-NMR spectrum of λ-carrageenan has been a matter of debate; the calculated chemical shifts of λ-carrageenan, together with all

Tab. 2 ^{13}C-NMR spectral assignment of different carrageenans

Carrageenan	*Unit*	*Chemical shift (ppm)*[a]					
		C-1	*C-2*	*C-3*	*C-4*	*C-5*	*C-6*
ι	G4S	103.1	69.9	77.3	72.5	75.3	61.8
	DA2S	92.4	75.3	78.7	78.9	77.5	70.3
κ	G4S	103.1	70.2	79.1	74.4	75.3	61.8
	DA	95.5	70.2	79.7	78.9	77.3	70.0
λ	G2S	103.9	77.4	75.8	64.6	74.2	61.7
	D2S,6S	92.0	74.8	69.5	80.3	68.7	68.1
ν	G4S	105.3	70.8	80.4	74.4	75.3	61.8
	D2S,6S	98.7	76.8	68.7	79.7	68.7	68.3
μ	G4S	105.3	70.8	78.9	74.4	75.3	61.8
	D6S	98.4	69.1	70.9	79.7	68.7	68.3

[a] Chemical shifts from Ciancia et al. (1993), Falshow and Furneaux (1994) and Stortz et al. (1994) and letter codes from Knutzsen et al. (1994).

other possible carrageenan structures, were published in 1992 (Stortz and Cerezo, 1992). Two years later, the chemical shifts of λ-carrageenan were published; however, some of the observed resonances were not assigned (Stortz et al., 1994). Table 2 gives an overview of the chemical-shift data of ^{13}C-NMR spectra of the major carrageenan types. Pyruvate acetal substitution of the carrageenan backbone results in the appearance of resonances at 25.5 ppm (methyl), 101.6 ppm (acetal), and 175.7 ppm (carboxyl) (Chiovitti et al., 1999). Floridean starch is a reserve polysaccharide in red seaweeds and is identified in carrageenan samples by resonances with chemical shifts of 72.9 and 102.5 ppm (Turquois et al., 1996). Commercial carrageenan samples are often blended with sucrose. This limits the composition determination, as the resonance of the C-1 carbon of α-D-glucose overlaps with that of C-1 of ι-carrageenan (DA2S at 92.4 ppm) (Turquois et al., 1996). Small molecules, such as sugar, can be removed by dialysis prior to NMR analysis.

^{1}H-NMR spectroscopy is much more sensitive than ^{13}C-NMR spectroscopy but shows

Tab. 3 ^{1}H-NMR chemical shifts of the α-anomeric protons of carrageenans

Carregeenan	*Unit*	*Chemical shift (ppm)*[a]
ι	DA2S	5.32
κ	DA	5.11
λ	D2S,6S	5.59
ν	D2S,6S	5.52
μ	D6S	5.26

[a] Chemical shifts from Chiancia et al. (1993) and Stortz et al. (1994) and letter codes from Knutsen et al. (1994).

significantly less chemical shift dispersion. Therefore, ^{1}H-NMR is mainly used for two different purposes. For the first purpose, the high sensitivity is used for a fast analysis of low concentrated samples. The quantification of the different carrageenan types in a sample is based on the fact that the α-anomeric protons give five resolved signals in the region from 5.1–5.7 ppm (see Table 3). The signals for the β-anomeric protons cannot be used for either identification or quantification of the diads. ^{1}H-NMR spectroscopy can in this way be used for routine analysis of commercial carrageenan samples, which are often blended with sugars and salts, for their total carrageenan content and molar ratio of the different carrageenan types.

In the second approach, oligosaccharides are prepared by enzymatic hydrolysis of the carrageenan samples by either κ- or ι-specific carrageenases (see Section 8.2). ^{1}H-NMR spectroscopy is used for its high sensitivity and possibility to resolve the detailed molecular structure of these carrageenan oligosaccharides. The higher- and lower-molecular-mass fractions were separated by precipitation methods and further fractionated by gel permeation chromatography (GPC) (see Section 6.1). The obtained fractions are then investigated by high field ^{1}H-NMR spectroscopy. The detailed information about the molecular structure of the different di-, tetra-, and oligosaccharides in the different fractions is used to deduce the structure of the original carrageenan sample (Knutsen, 1992; Knutsen and Grasdalen, 1992).

6.4 Chromatographic Analysis

Chromatographic techniques are applied to study different structural aspects of carrageenans, such as molecular mass distribution, sulfate ester content, monosaccharide composition, and glycosidic linkage analysis.

6.4.1 Molecular Mass Determination

The molecular mass of carrageenans is an important parameter of their rheological behavior (e.g., yield stress and gel strength) and, consequently, their application. Molecular mass distributions are normally determined by size exclusion chromatography (SEC) or GPC with multi-angle laser-light scattering (MALLS) detection (Hoffmann et al., 1996). The chromatographic system separates the carrageenans according to their hydrodynamic volume, giving the molecular mass distribution, whereas the light-scattering detector reveals the absolute molecular mass of each individual peak. However, several problems are associated with SEC or GPC, such as adsorption of the sample onto the column packing, degradation of the polymers due to the high shear forces within the column, and difficulties with separation of the sample with a high-molecular-mass or aggregated material, resulting from the fact that only dissolved material can be analyzed. The above problems are less severe in asymmetrical flow field-flow fractionation. Recently, field-flow fractionation with MALLS detection was used successfully for the molecular mass determination of κ-carrageenans (Viebke and Williams, 2000).

6.4.2 Sulfate Content

Sulfate groups are such an important structural substituent of the galactose backbone of carrageenans that a rapid and reproducible method for sulfate ester content is important. The officially used method for determining the sulfate content in carrageenans is based on the selective hydrolysis of the sulfate ester by acid and subsequent selective precipitation of the sulfate ions as barium sulfate. Barium sulfate is then measured by weighing or turbidimetry as described by the FAO/WHO Joint Expert Committee on Food Additives and most recently updated in 1992 (FAO, 1992). This method has disadvantages: it is laborious and requires gram quantities of sample. A promising development in this area is the use of HPLC to measure free sulfate that is liberated from carrageenans by acid hydrolysis. Typically, this analysis is performed on an HPLC system using an anion-exchange column with conductivity detection (Jol et al., 1999).

6.4.3 Monosaccharide Composition

The monosaccharide composition of carrageenans generally is measured to determine the ratio between galactose and anhydrogalactose. Normal acid hydrolysis of polysaccharides to obtain monosaccharides results in the conversion of anhydrogalactose to either galactose or degradation products, such as 5-hydroxymethylfurfural. The introduction of the reductive hydrolysis remedies this problem. Acid hydrolysis in the presence of a methylmorpholine-borane (MMB) complex produces the alditols of monosaccharides without destroying the 3,6-anhydro moiety. Acetylation of the alditols is necessary for the analysis with gas chromatography (GC) However, this derivatizing step is time-consuming and results in loss of accuracy. Recently, high-performance anion-exchange chromatography (HPAEC) was described as an accurate method for the direct analysis of alditols, obtained by reductive hydrolysis with MMB and trifluoroacetic acid (Jol et al., 1999). This method also reveals the presence of small amounts (below 5 mol%) of glucose and xylose in commercial carrageenan samples. The reductive hydrolysis also can be applied to seaweeds samples and, thus, facilitate chemotaxonomic studies.

6.4.4 Glycosidic Linkage Analysis

The glycosidic linkages in carrageenans can be revealed in principle by traditional methylation analysis. Methylation of carrageenans prior to hydrolysis modifies the unsubstituted hydroxyl groups. The major problem encountered with methylation analysis is the incomplete methylation of hydroxyl groups of sulfated polysaccharides. This problem can be eliminated by conversion of sulfate ester groups into the triethylammonium salts (Stevenson and Furneaux, 1991). Reductive hydrolysis of the methylated carrageenans results in the formation of the corresponding monosaccharide alditols that can be analyzed directly by HPAEC or by GC after acetylation. Information is obtained about the number and distribution of unsubstituted hydroxyl groups. Removal of the sulfate ester groups prior to methylation and hydrolysis reveals the number and position of hydroxyl groups that are involved in linkages between the different monosaccharide units.

7 Biosynthesis

Very little is known about the biosynthesis of carrageenans. The synthesis of the galactan backbone is believed to take place within the Golgi bodies in the cells (Bellion and Hamer, 1981). The sulfate esterification of the galactan backbone is catalyzed by sulfate transferases that occur in the Golgi dictyosomes, not in the cell wall (Craigie, 1990). The mechanisms and enzymes responsible for the introduction of other substituents, such as pyruvate and methoxy groups, await elucidation.

The (biological) precursors of the gel forming κ- and ι-carrageenan are called μ- and ν-carrageenan, respectively, and lack the 3,6-anhydro bridge. The formation of the 3,6-anhydro bridges is catalyzed by sulfohydrolases that are thought to act in the cell wall (Wong and Craigie, 1978). For ι-carrageenan the relationship between precursor and product is clear, but for κ-carrageenan this remains ambiguous because several κ/ι-hybrid carrageenan samples contain only ν-carrageenan as biological precursor (Amimi et al., 2001; VandeVelde et al., 2001). Bellion et al. (1983) report the presence of ν-carrageenan in samples of κ-carrageenan from *K. alvarazii*, known for its production of almost pure κ-carrageenan. Suggestions are made that ν-carrageenan is the common precursor for both κ- and ι-carrageenan. Further research is necessary to elucidate the different stages in the biosynthesis of the different carrageenans.

Recently, two sulfohydrolases from *Chondrus crispus* were purified to electrophoretic homogeneity and cloned (De Ruiter et al., 2000; Genicot et al., 2000). The two sulfohydrolases, sulfohydrolase I and sulfohydrolase II, exhibit the same substrate specificity and are indicated as ν-carrageenan 6-O-sulfohydrolases. Sulfohydrolase I releases sulfate without modifying the viscosity of the polymer. It is therefore likely that this enzyme acts randomly. In contrast, sulfohydrolase II is thought to remove sulfate from the ν-carrageenan processively, resulting in long uninterrupted chains of ι-carrageenan dimers. This results in a significant increase of the polysaccharide gelling behavior, indicating the formation of a helical structure.

8 Extracellular Biodegradation

Carrageenans are degraded by enzymes called carrageenases. Several marine bacteria produce κ- and/or ι-carrageenases that degrade κ- and ι-carrageenan, respectively. The structural biology and molecular evolution of carrageenases have been reviewed recently (Barbeyron et al., 2001), as well as their role in carrageenan biotechnology (De Ruiter and Rudolph, 1997).

8.1 Enzymology of Degradation

As far as we are aware, the first enzyme able to modify carrageenans was described in 1943 by the Japanese researcher Mori (1943), followed 12 years later by a Canadian group (Yaphe and Baxter, 1955). After that time, many carrageenan-modifying enzymes have been described, as listed in Table 1 of the review about carrageenan biotechnology (De Ruiter and Rudolph, 1997). A considerable number of those enzymes have been purified from marine bacteria such as *Alteromonas carrageenovora* (previously called *Pseudomonas carrageenovora*) and from different algae (e.g., *Gigartina* spp.).

The best-characterized enzyme is the κ-carrageenase isolated from *Alteromonas carrageenovora* (Bellion et al., 1982; Knutsen, 1992; Potin et al., 1995). This enzyme acts as an endo-hydrolase specifically hydrolyzing the β-glycosidic linkage between 3,6-anhydro-D-galactose and D-galactose, resulting in the formation of neocarrabiose oligosaccharide DA-G4S with 3-linked β-D-galactopyranose 4-sulfate as reducing end. The lytic mechanism and specificity were revealed by detailed analysis of the reaction products by ^{1}H-NMR (Knutsen and Grasdalen, 1992). ι-Carrageenases from *Alteromonas fortis* and *Zobellia galactanovorans* hydrolyze the β-1,4 linkages of ι-carrageenan (Barbeyron et al., 2001). The hydrolysis is a one-step nucleophilic substitution reaction and results in the inversion of the anomeric bond configuration. In contrast to κ-carrageenases (Potin et al., 1995), ι-carrageenases do not show transglycosylation activity (Barbeyron et al., 2001). To our knowledge, no enzymes have been reported that are able to cleave the glycosidic linkage of the resulting neocarrabiose dimers or that cleave the 3,6-anhydro-α-D-galactose linkage, resulting in a reducing end of 3,6-anhydro-galactose residues.

The crystallization and preliminary X-ray analysis are reported for the κ-carrageenase from *P. carrageenovora* and the ι-carrageenase from *A. fortis*. The structure of the κ-carrageenase has been resolved to 1.54 Å resolution (Michel et al., 2001). This was the first three-dimensional structure of a carrageenase. The active site of this enzyme is tunnel-shaped, indicating that the enzyme is able to degrade solid substrates.

8.2 Genetic Basis of Degradation

The gene encoding for the κ-carrageenase from *P. carrageenova* has been cloned from a genomic library of this bacterium (Barbeyron et al., 1994). Based on the amino acid sequence, the enzyme was classified in family 16 of the glycosyl hydrolases of the system designed by Dr. Bernard Henrissat (Potin et al., 1995). The family-16 glycoside hydrolases share two conserved glutamic acid residues and one aspartic acid residue in their active sites. The κ-carrageenase from *Zobellia galactanovorans*, whose gene has been cloned (Barbeyron et al., 1998a), belongs also to this family 16 of glycoside hydrolases. In contrast, the ι-carrageenases do not belong to the family 16 of glycoside hydrolases (Barbeyron et al., 2001). They are totally unrelated to the κ-carrageenases and belong to a new family of glycoside hydrolases.

9 Production

9.1 Seaweed Harvesting

The harvesting of carrageenan-containing seaweeds is labor-intensive, as the largest part is manually collected (Rudolph, 2000; Therkelsen, 1993). This is done in seaweed farming areas for seaweeds that have been thrown ashore after a storm and at low tides for *Iridaea* and *Gigartina* species. Hand- or drag-rakes are used to harvest *C. crispus*, which adheres to rocks. This raking from boats was developed in Canada to harvest fresh seaweed of a good quality. Floating seaweeds, e.g., *Furcellaria*, are easily harvested by trawling. Collection of seaweeds from the shore has its limitations. The amount that will reach the shore is limited, and seaweeds on the shore decompose fast when exposed to rain and sunlight.

9.2 Seaweed Farming

Harvesting of seaweeds from natural populations always contains the risk of overproduction and loss of the total population; therefore, much effort has been put into seaweed farming. Seaweed farming started almost 200 years ago in Japan. Fisherman stacked brush on the seashore to expand the settling grounds of *Porphyra*. Domestication of seaweed cultivars is important to remove the harvest pressure on target species in sensitive ecosystems. Scientific information about the seaweed life cycles allowed artificial seeding in the 1950s. Today, nearly a dozen seaweed taxa are cultivated commercially, which lowers the pressure on naturally occurring populations. An example of an important success in this field is in the Philippines, where Maxwell S. Doty pioneered the marine culture of *Eucheuma* species (Santelices, 1999). Seedlings of a variety that grows rapidly and that has a good resistance to diseases are tied to lines. The lines are placed at the inside of coral reefs, where the seedlings can grow without floating away. Plants are harvested when they reach a size of 0.8 to 1.2 kg (Rudolph, 2000). Today's production of *Eucheuma* and *Kappaphycus* cultivars counts for most of this country's carrageenan production without the need of harvesting natural resources.

Another approach to seaweed farming is the co-cultivation of seaweeds with intensive fish farming. Intensive fish and shrimp farming results in the release of nutrients that add to costal eutrophication. Seaweeds are successfully used as biofilters to remove these nutrients. Co-cultivation in tank systems or cultivation on ropes near a fish farm increases the seaweed production yield and reduces the amount of nutrients released by the intensive fish and shrimp farming. Studies on co-cultivation started in the mid-1970s and gained new interest at the end of the 20th century (Troell et al., 1999).

9.3 Manufacturing

Harvested seaweed is washed to remove sand and stones, followed by drying to preserve the quality of carrageenan and to reduce weight. In tropical regions, sun drying is mainly used, but in colder climates fuel-fired rotary air dryers are applied. After drying, seaweed is shipped to the production plants or warehouses where it is stored before use. For manufacturing plants that are located near the harvesting area, the use of wet seaweed is economically attractive because it reduces the drying costs.

To produce a certain carrageenan product with constant properties, different batches are combined. The quality of the final product is assured by detailed analysis of the raw material. The manufacturing processes for both refined and semi-refined carrageenans are drawn schematically in Figure 4.

In the refined carrageenan process, the seaweed is extracted under alkaline conditions. In this way, the 3,6-anhydro bridges are formed during the extraction step and a good quality of κ- and ι-carrageenan is obtained. The extraction is performed for several hours at a temperature near the boiling point of the alkaline solution

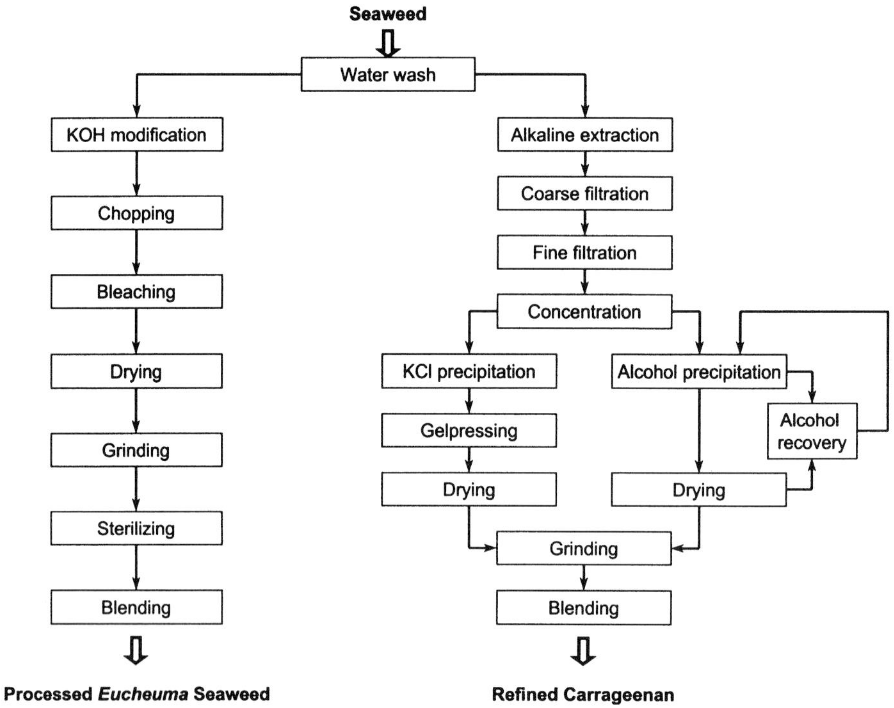

Fig. 4 Manufacturing process for processed *Eucheuma* seaweed and refined carrageenan.

(≥ 110°C). Different types of alkali can be used to manipulate the salt type of the carrageenan produced. Numerous patents describe different conditions to obtain carrageenans with special properties. For example, extraction at lower pH values or lower temperatures, resulting in an increasing amount of precursor units, is applied to obtain special carrageenan preparations (Hansen et al., 2000), which are typically used as an alternative to gelatin in specific applications.

The extract is subsequently filtered by multistage filtration. A coarse filter removes the residual seaweed material (mostly cellulose), and a fine filter is used to obtain a clear liquid. The diluted carrageenan solution is concentrated by multistage vacuum evaporators to a final carrageenan concentration of 3% to increase the yield in the precipitation step. The carrageenan is precipitated by the addition of isopropanol. The gelled fibrous material is pressed to remove the excess of solvent. The alcohol water mixture is distilled and the alcohol reused to reduce process costs. For κ-carrageenan an alternative precipitation protocol, also referred to as "gel press technology", based on the specific gelation of κ-carrageenan in the presence of potassium ions followed by removing the liquid phase, is possible. In this process the carrageenan solution is extruded into a concentrated KCl solution, in which the carrageen precipitates. The gelled material is subsequently pressed followed by a freeze/thaw cycle to remove the water.

The pressed fibers derived from both processes are dried and finally ground to an appropriate particle size. To obtain a product of consistent quality, extracted carrageenan from several batches are blended and/or mixed with salts or sugars. Different additives are introduced to obtain specific properties and to meet customer requirements.

In the manufacture of semi-refined carrageenan, the seaweed is treated with a potassium hydroxide solution at elevated temperature (70°C to 80°C). This temperature is high enough to catalyze the formation of the 3,6-anhydro bridges in both κ- and ι-carrageenan, but is not high enough to allow the carrageenan to be extracted. Most of the low-molecular-mass material is removed by the alkaline treatment. The modified seaweed is chopped and bleached to reduce the color of the final product. After washing, the material is dried and ground to the required particle size. Before blending and standardization, this product is sometimes sterilized to improve the microbiological quality of the final product.

9.4 Modified Carrageenan Functionalities

A continuous challenge for manufacturers of carrageenans is to make new, high value-added products and to widen the applications of carrageenan. An attractive market opportunity occurred in the 1990s when, as a result of "mad cow disease", the food industry started a search for gelatin alternatives for food applications. Although carrageenan cannot replace gelatin in all applications, some very good gelatin mimics are on the market based on ι-carrageenan for use in confectionery, e.g., in wine gums.

A promising development in this area is the production of engineered carrageenans with the help of polysaccharide modifying enzymes. In collaboration with CP Kelco, Dr. Bernard Kloareg and coworkers (CNRS Station Biologique, Roscoff, France) cloned two sulfohydrolases from *Chondrus crispus* (De Ruiter et al., 2000). These enzymes provide a new tool to produce ν/ι-hybrid carrageenans with special properties, which can be tailor-made for application in food industry.

Sulfated oligogalactans are recognized as valuable, active biocompounds for non-food applications, such as agriculture, cosmetics, or pharmaceuticals. For this reason, a variety of carrageenan modifying enzymes have been purified and cloned. Several carrageenases have been developed in Dr. Kloareg's lab for the selective hydrolysis of κ- or ι-carrageenan (Barbeyron et al., 2001). These enzymes can be used to produce carrageenan oligosaccharides with specific structures and properties. Moreover, the κ-carrageenase from *Pseudoalteromona carrageenovora* was shown to display endo-transglycosylation activity (Potin et al., 1995). This activity may be used to transfer various substituents onto κ-carrageenan oligosaccharides. The patent on this group of enzymes belongs to Goëmar Laboratories (Barbeyron et al., 1998b).

9.5 Current World Market

Carrageenan is the third most important hydrocolloid in the food area worldwide, after gelatin and starch. The total world production of seaweeds for carrageenan extraction is currently around 140 million kg dry weight ("as sold"), giving approximately 26 million kg of carrageenans with an estimated value of 310 million USD in 2000. The total carrageenan market showed a grow rate of 3% per year during the last few years.

9.6 Companies Producing Carrageenan

Today, six companies manufacture over 80% of the supply of carrageenans. Two types of companies can be distinguished: the traditional biopolymer-producing companies and the total ingredient suppliers, which manufacture their own carrageenan for the use in blends. The biopolymer producing companies include:

- FMC Corporation (USA; the former Marine Colloids Inc. known from several innovative patents in the sixties; www.fmcbiopolymer.com);
- CP Kelco (USA; formed in 2000 by merging the Copenhagen Pectin/Food Gums division of Hercules Incorporated and the Kelco Biopolymers group of Monsanto/Pharmacia Corporation; www.cpkelco.com); and
- Ceamsa (Spain; www.ceamsa.com).

The worldwide operating companies that, besides carrageenans, also manufacture a complete range of food ingredients, such as emulsifiers, flavors and so forth, include:

- Degussa (Germany; formerly SKW Biosystems; www.degussa.com or www.texturantsystems.com);
- Danisco (Denmark; www.danisco.com); and
- Quest International (Netherlands; part of ICI; www.questintl.com).

Because of the price erosion of carrageenan during the last 10 years, the use of carrageenan in blends increased at the expense of the traditional biopolymer companies.

10 Properties

10.1 Physical Properties

10.1.1 Solubility

Carrageenans are water-soluble, but their solubility depends on the type of carrageenan, temperature, pH, type, and concentration of counter ions and other solutes present. The most soluble carrageenan is λ-carrageenan, which lacks the more hydrophobic 3,6-anhydrogalactose and has three hydrophilic sulfate ester groups. λ-Carrageenan is readily water-soluble under most conditions. κ-Carrageenan is the most difficult carrageenan to dissolve. The sodium salt of κ-carrageenan is soluble in cold water, whereas the potassium salt of κ-carrageenan is soluble only in hot solutions. ι-Carrageenan has an intermediate solubility. In many practical systems, dispersions of carrageenans are used to add carrageenan to the process.

10.1.2 Coil-helix Transitions

κ-Carrageenan and ι-carrageenan can occur in two forms: the unstructured random coil conformation typically at elevated temperatures and the structured double helices, which are usually formed upon cooling. By cooling hot solutions of the carrageenans, the double helices are formed from the random coils at a certain temperature, called coil-helix transition temperature. This coil-helix transition temperature is an important parameter for the functional properties of carrageenans and depends heavily on intrinsic factors of the carrageenans, such as polydispersity, distribution of sulfate ester groups, amount of precursor units, and extrinsic conditions such as concentration and type of salts, pH, and cooling speed.

At temperatures above 60°C, κ-carrageenan is always in the random coil conformation, and, depending on the detailed chemical structure of the κ-carrageenan and the system, a typical coil-helix transition temperature in water and milk is approximately 38°C. In the presence of potassium ions, the double helices of κ-carrageenan cluster together into bigger units that are called "fibers" (Snoeren and Payens, 1976), "helix aggregates" (Morris et al., 1980), "superstrands" (Hermansson et al., 1991), or "superhelical rods" (Piculell, 1995). Electron microscopy revealed that the maximum size of the aggregates is about 20 nm (Hermansson et al., 1991). Potassium acts as intramolecular glue, forming electrostatic interactions with the sulfate esters and anhydrooxygen atom of κ-carrageenan (Figure 5) (Tako and Nakamura, 1986). The level of aggregation heavily determines the physical properties such as gel strength, syneresis, etc.

ι-Carrageenan also is always in the random coil conformation at temperatures above 60°C. Upon cooling, helices are formed, but these are very different from those of κ-carrageenan and are independent of the type of cation that is present. As far as is known in the literature, ι-carrageenan does not form big aggregates, as does κ-carrageenan, but under some conditions some aggregation of the helices has been observed (Morris et al., 1980). The coil-helix transition temperature of ι-carrageenan is also very dependent on intrinsic and extrinsic factors and in typical water or milk systems is approximately 45°C. The single helices of ι-carrageenan can be bridged by intermolecular calcium (Figure 6) (Tako et al., 1987).

λ-Carrageenan does not form helices and always is present in the random coil conformation. Therefore, λ-carrageenan cannot form gels.

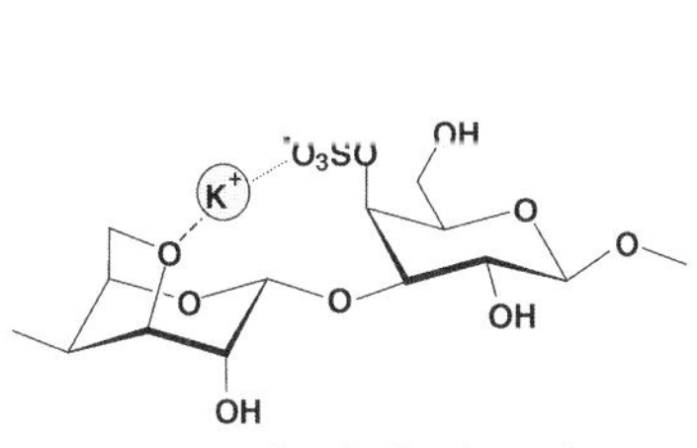

Fig. 5 Intramolecular binding of potassium ions to κ-carrageenan according to Tako and Nakamura (1986).

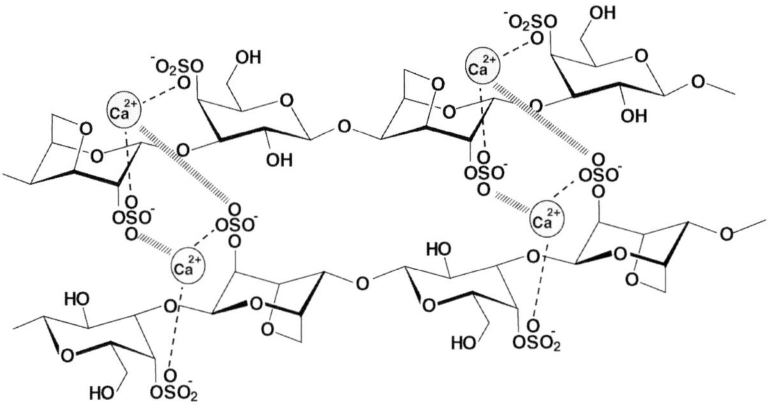

Fig. 6 Schematic representation of the intermolecular Ca^{2+}-bridges between ι-carrageenan helices as proposed by Tako et al. (1987).

10.1.3
Viscosity

Carrageenans above their coil-helix transition temperature typically form highly viscous solutions. In this situation, carrageenan molecules are random coils and are often highly extended because of the electrostatic repulsion of the negatively charged sulfate ester groups along the linear galactan chain. The viscosity increases almost exponentially with concentration. Addition of salts to a carrageenan solution reduces the viscosity because of shielding of the charges on the polymer. Viscosity increases with the molecular mass of the carrageenan, following the so-called Mark-Houwink equation,

$$[\eta] = K\,(M_w)^a \qquad (1)$$

where $[\eta]$ is the intrinsic viscosity and M_w is the weight-average molecular mass (strictly speaking, the viscosity-average molecular mass). The values for the constants K and a are dependent of the carrageenan type. The exponent a is close to unity, which indicates rigid, rod-like molecules.

10.1.4
Gelation

Gel formation is the most important feature of carrageenans. Only κ- and ι-carrageenan are able to form a gel, as the other carrageenans lack the essential $^1C_4$-conformation that results from the 3,6-anhydrobridge (see Section 3.2). Gels are formed upon cooling of a hot carrageenan solution. The gelling behavior of carrageenans is described in general in several textbooks (Guiseley et al., 1980; Imeson, 2000; Therkelsen, 1993) and in detail in review articles (Lahaye, 2001; Piculell, 1995).

In general terms, κ-carrageenan gels are hard, strong, and brittle gels that are freeze/thaw instable, whereas ι-carrageenan forms soft and weak gels that are freeze/thaw stable. The strength of both types of gels is controlled by the concentration of gelling cations. Both types of gels are thermoreversible, which means that the gels will melt when heated and form a gel again upon cooling. ι-Carrageenan forms gels in the presence of calcium, and the strength is related to the amount of calcium. The gelation of κ-carrageenan is specifically promoted by some monovalent cations

(K^+, Rb^+, Cs^+, and $NH4^+$). These cations promote the aggregation of κ-carrageenan double helices to form so-called aggregated "domains" or "superhelical rods" (see Section 10.1.2). This association of helical strains explains the hysteresis observed in optical rotation and rheological measurements of κ-carrageenan. The coil-to-helix transition of ι-carrageenan appears not to be cation-specific. The thermal conformational transition of ι-carrageenan shows no hysteresis, which suggests that there is little or no interhelical aggregation. Therefore, the gel forming in ι-carrageenan gels is assumed to take place at the helical level. The ι-carrageenan gels are thixotropic, which means that a gel, whose structure has been destroyed, will reform if left for an appropriate period of time without disturbance.

10.1.5 Synergism with Gums

κ-Carrageenans show synergism with galactomannan gums, especially with locust bean gum, less with konjac gum and guar gum. A synergistic effect means that the minimum gelling polymer concentration can be reduced even though locust bean gum is non-gelling alone. The mechanism of the interaction is assumed to proceed via alignment of the non-branched mannan backbone regions of the galactomannans with the aggregates of the double helices of κ-carrageenan (Dea and Morrison, 1975; Harding et al., 1995). Addition of galactomannans to κ-carrageenan gels changes the morphology from brittle, rigid gels to strong elastic gels with low syneresis.

Some starches are known to have synergism with ι-carrageenan; however, the detailed mechanism has not been revealed so far.

10.1.6 Interaction with Proteins

In the dairy industry, carrageenans are used for their specific synergistic interactions with milk proteins. In milk systems, five times lower carrageenan concentrations are necessary to obtain a gel compared to water systems.

In milk systems, carrageenan concentrations one-fifth of those required for water systems are necessary to obtain a gel. The synergistic effect is related to the interactions of the negatively charged carrageenans with the positively charged amino acids at the surface of casein micelles, especially between κ-carrageenan and κ-casein. Very low levels of 0.015% to 0.025% w/w of carrageenan are sufficient to prevent whey separation and to stabilize cocoa particles in chocolate milk.

10.2 Chemical Properties

Carrageenans are very sensitive to acid and oxidative breakdown. Cleavage of the glycosidic linkages increases with temperature and time. Acid-catalyzed hydrolysis occurs mainly at the $(1 \rightarrow 3)$ glycosidic linkage. The 3,6-anhydrogalactose-ring system favors the hydrolysis, whereas the presence of a sulfate ester group at the 2-position reduces hydrolysis (Therkelsen, 1993). The acid hydrolysis of κ-carrageenan is faster than that of ι-carrageenan. Depolymerization of carrageenans also could occur by acidification of the sulfate groups (releasing sulfuric acid) as a result of exhaustive dialysis against deionized water. Autohydrolysis also could occur during a lyophilization step normally following the dialysis or when carrageenans are stored for a long time at dry conditions. Carrageenans in the gelled state retain their associated cations and are protected from autohydrolysis (Hoffmann et al., 1996).

10.3 Safety

Seaweed products in general and carrageenans in particular have been used as food ingredient for centuries. Food-grade carrageenans, with a typical weight-average molecular mass above 100 kDa, have been demonstrated to be completely safe and nontoxic (see also Section 11.6) and therefore can be used in food in unlimited amounts. Food-grade carrageenan is not absorbed and there is no evidence that carrageenan is ulcerogenic in humans (Guiseley et al., 1980; Weiner 1991).

11 Applications

Overviews of carrageenan applications are given in almost every textbook dealing with carrageenans (Guiseley et al., 1980; Imeson, 2000; Rudolph, 2000; Therkelsen, 1993). The most important fields of carrageenan application are listed below.

11.1 Technical Applications

Water-based paints and inks can be thickened and stabilized with κ- or ι-carrageenan (0.15% to 0.25% w/w) to prevent settling of the pigment particles. Addition of carrageenan also improves the flow behavior of the paint. ι-Carrageenan (0.25% to 0.8% w/w) is used to suspend and stabilize insolutes in abrasive suspensions and ceramic glazes.

Beverage clarification is performed with all types of carrageenans, κ-carrageenan being the preferred one. Carrageenan added to beer or wine serves as a fining agent by complexation with proteins. Filtration or centrifugation removes insoluble particles. Semi-refined carrageenan or bleached and ground seaweed is preferred over refined carrageenans.

In oil-well drilling, carrageenans and other polysaccharides are used to increase the viscosity of drilling fluid. Increased viscosity enhances the carrying capacity of the fluid.

11.2 Medical Applications

In medical research, carrageenans are used in the so-called rat-paw edema assay for the screening of new drugs (Renn, 1997). Administration of carrageenan fractions directly into the blood system causes a systemic effect on the immune system. Carrageenan injections are used in animal models of inflammation. The rat-paw edema assay has led to the development of several anti-inflammatory drugs.

Today, carrageenans are recognized as agents that can prevent or inhibit development of sexually transferred viral infections (Neushul, 1990; Schaeffer and Krylov, 2000). Thus, carrageenans are considered promising new anti-HIV drugs. The mode of action of carrageenans (and other sulfated polysaccharides) is believed to be inhibition of the virus reverse transcriptase (Nakashima et al., 1987). Seaweed fucans are more active than carrageenans because of their higher degree of sulfatation and, therefore, are preferred for application in drugs.

11.3 Excipient Applications in Drugs

In the pharmaceutical industry, carrageenans are used as excipients because of their physical properties. ι-Carrageenan (0.1% to 0.5% w/w) is used to stabilize emulsions and suspensions of mineral oil and insoluble drug preparation. Complexes between carrageenan and drug molecules can be used for controlled release of pharmaceutical compounds. In tablet manufacture, carrageenan is used as ingredient as well as coating agent.

11.4 Personal Care and Household

One of the largest non-food applications of carrageenan is as viscosifier in high-quality toothpastes. Modern toothpaste is a complex mixture of inorganic and organic compounds suspended in a continuous phase, stabilized by a hydrocolloid. ι-Carrageenan (0.8% to 1.2% w/w) is added to prevent separation of the liquid phase and the abrasive. Addition of carrageenan offers texture, good stand-up, and thixotropic flow. The functional properties of carrageenan are influenced by interactions with the other ingredients, such as pigments, calcium carbonate, dicalcium phosphate, and silica. As carrageenans cannot be degraded easily by enzymes, the use of carrageenan relative to cellulose derivatives, such as carboxymethylcellulose (CMC) improves the stability of the paste during storage and use in areas where high temperature and humidity prevail.

λ-Carrageenan (0.1% to 1.0% w/w) is used in hand lotions and creams to provide slip and improve rubout. Both natural and low-molecular-mass carrageenans are used for the stabilization and thickening of shampoos.

Air-freshener gels are prepared from κ-carrageenans, combined with other gums and gelling salts (up to 2.5% w/w of gums). Fragrance oils and odor-absorbing compounds that are incorporated within the gel release uniformly from the gel surface as the gel dries down.

11.5 Agriculture

Carrageenan oligosaccharides, prepared by either acid hydrolysis or enzymatic degradation, are marketed by Goëmar Laboratories (Saint-Malo, France) as fertilizers and growth biostimulants. When applied at the instant of flowering, these compounds stimulate the nutrition and reproduction of numerous crops, resulting in better flower fertilization and fruit formation. Sulfated oligosaccharides are recognized to stimulate the defense mechanism of plants (Bouarab et al., 1999; Potin et al., 1999). Therefore, oligocarrageenans can be used as a natural growth-enhancing agent.

11.6 Food Application

In the U.S., carrageenan is generally recognized as safe (GRAS) by experts of the Food and Drug Administration (21 CFR 182.7255) and is approved as a food additive (21 CFR 172.620). In the EU, a difference is made between carrageenan and processed *Eucheuma* seaweed. They are approved as E407 and E407a, respectively, in the list of permitted emulsifiers, stabilizers, and thickening and gelling agents. The World Health Organization Joint Expert Committee of Food Additives has concluded that it is not necessary to specify an acceptable daily intake limit for carrageenan. These registrations made carrageenans applicable in food products. The use of carrageenan in food applications has recently been reviewed in the Handbook of Hydrocolloids (Imeson, 2000). The most important food applications of carrageenan are summarized in Table 4.

Tab. 4 Typical food applications for carrageenan[a]

Use	*Carrageenan type*	*Function*	*Use level (% w/w)*
Water dessert gels	Kappa + iota	Gelation	0.6–0.9
	Kappa + iota + LBG[b]		
Low-calorie gels	Kappa + iota	Gelation	0.5–1.0
Cooked flans	Kappa, kappa + iota	Gelation, mouthfeel	0.2–0.3
Cold prepared custards	Kappa, iota, lambda	Thicekning, gelation	0.2–0.3
Instant breakfasts	Lambda	Suspension, bodying	0.1–0.2
Chocolate milk	Kappa, lambda	Suspension, mouthfeel	0.01–0.03
Filled milk	Iota, lambda	Emulsifying	0.05
Dairy creams	Kappa, iota	Stabilization of the emulsion	0.01–0.05
Ice cream	Kappa + GG, LBG, XG	Whey prevention, emulsion stabilization, control meltdown	0.01–0.02
Whipped cream	Lambda	Stabilize overrun	0.5–0.15
Soy milk	Kappa + iota	Suspension, mouthfeel	0.02–0.04
Processed cheese	Kappa	Improve slicing, control melting	0.05–3.0
Canned and processed meats	Kappa	Moisture retention, slicing properties	0.6
Salad dressings	Iota	Stabilization of suspended herbs	0.3
Sauces	Kappa	Bodying	0.2–0.5
Pie fillings	Kappa	Reduced starch, lower burn-on	0.1–0.2
Tart glazing	Kappa + LBG	Gelation	0.7–1.0
Syrups	Kappa, lambda	Suspension, bodying	0.3–0.5
Beer or wine fining	Kappa	Complexation of proteins	?
Wine gums	Iota	Creating structure	?

[a] Data from (Guiseley et al. (1980), Imeson (2000), and Therkelsen (1993).[b] LBG = locust bean gum; GG = guar gum; XG = xanthan gum.

11.7 Other Applications

Chibata and coworkers pioneered the use of carrageenans for the immobilization of enzymes and microorganism (Chibata et al., 1979). Especially κ-carrageenan is used for the immobilization of microorganisms, plant cells, and enzymes (Iborra et al., 1997). For this application, the potassium-induced gelation of κ-carrageenan is used. Cells or enzymes are mixed with an aqueous solution of carrageenan, and beads are prepared by dripping this solution in a (cold) 0.3 M KCl solution, in which the κ-carrageenan solution immediately forms a gel. When the biomaterial is thermostable, the thermal reversibility of carrageenan can be used to prepare gels by cooling a hot solution. In this way, various shapes, e.g., cubes, beads, or membranes, of immobilized biocatalysts can be tailored for a particular application.

λ-Carrageenan is used as a chiral selector in capillary electrophoresis separations (Beck et al., 2000). With λ-carrageenan as the chiral selector, various racemic beta-blockers and tryptophan derivatives are successfully separated into their enantiomers.

12 Relevant Patents

Numerous patents describe the extraction/processing procedures for carrageenan manufacture as well as the specific application of carrageenans in particular areas. The patent of Goëmar Laboratories (France) describes the production of glycosylhydrolases, such as κ- and ι-carrageenase, for the biodegradation of carrageenans (Barbeyron et al., 1998b). The patent of CP Kelco (USA) describes the production of two sulfohydrolases from *Chondrus crispus* for the enzyme-catalyzed gelation of ν/ι-mixed carrageenans (De Ruiter et al., 2000). In addition, there is the development of a mild alkaline extraction process to prepare ι-carrageenan that contains a small fraction of ν-repeating units (around 5%) to obtain a ι-carrageenan preparation with enhanced functional properties (Hansen et al., 2000).

13 Current Problems and Limitations

Carrageenans have been used as healthy and natural products in food for centuries. However, reports of cecal and colonal ulceration induced by a highly degraded carrageenan resulted in a negative image of carrageenans. Unfortunately, the carrageenans used in these studies were significantly chemically modified compared to the regular carrageenan polymers used in food for so long; therefore, these results, scientifically speaking, cannot be extrapolated to food carrageenans. Intensive investigation into carrageenan's safety by the FDA resulted in the GRAS-status for food-grade carrageenan in 1976. Unfortunately, this could not remove the public opinion that carrageenan consumption has a negative association.

Further development of carrageenan technology is hampered by the fact that the whole industry has a slight overcapacity, which during the last 10 years resulted in price erosion. This limits the willingness of the traditional biopolymer industry to invest in research and development. While the total annual sales of pectin and carrageenan are almost equal, the number of patents and scientific articles on pectin is estimated to be at least a factor of 10 higher than on carrageenan.

14
Outlook and Perspectives

In order to improve process efficiency, a further increase in the production of semi-refined carrageenan (such as processed *Eucheuma* seaweed) in the country of harvesting will take place. Production in the country of harvesting reduces the transportation costs and the loss of material due to degradation on storage and drying. Development of better technologies for the drying and storage of carrageenan will increase the seaweed quality and will reduce the need for raw materials (Kapraun, 1999).

Because the exploitation of natural populations results in a decrease of resources, the domestication of seaweed cultivars is of greatest importance. The marine culture of commercial seaweed species will result in less environmental impact. The integration of seaweed production with other processes will become more important for the cultivation of seaweeds for carrageenan production. Co-cultivation of seaweeds with intensive marine culture system, e.g., salmon farming, reduces both the nutrient effluent of the salmon farm and the harvest of natural seaweed populations. In an integrated approach of carrageenan production and agriculture, the seaweed residues can be used as fertilizer.

15 References

Amimi, A., Mouradi, A., Givernaud, T., Chiadmi, N., Lahaye, M. (2001) Structural analysis of *Gigartina pistillata* carrageenans (Gigartinaceae, Rhodophyta), *Carbohydr. Res.* **333**, 271–279.

Barbeyron, T., Henrissat, B., Kloareg, B. (1994) The gene encoding the kappa-carrageenase of Alteromonas carrageenovora is related to beta-1,3-1,4-glucanases, *Gene* **139**, 105–109.

Barbeyron, T., Gerard, A., Potin, P., Henrissat, B., Kloareg, B. (1998a) The kappa-carrageenase of the marine bacterium *Cytophaga drobachiensis*. Structural and phylogenetic relationships within family-16 glycoside hydrolases, *Mol. Biol. Evol.* **15**, 528–537.

Barbeyron, T., Henrissat, B., Kloareg, B., Potin, P., Richard, C., Yvin, J. C. (1998b) Glycosylhydrolase genes and their use for producing enzymes for the biodegradation of carrageenans, World patent application: WO9815617.

Barbeyron, T., Flament, D., Michel, G., Potin, P., Kloareg, B. (2001) The sulphated-galactan hydrolases, agarases and carrageenases: structural biology and molecular evolution, *Cah. Biol. Mar.* **42**, 169–183.

Beck, G. M., Neau, S. H., Holder, A. J., Hemenway, J. N. (2000) Evaluation of quantitative structure property relationships necessary for enantioresolution with lambda- and sulfobutylether lambda-carrageenan in capillary electrophoresis, *Chirality* **12**, 688–696.

Bellion, C., Hamer, G. K. (1981) Analysis of kappa iota hybrid carrageenans with kappa-carrageenase, iota-carrageenase and ^{13}C NMR, in: *Proceedings of the Xth Int. Seaweed Symp.* (Levring, T., Ed.), Berlin: de Gruyter, 379–384.

Bellion, C., Hamer, G. K., Yaphe, W. (1982) The degradation of *Eucheuma spinosum* and *Eucheuma cottonii* carrageenans by ι-carrageenases and κ-carrageenases from marine bacteria, *Can. J. Microbiol.* **28**, 874–880.

Bellion, C., Brigand, G., Prome, J.-C., Bociek, D. W. S. (1983) Identification et caractérisation des précurseurs diologiques des carraghénanes par spectroscopie de RMN-13C, *Carbohdr. Res.* **119**, 31–48.

Bixler, H. J. (1996) Recent developments in manufacturing and marketing carrageenan, *Hydrobiologia* **326/327**, 35–57.

Bouarab, K., Potin, P., Correa, J., Kloareg, B. (1999) Sulfated oligosaccharides mediate the interaction between a marine red alga and its green algal pathogenic endophyte, *Plant Cell* **11**, 1635–1650.

Chibata, I., Tosa, T., Takata, I. (1979) Immobilized catalytically active substance and method of preparing the same, US Patent No. 4,138,292.

Chiovitti, A., Bacic, A., Kraft, G. T., Craik, D. J., Liao, M.-L. (1999) Pyruvated carrageenans from *Solieria robusta* and its adelphoparasite *Tikvahiella candida*, *Hydrobiologia* **398/399**, 401–409.

Chopin, T., Whalen, E. (1993) A new and rapid method for carrageenan identification by FT IR diffuse reflectance spectroscopy directly on dried, ground algal material, *Carbohydr. Res.* **246**, 51–59.

Ciancia, M., Matulewicz, M. C., Finch, P., Cerezo, A. S. (1993) Determination of the structures of cystocarpic carrageenans from *Gigartina skottsbergii* by methylation analysis and NMR spectroscopy, *Carbohydr. Res.* **238**, 241–248.

Correa Diaz, F., Aguilar Rosas, R., Aguilar-Rosas, L. E. (1990) Infrared analysis of eleven carrageenophytes form Baja California, Mexico, *Hydrobiologia* **204/205**, 609–614.

Craigie, J. S. (1990) Cell walls, in: *Biology of the Red Algae* (Cole, K. M., Sheath, R. G., Ed.), Cambridge: Cambridge University Press, 221–257.

Dea, I. C. M., Morrison, A. (1975) Chemistry and interactions of seed galactomannans, in: *Ad-*

vances in Carabohydrate Chemistry and Biochemistry (Tipson, R. S., Horton, D., Ed.), New York: Academic Press, 241–312.

De Ruiter, G. A., Rudolph, B. (1997) Carrageenan biotechnology, *Trends Food Sci. Technol.* **8**, 389–395.

De Ruiter, G., Richard, O., Rudolph, B., Genicot, S., Kloareg, B., Penninkhof, B., Potin, P. (2000) Sulfohydrolases, corresponding amino acid and nucleotide sequences, sulfohydrolase preparations, processes, and products thereof, World patent application: WO0068395.

Falshaw, R., Furneaux, R. (1994) Carrageenan from the tetrasporic stage of *Gigartina decipiens* (Gigartinaceae, Rhodophyta), *Carbohydr. Res.* **252**, 171–182.

FAO (1992) *Compendium of Food Additive Specifications.* Rome: FAO/WHO Joint Expert Committee on Food Additives, Food and Agricultural Organization of the United Nations.

Genicot, S., Richard, O., Crépineau, F., Boyen, C., Penninkhof, B., Potin, P., Rousvoal, S., Rudolph, B., Kloareg, B. (2000) Purification and cloning of polysaccharide-modifying enzymes from marine algae, in: *Polymerix 2000* (Conanec, R., Ed.), Centre de Biotechnologies en Bretagne, 157–163.

Gretz, M. R., Mollet, J.-C., Falshaw, R. (1997) Analysis of red algal extracellular matrix polysaccharides, in: *Techniques in Glycobiology* (Townsend, R. R., Hotchkiss, A. T., Ed.), New York: Marcel Dekker, 613–628.

Guiseley, K. B., Stanley, N. F., Whitehouse, P. A. (1980) Carrageenan, in: *Handbook of Water-Soluble Gums and Resins* (Davidson, R. L., Ed.), New York: McGraw-Hill, Chapter 5.

Hansen, J. H., Larsen, H., Groendal, J. (2000) Carrageenan compositions and methods for their production, U.S. patent application: US 6,063,915.

Harding, S. E., Hill, S. E., Mitchell, J. R. (1995) Biopolymer mixtures, Nottingham University Press, Nottingham.

Hermansson, A. M., Eriksson, E., Jordansson, E. (1991) Effects of potassium, sodium and calcium on the microstructure and rheological behavior of kappa-carrageenan gels, *Carbohydr. Polym.* **16**, 297–320.

Hoffmann, R. A., Russell, A. R., Gidley, M. J. (1996) Molecular weight distribution of carrageenans: characterisation of commercial stabilisers and effect of cation depletion on depolymerisation, in: *Gums and Stabilisers for the Food Industry 8* (Phillips, G. O., Williams, P. J., Wedlock, D. J., Ed.), Oxford: Oxford University Press, 137–148.

Iborra, J. L., Manjón, A., Cánovas, M. (1997) Immobilization in carrageenans, in: *Methods in Biotechnology* (Bickerstaff, G. F., Ed.), Totowa, NJ: Humana Press Inc, 53–60.

Imeson, A. P. (2000) Carrageenan, in: *Handbook of hydrocolloids* (Phillips, G. O., Williams, P. A., Ed.), Cambridge: Woodhead Publishing, 87–102.

Jol, C. N., Neiss, T. G., Penninkhof, B., Rudolph, B., De Ruiter, G. A. (1999) A novel high-performance anion-exchange chromatographic method for the analysis of carrageenans and agars containing 3,6-anhydrogalactose, *Anal. Biochem.* **268**, 213–222.

Kapraun, D. F. (1999) Red algal polysaccharides industry: economics and research status at the turn of the century, *Hydrobiologia* **398/399**, 7–14.

Kloareg, B., Quantrano, R. S. (1988) Structure of the cell walls of marine algae and ecophysiological functions of the matrix polysaccharides, *Oceanogr. Mar. Biol. Annu. Rev.* **26**, 259–315.

Knutsen S. H. (1992) Isolation and analysis of red algal galactans, University of Trondheim, Norway.

Knutsen, S. H., Grasdalen, H. (1992) Analysis of carrageenans by enzymic degradation gel, filtration and ^{1}H NMR spectroscopy, *Carbohydr. Polym.* **19**, 199–210.

Knutsen, S. H., Myslabodski, D. E., Larsen, B., Usov, A. I. (1994) A modified system of nomenclature for red algal galactans, *Bot. Mar.* **37**, 163–169.

Knutsen, S. H., Sletmoen, M., Kristensen, T., Barbeyron, T., Kloareg, B., Potin, P. (2001) A rapid method for the separation and analysis of carrageenan oligosaccharides released by *iota-* and *kappa*-carrageenase, *Carbohydr. Res.* **331**, 101–106.

Lahaye, M. (2001) Chemistry and physico-chemistry of phycocolloids, *Cah. Biol. Mar.* **42**, 137–157.

Michel, G., Chantalat, L., Duee, E., Barbeyron, T., Henrissat, B., Kloareg, B., Dideberg, O. (2001) The kappa-carrageenase of *P. carrageenovora* features a tunnel-shaped active site: A novel insight in the evolution of clan-B glycoside hydrolases, *Structure* **9**, 513–525.

Mori, T. (1943) The enzyme catalyzing the decomposition of mucilage of *Chondrus ocellatus* III. Purification, unit determination, and distribution of the enzyme, *J. Agric. Chem. Soc. Jpn.* **19**, 740–742.

Morris, E. R., Rees, D. A., Robinson, G. (1980) Cation-specific aggregation of carrageenan helices: domain model of polymer gel structure, *J. Mol. Biol.* **138**, 349–362.

Nakashima, H., Kido, Y., Kobayashi, N., Motoki, Y., Neushul, M., Yamamoto, N. (1987) Purification and characterization of an avian myeloblastosis and human immunodeficiency virus reverse transcriptase inhibitor, sulfated polysaccharides extracted from sea algae, *Antimicrob. Agents Chemother.* **31**, 1524–1528.

Neushul, M. (1990) Antiviral carbohydrates from marine red algae, *Hydrobiologica* **204/205**, 99–104.

O'Neill, A. N. (1955a) 3,6-Anhydro-D-galactose as a constituent of kappa-carrageenin, *J. Am. Chem. Soc.* **77**, 2837–2839.

O'Neill, A. N. (1955b) Derivatives of 4-O-beta-D-galactopyranosyl-3,6-anhydro-D-galactose form kappa-carrageenin, *J. Am. Chem. Soc.* **77**, 6324–6326.

Piculell, L. (1995) Gelling carrageenans, in: *Food Polysaccharides and Their Applications* (Stephan, A. M., Ed.), New York: Marcel Dekker, 205–244.

Potin, P., Richard, C., Barbeyron, T., Henrissat, B., Gey, C., Petillot, Y., Forest, E., Dideberg, O., Rochas, C., Kloareg, B. (1995) Processing and hydrolytic mechanism of the cgkA-encoded kappa-carrageenase of *Alteromonas carrageenovora*, *Eur. J. Biochem.* **228**, 971–975.

Potin, P., Bouarab, K., Kupper, F., Kloareg, B. (1999) Oligosaccharide recognition signals and defence reactions in marine plant-microbe interactions, *Curr. Opin. Microbiol.* **2**, 276–283.

Renn, D. (1997) Biotechnology and the red seaweed polysaccharide industry: status, needs and prospects, *Trends Biotechnol.* **15**, 9–14.

Roberts, M. A., Quemener, B. (1999) Measurement of carrageenans in food: challenges, progress, and trends in analysis, *Trends Food Sci. Technol.* **10**, 169–181.

Rudolph, B. (2000) Seaweed product: red algae of economic significance, in: *Marine and Freshwater Products Handbook* (Martin, R. E., Carter, E. P., Davis, L. M., Flich, G. J., Ed.), Boca Raton, FL: CRC Press, 515–529.

Santelices, B. (1999) A conceptual framework for marine agronomy, *Hydrobiologia* **398/399**, 15–23.

Schaeffer, D. J., Krylov, V. S. (2000) Anti-HIV activity of extracts and compounds from algae and cyanobacteria, *Ecotoxicol. Environ. Saf.* **45**, 208–227.

Smith, D. B., Cook, W. H. (1953) Fractionation of carrageenin, *Arch. Biochem. Biophys.* **232**, 232–233.

Smith, D. B., O'Neill, A. N., Perlin, A. S. (1955) Studies on the heterogeneity of carrageenin, *Can. J. Chem.* **32**, 1352–1360.

Snoeren, T. H. M., Payens, T. A. J. (1976) On the sol-gel transition in solutions of kappa-carrageenan, *Biochim. Biophys. Acta* **437**, 264–272.

Stancioff, D. J., Stanley, N. F. (1969) Infrared and chemical studies on algal polysaccharides, *Proc. Intl. Seaweed Symp.* **6**, 595–609.

Stevenson, T. T., Furneaux, R. H. (1991) Chemical methods for the analysis of sulphated galactans from red algae, *Carbohydr. Res.* **210**, 277–298.

Stortz, C. A., Cerezo, A. S. (1992) The ^{13}C NMR spectroscopy of carrageenans: calculation of chemical shifts and computer-aided structural determination, *Carbohydr. Polym.* **18**, 237–242.

Stortz, C. A., Bacon, C. E., Cherniak, R., Cerezo, A. S. (1994) High-field NMR spectroscopy of cystocarpic and tetrasporic carrageenans form *Iridaea undulosa*, *Carbohydr. Res.* **261**, 317–326.

Tako, M., Nakamura, S. (1986) Indicative evidence for a conformational transition in kappa-carrageenan from studies of viscosity-shear rate dependence, *Carbohydr. Res.* **155**, 200–205.

Tako, M., Nakamura, S., Kohda, Y. (1987) Indicative evidence for a conformational transitions in iota-carrageenan, *Carbohydr. Res.* **161**, 247–255.

Therkelsen, G. H. (1993) Carrageenan, in: *Industrial Gums: Polysaccharides and Their Derivatives* (Whistler, R. L., BeMiller, J. N., Ed.), San Diego, CA: Academic Press, 145–180.

Troell, M., Rönnbäck, P., Halling, C., Kautsky, N., Buschmann, A. (1999) Ecological engineering in aquaculture: use of seaweeds for removing nutrients from intensive marineculture, *J. Appl. Phycol.* **11**, 89–97.

Tseng, C. K. (1945) The terminology of seaweed colloids, *Science* **101**, 597–602.

Turquois, T., Acquistapace, S., Arce-Vera, F., Welti, D. H. (1996) Composition of carrageenan blends inferred from ^{13}C-NMR and infrared spectroscopy, *Carbohydr. Polym.* **31**, 269–278.

Usov, A. I., Yarotsky, S. V., Shashkov, A. S. (1980) ^{13}C-NMR spectroscopy of red algal galactans, *Biopolymers* **19**, 977–990.

Usov, A. I. (1998) Structural analysis of red seaweed galactans of agar and carrageenan groups, *Food Hydrocolloids* **12**, 301–308.

VandeVelde, F., Peppelman, H. A., Rollema, H. S., Tromp, R. H. (2001) On the structure of kappa/iota-hybrid carrageenans, *Carbohydr. Res.* **331**, 271–283.

Viebke, C., Williams, P. A. (2000) Determination of molecular mass distribution of kappa-carrageenan and xanthan using asymmetrical flow field-flow fractionation, *Food Hydrocolloids* **14**, 265–270.

Weiner, M. L. (1991) Toxicological properties of carrageenan, *Agents Actions* **32**, 46–51.

Wong, K. F., Craigie, J. S. (1978) Sulfohydrolase activity and carrageenan biosynthesis in *Chondrus crispus* (Rhodophyceae), *Plant Physiol.* **61**, 663–666.

Yaphe, W., Baxter, B. (1955) The enzymatic hydrolysis of carrageenan, *Appl. Microbiol.* **3**, 380–383.

11
Bacterial Cellulose

Prof. Dr. Eng. Stanislaw Bielecki[1], Dr. Eng. Alina Krystynowicz[2], Prof. Dr. Marianna Turkiewicz[3], Dr. Eng. Halina Kalinowska[4]

[1] Institute of Technical Biochemistry, Technical University of Lódz, Stefanowskiego 4/10, 90-924 Lódz, Poland; Tel: +48-4263-13442; Fax: +48-4263-402; E-mail: biochem@ck-sg.p.lodz.pl

[2] Institute of Technical Biochemistry, Technical University of Lódz, Stefanowskiego 4/10, 90-924 Lódz, Poland; Tel: +48-4263-13442; Fax: +48-4263-402; E-mail: biochem@ck-sg.p.lodz.pl

[3] Institute of Technical Biochemistry, Technical University of Lódz, Stefanowskiego 4/10, 90-924 Lódz, Poland; Tel: +48-4263-13442; Fax: +48-4263-402; E-mail: biochem@ck-sg.p.lodz.pl

[4] Institute of Technical Biochemistry, Technical University of Lódz, Stefanowskiego 4/10, 90-924 Lódz, Poland; Tel: +48-4263-13442; Fax: +48-4263-402; E-mail: biochem@ck-sg.p.lodz.pl

Biopolymers for Medical and Pharmaceutical Applications. Edited by A. Steinbüchel and R. H. Marchessault

ISBN: 3-527-31154-8

A-BC	bacterial cellulose from agitated culture
ATP	adenosine triphosphate
BC	bacterial cellulose
CBH	cellobiohydrolase
c-di-GMP	cyclic diguanosine monophosphate
Cel^-	cellulose-negative mutant
Cel6A, Cel7A	cellobiohydrolases belonging to 6A and 7A families, respectively
CM	carboxymethyl-
CS	cellulose synthase
CSL	corn steep liquor
D	aspartic acid

DMSO	dimethyl sulfoxide
DP	degree of polymerization
E	glutamic acid
FBP	fructose-1,6-biphosphate phosphatase
FK	fructokinase
Fru-bi-P	fructose-1,6-biphosphate
Fru-6-P	fructose-6-phosphate
G	guanine
GK	glucokinase
Glc	glucose
Glc-6(1)-P	glucose-6(1)-phosphate
G6PDH	glucose-6-phosphate dehydrogenase
H-S medium	Hestrin and Schramm medium (1954)
IS	insertion sequence
Lip	lipid
LP	UDPGPT: lipid pyrophosphate: UDPGlc- phosphotransferase
LPP	lipid pyrophosphate phosphohydrolase
Man	mannose
NMR	nuclear magnetic resonance
PC	plant cellulose
PDEA	phosphodiesterase A
PDEB	phosphodiesterase B
Pel^{-}	pellicle non-forming
1PFK	fructose-1-phosphate kinase
PGA	phosphogluconic acid
PGI	phosphoglucoisomerase
PMG	phosphoglucomutase
PTS	system of phosphotransferases
Q	glutamine
R	arginine
Rha	rhamnose
Rib	D-ribose
S	serine
S-BC	bacterial cellulose from static culture
TC	terminal complexe
U	uridine
UDP	uridine diphosphate
UDPGlc	uridine diphosphoglucose
UGP	pyrophosphorylase uridine diphosphoglucose
UMP	uridine monophosphate
v.v.m.	volume per volume per minute
W	tryptophan

1 Introduction

Cellulose is the most abundant biopolymer on earth, recognized as the major component of plant biomass, but also a representative of microbial extracellular polymers. Bacterial cellulose (BC) belongs to specific products of primary metabolism and is mainly a protective coating, whereas plant cellulose (PC) plays a structural role.

Cellulose is synthesized by bacteria belonging to the genera *Acetobacter, Rhizobium, Agrobacterium,* and *Sarcina* (Jonas and Farah, 1998). Its most efficient producers are Gram-negative, acetic acid bacteria *Acetobacter xylinum* (reclassified as *Gluconacetobacter xylinus,* Yamada et al., 1997; Yamada, 2000), which have been applied as model microorganisms for basic and applied studies on cellulose (Cannon and Anderson, 1991). Investigations have been focused on the mechanism of biopolymer synthesis, as well as on its structure and properties, which determine practical use (Legge, 1990; Ross et al., 1991). One of the most important features of BC is its chemical purity, which distinguishes this cellulose from that from plants, usually associated with hemicelluloses and lignin, removal of which is inherently difficult.

Because of the unique properties, resulting from the ultrafine reticulated structure, BC has found a multitude of applications in paper, textile, and food industries, and as a biomaterial in cosmetics and medicine (Ring et al., 1986). Wider application of this polysaccharide is obviously dependent on the scale of production and its cost. Therefore, basic studies run together with intensive research on strain improvement and production process development.

2 Historical Outline

Although synthesis of an extracellular gelatinous mat by *A. xylinum* was reported for the first time in 1886 by A. J. Brown, BC attracted more attention in the second half of the 20th century. Intensive studies on BC synthesis, using *A. xylinum* as a model bacterium, were started by Hestrin et al. (1947, 1954), who proved that resting and lyophilized *Acetobacter* cells synthesized cellulose in the presence of glucose and oxygen. Next, Colvin (1957) detected cellulose synthesis in samples containing cell-free extract of *A. xylinum,* glucose, and ATP. Further milestones in studies on BC synthesis, presented in this review, contributed to the elucidation of mechanisms governing not only the biogenesis of the bacterial polymer, but also that of plants, thus leading to the understanding of one of the most important processes in nature. The true historical outline is presented throughout all the paragraphs below, including the references.

3 Structure of BC

Cellulose is an unbranched polymer of β-1,4-linked glucopyranose residues. Extensive research on BC revealed that it is chemically identical to PC, but its macromolecular structure and properties differ from the latter (Figure 1). Nascent chains of BC aggregate to form subfibrils, which have a width of approximately 1.5 nm and belong to the thinnest naturally occurring fibers, comparable only to subelemental fibers of cellulose detected in the cambium of some plants and in quinee mucous (Kudlicka, 1989). BC subfibrils are crystallized into microfibrils (Jonas and Farah, 1998), these into bundles, and the latter into ribbons (Yamanaka et al.,

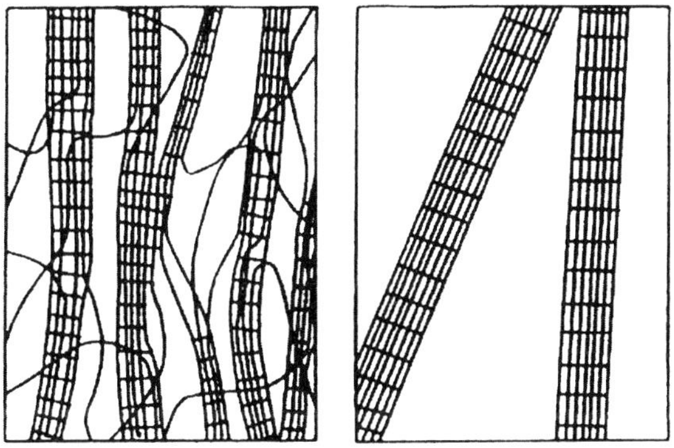

Fig. 1 Schematic model of BC microfibrils (right) drawn in comparison with the 'fringed micelles'; of PC fibrils (Iguchi et al., 2000; with kind permission).

2000). Dimensions of the ribbons are 3–4 (thickness) × 70–80 nm (width), according to Zaar (1977), 3.2 × 133 nm, according to Brown et al. (1976), or 4.1 × 117 nm, according to Yamanaka et al. (2000), whereas the width of cellulose fibers produced by pulping of birch or pine wood is two orders of magnitude larger ($1.4-4.0 \times 10^{-2}$ and $3.0-7.5 \times 10^{-2}$ mm, respectively). The ultrafine ribbons of microbial cellulose, the length of which ranges from 1 to 9 μm, form a dense reticulated structure (Figure 2), stabilized by extensive hydrogen bonding (Figure 3). BC is also distinguished from its plant counterpart by a high crystallinity index (above 60%) and different degree of polymerization (DP), usually between 2000 and 6000 (Jonas and Farah, 1998), but in some cases reaching even 16,000 or 20,000 (Watanabe et al., 1998b), whereas the average DP of plant polymer varies from 13,000 to 14,000 (Teeri, 1997).

Macroscopic morphology of BC strictly depends on culture conditions (Watanabe et al., 1998a; Yamanaka et al., 2000). In static conditions (Figure 4), bacteria accumulate cellulose mats (S-BC) on the surface of nutrient broth, at the oxygen-rich air–liquid interface. The subfibrils of cellulose are continuously extruded from linearly ordered pores at the surface of the bacterial cell, crystallized into microfibrils, and forced deeper into the growth medium. Therefore, the leather-like pellicle, supporting the population of *A. xylinum* cells, consists of overlapping and intertwisted cellulose rib-

a)

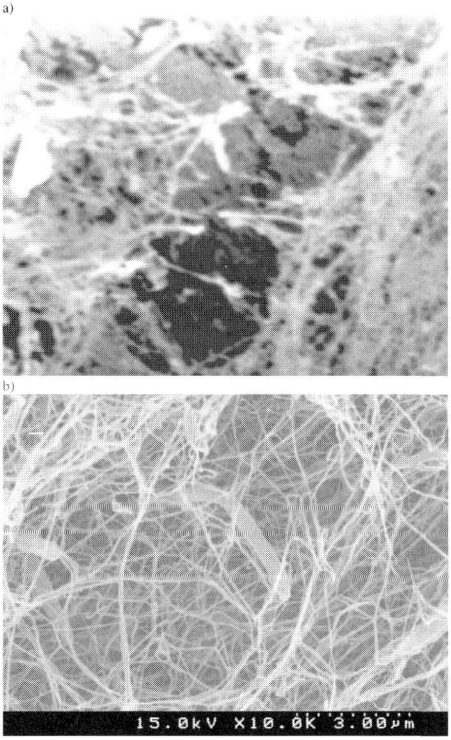

Fig. 2 Scanning electron microscopy images of BC membrane from static culture of *A. xylinum* (a) and bacterial cell with attached cellulose ribbons (b).

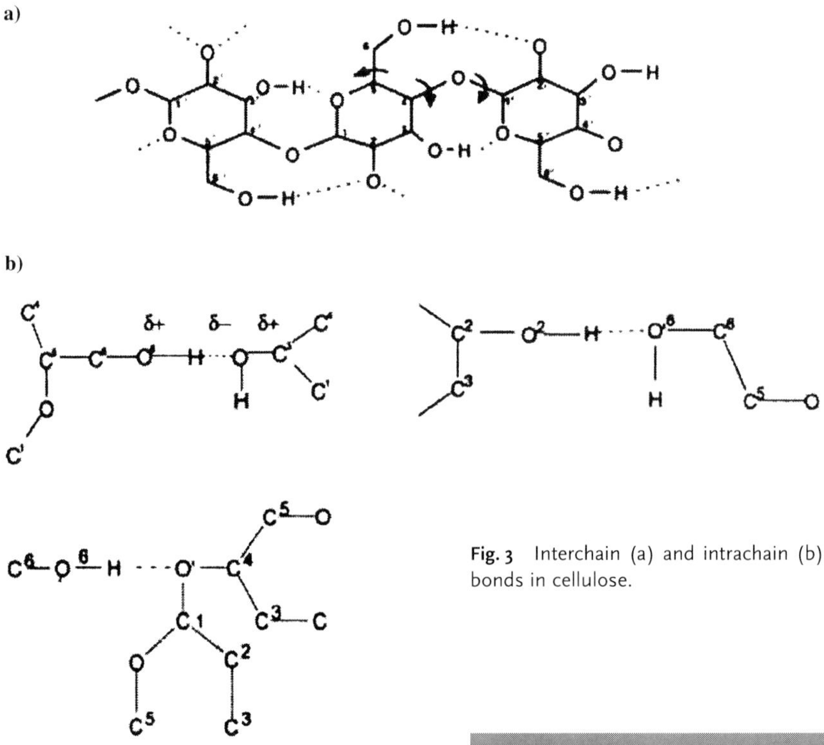

Fig. 3 Interchain (a) and intrachain (b) hydrogen bonds in cellulose.

bons, forming parallel but disorganized planes (Jonas and Farah, 1998). The adjacent S-BC strands branch and interconnect less frequently than these in BC produced in agitated culture (A-BC), in a form of irregular granules, stellate and fibrous strands, well-dispersed in culture broth (Figure 5) (Vandamme et al., 1998). The strands of reticulated A-BC interconnect to form a grid-like pattern, and have both roughly perpendicular and roughly parallel orientations (Watanabe et al., 1998a).

Differences in three-dimensional structure of A-BC and S-BC are noticeable in their scanning electron micrographs. The S-BC fibrils are more extended and piled above one another in a criss-crossing manner. Strands of A-BC are entangled and curved (Johnson and Neogi, 1989). Besides, they

Fig. 4 BC pellicle formed in static culture.

Fig. 5 BC pellets formed in agitated culture.

have a larger cross-sectional width (0.1–0.2 μm) than S-BC fibrils (usually 0.05–0.10 μm). Morphological differences between S-BC and A-BC contribute to varying degrees of crystallinity, different crystallite size and I_α cellulose content.

Two common crystalline forms of cellulose, designated as I and II, are distinguishable by X-ray, nuclear magnetic resonance (NMR), Raman spectroscopy, and infrared analysis (Johnson and Neogi, 1989). It is known that in the metastable cellulose I, which is synthesized by the majority of plants and also by *A. xylinum* in static culture, parallel β-1,4-glucan chains are arranged uniaxially, whereas β-1,4 glucan chains of cellulose II are arranged in a random manner. They are mostly antiparallel and linked with a larger number of hydrogen bonds that results in higher thermodynamic stability of the cellulose II.

A-BC has a lower crystallinity index and a smaller crystallite size than S-BC (Watanabe et al., 1998a). It was also observed that a significant portion of cellulose II occurred in BC synthesized in agitated culture. In nature, cellulose II is synthesized by only a few organisms (some algae, molds, and bacteria, such as *Sarcina ventriculi*) (Jonas and Farah, 1998); the industrial production of this kind of cellulose is based on chemical conversion of PC.

Using CP/MAS ^{13}C-NMR it is possible to reveal the presence of cellulose I_α and I_β – two distinct forms of cellulose (Watanabe et al., 1998a). These forms occur in algae-, bacteria-, and plant-derived cellulose. The latter one contains less I_α cellulose than BC (Johnson and Neogi, 1989). The irreversible crystal transformation from cellulose I_α to I_β shifts the X-ray and CP/MAS ^{13}C-NMR spectra because of the difference in the unit cell. S-BC contains more cellulose I_α than A-BC. It was reported that the difference in cellulose I_α content between A-BC and S-BC exceeded that in crystallinity index (Watanabe et al., 1998a), and the mass fraction of cellulose I_α was closely related to the crystallite size.

4 Chemical Analysis and Detection

For the detection of either crystalline or amorphous cellulose, several direct dyes, specific for the linear β-1,4-glucan, are used (Mondal and Kai, 2000). All of them are fluorescent brightening agents and form dye–cellulose complexes, stabilized by van der Waals and/or hydrogen bonding. One of these dyes is the fluorescent brightener Calcofluor. The direct dyes do not only enable visualization of cellulose chains, but also have been intensively applied for studies on nascent cellulose chains association and crystallization (see Section 7.3.2).

The weight-average DP of cellulose and the DP distribution are determined by high-performance gel-permeation chromatography (Watanabe et al., 1998a) of nitrated cellulose samples.

The differences between the reticulated structure of microbial cellulose, produced under agitated culture conditions, and the disorganized layered structure of cellulose pellicle, formed in static culture, are noticeable in scanning electron microscopy images (Johnson and Neogi, 1989).

To distinguish the parallel chain crystalline lattice of cellulose I from the antiparallel one of cellulose II, X-ray diffraction, Raman spectroscopy, infrared analysis, and NMR are applied (Johnson and Neogi, 1989). The crystallinity index and crystallite size are calculated based on X-ray diffraction measurements (Watanabe et al., 1998a).

Two distinct forms of cellulose I, i.e. cellulose I_α and I_β, are not distinguishable by X-ray diffraction, and therefore CP/MAS ^{13}C-NMR analysis, carried out on freeze-dried cellulose samples, has to be performed to determine their mass fractions (Watanabe et al., 1998a).

The physicochemical properties of cellulose such as water holding capacity, viscosity of disintegrated cellulose suspension, and the Young's modulus of dried sheets are determined using conventional methods (Watanabe et al., 1998a, Iguchi et al., 2000).

5
Occurrence

BC is synthesized by several bacterial genera, of which *Acetobacter* strains are best known. An overview of BC producers is presented in Table 1 (Jonas and Farah, 1998). The polymer structure depends on the organism, although the pathway of biosynthesis and mechanism of its regulation are probably common for the majority of BC-producing bacteria (Ross et al., 1991; Jonas and Farah, 1998).

Tab. 1 BC producers (Jonas and Farah, 1998, modified)

Genus	Cellulose structure
Acetobacter	extracellular pellicle composed of ribbons
Achromobacter	fibrils
Aerobacter	fibrils
Agrobacterium	short fibrils
Alcaligenes	fibrils
Pseudomonas	no distinct fibrils
Rhizobium	short fibrils
Sarcina	amorphous cellulose
Zoogloea	not well defined

A. xylinum (synonyms *A. aceti* ssp. xylinum, *A. xylinus*), which is the most efficient producer of cellulose, has been recently reclassified and included within the novel genus *Gluconacetobacter*, as *G. xylinus* (Yamada et al., 1998, 2000) together with some other species (*G. hansenii*, *G. europaeus*, *G. oboediens*, and *G. intermedius*).

6
Physiological Function

In natural habitats, the majority of bacteria synthesize extracellular polysaccharides, which form envelopes around the cells (Costeron, 1999). BC is an example of such a substance. Cells of cellulose-producing bacteria are entrapped in the polymer network, frequently supporting the population at the liquid–air interface (Wiliams and Cannon, 1989). Therefore, BC-forming strains can inhabit sewage (Jonas and Farah, 1998). The polymer matrix takes part in adhesion of the cells onto any accessible surface and facilitates nutrient supply, since their concentration in the polymer lattice is markedly enhanced due to its adsorptive properties, in comparison to the surrounding aqueous environment (Jonas and Farah, 1998; Costeron, 1999). Some authors suppose that cellulose synthesized by *A. xylinum* also plays a storage role and can be utilized by the starving microorganisms. Its decomposition would be then catalyzed by exo- and endoglucanases, the co-presence of which was detected in the culture broth of some cellulose-producing *A. xylinum* strains (Okamoto et al., 1994).

Because of the viscosity and hydrophilic properties of the cellulose layer, the resistance of producing bacterial cells against unfavorable changes (a decrease in water content, variations in pH, appearance of toxic substances, pathogenic organisms, etc.) in an habitat is increased, and they can further grow and develop inside the envelope. It was also found that cellulose placed over bacterial cells protects them from ultraviolet radiation. As much as 23% of the acetic acid bacteria cells covered with BC survived a 1h treatment with ultraviolet irradiation. Removal of the protective polysaccharide brought about a drastic decrease in their viability (3% only) (Ross et al., 1991).

7 Biosynthesis of BC

Synthesis of BC is a precisely and specifically regulated multi-step process, involving a large number of both individual enzymes and complexes of catalytic and regulatory proteins, whose supramolecular structure has not yet been well defined. The process includes the synthesis of uridine diphosphoglucose (UDPGlc), which is the cellulose precursor, followed by glucose polymerization into the β-1,4-glucan chain, and nascent chain association into characteristic ribbon-like structure, formed by hundreds or even thousands of individual cellulose chains. Pathways and mechanisms of UDPGlc synthesis are relatively well known, whereas molecular mechanisms of glucose polymerization into long and unbranched chains, their extrusion outside the cell, and self-assembly into fibrils require further elucidation.

Moreover, studies on BC synthesis may contribute to better understanding of PC biogenesis.

7.1 Synthesis of the Cellulose Precursor

Cellulose synthesized by *A. xylinum* is a final product of carbon metabolism, which depending on the physiological state of the cell involves either the pentose phosphate cycle or the Krebs cycle, coupled with gluconeogenesis (Figure 6) (Ross et al., 1991; Tonouchi et al., 1996). Glycolysis does not operate in acetic acid bacteria since they do not synthesize the crucial enzyme of this pathway – phosphofructose kinase (EC 2.7.1.56) (Ross et al., 1991). In *A. xylinum*, cellulose synthesis is tightly associated with catabolic processes of oxidation and consumes as much as 10% of energy derived from catabolic reactions (Weinhouse, 1977). BC production does not interfere with other anabolic processes, including protein synthesis (Ross et al., 1991).

A. xylinum converts various carbon compounds, such as hexoses, glycerol, dihydroxyacetone, pyruvate, and dicarboxylic acids, into cellulose, usually with about 50% efficiency. The latter compounds enter the Krebs cycle and due to oxalacetate decarboxylation to pyruvate undergo conversion to hexoses via gluconeogenesis, similarly to glycerol, dihydroxyacetone, and intermediates of the pentose phosphate cycle (Figure 6).

The direct cellulose precursor is UDPGlc, which is a product of a conventional pathway, common of many organisms, including plants, and involving glucose phosphoryla-

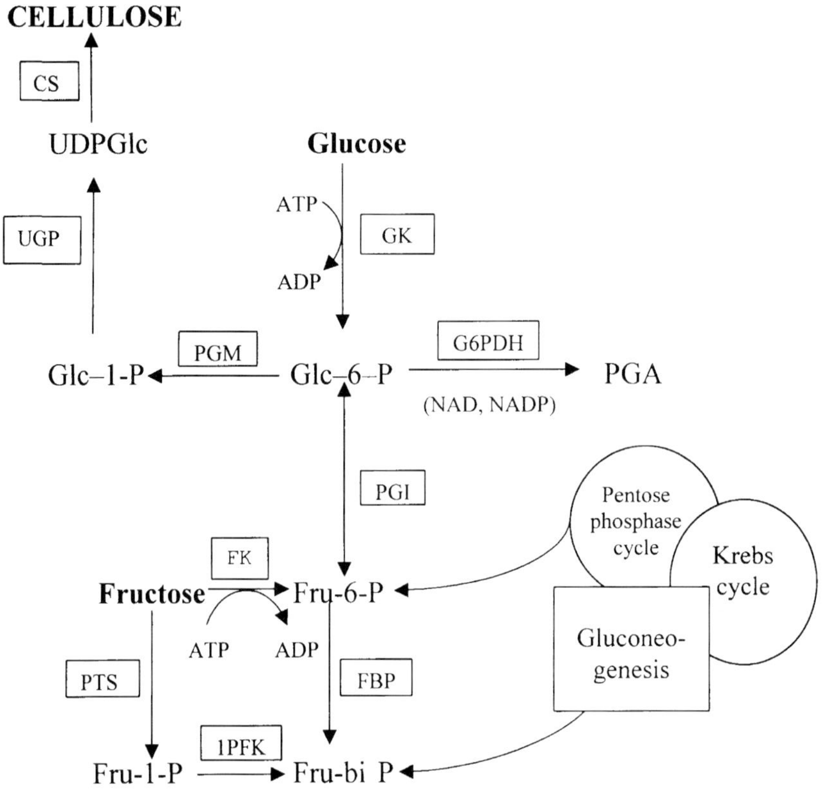

Fig. 6 Pathways of carbon metabolism in *A. xylinum*. CS, cellulose synthase (EC 2.4.2.12); FBP, fructose-1,6-biphosphate phosphatase (EC 3.1.3.11); FK, glucokinase (EC 2.7.1.2); G6PDH, glucose-6-phosphate dehydrogenase (EC 1.1.1.49); 1PFK, fructose-1-phosphate kinase (EC 2.7.1.56); PGI, phosphoglucoisomerase; PMG, phosphoglucomutase (EC 5.3.1.9); PTS, system of phosphotransferases; UGP, pyrophosphorylase UDPGlc (EC 2.7.7.9); Fru-bi-P, fructose-1,6-bi-phosphate; Fru-6-P, fructose-6-phosphate; Glc-6(1)-P, glucose-6(1)-phosphate; PGA, phosphogluconic acid; UDPGlc, uridine diphosphoglucose.

tion to glucose-6-phosphate (Glc-6-P), catalyzed by glucokinase, followed by isomerization of this intermediate to Glc-α-1-P, catalyzed by phosphoglucomutase, and conversion of the latter metabolite to UDPGlc by UDPGlc pyrophosphorylase. This last enzyme seems to be the crucial one involved in cellulose synthesis, since some phenotypic cellulose-negative mutants (Cel$^-$) are specifically deficient in this enzyme (Valla et al., 1989), though they display cellulose synthase (CS) activity, this was confirmed *in vitro* by means of observation of cellulose synthesis, catalyzed by cell-free extracts of Cel$^-$ strains (Saxena et al., 1989). Furthermore, the pyrophosphorylase activity varies between different *A. xylinum* strains and the highest activity was detected in the most effective cellulose producers, such as *A. xylinum* ssp. sucrofermentans BPR2001. The latter strain prefers fructose as a carbon source, displays high activity of phosphoglucoisomerase, and possesses a system of phosphotransferases, dependent on phosphoenolpyruvate. The system catalyzes conversion of fructose to fructose-1-phosphate and further to fructose-1,6-biphosphate (Figure 6).

7.2
Cellulose Synthase

Cellulose biosynthesis both in plants and in prokaryotes is catalyzed by the uridine diphosphate (UDP)-forming CS which is basically a processing 4-β-glycosyltransferase (EC 2.4.1.12, UDPGlc: 1,4-β-glucan 4-β-D-glucosyltransferase), since it transfers consecutive glucopyranose residues from UDPGlc to the newly formed polysaccharide chain and is all the time linked with this chain. Oligomeric CS complexes are frequently called terminal complexes (TCs). It is presumed that TCs are responsible first for β-1,4-glucan chains synthesis, extrusion through the outer membrane (if the globular, catalytic domain of CS is localized on the cytoplasmic side of the cell membrane), as well as association and crystallization into defined supramolecular structures, that follow the first two processes.

Cellulose synthase of *A. xylinum* is a typical membrane-anchored protein, having a molecular mass of 400–500 kDa (Lin and Brown, 1989), and is tightly bound to the cytoplasmic membrane. Because of this localization, purification of CS was extremely difficult, and isolation of the membrane fraction was necessary before CS solubilization and purification. Furthermore, *A. xylinum* CS appeared to be a very unstable protein (Lin and Brown, 1989). CS isolation from membranes was carried out using digitonin (Lin and Brown, 1989), or detergents (Triton X-100) and treatment with trypsin (Saxena et al., 1989), followed by CS entrapment on cellulose. According to Lin and Brown (1989), the purified CS preparation contained three different types of subunits, having molecular masses of 90, 67, and 54 kDa. Saxena et al. (1989) found only two types of polypeptides (83 and 93 kDa). The latter result seems to be more probable, since the mass of both subunits corresponds to the size of two genes *cesA* and *cesB*, detected in the CS operon by Saxena et al. (1990b, 1991), and the genes *bcsA* and *bcsB*, reported by Wong et al. (1990) and Ben-Bassat et al. (1993) for the same operon (see Section 7.4).

Photolabeling affinity studies indicate that the 83-kDa polypeptide is a catalytic subunit, displaying high affinity towards UDPGlc (Lin et al., 1990). According to the gene sequence, it contains 723 amino acid residues, of which the majority are hydrophobic. This subunit is probably synthesized as a proprotein, which contains a signal sequence, composed of 24 amino acid residues, which is cut off during the maturation process (Wong et al., 1990). The mature protein is anchored in the cell membrane by means of a transmembrane helix, close to the N-terminus of the polypeptide chain. The protein contains five more transmembrane helices, which can interact both with each other and with a large, catalytic site-comprising globular domain submerged in cytoplasma. Brown and Saxena (2000) claim that the catalytic subunit of *A. xylinum* CS operates in the same manner as processing glycosyltransferases of the second family, i.e. it catalyzes the direct glycoside bond synthesis in cellulose, assisted by simultaneous inversion of the configuration on the anomeric carbon, from the α-configuration in the UDPGlc, which is the monomer donor, to the β-configuration in the polysaccharide. A catalytically active aspartic acid residue (D), and short sequences DXD and QXXRW were found in the globular fragment of the numerous processing glycosyltransferases (Saxena and Brown, 1997). The same motives were detected in *A. xylinum* CS. This globular fragment has a two-domain character (i.e., comprises two domains A and B) in many of glycoside synthases (Saxena et al., 1995). The domain A includes the D residue, crucial for catalysis, and the DXD motive, which probably binds the nucleotide–sugar substrate. Participation of this motive in UDPGlc binding by *A. xylinum* CS was recently confirmed by Brown and Saxena (2000), who investigated enzyme mutants obtained using site-directed mutagenesis. It is not yet clear if the globular fragment of *A. xylinum* CS comprises one or two domains, although this second possibility is more probable. However, it is known that the catalytic subunit of *A. xylinum* CS is glycosylated. Two potential sites of glycosylation were found in its primary structure, deduced based on the gene sequence (Saxena et al., 1990a).

The 93-kDa polypeptide is tightly bound to the catalytic subunit. It contains the signal sequence close to its N-terminus and one transmembrane helix close to the C-terminus. The helix enables anchoring in the cell membrane. The polypeptide is probably a regulatory subunit, which interacts with the CS activator – cyclic diguanosine monophospahte (c-di-GMP) (see Section 7.5). The 93-kDa polypeptide does not combine with the 83-kDa subunit antibodies. However, it is not yet clear if CS is composed of these two subunits only, since this oligomeric complex plays multiple roles, i.e. β-1,4-glucan polymerization, extrusion of the subfibrils outside the cell, as well as self-assembly and crystallization of cellulose ribbons, composed of microfibrils. Therefore, the CS complex is probably comprised of other polypeptides involved in transmembrane pore formation. The analysis of microscopic images of the purified CS indicates that the enzyme tends to form tetrameric or octameric aggregates (Lin et al., 1989).

7.3 Mechanism of Biosynthesis

Synthesis of the metastable cellulose I allomorph, in *A. xylinum* and other cellulose-producing organisms, including plants, involves at least two intermediary steps, i.e. (1) polymerization of glucose molecules to the linear 1,4-β-glucan, and (2) assembly and crystallization of individual nascent polymer chains into supramolecular structures, characteristic for each cellulose-producing organism.

Assuming that the CS globular domain is localized on the cytoplasmic side of the cell membrane (see Section 7.2), the transfer of 1,4-β-glucan chains through the membrane, outside the cell is also required. All three steps are tightly coupled, and the rate of polymerization is limited by the rate of assembly and crystallization.

7.3.1 Mechanism of 1,4-β-Glucan Polymerization

Formation of BC is catalyzed by the CS complexes aligned linearly in the *A. xylinum* cytoplasmic membrane. Since these complexes can polymerize up to 200,000 molecules of Glc s^{-1} into the β-1,4-glucan chain (Ross et al., 1991), the process must run with high intensity. The mechanism of the reaction has not yet been definitely clarified. Currently, two different hypotheses for this mechanism in *A. xylinum* have been proposed.

The first of them, developed in Brown's laboratory (Brown, 1996; Brown and Saxena, 2000), assumes that 1,4-β-glucan polymerization does not involve a lipid intermediate, which transfers glucose from UDPGlc to the newly synthesized polymer chain. This hypothesis agrees with the fact that glycosyltransferases responsible for synthesis of unbranched homopolysaccharides are processing enzymes. The scheme of 1,4-β-glucan polymerization, proposed by Brown and his colleagues, is presented in Figure 7.

According to this model, there are three catalytic sites in the globular fragment of the CS catalytic subunit, similar to other proc-

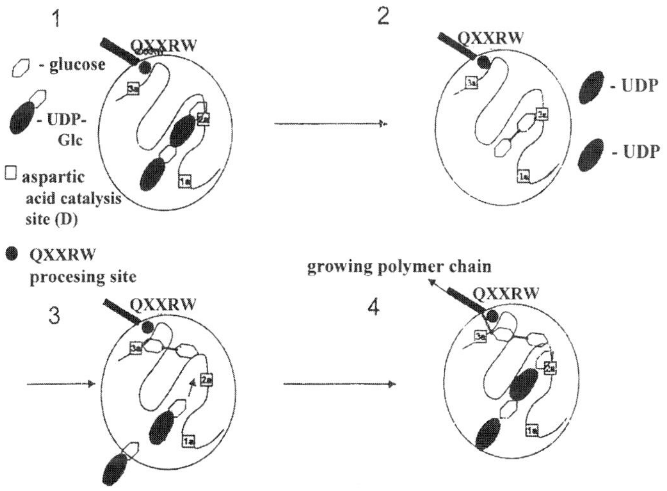

Fig. 7 Generalized concept for the polymerization reactions leading to β-1,4-glucan chain biosynthesis (Brown and Saxena, 2000; with kind permission).

essing glycosyltransferases. One of the catalytic sites (2a), comprising the DXD motive, binds two UDPGlc molecules (see Figure 7.1). The second catalytic site (1a), containing the crucial acidic aspartic acid (D) residue, catalyzes formation of the β-1,4-linkage between the two Glc residues docked in the pocket (see Figure 7.2), accompanied by releasing two UDP molecules. The third catalytic site (3a), containing the QXXRW motif, pulls the reducing end of the synthesized cellobiose (see Figure 7.3). The disaccharide leaves the area 1a–2a, which may bind two subsequent UDPGlc molecules (see Figure 7.4), and forms the second cellobiose molecule. In the next step, involving the QXXRW motif, the reducing end of the first cellobiose molecule is forced to the site of extrusion. At the same moment, the second cellobiose molecule is linked (with the aid of the D residue in the site 1a) to the nonreducing end of the first one, thus giving cellotetraose. The emptied 1a–2a area may bind two subsequent UDPGlc molecules, to repeat the cycle of reactions, attaching two more glucose residues to the chain. The simplified scheme of the proposed model of polymerization is depicted in Figure 8. According to this model, extrusion of the newly synthesized β-1,4-glucan chain starts from its reducing end.

Studies by Koyama et al. (1997) confirm this model. The authors proved that glucose residues are added to the nonreducing end of the polysaccharide and that reducing ends of nascent polymer chains are situated away from the cells. Furthermore, the torsion angle between two adjacent glucose residues in cellulose molecule is 180°, thus adding cellobiose residues (rather than single glucose moieties) to the growing chain favors maintaining the 2-fold screw axis of the β-1,4-glucan (Brown, 1996). Also Kuga and Brown (1988), who applied silver labeling, proved that cellulose chains were extruded outside the cell, starting from their reducing ends.

Han and Robyt (1998) who studied BC synthesis by the *A. xylinum* ATCC 10821 strain (resting cells and membrane preparations) using the ^{14}C pulse and chase reaction with D-glucose and UDPGlc, respectively,

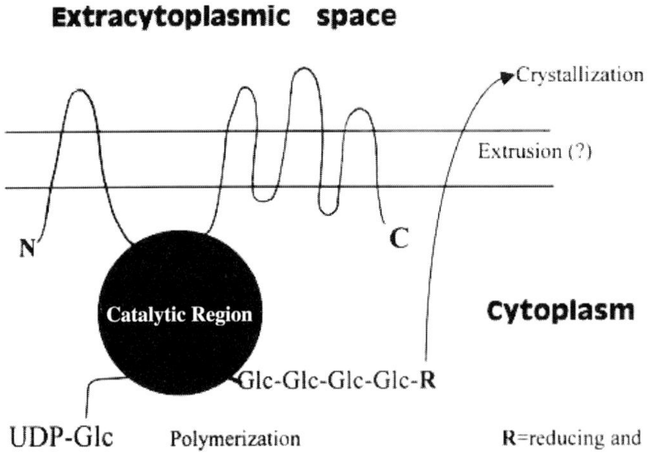

Fig. 8 Simplified scheme of BC biosynthesis according to the Brown's model (Brown and Saxena, 2000; with kind permission).

proposed a second, different molecular mechanism of this process. They believe that the consecutive residues of the activated glucopyranoses (from UDPGlc) are linked to the reducing end of the growing cellulose chain, because the ratio of D-[^{14}C]glucitol obtained from the extruded reducing end of the cellulose chain to D-[^{14}C]glucose was decreasing with time. Han and Robyt assumed that a lipid pyrophosphate with a polyisoprenoid character takes part in the biosynthesis. According to them, formation of the lipid intermediate during BC synthesis was confirmed by an extraction of as much as 33% of the pulsed ^{14}C label with a mixture of chloroform and methanol. Its quantitative extraction was impossible, since when the polysaccharide chain length exceeded 8–10 sugar moieties, the complex with lipid became insoluble in the extraction mixture.

Han and Robyt proposed that BC biosynthesis involved three enzymes embedded in the cytoplasmic membrane: CS, lipid pyrophosphate: UDPGlc phosphotransferase (LipPP: UDPGlc-PT), and lipid pyrophosphate phosphohydrolase (LPP). The reaction mechanism, called by the authors the insertion reaction, is presented in the Figure 9.

The first enzyme transfers Glc-α-1-P from UDPGlc onto the lipid monophosphate (Lip-P), thus giving the lipid pyrophosphate-α-D-Glc (LipPP-α-Glc) (reaction 1, Figure 9). The α configuration on the anomeric carbon, involved in the Glc phosphoester bond, remains the same as in the substrate molecule. The second product of this reaction is UMP (according to Brown's model, UDP is released).

In the next reaction (reaction 2, Figure 9), catalyzed by CS, the glucose residue is transferred from one LipPP-α-Glc molecule onto another one and the β-1,4-glycosidic linkage between the two glucose residues is formed, due to the attack of the C-4 hydroxyl group of one of them onto C-1 hydroxyl group of the second Glc (from the second LipPP-α-Glc).

In the third reaction (reaction 3, Figure 9), hydrolysis of the lipid pyrophosphate formed in the previous step occurs and

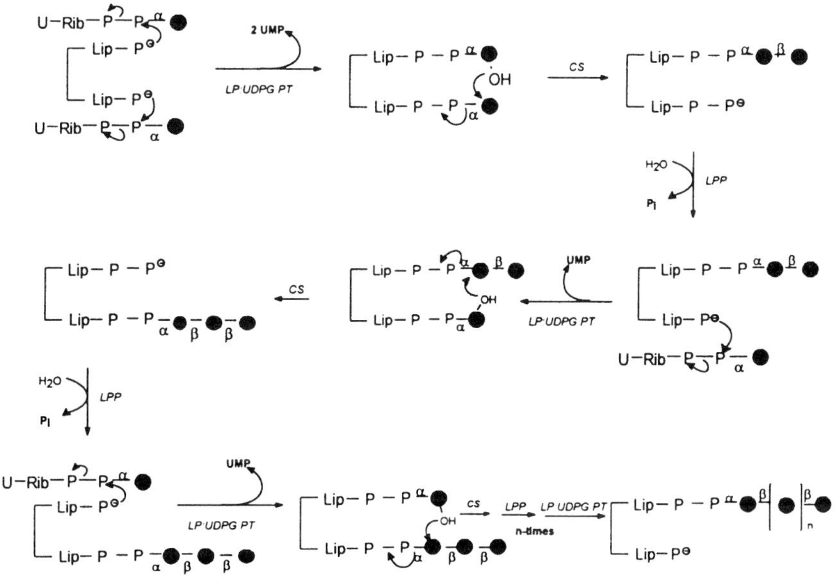

Fig. 9 Mechanism of BC biosynthesis involving lipid intermediates (Han and Robyt, 1998; with kind permission; modified). Lip, lipid; P, phosphoryl; Rib, D-ribose; U, uridine, •, glucosyl residue; LP: UDPGPT, lipid pyrophosphate: UDPGlc-phosphotransferase; CS; cellulose synthase; LPP, lipid pyrophosphate phosphohydrolase.

another Glc-α-1P from UDPGlc can be attached to the LipP, released in this reaction. The cycle of these three reactions is continued to give the β-1,4-glucan chain of an appropriate length. The mechanism includes the inversion of a configuration on the anomeric carbon in the Glc residue transferred either from one LipPP-α-Glc molecule to another, or to cellobiose, -triose, -tetraose, and subsequently to *n*-ose. An acceptor of the elongated chain of β-1,4-glucan is always a single α-D-Glc residue, linked with one of two LipPP molecules present in the polymerizing system. It means that the chain elongation occurs from its reducing end and nonreducing ends of nascent cellulose chains are situated away from the cells; this is contradictory to Brown's model. According to Han and Robyt's model, both the two cooperating LipPP molecules and the CS complex, which is not a processing glycosyltransferase, are embedded in the cytoplasmic membrane of the bacterial cell.

Participation of a lipid intermediate in cellulose biosynthesis was confirmed for *Agrobacterium tumefaciens* (Matthysse et al., 1995), as well as for synthesis of *Salmonella* O-antigen polysaccharide (Bray and Robbins, 1967) and *Xanthomonas campestris* xanthan (Ielpi et al., 1981). Furthermore, the lipid intermediate is probably also involved in the synthesis of acetan (De Lannino et al., 1988), which is a soluble polysaccharide produced together with cellulose by numerous *A. xylinum* strains (see Section 7.6). Further studies have to be carried out to elucidate whether the lipid intermediate plays in cellulose formation, the role postulated by Han and Robyt (1998). On the contrary, Brown's hypothesis (1996, 2000) is confirmed by cellulose synthesis *in vitro*, catalyzed by *A. xylinum* membrane preparations solubilized with digitonin (Lin et al., 1985; Bureau and Brown, 1987), and by polymer synthesis by purified CS subunits (Lin and Brown, 1989), which did not contain the lipid component.

According to both proposed hypotheses, BC biosynthesis does not require any primer, in agreement with earlier deductions (Canale-Parda and Wolte, 1964).

7.3.2
Assembly and Crystallization of Cellulose Chains

Polymerization of glucopyranose molecules to the β-1,4-glucan, whatever the mechanism, is the least complicated stage of cellulose synthesis, also in bacteria. The unique structure and properties of cellulose, which are dependent on its origin, result from the course of further stages: it means extrusion of the chains and their assembly outside the cell, giving supramolecular structures. Most important in these processes is the specific organization of multimeric complexes of CS, which are anchored in cytoplasmic membrane. In *A. xylinum* cells these complexes are ordered linearly along the long axis of the cell and one cell contains 50–80 TCs (Brown et al., 1976), whereas in vascular plants TCs form characteristic, 6-fold symmetrical rosettes (Brown and Saxena, 2000).

Stronger aeration, e.g., during agitated *A. xylinum* cultures, or the presence of certain substances, that cannot penetrate inside the cells, but can form competitive hydrogen bonds with the β-1,4-glucan chains (e.g., carboxymethyl-(CM)-cellulose, fluorescent brightener Calcofluor white), bring about significant changes in the supramolecular organization of cellulose chains, and instead of the ribbon-like polymer, i.e. the metastable cellulose I (Figure 10), which fibrils are colinear to CS complexes and gathered in a row, the second, thermodynamically more stable allomorph amorphous cellulose II is formed. This phenomenon is accompanied by disorganization of the TC' linear order. Discussing this problem, Brown and Saxena (2000) state that the conditions-dependent form of the final product may somehow determine the TC' arrangement in the cytoplasmic membrane, though it is believed that in *A. xylinum* these complexes have a stationary character, as opposed to some algae (Kudlicka, 1989). Worth mentioning is the well-known phenomenon of *A. xylinum* cell motion during BC synthesis, in the direction opposite to cellulose chain extrusion (Brown et al., 1976). Reversal movement of *A. xylinum* cells enabled calculation

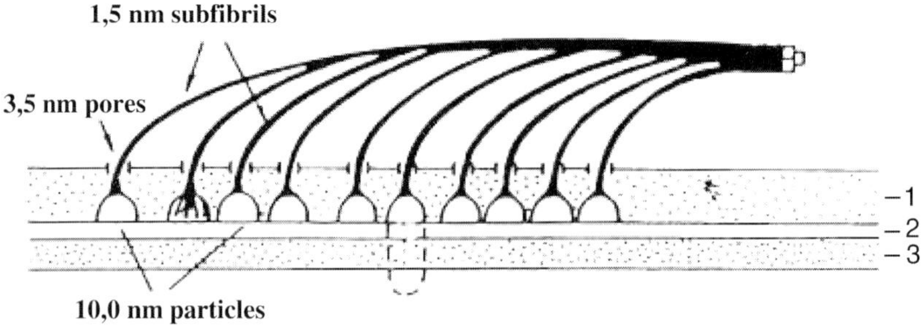

Fig. 10 Model of *A. xylinum* cellulose subfibrils formation: 1, lipopolysaccharide layer; 2, periplasmic space; 3, plasmalemma; 10 nm particles, CS subunits (Kudlicka, 1989; with kind permission).

of the typical chain elongation rate, which is equal to 2 μm min^{-1}, and corresponds to a polymerization of more than 10^8 glucose molecules into the β-1,4-glucan per hour and per single bacterial cell. Thus if forces associated with subfibrils assembly to bundles and ribbons, are strong enough to shift the motion of whole *A. xylinum* cells, these forces may as well influence the spatial arrangement of CS subunits, localized in the semi-liquid lipid bilayer of the membrane.

The spatial assembly of the β-1,4-glucan chains has a hierarchical character (Figure 10). In the first stage, 10–15 nascent β-1,4-glucan chains form a subfibril (also called a protomicrofibril), 1.5 nm in diameter, which is not a rod-like structure, but a left-handed triple helix (Ross et al., 1991). The subfibrils gather to form (also twisted) microfibrils composed of numerous parallel chains. The next structure in hierarchy is a bundle of microfibrils, followed by loosely wound ribbon, comprising about 1000 individual chains of the β-1,4-glucan (Haigler, 1985). In the presence of Calcofluor, which penetrates the outer membrane of the cell envelope, and forms complexes with protomicrofibrils, their aggregation to microfibrils is stopped. In the presence of much larger molecules such as CM-cellulose and xyloglucan (Yamamoto et al., 1996), which are too large to traverse the membrane, assembly of the bundles of microfibrils and ribbons is disrupted, thus indicating that formation of these two latter structures occurs exocellularly. Ross et al. (1991), who analyzed the mechanism of *A. xylinum* cellulose chain association, emphasized the important role of specific sites of adhesion in the inner and outer membranes of cell envelope, whose presence in bacterial cells was first reported by Bayer (1979). Their occurrence would enable an export of cellulose microfibrils without any interactions with the peptidoglycan gel, which fills the periplasmic space. Such interactions would prevent formation of correct hydrogen bonding between the protomicrofibrils.

The molecular organization of pores, through which *A. xylinum* cellulose microfibrils migrate outside the cell, remains unknown. The participation of expression products of the *bcsC* and *bcsD* genes, belonging to CS operon (see Section 7.4), in their formation cannot be excluded (Wong et al., 1990), all the more since these expression products appeared to be essential for cellulose synthesis *in vivo* (Ben-Bassat et al., 1993). Also the function of three proteins: CSAP20, 54, and 59 (CS-associated proteins), has to be scrutinized. They bind to cellulose together with the CS catalytic subunit, as revealed by experiments on CS purification (Benziman and Tal, 1995). It seems probable that isolation and purification of other components of the complex machinery responsible for cellulose synthesis, in concert with direct mutations of genes from *A. xylinum* CS operon, may help to elucidate the sequence of events during this process on the molecular level.

The process of assembly of nascent cellulose chains, not separated from parental *A. xylinum* cells, displays a unique and remarkable dynamics. Extrusion of these chains outside and their assembly into ordered structures is assisted by reversal of the motion of the cells. Throughout the whole process of the β-1,4-glucan synthesis and secretion, *A. xylinum* cell simultaneously turns around its own axis (uncoiling motion) (Brown et al., 1976). Also, this movement is driven by forming of twists of the ribbons, which are anchored in extrusion loci, localized on the side surface of the elongated cell. These twists are noticeable under the electron microscope.

7.4 Genetic Basis of Cellulose Biosynthesis

Cellulose biosynthesis involves several enzymes. Their function starts from the synthesis of the direct cellulose precursor UDPGlc, followed by polymerization of the monomer. It was found that the first reactions catalyzed by glucokinase, phosphoglucomutase and UDPGlc pyrophosphorylase, do not limit the final rate of BC synthesis, since *A. xylinum* synthesizes an excess of UDPGlc (Ben-Basat et al.,1993). The rate of BC synthesis is limited by diguanylate cyclase and oligomeric, regulatory complexes of CS. Therefore construction of more efficient cellulose-producing *A. xylinum* mutants, requires determination of organization of the genes coding for these enzymes. This has been studied by Wong et al. (1990) and Ben-Bassat et al. (1993), who proved that BC biosynthesis involves four coupled genes: *bcsA* (2261 bp), *bcsB* (2405 bp), *bcsC* (3956 bp), and *bcsD* (467 bp), forming the CS operon, which is 9217 bp in length, and is transcribed simultaneously as polycistron mRNA.

Although the function of the proteins encoded by each of these genes has not yet been precisely determined, some information on their role was derived from genetic complementation experiments (Wong et al., 1990). Results of these studies indicate that CS-deficient strains can be complemented by the gene *bcsB*, which codes for the 85-kDa protein (802 amino acid residues), and that *A. xylinum* mutants defective in both CS and diguanylate cyclase are complemented by the gene *bcsA*, coding for the 84-kDa protein (754 amino acid residues) (Ben-Bassat et al., 1993). Based on different mutations it was concluded that protein A takes part in the interaction of the CS complex with c-di-GMP, which is an allosteric activator of CS, earlier postulated by Saxena et al. (1991). The protein B is capable of binding UDPGlc and synthesizing β-1,4-glucan chains. A gene coding for the catalytic subunit (*cesA*) and analogous to the gene *bcsB* (Wong et al., 1990; Ben-Bassat et al., 1993) was also described by Saxena et al. (1990a), who one year later identified the gene coding for the regulatory subunit (*cesB*). However, Saxena et al. (1990a, 1991) suppose that the position of the first two genes in the CS operon is opposite and that the first gene codes for the CS catalytic subunit.

The role of the expression products of both *bcsC* and *bcsD* genes has not been determined so precisely. These genes probably also code for proteins (molecular mass of 141 and 17 kDa, respectively), anchored in the membrane, and both essential for cellulose synthesis *in vivo* and for protein secretion from the cells. Ben-Bassat et al. (1993) reported that splitting or deletion of the *bcsD* gene markedly reduced synthesis of cellulose. The presence of genes, which govern the assembly of cellulose chains, in the CS operon was also predicted by Saxena et al. (1990a) and Ross et al. (1991).

It is not clear if plasmid DNA participates in BC synthesis and, if so, what its role is. However it was proven that the composition and size of plasmids detected in 60% of the non-reverting Cel$^-$ mutants of the *A. xylinum* ATCC 10245 (currently 17005) strain, that were obtained by means of mutagenesis with *N*-methyl-*N*′-nitro-*N*-nitrosoguanidine, were markedly different from that of the wild strain. This observation suggests the plausible relationship between plasmid DNA structure and BC synthesis.

Mobile DNA fragments, including insertion sequences (IS), particularly widespread among prokaryotes, and 750–2500 bp in length (Galas and Chandler, 1989), are important factors of regulation of many genes. The ISs may activate or inactivate genes, and start DNA rearrangement such as deletions, inversions, and cointegrations. It was revealed that ISs participated in regulation of extracellular polysaccharides synthesis in *Pseudomonas atlantica* (Bartlett and Silverman, 1989), *Xanthomonas campestris* (Hotte et al., 1990), and *Zoogloea ramigera* (Easson et al., 1987). Their participation in BC synthesis regulation is also probable.

The best recognized insertion sequence in *A. xylinum* is the IS 1031 (950 bp, with known nucleotide sequence). Alteration in the IS 1031 profile was detected in the majority of Cel⁻ mutants, in comparison to the wild strain. Some recombinants contained two or more IS 1031 fragments. Coucheron (1991) observed even more significant changes in the IS pattern, pointing to DNA rearrangement and tried to obtain revertants by splitting off the additional ISs from the inactivated gene and obtained pseudorevertants, which produced a wax-like substance on the surface of nutrient broth, but the capability of cellulose synthesis was not restored. These experiments indicate that the presence of ISs contributes to the genetic instability of *A. xylinum*.

7.5 Regulation of Bacterial Cellulose Synthesis

Cyclo-3,6′:3′6 diguanosine monophosphate (c-di-GMP, see Figure 11) is a reversible allosteric activator of *A. xylinum* CS and plays a crucial role in regulation of the whole β-1,4-glucan biogenesis (Ross et al., 1987).

This compound binds to the enzyme regulatory subunit, and induces conformational changes that facilitate association of the CS protomers and lead to the enhancement of its reactivity (Ross et al., 1987).

The concentration of c-di-GMP in *A. xylinum* cells is regulated by 3 enzymes: diguanyl cyclase (CDG), phosphodiesterase A (PDEA) and phosphodiesterase B (PDEB)

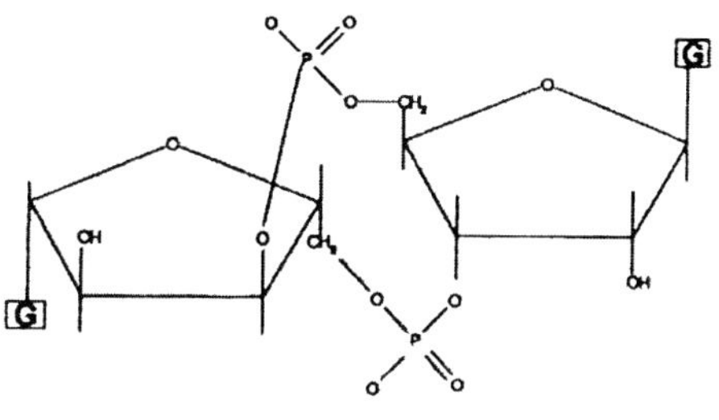

Fig. 11 Structure of c-di-GMP – the allosteric activator of *A. xylinum* CS.

(Figure 12) (Ross et al., 1987). PDEA and PDEB are anchored in the cytoplasmic membrane, and CDG has two forms. One of them is anchored in the cytoplasmic membrane and the second one operates in the cytoplasma (Ross et al., 1991). Recently, Weinhouse et al. (1997) reported on a new c-di-GMP-binding protein, which probably also participates in the intracellular regulation of free c-di-GMP concentration. Tal et al. (1998) proved that the cellular turnover of c-di-GMP in *A. xylinum* is controlled by three operons.

CDG, the key regulatory enzyme in the *A. xylinum* cellulose synthesis system, is composed of two polypeptide chains and encoded by two genes (Nichols and Singletary, 1998). CDG is activated by Mg^{2+} ions and specifically inhibited by saponin (Ohana et al., 1998), a glycosylated terpenoid. CDG converts two GTP molecules at first into the linear (pppGpG) and then into the cyclic (c-di-GMP) diguanosine monophosphate, which activates CS. PDEA splits active c-di-GMP into a linear inactive dimer (pGpG, di-GMP), further decomposed by PDEB into two molecules of 5′GMP (Figure 12). PDEA is inhibited by Ca^{2+} ions, which do not influence PDEB. Therefore, the Ca^{2+} concentration indirectly affects the rate of cellulose synthesis. High concentration of these ions enhances this rate, since the

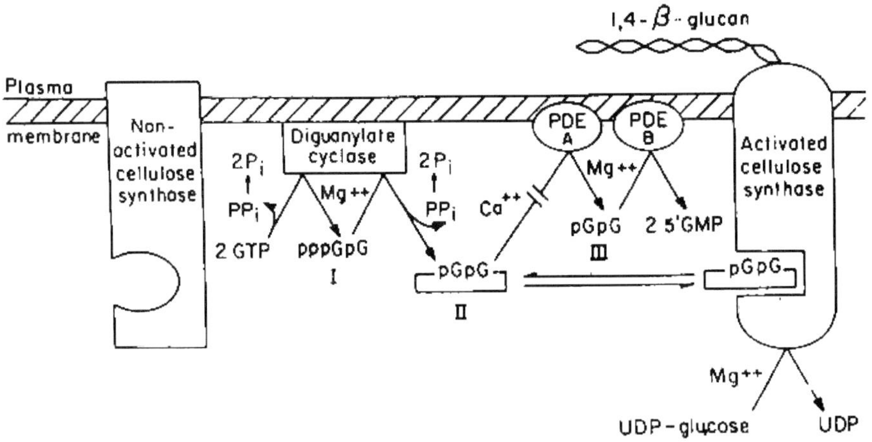

Fig. 12 Proposed model for cellulose synthesis in *A. xylinum*. For simplicity, the synthesis of a single β-1,4-glucan chain is depicted, although a more complex form of CS, polymerizing several chains simultaneously, might be the active enzyme unit in cellulose biogenesis (Ross et al., 1991; with kind permission). PDEA, phosphodiesterase A; PDEB, phosphodiesterase B; pppGpG, diphospho-di-GMP; pGpG (II), c-di-GMP (cyclic-di-GMP); pGpG, di-GMP (linear di-GMP).

conversion of c-di-GMP to the inactive linear form is inhibited.

It is believed that both the molecular background of the CS activation by c-di-GMP as well as the regulation of its synthesis, involving a positive influence of CDG, and a negative effect of PDEA and PDEB, have a unique character. To date, such a mechanism of cellulose synthesis regulation has only been detected in *A. xylinum*.

7.6
Soluble Polysaccharides Synthesized by *A. xylinum*

Apart from cellulose, some related soluble polysaccharides are also synthesized by *A. xylinum*. In 1970s, a soluble β-homoglucan was detected. Its main backbone was composed of β-1,4-linked glucose residues and every third glucose moiety of this chain was substituted with another glucose via β-1,2 linkage (Colvin and Leppard, 1977; Colvin et al., 1979). *A. xylinum* cellulose pellicle non-forming (Pel^-) strains synthesize α-glucan linked with the cell envelope (Dekker et al., 1977). Valla and Kjosbakken (1981) isolated a soluble polysaccharide containing Glc, Man, Rha, and glucuronic acid residues (molar ratio of 3:1:1:1, respectively) from the culture broth of the Cel^- strain. The related polymer, composed of the same moieties (altered molar ratio of 6:2:1:1, respectively), synthesized by another Cel^- strain, was detected by Minakami et al. (1984). Probably some of the residues in these polymers are acetylated (Tayama et al., 1985). Polysaccharides of this type have received the common name acetan (De Lannino et al., 1988). Moreover, some wild *A. xylinum* strains synthesize a similar soluble polysaccharide (Glc, Man, Rha, and glucuronic acid; 6:1:1:1, respectively) together with cellulose (Savidge and Colvin, 1985).

The question is whether cellulose-assisting soluble polysaccharides play any role in its biogenesis. It seems that they do not play any direct role, although kinetics of their biosynthesis are the same as that of cellulose (Marx-Figini and Pion, 1974) and, moreover,

UDPGlc was preferentially used for production of these soluble saccharides instead of the production of cellulose (Delmer, 1982). Some authors state that *in vivo* nascent cellulose microfibrils are coated with amorphous material (Leppard et al., 1975), which may be composed of the above-mentioned soluble polysaccharides. The absence of these soluble polymers in the liquid fraction of static culture broths and their presence in the cellulose mat confirms that assumption (Valla and Kjosbakken, 1981).

Much earlier, Ben-Hayyim and Ohad revealed in 1965 that CM-cellulose present in the medium, used for BC synthesis, resulted in incorporation of this soluble derivative of cellulose into microfibrils of the polymer. However, incorporation of the *A. xylinum* soluble exopolysaccharides into cellulose microfibrils has not been observed.

7.7 Role of Endo- and Exocellulases Synthesized by *A. xylinum*

The first reports on *A. xylinum* cellulases, detected independently by Okamoto et al. (1994) and Standal et al. (1994), included description of their genes. The first authors selected from *A. xylinum* IFO 3288 DNA gene libraries one gene coding for a 24-kDa protein (218 amino acid residues), which displayed CM cellulase activity. Standal et al. (1994) found another endocellulase gene in the *A. xylinum* ATCC 23769 Cel^- mutant, localized upstream the CS operon. They revealed that the loss of cellulase synthesis capability resulted from gene splitting. The conclusions of Tonouchi et al. (1997), who investigated the localization of the *A. xylinum* BPR 2001 endo-1,4-β-glucanase gene, were similar. The latter gene was localized upstream the CS operon, whereas the gene of the second cellulolytic enzyme, produced by this strain, i.e. an exo-1,4-β-glucosidase, was found downstream this operon.

A. xylinum BPR 2001 endo- and exocellulase were purified and characterized (Oikawa et al., 1997; Tahara et al., 1998). The studies of Tahara et al. (1997) also revealed that at pH 5.0, optimal for growth and BC synthesis, the activity of both cellulases is several times higher than at pH 4.0, at which the BC production is only slightly declined. It was also observed that at pH 5.0, the average DP was decreasing from 16,800 to 11,000 with the time of cultivation, whereas at pH 4.0 these changes were negligible. The cellulose obtained at pH 5.0 had inferior physical properties, i.e. a lower tensile strength (a lower Young's modulus value). These results and colocalization of genes of cellulases and CS operon, suggest that the endo-1,4-β-glucanases and exo-1,4-β-glucosidase are involved in *A. xylinum* cellulose biogenesis (Tahara, 1998).

More particularly, it relates to possible degradation of the nascent β-glucan chains, synthesized in the later phase of growth. Furthermore, the *A. xylinum* BPR 2001 exo-1,4-β-glucosidase displays both hydrolytic and transglycosylating activity (Tahara et al., 1998), similarly to other glycosidases, which do not cause inversion of configuration on the anomeric carbon. The latter activity may be responsible *in vivo* for changes in the average DP of the polymer; however, further research is necessary to explain this hypothesis. Tentative evidence for such a possibility might be the behavior of soybean cells cultured *in vitro*. The activity of their β-glucosidase, which also has transglycosylating activity, increases together with the length of the cells, since this enzyme probably participates in the transfer of glycosidic residues of hemicelluloses precursors to the growing terminus of the cell wall, in which intensive synthesis of these polysaccharides is taking place (Nari et al., 1983).

Whether *A. xylinum* cellulases are involved in releasing the energy stored in the form of cellulose, in periods of starvation, is still a question to answer, although some studies confirm this thesis (Okamoto et al., 1994). It is known that some *Sclerotium* endo-1,3-β-glucanases play such a role (Jones et al., 1974).

Current observations indicate that modification of the physical properties and DP of BC can be achieved either by using compounds that influence its biosynthesis, or by exploiting the activity and specificity of *A. xylinum* cellulases.

8 Biodegradation of BC

Independent of its origin, cellulose undergoes total biodegradation in nature. However, in comparison to PC, associated both with polymers susceptible to degradation, like hemicelluloses and pectin, and with lignin, which is the most resistant plant polymer, BC is relatively pure and *a priori* more susceptible to attack by cellulolytic enzymes, which are produced mainly by fungi and numerous bacterial species. In this respect, commercial application of BC is friendly for the environment.

Furthermore, the structure and properties of BC (large accessible surface area) make it the superior model substrate for studies on cellulases (1,4-β-glucan 4-glucanohydrolases, formerly endocellulases, EC 3.2.1.4) and cellobiohydrolases (cellulose 1,4-β-cellobiosidases, EC 3.2.1.91), that are the main components of cellulosomes, the specialized multienzymatic particles from cellulolytic bacteria, as well as fungal systems for cellulose decomposition.

Recently, Boisset et al. (1999) proved that *Clostridium thermocellum* cellulosomes completely decompose *A. xylinum* cellulose microfibrils faster than microcrystalline *Valonia ventricosa* cellulose. Ultrastructural observations of the hydrolysis process, using transmission electron microscopy, infrared spectroscopy, and X-ray diffraction analysis, indicated that the rapid hydrolysis of BC resulted from very efficient synergistic action of the various enzymic components, present in the cellulosome scaffolding structure. Further studies of Boisset et al. (2000) revealed that *Humicola insolvens* cellobiohydrolase Cel7A (previously CBH I) brought about thinning of dispersed BC ribbons, whereas the mixture of Cel6A (CBH II) and Cel7A (in the ratio of 2:1, respectively) cut the ribbons to shorter pieces, thus suggesting the partly endo-manner of CBH II attack. The phenomenon of low inherent endoactivity of some exoglucanases has been known for several years and explains more efficient synergistic action of cellobiohydrolases in comparison to endoglucanases (Teeri, 1997). According to current opinions, the system of multiple exo- and endocellulases, both in bacterial cellulosomes, and in fungal cellulolytic complexes, represents the perfectly balanced continuum of activity, providing efficient degradation of various cellulosic substrates.

BC susceptibility to cellulases was also observed by Samejima et al. (1997), who found that the mixture of *Trichoderma viride* CBH I and endoglucanase II, drastically disintegrated the twisted and bent ribbon-like structure of microfibril bundles to linear needle-like microcrystallites, and also caused rapid polymer fragmentation. Transformation of the coiled structure of BC ribbons to the needle-like one is driven by the remarkable twisting motion of the substrate, which is probably a result of a tension – released when the microfibrils are being decomposed to shorter fragments by cellulases. Samejima et al. (1997) also detected that the acid-treated BC, containing many microfibril aggregates, was not susceptible to attack of both enzymes.

Similar results were achieved by Srisodsuk et al. (1998), who digested microcrystalline BC with *Trichoderma reesei* CBH I and observed rapid solubilization of the polymer, but slow decrease in its DP. The endocellulase I from this organism attacked the substrate in the opposite manner. Both enzymes hydrolyzed cotton cellulose more slowly than BC, though cotton cellulose exhibits relatively high purity as compared to other plants.

One of the reasons of the susceptibility of BC to the attack of cellulases is its large accessible surface area, that even increases throughout hydrolysis and facilitates effective binding of cellulolytic enzymes, the crucial step in the process of degradation. It was concluded based on studies of Palonen et al. (1999), who found that the number of *T. reesei* CBH I and CBH II molecules linked to BC, increased during hydrolysis. Furthermore, the CBH II binding was markedly stronger (the adsorption/desorption of the enzyme from its substrate was assisted by 60–70% hysteresis), pointing to a distinctly different processing character of this enzyme.

Stalbrandt et al. (1998) compared the mode of attack of four *Cellulomonas fimi* endo- and exocellulases against microcrystalline BC and acid-swollen Avicel cellulose. The latter was decomposed more effectively by all the enzymes. In most cases 45–65% solubilization and a decrease in DP were observed, except for Cbh B, which caused 27% solubilization. A high degree of bacterial polymer solubilization (67%) was achieved for Cbh A, known as the processing enzyme. Results of Stalbrandt et al. (1998) proved that only the external surface of BC fibrils is accessible for *C. fimi* cellulases. Since Cbh A is the processing enzyme, it can remove external fibrils much faster than the other three cellulases and attack the deeper internal surface of the polymer. Therefore, the other enzymes (two nonprocessing endocellulases and Cbh B, which displays weaker processing properties) could not solubilize BC so effectively, whereas amorphous soluble cellulose was equally available to all of them.

Current results indicate that complete digestion of the highly crystalline bacterial polymer will not cause any problems, all the more since organisms responsible for BC biodegradation in natural environments produce a multitude of various endo- and exocellulases, whose activities complement each other. Furthermore, as opposed to cellulose of plant origin, which requires pretreatment to provide easier enzymatic conversion to sugars, BC can be directly digested by cellulases.

9 Biotechnological Production

To achieve high productivity and yields of BC and to reduce cost of its production, special emphasis should be given to the following aspects:

- development of screening methods providing selection of *A. xylinum* strains, which can efficiently produce cellulose from various inexpensive waste carbon sources – wild strains derived from the screening could be improved using genetic engineering methods;
- optimization of *A. xylinum* culture conditions (static or agitated culture, fermentor type), determining both a form (a pellicle or an amorphous gel) and properties (resilience, elasticity, mechanical strength, absorbency) of BC, which have to be tailored to the further polymer application;

- optimization of culture broth composition (carbon and nitrogen sources, biosynthesis stimulating compounds, microelements, etc.) and process conditions (pH, temperature, aeration).

These are discussed in the next sections.

9.1 Isolation from Natural Sources and Improvement of BC-producing Strains

One of the methods enabling selection of proper *A. xylinum* strains, is the screening for strains, which cannot oxidize glucose via gluconic acid to 2-, 5-, or 2,5-ketogluconate (Winkelman and Clark, 1984; Johnson and Neogi, 1989; De Wulf et al., 1996; Vandamme, 1998). In this respect, De Wulf et al. (1996) applied an agar nutrient medium containing Br^- and BrO_3^- ions, that combine at acidic pH to release molecular bromine, toxic to *A. xylinum* cells. Only those mutants, which do not convert glucose into gluconate or its derivatives, can survive in this medium. Vandamme et al. (1998) successfully used this method to isolate *A. xylinum* KJ33 strain, producing 3.3 g cellulose L^{-1}. *A. xylinum* strains, which do not metabolize glucose to gluconic acids, were also selected on $CaCO_3$ containing medium (Johnson and Neogi, 1989). Another approach to avoid conversion of glucose to organic acids, was screening for mutants defective in glucose-6-phosphate dehydrogenase. The mentioned increase in final concentration, volumetric productivity and total cellulose yield, was achieved in agitated cultures. For the same purpose, i.e. selection of the best cellulose producers, preparations of cellulases, added into the growth medium, have also been used (Brown, 1989c).

Mutation with chemical compounds, such as *N*-methyl-*N*'-nitro-*N*-nitrosoguanidine or ethylmethanesulphate, as well as with ultraviolet irradiation gave mutants that displayed higher cellulose productivity, reduced synthesis of the soluble polysaccharide acetane, and a much lower degree of glucose conversion into organic acids (Johnson and Noegi, 1989).

Screening of *A. xylinum* strains was also aimed at isolation of spontaneous or induced cellulose II-producing mutants. This cellulose is synthesized by bacteria that form smooth colonies and do not produce a pellicle on the surface of the nutrient broth. Instead, the polymer is dispersed in the whole volume of the medium.

Selection of the best BC producers has been also performed traditionally, by determination of the synthetic activity of a plethora of individual monocultures. For example, Toyosaki et al. (1995) isolated 2096 *Acetobacter* strains from a large number of plant samples (fruits, flowers) and 412 strains forming a mat on the surface of nutrient medium were subjected to further investigations. The procedure yielded *A. xylinum* ssp. sucrofermentans BPR2001 – the strain which efficiently synthesized BC in agitated culture.

9.1.1 Improvement of Cellulose-producing Strains by Genetic Engineering

Expression of genes coding for various carbohydrases from organisms other than *A. xylinum* enhances the scope of carbon sources available, that may increase BC yield and decrease the nutrient broth price. An example is the expression of sucrose phosphorylase (EC 2.4.1.7) gene from *Leuconostoc mesenteroides* in *Acetobacter* sp. (Tonouchi et al., 1998), that resulted in utilization of sucrose as a carbon source and increased cellulose production. In addition, Nakai et al. (1999) obtained double cellulose yield by means of expression of the mutant sucrose synthase (EC 2.4.1.13) gene from mung bean (*Vigna radiata*, Wilczek) in *A. xylinum* strain. In the mutant enzyme, the eleventh serine residue was replaced by glutamic acid (S11E); this caused higher affinity of the engineered enzyme toward sucrose and favored cleavage of this sugar for the synthesis of UDPGlc. Introduction of the mutant gene into *A. xylinum* not only changed sucrose metabolism in the recombinant strain, but also created a new metabolic pathway of direct UDPGlc synthesis (Figure 13). The energy driving this process is derived only from the cleavage of the glycosidic bond in sucrose without any other energy input required, which is the case when UDPGlc is synthesized via the conventional pathway. Furthermore, the UDP molecules resulting from glucose polymerization process can be directly recycled by the sucrose synthase activity, which increases the rate of the coupled reactions and a higher rate of cellulose synthesis is observed. It should be emphasized that higher plants have both systems of UDPGlc synthesis.

9.2 Fermentation Production

Growth on the surface of liquid media and synthesis of the gelatinous, leather-like mat, are natural properties of *A. xylinum* strains. Therefore static conditions seemed to be optimal for BC synthesis. However, the main drawbacks of static culture, such as the synthesis of the polymer only in the form of a sheet and relatively low productivity, have contributed to the development of new fermentation processes.

9.2.1 Carbon and Nitrogen Sources

In studies on factors affecting the BC production yield, much attention has been paid to carbon sources. Numerous mono-, di-, and polysaccharides, alcohols, organic acids, and other compounds were compared by Jonas and Farah (1998), who found out that the preferred carbon sources were D-arabitol and D-mannitol, which resulted in a 6.2- or 3.8-fold higher cellulose production, respectively, in comparison to glucose. Both sugar alcohols provided stabilization of the pH throughout the culture period, since they were not converted to gluconic acids.

Tonouchi et al. (1996), who used a strain of *A. xylinum* to obtain cellulose from glucose and fructose, found that fructose stimulated the activities of phosphoglucose isomerase and UDPGlc pyrophosphorylase, thus enhancing cellulose yield.

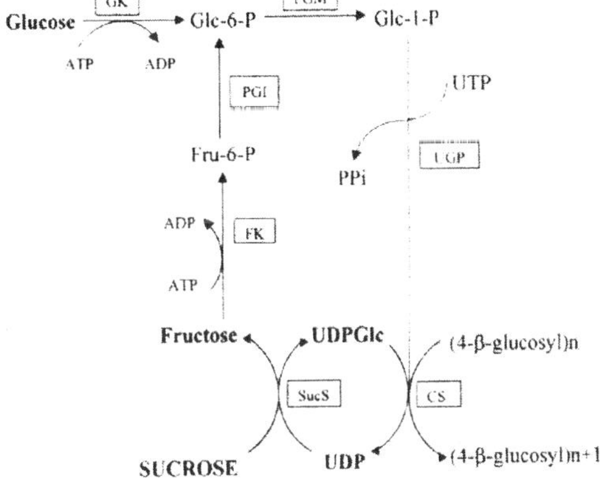

Fig. 13 UDPGlc biosynthesis pathways in a recombinant strain of *A. xylinum* containing the sucrose synthase (SucS) gene. Other abbreviations are the same as in Figure 6 (see Section 7.1).

When maltose was the sole carbon source in the nutrient medium (Masaoka, 1993) BC production was even 10 times lower than in glucose-containing medium and the polymer was markedly shorter (DP = 4000–5000) than in the presence of glucose (DP = 11,500).

Matsuoka et al. (1996) investigated BC synthesis by *A. xylinum* ssp. sucrofermentous BPR2001 in agitated culture and found out that the presence of lactate in the growth medium stimulated bacterial growth and enhanced the cellulose yield 4–5 times. The source of lactate is usually corn steep liquor (CSL), which is one of the main medium components applied for BC production, especially in agitated cultures. Contrary to glucose and fructose, lactate is not converted to UDPGlc, but is metabolized via pyruvate and oxalacetate in the Krebs cycle, being a source of energy for cellulose production. To provide lactate in the growth medium, cultivation of mixed cultures of lactic acid and acetic acid bacteria has been carried out. The best results gave strains of *Lactobacillus, Leuconostoc, Pediococcus,* and *Streptococcus.* The lactic and acetic acid bacteria were grown also in the presence of sucrose-hydrolyzing *Saccharomyces* yeast (β-fructofuranosidase producers). This principle provided a high BC yield, since after 14 days of agitated culture, as much as 8.1 g of cellulose L^{-1} was obtained, when in the absence of *Lactobacillus* strain, only 6.4 g L^{-1} (Seto et al., 1997).

One of the cellulose synthesis-stimulating compounds appeared to be ethanol (Naritomi et al., 1998). Continuous culture of *A. xylinum* in fructose-containing medium, revealed that 10 g ethanol L^{-1} enhanced a BC yield, but a concentration of 15 g L^{-1} prevented polymer synthesis. These results suggest that similarly to lactate, ethanol is also a source of energy (accumulated as ATP) and not a substrate for cellulose synthesis. ATP activates fructose kinase and inhibits glucose-6-phosphate dehydrogenase, thus halting conversion of 6-P-glucose to 6-phosphogluconate. It can be concluded that – due to these coupled reactions – the amount of fructose further isomerized to glucose-6-phosphate is increased.

Each cellulose-producing strain requires a specific complex nitrogen source, providing not only amino acids, but vitamins and mineral salts as well. These requirements are met with yeast extract, CSL as well as hydrolyzates of casein and other proteins. The preferred nitrogen sources are yeast extract and peptone, which are basic components of the model medium developed by Hestrin and Schramm (1954; H-S medium), applied in numerous studies on BC synthesis. However the most recommended nitrogen source for agitated cultures is CSL (Johnson and Noegi, 1989).

It was also found that a significant part of the expensive medium components, i.e., yeast extract and bactopeptone, can be replaced with CSL or even white cabbage juice. Also waste plant materials such as sugar beet molasses, spent liquors after glucose separation from starch hydrolyzates, as well as whey and some fermentation industry wastes (e.g., spent liquors after dextran precipitation with ethanol) were appropriate medium components (Krystynowicz et al., 2000).

Studies on the influence of vitamins on BC synthesis (Ishikawa et al., 1995, 1996b) revealed that the most stimulating ones were pyridoxine, nicotinic acid, *p*-aminobenzoic acid and biotin, and even CSL media should be fortified with these substances. Some other compounds, strongly stimulating cellulose production by *A. xylinum* strains, such as derivatives of choline, betaine, and fatty acids (salts and esters), were also identified (Hikawu et al., 1996).

High BC production (up to 25 g L^{-1}) can be achieved using optimum growth medium composition, designed by mathematical methods and computer analysis (Joris et al., 1990; Embuscado et al., 1994; Vandamme et al., 1998; Galas et al., 1999).

9.2.2
Effect of pH and Temperature

Analysis of the influence of pH on *A. xylinum* cellulose yield and properties, indicates that optimum pH depends on the strain, and usually varies between 4.0 and 7.0 (Johnson and Neogi, 1989; Galas et al., 1999). For instance, Ishikawa et al. (1996a) and Tahara et al. (1997), who applied two different *A. xylinum* strains, observed the highest polymer yield at pH 5.0. The same pH was found to be optimal by Krystynowicz et al. (1997). Comparison of adsorptive properties of the polymer obtained at different pH, learned that cellulose, accumulated in the S-H medium pH 4.8–6.0, displayed the highest water-binding capacity (Wlochowicz, 2001).

In addition to the pH of the nutrient broth, also the temperature influences BC yield and properties. In majority of reported experiments, the temperature ranged from 28 to 30 °C, and its variations caused changes of cellulose DP and water-binding capacity. For instance, BC synthesized at 30 °C had a lower DP (approximately 10,000) and a higher water-binding capacity (164%) in comparison to that produced at 25 and 35 °C (Wlochowicz, 2001).

9.2.3
Static and Agitated Cultures; Fermentor Types

Synthesis of BC is run either in static culture or in submerged conditions, providing proper agitation and aeration, necessary for medium homogeneity and effective mass transfer. The choice of culture conditions strictly depends on the polymer use and destination.

BC yield in static cultures is mostly dependent on the surface/volume ratio. Optimum surface/volume ratio protects from either too high (unnecessary) or too low aeration (cell growth and BC synthesis termination). Reported values of surface:volume ratio vary from 2.2 cm^{-1} (Joris et al., 1990) to 0.7 cm^{-1} (Krystynowicz et al., 1997). BC synthesis in static conditions can be achieved either in a one-step (medium inoculated with 5–10% cell suspension) or a two-step procedure. The latter one starts from agitated fermentation, followed by the static culture (Okiyama et al., 1992). Krystynowicz et al. (1997) modified the two-step procedure and applied two consecutive static cultures. Pellicles containing entrapped *A. xylinum* E_{25} cells, obtained in the first step (24 h), were cut to uniform pieces and used as an inoculum to start the next culture, run for 4–5 days. The method provided uniform bacterial growth and production of homogeneous pellicles.

The control of BC synthesis in static culture is very difficult since the pellicle limits access to the liquid medium. A particularly important parameter, which requires continuous control, is the pH. Accumulation of keto-gluconic acids in the culture broth brings about a decrease in pH far below the value that is optimum for growth and polysaccharide synthesis. Because conventional methods of pH adjustment cannot be used in static cultures, Vandamme et al. (1998) applied an *in situ* pH control via an optimized fermentation medium design, based on introducing acetic acid as an additional substrate for *Acetobacter* sp. LMG1518. Products of acetic acid catabolism counteracted the pH decrease caused by keto-gluconate formation, and provided a constant pH of 5.5 of the growth medium, throughout the whole process.

Fig. 14 *A. xylinum* cellulose formed on the surface of a roller in horizontal fermentor.

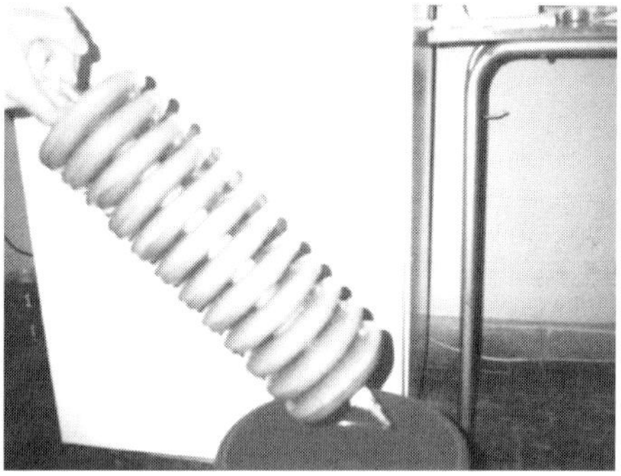

Fig. 15 *A. xylinum* cellulose deposited on disks in rotating disk fermentor.

Better control of BC synthesis can be achieved in special fermentors. Cellulose production in horizontal fermentors provides a combination of stationary and submerged cultures. The polymer is deposited on the surface of rollers or disks, rotating around the long axis (Figure 14). Part of their surface temporarily dips in the liquid medium or is above its surface (in the air). The advantages of this method include a larger polymer surface, synthesis of cellulose in a form of hollow fibers, different in diameter, as well as good process control, easy scale up, appropriate accessible surface for adhesion of bacteria and product deposition, better substrate utilization, higher rate of cellulose production, possibility of nutrient supply, additional aeration during the process, etc. (Sattler and Fiedler, 1990). Bungay and Serafica (1997) produced BC in a 1-L disk (12 cm in diameter) fermentor and revealed that optimum sugar (sucrose or glucose) concentration was 10 g L^{-1}, disks rotation rate 12 r.p.m., and constant pH 5.0. Krystynowicz et al. (1997) successfully applied a 11-L disk reactor containing 3 l of the H-S medium for BC synthesis by the *A. xylinum* E_{25} strain (Figure 15). The optimum rate of rotation of its 11 disks (12 cm in diameter each) appeared to be 3 r.p.m. As much as 4.2 g L^{-1} of dry cellulose was obtained after 7 d of growth.

BC production in submerged cultures usually requires common fermentors, equipped with some static parts (baffles, blades, etc.), enabling cell adhesion, since cellulose synthesis in the free aqueous phase usually drops. In analogy, various water-insoluble microparticles, such as sand, diatomaceous earth, or glass beads, added to the culture medium, enhanced BC productivity, since the biofilm formed by bacteria on the particles probably limited oxygen transfer and stopped glucose oxidation to gluconic acids (Vandamme et al., 1998).

Large-scale cellulose production, in fermentors with continuous agitation and aeration, encounters many problems, including spontaneous appearance of Cel^- mutants, which contribute to a decline in cellulose production. Optimized agitation and aeration prevent turbulence, which negatively effects cellulose polymerization and crystallization, thus reducing the polysaccharide yield. For instance, a rate of agitation equal to 60 r.p.m. and aeration of 0.6 volume per volume per minute (v.v.m.) were optimum for the *A. aceti* ssp. xylinum ATCC 2178 strain cultured in a 300-L fermentor for 45 h at 30 °C, and 10 g BC L^{-1} d^{-1} was obtained (Laboureur, 1988). Ben-Bassat et al. (1989) investigated BC production by the *A. xylinum* 1306–21 strain in 14-L Chemap fermentor, cultured in CSL medium containing glucose and CSL, 2% of each. The rate of agitation was equal to 900 r.p.m. and the dissolved oxygen concentration corresponded to 30% air saturation. The polymer yield amounted to 5.1 g L^{-1} d^{-1}. Chao et al. (2000) applied a 50-L internal loop airlift reactor for BC synthesis with the *A. xylinum* sp. BPR2001 strain. Aeration with oxygen-enriched air enlarged cellulose yield from 3.8 to 8 g L^{-1} after 67 h.

Production of BC in fermentors encounters similar agitation problems as cultivation of fungi or streptomycetes, since most of these organisms grow in pellet or filament form and their culture broths become non-Newtonian fluids. A high concentration of mycelium in the form of a dense suspension of diversely shaped particles limits agitation and gas transfer. Increasing the rate of agitation in order to improve aeration leads to damage to the product structure due to shearing forces. Bauer et al. (1992) took this aspect into consideration and introduced a polyacrylamide-protecting agent into the growth medium, applied for BC synthesis in fermentors. The protector reduced the shear damage and affected the specific productivity at high densities negatively since the cell growth rate was lower.

Accumulation of some metabolites can also affect production of cellulose. For instance, studies of Kouda et al. (1998) revealed that a high partial pressure of CO_2 negatively affected *A. xylinum* growth and reduced BC yield.

Laboureur (1998) developed a method for BC production by *A. xylinum* sp. ATCC 21780 strain in 300- and 500-L fermentors in a medium containing 5% sucrose, 0.05% yeast extract, 0.2% citric acid, nitrogen salts, Mg^{2+}, and phosphates. The medium was inoculated with 12% of inoculum, cultivated for 160 h. The process was run for 45 h at 30 °C and pH 4.8, and an aeration rate of 0.6 v.v.m. BC synthesis yield reached 18 g L^{-1} (10 g L^{-1} d^{-1}). Another strain, *Acetobacter* sp. ATCC 8303, was grown at 28 °C and pH 4.6, in a modified medium, containing 0.28% glucose, 0.07% maltose, 0.03% CSL, and 0.03% yeast extract. The aeration was 1 v.v.m. for the first 33 h and 0.5 v.v.m. for the last 12 h of the process, and the BC yield was 13 g L^{-1} d^{-1}.

Preparing an inoculum of appropriate cell density for large-volume fermentors is also a problem, mainly because the cells are entrapped in cellulose. To liberate the cells and increase their density, Brown (1989c) applied preparations of cellulases for partial cellulose hydrolysis. In the presence of these enzymes, the cell density reached 10^8 mL^{-1}, whereas in their absence it was only $1.12 \cdot 10^7$ mL^{-1}, after 30 h of growth.

In addition to the above-mentioned methods, the production of cellulose in the form of hollow fibrils of various diameters has also been tried (Yamanaka et al., 1990). Such material could be especially useful for the production of small-diameter blood vessels. The hollow BC fibril is obtained by growing the BC-producing bacteria on the inner and/or outer surface of an oxygen-permeable hollow carrier, produced from, for example, cellophane, Teflon, silicone, ceramics, etc.

9.2.4 Continuous Cultivation

Microbial cellulose can be also produced in a static continuous culture (Sakair et al., 1998). *A. xylinum* strain was cultured on trays, in the S-H medium, and after 2 d the pellicle produced on the surface was picked up, passed through a sodium dodecylsulfate bath to kill the cells, and set on a winding roller. The process was continued for a couple of weeks at a winding rate of 35 mm h^{-1} and fresh S-H medium was added into the trays every 8–12 h to keep its optimum level. A cellulose filament of more than 5 m was collected using this method, indicating its industrial application potential.

9.3 *In vitro* Biosynthesis

In vitro synthesis of cellulose has been one of the most difficult topics in the area of cellulose research. For the first time, cellulose was synthesized utilizing the 'reversed action' of cellulase as a catalyst, by Kobayashi et al. (1991), even though chemical synthetic methods have been applied before. Several monomers and catalysts have been used (Nakatsubo et al., 1989), but none of them gave the desired product, the stereoregular polymer of β-1,4-linked glucopyranoses. The idea of BC synthesis *in vitro* originated from experiments using *A. xylinum* cell-free extracts (Glaser, 1958), raw preparations of membranes (Colvin, 1980; Swissa et al., 1980; Aloni et al., 1982), and membranes solubilized with digitonin (Aloni et al., 1983). Such investigations proved that UDPGlc is the substrate for *A. xylinum* cellulose synthesis, c-di-GMP is the specific activator (Ross et al., 1987), and Ca^{2+} and Mg^{2+} ions are essential for this process (Swissa et al., 1980). Synthesis *in vitro*, carried out using *A. xylinum* membranes solubilized with digitonin (Lin et al., 1985), gave cellulose fibers 17 ± 2 Å in diameter. Their morphology and size resembled *A. xylinum* cellulose fibrils formed in the presence of factors disturbing crystallization of the nascent polymer (see Section 7.3.2). Further research, including X-ray diffraction analysis (Bureau and Brown, 1987) of cellulose produced by the *A. xylinum* membrane fractions, revealed that synthesis *in vitro* led to the amorphous cellulose II allomorph, whereas *in vivo* the highly crystalline cellulose I was obtained. Lin and Brown (1989) proved that purified CS also catalyzed cellulose synthesis *in vitro*. All these experiments were run on the microscale level and it is hard to believe today that commercial *in vitro* production of microbial polymer is possible, all the more since in the 1990s no progress in this area has been made.

9.4 Chemo-enzymatic Synthesis

Experiments on the chemical synthesis of cellulose have not given the expected results. Branched and low-molecular-weight glucans (Husemann and Muller, 1966) or polymers of β-1,4- and α-1,4-linked glucopyranoses (Micheel and Brodde, 1974) were obtained.

Successful experiments of Kobayashi et al. (1991) indicate that using coupled chemical and enzymatic methods may lead to the development of commercial *in vitro* synthesis of cellulose. The authors applied β-D-cellobiosyl fluoride, obtained by means of chemical synthesis, as a substrate and *T. reesei* cellulase as a catalyst in an aqueous–organic solvent system (a mixture of acetic buffer pH 5.0 and acetonitrile, 1:5 v/v) and achieved water-insoluble 'synthetic cellulose' having a DP > 22. X-ray and ^{13}C-NMR analyzes revealed that the product was the cellulose II. The authors stated that the DP of the product depends on reaction conditions, especially on the aqueous–organic solvent composition. They found that in the presence of higher concentrations of the substrate or acetonitrile, the cellulase – which acts as a glycosyltransferase – produces mainly soluble cellooligosaccharides (DP ≤ 8), which may also find numerous uses. Although these studies have not been continued, the method proposed by Kobayashi might be a promising alternative for total enzymatic *in vitro* cellulose production.

9.5
Production Processes Expected to be Applied in the Future

Recently, an original method of BC production on a large-scale was proposed by Nichols and Singletary (1998), who patented the idea of the construction of transgenic plants capable of this polymer synthesis. The authors intend to express three *A. xylinum* CS genes (*bcsA*, *bcsB*, and *bcsC*, see Section 7.4) and two *A. xylinum* diguanyl cyclase genes in storage tissues (roots, tubers, grains) of crops (potato, maize, oat, sorghum, millet, wheat, rice, sugarcane, etc.). They suppose to obtain masses of the pure polymer, easy and cheap to recover. According to their method, its production would be less expensive than by means of fermentation. The authors also emphasize the ecological aspect of their procedure, since an additional benefit of breeding of pure cellulose-producing transgenic plants would be a more economical use of forest resources.

9.6
Recovery and Purification

Microbial cellulose obtained via stationary or agitated culture is not completely pure and contains some impurities, such as culture broth components and whole *A. xylinum* cells. Prior to use in medicine, food production, or even the papermaking industry, all these impurities must be removed.

One of the most widely used purification methods is based on treatment of BC with a solution of hydroxides (mainly sodium and potassium), sodium chlorate or hypochlorate, H_2O_2, diluted acids, organic solvents, or hot water. The reagents can be used alone or in combination (Yamanaka et al., 1990). Immersing BC in such solutions (for 14–18 h, in some cases up to 24 h) at elevated temperature (55–65 °C) markedly reduces the number of cells and coloration degree.

Boiling in 2% NaOH solution after preliminary rinsing with running tap water was also reported (Yamanaka et al., 1989). Watanabe et al. (1998a) immersed the polymer in 0.1 NaOH at 80 °C for 20 min and then washed it with distilled water. Takai et al. (1997) treated BC with distilled water and 2% NaOH, and neutralized the mat with 2% acetic acid.

Krystynowicz et al. (1997) developed a procedure starting from washing crude BC in running tap water (overnight), followed by boiling in 1% NaOH solution for 2 h, washing in tap water to accomplish NaOH removal (1 d), neutralization with 5% acetic acid, and its removal with tap water. The final BC preparations obtained contained less than 3% protein, and were suitable for certain food and medical purposes.

Medical application of BC requires special procedures to remove bacterial cells and toxins, which can cause pyrogenic reactions. One of the most effective protocols begins with gentle pressing of the cellulose pellicle between absorbent sheets to expel about 80% of the liquid phase and then immersing the mat in 3% NaOH for 12 h. This procedure is repeated 3 times, and after that the pellicle is incubated in 3% HCl solution, pressed, and thoroughly washed in distilled water. The purified pellicle is sterilized in an autoclave or by ^{60}Co irradiation. It performs excellently as a wound dressing since it contains only 1–50 ng of lipopolysaccharide endotoxins, whereas BC purified using conventional methods usually contains 30 μg or more of these substances (Ring et al., 1986).

10
Properties

BC is an extremely insoluble, resilient, and elastic polymer, having high tensile strength. It has a reticulated structure, in which numerous ribbon-shaped fibrils are composed of highly crystalline and highly uniaxially oriented cellulose subfibrils. This three-dimensional structure, not found in plant-originating cellulose, brings about a higher crystallinity index (60–70%) of BC. Furthermore, fibers of the plant polymer are about 100 times thicker than BC microfibrils and therefore the bacterial polymer has an about 200 times larger accessible surface. In comparison to *A. xylinum* cellulose from agitated cultures, the polymer synthesized under static conditions has a higher DP (14,400 and 10,900, respectively), crystallinity index (71 and 63%, respectively), tensile strength (Young's modulus of 33.3 and 28.3, respectively), but lower water holding-capacity (45 and 170 g water g BC^{-1}, respectively) and suspension viscosity (0.04 and 0.52 Pa·s, respectively) in its disintegrated form (Watanabe et al., 1998a).

Microbial cellulose appears to be gelatinous, since its liquid component (usually water), present in voids among very fine ribbons, amounts to at least 95% by weight. The bacterial polysaccharide has a high water-binding capacity, but the majority of the water is not bound to the polymer and it can be squeezed out by gentle pressing. Drying of BC leads to paper-like sheets, having a thickness of 0.01–0.5 mm (Yamanaka et al., 1990; Krystynowicz et al., 1995, 1997) and good absorptive properties. In addition to a high Young's modulus, BC also displays high sonic velocity and, because of these unique mechanical properties, it can be applied as acoustic membranes (Vandamme et al., 1998).

The properties of BC can be modified either during its synthesis or when the culture is completed. Some compounds, like cellulose derivatives, sulfonic acids, alkylphosphates, or other polysaccharides (starch, dextran), introduced into the nutrient broth alter the macroscopic morphology, the tensile strength, the optical density, and the absorptive properties of the final product (Yamanaka et al., 2000). BC can also be combined with other substances, added either to wet or dried cellulose, thus giving composites of desired physicochemical properties (Yamanaka, 1990). The auxiliary materials used for this purpose comprise granules and fibers of various inorganic and organic compounds, such as alumina, glass, agar, alginates, carragenan, pullulan, dextran, polyacrylamide, heparin, polyhydroxylalcohols, gelatin, collagen, etc. They are combined with BC sheets by impregnation, lamination, or adsorption, or mixed with the disintegrated polymer. The composites are further subjected to a shaping treatment, thus giving various products.

Recently, Kim et al. (1999) reported an enzymatic method of BC modification using *L. mesenteroides* dextransucrase and alternansucrase. In their presence *A. xylinum* ATCC 10821 synthesized 'soluble cellulose', which was a glucan composed of 1,4-, 1,6-, and 1,3-linked monomers.

The basic BC properties are summarized in Table 2.

11
Applications

BC belongs to the generally recognized as safe (GRAS) polysaccharides and therefore it has already been put to a multitude of different uses. Commercial application of this polymer results from its unique properties and developments in effective technologies of production, based on growth of improved bacterial strains on cheap waste materials. The advantage of BC is its chem-

Tab. 2 Properties of BC

High purity
High degree of crystallinity
Greater surface area than that of conventional wood pulp
Sheet density from 300 to 900 kg m^{-3}
High tensile strength even at low sheet density (below 500 kg m^{-3})
High absorbency
High water-binding capacity
High elasticity, resilience, and durability
Nontoxicity
Metabolic inertness
Biocompatibility
Susceptibility to biodegradation
Good shape retention
Easy tailoring of physicochemical properties

ical purity and the absence of substances usually occurring in the plant polysaccharide, which requires laborious purification. In addition to the shape of BC sheets, their area and thickness can be tailored by means of culture conditions. Relatively easy BC modification during its biosynthesis enables regulation of such properties as molecular mass, elasticity, resilience, water-holding capacity, crystallinity index, etc. BC microfibrils may bind both low-molecular-weight substances and polymers, added for instance to the growth medium, thus giving novel commodities. BC can be also a raw material for further chemical modifications.

Based on the assumption that as much as 10,000 kg of the bacterial polymer can be obtained per year from static culture having 1 hectare surface area and only 600 kg of cotton is harvested in the same period of time from a field of the same area, the perspectives for a wider use of BC become more apparent (Kudlicka, 1989).

The main potential BC applications are summarized in the Table 3.

11.1 Technical Applications

Compared to PC sheets, BC has satisfactory tensile strength even at low sheet density (300–500 kg m^{-3}) (Johnson and Neogi,

Tab. 3 Applications of bacterial cellulose

Sector	*Application*
Cosmetics	Stabilizer of emulsions such as creams, tonics, nail conditioners, and polishes; component of artificial nails
Textile industry	Artificial skin and textiles; highly adsorptive materials
Tourism and sports	Sport clothes, tents, and camping equipment
Mining and refinery waste treatment	Spilt oils collecting sponges, materials for toxins adsorption, and recycling of minerals and oils
Sewage purification	Municipal sewage purification and water ultrafiltration
Broadcasting	Sensitive diaphragms for microphones and stereo headphones
Forestry	Artificial wood replacer, multi-layer plywood and heavy-duty containers
Paper industry	Specialty papers, archival documents repair, more durable banknotes, diapers, and napkins;
Machine industry	Car bodies, airplane parts, and sealing of cracks in rocket casings
Food production	Edible cellulose and 'nata de coco'
Medicine	Temporary artificial skin for therapy of decubitus, burns and ulcers; component of dental and arterial implants
Laboratory/research	Immobilization of proteins, chromatographic techniques, and medium component of *in vitro* tissue cultures

1989). Therefore, BC is an excellent component of papers, providing better mechanical properties. Microfibrils of the bacterial polymer form a great number of hydrogen bonds when the paper is subjected to drying, thus giving improved chemical adhesion and tensile strength. BC-containing paper not only shows better retention of solid additives, such as fillers and pigments, but is also more elastic, air permeable, resistant to tearing and bursting forces, and binds more water (Iguchi et al., 2000).

A beneficial effect such as an improved aging resistance was achieved by adding small amounts of BC to cotton fibers to obtain hand-made paper, used for old document repair. The paper displayed appropriate ink receptivity and specific snap (Krystynowicz et al., 1997). Paper containing 1% of BC meets the international standard ISO 9706:1994 for information and documentation papers, and also has a specific snap comparable to that of rag paper. It is possible to use BC for pressboard making, as well as production of paperboard used as an electroinsulating material and for bookbinding.

BC can also be used as a surface coating for specialty papers. For this purpose, a filtered suspension of BC (homogenized and mixed with other components) is added using special applicators, either within wet sheet formation, or to a partially or completely dried sheet. The coating improves properties such as gloss, brightness, smoothness, porosity, ink receptivity, and tensile strength. Other additives like starch, organic polymers, including CM-cellulose, organic, or inorganic pigments may also be used. Johnson and Neogi (1989) claim that paper coated with 3% of BC (solid matter) displays gloss properties and surface strength similar to rotogravure paper having 20% traditional coating. The authors also state that coating with a mixture of BC and CM-cellulose gives even better properties, since they act synergistically.

BC is also a valuable component of synthetic paper (Iguchi et al., 2000) since nonpolar polypropylene and polyethylene fibers, providing insulation, heat resistance, and fire-retarding properties, cannot form hydrogen bonds. The amount of wood pulp in this type of paper is usually from 20 to 50% to achieve good quality. Using BC enables us to decrease the amount of the additives without any effect on the synthetic paper properties.

BC also appeared to be a good binder in nonwoven fabric-like products (Yamanaka et al., 1990) commonly used in surgical drapes and gowns, and containing various hydrophilic and hydrophobic, natural and synthetic fibers, such as cellulose esters, polyolefin, nylon, acrylic glass, or metal fibers. Even small amounts of BC improve tensile and tear strength of the fabric, e.g. 10% of the bacterial polymer is equivalent to 20–30% of latex binder.

The scope of BC uses can be even wider since it is modifiable during synthesis (Brown, 1989a,b). For this purpose, CM-cellulose or copolymers of saccharides and dicarboxylic acids are added directly to the culture medium (Yamanaka et al., 1989). Cellulose obtained in the presence of CM-cellulose and dried with organic solvents has a resilient and elastic character, as well as higher water binding capacity (adsorbs faster more water). The optimum CM-cellulose concentrations range from 0.1 to 5% (w/v). The polarity of the organic solvent also influences the BC features. After treatment with acetone the polymer is elastic and rubber-like, whereas after drying with absolute ethanol, BC resembles leather.

Since BC is susceptible to enzymatic digestion, it is modified to obtain various composites of satisfactory biodegrability and strength. Their production is based either on chemical reaction between cellulose and the copolymer or on culturing the BC-producing strain in the copolymer-containing medium.

11.2 Medical Applications

A cellulose pad from a static culture is a ready-to-use, naturally 'woven' wound dressing material that meets the standards of modern wound dressings (Figure 16). It is sterilizable, biocompatible, porous, elastic, easy to handle and store, adsorbs exudation, provides optimum humidity, which is essential for fast wound healing, protects from secondary infection and mechanical injury, does not stick to the newly regenerated tissue, and alleviates pain by heat adsorption from burns. BC sheets are also excellent carriers for immobilization of medicinal preparations, which speed up the healing process.

Since BC pellicles can have various dimensions, it is relatively easy to produce dressings for extensive wounds. Because of recent problems with products of animal origin, collagen dressings can be replaced with BC ones. This thesis is additionally supported by undoubtedly positive results of clinical tests. For instance, the BC preparation, Prima Cel™, produced by Xylos Corp., according to the Rensselaer Polytechnic Institute (USA), has been applied as a wound dressing in clinical tests to heal ulcers. The results obtained were satisfactory, since after 8 weeks 54% of the patients recovered and the remaining ulcers were almost healed (Jonas and Farah, 1998). Other commercial preparations of *A. xylinum* cellulose, such as Biofil® and Bioprocess®, appeared to be excellent as skin transplants, and in the treatment of third-degree burns, ulcers, and decubitus. Another preparation, Gengiflex®, found application in recovering periodontal tissues (Jonas and Farah, 1998). Investigations on *A. xylinum* cellulose pads prepared by Krystynowicz et al. (2000) (Figure 17) indicate that general use as wound dressings is possible.

Results of investigations on BC hollow fibers use as artificial blood vessels and ureters are also promising (Yamanaka et al., 1990). Antithrombic BC property (blood

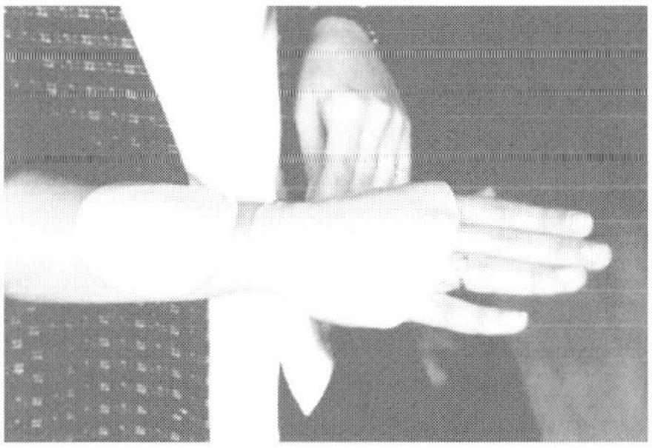

Fig. 16 Cellulose pellicle as a material for wound dressings.

a)

b)

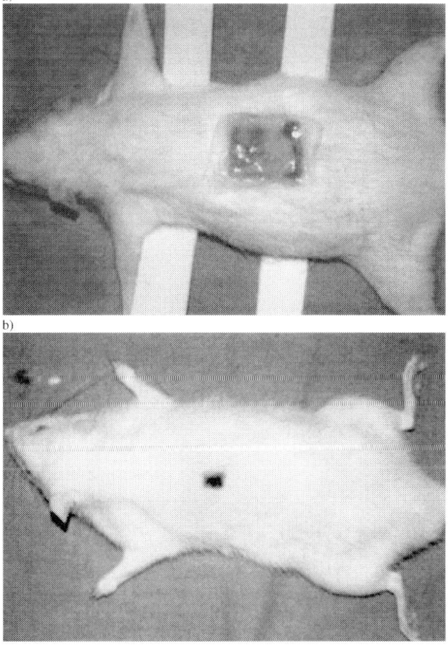

Fig. 17 Burns treated with a BC dressing, before (a) and after (b) healing (Krystynowicz et al., 2000).

compatibility) was evaluated by a test based on replacement of adult dog blood vessels (parts of the descending aorta and jugular vein) with the artificial counterpart made of BC. One month later the artificial vessel was removed and the state of adhesion of thrombi on its inner surface was examined. A good open state of the BC blood vessel was maintained.

Another successful experiment showing the application of biosynthetic cellulose was substitution of the dog dura matter in the brain (Jonas and Farah, 1998).

Because of its high tensile strength, elasticity, and permeability to liquids and gases, dried BC was applied as an additional membrane to protect immobilized glucose oxidase in biosensors used for assays of blood glucose levels. This BC membrane enhanced the electrode stability in 10 times diluted human blood solution, up to 200 h. Other commercial protecting membranes, such as cuprophan (AKZO, England), provided an electrode stability up to 30 h only. In undiluted human blood the biosensor coated with cuprophan was stable for 3–4 h, whereas the BC membrane prolonged its stability up to 24 h.

11.3 Food Applications

Chemically pure and metabolically inert BC has been applied as a noncaloric bulking and stabilizing agent in processed food. Similar to pure preparations of the plant polysaccharide and its derivatives, it is used for stabilization of foams, pectin and starch gels, emulsions, e.g. canned chocolate drinks and soups, texture modification, e.g. improvement of the consistency of pulps, enhancement of adhesion, replacement of lipids, including oils, and dietary fiber supplements (Ang and Miller, 1991; Kent et al., 1991; Krystynowicz et al., 1999).

The first successful commercial application of BC in food production is 'nata de coco' (Sutherland, 1998). It is a traditional dessert from the Philippines, prepared from coconut milk or coconut water with sucrose, which serves as a growth medium for BC-producing bacteria. Consumption of the pellicle is believed to protect against bowel cancer, artheriosclerosis, and coronary thrombosis, and prevents a sudden rise of glucose in the urine. Therefore 'nata de coco' is becoming increasingly popular, not only in Asia.

Another popular BC-containing food product is Chinese Kombucha (teakvass or tea-fungus), obtained by growing yeast and acetic acid bacteria on tea and sugar extract. The pellicle formed on the surface contains both cellulose and enzymes healthy for humans. Their abiotic activity is especially stimulating for the large bowel and the whole alimentary tract. Kombucha is believed to protect against certain cancers (Iguchi et al., 2000).

Results of studies carried out by the authors, who applied BC pellicles synthesized by *A. xylinum* E_{25} for wine and juice filtration, and for immobilization of polyphenols, are promising (Krystynowicz et al., 1999). Preparations of bioactive anthocyanins, enriched in dietary fiber, are excellent for functional food production. BC also appeared to be an attractive component of bakery products, since it plays a role of dietary fiber, is taste and odorless, and prolongs the shelf-life.

11.4 Miscellaneous Uses

The large accessible surface area, high durability, and superior adsorptive properties as well as possibility of modification by means of physical or chemical methods, means that BC can be applied as a carrier for immobilization of biocatalysts. Cellulose gels containing immobilized animal cells were used for their culture to produce interferon, interleukin-1, cytostatics, and monoclonal antibodies (Iguchi et al., 2000).

BC was also applied for adsorption of cells of *Gluconobacter oxydans, Acetobacter methanolyticus* and *Saccharomyces cerevisiae*. The immobilized strains appeared to be effective producers of gluconate (84–92% yield), dihydroxyacetone (90–98% yield), and ethanol (88–92% yield), respectively, and displayed better operational and thermal stability.

Purified BC can be a raw material for synthesis of cellulose acetate, nitrocellulose, CM-cellulose, hydroxymethylcellulose, methylcellulose, and hydroxycellulose (Yamanaka, 1990). If BC is produced in the presence of compounds that interfere with regular fibril assembly and which influence the β-1,4-glucan structure, such as CM-cellulose or other cellulose derivatives, and other carbohydrates (starch, dextran), sulfonates, and alkylphosphates, the resulting microbial polymer has novel, additive-dependent, and useful properties, including optical transparency or higher water-binding capacity, even after repeated soaking and drying.

The potential application of BC in the chemical, paper, and textile industries depends on its price and accessibility. To meet these demands, BC has to be produced by highly efficient strains, growing on cheap substrates, in sophisticated surface, solid-state, or submerged fermentors (Vandamme et al., 1998).

12
Patents

Growing rapidly, the interest in BC is reflected in a number of patents (annually about 20 since the 1980s) and publications (20–40 per year for the last 10 years) devoted to this unique polymer (Iguchi et al., 2000). Some basic patents concerning the biosynthesis, properties, and applications of BC are presented in Table 4. These patents that are cited in chapter are listed in the references.

13
Outlook and Perspectives

The first scientific paper reporting on an unusual substance formed by acetic acid bacteria and known for ages in many countries as 'vinegar plant' (Yamanaka et al., 1989) was published 115 years ago (Brown, 1886). Further studies revealed that the substance is a super-pure cellulose. Metabolic pathways and the complicated molecular machinery of the polysaccharide biosynthesis, as well as the intriguing dynamics of its nascent chains association into a structure with unique properties, has been elucidated. Although progress and limited commercialization of BC have undoubtedly taken place, the relevant biotechnology, competitive to modern industrial technologies for PC production, has not yet been developed.

So, what to do, to achieve success and accomplish commercialization of BC production? First of all, stable overproducer strains of the polymer have to be constructed, using recent achievements in molecular genetics and biology. These strains have to assimilate a wide range of carbon sources, as well as display lower tendency towards spontaneous mutation to Cell$^-$ mutants and effectively synthesize cellulose (10–15 g L^{-1} d^{-1}) under agitated culture conditions. Construction of new bioreactors for both stationary and submerged culture is necessary. The considerable reduction of BC production cost could be attained by replacing expensive nutrient media components

Tab. 4 Selected patents concerning BC

Patent number	*Patent holder*	*Inventors*	*Title*	*Date of publi-cation*	*Major claims*
WO 0125470	Novozymes A/S, Denmark	Herbert, W., Chanzy, H. D., Ernst, S., Schulein, M., Husum, T. L., Kongsbak, L.	Cellulose films for screening	2001	BC microfibril films containing fluorescein-labeled hemoglobin or galactomannan can be used to detect proteases and mannanases, respectively
WO 0105838	Pharmacia Corp., USA	Yang, Z. F., Sharma, S., Mohan, C., Kobzeff, J.	Process for drying reticulated bacterial cellulose without co-agents	2001	The reticulated *Acetobacter* cellulose subjected to dispersing in a solvent, e.g. hydrocarbon (hexane), aliphatic alcohol, and/or alkyl sulfoxide (DMSO), separation from the solvent, and drying, may be rehydrated to provide uniform dispersions having high viscosity
JP 11255806 A2	Bio-Polymer Research Co. Ltd, Japan	Watanabe, O.	Freeze drying method for microfibrous cellulose concentrate	1999	Freeze drying of *Acetobacter* cellulose gives dry microfibrous cellulose with good retention of its original properties after redispersing in water
WO 9940153	Monsanto	Smith, B. A., Colegrove, G. T., Rakitsky, W. G.	Acid-stable, cationic compatible cellulose compositions useful as rheology modifiers	1999	Cationic co-agents, acids, and/or cationic surfactants are used to form acid-stable *Acetobacter* cellulose compositions, which are useful as rheology modifiers
WO 9943748	Sony	Uryu, M., Tokura, K.	Biodegradable composite polymer material	1999	A new composite material has been obtained by drying the *A. xylinum* cellulose, its pulverization, and blending with a biodegradable polymer material
JP 0056669 A2	Canon Co., Japan	Minami, M., Mihara, C., Takeda, T., Kikuchi, Y.	Composite, for use in thermoformed articles, comprises cellulose and saccharide ester derivative	1999	A composite comprising BC and a saccharide ester can be used for production of biodegradable, thermoformed articles, which have improved processability, mechanical strength and flexibility
JP 11172115 A2	Ajinomoto Co. Inc., Japan	Suzuki, O., Kitamura, N., Matsumoto, R.	Bacterial cellulose-containing composite absorbents with high liquid absorption	1999	Dispersing highly water-absorbing polymer particles, dissolved in an organic solvent, in an aqueous solution of defibrillated BC, followed by organic solvent removal and partial drying, gives excellent and stable absorbents

Tab. 4 (cont.)

Patent number	Patent holder	Inventors	Title	Date of publi-cation	Major claims
JP 11246602 A2	Bio-Polymer Research Co. Ltd, Japan	Tahara, N., Hagamida, T., Miyashita, H., Watanabe, O.	Preservation of wet bacterial cellulose	1999	Wet BC can be preserved with alkyl sulfate salts or NaOH and/or KOH
JP 11187896 A2	Bio-Polymer Research Co. Ltd, Japan	Tabata, T., Toyosaki, H., Tsuchida, T., Yoshinaga, F.	A method for screening cellulose-producing bacteria using cellulase	1999	A rapid and convenient method for screening a large number of cellulose membrane-producing strains is presented; the metabolic peculiarities of the strains and conditions needed for them to produce cellulose pellicle are explained
JP 11092502 A2	Bio-Polymer Research Co. Ltd, Japan	Shoda, M., Kanno, Y., Koda, T., Yoshinaga, F.	Manufacture of bacterial cellulose under oxygen-enriched conditions	1999	Passing more than 21% oxygen-containing air through *A. xylinum* ssp. sucrofermentans culture in an air-lift fermentor, enables accumulation of 6.5 g L^{-1} of BC after 75 h
JP 11117120 A2	Toray Industries Inc., Japan	Hara, T., Amano, J.	Fibers made from blends of bacterial cellulose and polymers having flexible main backbone	1999	The blends contain polyvinyl alcohol-type polymers at weight ratio of 2–50%
JP 11181001 A2	Bio-Polymer Research Co. Ltd, Japan	Matsuoka, M., Toyosaki, H., Matsumura, T., Ougiya, H., Tsuchida, T., Yoshinaga, F.	Production of bacterial cellulose	1999	BC, useful for filler retention aids for paper-making, can be produced in agitated cultures of *Acetobacter* strains, in the presence of water-soluble polysaccharides, e.g. CM-cellulose
JP 11137163 A2	Shikishima Seipan Co. Ltd, Japan	Kondo, M., Yamada, M., Inoue, S.	Manufacture of bread from dough containing bacterial cellulose	1999	Addition of BC increases water absorption of dough, thus giving bread with high water-holding capacity
JP 11221072 A2	Bio-Polymer Research Co. Ltd, Japan	Ishikawa, A., Tsuchida, T., Yoshinaga, F.	Bacterial cellulose production enhancement	1999	An *A. xylinum* mutant having higher cellular levels of UDP, UTP, and UDPGlc as well as higher carbamoyl phosphate synthetase activity is an excellent cellulose producer

Tab. 4 (cont.)

Patent number	Patent holder	Inventors	Title	Date of publi-cation	Major claims
JP 11269797 A2	Toppan Printing Co. Ltd, Nakano Vinegar Co. Ltd, Japan	Yamawaki, K., Tomita, T., Harasawa, A., Kaminaga, J., Kawasaki, K., Matsuo, R., Sasaki, N., Fukagai, M., Tsukamoto, Y.	Impregnated paper with good water resistance and stiffness	1999	Paper impregnated with silane coupler-grafted BC derivatives is moisture resistant
PP 299907	Technical University of Lodz, Poland	Krystynowicz, A., Czaja, W., Bielecki, S.	Biosynthesis and application of bacterial cellulose	1999	Production of BC by wild and mutant *A. xylinum* strains, cultured in static or agitated cultures, and under various conditions as well as an influence of culture conditions on the cellulose properties are described
JP 11018758 A2	Bio-Polymer Research Co. Ltd, Japan	Yamamoto, T., Yano, H., Yoshinaga, F.	Horizontal type spinner culture vessel, having high oxygen-supplying efficiency	1999	BC can be produced in the spinner culture tank, equipped with mixing impellers, and providing high efficiency of oxygen supply
JP 10298204 A2	Ajinomoto Co. Inc., Japan	Ishikara, M., Yamanaka, S.	Bacterial cellulose with ribbon-like microfibril shape	1998	Cellulose fibrils having a short axis 10–100 nm and a long axis 160–1000 nm are produced extracellularly by cellulose-generating bacteria, e.g. *Acetobacter pasteurianus*, in a culture containing cell division inhibitor
US 005723764	Pioneer Hi-Bred International Inc., USA	Nichols, S. E., Singletary, G. W.	Cellulose synthesis in the storage tissue of transgenic plants	1998	Introducing the genes for cellulose biosynthesis from the species *A. xylinum* into a given plant provides a method of synthesizing cellulose in the storage tissue of transgenic plants
JP 10077302	Bio-Polymer Research Co. Ltd, Japan	Tabuchi, M., Watanabe, K., Morinaga, Y.	Solubilized bacterial cellulose and its compositions or composites for moldings and coatings.	1998	Cellulose synthesized by *A. xylinum* in stationary culture was solubilized by stirring in a mixture of DMSO and paraformaldehyde (25:5) at 100 °C for 3 h

Tab. 4 (cont.)

Patent number	*Patent holder*	*Inventors*	*Title*	*Date of publi-cation*	*Major claims*
WO 97/05271	Rensselaer Polytechnic Institute, USA	Bungay, H. R., Serafica, G.	Production of microbial cellulose using a rotating disc film bioreactor	1997	BC can be deposited by *A. xylinum*, cultured in a liquid medium inside the horizontal fermentor, on a plurality of disks, which rotate around the long axis of the fermentor and are partly submerged in the culture medium
JP 09025302 A2	Bio-Polymer Research Co. Ltd, Japan	Hioki, S., Watanabe, K., Ogya, H., Morinaga, Y.	Preparation of disintegrated bacterial cellulose for improved additive retention in paper manufacture	1997	BC which helps additive retention and causes no harm to freeness during paper manufacturing can be obtained by disintegration using a self-excited ultrasonic pulverizer
JP 09107892 A	Nakano Vinegar Co. Ltd, Japan	Furukawa, H., Maruyama, Y., Fukaya, M., Tsukamoto, Y., Kawamura, K.	Composition for stabilizing dispersion used in food	1997	A fine particulate (210 μm average particle diameter) BC, obtained by hydrolyzing *A. xylinum* cellulose with a mineral acid, is of low viscosity and is a sufficient dispersion stability agent in food products
WO 9744477	Bio-Polymer Research Co. Ltd, Japan	Naritomi, T., Kouda, T., Naritomi, M., Yano, H., Yoshinaga, F.	Continuous preparation of bacterial cellulose having a high production rate and yield	1997	BC obtained in continuous culture of *A. xylinum* (production rate at least 0.4 g L^{-1} h^{-1}) in a medium containing a substance enhancing the apparent substrate affinity to sugar (e.g. lactic acid) is useful as a thickener, humectant or stabilizer for production of food, cosmetics or paints, etc.
WO 9740135	Bio-Polymer Research Co. Ltd, Japan	Tsuchida, T., Tonouchi, N., Seto, A., Kojima, Y., Matsuoka, M., Yoshinaga, F.	Novel cellulose-producing bacteria and a process of producing it	1997	Production of cellulose with a new *A. xylinum* ssp. nonactoxidans, which lacks an ability to oxidize acetates and lactates, yields odor- and taste-less products, having excellent dispersability in water
WO 9712987	Bio-Polymer Research Co. Ltd, Japan	Kouda, T., Naritomi, T., Yano, H., Yoshinaga, F.	Production process for bacterial cellulose which is useful as material in various fields	1997	A new process for BC production involving maintaining a certain pressure inside a fermentor, reduces power required for agitation, and elevates production rate and yield
JP 09056392 A	Kikkoman Corp., Japan	Fukazawa, K., Imai, H., Kijima, T., Kikuchi, T.	Production of microorganism cellulose	1997	A cellulose pellicle synthesized by *A. xylinum* strain precultured in stationary conditions can be homogenized and used to inoculate the fresh culture medium

Tab. 4 (cont.)

Patent number	*Patent holder*	*Inventors*	*Title*	*Date of publi-cation*	*Major claims*
PL 171952 B1	Technical University of Lodz, Poland	Galas, E., Krystynowicz, A.	Method of obtaining bacterial cellulose	1997	A method of BC production in static culture using *A. xylinum* P23 strain is described
US 0824096	Bio-Polymer Research Co. Ltd, Japan	Kouda, T., Nagata, Y., Yano, H., Yoshinaga, F.	Production of bacterial cellulose through cultivation of cellulose-producing bacteria under specified conditions in aerated and agitated culture	1997	Cellulose produced by species of *Acetobacter, Agrobacterium, Rhizobium, Sarcina, Pseudomonas, Achromobacter, Alcaligenes, Aerobacter, Azotobacter,* and *Zoogloea,* in aerated and agitated cultures can be applied in production of food and cosmetics
JP 97–21905	Bio-Polymer Research Co. Ltd, Japan	Seto, H., Tsuchida, T., Yoshinaga, F.	Manufacture of bacterial cellulose by mixed culture of micro-organisms	1997	To provide lactate and split sucrose, cellulose-producing *A. xylinum* strain can be grown together with lactic acid bacteria and *Saccharomyces* yeast; their presence in the culture broth markedly enhances BC yield
JP 08127601 A	Bio-Polymer Research Co. Ltd, Japan	Hioki, S., Watabe, O., Hori, S., Morinaga, Y., Yoshinaga, F.	Freeness regulating agent	1996	Production of a freeness regulating agent, comprising a defiberized BC, synthesized by *A. xylinum* ssp. sucrofermentans is described
JP 96316922	Bio-Polymer Research Co. Ltd, Japan	Hikawu, S., Hiroshi, T., Takayasu, T., Yoshinaga, F.	Manufacture of bacterial cellulose by addition of cellulose formation stimulators	1996	Compounds such as derivatives of choline, betaine, and fatty acids (salts and esters) appeared to stimulate cellulose production by *A. xylinum* strains
JP 08056689 A	Bio-Polymer Research Co. Ltd, Japan	Seto, H., Tsuchida, T., Yoshinaga, F.	Production of bacterial cellulose	1996	To obtain an edible BC, excellent in aqueous dispersibility, useful for retaining the viscosity of foods, cosmetics, coatings, etc., and enrichment of foods, *Acetobacter* strains can be cultured in saponin-containing medium
JP 08033494 A	Bio-Polymer Research Co. Ltd, Japan	Tawara, N., Koda, T., Hagamida, T., Morinaga, Y., Yano, H.	Method for circulating continuous production and separation of bacterial cellulose	1996	Circulating a culture liquid containing *A. xylinum* cells between a culture and a separation apparatus enables separation of the produced cellulose; also, a flotation separator or an edge filter can be applied for this purpose

Tab. 4 (cont.)

Patent number	*Patent holder*	*Inventors*	*Title*	*Date of publi-cation*	*Major claims*
JP 08034802 A	Gun Ei Chemical Industries Co. Ltd, Japan	Hirooka, S., Hamano, T., Miyashita, Y., Hanaue, K., Yamazaki, K., Shiichi, K., Shiichi, F.	Bacterial cellulose, production thereof and processed product made therefrom	1996	*A. xylinum* can synthesize cellulose in culture broths containing difficult to ferment, branched oligosaccharides
JP 08000260 A	Bio-Polymer Research Co. Ltd, Japan	Ishikawa, A., Tsuchida, T., Yoshinaga, F.	Production of bacterial cellulose with pyrimidine analogue-resistant strain	1996	A method of production of BC, in a high yield and at a low cost, using a pyrimidine analog-resistant *Acetobacter* mutant is presented
JP 08325301 A	Bio-Polymer Research Co. Ltd, Japan	Hori, S., Watabe, O., Morinaga, Y., Yoshinaga, F.	Cellulose having high dispersibility and its production	1996	BC synthesized by *A. xylinum* ssp. sucrofermentans, subjected after harvesting to partial hydrolysis with HCl, yields a fraction exhibiting high birefringence
JP 08276126	Bio-Polymer Research Co. Ltd, Japan	Ogiya, H., Watabe, O., Shibata, A., Hioki, S., Morinaga, Y., Yoshinaga, F.	Emulsification stabilizer	1996	The method of production of the emulsification stabilizer containing BC obtained from agitated culture of *A. xylinum* ssp. sucrofermentans and having low index of crystallization, is presented
JP 08009965 A	Bio-Polymer Research Co. Ltd, Japan	Ishikawa, A., Tsuchida, T., Yoshinaga, F.	Production of bacterial cellulose using microbial strain resistant to inhibitor of DHO-dehydrogenase	1996	A method of efficient production of BC, using *Acetobacter* strains resistant either to an inhibitor of DHO-dehydrogenase or to DNP is presented
US 005382656	Weyerhauser Co., USA	Benziman, M., Tal, R.	Cellulose synthase associated proteins	1995	CS-associated proteins, which have molecular weights of 20, 54, and 59 kDa are not encoded by CS operon genes *bcsA*, *B*, *C*, or *D*
JP 07313181 A	Bio-Polymer Research Co. Ltd, Japan	Takemura, H., Tsuchida, T., Yoshinaga, F., Matsuoka, M.	Production of bacterial cellulose using PQQ-unproductive strain	1995	Pyrroloquinolinequinone-unproductive *Acetobacter* mutant strain enables high-yield production of BC
JP 07184675 A	Bio-Polymer Research Co. Ltd, Japan	Matsuoka, M., Tsuchida, T., Yoshinaga, F.	Production of bacterial cellulose	1995	To obtain BC useful for retaining the viscosity of foods, cosmetics, or coatings, *Acetobacter* strains can be cultured in a methionine-containing medium

Tab. 4 (cont.)

Patent number	*Patent holder*	*Inventors*	*Title*	*Date of publi-cation*	*Major claims*
JP 07268128	Fujitsuko Co. Ltd, Japan	Kiriyama, S., Fukui, H., Toda, T., Yamagishi, H.	Dried material of cellulose derived from microorganism and its production	1995	Water-soluble stabilizer (preferably glucose or gelatin) added to cellulose gel obtained by culturing *A. xylinum* strain before drying of the material provides excellent and stable physical properties
WO 95/32279	Bio-Polymer Research Co. Ltd, Japan	Tonouchi, N., Tsuchida, T., Yoshinaga, F., Horinouchi, S., Beppu, T.	Cellulose-producing bacterium transformed with gene coding for enzyme related to sucrose metabolism	1995	*A. xylinum* transformant with a gene coding for invertase accumulates cellulose in a sucrose-containing medium
JP 07184677 A	Bio-Polymer Research Co. Ltd, Japan	Seto, H., Tsuchida, T., Yoshinaga, F.	Production of bacterial cellulose	1995	High-yield production of an edible BC useful for retaining the viscosity of cosmetics or coatings can be achieved by culturing an *Acetobacter* strain in an invertase-added medium containing sucrose as a carbon source
JP 07039386 A	Bio-Polymer Research Co. Ltd, Japan	Matsuoka, M., Takemura, H., Tsuchida, T., Yoshinaga, F.	Production of bacterial cellulose	1995	BC, useful for retaining the viscosity of a food, a cosmetic, a coating, etc., and usable as a food additive, an emulsion stabilizer, etc., can be obtained by culturing *A. xylinum* in a culture medium containing a carboxylic acid (e.g. lactic acid) salt
JP 07274987 A	Bio-Polymer Research Co. Ltd, Japan	Toda K., Asakura, T.	Production of bacterial cellulose	1995	To obtain BC in high yield, the cellulose-producing *Acetobacter* strain is cultured in a cylindrical container and oxygen is fed through an oxygen-permeable membrane at the bottom
JP 06125780	Nakano Vinegar Co. Ltd, Japan	Fukaya, M., Okumura, H., Kawamura, K.	Production of cellulosic substance of microorganism	1994	A method used to improve production efficiency of cellulose by an *Acetobacter* strain is presented; the method is based on adding a protein having affinity to the cellulose to the culture medium
JP 06248594 A	Mitsubishi Paper Mills Ltd, Japan	Katsura, T., Okafuro, K.	Low-density paper having high smoothness	1994	Excellent, low-density paper having high smoothness contains BC and the broad-leaved pulp in the ratio of 1/99:1/1

Tab. 4 (cont.)

Patent number	*Patent holder*	*Inventors*	*Title*	*Date of publi-cation*	*Major claims*
JP 06206904	Shin Etsu Chemical Co. Ltd, Japan	Horii, F., Yamamoto, H.	Bacterial cellulose, its production and method for controlling crystal structure thereof	1994	Xanthan gum or sodium CM-cellulose added to the culture broth of *A. xylinum* enable control of the crystal structure of BC
JP 06113873 A	Nippon Paper Industries Co. Ltd, Japan	Samejima, K., Mamoto, K.	Production of microbial cellulose	1994	Adding a sulfite pulp waste liquor and/or its permeate from ultrafiltration into a culture medium of a cellulose-producing microorganism enhances the cellulose yield and lowers the cost of its production
WO 94/20626	Bio-Polymer Research Co. Ltd, Japan	Beppu, T., Tonouchi, N., Horinouchi, S., Tsuchida, T.	*Acetobacter*, plasmid originating therein, and shuttle vector constructed from said plasmid	1994	Genetic recombination of cellulose-producing *Acetobacter* strains is performed using an *Acetobacter* strain, its endogenous plasmid, and a shuttle vector constructed from the latter plasmid and a plasmid from *Escherichia coli*
JP 06001647 A	Shimizu Corp., Japan	Yano, H., Narutomi, T., Okamura, K., Kawai, T., Minami, S.	Concrete and coating material	1994	The concrete or coating material containing disaggregated BC displays better dispersibility of cement or pigment particles and an enhanced fluidity
US 005268274 A	Cetus Corp., USA	Ben-Bassat, A., Calhoon, R. D., Fear, A. L., Gelfand, D. H., Meade, J. H., Tal, R., Wong, H., Benziman, M.	Methods and nucleic acid sequences for the expression of the cellulose synthase operon	1993	Nucleic acid sequences encoding the BC synthase from *A. xylinum*, and methods for isolating the genes and their expression in hosts are presented
EP 0396344 A2	Ajinomoto Co. Inc., Japan	Yamanaka, S., Ono, E., Watanabe, K., Kusakabe, M., Suzuki, Y.	Hollow microbial cellulose, process for preparation thereof, and artificial blood vessel formed of said cellulose	1990	BC prepared by culturing a cellulose-producing strain on one or both surfaces of an oxygen-permeable hollow carrier is useful as a substitute for a blood vessel or another internal hollow organ; the cellulose can be impregnated with a medium, cured, and cut if necessary

Tab. 4 (cont.)

Patent number	*Patent holder*	*Inventors*	*Title*	*Date of publi-cation*	*Major claims*
US 004950597	University of Texas, USA	Saxena, I. M., Roberts, E. M., Brown, R. M.	Modification of cellulose normally synthesized by cellulose-producing micro-organisms	1990	Mutants of *A. xylinum* that do not form a pellicle in liquid culture and synthesize cellulose almost exclusively as the allomorph cellulose II, that arise spontaneously or by nitrosoguanidine mutagenesis are described and the cellulose they produce is characterized
JP 02182194 A	Asahi Chemical Industries Co. Ltd, Japan	Matsuda, Y., Kamiide, K.	Production of cellulose with acetic acid bacterium	1990	Acetic acid bacteria having a synchronized cell cycle enable efficient production of BC, excellent in water holding properties, tensile strength, and purity, and displaying a relatively low DP
WO 89/12107	Brown, R. M.	Brown, R. M.	Microbial cellulose as a building block resource for specialty products and processes thereof	1989	A novel process for manufacturing BC using different bacterial species belonging to *Acetobacter, Rhizobium, Agrobacterium*, and *Pseudomonas*, and production of various articles from this polymer are described
EP 0258038 A3	Brown, R. M.	Brown, R. M.	Use of cellulase preparations in the cultivation and use of cellulose-producing microorganisms	1989	To prepare an inoculum of appropriate cell density for large-volume fermentors, *A. xylinum* cells entrapped in cellulose can be liberated with cellulase preparation, which causes a partial cellulose hydrolysis
US 004863565	Weyerhauser Co., USA	Johnson, D. C., Neogi, A. M.	Sheeted products formed from reticulated microbial cellulose	1989	Strains of *Acetobacter* that are stable under agitated culture conditions and that exhibit reduced gluconic acids production, synthesize unique reticulated cellulose sheets, characterized by resistance to densification and great tensile strength
WO 89/11783	Brown, R. M.	Brown, R. M.	Microbial cellulose composites and processes for producing same	1989	Methods enabling production of various objects utilizing BC produced *in situ* or applied as a film are presented; a process for manufacturing currency from BC is described

Tab. 4 (cont.)

Patent number	*Patent holder*	*Inventors*	*Title*	*Date of publi-cation*	*Major claims*
EP 0323717 A3	ICI Plc, UK	Byrom, D.	Process for the production of microbial cellulose	1988	A process for the production of BC using a novel strain of the genus *Acetobacter* is described
EP 0 289993 A3	Weyerhaeuser Co., USA	Johnson, D. C., LeBlanc, H. A., Neogi, A. N.	Bacterial cellulose as surface treatment for fibrous web	1988	BC applied at relatively low concentrations, singularly or in combination, to at least one surface of a fibrous web gives excellent properties of gloss, smoothness, ink receptivity and holdout, and surface strength
WO 88/09381	Financial Union for Agricultural Development, France	Labourer, P. F.	Process for producing bacterial cellulose from material of plant origin	1988	Culturing of *A. xylinum* strain in a plant polysaccharide-containing medium enables efficient cellulose production
US 004655758	Johnson & Johnson products, Inc., USA	Ring, D. F., Nashed, W., Dow, T.	Microbial polysaccharide articles and methods of production	1987	After removal of excess liquid and bacterial cells, the cellulose pellicle can be impregnated and used for various purposes
WO 86/02095	Bio-Fill Industria e Comercio de Produtos Medico Hospitalares, Ltd, Brazil	Farah, L. F. X.	Process for the preparation of cellulose film, cellulose film produced thereby, artificial skin graft and its use	1986	BC film preparation, including optimal conditions of *A. xylinum* culturing, and methods of removing the formed film are described; the film appeared to be suitable for use as an artificial skin graft, a separating membrane, or artificial leather
EP 0200409 A3	Ajinomoto Co. Inc., Japan	Iguchi, M., Mitsuhashi, S., Ichimura, K., Nishi, Y., Uryu, M., Yamanake, S., Watanabe, K.	Molded material comprising bacteria-produced cellulose	1986	BC is an excellent component of molded materials having high dynamic strength as compared to conventional molded materials

with industrial wastes, rich in proper carbon sources, such as spent liquors from crystalline glucose production, etc. Simultaneously, a significant environmental benefit would be obtained. However, the major advantage of mass production of BC would be the protection of forests, which are presently disappearing at an alarming rate, thus leading to soil eutrophication and global climate changes.

BC has already been put to numerous uses, presented in this review. The polysaccharide can not only be replaced by some animal polymers (collagen), but also carriers of substances having a positive impact on human health (e.g. antioxidants and prebiotics). The usefulness of the bacterial polymer in medicine (wound, burn and ulcer dressing materials, component of implants) is not longer questioned. Moreover, cellulose granulates can be an excellent matrix for the immobilization of medicinal preparations. For example, if specific substances (receptors) are adsorbed on BC, the resulting molecules can be scavengers of either toxins or of the pathogenic microflora inhabiting the alimentary tract. Recently, unique nanocrystals ($30 \times 600-800$ nm) of BC have been obtained, derived from its commercial preparation Prima Cel^{TM} (Xylos). Selective modification (trimethyl silylation) of the surface of these nanocrystals while leaving their core intact has been achieved. Such modified crystals have great potential in several advanced technologies.

Deciphering of all the riddles of cellulose biosynthesis will lead to improvement and tailoring of BC supramolecular structure and properties; as a consequence, novel concepts for both its inexpensive production, and for its bulk and specialty applications will be developed.

14
References

Aloni, Y., Benziman, M. (1982) Intermediates of cellulose synthesis in *Acetobacter*, in: *Cellulose and Other Natural Polymers System* (Brown, R. M., Jr., Ed.), New York: Plenum Press, 341–361.

Aloni, Y., Cohen, Y., Benziman, M., Delmer, D. P. (1983) Solubilization of UDP-glucose: 1,4-β-glucan 4-β-D-glucosyl transferase (cellulose synthase) from *Acetobacter xylinum*, *J. Biol. Chem.* **258**, 4419–4423.

Ang, J. F., Miller, W. B. (1991) Multiple functions of powdered cellulose as a food ingredient, *J. Am. Ass. Cereal Chem.* **36**, 558–564.

Bartlett, D. H., Silverman, M. (1989) Nucleotide sequences of IS492, a novel insertion sequence causing variation of extracellular production in the marine bacterium *Pseudomonas atlantica*, *J. Bacteriol.* **171**, 1763–66.

Bauer, K., Codolington, K., Ben-Bassat, A. (1992) Methods for improving production of bacterial cellulose, *Abstr. Paper Am. Chem. Soc.*, **203** Meet., Pt 1, Biot. 94.

Bayer, M. E. (1979) The fusion sites between outer membrane and cytoplasmic membrane in bacteria: their role in membrane assembly and virus infection, in: *Bacterial Outer Membranes* (Inouye, M., Ed.), New York: John Wiley & Sons, 167–202.

Ben-Bassat, A., Bruner, R., Wong, H., Shoemaker, S., Aloni, Y. (1989) Production of bacterial cellulose by *Acetobacter*, *Abstr. Pap. Am. Chem. Soc.*, **198** Meet., MBTD20.

Ben-Bassat, A., Calhoon, R. D., Fear, A. L., Gelfand, D. H., Mead, J. H., Tal, R., Wong, H., Benziman, M. (1993) Methods and nucleic acid sequences for expression of the cellulose synthase operon, US patent 5 268 274.

Ben-Hayyim, G., Ohad, I. (1965) Synthesis of cellulose by *Acetobacter xylinum*; VIII. On the formation and orientation of bacterial cellulose fibrils in the presence of acidic polysaccharides, *J. Cell Biol.* **25**, 191–207.

Benziman, M., Tal, R. (1995) Cellulose synthase associated proteins. US patent 5 382 656.

Boisset, C., Chanzy, H., Henrissat, B., Lamed, R., Shoham, Y., Bayer, E. A. (1999) Digestion of crystalline cellulose substrates by the *Clostridium thermocellum* cellulosome: structural and morphological aspects, *Biochem. J.* **340**, 829–835.

Boisset, C., Fraschini, C., Schulein, M., Henrissat, B., Chanzy, H. (2000) Imaging the enzymatic digestion of bacterial cellulose ribbons reveals the endo character of the cellobiohydrolase Cel6A from *Humicola insolens* and its mode of synergy with cellobiohydrolase Cel7A, *Appl. Environ. Microbiol.* **66**, 1444–1452.

Brown, A. J. (1886) An acetic ferment which forms cellulose, *J. Chem. Soc.* **49**, 432–439.

Brown R. M., Jr. (1989a) Microbial cellulose composites and processes for producing same. WO 89/11783.

Brown, R. M., Jr. (1989b) Microbial cellulose as a building block resource for specialty products and processes thereof, WO 89/12107.

Brown, R. M., Jr. (1989c) Use of cellulase preparations in the cultivation and use of cellulose-producing microorganisms, European patent 0258038A3.

Brown, R. M., Jr. (1996) The biosynthesis of cellulose, *Pure Appl. Chem.* **A33**, 1345–1373.

Brown, R. M., Jr., Saxena, I. M. (2000) Cellulose biosynthesis: a model for understanding the assembly of biopolymers, *Plant Physiol. Biochem.* **38**, 57–60.

Brown, R. M., Jr., Willison, J. H. M., Richardson, C. L. (1976) Cellulose biosynthesis in *Acetobacter xylinum*: visualisation of the site of synthesis and direct measurement of the *in vivo* process, *Proc. Natl. Acad. Sci. USA* **73**, 4565–4569.

Bungay, H. R., Serafica, G. (1997) Production of microbial cellulose using a rotating disc film bioreactor, WO 97/05271.

Bureau, T. E., Brown, R. M., Jr. (1987) *In vitro* synthesis of cellulose II from a cytoplasmic membrane fraction of *Acetobacter xylinum*, *Proc. Natl. Acad. Sci. USA* **84**, 6985–6989.

Canale-Parda, E., Wolfe, R. S. (1964) Synthesis of cellulose by *Sarcina ventriculi*, *Biochim. Biophys. Acta* **82**, 403–405.

Cannon, R. E., Anderson, S. M. (1991) Biogenesis of bacterial cellulose, *Crit. Rev. Microbiol.* **17**, 435–439.

Chao, Y., Ishida, T., Sugano, Y., Shoda, M. (2000) Bacterial cellulose production by *Acetobacter xylinum* in a 50 L internal-loop airlift reactor, *Biotechnol. Bioeng.* **68**, 345–352.

Colvin, J. R. (1957) Formation of cellulose microfibrils in a homogenate of *Acetobacter xylinum*, *Arch. Biochem. Biophys.* **70**, 294–295.

Colvin, J. R. (1980) The biosynthesis of cellulose, in: *Plant Biochemistry* (Priess, J., Ed.), New York: Academic Press, 543–570, Vol. 3

Colvin, J. R., Leppard, G. G. (1977) The biosynthesis of cellulose by *Acetobacter xylinum* and *Acetobacter acetigenus*, *Can. J. Microbiol.* **23**, 701–709.

Colvin, J. R., Sowden, L. C., Daoust, V., Perry, M. (1979) Additional properties of a soluble polymer of glucose from cultures of *Acetobacter xylinum*, *Can. J. Biochem.* **57**, 1284–1288.

Costeron, J. W. (1999) The role of bacterial exopolysaccharides in nature and disease, *J. Ind. Microbiol. Biotechnol.* **22**, 551–563.

Coucheron, D. H. (1991) An *Acetobacter xylinum* insertion sequence element associated with inactivation of cellulose production, *J. Bacteriol.* **173**, 5723–2731.

Dekker, R. F. H., Rietschel, E. T., Sandermann, H. (1977) Isolation of α-glucan and lipopolysaccharide fractions from *Acetobacter xylinum*, *Arch. Microbiol.* **115**, 353–357.

De Lannino, N., Cuoso, R. O., Dankert, M. A. (1988) Lipid-linked intermediates and the synthesis of acetan in *A. xylinum*, *J. Gen. Microbiol.* **134**, 1731–1736.

Delmer, D. P. (1982) Biosynthesis of cellulose, *Adv. Carbohydr. Chem. Biochem.* **41**, 105–153.

De Wulf, P., Joris, K., Vandamme, E. J. (1996) Improved cellulose formation by an *Acetobacter xylinum* mutant limited in (keto)gluconate synthesis, *J. Chem. Tech. Biotechnol.* **67**, 376–380.

Easson, D. D., Jr., Sinskey, A. J., Peoples, O. P. (1987) Isolation of *Zoogloea ramigera* I-16 M exopolysaccharides biosynthetic genes and evidence for instability within this region, *J. Bacteriol.* **169**, 4518–4524.

Embuscado, M. E., Marks, J. S., Miller, J. N. (1994) Bacterial cellulose. II. Optimization of cellulose production by *Acetobacter xylinum* through response surface methodology, *Food Hydrocolloids* **8**, 419–430.

Galas, D. J., Chandler, M. (1989) Bacterial insertion sequences, in: *Mobile DNA* (Berg, D. E., Howe, M. M., Eds.) Washington, DC: ASM, 109–162.

Galas, E., Krystynowicz, A., Tarabasz-Szymanska, L., Pankiewicz, T., Rzyska, M. (1999) Optimization of the production of bacterial cellulose using multivariable linear regression analysis, *Acta Biotechnol.* **19**, 251–260.

Glaser, L. (1958) The synthesis of cellulose in cell-free extracts of *Acetobacter xylinum*, *J. Biol. Chem.* **232**, 627–636.

Haigler, C. H. (1985) The function and biogenesis of native cellulose, in: *Cellulose Chemistry and its Applications* (Nevel, R. P., Zeronian, S. H., Eds.), Chichester: Ellis Horwood, 30–83.

Han, N. S., Robyt, J. F. (1998) The mechanism of *Acetobacter xylinum* cellulose biosynthesis: direction of chain elongation and the role of lipid pyrophosphate intermediates in the cell membrane, *Carbohydr. Res.* **313**, 125–133.

Hestrin, S., Aschner, M., Mager J. (1947) Synthesis of cellulose by resting cells of *Acetobacter xylinum*, *Nature* **159**, 64–65.

Hestrin, S., Schramm, M. (1954) Synthesis of cellulose by *Acetobacter xylinum*, II. Preparation of freeze-dried cells capable of polymerizing glucose to cellulose, *Biochem. J.* **58**, 345–352.

Hikawu, S., Hiroshi, T., Takayasu, T., Yoshinaga, F. (1996) Manufacture of bacterial cellulose by addition of cellulose formation stimulators, Japanese patent 96316922.

Hotte, B., Roth-Arnold, I., Puhler, A., Simon, R. (1990) Cloning and analysis of a 35.3 kb DNA region involved in exopolysaccharide production by *Xanthomonas campestris*, *J. Bacteriol.* **172**, 2804–2807.

Husemann, E., Muller, G. J. M. (1966) Synthesis of unbranched polysaccharides, *Macromol. Chem.* **91**, 212–230.

Ielpi, L., Couso, R., Dankert, M. A. (1981) Lipid-linked intermediates in the biosynthesis of xanthan gum, *FEBS Lett.* **130**, 253–256.

Iguchi, M., Yamanaka, S., Budhioko, A. (2000) Bacterial cellulose – a masterpiece of nature's arts, *J. Mater. Sci.* **35**, 261–270.

Ishikawa, A., Matsuoka, M., Tsuchida, T., Yoshinaga, F. (1995) Increasing of bacterial cellulose production by sulfoguanidine-resistant mutants derived from *Acetobacter xylinum* subsp. sucrofermentans BPR2001, *Biosci. Biotechnol. Biochem.* **59**, 2259–2263.

Ishikawa, A., Tsuchida, T., Yoshinaga, F. (1996a) Production of bacterial cellulose using microbial strain resistant to inhibitor of DHO-dehydrogenase, Japanese patent 08009965A.

Ishikawa, A., Tsuchida, T., Yoshinaga, F. (1996b) Production of bacterial cellulose with pyrimidine analogue-resistant strain, Japanese patent 08000260.

Johnson, D. C., Neogi, A. N. (1989) Sheeted products formed from reticulated microbial cellulose, US patent 4 863 565.

Jonas, R., Farah, L. F. (1998) Production and application of microbial cellulose, *Polym. Degrad. Stabil.* **59**, 101–106.

Jones, D., Gordon, A. H., Bacon, J. S. D. (1974) β-1,3-Glucanases from *Sclerotium rolfsii*, *Biochem. J.* **140**, 47–55.

Joris, K., Billiet, F., Drieghe, S., Brachx, D., Vandamme, E. (1990) Microbial production of β-1,4-glucan, *Meded. Fac. Landbouwwet Rijksuniv. Gent* **55**, 1563–1566.

Kenji, K., Yukiko, M., Hidehi, L., Kunihiko, O. (1990) Effect of culture conditions of acetic acid bacteria on cellulose biosynthesis, *Br. Polym. J.* **22**, 167–171.

Kent, R. A., Stephens, R. S., Westland, J. A. (1991) Bacterial cellulose fiber provides an alternative for thickening and coating, *Food Technol.* **45**, 108.

Kim, D., Kim, Y. M., Park, D. H. (1999) Modification of *Acetobacter xylinum* bacterial cellulose using dextransucrase and alternansucrase, *J. Microbiol. Biotechnol.* **9**, 704–708.

Kobayashi, S., Kashiwa, K., Kawasaki, T., Shoda, S. (1991) Novel method for polysaccharide synthesis using an enzyme: the first *in vitro* synthesis of cellulose via nonbiosynthetic path utilising cellulase as catalyst, *J. Am. Chem. Soc.* **113**, 3079–3084.

Kouda, T., Naritomi, T., Yano, H., Yoshinaga, F. (1998) Inhibitory effect of carbon dioxide on bacterial cellulose production by *Acetobacter* in agitated culture, *J. Ferment. Bioeng.* **85**, 318–321.

Koyama, M., Helbert, W., Imai, T., Sugiyama, J., Henrissat, B. (1997) Parallel-up structure evidence the molecular directionality during biosynthesis of bacterial cellulose, *Proc. Natl. Acad. Sci. USA* **94**, 9091–9095.

Krystynowicz, A., Turkiewicz, M., Drynska, E., Galas, E. (1995) Bacterial cellulose – biosynthesis and application, *Biotechnologia* **30**, 120–132.

Krystynowicz, A., Galas, E., Pawlak, E. (1997) Method of bacterial cellulose production. Polish patent P-299907.

Krystynowicz, A., Czaja, W., Bielecki, S. (1999) Biosynthesis and application of bacterial cellulose, *Zywnosc* **3**, 22–33.

Krystynowicz, A., Czaja, W., Pomorski, L., Kolodziejczyk, M., Bielecki, S. (2000) The evaluation of usefulness of microbial cellulose as a wound dressing material, 14th Forum for Applied Biotechnology, Gent, Belgium, *Meded. Fac. Landbouwwet Rijksuniv. Gent*, Proceedings Part I, 213–220.

Kudlicka K. (1989) Terminal complexes in cellulose synthesis, *Postepy biologii komórki* **16**, 197–212 (abstract in English).

Kuga, G., Brown, R. M., Jr. (1988) Silver labeling of the reducing ends of bacterial cellulose, *Carbohydr. Res.* **180**, 345–350.

Laboureur, P. (1988) Process for producing bacterial cellulose from material of plant origin, WO 88/09381.

Legge R. L. (1990) Microbial cellulose as a specialty chemical, *Biotechnol. Adv.* **8**, 303–319.

Leppard G. G., Sowden, L. C., Ross, C. J. (1975) Nascent stage of cellulose biosynthesis, *Science* **189**, 1094–1095.

Lin, F. C., Brown, R. M., Jr. (1989) Purification of cellulose synthase from *Acetobacter xylinum*, in: *Cellulose and Wood Chemistry and Technology* (Schmerck, C., Ed.), New York: John Wiley & Sons, 473–492.

Lin, F. C., Brown, R. M., Jr., Cooper, J. B., Delmer, D. P. (1985) Synthesis of fibrils *in vitro* by a solubilized cellulose synthase from *Acetobacter xylinum*, *Science* **230**, 822–825.

Lin, F. C., Brown, R. M., Jr., Drake, R. R., Jr., Haley. B. E. (1990) Identification of the uridine-5'-diphosphoglucose (UDPGlc) binding subunit of cellulose synthase in *Acetobacter xylinum* using the photoaffinity probe 5-azido-UDPGlc, *J. Biol. Chem.* **265**, 4782–4784.

Marx-Figini, M., Pion, B. G. (1974) Kinetic investigations on biosynthesis of cellulose by *Acetobacter xylinum*, *Biochim. Biophys. Acta* **338**, 382–393.

Masaoka, S., Ohe, T., Sakota, N. (1993) Production of cellulose from glucose by *Acetobacter xylinum*, *J. Ferment. Bioeng.* **75**, 18–22.

Matsuoka, M., Tsuchida, T., Matsuchita, K., Adachi, O., Yoshinaga, F. (1996) A synthetic medium for bacterial cellulose production by *Acetobacter xylinum* subsp. sucrofermentans, *Biosci. Biotechnol. Biochem.* **60**, 575–579.

Matthyse, A., Thomas, D. I., White, A. R. (1995) Mechanism of cellulose synthesis in *Agrobacterium tumefaciens*, *J. Bacteriol.* **177**, 1076–1081.

Micheel, F., Brodde, O. E. (1974) Polymerization of 1,4-anhydro-2,3,6-tri-*O*-benzyl-α-D-glucopyranose, *Liebigs Ann. Chem.* **124**, 702–708.

Minakami, H., Entani, K., Tayama, S., Fujiyama, S., Masai, H. (1984) Isolation and characterization of a new polysaccharide-producing *Acetobacter* sp. *Agric. Biol. Chem.* **48**, 2405–2414.

Mondal, I. H., Kai, A. (2000) Control of the crystal structure of microbial cellulose during nascent stage, *J. Appl. Polym. Sci.***79**, 1726–1734.

Nakai, T., Tonouchi, N., Konishi, T., Kojima, Y., Tsuchida, T., Yoshinaga, F., Sakai, F., Hayashi, T. (1999) Enhancement of cellulose production by expression of sucrose synthase in *Acetobacter xylinum*, *Proc. Natl. Acad. Sci. USA* **96**, 14–18.

Nakatsubo, F., Takano, T., Kawada, T., Murakami, K. (1989) Toward the synthesis of cellulose: synthesis of cellooligosaccharides, in: *Cellulose: Structural and Functional Aspects* (Kennedy, J. F., Philips, G. O., Williams, P. A., Eds.), New York: Ellis Horwood, 201–206.

Nari, J., Noat, G., Richard, J., Franchini, E., Monstacas, A. M. (1983) Catalytic properties and tentative function of a cell wall β-glucosyltransferase from soybean cells cultured *in vitro*, *Plant Sci. Lett.* **28**, 313–320.

Naritomi, T., Kouda, T., Yano, H., Yoshinaga, F. (1998) Effect of ethanol on bacterial cellulose production from fructose in continuous culture, *J. Ferment Bioeng.* **85**, 598–603.

Nichols, S. E., Singletary, G. W. (1998) Cellulose synthesis in the storage tissue of transgenic plants, US patent 5 723 764.

Ohana, P., Delmer, D. P., Volman, G., Benziman, M. (1998) Glycosylated triterpenoid saponin: a specific inhibitor of diguanylate cyclase from *Acetobacter xylinum*, *Plant Cell Physiol.* **39**, 153–159.

Oikawa, T., Kamatani, N., Kaimura, T., Ameyama, M., Soda, K. (1997) Endo-β-glucanase from *A. xylinum* – purification and characterization, *Curr. Microbiol.* **34**, 309–313.

Okamoto, T., Yamano, S., Ikeaga, H., Nakamura, K. (1994) Cloning of the *A. xylinum* cellulase gene and its expression in *E. coli* and *Zymomonas mobilis*, *Appl. Microbiol. Biotechnol.* **42**, 563–568.

Okiyama, A., Shirae, H., Kano, H., Yamanaka, S. (1992) Two-stage fermentation process for cellulose production by *Acetobacter aceti*, *Food Hydrocolloids* **6**, 471–477.

Palonen, H., Tenkanen, M., Linder, M. (1999) Dynamic interaction of *Trichoderma reesei* cellobiohydrolases Cel6A and Cel7A and cellulose at equilibrium and during hydrolysis, *Appl. Environ. Microbiol.* **65**, 5229–5233.

Ring, D. F., Nashed, W., Dow, T. (1986) Liquid loaded pad for medical applications, US patent 4 588 400.

Ross, P., Weinhouse, H., Aloni, Y., Michaeli, D., Ohana, P., Mayer, R., Braun, S., de Vroom, E., van der Marel, G. A., van Boom, J. H., Benziman, M. (1987) Regulation of cellulose synthesis in *Acetobacter xylinum* by cyclic diguanylic acid, *Nature* **325**, 279–281.

Ross, P., Mayer, R., Benziman, M. (1991) Cellulose biosynthesis and function in bacteria, *Microbiol. Rev.* **55**, 35–58.

Sakair, N., Asamo, H., Ogawa, M., Nishi, N., Tokura, S. (1998) A method for direct harvest of bacterial cellulose filaments during continuous cultivation of *Acetobacter xylinum*, *Carbohydr. Polym.* **35**, 233–237.

Samejima, M., Sugiyama, J., Igarashi, K., Eriksson, K. E. L. (1997) Enzymatic hydrolysis of bacterial cellulose, *Carbohydr. Res.* **305**, 281–288.

Sattler, K., Fiedler, S. (1990) Production and application of bacterial cellulose. II. Cultivation in a rotating drum fermentor, *Zbl. Microbiol.* **145**, 247–252.

Savidge, R. A., Colvin, J. R. (1985) Production of cellulose and soluble polysaccharides by *Acetobacter xylinum*, *Can. J. Microbiol.* **31**, 1019–1025.

Saxena, I. M., Brown, R. M., Jr. (1989) Cellulose biosynthesis in *Acetobacter xylinum*: a genetic approach, in: *Cellulose and Wood Chemistry and Technology* (Schnerck, C., Ed.). New York: John Wiley & Sons, 537–557.

Saxena, I. M., Brown, R. M., Jr. (1997) Identification of cellulose synthase(s) in higher plants: Sequence analysis of processive β-glycosyltransferases with common motif 'D, D, D 35Q (RQ)XRW', *Cellulose* **4**, 33–49.

Saxena, I. M., Brown, R. M., Jr. (2000) Cellulose synthases and related proteins, *Curr. Opin. Plant Biol.* **3**, 523–531.

Saxena, I. M., Lin, F. C., Brown, R. M., Jr. (1990a) Cloning and sequencing of the cellulase synthase catalytic subunit gene of *Acetobacter xylinum*, *Plant Mol. Biol.* **15**, 673–683.

Saxena, I. M., Roberts, E. M., Brown, R. M., Jr. (1990b) Modification of cellulose normally synthesised by cellulose-producing microorganisms, US patent 4 950 597.

Saxena, I. M., Lin, F. C., Brown, R. M., Jr. (1991) Identification of a new gene in an operon for cellulose biosynthesis in *Acetobacter xylinum*, *Plant Mol. Biol.* **16**, 947–954.

Saxena, I. M., Brown, R. M., Jr., Fevre, M., Geremia, R. A., Henrissat, B. (1995) Multidomain architecture of β-glycosyl transferase: Implications for mechanism of action, *J. Bacteriol.* **177**, 1419–1424.

Seto, H., Tsuchida, T., Yoshinaga, F., Beppu, T., Horinouchi, S. (1996) Characterization of the biosynthetic pathway of cellulose from glucose and fructose in *Acetobacter xylinum*, *Biosci. Biotechnol. Biochem.* **60**, 1377–1379.

Seto, H., Tsuchida, T., Yoshinaga, F. (1997) Manufacture of bacterial cellulose by mixed culture of microorganisms, Japanese patent 9721905.

Srisodsuk, M., Kleman-Leyer, K., Keranen, S., Kirk, T. K., Teeri, T. T. (1998) Modes of action on cotton and bacterial cellulose of a homologous endoglucanase-exoglucanase pair from *Trichoderma reesei*, *Eur. J. Biochem.* **62**, 185–187.

Stalbrandt, H., Mansfield, S. D., Saddler, J. N., Kilburn, D. G., Warren, R. A. J., Gilkes, N. R. (1998) Analysis of molecular size of cellulose by recombinant *Cellulomonas fimi* β-1,4-glucanase, *Appl. Environ. Microbiol.* **64**, 2374–2379.

Standal, R., Iversen, T. G., Coucheron, D. H., Fjaervik, E., Blatny, J. M., Valla, S. (1994) A new gene required for cellulose production and gene encoding cellulolytic activity in *A. xylinum* are localized with *bcs* operon, *J. Bacteriol.* **176**, 665–672.

Sutherland, I. W. (1998) Novel and established applications of microbial polysaccharides, *TIBTECH* **16**, 41–46.

Swissa, M., Aloni, Y., Weinhouse, H., Benziman, M. (1980) Intermediary steps in cellulose synthesis in *Acetobacter xylinum*: studies with whole cells and cell-free preparation of the wild type and a cellulose-less mutant, *J. Bacteriol.* **143**, 1142–1150.

Tahara, N., Tabuchi, M., Watanabe, K., Yano, H., Morinaga, Y., Yoshinaga, F. (1997) Degree of polymerization of cellulose from *A. xylinum* BPR 2001 decreased by cellulase producing strain, *Biosci. Biotechnol. Biochem.* **61**, 1862–1865.

Tahara, N., Tonouchi, M., Yano, H., Yoshinaga, F. (1998) Purification and characterization of exo-1,4-β-glucosidase from *A. xylinum* BPR 2001, *J. Ferment. Bioeng.* **85**, 589–594.

Takai, M., Tsuta, Y., Watanabe, S. (1997) Biosynthesis of cellulose by *Acetobacter xylinum* and characterization of bacterial cellulose, *Polym. J.* **7**, 137–146.

Tal, R., Wong, H. C., Calhon, R., Gelfand, D., Fear, A. L., Volman, G., Mayer, R., Ross, P., Amikam, D., Weinhouse, H., Cohen, A., Sapir, S., Ohana, P., Benziman, M. (1998) Three *cdg* operons control cellular turnover of c-di-GMP in *Acetobacter xylinum*: genetic organization and occurrence of conserved domain in isozymes, *J. Bacteriol.* **180**, 4416–4425.

Tayama, K., Minakami, H., Entani, E., Fujiyama, S., Masai, H. (1985) Structure of an acidic polysaccharide from *Acetobacter*, sp. NBI 1022, *Agric. Biol. Chem.* **49**, 959–966.

Teeri, T. T. (1997) Crystalline cellulose degradation: new insight into the function of cellobiohydrolases, *TIBTECH* **15**, 160–167.

Tonouchi, N., Tsuchida, T., Yoshinaga, F., Beppu, T. (1996) Characterization of the biosynthetic pathway of cellulose from glucose and fructose in *Acetobacter xylinum*, *Biosci. Biotechnol. Biochem.* **60**, 1377–1379.

Tonouchi, N., Tahara, N., Kojima, Y., Nakai, T., Sakai, F., Hayashi, T., Tsuchida, T., Yoshinaga, F. (1997) A β-glucosidase gene downstream of the cellulase synthase operon in cellulase producing *Acetobacter*, *Biosci. Biotechnol. Biochem.* **61**, 1789–1790.

Tonouchi, N., Hirinouchi, S., Tsuchida, T., Yoshinaga, F. (1998) Increased cellulose production by *Acetobacter* after introducing the sucrose phosphorylase gene, *Biosci. Biotechnol. Biochem.* **62**, 1778–1780.

Toyosaki, H., Naritomi, T., Seto, A., Matsuoka, M., Tsuchida, T., Yoshinga, F. (1995) Screening of bacterial cellulose- producing *Acetobacter* strains suitable for agitated culture, *Biosci. Biotechnol. Biochem.* **59**, 1498–1502.

Valla, S., Kjosbakken, J. (1981) Isolation and characterization of a new extracellular polysaccharide from a cellulose-negative strain of *Acetobacter xylinum*, *Can. J. Microbiol.* **27**, 599–603.

Valla, S., Coucheron, D. H., Fjaervik, E., Kjosbakken, J., Weinhose, H., Ross, P., Amikam, D., Benziman, M. (1989) Cloning of a gene involved in cellulose biosynthesis in *Acetobacter xylinum*: complementation of cellulose-negative mutant by the UDPG pyrophosphorylase structure gene, *Mol. Gen. Genet.* **217**, 26–30.

Vandamme, E. J., De Baets, S., Vanbaelen, A., Joris, K., De Wulf P. (1998) Improved production of bacterial cellulose and its application potential, *J. Polymer Degrad. Stabil.* **59**, 93–99.

Watanabe, K., Tabuchi, M., Morinaga, Y., Yoshinaga, F. (1998a) Structural features and properties of bacterial cellulose produced in agitated culture, *Cellulose* **5**, 187–200.

Watanabe, K., Tabuchi, M., Ischikawa, M., Takemura, H., Tsuchida, T., Morinaga, Y. (1998b)

Acetobacter xylinum mutant with high cellulose productivity and ordered structure, *Biosci. Biotechnol. Biochem.* **62**, 1290–1292.

Weinhouse, R. (1977) Regulation of carbohydrate metabolism in *Acetobacter xylinum*, PhD thesis, Hebrew University of Jerusalem. Jerusalem, Israel.

Weinhouse, H., Sapir, S., Amikam, D., Shiko, Y., Volman, G., Ohana, P., Benziman, M. (1997) c-di-GMP-binding protein, a new factor regulating cellulose synthesis in *Acetobacter xylinum, FEBS Lett.* **416**, 207–211.

Wiliams, W. S., Cannon, R. E. (1989) Alternative environmental roles for cellulose produced by *Acetobacter xylinum, Appl. Environ. Microbiol.* **55**, 2448–2452.

Winkelman, J. W., Clark, D. P. (1984) Proton suicide: general method for direct selection of sugar transport and fermentation-defective mutants, *J. Bacteriol.* **11**, 687–690.

Wlochowicz, A. (2001) Personal communication.

Wong, H. C., Fear, A. L., Calhoon, R. D., Eichinger, G. M., Mayer, R., Amikam, D., Benziman, M., Gelfand, D. H., Meade, J. H., Emerick, A. W., Bruner, R., BenBassat, A., Tal, R. (1990) Genetic organization of cellulose synthase operon in *A. xylinum, Proc. Natl. Acad. Sci. USA* **87**, 8130–8134.

Yamada, Y., Hoshino, K., Ishikawa, T. (1997) The phylogeny of acetic acid bacteria based on the partial sequences of 16 S ribosomal RNA: the elevation of the subgenus *Gluconacetobacter* to the generic level, *Biosci. Biotechnol. Biochem.* **61**, 1244–51.

Yamada, Y., Hoshino, K., Ishikawa, T. (1998) *Gluconacetobacter corrig.* (*Gluconoacetobacter* [sic]), in: Validation of publication of new names and new combinations previously effectively published outside the IJSB, List no 64, *Int. J. Syst. Bacteriol.* **48**, 327–328.

Yamada, Y. (2000) Transfer *Acetobacter oboediens* and *A. intermedius* to the genus *Gluconacetobacter* as *G. oboediens* comb. nov. and *G. intermedius* comb. nov., *Int. J. System. Evolut. Microbiol.* **50**, 2225–2227.

Yamamoto, H., Horii, F., Hirai, A. (1996) *In situ* crystallization of bacterial cellulose. 2. Influence of different polymeric additives on the formation of celluloses I_{α} and I_{β} at the early stage of incubation, *Cellulose* **3**, 229–242.

Yamanaka, S., Watanabe, K., Kitamura, N., Iguchi, M., Mitsuhashi, S., Nishi, Y., Uryu, M. (1989) The structure and mechanical properties of sheets prepared from bacterial cellulose, *J. Mater. Sci.* **24**, 3141–3145.

Yamanaka, S., Watanabe, K., Suzuki, Y. (1990) Hollow microbial cellulose, process for preparation thereof, and artificial blood vessel formed of said cellulose, European patent 0396344A2.

Yamanaka, S., Ishihara, M., Sugiyama, J. (2000) Structural modification of bacterial cellulose, *Cellulose* **7**, 213–225.

Zaar, K. (1977) The biogenesis of cellulose by *Acetobacter xylinum, Cytobiologie* **16**, 1–15.

12
Chitin and Chitosan in Fungi

Prof. Dr. Martin G. Peter
University of Potsdam, Institute of Organic Chemistry and Structure Analysis, and Interdisciplinary Research Center for Biopolymers, Karl-Liebknecht-Str. 25, D-14476 Golm, Germany; Tel.: +49-331-977-5401; Fax: +49-331-977-5300; E-mail: peter@serv.chem.uni-potsdam.de

Biopolymers for Medical and Pharmaceutical Applications. Edited by A. Steinbüchel and R. H. Marchessault

ISBN: 3-527-31154-8

AA	amino acid residues
CDA	chitin deacetylase
CHS	chitin synthase gene
CS	chitin synthase
CSN	chitosanase gene
DP	degree of polymerization
ER	endoplasmic reticulum
F_A	mole fraction of *N*-acetylglucosamine residues
FMOC	9-fluorenylmethoxycarbonyl
FT-IR	Fourier transformation infrared spectroscopy
GlcN	2-amino-2-deoxy-D-glucopyranose, β-(1-4)-linked in chitin/chitosan
GlcNAc	2-acetamido-2-deoxy-D-glucopyranose, β-(1-4)-linked in chitin/chitosan
GlcNase	glucosaminidase
HPLC	high-performance liquid chromatography
M	molecular mass (daltons)
MALDI TOF MS	matrix-assisted laser desorption ionization time-of-flight mass spectrometry
M_v	viscosity average molecular mass

M_w	mass average of molecular mass
ORF	open reading frame
PITC	phenylisothiocyanate
UDP-GlcNAc	uridine diphospho-*N*-acetylglucosamine
WGA	wheat germ agglutinin

1 Introduction

The *aminoglucan* chitin (poly-GlcNAc) is widely distributed in nature, whereas the related polysaccharide chitosan (poly-GlcN) occurs in certain fungi only. Fungal chitin shows some special features, in particular with respect to chemical structure and biosynthesis. Thus, it appears appropriate to look at fungal chitin and chitosan in their own rights and to treat them in a separate chapter.

The basic principles of chitin and chitosan are treated elsewhere in this volume, and readers who are not familiar with these aminoglucans should consult also Chapter 15, this volume (Chitin and Chitosan from Animal Sources).

2 Chemical Structure

The chitin of fungi possesses principally the same structure as the chitin occurring in other organisms (see Chapter 15, this volume). However, a major difference results from the fact that fungal chitin is associated with other polysaccharides which do not occur in the exoskeleton of arthropods. Furthermore, the occurrence of chitosan is apparently restricted to fungi.

2.1 Chitin

The molecular mass of chitin in fungi is not known. However, it was estimated that bakers' yeast synthesizes rather uniform chains containing 120–170 GlcNAc monomer units which corresponds to ca. 24,000–34,500 Daltons (for a reference, see Valdivieso et al., 1999). This is roughly one order of magnitude lower than the estimated molecular mass of chitosan isolated from Mucoraceae (see below).

In *Saccharomyces cerevisiae*, terminal reducing ends of chitin chains are attached through β-(1,4)- or β-(1,2)-linkages to the nonreducing end of β-(1,3)-glucan branches which are linked to β-(1,6)-glucan. Attachment of chitin to glucan is catalyzed by chitin synthase 3 (Hartland et al., 1994). A mannoprotein is attached to β-(1,6)-glucan through a glycosylphosphatidylinositol anchor containing five α-linked mannosyl residues (Kollar et al., 1995, 1997). A mutant of *S. cerevisiae* with a reduced β-(1,3)-glucan content shows increased cross-linking of mannoproteins to chitin trough β-(1,6)-glucan (Kapteyn et al., 1997).

Likewise, the cell wall of *Aspergillus fumigatus* is highly complex, containing in the alkali-insoluble fraction a linear β-(1,3/1,4)-glucan, galactomannan, chitin, and β-(1,3)-glucan, whereas β-(1,6)-glucan is absent. The β-(1,3)-glucan shows 4% of β-(1,6)-branching. As in *S. cerevisiae*, chitin is β-(1,4)-linked to β-(1,3)-glucan (Fontaine et al., 2000). The content of GlcNAc in cell walls of

A. fumigatus in only 50% of that in *Aspergillus nidulans* (Guest and Momany, 2000).

Staining with fluorescent lectins reveals distinct distributions of mannoproteins, glucans and chitin in the cell wall of *Candida albicans* (Ruiz-Herrera et al., 1994). The cell wall of *Fusarium oxysporum* is composed of an outer layer of glycoproteins covering an inner layer of chitin and glucan (Schoffelmeer et al., 1999). Cell walls of the geophilic Dermatophytes *Microsporum fulvum* and *Epidermophyton stockdaleae* contain 4.0–6.5% of glucomannan and 44.2–71.0% of a glucan–chitin complex (Guarro et al., 1993).

2.2 Chitosan

Chitosans isolated from Mucorales typically show M_v in the range 4×10^5 to 1.2×10^6 Daltons and F_A values between 0.2 and 0.09. Amino acid analysis of chitosan prepared from *Aspergillus niger* reveals covalently bound arginine, serine, and proline (Lestan et al., 1993). The chitosan–glucan complexes from mycelia of *A. niger, Humicola lutea*, and *Fusarium moniliforme* contain 0.05–0.06% of amino acids, mostly as lysine and histidine (Velichkov and Sotirov, 1990).

Bacillus pumilus chitosanase is more effective than *Streptomyces griseus* chitinase in digestion of the cell wall of *Fusarium oxysporum*. Besides GlcN-GlcNAc as the main product, maltose is also observed as a degradation product, which indicates the presence of α-(1,4)-linked glucan (Fukamizo et al., 1996).

2.3 Polyphenolic Pigments

Cell walls or sporophore capsules of fungi contain often polyphenolic pigments which presumably function to enforce the polysaccharide–protein network by oxidative cross-linking or impregnation with a hydrophobic polymer. According to chemical logic, the polyphenols should be covalently bound to other biopolymers, though this has not been proven unequivocally. Precursors of fungal pigments are L-tyrosine and its hydroxylation product, 3,4-dihydroxy-phenylalanine (DOPA). Black and brown pigments, which are sometimes incorrectly named melanins, are often also derived from 1,8-dihydroxynaphthalene or from simple catechols (Prota, 1992). Several fungi use γ-glutaminyl-4-hydroxybenzene as a precursor (Pierce and Rast, 1995). Wood- inhabiting fungi, e.g., *Inonotus hispidus*, produce a phenolic biopolymer that is derived from the stryrylpyrone hispidin (for a general review of fungal pigments, see Gill and Steglich, 1987).

3 Occurrence

Chitin is widely distributed in fungi, occurring in Basidiomycetes, Ascomycetes, and Phycomycetes, where it is a component of the cell walls and structural membranes of mycelia, stalks, and spores. The amounts vary between traces and up to 45% of the organic fraction, the rest being mostly proteins, glucans and mannans (Roberts, 1992). However, not all fungi contain chitin, and the polymer may be absent in one species that is closely related to another. Variations in the amounts of chitin may depend on physiological parameters in natural environments as well as on the fermentation conditions in biotechnological processing or in cultures of fungi.

Chitin is the major component in primary septa between mother and daughter cells of *S. cerevisiae*, and also one of the main components of the hyaline outer wall of spores of four arbuscular mycorrhizal *Glo-*

mus species (Sbrana et al., 1995). Hyphal walls of the Oomycete *Pythium ultimum* contain cellulose and chitin, whereas only cellulose is present in another Oomycete, *Phythophthora parasitica*. Both polysaccharides are present in cell walls of the Ascomycetes *Ophiostoma ulmi* and *Colletotrichum lindemuthianum*, whereas the Ascomycete *Fusarium oxysporum* and the Basidiomycete *Rhizoctonia solani* contain only chitin (Cherif et al., 1993). The zoopathogenic fungi *Cryptococcus neoformans, Pityrosporum canis* and *Rhizopus oryzae* contain chitin, but not β-(1,3)-glucan (Nicholas et al., 1994).

The chitin of the cell wall of the white-rot fungus *Rigidoporus lignosus* is degraded by enzymes excreted as a defense response by the host cell, and therefore is not detectable during the process of infection (Nicole and Benhamou, 1991). The fungal sheaths of another white-rot fungus, *Phellinus noxius*, do not contain chitin (Nicole et al., 1995).

The mycelia, and the caps and stalks of fruiting bodies of four edible mushrooms, *Lentinus edodes, Lycophyllum shimeji, Pleurotussajor-caju*, and *Volvariella volvacea* contain chitin as a minor component (Cheung, 1996).

Chitosan occurs naturally in the Mucorales, in particular *Mucor, Absidia*, and *Rhizopus* species. There is apparently only one report on the presence of chitosan in a Basidiomycete, *Lentinus edodes* (Shiitake mushroom) (Crestini and Giovannozzi-Sermanni, 1996; Crestini et al., 1996).

Slime molds (Myxomycetes) and bacteria (Schizomycetes) are devoid of chitin.

4 Physiological Function

Chitin serves as a fibrous strengthening element responsible for cell wall rigidity. However, there are other functions of chitin and chitosan, as revealed by mutants bearing a defect in the complex machinery of chitin biosynthesis, intracellular trafficking of chitin synthases, or deposition of the polysaccharide in cell walls, although the morphology of a mutant may be indistinguishable from that of the wild-type. Thus, chsD disruptants of *Aspergillus nidulans* show excessive swelling and lysis of conidia in hypoosmotic media (Specht et al., 1996) and *S. cerevisiae* chs5Δ null mutants fail to undergo cell fusion in mating (Santos et al., 1997) (for details, see Section 6.1).

5 Chemical Analysis and Detection

Determination of fungal chitin in biological samples is important for estimating fungal biomass, e.g., in infected plant tissues. A comprehensive review of the most common methods was given by Muzzarelli (1997) (see also Chapter 15, this volume). Frequently, GlcN is quantified by colorimetric methods in hydrolyzates of alkali-resistant fractions to determine the amounts of chitin and chitosan (Plassard, 1997). GlcN was also determined in acid hydrolyzates by high-performance liquid chromatography (HPLC) of 9-fluorenylmethoxycarbonyl (FMOC)- or phenylisothiocyanate (PITC)-GlcN (Ekblad and Naesholm, 1996; Osswald et al., 1995), or by gas chromatography-mass spectrometry (GC-MS) techniques (Penman et al., 2000).

Localization of chitin in cell walls or spores of fungi is achieved by using dyes that intercalate with polysaccharides. Calcofluor white shows enhanced fluorescence when binding to β-(1,4)-glucans, such as chitin, chitosan and cellulose, whereas β-(1,3)-glucans are selectively stained with aniline blue (Nicholas et al., 1994). Various wheat germ agglutinin (WGA) labeling techniques in combination with fluorochromes or gold labeling are also described for the detection of chitin in fungi (Sbrana et al., 1995; Hu and Rijkenberg, 1998; Ekramoddoullah et al., 2000).

Fourier transform (FT) Raman spectroscopy of cell walls of fungi was applied to discriminate between different mixed species in culture media (Edwards et al., 1995).

6 Biosynthesis of Chitin and Chitosan

Chitin is biosynthesized in all chitinous fungi, including the relatively few investigated examples of Mucorales which contain chitosan. Chitin synthases as well as chitin deacetylases are reviewed in this section.

6.1 Chitin Synthases (CS)

In contrast to the situation in arthropods, many details of chitin biosynthesis are known in fungi. Most of the current knowledge is based on studies on baker's yeast, *S. cerevisiae*. The earlier literature was discussed comprehensively by Cabib (1987) and, since then, various aspects of chitin synthesis in fungi have been reviewed (Bulawa, 1993; Martinez and Gozalbo, 1994; Ruiz-Herrera and Xoconostle-Cazares, 1995; Bruyere et al., 1996; Cabib et al., 1996; Gooday, 1996; Merz et al., 1999a; Ruiz-Herrera and Martinez-Espinoza, 1999; Valdivieso et al., 1999; Karnezis et al., 2000).

6.1.1 Enzymology and Subcellular Localization of CS

The biosynthesis of chitin takes place vectorially in a membrane-bound protein complex. Chain elongation occurs by sequential transfer of GlcNAc from UDP-GlcNAc to the nonreducing end of the growing polymer. CS (chitin-(UDP-GlcNAc)-transferase, EC 2.4.1.16) belongs to family 2 glycosyltransferases which catalyze glycosyltransfer with inversion of the anomeric configuration (Coutinho and Henrissat, 1999). Further classification is based on sequence similarities or identities, and to date five CS classes have been assigned (cf. Table 1).

In general, β-glycosyltransferases, including cellulose and chitin synthases, have a highly conserved common motif 'D, D, D35Q(R,Q)XRW'. The second residue (R or Q) in the Q(R,Q)XRW sequence is probably involved in determining the degree of polymerization (DP) of the glucan chain (Saxena and Brown, 1997). Alignment with the deduced protein sequences of most known chitin synthase genes (CHS) reveals five to seven conserved domains (Xoconostle-Cazares et al., 1996).

CS is detected in membrane fractions and in chitosomes, the latter constituting small secretory vesicles (ca. 100S) which function as conveyors of CS to the cell surface (for a review, see Ruiz-Herrera and Martinez-Espinoza, 1999). Chitosome-membrane trafficking of CS was also observed in *Neurospora crassa* (Leal-Morales et al., 1994b). Solubilization of CS is achieved with detergents, such as digitonin, yielding a catalytically active 16S protein complex of molecular mass ca. 500 kDa. CS isolated from a microsomal fraction of *Absidia glauca* is a 30-kDa zymogenic polypeptide (Machida and Saito, 1993). A catalytically active CS was isolated as a 60-kDa polypeptide from 100S chitosomes of *Mucor rouxii* (Merz et al., 1999b).

In *S. cerevisiae*, CSI is more abundant than CSII, and both are localized in low- density chitosomes ($d = 1.15$ g mL^{-1}) as well as in high-density membrane fractions ($d = 1.21$ g mL^{-1}) (Leal-Morales et al., 1994a). Likewise, the transcripts from the *Ustilago maydis* gene, UmCHS1, appear to be present at a higher level than those from UmCHS2, and both transcripts appear to be more

Tab. 1 Chitin synthase genes from fungi (alphabetical listing by name of organism)

Organism	***Gene***	***Comments***	***Reference***
Agaricus bisporus	CHS1	CS class III; 2727 bp (ORF); 909 AA;	Sreenivasaprasad et al. (2000)
Ampelomyces quisqualis	AqCHSA	CS class I; 2786 bp; 910 AA	Weiss et al. (1996)
Aspergillus fumigatus	CHSD	CS-like; low but significant similarity to other CS	Mellado et al. (1996)
Aspergillus nidulans	CHSA CHSB	1013 AA 916 AA	Yanai et al. (1994)
A. nidulans	CHSD CHSE	CS class V and CS class IV; high sequence identity to ScCHS3 and CaCHS3	Specht et al. (1996)
Beauveria brongniartii	BbCHS1	Fragment; CS class II; 95.8% similarity with CHS2 of *Metarhizium anisopliae*	Nam et al. (1997)
Candida albicans	CaCHS1A	775 AA	Sudoh et al. (1995)
Fonsecaea pedrosoi	FpCHS1 FpCHS2 FpCHS3	600 bp and 366 bp; CS class I and II; homology to *S. cerevisiae* CS	Karuppayil et al. (1996)
Metarhizium anisopliae	MaCHS1 MaCHS2 MaCHS3	CS class I CS class I CS class III	Nam et al. (1998)
Mucor circinelloides	McCHS1	CS class TI; expressed during exponentially growing hyphal stage	Lopez-Matas et al. (2000)
Neurospora crassa	CHS2	Similar to CHS from other fungi	Din and Yarden (1994)
Paracoccidioides brasiliensis	CHS2	CS class II; 1043 AA	Nino-Vega et al. (1998)
Penicillium chrysogenum	PcCHS1 PcCHS2 PcCHS3 PcCHS4	CS class I CS class II CS class II CS class III	Namgung et al. (1996)
P. chrysogenum	CHS4	CS class III; 915 AA (ORF); close relationship between *P. chrysogenum* and *Aspergillus* CHS	Park et al. (2000)
Phialophora verrucosa	PvCHS1 PvCHS2 PvCHS3	CS class I and II; 614 bp CS class III; 366 bp; 88.2% similarity and 78.4% identity; with the *S. cerevisiae* enzyme	Peng et al. (1995)
Pyricularia oryzae	Fragment	340 bp; 86% homologous to *A. fumigatus* CHSE	Hwang et al. (1997)
Rhizopus oligosporus	CHS3	CS class IV, sequence similarity to CHS3 of *S. cerevisiae*; 46.7% identity with class IV CS of *N. crassa*	Motoyama et al. (1998)
Saccharomyces cerevisiae	CHS4	696 AA	Trilla et al. (1997)
S. cerevisiae	CHS5	671 AA	Santos et al. (1997)
S. cerevisiae	CHS6	See text	Ziman et al. (1998)
S. cerevisiae	CHS7	See text	Trilla et al. (1999)
Saprolegnia monoica	CHS2	Oomycetes and chitinous fungi have conserved CS	Mort-Bontemps et al. (1997)
Tuber borchii	–	CS class II; ca. 600 bp	Lanfranco et al. (1995)

Tab. 1 (cont.)

Organism	*Gene*	*Comments*	*Reference*
T. magnatum	TmCHS4	1230 AA; 62% homology to class IV CHS of *N. crassa*	Garnero et al. (2000)
Ustilago maydis	UmCHS1 UmCHS2	See text	Xoconostle-Cazares et al. (1996)
U. maydis	UmCHS5	Predicted CHS class IV; high similarity CHS3 from *S. cerevisiae* and *C. albicans*, CHS4 from *N. crassa*, CHSE from *A. nidulans*	Xoconostle-Cazares et al. (1997)
Wangiella dermatitidis	WdCHS4	High homology with CS class IV (Chs3p) of *S. cerevisiae*	Wang et al. (1999)

abundant in the mycelial form (Xoconostle-Cazares et al., 1996).

In *N. crassa*, CSII is compartmentalized in chitosomes which are abundant in the vicinity of the hyphal tip. Immunological studies have revealed that several peptides of microsomal membrane fractions react with a polyclonal antibody to CSII (Sietsma et al., 1996).

Though not proven conclusively *in vivo*, it appears that CSI and CSII are activated by proteolysis, occurring in a zymogenic form which contains a cytosolic amino-terminal region. The CS from *Saprolegnia monoica* is stimulated by digitonin and remains zymogenic after dissociation (Leal-Morales et al., 1997). Proteolytic activation of CSIII is observed in the presence of substrate (for discussion and references, see Merz et al., 1999a; Valdivieso et al., 1999).

As a rule, CSI and CSII are activated by divalent metal ions as co-factors, most commonly Mg^{2+} or Mn^{2+}. However, CSII of *S. cerevisiae* requires Co^{2+} rather than Mg^{2+} (Leal-Morales et al., 1994a).

In the cellulosic, nonfibrillar α-chitin-producing Oomycete *Saprolegnia monoica*, CS is found in high-density membrane components, but not in chitosomes (Leal-Morales et al., 1997).

6.1.2 Genetics of CS

The genetics of fungal CS are investigated very intensively, often using CHS from *S. cerevisiae* as hybridization probes, yielding useful information for taxonomy and phylogenetic relations, as well as the basis for the understanding of CS functions and their regulation (for reviews, see Bulawa, 1993; Ruiz-Herrera and Xoconostle-Cazares, 1995; Valdivieso et al., 1999). The variety of CHS is illustrated with a few recent examples in Table 1. A comprehensive listing of sequences is available, e.g., in the CAZy database (Coutinho and Henrissat, 1999; see also Campbell et al., 1997).

Much insight into the functions of CHS and their transcripts has been obtained from studies on the effects of gene deletion or disruption. Eight CHS have been described to date in *S. cerevisiae*. ScCHS1 and ScCHS2 are the structural genes for CSI and CSII, respectively, whereas the remaining genes are components of the CSIII complex (Valdivieso et al., 1999). Disruption of ScCHS2 or simultaneous disruption of ScCHS2 and ScCHS3, but not of ScCHS1 or ScCHS3, is lethal. However, a gene suppressing the lethality of disruption of ScCHS2 occurs in a *S. cerevisiae* strain which does not require ScCHS2 for viability. A mutant containing the suppressor and lacking ScCHS1 and ScCHS2 has normal amounts of chitin in its cell wall. Apparently, the suppressor gene encodes or controls the expression of CSIII (Baymiller and McCullough, 1993). Chs6p is required for anterograde transport of Chs3p from the chitosome to the plasma membrane (Ziman et al., 1998). The CHS7 gene is specifically involved in Chs3p export from the endoplasmic reticulum (ER) (Trilla et al., 1999).

S. cerevisiae CHS show significant homology to insect CHS and to bacterial and vertebrate hyaluronan synthase HAS genes (DeAngelis et al., 1994; Ibrahim et al., 2000; see also Chapter 15 in Volume 5 and Chapter 15 in this volume).

The CS of *C. albicans*, called CaChs1p, CaChs2p, and CaChs3p, are structurally and functionally analogous to the *S. cerevisiae* CS. CaChs1p is involved in septum formation and is required for the viability of *C. albicans*. Inhibition of CaChsp1 with RO-09-3143 (see below) causes cell death in the cachs2Δ null, but not in cachs3Δ null mutants (Sudoh et al., 2000).

In *U. maydis*, six CHS or fragments are presently known which could operate to compensate an eventual loss of one activity by the remaining enzymes, thus maintaining fungal viability. Umchs5 null mutants display significant reduction in growth rate, chitin content, and chitin synthase activity, especially in the mycelial form, and reduced virulence to corn plants (Xoconostle-Cazares et al., 1997).

Inactivation of the *N. crassa* CHS2 gene produces progeny which is indistinguishable from those of the wild-type, though a significant reduction in CS activity and increased sensitivity to the phosphatidylcholine biosynthesis inhibitor edifenphos were observed (Din and Yarden, 1994).

CHSB, but not CHSA, is essential for hyphal growth in *A. nidulans* (Yanai et al., 1994; Borgia et al., 1996). Chitin synthesized by the CHSD-encoded isoenzyme contributes to the rigidity of the walls of germinating conidia, of the subapical region of hyphae, and of conidiophore vesicles, but is not necessary for normal morphology of these cells. Hyphae from both, chsD and chsE disruptants contain ca. 60–70% of the chitin present in wild-type hyphae. The morphology and development of chsE disruptants are indistinguishable from those of wild-type cells (Specht et al., 1996).

An interesting feature of CHS is the presence of a N-terminal myosin motor-like sequence that has first been observed in the *A. nidulans* csmA gene which contains a large open reading frame (ORF) encoding a polypeptide of 1852 amino acids (Fujiwara et al., 1997; see also Zhang and Gurr, 2000). Apparently, the csmA transcript has important roles in polarized cell wall synthesis and maintenance of cell wall integrity (Horiuchi et al., 1999).

6.1.3
Regulation of CS

On the enzymatic level, CS of *Mucor rouxii* is allosterically activated by GlcNAc which shows cooperative binding (Horsch and Rast, 1993; Merz et al., 1999a).

Septum formation in mycelial fungi and yeasts, as well as apical growth of the hyphae of filamentous fungi requires a precisely regulated, complex interplay of CS and chitinases. In *S. cerevisiae*, CSI is involved in repair functions at the end of cytokinesis. CSII deposits a disk of chitin in the mother-bud neck, forming the primary septum at the end of mitosis, and CSIII synthesizes a ring of chitin at the onset of bud emergence. Genomic analysis reveals multigenic control of chitin synthesis (Valdivieso et al., 1999). Post-translational regulation, probably by activation of latent zymogenic forms, appears to be predominant for the three CS of

S. cerevisiae (Choi et al., 1994). Furthermore, Chs2p and Chs3p are spatially and timely regulated, involving also differential trafficking (Chuang and Schekman, 1996). The ScCHS4 gene which encodes a protein with no potential transmembrane domain regulates the catalytic activity of CSIII, as V_{max} is reduced in the enzyme of chs4 null mutants. In addition to the chitin defect, the chs4 mutant shows a severe defect in mating (Trilla et al., 1997). Chitin synthesis in *S. cerevisiae* is suppressed on the transcriptional level by the KNR4 gene (Martin et al., 1999).

Chitin synthesis in *S. cerevisiae* is also under control of the α mating factor (Martin et al., 1999; Santos and Snyder, 1997; for references, see also Valdivieso et al., 1999).

6.1.4
Inhibition of CS

Benzoylphenylureas and tunicamycin are not inhibitors of chitin synthesis in fungi, whereas the nucleotide analogous nikkomycins and polyoxins are highly effective (for reviews, see Cohen, 1993; Munro and Gow, 1995; Palli and Retnakaran, 1999; Zhang and Miller, 1999; Rast et al., 2000). However, Nikkomycin Z is not active against *S. cerevisiae*, as CSIII but not CSII is inhibited (Gaughran et al., 1994).

A potent inhibitor of CSI of *C. albicans* has recently been identified as 8-(6,6-dimethylaminohepta-2,4-diynyl)amino-4*H*-benz[1,4]-oxazin-3-one (RO-09-3143) (K_i for CaChs1p 0.55 nM) which arrests cell growth in wild-type yeast at MIC_{50} 0.27 μM (Sudoh et al., 2000).

A number of natural products isolated from plants or microorganisms show antifungal effects by inhibiting CS or functional components required for CS activity. Further review is beyond the scope of this article, however.

6.2
Glucan Transferase

The formation of branched glucan, glucan–glucan cross-links, and glucan–chitin cross-links in fungal walls involves the action of glucanases, glycosyltransferases, and transglycosylases. Glucosyltransferases from cell walls of *S. cerevisiae* and *C. albicans* were partly characterized. An activated intermediate is formed from a donor β-(1,3)-glucan by cleaving off a disaccharide (Goldman et al., 1995). A chitin–glucan-β-(1,4)-transferase which catalyzes formation of the linkage between the terminal reducing GlcNAc residue of chitin and the nonreducing Glc residue of β-(1,3)-glucan, as well as a potential use of the enzyme for assaying antifungal agents, are described in a patent (Kollar et al., 1996).

6.3
Chitin Deacetylase (CDA)

Enzymatic deacetylation of chitin by CDA [EC 3.5.1.41] is apparently restricted to fungi and bacteria (for reviews, see Kolodziejska et al., 1995; Tsigos et al., 2000). A few earlier reports on the occurrence of CDA in arthropods (Aruchami et al., 1986) have so far been neither confirmed nor disproved. A review of CDA is given in the following section. The selective deacetylation at the nonreducing end of $(GlcNAc)_n$ for Nod-factor biosynthesis as well as applications of deacetylases for the synthesis of partially acetylated chitooligosaccharides, and the deacetylation of small substrates for the terminal metabolism of chitin are excluded, as these topics are discussed in Chapter 15, this volume.

6.3.1
Enzymology of CDA

CDA deacetylates preferentially amorphous chitin of medium F_A (for details, see Chapter

15, this volume). CDA isolated from mycelial extracts of *M. rouxii* occurs as a high-mannose glycoprotein (ca. 30% carbohydrates) of molecular mass 75 kDa (by sodium dodecylsulfate-polyacrylamide gel electrophoresis; SDS-PAGE) or 80 kDa (by size-exclusion chromatography; SEC), respectively. The enzyme shows an optimum temperature of 50 °C and an optimum pH of 4.5. It requires at least four GlcNAc residues in sequence, and is inhibited by carboxylic acids, particularly acetic acid (Kafetzopoulos et al., 1993a). $(GlcNAc)_4$ and $(GlcNAc)_5$ are fully deacetylated, whereas higher GlcNAc oligomers always yield products carrying GlcNAc at the reducing end (Tsigos et al., 1999). Furthermore, the enzyme catalyzes the hydrolysis of acetamido groups in a processive fashion, starting at the nonreducing end. However, high M_v, partially *N*-acetylated chitosans reveal no correlation of K_m with F_A, though the relative rate increases with increasing F_A. CDA operates by a multiple-random-attack mechanism (Martinou et al., 1998).

Crude CDA extracted from mycelium of *M. rouxii* shows an optimum temperature of 50 °C and an optimum pH of 5.8 (substrate: chitosan). Ca^{2+}, Mn^{2+}, and Zn^{2+} do not affect activity. The enzyme is slightly activated by Co^{2+}, whereas EDTA, Fe^{2+}, and Fe^{3+} cause partial inhibition. The extracellular enzyme of the culture medium shows an optimum pH of 4.8 (Kolodziejska et al., 1999).

CDA of *Colletotrichum lindemuthianum*, molecular mass ca. 32 kDa, shows an optimum temperature of 60 °C and an optimum pH of 11.5 – 12.0. Kinetic analysis reveals K_m and k_{cat} values of 2.55 mM and 27.1 s^{-1} for glycol chitin, and 0.4 mM and 83.2 s^{-1} for $(GlcNAc)_5$, respectively (Tokuyasu et al., 1996). The reaction pattern of the CDA of *C. lindemuthianum* is distinct from that of *M. rouxii*, as the former recognizes a sequence of four monosaccharide residues which bind at sites –3 to +1. Sites –3 and –1 must be occupied by GlcNAc, whereas the residue bound at sites –2 and +1 can be either GlcNAc or GlcN. However, like *M. rouxii* CDA, the *C. lindemuthianum* enzyme operates by a random-attack, multiple-chain mechanism (Tokuyasu et al., 2000). (In the original paper, the binding sites are named –2, –1, 0, and +1.)

The CDA (M = 75 kDa) from *Absidia coerulea* is located in the periplasmic space. The enzyme is active on $(GlcNAc)_n$, n > 2, and glycolchitin, the optimum pH is 5.0 and the optimum temperature is 50 °C (substrate: glycol chitin). It is heat-stable and strongly inhibited by Fe^{3+} (Gao et al., 1995).

A heat-stable CDA was purified to homogeneity from autolyzed cultures of *Aspergillus nidulans*. The enzyme is a glycoprotein of pI 2.75, M_r 27,000 Daltons, optimum pH 7.0, and optimum temperature 50 °C. Substrates are glycol chitin, colloidal chitin, and $(GlcNAc)_n$ (Alfonso et al., 1995).

The CDA of *U. viciae-fabae* occur as five isoenzymes of molecular mass 48.1, 30.7, 25.2, 15.2, and 12.7 kDa. The enzyme is not thermostable, and shows an optimum pH of 5.5 – 6.0 (Deising and Siegrist, 1995).

6.3.2
CDA Genes

The cDNA to the *M. rouxii* mRNA encoding CDA shows significant similarity to bacterial *Rhizobium* NodB proteins (Kafetzopoulos et al., 1993b). *N,N'*-Diacetylchitobiose deacetylase from the bacterium *Vibrio alginolyticus* H-8 shows 37% identity with NodB and 28% identity with CDA from *M. rouxii*. This enzyme contains a domain which is 39% homologous to a chitin-binding domain of chitinase A from *Bacillus circulans* (Ohishi et al., 2000).

Two isoenzyme genes, CDA1 and CDA2, were identified in *S. cerevisiae* and functionally assigned to formation of the ascospore wall. Expression of the genes is restricted to a distinct time period during sporulation. Diploids disrupted for both genes sporulate as efficiently as wild-type cells, and the resulting mutant spores are viable under standard laboratory conditions. However, the mutant spores are relatively sensitive to hydrolytic enzymes, ether, and heat shock (Christodoulidou et al., 1996, 1999; Mishra et al., 1997).

The ORF of the CDA gene of *C. lindemuthianum* consists of a possible prepro-sequence of 27 amino acids at the N-terminus and of a mature CDA with a predicted molecular mass of 24.3 kDa, as was confirmed by MALDI-TOF MS. The activity of recombinant CDA produced in the culture medium of *E. coli* cells is enhanced two-fold by trypsin digestion (Tokuyasu et al., 1999a, b).

6.3.3 Regulation of CDA

Very little is known on the regulation of CDA. Enzymatic activity is enhanced in crude extracts of *M. rouxii* and *A. coerulea* by the addition of chitin of F_A 0.59 (Dunkel and Knorr, 1994). In rust fungi, CDA activity massively increases at the onset of penetration through the stomata of the host plant, as was described for a broad-bean rust *Uromyces viciae-fabae* (Deising et al., 1995) and the wheat stem rust *Puccinia graminis* f. sp. *tritici* (El-Gueddari and Moerschbacher, 2000). Deacetylation of fungal chitin to chitosan serves to evade recognition by the host and antifungal chitinases elicited during plant tissue colonization

7 Biodegradation

Chitin and chitosan are degraded by chitinases [EC 3.2.1.14] releasing GlcNAc as the reducing sugar unit, and by chitosanases [EC 3.2.1.132] releasing GlcN or GlcNAc, respectively. Most aspects of chitinases are discussed in Chapter 15, this volume, and only fungal chitinases are treated in this section, while chitosanases are covered in greater detail, including also enzymes from prokaryotes. Monosaccharide-producing enzymes, i.e., chitobiases and β-GlcN-ases, are briefly mentioned.

7.1 Chitinase

A sophisticated, strictly regulated interplay of chitin synthesis and degradation is necessary during apical growth, cell separation and branching, spore swelling and germination, and liberation of spores (for reviews, see Cabib et al., 1992, 1996; Gooday et al., 1992; Lorito, 1998; Herrera-Estrella and Chet, 1999).

7.1.1 Enzymology of Fungal Chitinases

Only one crystal structure of a fungal chitinase, i.e., an enzyme from *Coccidioides immitis* is known to date (Hollis et al., 2000). The enzyme shows transglycosylating exo-chitobiosidase activity, generating $(GlcNAc)_8$ from $(GlcNAc)_6$ (Fukamizo et al., 2001). All fungal chitinases described so far are retaining family 18 glycosidases.

Much of the current knowledge on fungal chitinases is obtained from mycoparasitic species, in particular *Trichoderma harzianum* (Gokul et al., 2000), which are of interest as antifungal agents (Deshpande, 1999) as well as sources for biotechnological production of chitinases (Margolles-Clark et al., 1996; Krishnamurthy et al., 1999; El-Katatny et al., 2000). Numerous patents describe various aspects of fungal chitinases and their applications.

The chitinolytic system of *T. harzianum* consists of at least four endochitinases (CHIT52, CHIT42, CHIT33, and CHIT31)

which differ in substrate specificity with respect to F_A, and of β-GlcNAc-ases (chitobiases) (Tronsmo et al., 1996). Another strain, *Trichoderma* sp. T6 produces a 46-kDa extracellular endochitinase (CHIT46) (Lima et al., 1999). A recombinant enzyme, ENCI from *T. harzianum*, which was expressed in *S. cerevisiae*, has a molecular mass of 44 kDa, pI 6.3, pH optimum 7.0, and temperature optimum 20 °C (Draborg et al., 1996).

Endochitinase activities of *Trichoderma* strains may range between 20 and 1265 units g^{-1} of dry mycelium. Disruption or overexpression of the *T. harzianum* CHT42 gene results in decreased or enhanced activity, respectively, of the strains against *R. solani* (Baek et al., 1999). A strain of *T. harzianum* overexpressing the 33-kDa chitinase was described (Limon et al., 1996). However, there is no clear-cut correlation between chitinase activity and mycoparasitic efficiency, since antifungal activity may also depend on other glucanases (Lorito et al., 1996; Turoczi et al., 1996; De la Cruz and Llobell, 1999). Some mycoparasitic fungi, e.g., the Basidiomycete *Schizophyllum commune*, produce barely detectable activities of extracellular chitinase, even when grown on chitin-containing substrates, and attack of other fungi is primarily effected by β-(1,3)-glucanase (Chiu and Tzean, 1995). Furthermore, the host may defend itself against the attack, as was shown in the case of *R. solani* which, among other enzymes, secretes an exochitinase that is potentially detrimental to *T. harzianum* (Bertagnolli et al., 1996).

Expression of the *Janthinobacterium lividum* chi69 chitinase in *Saccharomyces cerevisiae* DY-150 does apparently not result in deleterious effects on the yeast (Molloy and Burke, 1997).

Chitinases have been investigated in several other species. A complex chitinolytic system, consisting of β-GlcNAc-ase and at least seven chitinases, two of those being particulate, occurs in exponentially growing mycelia of *Mucor rouxii* (Rast et al., 1991). A cytosolic chitinase (molecular mass 20 kDa by SDS-PAGE) was partially purified from *Neurospora crassa*, and linear kinetics for enzyme activity were obtained using the substrate [^{3}H]chitin. The enzyme is sensitive to allosamidin (IC_{50} 1.6 μM, noncompetitive inhibition) (McNab and Glover, 1991). The entomopathogenic fungus *Metarhizium anisopliae* produces a 60-kDa chitinase, optimum pH 5.0, which shows both exo- and endo-activity (Kang et al., 1999). The acaropathogenic fungi *Hirsutella thompsonii* and *Hirsutella necatrix* excrete, besides a chitobiase, endochitinases of 162, 66, and 38 kDa, respectively (Chernin et al., 1997). The chitinase of the mycoparasitic and entomopathogenous Antarctic fungus, *Verticillium cfr. lecanii*, is a glycoprotein of 45 kDa, pI 4.9, which is especially adapted to low temperature, showing at 5 °C a four-fold higher activity than *T. harzianum* chitinase (Fenice et al., 1998b). A chitobiosidase of 40 kDa which also depolymerizes chitin, thus showing exo-/endo-activity, was isolated from the marine fungus *Corollospora maritima* Werdermann (Grant et al., 1996). An intracellular chitinase, showing optimum pH 5.0 and optimum temperature 50 °C occurs in *Candida albicans* (Mellor et al., 1994). The intracellular 43.5-kDa chitinase III of *Rhizopus oligosporus* shows a pH optimum of 6.0 (Takaya et al., 1998a). The anaerobic ruminal fungus *Piromyces communis* contains a membrane-bound endo-type chitinase (molecular mass 43.6 ± 1.2 kDa; pI 4.4; optimum pH 6.0 at 39 °C) which is similar to the extracellular chitinase of the organism (Sakurada et al., 1998).

Besides *Trichoderma*, some other species were investigated for biotechnological production of chitinases, such as *Penicillium janthinellum* P9 (Fenice et al., 1998a). The alkalophilic marine fungus *Beauveria bassiana* produces maximum chitinase activity at pH 9.2 (Suresh and Chandrasekaran, 1999).

7.1.2
Chitinase Genes

Chitinase genes of fungi have been reviewed (Lorito, 1998; Valdivieso et al., 1999). The chit42 cDNA of *Trichoderma harzianum* codes for a 42-kDa endochitinase containing 423 amino acids, including a signal peptide of 22 amino acids and a second peptide of 12 amino acids. The deduced amino acid sequence has putative catalytic, phosphorylation and glycosylation domains. Southern analysis shows that it is present as a single-copy gene (Garcia et al., 1994). A 1096-bp cDNA fragment was isolated from thalli of *T. harzianum*. The full-length 1554-bp cDNA sequence contains an ORF expressing a protein of 424 amino acids. The putative catalytic domain of the enzyme is highly conserved between endochitinase genes from *Trichoderma* and various bacteria (Hayes et al., 1994). Expression of a *T. harzianum* cDNA in *S. cerevisiae* revealed 11 full-length cDNAs encoding endochitinase I (ENCI). The 1473-bp chi1 cDNA encodes a 424 amino acid precursor protein, including both a signal sequence and a propeptide. The deduced ENCI amino acid sequence is homologous to other fungal and bacterial chitinases (Draborg et al., 1996).

The cts1 gene from *Coccidioides immitis* contains five introns and a 1281-bp ORF which translates a 427 amino acid protein of 47.4 kDa. The cts2 gene contains two introns (2580-bp ORF; 860 amino acid protein of 91.4 kDa). The deduced CTS1 protein showed greatest homology to *Aphanocladium album* and *T. harzianum* CTS (74 and 76%, respectively), whereas CTS2 showed greatest homology to the CTS of *S. cerevisiae* and *C. albicans* (47 and 51%, respectively). The putative N-terminal sequence of the mature CTS1 protein also showed 89% homology to the reported N-terminal sequence of a 48-kDa complement fixation antigen (CF-Ag) of *C. immitis* which has demonstrated chitinase activity. CTS1 and CTS2 of *C. immitis* are proposed to be members of two distinct classes of fungal chitinases (Pishko et al., 1995).

The deduced amino acid sequence of *Rhizopus oligosporus* chitinase III is similar to that of bacterial chitinases and chitinases from some mycoparasitic fungi. The enzyme lacks a N-terminal secretory signal sequence. Northern blot analysis reveals that chi3 is transcribed during hyphal growth, suggesting that the enzyme may function during morphogenesis (Takaya et al., 1998a). Additional references for fungal chitinase genes are given in Table 2.

7.1.3
Regulation of Chitinases

The regulation of chitinase activities in fungi has been reviewed (Lambais and Mehdy, 1995; Felse and Panda, 1999; Valdivieso et al., 1999).

In general, chitinolytic enzymes are induced and excreted during growth of fungi on chitin as the sole carbon source, as has been shown for example in *Trichoderma* (Ulhoa and Peberdy, 1993; Tronsmo et al., 1996) and several Basidiomycetes (Hodge et al., 1995). Glucose or GlcNAc usually repress chitinase activity, but the intracellular CHIT102 of *Trichoderma* is expressed at a low constitutive level when the fungus is grown on glucose (Haran et al., 1995). Other nutrient sources, e.g., alanine, also repress chitinase (St. Leger et al., 1993).

Expression of the *T. harzianum* chit42 mRNA is induced by chitin and chitin-containing cell walls (Garcia et al., 1994) and also by soluble chitooligosaccharides that are initially produced by the low constitutive activity of chitinolytic enzymes (Zeilinger

Tab. 2 Chitinase genes of fungi

Organism	***Gene***	***Comments***	***Reference***
Aspergillus nidulans	chiA	660 AA	Takaya et al. (1998b)
Candida albicans	CHT2	583 AA (60.8 kDa)	McCreath et al. (1995)
	CHT3	567 AA (60 kDa)	
Candida albicans	CHT1	416 AA (36% similarity Cht2 and Cht3)	McCreath et al. (1996)
Metarhizium anisopliae	–	58 kDa; 24.4–36.4% identity with other fungal chitinases, higher for the active site	Kang et al. (1998)
Metarhizium anisopliae	chit1	1521-bp ORF; 423 AA; 42 kDa; pI 5.8; 35 AA signal peptide	Bogo et al. (1998)
Trichoderma harzianum	chit33	1.2 kb cDNA; 321 AA; signal peptide of 19 AA; 43% identity with fungal and plant chitinases	Limon et al. (1995)
Trichoderma harzianum	chi1	424 AA; contains signal sequence and propeptide	Draborg et al. (1996)

et al., 1999). Transcription of the ech42 gene is strongly enhanced during direct interaction of the mycoparasite with *Botrytis cinerea* and also upon light-induced sporulation (Carsolio et al., 1994). Furthermore, expression of ech42 is induced by starvation (Mach et al., 1999). Enzyme production by *T. harzianum* Rifai is stimulated at pH 5.5–6.0, and by glucose or GlcNAc together with chitin (El-Katatny et al., 2000). The expression of the various β-GlcNAc-ases and endochitinases during mycoparasitism is affected in a specific manner by the host. When *T. harzianum* antagonizes *Rhizoctonia solani*, the initially appearing β-GlcNAc-ase is followed by three endochitinases (Haran et al., 1996).

The chit33 gene of *T. harzianum* is weakly expressed during growth on chitin and *Rhizoctonia solani* cell walls. GlcNAc and elevated temperature (40 °C) induce, whereas glucose or glycerol prevent, gene expression (Dana et al., 2001).

In *S. cerevisiae*, increased expression of the chitinase CTS1 gene is under control of the cell cycle-dependent transcriptional ACE2 regulator protein (Ace2p) which is localized in the nucleus in late M and early G_1 phases (King and Butler, 1998; O'Conallain et al., 1999). Acp2p is regulated by phosphorylation, involving protein kinase Cbk1p (Racki et al., 2000).

Expression of *Aspergillus nidulans* chiA increases during conidiophore development (Takaya et al., 1998b).

Acidic chitinases (pI 4.8) of 43.5 and 45 kDa appear during penetration of the entomopathogenic fungi, *Metarhizium anisopliae, M. flavoviride,* and *Beauveria bassiana* through insect cuticle, and increase after proteolytic degradation, which suggests that the release of the chitinase is dependent on the accessibility of the substrate (St. Leger et al., 1996).

7.2
Chitosanases

Chitosanases [EC 3.2.1.132] are family 46 inverting glycosidases, occurring in fungi, bacteria, and in plants. To date, no chitosanase has been reported from Arthropods, Vertebrates, or other phylae, although *Chlorella* virus genomes contain relevant genes (Lu et al., 1996; Yamada et al., 1997; Sun et al., 1999).

7.2.1
Structure and Mechanism of Chitosanases

Chitosanases share no sequence similarity with lysozymes or chitinases. However, the X-ray structure of *Streptomyces* N174 chitosanase (29.5 kDa) reveals an α-helical fold with a structural core that is similar to lysozymes and family 19 barley endochitinase (Marcotte et al., 1996). Cleavage of the glycosidic bond occurs by a S_N2-type mechanism with Glu22 as the catalytic acid residue and Asp40 activating the attacking water molecule (Marcotte et al., 1996; Robertus and Monzingo, 1999) which was also confirmed by site-directed mutagenesis (Boucher et al., 1995). An extraordinarily basic arginine residue (Arg205; $pK_a > 20.0$) is located in the catalytic site of the *Streptomyces* N174 enzyme in close proximity to Asp145 (pK_a –1.6), which is in turn close to Arg190 (pK_a 17.7), and these residues are important for stability and activity of the enzyme (Fukamizo et al., 2000). The crystal structure of a 29-kDa extracellular chitosanase from *Bacillus circulans* MH-K1 reveals an overall folding pattern similar to the chitosanase from *Streptomyces* sp. N174, though the two enzymes possess only 20% sequence homology (Saito et al., 1999). Alignment of the primary sequences of 14 chitosanases shows a common E-[DNQ]-x(8,17)-Y-x(7)-D-x-[RD]-[GP]-x-[TS]-x(3)-[AIVFLY]-G-x(5,11)-D sequence pattern (Tremblay et al., 2000).

7.2.2
Enzymology of Chitosanases

A survey of chitosanases is given in Table 3. Two synergistically acting extracellular chitosanases, both showing molecular mass of 31 kDa, are produced by *Bacillus lentus* (El-Din et al., 1995). Chitosanase B is activated by Ca^{2+}. The optimum pH is 6.0, 7.0, and 7.5 with chitosan, *M. rouxii* and *M. indicaeseudaticae* cell wall as substrates, respectively. K_m is lower with chitosan than with *Mucor* cell walls. *Acinetobacter* sp. CHB101 secretes two endo-type chitosanases (molecular mass 37 and 30 kDa), cleaving chitosan of F_A 0.1–0.3 or colloidal chitin and chitosan of F_A 0.3, respectively (Shimosaka et al., 1995). The 45-kDa chitosanase from *Bacillus* sp. GM44 produces $(GlcN)_n$, n = 3–8, as major end-products from chitosan (Choi et al., 1997). A 40-kDa chitosanase from the culture filtrate of *Aspergillus oryzae* IAM2660 hydrolyzes chitosan of F_A <0.3 and $(GlcN)_n$, n >5 (Zhang et al., 2000). Likewise, the endochitosanase from *Aspergillus* sp. Y2K degrades chitosan to yield $(GlcN)_n$, n = 3–5, and $(GlcN)_6$ is the smallest oligosaccharide substrate, which is cleaved into two moles of $(GlcN)_3$ (Cheng and Li, 2000). An increase of enzyme activity with decreasing F_A of chitosan has been observed in several cases. Several bacterial chitosanases show also β-(1,4)- or β-(1,3)-glucanase activity (Kusaoke et al., 1997; Pedraza-Reyes and Gutierrez-Corona, 1997; Mitsutomi et al., 1998).

According to substrate cleavage patterns, chitosanases are divided into three classes (Table 4). Class I enzymes cleave GlcN-GlcN or GlcNAc-GlcN glycosidic bonds. Thus, chitosanase from *Bacillus pumilus* BN-262 hydrolyzes chitosan (F_A 0.25–0.35), releasing hetero-oligosaccharides containing GlcN or GlcNAc at the reducing end, but exclusively GlcN at the nonreducing ends (Fukamizo et al., 1994). Class II chitosanases, e.g., from *Bacillus* sp. No. 7-M, cleave GlcN-GlcN bonds only, recognizing specifically a -$(GlcN)_3$- sequence at sites –2 to +1, whereas the residue at the +2 binding site can be either GlcN or GlcNAc, as was shown with a chitosan of F_A 0.35 and DP 20 (Izume et al., 1992; Vårum et al., 1996; Vårum and Smidsrød, 1997). The 40-kDa class III chitosanase from *Bacillus circulans* WL-12 shows similar substrate specificity as a chitosanase from *S. griseus* HUT6037, hydro-

Tab. 3 Properties of chitosanases

Organism	M [kDa]	pI	pH optimum	Temperature optimum [°C]	Further data	Reference
Fungi						
Aspergillus fumigatus KH-94	22.5	7.3	5.5	70– 80	Endo-enzyme, activated by Mn^{2+}	Kim et al. (1998)
Aspergillus sp. Y2K	25	8.4	6.5	65–70	Stable at pH 4–7.5/ 55 °C	Cheng and Li (2000)
Fusarium solani f. sp. *phaseoli* SUF386	36		5.6	40		Shimosaka et al. (1993)
Mucor rouxii autolyzed cultures	76 and 58	4.9 and 4.7	5.0	55 and 50		Alfonso et al. (1992)
Penicillium spinulosum			5	55	K_m 1.7 mg chitosan mL^{-1}; V_{max} 146 µmol GlcN mL^{-1} min^{-1}	Ak et al. (1998)
Actinomycetes						
Nocardioides strain K-01	27	8.8	5.0	55–60		Okajima et al. (1995)
Prokaryotes						
Amycolatopsis sp. CsO-2	27	8.8	5.3	55	Thermostable	Okajima et al. (1994)
Bacillus cereus	47	8.8	5.8	54		Piza et al. (2000)
Bacillus cereus S1	45		6	60	Stable at pH 6.0–11.0/40 °C	Kurakake et al. (2000)
Bacillus sp. GM44	45		5.0	70	Stable at pH 3.0–10.0/40 °C	Choi et al. (1997)
Bacillus sp. HW-002	46		5.5–6.0	65		Lee et al. (1996b)
Bacillus sp. KFB-C108	48		6.5	55	Thermostable; no activation by metal ions; major product is $(GlcN)_4$	Yoon et al. (1998)
Bacillus subtilis KH1	28	8.3				Omumasaba et al. (2000)
Matsuebacter chitosanotabidus 3001	34		4.0	30–40		Park et al. (1999)

lyzing GlcN-GlcN and GlcN-GlcNAc linkages (Mitsutomi et al., 1998).

The differences in specificity are explained by differences in the spatial arrangement of the substrate-binding cleft, which is more narrow in *Bacillus circulans* MH-K1 than in *Streptomyces* N174 chitosanase (Saito et al., 1999).

7.2.3 Chitosanases in Plants

Chitosan elicits chitosanases as defensive, antifungal enzymes in plants, as has been reported in a number of papers and patents. In some cases, the substrate specificity of the enzymes has not been unequivocally established, and it is not always clear whether they

Tab. 4 Cleavage patterns of chitosanases and degree of polymerization (DP) of oligosaccharides produced

Organism	*DP*	*Nonreducing end*	*Reducing end*	*Class*	*Reference*
Penicillium islandicum	2–3	GlcN	GlcNAc	–	Mitsutomi et al. (1996)
Bacillus pumilus BN-262	2–4	GlcN	GlcN/GlcNAc	I	Fukamizo et al. (1994)
Streptomyces N174	2–4	GlcN	GlcN/GlcNAc	I	Fukamizo et al. (1995)
Bacillus sp. 7-M	2–6	GlcN	GlcN	II	Izume et al. (1992)
S. griseus HUT 6037	2–6	GlcN/GlcNAc	GlcN	III	Mitsutomi and Ohtakara (1992)
B. circulans MH-K1	2–5	GlcN/GlcNAc	GlcN	III	Mitsutomi et al. (1996)
B. circulans WL-12	2–5	GlcN/GlcNAc	GlcN	III	Mitsutomi et al. (1998)

Tab. 5 Chitosanase genes (CSN)

Organism	*Comments*	*Reference*
Bacillus amyloliquefaciens UTK	278 AA ORF	Seki et al. (1997)
B. circulans	900-bp fragment; 259 AA ORF; contains signal peptide	Ando et al. (1992)
B. ehimensis	1.9-kbp fragment; 302 AA ORF; M = 31 kDa	Akiyama et al. (1999)
Bacillus sp. CK4	822-bp/242 AA ORF; 30 kDa; contains signal peptide; 76.6 (15.3, and 14.2% similarities to CSN from *B. subtilis*, *B. ehemensis*, and *B. circulans*, respectively	Yoon et al. (2000)
Burkholderia gladioli CHB101	355 AA ORF; 28-kDa protein	Shimosaka et al. (2000)
Fusarium solani f. sp. *phaseoli* SUF386	500-bp fragment; 304 AA ORF; contains 19 AA signal sequence	Shimosaka et al. (1996)
Matsuebacter chitosanotabidus 3001	391 AA ORF; contains signal peptide; no significant homology with other chitosanases	Park et al. (1999)
Streptomyces sp. N174	1.6-kb fragment; 238 AA ORF; contains signal peptide	Masson et al. (1994)

also cleave chitin or show lysozyme activity (cf. Osswald et al., 1993, 1994; Mayer et al., 1996).

7.2.4
Chitosanase Genes

A survey on chitosanase genes is given in Table 5. Expression of the chitosanase gene of the Actinomycete *Nocardioides* N106 in *S. lividans* yielded an enzyme of 29.5 kDa, pI 8.1, which shows 74.4% homology with the chitosanase gene from *Streptomyces* sp. N174 (Masson et al., 1995). A gene identified in the genome of *Bacillus subtilis* 168 encodes a chitosanase which is more similar to the chitosanase from *Streptomyces* N174 than to chitosanases from other *Bacillus* strains (Parro et al., 1997; Rivas et al., 2000).

7.2.5
Regulation of Chitosanases

Very little is known on the regulation of chitosanase genes. It appears that enzyme production is high when fungi are grown in media containing GlcN, GlcNAc, chitosan, or fungal cell walls. Chitosanase from *Bacillus cereus* sp. HW-002 is a constitutive enzyme which does not require an inducer

(Lee et al., 1996a). The gene of *Bacillus subtilis* is temporally regulated and is not subject to metabolic repression (Rivas et al., 2000).

7.3 Exo-β-D-glucosaminidases

Several β-GlcN-ases, i.e., enzymes releasing GlcN from the nonreducing end of chitosan or GlcN oligosaccharides, have been isolated recently. The GlcN-ase of *Trichoderma reesei* PC-3-7 is a retaining glycanase (molecular mass 93 kDa, optimum pH 4.0, optimum temperature 50 °C) which cleaves either GlcN-GlcN or GlcN-GlcNAc glycosidic bonds (Nogawa et al., 1998). *Aspergillus fumigatus* KH-94 produces a 108-kDa GlcN-ase, pI 4.8, optimum pH 4.5–5.5, optimum temperature 50–60 °C, which possesses transglycosylating activity. This enzyme is activated by Mn^{2+} (Kim et al., 1998).

The enzyme from *Aspergillus oryzae* IAM-2660 shows molecular mass 135 kDa (Zhang et al., 2000). A GlcN-ase of 110 kDa, pI 7.4, was purified from the culture filtrate of *Penicillium* sp. AF9-P-112 (Uchida et al., 1997).

8 Biotechnological Production

As yet, the production of pure chitin from fungal biomass has not been reported, as the aminoglucan is covalently liked to other glycans, mostly β-(1,3)-glucan. Furthermore, fungal biomass often contains pigments which are difficult to remove (see Section 3). However, the chitin–glucan complex is of considerable interest because of the biological properties of this material (see Section 9).

Though not yet technically feasible, production of chitosan from fungal biomass of Mucorales is discussed as an alternative to animal sources.

8.1 Screening for Chitosan Producer Strains

Screening of 125 strains selected from *Absidia, Actinomucor, Circinella, Mucor, Phycomyces, Rhizopus,* and *Zygorhychnus* for chitosan production resulted in identification of *Absidia butleri* HUT 1001 as the best producer (Shimahara et al., 1989) (Table 6). Out of 24 strains of *Absidia, Rhizopus,* and *Mucor* species, highest yields of chitosan were found with *Absidia coerulea* and *Absidia spinosa*. An increase in chitosan yield to 2.7 g L^{-1} was observed when glucose and $(NH_4)_2SO_4$ were added continuously to the culture medium. Addition of $CoCl_2$ or GlcNAc had no effect on chitosan production (Kuhlmann et al., 2000). Screening of *A. coerulea, Mucor rouxii, Gongronella butleri, Phycomyces blakesleeanus,* and *A. blakesleeana* for production of chitosan from mycelia gave best results for *A. coerulea*, yielding in 48 h at pH 5.5 up to 0.51 g L^{-1} chitosan (F_A 0.05–0.15; average $M_v = 4.5 \times 10^5$) in batch culture. The addition of 5 mg L^{-1} Co^{2+} increased the yield by 20%, and this could be further increased three-fold by fermentation in continuous culture. *M. rouxii* produced chitosan at a yield of ca. 0.3 g L^{-1}, and yields were even lower with the other species investigated (Rane and Hoover, 1993a). *Gongronella butleri* USDB 0201 produces the highest amount of chitosan when harvested at the late exponential growth phase. In terms of yield of chitosan per unit mycelia mass, *Cunninghamella echinulata* was the best strain among several fungi tested (Tan et al., 1996). Screening of 96 marine fungi afforded highest yields of chitosan from *Rhizopus oryzae* as determined by quantification of glucosamine (Yoshihara et al., 1996).

The Basidiomycete *Lentinus edodes* yields 6.8 g per kg fermentation medium of chitosan of F_A 0.55–0.11 when cultured under

Tab. 6 Chitosan production by *Mucorales*

Fungus	*Chitosan yield*	M_v	F_A	*Reference*
Absidia butleri HUT 1001	0.9 g L^{-1}	1.2×10^6	0.13	Shimahara et al. (1989)
Absidia coerulea	1.05 g L^{-1}	7×10^5	0.09	Kuhlmann et al. (2000)
Absidia orchidis	2.03 g L^{-1}	5.4×10^5	0.45	Jaworska and Szewczyk (1998)
Absidia spinosa	1.25 g L^{-1}	7×10^5	0.14	Kuhlmann et al. (2000)
Rhizopus oryzae	5.4% of dry matter	–	–	Yoshihara et al. (1996)

solid-state conditions on wheat straw (Crestini and Giovannozzi-Sermanni, 1996).

8.2 Isolation of Chitin and Chitosan from Fungal Biomass

Chitin is always obtained as a complex with glucans which, depending on the source of the material, also may contain insoluble melanins.

The protocols for isolation of chitosan from fungal biomass differ considerably from those used for preparation of chitosan from Crustacea, as demineralization is not necessary. Separation of the biomass from the culture by filtration or centrifugation is followed by disruption of the cells. Chitosan is extracted from *Absidia* by autoclaving the mycelia first in alkali (e.g., 1 M NaOH solution) at 121 °C for 15–30 min, followed by heating in diluted hydrochloric or 2% aqueous acetic acid (Rane and Hoover, 1993b; Hu et al., 1999).

The polymers may be subjected to various derivatizations, such as carboxymethylation, sulfonation or oxidation (cf. Chapter 15, this volume). Polyuronans were prepared from biomass of *Aspergillus niger, Trichoderma reesei*, and *Saprolegnia* sp. by oxidation with NaOCl and NaBr for 30 min at room temperature in the presence of 4-methoxy-2,2,6,6-tetramethylpiperidin-1-yloxyl (TEMPO), and various applications were suggested (Muzzarelli et al., 1999, 2000).

8.3 Production of CDA

The production of chitosan by enzymatic deacetylation is reviewed in Chapter 15, this volume. Here, we shall briefly consider approaches towards production of the enzyme CDA on a technical scale. Optimization of culture parameters was reported in a few cases. Extracts of *M. rouxii* mycelia show a maximum of specific activity after 12–96 h, with a maximum of total activity after 96 h of incubation in yeast extract-peptone-glucose (YPG) medium at pH 5. Digitonin (0.1 mg mL^{-1}) enhances the activity (Malesa-Ciecwierz et al., 1997). Growth of *A. coerulea* in a medium containing yeast extract is optimal at 30 °C, pH > 4.5 to neutral, and in the presence of a low concentration of detergent. Enzyme levels are significantly enhanced when chitin of F_A ca. 0.4 is included in the medium (Win et al., 2000).

The CDA gene from *M. rouxii* was cloned into the yeast *Pichia pastoris*. CDA gene expression resulted in efficient secretion of the enzyme into the medium and scale-up to a 30-L fermentor was reported (Jaspar-Versali and Clerisse, 1997).

9 Applications

Pure chitin and chitosan are obtained technically from Crustacea. Thus, most applications are developed with those polymers (see Chapter 15, this volume), and very few uses of chitin or chitosan from fungi are described in the literature. However, several interesting applications have been suggested for fungal biomass as well as for chitin–glucan complexes.

9.1 Adsorption of Coloring Matters

Adsorption of humic acid by *Rhizopus arrhizus* is generally favored at low pH and high ionic strength (Zhou and Banks, 1993). The biomass of the mucoralean fungus *Rhizomucor pusillus* was found to absorb 51% of color from a bleach plant effluent, whereas chitosan and a strong cation exchanger removed 34% and 4%, respectively (Christov et al., 1999). The alkali-treated, autoclaved and granulated biomass of *Rhizopus oryzae* 26668 proved to be most effective for removing the reactive dye Levafix Brilliant Red E-4B A (Gallagher et al., 1997).

9.2 Metal Ion Adsorption

Fungi or fungal biomass have been investigated as agents for the adsorption of metals, primarily from soil and from wastewater (Galli et al., 1994). The biomass of *A. nidulans* and *R. pusillus* showed a high capacity for rapid sorption of Cd^{2+}, as ca. 73% of the metal ions were taken up during 5 min at pH 3–5 (Plaza et al., 1996). Among the mycelia of six fungal species (*Rhizopus arrhizus, Mucor racemosus, Mycotypha africana, Aspergillus nidulans, A. niger,* and *Schizosaccharomyces pombe*), *R. arrhizus* showed the highest capacity for adsorption of Zn^{2+} (213 μmol g^{-1} dry weight). Low pH reduced Zn^{2+} sorption (Zhou, 1999). A material named "Mycoton" was developed from the hyphal walls of Basidiomycetes belonging to the Aphyllophorales (Gorovoj and Kosyakov, 1994). The complex contains hollow fibers of 3–5 μm in diameter and up to several millimeters in length, and is composed of 60–95% chitin, 5–35% glucans, and 0–10% melanins. Mycoton strongly adsorbs heavy metal ions, including radionuclides (Gorovoj and Kosyakov, 1994, 1997; Gorovoj and Petyuschenko, 2000).

9.3 Healthcare

The alkali-treated mycelium of *Mucor mucedo* and sporangiophores of *Phycomyces blakesleeanus* show proproliferant activity on human F1000 fibroblasts at low concentrations. The activity is correlated with the chitin content of these materials, which provides a matrix for the anchorage of fibroblasts and thus contributes to the granulation phase of the healing cascade (Chung et al., 1994). However, mycelia or sporangiophores of several other fungi showed either anti- or proproliferant effects on murine L929 fibroblasts. Apparently, the effects are modulated not only by chitin but also by H_2O_2 which is generated by the fungal materials (Chung et al., 1998). Dressings of Mycoton accelerate the healing of wounds, burns and ulcers (Gorovoj and Kosyakov, 1994; Gorovoj et al., 1997, 2000a), and the material also shows interesting beneficial effects in gastric infectious diseases (Gorovoj et al., 2000b) as well as in chronic hepatitis (Nakonechna et al., 2000). A membrane, named "Sacchachitin", has been developed as a skin substitute, and shows similar wound-healing effects as chitin. This material is prepared from the residue of the fruiting body of *Ganoderma tsugae* (Su et al., 1999).

10 Outlook

It is evident that fungal chitin and chitosan do not yet play a major role in applications of these polysaccharides. However, there is a great potential to produce chitin–glucan complexes and chitosan from the huge amounts of fungal biomass that is generated in various fermentation processes in the biotechnology industry.

Fungal deacetylases bear a great potential for environmentally friendly production of chitosan from crustacean waste, and thus will continue to attract research efforts in microbiology and biotechnological process engineering.

Future research on the biochemistry and molecular genetics of fungal chitin and chitosan biosynthesis and degradation will be directed to explore some fascinating possibilities for human welfare. Thus, unraveling the complex machinery for chitin biosynthesis and degradation in fungi with the tools of molecular biology will lead to novel approaches towards combating pathogenic fungi which infect humans, animals, and plants. Furthermore, in view of the vast biodiversity of fungi and the assumed large variety of presently unknown fungi, it is expected that new, hitherto unrecognized enzymes and proteins interacting with chitin and chitosan will be identified in the future, as more fungal strains and their genes are investigated.

Tab. 7 Some new patents on fungal chitin and chitosan

Patent number	*Patent holder*	*Inventors*	*Title of patent*	*Publication date (YYYYMMDD)*
WO 9307262	Biotechnology Trading Co., Inc., USA	Bouriotis, V., Kafetzopoulos, D., Vournakis, J. J., Martinou, A.	Chitin deacetylase and its purification from fungal mycelia	19930415
WO 9405778	Sandoz-Patent-GmbH, Germany; Sandoz Ltd.	Brodel, B. M. L., Debeire, P. V. J., Monsan, P. F. E., Paul, F. M. B., Touzel, J. P. M.	New products of depolymerized partially acetylated chitosan, new chitosanase, and new *Thermoactinomyces* strain for the production thereof	19940317
WO 9833934		Miller, M., Reeslev, M.	Method of selectively determining a fungal biomass	19980806
WO 9620717	American Cyanamid Company, USA	Kollar, R., Cabib, E., Silverman, S. J., Petrakova, E.	Compositions and methods for inhibiting fungal cell wall formation	19960711
WO 0001812	Cornell Research Foundation, Inc., USA	Harman, G. E., Lorito, M., Woo, S., Brants, A., Earle, E., Kubicek, C. P., Peterbauer, C. K., Tronsmo, A., Klemsdahl, S.	Fungal signal peptide and their use in secreting chitinolytic proteins from transgenic plants	20000113
WO 9946779	Nukem Nuklear G.m.b.H., Germany	Gorowi, L. F., Kojakow, V. N.	Adsorption means for radionuclides	19990916

11 Patents

Numerous patents deal with antifungal agents, using also chitinolytic enzymes and their genes from a variety of organisms, including fungi. These aspects are covered in Chapter 15, this volume. Some World patent applications are detailed in Table 7 which refer specifically to uses of fungi.

Index of fungi

Absidia blakesleeana
Absidia butleri
Absidia coerulea
Absidia glauca
Absidia nidulans
Absidia orchidis
Absidia spinosa
Actinomucor
Agaricus bisporus
Ampelomyces quisqualis
Aphanocladium album
Armillaria ostoyae
Aspergillus fumigatus
Aspergillus nidulans
Aspergillus niger
Beauveria bassiana
Beauveria brongniartii
Boletinus cavipes
Candida albicans
Circinella
Colletotrichum lindemuthianum
Corollospora maritima
Cryptococcus neoformans
Cunninghamella echinulata
Epidermophyton stockdaleae
Fonsecaea pedrosoi
Fusarium moniliforme
Fusarium oxysporum
Fusarium solani f. sp. *phaseoli* SUF386
Ganoderma tsugae
Glomus
Gongronella butleri
Heterobasidion annosum
Hirsutella necatrix
Hirsutella thompsonii
Humicola lutea
Inonotus hispidus
Lentinus edodes
Lycophyllum shimeji
Metarhizium anisopliae
Metarhizium flavoviride
Microsporum fulvum
Mucor circinelloides
Mucor mucedo
Mucor racemosus
Mucor rouxii
Mycothyph africana
Neurospora crassa
Nocardioides sp.
Ophiostoma ulmi
Paracoccidioides brasiliensis
Paxillus involutus
Penicillium chrysogenum
Penicillium islandicum
Penicillium janthinellum
Penicillium spinulosum
Phellinus noxius
Phialophora verrucosa
Phycomyces blakesleeanus
Phythophthora parasitica
Phytophthora cinnamomi
Piromyces communis
Pisolithus tinctorius
Pityrosporum canis
Pleurotus sajor-caju
Puccinia graminis f. sp. *tritici*
Pyricularia oryzae
Pythium dimorphum
Pythium ultimum
Rhizoctonia solani
Rhizomucor pusillus
Rhizopus arrhizus
Rhizopus oligosporus
Rhizopus oryzae
Rigidoporus lignosus
Saccharomyces cerevisiae
Saprolegnia monoica
Schizophyllum commune

Schizosaccharomyces pombe
Suillus variegatus
Trichoderma magnatum
Trichoderma harzianum
Tuber borchii
Uromyces viciae-fabae
Ustilago maydis
Verticillium cfr. *lecanii*
Volvariella volvacea
Wangiella dermatitidis
Zygorhychnus

Acknowledgements
The production of this chapter was, in part, supported by the Fonds der Chemischen Industrie.

12 References

Ak, O., Bakir, U., Guray, T. (1998) Production, purification and characterization of chitosanase from *Penicillium spinulosum*, *Biochem. Arch.* **14**, 221–225.

Akiyama, K., Fujita, T., Kuroshima, K.-I., Sakane, T., Yokota, A., Takata, R. (1999) Purification and gene cloning of a chitosanase from *Bacillus ehimensis* EAG1, *J. Biosci. Bioeng.* **87**, 383–385.

Alfonso, C., Martinez, M. J., Reyes, F. (1992) Purification and properties of two endochitosanases from *Mucor rouxii* implicated in its cell wall degradation, *FEMS Microbiol. Lett.* **95**, 2–3.

Alfonso, C., Nuero, O. M., Santamaria, F., Reyes, F. (1995) Purification of a heat-stable chitin deacetylase from *Aspergillus nidulans* and its role in cell wall degradation, *Curr. Microbiol.* **30**, 49–54.

Ando, A., Noguchi, K., Yanagi, M., Shinoyama, H., Kagawa, Y., Hirata, H., Yabuki, M., Fujii, T. (1992) Primary structure of chitosanase produced by *Bacillus circulans* MH-K1, *J. Gen. Appl. Microbiol.* **38**, 135–144.

Aruchami, M., Sundara Rajulu, G., Gowri, N. (1986) Distribution of chitin deacetylase in Arthropoda, in: *Chitin in Nature and Technology* (Muzzarelli, R., Jeniaux, C., Gooday, G. W., Eds.), New York: Plenum Press, 263–265.

Baek, J.-M., Howell, C. R., Kenerley, C. M. (1999) The role of an extracellular chitinase from *Trichoderma virens* Gv29-8 in the biocontrol of *Rhizoctonia solani*, *Curr. Genet.* **35**, 41–50.

Baymiller, J., McCullough, J. E. (1993) Identification of a *Saccharomyces cerevisiae* mutation that allows cells to grow without chitin synthase 1 or 2, *Curr. Genet.* **23**, 102–107.

Bertagnolli, B. L., Dalsoglio, F. K., Sinclair, J. B. (1996) Extracellular enzyme profiles of the fungal pathogen *Rhizoctonia solani* isolate 2B-12 and of two antagonists, *Bacillus megaterium* strain B153-2-2 and *Trichoderma harzianum* isolate Th008. 1. Possible correlations with inhibition of growth and biocontrol, *Physiol. Mol. Plant Pathol.* **48**, 145–160.

Bogo, M. R., Rota, C. A., Pinto, H., Jr., Ocampos, M., Correa, C. T., Vainstein, M. H., Schrank, A. (1998) A chitinase encoding gene (chit1 gene) from the entomopathogen *Metarhizium anisopliae*: isolation and characterization of genomic and full-length cDNA, *Curr. Microbiol.* **37**, 221–225.

Borgia, P. T., Iartchouk, N., Riggle, P. J., Winter, K. R., Koltin, Y., Bulawa, C. E. (1996) The chsB gene of *Aspergillus nidulans* is necessary for normal hyphal growth and development, *Fungal Genet. Biol.* **20**, 193–203.

Boucher, I., Fukamizo, T., Honda, Y., Willick, G. E., Neugebauer, W. A., Brzezinski, R. (1995) Site-directed mutagenesis of evolutionary conserved carboxylic amino acids in the chitosanase from *Streptomyces* sp. N174 reveals two residues essential for catalysis, *J. Biol. Chem.* **270**, 31077–31082.

Bruyere, T., Unger, C., Moesinger, E. (1996) Chitin synthase in phytopathogenic fungi, in: *Chitin Enzymology*, (Muzzarelli, R. A. A., Ed.), Grottammare: Atec, Vol. 2, 477–479.

Bulawa, C. E. (1993) Genetics and molecular biology of chitin synthesis in fungi, *Annu. Rev. Microbiol.* **47**, 505–534.

Cabib, E. (1987) The synthesis and degradation of chitin, *Adv. Enzymol.* **59**, 59–101.

Cabib, E., Silverman, S. J., Shaw, J. A. (1992) Chitinase and chitin synthase 1: counterbalancing activities in cell separation of *Saccharomyces cerevisiae*, *J. Gen. Microbiol.* **138**, 97–102.

Cabib, E., Shaw, J. A., Mol, P. C., Bowers, B., Choi, W. J. (1996) Chitin biosynthesis and morphogenetic processes, in: *Mycota* (Brambl, R., Marzluf, G. A., Eds.), Berlin: Springer-Verlag, Vol. 3, 243–267.

Campbell, J. A., Davies, G. J., Bulone, V., Henrissat, B. (1997) A classification of nucleotide-diphospho-sugar glycosyltransferases based on amino acid sequence similarities, *Biochem. J.* **326**, 929–942.
Carsolio, C., Gutierrez, A., Jimenez, B., Van, M. M., Herrera-Estrella, A. (1994) Characterization of ech-42, a *Trichoderma harzianum* endochitinase gene expressed during mycoparasitism, *Proc. Natl. Acad. Sci. USA* **91**, 10903–10907.
Cheng, C. Y., Li, Y. K. (2000) An *Aspergillus* chitosanase with potential for large-scale preparation of chitosan oligosaccharides, *Biotechnol. Appl. Biochem.* **32**, 197–203.
Cherif, M., Benhamou, N., Belanger, R. R. (1993) Occurrence of cellulose and chitin in the hyphal walls of *Pythium ultimum*: a comparative study with other plant pathogenic fungi, *Can. J. Microbiol.* **39**, 213–222.
Chernin, L., Gafni, A., Mozes-Koch, R., Gerson, U., Sztejnberg, A. (1997) Chitinolytic activity of the acaropathogenic fungi *Hirsutella thompsonii* and *Hirsutella necatrix*, *Can. J. Microbiol.* **43**, 440–446.
Cheung, P. C. K. (1996) Dietary fiber content and composition of some cultivated edible mushroom fruiting bodies and mycelia, *J. Agric. Food Chem.* **44**, 468–471.
Chiu, S. C., Tzean, S. S. (1995) Glucanolytic enzyme production by *Schizophyllum commune* Fr. during mycoparasitism, *Physiol. Mol. Plant Pathol.* **46**, 83–94.
Choi, W.-J., Santos, B., Durán, A., Cabib, E. (1994) Are yeast chitin synthases regulated at the transcriptional or the posttranslational level? *Mol. Cell. Biol.* **14**, 7685–7694.
Choi, Y. J., Kim, E. J., Kim, T. U., Shin, Y. C. (1997) An endo-chitosanase from *Bacillus* sp. GM44 that produces chitosan oligosaccharides with high degree of polymerization, *Adv. Chitin Sci.* **2**, 296–301.
Christodoulidou, A., Bouriotis, V., Thireos, G. (1996) Two sporulation-specific chitin deacetylase-encoding genes are required for the ascospore wall rigidity of *Saccharomyces cerevisiae*, *J. Biol. Chem.* **271**, 31420–31425.
Christodoulidou, A., Briza, P., Ellinger, A., Bouriotis, V. (1999) Yeast ascospore wall assembly requires two chitin deacetylase isozymes, *FEBS Lett.* **460**, 275–279.
Christov, L. P., Van Driessel, B., Du Plessis, C. A. (1999) Fungal biomass from *Rhizomucor pusillus* as adsorbent of chromophores from a bleach plant effluent, *Process Biochem.* **35**, 91–95.
Chuang, J. S., Schekman, R. W. (1996) Differential trafficking and timed localization of two chitin synthase proteins, Chs2p and Chs3p, *J. Cell Biol.* **135**, 597–610.
Chung, L. Y., Schmidt, R. J., Hamlyn, P. F., Sagar, B. F., Andrews, A. M., Turner, T. D. (1994) Biocompatibility of potential wound management products: fungal mycelia as a source of chitin/chitosan and their effect on the proliferation of human F1000 fibroblasts in culture, *J. Biomed. Mater. Res.* **28**, 463–469.
Chung, L. Y., Schmidt, R. J., Hamlyn, P. F., Sagar, B. F., Andrews, A. M., Turner, T. D. (1998) Biocompatibility of potential wound management products: hydrogen peroxide generation by fungal chitin/chitosans and their effects on the proliferation of murine L929 fibroblasts in culture, *J. Biomed. Mater. Res.* **39**, 300–307.
Cohen, E. (1993) Chitin synthesis and degradation as targets for pesticide, *Arch. Insect Biochem. Physiol.* **22**, 245–261.
Coutinho, P. M., Henrissat, B. (1999) Carbohydrate-Active Enzymes server at URL: http://afmb.cnrs-mrs.fr/~pedro/CAZY/db.html.
Crestini, C., Giovannozzi-Sermanni, G. (1996) Solid-state fermentation of *Lentinus edodes*: a new and efficient approach to chitosan production, in: *Chitin Enzymology*, (Muzzarelli, R. A. A., Ed.), Grottammare: Atec, Vol. 2, 595–600.
Crestini, C., Kovac, B., Giovannozzi-Sermanni, G. (1996) Production and isolation of chitosan by submerged and solid-state fermentation from *Lentinus elodes*, *Biotechnol. Bioeng.* **50**, 207–210.
Dana, M. D., Limon, M. C., Mejias, R., Mach, R. L., Benitez, T., Pintor-Toro, J. A., Kubicek, C. P. (2001) Regulation of chitinase 33 (chit33) gene expression in *Trichoderma harzianum*, *Curr. Genet.* **38**, 335–342.
DeAngelis, P. L., Yang, N., Weigel, P. H. (1994) The *Streptococcus pyogenes* hyaluronan synthase: sequence comparison and conservation among various group A strains, *Biochem. Biophys. Res. Commun.* **199**, 1–10.
Deising, H., Siegrist, J. (1995) Chitin deacetylase activity of the rust *Uromyces viciae-fabae* is controlled by fungal morphogenesis, *FEMS Microbiol. Lett.* **127**, 207–212.
Deising, H., Rauscher, M., Haug, M., Heiler, S. (1995) Differentiation and cell wall degrading enzymes in the obligately biotrophic rust fungus *Uromycesviciae-fabae*, *Can. J. Bot.* **73**, S624–S631.
De la Cruz, J., Llobell, A. (1999) Purification and properties of a basic endo-β-(1,6)-glucanase

(BGN16.1) from the antagonistic fungus *Trichoderma harzianum*, *Eur. J. Biochem.* **265**, 145–151.

Deshpande, M. V. (1999) Mycopesticide production by fermentation: potential and challenges, *Crit. Rev. Microbiol.* **25**, 229–243.

Din, A. B., Yarden, O. (1994) The *Neurospora crassa* chs-2 gene encodes a non-essential chitin synthase, *Microbiology* **140**, 2189–2197.

Draborg, H., Christgau, S., Halkier, T., Rasmussen, G., Dalboege, H., Kauppinen, S. (1996) Secretion of an enzymically active *Trichoderma harzianum* endochitinase by *Saccharomyces cerevisiae*, *Curr. Genet.* **29**, 404–409.

Dunkel, C., Knorr, D. (1994) Enhancement of chitin deacetylase activity in *Mucor rouxii* and *Absidia coerulea* with chitin and its detection with a non-radioactive substrate, *Food Biotechnol.* **8**, 67–74.

Edwards, H. G. M., Russell, N. C., Weinstein, R., Wynn-Williams, D. D. (1995) Fourier transform Raman spectroscopic study of fungi, *J. Raman Spectrosc.* **26**, 911–916.

Ekblad, A., Naesholm, T. (1996) Determination of chitin in fungi and mycorrhizal roots by an improved HPLC analysis of glucosamine, *Plant Soil* **178**, 29–35.

Ekramoddoullah, A. K. M., Jensen, G. D., Manning, L. E. (2000) Ultrastructural localization of chitin in the five spore stages of the blister rust fungus, *Cronartium ribicola*, *Mycol. Res.* **104**, 1384–1388.

El-Din, S. B., El-Tayeb, O. M., Hashem, A. M., Kansoh, A. A. (1995) Microbial mycolytic activity: VII. Purification and characterization of extracellular chitosanases from *Bacillus lentus*, *Al Azhar J. Microbiol.* **28**, 23–40.

El-Gueddari, N. E., Moerschbacher, B. M. (2000) A rust fungus turns chitin into chitosan upon plant tissue colonization to evade recognition by the host, *Adv. Chitin Sci.* **4**, 588–592.

El-Katatny, M. H., Somitsch, W., Robra, K. H., El-Katatny, M. S., Gubitz, G. M. (2000) Production of chitinase and β-(1-3)-glucanase by *Trichoderma harzianum* for control of the phytopathogenic fungus *Sclerotium rolfsii*, *Food Technol. Biotechnol.* **38**, 173–180.

Felse, P. A., Panda, T. (1999) Regulation and cloning of microbial chitinase genes, *Appl. Microbiol. Biotechnol.* **51**, 141–151.

Fenice, M., Leuba, J. L., Federici, F. (1998a) Chitinolytic enzyme activity of *Penicillium janthinellum* P9 in bench-top bioreactor, *J. Ferment. Bioeng.* **86**, 620–623.

Fenice, M., Selbmann, L., Di Giambattista, R., Federici, F. (1998b) Chitinolytic activity at low temperature of an Antarctic strain (A3) of *Verticillium lecanii*, *Res. Microbiol.* **149**, 289–300.

Fontaine, T., Simenel, C., Dubreucq, G., Adam, O., Delepierre, M., Lemoine, J., Vorgias, C. E., Diaquin, M., Latge, J. P. (2000) Molecular organization of the alkali-insoluble fraction of *Aspergillus fumigatus* cell wall, *J. Biol. Chem.* **275**, 27594–27607.

Fujiwara, M., Horiuchi, H., Ohta, A., Takagi, M. (1997) A novel fungal gene encoding chitin synthase with a myosin motor-like domain, *Biochem. Biophys. Res. Commun.* **236**, 75–78.

Fukamizo, T., Ohkawa, T., Ikeda, Y., Goto, S. (1994) Specificity of chitosanase from *Bacillus pumilus*, *Biochim. Biophys. Acta* **1205**, 183–188.

Fukamizo, T., Honda, Y., Goto, S., Boucher, I., Brzezinski, R. (1995) Reaction mechanism of chitosanase from *Streptomyces* sp. N174, *Biochem. J.* **311**, 377–383.

Fukamizo, T., Honda, Y., Toyoda, H., Ouchi, S., Goto, S. (1996) Chitinous component of the cell wall of *Fusarium oxysporum*, its structure deduced from chitosanase digestion, *Biosci. Biotechnol. Biochem.* **60**, 1705–1708.

Fukamizo, T., Juffer, A. H., Vogel, H. J., Honda, Y., Tremblay, H., Boucher, I., Neugebauer, W. A., Brzezinski, R. (2000) Theoretical calculation of pK(a) reveals an important role of Arg(205) in the activity and stability of *Streptomyces* sp. N174 chitosanase, *J. Biol. Chem.* **275**, 25633–25640.

Fukamizo, T., Sasaki, C., Schelp, E., Bortone, K., Robertus, J. D. (2001) Kinetic properties of chitinase-1 from the fungal pathogen *Coccidioides immitis*, *Biochemistry* **40**, 2448–2454.

Gallagher, K. A., Healy, M. G., Allen, S. J. (1997) Biosorption of synthetic dye and metal ions from aqueous effluents using fungal biomass, *Stud. Environ. Sci.* **66**, 27–50.

Galli, U., Schueepp, H., Brunold, C. (1994) Heavy metal binding by mycorrhizal fungi, *Physiol. Plant.* **92**, 364–368.

Gao, X.-D., Katsumoto, T., Onodera, K. (1995) Purification and characterization of chitin deacetylase from *Absidia coerulea*, *J. Biochem. (Tokyo)* 117, 257–263

Garcia, I., Lora, J. M., de la Cruz, J., Benitez, T., Llobell, A., Pintor-Toro, J. A. (1994) Cloning and characterization of a chitinase (CHIT42) cDNA from the mycoparasitic fungus *Trichoderma harzianum*, *Curr. Genet.* **27**, 83–89.

Garnero, L., Lazzari, B., Mainieri, D., Viotti, A., Bonfante, P. (2000) TMchs4, a class IV chitin synthase gene from the ectomycorrhizal *Tuber magnatum*, *Mycol. Res.* **104**, 703–707.

Gaughran, J. P., Lai, M. H., Kirsch, D. R., Silverman, S. J. (1994) Nikkomycin Z is a specific inhibitor of *Saccharomyces cerevisiae* chitin synthase isoenzyme Chs3 *in vitro* and *in vivo*, *J. Bacteriol.* **176**, 5857–5860.

Gill, M., Steglich, W. (1987) Pigments of fungi (Macromycetes), *Progr. Chem. Nat. Prod.* **51**, 1–317.

Gokul, B., Lee, J. H., Song, K. B., Rhee, S. K., Kim, C. H., Panda, T. (2000) Characterization and applications of chitinases from *Trichoderma harzianum* – A review, *Bioproc. Eng.* **23**, 691–694.

Goldman, R. C., Sullivan, P. A., Zakula, D., Capobianco, J. O. (1995) Kinetics of β-(1,3)-glucan interaction at the donor and acceptor sites of the fungal glucosyltransferase encoded by the BGL2 gene, *Eur. J. Biochem.* **227**, 372–378.

Gooday, G. W. (1996) Chitin and chitosan in fungi, in: *Chitin in Life Sciences* (Giraud-Guille, M.-M., Ed.), Lyon: J. André, 20–29.

Gooday, G. W., Zhu, W. Y., Odonnell, R. W. (1992) What are the roles of chitinases in the growing fungus, *FEMS Microbiol. Lett.* **100**, 1–3.

Gorovoj, L. F., Kosyakov, V. N. (1994) Mycoton – new chitin materials produced from fungi, in: *Chitin World* (Karnicki, Z. S., Wojtasz-Pajak, A., Brzeski, M. M., Bykowski, P. J., Eds.), Bremerhaven: Wirtschaftsverlag NW, 632–647.

Gorovoj, L., Kosyakov, V. (1997) Chitin and chitosan biosorbents for radionuclides and heavy metals, *Adv. Chitin Sci.* **2**, 858–863.

Gorovoj, L. F., Petyuschenko, A. P. (2000) Influence of medium pH on the biosorption of heavy metals by chitin-containing sorbent Mycoton, *Adv. Chitin Sci.* **4**, 310–314.

Gorovoj, L., Burdyukova, L., Zemskov, V., Prilutsky, A. (1997) Chitin health product “MYCOTON” produced from fungi, *Adv. Chitin Sci.* **2**, 648–655.

Gorovoj, L. F., Burdyukova, L. I., Zemskov, V. S., Prilutsky, A. I., Artamonov, V. S., Ivanyuta, S. O., Prilutskaya, A. B. (2000a) Chitin-containing materials Mycoton for wounds treatment, *Adv. Chitin Sci.* **4**, 68–74.

Gorovoj, L. F., Seniouk, O., Beketova, G., Savichuk, N., Amanbaeva, G. (2000b) The chitin-containing preparation Mycoton in a pediatric gastroenterology case, *Adv. Chitin Sci.* **4**, 280–286.

Grant, W. D., Atkinson, M., Burke, B., Molloy, C. (1996) Chitinolysis by the marine ascomycete *Corollospora maritima* Werdermann: purification and properties of a chitobiosidase, *Bot. Mar.* **39**, 177–186.

Guarro, J., Cano, J., Leal, J. A., Gomez-Miranda, B., Bernabe, M. (1993) Composition of the cell wall polysaccharides in some geophilic Dermatophytes, *Mycopathologia* **122**, 69–77.

Guest, G. M., Momany, M. (2000) Analysis of cell wall sugars in the pathogen *Aspergillus fumigatus* and the saprophyte *Aspergillus nidulans*, *Mycologia* **92**, 1047–1050.

Haran, S., Schickler, H., Oppenheim, A., Chet, I. (1995) New components of the chitinolytic system of *Trichoderma harzianum*, *Mycol. Res.* **99**, 441–446.

Haran, S., Schickler, H., Oppenheim, A., Chet, I. (1996) Differential expression of *Trichoderma harzianum* chitinases during mycoparasitism, *Phytopathology* **86**, 980–985.

Hartland, R. P., Vermeulen, C. A., Klis, F. M., Sietsma, J. H., Wessels, J. G. H. (1994) The linkage of (1-3)-β-glucan to chitin during cell wall assembly in *Saccharomyces cerevisiae*, *Yeast* **10**, 1591–1599.

Hayes, C. K., Klemsdal, S., Lorito, M., Di Pietro, A., Peterbauer, C., Nakas, J. P., Tronsmo, A., Harman, G. E. (1994) Isolation and sequence of an endochitinase-encoding gene from a cDNA library of *Trichoderma harzianum*, *Gene* **138**, 143–148.

Herrera-Estrella, A., Chet, I. (1999) Chitinases in biological control, *EXS 87* (Chitin and Chitinases), 171–184.

Hodge, A., Alexander, I. J., Gooday, G. W. (1995) Chitinolytic enzymes of pathogenic and ectomycorrhizal fungi, *Mycol. Res.* **99**, 935–941.

Hollis, T., Monzingo, A. F., Bortone, K., Ernst, S., Cox, R., Robertus, J. D. (2000) The X-ray structure of a chitinase from the pathogenic fungus *Coccidioides immitis*, *Protein Sci.* **9**, 544–551.

Horiuchi, H., Fujiwara, M., Yamashita, S., Ohta, A., Takagi, M. (1999) Proliferation of intrahyphal hyphae caused by disruption of csmA, which encodes a class V chitin synthase with a myosin motor-like domain in *Aspergillus nidulans*, *J. Bacteriol.* **181**, 3721–3729.

Horsch, M., Rast, D. M. (1993) Allosteric activation of chitin synthetase by N-acetylglucosamine: a mechanistic study, in: *Chitin Enzymology* (Muzzarelli, R. A. A., Ed.), Lyons: European Chitin Society, 47–56.

Hu, G. G., Rijkenberg, F. H. J. (1998) Cytochemistry of the interaction between wheat (*Triticum aestivum*) and *Puccinia recondita* f.sp. *tritici*. Localization of cellulose, chitin, glucose/mannose, galactose, and fucose, *J. Phytopathol.* **146**, 365–377.

Hu, K.-J., Yeung, K.-W., Ho, K.-P., Hu, J.-L. (1999) Rapid extraction of high-quality chitosan from mycelia of *Absidia glauca*, *J. Food Biochem.* **23**, 187–196.

Hwang, C.-W., Park, I.-C., Yeh, W.-H., Takagi, M., Ryu, J.-C. (1997) A partial nucleotide sequence of chitin synthase (CHS) gene from rice blast fungus, *Pyricularia oryzae* and its cloning, *J. Microbiol. Biotechnol.* **7**, 157–159.

Ibrahim, G. H., Smartt, C. T., Kiley, L. M., Christensen, B. M. (2000) Cloning and characterization of a chitin synthase cDNA from the mosquito *Aedes aegypti*, *Insect Biochem. Mol. Biol.* **30**, 1213–1222.

Izume, M., Nagae, S., Kawagishi, H., Mitsutomi, M., Ohtakara, A. J. (1992) Action pattern of *Bacillus* sp. No. 7-M chitosanase on partially N-acetylated chitosan, *Biosci. Biotechnol. Biochem.* **56**, 448–453.

Jaspar-Versali, M.-F., Clerisse, F. (1997) Expression and characterization of recombinant chitin deacetylase, *Adv. Chitin Sci.* **2**, 273–278.

Jaworska, M. M., Szewczyk, K. W. (1998) Chitosan from *Absidia orchidis*: influence of pH of culture medium, *Adv. Chitin Sci.* **3**, 489–494.

Kafetzopoulos, D., Martinou, A., Bouriotis, V. (1993a) Bioconversion of chitin to chitosan: purification and characterization of chitin deacetylase from *Mucor rouxii*, *Proc. Natl. Acad. Sci. USA* **90**, 2564–2568.

Kafetzopoulos, D., Thireos, G., Vournakis, J. N., Bouriotis, V. (1993b) The primary structure of a fungal chitin deacetylase reveals the function for two bacterial gene products, *Proc. Natl. Acad. Sci. USA* **90**, 8005–8008.

Kang, S. C., Park, S., Lee, D. G. (1998) Isolation and characterization of a chitinase cDNA from the entomopathogenic fungus, *Metarhizium anisopliae*, *FEMS Microbiol. Lett.* **165**, 267–271.

Kang, S. C., Park, S., Lee, D. G. (1999) Purification and characterization of a novel chitinase from the entomopathogenic fungus, *Metarhizium anisopliae*, *J. Invertebr. Pathol.* **73**, 276–281.

Kapteyn, J. C., Ram, A. F. J., Groos, E. M., Kollar, R., Montijn, R. C., van den Ende, H., Llobell, A., Cabib, E., Klis, F. M. (1997) Altered extent of crosslinking of β-1,6-glucosylated mannoproteins to chitin in *Saccharomyces cerevisiae* mutants with reduced cell wall β-1,3-glucan content, *J. Bacteriol.* **179**, 6279–6284.

Karnezis, T., McIntosh, M., Wardak, A. Z., Stanisich, V. A., Stone, B. A. (2000) The biosynthesis of β-glycans, *Trends Glycosci. Glycotechnol.* **12**, 211–227.

Karuppayil, S. M., Peng, M., Mendoza, L., Levins, T. A., Szaniszlo, P. J. (1996) Identification of the conserved coding sequences of three chitin synthase genes in *Fonsecaea pedrosoi*, *J. Med. Vet. Mycol.* **34**, 117–125.

Kim, S.-Y., Shon, D.-H., Lee, K.-H. (1998) Purification and characteristics of two types of chitosanases from *Aspergillus fumigatus* KH-94, *J. Microbiol. Biotechnol.* **8**, 568–574.

King, L., Butler, G. (1998) Ace2p, a regulator of CTS1 (chitinase) expression, affects pseudohyphal production in *Saccharomyces cerevisiae*, *Curr. Genet.* **34**, 183–191.

Kollar, R., Petrakovas, E., Ashwell, G., Robbins, P. W., Cabib, E. (1995) Architecture of the yeast cell wall. The linkage between chitin and β-(1-3)-glucan, *J. Biol. Chem.* **270**, 1170–1178.

Kollar, R., Cabib, E., Silverman, S. J., Petrakova, E. (1996) American Cyanamid Company, USA, Compositions and methods for inhibiting fungal cell wall formation, WO 9620717 [*Chem. Abstr.* **125**, 185857].

Kollar, R., Reinhold, B. B., Petrakova, E., Yeh, H. J. C., Ashwell, G., Drgonova, J., Kapteyn, J. C., Klis, F. M., Cabib, E. (1997) Architecture of the yeast cell wall. β-(1-6)-Glucan interconnects mannoprotein, β-(1-3)-glucan, and chitin, *J. Biol. Chem.* **272**, 17762–17775.

Kolodziejska, I., Wojtasz-Pajak, A., Sikorski, Z. E. (1995) Enzymic modification of chitin, *Biotechnologia*, 133–139.

Kolodziejska, I., Malesa-Ciecwierz, M., Lerska, A., Sikorski, Z. (1999) Properties of chitin deacetylase from crude extracts of *Mucor rouxii* mycelium, *J. Food Biochem.* **23**, 45–57.

Krishnamurthy, J., Samiyappan, R., Vidhyasekaran, P., Nakkeeran, S., Rajeswari, E., Raja, J. A. J., Balasubramanian, P. (1999) Efficacy of *Trichoderma* chitinases against *Rhizoctonia solani*, the rice sheath blight pathogen, *J. Biosci.* **24**, 207–213.

Kuhlmann, K., Czupala, A., Haunhorst, J., Weiss, A., Prasch, T., Schörken, U. (2000) Preparation and characterization of chitosan from *Mucorales*, *Adv. Chitin Sci.* **4**, 7–14.

Kurakake, M., You, S., Nakagawa, K., Sugihara, M., Komaki, T. (2000) Properties of chitosanase from *Bacillus cereus* S1, *Curr. Microbiol.* **40**, 6–9.

Kusaoke, H., Kimoto, H., Taketo, A. (1997) Total sequence of a bacterial gene encoding chitosanase-glucanase activities, *Adv. Chitin Sci.* **2**, 290–295.

Lambais, M. R., Mehdy, M. C. (1995) Differential expression of defense-related genes in arbuscular mycorrhiza, *Can. J. Bot.* **73**, S533–S540.

Lanfranco, L., Garnero, L., Delpero, M., Bonfante, P. (1995) Chitin synthase homologs in three ectomycorrhizal truffles, *FEMS Microbiol. Lett.* **134**, 109–114.

Leal-Morales, C. A., Bracker, C. E., Bartnicki-Garcia, S. (1994a) Subcellular localization, abundance and stability of chitin synthetases 1 and 2 from *Saccharomyces cerevisiae*, *Microbiology* **140**, 2207–2216.

Leal-Morales, C. A., Bracker, C. E., Bartnicki-Garcia, S. (1994b) Distribution of chitin synthetase and various membrane marker enzymes in chitosomes and other organelles of the slime mutant of *Neurospora crassa*, *Exp. Mycol.* **18**, 168–179.

Leal-Morales, C. A., Gay, L., Fevre, M., Bartnicki-Garcia, S. (1997) The properties and localization of *Saprolegnia monoica* chitin synthase differ from those of other fungi, *Microbiology* **7**, 2473–2483.

Lee, H.-W., Choi, J.-W., Han, D.-P., Lee, N.-W., Park, S.-L., Yi, D.-H. (1996a) Identification and production of constitutive chitosanase from *Bacillus* sp. HW-002, *J. Microbiol. Biotechnol.* **6**, 12–18.

Lee, H.-W., Choi, J.-W., Han, D.-P., Park, M.-J., Lee, N.-W., Yi, D.-H. (1996b) Purification and characteristics of chitosanase from *Bacillus* sp. HW-002, *J. Microbiol. Biotechnol.* **6**, 19–25.

Lestan, M., Pecavar, A., Lestan, D., Perdih, A. (1993) Amino acids in chitin-glucan complex of *Aspergillus niger*, *Amino Acids* **4**, 169–176.

Lima, L. H. C., De Marco, J. L., Ulhoa, C. J., Felix, C. R. (1999) Synthesis of a Trichoderma chitinase which affects the *Sclerotium rolfsii* and *Rhizoctonia solani* cell walls, *Folia Microbiol.* **44**, 45–49.

Limon, M. C., Lora, J. M., de la Cruz, J., Garcia, I., Llobell, A., Benitez, T., Pintor-Toro, J. A. (1995) Primary structure and expression pattern of the 33-kDa chitinase gene from the mycoparasitic fungus *Trichoderma harzianum*, *Curr. Genet.* **28**, 478–483.

Limon, M. C., Llobel, A., Pintor-Toro, J. A., Benitez, T. (1996) Overexpression of chitinase by Trichoderma harzianum strains used as biocontrol fungi, in: *Chitin Enzymology*, (Muzzarelli, R. A. A., Ed.), Grottammare: Atec, Vol. 2, 245–252.

Lopez-Matas, M. A., Eslava, A. P., Diaz-Minguez, J. M. (2000) Mcchs1, a member of a chitin synthase gene family in *Mucor circinelloides*, is differentially expressed during dimorphism, *Curr. Microbiol.* **40**, 169–175.

Lorito, M. (1998) Chitinolytic enzymes and their genes, in: *Trichodermaglio cladium*, (Harman, G. E., Kubicek, C. P., Eds.), London: Taylor & Francis, Vol. 2, 73–99.

Lorito, M., Woo, S. L., Donzelli, B., Scala, F. (1996) Synergistic, antifungal interactions of chitinolytic enzymes from fungi, bacteria and plants, in: *Chitin Enzymology*, (Muzzarelli, R. A. A., Ed.), Grottammare: Atec, Vol. 2, 157–164.

Lu, Z., Li, Y., Que, Q., Kutish, G. F., Rock, D. L., Van Etten, J. L. (1996) Analysis of 94 kb of the chlorella virus PBCV-1 330-kb genome: map positions 88 to 182, *Virology* **216**, 102–123.

Mach, R. L., Peterbauer, C. K., Payer, K., Jaksits, S., Woo, S. L., Zeilinger, S., Kullnig, C. M., Lorito, M., Kubicek, C. P. (1999) Expression of two major chitinase genes of *Trichoderma atroviride* (*T. harzianum* P1) is triggered by different regulatory signals, *Appl. Environ. Microbiol.* **65**, 1858–1863.

Machida, S., Saito, M. (1993) Purification and characterization of membrane-bound chitin synthase, *J. Biol. Chem.* **268**, 1702–1707.

Malesa-Ciecwierz, M., Kolodziejska, I., Krajka-Nanowska, R., Sikorski, Z. E. (1997) Influence of cultivation conditions on the activity of chitin deacetylase from *Mucor rouxii*, *Adv. Chitin Sci.* **2**, 266–272.

Marcotte, E. M., Monzingo, A. F., Ernst, S. R., Brzezinski, R., Robertus, J. D. (1996) X-ray structure of an anti-fungal chitosanase from *Streptomyces* N174, *Nature Struct. Biol.* **3**, 155–162.

Margolles-Clark, E., Hayes, C. K., Harman, G. E., Penttila, M. (1996) Improved production of *Trichoderma harzianum* endochitinase by expression in *Trichoderma reesei*, *Appl. Environ. Microbiol.* **62**, 2145–2151.

Martin, H., Dagkessamanskaia, A., Satchanska, G., Dallies, N., Francois, J. (1999) KNR4, a suppressor of *Saccharomyces cerevisiae* cwh mutants, is involved in the transcriptional control of chitin synthase genes, *Microbiology* **145**, 249–258.

Martinez, J. P., Gozalbo, D. (1994) Chitin synthetases in *Candida albicans*: a review on their subcellular distribution and biological function, *Microbiologia* **10**, 239–248.

Martinou, A., Bouriotis, V., Stokke, B. T., Vårum, K. M. (1998) Mode of action of chitin deacetylase from *Mucor rouxii* on partially N-acetylated chitosans, *Carbohydr. Res.* **311**, 71–78.

Masson, J.-Y., Denis, F., Brzezinski, R. (1994) Primary sequence of the chitosanase from *Streptomyces* sp. strain N174 and comparison with other endoglycosidases, *Gene* **140**, 103–107.

Masson, J.-Y., Boucher, I., Neugebauer, W. A., Ramotar, D., Brzezinski, R. (1995) A new chitosanase gene from a *Nocardioides* sp. is a third member of glycosyl hydrolase family 46, *Microbiology* **141**, 2629–2635.

Mayer, R. T., McCollum, T. G., Niedz, R. P., Hearn, C. J., McDonald, R. E., Berdis, E., Doostdar, H. (1996) Characterization of seven basic endochitinases isolated from cell cultures of *Citrus sinensis, Planta* **200**, 289–295.

McCreath, K. J., Specht, C. A., Robbins, P. W. (1995) Molecular cloning and characterization of chitinase genes from *Candida albicans, Proc. Natl. Acad. Sci. USA* **92**, 2544–2548.

McCreath, K. J., Specht, C. A., Liu, Y., Robbins, P. W. (1996) Molecular cloning of a third chitinase gene (CHT1) from *Candida albicans, Yeast* **12**, 501–504.

McNab, R., Glover, L. A. (1991) Inhibition of *Neurospora crassa* cytosolic chitinase by allosamidin, *FEMS Microbiol. Lett.* **82**, 79–82.

Mellado, E., Specht, C. A., Robbins, P. W., Holden, D. W. (1996) Cloning and characterization of chsD, a chitin synthase-like gene of *Aspergillus fumigatus, FEMS Microbiol. Lett.* **143**, 69–76.

Mellor, K. J., Nicholas, R. O., Adams, D. J. (1994) Purification and characterization of chitinase from *Candida albicans, FEMS Microbiol. Lett.* **119**, 111–118.

Merz, R. A., Horsch, M., Nyhlen, L. E., Rast, D. M. (1999a) Biochemistry of chitin synthase, *EXS 87* (Chitin and Chitinases), 9–37.

Merz, R. A., Horsch, M., Ruffner, H. P., Rast, D. M. (1999b) Interaction between chitosomes and concanavalin A, *Phytochemistry* **52**, 211–224.

Mishra, C., Semino, C. E., McCreath, K. J., de la Vega, H., Jones, B. J., Specht, C. A., Robbins, P. W. (1997) Cloning and expression of two chitin deacetylase genes of *Saccharomyces cerevisiae, Yeast* **13**, 327–336.

Mitsutomi, M., Ohtakara, A. (1992) Difference between microbial chitinase and chitosanase in the mode of action on partially N-acetylated chitosan, in: *Advances in Chitin and Chitosan* (Brine, C. J., Sandford, P. A., Zikakis, J. P., Eds.), London: Elsevier Applied Science, 304–313.

Mitsutomi, M., Ueda, M., Arai, M., Ando, A., Watanabe, T. (1996) Action patterns of microbial chitinases and chitosanases on partially N-acetylated chitosan, in: *Chitin Enzymology* (Muzzarelli, R. A. A., Ed.), Grottammare: Atec, Vol. 2, 273–284.

Mitsutomi, M., Isono, M., Uchiyama, A., Nikaidou, N., Ikegami, T., Watanabe, T. (1998) Chitosanase activity of the enzyme previously reported as β-(1-3/1-4)-glucanase from *Bacillus circulans* WL-12, *Biosci. Biotechnol. Biochem.* **62**, 2107–2114.

Molloy, C., Burke, B. (1997) Expression and secretion of *Janthinobacterium lividum* chitinase in *Saccharomyces cerevisiae, Biotechnol. Lett.* **19**, 1161–1164.

Mort-Bontemps, M., Gay, L., Fevre, M. (1997) CHS2, a chitin synthase gene from the oomycete *Saprolegnia monoica, Microbiology* **143**, 2009–2020.

Motoyama, T., Horiuchi, H., Ohta, A., Yamaguchi, I., Takagi, M. (1998) Isolation of a class IV chitin synthase gene from a zygomycete fungus, *Rhizopus oligosporus, FEMS Microbiol. Lett.* **169**, 1–8.

Munro, C. A., Gow, N. A. R. (1995) Chitin biosynthesis as a target for antifungals, in: *Antifungal Agents: Discovery, Mode, and Action* (Dixon, G. K., Copping, L. G., Hollomon, D. W., Eds.), Oxford: Bios Scientific Publishers, 161–171.

Muzzarelli, R. A. A. (1997) The determination of minute quantities of chitin in tissues, in: *Chitin Handbook* (Muzzarelli, R. A. A., Peter, M. G., Eds.), Grottammare: Atec, 15–25.

Muzzarelli, R. A. A., Muzzarelli, C., Cosani, A., Terbojevich, M. (1999) 6-Oxychitins, novel hyaluronan-like regiospecifically carboxylated chitins, *Carbohydr. Polym.* **39**, 361–367.

Muzzarelli, R. A. A., Miliani, M., Cartolari, M., Tarsi, R., Tosi, G., Muzzarelli, C. (2000) Polyuronans obtained by regiospecific oxidation of polysaccharides from *Aspergillus niger, Trichoderma reesei* and *Saprolegnia* sp., *Carbohydr. Polym.* **43**, 55–61.

Nakonechna, A. A., Drannik, G. N., Gorovoj, L. F., Kushko, L. J., Gorova, I. L. (2000) Clinicoimmunological efficiency of the chitin-containing drug Mycoton in complex treatment of a chronic hepatitis, *Adv. Chitin Sci.* **4**, 270–274.

Nam, J.-S., Lee, D. H., Park, H.-Y., Bae, K. S. (1997) Cloning and phylogenetic analysis of chitin synthase gene from entomopathogenic fungus, *Beauveria brongniartii, J. Microbiol.* **35**, 222–227.

Nam, J.-S., Lee, D.-H., Lee, K. H., Park, H.-M., Bae, K. S. (1998) Cloning and phylogenetic analysis of chitin synthase genes from the insect pathogenic fungus, *Metarhizium anisopliae* var. *anisopliae, FEMS Microbiol. Lett.* **159**, 77–84.

Namgung, J., Park, B. C., Lee, D. H., Bae, K. S., Park, H.-M. (1996) Cloning and characterization of chitin synthase gene fragments from *Penicillium chrysogenum, FEMS Microbiol. Lett.* **145**, 71–76.

Nicholas, R. O., Williams, D. W., Hunter, P. A. (1994) Investigation of the value of β-glucan-specific fluorochromes for predicting the β-glucan content of the cell walls of zoopathogenic fungi, *Mycol. Res.* **98**, 694–698.

Nicole, M. R., Benhamou, N. (1991) Ultrastructural localization of chitin in cell walls of *Rigidoporus lignosus*, the white-rot fungus of rubber tree roots, *Physiol. Mol. Plant Pathol.* **39**, 415–431.

Nicole, M., Chamberland, H., Rioux, D., Xixuan, X., Blanchette, R. A., Geiger, J. P., Ouellette, G. B. (1995) Wood degradation by *Phellinus noxius*: ultrastructure and cytochemistry, *Can. J. Microbiol.* **41**, 253–265.

Nino-Vega, G. A., Buurman, E. T., Gooday, G. W., San-Blas, G., Gow, N. A. R. (1998) Molecular cloning and sequencing of a chitin synthase gene (CHS2) of *Paracoccidioides brasiliensis*, *Yeast* **14**, 181–187.

Nogawa, M., Takahashi, H., Kashiwagi, A., Ohshima, K., Okada, H., Morikawa, Y. (1998) Purification and characterization of exo-β-D-glucosaminidase from a cellulolytic fungus, *Trichoderma reesei* PC-3-7, *Appl. Environ. Microbiol.* **64**, 890–895.

O'Conallain, C., Doolin, M. T., Taggart, C., Thornton, F., Butler, G. (1999) Regulated nuclear localization of the yeast transcription factor Ace2p controls expression of chitinase (CTS1) in *Saccharomyces cerevisiae*, *Mol. Gen. Genet.* **262**, 275–282.

Ohishi, K., Murase, K., Ohta, T., Etoh, H. (2000) Cloning and sequencing of the deacetylase gene from *Vibrio alginolyticus H-8*, *J. Biosci. Bioeng.* **90**, 561–563.

Okajima, S., Ando, A., Shinoyama, H., Fujii, T. (1994) Purification and characterization of an extracellular chitosanase produced by *Amycolatopsis* sp. CsO-2, *J. Ferment. Bioeng.* **77**, 617–620.

Okajima, S., Konouchi, T., Mikami, Y., Ando, A. (1995) Purification and some properties of a chitosanase of *Nocardioides* sp., *J. Gen. Appl. Microbiol.* **41**, 351–357.

Omumasaba, C. A., Yoshida, N., Sekiguchi, Y., Kariya, K., Ogawa, K. (2000) Purification and some properties of a novel chitosanase from *Bacillus subtilis* KH1, *J. Gen. Appl. Microbiol.* **46**, 19–27.

Osswald, W. F., Shapiro, J. P., Mcdonald, R. E., Niedz, R. P., Mayer, R. T. (1993) Some *Citrus* chitinases also possess chitosanase activities, *Experientia* **49**, 888–892.

Osswald, W. F., Shapiro, J. P., Doostdar, H., McDonald, R. E., Niedz, R. P., Nairn, C. J., Hearn, C. J., Mayer, R. T. (1994) Identification and characterization of acidic hydrolases with chitinase and chitosanase activities from sweet orange callus tissue, *Plant Cell Physiol.* **35**, 811–820.

Osswald, W. F., Jehle, J., Firl, J. (1995) Quantification of fungal infection in plant tissues by determining the glucosamine phenyl isothiocyanate derivative using HPLC techniques, *J. Plant Physiol.* **145**, 393–397.

Palli, S. R., Retnakaran, A. (1999) Molecular and biochemical aspects of chitin synthesis inhibition, *EXS 87* (Chitin and Chitinases), 85–98.

Park, J., Shimono, K., Ochiai, N., Shigeru, K., Kurita, M., Ohta, Y., Tanaka, K., Matsuda, H., Kawamukai, M. (1999) Purification, characterization, and gene analysis of a chitosanase (ChoA) from *Matsuebacter chitosanotabidus* 3001, *J. Bacteriol.* **181**, 6642–6649.

Park, Y. D., Lee, M. S., Kim, J. H., Namgung, J., Park, B. C., Bae, K. S., Park, H. M. (2000) Genomic organization of *Penicillium chrysogenum* chs4, a class III chitin synthase gene, *J. Microbiol.* **38**, 230–238.

Parro, V., Roman, M. S., Galindo, I., Purnelle, B., Bolotin, A., Sorokin, A., Mellado, R. P. (1997) A 23 911 bp region of the *Bacillus subtilis* genome comprising genes located upstream and downstream of the Iev operon, *Microbiology* **143**, 1321–1326.

Pedraza-Reyes, M., Gutierrez-Corona, F. (1997) The bifunctional enzyme chitosanase-cellulase produced by the gram-negative microorganism *Mycobacter* sp. AL-1 is highly similar to *Bacillus subtilis* endoglucanases, *Arch. Microbiol.* **168**, 321–327.

Peng, M., Karuppayil, S. M., Mendoza, L., Levins, T. A., Szaniszlo, P. J. (1995) Use of the polymerase chain reaction to identify coding sequences for chitin synthase isoenzymes in *Phialophora verrucosa*, *Curr. Genet.* **27**, 517–523.

Penman, D., Britton, G., Hardwick, K., Collin, H. A., Isaac, S. (2000) Chitin as a measure of biomass of *Crinipellis perniciosa*, causal agent of witches' broom disease of *Theobroma cacao*, *Mycol. Res.* **104**, 671–675.

Pierce, J. A., Rast, D. M. (1995) A comparison of native and synthetic mushroom melanins by Fourier-transform infrared spectroscopy, *Phytochemistry* **39**, 49–55.

Pishko, E. J., Kirkland, T. N., Cole, G. T. (1995) Isolation and characterization of two chitinase-encoding genes (cts1, cts2) from the fungus *Coccidioides immitis*, *Gene* **167**, 173–177.

Piza, F. A. T., Siloto, A. M. P., Carvalho, C. V., Franco, T. T. (2000) Extraction and purification of chitosanase from *Bacillus cereus*, *Adv. Chitin Sci.* **4**, 570–574.

Plassard, C. (1997) Assay of fungal chitin and estimation of mycelial biomass, in: *Chitin Handbook* (Muzzarelli, R. A. A., Peter, M. G., Eds.), Grottammare: Atec, 27–31.

Plaza, G., Lukasik, W., Ulfig, K. (1996) Sorption of cadmium by filamentous soil fungi, *Acta Microbiol. Pol.* **45**, 193–201.

Prota, G. (1992) *Melanins and Melanogenesis*, New York: Academic Press.

Racki, W. J., Becam, A. M., Nasr, F., Herbert, C. J. (2000) Cbk1p, a protein similar to the human myotonic dystrophy kinase, is essential for normal morphogenesis in *Saccharomyces cerevisiae*, *EMBO J.* **19**, 4524–4532.

Rane, K. D., Hoover, D. G. (1993a) Production of chitosan by fungi, *Food Biotechnol.* **7**, 11–33.

Rane, K. D., Hoover, D. G. (1993b) An evaluation of alkali and acid treatments for chitosan extraction from fungi, *Process Biochem.* **28**, 115–118.

Rast, D. M., Horsch, M., Furter, R., Gooday, G. W. (1991) A complex chitinolytic system in exponentially growing mycelium of *Mucor rouxii*: properties and function, *J. Gen. Microbiol.* **137**, 2797–2810.

Rast, D. M., Merz, R. A., Jeanguenat, A., Mösinger, E. (2000) Enzymes of chitin metabolism for the design of antifungals, *Adv. Chitin Sci.* **4**, 479–505.

Rivas, L. A., Parro, V., Moreno-Paz, M., Mellado, R. P. (2000) The *Bacillus subtilis* 168 csn gene encodes a chitosanase with similar properties to a *Streptomyces* enzyme, *Microbiology U.K.* **146**, 2929–2936.

Roberts, G. A. F. (1992) *Chitin Chemistry*, Houndmills: Macmillan.

Robertus, J. D., Monzingo, A. F. (1999) The structure and action of chitinases, *EXS 87* (Chitin and Chitinases), 125–135.

Ruiz-Herrera, J., Martinez-Espinoza, A. D. (1999) Chitin biosynthesis and structural organization in vivo, *EXS 87* (Chitin and Chitinases), 39–53.

Ruiz-Herrera, J., Xoconostle-Cazares, B. (1995) Molecular and genetic control of chitin biosynthesis in fungi, *Arch. Med. Res.* **26**, 315–321.

Ruiz-Herrera, J., Mormeneo, S., Vanaclocha, P., Font-de-Mora, J., Iranzo, M., Puertes, I., Sentandreu, R. (1994) Structural organization of the components of the cell wall from *Candida albicans*, *Microbiology* **140**, 1513–1523.

Saito, J.-I., Kita, A., Higuchi, Y., Nagata, Y., Ando, A., Miki, K. (1999) Crystal structure of chitosanase from *Bacillus circulans* MH-K1 at 1.6 Å resolution and its substrate recognition mechanism, *J. Biol. Chem.* **274**, 30818–30825.

Sakurada, M., Morgavi, D. P., Ushirone, N., Komatani, K., Tomita, Y., Onodera, R. (1998) Purification and characteristics of membrane-bound chitinase of anaerobic ruminal fungus *Piromyces communis* OTS1, *Curr. Microbiol.* **37**, 60–63.

Santos, B., Snyder, M. (1997) Targeting of chitin synthase 3 to polarized growth sites in yeast requires Chs5p and Myo2p, *J. Cell Biol.* **136**, 95–110.

Santos, B., Duran, A., Valdivieso, M. H. (1997) CHS5, a gene involved in chitin synthesis and mating in *Saccharomyces cerevisiae*, *Mol. Cell. Biol.* **17**, 2485–2496.

Saxena, I. M., Brown, R. M. (1997) Identification of cellulose synthase(s) in higher plants: sequence analysis of processive β-glycosyltransferases with the common motif 'D, D, D35Q(R,Q)XRW', *Cellulose* **4**, 33–49.

Sbrana, C., Avio, L., Giovannetti, M. (1995) The occurrence of calcofluor and lectin binding polysaccharides in the outer wall of arbuscular mycorrhizal fungal spores, *Mycol. Res.* **99**, 1249–1252.

Schoffelmeer, E. A. M., Klis, F. M., Sietsma, J. H., Cornelissen, B. J. C. (1999) The cell wall of *Fusarium oxysporum*, *Fungal Genet. Biol.* **27**, 275–282.

Seki, K., Kuriyama, H., Okuda, T., Uchida, Y. (1997) Molecular cloning of the gene encoding chitosanase from *Bacillus amyloliquefaciens* UTK, *Adv. Chitin Sci.* **2**, 284–289.

Shimahara, K., Takiguchi, Y., Kobayashi, T., Uda, K., Sannan, T. (1989) Screening of Mucoraceae strains suitable for chitosan production, in: *Chitin and Chitosan* (Skjåk-Bræk, G., Anthonsen, T., Sandford, P., Eds.), London: Elsevier Applied Science, 171–178.

Shimosaka, M., Nogawa, M., Ohno, Y., Okazaki, M. (1993) Chitosanase from the plant pathogenic fungus, *Fusarium solani* f. sp. *phaseoli* – purification and some properties, *Biosci. Biotechnol. Biochem.* **57**, 231–235.

Shimosaka, M., Nogawa, M., Wang, X.-Y., Kumehara, M., Okazaki, M. (1995) Production of two chitosanases from a chitosan-assimilating bacterium, *Acinetobacter* sp. strain CHB101, *Appl. Environ. Microbiol.* **61**, 438–442.

Shimosaka, M., Kumehara, M., Zhang, X.-Y., Nogawa, M., Okazaki, M. (1996) Cloning and characterization of a chitosanase gene from the plant pathogenic fungus *Fusarium solani*, *J. Ferment. Bioeng.* **82**, 426–431.

Shimosaka, M., Fukumori, Y., Zhang, X. Y., He, N. J., Kodaira, R., Okazaki, M. (2000) Molecular cloning and characterization of a chitosanase from the chitosanolytic bacterium *Burkholderia gladioli* strain CHB101, *Appl. Microbiol. Biotechnol.* **54**, 354–360.

Sietsma, J. H., Din, A. B., Ziv, V., Sjollema, K. A., Yarden, O. (1996) The localization of chitin synthase in membranous vesicles (chitosomes) in *Neurospora crassa*, *Microbiology* **142**, 1591–1596.

Specht, C. A., Liu, Y., Robbins, P. W., Bulawa, C. E., Iartchouk, N., Winter, K. R., Riggle, P. J., Rhodes, J. C., Dodge, C. L., Culp, D. W., Borgia, P. T. (1996) The chsD and chsE genes of *Aspergillus nidulans* and their roles in chitin synthesis, *Fungal Genet. Biol.* **20**, 153–167.

Sreenivasaprasad, S., Burton, K. S., Wood, D. A. (2000) Cloning and characterisation of a chitin synthase gene cDNA from the cultivated mushroom *Agaricus bisporus* and its expression during morphogenesis, *FEMS Microbiol. Lett.* **189**, 73–77.

St. Leger, R. J., Staples, R. C., Roberts, D. W. (1993) Entomopathogenic isolates of *Metarhizium anisopliae, Beauveria bassiana*, and *Aspergillus flavus* produce multiple extracellular chitinase isozymes, *J. Invertebr. Pathol.* **61**, 81–84.

St. Leger, R. J., Joshi, L., Bidochka, M. J., Rizzo, N. W., Roberts, D. W. (1996) Characterization and ultrastructural localization of chitinases from *Metarhizium anisopliae, M. flavoviride*, and *Beauveria bassiana* during fungal invasion of host (*Manduca sexta*) cuticle, *Appl. Environ. Microbiol.* **62**, 907–912.

Su, C. H., Sun, C. S., Juan, S. W., Ho, H. O., Hu, C. H., Sheu, M. T. (1999) Development of fungal mycelia as skin substitutes: effects on wound healing and fibroblast, *Biomaterials* **20**, 61–68.

Sudoh, M., Watanabe, M., Mio, T., Nagahashi, S., Yamada-Okabe, H., Takagi, M., Arisawa, M. (1995) Isolation of canCHS1A, a variant gene of *Candida albicans* chitin synthase, *Microbiology* **141**, 2673–2679.

Sudoh, M., Yamazaki, T., Masubuchi, K., Taniguchi, M., Shimma, N., Arisawa, M., Yamada-Okabe, H. (2000) Identification of a novel inhibitor specific to the fungal chitin synthase. Inhibition of chitin synthase 1 arrests the cell growth, but inhibition of chitin synthase 1 and 2 is lethal in the pathogenic fungus *Candida albicans*, *J. Biol. Chem.* **275**, 32901–32905.

Sun, L., Adams, B., Gurnon, J., Ye, Y., van Etten, J. (1999) Characterization of two chitinase genes and one chitosanase gene encoded by chlorella virus PBCV-1, *Virology* **263**, 376–387.

Suresh, P. V., Chandrasekaran, M. (1999) Impact of process parameters on chitinase production by an alkalophilic marine *Beauveria bassiana* in solid state fermentation, *Process Biochem.* **34**, 257–267.

Takaya, N., Yamazaki, D., Horiuchi, H., Ohta, A., Takagi, M. (1998a) Intracellular chitinase gene from *Rhizopus oligosporus*: molecular cloning and characterization, *Microbiology* **144**, 2647–2654.

Takaya, N., Yamazaki, D., Horiuchi, H., Ohta, A., Takagi, M. (1998b) Cloning and characterization of a chitinase-encoding gene (chiA) from *Aspergillus nidulans*, disruption of which decreases germination frequency and hyphal growth, *Biosci. Biotechnol. Biochem.* **62**, 60–65.

Tan, S. C., Tan, T. K., Wong, S. M., Khor, E. (1996) The chitosan yield of Zygomycetes at their optimum harvesting time, *Carbohydr. Polym.* **30**, 239–242.

Tokuyasu, K., Ohnishi-Kameyama, M., Hayashi, K. (1996) Purification and characterization of extracellular chitin deacetylase from *Colletotrichum lindemuthianum*, *Biosci. Biotechnol. Biochem.* **60**, 1598–1603.

Tokuyasu, K., Kaneko, S., Hayashi, K., Mori, Y. (1999a) Production of a recombinant chitin deacetylase in the culture medium of *Escherichia coli* cells, *FEBS Lett.* **458**, 23–26.

Tokuyasu, K., Ohnishi-Kameyama, M., Hayashi, K., Mori, Y. (1999b) Cloning and expression of chitin deacetylase gene from a deuteromycete, *Colletotrichum lindemuthianum*, *J. Biosci. Bioeng.* **87**, 418–423.

Tokuyasu, K., Mitsutomi, M., Yamaguchi, I., Hayashi, K., Mori, Y. (2000) Recognition of chitooligosaccharides and their N-acetyl groups by putative subsites of chitin deacetylase from a deuteromycete, *Colletotrichum lindemuthianum*, *Biochemistry* **39**, 8837–8843.

Tremblay, H., Blanchard, J., Brzezinski, R. (2000) A common molecular signature unifies the chitosanases belonging to families 46 and 80 of glycoside hydrolases, *Can. J. Microbiol.* **46**, 952–955.

Trilla, J. A., Cos, T., Duran, A., Roncero, C. (1997) Characterization of CHS4 (CAL2), a gene of *Saccharomyces cerevisiae* involved in chitin biosynthesis and allelic to SKT5 and CSD4, *Yeast* **13**, 795–807.

Trilla, J. A., Durán, A., Roncero, C. (1999) Chs7p, a new protein involved in the control of protein export from the endoplasmic reticulum that is specifically engaged in the regulation of chitin synthesis in *Saccharomyces cerevisiae*, *J. Cell Biol.* **145**, 1153–1163.

Tronsmo, A., Hjeljord, L., Klemsdal, S. S., Vårum, K. M., Nordtveit Hjerde, R., Harman, G. E. (1996) Chitinolytic enzymes from the biocontrol agent Trichoderma harzianum, in: *Chitin Enzymology* (Muzzarelli, R. A. A., Ed.), Grottammare: Atec, Vol. 2, 235–244.

Tsigos, I., Zydowicz, N., Martinou, A., Domard, A., Bouriotis, V. (1999) Mode of action of chitin deacetylase from *Mucor rouxii* on *N*-acetylchitooligosaccharides, *Eur. J. Biochem.* **261**, 698–705.

Tsigos, I., Martinou, A., Kafetzopoulos, D., Bouriotis, V. (2000) Chitin deacetylases: new, versatile tools in biotechnology, *Trends Biotechnol.* **18**, 305–312.

Turoczi, G., Fekete, C., Kerenyi, Z., Nagy, R., Pomazi, A., Hornok, L. (1996) Biological and molecular characterization of potential biocontrol strains of *Trichoderma*, *J. Basic Microbiol.* **36**, 63–72.

Uchida, Y., Takeda, H., Ohkuma, A., Seki, K. (1997) Purification and properties of exo-β-D-glucosaminidase from *Penicillium* sp. and its applications, *Adv. Chitin Sci.* **2**, 244–249.

Ulhoa, C. J., Peberdy, J. F. (1993) Effect of carbon sources on chitobiase production by *Trichoderma harzianum*, *Mycol. Res.* **97**, 45–48.

Valdivieso, M. H., Durán, A., Roncero, C. (1999) Chitin synthases in yeast and fungi, *EXS 87* (Chitin and Chitinases), 55–69.

Vårum, K. M., Smidsrød, O. (1997) Specificity in enzymic and chemical degradation of chitosans, *Adv. Chitin Sci.* **2**, 168–175.

Vårum, K. M., Holme, H. K., Izume, M., Stokke, B. T., Smidsrød, O. (1996) Determination of enzymic hydrolysis specificity of partially N-acetylated chitosans, *Biochim. Biophys. Acta* **1291**, 5–15.

Velichkov, A., Sotirov, N. (1990) Amino acid content in four samples of chitosan-glucan complex, *Dokl. Bolg. Akad. Nauk* **43**, 69–71 [*Chem. Abstr.* **115**, 227880d].

Wang, Z., Zheng, L., Hauser, M., Becker, J., Szaniszlo, P. (1999) WdChs4p, a homolog of chitin synthase 3 in *Saccharomyces cerevisiae*, alone cannot support growth of *Wangiella* (*Exophiala*) *dermatitidis* at the temperature of infection, *Infect. Immunity* **67**, 6619–6630.

Weiss, N., Sztejnberg, A., Yarden, O. (1996) The chsA gene, encoding a class-I chitin synthase from *Ampelomyces quisqualis*, *Gene* **168**, 99–102.

Win, N. N., Pengju, G., Stevens, W. F. (2000) Deacetylation of chitin by fungal enzymes, *Adv Chitin Sci.* **4**, 55–62.

Xoconostle-Cazares, B., Leon-Ramirez, C., Ruiz-Herrera, J. (1996) Two chitin synthase genes from *Ustilago maydis*, *Microbiology* **142**, 377–387.

Xoconostle-Cazares, B., Specht, C. A., Robbins, P. W., Liu, Y. L., Leon, C., Ruiz-Herrera, J. (1997) Umchs5, a gene coding for a class IV chitin synthase in *Ustilago maydis*, *Fungal Genet. Biol.* **22**, 199–208.

Yamada, T., Hiramatsu, S., Songsri, P., Fujie, M. (1997) Alternative expression of a chitosanase gene produces two different proteins in cells infected with chlorella virus CVK2, *Virology* **230**, 361–368.

Yanai, K., Kojima, N., Takaya, N., Horiuchi, H., Ohta, A., Takagi, M. (1994) Isolation and characterization of two chitin synthase genes from *Aspergillus nidulans*, *Biosci. Biotechnol. Biochem.* 5, 1828–1835.

Yoon, H.-G., Ha, S.-C., Lim, Y.-H., Cho, H.-Y. (1998) New thermostable chitosanase from *Bacillus* sp.: purification and characterization, *J. Microbiol. Biotechnol.* **8**, 449–454.

Yoon, H. G., Kim, H. Y., Lim, Y. H., Kim, H. K., Shin, D. H., Hong, B. S., Cho, H. Y. (2000) Thermostable chitosanase from *Bacillus* sp. strain CK4: cloning and expression of the gene and characterization of the enzyme, *Appl. Environ. Microbiol.* **66**, 3727–3734.

Yoshihara, K., Kubo, T., Hosokawa, J., Yokochi, T., Nakahara, T., Higashihara, T. (1996) Screening of marine fungi rich in chitinous material, *Shikoku Kogyo Gijutsu Kenkyusho Hokoku* **27**, 29–32 [*Chem. Abstr.* **125**, 109881].

Zeilinger, S., Galhaup, C., Payer, K., Woo, S. L., Mach, R. L., Fekete, C., Lorito, M., Kubicek, C. P. (1999) Chitinase gene expression during mycoparasitic interaction of *Trichoderma harzianum* with its host, *Fungal Genet. Biol.* **26**, 131–140.

Zhang, Z., Gurr, S. J. (2000) Walking into the unknown: a 'step down' PCR-based technique leading to the direct sequence analysis of flanking genomic DNA, *Gene* **253**, 145–150.

Zhang, D., Miller, M. J. (1999) Polyoxins and nikkomycins: progress in synthetic and biological studies, *Curr. Pharm. Des.* **5**, 73–99.

Zhang, X. Y., Dai, A. L., Zhang, X. K., Kuroiwa, K., Kodaira, R., Shimosaka, M., Okazaki, M. (2000) Purification and characterization of chitosanase and exo-β-D-glucosaminidase from a Koji Mold, *Aspergillus oryzae* IAM2660, *Biosci, Biotechnol Biochem.* **64**, 1896–1902.

Zhou, J. L. (1999) Zn biosorption by *Rhizopus arrhizus* and other fungi, *Appl. Microbiol. Biotechnol.* **51**, 686–693.

Zhou, J. L., Banks, C. J. (1993) Mechanism of humic acid color removal from natural waters by fungal biomass biosorption, *Chemosphere* **27**, 607–620.

Ziman, M., Chuang, J. S., Tsung, M., Hamamoto, S., Schekman, R. (1998) Chs6p-dependent anterograde transport of Chs3p from the chitosome to the plasma membrane in *Saccharomyces cerevisiae*, *Mol. Biol. Cell* **9**, 1565–1576.

13
Chitin and Chitosan from Animal Sources

Prof. Dr. Martin G. Peter
University of Potsdam, Institute of Organic Chemistry and Structure Analysis, and Interdisciplinary Research Center for Biopolymers, Karl-Liebknecht-Str. 25, D-14476 Golm, Germany; Tel.: +49-331-977-5401; Fax: +49-331-977-5300; E-mail: peter@serv.chem.uni-potsdam.de

Biopolymers for Medical and Pharmaceutical Applications. Edited by A. Steinbüchel and R. H. Marchessault

ISBN: 3-527-31154-8

CBP	chitin-binding protein
CDA	chitin deacetylase
CP/MAS NMR	cross-polarization/magic angle spinning nuclear magnetic resonance (solid-state)
CSA	10-camphorsulfonic acid
DA	degree of acetylation
DD	degree of deacetylation
DEAc	*N*,*N*-diethylacetamide

DMAc	*N,N*-dimethylacetamide
DMSO	dimethylsulfoxide
DP	degree of polymerization
DS	degree of substitution
ELISA	enzyme-linked immunosorbent assay
ESI MS	electrospray ionization mass spectrometry
F_A	mole fraction of *N*-acetylglucosamine residues
FAB MS	fast atom bombardment mass spectrometry
FT-IR	Fourier transformation infrared spectroscopy
Gal	D-galactopyranose
Glc	D-glucopyranose
GlcN	2-amino-2-deoxy-D-glucopyranose, β-(1-4)-linked in chitin/chitosan
GlcNAc	2-acetamido-2-deoxy-D-glucopyranose, β-(1-4)-linked in chitin/chitosan
GlcNAc-MU	4-methylumbelliferyl *N*-acetylglucosaminide
GlcNAc-*p*NP	*p*-nitrophenyl *N*-acetylglucosaminide
GlcNAc-*o*NP	*o*-nitrophenyl *N*-acetylglucosaminide
GlcNase	glucosaminidase
GPC	gel-permeation chromatography
HMW	high molecular weight
HP-GPC	high-performance gel-permeation chromatography
HPLC	high-performance liquid chromatography
HP-SEC	high-performance size-exclusion chromatography
HTS	high-throughput screening
IUB	International Union of Biochemistry
LHRH	luteinizing hormone releasing hormone
LMW	low molecular weight
Man	D-mannopyranose
MALDI TOF MS	matrix-assisted laser desorption ionization time-of-flight mass spectrometry
MALLS	multiple-angle laser light-scattering
MCCh	microcrystalline chitosan
M	molecular mass (daltons)
M_n	number average of molecular mass
M_v	viscosity average molecular mass
M_w	mass average of molecular mass
MS	mass spectrometry
MU	4-methylumbelliferyl
NMR	nuclear magnetic resonance
20-OH-E	20-hydroxyecdysone
PAL	phenylalanine ammonia lyase
PD MS	plasma desorption mass spectrometry
PEG	polyethylene glycol
$\langle R_g \rangle z$	radius of gyration
SEC	size-exclusion chromatography
SLS	static light scattering

TGF	transforming growth factor
THF	tetrahydrofuran
TIM	triose phosphate isomerase
p-TosOH	*p*-toluenesulfonic acid
UDP-GlcNAc	uridine diphospho-*N*-acetylglucosamine
WGA	wheat germ agglutinin

1 Introduction

Chitin and chitosan are aminoglucopyranans composed of *N*-acetylglucosamine (GlcNAc) and glucosamine (GlcN) residues. These polysaccharides are renewable resources which are currently being explored intensively by an increasing number of academic and industrial research groups. Indeed, an impressive number of applications – especially of chitosan and its derivatives – have been suggested in the literature, and several of those have been commercialized.

The chemistry and biochemistry of chitin is reviewed in this chapter. Although chitosan is a biopolymer that occurs naturally in several fungi, its technical production is based mostly on processing of chitinous resources from animals. Hence, the chemistry and applications of chitosan are included here, while the compound's biochemistry is discussed in more detail in Chapter 5 of this volume. Due to space limitations, only a small fraction of the extensive literature (over 14,000 references during the past decade, including approximately 3500 patents) will be considered. (Only in exceptional cases is literature available in Chinese or Japanese language cited in this review.)

2 Historical Outline

Chitin is a rather "old" molecule, with chemically detectable remains having been found in fossil insects from the Oligocene period (24.7 million years ago) (Stankiewicz et al., 1997). The history of chitin research began in 1811, when the French Professor of Natural History, H. Braconnot, published a paper entitled "Sur la nature des champignons" in *Ann. Chim. Phys.* (Paris). Prof. Braconnot described an alkali-insoluble material from higher fungi which he named "fungine" (for references, see Muzzarelli, 1977a; Roberts, 1992). Twelve years later, A. Odier isolated a similar fraction from insect exoskeletons which he named chitin (Greek: χιτων = tunic, armor). In 1859, C. Rouget showed that saponification of chitin produces chitosan, whilst in 1878 G. Ledderhose found that hydrolysis of chitin yields "glycosamine" and acetic acid. F. Tiemann demonstrated in 1886 that the phenylosazone of glucosamine was identical to the phenylosazones prepared two years earlier by E. Fischer from "dextrose" (= glucose) and laevulose (= fructose). F. Hoppe-Seyler reported in 1894 on the occurrence of chitin in crabs, scorpions, and spiders, whilst in 1903, E. Fischer and H. Leuchs obtained glucosamine and mannosamine from arabinose. W. V. Haworth and coworkers finally presented proof of the absolute configuration by synthesis of D-glucosamine in 1939. The first mention of a chitinase appeared in 1929, when P. Karrer, A. Hofmann, and G.

von Francois used "Schneckensaft" (an extract from *Helix pomatia*) to show that chitin from lobsters, *Homarus vulgaris*, showed essentially the same behavior as chitin from the edible mushroom, *Boletus edulis*, both yielding *N*-acetylglucosamine (GlcNAc). The first patents on chitin, together with several applications, were filed in 1935 by G. W. Rigby, and in 1936 by Du Pont de Nemours & Co.

The first textbook on chitin appeared in 1977 (Muzzarelli, 1977). (The original book is out of print; a facsimile copy is available from Franklin Book Company, Inc., 7804 Montogomery Avenue, Eliks Park, PA 19117, USA, FAX: +1-225-635-6155.) This was followed by a comprehensive presentation of chitin chemistry (Roberts, 1992). *Chitin Handbooks* are available in Japanese (Society for the Study of Chitin and Chitosan, 1991; Japanese Society for Chitin and Chitosan, 1995) and in English (Muzzarelli and Peter, 1997). Besides a growing number of publications in scientific journals, important reference sources are available in a series of conference proceedings, most of which are referred to throughout this chapter.

3 Structure of Chitin and Chitosan

It is usually understood that chitin (Chemical Abstracts Registry (CAS) 1398–61-4) is the polymer of β-1,4-linked *N*-acetylglucosamine (2-acetamido-2-deoxy-β-D-glucopyranose, GlcNAc) whereas chitosan (CAS 9012–76-4) is the corresponding polymer of glucosamine (GlcN). However, neither chitin nor chitosan are homopolymers, as both contain varying fractions of GlcNAc and GlcN residues (Figure 1). Roberts (1997) suggested indicating the mole fraction of GlcNAc residues by the "F_A value". Thus, the homopolymer of GlcNAc is chitin (F_A 1.0) and the homopolymer of GlcN is chitosan (F_A 0.0). Other (sometimes confusing) terms are used throughout the literature, such as degree of acetylation (DA), degree of deacetylation (DD or DDA), or residual degree of acetyl groups. The polymers may be distinguished by their solubility in 1% aqueous acetic acid. Chitin, containing ca. >40% GlcNAc residues (F_A >0.4) is insoluble, whereas soluble polymers are named chitosan.

Chitin from animals is highly acetylated (F_A >0.9); the average molecular weight of chitin in insect cuticles is estimated at 1–2×10^6, corresponding to a degree of polymerization (DP) of ca. 5000–10,000 (Hackman, 1987).

Chitosan is prepared from suitable chitinous raw materials, mostly by a sequence of deproteinization, demineralization, and chemical deacetylation procedures. The molecular weight of chitosan depends on the source of the biological material, as well as on the conditions of the deacetylation process (see Section 9).

3.1 Conformation in Solution

Light scattering of low-viscosity solutions of depolymerized chitin (DP ca. 40–50) in DMAc/5% LiCl ([η] <10 dL g^{-1}) indicate that the polysaccharide forms stable aggregates of rod-shaped, semirigid chains with a

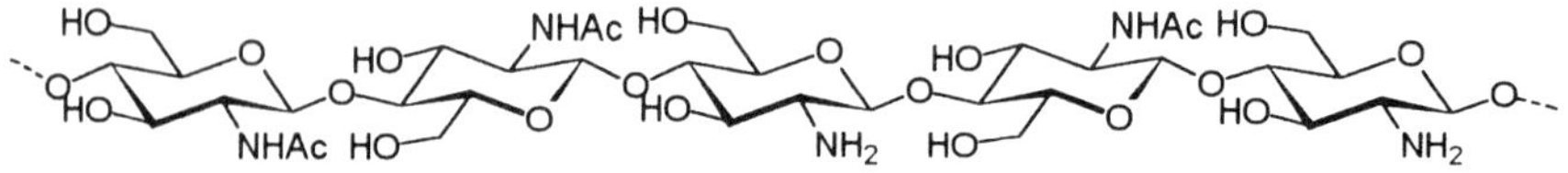

Fig. 1 Structure of chitin (F_A 0.60).

persistence length of 21.4 ± 1 nm, whereas higher-viscosity solutions contain molecularly dispersed chitin (Terbojevich et al., 1996).

It is generally accepted that the worm-like chain model applies to the conformation of chitosan, independent of F_A (Beri et al., 1993; Berth et al., 1998; Terbojevich et al., 1988). Due to the polyelectrolyte nature of chitosan, however, the data derived for conformational analysis are sensitive to M_w, polydispersity, ionic strength, polymer concentration, and the presence of hydrogen bond-breaking reagents such as urea. Thus, other models describing rod-shaped chains and increasing stiffness with increasing F_A, as well as random coil conformations have been suggested (Errington et al., 1993; Vårum et al., 1994; Tsaih and Chen, 1997, 1999; Chen and Tsaih, 2000; Signini et al., 2000).

3.2 Crystal Structures

Similar to cellulose, chitin occurs in three polymorphic forms, named α-, β-, and γ-chitin, which differ in the orientation of the polysaccharide chains. They are antiparallel in α-chitin, where intrachain hydrogen bonds are donated from C(3′)-OH to O(5), from C(6′)-OH to the acetamido carbonyl O(7), and from C(6)-OH to O(6′). Interchain hydrogen bonds exist between NH and acetamido carbonyl O(7′) and between C(6′)-OH and O(6). The unit cell dimensions of α-chitin are $a = 4.74$, $b = 10.32$, and $c = 18.86$ Å. In β-chitin, the chains are parallel whereas γ-chitin possesses two parallel chains in association with one antiparallel (for a comprehensive discussion and references, see Roberts, 1992; Ebert, 1993; Chanzy, 1997).

Crystalline chitosan exists in several polymorphs, all having an extended two-fold helical structure, but differing in packing density and water content. Intrachain hydrogen bonding occurs between C(3)-OH and O(5′). Complexes with acids or metal salts are grouped into type I, showing a 10.3 Å axial repeat, while type II has a 40 Å axial repeat (Ogawa et al., 2000; Okuyama et al., 2000).

4 Occurrence

Chitin is widely distributed in living organisms; its occurrence has been detected in fungi, algae, Protozoa, Cnidaria, Aschelmintes, Endoprocta, Bryozoa, Phoronida, Brachiopoda, Echiurida, Annelida, Mollusca, Onychophora, Arthropoda, Chaetognatha, Pognophora, and Tunicata (Muzzarelli, 1977; Roberts, 1992). Chitooligosaccharides, i.e. $(GlcNAc)_n$, $n > 2$, are synthesized in numerous organisms, including bacteria of the family Rhizobiaceae (Denarie et al., 1996), filaria (Nematoda) (Haslam et al., 1999), teleost fish (Jeuniaux et al., 1996), and even humans (Bakkers et al., 1999) (see also Section 7). *N,N′*-Diacetylchitobiose $(GlcNAc)_2$ is a normal constituent of the core structure of *N*-glycans of glycoproteins.

Chitin occurs mostly in the thermodynamically stable α-modification, whereas the metastable β-chitin is present in the shell of Inarticulata (Brachiopoda), the chaetae and gizzard cuticle of some Annelidae, the pen of *Loglio* and *Octopus* (Cephalopoda), and in the tubes of Pognophora. γ-Chitin is observed rarely, e.g. in the stalk cuticle of Inarticulata (Brachiopoda) and in the stomach cuticle of cephalopods (Muzzarelli, 1977; Roberts, 1992).

It is frequently stated that chitin is, after cellulose, the second most abundant polysaccharide on earth. Actually, chitin represents only a small fraction of the estimated 2.7×10^{11} tons of organic carbon in the biosphere, about 99% of which (i.e., 2.63×10^{11} tons) is located in plants, with 40% of that (i.e., ca. 1.1×10^{11} tons) being bound in cellulose (Ebert, 1993). The annual regeneration of cellulose by photosynthesis is estimated at 1.3×10^9 tons (an "average" tree synthesizes about 14 g of cellulose per day) (Falbe and Regitz, 1989). The synthesis of chitin by Crustacea (including the suborders Entomostraca and Malacostraca) in the hydrosphere alone is estimated at ca. 2.3×10^9 tons per year (Jeuniaux et al., 1993; Cauchie, 1997), some authors even citing a production of 10^{11} tons per annum. The chitin released into the seas by molting processes and dead chitinous organisms is called "marine snow" (for references, see Keyhani and Roseman, 1999). Cellulose is most abundant in terms of static occurrence, but chitin is most prominent in terms of metabolism dynamics, exceeding the annual regeneration of cellulose probably by roughly one order of magnitude!

The relative amounts of chitin vary considerably within the organisms and their chitinous organs or structures, ranging from traces to 80% of the organic fraction. Particularly rich sources of chitin are fungi, which contain up to 45% chitin (see Chapter 5, this volume), and arthropods. The organic fraction of insect cuticle contains 20–60% chitin, while that of decalcified crustacean cuticle contains up to 80% (Hackman, 1987).

5 Physiological Function

Chitin is often compared with cellulose: both are extracellular polysaccharides forming fibrous elements in biological composite materials. Thus, the most prominent role of chitin is apparent in the construction of the insect and crustacean exoskeleton which, depending on the degree of sclerotization (or mineralization, respectively), provides mechanical rigidity for the maintenance of morphology. There is a fundamental difference however in the function of extracellular structures in plants and animals in that the exoskeleton of the invertebrates also serves as attachments for tendons and muscles and, of course, this is the premise for mobility.

Chitin fibers are always embedded in a matrix of proteins which presumably are covalently linked to the polysaccharide. Although many proteins have been isolated and characterized, particularly from insect cuticles (Andersen et al., 1995), the extent of cross-linking as well as the exact chemical structure of the polysaccharide–polypeptide linkages are essentially unknown. With the possible exception of the chitin synthase complex (see Section 7), neither a glycosyltransferase nor a peptidyltransferase catalyzing the formation of covalent chitin–protein bonds in arthropods has yet been described. X-ray analysis of chitin–protein complexes from the ovipositor of the fly *Megarhyssa lunator* showed that the proteins form a sheath of a 6_1 helix of subunits around the chitin microfiber (Blackwell and Weigh, 1984).

The mechanical properties of the biological structures are clearly correlated with the packing of the chitin microfibrils, forming smectic, nematic, or cholesteric mesophases (Hackman, 1987). In most cuticles of insects, plywood-like stacking of the fibers yields isotropic materials. The chitin occurs in the form of microfibrils with diameters varying from 2.5 to 25 nm, and a length of ca. 0.36 μm (for references, see Vincent, 1990). The stiffness of the chitin fibers is estimated at 70–90 GPa, and that of the protein matrix at 120 MPa. Anisotropic materials such as the tendon of locusts show a parallel fiber orientation. The value of Young's modulus is 11 GPa in the longitudinal direction, and 0.15 GPa in the transverse direction. A helicoidal packing of chitin fibrils around the tracheae and other discontinuities in the cuticle yields materials that are isotropic in the plane (Vincent, 1990).

In addition to fiber orientation, the nature of the protein also determines the functional properties. Thus, the wing hinge of the locust and the intersegmental membranes of the integument of insects contain the rubber-like protein resilin (for a classic text on structure–function relationships in insects, see Chapman, 1971; for some recent articles on resilin-containing structures in insects, see Frazier et al., 1999; Haas et al., 2000; Neff et al., 2000).

The polysaccharide–protein complex of insects is further stabilized in soft cuticles with low molecular-weight phenolic compounds, in particular catecholamine derivatives, which are enzymatically oxidized and polymerized in sclerotized cuticles (Peter, 1993; Andersen et al., 1996). Evidence for the covalent coupling of polyphenols to chitin has been obtained by solid-state NMR spectroscopy (Kramer et al., 1995) and ESI mass spectrometry (Kerwin et al., 1999). In crustaceans, hardening of the exoskeleton occurs by mineralization, i.e., by the incorporation predominantly of calcium carbonate.

Besides occurring in the exoskeleton of arthropods, chitin in combination with proteins and further components serves as a fibrous element in many other biological structures, such as in the peritrophic membrane that forms the intestinal lining in insects (Tellam and Eisemann, 2000). The ovipositor of many insects and the mandibles of herbivores are examples of particularly hard structures, being reinforced by up to 10% of their dry weight with zinc or manganese ions (Vincent, 1990). The eggshells of nematodes and the cysts of the brine shrimp, *Artemia salina*, are protected against dehydration and mechanical damage by a chitin-containing envelope.

6 Detection of Chitin in Animals and Analysis of Chitin and Chitosan

Methods used to detect chitin in samples from animals are outlined below. Additional procedures are available for the analysis of chitin and chitosan in fungi (see Chapter 5, this volume). General methods of determining structural parameters and macromolecular properties of chitin are described in this section.

6.1 Detection of Chitin in Biological Samples

Earlier gravimetric methods as well as the determination of chitin by ion-exchange chromatography, using instrumentation for amino acid analysis, are not considered here.

Chitin can be detected with high sensitivity (nmol equivalents of GlcNAc or GlcN) by using an enzyme-linked immunosorbent assay (ELISA) which is based on the immobilization of chitin, chitosan or oligosaccharides on the surface of microtiter plates. Quantification is achieved either by measuring the binding of a polyclonal chitin antibody by means of IgG-peroxidase, or by competitive displacement of a known amount of antibody with the analytical sample (Buss et al., 1996). A similar procedure was developed for the quantification of oligomers of GlcNAc and GlcN, using a polyclonal oligosaccharide antibody (Kim et al., 2000).

Degradation of chitin by chitinase yields oligomers, mainly $(GlcNAc)_2$ which are further cleaved to GlcNAc by *N*-acetylhexosaminidase (see Section 8), and determination of chitin is based on the quantification of GlcNAc by chromatographic or colorimetric methods. Digestion of crude biological samples with chitinase yields GlcNAc from the fraction of the polysaccharide that is accessible to the enzyme ("free chitin"), whereas digestion of purified chitin from the same specimen indicates the total chitin. The difference is then calculated as "bound chitin" (Jeuniaux and Voss-Foucard, 1997).

The lectin wheatgerm agglutinin (WGA) possesses a high-affinity binding site for oligomers of GlcNAc (DP > 3) and chitin. Application of WGA-gold labeling in combination with electron microscopy is a powerful tool for the ultrastructural localization of chitin in various organisms (Neuhaus et al., 1997; Peters and Latka, 1997), including fungi (see Chapter 5, this volume). Other methods which are mostly used to detect chitin in fungi include staining by means of WGA-fluorescein or Calcofluor White.

6.2
Determination of F_A

Many methods for the determination of F_A of chitin and chitosan are described in the literature, including infrared (IR), CP/MAS ^{13}C-NMR, first-derivative-UV, and CD spectroscopy, potentiometric and dye adsorption titration, pyrolysis-gas chromatography, quantification of acetic acid in hydrolyzates by liquid chromatography, thermal and elemental analysis (Roberts, 1992; Muzzarelli et al., 1997b). Only the most commonly applied methods are considered here. It should be mentioned that each method has its advantages or disadvantages, depending on F_A, but calibration with an independent method is recommended.

6.2.1
IR Spectroscopy

The variables, i.e., the intensities of the amide absorptions at 1660 and 1630 cm^{-1} ($\nu_{C=O}$, amide I, two types of H-bonds) and at 1550 (δ_{NH}; amide II) cm^{-1} relative to the invariables, i.e., absorption at 3450 (ν_{OH}), 2950–2880 (ν_{CH}), or 1150 and 1020–1100 ($\nu_{C\text{-}O}$, asymm. and symm.) cm^{-1} (KBr) are calculated by various methods, differing in selection of the absorption band and definition of the baseline of the IR spectrum (Roberts, 1992; Shigemasa et al., 1996a,b; Muzzarelli et al., 1997b; Ratajska et al., 1997; Struszczyk et al., 1997; Duarte et al., 2000; Struszczyk et al., 2000; Brugnerotto et al., 2001). In the author's laboratory, the intensities of the amide I band and ν_{OH} are used for calculation of F_A from A_{1650}/A_{3450} (Figure 2) (Struszczyk et al., 1997).

The absorption bands selected may be obscured by ν_{OH} at ca. 3480, ν_{NH} at 3269, δ_{OH} at ca. 1640, and, particularly in low F_A samples, by NH_2 at 1597 cm^{-1}. Absorptions in the fingerprint regions are very sensitive to differences in structure (Duarte et al., 2000).

6.2.2
NMR Spectroscopy

Solid-state NMR is the method of choice for analysis of insoluble samples, usually requiring amounts of about 200–300 mg. Soluble chitosans of low to medium DP ($M \leq 25{,}000$ Daltons) may also be analyzed in aqueous systems using D_2O as the solvent.

Solid-state ^{13}C-NMR

CP/MAS ^{13}C-NMR (natural ^{13}C abundance) is useful to determine the degree of acetylation of chitin of $F_A > 0.05$. It is also suitable for the analysis of chitin in highly complex biological matrices such as sclerotized insect cuticle which contains also proteins and polyphenols (Grün et al., 1984; Kramer et al., 1995; Zhang et al., 2000). The principles and application of CP/MAS ^{13}C-NMR for chitin analysis are explained by Ebert and Fink (1997). Representative spectra are depicted in Figure 3 and Table 1.

Besides determination of F_A, high-resolution CP/MAS ^{13}C-NMR is used for analysis

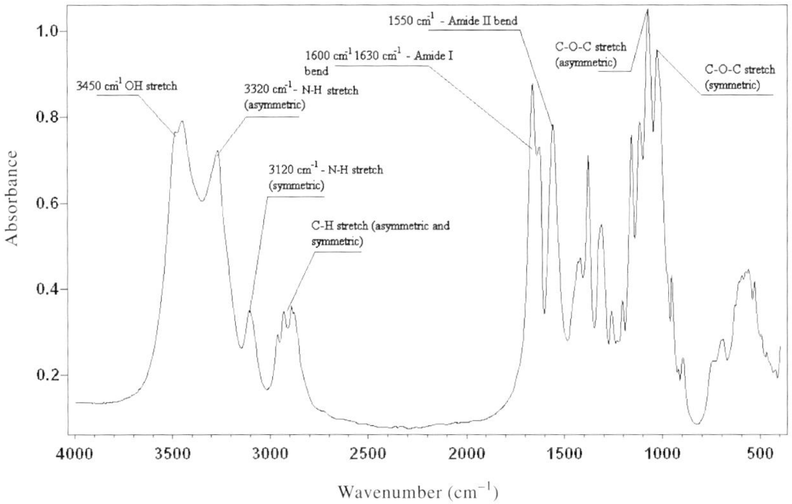

Fig. 2 FT-IR spectrum (KBr) of chitin from *Pandalus borealis* (Struszczyk et al., 1997).

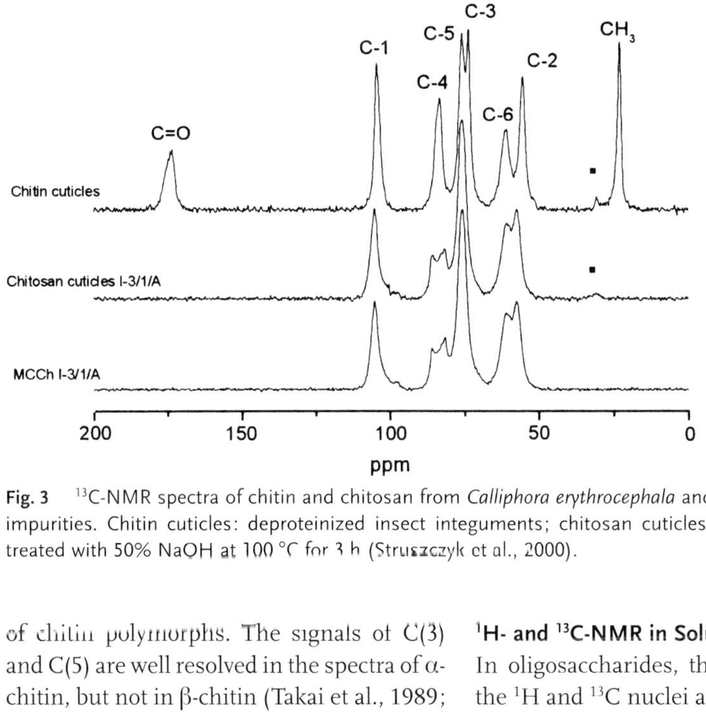

Fig. 3 ^{13}C-NMR spectra of chitin and chitosan from *Calliphora erythrocephala* and of MCCh. ■: Signals of impurities. Chitin cuticles: deproteinized insect integuments; chitosan cuticles I-3/1/A: chitin cuticles treated with 50% NaOH at 100 °C for 3 h (Struszczyk et al., 2000).

of chitin polymorphs. The signals of C(3) and C(5) are well resolved in the spectra of α-chitin, but not in β-chitin (Takai et al., 1989; Vincendon et al., 1989). The duplication of the C(4) signal observed in CP/MAS ^{13}C-NMR spectra of chitosan is explained by the presence of type I and II polymorphs (Takai et al., 1989).

^{1}H- and ^{13}C-NMR in Solution

In oligosaccharides, the chemical shifts of the ^{1}H and ^{13}C nuclei are influenced significantly by the next-neighbor sugar residue. High-resolution ^{1}H- and ^{13}C-NMR spectroscopy of partially acetylated chitosan ($F_A < 0.7$) yields statistical information about the occurrence of GlcNAc and GlcN sequen-

Tab. 1 ^{13}C Chemical shifts of solid chitin and chitosan. (External standard: adamantane or benzene; for references, see Ebert and Fink, 1997.)

	C=O	C-1	C-4	C-5	C-3	C-6	C-2	CH_3
α-Chitin								
Crab shell	173.7	104.5	83.2	76.0	73.6	61.0	55.4	23.1
Shrimp shell	173.7	104.4	83.3	75.8	73.6	61.4	55.5	23.1
β-Chitin								
Tevnia tube (dried)	175.5	105.3	84.4	75.4	73.1	59.8	55.2	22.7
	176.4						56.0	
Squid pen	174.8	104.8	83.7	75.4	74.5	59.3	56.0	23.3
						61.0		24.1
Chitosan								
Crab shell		105.0	85.6	75.5		60.3	56.4	
			81.1					
Shrimp shell		105.2	86.1	75.6		61.0	57.2	
			81.2					

ces in diads and triads, as well as about the nature of the sugar at the reducing and nonreducing ends. HMW chitosan must be depolymerized partially, e.g., by acid-catalyzed limited hydrolysis or by treatment with nitrous acid (cf. Section 10) (Vårum et al., 1991a, b). The ^{1}H chemical shifts of $(GlcNAc)_2$, $(GlcNAc)_3$, and the *O*-acetates, as well as the ^{13}C shifts of the sugar carbon atoms in diads and triads of different GlcNAc and GlcN sequences were tabulated (Schanzenbach and Peter, 1997).

6.2.3 Titration Methods

Titration of chitosan is useful for analysis of low F_A samples. The pK_a of chitosan falls between 6.2 and 6.4. Potentiometry or conductometry is used for titration of chitosan hydrochloride with NaOH (Raymond et al., 1993; Koetz and Kosmella, 1997) or of chitosan acetate with perchloric acid in glacial acetic acid/anhydrous 1,4-dioxan (Bodek, 1994, 1995). The F_A is calculated from the inflexion points of the potentiometric respectively from the section indicating dissociation of the ammonium groups in the conductometric titration curve.

For polyelectrolyte titration, sodium polystyrenesulfonate is the polyanion of choice. The end-point is determined by turbidity measurements (Koetz and Kosmella, 1997) or by toluidine blue adsorption (Park et al., 1995). The latter method requires a minimum of four sequential GlcN residues (Hattori et al., 1999).

6.3 Mass Spectrometry of Chitin and Chitosan Oligosaccharides

Although various polymers are conveniently analyzed using soft ionization techniques, few reports are available on the investigation of polysaccharides (Mohr et al., 1995). Chitin and chitosan are difficult to analyze by MS, due to the insolubility of chitin, and with regard to the polycationic nature of chitosan. The highest oligomer identified in a mixture of oligosaccharides prepared from fully deacetylated chitin ($F_A < 0.02$) is $(GlcN)_{49}$ (calculated mass: 7932 $[M + Na]^{+}$) (Figure 4) (Letzel et al., 2000).

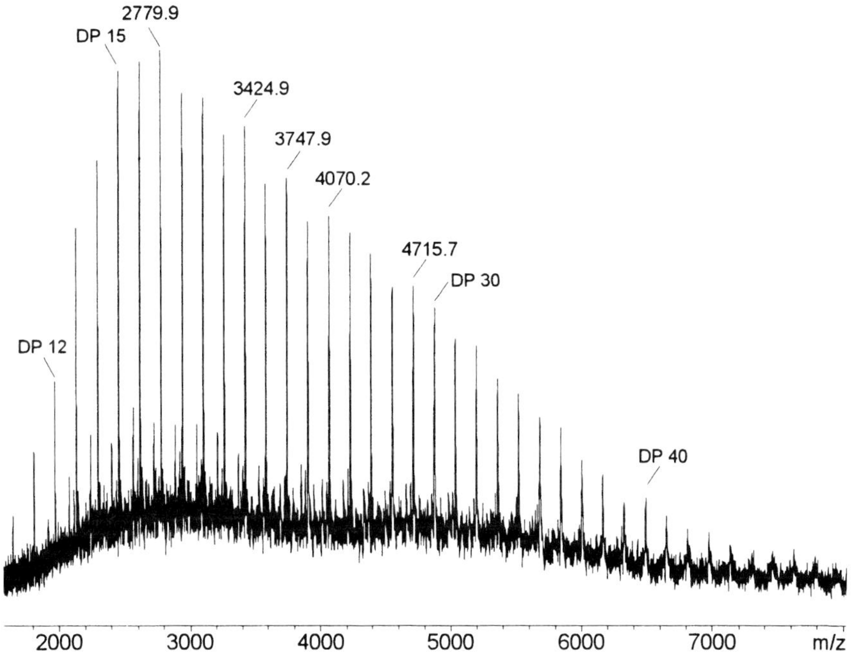

Fig. 4 MALDI-TOF MS (matrix: THAP) of GlcN oligomers obtained by HCl cleavage of chitosan ($F_A < 0.02$). The labeled peaks refer to $[M+Na]^+$ ions of the $(GlcN)_n$ series (Letzel et al., 2000).

Mass spectrometry of chitooligosaccharides is frequently used to determine the products of enzymatic degradation or deacetylation of chitin and chitosan (for a survey of MS methods, see Haebel et al., 1997). Fully acetylated GlcNAc oligomers of DP ≤ 7 were analyzed by FAB (Bosso and Domard, 1992), PD (Lopatin et al., 1995), or MALDI TOF MS (Akiyama et al., 1995a). The fragmentation of GlcN oligomers of DP ≤ 9 was investigated by positive and negative ion mode FAB MS (Bosso and Domard, 1992). MALDI TOF mass spectra of partially acetylated chitooligosaccharides containing GlcNAc and GlcN were measured for pure compounds of DP ≤ 6 (Akiyama et al., 1995b) and of mixtures containing up to DP 17 which were obtained by depolymerization of chitosan ($F_A = 0.24$) with an enzyme cocktail (Zhang et al., 1999) (see Section 8). The spectra of higher partially acetylated oligomers become increasingly complex due to the chemical inhomogeneity of the sample.

6.4
Macromolecular Characterization of Chitin and Chitosan

The most commonly used methods for determination of M_v, M_w, and M_n are viscosimetry, light scattering (SLS and MALLS), and GPC (SEC) (Terbojevich and Cosani, 1997).

6.4.1
Viscosimetry

The relationship between M_v and intrinsic viscosity $[\eta]$ is expressed by the Mark-Houwink-Kuhn-Sakurada equation, $[\eta] = KM^a$, where the viscosity parameters K and a depend on the polymer, the temperature,

Tab. 2 Viscosity parameters of chitosan (for references and discussion, see Roberts, 1992)

M_v	Solvent	K	a
113,000–492,000	0.2 M HOAc, 0.1 M NaCl, 4 M urea	8.93×10^{-4}	0.71
90,000–1,140,000	0.1 M HOAc, 0.2 M NaCl	1.81×10^{-3}	0.93
13,000–135,000	0.33 M HOAc, 0.3 M NaCl	3.41×10^{-3}	1.02

the solvent, and the salt concentration. The values found for chitin in DMAc-5% LiCl solution are $K=2.2\times10^{-4}\,\mathrm{dL\,g^{-1}}$, $a=0.88$ (Terbojevich et al., 1996). The viscosity parameters used widely for chitosan are listed in Table 2 (Roberts, 1992).

6.4.2 Chromatography

Separation of GlcN oligomers of low DP (<50) is achieved by size-exclusion chromatography (SEC) on polyacrylamide gels, such as Biogel P (Domard and Cartier, 1991). However, the method is not universally applicable to chitosan separation because higher oligomers (DP >50) cannot be separated, even by the use of larger pore-size gels. The apparent pK_a of chitosan decreases with increasing DP, and strong interactions can cause strong adsorption of the polymer onto the stationary phase (Domard and Cartier, 1991). Irreversible adsorption may also occur with lower GlcN oligomers on some lots of Biogel (M. G. Peter, unpublished observations). The supplier does not guarantee the specific applicability of these materials, and it is recommended that the suitability of each lot of Biogel be tested before setting up columns for preparative separations, even of low-DP GlcN oligomers.

HP-SEC is conveniently performed with a series of TSK-60 and/or TSK-50 columns (Beri et al., 1993; Terbojevich and Cosani, 1997). Calibration of the chromatographic system with dextrans may provide erroneous data, as the M_w of chitosan is overestimated, especially when small quantities of the polymers are injected. HP-SEC in combination with MALLS provides more reliable values of polydispersity.

Lower oligosaccharides of mixed acetylation patterns are separated by ion exchange on CM Sephadex C-25 columns (Mitsutomi et al., 1996). For chitosan, Sepharose CL-2B is suitable for determination of polydispersity in open column systems (Berth et al., 1998).

7 Biosynthesis of Chitin in Animals

All chitin-containing organisms must synthesize the polysaccharide. In addition, chitin synthase homologous genes occur in bacteria and in vertebrates, including humans. This section focuses on chitin synthesis, mostly in insects; chitin synthase-like activities in bacteria and vertebrates are also briefly reviewed.

7.1 Synthesis of Substrates for the Polymerizing Enzyme

An overview of the enzymatic transformations involved in chitin biosynthesis is given in Figure 5. Trehalose [α-D-Glc(1-1)α-D-Glc] is the primary storage form of glucose in many organisms, e.g., fungi, lichens, algae, bacteria, mosses, and arthropods. Enzymatic hydrolysis by the α-glycosidase trehalase [EC 3.2.1.28] gives glucose (Figure 5, step 1). Apparently, trehalose does not occur in vertebrates. Trehalase has only rarely been

Step #	Transformation	Enzyme
	Trehalose	
1	↓	Trehalase
	Glucose	
2	↓	Hexokinase
	Glucose-6-phosphate	
3	↓	Glucosephosphate isomerase
	Fructose-6-phosphate	
4	↓	Glutamine fructose-6-phosphate aminotransferase
	Glucosamine-6-phosphate	
5	↓	Glucosamine 6-phosphate-*N*-acetyltransferase
	N-Acetylglucosamine-6-phosphate	
6	↓	Phosphoacetylglucosamine mutase
	N-Acetylglucosamine-1-phosphate	
7	↓	Uridine-diphosphate-*N*-acetylglucosamine pyrophosphorylase
	UDP-*N*-Acetyl-glucosamine	
8	↓	Chitin synthase
	Chitin	

Fig. 5 Biochemical pathway of chitin biosynthesis.

found in higher organisms, and therefore, the enzyme appears as an attractive target for the development of pesticides (for a review on the chemistry and biological activity of trehalase inhibitors and analogues, see Berecibar et al., 1999). Steps 2–7 in Figure 5, i.e., the transformation of Glc into UDP-GlcNAc are most common biochemical reactions.

7.2 Enzymology of Chitin Synthase

The enzymology of chitin synthesis in arthropods is fairly well understood. However, the ultrastructural organization as well as the mechanism of chitin synthase [chitin-(UDP-GlcNAc)-transferase, EC 2.4.1.16] in animals are essentially unknown. Chitin synthase is located in membrane fractions, most likely as a component of a multiprotein complex, and activity is lost during attempts to purify the enzyme (for reviews on chitin synthesis in arthropods, covering also aspects of chitin degradation and regulation of synthesis, see Kramer and Koga, 1986; Spindler and Spindler-Barth, 1996; Londershausen et al., 1997).

In contrast to the large number of studies on arthropods, the investigation of chitin synthesis in lower organisms is limited to incorporation studies of radiolabeled precursors and/or histochemical evidence in the ciliated protozoon *Eufolliculina uhligi* (Schermuly et al., 1996; Mulisch and Schermuly, 1997) and the hydrothermal vent worm *Riftia pachytila* (Vestimentifera) (Shillito et al., 1993). Both organisms produce β-chitin.

7.2.1 Assays of Chitin Biosynthesis

Chitin synthesis is studied in tissue or cell homogenates, or in microsomal fractions. The incorporation of radiolabeled GlcNAc, GlcN, or UDP-GlcNAc into the alkali-resistant fraction yields the amount of chitin synthesized (Spindler-Barth, 1997). The presence of chitin is verified by acid hydrolysis and/or by digestion with chitinase, followed by thin-layer chromatography (TLC) analysis of the products. Also, ELISA or staining with WGA or Calcofluor White is used to assay for the presence of chitin (see Section 6). Chitin synthesis may be enhanced in crude homogenates when a chitinase inhibitor is added to the assay mixture (Peter and Schweikart, 1990).

7.2.2
Polymerization of GlcNAc

Assembly of the chitin chain occurs by transfer of GlcNAc from UDP-GlcNAc, most likely to the reducing end of the growing chain. The pH optimum of chitin synthesis in cell-free systems is ca. 7.0–8.0, and the K_m values for UDP-GlcNAc are in the range of 0.1 to 1 mM (Kramer and Koga, 1986). Chitin synthesis shows an apparent K_m of 4.1 mM for GlcNAc in homogenates of embryonic tissues of the tick, *Boophilus microplus* (Londershausen et al., 1993). Reports on the initialization of the polymerization reaction by transfer of GlcNAc oligomers from dolichol-P-$(GlcNAc)_n$ to "chitinoproteins", involving the Golgi apparatus, are limited to the crustacean species *Artemia salina* (Horst, 1986; Horst and Walker, 1993), and await confirmation by analogous studies in other species as well as rigorous identification of the intermediary lipid-linked chitooligosaccharides. Tunicamycin, which inhibits (UDP-GlcNAc):dolichyl-P-GlcNAc-1-P transferase, apparently interferes strongly with chitin synthesis in intact tissues, but much less in subcellular microsomal fractions (for discussion and references, see Kramer and Koga, 1986; Londershausen et al., 1997).

Reports on the occurrence of sulfated "chitinoproteins" in *Drosophila* and other insect species, yielding $(GlcNAc)_2$ and $(GlcNAc)_3$ upon chitinase digestion (Kramerov et al., 1986, 1990), should be critically re-evaluated by application of appropriate methods of structure analysis, especially as *O*-linked GlcNAc occur in many nuclear and cytoplasmic proteins (for a review, see Haltiwanger, 2000). On the other hand, chitooligosaccharides appear to be involved in hyaluronan biosynthesis (see below).

7.2.3
Translocation and Finishing of the Polymer

Translocation of the polymer may occur by exocytosis and/or extrusion, the exocytosis being mediated by GTP binding proteins. Indirect supportive evidence for exocytosis is provided by the finding that an analogue of GTP, γ-S-GTP, stimulates chitin synthesis (Londershausen et al., 1997). On the other hand, extrusion of the polysaccharide chain ensues when the chitin synthase complex is located in the plasma membrane. It is of interest to note that fungal chitin synthase genes contain a myosin motor-like sequence (see Chapter 5, this volume).

With respect to the ultimate steps of chitin synthesis, two intriguing questions remain unanswered to date:

- The mechanism of crystallization of α-chitin is not understood, though several models have been proposed. When the polymer grows into the lumen of a vesicle, the chains could be aligned in an antiparallel fashion. However, crystallization could also occur at overlapping ends, and additional processes would be required to achieve higher degrees of crystallinity.
- The mechanism accounting for the presence of a small but significant number of GlcN residues in the polymer is unknown. The most simple hypothesis would be that GlcN residues in animal chitin are artifacts, resulting from the isolation process. The occurrence of a chitin deacetylase (an earlier report on the enzymatic deacetyla-

tion of chitin in arthropods was apparently not followed up by further studies) or of a chitin-polypeptide acyltransferase in animals has not yet been described. However, as GlcN is incorporated into chitin via GlcNAc, a heteropolymer could result when the enzymes transforming GlcN into UDP-GlcNAc and catalyzing transfer of UDP-GlcNAc to chitin are not absolutely substrate-specific.

7.2.4 Inhibition of Chitin Synthesis

A large variety of chitin synthesis inhibitors have been described as potential pesticides (for reviews, see Wright and Retnakaran, 1987; Spindler et al., 1990; Cohen, 1993; Palli and Retnakaran, 1999). The most important, commercially exploited inhibitors of chitin synthesis are benzoylphenyl ureas (Wright and Retnakaran, 1987; Palli and Retnakaran, 1999). The mechanism of inhibition remains unknown, though several sites of action have been proposed (Londershausen et al., 1997). Analogues of UDP-GlcNAc, i.e., polyoxins and nikkomycins, are effective in insects (Cohen, 1987; Tellam et al., 2000), but have not found any practical application. Several compounds inhibit chitin synthesis nonspecifically by disrupting membrane organization (Londershausen et al., 1997). As mentioned above, data on the inhibition of chitin synthesis by tunicamycin are at variance. Other natural products possessing general toxicity sometimes inhibit chitin synthesis quite efficiently, e.g., the naphthoquinone plumbagin which shows IC_{50} = 10 μM in *Chiromomus tentans* cell line preparations and IC_{50} = 6 μM in tick embryonic tissue homogenates (Londershausen et al., 1993).

7.3 Genetic Basis of Chitin Synthesis

Various chitin synthase and chitin synthase-like genes have been identified in fungi (see Chapter 5, this volume), bacteria, and vertebrates (see below). In contrast, only two putative chitin synthase genes of arthropods have been cloned to date, i.e., chitin synthase encoding cDNA from the mosquito *Aedes aegypti* (Ibrahim et al., 2000) and chitin synthase-like protein, LcCS-1, from larvae of the fly *Lucilia cuprina* (Tellam et al., 2000). Chitin synthase is a single copy gene, at least in *A. aegypti*. The sequences show high homology to related genes from the nematode *Caenorhabditis elegans* and lower (but significant) similarity to yeast, *Saccharomyces cerevisiae* chitin synthases, with stronger conservation centered in the active sites of the yeast enzymes. The cDNA of *A. aegypti* is 3.5 kb in length with an open reading frame of 2.6 kb, encoding a protein of 865 amino acids with a predicted M = 99,500 daltons. The cDNA of *L. cuprina* is 5757 bp in length, coding for a protein of 1592 amino acids, and M = 180,717 Daltons. LcCS-1 contains 15 to 18 potential transmembrane segments, indicative of an integral membrane protein. Highly related genomic sequences were demonstrated in three insect orders, in one arachnid and in the nematode *C. elegans* (Tellam et al., 2000).

7.3.1 Chitin Synthase-like Genes in Bacteria and in Vertebrates

The products of chitin synthase activity in Rhizobia and in vertebrates are chitin oligosaccharides of DP 4–6, which in the bacteria are modified via a variety of reactions to yield lipochitooligosaccharides (Nod factors) (Figure 6) (for reviews, see Denarie et al., 1996; Bakkers et al., 1999).

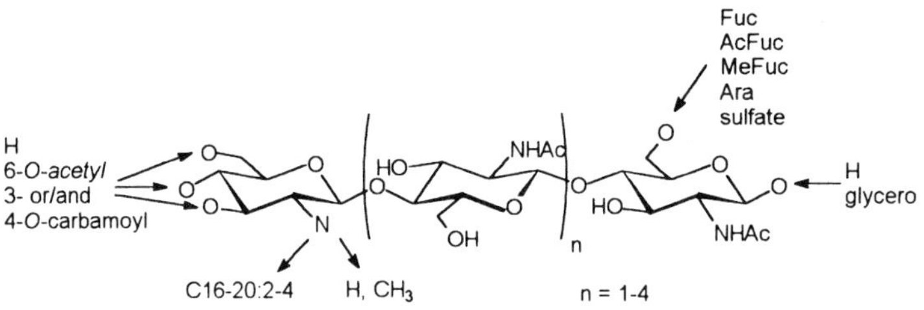

Fig. 6 Structures of *Rhizobium* Nod factors.

Nod factors stimulate nodule formation in leguminous plants, their synthesis being initiated by a flavonoid signal released by the plant. Assembly of the chitooligosaccharide is catalyzed by NodC, transferring GlcNAc from UDP-GlcNAc to the nonreducing end of free GlcNAc and the growing oligosaccharide chain (Kamst et al., 1999).

A developmentally regulated gene of *Xenopus*, DG42, which is expressed between the late midblastula and neurulation stages of embryonic development, was shown to synthesize chitooligosaccharides (Semino and Robbins, 1995). Analogous genes occur also in zebrafish and mouse (Semino et al., 1996). Interference with chitooligosaccharide synthesis or their modification by the fucosyltransferase NodZ in developing embryos of zebrafish leads to severe defects in trunk and tail development (Bakkers et al., 1997). On the other hand, in the presence of UDP-GlcNAc and UDP-GlcA, DG42 synthesizes hyaluronan and it was suggested that chitooligosaccharides act as primers for hyaluronan biosynthesis (Meyer and Kreil, 1996) (for comments, see Varki, 1996). Indeed, recombinant mouse hyaluronan synthase *has*1 incorporates UDP-GlcNAc into chitooligosaccharides (Yoshida et al., 2000). DG42 and various hyaluronan synthase genes share >70% sequence homology and significant homology with the chitin synthases NodC from Rhizobia and *chs*3 from *S. cerevisiae* (Bakkers et al., 1999; Spicer and McDonald, 1998).

7.4 Regulation of Chitin Synthesis

A wealth of information is available on metabolic, translational, and transcriptional regulation of chitin synthesis in arthropods. Apparently, regulation of chitin synthesis has not been investigated in lower chitinous organisms.

7.4.1 At the Enzymatic Level

Proteolytic activation of chitin synthesis seems to be required for efficient *in vitro* incorporation of sugar precursors into the polysaccharide (Kramer and Koga, 1986; Ludwig et al., 1991).

In contrast to fungal chitin synthase (see Chapter 5, this volume), the enzyme from an insect cell line of *Chironomus tentans* is not allosterically regulated by GlcN or GlcNAc, but is stimulated by Mg^{2+} and inhibited by UDP, UMP, and UTP (Ludwig et al., 1991).

7.4.2 At the Translational Level

In-situ hybridization studies of midgut samples of *A. aegypti* revealed that chitin synthase mRNA increases following a blood meal. Localization at the periphery of the epithelial cells facing the midgut lumen underlines the role of these cells in formation of the chitin-containing peritrophic membrane (Ibrahim et al., 2000).

7.4.3
At the Transcriptional Level

In arthropod cuticle, the activity of chitin synthesis is correlated with molting (ecdysis) and differentiation. Both processes are strictly regulated by an array of hormones, *inter alia* prothoracicotrophic hormone, juvenile hormones and ecdysteroids, the latter regulating also chitin synthesis. Hormones in general induce expression of transcription factors which act in various ways on other functions (for reviews of ecdysteroid regulation of chitin synthesis, see Marks and Ward, 1987; Palli and Retnakaran, 1999).

Molting in insects is initiated by a rise in the titer of 20-hydroxyecdysone (20-OH-E) which also induces or suppresses chitin synthesis, depending critically on hormone concentration as well as on changes in hormone titers (Spindler and Spindler-Barth, 1996). In the pupal-adult development of holometabolous insects, high concentrations of 20-OH-E inhibit chitin synthesis during differentiation of organs in imaginal discs. A decline in hormone titer induces formation of the epicuticle, which is devoid of chitin, whilst formation of the cuticle requires low basal titers of 20-OH-E. In larval molting, high ecdysterone titers induce chitin synthesis. A high permanent ecdysteroid titer also inhibits chitin synthesis in an insect cell line from *C. tentans* (Spindler-Barth, 1993).

8
Biodegradation

The enormous amounts of chitin synthesized by living organisms do not accumulate in the biosphere, because the polysaccharide is degraded by microorganisms which use chitin as an energy source. In addition, chitin degradation plays an essential role during the metamorphosis of arthropods, as well as in a variety of developmentally regulated processes in many organisms, including vertebrates.

8.1
Enzymology of Chitin Degradation

A wide variety of enzymes is involved in chitin and chitosan degradation, including endo- and exochitinases, chitosanases, chitodextrinases, chitotriosidases, chitobiosidases, chitobiases, *N*-acetylhexosaminidases, glucosaminidases, and, finally, enzymes of the terminal metabolism of GlcNAc and GlcN. This section focuses on chitinases, excluding enzymes from fungi (see Chapter 5, this volume). An overview of the enzymes involved in chitin degradation is given in Table 3. Lysozymes and enzymes of the terminal metabolism of $(GlcNAc)_2$ are mentioned briefly.

8.1.1
Chitinases: An Overview

There is a wide diversity of chitinases that differ in structure, catalytic mechanism, and substrate specificity. Chitinases, besides being of academic interest, possess a great potential for many applications, such as plant protection in agriculture, and processing in the food industry, as well as for the biotechnological production of chitooligosaccharides, which show intriguing biological activities (see Section 10). Various aspects of these enzymes have been reviewed (Beintema, 1994; Schrempf, 1996; Kramer and Muthukrishnan, 1997; Keyhani and Roseman, 1999; Koga et al., 1999; Robertus and Monzingo, 1999; Patil et al., 2000).

Nomenclature, Substrate Specificities and Cleavage Patterns of Chitinolytic Enzymes

Chitinases [EC 3.1.14.1] hydrolyze glycosidic bonds of $(GlcNAc)_n$, $n > 2$. Based on sequence similarities, chitinases are classified

Tab. 3 Overview of enzymes of chitin and chitosan degradation

Enzyme	*EC no.*	*Family*[a]	*Substrate*	*Occurrence*
Chitinase	3.2.1.14	18	Chitin	Widely
Chitinase	3.2.1.14	19	Chitin	Plants, *Streptomyces*
Chitodextrinase	3.2.1.–	18	$(GlcNAc)_n$	*Vibrio furnissii*
Lysozyme	3.2.1.17	22–25		Widely
di-*N*-Acetylchitobiase[b]	3.2.1.52	20	$(GlcNAc)_2$	Widely
β-*N*-Acetylglucosaminidase (exo)	3.2.1.52	20	$(GlcNAc)_n$	Widely
Chitin deacetylase	3.5.1.41	–	Chitin	Fungi
Chitosanase	3.2.1.132	46	Chitosan	Bacteria, fungi
β-Glucosaminidase	–	–	4′-terminal GlcN	Fungi
GlcNAc-6-P-deacetylase	3.5.1.25	–		Bacteria, viruses

[a] For classification based on sequence similarities of carbohydrate active enzymes, see Henrissat (1999) and a URL (Coutinho and Henrissat, 1999) which also provides links to additional relevant databases. [b] Former EC 3.2.1.30 was deleted and chitobiase is now hexosaminidase, EC 3.2.1.52.

as family 18 or family 19 glycosidases (Coutinho and Henrissat, 1999; Henrissat, 1999). Family 18 plant chitinases are further divided into classes III and V, and family 19 chitinases are grouped into classes I, II, IV, VI enzymes, according to domain structures and cellular location (for reviews, see Terwisscha van Scheltinga, 1997; Patil et al., 2000). A revised nomenclature dividing chitinase genes into four families has been suggested (Neuhaus et al., 1996). In addition, chitinases are assigned as endo- or exo-enzymes, as chitobiosidases, chitotriosidases, chitodextrinases, acidic and basic enzymes, or by molecular mass. In fact, the situation is becoming increasingly complex (if not confusing) as an increasing number of genes and enzymes has become available, while standard rules for nomenclature are lacking.

The substrate specificities of chitinases differ considerably, as illustrated by a few recent examples in Table 4. In most cases, the smallest substrate cleaved is $(GlcNAc)_3$ and the end-products of chitin hydrolysis are GlcNAc and $(GlcNAc)_2$. *Plasmodium* chitinases do not cleave $(GlcNAc)_3$ (Vinetz et al., 1999, 2000), and the smallest substrate cleaved by the plant chitinase hevamine is $(GlcNAc)_5$ (Terwisscha van Scheltinga et al., 1995; Bokma et al., 2000).

Chitodextrinase is different from chitinase, as it cleaves chitooligosacharides but not $(GlcNAc)_2$ or chitin (Bassler et al., 1991).

The substrate-binding characteristics of chitinases and cleavage patterns of partially acetylated chitosans are determined by NMR spectroscopy (Stokke et al., 1995; Vårum and Smidsrød, 1997) or by enzymatic sequencing of the products (Mitsutomi et al., 1996).

Chitinase Assays

A great variety of assays for chitinase activity has been described in the literature, using different experimental set-ups and substrates (for a description of the most common methods, see Spindler, 1997). Selection of the appropriate substrate is crucial when discrimination between chitinase, chitodextrinase, and *N*-acetylglucosaminidase is desired, and additional investigations may be necessary when the function of a particular enzyme is investigated. Different kinetics, pH and temperature optima for oligosaccharide versus polymeric substrates have been reported for various chitinases (Brurberg et al., 1996; for additional references, see Keyhani and Roseman, 1999):

Tab. 4 Substrates and cleavage patterns of chitinases

Organism	*Substrate(s)*	*Results*	*References*
Human			
Homo sapiens, granulocyte-rich homogenates	[^{3}H] chitin	Release of 7.2 nmol GlcNAc min^{-1} mg $protein^{-1}$	Escott and Adams (1995)
Insects			
Bombyx mori, larvae 65 and 88 kDa chitinases	$(GlcNAc)_3$ →	GlcNAc plus $(GlcNAc)_2$	Koga et al. (1997)
	$(GlcNAc)_4$ →	2 $(GlcNAc)_2$	
	$(GlcNAc)_5$ →	$(GlcNAc)_2$ plus $(GlcNAc)_3$	
	$(GlcNAc)_6$ →	$(GlcNAc)_2$ plus $(GlcNAc)_4$ (65%), and 2 $(GlcNAc)_3$ (35%)	
Molluscs			
Todarodes pacificus (squid)	glycol chitin		
Endochitinase	$(GlcNAc)_4$ →	2 $(GlcNAc)_2$	Matsumiya and Mochizuki (1997)
	$(GlcNAc)_5$ →	$(GlcNAc)_2$ plus $(GlcNAc)_3$	
	$(GlcNAc)_6$ →	$(GlcNAc)_2$ plus $(GlcNAc)_4$ (84%) and 2 $(GlcNAc)_3$ (16%)	
Plants			
Dioscorea opposita (yam) chitinase A	glycol chitin; $(GlcNAc)_n$, $n=2-6$	Endo/random, released the monosaccharide from all hydrolyzed oligosaccharides; prefers oligosaccharides with longer chain lengths	Arakane and Koga (1999)
Rehmannia glutinosa, leaves: chitinase P2	$(GlcNAc)_{4\text{-}6}$ chitin	$(GlcNAc)_2$	Lee et al. (1999a)
Hordeum vulgare	$(GlcNAc)_6$	2 $(GlcNAc)_3$, and $(GlcNAc)_4$ plus $(GlcNAc)_2$; $(GlcNAc)_3$-MU: K_m 33 μM, k_{cat} 0.33 min^{-1}; $(GlcNAc)_4$: K_m 3 μM, k_{cat} 35 min^{-1}; K_d for $(GlcNAc)_n$: n = 2: 43, n = 3: 19, n = 4: 6 μM; Bi-bi kinetics, with $(GlcNAc)_4$ and H_2O as substrates and $(GlcNAc)_2$ as product	Hollis et al. (1997)
Bacteria			
Serratia marcescens ChiA; endo/exochitinase	$(GlcNAc)_n$; chitin	GlcNAc plus $(GlcNAc)_2$; shows higher specific activity towards chitin than ChiB	Brurberg et al. (1996)
	$(GlcNAc)_3$-MU →	GlcNAc plus $(GlcNAc)_2$ plus MU	
S. marcescens ChiB	$(GlcNAc)_n$; chitin	$(GlcNAc)_2$ plus GlcNAc; synergistic effects upon combining ChiA and ChiB	Brurberg et al. (1996)
Endo/exochitinase	$(GlcNAc)_3$-MU →	GlcNAc-MU plus $(GlcNAc)_2$	
Streptomyces griseus HUT 6037, 49 kDa chitinase	chitosan F_A 0.46	$(GlcNAc/GlcN)_n$, always GlcNAc at reducing end; partially *N*-acetylated chitosan is hydrolyzed more easily than colloidal chitin	Tanabe et al. (2000)

Tab. 4 (cont.)

Organism	*Substrate(s)*	*Results*	*References*
Protozoa			
Plasmodium falciparum PfCHT1; endochitinase	glycol chitin		
	$(GlcNAc)_3$	Is not hydrolyzed	Vinetz et al. (1999)
	$(GlcNAc)_4$ →	2 $(GlcNAc)_2$	
	$(GlcNAc)_5$ →	$(GlcNAc)_2$ plus $(GlcNAc)_3$	
	$(GlcNAc)_6$ →	$(GlcNAc)_3$ (major), and $(GlcNAc)_2$ plus $(GlcNAc)_4$ (minor)	
Plasmodium gallinaceum PgCHT1; endochitinase		Similar to PfCHT1	Vinetz et al. (2000)

- Radiolabeled colloidal chitin is prepared by acetylation of chitosan, preferentially with [^{3}H]acetic anhydride. Dye-modified chitin derivatives (Chitin Red or Chitin Blue), suitable for photometric assays, are available commercially. In a recently described plate assay, microorganisms are grown on plates prepared from a suspension of colloidal chitin dyed with Remazol Brilliant Red in agar. The presence of chitinase is detected through observation of a clearing zone (Draborg et al., 1997). Other frequently used polymer substrates are 6-*O*-hydroxyethylchitin or 6-*O*-acetoxyethylchitin. Differences in the hydrolysis rates were observed, depending on the degree of substitution and source of the chitin used for derivatization (Koga and Kramer, 1983; Inui et al., 1996).
- The use of chromogenic or fluorogenic oligosaccharide derivatives, i.e., $(GlcNAc)_n$-*p*NP, $(GlcNAc)_n$-*o*NP, or $(GlcNAc)_n$-MU, respectively, is probably the most widely used assay for quantification of hydrolytic activities of enzymes specified by EC 3.2.1.14, EC 3.2.1.52, and EC 3.2.1.17. Advantages are high sensitivity, especially with fluorogenic 4-methylumbelliferyl (MU) derivatives, the simple protocol, and suitability for high-throughput screening (HTS) on microtiter plates. The assay is also useful to analyze cleavage patterns of chitinases and lysozymes, though care must be taken to differentiate chitinases with respect to substrate specificity and function. Uncritical use of these substrate my lead to invalid conclusions and contribute further to the confusion about the nomenclature of chitinases (Keyhani and Roseman, 1999).
- HPLC assays are used to detect $(GlcNAc)_n$ of DP ≤ 7. Amide-80 columns allow the analysis of substrate cleavage patterns as well as the anomeric ratios of free oligosaccharides (Terwisscha van Scheltinga et al., 1995; Koga et al., 1998). Partially acetylated oligomers are not eluted from the column. Fully acetylated oligomers – but not anomers – may also be separated on amino phases (Mitsutomi et al., 1996). Reductive amination with *p*-aminoethylbenzoate of an oligosaccharide mixture followed by HPLC on a C_{18} column was used to determine chitinase kinetics. This method is not suitable for analysis of anomers, and the derivatives of $(GlcNAc)_4$ and $(GlcNAc)_5$ are not separated (Bokma et al., 2000).

Tab. 5 Properties of chitinases

Organism	*M [kDa]*	*pI*	*pH opt.*	*Temperature optimum [°C]*	*Comments*	*References*
Human						
Homo sapiens, granulocyte-rich homogenates	48 and 56				Partial inhibition by allosamidin at 9 μM; low levels in serum	Escott and Adams (1995)
Homo sapiens, plasma, spleen	39 and 50	8.0; 7.2			Both enzymes have identical N-terminal sequence	Renkema et al. (1995)
Insects						
Bombyx mori, larvae	65 and 88		5.5 6.5		pH optima 5.5 and 6.5 with $(GlcNAc)_5$ and pH 4–10 with glycolchitin; substrate inhibition with small $(GlcNAc)_n$; 88 kDa chitinase is four-fold times more active with glycol chitin than 65 kDa enzyme	Koga et al. (1997)
Molluscs						
Todarodes pacificus (squid), liver	38	8.3	1.5 and 8.5	50		Matsumiya and Mochizuki (1997)
Plants						
Dioscorea opposita (yam) chitinase A	28	3.6	4.0	60	Inhibited 76% by 55 μM allosamidin	Arakane and Koga (1999)
Rehmannia glutinosa, leaves, chitinase P2	28.6	8.46	5.0	60	Basic exochitinase	Lee et al. (1999a)
Hordeum vulgare endochitinase					pK_a 3.9 (Glu89) and 6.9 (Glu67)	Hollis et al. (1997)
Bacteria						
Serratia marcescens ChiA	58.5		5.0–6.0	50–60		Brurberg et al. (1996)
Serratia marcescens ChiB	55.4		5.0–6.0	50–60		Brurberg et al. (1995, 1996)
Streptomyces erythraeus	30.9				290 AA residues, two disulfide bridges	Kamei et al. (1989)
Streptomyces griseus HUT 6037	49	7.3			N-terminal amino acid sequence fibronectin type III-like sequence	Tanabe et al. (2000)
Streptomyces thermoviolaceus OPC-520	30	3.8	4.0	60	Also 25 and 40 kDa chitinases found	Tsujibo et al. (2000a)
Arthrobacter sp. NHB-10	30	6.8	5.0	45	Internal sequences AGPQLLTGYY and IGGVMT identified	Okazaki et al. (1999)

Tab. 5 (cont.)

Organism	M [kDa]	pI	pH opt.	Temperature optimum [°C]	Comments	References
Protozoa						
Plasmodium falciparum PfCHT1	39		5.0		Inhibited by allosamidin, IC_{50} 0.04 μM (pH > 6.0)	Vinetz et al. (1999)
Plasmodium gallinaceum PgCHT1	60		5.0		Inhibited by allosamidin, IC_{50} 12 μM (pH > 6.0)	Vinetz et al. (2000)

8.1.2
Occurrence and Functions of Chitinases

The CAZy database (Coutinho and Henrissat, 1999) contains (at the time of writing) sequences or fragments with 247 entries for family 18 chitinases, and 220 entries for family 19 chitinases. In addition, numerous related sequences of unknown or undetermined enzymatic activities, including also lectins, are listed. Family 18 chitinases are widely distributed in mammals, arthropods, plants, fungi, bacteria, and viruses. The family 19 enzymes occur typically in plants, but have also been observed in bacteria of the genus *Streptomyces* (Watanabe et al., 1999; Tsujibo et al., 2000c; for reviews, see Koga et al., 1999; Robertus and Monzingo, 1999). Some recent data on the occurrence and properties of chitinases are listed in Table 5.

Chitinase activity has also been discovered in human serum (Overdijk and Van Steijn, 1994), and the enzyme as well as its production by cloning and use as an antifungal has been patented (Aerts, 1996). Guinea pigs infected with *Aspergillus fumigatus* show elevated levels of the enzyme, and it is thought that chitinase is part of the mammalian defense against fungi (Overdijk et al., 1996, 1999). However, in ca. 6% of Caucasian people, the enzyme shows pseudodeficiency (i.e., very low activity without apparent symptoms) (Eiberg and Dentandt, 1997). Apparently, there are other, yet unknown functions of chitinases. A several hundred-fold increase of chitinase (chitotriosidase) was found in lipid-loaden macrophages and in plasma of symptomatic Gaucher disease patients (Renkema et al., 1995; Aerts et al., 1996). Furthermore, chitotriosidase enzyme activity is elevated up to 55-fold in extracts of atherosclerotic tissue, and it is postulated that chitinase plays a role in cell migration and tissue remodeling during atherogenesis (Renkema et al., 1998; Bleau et al., 1999; Boot et al., 1999). The binding domain of human chitinase was shown to recognize specifically chitin and several yeasts and fungi (Tjoelker et al., 2000).

Other occurrences of chitinases in vertebrates include the intestine of cod, *Gadus morhua* (Danulat, 1986), and a variety of other fishes (Matsumiya and Mochizuki, 1996), their function clearly being to digest chitinous food components.

Chitinases have important functions in the regulation of growth and development of many organisms which often contain several synergistically acting chitinases.

Insects digest part of the cuticle during molting, and therefore inhibition of chitinases is one of many the strategies for the development of new pesticides (for reviews, see Kramer and Koga, 1986; Kramer and Muthukrishnan, 1997). Two chitinase isoenzymes were purified from fifth-instar larvae of *Bombyx mori*, showing similar characteristics as the chitinases from *Manduca sexta* (Koga et al., 1997). They are most likely involved in the initial and intermediate stages of chitin degradation during molting.

In plants, the most important function of chitinases is to act as phytoalexins in defense against fungal infections. A great variety of chitinase isoenzymes are known. In yam tubers (*Dioscorea opposita*), chitinases H1 and H2 are family 18 enzymes, whereas chitinases E and F belong to family 19, as revealed by immunological properties of the isoenzymes (Arakane et al., 2000). In contrast to yam chitinase G, chitinases E, F, and H1 show high lytic activity against the plant pathogen, *Fusarium oxysporum*. Chitinases E, F and H1 show two optimum pH ranges of 3–4 and 7.5–9 towards glycolchitin.

Many bacteria metabolize chitin ultimately to CO_2, NH_4^+, and H_2O (Keyhani and Roseman, 1999). Chitinase-producing soil bacteria are active against fungi and nematodes (Cohen-Kupiec and Chet, 1998). The marine bacterium, *Vibrio furnissii*, shows chemotaxis to GlcNAc oligomers and produces at least two extracellular chitinases – one to provide the nutrient $(GlcNAc)_2$, and the other to release oligosaccharides for chemotaxis from a remote source of chitin. *Serratia marcescens* produces at least three family 18 extracellular chitinases (ChiA, ChiB, ChiC), but ChiB is also located in the periplasm (Brurberg et al., 1996; Watanabe et al., 1997). *Streptomyces* produce both, family 18 and family 19 chitinases (Schrempf et al., 1993; Saito et al., 1999; Watanabe et al., 1999).

A number of serious human and animal infectious diseases are caused by protozoan and metazoan parasites, and evidence is accumulating rapidly that chitinases play crucial roles in these infectious mechanisms. Thus, during the development of the malarial parasite, *Plasmodium*, ookinetes migrate from the intestine of the mosquito host into the hemolymph, traversing the chitinous peritrophic membrane (Shahabuddin and Kaslow, 1994; Ramasamy et al., 1997), and family 18 chitinases occur in the human malarial parasite *Plasmodium falciparum* (Vinetz et al., 1999) and in *P. gallinaceum* (Vinetz et al., 2000). Other parasites producing chitinases include *Leishmania*, *Trypanosoma*, Filaria, and Entamoeba (for references, see Wu et al., 1996; De la Vega et al., 1997; Shakarian and Dwyer, 1998; Shahabuddin and Vinetz, 1999; Shakarian and Dwyer, 2000).

8.1.3
Structures and Mechanisms of Chitinases

Crystal structures are presently known of five family 18 chitinases (Table 6). The common motif of the catalytic domain is a $(\beta\alpha)_8$-barrel (TIM-barrel; Figure 7). The solution structure of the catalytic domain of *Bacillus circulans* WL-12 chitinase A1 was also resolved by NMR spectroscopy (Ikegami et al., 2000).

In addition to the catalytic domain, many chitinases contain chitin-binding and/or FnIII type domains which are important for recognition of the polysaccharidic substrate and possibly also assist in substrate binding (see also Section 8.2). Deletion of

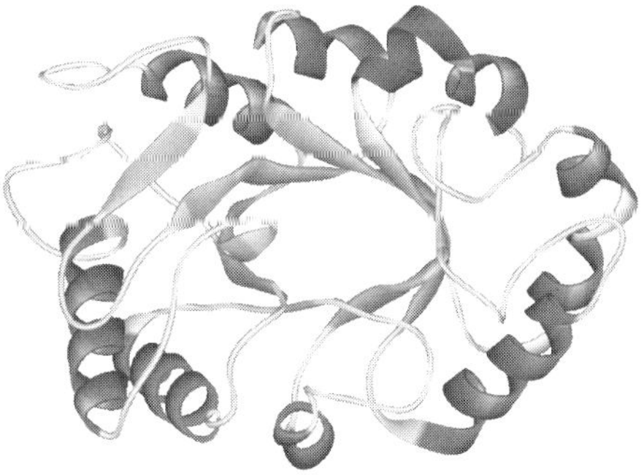

Fig. 7 Ribbon display of the family 18 chitinase hevamine (Terwisscha van Scheltinga et al., 1996). PDB code 2HVM; display by WebLab viewer.

Tab. 6 Known crystal structures of family 18 chitinases

Genus	*Enzyme*	*M [kDa]*	*Catalytic residue*	*Reference*
Plants	*Hevea brasiliensis*, hevamine	29	Glu127	Terwisscha van Scheltinga et al. (1996)
Bacteria	*Serratia marcescens*, ChiA	58.7	Glu315	Perrakis et al. (1994)
Bacteria	*Serratia marcescens*, ChiB	55.5	Glu144	van Aalten et al. (2000)
Bacteria	*Bacillus circulans* WL-12, ChiA1	74	Glu204	Matsumoto et al. (1999)
Fungi	*Coccidioides immitis*, CiX1	47.4	Glu171	Hollis et al. (2000)

the chitin-binding domain does not abolish catalytic activity towards smaller oligosaccharides, but does diminish or abolish the hydrolysis of chitin (Vorgias, 1997; Hashimoto et al., 2000; Tjoelker et al., 2000).

Serratia marcescens ChiA is a family 18 endo/exo-chitinase that possesses a deep active site groove composed of six sugar-binding subsites. The chitin-binding FnIII type domain of ChiA extends the substrate binding cleft towards the nonreducing side of chitin. Release of chitobiose must therefore occur from the reducing end of the substrate. In contrast, *S. marcescens* ChiB possesses a relatively short binding pocket, and the chitin-binding domain is directed towards the reducing end of the substrate. Therefore, chitobiose/chitotriose are released from the nonreducing end of chitin (van Aalten et al., 2000). Whereas both ChiA and ChiB from *S. marcescens* each have considerable exo-activity, the plant chitinase hevamine has a much more open active site and appears to be an endochitinase.

Family 18 chitinases hydrolyze the substrate with retention of configuration at the anomeric center, with the acetamido group providing anchimeric assistance. Binding of the substrate and protonation of the glycosidic bond leads to distortion of the sugar ring at the –1 binding site to a boat-like conformation, and to an approximately perpendicular orientation of the planes of sugar rings at the – and the + binding sites as well as a co-linear orientation of the leaving glycosidic oxygen and the incoming acetamido carbonyl oxygen, leading to an oxazolinium ion intermediate (Tews et al., 1997; Brameld and Goddard, 1998). Substrate-assisted catalysis is also evident from kinetic data. Chitinase A1 from *Bacillus circulans* WL-12 cleaves MU from $(GlcNAc)_2$-MU, GlcN-GlcNAc-MU, and GlcNAc-GlcN-MU, showing k_{cat}/K_m values of 145.3, 8.3, and 0.1 $s^{-1} M^{-1}$, respectively. The fully deacetylated disaccharide glycoside $(GlcN)_2$-MU is not hydrolyzed at all (Honda et al., 2000b).

Family 19 chitinases follow the classical mechanism of inverting glycosidases, cleaving the glycosidic bond by direct displacement of the leaving glycosidic oxygen with a water molecule. To date, two crystal structures of plant family 19 chitinases, i.e., from jack bean (Hahn et al., 2000) and from *Hordeum vulgare* (Hart et al., 1995), are known. The X-ray structure of a chitinase from barley, *H. vulgare*, shows a lysozyme type $\alpha+\beta$ fold, which is located within the structure, composed of a β-sheet, 10 α-helices, and three disulfide bonds. The catalytic site contains two glutamic acid residues where Glu67 acts as proton donor and Glu89 acts as a base coordinating the attacking water molecule (Robertus and Monzingo, 1999). Jack bean chitinase shows analogous structural features (Hahn et al., 2000).

Glycosidases are further distinguished with respect to the direction of the laterally occurring protonation as *syn* or *anti* protonators (for a review, see Heightman and Vasella, 1999).

8.1.4
Inhibition of Chitinases

The most potent, well-known chitinase inhibitors are allosamidin and congeners, including several synthetic and semi-synthetic analogues (for reviews, see Berecibar et al., 1999; Koga et al., 1999; Spindler and Spindler-Barth, 1999; Rast et al., 2000). Although not all chitinases which are inhibited by allosamidin are classified, it is thought that the compound is specific for family 18 chitinases. Remarkable species selectivity is observed, with IC_{50} values of 0.32 µM for insect (*Lucilia cuprina*), 0.69 µM for tick (*Boophilus microplus*), and 0.048 µM for nematode (*Haemonchus contortus*) chitinases (Londershausen et al., 1996). Allosamidin inhibits the family 18 plant chitinase hevamine with a rather high K_i of 3.1 µM (Bokma et al., 2000), and a family 19 plant class I chitinase from *Phaseolus vulgaris* with IC_{50} 1 µM (Londershausen et al., 1996). A number of chitooligosaccharide-derived glycosylamides were found to yield different inhibition of ChiA and ChiB from *S. marcescens* (Rottmann et al., 1999, 2000). Although the IC_{50} values are rather high (in the range of 0.2 to 0.5 mM), the results suggest that the number of sugar residues is crucial for binding in the case of ChiA, whereas the hydrophobicity of the aglycone contributes to affinity towards ChiB. A mixture of partially acetylated chitooligosaccharides strongly inhibits ChiB of *S. marcescens* (Letzel et al., 2000).

Several nonsugar chitinase inhibitors have been described recently. The fungal cyclic peptides, argifin and argadin, inhibit the insect *Lucilia cuprina* chitinase with IC_{50} 3.7 µM and 150 nM respectively at 37 °C and 0.1 µM and 3.4 nM respectively at 10 °C (Arai et al., 2000; Shiomi et al., 2000). Styloguanidins were isolated from a marine sponge, *Stylotella aurantium*, and shown to inhibit the molting of cyprid larvae of barnacles (Kato et al., 1995).

8.1.5
Lysozymes

Lysozymes [EC 3.2.1.17] hydrolyze *N*-acetylmuramic acid. Although not classified as chitinases, theses enzymes show also chitinase [EC 3.2.1.14] activity. Hydrolysis of depolymerized chitosan of DP 30 and F_A 0.68, with hen egg-white lysozyme revealed that substrate binding requires at least three GlcNAc residues in sequence, and that cleavage occurs exclusively between two GlcNAc residues (Vårum et al., 1996; Vårum and Smidsrød, 1997). Chitosan of $F_A < 0.16$ is not hydrolyzed by lysozyme (Tokura et al., 1984). Many plant chitinases also show lysozyme activity (Beintema and Terwisscha van Scheltinga, 1996); however, it was proven that, at least for hevamine, hydrolysis of *N*-acetylmuramic acid occurs with release of GlcNAc as the reducing sugar and therefore, the enzyme is not a lysozyme (Bokma et al., 1997).

Interactions of lysozyme with partially deacetylated chitosan were studied by NMR spectroscopy (Kristiansen et al., 1998), with $(GlcNAc)_6$ by mass spectrometry (He et al., 1999), and with $(GlcNAc)_n$ or MU derivatives by X-ray crystallography (Harata and Muraki, 1995; Karlsen and Hough, 1995; Kopacek et al., 1999).

8.1.6
Chitin-binding Proteins and Lectins

A large number of chitin-binding proteins (CBPs) and lectins are known that show no enzymatic activity. A characteristic feature of chitin-binding domains and of CBPs is a tryptophan-rich sequence. The CBPs from *Vibrio* species are expressed constitutively or are induced by chitin or higher chitooligosaccharides. Their function is to mediate adhesion of the bacteria to chitin and to immobilize larger chitin fragments on the cell membrane (Keyhani and Roseman, 1999). The CBPs of *Streptomyces olivaceoviridis* are small proteins which specifically bind α-chitin, but not β-chitin (Schrempf, 1999).

Chitin-specific plant lectins were reviewed recently (Beintema, 1994; Van Damme et al., 1998). Sequence homologies and X-ray structures of narbonin (Hennig et al., 1992) and concanavalin B (Hennig et al., 1995) revealed a catalytically inactive $(\beta\alpha)_8$ barrel where a glutamine is in the position of the catalytic glutamic acid residue of a homologous chitinase. The three-dimensional (3D) structure of the plant lectin hevein from *Hevea brasiliensis* is also related to family 18 chitinases, as shown by NMR spectroscopy (Asensio et al., 1995, 1998; Poveda et al., 1997). A lectin related to hevein occurs in the stinging nettle, *Urtica dioica* and 3D structures of complexes with NAG oligomers were determined by X-ray crystallography (Harata and Muraki, 2000; Saul et al., 2000).

8.1.7 Transmembrane Transport and Intracellular Degradation of Chitooligosaccharides

The catabolism of chitin by bacteria which use the polysaccharide as an energy source depends ultimately on the uptake of low molecular-weight precursors. Thus, a variety of proteins mediating transport and hydrolysis of oligosaccharides as well as metabolizing GlcNAc and GlcN are required.

Chitoporins

In Gram-negative bacteria, the oligosaccharides generated from chitin by extracellular chitinases are transported via porins into the periplasm where they are further hydrolyzed by chitodextrinase and β-*N*-acetylglucosaminidase to GlcNAc and $(GlcNAc)_2$ (for a review, see Keyhani and Roseman, 1999). A specific chitoporin, M = 40 kDa, which is inducible by $(GlcNAc)_n$, n = 2–6, occurs in the outer membrane of *Vibrio furnissii*. The protein was cloned and sequenced recently (Keyhani et al., 2000).

The Final Steps

The final steps in the degradation of chitin in Gram-negative bacteria encompass transport of GlcNAc and $(GlcNAc)_2$ into the cytoplasm (Keyhani and Roseman, 1999). A periplasmic *N*-acetylglucosaminidase cleaves oligosaccharides specifically from the nonreducing terminal. Translocation of GlcNAc is driven by the phosphoenolpyruvate:glycose phosphotransferase system, yielding GlcNAc-6-P as the product, whereas $(GlcNAc)_2$ is transported unchanged or may also become phosphorylated (Park et al., 2000). Additional enzymes of yet unknown function include a chitobiase which is incapable of hydrolyzing GlcNAc-*p*NP-or GlcNAc-MU but which hydrolyzes $(GlcNAc)_2$, and a β-*N*-acetylhexosaminidase, which hydrolyzes GlcNAc-*p*NP and GlcNAc-MU, but not chitin or chitin oligosaccharides. A detailed discussion is beyond the scope of this chapter, and the reader is referred to the review by Keyhani and Roseman (1999).

The chitobiase (hexosaminidase) of *S. marcescens* is a family 20 glycosidase observing a retaining mechanism of hydrolysis *via* the oxazoline intermediate (Drouillard et al., 1997). The crystal structure of this enzyme has been resolved (Tews et al., 1996).

N-Acetyl-β-D-hexosaminidases of arthropods occur in large varieties (Kramer and Koga, 1986). Three enzymes were characterized from the brine shrimp, *Artemia salina*, showing K_m 0.16–0.72 mM for GlcNAc-*p*NP, and M = 83, 110, and 56 kDa respectively (Spindler and Funke-Höpfner, 1991). The *N*-acetyl-β-D-hexosaminidases from the liver of the prawn *Penaeus japonicus* are acidic enzymes with *p*I < 3.2, optimum pH 5.0–5.5, and optimum temperature 50 °C, M = 64 and 110 kDa, respectively, preferring shorter oligosaccharides from which GlcNAc is cleaved from the non-reducing end (Koga et al., 1996). *Alteromonas* contains an outer-membrane β-*N*-acetylglucosaminidase which was characterized by gene cloning (Tsujibo et al., 2000b). Finally, deacetylation of substrates may occur on the level of GlcNAc, GlcNAc-6-P, and/or $(GlcNAc)_2$-P (Yamano et al., 1996).

8.2 Chitinase Genes

Chitinase genes from a variety of organisms have been cloned and sequenced, as reviewed on several occasions (Graham and Sticklen, 1994; Kramer and Muthukrishnan, 1997; Neuhaus, 1999; Saito et al., 1999; Gokul et al., 2000; Patil et al., 2000), and a few examples are listed in Table 7.

Many chitinases are composed of a catalytic domain and one or more chitin binding and/or FnIII type domains (Figure 8). The catalytic domain of family 18 chitinases contains the highly conserved motif DXXDXDXE (Synstad et al., 2000). Analysis of gene structures often reveals the presence of a signal sequence of 21–23 amino acid residues.

Human chitinase (chitotriosidase) contains a C-terminal chitin-binding domain of at least 49 amino acid residues, including six cysteines accounting for three disulfide bridges which are essential for chitin binding (Tjoelker et al., 2000). The gene is located on chromosome 1q (Eiberg and Dentandt, 1997). Cloning of a cDNA encoding human macrophage chitotriosidase confirmed assignment to family 18 chitinases. Expression is strongly regulated as the mRNA occurs at a late stage of differentiation of monocytes to activated macrophages (Boot et al., 1995). Several mammalian proteins of yet unknown function, including human articular cartilage chondrocyte protein YKL-39, YKL-40, oviductal glycoprotein, and macrophage YM-1, share significant sequence homologies with $(\beta\alpha)_8$ family 18 chitinases (Hu et al., 1996; Jin et al., 1998; Renkema et al., 1998; Bleau et al., 1999).

Insect chitinase genes have been cloned in a few cases, i.e., the tobacco hornworm, *Manduca sexta* (Kramer et al., 1993; Choi et al., 1997), the silkworm *Bombyx mori* (Kim et al., 1998), and *Aedes, Anopheles,* and *Drosophila* species (De la Vega et al., 1998). Typically, these genes code for N-terminal signal sequences and for chitin-

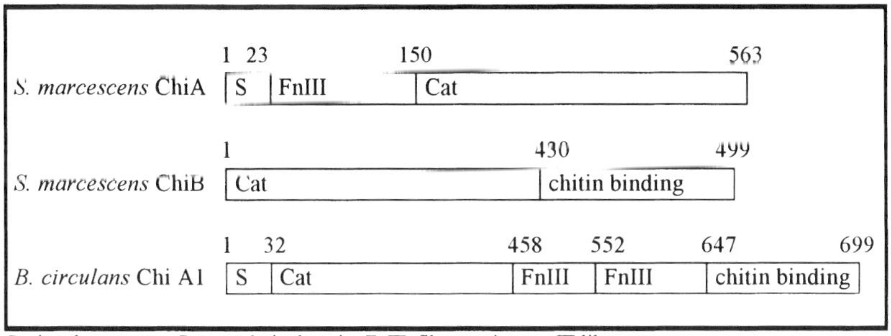

S: signal sequence, Cat: catalytic domain, FnIII: fibronectin type III like sequence

Fig. 8 Examples of typical domain organizations of family 18 chitinases. The numbers indicate the positions of amino acids in the sequence.

Tab. 7 Chitinase genes

Organism	*No. of amino acids*	*Comments*	*References*
Human			
Homo sapiens	466	N-terminal signal sequence	Aerts et al. (1996)
Insects			
Bombyx mori	565	75% homology with *M. sexta* chitinase; contains signal peptide; N-glycosylated	Kim et al. (1998)
Hyphantria cunea	553	77–80% homology with *M. sexta* chitinase; contains signal peptide; N-glycosylated	Kim et al. (1998)
Bacteria			
Serratia marcescens chitinase A	563	N-terminal signal peptide of 23 residues; domain structure: see Figure 8	Brurberg et al. (1994)
Serratia marcescens chitinase B	499	Does not contain a signal peptide; domain structure: see Figure 8	Brurberg et al. (1995)
Streptomyces thermoviolaceus; thermostable chitinase Chi30	347	Catalytic domain only; similarity with ChiA from *S. coelicolor* and ChiA from *S. lividans*; 12-bp direct repeat sequence involved probably in induction by chitin and repression by Glc	Tsujibo et al. (2000a)
Protozoa			
Plasmodium falciparum PfCHT1	378	N-terminal signal sequence; lacks proenzyme and chitin-binding domains	Vinetz et al. (1999)
Plasmodium gallinaceum PgCHT1	587	N-terminal signal sequence; C-terminal chitin binding domain; a proenzyme form was identified	Vinetz et al. (2000)

binding domains, besides the catalytic domain of typically family 18 chitinases. An exception is the chitinase from a cell line from *Chironomus tentans* which is devoid of a chitin-binding domain (Feix et al., 2000).

In plants, chitinases belong to the PR-genes, and degrade chitin as a component of fungal cell walls. Rapid evolution in plant chitinases shows that nonsynonymous substitution rates in plant class I chitinase often exceed synonymous rates in the plant genus (Bishop et al., 2000). The chitinases from bulbs of genus *Tulipa* TBC-1 and TBC-2 each consist of 275 amino acid residues with M = 30,825 and 30,863 daltons, respectively (Yamagami and Ishiguro, 1998). The enzymes share 90% and 63% identity respectively with *Gladiolus* chitinase A. The chitinase A (M = 22,391 daltons) from leaves of pokeweed (*Phytolacca americana*) does not contain a chitin-binding domain (Yamagami et al., 1998).

The gene encoding ChiB from *S. marcescens* does not contain a signal peptide, and the enzyme is exported to the periplasm without processing (Brurberg et al., 1995). Also, family 18 ChiC1 of *S. marcescens* lacks a signal sequence (Suzuki et al., 1999). Deletion of the C-terminal domain reduces the hydrolytic activity towards powdered chitin and regenerated chitin, but not towards colloidal chitin and glycol chitin, illustrating the importance of the chitin-binding domain for efficient hydrolysis of crystalline chitin. ChiC2 corresponds to the catalytic domain of ChiC1, and is probably generated by proteolytic removal of the FnIII-like and chitin-binding domains. Phylogenetic analysis shows that family 18 chitinases can be clustered in three subfamilies which have diverged at an early stage of bacterial chitinase evolution (Suzuki et al., 1999). *S. marcescens* chitinase C1 is found in one subfamily, whereas chitinases A and B of the same bacterium belong to another subfamily. Chitinase ChiA1 of *Bacillus circulans* WL-12 contains an N-terminal catalytic domain, two FnIII type domains, and a C-terminal chitin-binding domain (Hashimoto et al., 2000).

A high multiplicity of family 18 and 19 chitinase genes exists in *Streptomyces*, and proteolytic cleavage of the primary gene products contributes to the diversity of chitinases in these bacteria (Saito et al., 1999). The family 19 chitinase genes, chi35 and chi25 of *Streptomyces thermoviolaceus* OPC-520, are arranged in tandem (Tsujibo et al., 2000c). Alignment of the deduced amino acid sequences reveals that Chi35 has a N-terminal polysaccharide-binding domain, similar to xylanase but not yet observed in chitinase, containing three reiterated amino acid sequences starting from C-L-D and ending with W, and a catalytic domain. Chi25 encodes only a catalytic domain.

The chitinases of *Plasmodium* differ in domain structure. In contrast to PgCHT1, PfCHT1 lacks proenzyme and chitin-binding domains. (Vinetz et al., 1999, 2000).

8.3
Regulation of Degradation

In insects, chitinase digests chitin in non-sclerotized parts of the cuticle during molting. Induction of the integumental enzyme by 20-OH-E was observed in fifth instar larvae of the tobacco hornworm, *Manduca sexta* (Fukamizo and Kramer, 1987), in the silkworm, *Bombyx mori* (Koga et al., 1991), and in insect cell lines of *Chironomus tentans* (Spindler-Barth, 1993). Gene expression is up-regulated during the molting process, larval–pupal transformation and pupal–adult transformation.

In plants, different ethylene-inducible genes are involved in ripening of fruit, senescence of flower petals, or during pathogen infection (Deikman, 1997; Ohme-Takagi et al., 2000). The ethylene-responsive element was identified as the promoter GCC box (AGCCGCC), and transcription factors respond to extracellular signals to modulate GCC box-mediated gene expression. Transcription of class I chitinase genes is enhanced in response to subnanomolar concentrations of $(GlcNAc)_n$, $n > 5$. The elicitor signal evokes also phytoalexin production and the expression of elicitor-responsive genes is probably transmitted through a 75 kDa high-affinity binding plasma membrane protein. However, the induction of chitinase expression by *N*-acetylchitooligosaccharides may require protein phosphorylation, but not *de novo* protein synthesis (Nishizawa et al., 1999).

In bacteria, chitinase synthesis is induced by chitin and repressed by the presence of glucose (for a review, see Keyhani and Roseman, 1999). The promoter region of most of the *Streptomyces* chitinase genes contains a pair of 12 bp direct repeat sequences, which apparently is involved in both chitin induction and glucose repression (Saito et al., 1999; Tsujibo et al., 2000a). A pleiotropic regulatory gene, *reg*1, was identified in *Streptomyces lividans*, which controlled both amylase and chitinase expression (Nguyen et al., 1997).

9
Production of Chitin and Chitosan

Chemical as well as biotechnological procedures are currently under investigation for chitin and chitosan production. In practical terms, chemical processing of the waste fraction of the shellfish industry is the most important source of these polysaccharides, but biotechnological procedures for deproteinization, demineralization, and deacetylation by fermentation or enzymatic reactions are being evaluated in a number of research laboratories. To date, the investigation of additional sources such as insects (Struszczyk et al., 2000; Zhang et al., 2000) and fungi (see Chapter 5, this volume) has not provided any viable technical process for chitin and chitosan production.

9.1
Isolation of Chitin and Chitosan from Shellfish Waste

The primary sources of α-chitin and chitosan are waste fractions from crabs, prawns, shrimps, and freshwater crayfish which are harvested for human food consumption, but never for the sole purpose of producing chitin or chitosan. Chitin and chitosan are produced by chemical processes suitable for removing proteins, minerals, pigments, and varying fractions of the *N*-acetyl groups, respectively. As a rule, the yield of chitin from crustaceans is in the order of 1–2% of the raw material wet weight.

9.1.1
Resources

The waste generated globally from processing of crustaceans is estimated at 1.44×10^6 tons per annum on a dry matter basis, and the potential for chitin production is approximately 2×10^5 tons (for references, see Roberts, 1992; Shahidi and Synowiecki, 1992). Some data collected from recent publications are listed in Table 8.

The estimated resources of the Northern Atlantic deep-water shrimp, *Pandalus borealis*, in the Barents Sea and the Spitzbergen area alone are ca. 161,000 tons, 25% of which are landed in Norway. Additional resources of *Pandalus*, utilized predominantly in Canada, Greenland, and Iceland, occur in the North-West Atlantic. Shrimp shells and heads yield about 40% of the processed material, most of which is either disposed as waste (thereby causing serious pollution

Tab. 8 Potential resources for chitin and chitosan production

Country/year	*Source*	*Annual catch [tons]*	*References*
India, 1997	Various shellfish	80,000 (waste)	See Kumar (2000)
Japan, 1996	Krill, *Euphausia superba*	60,500	Krasavtsev (2000)
Mexico, 1998	Prawn, *Penaeus monodon*	80,000	Cira et al. (2000)
Norway, 1993	Shrimp, *Pandalus borealis*	40,000	Stenberg and Wachter (1996)
Panama, 1996	Krill, *Euphausia superba*	495	Krasavtsev (2000)
Poland, 1996	Krill, *Euphausia superba*	20,600	Krasavtsev (2000)
Russia, 1993	Shrimp, *Pandalus borealis*	20,000	Stenberg and Wachter (1996)
Tasmania, 1992	Lobster, *Jasus edwardsii* and *J. verreouxi*;	1907	Kow (1994)
	Crab, *Pseudocarcinus gigas*;	76	
	Squid, *Nototodarus gouldi*	14	
Thailand, 1997	Shrimp, *Penaeus monodon*	200,000 (waste)	Rao and Stevens (1997)
Ukraine, 1996	Krill, *Euphausia superba*	13,400	Krasavtsev (2000)

problems) or processed to shrimp meal to be used as a feed additive for aquacultured salmon. The production of chitosan from *Pandalus* waste has been established in some countries along the North-East Atlantic coast line, including Norway and Iceland (Stenberg and Wachter, 1996).

The primary source for chitin and chitosan production in Japan are shells from limbs of the Red Crab, *Chionoecetes japonicas*, which yield chitosan at about 20–30% of dry matter (Hirano, 1989).

The Antarctic krill, *Euphausia superba*, is caught on a large scale by fishing fleets, mainly of Japan, Panama, Poland, and Ukraine. Estimates of the resources vary between 50×10^6 and 1000×10^6 tons annually (Peter and Köhler, 1996; Krasavtsev, 2000). The actual world catch of Antarctic krill is in the order of 100,000 tons, which is $< 1\%$ of the permissible amount of $15-16 \times 10^6$ tons. Approximately 35,000 tons are harvested by Polish and Ukrainian fleets (Krasavtsev, 2000). However, krill is apparently not used for the production of chitin or chitosan on a technical scale.

9.1.2
Chemical Processes

The technology of chitin and chitosan production has been reviewed (Muzzarelli, 1977; Roberts, 1992; No and Meyers, 1997); however, only the basic principles of the process are outlined in this section.

The procedure for isolation of chitin from crustacean waste consists principally of four steps (Table 9). The concentrations of reagents, temperatures and reaction times vary widely, depending on the source of the starting material and the desired specifications of the final product, i.e., mainly M_v and F_A of chitosan. There is an inverse correlation between M_v and F_A versus reagent concentration as well as reaction temperature and reaction time during deacetylation (Wojtasz-Pajak et al., 1994; Roberts, 1997; Struszczyk et al., 1998; Ng et al., 2000; Roberts and Wood, 2000; Struszczyk et al., 2000).

A modified process consisting of an initial demineralization of crustacean shells, followed by deproteinization and a second demineralization, before deacetylation yields chitosan of M_v 900,000–1,000,000 and F_A 0.2–0.12, with an ash content $<0.1\%$ (Wachter et al., 1995). Mechanical shear does not significantly affect M_v and F_A (Rege and Block, 1999).

Insect chitin has a lower crystallinity than crustacean chitin, and relatively mild conditions for deacetylation yield chitosans of high M_v and low F_A (Struszczyk et al., 2000; Zhang et al., 2000). β-Chitin was isolated from pens of the squid, *Loglio vulgaris*, and used as starting material for the isolation of chitosan of $M = 500{,}000$ daltons (Tolaimate et al., 2000).

9.1.3
Fermentation Processes

Fermentation of prawn waste with *Lactobacillus* is performed under cofermentation

Tab. 9 Principal steps of chitin and chitosan production (for details, see No and Meyers, 1997; Roberts, 1992)

Step	*Reagent*	*Temperature [°C]*	*Time [h]*
Deproteinization	0.5–15 % NaOH	25–100 °C	0.5–72
Demineralization	2.5–8%	15–30 °C	0.5–48
Decoloration	Various organic solvents; NaOCl, H_2O_2	20–30 °C	Washing, 1.0
Deacetylation	39–60% NaOH	60–150 °C	0.5–144

with additional carbon sources, such as sugar cane, lactose and whey powder (Rao and Stevens, 1997; Cira et al., 2000). The process removes up to 90% of the protein and 80% of the calcium carbonate and calcium phosphate. The crude chitin from silage is treated with acid and alkali to complete the removal of minerals and proteins.

Acetic acid fermentation using *Lactococcus lactis* or *Clostridium formicoaceticum* and a chitinous fraction of crawfish, *Programbarum clarkii*, containing ca. 18% protein and 52% ash, affords chitin of ca. 95% purity, containing ca. 1% protein and 2% ash (Bautista et al., 2000).

Screening of proteolytic microorganisms including *Pseudomonas maltophilia*, *Bacillus subtilis*, *Streptococcus faecium*, *Pediococcus pentosaseus*, and *Aspergillus oryzae* resulted in 82% deproteinization of demineralized prawn shell when an inoculum containing four strains (*B. subtilis*, *S. faecium*, *P. pentosaseus*, and *A. oryzae*) is used (Bustos and Healy, 1994).

9.1.4 Enzymatic Deacetylation of Chitin

Chitin deacetylases (CDA) occur in bacteria and fungi, and the biochemistry of the enzyme is described in Chapter 5, this volume. In this section, approaches towards the utilization of CDA for preparation of chitosan are briefly mentioned, though a review of CDA was published recently (Tsigos et al., 2000). The extracellular CDAs of *Mucor rouxii*, *Absidia glauca* and *A. coerulea* species are effective on amorphous chitin of medium F_A, whereas data on the specificity of CDA of *Colletotrichum* are at variance (Martinou et al., 1997; Win et al., 2000). The *Mucor* enzyme deacetylates chitin of F_A 0.42 or 0.28 to yield, within 6 to 13 h, chitosans of F_A 0.02 or 0.03, respectively. Crystalline α-chitin from shrimps or crabs reacts very slowly, with $<1\%$ deacetylation occurring, while amorphous crustacean chitin undergoes ca. 10% deacetylation (Martinou et al., 1997). Although the process has been demonstrated on a laboratory scale, it is not yet technically feasible, though the optimization of enzyme production is currently under investigation (Win et al., 2000).

9.2 Preparation of Low Molecular-weight Chitin, Chitosan, and of Chitooligosaccharides

Oligomers of GlcNAc, GlcN, as well as chitooligosaccharides composed of both monosaccharides are of increasing interest because of their remarkable biological activities (see Section 10). Principally, these compounds can be obtained by enzymatic or chemical depolymerization of chitin of any F_A, or by enzymatic, chemical or chemo-enzymatic synthesis.

9.2.1 Depolymerization of Chitin and Chitosan

The traditional procedures of hydrolysis of chitin by means of hydrochloric acid, hydrogen fluoride, or sulfuric acid/acetic anhydride are discussed elsewhere (Defaye et al., 1989; Schanzenbach et al., 1997; Scheel and Thiem, 1997). Other methods for depolymerization of chitin or chitosan include γ-irradiation (Lim et al., 1998), ultrasonic irradiation (Takiguchi et al., 1999), or reaction with hydroxyl radicals (Tanioka et al., 1996). Depolymerization of chitosan by diazotization of amino groups with nitrous acid results in the formation of 2,5-anhydro-D-mannose units at the cleaved glycosidic bond (Allan and Peyron, 1997). Treatment with sodium borohydride then yields the corresponding anhydromannitol (Figure 9).

Low molecular-weight chitosan of M_v 11,000 to 436,000 and polydispersities between 2.01 and 4.16 is obtained by acid and alkali treatment of β-chitin (Shimojoh et al., 1998).

Enzymatic depolymerization of partially acetylated chitin is achieved by chitinases, chitosanases, and by lysozymes (cf. Section 8). Chitooligosaccharides of DP 3–6 are produced by degradation of chitosan (F_A 0.11) in an ultrafiltration membrane reactor using chitosanase from *Bacillus pumilus* (Jeon and Kim, 2000).

Besides lysozymes and chitinases, hemicellulase, cellulase, α-amylase, papain, tannase, and lipases can also be used for depolymerization of chitin and for the preparation of chitooligosaccharides (Pantaleone and Yalpani, 1993; Muzzarelli et al., 1994, 1995; Yalpani and Pantaleone, 1994; Muzzarelli, 1997a; Zhang et al., 1999). Generally, high enzyme concentrations and rather long reaction times are required, however.

9.2.2
Synthesis of Chitooligosaccharides

The synthesis of β-(1-6)-linked GlcN or GlcNAc as well as attachment of GlcNAc to glycans containing for example Man or Gal residues is not considered here. An impressive number of outstanding chemical investigations have been published on the synthesis of $(GlcN)_n$, n $\leq$ 12 (Kuyama and Ogawa, 1997) and on lipochitooligosaccharides (Nod factors) (for a recent reference, see Demont-Caulet et al., 1999). Polycondensation of *O*-benzyl-protected ethylthio-*N*-acetylglucosaminide affords *O*-benzyl oligomers of GlcNAc of DP $\leq$12 (Hashimoto et al., 1989).

Oligosaccharides of β-(1-4)-linked GlcN and GlcNAc are most conveniently synthesized by chemoenzymatic approaches using enzymatic transglycosylations and selective *N*-acetylation or *N*-deacetylation reactions. A summary of products and methods is provided in Table 10.

9.2.3
Abiotic Synthesis of Chitin

The synthesis of chitin of M_v 46,000 has recently been achieved using an elegant chemoenzymatic approach with the intermediate of the enzymatic hydrolysis of chitin, i.e., the oxazoline which is readily prepared from $(GlcNAc)_2$ (Figure 10) (Kobayashi et al., 1996). X-ray diffraction and ^{13}C-CP/MAS NMR revealed the product to be α-chitin (Kiyosada et al., 1998).

A hyperbranched chitin analogous dendrimer of M_w 4.5×10^5 daltons, consisting of β-(1-3/1-4)-linked GlcNAc, is obtained by camphorsulfonic acid-catalyzed polycondensation of the 1,2-oxazoline derivative of 6-*O*-tosyl-GlcNAc (Kadokawa et al., 1998).

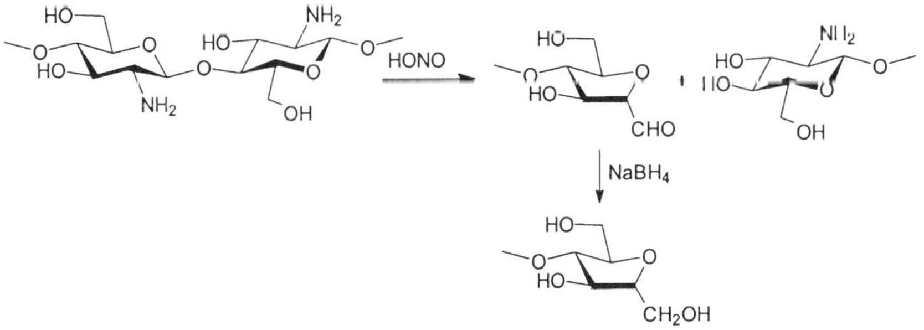

Fig. 9 Depolymerization of chitosan with nitrous acid.

Tab. 10 Chemoenzymatic preparation of chitin and chitosan oligosaccharides.

Product	*Method*	*Reference*
(GlcNAc/GlcN)$_n$, n = 4–12	Transglycosylation with lysozyme; substrates: $(GlcNAc)_3$ and/or $(GlcN\text{-}COCH_2Cl)_3$	Akiyama et al. (1995a)
$(GlcNAc)_3$- and $(GlcNAc)_4$-6-*O*-sulfate	NodST and Arylsulfotransferase IV as PAPS regenerating system	Burkart et al. (1999)
GlcNAc-GlcN	*N*-Acetylation of $(GlcN)_2$ by reverse hydrolysis with chitin deacetylase from *C. lindemuthianum*	Tokuyasu et al. (1999a)
GlcNAc-GlcNAc-GlcNAc-GlcN	*N*-Acetylation of $(GlcN)_4$ by reverse hydrolysis with chitin deacetylase from *C. lindemuthianum*	Tokuyasu et al. (2000)
$(GlcNAc)_5$; $(GlcN)(GlcNAc)_4$	Synthesis with NodC, N-terminal deacetylation with NodBC, expressed in recombinant *E. coli*	Samain et al. (1997)
GlcN-GlcNAc-MU; GlcNAc-GlcN-MU; $(GlcN)_2$-MU	*N*-Deacetylation with chitin deacetylase from *C. lindemuthianum*	Honda et al. (2000a)
$(GlcNAc)_2$	Enzymatic transfer of GlcNAc-oxazoline to GlcNAc	Kobayashi et al. (1997)
$(GlcNAc)_2$-*p*NP	Transglycosylation of GlcNAc-*p*NP with hexosaminidase	Kubisch et al. (1999)
GlcN-GlcNAc-*p*NP	Enzymatic deacetylation of $(GlcNAc)_2$-*p*NP	Tokuyasu et al. (1999b)
$(GlcNAc)_n$, n = 3–6	*N*-Acetylation of oligomers produced by chitosanase digestion of chitosan	Izume et al. (1992)
$(GlcNAc)_n$, n = 3–8	Transglycosylation of $(GlcNAc)_n$, n = 2–7 with hexosaminidase	Dvorakova et al. (2001)

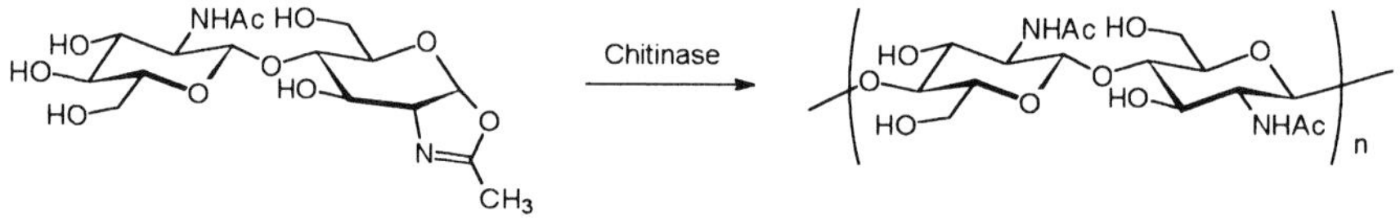

Fig. 10 Chemoenzymatic synthesis of chitin (Kobayashi et al., 1996).

9.3 Current World Market and Economics

The commercial aspects of chitin and chitosan have been detailed recently (Technical-Insights, 1998). Bulk prices for chitosan vary between approximately US$ 10 and US$ 150 per kg, depending primarily on the product specifications, though clearly the cost of highly pure, pyrogen-free chitosan suitable for high-value applications is significantly higher.

Reliable data relating to the current world production of chitin and chitosan are not available. In 1988, the total capacity of chitin production by 15 Japanese companies was reported as 2000 tons, but with an actual output of 700 tons. This was used in the production of chitosan (500 tons), glucosamine and oligosaccharides (60 tons), and for other purposes (40 tons), leaving an overproduction of 100 tons of chitin (Hirano, 1989). Since then, the demand for chitin and chitosan has grown remarkably, and production sites for both compounds have been established in many countries worldwide. Based on information received by the author from various sources, some companies are planning to increase production capacities for chitosan to several 1000 tons per annum, and so it is probably safe to estimate that the global annual production of

Tab. 11 Some companies offering chitin, chitosan, and related products (alphabetical listing)

Company	*Product*	*Contact*
BioPrawn AS, Tromsø, Norway / Henkel KGaA, Düsseldorf, Germany	Chitosan for use in cosmetics and pharmaceuticals	
Chito-Bios, Ancona, Italy	*N*-Carboxyisobutyl chitosan	Via Torresi, 8; I-60128 Ancona, Italy. Ph.: +39 (0) 71 89 39 58
ChitoGenics, Halifax, Canada	Chitin, chitosan, *N,O*-carboxymethylchitosan	FAX: +1-902-466 6889
Genis hf, Iceland	Chitosan, Oligosaccharides	http://www.genis.is
Heppe GmbH, Halle-Queis, Germany	Chitosan; R&D contracts	http://www.biolog-heppe.de
Katakura Chikkarin Co., Ltd., Tokyo, Japan	Chitin, chitosan	
Kate International	Chitin, chitosan	http://www.kateinternational.com
Kitto Life Co., Seoul, Korea	Chitosan, oligosaccharides, glucosamine	http://www.kittolife.co.kr
Korea Chitosan CO., Ltd.; Youngdeok Chitosan Co., Ltd., Korea	Chitin / chitosan fiber for medical / textile use, water soluble chitosan, derivatives of chitin / chitosan,	http://www.koreachitosan.com
Micromod Gmbh, Germany	Chitosan micro- and nanocapsules	http://www.micro-mod.de/contact.htm
Primex Ingredients ASA, Norway	Chitosan for food and cosmetics	http://www.primex.no
Pronova, Norway	Various chitosans of low to high viscosity / Mw; drug master files are expected	www.pronova.com
Seafresh Chitosan, Thailand	Chitosan	seafresh@samarth.co.th
SONAT, JSC, Russia	Chitosan, intended for use in food industry, as a food ingredient and emulgator	chitosan@col.ru

chitin and chitosan currently ranges from 3000–10,000 tons. (A recent report of >100 billion tons of chitosan being produced in Japan alone is clearly incorrect.) It appears that approximately 1% of the shellfish waste produced globally is actually utilized for chitin and chitosan production.

It has been suggested that chitin and derivatives might become competitors of cellulose or synthetic polymers, but this appears to be unrealistic for reasons of cost and availability. Cellulose is available in amounts exceeding those of chitin by several orders of magnitude. However, the unique properties of chitosan might lead to a considerable number of high-added value applications, in addition to the bulk use of cheap chitosan in some technical processes (see Section 11).

9.4 Companies Producing the Polymer

Companies which produce chitin, chitosan, and/or related products are listed in Table 11, this list having been compiled from various sources in recent years. The list is most likely incomplete, as not all producers are listed on the Internet, nor do they attend chitin conferences. However, some companies may recently have ceased production. Additional addresses of companies and research institutions are listed in a commercial reference source (Technical-Insights, 1998; the Table of Contents is available at the URL cited).

10 Properties of Chitin and Chitosan

Some of the remarkable physico-chemical and biological properties of chitin and chitosan have been referred to in previous sections of this chapter. Here, attention will be focused on those properties that are especially important with respect to the development of practical applications of the polysaccharides (see Section 11).

10.1 Physico-chemical Properties

The latest comprehensive compilation of data on the physical and chemical properties of chitin and chitosan appeared some ten years ago (Roberts, 1992; see also Muzzarelli, 1977). Some of the fundamental parameters are summarized in Table 12.

Water-soluble chitin is obtained by steeping α-chitin in concentrated aqueous NaOH, followed by the addition of crushed ice, thereby reducing the NaOH concentration to 10%; the solution is then maintained at room temperature for 3 h. The resulting water-soluble product is actually chitin of F_A 0.45–0.55. β-Chitin swells readily in water, and is also soluble in formic acid. A recently described organic solvent for chitosan consists of DMSO and *p*-toluenesulfonic acid (*p*-TosOH) or 10-camphorsulfonic acid (CSA) (Sashiwa et al., 2000).

Microcrystalline chitosan (MCCh) is prepared by limited hydrolysis of chitosan, followed by precipitation with alkali under high shear (Struszczyk, 1987). This material possesses interesting properties, for example a high water retention value (Table 13) (Struszczyk and Kivekäs, 1992).

10.2 Materials Produced from Chitin or Chitosan

A variety of materials can be prepared from chitosan, including films, semipermeable membranes, fibers, micro- and nanocapsules, as well as composites with inorganic components.

Tab. 12 Chemical properties of chitin and chitosan (Roberts, 1992)

Parameter	*α-Chitin*	*Chitosan*
M	$1-2\times10^6$ Da (estimated)	Varying widely; c.f. Section 9
Solubility	5% LiCl in DMAc[a]; hexafluoroacetone; hexafluoro-2-propanol; 12 M cold HCl[b]	Dilute aqueous organic or mineral acids
pK_a	–	6.5, decreasing with increasing DP

[a] Similar solvents are 5% LiCl in DEAc or *N*-methyl-2-pyrrolidone.
[b] Chitin is slowly depolymerized in cold concentrated HCl.

Tab. 13 Properties of microcrystalline chitosan (MCCh) (Struszczyk and Kivekäs, 1992)

Parameter	*Hydrogel*	*Dry powder*
M_w	10^4–10^6	10^4–10^6
Water retention (%)	500–5000	200–800
F_A	<0.65	<0.65
Crystallinity (%)	–	≤95
Polymer content (%)	0.1–10.0	85–95
pH	≥7	6.5–7.5

10.2.1
Films, Membranes, and Fibers

Chitosan forms films when diluted aqueous solutions of the polymer (<2%) in for example 1% acetic or lactic acid are allowed to dry, using either solvent-casting or surface-coating techniques. The tensile strengths of chitosan films fall in the range of 38 to 66 MPa, which is approximately twice the tensile strength of polyethylene (Remunan-Lopez and Bodmeier, 1996; Park et al., 1999). Trimethylsilyl chitin provides films from solutions in organic solvents (Kurita et al., 1999). Membranes formed from chitosan are less permeable to oxygen, nitrogen and carbon dioxide than cellulose acetate membranes. Many blends have been developed in order to modify the properties of films and other materials prepared from chitosan. Blend films of bacterial poly(3-hydroxybutyric acid) (PHB) with chitin or chitosan show enhanced biodegradability as compared with materials which contain a single component (Ikejima and Inoue, 2000).

Fibers are manufactured by wet spinning of appropriate solutions, applying principally the technologies used in the cellulose industry (Rathke and Hudson, 1994). Treatment of materials prepared from chitosan with acetic anhydride affords the corresponding insoluble chitin materials.

10.2.2
Polyelectrolyte Complexes

Chitosan, as a polycation, forms polyelectrolyte complexes with poylanions such as dextran sulfate (Sakiyama et al., 1999), chondroitin sulfate, and hyaluronan (Denuziere et al., 1996). Microcapsules derived from chitin as the only precursor are prepared from chitosan and 6-oxychitin (Muzzarelli, 2000a; Muzzarelli et al., 2000a) or chitosan sulfate (Holme and Perlin, 1997; Voigt et al., 1999). The mechanical strength and permeability of the capsules is optimized by cross-linking, using glutaraldehyde or other bifunctional reagents.

10.2.3
Lipid-binding Properties

Chitosan is well known for its lipid-binding properties, including fatty acids, di- and triglycerides, and steroids. The mechanism of adsorption is not fully understood, though the electrostatic interactions with carboxylates and emulsifying properties are well documented. Chitosan interacts with fatty acids, leading to flocculation or dispersion of the system (Domard and Demarger-Andre, 1994). Columns of chitosan, when percolated with olive oil, retain the oil at up to twice the weight of the chitosan, and the retained fraction is enriched in phytosteroids (Muzzarelli et al., 2000b). Stable emulsions with sunflower oil are formed in 0.2–2% solutions of chitosan (F_A 0.05–0.25) in diluted hydrochloric acid at weight ratios of ca. 1–5 parts chitosan to 1 part oil (Rodriguez et al., 2000).

10.3
Chemistry of Chitin and Chitosan

Chitosan possesses unique chemical functionalities, as the biopolymer contains primary and secondary hydroxy groups as well as primary amino groups of comparatively low pK_a. Thus, numerous derivatives and

Tab. 14 Derivatives of chitin and chitosan

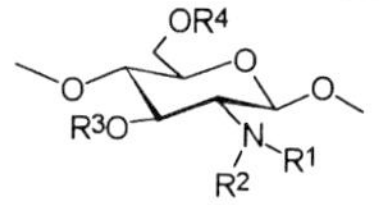

Derivative	R^1; R^2	R^3	R^4	*References*
Triphenylsilylchitin	H, Ac	Ph_3Si	Ph_3Si	Vincendon (1997)
N-Acylchitosan	H, acyl	H	H	Hirano (1997)
N-Alkylchitosan	H, alkyl	H	H	Muzzarelli and Tanfani (1985)
Quaternary chitosan salt	$(R_3)^+$	H	H	Lang et al. (1997b)
N-Alkylidenechitosan	=CH-alkyl	H	H	Hirano (1997)
N-Arylidenechitosan	=CH-aryl	H	H	Hirano (1997)
N- or *O*-Carboxymethylchitosan	CH_2COO^-	H, CH_2COO^-	H, CH_2COO^-	Muzzarelli (1988)
N,N-Dicarboxymethylchitosan	CH_2COO^-	H	H	Muzzarelli et al. (1997c)
N-or *O*-Sulfonyl chitosan	H, SO_3^-	H	H	Holme and Perlin (1997); Nishimura et al. (1998)
N- or *O*-Hydroxyalkylchitosan	alkyl-OH	H	H or alkyl-OH	Lang et al. (1997a)

procedures for their preparation have been described in the literature.

10.3.1
Reactivity

α-Chitin, because of its insolubility, is rarely subjected to chemical reactions, except for the preparation of chitosan by deacetylation, though techniques utilizing highly reactive regenerated chitin by precipitation from phosphorous acid have been described (Vincendon, 1997). β-Chitin has relatively high reactivity (Kurita et al., 1994a).

10.3.2
Derivatives

The derivatives prepared from chitin or chitosan are numerous, and only a very limited selection of recently produced examples is considered here (Table 14) (for reviews, see Hirano, 1989; Lang, 1995; Kurita, 1997; Kurita et al., 2000).

Triphenylsilylchitin of DS 1.5 containing predominantly 6-*O*-silyl groups is soluble in dioxan, tetrahydrofuran (THF), chloroform, and toluene, forming low-viscosity solutions. The polymer is stable in water and thermostable up to 250 °C (Vincendon, 1997). Likewise, 3,6-di-*O*-trimethylsilyl chitin of DS 2.0, prepared from β-chitin is soluble in organic solvents and can be modified further to give for example 6-*O*-tosylated or 6-*O*-glycosylated chitin, containing α-Man or β-GlcNAc branches (Kurita et al., 1994b, 1999). 6-*O*-Tosylchitin is used for further modifications, such as preparation of 6-mercaptochitin (Kurita et al., 1997). *N*-Phthaloyl-chitosan is another organosoluble precursor for the regioselective introduction of various branches into chitin (Kurita, 1997; Kurita et al., 2000).

N-Acyl derivatives include various fatty acid and amino acid amides. The reaction of solid *N*-α-chloropropionyl or *N*-chloroacetyl chitosan with ammonia or aromatic amines affords the corresponding *N*-alanyl and *N*-(*N'*-aryl)glycyl derivatives of chitosan, respectively (Hirano et al., 1992).

Carboxymethyl chitins and chitosans have long been known (Muzzarelli, 1988). Substitution occurs preferentially at the 6-OH position, yielding betaines which are soluble at acidic or basic pH values, but which form gels at pH ca. 7. Depending on the reaction conditions, *N*-, 6-*O*- or 3,6-di-*O*-, or *N,N*-dicarboxymethylchitosans are obtained.

The oxidation of chitin or fungal chitin–glucan complexes by means of NaOCl in the presence of 2,2,6,6-tetramethylpiperidine-*N*-oxide and NaBr gives 6-oxychitin (Figure 11). This was suggested as a substitutive for hyaluronan (Muzzarelli et al., 1999) and, indeed, possesses interesting biological properties (see below). Bromination of *N*-acetylchitosan followed by reduction with $NaBH_4$ affords 6-deoxychitin (see Figure 11) (Zhang et al., 1997).

10.3.3 Hybrid Polymers

Copolymers containing alternating chitooligosacharides and poly(propylene glycol) are resistant to chitinase digestion (Kadokawa et al., 1995). Reaction of chitosan with PEG-aldehyde followed by *N*-acetylation produces a polymer which is soluble in water and in phosphate buffer (Sugimoto et al., 1998). *N*-Ethyl-*N'*-(dimethylaminopropyl)carbodiimide is used to prepare chitosan esters at neutral pH, whereas dialkylpyrocarbonate and cyanuryl dichloride are preferably used under weakly basic conditions (Aiba, 1993). Graft copolymerization of 2-methyl-2-oxazoline onto tosyl- or iodo-chitins produces poly(*N*-acetylethyleneimine) side chains (Kurita et al., 1996a). Likewise, mercapto-chitin is used for graft copolymerization of methyl methacrylate or styrene onto chitin (Kurita et al., 1996b,c, 2000).

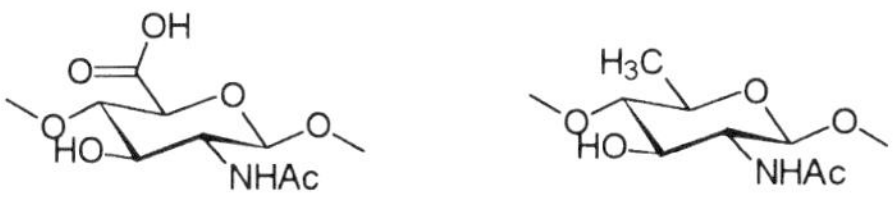

Fig. 11 Structures of 6-oxychitin and 6-deoxychitin.

Chitosan is a suitable carrier for peptide synthesis, as shown by preparation of protected precursors of the putative *Ras* protein farnesyl transferase inhibitors, i.e., Fmoc-Cys-Aib-Aib-Met-NH_2, Fmoc-Cys-Aib-Aib-Met-OH, and Fmoc-Cys-Val-Gly-Met-NH_2. Different cleavable linkers such as 4-(2',4'-dimethoxyphenyl-Fmoc-aminomethyl)-phenoxy-acetic acid, 4-hydroxy-methyl-3-methoxyphenoxy-acetic acid, or 4-methylene-3-methoxy-phenol were grafted onto chitosan prior to peptide synthesis (Neugebauer et al., 2000).

10.4 Biological Properties

Although the outstanding biological activities of chitin, chitosan and chitooligosaccharides were discovered some 15 years ago, some novel aspects are reviewed in this section.

10.4.1 Biological Activities in Mammalian Systems

The most spectacular results were observed when chitosan or derivatives were assayed in vertebrates for acceleration of bone regeneration and wound healing. Other biological effects include antimetastatic and antiviral activities of both chitooligosaccharides and chitosan derivatives.

Bone Regeneration and Wound Healing

Chitosan, *N,N*-dicarboxymethylchitosan, and 6-oxychitin show a remarkable acceleration effect on bone regeneration (Muzzarelli, 2000a) and wound healing (Stone et al., 2000). The molecular mechanisms of these effects are unknown, but chitosan-coated surfaces promote the growth of human osteoblasts and chondrocytes and propagate the expression of extracellular matrix proteins (Klokkevold et al., 1997b; Lahiji et al., 2000). Chitosan films promote the growth of keratinocytes, but not of fibroblasts which adhere to the surface of the films almost twice as strongly as do keratinocytes (Chatelet et al., 2001). Moreover, as discussed in Section 7.3, chitooligosaccharides ultimately are primers for hyaluronan biosynthesis (Varki, 1996; Muzzarelli, 1997b; Muzzarelli et al., 1997a), whilst GlcN is both substrate for, and regulator of, glycosaminoglycan biosynthesis (Rubin et al., 2000). Chitosan also inhibits the production of nitric oxide (NO) in activated macrophages, thus reducing cytotoxicity in cell proliferation during inflammation processes in wound healing (Hwang et al., 2000).

Activation of complement C3 and C5, as well as stimulation of fibroblasts to produce interleukin-8 are important events in wound healing acceleration by chitin and chitosan (Minami et al., 1997). Complement C3 activation by chitosan decreases with decreasing F_A, but is not seen with low-M_w chitosans (Suzuki et al., 2000). An investigation of the effects of chitin and derivatives on proliferation of fibroblasts *in vitro* revealed that the biological response is mediated through additional factors, including the composition of the culture medium (Mori et al., 1997).

Other Effects

Although $(GlcNAc)_6$ shows significant antimetastatic effects in mice bearing Lewis lung carcinoma (Suzuki et al., 1986, 1989), this issue does not appear to have been investigated with appropriate clinical studies.

Sulfation of 6-*O*-tritylchitosan at N-2 and O-3 affords potent antiretroviral agents which inhibit the infection of HIV-1 to T lymphocytes at 0.28 μg mL^{-1}, without significant cytotoxicity, whilst 6-*O*-sulfochitin strongly inhibits blood coagulation (Nishimura et al., 1998).

10.4.2 Antimicrobial Activity

GlcN oligomers of DP ≥ 30 possess antimicrobial activity against a number of Gram-negative, Gram-positive and lactic acid bacteria, whereas low-DP oligomers are ineffective (Ueno et al., 1997; Jeon et al., 2001). Apparently, the smaller chitosan oligomers serve as nutrients for bacteria, whereas the higher oligomers are toxic by virtue of their charge-mediated adhesion to the cell membrane, which in turn prevents the uptake of nutrients through the cell wall (Tokura et al., 1997).

10.4.3 Elicitor Activity in Plants

Besides induction of cell division in the cortex of Leguminosae by Nod factors (see Section 7), the most prominent activity of chitin and/or chitosan in plants is the elicitation of defense reactions, i.e., the induction of chitinases or of metabolites of the shikimate pathway, including lignin.

Partially *N*-acetylated, polymeric chitosans elicit phenylalanine-ammonia-lyase (PAL) and peroxidase activities in wheat (*Triticum aestivum*) leaves. Maximum activity is observed with chitosans of intermediate F_A values, whereas $(GlcNAc)_n$, $n \geq 7$, elicits peroxidase. GlcN oligomers are not active as elicitors. All chitosans (but not the chitin oligomers) induce the deposition of lignin (Vander et al., 1998). Partially acetylated (F_A 0.33–0.52) chitooligosaccharides of DP 5–7 elicit pisatin in pea epicotyls (Akiyama et al., 1995b).

Chitosan stimulates the production of secondary metabolites in plant cell cultures, as was shown for anthraquinone by *Morinda citrifolia* and amaranthin by *Chenopodium rubrum*, presumably by elicitation and membrane permeabilization (Doernenburg and Knorr, 1996).

Chitin-oligosaccharides induce an extracellular class III chitinase and PAL in a rice suspension culture. The maximum activity is observed with $(GlcNAc)_6$ at 1 µg mL^{-1}, whereas lower GlcNAc and GlcN oligosaccharides are less active (Inui et al., 1997). The induction of anti-plant pathogen-specific chitinases by various elicitors, including chitin, chitosan and chitooligosaccharides has been demonstrated in grass and yam (Koga, 1996).

At the physiological level, GlcNAc oligomers of DP ≥ 4 cause a transient alkalinization of the culture medium of suspension-cultured tomato cells at 1.4 nM for whole cells and 23 nM for microsomal membranes, whereas chitosan oligomers are inactive (Baureithel et al., 1994).

11 Applications of Chitin and Chitosan

The great potential for applications of chitin and chitosan is reflected by the coexistence of approximately 3500 patents or patent applications, in addition to a much higher number of scientific articles which have appeared in the literature during the past decade. Many of the claims suggest uses for chitin and chitosan where these biopolymers actually replace existing synthetic or natural polymers, irrespective of economical aspects (for general reviews, see Lang, 1995; Peter, 1995; Daly and Macossay, 1997; Kurita, 1997; Kumar, 2000). The most frequently cited advantages of chitosan are seen in the unique physico-chemical and biological properties of the polymer, including antimicrobial activity and biodegradability. A survey of applications suggested over the years is listed in Table 15. The commercial exploration of chitosan has been particularly successful in water engineering, the manufacture of textiles, and as ingredients in cosmetics and dietary health food products.

11.1 Technical Applications

Numerous technical applications of chitosan have been suggested, the most intensively studied topics being the removal of proteins, metal ions, and dyes from various factory effluents, and applications in the textile and paper industry.

11.1.1 Waste-water Engineering

As chitosan interacts with negatively charged proteins, it is not surprising that applications for the flocculation of proteins from waste waters have been reported (for example, Guerrero et al., 1998; Savant and Torres, 2000). The recovered chitin–protein complexes can then be used for animal feed.

The metal ion-complexing properties of chitosan and derivatives are also well documented (Guibal et al., 1997; Domard and Piron, 2000), and a compilation of data is given in Table 16 (Peter, 1995). Possible applications are seen for example in the galvanizing industry, or in the purification of effluents from nuclear power plants.

Chitosan possesses a high affinity for a wide range of dyes, in particular negatively charged compounds. The sorption characteristics are influenced primarily by the pH of the solution, and also by the presence of other contaminants. Applications are envisaged in the purification of waste water from the textile industry (Kumar, 2000).

Chitosan pervaporation membranes are used to separate ethanol or DMSO from water (Kurita, 1997).

Tab. 15 Summary of applications of chitosan

Application	*Properties of chitosan utilized*
Technical	
Water engineering: adsorption of metal ions and dyes; flocculation of proteins; separation of organic solvents	Polycation; metal ion complexation; biodegradability
Textiles, fibers, nonwoven fabrics, leather	Polycation; film formation ; antibacterial properties
Paper coating	Complex formation with polyanions and polysaccharides
Biotechnology: enzyme immobilization; plant culture medium supplement; cell encapsulation; protein purification	Chemical functionality; polyelectrolyte complexes
Medicine and healthcare	
Lowering of serum lipids	Polycation; lipid complexation
Bone regeneration; treatment of rheumatoid diseases	Osteoconductivity; GAG synthesis regulation
Vascular medicine and surgery	Tissue adhesion; hemostatic
Wound care; artificial skin; hemostasis	Antibacterial; biological activity on cells
Cosmetics	
Skin moisturizing ingredient; hair shampoos; hair styling; dentrifices	Gel and film formation; antibacterial
Agriculture	
Plant growth regulators; elicitors of plant defense; seed conservation; soil fertilizer; anti-fungals and anti-nematodals	Plant growth regulator; regulation of resistance proteins; stimulation of chitinase-producing soil bacteria
Food	
Health ingredient, dietary fiber; foam stabilizer; preservative; clarification of beverages; packaging materials	Viscosity; polycation; bacteriostatic activity; film formation

11.1.2 Fibers, Textiles, and Nonwoven Fabrics

A comprehensive review on fibers and films made from chitin and chitosan was published by Rathke and Hudson (1994) (see also Struszczyk, 1997a; Kumar, 2000). Solvents used for the spinning of chitin fibers are chosen according to the solubility of the polymer in LiCl/DMAc or halogenated organic solvents. Alternatively, an aqueous alkaline solution of sodium xanthate can be used, and *N*-propionylchitosan fibers are obtained in the same way. *N*-Acylchitosan fibers containing higher fatty acid residues are obtained by *N*-acylation of chitosan fibers; these may be prepared by spinning a 3% solution of chitosan in 2% acetic acid into an aqueous 10% NaOH solution containing 30% sodium acetate (Hirano et al., 2000). The mechanical properties of chitosan fibers (Table 17) are modulated by stretching or by immersing the fibers in phosphate- and phthalate-containing solutions (Knaul et al., 1999).

Chitosan–cellulose blend fibers are generated from mixtures of the corresponding xanthates or of chitosan salts and cellulose xanthate (Hirano and Midorikawa, 1998). Conventional textile fibers may also be coated with chitosan (Struszczyk, 1997b) and cross-linked, for example by periodate activation of the cellulose (Liu et al., 2001).

Chitosan-containing textiles, which are manufactured by several companies, possess

Tab. 16 Binding capacities of polymers for metal ions (mEq g^{-1}) (Peter, 1995)

Polymer	*Method*	*Mg^{2+}*	*Ca^{2+}*	*Sr^{2+}*	*Ba^{2+}*	*Mn^{2+}*	*Ni^{2+}*	*Cu^{2+}*	*Cd^{2+}*	*Pb^{2+}*	*Zn^{2+}*	*Hg^{2+}*	*Co^{2+}*	*Cr^{2+}*	*Fe^{2+}*	*Ag^{+}*	*Au^{3+}*	*Pt^{4+}*	*Pd^{2+}*
Chitosan (F_A 0.45)	A	0.3	0.8	1.5	1.1	1.1	3.5	5.3	6.5		5.5								
Chitosan (F_A 0.03)	A	0.5	0.4	0.6	0.8	0.5	2.3	4.8	4.9		3.2								
Chitosan (F_A not specified)	B					1.44	3.15	3.12	2.78	3.97	3.70	5.60	2.47	0.46	1.18	3.26	5.84	4.52	6.28
p-Aminostyrene	B						0.02	1.31	0.10	0.17	0.52	5.70	0.07	0.03	0.05	1.98	5.75	3.82	0.87
6-*O*-CM-chitin	C	2.08	2.50	2.12	2.15	2.22	2.22		2.42	2.13									
3,6-di-*O*-CM-chitin	C	4.0	3.3*	2.8	10.1+	3.4	4.1+	3.1+	9.2+	8.5+									

Method A: 0.04 M metal ion; pH 7.4; 20 h; 30 °C. Method B: 1 g polymer; 0.02 – 0.4 mM metal chloride or nitrate; 24 h. Method C: not specified.6-*O*-CM-chitin = 6-*O*-carboxymethylchitin; 3,6-di-*O*-CM-chitin = 3,6-di-*O*-carboxymethylchitin. * Gel formation or precipitation. + Ca^{2+} can be desorbed by means of EDTA.

Tab. 17 Some properties of chitosan fibers (Struszczyk, 1997a)

Property	*Value*
Titer [dtex]	1.5 – 3.0
Tenacity in standard conditions [cN/tex]	10 – 15
Tenacity in wet conditions [cN/tex]	3 – 7
Loop tenacity [cN/tex]	3 – 7
Elongation in standard conditions [%]	10 – 20
Water retention [%]	≤ 250

antimicrobial properties and so are especially useful in the manufacture of underwear and sanitary products. Nonwoven fabrics may also be used as wound dressings (see below).

11.1.3
Paper Technology

Paper coated with chitosan shows improved breaking strength, burst resistance or tearing strength as compared with noncoated paper (Wieczorek and Mucha, 1997). Techniques for the manufacture of this include casting after addition of chitosan to paper pulp, as well as coating of paper sheets with chitosan.

11.1.4
Biotechnology

Chitosan is used as a stationary phase for the chromatography of various biomolecules, e.g., proteins and saccharides. Porous beads are manufactured from chitosan by cross-linking, and are available commercially in Japan as Chitopearl® with a range of specifications, including chemically modified materials, e.g., for chiral recognition, ion-exchange, gel-permeation, or affinity chromatography, as well as for the immobilization of cells, enzymes, antibodies, or other polypeptides.

Microbial and mammalian cells are also immobilized by encapsulation, often using alginates as the negatively charged polymer component, with possible applications in biotechnology, medicine, and pharmacy (Zielinski and Aebischer, 1994; Vogt et al., 1996; Quong et al., 1997; Bartkowiak and Hunkeler, 2000).

11.2 Medicine and Healthcare

The low toxicity, antimicrobial activity and physico-chemical functionality of chitosan offer a great potential for medical applications, as documented in several reviews (Muzzarelli, 1997b; Muzzarelli et al., 1997a; Paul and Sharma, 2000; see also Muzzarelli, 2000b). Chitosan is currently not registered as a drug for the treatment of diseases, but GlcN – either as the hydrochloride or the sulfate – is used widely for the relief of pain in musculoskeletal rheumatoid diseases, including arthrosis, arthritis, or osteoporosis. GlcN is available in some European countries by prescription, but is sold elsewhere (e.g., the USA) as an over-the-counter (OTC) pharmaceutical. GlcN sulfate is prepared by the hydrolysis of chitin with hydrochloric acid, followed by ion exchange. The starting material for GlcN production is chitin, and significant amounts of the available resources of this polymer are utilized to satisfy the huge market resulting from applications in human and veterinary medicine.

11.2.1 Treatment of Obesity and Hyperlipidemia

Although reports on the hypocholesterolemic and hypolipidemic activity of chitosan in animals first appeared some 25 years ago, a comprehensive discussion of the hypolipidemic action (including possible mechanisms) and drug formulations was published only recently (Furda, 2000).

It has been suggested that chitosan binds to bile acids, thereby removing them from the enterohepatic circulation, and showing similar effects as other polycations such as cholestyramine or cholestipol. Oral administration of chitosan to mice showed that fatty and bile acids, cholesterol and triglycerides which bind to chitosan and are not absorbed, resulting in hypolipidemic and anti-atherogenic effects (Muzzarelli, 1998; Ormrod et al., 1998; Han et al., 1999). In rats, cholesterol resorption in the intestine was reduced, while fat and bile acid excretion was enhanced after feeding chitosan (Gallaher et al., 2000). On the other hand, a hypocholesterolemic action of chitosan was absent in rabbits with high serum cholesterol levels, though enhanced excretion of sterols and bile acids were noted (Hirano and Akiyama, 1995)

Chitosan at a dose of 3–6 g per day to humans was reported to lower total serum cholesterol, increase high-density lipoprotein (HDL), and enhance the excretion of primary bile acids, cholic acid and chenodeoxycholic acid (Maezaki et al., 1993). Derivatives of chitosan bearing a tertiary or quaternary amino group show enhanced binding capacity for glycocholate (Table 18) (Lee et al., 1999b).

A clinical study showed that 66% of a moderately obese population responded to a dose of 2.25 g per day chitosan over 12 weeks, with a mean weight reduction of ca. 4 kg (Thom and Wadstein, 2000). Chitosan

Tab. 18 Binding capacity of polymers for glycocholic acid (Lee et al., 1999b)

Polymer	*Binding capacity [mmol g^{-1}]*
Chitosan	1.42
DEAE-chitosan	3.12
Quaternized DEAE-chitosan	4.06
Cholestyramine	2.78

in combination with a hypocaloric diet, with or without additional fibers and micronutrients, was reported also to reduce body weight, serum triglycerides, total and low-density lipoprotein (LDL) cholesterol, but to increase in HDL cholesterol (Colombo and Sciutto, 1996; Girola et al., 1996; Macchi, 1996; Veneroni et al., 1996a,b). Chitooligosaccharides appeared to have similar effects, in addition to lowering blood pressure (Kawasaki et al., 1998).

Some critical issues were raised, however. For example, Ernst and Pittler (1998) and other investigators reported no significant differences in body weight reduction or serum lipid levels when a normal diet was administered with or without chitosan (Pittler et al., 1999; Stern et al., 2000), including microcrystalline chitosan (MCCh) (Wuolijoki et al., 1999). Possible side effects, such as impairment of the resorption of minerals or lipid-soluble vitamins were not observed in all studies. Vitamin K levels were increased from 0.33 (placebo) to 0.56 ng mL^{-1} (Pittler et al., 1999).

When administered orally to dogs, β-chitin was not degraded during passage through the intestine, whereas chitosan ($M = 4 \times 10^5$; particle size ca. 0.5 – 3 mm) underwent transformation into a gel in the stomach, thereby losing ca. 15% of its weight. Further degradation of chitosan was found to take place in the large intestine, and < 10% was recovered in the feces (Okamoto et al., 2001). The DP of the degradation products was not determined. Similar results were obtained in rabbits and chickens, whereas the reverse situation appears to exist in sheep (for references, see Okamoto et al., 2001). Feeding of [^{14}C]chitosan to rats resulted in the distribution of radioactivity among all tissues (Nishimura et al., 1997).

11.2.2 Bone Regeneration and Prosthetic Implants

The osteoconductive properties of chitosan and some derivatives are described in Section 10.3.1. 6-Oxychitin was shown to promote reconstruction of the osteo-architectural morphology in surgical lesions in rat condylus, even though healing was slower when compared with *N,N*-dicarboxymethylchitosan (Mattioli-Belmonte et al., 1999). Complete healing was observed with *N,N*-dicarboxymethylchitosan within 3 weeks. A complex of *N,N*-dicarboxymethylchitosan with calcium acetate and Na_2HPO_4 promoted bone mineralization in osteogenesis, leading to complete healing of otherwise nonhealing surgical defects in sheep (Muzzarelli et al., 1998).

A mixture of chitosan, hydroxyapatite granules, ZnO, and CaO, pH 7.4, was used to prepare a self-hardening paste that produced much less heat on setting as compared with conventional poly(methyl methacrylate) bone cement (Maruyama and Ito, 1996).

The coating of prosthetic materials made from ceramics or titanium with chitosan showed improved biocompatibility and facilitated the attachment of fibroblasts and osteoblasts, thus promoting bone healing and integration of the implant (Muzzarelli, 2000a; Pittermann et al., 1997b).

11.2.3 Vascular Medicine and Surgery

N-Acylchitosans are blood-compatible and show anticoagulative effects, especially in the case of *N*-hexanoylchitosan of DS 0.2 – 0.5, which is biodegradable by lysozyme (Lee et al., 1995). In contrast, chitosans of F_A 0.3 – 0.57 show hemostatic activity (Klokkevold et al., 1997a; Sugamori et al., 2000). MCCh was found to seal arterial puncture sites when applied after arterial catheterization via an arterial sheath (Hoekstra et al., 1998).

Photocross-linking by UV irradiation of chitosan containing *p*-azidobenzamide and lactobionic acid amide side chains produces an insoluble hydrogel within 60 s. The binding strength of the noncytotoxic hydrogel is similar to that of fibrin glue, and the material seals against air leakage from pinholes on isolated small intestine, aorta, and also from incisions on isolated trachea (Ono et al., 2000).

11.2.4
Wound Care and Artificial Skin

Spectacular results on wound healing have been observed when chitosan was applied to damaged skin in the form of bandages made from nonwoven tissues, or sponges, films and membranes (Muzzarelli, 1997b; Muzzarelli et al., 1997a). MCCh was shown to absorb and retain blood, to induce the release of substances involved in platelet activation, such as β-thromboglobulin and platelet factor 4, and to possess excellent hemoagglutinative activity in dogs (Sugamori et al., 2000). Chitosan, as a semipermeable biological dressing, maintained a sterile wound exudate beneath a dry scab, preventing dehydration and contamination of the wound, facilitating rapid wound re-epithelialization, including regeneration of nerves within a vascular dermis (Stone et al., 2000). Various formulations of chitosan with other biopolymers, such as collagen, hyaluronan, or chondroitin sulfate have been described (Denuziere et al., 1996; Choi et al., 2001).

11.2.5
Other Medical Applications

Chitosan–iron complexes have been investigated for the treatment of hyperphosphatemia (Baxter et al., 2000). Chitosan immobilized on poly(L-lysine) binds bilirubin, and is suggested as an adsorbent for hematoperfusion in hyperbilirubinemia (Chandy and Sharma, 1992). Contrast agents used in X-ray, ultrasound or NMR imaging diagnostics have been described (mostly in the patent literature) in considerable variety as formulations containing chitosan or derivatives.

11.2.6
Toxicology of Chitin and Chitosan

Chitin is a normal constituent of seafood and mushrooms, both of which are considered to be nontoxic and safe for human consumption, except for a very small percentage of the population that is allergic to seafood products due to the presence of proteinaceous impurities. However, hypersensitization is most likely due to protein components rather than to the polysaccharide. Intravenous or peritoneal application of chitin is not possible, and has actually never been suggested.

Earlier studies published some 20 years ago indicated a very low toxicity of chitosan in mice (for references, see Hwang and Damodaran, 1995). A recent extensive toxicological study with PROTASAN™ UP (ultrapure; F_A 0.17), including various application routes as well as hypersensitization, mutagenicity, and cytotoxicity assays,

Tab. 19 Toxicology of chitosan glutamate (PROTASAN™ UP) in rats (Dornish et al., 1997a)

Application	*No toxic effects seen at*	*Application for*
Oral	up to 600 mg kg^{-1} per day	13 weeks
Intravenous	up to 25 mg kg^{-1}	Once
Intraperitoneal	bolus injection, up to 500 mg kg^{-1}	7 days
Nasal mucosa	up to 3 mg per rat per day	7 days

showed that chitosan glutamate is essentially nontoxic in rats (Table 19) (Dornish et al., 1997a). The testing of cosmetic preparations of Hydagen® CMF on the skin of volunteers revealed favorable properties of high-M_v chitosan when erythema and squamations were evaluated (Pittermann et al., 1997a).

Data on the toxicological properties of carboxymethylchitosans are rare. The LD_{50} was reported as > 5 g kg^{-1} in rats (Liu et al., 1997).

11.2.7 Regulatory Aspects

As mentioned earlier, chitosan has not been registered for drug use, for a number of reasons, including:

- Lack of standards: apparently, no obligatory protocol for defining the chemical, physico-chemical, and purity specifications as well as other parameters has been established (Table 20).
- Scarcity of data on efficacy: although numerous spectacular results have been detailed in the literature, some are apparently at variance, and the results of appropriate clinical studies are extremely rare. However, this situation may soon change as some companies are clearly investigating drug master files.
- Lack of long-term toxicological and safety data. The use of chitosan as a pharmaceutical excipient is currently hampered by a lack of appropriate preclinical testing and assessment of the compound's safety (Baldrick, 2000).

11.3 Pharmaceutical Applications

The most intensively studied topics in pharmaceutical applications are enhanced transmucosal drug delivery, sustained drug release, and drug targeting. Again, the polycationic nature of chitosan as well as its biodegradability are the most important features of chitosan with regard to interactions with negatively charged tissue surfaces, as well as the development of slow-release formulations, including micro- and nanocapsular systems.

Tab. 20 Specifications of chitosan for pharmaceutical-grade products

Parameter	*Specification*	*Reference*[a]
Viscosity	Should be specified	A, B, C
F_A	Should be specified	A, B, C
Acid content	Should be specified	A
Moisture	Should be specified	A, B, C
Solubility, turbidity	$> 99.9\%$	A, B, C
Heavy metals	< 27 ppm	A, B
Endotoxin	< 625 EU g^{-1}	A
Microbiology	< 1 c.f.u. g^{-1}	A
Color	Colorless, grayish, cream, or pink	B, C
Ash	$< 3\%$	B, C
Appearance	Powder, flakes	B
Grain size	Less than 25% > 2 mm	B
Chromatographic analysis	Suggested	C
Spectroscopic analysis	Suggested	C
Chemical analysis; I_2, $KMnO_4$	Suggested	C
Protein/amino acids	Suggested	C

[a] A = Dornish et al. (1997); B = Wojtasz-Pajak et al. (1994); C = Stevens (2001).

11.3.1
Transmucosal Drug Delivery

Chitosan is strongly mucoadhesive (Deacon et al., 1999; Harding et al., 1999). Furthermore, chitosans of DP > 50 and $F_A < 0.4$ have been shown to induce the opening of tight junctions in confluent monolayers of Caco-2 cells (Holme et al., 2000). An increased transport of polar drugs across epithelial surfaces was observed when insulin was applied with chitosan glutamate (PROTASAN™) to the nasal epithelia of rats, with little or no toxicity to the nasal mucosa being observed (Dornish et al., 1997b). Hydrocortisone or transforming growth factor β (TGF-β) was applied with chitosan lactate to porcine buccal mucosa, and this resulted in a six-fold enhancement of tissue permeability (Senel et al., 2000a,b). Peptide delivery through the intestinal epithelia has also been demonstrated in Caco-2 cells with the luteinizing hormone-releasing hormone (LHRH) analogue buserelin, using trimethylchitosonium chloride (DS 0.4 or 0.6) (Thanou et al., 2000).

By contrast, the screening of cationic compounds as absorption enhancers for nasal drug delivery has led to the conclusion that poly(L-arginine) is superior to chitosan in terms of efficiency and side effects when fluorescein isothiocyanate-labeled dextran was used as a model drug in rats (Natsume et al., 1999).

11.3.2
Sustained-release Formulations

In addition to peptides (for a review, see Bernkop-Schnurch, 2000), a variety of other drugs, e.g., nifedipine, diclofenac, and propranolol, have been investigated for sustained oral, buccal, transdermal, or colon-specific delivery from beads, granules, membranes, films, tablets, or microcapsules, manufactured from chitosan, often in combination with other polymers (for some recent reviews, see Zecchi et al., 1997; Felt et al., 1998; Illum, 1998; Davis, 2000; Gupta and Kumar, 2000; Kumar, 2000; Paul and Sharma, 2000).

$(GlcNAc)_n$ has been investigated as a slow-release form in the treatment of osteoarthritis (Rubin et al., 2000).

11.4
Cosmetics

Applications of chitosan are based on the film-forming properties on negatively charged surfaces, and on the formation of hydrogels. Hair conditioners, shampoos, and styling products containing either chitosan or derivatives such as carboxymethyl- and hydroxyalkylchitosans as well as quaternary ammonium salts have long been used (Lang, 1995), while numerous other uses of chitosan have been suggested in skin lotions and creams, soaps, and dentifrices. A chitosan-containing soap is currently marketed in Korea.

11.5
Agriculture

Chitosan can be used as an antimicrobial film for fruit and seed coating, resulting in higher crop yields. Application of chitosan to the soil stimulates growth of the chitinase-producing bacteria that mediate antifungal and anti-nematodal effects (Gooday, 1999; Herrera-Estrella and Chet, 1999). The elicitor activities for plant defense reactions and the stimulation of nodulation were discussed in Sections 8 and 9.

11.6
Food

Chitosan, though not registered as a drug, is successfully marketed as a health food ingredient and dietary fiber by several companies, predominantly in the Far East, North America, and in some European countries (Muzzarelli, 1996; Yalpani, 1997; Shahidi et al., 1999).

Applications of chitosan in food processing technology include uses as a foam stabilizer and emulsifier, as a preservative agent, and as a flocculent of tannins and proteins in beverages. A review of these topics is beyond the scope of this chapter, however.

12 Current Problems and Limits

The appearance of the book *Chitin* almost a quarter of a century ago (Muzzarelli, 1977) marked the beginning of a world-wide systematic exploitation of chitin and chitosan. Whilst it is likely that all major chemical and pharmaceutical companies have filed patents on the production and uses of biopolymers, the question remains as to why chitin and chitosan have not yet acquired a position in the large, high value-added markets. It is possible that chitosan has remained underinvestigated, even though the routes of production of both chitin and chitosan are well established and the chemistry of the polymers is reasonably well understood.

In addition to the relatively high price of good quality chitosan, practical utilization is slowed by regulatory matters (see Section 11). Other problems are sometimes thought to arise from limitations of the availability of raw materials, which depends not only on the population dynamics of marine organisms, but also on political and ecological issues. However, in view of the vast amounts of shellfish waste produced globally, it is envisaged that huge resources are always available, even when the catch of a particular crustacean species might be restricted. To secure the resources initially is a matter of developing suitable logistics. The fishing and seafood processing industries are most likely not aware of the potential value of shellfish waste, and much of this is simply returned either to the sea or to ecologically questionable terrestrial landfill sites. Another major problem arises from the nature of the biological material itself, as it rapidly decomposes and must therefore be processed with minimal delay after peeling of the crustacean catch. Nonetheless, it appears that these problems can be solved, as demonstrated by the fact that several companies now produce these polymers successfully.

The comment has often been made that the production of chitosan leads to environmental problems, mainly because of the harsh conditions involved such as high temperatures, aggressive acid and alkali solutions, and the large volumes of fresh water required. However, similar concerns might be expressed with regard to the production of cellulose.

13 Outlook and Perspectives

It is expected that the global demand for chitosan will increase dramatically, once the polysaccharide has been registered for use as a food additive and pharmaceutical excipient, and for certain medical and agricultural applications. Moreover, it is possible that this might happen in the near future. Subsequently, the production of chitin and chitosan will depend for many years essentially on marine and freshwater crustacean resources, though biotechnological processes to remove most of the protein and mineral fractions show great promise. However, it is unlikely that the development of alternative routes such as enzymatic deacetylation, or the use of either insects, fungi, or synthetic approaches will become practicable in the near future.

Often, when the subject of chitin is raised, it is the exoskeleton of insects and crabs that comes first to mind, with chitosan remaining something of an unknown quantity. However, there is an increasing awareness of the fascinating properties of chitosan and its vast potential in high value-added applications. Consequently, as our knowledge of chitosan develops, then this biopolymer – which is clearly much more than simply an "interesting skeletal material" – is likely to be exploited to a much greater extent.

14
Relevant Patents for Isolation, Production and Applications of Chitin and Chitosan

A patent search in *Chemical Abstracts* on chitin or chitosan related citations yielded a total of 3527 references containing relevant terms either in the title and/or in the abstract (date of patent search: May 25, 2001). The number of references is rapidly growing, averaging presently ca. 250 *per annum* (Figure 12).

A summary of the topics covered in the patent literature is given in Table 21.

Tables 22–34 refer to details. The citations are sorted alphabetically by the first inventor. Obviously, it is neither practical nor useful to evaluate all patents and thus, a rather narrow selection is made from the International (World) patents or applications which were published after the year 1990. Original texts were reviewed in a few cases only, and most of the information is retrieved from the titles and abstracts. Patents describing uses of these polysaccharides as substitutes for other biopolymers are generally excluded.

This survey should not be used as a support for future patent applications. It cannot not replace a literature search which must always be done when a new patent should be considered. Indeed, the selection is very incomplete, covering less than 10% of the relevant literature. However, we believe that it is representatively illustrating the various activities going on in the exciting field of research and applications of chitin and chitosan.

Acknowledgements
The investigations performed in the author's laboratory were supported by the Deutsche Forschungsgemeinschaft, by EU grant no. BIO4-CT-960670, and by the Fonds der Chemischen Industrie. The author's thanks are also extended to H.-M. Cauchie (Luxembourg), V. G. H Eijsink (Ås), J.-P. Thomé (Liège), and J. F. V. Vincent (Bath) for their valuable comments on different sections of the manuscript.

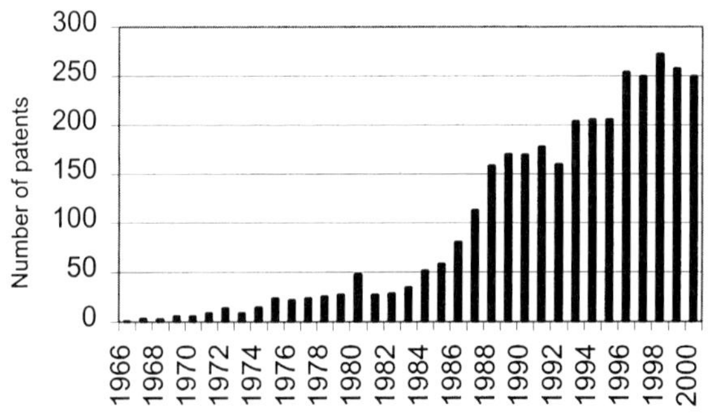

Fig. 12 Annual number of chitin or chitosan related patents or patent applications, 1966–2000.

Tab. 21 Summary of the main areas and topics of chitin or chitosan relevant patents cited in Tables 22–34

Field	*Table No.*	*Number of patents cited*	*Selected topics*
Agrochemical bioregulators	22	14	Antimicrobial chitinolytic enzymes and chitin binding proteins Chitosan as plant growth regulator Chitosan in release formulations for antimicrobial agents
Biochemical genetics	23	16	Chitinases as antifreeze proteins in plants Construction of transgenic plants expressing chitinases Genes and cloning of chitinases, including human chitinase Uses of cloned chitinases and chitin binding proteins as antifungal agents Yeast chitin synthase
Biochemical methods	24	6	Chitinases or chitin binding proteins as diagnostic or analytical tools Chitosan materials for cell culture
Enzymes	25	8	Chitin deacetylases, including c-DANN's Enzymes as antimicrobial agents, as therapeutic and analytical tools Enzymes for production of chitooligosaccharides Uses of chitosan for immobilization of enzymes
Cosmetics and ingredients	26	11	Liquid cleansers Products for skin and hair care, including microcapsule formulations Sunscreen formulations Toothpastes and oral hygiene
Industrial biotechnology	27	4	Cell adhesive surfaces Chitosan beads as carriers for proteins Manufacture of chitin and chitosan using fungal cultures Production of *N*-acetylglucosamine from chitin Use of chitosan for improved production of natural products in plant tissue cell culture
Food and feed chemistry	28	5	Chitosan for use in weight loss programmes Gels and emulsions for food processing Ingredient in food preservatives Protective coating of fruits, vegetables and other food Stabilization and formulation of vitamins
Immunochemistry	29	3	Chitinase or chitin binding proteins for diagnostics to detect fungal infections Chitosan as adjuvant in influenza vaccines Immunstimulation and -potentiation by chitosan
Polysaccharide chemistry and processes	30	38	Applications in drug delivery or targeting systems, wound dressings, magnetic supports, coatings Carboxymethylchitosans, chitin-glucan complexes Chitin or derivatives in form of hydrogels, films, fibers, foams, beads, porous particles, micro- and nanocapsules Cyclodextrin derivatives Manufacture of chitooligosaccharides Methylpyrrolidinone chitosans *N*-Halochitosans for use as flocculants Processes and technologies for production of chitosan Processes for crosslinking of chitosan Quaternary ammonium salts and their hypocholesterolemic properties

Tab. 21 (cont.)

Field	*Table No.*	*Number of patents cited*	*Selected topics*
			Various Derivatives for uses in cosmetics, transfection of cells, synthetic bone and teeth
Pharmaceuticals and pharmacology	31	100	Adjuvants for vaccines
			Antimicrobial non-woven fabrics
			Antiviral, antifungal, or antiinflammatory formulations
			Artificial tears for treatment of dry-eye syndrome
			Biodegradable cationic bioadhesives
			Bone substitutes
			N,O-Carboxymethylchitosan for prevention of postsurgical tissue adhesion
			Chitosan for prevention of infections
			Chitosan-drug conjugates
			Compositions and formulations for treatment of rheumatoid diseases
			Detection and monitoring of fungal infections
			Doxorubicin-chitooligosaccharide conjugates
			Drug targeting and enhancement of resorption in the intestinal tract
			Enhancers for transmucosal drug delivery
			Foam scaffolds for cell entrappment and growth supports for encapsulated cells
			Formulations for reducing skin irritation
			Hygiene products
			Imageing materials
			Immunostimulants for tumor therapy
			Implantable polymers, encapsulation of cells for implantation
			Liquid and odour absorbent materials for medical and pharmaceutical uses
			Manufacture of UV radiation absorbers
			Nanoparticles for drug delivery
			Osteoinductive materials
			pH Controlled gel formation
			Promotion of tissue repair and tissue engineering
			Reduction of fat absorption from food
			Sustained drug release formulations
			Thromboresistant surfaces
			Treatment of fungal infections with chitin synthesis inhibitors
			Treatment of periodontic diseases
			Wound dressings
Textiles and Paper	32	11	Modified fibers
			Pulp and paper additives
			Textile finishing
Water purification	33	6	Adsorption of anions, dyes, metal ions, and organochemicals
Various applications	34		Anticorrosives
			Filtration media
			Liquid absorbers
			Metal ion adsorbers
			Packaging materials
			Semipermeable membranes, also for pervaporation

Tab. 22 Agrochemical bioregulators

Patent number	Patent holder	Inventors	Title of patent	Publication date (YYYYMMDD)
WO 0032041	Safescience Inc., USA	Ben-Shalom, N., Pinto, R.	Chitosan metal complexes as agents for controlling microbial diseases in crops	20000608
WO 9942594	Cornell Research Foundation, Inc., USA	Broadway, R. M., Harman, G. E.	Fungus and insect control with chitinolytic enzymes from *Streptomyces albidoflavus*	19990826
WO 9411511	Zeneca Ltd. UK	Broekaert, W. F., Cammue, B. P. A., Osborn, R. W., Rees, S. B.	Antimicrobial plant proteins and their purification	19940526
WO 0051428	Virbac S.A., France	Broussaud, O., Derrieu, G., Jouan Daniel, R., Pougnas, J.-I.	Microparticulate or nanoparticulate vectors for active agents in aquatic media	20000908
WO 9732973	Cornell Research Foundation, Inc., USA	Harman, G. E., Lorito, M., Di Pietro, A., Hayes, C. K., Scala, F., Kubicek, C. P.	Combinations of fungal cell wall degrading enzyme synergistic use with fungicides, transgenic plants, and agricultural and therapeutic uses	19970912
WO 9832335	DCV, Inc., USA.	Heinsohn, G. E., Bjornson, A. S.	Chitosan salts as crop yield enhancers	19980730
WO 9639824	Chitogenics, Inc. USA	Henderson, S. E., Curran, D. T., Elson, C. M.	Negatively-charged chitosan derivative semiochemical delivery system	19961219
WO 0024260	Université Pierre et Marie Curie (Paris VI), France	Khoury, C., Minier, M., Le Goffic, F., Van Huynh, N.	Glycolytic enzyme-based fungicidal composition	20000504
WO 9708944	Kansas State University Research Foundation, USA	Kramer, K. J., Muthukrishnan, S., Choi, H. K., Corpuz, L., Gopalakrishnan, B.	A chitinase of *Manduca sexta* and a cDNA encoding it and their use in the preparation of insect resistant plants	19970313
WO 9942488	Vanson, Inc. USA	Nichols, E. J.	Composition and method for reducing transpiration in plants	19990826
WO 9403062	Novasso OY, Finland	Struszczyk, H., Kivekaes, O.	Seed encrusting with microcrystalline chitosan	19940217
WO 0119187	Instytut Wlokien Chemicznych, Poland; Instytut Ochrony Roslin; Instytut Sadownictwa i Kwiaciarstwa	Struszczyk, H., Niekraszewicz, A., Wisniewska-Wrona, M., Urbanowski, A., Pospieszny, H., Orlikowski, L., Wojdyla, A., Skrzypczak, C.	Chitosan gel for protecting plants against diseases	20010322
WO 9945784	Oji Paper Co., Ltd., Japan	Takahashi, T.	Antimicrobial agents containing Eucalyptus leaf extracts and chitosan	19990916
WO 9709879	Bioestimulantes Organicos, Lda., Portugal	Villanueva, J., Valenzuela, P.	Chitosan formulation to increase resistance of plants to pathogenic agents and environmental stress	19970320

Tab. 23 Biochemical genetics

Patent number	*Patent holder*	*Inventors*	*Title of patent*	*Publication date (YYYYMMDD)*
WO 0009729	Pioneer Hi-Bred International, Inc., USA	Dhugga, K. S., Anderson, P. C., Nichols, S. E.	Expression of chitin synthase and chitin deacetylase genes in plants to alter the cell wall for industrial uses and improved disease resistance	20000224
WO 9201792	SANOFI S. A., France; Societe Nationale Elf Aquitaine	Dubois, M., Grison, R., Leguay, J. J., Pignard, A., Toppan, A.	Chimeric endochitinase gene and transgenic, pathogen-resistant plants expressing it	19920206
WO 0024874	Johns Hopkins University, USA	Fomenkov, A., Keyhani, N. O., Roseman, S.	Polynucleotide and polypeptide of *Vibrio furnissii* gene chiE extracellular chitinase, sequences and biological uses thereof	20000504
WO 9747752	Icos Corp., USA	Gray, P. W.	Human chitinase, its cDNA sequence and cloning, and its use in antifungal compositions	19971218
WO 9946390	Icos Corp., USA	Gray, P. W., Tjoelker, L. W.	Chitinase chitin-binding fragments and their use in chitin determination and as anti-fungal agents	19990916
WO 9424288	Cornell Research Foundation, Inc., USA	Harman, G. E., Tronsmo, A., Hayes, C. K., Lorito, M.	Gene encoding for endochitinase	19941027
WO 9906565	Ice Biotech Inc., Canada	Hew, C., Xiong, F., Moffatt, B., Griffith, M.	Cloning expression systems and cDNA sequences encoding antifreeze proteins from winter rye	19990211
WO 9736917	Smith-Kline Beecham Corp., USA; Human Genome Sciences, Inc.	Kirkpatrick, R. B., Thotakura, N. R., Ni, J., Gentz, R., Chopra, A.	Human chitotriosidase and its cDNA, detection of altered chitotriosidase expression, and diagnosis of tissue remodeling disorder	19971009
WO 9716540	Chemgenics Pharmaceuticals, Inc., USA	Koltin, Y., Riggle, P., Gavrias, V., Bulawa, C., Winter, K.	Yeast chitin synthase 1 gene sequence, recombinant vector and auxotrophic host cell, and method for determining growth-associated proteins and genetic elements	19970509
WO 0116353	United States of America, as Represented by the Secretary of Agriculture, USA; Novo Nordisk Biotech, Inc.	Okubara, P. A., Blechl, A. E., Hohn, T. M., Berka, R. M.	Nucleic acid sequences encoding cell wall-degrading enzymes and their use to engineer plant resistance to *Fusarium* and other pathogens	20010308
WO 0109283	The Regents of the University of California, USA; The	Ronald, P., He, Z., Chory, J., Lamb, C., Li, J.	DNA constructs encoding chimeric plant RRK receptors (Bri1::Xa21 and Hevein::Xa21), and	20010208

Tab. 23 (cont.)

Patent number	*Patent holder*	*Inventors*	*Title of patent*	*Publication date (YYYYMMDD)*
	Salk Institute for Biological Studies		their use in production of transgenic plants	
WO 9625424	Johns Hopkins University, USA	Roseman, S., Bassler, B., Keyhani, N., Chitlaru, E., Rowe, C., Yu, C.	Bacterial catabolism of chitin	19960822
WO 0056908	Pioneer Hi-Bred International, Inc., USA	Simmons, C. R., Yalpani, N.	Maize chitinases and cDNAs and method of modulating chitinase activity in plants	20000928
WO 0102577	Provalis UK Ltd., UK	Smith, C. J., Thompson, S. E., Smith, M. W., Peek, K., Sizer, P. J. H., Wilkinson, M. C.	*Pseudomonas aeruginosa* antigens related to chitinase and groEL genes and their use in vaccine preparation	20010111
WO 9829542	Japan Tobacco Inc., Japan	Takakura, Y., Inoue, T., Saito, H., Ito, T.	Cloning of gene for chitinase of rice, characterization of floral organ-specific promoter of the monocotyledonous plant, and use for breeding pathogen-resistant plant	19980709
WO 9961635	Mycogen Corp., USA	Yenofsky, R. L., Fine, M., Rangan, T. S., Anderson, D. M.	Cotton cells, plants, and seeds genetically engineered to express insecticidal and fungicidal chitin binding proteins (lectins)	19991202

Tab. 24 Biochemical methods

Patent number	*Patent holder*	*Inventors*	*Title of patent*	*Publication date (YYYYMMDD)*
WO 9925390	Albert Einstein College of Medicine of Yeshiva University, USA	Burns, E. R., Wittner, M., Faskowitz, F.	A method for detecting chitin-containing organisms	19990527
WO 0047716	Advanced Medical Solutions Ltd., UK	Hamilton, D. W., Ives, C. L., Middleton, I. P., Rosetto, C.	Fibers for culturing eukaryotic cells	20000817
WO 9735037	Smith-Kline Beecham Corp., USA	Kirkpatrick, R. B.	Tissue remodeling proteins	19970925
WO 9802742	Board of Supervisors of Louisiana State University, USA	Laine, R. A., Lo, W. C. J.	Diagnosis of fungal infections with a chitinase	19980122
WO 9851712	National Institute of Sericultural and Entomological Science Ministry, Japan	Tsukada, M., Shirata, A., Hayasaka, S.	Chitin beads, chitosan beads, processes for producing these beads, carriers made of these beads, and processes for producing microsporidian spores	19981119
WO 9217786	SRI International, USA	Tuse, D., Dousman, L.	Assay device and enzymic-spectrophotometric method of detecting chitin	19921015

Tab. 25 Enzymes

Patent number	*Patent holder*	*Inventors*	*Title of patent*	*Publication date (YYYYMMDD)*
WO 9640940	Universiteit Van Amsterdam, The Netherlands	Aerts, J. M. F.	A human macrophage chitinase for use in the treatment or prophylaxis of infection by chitin-containing pathogens and manufacture of the enzyme by expression of the cloned gene	19961219
WO 9402598	Cornell Research Foundation, Inc., USA	Harman, G. E., Broadway, R. M., Tronsmo, A., Lorito, M., Hayes, C. K., Di, P. A.	Chitobiase and endochitinase of *Trichoderma harzianum* and their use in control of plant pathogenic fungi	19940203
WO 9958650	Biomarin Pharmaceuticals, USA	Klock, J. C., Mishra, C., Starr, C. M.	Lytic enzymes from *Trichoderma useful* for treating fungal infections	19991118
WO 9854304	Institut für Pflanzengenetik und Kulturpflanzenfor-schung, Germany	Schlesier, B., Wolf, A., Koch, G., Horstmann, C., Konig, S., Gillandt, R.-m.	Chitinase of *Canavalia ensiformis* seed and its use in preparation of *N*-acetyl-D-glucosamine from chitin	19981203
WO 9413815	Institute for Molecular Biology and Biotechnology/Forth, Greece	Thireos, G., Kafetzopoulos, D.	Microbial chitin deacetylases and the genes encoding them	19940623
WO 9853053	Foundation for Research and Technology, Hellas, Greece	Thireos, G., Kafetzopoulos, D.	Cloning of cDNA for an arthropod chitin synthase and its use for developing chitin synthase inhibitors	19981126
WO 9806859	Human Genome Sciences, Inc., USA	Thotakura, N. R., Chopra, A., Choi, G. H., Genzt, R. L., Rosen, C. A.	Human chitinase α and chitinase α-2 and their cDNA sequences and therapeutic applications	19980219
WO 0073488	Board of Regents of the University of Texas System, USA	Vinetz, J. M.	*Plasmodium* gene CHT1 chitinases and cDNAs and methods for preventing malaria transmission by mosquitoes	20001207

Tab. 26 Cosmetics and ingredients

Patent number	*Patent holder*	*Inventors*	*Title of patent*	*Publication date (YYYYMMDD)*
WO 0060038	Laboratoire Medidom S.A., Switzerland	Cantoro, A.	Viscosity-enhanced ophthalmic solutions having detergent action and their use on contact lenses	20001012
WO 9219218	Procter and Gamble Co., USA	Cooke, R. J., Date, R. F., Barrington, N. G., Spengler, E. G.	Carboxymethylchitin-containing cosmetic composition for skin and hair	19921112
WO 9822210	Merck S.A., France	Grandmontagne, B., Marchio, F.	Chitin or chitin derivative microcapsules containing a hydrophobic sunscreen	19980528
WO 0054738	Cognis Deutschland GmbH, Germany; Biotec ASA	Griesbach, U., Wachter, R., Ansmann, A., Fabry, B., Eisfeld, W., Engstad, R. E.	Cosmetic preparations containing chitosans and β-(1-3)-glucans	20000921
WO 9525752	Kao Corp., Japan	Hasebe, Y., Sawada, M., Furukawa, M., Nakayama, T., Kodama, K., Ito, Y., Nakamura, G., Fukumoto, Y.	Fine porous polysaccharide particles and their use in cosmetics	19950928
WO 9817245	Henkel KGaA, Germany	Heilemann, A., Holzer, J., Horlacher, P., Sander, A., Wachter, R.	Collagen-free cosmetic preparations obtained from crosslinked chitosan hydrogels	19980430
WO 0047177	Cognis Deutschland GmbH, Germany	Kropf, C., Fabry, B., Foerster, T., Wachter, R., Reil, S., Panzer, C.	Use of nanoscale chitosans and/or chitosan derivatives	20000817
WO 9530403	Larm, O., Sweden	Larm, O.	Dentifrices containing chitosan and sulfated polysaccharides for improvement of oral hygiene	19951116
WO 0025734	Cognis Deutschland GmbH, Germany	Panzer, C., Tesmann, H., Wachter, R.	Ethanolic cosmetic preparations containing chitosan	20000511
WO 9413774	Allergan, Inc., USA	Powell, C. H., Karageozian, H., Currie, J. P.	Compositions and methods for cleaning hydrophilic contact lenses	19940623
WO 9629979	Unilever Plc, UK	Tsaur, L. S., Shen, S., Jobling, M., Aronson, M. P.	Liquid cleansers comprising hydrogel dispersions	19961003

Tab. 27 Industrial biotechnology

Patent number	*Patent holder*	*Inventors*	*Title of patent*	*Publication date (YYYYMMDD)*
WO 9317121	Phyton Catalytic, Inc., USA; Prince, C. L.; Schubmehl, B. F.; Kane, E. J.; Roach, B.	Bringi, V., Kadkade, P. G.	Increased yields of taxol and taxanes from cell cultures of Taxus species	19930902
WO 9731121	University of British Columbia, Canada	Haynes, C. A., Aloise, P., Creagh, A. L.	Process for producing *N*-acetyl-D-glucosamine	19970828
WO 9409148	Toray Industries, Inc., Japan	Miwa, K., Fukuyama, M., Uchiyama, T.	Process for producing major histocompatibility antigen class II (MHC) protein and immobilized MHC II or fragments	19940428
WO 9314216	Nakamura, Tomotaka, Japan	Nakamura, K., Yamada, C.	Manufacture of chitin and chitosan with dermatophyte	19930722

Tab. 28 Food and feed chemistry.

Patent number	*Patent holder*	*Inventors*	*Title of patent*	*Publication date (YYYYMMDD)*
WO 0004086	Kemestrie Inc., Canada	Chornet, E., Dumitriu, S.	Polyionic hydrogels based on xanthan and chitosan for stabilization and controlled release of vitamins	20000127
WO 0005974	E-Nutriceuticals, Inc., USA	Hardinge-Lyme, N.	Fat-absorbing chitosan-containing liquid compositions for use in weight-loss programs	20000210
WO 0103511	CH20 Incorp., USA	Iverson, C. E., Ager, S. P.	Method of coating food products, and a coating composition	20010118
WO 9748291	Fuji Oil Co., Ltd., Japan	Nakamura, A., Hattori, M., Maeda, H.	Food preservatives	19971224
WO 9837775	Nestle S.A., Switzerland	Popper, G., Ekstedt, S., Walkenstrom, P., Hermansson, A.-M.	Gelled emulsion products containing chitosan	19980903

Tab. 29 Immunochemistry.

Patent number	*Patent holder*	*Inventors*	*Title of patent*	*Publication date (YYYYMMDD)*
WO 0123430	Icos Corp., USA	Allison, D. S., Dietsch, G. N., Gray, P. W., Shaw, K. D., Steiner, B. H.	Chitinase immunoglobulin fusion products	20010405
WO 9716208	Medeva Holdings B.V., The Netherlands	Chatfield, S. N.	Influenza vaccine compositions	19970509
WO 9609805	Zonagen, Inc., USA	Podolski, J. S., Hsu, K., Bhogal, B., Singh, G.	Chitosan induced immunopotentiation	19960404

Tab. 30 Polysacchride chemistry and processes.

Patent number	*Patent holder*	*Inventors*	*Title of patent*	*Publication date (YYYYMMDD)*
WO 9964470	Bioeffect A/S, Norway	Arntsen, D.	An integrated plant and method for automatic multistage production of chitosan	19991216
WO 0024490	Cognis Deutschland Gmbh, Germany	Blum, S., Horlacher, P., Trius, A., Wagemans, P., Weitkemper, N., Albiez, W.	Continuous extraction of natural substances and production of chitin and chitosan	20000504
WO 9626965	Akzo Nobel N.V., The Netherlands	Bredehorst, R., Pomato, N., Scheel, O., Thiem, J.	High-yield manufacture of water-soluble dimeric to decameric chitin oligomers	19960906
WO 9206136	Pfizer Inc., USA	Chandler, C. E., Curatolo, W. J.	Hypocholesterolemic quaternary ammonium salts derived from chitosan	19920416
WO 9208742	Du Pont de Nemours, E. I., and Co., USA	Dingilian, E. O., Heinsohn, G. E.	*N*-halochitosans and their preparation and uses	19920529
WO 9641818	Drohan, W. N., et al., USA	Drohan, W. N., Macphee, M. J., Miekka, S. I., Singh, M., Elson, C., Taylor, J.	Manufacture and use of supplemented chitin hydrogels	19961227
WO 9400512	Albany International Corp., USA	Eagles, D. B., Bakis, G., Jeffery, A. B., Mermingis, C., Hagoort, T. H.	Method of producing polysaccharide foams	19940106
WO 9806489	Wolff Walsrode AG, Germany	Engelhardt, J., Koch, W., Pannek, J.-B., Vorlop, K.-D., Patel, A. V.	Polymeric immobilizate for encapsulation of useful materials	19980219

Tab. 30 (cont.)

Patent number	*Patent holder*	*Inventors*	*Title of patent*	*Publication date (YYYYMMDD)*
WO 9921648	Everest-Todd Research & Development Ltd., UK	Everest-Todd, S.	Formulation of oils	19990506
WO 9730092	Innovative Technologies Ltd., UK	Gilding, D. K., Al-Lamee, K. G.	Biocompatible hydrogels made from crosslinked chitosan and galactomannan-type or/and pectic substances and their manufacture and uses	19970821
WO 9745453	Virginia Tech Intellectual Properties, Inc., USA	Glasser, W. G., Jain, R. K.	Manufacture of ester-cross-linked chitosan hydrogels as support material and quaternized, cyclodextrin-modified chitosan	19971204
WO 9900452	Virginia Tech Intellectual Properties, Inc., USA	Glasser, W. G., Jain, R. K.	Method of making magnetic, crosslinked chitosan support materials and products thereof	19990107
WO 8606082	Matcon Raadgwende Ingenioerfirma A/S, Denmark	Joensen, J. O.	Recovering chitin from materials in which chitin occurs together with or connected to proteinaceous substances	19861023
WO 9938893	Matcon Radgivende Ingeniorfirma A/S, Denmark	Joensen, J. O.	Chitosan production by deacetylation of chitin with reuse of the deacetylating hydroxide solution	19990805
WO 9808877	Wakunaga Seiyaku Kabushiki Kaisha, Japan	Kakimoto, M., Ushijima, M., Kasuga, S., Shiraishi, S., Itakura, Y.	Chitosan derivatives, their manufacture and uses in treatment of constipation	19980305
WO 9742226	Nippon Suisan Kaisha, Ltd., Japan	Kawahara, H., Jinno, S., Okita, Y.	Chitin derivatives having lower carboxyalkyl groups as hydrophilic substituents and hydrophobic substituents, and high-molecular micellar supports and aqueous compositions thereof	19971113
WO 9817693	Transgene S.A., France	Kolbe, H.	Chitosan compositions for transferring therapeutic agents into host cells	19980430
WO 0132751	Cognis Deutschland GmbH, Germany	Kropf, C., Reil, S.	Method for producing nanoparticulate chitosans or chitosan derivatives	20010510
WO 0011038	Kumar, G., Payne, G. F., USA	Kumar, G., Payne, G. F.	Modified chitosan polymers and enzymic methods for their production	20000302

Tab. 30 (cont.)

Patent number	*Patent holder*	*Inventors*	*Title of patent*	*Publication date (YYYYMMDD)*
WO 9948809	Teknimed S.A., France	Lacout, J.-L., Hatim, Z., Frache-Botton, M.	Preparation and *in situ* hardening of hydroxyapatite formulations for biomaterial for synthetic bone and teeth	19990930
WO 9961482	Trustees of Tufts College, USA; University of Massachusetts Lowell; United States Dept. of the Army	Lee, J. W., Yeomans, W. G., Allen, A. L., Deng, F., Gross, R. A., Kaplan, D. L.	Production of chitosan- and chitin-like polymers	19991202
WO 9209635	Merck Patent GmbH, Germany	Muzzarelli, R.	Methylpyrrolidinone-substituted chitosans, production and uses thereof	19920611
WO 0123398	Carbion Oy, Finland	Natunen, J.	Preparation of acetamido-deoxy fucosylated oligosaccharides via enzymic glycosidation reaction	20010405
WO 9729132	Merck Patent GmbH, Germany	Nies, B.	Crosslinked products of amino-group-containing biopolymers	19970814
WO 9533773	Vitaphore Corp., USA	Prior, J. J., Yamamoto, R. K., Brode, G. L.	Lysozyme-degradable biomaterials and their uses	19951214
WO 9306136	Fidia S.P.A., Italy	Romeo, A., Toffano, G., Callegaro, L.	Chitosan-based carboxylic polysaccharide derivatives	19930401
WO 0104207	Cognis Deutschland GmbH, Germany	Schafer, G., Holzer, J., Sander, A., Heilemann, A.	Chitosan preparations free from crosslinking agents	20010118
WO 9105808	Firextra Oy, Finland	Struszczyk, H., Kivekaes, O.	Method and apparatus for manufacture of chitosan and other products from shells of organisms, especially marine organisms	19910502
WO 9100298	Firextra Oy, Finland; Instytut Wlokien Chemicznych	Struszczyk, H., Niekraszewicz, A., Wrzesniewska-Tosik, K., Koch, S., Kivekas, O.	Method for continuous manufacture of microcrystalline chitosan	19910110
WO 9901480	National Starch and Chemical Investment Holding Corp., USA	Sugimoto, M., Shigemasa, Y.	Hydrophilic chitin derivatives and manufacture methods	19990114
WO 9901479	National Starch and Chemical Investment Holding Corp., USA	Sugimoto, M., Shigemasa, Y.	Chitosan derivatives and manufacture methods	19990114

Tab. 30 (cont.)

Patent number	*Patent holder*	*Inventors*	*Title of patent*	*Publication date (YYYYMMDD)*
WO 9625437	Abion Beteiligungs- und Verwaltungsgesellschaft mbH, Germany	Teslenko, A.	Preparation of chitosan-glucan complexes and their uses	19960822
WO 0014155	University of Strathclyde, UK	Uchegbu, I. F.	Erodible solid hydrogels for delivery of biologically active materials	20000316
WO 9944710	Virginia Tech Intellectual Properties, Inc., USA	Velander, W. H., Van Cott, K. E., Van Tassell, R.	Inside-out crosslinked and commercial-scale hydrogels, and sub-macromolecular selective purification using the hydrogels	19990910
WO 9803523	Villalva Basabe, J., Spain	Villalva Basabe, J.	Process for obtaining [2,1-d]-2-oxazolines of amino sugars from chitin or its derivatives	19980129
WO 0068273	Cognis Deutschland GmbH, Germany	Wachter, R., Griesbach, U., Horlacher, P.	Collagen-free compositions for cosmetics	20001116
WO 9840411	Henkel KGaA, Germany	Wachter, R., Priebe, C., Ritter, W., Skwiercz, M.	Crosslinking of chitosans	19980917
WO 9707139	Ciba-Geigy A.-G., Switzerland	Yen, S.-F., Sou, M.	Process for sterilizing solutions of chitosan or its derivatives	19970227

Tab. 31 Pharmaceuticals and pharmacology

Patent number	*Patent holder*	*Inventors*	*Title of patent*	*Publication date (YYYYMMDD)*
WO 9407999	Brown University Research Foundation, USA	Aebischer, P., Zielinski, B. A.	Chitosan matrixes for encapsulated cells for implantation	19940414
WO 9804244	Universidade de Santiago de Compostela, Spain	Alonso Fernandez, M. J., Calvo Salve, P., Remunan Lopes, C., Vila Jato, J. L.	Application of nanoparticles based on hydrophilic polymers as pharmaceutical forms	19980205
WO 0056362	The Secretary of State for Defence, UK	Alpar, H. O., Eyles, J. E., Somavarapu, S., Williamson, E. D., Baillie, L. W. J.	Immunostimulants comprising polycationic carbohydrates	20000928
WO 0050090	DCV, Inc., USA	Angerer, J. D., Cyron, D. M., Iyer, S., Jerrell, T. A.	Dry acid-chitosan complexes	20000831
WO 9919003	Battelle Memorial Institute, USA	Armstrong, B. L., Campbell, A. A., Gutowska, A., Song, L.	Implantable polymer/ceramic composites	19990422
WO 9620730	Astra AB, Sweden	Artursson, P., Schipper, N.	Chitosan polymer with specific degree of acetylation as drug absorption enhancer	19960711
WO 9211844	Enzytech, Inc., USA	Auer, H. E., Brown, L. R., Gross, A.	Stabilization of proteins by cationic biopolymers	19920723
WO 0009123	F. Hoffmann-La Roche A.-G., Switzerland	Bailly, J., Fleury, A., Hadvary, P., Lengsfeld, H., Steffen, H.	Pharmaceutical compositions containing lipase inhibitors and chitosan	20000224
WO 9313137	Howmedica Inc., USA	Barry, J. J., Higham, P. A., Mann, N. N.	Process for insolubilizing *N*-carboxyalkyl derivatives of chitosan	19930708
WO 0001373	Ecole Polytechnique Federale De Lausanne, Switzerland; Kiddle, Simon	Bartkowiak, A., Hunkeler, D.	Materials and methods relating to encapsulation by formation of complex reaction between oppositely charged polymers	20000113
WO 9948480	Aventis Research & Technologies GmbH & Co. KG, Germany	Bayer, U.	Method for the production of microcapsules	19990930
WO 9948479	Aventis Research & Technologies GmbH & Co. KG, Germany	Bayer, U., Hahn, B., Majeres, A.	Slow release microcapsules	19990930

Tab. 31 (cont.)

Patent number	Patent holder	Inventors	Title of patent	Publication date (YYYYMMDD)
WO 9503786	Fidia Advanced Biopolymers SRL, Italy	Benedetti, L., Callegaro, L.	Pharmaceutical compositions for topical use containing hyaluronic acid and its derivatives	19950209
WO 0100246	Shearwater Polymers, Incorp., USA	Bentley, M. D., Zhao, X.	Hydrogels derived from chitosan and poly(ethylene glycol) or related polymers	20010104
WO 0122973	Medicarb AB, Sweden	Bergstrom, T., Trybala, E., Back, M., Larm, O.	The use of a positively charged carbohydrate polymer such as chitosan for the prevention of infection	20010405
WO 0057931	Paul Hartmann A.-G., Germany	Bitterhof, A.	Breast-milk absorbent pad containing chitosan to improve performance	20001005
WO 9961079	Kimberly-Clark Worldwide, Inc., USA	Boney, L. C., Borders, R. A., Di Luccio, R. C., Kepner, E. S., Yahiaoui, A.	Enhanced odor absorption by natural and synthetic polymers	19991202
WO 0126635	Elan Pharma International Ltd., Ireland; McGurk, S. L.	Bosch, H. W., Cooper, E. R.	Bioadhesive nanoparticulate compositions having cationic surface stabilizers	20010419
WO 9209636	Baker Cummins Dermatologicals, Inc., USA	Brode, G. L., Adams, N. W., Figueroa, R.	Skin protective compositions and method of inhibiting skin irritation	19920611
WO 9523235	Myco Pharmaceuticals Inc., USA	Bulawa, C. E.	Methods for identifying inhibitors of fungal pathogenicity and use in pharmacologically and agriculturally useful fungicide analysis and insecticide analysis	19950831
WO 9610421	Medeva Holdings B.V., The Netherlands	Chatfield, S. N.	Vaccine compositions	19960411
WO 9907416	Bio-Syntech Ltd., Canada	Chenite, A., Chaput, C., Combes, C., Jalal, F., Selmani, A.	Temperature-controlled pH-dependant formation of ionic polysaccharide gels	19990218
WO 9613253	Dong Kook Pharmaceutical Co., Ltd., S. Korea	Chung, C. P., Lee, S. J.	Biodegradable sustained release preparation for treating periodontitis	19960509
WO 9901498	Danbiosyst UK Ltd., UK	Davis, S. S., Lin, W., Bignotti, F., Ferruti, P.	Conjugate of polyethylene glycol and chitosan	19990114

Tab. 31 (cont.)

Patent number	*Patent holder*	*Inventors*	*Title of patent*	*Publication date (YYYYMMDD)*
WO 9426788	Development Biotechnological Processes S.N.C., Italy; IMS-International Medical Servica S.R.L	De Rosa, A., Rossi, A., Affaitati, P.	Process for the preparation of iodinated biopolymers having disinfectant and cicatrizing activity, and the iodinated biopolymers obtainable thereby	19941124
WO 0056275	Virbac, France	Derrieu, G., Pougnas, J.-I.	Pharmaceutical compositions based on chondroitin sulphate and chitosan for preventing or treating rheumatic disorders by systemic administration	20000928
WO 9909962	Reckitt & Colman Products Ltd., UK	Dettmar, P. W., Jolliffe, I. G., Skaugrud, O.	In situ formation of bioadhesive polymeric material	19990304
WO 0107486	Polygene Ltd., Israel	Domb, A. J.	A biodegradable polycation composition for delivery of an anionic macromolecule in gene therapy	20010201
WO 9932697	Kimberly-Clark Worldwide, Inc., USA	Dutkiewicz, J., Bevernitz, K. J., Huard, L. S., Qin, J., Sun, T., Wallajapet, P. R. R.	Antimicrobial structures based on polymer nonwoven fabrics for diapers	19990701
WO 9635433	Chitogenics, Inc., USA	Elson, C. M.	*N,O*-carboxymethylchitosan for prevention of surgical adhesions	19961114
WO 0071136	Chitogenics, Inc., USA	Elson, C. M., Lee, T. D. G.	Adhesive *N,O*-carboxymethyl chitosan coatings which inhibit attachment of substrate-dependent cells and proteins	20001130
WO 0015192	Zonagen, Inc., USA	Fontenot, G. K.	Methods and materials related to bioadhesive contraceptive gels	20000323
WO 9528158	Janssen Pharmaceutica N.V., Belgium	Francois, M. K. J., Embrechts, R. C. A., Illum, L.	Intranasal antimigraine composition	19951026
WO 9729760	Furda, Ivan, USA	Furda, I.	Multifunctional fat absorption and blood cholesterol reducing formulation containing chitosan and nicotinic acid	19970821
WO 9902252	Norsk Hydro ASA, Norway	Gaserod, O., Skaugrud, O., Dettmar, P., Sjak-Braek, G., Jolliffe, I.	Preparation of capsules based on ionic polysaccharides	19990121
WO 9602260	Medicarb AB, Sweden	Gouda, I., Larm, O.	Wound healing agents comprising polysaccharides conjugates with chitosan	19960201

Tab. 31 (cont.)

Patent number	*Patent holder*	*Inventors*	*Title of patent*	*Publication date (YYYYMMDD)*
WO 0030609	Laboratoire Mecidom S.A., Switzerland	Gurny, R., Felt, O.	Aqueous ophthalmic formulations comprising chitosan	20000602
WO 0125321	Alvito Biotechnologie GmbH, Germany	Hahnemann, B., Pahmeier, A., Loetzbeyer, T.	Three-dimensional matrix for producing cell transplants	20010412
WO 9608149	USA	Haimovich, B., Freeman, A., Greco, R.	Thromboresistant surface treatment for biomaterials	19960321
WO 9725994	Hansson, H.-A., Sweden	Hansson, H.-A.	Control of healing process	19970724
WO 9602258	Astra AB, Sweden	Hansson, H.-a., Johansson-Ruden, G., Larm, O.	Immobilized polysaccharide as anti-adhesion agent in wound healing	19960201
WO 9602259	Astra AB, Sweden	Hansson, H.-A., Johansson-Ruden, G., Larm, O.	Hard tissue stimulating agent comprising chitosan and a polysaccharide immobilized thereto	19960201
WO 9707802	Shaman Pharmaceuticals, Inc., USA	Hector, R. F., Sabouni, A.	Methods and compositions for treating fungal infections in mammals	19970306
WO 9855149	Shionogi & Co., Ltd., Japan	Horie, K., Masuda, K., Sakagami, M.	Use of drug carriers for producing lymph node migrating drugs	19981210
WO 9531184	Vaccine Technologies Pty. Ltd., Australia	Husband, A., Kingston, D.	Film-coated microparticles for bioactive molecule delivery	19951123
WO 9315737	Danbiosyst UK Ltd., UK	Illum, L.	Compositions for nasal administration containing polar metabolites of opioid analgesics	19930819
WO 9720576	Danbiosyst UK Ltd., UK	Illum, L.	Vaccine compositions for intranasal administration containing chitosan as adjuvant	19970612
WO 9801160	Danbiosyst UK Ltd., UK	Illum, L.	Chitosan compositions suitable for delivery of genes to epithelial cells	19980115
WO 9203480	Drug Delivery System Institute, Ltd., Japan	Inoue, K., Ito, T., Okuno, S., Aono, K.	Preparation of *N*-acetyl-*O*-(carboxymethyl)chitosan derivatives for drug targeting	19920305
WO 9833521	Korea Institute of Science and Technology, S. Korea	Jeong, S. Y., Kwon, I.-C., Kim, Y.-H., Seong, S.-Y.	Pneumococcal polysaccharide conjugate with cholera toxin B subunit for use in vaccines	19980806

Tab. 31 (cont.)

Patent number	*Patent holder*	*Inventors*	*Title of patent*	*Publication date (YYYYMMDD)*
WO 9826788	Noviscens AB, Sweden	Johansson, B., Niklasson, B.	Medical composition and use thereof for the manufacture of a topical barrier formulation, a UV-radiation absorbing formulation, or an antiviral, antifungal, or antiinflammatory formulation	19980625
WO 9916478	Metagen, Llc, USA	Johnson, J. R., Marx, J. G., Johnson, W. D.	Bone substitutes comprising rigid framework filled with osteoconductive materials and resilient polymers	19990408
WO 9822114	Dumex-Alpharma A/S, Denmark	Jorgensen, T., Moss, J., Nicolajsen, H. V., Nielsen, L. S.	Method and carbohydrate composition for promoting tissue repair	19980528
WO 0050507	Cognis Corporation, USA	Kraechter, H. U., Wachter, R.	Latex products with reduced hypersensitivity	20000831
WO 0074720	Mochida Pharmaceutical Co., Ltd., Japan	Kudo, Y., Ueshima, H., Sakai, K.	System for release in lower digestive tract	20001214
WO 9805341	Medicarb AB, Sweden	Larm, O., Back, M., Bergstrom, T.	The use of heparin or heparan sulfate in combination with chitosan for the prevention or treatment of infections caused by herpes virus	19980212
WO 9819718	Challenge Bioproducts Co., Ltd., Taiwan; Lee, H. L., Lin, C.-k., Sung, H.-w.	Lee, T. C.-j., Sung, H.-w.	Chemical modification of biomedical materials with genipin	19980514
WO 9805304	Cytotherapeutics Inc., USA	Li, R., Hazlett, T. F.	Biocompatible devices with foam scaffolds	19980212
WO 0006210	Minnesota Mining and Manufacturing Company, USA	Lyon, K. R., Rock, M. M., Jr.	Method for the manufacture of antimicrobial articles	20000210
WO 9927960	Medeva Europe Ltd., UK	Makin, J. C., Bacon, A. D.	Vaccine compositions for mucosal administration comprising chitosan	19990610
WO 0100792	Marchosky, J. A., USA	Marchosky, J. A.	Compositions containing growth factors and methods for forming and strengthening bone	20010104
WO 9922739	Taisho Pharmaceutical Co., Ltd., Japan	Miwa, A., Yamahira, T., Ishibe, A., Itai, S.	Medicinal steroid compound composition	19990514

Tab. 31 (cont.)

Patent number	*Patent holder*	*Inventors*	*Title of patent*	*Publication date (YYYYMMDD)*
WO 9742975	Genemedicine, Inc., USA	Mumper, R. J., Rolland, A.	Chitosan-related compositions and methods for delivery of nucleic acids into a cell	19971120
WO 9919366	Abiogen Pharma SRL, Italy	Muzzarelli, R., Rosini, S., Trasciatti, S.	Chelates of chitosans and alkaline-earth metal insoluble salts and the use thereof as medicaments useful in osteogenesis	19990422
WO 9921566	Rexall Sundown, Inc., USA	Myers, A. E., Priddy, M. R.	Dietary compositions with lipid binding properties for weight management and serum lipid reduction	19990506
WO 0124840	Coloplast A/S, Denmark	Nielsen, B.	Wound care device based on chitosan fibers	20010412
WO 9947162	Wound Healing of Oklahoma, Inc., USA	Nordquist, R. E., Chen, W. R., Carubelli, R.	Laser/sensitizer-assisted immunotherapy of cancer	19990923
WO 0115716	Medex Scientific (UK) Ltd., UK	Oben, J. E.	Plant extracts, chitosan derivatives and antioxidant vitamins for treatment of weight-related disorders	20010308
WO 9747323	Zonagen, Inc., USA	Podolski, J. S., Hsu, K. T., Singh, G.	Chitosan drug delivery system	19971218
WO 9842374	Zonagen, Inc., USA	Podolski, J. S., Martinez, M. L.	Chitosan-induced immunopotentiation	19981001
WO 9613284	Innovative Technologies Ltd., UK	Qin, Y., Gilding, K. D.	Wound treatment composition	19960509
WO 9613282	Innovative Technologies Ltd., UK	Qin, Y., Gilding, K. D.	Absorbent wound dressing	19960509
WO 9706782	Ciba-Geigy A.-G., Switzerland	Reed, K. W., Yen, S.-F.	Ophthalmic pharmaceuticals containing *O*-carboxyalkyl chitosan	19970227
WO 0124795	SKW Trostberg A.-G., Germany	Schuhbauer, H., Pischel, I., Bernkop-Schnuerch, A.	Delayed-release dosage form containing α-lipoic acid (derivatives)	20010412
WO 9965521	Zonagen, Inc., USA	Seid, C. A., Singh, G.	Prostate-associated antigen composition with chitosan metal chelate for the treatment of prostatic carcinoma	19991223
WO 0021567	Societe De Conseils De Recherches Et D'applications Scientifiques S.A., France	Shalaby, S. W., Jackson, S. A., Ignatious, F. X., Moreau, J.-P., Russell, R. M.	Ionic molecular conjugates of N-acylated derivatives of poly(2-amino-2-deoxy-D-glucose) and polypeptides	20000420
WO 0110467	Dainippon Pharmaceutical Co., Ltd., Japan	Shimono, N., Mori, M., Higashi, Y.	Solid preparations containing chitosan powder and process for producing the same	20010215

Tab. 31 (cont.)

Patent number	*Patent holder*	*Inventors*	*Title of patent*	*Publication date (YYYYMMDD)*
WO 0015766	Oncopharmaceutical, Inc., USA	Singh, S. S.	Treatment of oncologic tumors with an injectable formulation of a Golgi apparatus disturbing agent	20000323
WO 0030658	The University of Akron, USA	Smith, D. J., Serhatkulu, S.	Chitosan-based nitric oxide donor compositions	20000602
WO 0016817	Korea Atomic Energy Research Institute, S. Korea	Son, Y.-S., Youn, Y.-H., Hong, S.-I., Lee, S.-H., Gin, Y.-J.	Dermal scaffold using neutralized chitosan sponge or neutralized chitosan/collagen mixed sponge	20000330
WO 9324112	Clover Consolidated, Ltd., Switzerland	Soon, S. P., Heintz, R. A.	Microencapsulation of cells for transplantation	19931209
WO 0024785	Novasso Oy, Finland	Struszczyk, H., Pomoell, H., Wulff, M., Saynatjoki, E., Ylitalo, P., Wuolijoki, E., Hakli, H.	Chitosan-based pharmaceuticals for reduction of cholesterol and lipid contents	20000504
WO 9947186	University of Pittsburgh, USA; Wayne State University	Suh, J.-K., Matthew, H., Fu, F. H., Sechriest, F. V., Manson, T. T.	Chitosan-based composite materials containing glycosaminoglycan for cartilage repair	19990923
WO 9702125	FMC Corp., USA	Thomas, W. R., Pfeffer, H. A., Guiliano, B. A., Sewall, C. J., Tomko, S.	Process for making biopolymer gel microbeads	19970123
WO 9811903	Chicura Cooperatie U.A., The Netherlands	Timmermans, C.	Chitosan derivatives, and preparation thereof, for treating malignant tumors	19980326
WO 9945900	Kabushiki Kaisha Soken, Japan	Tokuyama, T., Jo, M.	Topical compositions for improving skin conditions	19990916
WO 9842755	University of Strathclyde, UK	Uchegbu, I. F.	Particulate drug carriers	19981001
WO 9205791	Dana-Farber Cancer Institute, Inc., USA	Vercellotti, S. V., Ruprecht, R. M., Gama Sosa, M. A.	Inhibition of viral replication with chitosan derivatives	19920416
WO 9843609	Henkel KGaA, Germany	Wachter, R., Horlacher, P.	Chitosan microspheres	19981008
WO 9930751	Kimberly-Clark Worldwide, Inc., USA	Wallajapet, P. R. R., Romans-Hess, A. Y., Ta, E. T., Qin, J.	Absorbent structure having balanced pH profile	19990624

Tab. 31 (cont.)

Patent number	*Patent holder*	*Inventors*	*Title of patent*	*Publication date (YYYYMMDD)*
WO 9312875	Vanderbilt University, USA	Wang, T. G. J.	Method and apparatus for producing uniform polymeric spheres	19930708
WO 9936075	Danbiosyst UK Ltd., UK	Watts, P. J., Davis, S. S.	Oral oil sorbing compositions for treatment of obesity	19990722
WO 9605810	Danbiosyst UK Ltd., UK	Watts, P. J., Illum, L.	Drug delivery composition containing chitosan or its derivative having a defined zeta-potential	19960229
WO 9830207	Danbiosyst UK Ltd., UK	Watts, P. J., Illum, L.	Chitosan-gelatin A microparticles	19980716
WO 9603142	Danbiosyst UK Ltd., UK	Watts, P.J., Illum, L.	Drug delivery composition for the nasal administration of antiviral agents	19960208
WO 9823275	Henkel KGaA,	Weitkemper, N., Fabry, B.	Use of mixtures of phytostanols and tocopherols for the production of hypocholesteremic agents	19980604
WO 9404135	Teikoku Seiyaku Kabushiki Kaisha, Japan	Yamada, A., Wato, T., Uchida, N., Kadoriku, M., Takama, S., Inamoto, Y.	Oral preparation for release in lower digestive tract	19940303
WO 9741894	Taisho Pharmaceutical Co., Ltd., Japan	Yamahira, T., Miwa, A., Ishibe, A., Itai, S.	Micellar aqueous composition and method for solubilizing hydrophobic drug	19971113
WO 9610426	Yokota, S., et al., Japan	Yokota, S., Shimokawa, S., Sonohara, R., Okada, A., Takahashi, K.	Osteoplastic graft	19960411
WO 0027889	Netech Inc., Japan	Yura, H., Saito, Y., Ishihara, M., Ono, K., Saeki, S.	Functional chitosan derivatives	20000518

Tab. 32 Textiles and paper

Patent number	*Patent holder*	*Inventors*	*Title of patent*	*Publication date (YYYYMMDD)*
WO 0049219	Foxwood Research Limited, UK	Blowes, P. C., Taylor, A. J., Roberts, G., Wood, F.	Substrates with biocidal properties and process for making them	20000824
WO 9851709	Hong, Y. K., S. Korea	Hong, Y. K.	Aqueous cellulose solution containing chitosan or alginic acid or derivatives and rayon fiber produced therefrom	19981119
WO 9904083	Consejo Superior De Investigaciones Cientificas, Spain	Julia Ferres, R., Erra Serrabasa, P., Diez Tascon, J. M.	Shrinkproofing process for wool	19990128
WO 0134897	Moritoshi Co., Ltd., Japan	Mori, T.	Fiber-treating agent and fiber	20010517
WO 9109163	Kemira Oy, Finland	Nousiainen, P., Struszczyk, H.	Microcrystalline chitosan-modified viscose fibers and method for their manufacture	19910627
WO 9812369	Mitsubishi Rayon Co., Ltd., Japan	Ohnishi, H., Nishihara, Y., Hosokawa, H., Oishi, S., Iwamoto, M., Fujii, Y., Itoh, H., Ohsuga, N.	Chitosan-containing acrylic fibers and process for preparing the same	19980326
WO 0046441	Gunze Limited, Japan	Ozawa, N., Akieda, S., Mukai, H.	Deodorizing fibers and production process	20000810
WO 9511925	Kimberly-Clark Corp., USA	Qin, J., Gross, J. R., Mui, W. J., Ning, X., Schroeder, W. Z., Sun, T.	Crosslinked polysaccharides having enhanced absorbency and process for their preparation	19950504
WO 9830752	Du Pont De Nemours and Co., USA	Ramachandran, S.	Chitosan-coated pulp, a paper using the pulp, and a process for making them	19980716
WO 9409192	Novasso OY, Finland	Struszczyk, H., Nousiainen, P., Kivekaes, O., Niekraszewicz, A.	Modified viscose fibers and their manufacture	19940428
WO 9842818	Procter & Gamble, USA	Surutzidis, A., Heist, B. M., Leblanc, M. J.	Laundry additive particle having multiple surface coatings	19981001

Tab. 33 Water purification

Patent number	*Patent holder*	*Inventors*	*Title of patent*	*Publication date (YYYYMMDD)*
WO 9937584	Fountainhead Technologies, Inc., USA	Denkewicz, R. P., Jr., Senderov, E. E., Grenier, J. W.	Biocidal compositions for treating water	19990729
WO 9202460	Archaeus Technology Group Ltd., UK	Harris, R., Jacques, A. M., Buchan, A., Brown, M.	Decolorization of water	19920220
WO 9533690	Vanson L.P., USA	Heinsohn, G. E.	Removal of phosphates from aqueous media	19951214
WO 9817386	I.N.P. - Industrial Natural Products SRL, Italy	Pifferi, P., Spagna, G., Manenti, I.	Method for removing pesticides and/or phytodrugs from liquids using cellulose, chitosan and pectoligníncellulosic material derivatives	19980430
WO 9002708	Firextra Oy, Finland	Struszczyk, H., Kivekaes, O., Lindberg, J. J., Riekkola, M. L., Holopainen, K.	Method for purification of waste aqueous medium	19900322
WO 9737008	Yissum Research Development Company of the Hebrew University of Jerusalem, Israel	Van Rijn, J., Nussinovitch, A., Tal, J.	Bacterial immobilization in polymeric beads for nitrate removal from aquarium water	19971009

Tab. 34 Various applications

Patent number	*Patent holder*	*Inventors*	*Title of patent*	*Publication date (YYYYMMDD)*
WO 9723390	Tetra Laval Holdings & Finance SA, Switzerland	Berlin, M.	A laminated packaging material, its production, and packaging containers having gas barrier properties even in humid environments	19970703
WO 9325716	EMC Services, France	Durecu, S., Berthelin, J., Thauront, J.	Detoxication of combustion residues by removing soluble toxic compounds and by binding and concentrating the compounds obtained from the treatment solutions	19931223
WO 9702077	Kimberly-Clark Corp., USA	Gillberg-La, F. G. E., Turkevich, L. A., Kiick-Fischer, K. L.	Surface-modified fibrous material as filtration medium	19970123
WO 9822540	Merck S.A., France	Grandmontagne, B., Brousse, B.	Coating of mineral pigment microparticles with chitin	19980528
WO 0053295	University of Waterloo, Canada	Huang, R. Y. M., Pal, R., Moon, G. Y.	Two-layer composite pervaporation and reverse-osmosis membranes consisting of alginic acid derivative and chitosan	20000914
WO 9962974	Kao Corp., Japan	Kuwahara, K., Hasebe, Y., Akaogi, A., Yasumura, T., Takahashi, T., Tachizawa, O., Terada, E.	Core-shell polymer emulsions and process their manufacture	19991209
WO 9824832	Kimberly-Clark Worldwide, Inc., USA	Qin, J., Wallajapet, P. R. R.	Absorbent composition for disposable absorbent sheets	19980611
WO 9114499	Toray Industries, Inc., Japan	Shiro, K., Himeshima, Y., Yamada, S., Watanabe, T., Uemura, T., Kurihara, M.	Polythiophenylene-polysulfone composite membranes with active layers	19911003
WO 0121854	Potchefstroom University for Christian Higher Education, S. Africa	Vorster, S. W., Waanders, F. B., Geldenhuys, A. J.	Chitosan-based corrosion inhibitor for steel in aqueous acidic media	20010329

15 References

Aerts, J. M. F. (1996) Universiteit Van Amsterdam, Netherlands; Aerts, Johannes Maria Franciscus Gerardus, A human macrophage chitinase for use in the treatment or prophylaxis of infection by chitin-containing pathogens and manufacture of the enzyme by expression of the cloned gene, WO 9640940 [*Chem. Abstr.* **126**, 128715].

Aerts, J. M. F. G., Boot, R. G., Renkema, G. H., van Weely, S., Hollak, C. E. M., Donker-Koopman, W., Strijland, A., Verhoek, M. (1996) Chitotriosidase: a human macrophage chitinase that is a marker for Gaucher disease manifestation, in: *Chitin Enzymology*, (Muzzarelli, R. A. A., Ed.), Grottammare: Atec, 3–10, vol. 2.

Aiba, S. (1993) Studies on chitosan. 5. Reactivity of partially *N*-acetylated chitosan in aqueous media, *Makromol. Chem.* **194**, 65–75.

Akiyama, K., Kawazu, K., Kobayashi, A. (1995a) A novel method for chemo-enzymic synthesis of elicitor-active chitosan oligomers and partially *N*-deacetylated chitin oligomers using *N*-acylated chitotrioses as substrates in a lysozyme-catalyzed transglycosylation reaction system, *Carbohydr. Res.* **279**, 151–160.

Akiyama, K., Kawazu, K., Kobayashi, A. (1995b) Partially *N*-deacetylated chitin oligomers (pentamer to heptamer) are potential elicitors for (+)-pisatin induction in pea epicotyls, *Z. Naturforsch.* **50C**, 391–397.

Allan, G. G., Peyron, M. (1997) Depolymerization of chitosan by means of nitrous acid, in: *Chitin Handbook* (Muzzarelli, R. A. A., Peter, M. G., Eds.), pp. 175–180. Grottammare: Atec.

Andersen, S. O., Hojrup, P., Roepstorff, P. (1995) Insect cuticular proteins, *Insect Biochem. Mol. Biol.* **25**, 153–176.

Andersen, S. O., Peter, M. G., Roepstorff, P. (1996) Cuticular sclerotization in insects, *Comp. Biochem. Physiol.* **B113**, 689–705.

Arai, N., Shiomi, K., Yamaguchi, Y., Masuma, R., Iwai, Y., Turberg, A., Kolbl, H., Omura, S. (2000) Argadin, a new chitinase inhibitor, produced by *Clonostachys* sp FO-7314, *Chem. Pharm. Bull.* **48**, 1442–1446.

Arakane, Y., Koga, D. (1999) Purification and characterization of a novel chitinase isozyme from yam tuber, *Biosci. Biotechnol. Biochem.* **63**, 1895–1901.

Arakane, Y., Hoshika, H., Kawashima, N., Fujiya-Tsujimoto, C., Sasaki, Y., Koga, D. (2000) Comparison of chitinase isozymes from yam tuber – Enzymatic factor controlling the lytic activity of chitinases, *Biosci. Biotechnol. Biochem.* **64**, 723–730.

Asensio, J. L., Canada, F. J., Bruix, M., Rodriguez-Romero, A., Jimenez-Barbero, J. (1995) The interaction of hevein with *N*-acetylglucosamine-containing oligosaccharides. Solution structure of hevein complexed to chitobiose, *Eur. J. Biochem.* **230**, 621–633.

Asensio, J. L., Canada, F. J., Bruix, M., Gonzalez, C., Khiar, N., Rodriguez-Romero, A., Jimenez-Barbero, J. (1998) NMR investigations of protein-carbohydrate interactions: refined three-dimensional structure of the complex between hevein and methyl beta-chitobioside, *Glycobiology* **8**, 569–577.

Bakkers, J., Semino, C. E., Stroband, H., Kijne, J. W., Robbins, P. W., Spaink, H. P. (1997) An important developmental role for oligosaccharides during early embryogenesis of cyprinid fish, *Proc. Natl. Acad. Sci. USA* **94**, 7982–7986.

Bakkers, J., Kijne, J. W., Spaink, H. P. (1999) Function of chitin oligosaccharides in plant and animal development, *EXS* **87** (Chitin and Chitinases), 71–83.

Baldrick, P. (2000) Pharmaceutical excipient development: The need for preclinical guidance, *Regul. Toxicol. Pharmacol.* **32**, 210–218.

Bartkowiak, A., Hunkeler, D. (2000) Alginate-oligochitosan microcapsules. II. Control of mechanical resistance and permeability of the membrane, *Chem. Mater.* **12**, 206–212.

Bassler, B. L., Yu, C., Lee, Y. C., Roseman, S. (1991) Chitin utilization by marine bacteria, *J. Biol. Chem.* **266**, 24276–24286.

Baureithel, K., Felix, G., Boller, T. (1994) Specific, high affinity binding of chitin fragments to tomato cells and membranes. Competitive inhibition of binding by derivatives of chitooligosaccharides and a NOD factor of *Rhizobium*, *J. Biol. Chem.* **269**, 17931–17938.

Bautista, J., Cremades, O., Corpas, R., Ramos, R., Iglesias, F., Vega, J., Fontiveros, E., Perales, J., Parrados, J., Millan, F. (2000) Preparation of chitin by acetic acid fermentation, *Adv. Chitin Sci.* **4**, 28–33.

Baxter, J., Shimizu, F., Takiguchi, Y., Wada, M., Yamaguchi, T. (2000) Effect of iron(III) chitosan intake on the reduction of serum phosphorus in rats, *J. Pharm. Pharmacol.* **52**, 863–874.

Beintema, J. J. (1994) Structural features of plant chitinases and chitin-binding proteins, *FEBS Lett.* **350**, 159–163.

Beintema, J. J., Terwisscha van Scheltinga, A. C. (1996) Plant lysozymes, *EXS* **75** (Lysozymes: Model Enzymes in Biochemistry and Biology), 75–86.

Berecibar, A., Grandjean, C., Siriwardena, A. (1999) Synthesis and biological activity of natural aminocyclopentitol glycosidase inhibitors: mannostatins, trehazolin, allosamidins and their analogues, *Chem. Rev.* **99**, 779–844.

Beri, R. G., Walker, J., Reese, E. T., Rollings, J. E. (1993) Characterization of chitosans via coupled size-exclusion chromatography and multiple-angle laser light-scattering technique, *Carbohydr. Res.* **238**, 11–26.

Bernkop-Schnurch, A. (2000) Chitosan and its derivatives: potential excipients for peroral peptide delivery systems, *Int. J. Pharm.* **194**, 1–13.

Berth, G., Dautzenberg, H., Peter, M. G. (1998) Physicochemical characterization of chitosans varying in degree of acetylation, *Carbohydr. Polym.* **36**, 205–216.

Bishop, J. G., Dean, A. M., Mitchell-Olds, T. (2000) Rapid evolution in plant chitinases: Molecular targets of selection in plant-pathogen coevolution, *Proc. Natl. Acad. Sci. USA* **97**, 5322–5327.

Blackwell, J., Weigh, M. A. (1984) The structure of chitin–protein complexes, in: *Chitin, Chitosan, and Related Enzymes* (Zikakis, J. P., Ed.), New York: Academic Press, 257–272.

Bleau, G., Massicotte, F., Merlen, Y., Boisvert, C. (1999) Mammalian chitinase-like proteins, *EXS* **87** (Chitin and Chitinases), 211–221.

Bodek, K. H. (1994) Potentiometric method for determination of the degree of deacetylation of chitosan, in: *Chitin World* (Karnicki, Z. S., Wojtasz-Pajak, A., Brzeski, M. M., Bykowski, P. J., Eds.), Bremerhaven: Wirtschaftsverlag NW, 456–461.

Bodek, K. H. (1995) Evaluation of the potentiometric titration in anhydrous medium for determination of the chitosan deacylation degree, *Acta Pol. Pharm.* **52**, 337–343.

Bokma, E., van Koningsveld, G. A., Jeronimus-Stratingh, M., Beintema, J. J. (1997) Hevamine, a chitinase from the rubber tree Hevea brasiliensis, cleaves peptidoglycan between the C-1 of *N*-acetylglucosamine and C-4 of *N*-acetylmuramic acid and therefore is not a lysozyme, *FEBS Lett.* **411**, 161–163.

Bokma, E., Barends, T., van Scheltinga, A. C. T., Dijkstra, B. W., Beintema, J. J. (2000) Enzyme kinetics of hevamine, a chitinase from the rubber tree *Hevea brasiliensis*, *FEBS Lett.* **478**, 119–122.

Boot, R. G., Renkema, G. H., Strijland, A., van Zonneveld, A. J., Aerts, J. M. F. G. (1995) Cloning of a cDNA encoding chitotriosidase, a human chitinase produced by macrophages, *J. Biol. Chem.* **270**, 26252–26256.

Boot, R. G., Van Achterberg, T. A. E., Van Aken, B. E., Renkema, G. H., Jacobs, M. J. H. M., Aerts, J. M. F. G., De Vries, C. J. M. (1999) Strong induction of members of the chitinase family of proteins in atherosclerosis: chitotriosidase and human cartilage gp-39 expressed in lesion macrophages, *Arterioscler. Thromb. Vasc. Biol.* **19**, 687–694.

Bosso, C., Domard, A. (1992) Characterization of glucosamine and *N*-acetylglucosamine oligomers by fast atom bombardment mass spectrometry, *Org. Mass Spectrom.* **27**, 799–806.

Brameld, K. A., Goddard, W. A. (1998) Substrate distortion to a boat conformation at subsite-1 is critical in the mechanism of family 18 chitinases, *J. Am. Chem. Soc.* **120**, 3571–3580.

Brugnerotto, J., Lizardi, J., Goycoolea, F. M., Arguelles-Monal, W., Desbrieres, J., Rinaudo, M. (2001) An infrared investigation in relation with chitin and chitosan characterization, *Polymer* **42**, 3569–3580.

Brurberg, M. B., Eijsink, V. G. H., Nes, I. F. (1994) Characterization of a chitinase gene (chiA) from *Serratia marcescens* BJL200 and one-step purification of the gene product, *FEMS Microbiol. Lett.* **124**, 399–404.

Brurberg, M. B., Eijsink, V. G. H., Haandrikman, A. J., Venema, G., Nes, I. F. (1995) Chitinase B from *Serratia marcescens* BJL200 is exported to the periplasm without processing, *Microbiology* **141**, 123–131.

Brurberg, M. B., Nes, I. F., Eijsink, V. G. H. (1996) Comparative studies of chitinases A and B from *Serratia marcescens*, *Microbiology* **142**, 1581–1589.

Burkart, M. D., Izumi, M., Wong, C. H. (1999) Enzymatic regeneration of 3′-phosphoadenosine-5′-phosphosulfate using aryl sulfotransferase for the preparative enzymatic synthesis of sulfated carbohydrates, *Angew. Chem. Int. Edn.* **38**, 2747–2750.

Buss, U., Vårum, K. M., Peter, M. G., Spindler-Barth, M. (1996) ELISA for quantitation of chitin, chitosan and related compounds, *Adv. Chitin Sci.* **1**, 254–261.

Bustos, R. O., Healy, M. G. (1994) Microbial extraction of chitin from prawn waste, in: *Chitin World* (Karnicki, Z. S., Wojtasz-Pajak, A., Brzeski, M. M., Bykowski, P. J., Eds.), Bremerhaven: Wirtschaftsverlag NW, 15–25.

Cauchie, H.-M. (1997) An attempt to estimate crustacean chitin production in the hydrosphere, *Adv. Chitin Sci.* **2**, 32–39.

Chandy, T., Sharma, P. (1992) Polylysine-immobilized chitosan beads as adsorbents for bilirubin, *Artif. Organs* **16**, 568–576.

Chanzy, H. (1997) Chitin crystals, *Adv. Chitin Sci.* **2**, 11–21.

Chapman, R. F. (1971) *The Insects, Structure and Function*, London: The English Universities Press.

Chatelet, C., Damour, O., Domard, A. (2001) Influence of the degree of acetylation on some biological properties of chitosan films, *Biomaterials* **22**, 261–268.

Chen, R. H., Tsaih, M. L. (2000) Urea-induced conformational changes of chitosan molecules and the shift of break point of Mark-Houwink equation by increasing urea concentration, *J. Appl. Polym. Sci.* **75**, 452–457.

Choi, H. K., Choi, K. H., Kramer, K. J., Muthukrishnan, S. (1997) Isolation and characterization of a genomic clone for the gene for an insect molting enzyme, chitinase, *Insect Biochem. Mol. Biol.* **27**, 37–47.

Choi, Y. S., Lee, S. B., Hong, S. R., Lee, Y. M., Song, K. W., Park, M. H. (2001) Studies on gelatin-based sponges. Part III: A comparative study of cross-linked gelatin/alginate, gelatin/hyaluronate and chitosan/hyaluronate sponges and their application as a wound dressing in full-thickness skin defect of rat, *J. Mater. Sci., Mater. Med.* **12**, 67–73.

Cira, L. A., Huerta, S., Guerrero, I., Rosas, R., Hall, G. M., Shirai, K. (2000) Scaling up of lactic acid fermentation of prawn wastes in packed-bed column reactor for chitin recovery, *Adv. Chitin Sci.* **4**, 21–27.

Cohen, E. (1987) Interference with chitin biosynthesis in insects, in: *Chitin and Benzoylphenyl Ureas* (Wright, J. E., Retnakaran, A., Eds.), Dordrecht: W. Junk, 43–73.

Cohen, E. (1993) Chitin synthesis and degradation as targets for pesticide, *Arch. Insect Biochem. Physiol.* **22**, 245–261.

Cohen-Kupiec, R., Chet, I. (1998) The molecular biology of chitin digestion, *Curr. Opin. Biotechnol.* **9**, 270–277.

Colombo, P., Sciutto, A. M. (1996) Nutritional aspects of chitosan employment in hypocaloric diet, *Acta Toxicol. Ther.* **17**, 287–302.

Coutinho, P. M., Henrissat, B. (1999) Carbohydrate-Active Enzymes server at URL: http://afmb.cnrs-mrs.fr/~pedro/CAZY/db.html.

Daly, W. H., Macossay, J. (1997) An overview of chitin and derivatives for biodegradable materials applications, *Fib. Text. East. Eur.* **5**, 22–27.

Danulat, E. (1986) The effects of various diets on chitinase and beta-glucosidase activities and the condition of cod, *Gadus morhua* (L.), *J. Fish Biol.* **28**, 191–197.

Davis, S. S. (2000) Drug delivery systems, *Interdiscipl. Sci. Rev.* **25**, 175–183.

De la Vega, H., Specht, C. A., Semino, C. E., Robbins, P. W., Eichinger, D., Caplivski, D., Ghosh, S., Samuelson, J. (1997) Cloning and expression of chitinases of Entamoebae, *Mol. Biochem. Parasitol.* **85**, 139–147.

De la Vega, H., Specht, C. A., Liu, Y., Robbins, P. W. (1998) Chitinases are a multi-gene family in *Aedes*, *Anopheles* and *Drosophila*, *Insect Mol. Biol.* **7**, 233–239.

Deacon, M. P., Davis, S. S., White, R. J., Nordman, H., Carlstedt, I., Errington, N., Rowe, A. J., Harding, S. E. (1999) Are chitosan–mucin interactions specific to different regions of the stomach? Velocity ultracentrifugation offers a clue, *Carbohydr. Polym.* **38**, 235–238.

Defaye, J., Gadelle, A., Pedersen, C. (1989) Chitin and chitosan oligosaccharides, in: *Chitin and Chitosan* (Skjåk-Bræk, G., Anthonsen, T., Sandford, P., Eds.), London: Elsevier Applied Science, 415–429.

Deikman, J. (1997) Molecular mechanisms of ethylene regulation of gene transcription, *Physiol. Plantarum* **100**, 561–566.

Demont-Caulet, N., Maillet, F., Tailler, D., Jacquinet, J.-C., Prome, J.-C., Nicolaou, K. C., Truchet, G., Beau, J.-M., Denarie, J. (1999) Nodule-inducing activity of synthetic *Sinorhizobium meliloti* nodulation factors and related lipo-chitooligosaccharides on alfalfa. Importance of the acyl chain structure, *Plant Physiol.* **120**, 83–92.

Denarie, J., Debelle, F., Prome, J.-C. (1996) Rhizobium lipo-chitooligosaccharide nodulation factors: signaling molecules mediating recognition and morphogenesis, *Annu. Rev. Biochem.* **65**, 503–535.

Denuziere, A., Ferrier, D., Domard, A. (1996) Chitosan–chondroitin sulfate and chitosan–hyaluronate polyelectrolyte complexes. Physicochemical aspects, *Carbohydr. Polym.* **29**, 317–323.

Doernenburg, H., Knorr, D. (1996) Chitosan for elicitation, permeabilization and immobilization of plant cell cultures, *Bioforum* **19**, 52–62.

Domard, A., Cartier, N. (1991) Glucosamine oligomers: 3. Study of the elution process on polyacrylamide gels, application to other polyamine oligomer series, *Polym. Commun.* **32**, 116–119.

Domard, A., Demarger-Andre, S. (1994) Chitosan behaviors in fatty acid dispersions, *Macromol. Rep.* **A31**, 849–856.

Domard, A., Piron, E. (2000) Recent approach of metal binding by chitosan and derivatives, *Adv. Chitin Sci.* **4**, 295–301.

Dornish, M., Hagen, A., Hansson, E., Pecheur, C., Verdier, F., Skaugrud, O. (1997a) Safety of Protasan: ultrapure chitosan salts for biomedical and pharmaceutical use, *Adv. Chitin Sci.* **2**, 664–670.

Dornish, M., Skaugrud, O., Illum, L., Davis, S. S. (1997b) Nasal drug delivery with Protasan, *Adv. Chitin Sci.* **2**, 694–697.

Draborg, H., Kauppinen, S., Christgau, S., Dalboge, H. (1997) A sensitive plate assay for the detection and expression cloning of endochitinase from *Trichoderma harzianum*, in: *Chitin Handbook* (Muzzarelli, R. A. A., Peter, M. G., Eds.), Grottammare: Atec, 261–266.

Drouillard, S., Armand, S., Davies, G. J., Vorgias, C. E., Henrissat, B. (1997) *Serratia marcescens* chitobiase is a retaining glycosidase utilizing substrate acetamido group participation, *Biochem. J.* **328**, 945–949.

Duarte, M. L., M.C. Ferreira, M. C., Marvão, M. R. (2000) A statistical evaluation of IR spectroscopic methods to determine the degree of acetylation of α-chitin and chitosan, *Adv. Chitin Sci.* **4**, 367–374.

Dvorakova, J., Schmidt, D., Hunkova, Z., Thiem, J., Kren, V. (2001) Enzymatic rearrangements of chitin hydrolysates with β-*N*-acetylhexosaminidase from *Aspergillus oryzae*, *J. Mol. Catal. B: Enzymatic* **11**, 225–232.

Ebert, A., Fink, H.-P. (1997) Solid-state NMR spectroscopy of chitin and chitosan, in: *Chitin Handbook* (Muzzarelli, R. A. A., Peter, M. G., Eds.), Grottammare: Atec, 137–143.

Ebert, G. (1993) *Biopolymere*, Stuttgart: Teubner.

Eiberg, H., Dentandt, W. R. (1997) Assignment of human plasma methylumbelliferyl-tetra-*N*-acetylchitotetraose hydrolase or chitinase to chromosome 1q by a linkage study, *Hum. Genet.* **101**, 205–207.

Ernst, E., Pittler, M. H. (1998) Chitosan as a treatment for body weight reduction? A meta-analysis, *Perfusion* **11**, 461–462.

Errington, N., Harding, S. E., Vårum, K. M., Illum, L. (1993) Hydrodynamic characterization of chitosans varying in degree of acetylation, *Int. J. Biol. Macromol.* **15**, 113–117.

Escott, G. M., Adams, D. J. (1995) Chitinase activity in human serum and leukocytes, *Infect. Immun.* **63**, 4770–4773.

Falbe, J., Regitz, M. (Eds.) (1989) *Römpp Chemie Lexikon*, Stuttgart: Thieme, 613.

Feix, M., Glоggler, S., Londershausen, M., Weidemann, W., Spindler, K. D., Spindler-Barth, M. (2000) A cDNA encoding a chitinase from the epithelial cell line of *Chironomus tentans* (Insecta, Diptera) and its functional expression, *Arch. Insect Biochem. Physiol.* **45**, 24–36.

Felt, O., Buri, P., Gurny, R. (1998) Chitosan: a unique polysaccharide for drug delivery, *Drug Dev. Ind. Pharm.* **24**, 979–993.

Frazier, S. F., Larsen, G. S., Neff, D., Quimby, L., Carney, M., DiCaprio, R. A., Zill, S. N. (1999) Elasticity and movements of the cockroach tarsus in walking, *J. Comp. Physiol.* **A185**, 157–172.

Fukamizo, T., Kramer, K. J. (1987) Effect of 20-hydroxyecdysone on chitinase and β-*N*-acetylglucosaminidase during the larval-pupal transformation of *Manduca sexta*, *Insect Biochem. Mol. Biol.* **17**, 547–550.

Furda, I. (2000) Reduction of absorption of dietary lipids and cholesterol by chitosan and its derivatives and special formulations, in: *Chitosan* per os, *from Dietary Supplement to Drug Carrier* (Muzzarelli, R. A. A., Ed.), Grottamare: Atec, 41–63.

Gallaher, C. M., Munion, J., Hesslink, R., Wise, J., Gallaher, D. D. (2000) Cholesterol reduction by glucomannan and chitosan is mediated by changes in cholesterol absorption and bile acid and fat excretion in rats, *J. Nutr.* **130**, 2753–2759.

Girola, M., De Bernardi, M., Contos, S., Tripodi, S., Ventura, P., Guarino, C., Marletta, M. (1996) Dose effect in lipid-lowering activity of a new dietary integrator (chitosan, *Garcinia cambogia* extract and chrome), *Acta Toxicol. Ther.* **17**, 25–40.

Gokul, B., Lee, J. H., Song, K. B., Rhee, S. K., Kim, C. H., Panda, T. (2000) Characterization and applications of chitinases from *Trichoderma harzianum* – A review, *Bioproc. Eng.* **23**, 691–694.

Gooday, G. W. (1999) Aggressive and defensive roles for chitinases, *EXS* **87** (Chitin and Chitinases), 157–169.

Graham, L. S., Sticklen, M. B. (1994) Plant chitinases, *Can. J. Bot.* **72**, 1057–1083.

Grün, L., Förster, H., Peter, M. G. (1984) CP/MAS-^{13}C-NMR spectra of sclerotized insect cuticle and of chitin, *Angew. Chem. Int. Edn.* **23**, 638–639.

Guerrero, L., Omil, F., Mendez, R., Lema, J. M. (1998) Protein recovery during the overall treatment of wastewaters from fish-meal factories, *Bioresour. Technol.* **63**, 221–229.

Guibal, E., Milot, C., Roussy, J. (1997) Chitosan gel beads for metal ion recovery, in: *Chitin Handbook* (Muzzarelli, R. A. A., Peter, M. G., Eds.), Grottammare: Atec, 423–429.

Gupta, K. C., Kumar, M. N. V. R. (2000) Trends in controlled drug release formulations using chitin and chitosan, *J. Sci. Ind. Res.* **59**, 201–213.

Haas, F., Gorb, S., Blickhan, R. (2000) The function of resilin in beetle wings, *Proc. R. Soc. London, Ser. B, Biol. Sci.* **267**, 1375–1381.

Hackman, R. H. (1987) Chitin and the fine structure of insect cuticles, in: *Chitin and Benzoylphenyl Ureas* (Wright, J. E., Retnakaran, A., Eds.), Dordrecht: W. Junk, 1–32.

Haebel, S., Peter-Katalinic, J., Peter, M. G. (1997) Mass spectrometry of chito-oligosaccharides, in: *Chitin Handbook* (Muzzarelli, R. A. A., Peter, M G., Eds.), Grottammare: Atec, 205–214.

Hahn, M., Hennig, M., Schlesier, B., Höhne, W. (2000) Structure of jack bean chitinase, *Acta Crystallogr.* **D56**, 1096–1099.

Haltiwanger, R. S. (2000) Structures and function of nuclear and cytoplasmic glycoproteins, in: *Carbohydrates in Chemistry and Biology* (Ernst, B., Hart, G. W., Sinay, P., Eds.), Weinheim: Wiley-VCH, 651–667, vol. 4.

Han, L. K., Kimura, Y., Okuda, H. (1999) Reduction in fat storage during chitin-chitosan treatment in mice fed a high-fat diet, *Int. J. Obes.* **23**, 174–179.

Harata, K., Muraki, M. (1995) X-ray structure of turkey egg lysozyme complex with di-*N*-acetylchitobiose. Recognition and binding of α-anomeric form, *Acta Crystallogr.* **D51**, 718–724.

Harata, K., Muraki, M. (2000) Crystal structures of *Urtica dioica* agglutinin and its complex with tri-*N*-acetylchitotriose, *J. Mol. Biol.* **297**, 673–681.

Harding, S. E., Davis, S. S., Deacon, M. P., Fiebrig, I. (1999) Biopolymer mucoadhesives, *Biotechnol. Genet. Engin. Rev.* **16**, 41–86.

Hart, P. J., Pfluger, H. D., Monzingo, A. F., Hollis, T., Robertus, J. D. (1995) The refined crystal structure of an endochitinase from *Hordeum vulgare* L. seeds at 1.8 Å resolution, *J. Mol. Biol.* **248**, 402–413.

Hashimoto, H., Abe, Y., Horito, S., Yoshimura, J. (1989) Synthesis of chitooligosaccharide derivatives by condensation polymerization, *J. Carbohydr. Chem.* **8**, 307–311.

Hashimoto, M., Ikegami, T., Seino, S., Ohuchi, N., Fukada, H., Sugiyama, J., Shirakawa, M., Watanabe, T. (2000) Expression and characterization of the chitin-binding domain of chitinase A1 from *Bacillus circulans* WL-12, *J. Bacteriol.* **182**, 3045–3054.

Haslam, S. M., Houston, K. M., Harnett, W., Reason, A. J., Morris, H. R., Dell, A. (1999) Structural studies of *N*-glycans of filarial parasites. Conservation of phosphorylcholine-substituted glycans among species and discovery of novel chito-oligomers, *J. Biol. Chem.* **274**, 20953–20960.

Hattori, T., Katai, K., Kato, M., Izume, M., Mizuta, Y. (1999) Colloidal titration of chitosan and critical unit of chitosan to the potentiometric colloidal titration with poly(vinyl sulfate) using Toluidine Blue as indicator, *Bull. Chem. Soc. Jpn.* **72**, 37–41.

He, F., Ramirez, J., Lebrilla, C. B. (1999) Evidence for enzymatic activity in the absence of solvent in gas-phase complexes of lysozyme and oligosaccharides, *Int. J. Mass Spectrometry* **193**, 103–114.

Heightman, T. D., Vasella, A. T. (1999) Recent insights into inhibition, structure, and mechanism of configuration-retaining glycosidases, *Angew. Chem. Int. Edn.* **38**, 750–770.

Hennig, M., Schlesier, B., Dauter, Z., Pfeffer, S., Betzel, C., Höhne, W. E., Wilson, W. S. (1992) A TIM barrel protein without enzymatic activity? Crystal-structure of narbonin at 1.8 Å resolution, *FEBS Lett.* **306**, 80–84.

Hennig, M., Jansonius, J. N., Terwisscha van Scheltinga, A. C., Dijkstra, B. W., Schlesier, B. (1995) Crystal structure of concanavalin B at 1.65 Å resolution. An inactivated chitinase from seeds of *Canavalia ensiformis*, *J. Mol. Biol.* **254**, 237–246.

Henrissat, B. (1999) Classification of chitinases modules, *EXS* **87** (Chitin and Chitinases), 137–156.

Herrera-Estrella, A., Chet, I. (1999) Chitinases in biological control, *EXS* **87** (Chitin and Chitinases), 171–184.

Hirano, S. (1989) Production and application of chitin and chitosan in Japan, in: *Chitin and Chitosan* (Skjåk-Bræk, G., Anthonsen, T., Sandford, P., Eds.), London: Elsevier Applied Science, 37–43.

Hirano, S. (1997) *N*-Acyl-, *N*-arylidene- and *N*-alkylidene chitosans and their hydrogels, in: *Chitin Handbook* (Muzzarelli, R. A. A., Peter, M. G., Eds.), Grottammare: Atec, 71–75.

Hirano, S., Akiyama, Y. (1995) Absence of a hypocholesterolemic action of chitosan in high-serum-cholesterol rabbits, *J. Sci. Food Agric.* **69**, 91–94.

Hirano, S., Midorikawa, T. (1998) Novel method for the preparation of *N*-acylchitosan fiber and *N*-acylchitosan-cellulose fiber, *Biomaterials* **19**, 293–297.

Hirano, S., Sakaguchi, T., Kuramitsu, K. (1992) *N*-Alanyl and some *N*-(*N'*-aryl)glycyl derivatives of chitosan, *Carbohydr. Polym.* **19**, 135–138.

Hirano, S., Zhang, M., Chung, B., Kim, S. (2000) The *N*-acylation of chitosan fibre and the *N*-deacetylation of chitin fibre and chitin-cellulose blended fibre at a solid state, *Carbohydr. Polym.* **41**, 175–179.

Hoekstra, A., Struszczyk, H., Kivekäs, O. (1998) Percutaneous microcrystalline chitosan application for sealing arterial puncture sites, *Biomaterials* **19**, 1467–1471.

Hollis, T., Honda, Y., Fukamizo, T., Marcotte, E., Day, P. J., Robertus, J. D. (1997) Kinetic analysis of barley chitinase, *Arch. Biochem. Biophys.* **344**, 335–342.

Hollis, T., Monzingo, A. F., Bortone, K., Ernst, S., Cox, R., Robertus, J. D. (2000) The X-ray structure of a chitinase from the pathogenic fungus *Coccidioides immitis*, *Protein Sci.* **9**, 544–551.

Holme, H. K., Hagen, A., Dornish, M. (2000) Influence of chitosans on permeability of human intestinal epithelial (Caco-2) cells: the effect of molecular weight, degree of deacetylation and exposure time, *Adv. Chitin Sci.* **4**, 259–265.

Holme, K. R., Perlin, A. S. (1997) Chitosan *N*-sulfate. A water-soluble polyelectrolyte, *Carbohydr. Res.* **302**, 7–12.

Honda, Y., Tanimori, S., Kirihata, M., Kaneko, S., Tokuyasu, K., Hashimoto, M., Watanabe, T. (2000a) Chemo- and enzymatic synthesis of partially and fully *N*-deacetylated 4-methylumbelliferyl chitobiosides: fluorogenic substrates for chitinase, *Bioorg. Med. Chem. Lett.* **10**, 827–829.

Honda, Y., Tanimori, S., Kirihata, M., Kaneko, S., Tokuyasu, K., Hashimoto, M., Watanabe, T., Fukamizo, T. (2000b) Kinetic analysis of the reaction catalyzed by chitinase A1 from *Bacillus circulans* WL-12 toward the novel substrates, partially *N*-deacetylated 4-methylumbelliferyl chitobiosides, *FEBS Lett.* **476**, 194–197.

Horst, M. N. (1986) Lipid-linked intermediates in crustacean chitin synthesis, in: *Chitin in Nature and Technology* (Muzzarelli, R. A. A., Jeuniaux, C., Gooday, G. W., Eds.), New York: Plenum Press, 45–52.

Horst, M. N., Walker, A. N. (1993) Crustacean chitin synthesis and the role of the Golgi apparatus, in: *Chitin Enzymology* (Muzzarelli, R. A. A., Ed.), Lyon: European Chitin Society, 109–118.

Hu, B., Trinh, K., Figueira, W. F., Price, P. A. (1996) Isolation and sequence of a novel human chondrocyte protein related to mammalian members of the chitinase protein family, *J. Biol. Chem.* **271**, 19415–19420.

Hwang, D.-C., Damodaran, S. (1995) Selective precipitation and removal of lipids from cheese whey using chitosan, *J. Agric. Food Chem.* **43**, 33–37.

Hwang, S. M., Chen, C. Y., Chen, S. S., Chen, J. C. (2000) Chitinous materials inhibit nitric oxide production by activated RAW 264.7 macrophages, *Biochem. Biophys. Res. Commun.* **271**, 229–233.

Ibrahim, G. H., Smartt, C. T., Kiley, L. M., Christensen, B. M. (2000) Cloning and characterization of a chitin synthase cDNA from the mosquito *Aedes aegypti*, *Insect Biochem. Mol. Biol.* **30**, 1213–1222.

Ikegami, T., Okada, T., Hashimoto, M., Seino, S., Watanabe, T., Shirakawa, M. (2000) Solution structure of the chitin-binding domain of *Bacillus circulans* WL-12 chitinase A1, *J. Biol. Chem.* **275**, 13654–13661.

Ikejima, T., Inoue, Y. (2000) Crystallization behavior and environmental biodegradability of the blend films of poly(3-hydroxybutyric acid) with chitin and chitosan, *Carbohydr. Polym.* **41**, 351–356.

Illum, L. (1998) Chitosan and its use as a pharmaceutical excipient, *Pharm. Res.* **15**, 1326–1331.

Inui, H., Yoshida, M., Hirano, S. (1996) Effects of a 6-*O*-hydroxyethyl group on the hydrolysis of 6-*O*-hydroxyethylchitin (glycolchitin) by chitinase, *Biosci. Biotechnol. Biochem.* **60**, 1886–1887.

Inui, H., Yamaguchi, Y., Hirano, S. (1997) Elicitor actions of *N*-acetylchitooligosaccharides and laminarioligosaccharides for chitinase and L-phenylalanine ammonia-lyase induction in rice suspension culture, *Biosci. Biotechnol. Biochem.* **61**, 975–978.

Izume, M., Nagae, S., Kawagishi, H., Ohtakara, A. (1992) Preparation of normal-acetylchitooligosaccharides from enzymatic hydrolyzates of chitosan, *Biosci. Biotechnol. Biochem.* **56**, 1327–1328.

Japanese Society for Chitin and Chitosan (1995) *Kichin, Kitosan Handobukku* (*Chitin and Chitosan Handbook*), Tokyo: Gihodo Shuppan. [*Chem. Abstr.* **122**, 291440].

Jeon, Y. J., Kim, S. K. (2000) Production of chitooligosaccharides using an ultrafiltration membrane reactor and their antibacterial activity, *Carbohydr. Polym.* **41**, 133–141.

Jeon, Y. J., Park, P. J., Kim, S. K. (2001) Antimicrobial effect of chitooligosaccharides produced by bioreactor, *Carbohydr. Polym.* **44**, 71–76.

Jeuniaux, C., Voss-Foucard, M. F. (1997) A specific enzymatic method for the quantitative estimation of chitin, in: *Chitin Handbook* (Muzzarelli, R. A. A., Peter, M. G., Eds.), Grottammare: Atec, 3–7.

Jeuniaux, C., Voss-Foucart, M. F., Bussers, J.-C. (1993) La production de chitine par les crustacés dans les écosystèmes marins, *Aquat. Living Resour.* **6**, 331–341.

Jeuniaux, C., Compere, P., Toussaint, C., Decloux, N., Voss-Foucart, M.-F. (1996) Confirmation of the presence of chitin in fin cuticles of some Blennidae (Teleostei) by enzymic and cytochemical methods, *Adv. Chitin Sci.* **1**, 18–25.

Jin, H. M., Copeland, N. G., Gilbert, D. J., Jenkins, N. A., Kirkpatrick, R. B., Rosenberg, M. (1998) Genetic characterization of the murine Ym1 gene and identification of a cluster of highly homologous genes, *Genomics* **54**, 316–322.

Kadokawa, J.-I., Yamashita, K., Karasu, M., Tagaya, H., Chiba, K. (1995) Preparation and enzymic hydrolysis of block copolymer consisting of oligochitin and poly(propylene glycol), *J. Macromol. Sci., Pure Appl. Chem.* **A32**, 1273–1280.

Kadokawa, J., Sato, M., Karasu, M., Tagaya, H., Chiba, K. (1998) Synthesis of hyperbranched aminopolysaccharides, *Angew. Chem. Int. Edn.* **37**, 2373–2376.

Kamei, K., Yamamura, Y., Hara, S., Ikenaka, T. (1989) Amino acid sequence of chitinase from *Streptomyces erythraeus*, *J. Biochem.* **105**, 979–985.

Kamst, E., Bakkers, J., Quaedvlieg, N. E. M., Pilling, J., Kijne, J. W., Lugtenberg, B. J. J., Spaink, H. P. (1999) Chitin oligosaccharide synthesis by *Rhizobia* and zebrafish embryos starts by glycosyl transfer to O-4 of the reducing-terminal residue, *Biochemistry* **38**, 4045–4052.

Karlsen, S., Hough, E. (1995) Crystal structures of three complexes between chito-oligosaccharides and lysozyme from the rainbow trout. How distorted is the NAG sugar in site D? *Acta Crystallogr.* **D51**, 962–978.

Kato, T., Shizuri, Y., Izumida, H., Yokoyama, A., Endo, M. (1995) Styloguanidines, new chitinase inhibitors from the marine sponge *Stylotella aurantium*, *Tetrahedron Lett.* **36**, 2133–2136.

Kawasaki, T., Kawasaki, M., Notomi, A., Itoh, K., Ikeyama, N. (1998) Effects of chitin-oligosaccharides on blood pressure, lipid and glucose metabolism in clinically healthy subjects – a randomized single-blind placebo-controlled crossover trial, *Kichin, Kitosan Kenkyu* **4**, 316–324 [*Chem. Abstr.* **130**, 60903].

Kerwin, J. L., Whitney, D. L., Sheikh, A. (1999) Mass spectrometric profiling of glucosamine, glucosamine polymers and their catecholamine adducts. Model reactions and cuticular hydrolysates of *Toxorhynchites amboinensis* (Culicidae) pupae, *Insect Biochem. Mol. Biol.* **29**, 599–607.

Keyhani, N. O., Roseman, S. (1999) Physiological aspects of chitin catabolism in marine bacteria, *Biochim. Biophys. Acta* **1473**, 108–122.

Keyhani, N. O., Li, X. B., Roseman, S. (2000) Chitin catabolism in the marine bacterium *Vibrio furnissii* – Identification and molecular cloning of a chitoporin, *J. Biol. Chem.* **275**, 33068–33076.

Kim, M. G., Shin, S. W., Bae, K. S., Kim, S. C., Park, H. Y. (1998) Molecular cloning of chitinase cDNAs from the silkworm, *Bombyx mori* and the fall webworm, *Hyphantria cunea*, *Insect Biochem. Mol. Biol.* **28**, 163–171.

Kim, S. Y., Shon, D. H., Lee, K. H. (2000) Enzyme-linked immunosorbent assay for detection of chitooligosaccharides, *Biosci. Biotechnol. Biochem.* **64**, 696–701.

Kiyosada, T., Shoda, S., Kobayashi, S. (1998) Enzymic ring-opening polyaddition for chitin synthesis. A cationic mechanism in basic solution? *Macromol. Symp.* **132**, 415–420.

Klokkevold, P., Fukayama, H., Sung, E. (1997a) The effect of chitosan on hemostasis: current work and review of the literature, *Adv. Chitin Sci.* **2**, 698–704.

Klokkevold, P., Redd, M., Salamati, A., Kim, J., Nishimura, R. (1997b) The effect of chitosan on guided bone regeneration: a pilot study in the rabbit, *Adv. Chitin Sci.* **2**, 656–663.

Knaul, J. Z., Hudson, S. M., Creber, K. A. M. (1999) Improved mechanical properties of chitosan fibers, *J. Appl. Polym. Sci.* **72**, 1721–1732.

Kobayashi, S., Kiyosada, T., Shoda, S. (1996) Synthesis of artificial chitin: irreversible catalytic behavior of a glycosyl hydrolase through a transition state analogue substrate, *J. Am. Chem. Soc.* **118**, 13113–13114.

Kobayashi, S., Kiyosada, T., Shoda, S. (1997) A novel method for synthesis of chitobiose via enzymic glycosylation using a sugar oxazoline as glycosyl donor, *Tetrahedron Lett.* **38**, 2111–2112.

Koetz, J., Kosmella, S. (1997) Polyelectrolyte complex formation with chitosan, *Adv. Chitin Sci.* **2**, 476–483.

Koga, D. (1996) Induction of chitinase in fine bentgrass and yam by various elicitors, *Front. Biomed. Biotechnol.* **3**, 231–241.

Koga, D., Kramer, K. J. (1983) Hydrolysis of glycol chitin by chitinolytic enzymes, *Comp. Biochem. Physiol.* **76B**, 291–293.

Koga, D., Funakoshi, T., Fujimoto, H., Kuwano, E., Eto, M., Ide, A. (1991) Effects of 20-hydroxyecdysone and KK-42 on chitinase and β-*N*-acetylglucosaminidase during the larval-pupal transformation of *Bombyx mori*, *Insect Biochem.* **21**, 277–284.

Koga, D., Hoshika, H., Matsushita, M., Tanaka, A., Ide, A., Kono, M. (1996) Purification and characterization of β-*N*-acetylhexosaminidase from the liver of a prawn, *Penaeus japonicus*, *Biosci. Biotechnol. Biochem.* **60**, 194–199.

Koga, D., Sasaki, Y., Uchiumi, Y., Hirai, N., Arakane, Y., Nagamatsu, Y. (1997) Purification and characterization of *Bombyx mori* chitinases, *Insect Biochem. Mol. Biol.* **27**, 757–767.

Koga, D., Yoshioka, T., Arakane, Y. (1998) HPLC analysis of anomeric formation and cleavage pattern by chitinolytic enzymes, *Biosci. Biotechnol. Biochem.* **62**, 1643–1646.

Koga, D., Mitsutomi, M., Kono, M., Matsumiya, M. (1999) Biochemistry of chitinases, *EXS* **87** (Chitin and Chitinases), 111–123.

Kopacek, P., Vogt, R., Jindrak, L., Weise, C., Safarik, I. (1999) Purification and characterization of the lysozyme from the gut of the soft tick *Ornithodoros moubata*, *Insect Biochem. Mol. Biol.* **29**, 989–997.

Kow, F. (1994) Seafood offal available as raw material for chitin and chitosan production in Tasmania, in: *Chitin World* (Karnicki, Z. S., Wojtasz-Pajak, A., Brzeski, M. M., Bykowski, P. J., Eds.), Bremerhaven: Wirtschaftsverlag NW, 26–29.

Kramer, K. J., Koga, D. (1986) Insect chitin: physical state, synthesis, degradation, and metabolic regulation, *Insect Biochem.* **16**, 851–877.

Kramer, K. J., Muthukrishnan, S. (1997) Insect chitinases: molecular biology and potential use as biopesticides, *Insect Biochem. Mol. Biol.* **27**, 887–900.

Kramer, K. J., Corpuz, L., Choi, H. K., Muthukrishnan, S. (1993) Sequence of a cDNA and expression of the gene encoding epidermal and gut chitinases of *Manduca sexta*, *Insect Biochem. Mol. Biol.* **23**, 691–701.

Kramer, K. J., Hopkins, T. L., Schaefer, J. (1995) Applications of solid-state NMR to the analysis of insect sclerotized structures, *Insect Biochem. Mol. Biol.* **25**, 1067–1080.

Kramerov, A. A., Mukha, D. V., Metakovsky, E. V., Gvozdev, V. A. (1986) Glycoproteins containing sulfated chitin-like carbohydrate moiety, *Insect Biochem.* **16**, 417–432.

Kramerov, A. A., Rozovsky, Y. M., Baikova, N. A., Gvozdev, V. A. (1990) Cognate chitinoproteins are detected during *Drosophila melanogaster* development and in cell cultures from different insect species, *Insect Biochem.* **20**, 769–775.

Krasavtsev, V. E. (2000) Krill as a promising material for the production of chitin in Europe, *Adv. Chitin Sci.* **4**, 1–3.

Kristiansen, A., Vårum, K. M., Grasdalen, H. (1998) Quantitative studies of the non-productive binding of lysozyme to partially *N*-acetylated chitosans. Binding of large ligands to a one-dimensional binary lattice studied by a modified McGhee and Von Hippel model, *Biochim. Biophys. Acta* **1425**, 137–150.

Kubisch, J., Weignerova, L., Kotter, S., Lindhorst, T. K., Sedmera, P., Kren, V. (1999) Enzymatic synthesis of *p*-nitrophenyl β-chitobioside, *J. Carbohydr. Chem.* **18**, 975–984.

Kumar, M. N. V. R. (2000) A review of chitin and chitosan applications, *React. Funct. Polym.* **46**, 1–27.

Kurita, K. (1997) Chitin and chitosan derivatives, in: *Desk Reference of Functional Polymers, Synthesis and Applications* (Arshady, R., Ed.), Washington, DC: American Chemical Society, 239–259.

Kurita, K., Ishii, S., Tomita, K., Nishimura, S. I., Shimoda, K. (1994a) Reactivity characteristics of squid β-chitin as compared with those of shrimp chitin: high potentials of squid chitin as a starting material for facile chemical modifications, *J. Polym. Sci., Part A: Polym. Chem.* **32**, 1027–1032.

Kurita, K., Kobayashi, M., Munakata, T., Ishii, S., Nishimura, S.-I. (1994b) Synthesis of non-natural branched polysaccharides. Regioselective introduction of α-mannoside branches into chitin, *Chem. Lett.*, 2063–2066.

Kurita, K., Hashimoto, S., Ishi, S., Mori, T., Nishimura, S.-I. (1996a) Efficient graft copolymerization of 2-methyl-2-oxazoline onto tosyl- and iodo-chitins in solution, *Polym. J.* **28**, 686–689.

Kurita, K., Hashimoto, S., Ishii, S., Mori, T. (1996b) Chitin/poly(methyl methacrylate) hybrid materials. Efficient graft copolymerization of methyl methacrylate onto mercapto-chitin, *Polym. Bull.* **36**, 681–686.

Kurita, K., Hashimoto, S., Yoshino, H., Ishii, S., Nishimura, S.-I. (1996c) Preparation of chitin/polystyrene hybrid materials by efficient graft copolymerization based on mercaptochitin, *Macromolecules* **29**, 1939–1942.

Kurita, K., Yoshino, H., Nishimura, S.-I., Ishii, S., Mori, T., Nishiyama, Y. (1997) Mercapto-chitins: a new type of supports for effective immobilization of acid phosphatase, *Carbohydr. Polym.* **32**, 171–175.

Kurita, K., Hirakawa, M., Nishiyama, Y. (1999) Silylated chitin: a new organo-soluble precursor for facile modifications and film casting, *Chem. Lett.*, 771–772.

Kurita, K., Inoue, M., Nishiyama, Y. (2000) Graft copolymerization of methyl methacrylate onto mercapto-chitin, *Adv. Chitin Sci.* **4**, 417–421.

Kuyama, H., Ogawa, T. (1997) Synthesis of chitosan oligomers, in: *Chitin Handbook* (Muzzarelli, R. A. A., Peter, M. G., Eds.), Grottammare: Atec, 181–189.

Lahiji, A., Sohrabi, A., Hungerford, D. S., Frondoza, C. G. (2000) Chitosan supports the expression of extracellular matrix proteins in human osteoblasts and chondrocytes, *J. Biomed. Mater. Res.* **51**, 586–595.

Lang, G. (1995) Chitosan derivatives – preparation and potential uses, in: *Chitin Chitosan* (Zakaria, M. B., Wan-Muda, W. M., Pauzi, A., Eds.), Bangi: Penerbit Universiti Kebangsaan Malaysia, 109–118.

Lang, G., Maresch, G., Birkel, S. (1997a) Hydroxyalkyl chitosans, in: *Chitin Handbook* (Muzzarelli, R. A. A., Peter, M. G., Eds.), Grottammare: Atec, 61–66.

Lang, G., Maresch, G., Birkel, S. (1997b) Quaternary chitosan salts, in: *Chitin Handbook* (Muzzarelli, R. A. A., Peter, M. G., Eds.), Grottammare: Atec, 67–70.

Lee, E. A., Pan, C. H., Son, J. M., Kim, S. I. (1999a) Isolation and characterization of basic exochitinase from leaf extract of *Rehmannia glutinosa*, *Biosci. Biotechnol. Biochem.* **63**, 1781–1783.

Lee, J. K., Kim, S. U., Kim, J. H. (1999b) Modification of chitosan to improve its hypocholesterolemic capacity, *Biosci. Biotechnol. Biochem.* **63**, 833–839.

Lee, K. Y., Ha, W. S., Park, W. H. (1995) Blood compatibility and biodegradability of partially *N*-acylated chitosan derivatives, *Biomaterials* **16**, 1211–1216.

Letzel, M. C., Synstad, B., Eijsink, V. G. H., Peter-Katalinic, J., Peter, M. G. (2000) Libraries of chitooligosaccharides of mixed acetylation patterns and their interactions with chitinases, *Adv. Chitin Sci.* **4**, 545–552.

Lim, L.-Y., Khor, E., Koo, O. (1998) γ-Irradiation of chitosan, *J. Biomed. Mater. Res.* **43**, 282–290.

Liu, W., Chen, X., Zhang, X., Liu, C., Cong, R., Jin, L., Lang, G., Wang, S. (1997) Study on toxicity of CM-chitosan, *Zhongguo Haiyang Yaowu* **16**, 17–19 [*Chem. Abstr.* **129**, 50606].

Liu, X. D., Nishi, N., Tokura, S., Sakairi, N. (2001) Chitosan coated cotton fiber: preparation and physical properties, *Carbohydr. Polym.* **44**, 233–238.

Londershausen, M., Tuberg, A., Buss, U., Spindler-Barth, M., Spindler, K. D. (1993) Comparison of chitin synthesis from an insect cell line and embryonic tick tissues, in: *Chitin Enzymology* (Muzzarelli, R. A. A., Ed.), Lyon: European Chitin Society, 101–108.

Londershausen, M., Turberg, A., Bieseler, B., Lennartz, M., Peter, M. G. (1996) Characterization and inhibitor studies of chitinases from a parasitic blowfly (*Lucilia cuprina*), a tick (*Boophilus microplus*), an intestinal nematode (*Haemonchus contortus*) and a bean (*Phaseolus vulgaris*), *Pest. Sci.* **48**, 305–314.

Londershausen, M., Turberg, A., Ludwig, M., Hirsch, B., Spindler-Barth, M. (1997) Properties of insect chitin synthase: effects of inhibitors, γ-S-GTP, and compounds influencing membrane lipids, in: *Applications of Chitin and Chitosan* (Goosen, M. F. A., Ed.), Lancaster: Technomic, 155–170.

Lopatin, S. A., Ilyin, M. M., Pustobaev, V. N., Bezchetnikova, Z. A., Varlamov, V. P., Davankov, V. A. (1995) Mass-spectrometric analysis of *N*-acetylchitooligosaccharides prepared through enzymic hydrolysis of chitosan, *Anal. Biochem.* **227**, 285–288.

Ludwig, M., Spindler-Barth, M., Spindler, K. D. (1991) Properties of chitin synthase in homogenates from *Chironomus* cells, *Arch. Insect Biochem. Physiol.* 18, 251–263.

Macchi, G. (1996) A new approach to the treatment of obesity: chitosan's effects on body weight reduction and plasma cholesterol's levels, *Acta Toxicol. Ther.* **17**, 303–320.
Maezaki, Y., Tsuji, K., Nakagawa, Y., Kawai, Y., Akimoto, M., Tsugita, T., Takekawa, W., Terada, A., Hara, H., Mitsuoka, T. (1993) Hypocholesterolemic effect of chitosan in adult males, *Biosci. Biotechnol. Biochem.* **57**, 1439–1444.
Marks, E. P., Ward, G. B. (1987) Regulation of chitin synthesis: mechanisms and methods, in: *Chitin and Benzoylphenyl Ureas* (Wright, J. E., Retnakaran, A., Eds.), Dordrecht: W. Junk, 33–42.
Martinou, A., Tsigos, I., Bouriotis, V. (1997) Preparation of chitosan by enzymatic deacetylation, in: *Chitin Handbook* (Muzzarelli, R. A. A., Peter, M. G., Eds.), pp. 501–505. Grottammare: Atec.
Maruyama, M., Ito, M. (1996) In vitro properties of a chitosan-bonded self-hardening paste with hydroxyapatite granules, *J. Biomed. Mater. Res.* **32**, 527–532.
Matsumiya, M., Mochizuki, A. (1996) Distribution of chitinase and β-*N*-acetylhexosaminidase in the organs of several fishes, *Fish Sci.* **62**, 150–151.
Matsumiya, M., Mochizuki, A. (1997) Purification and characterization of chitinase from the liver of Japanese common squid *Todarodes pacificus*, *Fish Sci.* **63**, 409–413.
Matsumoto, T., Nonaka, T., Hashimoto, M., Watanabe, T., Mitsui, Y. (1999) Three-dimensional structure of the catalytic domain of chitinase A1 from *Bacillus circulans* WL-12 at a very high resolution, *Proc. Jpn. Acad., Ser. B, Phys. Biol. Sci.* **75**, 269–274.
Mattioli-Belmonte, M., Nicoli-Aldini, N., De Benedittis, A., Sgarbi, G., Amati, S., Fini, M., Biagini, G., Muzzarelli, R. A. A. (1999) Morphological study of bone regeneration in the presence of 6-oxychitin, *Carbohydr. Polym.* **40**, 23–27.
Meyer, M. F., Kreil, G. (1996) Cells expressing the DG42 gene from early *Xenopus* embryos synthesize hyaluronan, *Proc. Natl. Acad. Sci. USA* **93**, 4543–4547.
Minami, S., Okamoto, Y., Mori, T., Fujinaga, T., Shigemasa, Y. (1997) Mechanism of wound healing acceleration by chitin and chitosan, *Adv. Chitin Sci.* **2**, 633–639.
Mitsutomi, M., Ueda, M., Arai, M., Ando, A., Watanabe, T. (1996) Action patterns of microbial chitinases and chitosanases on partially *N*-acetylated chitosan, in: *Chitin Enzymology*, (Muzzarelli, R. A. A., Ed.), Grottammare: Atec, 272–284, vol. 2.
Mohr, M. D., Boernsen, K. O., Widmer, H. M. (1995) Matrix-assisted laser desorption/ionization mass spectrometry: improved matrix for oligosaccharides, *Rapid Commun. Mass Spectrom.* **9**, 809–814.
Mori, T., Okumura, M., Matsuura, M., Ueno, K., Tokura, S., Okamoto, Y., Minami, S., Fujinaga, T. (1997) Effects of chitin and its derivatives on the proliferation and cytokine production of fibroblasts in vitro, *Biomaterials* **18**, 947–951.
Mulisch, M., Schermuly, G. (1997) Chitin synthesis in lower organisms, in: *Chitin Handbook* (Muzzarelli, R. A. A., Peter, M. G., Eds.), Grottammare: Atec, 337–344.
Muzzarelli, R. A. A. (1977) *Chitin*, Oxford: Pergamon Press.
Muzzarelli, R. A. A. (1988) Carboxymethylated chitins and chitosans, *Carbohydr. Polym.* **8**, 1–21.
Muzzarelli, R. A. A. (1996) Chitosan-based dietary foods, *Carbohydr. Polym.* **29**, 309–316.
Muzzarelli, R. A. A. (1997a) Depolymerization of chitins and chitosans with hemicellulase, lysozyme, papain and lipases, in: *Chitin Handbook* (Muzzarelli, R. A. A., Peter, M. G., Eds.), Grottammare: Atec, 153–163.
Muzzarelli, R. A. A. (1997b) Human enzymic activities related to the therapeutic administration of chitin derivatives, *Cell. Mol. Life Sci.* **53**, 131–140.
Muzzarelli, R. A. A. (1998) Management of hypercholesterolemia and overweight by oral administration of chitosans, *Biomed. Health Res.* **16**, 135–142.
Muzzarelli, R. A. A. (2000a) Chemical and preclinical studies on 6-oxychitin, *Adv. Chitin Sci.* **4**, 171–175.
Muzzarelli, R. A. A. (Ed.) (2000b) *Chitosan* per os, *from Dietary Supplement to Drug Carrier*, Grottamare: Atec.
Muzzarelli, R. A. A., Peter, M. G. (Eds.) (1997) *Chitin Handbook*, Grottammare: Atec.
Muzzarelli, R. A. A., Tanfani, F. (1985) The *N*-permethylation of chitosan and the preparation of *N*-trimethyl chitosan iodide, *Carbohydr. Polym.* **5**, 297–307.
Muzzarelli, R. A. A., Tomasetti, M., Ilari, P. (1994) Depolymerization of chitosan with the aid of papain, *Enzyme Microb. Technol.* **16**, 110–114.
Muzzarelli, R. A. A., Xia, W., Tomasetti, M., Ilari, P. (1995) Depolymerization of chitosan and substituted chitosans with the aid of a wheat germ lipase preparation, *Enzyme Microb. Technol.* **17**, 541–545.
Muzzarelli, R. A. A., Mattioli-Belmonte, M., Muzzarelli, B., Mattei, G., Fini, M., Biagini, G. (1997a) Medical and veterinary applications of chitin and chitosan, *Adv. Chitin Sci.* **2**, 580–589.

Muzzarelli, R. A. A., Rochetti, R., Stanic, V., Weckx, M. (1997b) Methods for the determination of the degree of acetylation of chitin and chitosan, in: *Chitin Handbook* (Muzzarelli, R. A. A., Peter, M. G., Eds.), Grottammare: Atec, 109–119.

Muzzarelli, R. A. A., Rosini, S., Trasciatti, S. (1997c) Abiogen Pharma S.R.L., Italy, Chelates of chitosans and alkaline-earth metal insoluble salts and the use thereof as medicaments useful in osteogenesis, WO 9919366 [*Chem. Abstr.* **130**, 301742].

Muzzarelli, R. A. A., Ramos, V., Stanic, V., Dubini, B., Mattioli-Belmonte, M., Tosi, G., Giardino, R. (1998) Osteogenesis promoted by calcium phosphate-*N,N*-dicarboxymethyl chitosan, *Carbohydr. Polym.* **36**, 267–276.

Muzzarelli, R. A. A., Muzzarelli, C., Cosani, A., Terbojevich, M. (1999) 6-Oxychitins, novel hyaluronan-like regiospecifically carboxylated chitins, *Carbohydr. Polym.* **39**, 361–367.

Muzzarelli, R. A. A., Miliani, M., Cartolari, M., Genta, I., Perugini, P., Modena, T., Pavanetto, F., Conti, B. (2000a) Oxychitin-chitosan microcapsules for pharmaceutical use, *STP Pharm. Sci.* **10**, 51–56.

Muzzarelli, R. A. A., Frega, N., Miliani, M., Cartolari, M. (2000b) Interactions of chitin, chitosan, *N*-laurylchitosan, and *N*-dimethylaminopropyl chitosan with olive oil, *Adv. Chitin Sci.* **4**, 275–279.

Natsume, H., Iwata, S., Ohtake, K., Miyamoto, M., Yamaguchi, M., Hosoya, K.-I., Kobayashi, D., Sugibayashi, K., Morimoto, Y. (1999) Screening of cationic compounds as an absorption enhancer for nasal drug delivery, *Int. J. Pharm.* **185**, 1–12.

Neff, D., Frazier, S. F., Quimby, L., Wang, R. T., Zill, S. (2000) Identification of resilin in the leg of cockroach, *Periplaneta americana*: confirmation by a simple method using pH dependence of UV fluorescence, *Arthropod Struct. Devel.* **29**, 75–83.

Neugebauer, W. A., D'Orléans-Juste, P., Bkaily, G. (2000) Peptide synthesis on chitosan/chitin, *Adv. Chitin Sci.* **4**, 411–416.

Neuhaus, B., Bresciani, J., Peters, W. (1997) Ultrastructure of the pharyngeal cuticle and lectin labelling with wheat germ agglutinin–gold conjugate indicating chitin in the pharyngeal cuticle of *Oesophagostomum dentatum* (Strongylida, Nematoda), *Acta Zool.* **78**, 205–213.

Neuhaus, J.-M. (1999) Plant chitinases (PR-3, PR-4, PR-8, PR-11), in: *Pathogen-related Proteins in Plants* (Datta, S. K., Muthukrishnan, S., Eds.), Boca Raton, FL: CRC Press, 77–105.

Neuhaus, J.-M., Fritig, B., Linthorst, H. J. M., Meins, F., Jr., Mikkelsen, J. D., Ryals, J. (1996) A revised nomenclature for chitinase genes, *Plant Mol. Biol. Rep.* **14**, 102–104.

Ng, C.-H., Chandrkrachang, S., Stevens, W. F. (2000) Effect of the rate of deacetylation on the physico-chemical properties of cuttlefish chitin, *Adv. Chitin Sci.* **4**, 50–54.

Nguyen, J., Francou, F., Virolle, M.-J., Guerineau, M. (1997) Amylase and chitinase genes in *Streptomyces lividans* are regulated by reg1, a pleiotropic regulatory gene, *J. Bacteriol.* **179**, 6383–6390.

Nishimura, S., Kai, H., Shinada, K., Yoshida, T., Tokura, S., Kurita, K., Nakashima, H., Yamamoto, N., Uryu, T. (1998) Regioselective syntheses of sulfated polysaccharides: specific anti-HIV-1 activity of novel chitin sulfates, *Carbohydr. Res.* **306**, 427–433.

Nishimura, Y., Watanabe, Y., Hong, J. M., Takeda, H., Wada, M., Yukawa, M. (1997) Intestinal absorption of ^{14}C-chitosan in rats, *Kichin, Kitosan Kenkyu* **3**, 55–61 [*Chem. Abstr.* **126**, 340532].

Nishizawa, Y., Kawakami, A., Hibi, T., He, D.-Y., Shibuya, N., Minami, E. (1999) Regulation of the chitinase gene expression in suspension-cultured rice cells by *N*-acetylchitooligosaccharides: differences in the signal transduction pathways leading to the activation of elicitor-responsive genes, *Plant Mol. Biol.* **39**, 907–914.

No, K. H., Meyers, S. P. (1997) Preparation of chitin and chitosan, in: *Chitin Handbook* (Muzzarelli, R. A. A., Peter, M. G., Eds.), Grottammare: Atec, 475–489.

Ogawa, K., Kawada, J., Yui, T., Okuyama, K. (2000) Crystalline behavior of chitosan, *Adv. Chitin Sci.* **4**, 324–329.

Ohme-Takagi, M., Suzuki, K., Shinshi, H. (2000) Regulation of ethylene-induced transcription of defense genes, *Plant Cell Physiol.* **41**, 1187–1192.

Okamoto, Y., Nose, M., Miyatake, K., Sekine, J., Oura, R., Shigemasa, Y., Minami, S. (2001) Physical changes of chitin and chitosan in canine gastrointestinal tract, *Carbohydr. Polym.* **44**, 211–215.

Okazaki, K., Kawabata, T., Nakano, M., Hayakawa, S. (1999) Purification and properties of chitinase from *Arthrobacter* sp. NHB-10, *Biosci. Biotechnol. Biochem.* **63**, 1644–1646.

Okuyama, K., Noguchi, K., Kanenari, M., Egawa, T., Osawa, K., Ogawa, K. (2000) Structural diversity of chitosan and its complexes, *Carbohydr. Polym.* **41**, 237–247.

Ono, K., Saito, Y., Yura, H., Ishikawa, K., Kurita, A., Akaike, T., Ishihara, M. (2000) Photocrosslinkable chitosan as a biological adhesive, *J. Biomed. Mater. Res.* **49**, 289–295.

Ormrod, D. J., Holmes, C. C., Miller, T. E. (1998) Dietary chitosan inhibits hypercholesterolemia and atherogenesis in the apolipoprotein E-deficient mouse model of atherosclerosis, *Atherosclerosis* **138**, 329–334.

Overdijk, B., Van Steijn, G. J. (1994) Human serum contains a chitinase: identification of an enzyme, formerly described as 4-methylumbelliferyl-tetra-*N*-acetylchitotetraoside (MU-TACT hydrolase), *Glycobiology* **4**, 797–803.

Overdijk, B., Van Steijn, G. J., Odds, F. C. (1996) Chitinase levels in guinea pig blood are increased after systemic infection with *Aspergillus fumigatus*, *Glycobiology* **6**, 627–634.

Overdijk, B., Van Steijn, G. J., Odds, F. C. (1999) Distribution of chitinase in guinea pig tissues and increases in levels of this enzyme after systemic infection with *Aspergillus fumigatus*, *Microbiology* **145**, 259–269.

Palli, S. R., Retnakaran, A. (1999) Molecular and biochemical aspects of chitin synthesis inhibition, *EXS* **87** (Chitin and Chitinases), 85–98.

Pantaleone, D., Yalpani, M. (1993) Unusual susceptibility of aminoglycans to enzymic hydrolysis, *Front. Biomed. Biotechnol.* **1**, 44–51.

Park, H. J., Jung, S. T., Song, J. J., Kang, S. G., Vergano, P. J., Testin, R. F. (1999) Mechanical and barrier properties of chitosan-based biopolymer film, *Kichin, Kitosan Kenkyu* **5**, 19–26 [*Chem. Abstr.* **131**, 20503].

Park, J. K., Keyhani, N. O., Roseman, S. (2000) Chitin catabolism in the marine bacterium *Vibrio furnissii* – identification, molecular cloning, and characterization of a *N,N'*-diacetylchitobiose phosphorylase, *J. Biol. Chem.* **275**, 33077–33083.

Park, R.-D., Cho, Y.-Y., Kim, K.-Y., Bom, H.-S., Oh, C.-S., Lee, H.-C. (1995) Adsorption of Toluidine Blue O onto chitosan, *Han'guk Nonghwa Hakhoechi* **38**, 447–451 [*Chem. Abstr.* **124**, 225723].

Patil, R. S., Ghormade, V., Deshpande, M. V. (2000) Chitinolytic enzymes: an exploration, *Enzyme Microb. Technol.* **26**, 473–483.

Paul, W., Sharma, C. P. (2000) Chitosan, a drug carrier for the 21st century: a review, *STP Pharm. Sci.* **10**, 5–22.

Perrakis, A., Tews, I., Dauter, Z., Oppenheim, A. B., Chet, I., Wilson, K. S., Vorgias, C. E. (1994) Crystal structure of a bacterial chitinase at 2.3 Å resolution, *Structure* **2**, 1169–1180.

Peter, M. G. (1993) Die molekulare Architektur des Exoskeletts von Insekten, *Chem. uns. Zeit* **27**, 189–197.

Peter, M. G. (1995) Applications and environmental aspects of chitin and chitosan, *J. Macromol. Sci., Pure Appl. Chem.* **A32**, 629–640.

Peter, M. G., Köhler, L. (1996) Chitin und Chitosan: Nachwachsende Rohstoffe aus dem Meer, in: *Perspektiven nachwachsender Rohstoffe in der Chemie* (Eierdanz, H., Ed.), Weinheim: VCH, 328–331.

Peter, M. G., Schweikart, F. (1990) Chitin biosynthesis enhancement by the endochitinase inhibitor allosamidin, *Biol. Chem. Hoppe Seyler* **371**, 471–473.

Peters, W., Latka, I. (1997) Wheat Germ Agglutinin-gold labelling, in: *Chitin Handbook* (Muzzarelli, R. A. A., Peter, M. G., Eds.), Grottammare: Atec, 33–40.

Pittermann, W., Hörner, V., Wachter, R. (1997a) Efficiency of high molecular weight chitosan in skin care applications, in: *Chitin Handbook* (Muzzarelli, R. A. A., Peter, M. G., Eds.), Grottammare: Atec, 361–372.

Pittermann, W., Wachter, R., Hörner, V. (1997b) Henkel K.G.a.A., Germany, Use of chitosan and/or chitosan derivatives for surface coating of implants and medical instruments, DE 19724869 [*Chem. Abstr.* **130**, 71621].

Pittler, M. H., Abbot, N. C., Harkness, E. F., Ernst, E. (1999) Randomized, double-blind trial of chitosan for body weight reduction, *Eur. J. Clin. Nutr.* **53**, 379–381.

Poveda, A., Asensio, J. L., Espinosa, J. F., Martin Pastor, M., Canada, J., Jimenez Barbero, J. (1997) Applications of nuclear magnetic resonance spectroscopy and molecular modeling to the study of protein–carbohydrate interactions, *J. Mol. Graph. Model.* **15**, 9–17.

Quong, D., Groboillot, A., Darling, G. D., Poncelet, D., Neufeld, R. J. (1997) Microencapsulation with cross-linked chitosan membranes, in: *Chitin Handbook* (Muzzarelli, R. A. A., Peter, M. G., Eds.), Grottammare: Atec, 405–410.

Ramasamy, M. S., Kulasekera, R., Wanniarachchi, I. C., Srikrishnaraj, K. A., Ramasamy, R. (1997) Interactions of human malaria parasites, *Plasmodium vivax* and *P. falciparum*, with the midgut of *Anopheles mosquitoes*, *Med. Vet. Entomol.* **11**, 290–296.

Rao, M. S., Stevens, W. F. (1997) Processing parameters in scale-up of *Lactobacillus* fermentation of shrimp biowaste, *Adv. Chitin Sci.* **2**, 88–93.

Rast, D. M., Merz, R. A., Jeanguenat, A., Mösinger, E. (2000) Enzymes of chitin metabolism for the design of antifungals, *Adv. Chitin Sci.* **4**, 479–505.

Ratajska, M., Struszczyk, M. H., Boryniec, S., Peter, M. G., Loth, F. (1997) The degree of deacetylation of chitosan: optimization of the IR method, *Polimery* **42**, 572–575.

Rathke, T. D., Hudson, S. M. (1994) Review of chitin and chitosan as fiber and film formers, *J. Macromol. Sci., Rev. Macromol. Chem. Phys.* **C34**, 375–437.

Raymond, L., Morin, F. G., Marchessault, R. H. (1993) Degree of deacetylation of chitosan using conductometric titration and solid-state NMR, *Carbohydr. Res.* **246**, 331–336.

Rege, P. R., Block, L. H. (1999) Chitosan processing: influence of process parameters during acidic and alkaline hydrolysis and effect of the processing sequence on the resultant chitosan's properties, *Carbohydr. Res.* **321**, 235–245.

Remunan-Lopez, C., Bodmeier, R. (1996) Mechanical and water vapor transmission properties of polysaccharide films, *Drug Dev. Ind. Pharm.* **22**, 1201–1209.

Renkema, G. H., Boot, R. G., Muijsers, A. O., Donker-Koopman, W. E., Aerts, J. M. F. G. (1995) Purification and characterization of human chitotriosidase, a novel member of the chitinase family of proteins, *J. Biol. Chem.* **270**, 2198–2202.

Renkema, G. H., Boot, R. G., Au, F. L., Donker-Koopman, W. E., Strijland, A., Muijsers, A. O., Hrebicek, M., Aerts, J. M. F. G. (1998) Chitotriosidase, a chitinase, and the 39-kDa human cartilage glycoprotein, a chitin-binding lectin, are homologs of family 18 glycosyl hydrolases secreted by human macrophages, *Eur. J. Biochem.* **251**, 504–509.

Roberts, G. A. F. (1992) *Chitin Chemistry*, Houndmills: Macmillan.

Roberts, G. A. F. (1997) Chitosan production routes and their role in determining the structure and properties of the product, *Adv. Chitin Sci.* **2**, 22–31.

Roberts, G. A. F., Wood, F. A. (2000) Inter-source reproducibility of the chitin deacetylation process, *Adv. Chitin Sci.* **4**, 34–39.

Robertus, J. D., Monzingo, A. F. (1999) The structure and action of chitinases, *EXS* **87** (Chitin and Chitinases), 125–135.

Rodriguez, M. S., Albertengo, L. A., Agulló, E. (2000) Chitosan emulsification properties, *Adv. Chitin Sci.* **4**, 382–388.

Rottmann, A., Synstad, B., Eijsink, V. G. H., Peter, M. G. (1999) Synthesis of *N*-acetylglucosaminyl and diacetylchitobiosyl amides of heterocyclic carboxylic acids as potential chitinase inhibitors, *Eur. J. Org. Chem.*, 2293–2297.

Rottmann, A., Synstad, B., Thiele, G., Schanzenbach, D., Eijsink, V. G. H., Peter, M. G. (2000) Approaches towards the design of new chitinase inhibitors, *Adv. Chitin Sci.* **4**, 553–557.

Rubin, B. R., Talent, J. M., Pertusi, R. M., Forman, M. D., Gracy, R. W. (2000) Oral polymeric *N*-acetyl-D-glucosamine as potential treatment for patients with osteoarthritis, *Adv. Chitin Sci.* **4**, 266–269.

Saito, A., Fujii, T., Miyashita, K. (1999) Chitinase system in *Streptomyces*, *Actinomycetologica* **13**, 1–10.

Sakiyama, T., Takata, H., Kikuchi, M., Nakanishi, K. (1999) Polyelectrolyte complex gel with high pH-sensitivity prepared from dextran sulfate and chitosan, *J. Appl. Polym. Sci.* **73**, 2227–2233.

Samain, E., Drouillard, S., Heyraud, A., Driguez, H., Geremia, R. A. (1997) Gram-scale synthesis of recombinant chitooligosaccharides in *Escherichia coli*, *Carbohydr. Res.* **302**, 35–42.

Sashiwa, H., Shigemasa, Y., Roy, R. (2000) Dissolution of chitosan in dimethyl sulfoxide by salt formation, *Chem. Lett.*, 596–597.

Saul, F. A., Rovira, P., Boulot, G., Van Damme, E. J. M., Peumans, W. J., Truffa-Bachi, P., Bentley, G. A. (2000) Crystal structure of *Urtica dioica* agglutinin, a superantigen presented by MHC molecules of class I and class II, *Struct. Fold. Design* **8**, 593–603.

Savant, V. D., Torres, J. A. (2000) Chitosan-based coagulating agents for treatment of cheddar cheese whey, *Biotechnol. Progr.* **16**, 1091–1097.

Schanzenbach, D., Matern, C., Peter, M. G. (1997) Cleavage of chitin by means of sulfuric acid/acetic anhydride, in: *Chitin Handbook* (Muzzarelli, R. A. A., Peter, M. G., Eds.), Grottammare: Atec, 171–174.

Schanzenbach, D., Peter, M. G. (1997) NMR spectroscopy of chito-oligosaccharides, in: *Chitin Handbook* (Muzzarelli, R. A. A., Peter, M. G., Eds.), Grottammare: Atec, 199–204.

Scheel, O., Thiem, J. (1997) Cleavage of chitin by means of aqueous hydrochloric acid and isolation of chito-oligosaccharides, in: *Chitin Handbook* (Muzzarelli, R. A. A., Peter, M. G., Eds.), Grottammare: Atec, 165–170.

Schermuly, G., Markmann-Mulisch, U., Mulisch, M. (1996) In vivo studies on the pathway of chitin synthesis in the ciliated protozoon *Eufolliculina uhligi*, *Adv. Chitin Sci.* **1**, 10–17.

Schrempf, H. (1996) The chitinolytic system of *Streptomycetes*, *Adv. Chitin Sci.* **1**, 123–128.

Schrempf, H. (1999) Characteristics of chitin-binding proteins from *Streptomycetes*, *EXS* **87** (Chitin and Chitinases), 99–108.

Schrempf, H., Blaak, H., Schnellmann, J., Stock, S. (1993) Chitinases from *Streptomyces olivaceoviridis*, *DECHEMA Monogr.* **129**, 261–271.

Semino, C. E., Robbins, P. W. (1995) Synthesis of "Nod"-like chitin oligosaccharides by the *Xenopus* developmental protein DG42, *Proc. Natl. Acad. Sci. USA* **92**, 3498–3501.

Semino, C. E., Specht, C. A., Raimondi, A., Robbins, P. W. (1996) Homologs of the *Xenopus* developmental gene DG42 are present in zebrafish and mouse and are involved in the synthesis of Nod-like chitin oligosaccharides during early embryogenesis, *Proc. Natl. Acad. Sci. USA* **93**, 4548–4553.

Senel, S., Kremer, M. J., Kas, S., Wertz, P. W., Hincal, A. A., Squier, C. A. (2000a) Effect of chitosan in enhancing drug delivery across buccal mucosa, *Adv. Chitin Sci.* **4**, 254–258.

Senel, S., Kremer, M. J., Kas, S., Wertz, P. W., Hincal, A. A., Squier, C. A. (2000b) Enhancing effect of chitosan on peptide drug delivery across buccal mucosa, *Biomaterials* **21**, 2067–2071.

Shahabuddin, M., Kaslow, D. C. (1994) *Plasmodium*: parasite chitinase and its role in malaria transmission, *Exp. Parasitol.* **79**, 85–88.

Shahabuddin, M., Vinetz, J. M. (1999) Chitinases of human parasites and their implications as antiparasitic targets, *EXS* **87** (Chitin and Chitinases), 223–234.

Shahidi, F., Synowiecki, J. (1992) Quality and compositional characteristics of Newfoundland shellfish processing discards, in: *Advances in Chitin and Chitosan* (Brine, C. J., Sandford, P. A., Zikakis, J. P., Eds.), London: Elsevier Applied Science, 617–626.

Shahidi, F., Arachchi, J. K. V., Jeon, Y.-J. (1999) Food applications of chitin and chitosans, *Trends Food Sci. Technol.* **10**, 37–51.

Shakarian, A. M., Dwyer, D. M. (1998) The Ld Cht1 gene encodes the secretory chitinase of the human pathogen *Leishmania donovani*, *Gene* **208**, 315–322.

Shakarian, A. M., Dwyer, D. M. (2000) Pathogenic *Leishmania* secrete antigenically related chitinases which are encoded by a highly conserved gene locus, *Exp. Parasitol.* **94**, 238–242.

Shigemasa, Y., Matsuura, H., Sashiwa, H., Saimoto, H. (1996a) An improved IR spectrometric determination of degree of deacetylation of chitin, *Adv. Chitin Sci.* **1**, 204–209.

Shigemasa, Y., Matsuura, H., Sashiwa, H., Saimoto, H. (1996b) Evaluation of different absorbance ratios from infrared spectroscopy for analyzing the degree of deacetylation in chitin, *Int. J. Biol. Macromol.* **18**, 237–242.

Shillito, B., Lechaire, J. P., Goffinet, G., Gaill, F. (1993) The chitin secreting microvilli of hydrothermal vent worms, in: *Chitin Enzymology* (Muzzarelli, R. A. A., Ed.), Lyon: European Chitin Society, 129–136.

Shimojoh, M., Fukushima, K., Kurita, K. (1998) Low-molecular-weight chitosans derived from β-chitin: preparation, molecular characteristics and aggregation activity, *Carbohydr. Polym.* **35**, 223–231.

Shiomi, K., Arai, N., Iwai, Y., Turberg, A., Kolbl, H., Omura, S. (2000) Structure of argifin, a new chitinase inhibitor produced by *Gliocladium* sp., *Tetrahedron Lett.* **41**, 2141–2143.

Signini, R., Desbrieres, J., Campana, S. P. (2000) On the stiffness of chitosan hydrochloride in acid-free aqueous solutions, *Carbohydr. Polym.* **43**, 351–357.

Society for the Study of Chitin and Chitosan (1991) *Kichin, Kitosan Jikken Manyuaru* (*Manual for Chitin and Chitosan Experiments*), Tokyo: Gihodo Shuppan. [*Chem. Abstr.* **117**, 107775].

Spicer, A. P., McDonald, J. A. (1998) Characterization and molecular evolution of a vertebrate hyaluronan synthase gene family, *J. Biol. Chem.* **273**, 1923–1932.

Spindler, K. D. (1997) Chitinase and chitosanase assays, in: *Chitin Handbook* (Muzzarelli, R. A. A., Peter, M. G., Eds.), Grottammare: Atec, 229–235.

Spindler, K. D., Funke-Höpfner, B. (1991) *N*-Acetyl-β-D-hexosaminidases of the brine shrimp *Artemia*: partial purification and characterization, *Z. Naturforsch.* **46C**, 781–788.

Spindler, K.-D., Spindler-Barth, M. (1996) Chitin degradation and synthesis in arthropods, in: *Chitin in Life Sciences* (Giraud-Guille, M. M., Ed.), Lyon: André, 41–52.

Spindler, K.-D., Spindler-Barth, M. (1999) Inhibitors of chitinases, *EXS* **87** (Chitin and Chitinases), 201–209.

Spindler, K. D., Spindler-Barth, M., Londershausen, M. (1990) Chitin metabolism: a target for drugs against parasites, *Parasitol. Res.* **76**, 283–288.

Spindler-Barth, M. (1993) Hormonal regulation of chitin metabolism in insect cell lines, in: *Chitin Enzymology* (Muzzarelli, R. A. A., Ed.), Lyon: European Chitin Society, 75–82.

Spindler-Barth, M. (1997) Quantitative determination of chitin biosynthesis, in: *Chitin Handbook* (Muzzarelli, R. A. A., Peter, M. G., Eds.), Grottammare: Atec, 331–335.

Stankiewicz, B. A., Briggs, D. E. G., Evershed, R. P., Flannery, M. B., Wuttke, M. (1997) Preservation of chitin in 25-million-year-old fossils, *Science* **276**, 1541–1543.

Stenberg, E., Wachter, R. (1996) Potential of the northern shrimp (*Pandalus borealis*) as a source of chitosan in Norway, *Adv. Chitin Sci.* **1**, 166–172.

Stern, J. S., Gades, M. D., Halsted, C. H. (2000) Chitosan does not block fat absorption in men fed a high fat diet, *Obesity Res.* **8**, B94.

Stevens, W. F. (2001) Proceedings, 8th ICCC, Yamaguchi, Japan, September 20–23, 2000, in press, and personal communication.

Stokke, B. T., Vårum, K. M., Holme, H. K., Hjerde, R. J. N., Smidsrød, O. (1995) Sequence specificities for lysozyme depolymerization of partially *N*-acetylated chitosans, *Can. J. Chem.* **73**, 1972–1981.

Stone, C. A., Wright, H., Clarke, T., Powell, R., Devaraj, V. S. (2000) Healing at skin graft donor sites dressed with chitosan, *Br. J. Plast. Surgery* **53**, 601–606.

Struszczyk, H. (1987) Microcrystalline chitosan. I. Preparation and properties of microcrystalline chitosan, *J. Appl. Polym. Sci.* **33**, 177–189.

Struszczyk, H. (1997a) Preparation of chitosan fibers, in: *Chitin Handbook* (Muzzarelli, R. A. A., Peter, M. G., Eds.), Grottammare: Atec, 437–440.

Struszczyk, H. (1997b) Coating of textile fibers with chitosan, in: *Chitin Handbook* (Muzzarelli, R. A. A., Peter, M. G., Eds.), Grottammare: Atec, 441–444.

Struszczyk, H., Kivekäs, O. (1992) Recent developments in microcrystalline chitosan applications, in: *Advances in Chitin and Chitosan* (Brine, C. J., Sandford, P. A., Zikakis, J. P., Eds.), London: Elsevier Applied Science, 549–555.

Struszczyk, M. H., Loth, F., Peter, M. G. (1997) Analysis of degree of deacetylation in chitosans from various sources, *Adv. Chitin Sci.* **2**, 71–77.

Struszczyk, M. H., Loth, F., Köhler, L. A., Peter, M. G. (1998) Characterization of chitosan, in: *Carbohydrates as Organic Raw Materials*, (Praznik, W., Huber, A., Eds.), Vienna: WUV Universitätsverlag, 110–117, vol. 4.

Struszczyk, M. H., Hahlweg, R., Peter, M. G. (2000) Comparative analysis of chitosans from insects and Crustacea, *Adv. Chitin Sci.* **4**, 40–49.

Sugamori, T., Iwase, H., Maeda, M., Inoue, Y., Kurosawa, H. (2000) Local hemostatic effects of microcrystalline partially deacetylated chitin hydrochloride, *J. Biomed. Mater. Res.* **49**, 225–232.

Sugimoto, M., Morimoto, M., Sashiwa, H., Saimoto, H., Shigemasa, Y. (1998) Preparation and characterization of water-soluble chitin and chitosan derivatives, *Carbohydr. Polym.* **36**, 49–59.

Suzuki, K., Mikami, T., Okawa, Y., Tokoro, A., Suzuki, S., Suzuki, M. (1986) Antitumor effect of hexa-*N*-acetylchitohexaose and chitohexaose, *Carbohydr. Res.* **151**, 403–408.

Suzuki, K., Taiyoji, M., Sugawara, N., Nikaidou, N., Henrissat, B., Watanabe, T. (1999) The third chitinase gene (ChiC) of *Serratia marcescens* 2170 and the relationship of its product to other bacterial chitinases, *Biochem. J.* **343**, 587–596.

Suzuki, S., Matsumoto, T., Tsukada, K., Aizawa, K., Suzuki, M. (1989) Antimetastatic effects of *N*-acetyl chitohexaose on mouse-bearing Lewis lung carcinoma, in: *Chitin and Chitosan* (Skjåk-Bræk, G., Anthonsen, T., Sandford, P., Eds.), London: Elsevier Applied Science, 707–711.

Suzuki, Y., Okamoto, Y., Morimoto, M., Sashiwa, H., Saimoto, H., Tanioka, S., Shigemasa, Y., Minami, S. (2000) Influence of physico-chemical properties of chitin and chitosan on complement activation, *Carbohydr. Polym.* **42**, 307–310.

Synstad, B., Gaseidnes, S., Vriend, G., Nielsen, J.-E., Eijsink, V. G. H. (2000) On the contribution of conserved acidic residues to catalytic activity of chitinase B from *Serratia marcescens*, *Adv. Chitin Sci.* **4**, 524–529.

Takai, M., Shimizu, Y., Hayashi, J., Uraki, Y., Tokura, S. (1989) NMR and X-ray studies of chitin and chitosan in solid state, in: *Chitin and Chitosan* (Skjåk-Bræk, G., Anthonsen, T., Sandford, P., Eds.), London: Elsevier Applied Science, 431–436.

Takiguchi, Y., Sakamaki, Y., Yamaguchi, T. (1999) Depolymerization of chitosan by ultrasonic irradiation, *Kichin, Kitosan Kenkyu* **5**, 75–79 [*Chem. Abstr.* **131**, 130184].

Tanabe, T., Kawase, T., Watanabe, T., Uchida, Y., Mitsutomi, M. (2000) Purification and characterization of a 49-kDa chitinase from *Streptomyces griseus* HUT 6037, *J. Biosci. Bioeng.* **89**, 27–32.

Tanioka, S., Matsui, Y., Irie, T., Tanigawa, T., Tanaka, Y., Shibata, H., Sawa, Y., Kono, Y. (1996) Oxidative depolymerization of chitosan by hydroxyl radical, *Biosci. Biotechnol. Biochem.* **60**, 2001–2004.

Technical-Insights (1998) Chitin and chitosan: an expanding range of markets await exploitation, 3rd edn. URL: www.wiley/com/technical insights/reporttocs/chitin3.html.

Tellam, R. L., Eisemann, C. (2000) Chitin is only a minor component of the peritrophic matrix from larvae of *Lucilia cuprina*, *Insect Biochem. Mol. Biol.* **30**, 1189–1201.

Tellam, R. L., Vuocolo, T., Johnson, S. E., Jarmey, J., Pearson, R. D. (2000) Insect chitin synthase – cDNA sequence, gene organization and expression, *Eur. J. Biochem.* **267**, 6025–6042.

Terbojevich, M., Cosani, A. (1997) Molecular weight determination of chitin and chitosan, in: *Chitin Handbook* (Muzzarelli, R. A. A., Peter, M. G., Eds.), Grottammare: Atec, 87–101.

Terbojevich, M., Carraro, C., Cosani, A. (1988) Solution studies of the chitin-lithium chloride-*N,N*-di-methylacetamide system, *Carbohydr. Res.* **180**, 73–86.

Terbojevich, M., Cosani, A., Bianchi, E., Marsano, E. (1996) Solution behavior of chitin in dimethylacetamide/lithium chloride, *Adv. Chitin Sci.* **1**, 333–339.

Terwisscha van Scheltinga, A. C. (1997) *Structure and mechanism of the plant chitinase hevamine*, PhD Thesis, University of Groningen.

Terwisscha van Scheltinga, A. C., Armand, S., Kalk, K. H., Isogai, A., Henrissat, B., Dijkstra, B. W. (1995) Stereochemistry of chitin hydrolysis by a plant chitinase/lysozyme and x-ray structure of a complex with allosamidin. Evidence for substrate assisted catalysis, *Biochemistry* **34**, 15619–15623.

Terwisscha van Scheltinga, A. C., Hennig, M., Dijkstra, B. W. (1996) The 1.8 Å resolution structure of hevamine, a plant chitinase/lysozyme, and analysis of the conserved sequence and structure motifs of glycosyl hydrolase family 18, *J. Mol. Biol.* **262**, 243–257.

Tews, I., Perrakis, A., Oppenheim, A., Dauter, Z., Wilson, K. S., Vorgias, C. E. (1996) Bacterial chitobiase structure provides insight into catalytic mechanism and the basis of Tay-Sachs disease, *Nature Struct. Biol.* **3**, 638–648.

Tews, I., Terwisscha van Scheltinga, A. C., Perrakis, A., Wilson, K. S., Dijkstra, B. W. (1997) Substrate-assisted catalysis unifies two families of chitinolytic enzymes, *J. Am. Chem. Soc.* **119**, 7954–7959.

Thanou, M., Verhoef, J. C., Florea, B. I., Junginger, H. E. (2000) Chitosan derivatives as intestinal penetration enhancers of the peptide drug buserelin *in vivo* and *in vitro*, *Adv. Chitin Sci.* **4**, 244–249.

Thom, E., Wadstein, J. (2000) Chitosan in weight reduction: results from a large scale consumer study, *Adv. Chitin Sci.* **4**, 229–232.

Tjoelker, L. W., Gosting, L., Frey, S., Hunter, C. L., Le Trong, H., Steiner, B., Brammer, H., Gray, P. M. (2000) Structural and functional definition of the human chitinase chitin-binding domain, *J. Biol. Chem.* **275**, 514–520.

Tokura, S., Nishi, N., Nishimura, S.-I., Ikeuchi, Y., Azuma, I., Nishimura, K. (1984) Physicochemical, biochemical, and biological properties of chitin derivatives, in: *Chitin, Chitosan, and Related Enzymes* (Zikakis, J. P., Ed.), New York: Academic Press, 303–325.

Tokura, S., Ueno, K., Miyazaki, S., Nishi, N. (1997) Molecular weight-dependent antimicrobial activity by chitosan, *Macromol. Symp.* **120**, 1–9.

Tokuyasu, K., Ono, H., Hayashi, K., Mori, Y. (1999a) Reverse hydrolysis reaction of chitin deacetylase and enzymatic synthesis of β-D-GlcNAc-(1-4)-GlcN from chitobiose, *Carbohydr. Res.* **322**, 26–31.

Tokuyasu, K., Ono, H., Kitagawa, Y., Ohnishi-Kameyama, M., Hayashi, K., Mori, Y. (1999b) Selective *N*-deacetylation of *p*-nitrophenyl *N,N'*-diacetyl-β-chitobioside and its use to differentiate the action of two types of chitinases, *Carbohydr. Res.* **316**, 173–178.

Tokuyasu, K., Ono, H., Mitsutomi, M., Hayashi, K., Mori, Y. (2000) Synthesis of a chitosan tetramer derivative, β-D-GlcNAc-(1-4)-β-D-GlcNAc-(1-4)-β-D-GlcNAc-(1-4)-D-GlcN through a partial *N*-acetylation reaction by chitin deacetylase, *Carbohydr. Res.* **325**, 211–215.

Tolaimate, A., Desbrieres, J., Rhazi, M., Alagui, A., Vincendon, M., Vottero, P. (2000) On the influence of deacetylation process on the physicochemical characteristics of chitosan from squid chitin, *Polymer* **41**, 2463–2469.

Tsaih, M. L., Chen, R. H. (1997) Effect of molecular weight and urea on the conformation of chitosan molecules in dilute solutions, *Int. J. Biol. Macromol.* **20**, 233–240.

Tsaih, M. L., Chen, R. H. (1999) Effects of ionic strength and pH on the diffusion coefficients and conformation of chitosans molecule in solution, *J. Appl. Polym. Sci.* **73**, 2041–2050.

Tsigos, I., Martinou, A., Kafetzopoulos, D., Bouriotis, V. (2000) Chitin deacetylases: new, versatile tools in biotechnology, *Trends Biotechnol.* **18**, 305–312.

Tsujibo, H., Hatano, N., Endo, H., Miyamoto, K., Inamori, Y. (2000a) Purification and characterization of a thermostable chitinase from *Streptomyces thermoviolaceus* OPC-520 and cloning of the encoding gene, *Biosci. Biotechnol. Biochem.* **64**, 96–102.

Tsujibo, H., Miyamoto, J., Kondo, N., Miyamoto, K., Baba, N., Inamori, Y. (2000b) Molecular cloning of the gene encoding an outer-membrane-associated β-*N*-acetylglucosaminidase involved in chitin degradation system of *Alteromonas* sp. strain O-7, *Biosci. Biotechnol. Biochem.* **64**, 2512–2516.

Tsujibo, H., Okamoto, T., Hatano, N., Miyamoto, K., Watanabe, T., Mitsutomi, M., Inamori, Y. (2000c) Family 19 chitinases from *Streptomyces thermoviolaceus* OPC-520: molecular cloning and characterization, *Biosci. Biotechnol. Biochem.* **64**, 2445–2453.

Ueno, K., Yamaguchi, T., Sakairi, N., Nishi, N., Tokura, S. (1997) Antimicrobial activity by fractionated chitosan oligomers, *Adv. Chitin Sci.* **2**, 156–161.

van Aalten, D. M. F., Synstad, B., Brurberg, M. B., Hough, E., Riise, B. W., Eijsink, V. G. H., Wierenga, R. K. (2000) Structure of a two-domain chitotriosidase from *Serratia marcescens* at 1.9 Å resolution, *Proc. Natl. Acad. Sci. USA* **97**, 5842–5847.

Van Damme, E. J. M., Peumans, W. J., Barre, A., Rouge, P. (1998) Plant lectins: a composite of several distinct families of structurally and evolutionary related proteins with diverse biological roles, *Crit. Rev. Plant Sci.* **17**, 575–692.

Vander, P., Vårum, K. M., Domard, A., El Gueddari, N. E., Moerschbacher, B. M. (1998) Comparison of the ability of partially *N*-acetylated chitosans and chitooligosaccharides to elicit resistance reactions in wheat leaves, *Plant Physiol.* **118**, 1353–1359.

Varki, A. (1996) Does DG42 synthesize hyaluronan or chitin ?: a controversy about oligosaccharides in vertebrate development, *Proc. Natl. Acad. Sci. USA* **93**, 4523–4525.

Vårum, K. M., Smidsrød, O. (1997) Specificity in enzymic and chemical degradation of chitosans, *Adv. Chitin Sci.* **2**, 168–175.

Vårum, K. M., Anthonsen, M. W., Grasdalen, H., Smidsrød, O. (1991a) Determination of degree of *N*-acetylation and the distribution of *N*-acetyl groups in partially *N*-deacetylated chitins (chitosans) by high field NMR spectroscopy, *Carbohydr. Res.* **211**, 17–23.

Vårum, K. M., Anthonsen, M. W., Grasdalen, H., Smidsrød, O. (1991b) ^{13}C-NMR studies of the acetylation sequences in partially *N*-deacetylated chitins (chitosans), *Carbohydr. Res.* **217**, 19–27.

Vårum, K. M., Anthonsen, M. W., Nordtveit, R. J., Ottoy, M. H., Smidsrød, O. (1994) Structure–property relationships in chitosans, in: *Chitin World* (Karnicki, Z. S., Wojtasz-Pajak, A., Brzeski, M. M., Bykowski, P. J., Eds.), Bremerhaven: Wirtschaftsverlag NW, 166–174.

Vårum, K. M., Holme, H. K., Izume, M., Stokke, B. T., Smidsrød, O. (1996) Determination of enzymic hydrolysis specificity of partially *N*-acetylated chitosans, *Biochim. Biophys. Acta* **1291**, 5–15.

Veneroni, G., Veneroni, F., Contos, S., Tripodi, S., De Bernardi, M., Guarino, C., Marletta, M. (1996a) Effect of a new chitosan dietary integrator and hypocaloric diet on hyperlipidemia and overweight in obese patients, *Acta Toxicol. Ther.* **17**, 53–70.

Veneroni, G., Veneroni, F., Contos, S., Tripodi, S., De Bernardi, M., Guarino, C., Marletta, M. (1996b) Effect of a new chitosan on hyperlipidemia and overweight in obese patients, in: *Chitin Enzymology* (Muzzarelli, R. A. A., Ed.), Grottammare: Atec, 63–67, vol. 2.

Vincendon, M. (1997) Triphenylsilylchitin: a new chitin derivative soluble in organic solvents, *Adv. Chitin Sci.* **2**, 328–333.

Vincendon, M., Roux, J. C., Chanzy, H., Tanner, S., Belton, P. (1989) Solid-state CP/MAS ^{13}C NMR analysis of the crystalline chitin polymorphs, in: *Chitin and Chitosan* (Skjåk-Bræk, G., Anthonsen, T., Sandford, P., Eds.), London: Elsevier Applied Science, 437–438.

Vincent, J. F. V. (1990) *Structural Biomaterials*, Princeton: Princeton University Press.

Vinetz, J. M., Dave, S. K., Specht, C. A., Brameld, K. A., Xu, B., Hayward, R., Fidock, D. A. (1999) The chitinase PfCHT1 from the human malaria parasite *Plasmodium falciparum* lacks proenzyme and chitin-binding domains and displays unique substrate preferences, *Proc. Natl. Acad. Sci. USA* **96**, 14061–14066.

Vinetz, J. M., Valenzuela, J. G., Specht, C. A., Aravind, L., Langer, R. C., Ribeiro, J. M. C., Kaslow, D. C. (2000) Chitinases of the avian malaria parasite *Plasmodium gallinaceum*, a class of enzymes necessary for parasite invasion of the mosquito midgut, *J. Biol. Chem.* **275**, 10331–10341.

Vogt, W., Bachen, M., Gaumann, A., Jacob, B., Laue, C., Pommersheim, R., Schrezenmeir, J. (1996) Immobilization of enzymes and living cells by multilayer microcapsules, *DECHEMA Monogr.* **132**, 195–204.

Voigt, A., Lichtenfeld, H., Sukhorukov, G. B., Zastrow, H., Donath, E., Baeumler, H., Moehwald, H. (1999) Membrane filtration for microencapsulation and microcapsules fabrication by layer-by-layer polyelectrolyte adsorption, *Ind. Eng. Chem. Res.* **38**, 4037–4043.

Vorgias, C. E. (1997) Structural basis of chitin hydrolysis in bacteria, *Adv. Chitin Sci.* **2**, 176–187.

Wachter, R., Tesmann, H., Svenning, R., Olsen, R., Stenberg, E. (1995) Henkel KGaA, Germany, Kationische Biopolymere, EP 0737 211.

Watanabe, T., Kimura, K., Sumiya, T., Nikaidou, N., Suzuki, K., Suzuki, M., Taiyoji, M., Ferrer, S., Regue, M. (1997) Genetic analysis of the chitinase system of *Serratia marcescens* 2170, *J. Bacteriol.* **179**, 7111–7117.

Watanabe, T., Kanai, R., Kawase, T., Tanabe, T., Mitsutomi, M., Sakuda, S., Miyashita, K. (1999) Family 19 chitinases of *Streptomyces* species: characterization and distribution, *Microbiology UK* **145**, 3353–3363.

Wieczorek, A., Mucha, M. (1997) Application of chitin derivatives and their composites to biodegradable paper coatings, *Adv. Chitin Sci.* **2**, 890–896.

Win, N. N., Pengju, G., Stevens, W. F. (2000) Deacetylation of chitin by fungal enzymes, *Adv. Chitin Sci.* **4**, 55–62.

Wojtasz-Pajak, A., Brzeski, M. M., Malesa-Ciecwierz, M. (1994) Regulatory aspects of chitosan and its practical applications, in: *Chitin World* (Karnicki, Z. S., Wojtasz-Pajak, A., Brzeski, M. M., Bykowski, P. J., Eds.), Bremerhaven: Wirtschaftsverlag NW, 435–445.

Wright, J. E., Retnakaran, A. (Eds.) (1987) *Chitin and Benzoylphenyl Ureas*, Dordrecht: W. Junk.

Wu, Y., Adam, R., Williams, S. A., Bianco, A. E. (1996) Chitinase genes expressed by infective larvae of the filarial nematodes, *Acanthocheilonema viteae* and *Onchocerca volvulus*, *Mol. Biochem. Parasitol.* **75**, 207–219.

Wuolijoki, E., Hirvela, T., Ylitalo, P. (1999) Decrease in serum LDL cholesterol with microcrystalline chitosan, *Methods Find. Exp. Clin. Pharmacol.* **21**, 357–361.

Yalpani, M. (1997) Nutraceuticals. A materials-based perspective, *Front. Foods Food Ingredients* **2**, 53–70.

Yalpani, M., Pantaleone, D. (1994) An examination of the unusual susceptibilities of aminoglycans to enzymic hydrolysis, *Carbohydr. Res.* **256**, 159–175.

Yamagami, T., Ishiguro, M. (1998) Complete amino acid sequences of chitinase-1 and -2 from bulbs of genus *Tulipa*, *Biosci. Biotechnol. Biochem.* **62**, 1253–1257.

Yamagami, T., Tanigawa, M., Ishiguro, M., Funatsu, G. (1998) Complete amino acid sequence of chitinase-A from leaves of pokeweed (*Phytolacca americana*), *Biosci. Biotechnol. Biochem.* **62**, 825–828.

Yamano, N., Matsushita, Y., Kamada, Y., Fujishima, S., Arita, M. (1996) Purification and characterization of *N*-acetylglucosamine 6-phosphate deacetylase with activity against *N*-acetylglucosamine from *Vibrio cholerae* Non-O1, *Biosci. Biotechnol. Biochem.* **60**, 1320–1323.

Yoshida, M., Itano, N., Yamada, Y., Kimata, K. (2000) In vitro synthesis of hyaluronan by a single protein derived from mouse HAS1 gene and characterization of amino acid residues essential for the activity, *J. Biol. Chem.* **275**, 497–506.

Zecchi, V., Aiedeh, K., Orienti, I. (1997) Controlled drug delivery systems, in: *Chitin Handbook* (Muzzarelli, R. A. A., Peter, M. G., Eds.), Grottammare: Atec, 397–404.

Zhang, H., Du, Y. G., Yu, X. J., Mitsutomi, M., Aiba, S. (1999) Preparation of chitooligosaccharides from chitosan by a complex enzyme, *Carbohydr. Res.* **320**, 257–260.

Zhang, M., Inui, H., Hirano, S. (1997) A facile method for the preparation of 6-deoxy derivatives of chitin, *J. Carbohydr. Chem.* **16**, 673–679.

Zhang, M., Haga, A., Sekiguchi, H., Hirano, S. (2000) Structure of insect chitin isolated from beetle larva cuticle and silkworm (*Bombyx mori*) pupa exuvia, *Int. J. Biol. Macromol.* **27**, 99–105.

Zielinski, B. A., Aebischer, P. (1994) Chitosan as a matrix for mammalian cell encapsulation, *Biomaterials* **15**, 1049–1056.

14
Curdlan

Dr. In-Young Lee
Korea Research Institute of Bioscience and Biotechnology and DawMaJin Biotech Corp., P.O. Box 115, Daeduk Valley, Daejon 305-600, Korea; Tel.: +82-428620722; Fax: +82-428620702; E-mail: leeiy@dmj-biotech.com

Biopolymers for Medical and Pharmaceutical Applications. Edited by A. Steinbüchel and R. H. Marchessault

ISBN: 3-527-31154-8

AIDS	acquired immunodeficiency syndrome
AMP	adenosine 5′-monophosphate
ATCC	American Type Culture Collection
DMSO	dimethylsulfoxide
DP	degree of polymerization
FDA	Food and Drug Administration
HIV	human immunodeficiency virus
MNNG (also NTG)	*N*-methyl-*N*-nitro-*N*-nitrosoguanidine
NMR	nuclear magnetic resonance
TEM	transmission electron microscopy
UDP-glucose	uridine 5′-diphosphate glucose
UMP	uridine 5′-monophosphate

1 Introduction

Curdlan, an insoluble microbial exopolymer is composed almost exclusively of β-(1,3)-glucosidic linkages. One of the unique features of curdlan is that aqueous suspensions can be thermally induced to produce high-set gels, which will not return to the liquid state upon reheating (Harada et al., 1968), and this has attracted the attention of the food industry. In addition to this, curdlan offers many health benefits, as the beta-glucan family is well known among the scientific community to have immunestimulatory effects. Many informative reviews have been produced on this subject (Harada, 1977; Harada et al., 1993); however, apart from offering a brief review of β-(1,3)-glucan, the present chapter article provides an updated overview of the production, properties, and application of curdlan.

2 Historical Outline

Curdlan was discovered in 1966 by Professor Harada and coworkers, and given its name because of its ability to "curdle" when heated (Harada et al., 1966). At this time, Harada and his colleagues were working on the identification of organisms capable of utilizing petrochemical materials, and isolated *Alcaligenes faecalis* var. *myxogenes* 10C3 from soil. This organism was found to be capable of growing on a medium containing 10% ethylene glycol as the sole carbon source

(Harada et al., 1965), and also produced a new β-(1,3)-glucan that contained about 10% succinic acid, and which was named succinoglucan (Harada 1965; Harada and Yoshimura, 1965). They were also able to derive a spontaneous mutant that mainly produced a water-insoluble neutral polysaccharide, β-(1,3)-glucan, and which did not contain succinoglucan.

Scientists at Takeda Chemical Industries Ltd. (Osaka, Japan) have played a pioneering role in both the research and development of curdlan. Thus, as early as 1989, curdlan was approved and commercialized for food usage in Korea, Taiwan, and Japan. Upon obtaining approval in December 1996, Pureglucan™ – the tradename of curdlan – was launched in the US market as a formulation aid, processing aid, stabilizer, and thickener or texture modifier for food use (Spicer et al., 1999). No evidence of any toxicity nor carcinogenicity of Pureglucan has been observed.

3 Structure

3.1 Beta-1,3-glucan

Both chemical and enzymatic analyses have confirmed that curdlan is a homopolymer of D-glucose linked in β-(1,3) fashion (Saito et al., 1968) (Figure 1). Curdlan has an average degree of polymerization (DP) of approximately 450, and is unbranched (Naganishi et al., 1976). Nakata et al. (1998) reported that the average molecular weight of curdlan in 0.3 N NaOH is in the range of 5.3×10^4 to 2.0×10^6 daltons. Within the class of polysaccharides classified as β-(1,3)-glucans, there are a number of structural variants. The sources of glucans and their structural differences are listed in Table 1. Mycelial fungi are an abundant source of β-(1,3)-glucans; grifolan, which is produced from *Grifola frondosa* and stimulates cytokine production from macrophages, is a β-(1,3)-glucan with a molecular weight of $> 4.5 \times 10^5$ daltons (Okazaki et al., 1995), while lentinan, from *Lentinus edodes*, has a molecular weight of 5×10^5 daltons and two glucose branches for every five β-(1,3)-glucosyl units in the backbone (Jong and Birmingham, 1993). The structure of schizophyllan is very similar to that of lentinan, but it has one glucose branch for every third glucose in the β-1,3-backbone and its molecular weight is 4.5×10^5 daltons (Misaki et al., 1993). Scleroglucan from *Sclerotium rolfsii* has one glucose branch for every third glucose unit (Farina et al., 2001), whereas SSG from *Sclerotinia sclerotiorum* is a highly branched β-(1,3)-glucan (Sakurai et al., 1991). Pachyman, from *Poria cocos*, has an average of 3.2 branch points per molecule of β-(1,3)-glucan (Okuyama et al., 1996; Zhang et al., 1997), whilst krestin is a protein-linked β-glucan with a molecular weight of c. 100,000 daltons, which can be extracted from the mycelia of *Coriolus versicolor* (Azuma, 1987). β-(1,3)-Glucan is also present in the inner cell wall of the bakers' yeast *Saccha-*

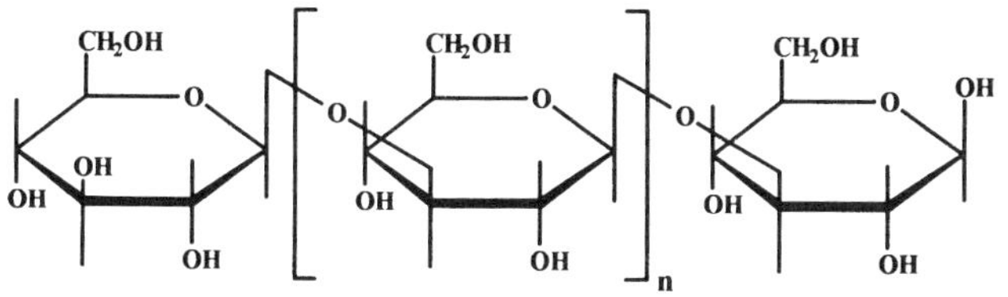

Fig. 1 Structure of curdlan.

Tab. 1 A variety of glucans having β-1,3 linkage in their backbones

Source	*Branch*	M_w	*Reference*
Bacteria			
Curdlan (*Agrobacterium sp. Alcaligens sp.*)	Exclusively β-(1,3)-glucosidic linkages	5.3×10^4 -2.0×10^6	Nakata et al. (1998)
Fungi			
Grifolan (*Grifola frondosa*)	Branched β-1,3-gulcan	4.5×10^5	Okazaki et al. (1995)
Lentinan (*Lentinus eeodes*)	Two glucose branches for every five glucose unit	5×10^5	Jong and Birmingham (1993)
Schizophyllan (*Schizophyllum commune*)	One glucose branch for every third glucose unit	4.5×10^5	Misaki et al. (1993)
Scleroglucan (*Sclerotium glucanum*)	One glucose branch for every third glucose unit	$1.6-5.0\times10^6$	Farina et al. (2001)
SSG (*Sclerotinia sclerotiorum*)	Highly branched β-1,3-glucan	$2\times10^5-2\times10^6$	Okazaki et al. (1995)
Pachyman (*Poria cocos*)	Several β-1,6-linked branch points per molecule	2.06×10^4, 8.93×10^4	Zhang et al. (1997)
Krestin (*Coriolus versicolor*)	β-1,3-glucan	1.0×10^6	Azuma (1987)
Yeast (*Saccharomyces cerevisiae*)			
Soluble glucan	β(1,6) linkage to β-(1,3) backbone	2×10^5–2×10^6	Janusz et al. (1986)
Insoluble glucan	β-(1,6) linkage to β-(1,3) backbone	3.53×10^4, 4.57×10^6	Williams et al. (1994)
Brown algea			
Laminarin (*Laminaria digitata*)	β-1,3-glucan and β-1,6-glucan		Read et al. (1996)

romyces cerevisiae to support the structural strength of its cell wall. Unlike lentinan, schizophyllan and scleroglucan, the side branches of yeast β-(1,3)-glucans are chains of glucose molecules, and not single glucose residues. Depending on the extraction procedure used and their subsequent treatment, the yeast glucans may be either particulate water-insoluble or water-soluble macromolecules.

3.2 Conformation in Solution

Many researchers have investigated the molecular structures of curdlan in aqueous system. Three conformers of soluble curdlan have been reported, including single-helix, triple-helix, and random coil. Ogawa et al. (1972) studied the conformational behavior of curdlan in alkaline solution by measuring the optical rotatory dispersion, intrinsic viscosity and flow birefringence. At low concentrations of sodium hydroxide, curdlan has a helical (ordered) conformation, but a significant conformational change occurs at a NaOH concentration of 0.19–0.24 N NaOH. In alkaline solution >0.2 N NaOH, curdlan is completely soluble and exists as random coils, but upon neutralization the polymer adopts an 'ordered state', which is composed of a mixture of single and triple helices. A ^{13}C-NMR study supported this finding (Saito et al., 1977). Increasing the salt concentration shifts the point of conformational transition to a higher alkali concentration (Ogawa et al., 1973a), and addition of nonsolvents such as 2-chloroethanol, dioxan or water to dimethylsulfoxide (DMSO) solution also changes the conformation of curdlan to a rigid, ordered structure (Ogawa et al., 1973b). These workers also showed that the optical rotation was dependent upon the DP of the curdlan in 0.1 N sodium hydroxide (Ogawa et al., 1973c), and concluded that the content of the ordered form increases with DP until becoming constant at DP values of about 200. Electron microscopic comparison of the molecular structures of curdlan with different DPs showed that only curdlan with higher DP can form a gel when heated (Koreeda et al., 1974).

The conversion between triple-helix and single-helix conformers is mediated by different chemical or physical treatments. Treatment of the triple-helix schizophyllan with NaOH has been used to prepare single helix-rich forms (Ohno et al., 1995). Young et al. (2000) proposed a transition mechanism after an investigation using fluorescence resonance energy transfer spectroscopy, which showed that a partially opened triple-helix conformer was formed on treatment with NaOH, and that increasing degrees of strand opening were associated with increasing concentrations of NaOH. After neutralizing the NaOH, the partially opened conformers gradually reverted to the triple-helix.

3.3 Gel Structure

Some clarification of the fine structure of dispersed molecules and networks is necessary to understand the viscoelastic properties of curdlan, which forms two distinct types of gel. For both gels – described as low-set and high-set gels – transmission electron microscopy (TEM) showed them to be composed of three curdlan molecules that are associated to form a triple helix. Tada et al. (1997) proposed a mechanism of formation of the low-set gel using static light-scattering measurements. The molecular associates are formed at a NaOH concentration of 0.01–0.1 N at 25 °C, and this association progresses with as the NaOH concentration decreases. Consequently, the average molecular weight for the molecular associate in 0.01 N NaOH is higher than that in 0.1 N NaOH aqueous solution. The molecular associates consist of a dense core and hydrophilic surface at low NaOH concentrations. By contrast, Kasai and Harada (1980) proposed an annealing model to form a high-set, resilient gel upon heating to >80 °C. This annealing is associated with the irreversible loss of water, and resulted in a more tightly coiled triple helix. The structure crystallizes as a triplex of right-handed, six-fold helical chains in a hexagonal unit cell with a fiber repeating length of 18.78 Å (Chuah et al., 1983). Further removal of water from this structure, by drying under vacuum, results in further tightening of the six-fold triple helix and a decrease in the fiber period to only 5.87 Å (Deslandes et al., 1980).

4 Occurrence

As described in Section 2, β-(1-3)-D-glucans are present in a variety of living systems, including fungi, yeasts, algae, bacteria and higher plants. However, until now only bacteria belonging to the *Alcaligenes* and *Agrobacterium* species have been reported to produce the linear β-(1,3)-glucan type of homopolymer, curdlan. Figure 2 illustrates

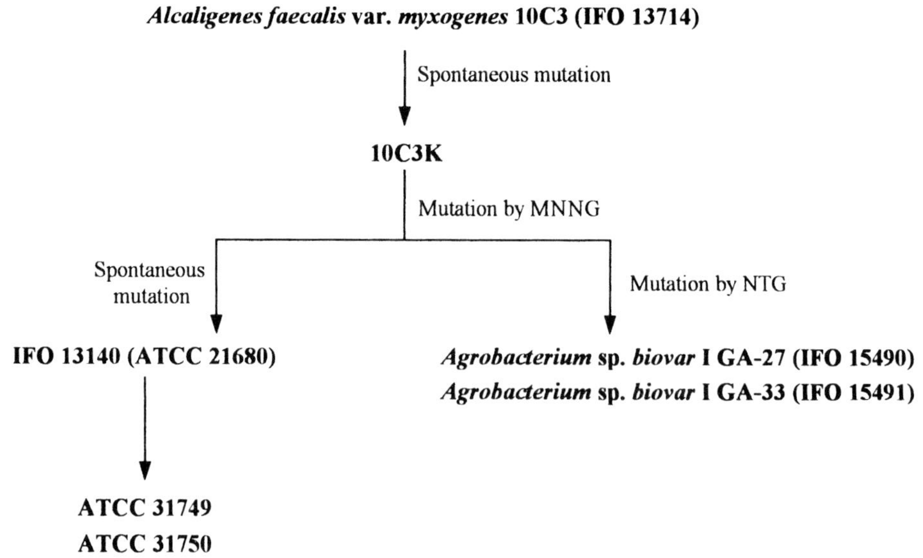

Fig. 2 Lineage of representative strains for curdlan production.

the lineage of the curdlan-producing strains since Harada et al. first isolated *Alcaligenes faecalis* var. *myxogenes* 10C3 during the screening of soil bacteria capable of metabolizing various petroleum fractions (Harada et al., 1965). The parent strain produced two different types of exopolysaccharide; a water-insoluble neutral homoglucan called 'curdlan', and a water-soluble acidic heteroglucan containing about 10% succinic acid, and referred to as 'succinoglucan' (Harada, 1965; Harada and Yoshimura 1965). Moreover, a mutant strain 10C3K was isolated from a stock culture of 10C3, which produced only curdlan. Strain 10C3K is a spontaneous mutant with a stable ability to produce exocellular polysaccharide. By inducing mutagenesis with *N*-methyl-*N*-nitro-*N*-nitrosoguanidine (MNNG), Takeda Chemical Industries Ltd. later isolated a uracil auxotrophic mutant from strain 10C3K, which was named *Alcaligenes faecalis* var. *myxogenes* IFO 13140 (ATCC 21680) and had improved gel-forming β-(1,3)-glucan-producing ability. Phillips and Lawford (1983) isolated a mutant strain from strain ATCC 21680 in a nitrogen-limited chemostat culture (the accession number was ATCC 31749). Unlike its auxotrophic parent strain, ATCC 31749 does not require uracil for its growth, and is not a revertant as it can be distinguished from 10C3K by its inability to hydrolyze starch and its ability to grow on citric acid as sole carbon source. ATCC 31750, which arose as a spontaneous variant of the parent ATCC 31749, produces only the water-insoluble curdlan-type glucan, while the parent strain produces both soluble and insoluble polysaccharides. All of these strains, which were formerly regarded as *Alcaligenes* species, have now been taxonomically reclassified as *Agrobacterium* species (IFO Research Communications, Vol. 15, pp. 57–75 (1991)). Takeda Chemical Industries Ltd. derived further mutant strains from strain 10C3K, which reduced the activity of the enzyme, phosphoenol pyruvic acid carboxykinase (Kanegae et al., 1996).

Naganishi et al. (1974) examined the occurrence of curdlan-type polysaccharides in microorganisms by using the water-soluble dye aniline blue, with which curdlan forms a blue complex. It was also shown that the rate of color complex formation was dependent both on the polymer concentration and DP; hence these findings provided an excellent tool for the screening of curdlan-producing bacteria. Naganishi et al. (1976) tested 687 strains of different genus of bacteria using the aniline blue staining technique. Among those examined, some strains of *Alcaligenes* and *Agrobacterium* species turned blue on agar plates containing aniline blue, and these have been used widely in the production of curdlan-type polysaccharides. Some strains of *Bacillus* formed blue complexes with aniline blue, but their polymeric constitution has not yet been studied.

5 Biosynthesis

Sutherland (1977, 1993) generalized the biosynthesis of extracellular polysaccharides into three major steps: (1) substrate uptake; (2) intracellular formation of polysaccharide; and (3) extrusion from cell. A metabolic pathway for exopolysaccharide biosynthesis is shown schematically in Figure 3. First, a carbohydrate substrate enters the cell by active transport and group translocation involving substrate phosphorylation. The substrate is then directed along either catabolic pathways, or those leading to polysaccharide synthesis. UDP-glucose, a key precursor, is synthesized by the UDP-glucose pyrophosphorylase-induced conversion of glucose-1-phosphate to UDP-glucose. Subsequently, polymer construction occurs together with the transfer of monosaccharides from UDP-glucose to a carrier lipid.

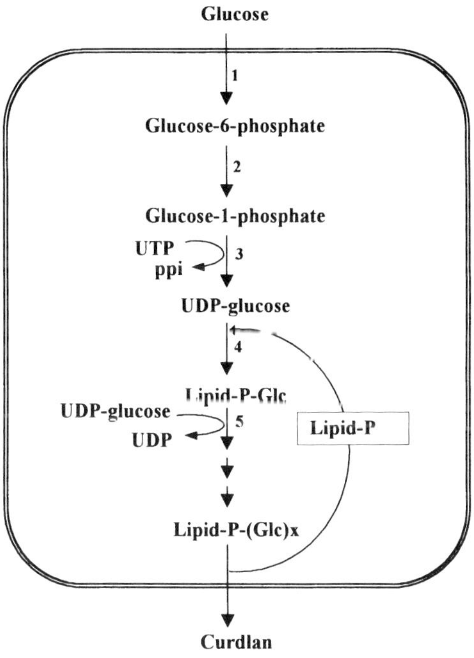

Fig. 3 Metabolic pathway for the synthesis of curdlan. 1, hexokinase; 2, phosphoglucomutase; 3, UDP-glucose pyrophosphorylase; 4, transferase; 5, polymerase. Lipid-P represents isoprenoid lipid phosphate.

After further chain elongation, the polymer is extruded from the cells.

Nitrogen-limited culture has also been employed for the production of curdlan, and has been generally explained by the roles of the carrier lipids (Sutherland, 1977; Ielpi et al., 1993). The availability of isoprenoid lipid may provide a way of regulating polysaccharide synthesis. Since curdlan biosynthesis takes place most extensively after cell growth has been stopped due to nitrogen exhaustion, isoprenoid lipids would be more available for carrying oligosaccharides instead of cellular liposaccharide and peptidoglycan.

The synthesis of precursor molecules is also of considerable importance in polysaccharide synthesis, in terms of the metabolic driving force. UDP-glucose serves as an activated precursor for glycosyl moieties in the synthesis of curdlan, a homopolysaccharide composed exclusively of β-1,3-linked glucose residues. In addition, cellular nucleotides not only play an important role in the synthesis of sugar nucleotides, but also have a widespread regulatory potential in cellular metabolism. Kim et al. (1999) examined the change of intracellular nucleotide levels and their stimulatory effects on curdlan synthesis in *Agrobacterium* species under different culture conditions. Under nitrogen-limited conditions where curdlan synthesis was stimulated, intracellular levels of UMP and AMP were at least twice as high as those occurring under nitrogen-sufficient conditions, though UDP-glucose levels were similar. The time profiles of curdlan synthesis and cellular nucleotide levels showed that curdlan synthesis is positively related with intracellular levels of UMP and AMP. *In vitro* enzyme reactions involved in the synthesis of UDP-glucose showed that a higher UMP concentration promotes the synthesis of UDP-glucose, while AMP neither inhibits nor facilitates the activity of UDP-glucose pyrophosphorylase. The addition of UMP to the medium also increased curdlan synthesis. From these results, these workers concluded that the higher intracellular UMP levels caused by nitrogen limitation enhance the metabolic flux of curdlan synthesis by promoting cellular UDP-glucose synthesis.

Kai et al. (1993) reported a study of the biosynthetic pathway of curdlan using ^{13}C-labeled glucose in *Agrobacterium* sp. ATCC 31749. By analyzing the labeled products, the biosynthesis of curdlan was interpreted as involving five routes: direct polymerization from glucose; rearrangement; isomerization of cleaved trioses; from fructose-6-phosphate; and from fructose fragments produced in various pathways of glycolysis. However, it was noted that more than 60% of curdlan is synthesized by direct polymerization, and that curdlan biosynthesis via glycolysis is comparatively low. This analysis also indicated that glycolysis occurs mainly via the pentose cycle and the Entner–Doudoroff pathway rather than the Embden–Meyerhof pathway.

6 Molecular Genetics

Little is known about the molecular genetics of bacterial curdlan biosynthesis, while there is growing information about the genes required for β-(1,3)-glucan synthesis in yeasts and filamentous fungi. Recently, Stasinopoulos et al. (1999) cloned genes that were essential for the production of curdlan and, by using comparative sequence analysis, identified them as putative curdlan synthase genes. Further genetic investigations will open up new avenues for curdlan synthesis, and these will doubtlessly be exploited to produce curdlan in higher yields.

7 Production

7.1 Carbon Source

Many crucial factors that affect curdlan production, including carbon, nitrogen, phosphate, oxygen supply, and pH have been investigated. High productivity using cheap carbon sources is important for the industrial production of curdlan. Lee, I.-Y. et al. (1997) reported that maltose and sucrose were efficient carbon sources for the production of curdlan by a strain of *Agrobacterium* species, with maximal production (60 g L^{-1}) being obtained from sucrose, with a productivity of 0.5 g L^{-1} h^{-1} when nitrogen was limited at a cell concentration of 16 g L^{-1}. Molasses, which contains large amounts of sucrose, might also be the substrate of choice, with up to 42 g L^{-1} of curdlan, at a yield of 0.35 g curdlan per gram total sugar, being obtained in 5-day cultivation. Sucrose is a less expensive substrate than glucose, and as sugar beet or sugar cane molasses are cheap byproducts widely available from the sugar industry, they are very attractive carbon sources from an economic point of view.

7.2 Nitrogen Effect

As previously described, relatively few strains of *Agrobacterium* and *Alcaligenes* species are known to produce curdlan. In such strains, curdlan production is associated with the poststationary phase of nitrogen depletion; thus, the operation involves an initial production of biomass, which is followed by curdlan production. Therefore, it is important to determine the initial concentration of the nitrogen source because it provides the limiting factor for cell growth during batch fermentation. Kim, M.-K. et al. (2000) reported that the cell growth rate decreased as the ammonium concentration increased. However, since higher cell concentrations produce more curdlan, an optimal ammonium concentration should be determined to provide an appropriate cell concentration while minimizing the inhibitory effect of the ammonium ion.

7.3 Oxygen Supply

Curdlan-producing stains are highly aerobic, and an adequate oxygen supply is therefore a key factor in production. Since curdlan is insoluble in water, the fermentation broth is of relatively low viscosity, and there is little resistance to oxygen transfer from gas to the liquid. However, a layer of insoluble exopolymer surrounding the cell mass offers resistance to oxygen transfer from the liquid into the cell, and therefore a high dissolved oxygen concentration is required for maximal productivity. Shake-flask fermentation results have shown that the specific production rate decreases as the volume of medium is increased, indicating that these cultures are limited by the relatively low oxygen transfer capacity of the system. Several investigations have been made into developing the process for curdlan production, especially with respect to reactor design (Lawford et al., 1986; Lawford and Rousseau, 1991, 1992). These workers employed two different types of impeller: a radial-flow, flat-blade impeller; and an axial-flow impeller. The radial-flow impeller was effective at providing high oxygen transfer rates to increase the production of curdlan, but the high shear characteristics of this design yielded a product of inferior quality in terms of tensile strength of the thermally induced gel. An axial-flow impeller typically produces less shear and more pumping. High volumetric oxygen transfer can be achieved by low shear designs equipped with sparging devices, which consist of microporous materials through which oxygen-enriched air is dispersed. The maximal specific production rate was 90 mg per g cells h^{-1} when 30% oxygen-enriched air was supplied in the low-shear system.

7.4 Phosphate Effect

Phosphate concentration must also be considered because it significantly influences cell growth and product formation. The production of rhamnose-containing polysaccharide by a *Klebsiella* strain was enhanced by a reduction in the phosphate content of the medium (Farres et al., 1997). In contrast, a sufficiency of phosphate resulted in good alginate yields in *Pseudomonas* strains and showed growth-associated production (Conti et al., 1994). Thus, the effect of phosphate on the production of polysaccharides is variable. Kim, M.-K. et al. (2000) investigated the influence of inorganic phosphate concentration on the production of curdlan by *Agrobacterium* species. Under nitrogen-limited conditions which allow curdlan production, the concentration of phosphate remains constant as it is not further utilized for cell growth. The optimal residual phosphate concentration for curdlan production was in the range 0.1–0.5 g L^{-1}. Relatively low concentrations appeared to be optimal for curdlan production, although without phosphate, curdlan production was extremely low. However, on increasing the cell phosphate concentration from 0.42 to 1.68 g L^{-1}, curdlan production increased from 0.44 to 2.80 g L^{-1}. Moreover, the optimal phosphate concentration range was not dependent upon cell concentration, and the specific production rate was about 70 mg curdlan per g cells h^{-1}, irrespective of cell concentration.

7.5 pH Effect

The pH of the culture is one of the most important factors because it significantly influences rates of both cell growth and product formation. High viscosity of the culture broth is often a critical problem in polysaccharide production. However, the viscosity problem can be obviated by operating the fermentation at a slightly acidic pH, since curdlan is insoluble under these conditions. Moreover, there appears to be more than one single optimal pH because fermentation of the culture for curdlan production is divided into two phases – the cell growth phase and the curdlan production phase. Lee, J.-H. et al. (1999) sought an optimal pH profile to maximize curdlan production in a batch fermentation of *Agrobacterium* species. The cell growth rate was maximal at pH 7.0, while curdlan production was maximal at pH 5.5. The pH profile provided a strategy to shift the culture pH from the optimal growth condition (pH 7.0) to the optimal production one (pH 5.5) at the time of ammonium exhaustion. By adopting the optimal pH profile in a batch process, these workers obtained a significant improvement in curdlan production (64 g L^{-1}) compared with that obtained using a constant-pH operation (36 g L^{-1}).

7.6 Batch Production

A high curdlan production was attempted by employing the optimal operation strategy with an *Agrobacterium* strain (Lee, I.-Y. et al., 1997, 1999a; Lee, J.-H. et al., 1999; Kim, M.-K. et al., 2000) that was tolerant of high concentrations of sucrose. This made batch operation possible, with an initial sucrose concentration of 140 g L^{-1}. An initial ammonium concentration 0.8 g L^{-1} was chosen, which produced 6.4 g L^{-1} of cells, the cell yield from ammonium being 8.0 g cells per g ammonium. A typical batch fermentation profile in a 300-L stirred tank reactor is shown in Figure 4. To produce a volumetric oxygen transfer coefficient of 146.5 h^{-1}, the agitation speed was set at 200 r.p.m. over the whole fermentation period, with an inner pressure of 0.2 bar. When the ammonium was exhausted in the culture broth at 20 h, the cell concentration had reached 6.8 g L^{-1}. The pH was controlled at 7.0 with 4 N NaOH/KOH at the cell growing stage. The culture pH was then shifted from 7.0 to 5.5 by adding 3 N HCl at the time of nitrogen limitation. Aeration rates were maintained at 0.5 vvm (volume $volume^{-1}$ min^{-1}). Curdlan production began from the onset of nitrogen exhaustion, and a maximum concentration of 58 g L^{-1} was obtained in 120 h cultivation. The dissolved oxygen level fell rapidly during cell growth, and increased immediately after nitrogen exhaustion, as cell growth ceased. Subsequently, as the curdlan concentration increased, the dissolved oxygen level became limited due to the increased viscosity of the culture broth.

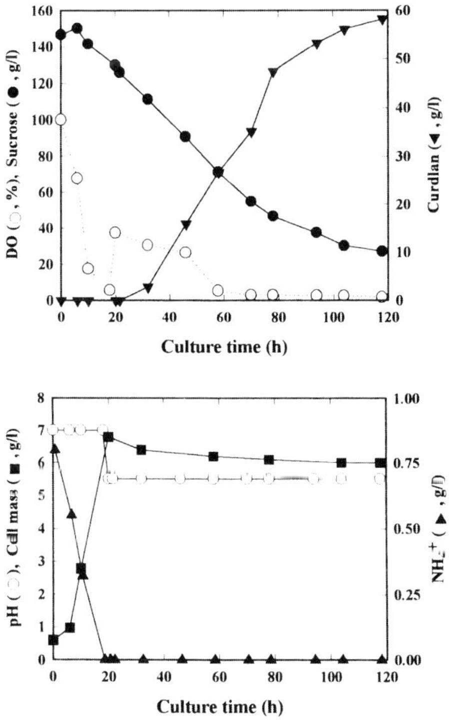

Fig. 4 Batch production of curdlan with *Agrobacterium sp.* ATCC 31750 in a 300-L jar fermenter (Lee, I.-Y. et al., 1999a).

7.7 Continuous Production

Continuous production of curdlan was attempted in a two-stage continuous process (Phillips and Lawford, 1983). In the first stage, the curdlan-producing strain *Alcaligenes* sp. ATCC 31749 was grown aerobically in a medium containing carbon and nitrogen. However, the amount of nitrogen in the first stage was so limited that the effluent contained substantially no inorganic nitrogen. The effluent was fed into the second stage in a constant-volume fermenter, where it was mixed with a nitrogen-free medium in order to induce curdlan production without further cell growth. Curdlan production was 7 g L^{-1} at a dilution rate of 0.02 h^{-1} in a steady-state culture, providing a curdlan productivity of 0.14 g L^{-1} h^{-1}.

7.8 Isolation Process

The recovery procedure is based on the conformational transition which occurs when the concentration of alkali exceeds 0.2 N (Ogawa et al., 1972). Under alkaline conditions, the biomass can be separated from the dissolved curdlan, which remains in the supernatant. Upon neutralization of the alkaline supernatant, the polymer forms an insoluble gel that can be recovered by centrifugation, the polymer being subsequently washed free of contaminating salt. Procedures for preserving curdlan in the dry state include both dehydration with organic solvents and spray drying.

8 Properties

8.1 Gel Formation

Curdlan is water-insoluble but soluble in alkali, and forms two-types of gels: high-set or low-set gels. The high-set gel is formed upon heating to around 80 °C, depending on its concentration in water (Maeda et al., 1967). The high-set curdlan gels are thermally irreversible and, unlike agar gel, further heating does not cause them to revert to the liquid state; they are also very elastic and resilient, whereas agar gels are brittle and relatively fragile. Curdlan forms a low-set reversible gel when its suspension is heated to 55 – 65 °C and then cooled. When a low-set gel is heated to 80 °C, however, it turns into a high-set gel. At the same curdlan concentration, a low-set gel has a weaker strength than a high-set gel. Changes in external factors such as pH, temperature, and ionic strength greatly affect gelation ability. Curdlan also forms a reversible gel when its alkaline solution is neutralized, and the addition of calcium or magnesium ions to a weakly alkaline solution of curdlan produces a gel with a bridged structure (Aizawa et al., 1974). Other methods to prepare curdlan gels include cooling DMSO solutions, or dialyzing its alkaline or DMSO solutions in water (Ogawa et al., 1972). The strength of the curdlan gel increases with temperature, concentration, and heating time (Maeda et al., 1967; Takeda technical report, 1997) (Figure 5). Curdlan, unlike other gelling agents, has a unique ability to form gels over a wide range of pH values (3.0 – 10.0).

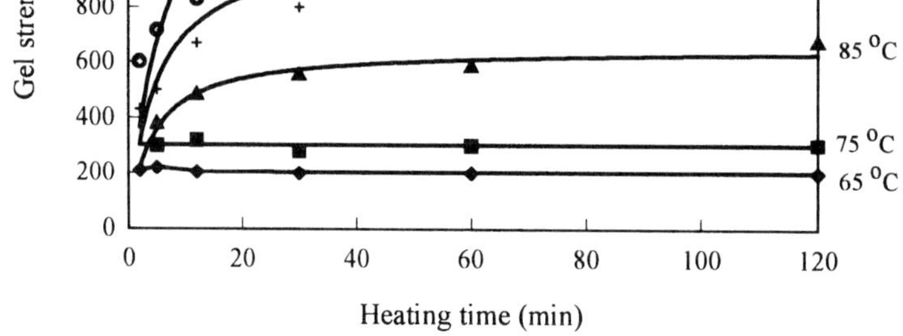

Fig. 5 Effect of heating temperature and time on curdlan gel strength (2% gel) (Takeda technical report, 1997).

8.2 Immunestimulatory Activity

β-(1,3)-Glucan has been found to be effective in stimulating macrophages (leukocytes or white blood cells). Macrophages form the immune system's first line of defense against foreign invaders, and can recognize and kill tumor cells, remove foreign debris resulting from oxidative and radiation damage, speed up the recovery of damage tissue, further activate cytokines, and initiate an immune cascade system to mobilize B and T cells (Di Luzio et al., 1979; Di Luzio, 1983; Suzuki et al., 1992; Hadden, 1993). Curdlan is a bacterial polysaccharide composed entirely of β-(1,3)-D-glucosidic linkages, and has been reported to show antitumor activity (Sasaki and Takasuka, 1976; Yoshioka et al., 1985). However, differences in biological activities seem to be dependent upon the degree of branching, molecular conformation, and molecular weight. Sasaki et al. (1978) examined the effect of chain length on antitumor activity by using different molecular weight glucans, which were obtained by the acid hydrolysis of curdlan. The results show that insoluble glucans with a number-average DP >50 have strong antitumor activity. As described in Section 2, β-glucans exist in three conformers: single-helix, triple-helix, and random coil. Among these, the ordered (helical) conformations are considered to be biologically active forms (Bohn and BeMiller, 1995). Some researchers reported that the triple-helical structure of schizophyllan, a β-glucan obtained from *Schizophyllum commune,* is essential for its anti-tumor activity (Yanaki et al., 1983; Kojima et al., 1986). However, other studies have suggested that the single-helix is more potent in this respect (Saito et al., 1991; Aketagawa et al., 1993). Ohno et al. (1995, 1996) showed that both the triple- and single-helix conformers of schizopyllan are active against tumors and leukopenia. In addition, the triple-helix conformer had a significant antagonistic effect upon zymosan-mediated hydrogen peroxide synthesis in peritoneal macrophages, whereas the single-helix conformer showed strong activity on the synthesis of tumor necrosis factor, nitric oxide, and hydrogen peroxide. It remains unclear as to which of the helical conformers is the most active, but it is likely that the structure–activity relationship of the β-glucan-mediated immunopharmacological activities vary, and are dependent upon the assay systems adopted.

9 Applications

9.1 Food Applications

Since the time of curdlan's original discovery, it has been mainly proposed for use in the food industry as a possible extender or substitute for natural plant gums in food preparations requiring a thickening or bodying additive. The food applications of curdlan are summarized in Table 2. These food additives and essential ingredients are generally divided in terms of how much curdlan is used; food additives require less than 1% curdlan, and essential ingredients more than 1%. Curdlan is useful as a gelling material to improve the textural quality, water-holding capacity and thermal stability of various foods. The polymer can be added during the production process, before heating, either as a powder, or as a suspension or slurry in either water or aqueous alcohol. The foods employing these functions include soy-bean curd (tofu), sweet bean paste jelly, boiled fish paste, noodles, sausages, jellies, and jams (Masayuki and Yukihiro, 1990; Taguchi et al., 1991; Ken and Akirou, 1992; Masanori et al., 1992; Shunsuke and Masatoshi, 1992; Hideaki et al., 1993; Hiroki and Etsuko, 1993; Masahiro and Masaru, 1993; Masatoshi et al., 1993; Yukihiro and Takahiro, 1993; Yasuhiro, 1994; Masaaki and Takaaki, 1995; Masataka et al., 1997; Akihiro and Takaaki, 1998; Hiroshi et al., 1999; Kazushi, 1999; Masao and Toshimi, 1999; Takaaki, 1999). Various tests have shown that this polymer is safe. The gel of curdlan has properties intermediate between the brittleness of agar gel and the elasticity of gelatin (Kimura et al., 1973). Furthermore, since the β-(1,3)-glucan polymer is not readily degraded by human digestive enzymes, it offers the possibility of new calorie-reduced products, such as those often referred to as "dietetic" foods. The fact that curdlan set-gels are known efficiently to absorb high concentrations of sugars from syrups and are relatively resistant to syneresis, suggests a use in sweet jellies and various other dessert-type foods. The resistance of curdlan gels to degradation by freezing and thawing also indicates potential in frozen food products. The pseudoplastic flow behavior of curdlan-containing fluids points to a possible role as thickeners and stabilizers in liquefied foods, such as salad dressings and spreads. Curdlan is also used to produce an edible casting film for foods with other water-soluble polymeric substances having heat-sealability (Akira and Atsushi, 1989).

9.2
Pharmaceutical Applications

Curdlan sulfate having a β-(1,3)-glucan backbone showed high anti-AIDS (acquired immunodeficiency syndrome) virus activity, with few adverse side effects; therefore, curdlan sulfate has a growing potential as an anti-AIDS drug (Jagodzinski et al., 1994; Takeda-Hirokawa et al., 1997). The Phase I/II trials (toxicity testing) of curdlan sulfate for human immunodeficiency virus (HIV) carriers have been carried out in the United States under the auspices of the Food and Drugs Administration (FDA) since 1992. A sulfuric acid ester of curdlan low molecular polymer, with an average DP ranging from 10 to 400 was found to be an active antiretrovirus ingredient useful for preventing and treating infection because of its high water solubility, low toxicity and relatively high inhibitory activity upon human retrovirus multiplication (Takashi and Junji, 1991). Mikio et al. (1995) disclosed that a

Tab. 2 Application of curdlan as food additives and essential ingredient (Spicer et al., 1999)

Function	*Objective foods*
Essential ingredient	
Gelling agent	Dessert, jelly, pudding, dry mixes
Bulking agent	Dietetic foods, diabetic foods
Low-calorie foods	Edible film, edible fiber casings
Food additive	
Improving viscoelasticity	Noodle, hamburger, sausage
Binding agent	Hamburger, starch jelly
Water-holding agent	Noodle, sausage, ham, starch jelly
Prevention of deterioration	Frozen egg products
Masking of malodors or aromas	Boiled rice
Retention of shape	Starch jelly, dry desert mixes
Thickeners and stabilizers	Salad dressing, low-calorie foods, frozen foods
Coating agent	Flavors

curdlan derivative modified by reaction with glycidol developed excellent antiviral activity whilst showing extremely low toxicity. Evans et al. (1998) subsequently reported that the low toxicity of curdlan and its marked anti-invasion activity on merozoites makes it a potential auxiliary treatment for severe malaria. Moreover, linear β-(1,3)-glucan sulfate has potential as a blood coagulation inhibitor useful for both the prevention and treatment of thrombosis, because linear β-(1,3)-glucan sulfate has been found to have a strong inhibitory action on blood coagulation (Tsuneo, 1990). Curdlan is also used as a sustained release suppository containing drug-active components, such as indomethacin, diclofenac sodium and ibuprofen (Toshiko, 1992). Kanke et al. (1992) investigated the *in vitro* release of a curdlan tablet containing theophylline; drug release from the tablet was the lowest tested, and was unaffected by pH and various other ions. Kim, B.-S. et al. (2000) proposed that hydroxyethyl derivatives of curdlan can be used as protein drug delivery vehicles. A curdlan hydrolysate with a number average molecular weight ranging from 340 to 4000 was used as an immunoactivator and did not cause any adverse side effects (Masahiro et al., 1998a). Hiroshi et al. (1997) have also reported that linear or branched β-(1,3)-glucans such as lentinan and curdlan are effective in the treatment of dementia.

9.3 Agricultural Applications

Fumio et al. (1989) reported that polysaccharides with the β-(1,3)-glucan structure, such as lentinan, schizophyllan, curdlan or laminarin are useful for the multiplication of *Bifidobacterium* bacteria and, when given to animals in the food, also inhibit the intestinal putrefying bacteria, thereby preventing the animals from aging. In addition, curdlan is incorporated into fish feed mixtures to improve immune activity (Masahiro et al., 1998b).

9.4 Other Industrial Applications

Curdlan has been studied as a support for immobilized enzymes (Murooka et al., 1977). For example, in the immobilization of an enzyme or a mold using curdlan gel, the curdlan is gelatinized by heating an aqueous suspension containing curdlan, an enzyme or a mold (Toshio, 1986). Other potential industrial applications include the use of curdlan films and fibers which are water-insoluble, biodegradable and impermeable to oxygen. Curdlan is also used to prevent nozzle clogging, and the electrostatic damage of electronic components when used as a seal in an ink-jet recoding head (Yoichi et al., 1996). Furthermore, curdlan functions as a concrete admixture to produce concrete with high fluidity but low separating properties (Toshiyuki et al., 1995; Lee, I.-Y. et al., 1999b). Applications of curdlan other than in the food industry are shown in Figure 6.

10 Outlook and Perspectives

Curdlan has been well accepted in the meat processing and noodle industries for its extremely unusual property of forming gels that are resistant to freezing or retorting. Takeda has been marketing this innovative product in Japan for several years, and sales are promising. Indeed, production has continued to grow by more than 10% each year during the past eight years such that, in recent years, Takeda Chemical Industries, Ltd. has produced 600–700 tonnes of curdlan annually.

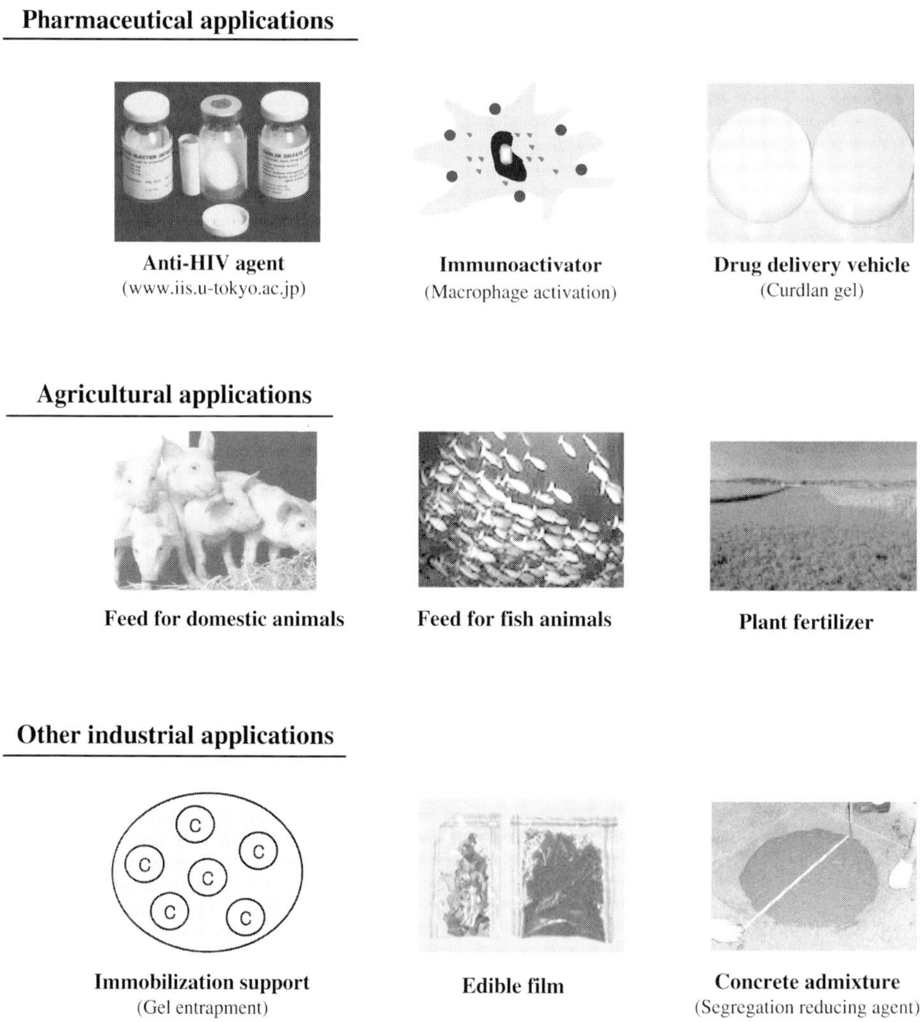

Fig. 6 Applications of curdlan in fields other than the food industry.

In addition, it should be noted that a variety of glucans with the β-1,3 linkage have immunestimulatory effects. Krestin, a β-glucan extracted from the mycelium of the basidiomycete *Coriolus versicolor*, has a general immunestimulatory effect in humans and has also been used as an anticancer agent (Kamisato and Nowakowski, 1988). Lentinan from *Lentinus edodes* is another glucan which inhibits tumor growth and increases resistance to infections by bacteria, viruses, and parasites (Jong and Birmingham, 1993). Schizophyllan (Numasaki et al., 1990) from *Schizophyllum commune* has the property of inducing the formation of cytotoxic macrophages and generating antitumor activity in the human body. In their preparation, these fungal glucans require extensive extraction operations to be performed in order to obtain comparatively small amounts of pure glucans, which makes the process very expensive. By contrast, curdlan as an exotype of β-(1,3)-glucan can be readily produced on a large scale by a submerged bacterial culture. Thus, if the material is correctly designed on a molecular

Tab. 3 Patent holders and publication year

Company	***Year of publication***													
	1988	*1989*	*1990*	*1991*	*1992*	*1993*	*1994*	*1995*	*1996*	*1997*	*1998*	*1999*	*2000*	***Total***
Ajinomoto	2		1							2				5
Asai							1							2
Chem Reizou	1													1
Dainippon	1	1							1	1	1			5
Daiichi		1												1
Daito										1				1
Endo Akira	1													1
Ezaki Glico														1
Fuji Oil	1				2	1							1	5
Hinoshoku												1	1	2
House foods	1							1						2
Hodaka						1								1
Ina						1					1		4	6
Japan Organo						2		2	2	7	2	2	2	19
Kanai Masako												1		1
Kanebo					2			1			1			4
Kanegafuchi				1			1				1			3
Kiteii					1									1
Kibun foods									1					1
Kyokuto				1										1
Meiji			2		1					1	1			5
Morigana												1		1
Nagai				1	1									2
Nikken									1					1
Nippon					2	1							1	4
Nitta Gelatin							1							1
Nokyo									1					1
Okada					2									2
Sanei		4		1			1		1					7
Sanyo	1													1
Shiseido							1		2					3
Snow Brand					1						2	1	2	6
Sumisho												1		1
Takeda		1	12	8	6	5	1	5	3	5	9	3	4	62
Taiyo Kagaku							1							1
Tsubuki								1						1
Tsukishima						1								1
Unie Colloid							2							2
Wako	5													5
Yamamoto										1				1
Total	13	7	15	12	18	12	8	13	9	19	19	11	15	171

basis to enhance its immunestimulatory activity, there is a large potential market in both the nutraceutical and pharmaceutical industries.

11 Patents

The Japanese patent literature between 1988 and 2000 was reviewed, as most of the patents relating to the application of curdlan were first filed in Japan. The companies holding patents, together with the annual number of patents filed, are listed in Table 3. Takeda and Japan Organo each hold about one-half of total 171 patents filed, indicating their major role in the development of curdlan application. It should also be noted that 122 patents relate to the application of curdlan as food additives or ingredients. The patents described in Section 8 are listed in Table 4.

Acknowledgments
The author thanks Mi-Kyoung Kim for her help in the preparation of the manuscript.

Tab. 4 A list of the patents of curdlan application

Patent no.	*Inventor(s)*	*Title*	*Year*
JP-61015688	Toshio, O.	Immobilized enzyme, or the like and preparation thereof	1986
JP-01289457	Akira, K., Atsushi, I.	Edible film	1989
JP-01137990A	Fumio, I., Kimikazu, I., Satoshi, S.	Polysaccharide having activity for multiplying *Bifidobacterium*	1989
JP-02124902	Tsuneo, A., Koichi, K., Junji, K.	Blood coagulation inhibitor	1990
JP-02249466	Masayuki, T., Yukihiro, N.	Noodle made of rice powder and producing method thereof	1990
JP-03218317	Takashi, T., Junji, K.	Anti-retrovirus agent	1991
JP-03157401	Taguchi, T., Hiroshi, K., Yukihiro, N.	Production of linear gel	1991
JP-04210507	Masanori, T., Tetsuya, T., Yukihiro, N.	Noodles	1992
JP-04330256	Shunsuke, O., Masatoshi, S.	Production of uncooked Chinese noodle and boiled Chinese noodle	1992
JP-04197148	Ken, O., Akirou, M.	Soybean protein-containing solid food	1992
JP-04074115	Toshiko, S.	Sustained release suppository	1992
JP-05184310	Masatoshi, H., Masami, F., Kenichi, H., Hiromi, K.	Devil's tongue for frozen food and its production	1993
JP-05207859	Hideaki, Y., Chieko, M., Takeshi, A.	Ganmodoki and its production	1993
JP-05023143	Yukihiro, N., Takahiro, F.	Ground fish meat or fish or cattle paste product and composition therefor	1993
JP-05076290	Hiroki, O., Etsuko, M.	Noodle-like soybean protein-containing food	1993
JP-05288909	Masahiro, K., Masaru, N.	Preparation of processed edible meat	1993
JP-06335370	Yasuhiro, S.	Formed jelly-containing liquid food and its production	1994
JP-07010624	Toshiyuki, U., Yoshio, T., Minoru, Y.	Additive for concrete	1995

Tab. 4 (cont.)

Patent no.	***Inventor(s)***	***Title***	***Year***
JP-07274876	Masaaki, A., Takaaki, I.	Quality improver for wheat flour product and wheat flour product having improved texture	1995
JP-07228601	Mikio, K., Yoshiro, O., Hirotomo, O., Yoshikazu, K., Hiroshi, I., Hisashi, M., Takeshi, K.	Water soluble β-1,3-glucan derivative and antiviral agent containing the derivative	1995
JP-0818787	Yoichi, T., Hiroshi, S., Makiko, K.	Biodegradable component for protecting ink jet recording head	1996
USP-5508191	Kanegae, Y. Yutani, A., Nakatsui, I.	Mutant strains of *Agrobacterium* for producing beta-1,3-glucan	1996
JP-09075017	Masataka, M., Hideo, Y., Yukihiro, N.	Devil's-tongue jelly and its production	1997
Patent number	Inventors	Title	Year
JP-09255579	Hiroshi, S., Nobuyoshi, N., Yutaro, K., Toru, M.	Medicine for treating dementia	1997
JP-10194977	Masahiro, K., Masaaki, K., Shinji, M., Hiroaki, K.	Immunoactivator	1998a
JP-10313794	Masahiro, K., Mikitomo, A., Akitomo, A., Tomonori, Y.	Feed composition for fish kind cultivation	1998b
JP-10014541	Akihiro, S., Takaaki, I.	Quality improving agent for fishery paste product and production of fishery paste product	1998
JP-11187819	Hiroshi, Y., Yuichi, S., Norio, I., Nana, I.	Frozen dessert food and its production	1999
JP-11075726	Masao, K., Toshimi, T.	Jelly-like food	1999
JP-11178533	Kazushi, M.	Production of bean curd including ingredient	1999
JP-11056246	Takaaki, I.	Quality improver for bean jam product and production of bean jam product	1999

12
References

Aizawa, M., Takahashi, M., Suzuki, S. (1974) Gel formation of curdlan-type polysaccharide in DMSO-H_2O mixed solvents, *Chem. Lett.* 193–196.

Aketagawa, J., Tanaka, S., Tamura, H., Shibata, Y., Saito, H. (1993) Activation of *Limulus* coagulation factor G by several (1→3)-β-D-glucans: comparison of the potency of glucans with identical degree of polymerization but different conformations, *J. Biochem.* **113**, 683–686.

Akihiro, S., Takaaki, I. (1998) Quality improving agent for fishery paste product and production of fishery paste product. Japanese patent 10014541.

Akira, K., Atsushi, I. (1989) Edible film. Japanese patent 01289457.

Azuma, I. (1987) Development of immunostimulants in Japan, in: *Immunostimulants: Now and Tomorrow* (Azuma, I., Jolles, G., Eds.), Tokyo: Japan Sci. Soc. Press/Berlin: Springer-Verlag, 41–45.

Bohn, J. A., BeMiller, J. N. (1995) (1→3)-β-D-glucans as biological response modifiers: a review of structure–functional activity relationships, *Carbohydr. Res.* **28**, 3–14.

Chuah, C. T., Sarko, A., Deslandes, Y., Marchessault, R. H. (1983) Triple-helical crystalline structure of curdlan and paramylon hydrates, *Macromolecules* **16**, 1375–1382.

Conti, E., Flaibani, A., O'Regan, M., Sutherland, I. W. (1994) Alginate from *Pseudomonas fluorescens* and *P. putida*: production and properties, *Microbiology* **140**, 1125–1132.

Deslandes, Y., Marchessault, R. H., Sarko, A. (1980) Triple-helical structure of (1→3)-β-D-glucan, *Macromolecules* **13**, 1466–1471.

Di Luzio, N. R. (1983) Immunopharmacology of glucan: a broad spectrum enhancer of host defence mechanisms, *Trends Pharmacol. Sci.* **4**, 344–347.

Di Luzio, N. R., Williams, D. L., McNamee, R. B., Edwards, B. F., Kitahama, A. (1979) Comparative tumor-inhibitory and anti-bacterial activity of soluble and particulate glucan, *Int. J. Cancer* **24**, 773–779.

Evans, S. G., Morrison, D., Kaneko, Y., Havlik, I. (1998) The effect of curdlan sulfate on development *in vitro* of *Plasmodium falciparum*, *Trans. R. Soc. Trop. Med. Hyg.* **92**, 87–89.

Farina, J. I., Sineriz, F., Molina, O. E., Perotti, N. I. (2001) Isolation and physico-chemical characterization of soluble scleroglucan from *Sclerotium rolfsii*. Rheological properties, molecular weight and conformational characteristics, *Carbohydr. Res.* **44**, 41–50.

Farres, J., Caminal, G., Lopez-Santin, J. (1997) Influence of phosphate on rhamnose-containing exopolysaccharide rheology and production by *Klebsiella* I-714, *Appl. Microbiol. Biotechnol.* **48**, 522–527.

Fumio, I., Kimikazu, I., Satoshi, S. (1989) Polysaccharide having activity for multiplying *Bifidobacterium*. Japanese patent 01137990A.

Hadden, J. W. (1993) Immunostimulants, *Immunology Today* **14**, 275–280.

Harada, T. (1965) Succinoglucan 10C3: a new acidic polysaccharide of *Alcaligenes faecalis* var. *myxogenes*, *Arch. Biochem. Biophys.* **112**, 65–69.

Harada, T. (1977) Production, properties, and application of curdlan, in: *Extracellular Microbial Polysaccharides* (Sanford, P. A. Laskin, A., Eds.), Washington, DC: American Chemical Society, 265–283.

Harada, T., Yoshimura, T. (1965) Rheological properties of succinoglucan 10C3 from *Alcaligenes faecalis* var. *myxogenes*, *Agr. Biol. Chem.* **29**, 1027–1032.

Harada, T., Yoshimura, T., Hidaka, H., Koreeda, A. (1965) Production of a new acidic polysaccharide,

succinoglucan by *Alcaligenes faecalis* var. *myxogenes, Agr. Biol. Chem.* **29**, 757–762.
Harada, T., Masada, M., Fujimori, K., Maeda, I. (1966) Production of a firm, resilient gel-forming polysaccharide by a mutant of *Alcaligenes faecalis* var. *myxogenes* 10C3, *Agr. Biol. Chem.* **30**, 196–198.
Harada, T., Misaki, A., Saito, H. (1968) Curdlan: a bacterial gel-forming β-1,3-glucan, *Arch. Biochem. Biophys.* **124**, 292–298.
Harada, T., Terasaki, M., Harada, A. (1993) Curdlan, in: *Industrial Gums* (Whistler, R. L., BeMiller, J. N., Eds.), San Diego, CA: Academic Press, Inc., 427–445.
Hideaki, Y., Chieko, M., Takeshi, A. (1993) Ganmodoki and its production. Japanese patent 05207859.
Hiroki, O., Etsuko, M. (1993) Noodle-like soybean protein-containing food. Japanese patent 05076290.
Hiroshi, S., Nobuyoshi, N., Yutaro, K., Toru, M. (1997) Medicine for treating dementia. Japanese patent 09255579.
Hiroshi, Y., Yuichi, S., Norio, I., Nana, I. (1999) Frozen dessert food and its production. Japanese patent 11187819.
Ielpi, L., Couso, R. O., Dankert, M. A. (1993) Sequential assembly and polymerization of the polyphenol-linked pentasaccharide repeating unit of the xanthan polysaccharide in *Xanthomonas campestris, J. Bacteriol.* **175**, 2490–2500.
Jagodzinski, P. P., Wiaderkiewicz, R., Kurzawski, G., Kloczewiak, M., Nakashima, H., Hyjek, E., Yamamoto, N., Uryu, T., Kaneko, Y., Posner, M. R., Kozbor, D. (1994) Mechanism of the inhibitory effect of curdlan sulfate on HIV-1 infection *in vitro, Virology* **202**, 735–745.
Janusz, M. J., Austen, K. F., Czop, J. K. (1986) Isolation of soluble yeast β-glucans that inhibit human monocyte phagocytosis mediated by β-glucan receptors, *J. Immunol.* **137**, 3270–3276.
Jong, S. C., Birmingham, J. M. (1993) Medicinal and therapeutic value of the Shiitake mushroom, *Adv. Appl. Microbiol.* **39**, 153–184.
Kai, A., Ishino, T., Arashida, T., Hatanaka, K., Akaike, Y., Matsuzaki, K., Kaneko, Y., Mimura, T. (1993) Biosynthesis of curdlan from culture media containing ^{13}C-labeled glucose as the carbon source, *Carbohydr. Res.* **240**, 153–159.
Kamisato, J. K., Nowakowski, M. (1988) Morphological and biochemical alterations of macrophages produced by a glucan, PSK, *Immunopharmacology* **16**, 88–96.
Kanegae, Y. Yutani, A., Nakatsui, I. (1996) Mutant strains of *Agrobacterium* for producing beta-1,3-glucan. US patent 5508191.
Kanke, M., Koda, K., Koda, Y., Katayama, H. (1992) Application of curdlan to controlled drug delivery. I. The preparation and evaluation of theophylline-containing curdlan tablets, *Pharmaceut. Res.* **9**, 414–418.
Kasai, N., Harada, T. (1980) Ultrastructure of curdlan. in: *Fiber Diffraction Methods,* (French, A. D., Gardner, K. H., Eds.), Washington, DC: American Chemical Society Symposium Series, 363–383.
Kazushi, M. (1999) Production of bean curd including ingredient. Japanese patent 11178533.
Ken, O., Akirou, M. (1992) Soybean protein-containing solid food. Japanese patent 04197148.
Kim, B.-S., Jung, I.-D., Kim, J.-S., Lee, J.-H., Lee, I.-Y., Lee, K.-B. (2000) Curdlan gels as protein drug delivery vehicles, *Biotechnol. Lett.* **22**, 1127–1130.
Kim, M.-K., Lee, I.-Y., Ko, J.-H., Rhee, Y.-H., Park, Y.-H., (1999) Higher intracellular levels of uridine monophosphate under nitrogen-limited conditions enhance metabolic flux of curdlan synthesis in *Agrobacterium* species, *Biotechnol. Bioeng.* **62**, 317–323.
Kim, M.-K., Lee, I.-Y., Lee, J.-H, Kim, K.-T., Rhee, Y.-H., Park, Y.-H. (2000) Residual phosphate concentration under nitrogen-limiting conditions regulates curdlan production in *Agrobacterium* species, *J. Ind. Microbiol. Biotechnol.* **25**, 180–183.
Kimura, H., Moritaka, S., Misaki, M. (1973) Polysaccharide 13140: a new thermo-gelable polysaccharide, *J. Food Sci.* **38**, 668–670.
Kojima, T., Tabata, K., Itoh, W., Yanaki, T. (1986) Molecular weight dependence of the antitumor activity of schizophyllan, *Agr. Biol. Chem.* **50**, 231–232.
Koreeda, A., Harada, T., Ogawa, K., Sato, S. Kasai, N. (1974) Study of the ultrastructure of gel-forming (1 → 3)-β-D-glucan (curdlan-type polysaccharide) by electron microscopy, *Carbohydr. Res.* **33**, 396–399.
Lawford, H. G., Rousseau, J. D. (1991) Bioreactor design considerations in the production of high-quality microbial exopolysaccharide, *Appl. Biochem. Biotechnol.* **28/29**, 667–684.
Lawford, H. G., Rousseau, J. D. (1992) Production of β-1,3-glucan exopolysaccharide in low shear systems, *Appl. Biochem. Biotechnol.* **34/35**, 597–612.
Lawford, H., Keenan, J., Phillips, K., Orts, W. (1986) Influence of bioreactor design on the rate and amount of curdlan-type exopolysaccharide production by *Alcaligenes faecalis, Biotechnol. Lett.* **8**, 145–150.
Lee, J.-H, Lee, I.-Y., Kim, M.-K., Park, Y.-H. (1999) Optimal pH control of batch processes for

production of curdlan by *Agrobacterium* species, *J. Ind. Microbiol. Biotechnol.* **23**, 143–148.

Lee, I.-Y., Seo, W.-T., Kim, K.-G., Kim, M.-K., Park, C.-S., Park, Y.-H. (1997) Production of curdlan using sucrose or sugar cane molasses by two-step fed-batch cultivation of *Agrobacterium* species, *J. Ind. Microbiol. Biotechnol.* **18**, 255–259.

Lee, I.-Y, Kim, M.-K., Lee, J.-H., Seo, W.-T., Jung, J.-K., Lee, H.-W., Park, Y.-H., (1999a) Influence of agitation speed on production of curdlan by *Agrobacterium* species, *Bioprocess Eng.* **20**, 283–287.

Lee, I.-Y., Kim, S.-W., Lee, J.-H., Kim, M.-K., Cho, I.-S., Park, Y.-H. (1999b) A high viscosity of curdlan at alkaline pH increases segregational resistance of concrete, *Korean J. Biotechnol. Bioeng.* **14**, 1–5.

Maeda, I., Saito, H., Masda, M., Misaki, A., Harada, T. (1967) Properties of gels formed by heat treatment of curdlan, a bacterial β-1,3 glucan, *Agr. Biol. Chem.* **31**, 1184–1188.

Masaaki, A., Takaaki, I. (1995) Quality improver for wheat flour product and wheat flour product having improved texture. Japanese patent 07274876.

Masahiro, K., Masaru, N. (1993) Preparation of processed edible meat. Japanese patent 05288909.

Masahiro, K., Masaaki, K., Shinji, M., Hiroaki, K. (1998a) Immunoactivator. Japanese patent 10194977.

Masahiro, K., Mikitomo, A., Akitomo, A., Tomonori, Y. (1998b) Feed composition for fish kind cultivation. Japanese patent 10313794.

Masanori, T., Tetsuya, T., Yukihiro, N. (1992) Noodles. Japanese patent 04210507.

Masao, K., Toshimi, T. (1999) Jelly-like food. Japanese patent 11075726.

Masataka, M., Hideo, Y., Yukihiro, N. (1997) Devil's-tongue jelly and its production. Japanese patent 09075017.

Masatoshi, H., Masami, F., Kenichi, H., Hiromi, K. (1993) Devil's tongue for frozen food and its production. Japanese patent 05184310.

Masayuki, T., Yukihiro, N. (1990) Noodle made of rice powder and producing method thereof. Japanese patent 02249466.

Mikio, K., Yoshiro, O., Hirotomo, O., Yoshikazu, K., Hiroshi, I., Hisashi, M., Takeshi, K. (1995) Water soluble β-1,3-glucan derivative and antiviral agent containing the derivative. Japanese patent 07228601.

Misaki, A., Kishida, E., Kakuta, M., Tabata, K. (1993) Antitumor fungal (1 → 3)-β-D-glucans: structural diversity and effects of chemical modification. In: *Carbohydrates and Carbohydrate Polymers* (Yalpani, M., Ed.), Mount Prospect, IL: ATL Press, 116–129.

Murooka, Y., Yamada, T., Harada, T. (1977) Affinity chromatography of *Klebsiella* arylsulfatase on tyrosyl-hexamethylenediamine-β-1,3-glucan and immunoadsorbent, *Biochim. Biophys. Acta* **485**, 134–140.

Naganishi, I., Kimura, K., Kusui, S., Yamazaki, E. (1974) Complex formation of gel-forming bacterial (1 → 3)-β-D-glucan (curdlan type polysaccharide) with dyes in aqueous solution, *Carbohydr. Res.* **32**, 47–52.

Naganishi, I., Kimura, K., Suzuki, T., Ishikawa, M., Banno, I., Sakene, T., Harada, T. (1976) Demonstration of curdlan-type polysaccharide and some other β-1,3-glucan in microorganisms with aniline blue, *J. Gen. Appl. Microbiol.* **22**, 1–11.

Nakata, M., Kawaguchi, T., Kodama, Y., Konno, A. (1998) Characterization of curdlan in aqueous sodium hydroxide, *Polymer* **39**, 1475–1481.

Numasaki, Y., Kikuchi, M., Sugiyama, Y., Ohba, Y. (1990) A glucan Sizofiran: T cell adjuvant property and antitumor and cytotoxic macrophage inducing activities, *Oyo Yakuri Pharmacometrics* **39**, 39–48.

Ogawa, K., Watanabe, T., Tsurugi, J., Ono, S. (1972) Conformational behavior of a gel-forming (1 → 3)-β-D-glucan in alkaline solution, *Carbohydr. Res.* **23**, 399–405.

Ogawa, K., Tsurugi, J., Watanabe, T. (1973a) Effect of salt on the conformation of gel-forming β-1,3-D-glucan in alkaline solution, *Chem. Lett.* 95–98.

Ogawa, K., Miyagi, M., Fukumoto, T., Watanabe, T. (1973b) Effect of 2-chloroethanol, dioxane, or water on the conformation of a gel-forming β-1,3-glucan in DMSO, *Chem. Lett.* 943–946.

Ogawa, K., Tsurugi, J., Watanabe, T. (1973c) The dependence of the conformation of a (1 → 3)-β-D-glucan on chain length in alkaline solution, *Carbohydr. Res.* **29**, 397–403.

Ohno, N., Miura, N. N., Chiba, N., Adachi, Y., Yadomae, T. (1995) Comparison of the immunopharmacological activities of triple and single-helical schizophyllan in mice, *Biol. Pharm. Bull.* **18**, 1242–1247.

Ohno, N., Hashimoto, T., Adachi, Y., Yadomae, T. (1996) Conformation dependency of nitric oxide synthesis of murine peritoneal macrophages by β-glucans *in vitro*, *Immunol. Lett.* **52**, 1–7.

Okazaki, M., Adachi, Y., Ohno, N., Yadomae, T. (1995) Structure–activity relationship of (1 → 3)-β-D-glucans in the induction of cytokine production from macrophages, *in vitro, Biol. Pharm. Bull.* **18**, 1320–1327.

Okuyama, K., Obata, Y., Noguchi, K., Kusaba, T., Ito, Y., Ohno, S. (1996) Single structure of curdlan triacetate, *Biopolymers* **38**, 557–566.

Phillips, K. R., Lawford, H. G. (1983) Curdlan: Its properties and production in batch and continuous fermentations, in: *Progress Industrial Microbiology* (Bushell, D. E., Ed.), Amsterdam: Elsevier Scientific Publishing Co., 201–229.

Read, S. M., Currie, G., Bacic, A. (1996) Analysis of the structural heterogeneity of laminarin by electrospray-ionisation-mass spectrometry, *Carbohydr. Res.* **281**, 187–201.

Saito, H., Misaki, A., Harada, T. (1968) Comparison of structure of curdlan and pachyman, *Agr. Biol. Chem.* **32**, 1261–1269.

Saito, H., Ohki, T., Sasaki, T. (1977) A ^{13}C Nuclear Magnetic Resonance study of gel-forming (1 → 3)-β-D-glucans. Evidence of the presence of single-helical conformation in a resilient gel of a curdlan-type polysaccharide 13140 from *Alcaligenes faecalis* var. *myxogenes* IFO 13140, *Biochemistry* **16**, 908–914.

Saito, H., Yoshioka, Y., Uehara, N., Aketagawa, J., Tanaka, S., Shibata, Y. (1991) Relationship between conformation and biological response for (1 → 3)-β-D-glucans in the activation of coagulation factor G from limulus amebocyte lysate and host-mediated antitumor activity. Demonstration of single-helix conformation as a stimulant, *Carbohydr. Res.* **217**, 181–190.

Sakurai, T., Suzuki, I., Kinoshita, A., Oikawa, S., Masuda, A., Ohsawa, M., Tadomae, T. (1991) Effect of intraperitoneally administered β-1,3-glucan, SSG, obtained from *Sclerotinia sclerotiorum* IFO 9395 on the functions of murine alveolar macrophages, *Chem. Pharm. Bull.* **39**, 214–217.

Sasaki, T., Takasuka, N. (1976) Further study of the structure of lentinan, an anti-tumor polysaccharide from *Lentinus edodes, Carbohydr. Res.* **17**, 99–104.

Sasaki, T., Abiko, N., Sugino, Y., Nitta, K. (1978) Dependence on chain length of antitumor activity of (1 → 3)-β-D-glucan from *Alcaligenes faecalis* var. *myxogenes,* IFO 13140, and its acid-degraded products, *Cancer Res.* **38**, 379–383.

Shunsuke, O., Masatoshi, S. (1992) Production of uncooked Chinese noodle and boiled Chinese noodle. Japanese patent 04330256.

Spicer, E. J. F., Goldenthal, E. I., Ikeda, T. (1999) A toxicological assessment of curdlan. *Fd. Chem. Toxicol.* **37**, 455–479.

Stasinopoulos, S. J., Fisher, P. R., Stone, B. A., Stanisich, V. A. (1999) Detection of two loci involved in (1 → 3)-β-glucan (curdlan) biosynthesis by *Agrobacterium* sp. ATCC 31749, and comparative sequence analysis of the putative curdlan synthase gene, *Glycobiology* **9**, 21–31.

Sutherland, I. W. (1977) Microbial exopolysaccharide synthesis, in: *Extracellular Microbial Polysaccharides* (Sanford, P. A., Laskin, A., Eds.), Washington, DC: American Chemical Society, 40–57.

Sutherland, I. W. (1993) Biosynthesis of extracellular polysaccharides, in: *Industrial Gums* (Whistler, R. L., BeMiller, J. N., Eds.), San Diego, CA: Academic Press, Inc., 69–85.

Suzuki, T., Ohno, N., Saito, K., Yamdomae, T. (1992) Activation of the complement system by (1,3)-β-D-glucan having different degrees of branching and different ultrastructures, *J. Pharmacobiodyn.* **15**, 277–285.

Tada, T., Matsumoto, T., Masuda, T. (1997) Influence of alkaline concentration on molecular association structure and viscoelastic properties of curdlan aqueous systems, *Biopolymers* **42**, 479–487.

Taguchi, T., Hiroshi, K., Yukihiro, N. (1991) Production of linear gel. Japanese patent 03157401.

Takaaki, I. (1999) Quality improver for bean jam product and production of bean jam product. Japanese patent 11056246.

Takashi, T., Junji, K. (1991) Anti-retrovirus agent. Japanese patent 03218317.

Takeda technical report. (1997) Pureglucan: basic properties and food applications. Takeda Chemical Industries, Ltd. Japan.

Takeda-Hirokawa, N., Neoh, L. P., Akimoto, H., Kaneko, H., Hishikawa, T., Sekigawa, I., Hashimoto, H., Hirose, S.-I., Murakami, T., Yamamoto, N., Mimura, T., Kaneko, Y. (1997) Role of curdlan sulfate in the binding of HIV-1 gp120 to CD4 molecules and the production of gp120-mediated THF-α, *Microbiol. Immunol.* **41**, 741–745.

Toshio, O. (1986) Immobilized enzyme, or the like and preparation thereof. Japanese patent 61015688.

Toshiko, S. (1992) Sustained release suppository. Japanese patent 04074115.

Toshiyuki, U., Yoshio, T., Minoru, Y. (1995) Additive for concrete. Japanese patent 07010624.

Tsuneo, A., Koichi, K., Junji, K. (1990) Blood coagulation inhibitor. Japanese patent 02124902.

Williams, D. L., Pretus, H. A., Ensley, H. E., Browder, I. W. (1994) Molecular weight analysis of a water-insoluble, yeast-derived (1 → 3)-β-D-

glucan by organic-phase size-exclusion chromatography, *Carbohydr. Res.* **253**, 293–298.
Yanaki, T., Ito, W., Tabata, K., Kojima, T., Norysuye, T., Takano, N., Fujita, H. (1983) Correlation between the antitumor activity of a polysaccharide schizophyllan and its triple-helical conformation in dilute aqueous solution, *Biophys. Chem.* **17**, 337–342.
Yasuhiro, S. (1994) Formed jelly-containing liquid food and its production. Japanese patent 06335370.
Yoichi, T., Hiroshi, S., Makiko, K. (1996) Biodegradable component for protecting ink jet recording head. Japanese patent 0818787.
Yoshioka, Y., Tabeta, R., Saito, R., Uehara, N., Fukuoka, F. (1985) Antitumor polysaccharides from *P. ostreatus* (Fr.) Quel: isolation and structure of a beta-glucan, *Carbohydr. Res.* **140**, 93–100.
Young, S.-H., Dong, W.-J., Jacobs, R. R. (2000) Observation of a partially opened triple-helix conformation in (1 → 3)-β-glucan by fluorescence resonance energy transfer spectroscopy, *J. Biol. Chem.* **275**, 11874–11879.
Yukihiro, N., Takahiro, F. (1993) Ground fish meat or fish or cattle paste product and composition therefor. Japanese patent 05023143.
Zhang, L., Ding, Q., Zhang, P., Zhu, R., Zhou, Y. (1997) Molecular weight and aggregation behaviour in solution of β-D-glucan from *Poria cocos* sclerotium, *Carbohydr. Res.* **303**, 193–197.

15
Dextran

Dr. Timothy D. Leathers
Fermentation Biochemistry Research Unit, National Center for Agricultural Utilization Research, Agricultural Research Service, United States Department of Agriculture;1815 N. University St.; Peoria, IL, 61604, USA; Tel. +1-309-681-6377; Fax: +1-309-681-6427; E-mail: leathetd@ncaur.usda.gov

Biopolymers for Medical and Pharmaceutical Applications. Edited by A. Steinbüchel and R. H. Marchessault

ISBN: 3-527-31154-8

ATP	adenosine 5′-triphosphate
CM	carboxymethyl
DEAE	diethylaminoethyl
EC	enzyme commission
FTIR	fourier-transform infrared
GRAS	Generally Regarded as Safe
HIV	human immunodeficiency virus
IU	international units
K_m	Michaelis-Menten constant
MRI	magnetic resonance imaging
NMR	nuclear magnetic resonance
QAE	diethyl(2-hydroxypropyl)aminoethyl
SP	sulfopropyl

Names are necessary to report factually on available data; however, the USDA neither guarantees nor warrants the standard of the product, and the use of the name by USDA implies no approval of the product to the exclusion of others that also may be suitable.

1 Introduction

Dextrans are defined as homopolysaccharides of glucose that feature a substantial number of consecutive α-(1 → 6) linkages in their major chains, usually more than 50% of total linkages. These α-D-glucans also possess side chains stemming from α-(1 → 2), α-(1 → 3), or α-(1 → 4) branch linkages. The exact structure of each type of dextran depends on its specific microbial strain of origin. Dextrans are produced by certain lactic acid bacteria, particularly strains of *Streptococcus* species and *Leuconostoc mesenteroides*. Dextrans from oral *Streptococcus* species are of clinical interest as components of dental plaque. Dextrans from *L. mesenteroides* are of commercial interest, primarily as specialty chemicals for clinical, pharmaceutical, research, and industrial uses. Numerous reviews have appeared on dextrans, including those by Evans and Hibbert (1946), Neely (1960), Jeanes (1966, 1978), Murphy and Whistler (1973), Sidebotham (1974), Walker (1978), Alsop (1983), Robyt (1986, 1992, 1995), de Belder (1990, 1993), and Cote and Ahlgren (1995).

2 Historical Outline

Because dextrans are formed from sucrose, they have long been known as troublesome contaminants of food products and sugar refineries. Pasteur made an early applied study of dextran formation in wine, proving that this phenomenon was caused by microbial activity (Pasteur, 1861). Scheibler (1874) determined that dextran was a carbohydrate of the empirical formula $(C_6H_{10}O_6)_n$ having a positive optical rotation and thus coined the term "dextran". Van Tieghem (1878) identified a dextran-forming bacterium and named it *Leuconostoc mesenteroides*. Beijerinck (1912) investigated the phenomenon, and Hehre (1941) demonstrated dextran synthesis by a cell-free culture filtrate. Scientific interest in dextrans was stimulated by studies suggesting its value as a blood-plasma volume expander (Gronwall and Ingelman, 1945, 1948). In the late 1940s, an extensive research program on dextrans was initiated at the Northern Regional Research Laboratory (now the National Center for Agricultural Utilization Research) of the Agricultural Research Service, U.S. Department of Agriculture, in Peoria, Illinois. Numerous dextran-producing strains were characterized, including *L. mesenteroides* strain NRRL B-512F used today for commercial dextran production in North America and Western Europe.

3 Chemical Structure

Within the general definition of dextran as a glucan in which α-(1 → 6) linkages predominate, chemical structures vary considerably as a function of the specific microbial strain of origin. The article of commerce is the product of a single strain of *L. mesenteroides*, NRRL B-512F. As shown in Figure 1, dextran from this strain features α-(1 → 6) linkages in the main chains with a relatively low level (about 5%) of α-(1 → 3) branch linkages (Van Cleve et al., 1956; Jeanes et al., 1954; Slodki et al., 1986). Larm et al. (1971) estimated that 40% of these side chains are one subunit long and 45% are two subunits long. The remaining side chains are probably greater than 30 subunits long, and branches appear to be distributed randomly (Bovey, 1959; Covacevich and Richards, 1977; Taylor et al., 1985; Kuge et al., 1987).

Dextrans are produced by numerous additional strains of bacteria, and the structures of these dextrans are diverse. In the classic study of Jeanes et al. (1954), 96 strains of *Leuconostoc* and *Streptococcus* were surveyed for the formation of polysaccharides from sucrose. In this study it was reported that α-(1 → 6) linkages in dextran varied from 50% to 97% of total linkages. The balance represented α-(1 → 2), α-(1 → 3), or α-(1 → 4) linkages, usually at branch points. Several isolates produced more than one type of polysaccharide, which were named on the basis of their greater (fraction *S*, for soluble) or lesser (fraction *L*) degree of solubility in water–ethanol mixtures.

Leuconostoc citreum strain NRRL B-742 (formerly *L. mesenteroides*; Takahashi et al., 1992) produces a fraction *L* dextran that contains approximately 15% α-(1 → 4) branch linkages and a fraction *S* dextran that contains a variable (30% to 45%) percentage of α-(1 → 3) branches (Jeanes et al., 1954; Seymour et al., 1979a; Côté and Robyt, 1983; Slodki et al., 1986). Dextran from *L. mesenteroides* strain NRRL B-1299 has a high percentage (27% to 35%) of α-(1 → 2) linked single-glucose branches (Jeanes et al., 1954; Kobayashi and Matsuda, 1977; Slodki et al., 1986). Soluble dextran produced by *Streptococcus sobrinus* strain 6715 (formerly *Streptococcus mutans* strain 6715) appears to be structurally similar to the *S* dextran from *L. citreum* strain NRRL B-742 in that it is an α-(1 → 6) glucan with a relatively high percentage of α-(1 → 3) branch linkages (Shimamura et al., 1982). Many strains of oral *Streptococcus* species, and some strains of *L. mesenteroides*, also make α-D-glucans containing linear sequences of consecutive α-(1 → 3) linkages. These low solubility glucans were once proposed to be "Class 3" dextrans (Seymour and Knapp, 1980) but now are considered to be forms of mutan, an important component of dental plaque.

L. mesenteroides strains NRRL B-1355, NRLL B-1498, and NRRL B-1501 produce an *S* fraction glucan with a unique backbone structure of regularly alternating α-(1 → 3) and α-(1 → 6) linkages (Côté and Robyt, 1982; Misaki et al., 1980; Seymour and

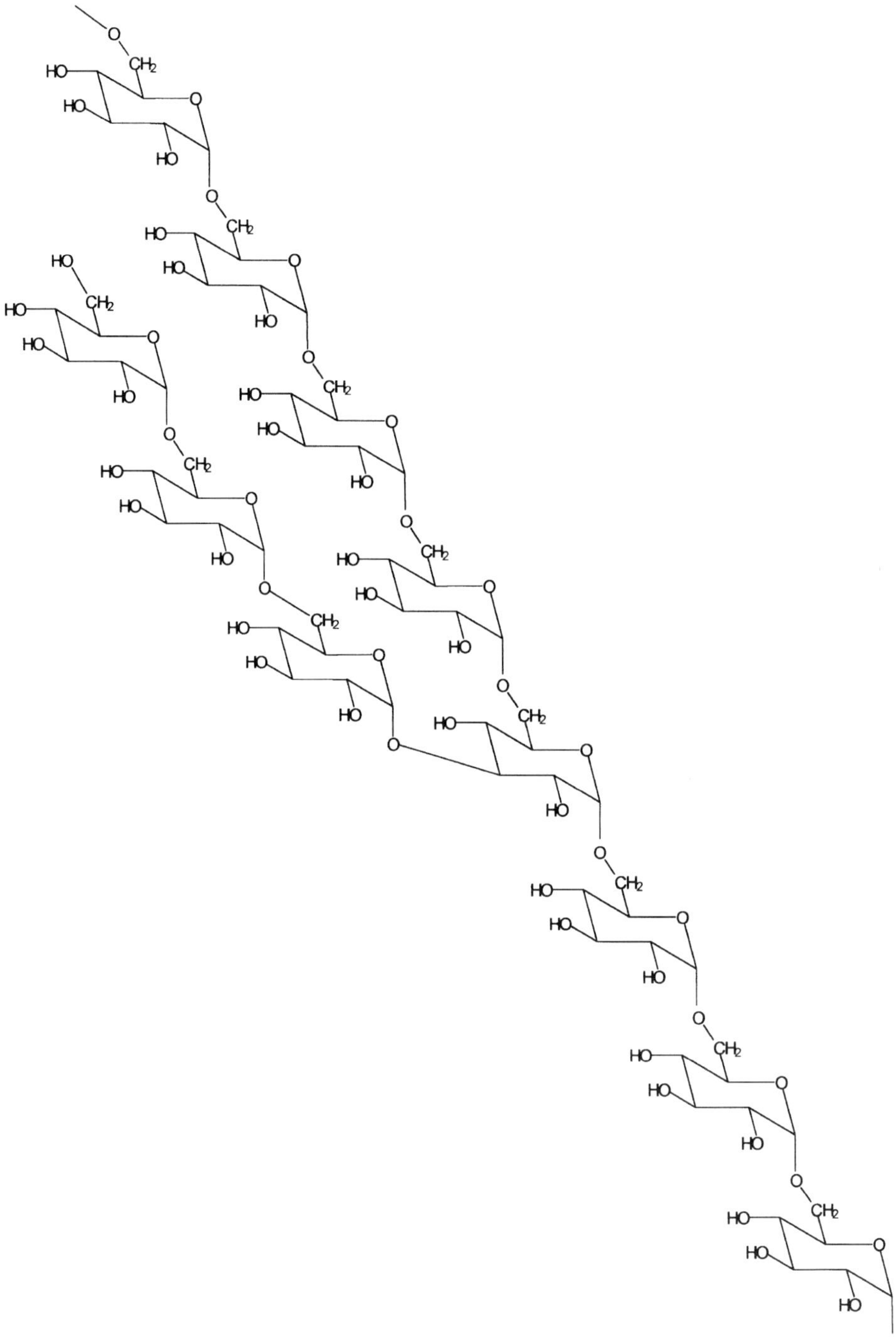

Fig. 1 Chemical structure of a representative portion of dextran from *Leuconostoc mesenteroides* strain NRRL B-512F. Figure courtesy of Dr. Gregory L. Côté.

Knapp, 1980). Because this polysaccharide does not contain significant regions of contiguous α-(1 → 6) linkages, it is not considered to be a true dextran and has been named alternan (Côté and Robyt, 1982). Alternan has distinctive properties of significant basic and applied interest, and it is the subject of Chapter 13, this volume. A mutant derivative of strain NRRL B-1355 recently was reported to produce a third polysaccharide, an insoluble α-D-glucan containing linear (1 → 3) and (1 → 6) linkages with (1 → 2) and (1 → 3) branch points (Smith et al., 1998; Côté et al., 1999).

4 Physiological Function

Oral *Streptococcus* species produce both dextran and mutan, an insoluble glucan in which α-(1 → 3) linkages predominate. These glucans comprise the matrix of dental plaque, which provides an environment for the proliferation of these bacteria (Hamada and Slade, 1980; Shimamura et al., 1982; Loesche, 1986). The physiological function of dextrans produced by *L. mesenteroides* is unknown. Since dextran-producing bacteria do not break down the polymers, dextrans presumably do not serve as storage materials. It is possible that these polysaccharides serve to protect cells from dessication or help them adhere to environmental substrates.

5 Chemical Analyses

General methods applicable to polysaccharides and polyglucans may be used to detect and quantitate dextrans. More specifically, dextrans show high positive specific optical rotation and exhibit characteristic infrared absorption bands at about 917, 840, and 768 cm^{-1} (Barker et al., 1956; Jeanes, 1966). Since dextrans are homopolysaccharides of glucose subunits, many studies have focused on the analysis of specific linkage patterns. Methods have included periodate analysis (Rankin and Jeanes, 1954; Dimler et al., 1955), methylation analysis (Van Cleve et al., 1956; Lindberg and Svensson, 1968; Seymour et al., 1977; Jeanes and Seymour, 1979; Seymour et al., 1979a; Slodki et al., 1986), nuclear magnetic resonance (NMR) spectroscopy (Seymour et al., 1976; Seymour et al., 1979b; Cheetham et al., 1991), and Fourier-transform infrared (FTIR) spectroscopy (Seymour and Julian, 1979). Enzymes that attack dextran in a specific fashion also have been exploited to reveal useful structural information (Covacevich and Richards, 1977; Sawai et al., 1978; Taylor et al., 1985; Pearce et al., 1990).

6 Occurrence

Several lactic acid bacteria have been reported to produce dextrans, principally including *Streptococcus* species and *L. mesenteroides* (Jeanes, 1966; Sidebotham, 1974; Cerning, 1990). *Leuconostoc* and *Streptococcus* are related genera, both composed of Gram-positive, facultatively anaerobic cocci. *L. mesenteroides* (incorporating as subspecies the former species *L. dextranicum* and *L. cremoris*) generally is found on plant materials, particularly on mature or harvested crops, and often plays a role in spoilage (Holzapfel and Schillinger, 1992; Stiles and Holzapfel, 1997). Because sucrose is the natural substrate for dextran synthesis, contamination problems are most evident in food products containing this sugar. *L. mesenteroides* can cause significant problems in sugar (particularly cane) refineries, where dextrans can clog filters and inhibit sugar crystallization (Jeanes, 1977). *Leuconostoc* species are considered Generally Regarded as Safe (GRAS) organisms because of their common appearance in natural fermented foods. In fact, certain strains are valued as starter cultures for buttermilk, cheese, and other dairy products (Holzapfel and Schillinger, 1992). *L. mesenteroides* also plays a role in the production of sauerkraut and other fermented vegetables. Recently, a strain of *L. mesenteroides* that produces both dextran and the related glucan alternan was isolated from the traditional fermented food Kim-Chi (Jung et al., 1999). Oral *Streptococcus* species produce dextran and the insoluble glucan mutan as components of dental plaque (Hamada and Slade, 1980; Shimamura et al., 1982). In addition, certain strains of the ubiquitous Gram-negative bacterium *Gluconobacter oxydans* (formerly *Acetobacter capsulatus*) produce dextran from starch-derived dextrins (Hehre and Hamilton, 1951; Kooi, 1958; Yamamoto et al., 1993a,b; Mountzouris et al., 1999).

7
Biosynthesis

Dextrans are produced extracellularly by secreted enzymes commonly referred to as glucansucrases or, more specifically, dextransucrases. These enzymes are glycosyltransferases (EC 2.4.1.5) that catalyze the transfer of D-glucopyranosyl subunits from sucrose to dextrans. Fructose is released and consumed by growing cells, if they are present. No adenosine 5′-triphosphate (ATP) or cofactors are required for these reactions, as the enzymes utilize energy available in the glycosidic bond between glucose and fructose. Glucansucrase synthesis in wild-type strains of *L. mesenteroides* is induced by growth on sucrose, while *Streptococcus* species produce this enzyme constitutively. Dextransucrase production by *L. mesenteroides* strain NRRL B-512F appears to be regulated at the transcriptional level (Quirasco et al., 1999).

Characterizations of dextransucrases have been complicated by the appearance of multiple enzyme species in culture fluids. Many of these species appear to be proteolytic-processing or degradation products, although some retain activity (Sanchez-Gonzalez et al., 1999). However, strains that produce more than one type of glucan appear to produce a separate glucansucrase for each. Furthermore, enzymes may exist as aggregates and typically are associated with their polysaccharide products, making purifications difficult. Despite these problems, a number of glucansucrases have been studied. Dextransucrase from *L. mesenteroides* strain NRRL B-512F has been purified and characterized (Robyt and Walseth, 1979; Kobayashi and Matsuda, 1980; Paul et al., 1984; Miller et al., 1986; Fu and Robyt, 1990; Kitaoka and Robyt, 1998a; Kim and Kim, 1999). The enzyme appears to have an initial molecular mass of 170 kDa, a pI value of 4.1, and a Michaelis-Menten constant (K_m) for sucrose of approximately 12 to 16 mM. Optimal reaction conditions are pH 5.0 to 5.5 and 30 °C. Kim and Kim (1999) reported a specific activity of up to 250 IU mg^{-1} protein for highly purified enzyme. Low levels of calcium are necessary for optimal enzyme production and activity. Various forms of dextransucrase have been described from *L. mesenteroides* strain NRRL B-1299, with molecular masses of 48 to 79 kDa, temperature optima of 35 °C to 45 °C, pH optima of 5.0 to 6.5, and K_m values for sucrose of 13 to 30 mM (Kobayashi and Matsuda, 1975, 1976; Dols et al., 1997). Multiple forms of dextransucrase have been described from *Streptococcus* species, having molecular masses of 94 to 170 kDa, pI values of approximately 4.0, temperature optima of

34 °C to 42 °C, pH optima of 5.0 to 5.7, and K_m values for sucrose of 2 to 9 mM (Chludzinski et al., 1974; Fukui et al., 1974; Shimamura et al., 1982; Furuta et al., 1985; McCabe, 1985). In addition, dextrandextrinase (also called dextrin dextranase) has been purified and characterized from *G. oxydans* strain ATTC 11894 (Yamamoto et al., 1992; Suzuki et al., 1999).

Robyt and colleagues have developed a model for the reaction mechanism of dextransucrase from *L. mesenteroides* strain NRRL B-512F (Robyt et al., 1974; Robyt, 1992, 1995; Su and Robyt, 1994). According to this model, two nucleophilic reaction sites exist in the catalytic domain of the enzyme. Sucrose is hydrolyzed at one or both sites, and the glucosyl residues are bound covalently to the enzyme in high-energy bonds conserved from sucrose. A dextran chain grows by successive glucosyl insertions between the enzyme and the reducing end of the chain, which remains bound to the enzyme. Branches are formed when glucosyl units or dextran chains are transferred to secondary hydroxyl positions on the dextran chains (Robyt and Taniguchi, 1976). Termination of chain extension occurs by transfer to an acceptor molecule.

Sucrose is strongly preferred as the glucosyl donor, although other natural and synthetic donors have been identified (Hehre and Suzuki, 1966; Binder and Robyt, 1983). However, a number of sugars and derivatives may function as alternative acceptors, including maltose, isomaltose, nigerose, α-methyl glucoside, and others (Robyt and Taniguchi, 1976; Robyt and Walseth, 1978; Robyt and Eklund, 1983; Fu et al., 1990). These acceptor reactions can be utilized to produce dextrans of lower average molecular weights, including clinical dextrans (Koepsell et al., 1955; Tsuchiya et al., 1955; Remaud et al., 1991; Robyt, 1992) and oligosaccharides of interest (Pelenc et al., 1991; Remaud et al., 1992; Remaud-Simeon et al., 1994; Dols et al., 1999). Maltose, the most effective alternative acceptor, accepts a glucosyl residue to form the trisaccharide panose (Killey et al., 1955; Heincke et al., 1999). The acceptor reaction with fructose, the natural co-product of dextran synthesis, has been studied for production of the disaccharide leucrose, a potential alternative sweetener and substrate for industrial conversions (Stodola et al., 1956; Swengers, 1991; Reh et al., 1996; Heincke et al., 1999). A minor product of this reaction is isomaltulose, also known as palatinose, likewise of interest as an alternative sweetener (Sharpe et al., 1960; Takazoe, 1989).

Robyt and Martin (1983) found evidence that a similar reaction mechanism exists for glucansucrases from *S. sobrinus* strain 6715. Alternative models for the glucansucrase reaction mechanism have been reviewed (Monchois et al., 1999). Because the glucansucrases from *Leuconostoc* and *Streptococcus* species appear to be closely related on a molecular level, it seems likely that they share a common reaction mechanism. If so, differences among the polymer structures might be determined by subtle differences in the stereochemistry of the reaction sites.

8 Genetics and Molecular Biology

L. mesenteroides strain NRRL B-512F, used for commercial production of dextran, has been described as a laboratory "substrain" that supplanted natural isolate NRRL B-512 in 1950 (Van Cleve et al., 1956). A dextransucrase hyperproducer mutant of NRRL B-512F was isolated as NRRL B-512FM (Miller and Robyt, 1984). Wild-type strains of *L. mesenteroides* form glucansucrases only when cultured on sucrose, and further mutations were obtained that allowed strain

B-512FMC to produce dextransucrase constitutively (Kim and Robyt, 1994). Further improvements in enzyme productivity were obtained through additional rounds of mutagenesis (Kim et al., 1997; Kitaoka and Robyt, 1998b). Although commercial dextran currently is produced by a fermentative process, such strains would be particularly valuable for dextran production by an enzymatic process. Similar glucansucrase mutants have been obtained for other strains of *Leuconostoc,* including NRRL B-742, NRRL B-1142, NRRL B-1299, and NRRL B-1355 (Kim and Robyt, 1994, 1995a,b, 1996; Kitaoka and Robyt, 1998b). Dextransucrase-deficient mutants of strain NRRL B-1355 also have been isolated for improved production of alternan (Smith et al., 1994; Leathers et al., 1995, 1997, 1998). Alternan is the subject of Chapter 13, this volume.

Because of clinical interest in developing anti-caries vaccines, a number of glucansucrase genes have been cloned from oral *Streptococcus* species (Shiroza et al., 1987; Ueda et al., 1988; Honda et al., 1990). Glucansucrase genes have been cloned and sequenced from *L. mesenteroides* NRRL B-512F (Wilke-Douglas et al., 1989; Bhatnagar and Singh, 1999; Arguello-Morales et al., 2000a; Funane et al., 2000; Ryu et al., 2000), *L. mesenteroides* strain NRRL B-1299 (Monchois et al., 1996, 1998), *L. citreum* strain NRRL B-742 (Kim et al., 2000), and *L. mesenteroides* strain NRRL B-1355 (Arguello-Morales et al., 2000b; Kossman et al., 2000). Interestingly, some of these genes apparently do not specify enzymes normally secreted *in vivo*. Glucansucrase genes appear to be closely related and exhibit a common organizational structure, with a conserved N-terminal catalytic domain and a C-terminal glucan-binding domain that contains a series of direct tandem repeat sequences (Monchois et al., 1999; Remaud-Simeon et al., 2000). Based on site-directed mutageneses and consensus sequences, potentially important catalytic sites have been proposed (Monchois et al., 1997; Arguello-Morales, 2000b; Monchois et al., 2000; Remaud-Simeon et al., 2000). On a broader scale, glucansucrases resemble enzymes in glycosyl hydrolase family 13, which includes α-amylases (Fujiwara et al., 1998; Janecek et al., 2000; Remaud-Simeon et al., 2000).

9 Biodegradation

A variety of fungi produce dextranases, including *Aspergillus* species (Carlson and Carlson, 1955b; Hiraoka et al., 1972), *Chaetomium gracile* (Hattori et al., 1981), *Fusarium* species (Simonson and Liberta, 1975; Shimizu et al., 1998), *Lipomyces starkeyi* (Webb and Spencer-Martins, 1983; Koenig and Day, 1988), *Paecilomyces lilacinus* (Lee and Fox, 1985; Sun et al., 1988; Galvez-Mariscal and Lopez-Munguia, 1991), and *Penicillium* species (Tsuchiya et al., 1956; Chaiet et al., 1970). These enzymes are endodextranases with specificity for internal α-(1→6) linkages, and they produce mainly isomaltose or isomaltotriose from dextran. Dextranases from *C. gracile* and *Penicillium* sp. are produced commercially and used for treatment of dextran contamination problems in sugar processing (Godfrey, 1983). Endodextranases have shown potential for the enzymatic production of specific molecular weight fractions of dextran (Carlson and Carlson, 1955a; Corman and Tsuchiya, 1957; Novak and Stoycos, 1958; Day and Kim, 1992; Kim and Day, 1995; Kim and Robyt, 1996). These enzymes also have been tested for the treatment of dental plaque (Fitzgerald et al., 1968; Caldwell et al., 1971), although the more highly branched dextrans are far less susceptible to endodextranase digestion. Limit endodextranase digestion of the branched dextran from *L. citreum* strain NRRL B-742 produces an interesting branched fraction with rheological characteristics similar to those of polydextrose (Cote et al., 1997).

Dextranases also have been reported from a number of bacteria. *Arthrobacter globiformis* produces isomaltodextranase, an exodextranase that successively releases isomaltose from the non-reducing ends of dextrans and oligosaccharides (Torii et al., 1976; Okada et al., 1988). This enzyme recognizes not only α-(1 → 6) linkages but also α-(1 → 2), α-(1 → 3), and α-(1 → 4) linkages. Unlike endodextranases, isomaltodextranase is able to partially hydrolyze alternan, producing an interesting limit alternan (Sawai et al., 1978; Cote, 1992). An isomaltodextranase from *Actinomadura* sp. exhibits slightly different specificities (Sawai et al., 1981). A dextran α-(1 → 2) debranching enzyme also has been described from a *Flavobacterium* sp. (Mitsuishi et al., 1979). Dextran from *L. mesenteroides* strain NRRL B-512F is degraded by intestinal bacteria and enzymes in mammalian tissues other than blood (Sery and Hehre, 1956; Fischer and Stein, 1960). Intravenously administered clinical dextrans are metabolized slowly and completely in the body.

10 Production

To date, commercial production of dextran has employed primarily simple batch fermentation methods, using live cultures grown on sucrose. Methods and conditions for dextran fermentation have been detailed (Tarr and Hibbert, 1931; Hehre et al., 1959; Jeanes, 1965b, 1966; Alsop, 1983; de Belder, 1993). *L. mesenteroides* is a fastidious organism, and its special nutritional requirements include glutamic acid, valine, biotin, nicotinic acid, thiamine, and pantothenic acid (Holzapfel and Schillinger, 1992). In dextran production, these needs are met by combinations of complex medium components, such as yeast extract, corn steep liquor, casamino acids, malt extract, peptone, and tryptone. Sucrose serves as a carbon source, inducer of dextransucrase, and substrate for dextran production. Low levels of calcium (e.g., 0.005%) are necessary for optimal enzyme and dextran yields, and other basal salts, including a source of phosphate, complete the medium. Operative production factors include initial pH (typically pH 6.7 to 7.2), temperature (about 25 °C), initial sucrose concentration (usually 2%), and time (usually 24 to 48 h). Dextran branching appears to increase at elevated temperatures (Sabatie et al., 1988). High levels of sucrose (10% to 50%) reduce the yield of high-molecular-weight dextran, and this observation has been exploited for the production of intermediate sized dextran (Tsuchiya et al., 1955; Alsop, 1983). The organism is facultatively anaerobic or microaerophilic, and fermentations are not aerated. During the first 20 h of fermentation, culture pH falls to approximately 5.0 because of the formation of organic acids, favorably near the optimal pH of dextran sucrase. Dextran may be recovered by precipitation with solvents, particularly alcohols (Hehre et al., 1959; Jeanes, 1965b).

It has long been recognized that dextran also can be produced enzymatically, using cell-free culture supernatants that contain dextransucrase (Hehre, 1941; Tsuchiya and Koepsell, 1954; Hellman et al., 1955; Behrens and Ringpfeil, 1962; Jeanes, 1965a). Accordingly, improved dextransucrase production and purification methods have been developed (Lawford et al., 1979; Paul et al., 1984; Miller et al., 1986; Fu and Robyt, 1990; Kim and Kim, 1999). Glucansucrases also have been immobilized, although this approach may be most useful for production of oligosaccharides (Kaboli and Reilly, 1980; Monsan et al., 1987; Cote and Ahlgren, 1994; Reh et al., 1996; Alcalde et al., 1999). Enzymatic synthesis offers advantages of product molecular weight and quality control, as well as the benefit of obtaining fructose as a valuable co-product. However, this approach has been largely ignored for commercial production, presumably for economic reasons. Dextran production from maltodextrins, using dextran-dextrinase from *Gluconobacter oxydans*, also has attracted interest (Hehre and Hamilton, 1951; Kooi, 1958; Yamamoto et al., 1993a,b; Mountzouris et al., 1999).

Clinical dextran fractions are produced primarily by simple methods of partial acid hydrolysis followed by differential fractionation in solvents (Wolff et al., 1955; Gronwall, 1957; de Belder, 1990). Attractive alternative methods to produce these fractions include the use of dextranases (Carlson and Carlson, 1955a; Corman and Tsuchiya, 1957; Novak and Stoycos, 1958; Day and Kim, 1992; Kim and Day, 1995; Kim and Robyt, 1996) and chain-terminating acceptor reactions (Koepsell et al., 1955; Tsuchiya et al., 1955; Remaud et al., 1991; Robyt, 1992).

Dextran has been produced commercially for many years and by a number of companies, including Dextran Products, Ltd., Toronto, Canada; Pfeifer und Langen, Dormagen, Germany; Pharmachem Corp., Bethlehem, Pennsylvania, USA; and Pharmacia, Uppsala, Sweden. Annual world production was recently estimated at 2000 tons per year (Vandamme et al., 1996). The wholesale price of dextran varies, but recently it has been near 3 USD per pound.

11
Properties and Applications

Purified dextrans are white, tasteless solids. Other physical and chemical properties vary depending on the specific chemical structure, which is determined by the microbial strain of origin and method of production. Dextrans with the highest percentages of α-$(1\rightarrow6)$ linkages are generally the most soluble in water. Dextran from *L. mesenteroides* strain NRRL B-512F is freely soluble in water and other solvents, including 6 M urea, 2 M glycine, formamide, glycerol, etc. (Jeanes, 1966; de Belder, 1990). Dextran solutions behave as Newtonian fluids, and their viscosity is a function of concentration, temperature, and average molecular weight (Granath, 1958; Gekko and Noguchi, 1971; Carrasco et al., 1989). Native dextran is polydisperse and typically of high average molecular weight (generally between 10^6 and 10^9 daltons). However, many of the current applications for dextran depend on the convenience with which it can be broken down to fractions of specific weight ranges. The relative linearity of dextran from strain NRRL B-512F is crucial for the production of such fractions. Dextrans exhibit characteristic serological reactions, apparently related to their molecular weight and degree of branching (Gronwall, 1957; Kabat and Bezer, 1958; Jeanes, 1986). However, intravenously administered clinical dextrans are of relatively low antigenicity, although individuals can exhibit hypersensitivity. The pharmacological properties of clinical dextrans have been reviewed recently (de Belder, 1996). Free hydroxyl groups in dextran are potential targets for chemical derivatizations, and dextran from strain NRRL B-512F is particularly suitable for these reactions because of its low level of branch linkages.

A number of bulk chemical applications have been demonstrated for dextran, including uses in oil-drilling operations, agriculture, food products, and the manufacture of photographic films and other products (Murphy and Whistler, 1973; Alsop, 1983; Glicksman, 1983). It should be noted that dextrans are not explicitly approved as food additives in the United States or Europe, although *L. mesenteroides* is a GRAS organism commonly found in fermented foods. Currently, dextran and dextran derivatives are used primarily as specialty chemicals in clinical, pharmaceutical, research, and industrial applications (Yalpani, 1986; de Belder, 1996; Vandamme et al., 1996). Early work by Gronwall and Ingelman (1945, 1948) established the potential of using a hydrolyzed dextran fraction as a blood-plasma volume expander (Figure 2). Clinical dextrans used today are Dextran 40 and Dextran 70, which are 40,000 and 70,000 dalton average molecular weight fractions, respectively. Dextrans are less expensive than the albumins and starch derivatives also used in plasma therapies (Lilley and Aucker, 1999). These colloids essentially replace normal blood proteins in providing osmotic pressure to pull fluid from the interstitial space into the plasma. This treatment is useful to prevent shock from hemorrhage, burns, surgery, or trauma and to reduce the risk of thrombosis and embolisms. Dextran 40 also improves blood flow and inhibits the aggregation of erythrocytes (de Belder, 1996).

Iron dextran is a colloidal preparation used especially in veterinary medicine for the treatment of anemia, particularly in newborn piglets. Special iron dextran preparations also have been developed to enhance magnetic resonance imaging (MRI) techniques (de Belder, 1996). Dextran sulfate has been used as a substitute for heparin in anticoagulant therapy, and, more recently, it is being studied as an antiviral agent, particularly in the treatment of human immunodeficiency virus (HIV) (Mitsuya et al., 1988; Piret et al., 2000). Dextran can be crosslinked by epichlorohydrin (Flodin and Porath, 1961) to form beads (Sephadex®) that have become widely used in

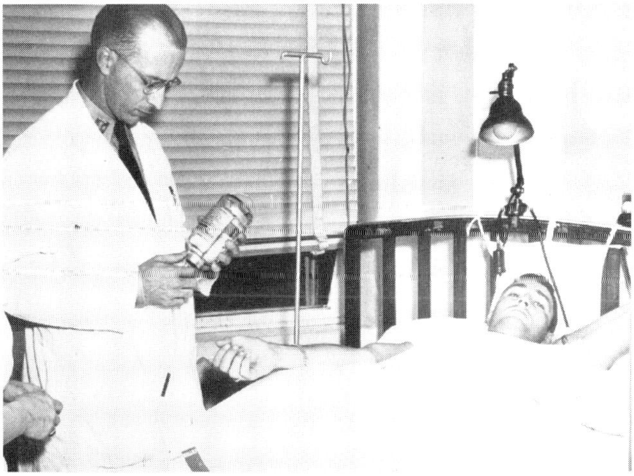

Fig. 2 Administration of dextran to a soldier at Walter Reed General Hospital, 1952. U.S. Dept. of Agriculture photograph.

research and industry for separations based on gel filtration. Anion and cation exchange resins based on Sephadex derivatives are widely used, including carboxymethyl (CM) Sephadex, diethylaminoethyl (DEAE) Sephadex, diethyl(2-hydroxypropyl)aminoethyl (QAE) Sephadex, and sulfopropyl (SP) Sephadex (Figure 3). Dextran is also an important component of many aqueous two-phase extraction systems, usually used in conjunction with polyethylene glycol (Tjerneld, 1992; Sinha et al., 2000).

Recently, oligosaccharides have received a great deal of attention as potential prebiotic compounds in food products, animal feeds, and cosmetics (Hidaka and Hirayama, 1991; Kohmoto et al., 1991; Monsan and Paul, 1995; Lamothe et al., 1996; Monsan et al., 2000). In contrast to clinical dextran preparations, prebiotic oligosaccharides must be resistant to digestion and preferentially utilized by beneficial bifidobacteria and lactic acid bacteria in the intestinal or skin microflora. Accordingly, dextran oligosaccharides of interest as prebiotics include the more branched varieties, containing α-(1 → 3) linkages from *L. citreum* strain NRRL B-742 (Remaud et al., 1992), α-(1 → 2) linkages from *L. mesenteroides* strain NRRL B-1299 (Remaud-Simeon et al., 1994; Dols et al., 1999), or the alternating α-(1 → 3) and α-(1 → 6) linkages from *L. mesenteroides* strain NRRL B-1355 (Pelenc et al., 1991).

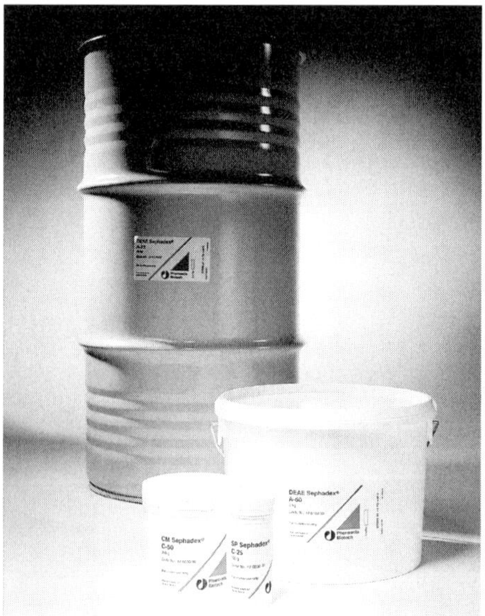

Fig. 3 Sephadex ion-exchange media are used widely for process scale applications. Photograph courtesy of Amersham Pharmacia Biotech, Inc.

12 Patents

Numerous patents claim methods for the production of dextran and dextran derivatives. The following examples, also summarized in Table 1, are illustrative. An early patent by Gronwall and Ingelman (1948) suggested that dextran might be useful as a blood-plasma volume expander. Hehre et al. (1959) described methods for dextran production by fermentation. Enzymatic production of dextran using dextransucrase also has been claimed (Tsuchiya and Koepsell, 1954; Hellman et al., 1955; Behrens and Ringpfeil, 1962). Wolff et al. (1955) described partial acid hydrolysis and fractionation of dextran for clinical applications. Alternative methods for the production of clinical dextrans include the use of dextranases (Carlson and Carlson, 1955a; Corman and Tsuchiya, 1957; Novak and Stoycos, 1958; Day and Kim, 1992) and chain-terminating acceptor reactions (Koepsell et al., 1955). Flodin and Porath (1961) described the cross-linking of dextran to form beads (Sephadex®) useful for gel filtration. Methods have been patented for the production of therapeutic iron dextrans (London and Twigg, 1958; Herb, 1979) and dextran sulfate (Morii et al., 1964; Usher, 1989). Recently, dextran oligosaccharides have garnered interest as potential prebiotic compounds (Lamothe et al., 1996).

Tab. 1 Selected patents related to dextran

Patent number	*Holder*	*Inventors*	*Title*	*Date*
U.S. Patent 2,437, 518	Pharmacia AB, Sweden	A. Gronwall, B. Ingelman	Manufacture of infusion and injection fluids	1948
U.S. Patent 2,686,147	U.S. Dept. Agriculture	H. M. Tsuchiya, H. J. Koepsell	Production of dextransucrase	1954
U.S. Patent 2,709,150	Enzmatic Chemicals, Inc., Delaware, USA	V. W. Carlson, W. W. Carlson	Method of producing dextran material by bacteriological and enzymatic action	1955
U.S. Patent 2,712,007	U.S. Dept. Agriculture	I. A. Wolff, R. L. Mellies, C. E. Rist	Fractionation of dextran products	1955
U.S. Patent 2,726,190	U.S. Dept. Agriculture	H. J. Koepsell, N. N. Hellman, H. M. Tsuchiya	Modification of dextran synthesis by means of alternate glucosyl acceptors	1955
U.S. Patent 2,726,985	U.S. Dept. Agriculture	N. N Hellman, H. M. Tsuchiya, S. P. Rogovin, R. W. Jackson, F. R. Senti	Controlled enzymatic synthesis of dextran	1955
U.S. Patent 2,841,578	The Commonwealth Engineering Co. of Ohio	L. J. Novak, G. S. Stoycos	Method for producing clinical dextran	1958
U.S. Patent 2,776,925	U.S. Dept. Agriculture	J. Corman, H. M. Tsuchiya.	Enzymic production of dextran of intermediate molecular weights	1957
U.S. Patent 2,820, 740	Benger Laboratories Ltd., England	E. London, G. D. Twigg	Therapeutic preparation of iron	1958
U.S. Patent 2,906,669	U.S. Dept. Agriculture	E. J. Hehre, H. M. Tsuchiya, N. N. Hellman, F. R. Senti	Production of dextran	1959
U.S. Patent 3,002,823	Pharmacia AB, Sweden	P. G. M. Flodin, J. O. Porath	Process of separating materials having different molecular weights and dimensions	1961
U.S. Patent 3,044,940	VEB Serum-Werk Bernburg, Germany	U. Behrens, M. Ringpfeil	Process for enzymatic synthesis of dextran	1962
U.S. Patent 3,141,014	Meito Sangyo Kabushiki Kaishu, Japan	E. Morii, K. Iwata, H. Kokkoku	Sodium and potassium salts of the dextran sulfate acid ester having substantially no anticoagulant activity but having lipolytic activity and the method of preparation thereof	1964
U.S. Patent 4,180,567	Pharmachem Corp., USA	J. R. Herb	Iron preparations and methods of making and administering the same	1979

Tab. 1 (cont.)

Patent number	*Holder*	*Inventors*	*Title*	*Date*
U.S. Patent 4,855,416	Polydex Pharmaceuticals, Ltd., The Bahamas	T. C. Usher	Method for the manufacture of dextran sulfate and salts thereof	1989
U.S. Patent 5,229,277	Louisiana State Univ.	D. F. Day, D. Kim	Process for the production of dextran polymers of controlled molecular size and molecular size distributions	1992
U.S. Patent 5,518,733.	Bioeurope, France	J.-P. Lamothe, Y. G. Marchenay, P. F. Monsan, F. M. B. Paul, V. Pelenc	Cosmetic compositions containing oligosaccharides	1996

13 Outlook and Perspectives

Advances in the molecular biology of glucansucrases promise not only to resolve fundamental questions concerning enzyme structure, function, and regulation but also to open new avenues for dextran applications. Recombinant organisms that overproduce dextransucrases may reduce the cost of dextran production by enzymatic synthesis, making dextrans more competitive for bulk chemical applications. Alternatively, dextrans might be produced in transgenic crops, as has been demonstrated recently for fructans (Caimi et al., 1996; Pilon-Smits et al., 1996; Sevenier et al., 1998) and *Streptococcus* glucans (Nichols, 2000a,b,c). Novel dextransucrases might be created by site-directed mutagenesis, chimeric recombination, or shuffling of dextransucrase genes. At the same time, dextran oligosaccharides appear to have considerable potential to find new and expanded markets as prebiotic supplements in foods, cosmetics, and animal feeds.

14
References

Alcalde, M., Plou, F. J., Gomez de Segura, A., Remaud-Simeon, M. Willemot, R. M., Monsan, P., Ballesteros, A. (1999) Immobilization of native and dextran-free dextransucrases from *Leuconostoc mesenteroides* NRRL B-512F for the synthesis of glucooligosaccharides, *Biotechnol. Tech.* **13**, 749–755.

Alsop, R. M. (1983) Industrial production of dextrans, in: *Progress in Industrial Microbiology*, (Bushell, M. E., Ed.), London: Elsevier, 1–44, Vol. 18.

Arguello-Morales, M. A., Remaud-Simeon, M., Pizzut, S., Sarcabal, P., Willemot, R.-M., Monsan, P. (2000a) *Leuconostoc mesenteroides* NRRL B-1355 *dsrC* gene for dextransucrase. GenBank Accession No. AJ250172.

Arguello-Morales, M. A., Remaud-Simeon, M., Pizzut, S., Sarcabal, P., Willemot, R.-M., Monsan, P. (2000b) Sequence analysis of the gene encoding alternansucrase, a sucrose glucosyltransferase from *Leuconostoc mesenteroides* NRRL B-1355. *FEMS Microbiol. Lett.* **182**, 81–85.

Barker, S. A., Bourne, E. J., Whiffen, D. H. (1956) Use of infrared analysis in the determination of carbohydrate structure, in: *Methods of Biochemical Analysis*, (Glick, D., Ed.), New York: Interscience Publishers, Inc., 213–245, Vol. 3.

Behrens, U., Ringpfeil, M. (1962) Process for enzymatic synthesis of dextran. U.S. Patent 3,044,940.

Beijerinck, M. W. K. (1912) Mucilaginous substances of the cell wall produced from cane sugar by bacteria. *Folia Microbiol.* **1**, 377.

Bhatnagar, R., Singh, D. K. S. (1999) Cloning and characterization of dextransucrase gene from *Leuconostoc mesenteroides* NRRL B-512F. GenBank Accession No. U81374.

Binder, T. P., Robyt, J. F. (1983) *p*-nitrophenyl α-D-glucopyranoside, a new substrate for glucansucrases, *Carbohydr. Res.* **124**, 287–299.

Bovery, F. A. (1959) Enzymatic polymerization. I. Molecular weight and branching during the formation of dextran, *J. Polymer Sci.* **35**, 167–182.

Caimi, P. G., McCole, L. M., Klein, T. M., Kerr, P. S. (1996) Fructan accumulation and sucrose metabolism in transgenic maize endosperm expressing a *Bacillus amyloliquefaciens SacB* gene, *Plant Physiol.* **110**, 355–363.

Caldwell, R. C., Sandham, H. J., Mann, W. V., Finn, S. B., Formicola, A. J. (1971) The effect of a dextranase mouthwash on dental plaque in young adults and children, *J. Amer. Dent. Assoc.* **82**, 124–131.

Carlson, V. W., Carlson, W. W. (1955a) Method of producing dextran material by bacteriological and enzymatic action. U.S. Patent 2,709,150.

Carlson, V. W., Carlson, W. W. (1955b) Production of endodextranase by *Aspergillus wentii*. U.S. Patent 2,716,084.

Carrasco, F., Chornet, E., Overend, R. P., Costa, J. (1989) A generalized correlation for the viscosity of dextrans in aqueous solutions as a function of temperature, concentration, and molecular weight at low shear rates, *J. Appl. Polymer Sci.* **37**, 2087–2098.

Cerning, J. (1990) Exocellular polysaccharides produced by lactic acid bacteria, *FEMS Microbiol. Rev.* **87**,113–130.

Chaiet, L., Kempf, A. J., Harman, R., Kaczka, E., Weston, R., Nollstadt, K., Wolf, F. J. (1970) Isolation of a pure dextranase from *Penicillium funiculosum*, *Appl. Microbiol.* **20**, 421–426.

Cheetham, N. W. H., Fiala-Beer, E., Walker, G. J. (1991) Dextran structural details from high-field proton NMR spectroscopy, *Carbohydr. Polymers* **14**, 149–158.

Chludzinski, A. M., Germaine, G. R., Schachtele, C. F. (1974) Purification and properties of dextran-

sucrase from *Streptococcus mutans*, *J. Bacteriol.* **118**, 1–7.
Corman, J., Tsuchiya, H. M. (1957) Enzymic production of dextran of intermediate molecular weights. U.S. Patent 2,776,925.
Côté, G. L. (1992) Low-viscosity α-D-glucan fractions derived from sucrose which are resistant to enzymatic digestion, *Carbohydr. Polym.* **19**, 249–252.
Côté, G. L., Ahlgren, J. A. (1994) Production, isolation, and immobilization of alternansucrase. Amer. Chem. Soc. 207 Natl. Meeting, Abstract CARB#12.
Côté, G. L., Ahlgren, J. A. (1995) Microbial polysaccharides, in: *Kirk-Othmer Encyclopedia of Chemical Technology* (Kroschvitz, J. I., Howe-Grant, M., Eds), New York: John Wiley & Sons, Inc., 578–612, 4th Ed., Vol. 16.
Côté, G. L., Robyt, J. F. (1982) Isolation and partial characterization of an extracellular glucansucrase from *L. mesenteroides* NRRL B-1355 that synthesizes an alternating (1 → 6), (1 → 3)-α-D-glucan, *Carbohyd. Res.* **101**, 57–74.
Côté, G. L., Robyt, J. F. (1983) The formation of α-D-(1 → 3) branch linkages by an exocellular glucansucrase from *Leuconostoc mesenteroides* NRRL B-742, *Carbohydr. Res.* **119**, 141–156.
Côté, G. L., Leathers, T. D., Ahlgren, J. A., Wyckoff, H. A., Hayman, G. T., Biely, P. (1997) Alternan and highly branched limit dextrans: Low-viscosity polysaccharides as potential new food ingredients, in: *Chemistry of Novel Foods* (Spanier, A. M., Tamura, M., Okai, H., Mills, O., Eds.), Carol Stream, IL: Allured Publishing Corp., 95–110.
Côté, G. L., Ahlgren, J. A., Smith, M. R. (1999) Some structural features of an insolube α-D-glucan from a mutant strain of *Leuconostoc mesenteroides* NRRL B-1355, *J. Ind. Microbiol. Biotechnol.* **23**, 656–660.
Covacevich, M. T., Richards, G. N. (1977) Frequency and distribution of branching in a dextran: an enzymic method, *Carbohydr. Res.* **54**, 311–315.
Day, D. F., Kim, D. (1992) Process for the production of dextran polymers of controlled molecular size and molecular size distributions. U.S. Patent 5,229,277.
de Belder, A. N. (1990) *Dextran*. Uppsala, Sweden: Pharmacia.
de Belder, A. N. (1993) Dextran, in: *Industrial Gums. Polysaccharides and Their Derivatives*, Third Edition (Whistler, R. L., BeMiller, J. N., Eds.), San Diego, CA: Academic Press, 399–425.
de Belder, A. N. (1996) Medical applications of dextran and its derivatives, in: *Polysaccharides. Medical Applications* (Dumitriu, S., Ed.), New York: Marcel Dekker, 505–523.
Dimler, R. J., Wolff, I. A., Sloan, J. W., Rist, C. E. (1955) Interpretation of periodate oxidation data on degraded dextran, *J. Am. Chem. Soc.* **77**, 6568–6573.
Dols, M., Remaud-Simeon, M., Willemot, R.-M., Vignon, M., Monsan, P. F. (1997) Characterization of dextransucrases from *Leuconostoc mesenteroides* NRRL B-1299, *Appl. Biochem. Biotechnol.* **62**, 47–59.
Dols, M., Remaud-Simeon, M., Willemot, R.-M., Demuth, B., Joerdening, H.-J., Buchholz, K., Monsan, P. (1999) Kinetic modeling of oligosaccharide synthesis catalyzed by *Leuconostoc mesenteroides* NRRL B-1299 dextransucrase, *Biotechnol. Bioeng.* **63**, 308–315.
Evans, T. H., Hibbert, H. (1946) Bacterial polysaccharides, in: *Adv. Carbohydr. Chem.* (Pigman, W. W., Wolfrom, M. L., Eds.). New York: Academic Press, 203–233, Vol. 2.
Fischer, E. H., Stein, E. A. (1960) Cleavage of O- and S-glycosidic bonds (survey), in: *The Enzymes* (Boyer, P. D., Lardy, H., Myrback, K., Eds.), New York: Academic Press, 301–312, Vol. 4.
Fitzgerald, R. J., Spinell, D. M., Stoudt, T. H. (1968) Enzymatic removal of artificial plaques, *Arch. Oral Biol.* **13**, 125–128.
Flodin, P. G. M., Porath, J. O. (1961) Process of separating materials having different molecular weights and dimensions. U.S. Patent 3,002,823.
Fu, D., Robyt, J. F. (1990) A facile purification of *Leuconostoc mesenteroides* B- 512FM dextransucrase, *Prep. Biochem.* **20**, 93–106.
Fu, D., Slodki, M. E., Robyt, J. F. (1990) Specificity of acceptor binding to *Leuconostoc mesenteroides* B512F dextransucrase: binding and acceptor-product structure of α-methyl-D-glucopyranoside analogs modified at C-2, C-3, and C-4 by inversion of the hydroxyl and by replacement of the hydroxyl with hydrogen, *Arch. Biochem. Biophys.* **276**, 460–465.
Fujiwara, T. Terao, Y., Hoshino, T., Kawabata, S., Ooshima, T., Sobue, S., Kimura, S., Hamada, S. (1998) Molecular analyses of glucosyltransferase genes among strains of *Streptococcus mutans*, *FEMS Microbiol. Lett.* **161**, 331–336.
Fukui, K., Fukui, Y., Moriyama, T. (1974) Purification and properties of dextransucrase and invertase from *Streptococcus mutans*, *J. Bacteriol.* **118**, 796–804.
Funane, K., Mizuno, K., Takahara, H., Kobayashi, M. (2000) Gene encoding a dextransucrase-like

protein in *Leuconostoc mesenteroides* NRRL B-512F, *Biosci. Biotechnol. Biochem.* **64**, 29–38.
Furuta, T., Koga, T., Nisizawa, T., Okahashi, N., Hamada, S. (1985) Purification and characterization of glucosyltransferases from *Streptococcus mutans* 6715, *J. Gen. Microbiol.* **131**, 285–293.
Galvez-Mariscal, A., Lopez-Munguia, A. (1991) Production and characterization of a dextranase from an isolated *Paecilomyces lilacinus* strain, *Appl. Microbiol. Biotechnol.* **36**, 327–331.
Gekko, K., Noguchi, H. (1971) Physicochemical studies of oligodextran. I. Molecular weight dependence of intrinsic viscosity, partial specific compressibility and hydrated water, *Biopolymers* **10**, 1513–1524.
Glicksman, M. (1983) Dextran, in: *Food Hydrocolloids* (Glicksman, M., Ed.), Boca Raton: CRC Press 157–166.
Godfrey, T. (1983) Dextranase and sugar processing, in: *Industrial Enzymology. The Application of Enzymes in Industry* (Godfrey, T., Reichelt, T., Eds.), New York: Nature Press, 422–424.
Granath, K. A. (1958) Solution properties of branched dextrans, *J. Colloid Sci.* **13**, 308–328.
Gronwall, A. (1957) *Dextran and Its Use in Colloidal Infusion Solutions.* Stockholm: Almqvist & Wiksell.
Gronwall, A., Ingelman, B. (1945) Dextran as a substitute for plasma, *Nature* **155**, 45.
Gronwall, A., Ingelman, B. (1948) Manufacture of infusion and injection fluids. U.S. Patent 2,437, 518.
Hamada, S., Slade, H. D. (1980) Biology, immunology, and cariogenicity of *Streptococcus mutans, Microbiol. Rev.* **44**, 331–384.
Hattori, A., Ishibashi, K., Minato, S. (1981) The purification and characterization of the dextranase from *Chaetomium gracile, Agric. Biol. Chem.* **45**, 2409–2416.
Hehre, E. J. (1941) Production from sucrose of a serologically reactive polysaccharide by a sterile bacterial extract, *Science* **93**, 237–238.
Hehre, E. J., Hamilton, D. M. (1951) The biological synthesis of dextran from dextrins, *J. Biol. Chem.* **192**, 161–174.
Hehre, E. J., Tsuchiya, H. M., Hellman, N. N., Senti, F. R. (1959) Production of dextran. U.S. Patent 2,906,669.
Hehre, E. J., Suzuki, H. (1966) New reactions of dextransucrase: α-D-glucosyl transfers to and from the anomeric sites of lactulose and fructose, *Arch. Biochem. Biophys.* **113**, 675–683.
Heincke, K., Demuth, B., Jordening, H.-J., Buchholz, K. (1999) Kinetics of the dextransucrase acceptor reaction with maltose - experimental results and modeling, *Enzyme Microbial Technol.* **24**, 523–534.
Herb, J. R. (1979) Iron preparations and methods of making and administering the same. U.S. Patent 4,180,567.
Hellman, N. N., Tsuchiya, H. M., Rogovin, S. P., Jackson, R. W., Senti, F. R. (1955) Controlled enzymatic synthesis of dextran. U.S. Patent 2,726,985.
Hidaka, H., Hirayama, M. (1991) Useful characteristics and commercial applications of fructo-oligosaccharides, *Biochem. Soc. Trans.* **19**, 561–565.
Hiraoka, N., Fukumoto, J., Tsuru, D. (1972) Studies on mold dextranases: III. Purification and some enzymatic properties of *Aspergillus carneus* dextranase, *J. Biochem.* **71**, 57–64.
Holzapfel, W. H., Schillinger, U. (1992) The genus *Leuconostoc,* in: *The Procaryotes,* 2nd Ed., (Ballows, A., Truper, H. G., Dworkin, M., Harder, W., Schleifer, K.-H., Eds.), New York: Springer-Verlag, 1508–1534, Vol. 2.
Honda, O., Kato, C., Kuramitsu, H. K. (1990) Nucleotide sequence of the *Streptococcus mutans gtfD* gene encoding the glucosyltransferase-S enzyme, *J. Gen. Microbiol.* **136**, 2099–2105.
Janecek, S., Svensson, B, Russell, R. R. B. (2000) Location of repeat elements in glucansucrases of *Leuconostoc* and *Streptococcus* species, *FEMS Microbiol. Lett.* **192**, 53–57.
Jeanes, A. (1965a) Dextrans. Preparation of a water soluble dextran by enzymic synthesis, in: *Methods in Carbohydrate Chemistry* (Whistler, R. L., BeMiller, J. N., Eds.), New York: Academic Press, 127–132, Vol. 5.
Jeanes, A. (1965b) Dextrans. Preparation of dextrans from growing *Leuconostoc* cultures, in: *Methods in Carbohydrate Chemistry* (Whistler, R. L., BeMiller, J. N., Eds.), New York: Academic Press, 118–126, Vol. 5.
Jeanes, A. (1966) Dextran, in: *Encyclopedia of Polymer Science and Engineering,* (Mark, H. F.; Bikales, N. M.; Overberger, C. G.; Menges, G.; Kroschwitz, J. I., Eds.), New York: John Wiley & Sons, 752–767, Vol. 4.
Jeanes, A. (1977) Dextrans and pullulans: industrially significant α-D-glucans, in: *ACS Symp. Series No. 45, Extracellular Microbial Polysaccharides* (Sandford, P. A., Laskin, A., Eds.), Washington, D. C.: American Chemical Society, 284–298.
Jeanes, A. (1978) *Dextran Bibliography.* Washington, D. C.: U. S. Dept. Agriculture.

Jeanes, A. (1986) Immunochemical and related interactions with dextrans reviewed in terms of improved structural information. *Mol. Immun.* **23**, 999–1028.

Jeanes, A., Seymour, F. R. (1979) The α-D-glucopyranosidic linkages of dextrans: comparison of percentages from structural analysis by periodate oxidation and by methylation, *Carbohydr. Res.* **74**, 31–40.

Jeanes, A., Haynes, W. C., Wilham, C. A., Rankin, J. C., Melvin, E. H., Austin, M. J., Cluskey, J. E., Fisher, B. E., Tsuchiya, H. M., Rist, C. E. (1954) Characterization and classification of dextrans from ninety-six strains of bacteria, *J. Amer. Chem. Soc.* **76**, 5041–5052.

Jung, H-K., Kim, K-N., Lee, H-S., Jung, S-H. (1999) Production of alternan by *Leuconostoc mesenteroides* CBI-110, *Kor. J. Appl. Microbiol. Biotechnol.* **27**, 35–40.

Kabat, E. A., Bezer, A. E. (1958) The effect of variation in molecular weight on the antigenicity of dextran in man, *Arch. Biochem. Biophys.* **78**, 306–318.

Kaboli, H., Reilly, P. J. (1980) Immobilization and properties of *Leuconostoc mesenteroides* dextransucrase, *Biotechnol. Bioeng.* **22**, 1055–1069.

Killey, M., Dimler, R. J., Cluskey, J. E. (1955) Preparation of panose by the action of NRRL B-512 dextransucrase on a sucrose-maltose mixture, *J. Amer. Chem. Soc.* **77**, 3315–3318.

Kim, D., Day, D. F. (1995) Isolation of a dextranase constitutive mutant of *Lipomyces starkeyi* and its use for the production of clinical size dextran, *Lett. Appl. Microbiol.* **20**, 268–270.

Kim, D., Kim, D-W. (1999) Facile purification and characterization of dextransucrase from *Leuconostoc mesenteroides* B-512FMCM, *J. Microbiol. Biotechnol.* **9**, 219–222.

Kim, D., Robyt, J. F. (1994) Production and selection of mutants of *Leuconostoc mesenteroides* constitutive for glucansucrases, *Enzyme Microb. Technol.* **16**, 659–664.

Kim, D., Robyt, J. F. (1995a) Dextransucrase constitutive mutants of *Leuconostoc mesenteroides* B-1299, *Enzyme Microb. Technol.* **17**, 1050–1056.

Kim, D., Robyt, J. F. (1995b) Production, selection, and characteristics of mutants of *Leuconostoc mesenteroides* B-742 constitutive for dextransucrases, *Enzyme Microb. Technol.* **17**, 689–695.

Kim, D., Robyt, J. F. (1996) Properties and uses of dextransucrases elaborated by a new class of *Leuconostoc mesenteroides* mutants, *Prog. Biotechnol.* **12**, 125–144.

Kim, D., Kim, D-W., Lee, J-H., Park, K-H., Day, L. M., Day, D. F. (1997) Development of constitutive dextransucrase hyper-producing mutants of *Leuconostoc mesenteroides* using the synchrotron radiation in the 70–1000 eV region, *Biotechnol. Tech.* **11**, 319–321.

Kim, H., Kim, D., Ryu, W-H., Robyt, J. F. (2000) Cloning and sequencing of the α-1 → 6 dextransucrase gene from *Leuconostoc mesenteroides* B-742CB, *J. Microbiol. Biotechnol.* **10**, 559–563.

Kitaoka, M., Robyt, J. F. (1998a) Large-scale preparation of highly purified dextransucrase from a high-producing constitutive mutant of *Leuconostoc mesenteroides* B-512FMC, *Enzyme Microb. Technol.* **23**, 386–391.

Kitaoka, M., Robyt, J. F. (1998b) Use of a microtiter plate screening method for obtaining *Leuconostoc mesenteroides* mutants constitutive for glucansucrase, *Enzyme Microb. Technol.* **22**, 527–531.

Kobayashi, M., Matsuda, K. (1975) Purification and characterization of two activities of the intracellular dextransucrase from *Leuconostoc mesenteroides* NRRL B-1299, *Biochim. Biophys. Acta* **397**, 69–79.

Kobayashi, M., Matsuda, K. (1976) Purification and properties of the extracellular dextransucrase from *Leuconostoc mesenteroides* NRRL B-1299, *J. Biochem.* **79**, 1301–1308.

Kobayashi, M., Matsuda, K. (1977) Structural characteristics of dextrans synthesized by dextransucrases from *Leuconostoc mesenteroides* NRRL B-1299, *Agric. Biol. Chem.* **41**, 1931–1937.

Kobayashi, M., Matsuda, K. (1980) Characterization of the multiple forms and main component of dextransucrase from *Leuconostoc mesenteroides* NRRL B-512F, *Biochim. Biophys. Acta* **614**, 46–62.

Koenig, D. W., Day, D. F. (1988) Production of dextranase by *Lipomyces starkeyi*, *Biotechnol. Lett.* **10**, 117–122.

Koepsell, H. J., Hellman, N. N., Tsuchiya, H. M. (1955) Modification of dextran synthesis by means of alternate glucosyl acceptors. U.S. Patent 2,726,190.

Kohmoto, T., Fukui, F., Takaku, H., Mitsuoka, T. (1991) Dose-response test of isomaltooligosaccharides for increasing fecal bifidobacteria, *Agric. Biol. Chem.* **55**, 2157–2159.

Kooi, E. R. (1958) Production of dextran-dextrinase. U.S. Patent 2,833,695.

Kossman, J., Welsh, T., Quanz, M., Knuth, K. (2000) Nucleic acid molecules encoding alternansucrase. PCT Patent WO00/47727.

Kuge, T., Kobayashi, K., Kitamura, S., Tanahashi, H. (1987) Degrees of long-chain branching in dextrans, *Carbohydr. Res.* **160**, 205–214.

Lamothe, J.-P., Marchenay, Y. G., Monsan, P. F., Paul, F. M. B., Pelenc, V. (1996) Cosmetic compositions containing oligosaccharides. U.S. Patent 5,518,733.

Larm, O., Lindberg, B., Svensson, S. (1971) Studies on the length of the side chains of the dextran elaborated by *Leuconostoc mesenteroides* NRRL B-512, *Carbohydr. Res.* **20**, 39–48.

Lawford, G. R., Kligerman, A., Williams, T. (1979) Dextran biosynthesis and dextransucrase production by continuous culture of *Leuconostoc mesenteroides, Biotechnol. Bioeng.* **21**, 1121–1131.

Leathers, T. D., Hayman, G. T., Cote, G. L. (1995) Rapid screening of *Leuconostoc mesenteroides* mutants for elevated proportions of alternan to dextran, *Curr. Microbiol.* **31**, 19–22.

Leathers, T. D., Hayman, G. T., Cote, G. L. (1997) Microorganism strains that produce a high proportion of alternan to dextran. U.S. Patent 5,702,942.

Leathers, T. D., Hayman, G. T., Cote, G. L. (1998) Rapid screening method to select microorganism strains that produce a high proportion of alternan to dextran. U.S. Patent 5,789,209.

Lee, J. M., Fox, P. F. (1985) Purification and characterization of *Paecilomyces lilacinus* dextranase, *Enzyme Microb. Technol.* **7**, 573–577.

Lilley, L. L., Aucker, R. S. (1999) Fluids and electrolytes, in: *Pharmacology and the Nursing Process.* St. Louis, MO: Mosby, Inc., 335–348.

Lindberg, B., Svensson, S. (1968) Structural studies on dextran from *Leuconostoc mesenteroides* NRRL B-512, *Acta Chem. Scand.* **22**, 1907–1912.

Loesche, W. J. (1986) Role of *Streptococcus mutans* in human dental decay, *Microbiol. Rev.* **50**, 353–380.

London, E., Twigg, G. D. (1958) Therapeutic preparation of iron. U.S. Patent 2,820, 740.

McCabe, M. M. (1985) Purification and characterization of a primer-independent glucosyltransferase from *Streptococcus mutans* 6715-13 mutant 27, *Infect. Immun.* **50**, 771–777.

Miller, A. W., Robyt, J. F. (1984) Stabilization of dextransucrase from *Leuconostoc mesenteroides* NRRL B-512F by nonionic detergents, poly(ethylene glycol) and high-molecular-weight dextran, *Biochim. Biophys. Acta* **785**, 89–96.

Miller, A. W., Eklund, S. H., Robyt, J. F. (1986) Milligram to gram scale purification and characterization of dextransucrase from *Leuconostoc mesenteroides* NRRL B-512F, *Carbohydr. Res.* **147**, 119–133.

Misaki, A., Torii, M., Sawai, T., Goldstein, I. J. (1980) Structure of the dextran of *Leuconostoc mesenteroides* B-1355, *Carbohydr. Res.* **84**, 273–285.

Mitsuishi, Y., Kobayashi, M., Matsuda, K. (1979) Dextran α-1, 2 debranching enzyme from *Flavobacterium* sp. M-73: its production and purification, *Agric. Biol. Chem.* **43**, 2283–2290.

Mitsuya, H., Looney, D. J., Kuno, S., Ueno, R., Wong-Staal, F., Broder, S. (1988) Dextran sulfate suppression of viruses in the HIV family: inhibition of virion binding to $CD4^+$ cells, *Science* **240**, 646–649.

Monchois, V., Willemot, R.-M., Remaud-Simeon, M., Croux, C., Monsan, P. (1996) Cloning and sequencing of a gene coding for a novel dextransucrase from *Leuconostoc mesenteroides* NRRL B-1299 synthesizing only a α(1-6) and α(1-3) linkages, *Gene* **182**, 23–32.

Monchois, V., Remaud-Simeon, M., Russell, R. R. B., Monsan, P., Willemot, R.-M. (1997) Characterization of *Leuconostoc mesenteroides* NRRL B512F dextransucrase (DSRS) and identification of amino-acid residues playing a key role in enzyme activity, *Appl. Microbiol. Biotechnol.* **48**, 465–472.

Monchois, V., Remaud-Simeon, M., Monsan, P., Willemot, R.-M. (1998) Cloning and sequencing of a gene coding for an extracellular dextransucrase (DSRB) from *Leuconostoc mesenteroides* NRRL B-1299 synthesizing only α(1-6) glucan, *FEMS Microbiol. Lett.* **159**, 307–315.

Monchois, V., Willemot, R.-M., Monsan, P. (1999) Glucansucrases: mechanism of action and structure-function relationships, *FEMS Microbiol. Rev.* **23**, 131–151.

Monchois, V., Vignon, M., Russell, R. R. B. (2000) Mutagenesis of asp-569 of glucosyltransferase I glucansucrase modulates glucan and oligosaccharide synthesis, *Appl. Environ. Microbiol.* **66**, 1923–1927.

Monsan, P., Paul, F. (1995) Enzymatic synthesis of oligosaccharides, *FEMS Microbiol. Rev.* **16**, 187–192.

Monsan, P., Paul, F., Auriol, D., Lopez, A. (1987) Dextran synthesis using immobilized *Leuconostoc mesenteroides* dextransucrase, in: *Methods in Enzymology: Immobilized Enzymes and Cells* (Mosbach, K., Ed.), Orlando, FL: Academic Press, Inc, 239–254, Vol. 136.

Monsan, P., Potocki de Montalk, G., Sarcabal, P., Remaud-Simeon, M., Willemont, R.-M. (2000) Glucansucrases: efficient tools for the synthesis of oligosaccharides of nutritional interest, in: *Food Biotechnology* (Bielecki, S., Tramper, J., Polak, J., Eds.), Amsterdam: Elsevier Science B. V., 115–122.

Morii, E., Iwata, K., Kokkoku, H. (1964) Sodium and potassium salts of the dextran sulphate acid ester having substantially no anticoagulant activity but having lipolytic activity and the method of preparation thereof. U.S. Patent 3,141,014.

Mountzouris, K. C., Gilmour, S. G., Jay, A. J., Rastall, R. A. (1999) A study of dextran production from maltodextrin by cell suspensions of *Gluconobacter oxydans* NCIB 4943, *J. Appl. Microbiol.* **87**, 546–556.

Murphy, P.T., Whistler, R. L. (1973) Dextrans, in *Industrial Gums,* Second Ed. (Whistler, R. L., BeMiller, J. N., Eds.), New York: Academic Press, 513–542.

Neely, W. B. (1960) Dextran: structure and synthesis, in: *Adv. Carbohydr. Chem.* (Wolfrom, M. L., Tipson, R. S., Eds.), New York: Academic Press, 341–369, Vol. 15.

Nichols, S. E. (2000a) Plant cells and plants transformed with *Streptococcus mutans* gene encoding glucosyltransferase C enzyme. U.S. Patent 6,127,603.

Nichols, S. E. (2000b) Plant cells and plants transformed with *Streptococcus mutans* genes encoding wild-type or mutant glucosyltransferase B enzymes. U.S. Patent 6,087,559.

Nichols, S. E. (2000c) Plant cells and plants transformed with *Streptococcus mutans* genes encoding wild-type or mutant glucosyltransferase D enzymes. U.S. Patent 6,127,602.

Novak, L. J., Stoycos, G. S. (1958) Method for producing clinical dextran. U.S. Patent 2,841,578.

Okada, G., Takayanagi, T., Sawai, T. (1988) Improved purification and further characterization of an isomaltodextranase from *Arthrobacter globiformis* T6, *Agric. Biol. Chem.* **52**, 495–501.

Pasteur, L. (1861) On the viscous fermentation and the butyrous fermentation, *Bull. Soc. Chim. Paris,* 30–31.

Paul, F., Auriol, D., Oriol, E. Monsan, P. (1984) Production and purification of dextransucrase from *Leuconostoc mesenteroides* NRRL B-512(F). *Ann. N. Y. Acad. Sci.* **434**, 267–270.

Pearce, B. J., Walker, G. J., Slodki, M. E., Schuerch, C. (1990) Enzymic and methylation analysis of dextrans and (1-3)-α-D-glucans, *Carbohydr. Res.* **203**, 229–246.

Pelenc, V., Lopez-Munguia, A., Remaud, M., Biton, J., Michel, J. M., Paul, F., Monsan, P. (1991) Enzymatic synthesis of oligoalternans, *Sci. Aliments* **11**, 465–476.

Pilon-Smits, E. A. H., Ebskamp, M. J. M., Jeuken, M. J. W., van der Meer, I. M., Visser, R. G. F., Weisbeek, P. J., Smeekens, S. C. M. (1996) Microbial fructan production in transgenic potato plants and tubers, *Ind. Crops Products* **5**, 35–46.

Piret, J., Lamontagne, J., Bestman-Smith, J., Roy, S., Gourde, P., Desormeaux, A., Omar, R. F., Juhasz, J., Bergeron, M. G. (2000) *In vitro* and *in vivo* evaluations of sodium lauryl sulfate and dextran sulfate as microbicides against herpes simplex and human immunodeficiency viruses, *J. Clin. Microbiol.* **38**, 110–119.

Quirasco, M., Lopez-Munguia, A., Remaud-Simeon, M., Monsan, P., Farres, A. (1999) Induction and transcriptional studies of the dextransucrase gene in *Leuconostoc mesenteroides* NRRL B-512F, *Appl. Environ. Microbiol.* **65**, 5504–5509.

Rankin, J. C., Jeanes, A. (1954) Evaluation of periodate oxidation method for structural analysis of dextrans, *J. Am. Chem. Soc.* **76**, 4435–4441.

Reh, K.-D., Noll-Borchers, M., Buchholz, K. (1996) Productivity of immobilized dextransucrase for leucrose formation, *Enzyme Microb. Technol.* **19**, 518–524.

Remaud, M., Paul, F., Monsan, P., Heyraud, A., Rinaudo, M. (1991) Molecular weight characterization and structural properties of controlled molecular weight dextrans synthesized by acceptor reaction using highly purified dextransucrase, *J. Carbohydr. Chem.* **10**, 861–876.

Remaud, M., Paul, F., Monsan, P. (1992) Characterization of α-(1 → 3) branched oligosaccharides synthesized by acceptor reaction with the extracellular glucosyltransferases from *L. mesenteroides* NRRL B-742, *J. Carbohydr. Chem.* **11**, 359–378.

Remaud-Simeon, M., Lopez-Munguia, A., Pelenc, V., Paul, F., Monsan, P. (1994) Production and use of glucosyltransferases from *Leuconostoc mesenteroides* NRRL B-1299 for the synthesis of oligosaccharides containing α-(1 → 2) linkages, *Appl. Biochem. Biotechnol.* **44**, 101–117.

Remaud-Simeon, M., Willemot, R.-M., Sarcabal, P., Potocki de Montalk, G., Monsan, P. (2000) Glucansucrases: molecular engineering and oligosaccharide synthesis, *J. Mol. Catalysis B: Enzymatic* **10**, 117–128.

Robyt, J. F. (1986) Dextran, in: *Encyclopedia of Polymer Science and Engineering,* (Mark, H. F., Gaylord, N. G., Bikales, N. M., Eds.), New York: John Wiley & Sons, 752–767, Vol. 4.

Robyt, J. F. (1992) Structure, biosynthesis, and uses of nonstarch polysaccharides: dextran, alternan, pullulan, and algin, in: *Developments in Biochemistry and Biophysics* (Alexander, R. J., Zobel, H. F., Eds.), St. Paul: Amer. Assoc. Cereal Chemists, 261–292.

Robyt, J. F. (1995) Mechanisms in the glucansucrase synthesis of polysaccharides and oligosaccharides from sucrose, in: *Adv. Carbohydr. Chem. Biochem.* (Horton, D., Ed.), San Diego: Academic Press, 133–168, Vol. 51.

Robyt, J. F., Eklund, S. H. (1983) Relative, quantitative effects of acceptors in the reaction of *Leuconostoc mesenteroides* B-512F dextransucrase, *Carbohydr. Res.* **121**, 279–286.

Robyt, J. F., Martin, P. J. (1983) Mechanism of synthesis of D-glucans by D-glucosyltransferases from *Streptococcus mutans* 6715, *Carbohydr. Res.* **113**, 301–315.

Robyt, J. F., Taniguchi, H. (1976) The mechanism of dextransucrase action. Biosynthesis of branch linkages by acceptor reactions with dextran, *Arch. Biochem. Biophys.* **174**, 129–135.

Robyt, J. F., Walseth, T. F. (1978) The mechanism of acceptor reactions of *Leuconostoc mesenteroides* B-512F dextransucrase, *Carbohydr. Res.* **61**, 433–445.

Robyt, J. F., Walseth, T. F. (1979) Production, purification and properties of dextransucrase from *Leuconostoc mesenteroides* NRRL B-512F, *Carbohydr. Res.* **68**, 95–111.

Robyt, J. F., Kimble, B. K., Walseth, T. F. (1974) The mechanism of dextransucrase action. Direction of dextran biosynthesis, *Arch. Biochem. Biophys.* **165**, 634–640.

Ryu, H-J., Kim, D., Kim, D-W., Moon, Y-Y., Robyt, J. F. (2000) Cloning of a dextransucrase gene (*fmcmds*) from a constitutive dextransucrase hyper-producing *Leuconostoc mesenteroides* B-512FMCM developed using VUV, *Biotechnol. Lett.* **22**, 421–425.

Sabatie, J., Choplin, L., Moan, M., Doublier, J. L., Paul, F., Monsan, P. (1988) The effect of synthesis temperature on the structure of dextran NRRL B 512F, *Carbohydr. Polymers* **9**, 87–101.

Sanchez-Gonzalez, M., Alagon, A., Rodriguez-Sotres, R., Lopez-Munguia, A. (1999) Proteolytic processing of dextransucrase of *Leuconostoc mesenteroides*, *FEMS Microbiol. Lett.* **181**, 25–30.

Sawai, T., Tohyama, T., Natsume, T. (1978) Hydrolysis of fourteen native dextrans by *Arthrobacter* isomaltodextranase and correlation with dextran structure, *Carbohydr. Res.* **66**, 195–205.

Sawai, T., Ohara, S., Ichimi, Y., Okaji, S., Hisada, K., Fukaya, N. (1981) Purification and some properties of the isomaltodextranase of *Actinomadura* strain R10 and comparison with that of *Arthrobacter globiformis* T6, *Carbohydr. Res.* **89**, 289–299.

Scheibler, C. (1874) Investigation on the nature of the gelatinous excretion (so-called frog's spawn) which is observed in production of beet-sugar juices, *Z. Ver. Dtsch. Zucker-Ind.* **24**, 309–335.

Sery, T. W., Hehre, E. J. (1956) Degradation of dextrans by enzymes of intestinal bacteria, *J. Bacteriol.* **71**, 373–380.

Sevenier, R., Hall, R. D., van der Meer, I. M., Hakkert, H. J. C., van Tunen, A. J., Koops, A. J. (1998) High level fructan accumulation in a transgenic sugar beet, *Nature Biotechnol.* **16**, 843–846.

Seymour, F. R., Julian, R. L. (1979) Fourier-transform, infrared difference-spectrometry for structural analysis of dextrans, *Carbohydr. Res.* **74**, 63–75.

Seymour, F. R., Knapp, R. D. (1980) Unusual dextrans: 13. Structural analysis of dextrans from strains of *Leuconostoc* and related genera, that contain 3-O-α-D-glucosylated α-D-glucopyranosyl residues at the branch points, or in consecutive linear position, *Carbohyd. Res.* **81**, 105–129.

Seymour, F. R., Knapp, R. D., Bishop, S. H. (1976) Determination of the structure of dextran by ^{13}C-nuclear magnetic resonance spectroscopy, *Carbohydr. Res.* **51**, 179–194.

Seymour, F. R., Slodki, M. E., Plattner, R. D., Jeanes, A. (1977) Six unusual dextrans: methylation structural analysis by combined G. L. C.-M. S. of per-O-acetylaldononitriles, *Carbohydr. Res.* **53**, 153–166.

Seymour, F. R., Chen, E. C. M., Bishop, S. H. (1979a) Methylation structural analysis of unusual dextrans by combined gas-liquid chromatography-mass spectrometry, *Carbohydr. Res.* **68**, 113–121.

Seymour, F. R., Knapp, R. D., Bishop, S. H. (1979b) Correlation of the structure of dextrans to their ^{1}H-NMR spectra, *Carbohydr. Res.* **74**, 77–92.

Sharpe, E. S., Stodola, F. H., Koepsell, H. J. (1960) Formation of isomaltulose in enzymatic dextran synthesis, *J. Org. Chem.* **25**, 1062–1063.

Shimamura, A., Tsumori, H., Mukasa, H. (1982) Purification and properties of *Streptococcus mutans* extracellular glucosyltransferase, *Biochim, Biophys. Acta* **702**, 72–80.

Shimizu, E., Unno, T., Ohba, M., Okada, G. (1998) Purification and characterization of an isomaltotriose-producing *endo*-dextranase from a *Fusarium* sp., *Biosci. Biotechnol. Biochem.* **62**, 117–122.

Shiroza, T., Ueda, S., Kuramitsu, H. K. (1987) Sequence analysis of the *gftB* gene from *Streptococcus mutans*, *J. Bacteriol.* **169**, 4263–4270.

Sidebotham, R. L. (1974) Dextrans, in: *Adv. Carbohydr. Chem. Biochem.* (Tipson, R. S., Horton, D., Eds.), New York: Academic Press, 371–444, Vol. 30.

Simonson, L. G., Liberta, A. E. (1975) New sources of fungal dextranase. *Mycologia* **4**, 845–851.

Sinha, J., Dey, P. K., Panda, T. (2000) Aqueous two-phase: the system of choice for extractive fermentations, *Appl. Microbiol. Biotechnol.* **54**, 476–486.

Slodki, M. E., England, R. E., Plattner, R. D., Dick Jr., W. E. (1986) Methylation analyses of NRRL dextrans by capillary gas-liquid chromatography, *Carbohyd. Res.* **156**,199–206.

Smith, M. R., Zahnley, J., Goodman, N. (1994) Glucosyltransferase mutants of *Leuconostoc mesenteroides* NRRL B-1355, *Appl. Environ. Microbiol.* **60**, 2723–2731.

Smith, M. R., Zahnley, J. C., Wong, R. Y., Lundin, R. E., Ahlgren, J. A. (1998) A mutant strain of *Leuconostoc mesenteroides* B-1355 producing a glucosyltransferase synthesizing $\alpha(1 \rightarrow 2)$ glucosidic linkages, *J. Ind. Microbiol. Biotechnol.* **21**, 37–45.

Stiles, M. E., Holzapfel, W. H. (1997) Lactic acid bacteria of foods and their current taxonomy, *Int. J. Food Microbiol.* **36**, 1–29.

Stodola, F. H., Sharpe, E. S., Koepsell, H. J. (1956) The preparation, properties and structure of the disaccharide leucrose, *J. Amer. Chem. Soc.* **78**, 2514–2518.

Su, D., Robyt, J. F. (1994) Determination of the number of sucrose and acceptor binding sites for *Leuconostoc mesenteroides* B-512FM dextransucrase, and the confirmation of the two-site mechanism for dextran synthesis, *Arch. Biochem. Biophys.* **308**, 471–476.

Sun, J., Cheng, X., Zhang, Y., Yan, Z., Zhang, S. (1988) A strain of *Paecilomyces lilacinus* producing high quality dextranase, *Ann. N. Y. Acad. Sci.* **542**, 192–194.

Suzuki, M., Unno, T., Okada, G. (1999) Simple purification and characterization of an extracellular dextrin dextranase from *Acetobacter capsulatum* ATTC 11894, *J. Appl. Glycosci.* **46**, 469–473.

Swengers, D. (1991) Leucrose, a ketodisaccharide of industrial design, in: *Carbohydrates as Organic Raw Materials* (Lichtenthaler, F. W., Ed.), Weinheim, Germany: VCH, 183–195.

Takahashi, M., Okada, S., Uchimura, T., Kozaki, M. (1992) *Leuconostoc amelibiosum* Schillinger, Holzapfel, and Kandler 1989 is a later subjective synonym of *Leuconostoc citreum* Farrow, Facklam, and Collins 1989, *Int. J. Syst. Bacteriol.* **42**, 649–651.

Takazoe, I. (1989) Palatinose - an isomeric alternative to sucrose, in: *Progress in Sweeteners* (Grenby, T. H., Ed.), New York: Elsevier Science, 143–167.

Tarr, H. L. A., Hibbert, H. (1931) Studies on reactions relating to carbohydrates and polysaccharides. XXXVII. The formation of dextran by *Leuconostoc mesenteroides, Can. J. Res.* **5**, 414–427.

Taylor, C., Cheetham, N. W. H., Walker, G. J. (1985) Application of high-performance liquid chromatography to a study of branching dextrans, *Carbohydr. Res.* **137**, 1–12.

Tjerneld, F. (1992) Aqueous two-phase partitioning on an industrial scale, in: *Poly(Ethylene Glycol) Chemistry: Biotechnical and Biomedical Applications* (Harris, J. M., Ed.), New York: Plenum Press, 85–102.

Torii, M., Sakakibara, K., Misaki, A., Sawai, T. (1976) Degradation of alpha-linked D-gluco-oligosaccharides and dextrans by an isomaltodextranase preparation from *Arthrobacter globiformis* T6, *Biochem. Biophys. Res. Comm.* **70**, 459–464.

Tsuchiya, H. M., Koepsell, H. J. (1954) Production of dextransucrase. U.S. Patent 2,686,147.

Tsuchiya, H. M., Hellman, N. N., Koepsell, H. J., Corman, J., Stringer, C. S., Rogovin, S. P., Bogard, M. O., Bryant, G., Feger, V. H., Hoffman, C. A., Senti, F. R., Jackson, R. W. (1955) Factors affecting molecular weight of enzymatically synthesized dextran, *J. Amer. Chem. Soc.* **77**, 2412–2419.

Tsuchiya, H. M., Jeanes, A., Bricker, H. M., Wilham, C. A. (1956) Production of dextranase. U.S. Patent 2,742,399.

Ueda, S., Shiroza, R., Kuramitsu, H. K. (1988) Sequence analysis of the *gtfC* gene from *Streptococcus mutans* GS-5, *Gene* **69**, 101–109.

Usher, T. C. (1989) Method for the manufacture of dextran sulfate and salts thereof. U.S. Patent 4,855,416.

Van Cleve, J. W., Schaefer, W. C., Rist, C. E. (1956) The structure of NRRL B-512 dextran. Methylation studies, *J. Amer. Chem. Soc.* **78**, 4435–4438.

Vandamme, E. J., Bruggeman, G., De Baets, S., Vanhooren, P. T. (1996) Useful polymers of microbial origin, *Agro-Food-Industry Hi-Tech* **Sept./Oct.**, 21–25.

Van Tieghem, P. (1878) On sugar-mill gum, *Ann. Sci. Nat. Bot. Biol. Veg.* **7**, 180–203.

Walker, G. J. (1978) Dextrans, in: *International Review of Biochemistry. Biochemistry of Carbohydrates II*, (Manners, D. J., Ed.), Baltimore, MD: University Park Press, 75–125, Vol. 16.

Webb, E., Spencer-Martins, I. (1983) Extracellular endodextranase from the yeast *Lipomyces starkeyi*, *Can. J. Microbiol.* **29**, 1092–1095.

Wilke-Douglas, M., Perchorowicz, J. T., Houck C. M., Thomas, B. R. (1989) Methods and compositions for altering physical characteristics of fruit and fruit products. PCT Patent WO89/12386.

Wolff, I. A., Mellies, R. L., Rist, C. E. (1955) Fractionation of dextran products. U.S. Patent 2,712,007.

Yalpani, M. (1986) Preparation and applications of dextran-derived products in biotechnology and related areas, *CRC Crit. Rev. Biotechnol.* **3**, 375–421.

Yamamoto, K., Yoshikawa, K., Kitahata, S., Okada, S. (1992) Purification and some properties of dextrin dextranase from *Acetobacter capsulatus* ATTC 11894, *Biosci. Biotechnol. Biochem.* **56**, 169–173.

Yamamoto, K., Yoshikawa, K., Okada, S. (1993a) Dextran synthesis from reduced maltooligosaccharides by dextrin dextranase from *Acetobacter capsulatus* ATTC 11894, *Biosci. Biotechnol. Biochem.* **57**, 136–137.

Yamamoto, K., Yoshikawa, K., Okada, S. (1993b) Effective dextran production from starch by dextrin dextranase with debranching enzyme, *J. Ferment. Bioeng.* **76**, 411–413.

16
Cell-Wall β-Glucans of *Saccharomyces cerevisiae*

Gerrit J. P. Dijkgraaf[1], Dr. Huijuan Li[2], Dr. Howard Bussey[3]
[1] Department of Biology, McGill University, Stewart Biology Building, 1205 Dr. Penfield Ave., Montreal, Quebec, Canada H3A 1B1; Tel.: 1-514-398-6439; Fax: 1-514-398-2595; E-mail: gdijkg@po-box.mcgill.ca
[2] Department of Biology, McGill University, Stewart Biology Building, 1205 Dr. Penfield Ave., Montreal, Quebec, Canada H3A 1B1; Tel.: 1-514-398-6439; Fax: 1-514-398-2595; E-mail: hli6@po-box.mcgill.ca
[3] Department of Biology, McGill University, 1205 Dr. Penfield Ave., Montreal, Quebec, Canada H3A 1B1; Tel.: 1-514-398-6439; Fax: 1-514-398-8051; E-mail: hbusse@po-box.mcgill.ca

Biopolymers for Medical and Pharmaceutical Applications. Edited by A. Steinbüchel and R. H. Marchessault

ISBN: 3-527-31154-8

aa	amino acid residue(s)
ADP	adenosine diphosphate
ATP	adenosine triphosphate
bp	base pair
CFW	calcofluor white
Con A	concanavalin A
DP	degree of polymerization
EDTA	ethylene diamine tetra-acetic acid
ER	endoplasmic reticulum
GAP	GTPase-activating protein
GEF	guanine-nucleotide exchange factor
GGTase I	geranylgeranyl transferase type I
Glc	glucose
GPI	glycosylphosphatidylinositol
G protein	GTP-binding protein
GS	glucan synthase
GTP	guanosine triphosphate
GTPase	guanosine triphosphatase
GTP[γS]	guanosine 5′[γ-thio]triphosphate
K_m	Michaelis constant
MAR	matrix-association region
NMR	nuclear magnetic resonance
PHS	phytosphingosine
SDS	sodium dodecyl sulfate
TRAPP	transport protein particle
UDP	uridine diphosphate
UGGT	UDP-glucose:glycoprotein glucosyltransferase
V_{max}	maximal velocity
X-Gal	5-bromo-4-chloro-3-indolyl-β-D-galactoside
YPD	yeast extract peptone dextrose

1 Introduction

Current understanding of the architecture and biosynthesis of yeast cell walls is based on an integration of chemical, biochemical, genetic and cell biological data obtained over more than a century. The cell wall of *Saccharomyces cerevisiae* represents some 30% of the cell dry weight, and its main components are glycoproteins, glucans and a small amount of chitin. The exact composition of the *S. cerevisiae* cell wall varies considerably with nutrient availability and the developmental program (i.e., vegetative growth, mating, sporulation, or pseudohyphal filamentation). The cell wall ensures osmotic stability, acts as a filter for large molecules, and retains enzymes required for nutrient uptake. Yeast species differ in the relative abundance and chemical linkage of their polymers, but their walls share a similar architecture and appear as bilayered structures when examined by electron microscopy. The dense outer layer consists predominantly of glycoproteins, which mediate wall permeability, flocculence, sexual agglutinability and pathogenicity (see Chapter 4 of this volume and references therein). The amorphous inner layer consists of glucans and chitin, with the latter being present close to the plasma membrane in lateral walls. This inner layer is thought to be responsible for the mechanical strength of the cell wall, since removal of the outer glycoprotein layer does not affect cell shape. Various kinds of bud and fission scars resulting from vegetative reproduction are the only differentiated regions of yeast cell walls. The chitin-rich bud scars of *S. cerevisiae* are easily visualized with the fluorescent dye calcofluor white (CFW; see Chapter 5 of this volume and references therein). We have gained considerable insight in the structure and biosynthesis of all these components, but the cross-linking reactions required to assemble the cell wall and the order in which they occur remain poorly understood. The yeast cell wall determines cell shape and has been studied as an example of morphogenesis, since its assembly is highly regulated and requires de-novo synthesis, delivery and remodeling of wall components at a specific time and place during the yeast cell cycle. *S. cerevisiae* also provides a genetically amenable model system to study wall biosynthesis with applications in pathogenic fungi. Fungal infections have become a significant cause of mortality over the past two decades due to an increase in the number of immunocompromised individuals and a higher incidence of drug-resistant strains. The fungal cell wall is perceived as a good target for novel therapeutic agents, since it is an essential organelle that is absent from humans. This chapter will focus on the molecular structure and biosynthesis of glucans in the budding yeast, *S. cerevisiae*. Where applicable, studies on other fungi are mentioned to emphasize the highly conserved nature of their polysaccharides. Structural work on yeast glucans has previously been summarized by Phaff (1971), Duffus et al. (1982) and Shahinian and Bussey (2000), and in more general reviews on yeast cell-wall biosynthesis by Cabib et al. (1982), Klis (1994), Cid et al. (1995) and Orlean (1997).

2 Historical Outline

The literature on yeast β-glucan is far too extensive to be comprehensively reviewed here. Hence, it was necessary to be selective and to focus on general progress in cell-wall research as well as on two important breakthroughs in yeast β-glucan biology.

2.1 Methodological Advances in Cell-Wall Research

The yeast cell wall was the subject of continued investigation during the first half of the twentieth century, but progress was slow due to the complex nature and relative insolubility of the cell-wall polysaccharides, which hampered purification and structural analysis. The harsh chemical treatments initially used to fractionate whole cells caused extensive degradation of wall polymers, and were uninformative with respect to cross-links between individual wall components, as it was unclear what chemical bonds they destroyed. In addition to improvements in methods of structural analysis, three procedures were developed during the 1950s that are still in use today. The first procedure was the isolation of cell walls by mechanical disruption of yeast cells, and involved either high-speed shaking in the presence of glass beads (Merkenschlager et al., 1957) or use of a modified French press (Simpson et al., 1963). The second procedure was the use of enzyme preparations for cell-wall digestion. Giaja (1914) was the first to observe that gastric juice of the garden snail, *Helix pomatia*, had the ability to lyse yeast cells, and this crude enzyme mixture was later used by Eddy and Williamson (1957) in conjunction with osmotic stabilizers to study spheroplast formation. Johnston and Mortimer (1959) introduced digestion of asci with this lytic mixture to allow isolation of ascospores, and in doing so opened up the field of yeast genetics. Snail digestive fluid is isolated from a small vesicle of the alimentary canal, and contains over 30 different enzymatic activities, though it is low in proteolytic activity (Holden and Tracey, 1950). Difficulties in obtaining large quantities of this material, and its complex composition and variable potency, made investigators look for alternative sources of yeast lytic enzymes. Tanaka and Phaff (1965) isolated several microorganisms from soil that were able to grow on mineral medium with autoclaved yeast cells as the carbon source, and one of their isolates – *Bacillus circulans* strain WL-12–produced a lytic mixture of endo-β-1,3- and endo-β-1,6-glucanases. Anderson and Millbank (1966) fractionated snail digestive fluid and also observed that these glucanase activities were sufficient for yeast cell lysis. Considerable variation in lysis susceptibility to both snail digestive fluid and microbial enzymes were seen between *S. cerevisiae* strains as well as between growth phases for a given strain. Pretreatment with either sulfhydryl reagents such as 2-mercaptoethanol (Duell et al., 1964) or a phosphomannanase (McLellan and Lampen, 1968) greatly improved lysis efficiency. The use of purified enzymes gave information on how wall components were retained in the cell wall, and contributed significantly to the elucidation of glucan structures, by confirming polymer configuration and linkage type. The third procedure involved development of electron microscopic techniques to study the morphology and regeneration of cell walls. Northcote and Horne (1952) and Houwink and Kreger (1953) exposed and described bud and birth scars after boiling cell walls in alkali and acid, respectively. They observed that cell walls have a smooth outside and a more fibrillar inner side, but that the bilayered structure of the wall was most obvious in ultrathin sections of whole yeast cells (Agar and Douglas, 1955). Mundkur (1960) demonstrated that mannan formed the outer layer of the wall. He treated cell walls with periodate and stained the aldehyde that was formed by the cleavage of unsubstituted hydroxyls at C-3 and C-4 positions with leucofuchsin. Yeast mannan has an abundance of 1,2- and 1,6-linkages, and stained much more intensely with this cytochemical technique than yeast glucan, which is mainly 1,3-linked. Kopecká et al. (1974) exposed the wall inner fibrillar layer by digesting cell walls with either purified endo-β-1,3- or endo-β-1,6-glucanase from *B. circulans* and concluded that this layer was mainly β-1,3-glucan, as it was completely degraded by an endo-β-1,3-glucanase from *Schizosaccharomyces versatilis*. Horisberger and Vonlanthen (1977) confirmed that mannan formed the outer layer of the wall by immunogold labeling, and demonstrated that chitin was present close to the plasma membrane in the lateral wall by labeling with wheatgerm agglutinin coupled to gold particles. Necas (1956) prepared *S. cerevisiae* spheroplasts by autolysis and found that a small fraction survived on nutrient agar. They increased in volume, became multinucleated, developed a wall, and started to bud after approximately three days. Eddy and Williamson (1959) greatly improved spheroplast formation by using snail gut enzyme and osmotic stabilizers. They observed only partial resynthesis in liquid media, with the newly formed walls being highly aberrant and unable to regenerate viable cells. *Saccharomyces* spheroplasts were found less suitable for cell-wall regeneration studies, as they need to be embedded in either 30% gelatin (Necas, 1961) or 2% agar (Svoboda, 1966) for regeneration to occur. *Candida albicans* and *Schizosaccharomyces pombe* regenerate more readily in liquid media and are, therefore, most used in cell-wall regeneration studies (see Osumi et al., 1998, for some spectacular photographs). We describe in the next two sections how researchers grappled with the puzzling structure of yeast glucan.

2.2 Elucidating the Molecular Structure of Yeast Glucans

Budding yeast contains both β-1,3- and β-1,6-linked glucan polymers, but it took over 75 years after the initial work to establish this. Salkowski (1894) was the first to isolate two polysaccharide fractions from yeast: an alkali-soluble fraction mainly composed of mannose which he termed "yeast gum", and an alkali-insoluble fraction mainly composed of glucose which he termed "yeast cellulose". Zechmeister and Tóth (1934) re-investigated the latter substance and found it to be different from cellulose: it failed to give the characteristic blue coloration with an iodine solution in potassium iodide and treatment with strong acid, and it did not yield cellobiose when treated with acetolyzing reagents. Methylation and subsequent hydrolysis of the product gave 2,4,6-tri-*O*-methyl-D-glucose indicating, for that time, a rather unusual 1,3-glucosidic linkage. In order to avoid confusion with real cellulose, Zechmeister and Tóth called the material "yeast polyose". Hassid et al. (1941) confirmed the presence of 1,3 linkages by methylation analysis, but failed to detect end groups in the form of tetramethylglucose and concluded that the molecule is probably of the closed chain type. Low specific rotations of acetylated and methylated derivatives combined with an upward rotation during hydrolysis suggested a predominance of β-linkages. Barry and Dillon (1943) first called the insoluble polysaccharide "yeast glucan" and provided further evidence for the 1,3 linkage by demonstrating a lack of oxidation by periodic acid, except for one terminal residue per 28 glucose units. They also confirmed the β-configuration of this linkage by showing hydrolysis with the β-glucosidase emulsin, but not with the α-glucosidase takadiastase.

In these early studies, glucan was isolated by drastic treatment of whole yeast cells with hot dilute alkali to remove mannan, followed by extraction with hot dilute hydrochloric acid to remove glycogen. These treatments degraded the glucan molecules, and the majority of cell-wall glucan was discarded in these extractions. A kilogram of pressed baker's yeast typically yielded only a few grams of an insoluble residue that was further washed with water, ethanol and ether. Bell and Northcote (1950) tried to avoid glucan degradation by extracting glycogen with 0.5 M acetic acid, and this procedure enabled them to obtain evidence for a highly branched polysaccharide of high molecular weight. By excluding oxygen during methylation analysis of the polysaccharide, they obtained evidence indicating chain units of nine β-1,3-linked glucose residues with β-1,2-interchain links. Peat et al. (1958a) used the same protocol to purify glucan, but autoclaved the residual material after acetic acid extraction in order to remove glycogen completely. Oligosaccharides of the laminaribiose series were obtained in quantity by acid hydrolysis of their glucan material, but sophorose was absent, ruling out the β-1,2-interchain link proposed by Bell and Northcote (1950). Instead, β-1,6-linkages were demonstrated for the first time based on the occurrence of oligosaccharides of the gentiobiose series. It was shown by periodate oxidation that approximately 10% of the glucose residues were in this category, and Peat et al. (1958b) confirmed this number by toluene-*p*-sulfonylation. However, they did not identify the trisaccharide 3,6-di-*O*-β-glucosylglucose among the hydrolysis products, and therefore proposed that yeast glucan was a linear molecule in which β-1,3- and β-1,6-linkages occur at random or in sequences. Manners and Patterson (1966) analyzed yeast glucan by methylation analysis and found evidence for: (1) a branched glucan molecule in the form of 2,4-di-*O*-methyl-D-glucose, representing glucose triply-linked at the C-1, C-3 and C-6 positions; and (2) the presence of β-1,6-linkages in the form of 2,3,4-tri-*O*-methyl-D-glucose. These authors assumed just one polysaccharide type, and proposed that the degree of substitution of the main chain as well as the length of side chains may vary with the conditions used to prepare the glucan sample. An alternative explanation for these results was offered by Bacon et al. (1969), who observed that extraction of alkali-treated yeast with hot dilute acetic acid and autoclaving in water released both glycogen and a water-soluble β-1,6-glucan. For the first time, yeast glucan appeared heterogeneous in nature and consisted of a mixture of two polysaccharides, with β-1,3-glucan being the major and β-1,6-glucan being a minor component. Although the β-1,6-glucan was not studied in detail, its degree of removal depended on the thoroughness with which the alkali-insoluble glucan was treated with acetic acid and water. Subsequently, these two alkali-insoluble glucan polymers were purified and characterized by Manners et al. (1973a,b; see Section 3).

2.3
Alkali-Insolubility of Glucan is Mediated by Cross-Linking to Chitin

The glucan discussed above is alkali-insoluble, but part of the glucan exists in an alkali-soluble form that can be separated from mannan by either: (1) addition of Fehling's solution to the alkali extract, which leads to the precipitation of mannan as its copper complex (Haworth et al., 1937); or (2) neutralization of the alkali extract, which leads to gel formation of the glucan (Eddy and Woodhead, 1968). Roelofsen (1953) was the first to report solubilization of glucan when boiling baker's yeast in dilute alkali. Kessler and Nickerson (1959) extracted *S. cerevisiae* and *C. albicans* with alkali at room temperature and were able to confirm the above observation, although considerable amounts of glucose were found to be associated with what they refer to as glucomannan–protein complexes. Eddy and Woodhead (1968) purified the alkali-soluble glucan from *S. carlsbergensis*, which represented about 20% of the wall weight. These authors carefully suggested the possibility that this material could be a metabolic precursor of the insoluble glucan. Subsequent structural analysis by Fleet and Manners (1976, 1977) indicated that this material was indeed very similar to the insoluble glucan. However, the biological relationship between the soluble and insoluble glucan remained unclear, as did the basis for their differential solubility. Houwink and Kreger (1953) made all glucan alkali-soluble by prolonged boiling of yeast cell walls in 2% hydrochloric acid, but did not appreciate that this procedure depolymerized cell-wall chitin. The first evidence that a covalent linkage between chitin and β-glucan affects β-glucan solubility was provided by Sietsma and Wessels (1979), who removed chitin from *Schizophyllum commune* walls with either chitinase or nitrous acid and observed that all β-glucan became soluble in water and alkali. Similar results were obtained for *S. cerevisiae* by Mol and Wessels (1987), and the actual linkage between chitin and glucan was demonstrated by Kollár et al. (1995) (see also Section 3). Katohda et al. (1976) separated small daughter cells from larger mother cells by fractional centrifugation and compared the composition of their cell walls. Both had a similar glucose content, but the young cells were found to have eight-fold less insoluble glucan than their older mothers. However, direct evidence for a precursor–product relationship between the two glucan populations was first established in regenerating *C. albicans* spheroplasts by Elorza et al. (1987), and later for *S. cerevisiae* by Hartland et al. (1994). Cells were pulse-labeled with radioactive glucose in both studies, and it was shown that the amount of alkali-soluble glucan decreased during the chase period with a concomitant increase in alkali-insoluble glucan.

3
Structural Studies on Yeast Glucan

The glucan content of *S. cerevisiae* cell walls is approximately 50–60% by weight, with roughly equal amounts being alkali-soluble and alkali-insoluble (Nguyen et al., 1998). A detailed chemical analysis of both glucan populations was performed in the 1970s, yet the interconnections between glucans and the other cell-wall components have only recently been investigated. As we discuss the chemical structure of yeast glucan, the reader should realize that any proposed glucan structure represents an *average* molecule, whose size and composition may vary with different strains of yeast, its conditions of growth and the methods used for the preparation of cell walls and their fractionation. The chemical fractionation scheme used by Manners et al. (1973a,b) for the purification of the two alkali-insoluble glucan polymers, and the enzymatic fractionation scheme used by Kollár et al. (1995, 1997) for the isolation of cross-link containing cell-wall fragments are shown in Figure 1. These structural studies became possible after certain key observations, which were crucial for both the experimental design and correct interpretation of the analytical results. Bacon et al. (1969) proposed that yeast glucan was a mixture of a major β-1,3- and a minor β-1,6-glucan component, and this suggestion formed the basis for the structural glucan

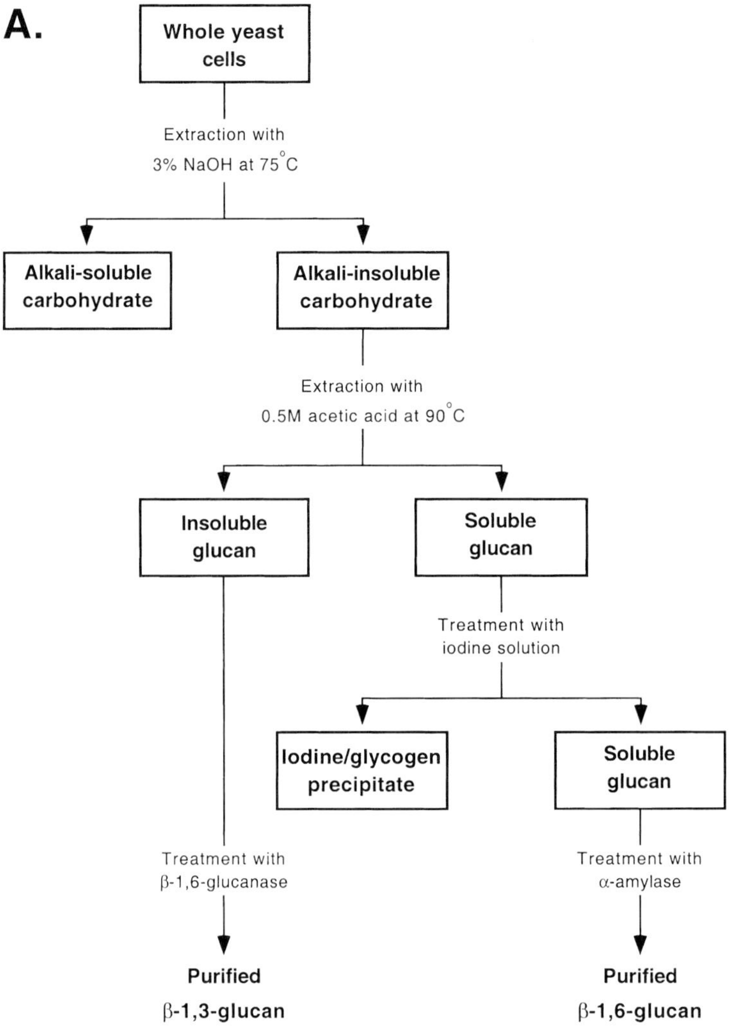

analysis by Manners et al. (1973a, b). Their findings indicated that 85% of the alkali-insoluble glucan was β-1,3-linked, contained 3% β-1,6-glucosidic interchain linkages, and had a degree of polymerization (DP) of 1450 ± 150 equivalent to a molecular weight of 240 kDa. It was unclear from their structural analysis whether single or multiple branching was present and whether this molecule had a laminated, comb or tree-type structure. The authors suggested that the low degree of branching could allow for alignment of linear chain segments and their packing into helices, to give macromolecular structures with a substantial degree of rigidity and insolubility in water. The remaining 15% of the alkali-insoluble glucan was β-1,6-linked, highly branched, contained about 19% β-1,3-glucosidic linkages and had a DP value of 141 ± 10 equivalent to a molecular weight of 22 kDa. Fleet and Manners (1976) used milder conditions for the purification of

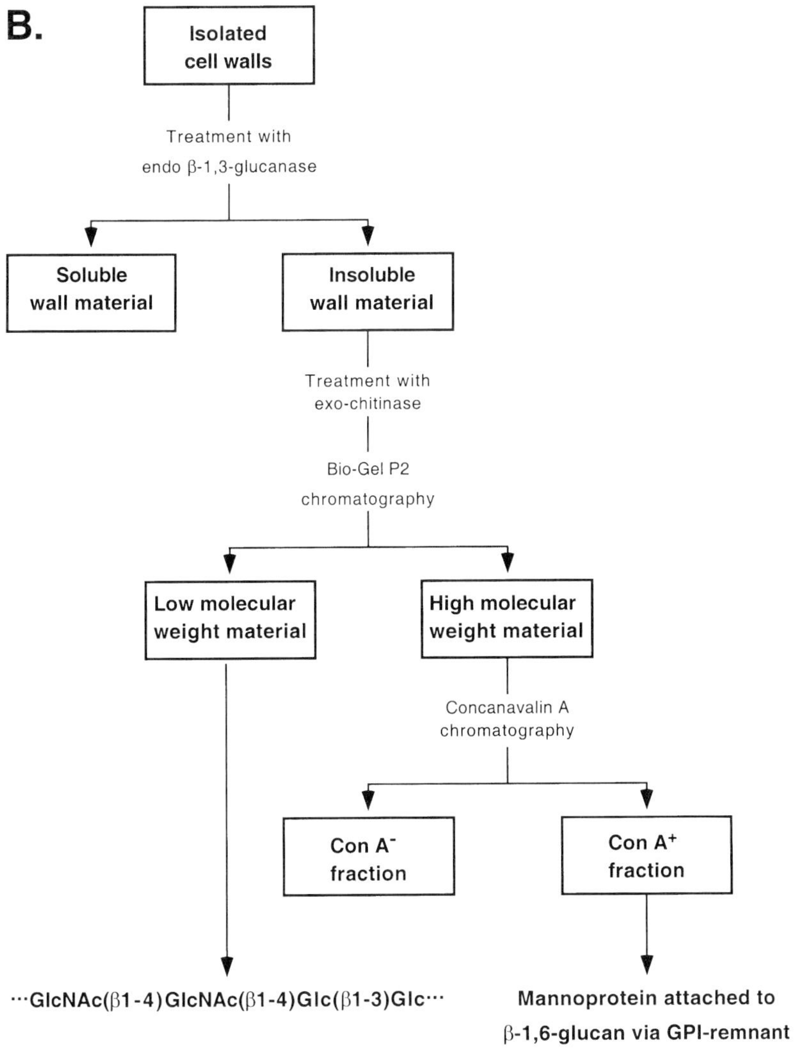

Fig. 1 Fractionation schemes for alkali-insoluble glucan purification and cross-link analysis. (A) The method used by Manners et al. (1973a,b) for the purification of alkali-insoluble glucans: starting with 2 kg of compressed yeast, 2.0 g of β-1,6-glucan and 6.9 g of β-1,3-glucan were obtained. (B) The approach used by Kollár et al. (1995, 1997) for the isolation of cross-link-containing wall fragments: cell walls were isolated from 10 g of yeast, and nanomolar amounts of material were obtained after enzymatic digestion and separation on the Bio-Gel P2 sizing column.

alkali-soluble glucan, since other investigators had reported glucan degradation by alkali at elevated temperatures. They extracted isolated cell walls with dilute sodium hydroxide at 4°C under nitrogen for six days and precipitated the alkali-soluble glucan by neutralization. The resulting material appeared homogeneous and was very similar to the mixed insoluble glucan: it contained 80–85% β-1,3-linkages, 8–12% β-1,6-linkages,

3–4% branched residues, and had a molecular weight of ~250 kDa. However, it contained an additional mannan component that could be released by digestion with an endo-β-1,6-glucanase (Fleet and Manners, 1977). Mol and Wessels (1987) provided strong evidence for a covalent linkage between chitin and β-1,3-glucan in *S. cerevisiae*, and Kollár et al. (1995) investigated this linkage. They used an approach similar to Surarit et al. (1988), who digested *C. albicans* cell walls with both β-1,3-glucanase and chitinase, and then separated the resulting wall fragments by gel filtration. Kollár et al. (1995) radiolabeled the β-1,3-glucanase-resistant material by reduction with sodium borotitride to facilitate detection and purification of wall fragments after the enzymatic hydrolysis of chitin. The chitinase- solubilized material was applied to a Bio-Gel P2 column and tritium-labeled material was detected in both the early void volume fraction as well as later peaks. The structure of the smaller oligosaccharides was further analyzed by NMR spectroscopy, and a β-1,4-linkage between the reducing end of a chitin chain and the nonreducing end of a β-1,3-glucan chain was demonstrated. The high molecular-weight material of the void volume fraction represented 4% of the total cell-wall carbohydrate, and was further fractionated by concanavalin A (Con A) chromatography in a subsequent study (Kollár et al., 1997). Analysis of the Con A$^+$ material indicated that a cell-wall complex had been isolated, in which all four wall components were linked together. The β-1,6-glucan polymer was shown to interconnect mannoproteins with chitin and β-1,3-glucan via its attachment to a remnant of the C-terminal glycosylphosphatidylinositol (GPI) anchor. The GPI remnant consisted of five α-linked mannosyl residues and lacked both the glucosamine residue as well as the inositol phospholipid. Fujii et al. (1999) obtained similar results for the attachment point of the cell wall protein Tip1p, but neither study elucidated the chemical nature of this linkage. The reducing terminus of chitin was found to be directly attached to β-1,3-linked branches of the β-1,6-glucan polymer, most probably via a β-1,4-linkage as was previously observed for the β-1,3-glucan polymer, although a β-1,2-linkage could not be excluded. The reducing end of β-1,6-glucan was connected to the nonreducing terminal glucose residue of β-1,3-glucan through a linkage that remains to be established. The Con A$^-$ material lacked mannoprotein and consisted mainly of β-1,6-glucan, with a few chitin and β-1,3-glucan stubs. This material was quite heterogeneous in size and had an average DP value of 300–350 residues, which was considerably larger than the 140 glucose units reported by Manners et al. (1973b) for β-1,6-glucan isolated by alkali and acetic acid extraction. We have included this structural information in Figure 2, which shows all four cell-wall components linked together in a module that has been proposed by Kapteyn et al. (1996) and others to serve as a "building block" of the yeast cell wall.

4
Enzymology of Glucan Synthesis

Although both β-1,6- and β-1,3-linked glucan polymers can be purified from *in vivo* cell-wall fractions, only the synthesis of β-1,3-glucan has been achieved *in vitro*. The glucan synthase (GS) is a plasma-membrane-localized enzyme complex that transfers glucose from UDP-Glc to the nonreducing end of a growing β-1,3-glucan chain (UDP-glucose: 1,3-β-D-glucan 3-β-D-glucosyltransferase, EC 2.4.1.34). The reducing ends of these glucan chains are labeled when ^{14}C-UDP-Glc is used

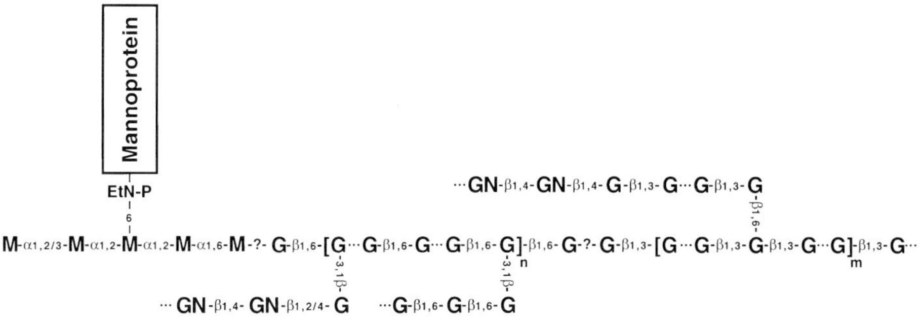

Fig. 2 Schematic partial structure of yeast glucan with cross-links to other cell-wall components. A segment of a β-1,6-glucan chain with two β-1,3-linked branch points attached to respectively, another β-1,6-glucan chain, and the reducing end of a chitin chain is shown between the $[\,]_n$ brackets. A segment of a β-1,3-glucan chain with a β-1,6-linked branch point attached to another β-1,3-glucan chain, whose nonreducing end is connected to the reducing terminus of a chitin chain via a β-1,4-linkage, is shown between the $[\,]_m$ brackets. The linkage between the reducing end of the mannose residue of the GPI remnant and the nonreducing terminal glucose residue of the β-1,6-glucan chain, and the linkage between the reducing end of the β-1,6-glucan chain and the nonreducing terminus of the β-1,3-glucan chain, are both unknown and are indicated with a question mark (?). G: glucose; M: mannose; GN: *N*-acetylglucosamine; EtN-P: phosphoethanolamine; ... denotes more sugar residues with the same linkage.

as a donor, suggesting that their formation occurs *de novo* and does not require an endogenous primer (Shematek et al., 1980). Stoichiometric amounts of UDP are liberated with each transferred glucose and the reaction can therefore be written as:

$$2n\ \text{UDP-}\alpha\text{-D-Glc} \rightarrow 2n\ \text{UDP} + [\text{Glc-}\beta1,3\text{-Glc}]_n$$

The reaction product was analyzed enzymatically and chemically and was found to be a linear β-1,3-glucan of variable chain length: Shematek et al. (1980) used a combination of reduction with sodium borohydride and acid hydrolysis and calculated an average chain length of 60 residues, while Douglas et al. (1994b) established a length of >2000 glucose units by gel filtration. The latter value more closely resembles the size of the *in vivo* polymer and is probably more accurate; the reaction product of Shematek et al. (1980) may have been partially degraded by (contaminating) snail enzymes that were used during enzyme preparation. Assay conditions for GS activity using cell-free extracts as well as permeabilized cells have been described. The preferred method for cell lysis appears to be high-speed shaking in the presence of glass beads, as microsomal fractions obtained by osmotic lysis of spheroplasts generated by enzymatic digestion with lytic enzymes not only produce glucan chains with a much smaller average length, but also tend to lose their *in vitro* GS activity (Szaniszlo et al., 1985). Crude cell lysates display reasonable GS activity, but have lower specific activity than microsomal fractions due to a higher protein content. UDP-Glc has been identified as the sole substrate of the glucan synthase, and this sugar nucleotide does not diffuse through intact membranes. Whole cells can be permeabilized with either inorganic solvents or sorbitol followed by osmotic shock, and these *in situ* GS assay systems are more suitable for large-scale screening of potential anti-fungal compounds inhibiting glucan synthesis (Frost et al., 1994; Crotti et al., 2001). Exponentially growing cells display the highest GS activity and are most often used in assays: GS activity gradually decreases as cells age, and extracts of stationary phase cells incorporate considerable amounts of glucose into glycogen (López-Romero and Ruiz-Herrera, 1977). The latter can also be prevented by either including α-amylase in the reaction mixture or by working in a strain background lacking glycogen synthase activity (Shematek et al., 1980). The GS is most active at pH 7.5 and 37°C, and many activators have been shown to improve the stability and/or activity of this enzyme complex, including triphosphate nucleotides, sodium fluoride, mild detergents, bovine serum albumin, β-lactoglobulin, glycerol, and EDTA. Their proposed mechanisms of action are discussed in detail by Ruiz-Herrera (1992), but except for guanosine triphosphate (GTP) and its analogs, their positive influence on the *in vitro* catalytic activity remains empirical. The GS is inhibited by UDP and is very sensitive to perturbations in the membrane environment caused by lipophilic compounds (Shematek et al., 1980; Ko et al., 1994). This latter property has been exploited in the development of antifungal drugs targeting the glucan synthases of various fungi (see Section 9). Kang and Cabib (1986) dissociated the GS activity by high salt and detergent treatment into two proteinaceous components that were inactive alone, but mixing the two fractions in the presence of GTP or its nonhydrolyzable analog guanosine 5′-[γ-thio]triphosphate (GTP[γS]) reconstituted the enzymatic activity. The membrane fraction was partially rescued from heat inactivation by the substrate UDP-Glc, and was found to contain the catalytic subunit. The active component in the soluble fraction bound GTP[γS] and was specifically protected from both heat- and EDTA-induced inactivation by this nucleotide. Szaniszlo et al. (1985) demonstrated that GTP stimulation was a common property of fungal glucan synthases, and Kang and Cabib (1986) provided further evidence for a highly conserved mechanism of glucan synthesis, as the soluble and membrane fractions from different fungi were interchangeable. The soluble fraction was further fractionated by Mono Q and Sephacryl S-300 chromatography and a 20-kDa GTP-binding protein was identified by photoaffinity labeling with [γ-^{32}P]8-azido-GTP (Mol et al., 1994). This suggested that the GS was regulated by a G-protein belonging to the Ras superfamily; Rho proteins were the most likely candidates, as defects in this Ras subfamily resulted in the lysis of small budded cells. Drgonová et al. (1996) and Qadota et al. (1996) subsequently identified Rho1p as the regulatory subunit of the β-1,3-glucan synthase, and their results are discussed in the next section. The observation that fungal glucan synthases exist as multimeric complexes containing both integral membrane components as well as a peripherally bound GTP binding regulatory subunit provided an explanation for the rapid loss in activity during its solubilization from the membrane, which had prevented glucan synthase purification. Inoue et al. (1995) partially purified the β-1,3-glucan synthase by product entrapment, and observed enrichment of a 200-kDa protein in parallel with an increase in specific activity. These authors cloned the gene encoding the putative catalytic subunit by reverse genetics, and their results are also discussed in the next section.

In the next two sections, we will discuss gene products involved in the biosynthesis of β-glucan polymers, but the reader is referred to Orlean (1997) for a comprehensive review of UDP-glucose synthesis. This nucleotide-activated monosaccharide serves as a direct sugar donor for the synthesis of both polysaccharides.

5 Gene Products Involved in β-1,3-Glucan Biosynthesis

Most of the gene products discussed in this section are shown in Figure 3.

5.1 Fks1p and Fks2p

The *S. cerevisiae FKS1* and *FKS2* genes encode a pair of 88% identical integral membrane proteins with 16 predicted transmembrane domains (Inoue et al., 1995). Together, Fks1p and Fks2p are essential components of the GS, and closely related proteins have been identified in 11 fungi and

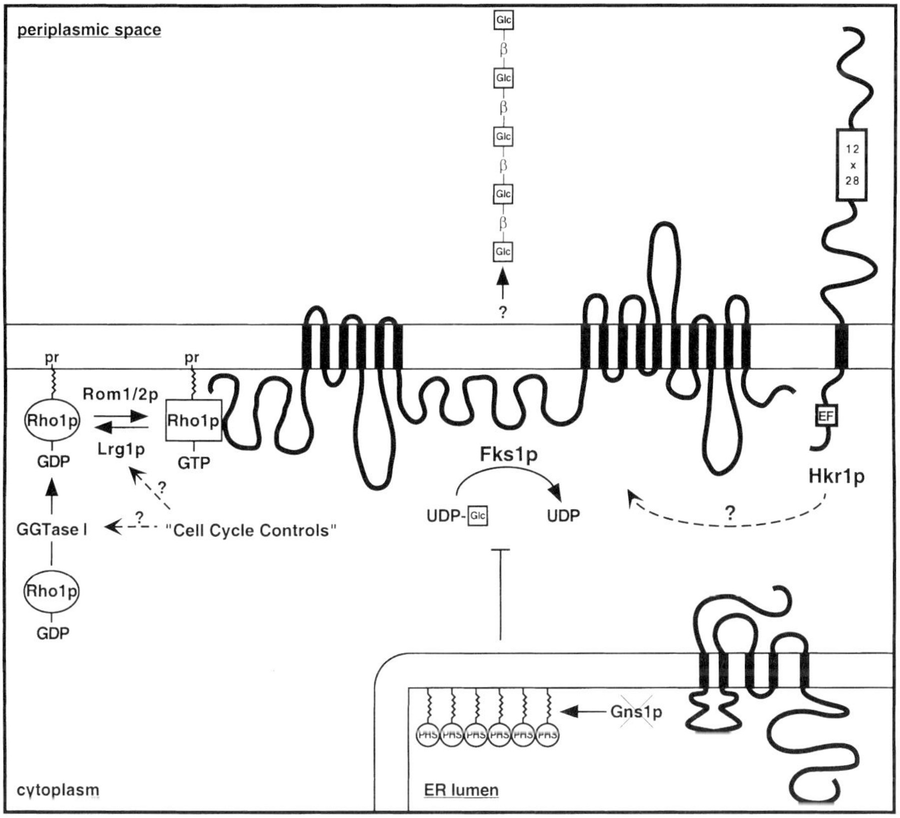

Fig. 3 Schematic representation of gene products involved in β-1,3-glucan biosynthesis. Membrane proteins are shown as solid black lines with black bars as transmembrane domains. The membrane topology of Fks1p and Gns1p was adapted from Douglas et al. (1994b) and El-Sherbeini and Clemas (1995b), respectively. An ↓ depicts a positive effect and a ⊥ depicts an inhibitory effect. Known events are indicated by a solid line, while dashed lines are used for hypothetical interactions. Nucleotide exchange on Rho1p induces a conformational change, which is depicted by an oval Rho1p in the GDP-bound form and a rectangular Rho1p in the GTP-bound form. The crossed-out Gns1p indicates a nonfunctional *gns1* allele, and phytosphingosine (PHS) accumulation is shown on the luminal side of the ER, but may occur on the cytosolic side. Pr: prenyl group. For explanations, see text.

four plants (http://afmb.cnrs-mrs.fr/~pedro/CAZY/gtf 48.html), suggesting a conserved role for these proteins in the synthesis of β-1,3-glucan. Inoue et al. (1995) cloned *FKS1* by reverse genetics, but several *fks1* alleles were identified in genetic screens for mutants that: (1) were resistant to glucan synthase inhibitors (Douglas et al., 1994a; Castro et al., 1995); (2) depended on chitin synthesis for cell integrity (El-Sherbeini and Clemas, 1995a; Ram et al., 1995); or (3) required calcineurin for viability (Parent et al., 1993; Garrett-Engele et al., 1995). *FKS2* was cloned by copurification of Fks2p with Fks1p and by low stringency hybridization of genomic DNA with a *FKS1* probe (Inoue et al., 1995; Mazur et al., 1995). The latter approach was also successfully used for the cloning of most fungal homologs. The yeast genome sequencing project revealed an hypothetical open reading frame (YMR306w) encoding a protein with 55% identity to Fks1p. This protein is referred to as Fks3p, even though it has not been implicated in β-1,3-glucan synthesis and remains of unknown function (Dijkgraaf et al., 2002).

While haploid *fks2Δ* mutants show no obvious phenotype, *fks1Δ* mutants are slow growing and have a 50% reduction in cell-wall β-glucan content that is accompanied by a 40% increase in mannan and a 10-fold increase in chitin (Dallies et al., 1998; Ram et al., 1998). The higher cell-wall mannan content is at least partly achieved by transcriptional up-regulation of the GPI-dependent cell-wall proteins Cwp1p, Pst1p, and Crh1p (Ram et al., 1998; Terashima et al., 2000). The higher chitin content is mediated by an increase in chitin synthase III activity, and is crucial for *fks1Δ* cell integrity, as *fks1Δ chs3Δ* mutants are inviable (Osmond et al., 1999). Chitin deposition is no longer restricted to the mother cell wall and septum in *fks1* mutants, but can now also be seen in the wall of the growing bud (García-Rodriquez et al., 2000). Many cell-wall mutants rely on this compensatory mechanism, which appears to involve Chs6p- independent delivery of Chs3p to the bud tip (Osmond et al., 1999; García-Rodriquez et al., 2000).

Fks1p and Fks2p are believed to be differentially regulated, but partially redundant catalytic subunits of the GS: (1) both Fks1p and Fks2p copurify with β-1,3-GS activity during successive rounds of product entrapment (Inoue et al., 1995); (2) *fks1Δ* mutants have reduced *in vitro* GS activity and reduced cell-wall β-1,3-glucan levels *in vivo* (Douglas et al., 1994b; Ram et al., 1998); and (3) the GS activity of yeast extracts can be immunoprecipitated and immunodepleted with Fksp-specific antibodies (Inoue et al., 1995; Mazur et al., 1995). While these proteins are required for normal GS activity, their precise function remains unknown. Both Fks1p and Fks2p lack clear similarities with other enzymes involved in the formation of glucan polymers, including bacterial curdlan (β-1,3-glucan) synthases (Stasinopoulos et al., 1999). The UDP-glucose binding site R/K-X-G-G found in glycogen synthases of several species (Farkas et al., 1990) is absent in Fks1p and Fks2p, and none of the Fks proteins thus far implicated in glucan synthesis has been demonstrated to bind this sugar nucleotide. Furthermore, even the most purified GS preparations contain other proteins, and no GS inhibitor has been shown to bind Fks1p or Fks2p (Kelly et al., 1996). In fact, a cross-linking study with a photoactivatable echinocandin analog with antifungal activity identified two major proteins of 40 and 18 kDa in *C. albicans* microsomes (Radding et al., 1998). Fks1p and Fks2p are predicted to have two large hydrophobic domains (with multiple transmembrane segments) preceded by large cytoplasmic hydrophilic domains, and this membrane topology is reminiscent of several classes of bacterial and eukaryotic transport proteins, although neither Fks protein has an ATP-binding cassette (Douglas et al., 1994b). The reaction mechanism of β-1,3-glucan synthesis has not been elucidated, and current evidence cannot distinguish between a glucan *synthase* and a glucan *transport* role for these proteins. Abe et al. (2002) performed a mutational analysis of Fks1p and isolated ten temperature-sensitive alleles, which were introduced in a *fks1*Δ *fks2*Δ background so that glucan synthesis was dependent on them. Biochemical and phenotypic analysis of these *fks1*$_{ts}$ *fks2*Δ mutants indicated that the central cytoplasmic loop may contain the catalytic site, as point mutations in this domain reduced GS activity *in vitro* and *in vivo*. These mutants arrested with a tiny bud-like projection at the restrictive temperature that failed to stain with the β-1,3-glucan-specific fluorescent dye, aniline blue. Dijkgraaf et al. (2002) examined the cell-wall polymer levels of these conditional mutants grown at the permissive temperature and observed differential reductions in β-1,3- and β-1,6-glucan levels. Their unexpected observation is difficult to reconcile with the proposed role of Fks1p and Fks2p as catalytic subunits of the GS. One interpretation is that β-1,3-glucan may serve as an acceptor for β-1,6-glucan chains at the cell surface, and the synthesis of the β-1,6-glucan polymer may depend on the availability and/or structure of this acceptor. However, current knowledge suggests that β-1,6-glucan chains are first attached to a remnant of the GPI anchor of cell-wall proteins and these β-1,6-glucosylated mannoproteins are subsequently cross-linked to the β-1,3-glucan network (Kapteyn et al., 1997). Alternatively, Fks1p and Fks2p may directly contribute to both β-1,3- and β-1,6-glucan synthesis as catalytic subunits of a dual-specificity GS or as glucan transporters. Two different glycosyltransferase activities have been demonstrated to exist within one polypeptide for the alternating addition of two different sugars to a growing polysaccharide chain (Jing and DeAngelis, 2000), but there is no precedent for one enzyme with two processive β-glucosyltransferase activities synthesizing two differently linked glucan polymers. In short, the exact role of Fks1p and Fks2p in β-glucan synthesis is still unknown.

Although Fks1p and Fks2p are differentially regulated at the transcriptional level, both proteins have overlapping roles as an *fks1*Δ *fks2*Δ mutant is lethal and cannot be rescued by osmotic stabilization of the medium (Mazur et al., 1995). *FKS1* is preferentially expressed during vegetative growth and its transcription is cell cycle-regulated, with mRNA levels peaking in late G_1/S-phase, consistent with a requirement for β-1,3-glucan synthesis during bud emergence (Ram et al., 1995). Multiple signaling pathways are involved in the transcriptional regulation of *FKS2* during a wide variety of stress conditions (Mazur et al., 1995; Zhao et al., 1998). *FKS2* expression is induced during entry into stationary phase and sporulation – two processes which are mediated by nutritional starvation. The signaling pathway involved in this up-regulation is still unknown, but Fks2p is essential for sporulation as homozygous *fks2*Δ diploids fail to sporulate (Mazur et al., 1995). *FKS2* mRNA also accumulates during growth on poor carbon sources, but this is a response to glucose depletion and is controlled by the protein kinase Snf1p and the transcriptional repressor Mig1p (Zhao et al., 1998). Induction of *FKS2* expression in response to mating pheromone, high extracellular Ca^{2+} or null mutations in *FKS1* requires the Ca^{2+}/calmodulin-dependent protein phosphatase calcineurin (Mazur et al., 1995). A detailed analysis of the *FKS2* promoter revealed a 24-bp calcineurin-dependent response element, recognized by the transcription factor Crz1p (Stathopolous and Cyert, 1997). Translocation of Crz1p from the cytosol to the nucleus requires dephosphorylation by calcineurin, which is necessary for binding of Crz1p to the karyopherin Nmd5p (Stathopoulos-Gerontides et al., 1999; Polizotto and Cyert, 2001). This calcineurin signaling pathway functions in parallel with the *PKC1* cell integrity pathway to induce *FKS2* expression during thermal stress: the early stages of this response are calcineurin-dependent, while persistent expression of *FKS2* at elevated temperature requires the MAPK pathway (Zhao et al., 1998). Induction of *FKS2* is usually accompanied by a decrease in *FKS1* expression, but it is unclear why cells exchange these GS components in response to stress. One explanation for this phenomenon is that Fks1p and Fks2p are functionally distinct. Douglas et al. (1994b) analyzed the *in vitro* reaction product of wild-type and *fks1*Δ microsomes by gel filtration, and found the synthesized β-1,3-glucan to have the same DP (> 2000). Thus, it seems that at least *in vitro*, Fks1p and Fks2p produce an identical product. However, Fks2p is not fully competent to substitute for Fks1p *in vivo*. A low-copy number plasmid containing *FKS2* behind the Fks1 promoter rescues both the slow growth and CFW-hypersensitivity of an *fks1*Δ mutant, but fails to restore the K1 killer toxin sensitivity (G. Dijkgraaf, unpublished observation). Elucidating the reaction mechanism of β-1,3-glucan synthesis should provide further insight into the role of Fks proteins.

5.2
Rho1p

The regulatory subunit Rho1p was identified by a genetic approach. An inability to switch on glucan synthesis during bud emergence would be lethal, but it was reasoned that conditional mutants may give rise to osmotically fragile buds that lyze at the restrictive temperature. The only known guanosine triphosphatases (GTPases) showing these phenotypes were mutant alleles of the *RHO* gene family (Matsui and Toh-e, 1992; Yamochi et al., 1994). Qadota et al. (1996), Drgonová et al. (1996) and Mazur and Baginsky (1996) all provided evidence for Rho1p being the regulatory subunit of the GS complex: (1) membranes of $rho1_{ts}$ mutants displayed strongly reduced GS activity even at permissive temperatures, and showed no stimulation by GTP; (2) GS activity could be restored in a GTP-dependent manner by addition of recombinant Rho1p; (3) addition of recombinant constitutively active (Q68H) Rho1p abolished the GTP requirement for GS activity; (4) Rho1p-specific ADP-ribosylation by *Clostridium botulinum* C3 exoenzyme inhibited GS activity; and (4) Rho1p copurified, coimmunoprecipitated and colocalized with Fks1p. The interaction of active Rho1p with the GS complex is required for polysaccharide biosynthesis, and seems to stabilize Fks1p and Fks2p (Inoue et al., 1999). Rho1p in its GTP-bound form binds and activates several other effectors and thus regulates β-1,3-glucan synthesis at multiple levels. Rho1p mediates the de - and repolarization of Fks1p during cell-wall stress via coordination of the actin cytoskeleton through the yeast protein kinase C homolog Pkc1p, presumably to help repair cell-wall damage (Helliwell et al., 1998; Delley and Hall, 1999). Pkc1p also controls transcription of a number of cell-wall biosynthetic genes, including *FKS1* and *FKS2*, via signaling through the *MPK1* cell-wall integrity pathway (Igual et al., 1996; Zhao et al., 1998). Finally, transport of Fks1p and Fks2p to the cell surface appears to occur along the secretory pathway (Lee et al., 1999). Delivery of these proteins to sites of polarized growth is likely to depend upon the actin cytoskeleton as well as the exocyst complex, a multiprotein complex required for vesicle targeting and docking at plasma membrane regions of active cell-surface expansion. Rho1p is required for the polarized localization of the exocyst complex and directly interacts with Sec3p, a member of this complex that is thought to be a spatial landmark for polarized exocytosis (Guo et al., 2001).

Like other small G-proteins, Rho1p acts as a molecular switch activating effectors in the GTP-bound state, but doing so only transiently as it moves to an inactive GDP-bound state. These changes in Rho1p state are regulated by other proteins: guanine-nucleotide exchange factors (GEFs) promote the exchange of GDP for GTP and function as activators, while GTPase-activating proteins (GAPs) enhance the intrinsic GTPase activity and function as negative regulators by stimulating the transition from a GTP- to a GDP-bound state (for a review, see Cabib et al., 1998). Rho1p is activated by Rom1p and Rom2p, a pair of homologous GEFs with overlapping functions. Like *rho1*Δ cells, *rom1*Δ *rom2*Δ double disruptants arrest with a small bud that is prone to lyze (Ozaki et al., 1996). These GEFs are regulated themselves via direct interaction with the cytoplasmic tails of the putative cell-wall sensors Wsc1p and Mid2p (Philip and Levin, 2001) and indirectly by the yeast phosphatidylinositol kinase homolog Tor2p (Schmidt et al., 1997). The latter molecule only activates the Rho1p–Pkc1p effector complex involved in cell cycle-dependent organization of the actin cytoskeleton (Helliwell et al., 1998) and it remains to be established whether Wsc1p and/or Mid2p are required for activation of the Rho1p–Fks1/2p effector complex. Lrg1p appears to be the only member among eight potential RhoGAPs to negatively regulate GS activity, as no RhoGAP mutant except for *lrg1*Δ restored the impaired β-1,3-glucan synthesis of *fks1-1154 fks2*Δ and *rho1-2* mutants (Watanabe et al., 2001). Bem2p and Sac7p have also been shown to possess GAP activity towards Rho1p (Peterson et al., 1994; Schmidt et al., 1997). However, these GAPs appear to negatively regulate the *PKC1* cell integrity pathway, as both *bem2* and *sac7*Δ mutants display elevated levels of phosphorylated Mpk1p at 30°C, while *lrg1*Δ cells do not (Watanabe et al., 2001). Thus, different RhoGAPs may control the activity of distinct Rho1p–effector complexes. The upstream factors of these GAPs and how GAP activities are controlled remains unknown. Rom2p and Lrg1p localize to sites of polarized growth, consistent with their role in regulating GS activity (Manning et al., 1997; Watanabe et al., 2001). Prenylation by geranylgeranyl transferase type I (GGTase I) is a third level of Rho1p regulation. The GGTase I of *S. cerevisiae* is a heterodimer composed of a catalytic α subunit encoded by *RAM2* and a protein substrate specificity β subunit encoded by *CAL1/CDC43* (Ohya et al., 1996). The latter protein recognizes a C-terminal CAAX motif (C is cysteine, A is aliphatic amino acid, and X is preferably leucine), of which the cysteine residue is the site of geranylgeranylation. Three observations suggest that lipid modification of Rho1p is essential for assembly and activation of the GS: (1) the *cal1-1* mutant is defective in modifying Rho1p and exhibits a dramatic reduction in GS activity; (2) unmodified active (Q68L C206S) Rho1p failed to reconstitute GS activity of *rho1-3* membranes; and (3) this recombinant protein was also unable to interact with Fks1p in a ligand overlay assay (Inoue et al., 1999). So far, Fks1p is the only effector that relies on this modification for interaction with Rho1p: Pkc1p, Bni1p, Skn7p and Sec3p readily interact with unmodified active Rho1p in a yeast two-hybrid assay (Nonaka et al., 1995;

Evangelista et al., 1997; Alberts et al., 1998; Guo et al., 2001). Nevertheless, it remains possible that prenylation contributes to efficient binding and activation of these effectors by Rho1p *in vivo*. Thus, the C-20 geranylgeranyl group not only targets Rho1p to membranes, but also enables this GTPase to interact with Fks1p and to activate the GS.

In summary, Rho1p acts as a master regulator of morphogenesis through at least three essential effectors; Pkc1p, Sec3p and Fks1/2p. β-1,3-Glucan synthesis is just one of several processes required for polarized growth that is under the control of this GTPase. Hyperactive Rho1p is toxic, and its activity is therefore tightly regulated at multiple levels. Although several gene products involved in the regulation of Rho1p and its effectors have been identified, we still lack a global view of how these molecules coordinate their activities during the cell cycle to promote β-glucan synthesis and polarized growth.

5.3 Gns1p

GNS1 (glucan synthesis 1) encodes a 40-kDa endoplasmic reticular protein with five predicted transmembrane domains that is required for fatty acid elongation and sphingolipid biosynthesis (Oh et al., 1997; David et al., 1998). Sphingolipids contain ceramide as a hydrophobic moiety and are highly concentrated in the plasma membrane, where they comprise roughly 40% of the inositol-containing lipid species and contribute to a number of critical membrane functions (for a review, see Dickson and Lester, 1999). The sphingolipid composition of the plasma membrane is an important factor in the susceptibility of *S. cerevisiae* to a wide variety of lipid-like antifungal compounds. *GNS1* as well as other sphingolipid biosynthetic genes have been cloned by complementation of mutants that were resistant to: (1) post-squalene sterol biosynthesis inhibitors such as the morpholine derivative, fenpropimorph (Lorenz and Parks, 1991) and the sterol isomerase inhibitor, SR31747 (Silve et al., 1996); (2) the β-1,3-glucan synthase inhibitor and pneumocandin B_0 analog, L-733,560 (El-Sherbeini and Clemas, 1995b); and (3) the cyclic lipodepsinonapeptide, syringomycin E (Stock et al., 2000). El-Sherbeini and Clemas (1995b) observed that *gns1Δ* mutants have only 10% of the wild-type *in vitro* GS activity with the mutant enzyme defective in its insoluble membrane component, but the physiological function of Gns1p remained obscure as these mutants had no apparent reduction in cell-wall β-glucan. Oh et al. (1997) noted the homology between Gns1p and Elo1p, a protein responsible for the conversion of C_{14} to C_{16} fatty acids, and implicated Gns1p in elongation of fatty acids up to 24 carbons. The endoplasmic reticulum (ER) was one of the proposed sites of sphingolipid biosynthesis, and David et al. (1998) localized Gns1p to this organelle. These results exclude the possibility of Gns1p being a subunit of the GS complex, but provide no explanation for the impaired *in vitro* GS activity of *gns1* mutants. Abe et al. (2001) performed a kinetic analysis of the *in vitro* GS activity of *gns1Δ* membrane fractions and observed that the mutant V_{max} value was one-fifth of a wild-type, with no change in the K_m value. However, the Fks1p and Rho1p levels of *gns1Δ* membranes were indistinguishable from wild-type, and both proteins exhibited a normal localization pattern in *gns1Δ* cells. This suggested that an inhibitor of GS activity accumulated in the membrane fraction of this mutant. Oh et al. (1997) observed accumulation of phytosphingosine (PHS) in *gns1Δ* mutants, and Abe et al. (2001) showed that this intermediate-length fatty acid precursor is a potent noncompetitive inhibitor of yeast GS activity *in vivo*. Although the majority of PHS accumulated in the ER membrane, PHS containing microdomains in the plasma membrane may prevent prenylated Rho1p from interacting with Fks1p or, alternatively, may alter the physico-chemical state of the lipid bilayer, thereby causing inactivation of the GS.

5.4
Knr4p

The role of Smi1p/Knr4p in β-1,3-glucan biosynthesis is less well understood. Fishel et al. (1993) originally cloned the *SMI1* gene by complementation of a mutant displaying temperature-sensitive induction of a matrix-association region (MAR) -inhibited reporter gene (*SMI*: suppression of MAR inhibition). These authors used β-galactosidase as a reporter and screened for conditional mutants that turned blue on X-GAL plates at 37°C. Failure to compare β-galactosidase activity of wild-type and mutant strains grown at the restrictive temperature led to an inability to distinguish between reporter gene induction, and altered permeability for the chromophore. Hong et al. (1994a,b) cloned the identical *KNR4* gene by complementation of a *Hansenula mrakii* K9 killer toxin-resistant mutant (*KNR*: killer nine resistant), and convincingly demonstrated that *knr4*Δ mutants have a highly permeable cell wall, offering an explanation why *SMI1*/*KNR4* was isolated in the color enhancement screen of Fishel et al. (1993). Several additional wall-related phenotypes have been observed for this mutant, including: (1) an inability to grow on YPD + 0.1% SDS; (2) an increased resistance towards Zymolyase digestion; and (3) CFW hypersensitivity (Hong et al., 1994a; Martin et al., 1999). Compared to wild-type, *knr4*Δ mutants have strongly reduced *in vitro* GS activity and incorporate two-fold less 3-[^{3}H]-D-glucose into β-1,3-glucan and four-fold less into β-1,6-glucan in a cell-wall labeling assay, leading to an overall 33% reduction in cell-wall β-glucans (Hong et al., 1994a,b). The *Neurospora crassa* GS-1 protein is related to Knr4p and *gs-1* null mutants are also defective in cell-wall formation and GS activity (Enderlin and Selitrennikoff, 1994). Knr4p was localized to cytoplasmic patches near the presumptive bud site in unbudded cells, and at the incipient bud site during bud emergence (Martin et al., 1999). Overexpression of *KNR4* suppressed several cell-wall mutants and reduced mRNA levels of the chitin synthases *CHS1*, *CHS2*, and *CHS3* (Martin et al., 1999). Since Knr4p can interact with the transcription factors Bas1p (Uetz et al., 2000) and Cin5p/Yap4p (Dagkessamanskaia et al., 2001) in a yeast two-hybrid system, it may have an indirect role in cell-wall biosynthesis by affecting the expression of biosynthetic genes. Intriguingly, preliminary observations by Martin-Yken et al. (2001) suggest that Knr4p physically interacts with Slt2p and is required for its kinase activity. This MAP kinase of the *PKC1* cell integrity pathway participates in many biological processes, and regulates the activity of the transcription factor Rlm1p (Watanabe et al., 1997) and the SBF transcriptional complex (Madden et al., 1997), both of which are implicated in the transcription of cell-wall biosynthetic genes.

5.5
Hkr1p

Kasahara et al. (1994) identified *HKR1* as a gene whose overproduction conferred resistance towards the *H. mrakii* HM-1 killer toxin, a small protein that interferes with β-1,3-glucan synthesis *in vivo*. *HKR1* encodes a serine- and threonine-rich protein of 1802 aa with a transmembrane domain towards its C terminus (aa 1486–1506) and a predicted type I topology (Kasahara et al., 1994). Hkr1p is not detectable without overexpression, and mainly localizes to the cell surface, suggesting that it may have a relatively small cytoplasmic tail of 292 aa and a large periplasmic domain of 1464 aa after removal of the signal sequence (Yabe et al., 1996). An EF hand consensus sequence for calcium-binding is present in the putative cytoplasmic tail, and the presumed periplasmic domain contains twelve 28-amino acid repeats, but the functional relevance of either motif for Hkr1p remains to be established. Disruption of *HKR1* in the RAY3A-D strain background was lethal, and hampered investigation of Hkr1p's involvement in β-1,3-glucan biosynthesis (Kasahara et al., 1994). Yabe et al. (1996) obtained a viable *hkr1* mutant allele by inserting the *LEU2* gene into an internal *Bam*HI site of *HKR1*, and this allele presumably made a truncated protein, Hkr1pΔC, lacking the transmembrane domain and cytoplasmic tail. The *hkr1*ΔC mutant was not affected in growth or HM-1 killer toxin sensitivity, but displayed one-third of wild-type *in vitro* GS activity and a 33% reduction in cell-wall β-glucan content. Overexpression of an N-terminal truncated version of Hkr1p, $HKR1^{tr}$ consisting of the last 665 aa, made cells more resistant to HM-1 killer toxin than a full- length Hkr1p, and doubled the wall β-glucan content (Kasahara et al., 1994). These results implicate Hkr1p in β-1,3-glucan biosynthesis, and demonstrate that its putative cytoplasmic tail with the Ca^{2+}-binding motif is required for wild-type levels of β-1,3-glucan synthesis. The *hkr1*ΔC mutant also had an aberrant budding pattern, and the authors proposed an additional role for Hkr1p in bud site selection, though this phenotype is likely an indirect consequence of the wall defect, as many wall mutants display aberrant budding patterns (Ni and Snyder, 2001). Future studies to elucidate Hkr1p function may benefit from the viability of *hkr1*Δ cells in the strain background of the yeast genome deletion project (Winzeler et al., 1999).

6 Gene Products Involved in β-1,6-Glucan Biosynthesis

Genetic screens for CFW hypersensitivity (Ram et al., 1994) and K1 killer toxin resistance (Brown et al., 1993) have identified most genes presently implicated in β-1,6-glucan biosynthesis. The K1 killer toxin is a pore-forming protein that binds to a β-1,6-glucan-containing cell-wall receptor and subsequently kills sensitive yeast (reviewed by Bussey, 1991). Thus, K1 killer toxin resistance is often associated with β-1,6-glucan defects, while a broad range of cell-wall mutants are CFW-hypersensitive. Deletion of *KRE* (killer toxin resistant) genes causes variable reductions in β-1,6-glucan, and their gene products have been localized along the secretory pathway and at the cell surface, suggesting that the biosynthesis of this polymer requires multiple intracellular and cell-surface events. We will restrict ourselves to a brief overview of the gene products involved in this process, as this subject was recently reviewed in detail by Shahinian and Bussey (2000).

6.1 Endoplasmic Reticulum

Kre5p, Cwh41p, Rot2p, and Cne1p are involved in β-1,6-glucan biosynthesis in *S. cerevisiae*, but related proteins have been implicated in ER "quality control" of *N*-glycoprotein folding in other systems. A key player of this folding apparatus is UGGT (UDP-glucose:glycoprotein glucosyltransferase), which recognizes misfolded proteins and adds a glucose residue on their *N*-glycosyl chains. This enables binding of the chaperone calnexin and facilitates another round of protein folding, after which glucosidase II removes the glucose residue so that correctly folded proteins can leave the ER (reviewed by Parodi, 2000). A *kre5Δ* mutant has no detectable β-1,6-glucan, grows extremely slowly, and is lethal in some backgrounds (Meaden et al., 1990; Shahinian et al., 1998). Although Kre5p has limited but significant similarity with UGGT enzymes (Parker et al., 1995), UGGT activity has not been detected in *S. cerevisiae* (Fernández et al., 1994). In addition, constitutive monoglucosylation of *N*-glycoproteins does not relieve the essential requirement for Kre5p, arguing that at least the essential role of Kre5p is distinct (Shahinian et al., 1998). Cwh41p and Rot2p are yeast glucosidase I and II, which trim the three glucose residues from *N*-chains following their transfer to proteins (Trombetta et al., 1996; Romero et al., 1997). Deletion of either gene decreased β-1,6-glucan levels, but growth and incorporation of the wall protein α-agglutinin were unaffected. The β-1,6-glucan defect was attributed to the persistent presence of *N*-chain glucose residues, with three or two residues causing a 50% reduction and one residue decreasing polymer amounts by 25% (Jiang et al., 1996; Shahinian et al., 1998). The glucose residues may interfere with the attachment of β-1,6-glucan to wall protein *N*-chains at the cell surface, though such a modification remains to be established. *CNE1* encodes the *S. cerevisiae* calnexin homolog (Parlati et al., 1995) and *cne1Δ* cells have a modest β-1,6-glucan defect. However, Cne1p has a significant role in β-1,6-glucan biosynthesis, as *cne1Δ cwh41Δ*, *cne1Δ rot2Δ* and *cne1Δ kre6Δ* double mutants display a more severe β-1,6-glucan reduction than either single mutant (Shahinian et al., 1998). Based on these observations, Shahinian et al. (1998) proposed that cell-wall proteins may receive two β-1,6-glucan modifications at the cell surface: (1) a major one being the attachment of β-1,6-glucan to GPI-remnants, which is required for growth and cell-wall protein incorporation; and (2) a minor one involving *N*-chain attachment, which appears to be less important for viability. In this model, Kre5p would be essential for both biosynthetic pathways, as *kre5Δ* mutants have virtually no cell-wall β-1,6-glucan. Cwh41p and Rot2p would indirectly contribute to proper *N*-chain modification, while Cne1p could be involved in either pathway.

6.2
Golgi

Kre6p and Skn1p are two related type II Golgi membrane proteins. A *kre6Δ* mutant grows more slowly, has a 50% reduction in β-1,6-glucan, and is defective in α-agglutinin incorporation (Roemer and Bussey, 1991; Lu et al., 1995). *SKN1* was identified as a multicopy suppressor of a *kre6Δ* mutant, and disruption of this gene is synthetically lethal with *kre6Δ* in some strain backgrounds, though *skn1Δ* cells lack obvious phenotypes (Roemer et al., 1993). Both genes are differentially regulated at the transcriptional level, with *KRE6* mRNA peaking at the G_1/S boundary and *SKN1* mRNA in the M phase (Igual et al., 1996; Spellman et al., 1998), consistent with these redundant proteins being involved in β 1,6-glucan biosynthesis at different stages of the cell cycle. The Kre6p and Skn1p luminal domains share 86% identity, and have significant similarities to family 16 glycoside hydrolases, suggesting that they may function as glucosyl hydrolases or transglucosylases (Roemer et al., 1994; Montijn et al., 1999). The cytoplasmic tails of Kre6p and Skn1p are far less similar, the N-terminal

137 aa of Kre6p being only 32% identical to the Skn1p counterpart. Replacement of the Kre6p cytoplasmic tail with the one of Skn1p fails to complement the K1 killer toxin resistance of a *kre6*Δ mutant, and deletion of this domain leads to Kre6p mislocalization, indicating that their cytoplasmic tails are required for full function (Li et al., 2002).

6.3 Cytoplasm and Cell Surface

Kre11p was originally identified as a putative cytoplasmic protein involved in β-1,6-glucan biosynthesis, and has recently been purified as a TRAPP II component (Brown et al., 1993; Sacher et al., 2001). TRAPP (transport protein particle) is a conserved protein complex required at multiple stages of membrane traffic: TRAPP I functions in ER to Golgi transport, while TRAPP II is required for traffic within the Golgi and/or exit from the Golgi (Sacher et al., 2001). Thus, it seems likely that Kre11p is involved in trafficking glucan biosynthetic components through the late secretory pathway. *KRE1* encodes a GPI-anchored plasma membrane protein, and two observations are consistent with a late role for Kre1p in β-1,6-glucan assembly: (1) that *kre1*Δ mutants have β-1,6-glucan defects in terms of quantity and polymer size (Boone et al., 1990); and (2) that Kre1p is more abundant at the mother cell surface of budding cells (Roemer and Bussey, 1995). Kre9p and Knh1p are a pair of 46% identical *O*-glycoproteins with unknown cellular localization, as they can only be found in the growth medium when overexpressed (Dijkgraaf et al., 1996). Disruption of *KRE9* results in super slow growth and reduces β-1,6-glucan levels by 80%, though *kre9*Δ mutants grow slightly better on nonglucose-containing medium (Brown and Bussey, 1993). *KNH1* overexpression can suppress these *kre9*Δ phenotypes and *kre9*Δ *knh1*Δ mutants are inviable, consistent with Knh1p making the residual β-1,6-glucan in *kre9*Δ cells. The fact that *KNH1* transcription is up-regulated when cells are grown on galactose-containing media offers an explanation for the carbon source-dependent growth of *kre9*Δ mutants, and supports the notion that these duplicated enzymatic activities may be involved in β-1,6-glucan biosynthesis under different environmental conditions (Dijkgraaf et al., 1996).

Figure 4 shows the (putative) cellular location of these gene products. It is still

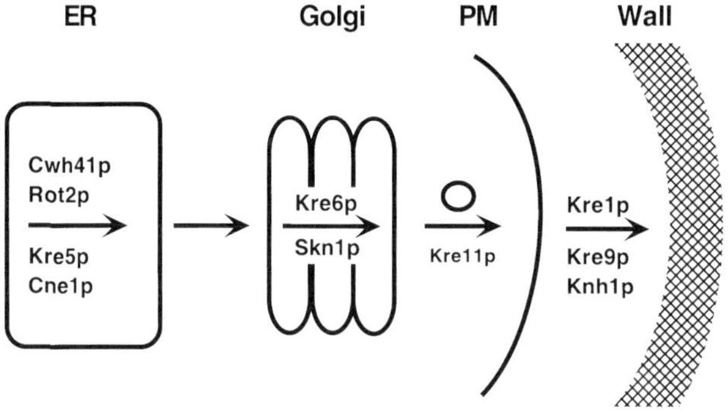

Fig. 4 Cellular location of gene products involved in β-1,6-glucan biosynthesis. The precise location of Kre9p and Knh1p is still unknown. Kre11p is shown in vesicular transport from the Golgi, but may also participate in inter-Golgi traffic. See text for details. PM: plasma membrane.

unclear how they contribute to the synthesis and assembly of β-1,6-glucan, as only Cwh41p and Rot2p have known enzymatic activities. Shahinian and Bussey (2000) proposed two models for β-1,6-glucan biosynthesis, which differ in the role of the intracellular *KRE* gene products, although both support the idea that the majority of β-1,6-glucan is synthesized at the cell surface by an yet-to-be-identified synthase. One model suggests that Kre proteins are involved in generating some kind of a protein-bound acceptor, to which β-1,6-glucan becomes attached at the cell surface. An immunocytochemical study by Montijn et al. (1999) provides indirect evidence for this model, as they failed to detect intracellular β-1,6-glucan even though this polymer was readily demonstrated in the *S. cerevisiae* cell wall. The other model suggests that Kre proteins initiate β-1,6-glucan synthesis, by making a protein-bound β-1,6-glucan precursor that is elongated at the cell surface. A recent *in situ* β-1,6-glucan localization study in *S. pombe* supports this model, as β-1,6-glucan-specific antibodies coupled to gold particles were found to be associated with the Golgi apparatus, small vesicles and cell wall (Humbel et al., 2001). Further investigations are required to establish whether any of these models is correct.

7 Glucan Remodeling and Cross-Linking

Individual β-1,3- and β-1,6-glucan chains are organized at the cell surface by cross-linking events, a poorly understood process that gives rise to a mature cell wall. Mouyna et al. (2000) proposed a β-1,3-glucan maturation model: linear β-1,3-glucan chains are branched through β-1,6-linkages, followed by β-1,3-glucan side chain elongation and cross-linking to the other cell-wall components. The β-1,6-glucan polymer is thought to be attached to the GPI-remnant of wall proteins, and these β-1,6-glucosylated modules are next covalently incorporated into the chitin/β-1,3-glucan network (Kollár et al., 1997; Ram et al., 1998). Glucan remodeling is required during growth, budding, mating, sporulation, and fermentation and involves glucanases and glucanosyltransferases localized at the cell surface (Cappellaro et al., 1998; Rodríguez-Peña et al., 2000).

7.1 Glucanases

Sixteen genes comprise the Exg1p, Bgl2p, Scw3p, and Crh1p protein families (see Cappellaro et al., 1998 for a family tree), encoding mostly predicted glucan hydrolases, and with Exg1p and Bgl2p having been shown to possess glucan 1,3-β-glucosidase activity (EC 3.2.1.58). The Exg1p protein family includes Exg1p, Exg2p, Spr1p, and Ybr056p. Exg1p is a soluble exo-β-glucanase active on both β-1,3- and β-1,6-glucan, as *EXG1* overproduction confers K1 killer toxin resistance and reduces cell-wall β-1,6-glucan levels (Jiang et al., 1995; Larriba et al., 1995). Homozygous *spr1*Δ diploids sporulate more slowly and lack exo-β-1,3-glucanase activity in a whole-cell assay (Muthukumar et al., 1993). A sporulation-specific role for Spr1p is further supported by a 25-fold increase in *SPR1* expression during late sporulation (Chu et al., 1998). *EXG2* encodes a putative GPI anchored protein, suggesting that Exg2p is the only plasma membrane-associated member of this family (Caro et al., 1997). Bgl2p is the best characterized member of the Bgl2p-Scw4p-Scw10p-Scw11p family of soluble glucanases (Cappellaro et al., 1998). This molecule binds to β-glucans and chitin, has endo-β-1,3-glucanase activity, and can cross-link β-1,3-glucan chains via a β-1,6-linkage *in vitro* (Klebl and

Tanner, 1989; Goldman et al., 1995). Scw3p is a soluble cell-wall protein required for timely separation of mother and daughter cell after the completion of the cell cycle (Mouassite et al., 2000). The other Scw3p family members Sim1p, Uth1p, Nca3p, and Ymr244p have been implicated in multiple cellular processes, including DNA replication, aging and mitochondrial biogenesis (Mouassite et al., 2000). It is unclear whether their null mutant phenotypes are an indirect effect of impaired cell wall assembly, as none of these proteins has been localized. The nonessential Crh1p family consists of the GPI-anchored Crh1p and Crh2p, and the less well-characterized membrane protein Crr1p (Caro et al., 1997; Rodríguez-Peña et al., 2000). A *crh1Δ crh2Δ* mutant has increased amounts of alkali-soluble glucan compared to wild-type cells, consistent with a role for Crh1p and Crh2p in glucan cross-linking. The members of this family show distinct patterns of transcriptional regulation: *CRH1* expression peaks at the G_1/S and M/G_1 boundaries, *CRR1* at the G_2/M boundary, and both genes are also induced during sporulation, while *CRH2* transcripts remain constant throughout the cell cycle (Rodríguez-Peña et al., 2000). Accordingly, a Crh1p-GFP fusion protein localized to the incipient bud site, around the septum in later stages of budding and in ascospore envelopes, while a Crh2p-GFP fusion protein localized mainly at the bud neck throughout the budding cycle (Rodríguez-Peña et al., 2000).

7.2 Glucanosyltransferases

GAS1 encodes a 1,3-β-glucanosyltransferase attached to the plasma membrane via a GPI-anchor (Mouyna et al., 2000). Four additional *GAS* homologs are encoded in the *S. cerevisiae* genome, but only Gas1p has been implicated in cell-wall assembly (reviewed by Popolo and Vai, 1999). A *gas1Δ* mutant has a more spherical morphology, and releases β-1,3-glucan and β-1,6-glucosylated wall proteins into the medium (Ram et al., 1998). This mutant is able to maintain partial cell-wall integrity by increasing chitin deposition and by alternative cross-linking of β-1,6-glucosylated wall proteins to chitin (Kapteyn et al., 1997; Valdivieso et al., 2000). Although Mouyna et al. (2000) recently demonstrated that Gas1p has 1,3-β-glucanosyltransferase activity *in vitro*, we still do not know how this molecule contributes to cell-wall construction *in vivo*.

8 Applications of Yeast β-Glucans

There is considerable commercial interest in yeast β-glucans (reviewed by Sutherland, 1998). The pharmaceutical industry is exploring their therapeutic potential as a biomedical agent to prevent infections and cancer (Williams et al., 1990; Jamas et al., 1996). These polymers can stimulate the antimicrobial and antitumor activity of the immune system in both particulate and soluble form by binding to receptors on macrophages as well as other white blood cells and activating them. Complement receptor 3 and Dectin-1 have been identified as β-glucan receptors, but the molecular mechanisms that underlie these immune responses are still not well understood (Ross et al., 1999; Willment et al., 2001). The cosmetics industry may exploit this immunomodulating activity to revitalize skin by including β-glucan in cosmetic products (Donzis, 1993). Other potential medical applications include: (1) a cholesterol-lowering effect (Robbins and Seeley, 1977); (2) use as a temporary skin substitute to promote wound healing (Williams et al.,

1990); (3) use as therapeutic contact lenses for treatment of eye infections after ophthalmic surgery (Erwin, 1990); and (4) use as a drug delivery system (Jamas et al., 1991). The food industry is mainly interested in yeast β-glucans to improve the texture of foods by using them as thickening agents, fat substitutes and/or as sources of dietary fiber (Nguyen et al., 1998). Yeast β-glucans will have to compete with the bacterial β-1,3-glucan, curdlan, for most of these food applications. The latter polymer has a relative low molecular weight and is produced as an exopolysaccharide by *Agrobacterium* and *Rhizobium* species (Sutherland, 1998).

9
Glucan Synthase Inhibitors

β-Glucans are absent from humans, so their biosynthesis has long been seen as a promising target area for therapeutic antifungal agents. Papulacandin B was the first of the "candins" shown to affect β-glucan synthesis in *S. cerevisiae* and *C. albicans* (Baguley et al., 1979). Subsequently, the glycolipid papulacandins and the lipopeptide echinocandins and pneumocandins were found to inhibit β-1,3-glucan synthase activity *in vitro* (for a review, see Georgopapadakou and Tkacz, 1995). More recently, Onishi et al. (2000) screened for compounds inhibiting *in vitro* β-1,3-glucan synthase activity and found a new class of agents, the acidic terpenoids. Application of semisynthetic medicinal chemistry to the candins has led to the development of several compounds with therapeutic potential (Georgopapadakou and Tkacz, 1995). The pneumocandin B analog caspofungin acetate (Merck MK-0991, Cancidas™) is one such compound which has recently been approved for clinical use in the USA for treatment of invasive aspergillosis in patients who are refractory or intolerant to other therapies (http://www.merck.com/product/usa/cancidas/hcp/home.html). The precise mode of action of these inhibitory compounds on the β-1,3-glucan synthase remains to be elucidated, but our growing knowledge about the biosynthetic components of fungal β-glucans promises to assist mechanism of action studies, and to allow screens for new β-glucan inhibitors.

10
Outlook and Perspectives

Knowledge of *S. cerevisiae* β-glucans has grown over the past three decades. Chemical approaches have demonstrated the existence of two differently linked β-glucans, and their molecular structures as well as several cross-links with other cell-wall macromolecules have been elucidated. In addition, an *in vitro* synthesis assay has been developed for one β-glucan type, and molecular genetics has discovered many genes involved in β-glucan biosynthesis. It is also apparent that the synthesis of these wall polymers is under the control of various intracellular signaling pathways to coordinate wall biogenesis with the cell cycle and to ensure cell integrity during cell-wall stress. However, a detailed understanding of β-glucan synthesis is lacking: the biochemical activity of most of the implicated gene products remains to be established, and many more genes are likely involved in these processes.

The application of functional genomics to *S. cerevisiae* research should assist in the identification of new genes involved in β-glucan synthesis. A yeast disruption set (Winzeler et al., 1999) is available for nearly all individual genes and should enable researchers to assign roles to previously uncharacterized genes via genomic-scale cell-wall phenotypic analyses. For instance,

Tab. 1 List of yeast β-glucan patents*

US Patent no.	Patent holder	Inventors	Date of publication	Title of patent
5,504,079	Alpha-Beta Technology, Inc.	Jamas, S., Easson, D. D., Jr., Ostroff, G. R.	1996	Method for immune system activation by administration of a β(1,3)glucan, which is produced by *Saccharomyces cerevisiae* strain R4
5,223,491	Donzis, B. A.	Donzis, B. A.	1993	Method for revitalizing skin by applying topically water insoluble glucan
5,032,401	Alpha-Beta Technology, Inc.	Jamas, S., Ostroff, G. R., Easson, D. D., Jr.	1991	Glucan drug delivery system and adjuvant
4,946,450	Biosource Genetics Corporation	Erwin, R. L.	1990	Glucan/collagen therapeutic eye shields
4,975,421.	Bioglucan, LP.	Williams, D. L., Browder, I. W., Diluzio, deceased, Nicholas, R., Diluzio, N. M. representative of Estate	1990	Soluble phosphorylated glucan: methods and composition for wound healing

* Full text and major claims can be accessed at http://www.uspto.gov/main/patents.htm.

our laboratory has systematically screened this null mutant collection and identified mutants in over 100 genes with altered K1 killer toxin sensitivity (N. Pagé, personal communication). Dallies et al. (1998) have developed a simple method for quantitative determination of monosaccharides released from isolated cell walls by acid hydrolysis. Their procedure is conducive to large-scale cell-wall analyses and can be used as a primary screen to measure the cell-wall glucose content of potential β-glucan mutants, though alterations should be confirmed by a secondary screen to distinguish between the two types of β-glucan. As both β-glucan polymers are indispensable for yeast viability, certain key genes involved in β-glucan synthesis may be essential and could have been missed in traditional screens using haploid strains. The gene disruption set has revealed all essential yeast genes, and their involvement in β-glucan biosynthesis can be tested by using any of the following three strategies to circumvent the lethality problem: (1) osmotic stabilization of the medium; (2) analysis of heterozygous diploid strains for haploinsufficient phenotypes (Giaever et al., 1999); and (3) preparation of conditional alleles. Changes in gene expression associated with a cell-wall perturbation caused by a particular mutation or drug can be monitored with transcriptional profiling. Global gene expression studies have already implicated hundreds of previously uncharacterized genes in diverse cellular processes, including cell-wall biosynthesis (Hughes et al., 2000). This technique may help identify new genes involved in β-glucan biosynthesis by matching the expression profile of an uncharacterized mutant with those of known β-glucan mutants. We also anticipate that functional genomics will help uncover many interactions among these genes. The yeast disruption set and robotic technologies will make systematic synthetic lethal screens and multicopy suppression analysis possible. Moreover, several large-scale protein–protein interaction studies have already been reported (Uetz et al., 2000; Ito et al., 2001), and novel applications for the null mutant collection will undoubtedly emerge.

A major challenge is to establish the biochemical function of all gene products implicated in β-glucan synthesis. Thus far, only a few enzymes involved in late steps of β-glucan assembly have been assigned catalytic activities, as glucan degradation and cross-linking are relatively straight forward to demonstrate *in vitro*. However, the early intracellular events required for proper biosynthesis of β-1,6-glucan remain largely unknown. Biosynthetic intermediates do not appear to accumulate in *kre*Δ backgrounds, and the lack of an *in vitro* assay has further complicated functional analysis of the *KRE* gene products. An *in vitro* assay opens the door to biochemistry, an indispensable discipline for the study of fungal β-glucan synthesis, allowing insights into the precise role of Fks proteins and the mechanism of the glucan synthesis reaction.

Much work remains to be done in order to obtain a global view of β-glucan synthesis in *S. cerevisiae*, but many of the tools are in place to permit progress. It will be interesting to see whether this knowledge is informative for plant polysaccharide synthesis.

Acknowledgments

H. Li is a Fonds de la Recherche en Santé du Québec postdoctoral fellow, and β-glucan research from our laboratory has been supported by the Natural Sciences and Engineering Research Council of Canada.

11
References

Abe, M., Nishida, I., Minemura, M., Qadota, H., Seyama, Y., Watanabe, T., Ohya, Y. (2001) Yeast 1,3-β-glucan synthase activity is inhibited by phytosphingosine localized to the endoplasmic reticulum, *J. Biol. Chem.* **276**, 26923–26930.

Abe, M., Minemura, M., Utsugi, T., Sekiya-Kawasaki, M., Hirata, A., Qadota, H., Morishita, K., Watanabe, T., Ohya, Y. (2002) Multiple functional domains of yeast 1,3-β-glucan synthase revealed by mutational analysis (in preparation).

Agar, H. D., Douglas, H. C. (1955) Studies of budding and cell wall structure of yeast, *J. Bacteriol.* **70**, 427–434.

Alberts, A. S., Bouquin, N., Johnston, L. H., Treisman, R. (1998) Analysis of RhoA-binding proteins reveals an interaction domain conserved in heterotrimeric G protein beta subunits and the yeast response regulator protein Skn7, *J. Biol. Chem.* **273**, 8616–8622.

Anderson, F. B., Millbank, J. W. (1966) Protoplast formation and yeast cell-wall structure. The action of the enzymes of the snail, *Helix pomatia*, *Biochem. J.* **99**, 682–687.

Bacon, J. S. D., Farmer, V. C., Jones, D., Taylor, I. F. (1969) The glucan components of the cell wall of baker's yeast (*Saccharomyces cerevisiae*) considered in relation to its ultrastructure, *Biochem. J.* **114**, 557–567.

Baguley, B. C., Rommele, G., Gruner, J., Wehrli, W. (1979) Papulacandin B: an inhibitor of glucan synthesis in yeast spheroplasts, *Eur. J. Biochem.* **97**, 245–351.

Barry, V. C., Dillon, T. (1943) On the glucan of the yeast membrane, *Proc. R. Ir. Acad., Sect. B* **49**, 177–185.

Bell, D. J., Northcote, D. H. (1950) The structure of a cell-wall polysaccharide of baker's yeast, *J. Chem. Soc.* 1944–1947.

Boone, C., Sommer, S. S., Hensel, A., Bussey, H. (1990) Yeast *KRE* genes provide evidence for a pathway of cell wall β-glucan assembly, *J. Cell Biol.* **110**, 1833–1843.

Brown, J. L., Bussey, H. (1993) The yeast *KRE9* gene encodes an O glycoprotein involved in cell surface β-glucan assembly, *Mol. Cell. Biol.* **13**, 6346–6356.

Brown, J. L., Kossaczka, Z., Jiang, B., Bussey, H. (1993) A mutational analysis of killer toxin resistance in *Saccharomyces cerevisiae* identifies new genes involved in cell wall (1 → 6)-β-glucan synthesis, *Genetics* **133**, 837–849.

Bussey, H. (1991) K1 killer toxin, a pore-forming protein from yeast, *Mol. Microbiol.* **5**, 2339–2343.

Cabib, E., Roberts, R., Bowers, B. (1982) Synthesis of the yeast cell wall and its regulation, *Annu. Rev. Biochem.* **51**, 763–793.

Cabib, E., Drgonová, J., Drgon, T. (1998) Role of small G proteins in yeast cell polarization and wall biosynthesis, *Annu. Rev. Biochem.* **67**, 307–333.

Cappellaro, C., Mrsa, V., Tanner, W. (1998) New potential cell wall glucanases of *Saccharomyces cerevisiae* and their involvement in mating, *J. Bacteriol.* **180**, 5030–5037.

Caro, L. H. P., Tettelin, H., Vossen, J. H., Ram, A. F. J., van den Ende, H., Klis, F. M. (1997) *In silicio* identification of glycosyl-phosphatidylinositol-anchored plasma-membrane and cell wall proteins of *Saccharomyces cerevisiae*, *Yeast* **13**, 1477–1489.

Castro, C., Ribas, J. C., Valdivieso, M. H., Varona, R., del Rey, F., Duran, A. (1995) Papulacandin B resistance in budding and fission yeasts: isolation and characterization of a gene involved in (1,3)β-D-glucan synthesis in *Saccharomyces cerevisiae*, *J. Bacteriol.* **177**, 5732–5739.

Chu, S., DeRisi, J., Eisen, M., Mulholland, J., Botstein, D., Brown, P. O., Herskowitz, I. (1998)

The transcriptional program of sporulation in budding yeast, *Science* **282**, 699–705.

Cid, V. J., Duran, A., del Rey, F., Snyder, M. P., Nombela, C., Sanchez, M. (1995) Molecular basis of cell integrity and morphogenesis in *Saccharomyces cerevisiae*, *Microbiol. Rev.* **59**, 345–386.

Crotti, L. B., Drgon, T., Cabib, E. (2001) Yeast cell permeabilization by osmotic shock allows determination of enzymatic activities *in situ*, *Anal. Biochem.* **292**, 8–16.

Dagkessamanskaia, A., Martin-Yken, H., Basmaji, F., Briza, P., Francois, J. (2001) Interaction of Knr4 protein, a protein involved in cell wall synthesis, with tyrosine tRNA synthetase encoded by *TYS1* in *Saccharomyces cerevisiae*, *FEMS Microbiol. Lett.* **200**, 53–58.

Dallies, N., François, J., Paquet, V. (1998) A new method for quantitative determination of polysaccharides in the yeast cell wall. Application to the cell wall defective mutants of *Saccharomyces cerevisiae*, *Yeast* **14**, 1297–1306.

David, D., Sundarababu, S., Gerst, J. E. (1998) Involvement of long chain fatty acid elongation in the trafficking of secretory vesicles in yeast, *J. Cell Biol.* **143**, 1167–1182.

Delley, P.-A., Hall, M. N. (1999) Cell wall stress depolarizes cell growth via hyperactivation of RHO1, *J. Cell Biol.* **147**, 163–174.

Dickson, R. C., Lester, R. L. (1999) Metabolism and selected functions of sphingolipids in the yeast *Saccharomyces cerevisiae*, *Biochim. Biophys. Acta* **1438**, 305–321.

Dijkgraaf, G. J. P., Brown, J. L., Bussey, H. (1996) The *KNH1* gene of *Saccharomyces cerevisiae* is a functional homolog of *KRE9*, *Yeast* **12**, 683–692.

Dijkgraaf, G. J. P., Abe, M., Ohya, Y., Bussey, H. (2002) Mutations in Fks1p affect the cell wall content of β-1,3- and β-1,6-glucan in *Saccharomyces cerevisiae*, yeast in press.

Donzis, B. A. (1993) Method for revitalizing skin by applying topically water insoluble glucan, US Patent 5, 223, 491.

Douglas, C. M., Marrinan, J. A., Li, W., Kurtz, M. B. (1994a) A *Saccharomyces cerevisiae* mutant with echinocandin-resistant 1,3 β D glucan synthase, *J. Bacteriol.* **176**, 5686–5696.

Douglas, C. M., Foor, F., Marrinan, J. A., Morin, N., Nielsen, J. B., Dahl, A. M., Mazur, P., Baginsky, W., Li, W., El-Sherbeini, M., Clemas, J. A., Mandala, S. M., Frommer, B. R., Kurtz, M. B. (1994b) The *Saccharomyces cerevisiae FKS1* (*ETG1*) gene encodes an integral membrane protein which is a subunit of 1,3-β-D-glucan synthase, *Proc. Natl. Acad. Sci. USA* **91**, 12907–12911.

Drgonová, J., Drgon, T., Tanaka, K., Kollár, R., Chen, G.-C., Ford, R. A., Chan, C. S. M., Takai, Y., Cabib, E. (1996) Rho1p, a yeast protein at the interface between cell polarization and morphogenesis, *Science* **272**, 277–279.

Duell, E. A., Inoue, S., Utter, M. F. (1964) Isolation and properties of intact mitochondria from spheroplasts of yeast, *J. Bacteriol.* **88**, 1762–1773.

Duffus, J. H., Levi, C., Manners, D. J. (1982) Yeast cell-wall glucans, *Adv. Microb. Physiol.* **23**, 151–181.

Eddy, A. A., Williamson, D. H. (1957) A method of isolating protoplasts from yeast, *Nature* **179**, 1252–1253.

Eddy, A. A., Williamson, D. H. (1959) Formation of aberrant cell walls and of spores by the growing yeast protoplast, *Nature* **183**, 1101–1104.

Eddy, A. A., Woodhead, J. S. (1968) An alkali-soluble glucan fraction from the cell walls of the yeast *Saccharomyces carlsbergensis*, *FEBS Lett.* **1**, 67–68.

Elorza, M. V., Murgui, A., Rico, H., Miragall, F., Sentandreu, R. (1987) Formation of a new cell wall by protoplasts of *Candida albicans*: effect of papulacandin B, tunicamycin and nikkomycin, *J. Gen. Microbiol.* **133**, 2315–2325.

El-Sherbeini, M., Clemas, J. A. (1995a) Nikkomycin Z supersensitivity of an echinocandin-resistant mutant of *Saccharomyces cerevisiae*, *Antimicrob. Agents Chemother.* **39**, 200–207.

El-Sherbeini, M., Clemas, J. A. (1995b) Cloning and characterization of *GNS1*: a *Saccharomyces cerevisiae* gene involved in synthesis of 1,3-β-glucan in vitro, *J. Bacteriol.* **177**, 3227–3234.

Enderlin, C. S., Selitrennikoff, C. P. (1994) Cloning and characterization of a *Neurospora crassa* gene required for (1,3)β-glucan synthase activity and cell wall formation, *Proc. Natl. Acad. Sci. USA* **91**, 9500–9504.

Erwin, R. L. (1990) Glucan/collagen therapeutic eye shields, US Patent 4, 946, 450.

Evangelista, M., Blundell, K., Longtine, M. S., Chow, C. J., Adames, N., Pringle, J. R., Peter, M., Boone, C. (1997) Bni1p, a yeast formin linking Cdc42p and the actin cytoskeleton during polarized morphogenesis, *Science* **272**, 118–122.

Farkas, I., Hardy, T. A., Depaoli-Roach, A. A., Roach, P. (1990) Isolation of the *GSY1* gene encoding yeast glycogen synthase and evidence for the existence of a second gene, *J. Biol. Chem.* **265**, 20879–20886.

Fernández, F. S., Trombetta, S. E., Hellman, U., Parodi, A. J. (1994) Purification to homogeneity of UDP-glucose:glycoprotein glucosyltransferase from *Schizosaccharomyces pombe* and apparent absence of the enzyme from *Saccharomyces cerevisiae, J. Biol. Chem.* **269**, 30701–30706.

Fishel, B., Sperry, A. O., Garrard, W. T. (1993) Yeast calmodulin and a conserved nuclear protein participate in the *in vivo* binding of a matrix association region, *Proc. Natl. Acad. Sci. USA* **90**, 5623–5627.

Fleet, G. H., Manners, D. J. (1976) Isolation and composition of an alkali-soluble glucan from the cell walls of *Saccharomyces cerevisiae, J. Gen. Microbiol.* **94**, 180–192.

Fleet, G. H., Manners, D. J. (1977) The enzymic degradation of an alkali-soluble glucan from the cell walls of *Saccharomyces cerevisiae, J. Gen. Microbiol.* **98**, 315–327.

Frost, D. J., Brandt, K., Capobianco, J., Goldman, R. (1994) Characterization of (1,3)-β-glucan synthase in *Candida albicans*: microsomal assay from the yeast or mycelial morphological forms and a permeabilized whole-cell assay, *Microbiology* **140**, 2239–2246.

Fujii, T., Shimoi, H., Iimura, Y. (1999) Structure of the glucan-binding sugar chain of Tip1p, a cell wall protein of *Saccharomyces cerevisiae, Biochim. Biophys. Acta* **1427**, 133–144.

García-Rodriguez, L. J., Trilla, J. A., Castro, C., Valdivieso, M. H., Durán, A., Roncero, C. (2000) Characterization of the chitin biosynthesis process as a compensatory mechanism in the *fks1* mutant of *Saccharomyces cerevisiae, FEBS Lett.* **478**, 84–88.

Garrett-Engele, P., Moilanen, B., Cyert, M. S. (1995) Calcineurin, the Ca^{2+}/calmodulin-dependent protein phosphatase, is essential in yeast mutants with cell integrity defects and in mutants that lack a functional vacuolar H^+-ATPase, *Mol. Cell. Biol.* **15**, 4103–4114.

Georgopapadakou, N. H., Tkacz, J. S. (1995) The fungal cell wall as a drug target, *Trends Microbiol.* **3**, 98–104.

Giaever, G., Shoemaker, D. D., Jones, T. W., Liang, H., Winzeler, E. A., Astromoff, A., Davis, R. (1999) Genomic profiling of drug sensitivities via induced haploinsufficiency, *Nature Genet.* **21**, 278–283.

Giaja, J. (1914) Sur l'action de quelques ferments sur les hydrates de carbone de la levure, *C. R. Soc. Biol.* **77**, 2–4.

Goldman, R. C., Sullivan, P. A., Zakula, D., Capobianco, J. O. (1995) Kinetics of β-1,3 glucan interaction at the donor and acceptor sites of the fungal glucosyltransferase encoded by the *BGL2* gene, *Eur. J. Biochem.* **227**, 372–378.

Guo, W. Tamanoi, F., Novick, P. (2001) Spatial regulation of the exocyst complex by Rho1 GTPase, *Nature Cell Biol.* **3**, 353–360.

Hartland, R. P., Vermeulen, C. A., Klis, F. M., Sietsma, J. H., Wessels, J. G. H. (1994) The linkage of (1-3)-β-glucan to chitin during cell wall assembly in *Saccharomyces cerevisiae, Yeast* **10**, 1591–1599.

Hassid, W. Z., Joslyn, M. A., McCready, R. M. (1941) The molecular constitution of an insoluble polysaccharide from yeast, *Saccharomyces cerevisiae, J. Am. Chem. Soc.* **63**, 295–298.

Haworth, W. N., Hirst, E. L., Isherwood, F. A. (1937) Yeast mannan, *J. Chem. Soc.* 784–791.

Helliwell, S. B., Schmidt, A., Ohya, Y., Hall, M. N. (1998) The Rho1 effector Pkc1, but not Bni1, mediates signalling from Tor2 to the actin cytoskeleton, *Curr. Biol.* **8**, 1211–1214.

Holden, M., Tracey, M. V. (1950) A study of enzymes that can break down tobacco-leaf components, *Biochem. J.* **47**, 407–414.

Hong, Z., Mann, P., Brown, N. H., Tran, L. E., Shaw, K. J., Hare R. S., DiDomenico, B. (1994a) Cloning and characterization of *KNR4*, a yeast gene involved in (1,3)-β-glucan synthesis, *Mol. Cell. Biol.* **14**, 1017–1025.

Hong, Z., Mann, P., Shaw, K. J., DiDomenico, B. (1994b) Analysis of β-glucans and chitin in a *Saccharomyces cerevisiae* cell wall mutant using high-performance liquid chromatography, *Yeast* **10**, 1083–1092.

Horisberger, M., Vonlanthen, M. (1977) Localization of mannan and chitin on thin sections of budding yeasts with gold markers, *Arch. Microbiol.* **115**, 1–7.

Houwink, A. L., Kreger, D. R. (1953). Observations on the cell wall of yeast, *Antonie van Leeuwenhoek* **19**, 1–24.

Hughes, T. R., Marton, M. J., Jones, A. R., Roberts, C. J., Stoughton, R. et al. (2000) Functional discovery via a compendium of expression profiles, *Cell* **102**, 109–126.

Humbel, B. M., Konomi, M., Takagi, T., Kamasawa, N., Ishijima, S. A., Osumi, M. (2001) *In situ* localization of β-glucans in the cell wall of *Schizosaccharomyces pombe, Yeast* **18**, 433–444.

Igual, J. C., Johnson, A. L., Johnston, L. H. (1996) Coordinated regulation of gene expression by the cell cycle transcription factor Swi4 and the protein kinase C MAP kinase pathway for yeast cell integrity, *EMBO J.* **15**, 5001–5013.

Inoue, S. B., Takewaki, N., Takasuka, T., Mio, T., Adachi, M., Fujii, Y., Miyamoto, C., Arisawa, M., Furuichi, Y., Watanabe, T. (1995) Characterization and gene cloning of 1,3-β-D-glucan synthase from *Saccharomyces cerevisiae*, *Eur. J. Biochem.* **231**, 845–854.

Inoue, S. B., Qadota, H., Arisawa, M., Watanabe, T., Ohya, Y. (1999) Prenylation of Rho1p is required for activation of yeast 1,3-β-glucan synthase, *J. Biol. Chem.* **274**, 38119–38124.

Ito, T., Chiba, T., Ozawa, R., Yoshida, M., Hattori, M., Sakaki, Y. (2001) A comprehensive two-hybrid analysis to explore the yeast protein interactome, *Proc. Natl. Acad. Sci. USA* **98**, 4569–4574.

Jamas, S., Ostroff, G. R., Easson, D. D., Jr. (1991) Glucan drug delivery system and adjuvant, US Patent 5, 032, 401.

Jamas, S., Easson, D. D., Jr., Ostroff, G. R. (1996) Method for immune system activation by administration of a β(1,3)glucan, which is produced by *Saccharomyces cerevisiae* strain R4, US Patent 5, 504, 079.

Jiang, B., Ram, A. F. J., Sheraton, J., Klis, F. M., Bussey, H. (1995) Regulation of cell wall β-glucan assembly: *PTC1* negatively affects *PBS2* action in a pathway that includes modulation of *EXG1* transcription, *Mol. Gen. Genet.* **248**, 260–269.

Jiang, B., Sheraton, J., Ram, A. F. J., Dijkgraaf, G. J. P., Klis, F. M., Bussey H. (1996) *CWH41* encodes a novel endoplasmic reticulum membrane *N*-glycoprotein involved in β1,6-glucan assembly, *J. Bacteriol.* **178**, 1162–1171.

Jing, W., DeAngelis, P. I. (2000) Dissection of the two transferase activities of the *Pasteurella multocida* hyaluronan synthase: two active sites exist in one polypeptide, *Glycobiology* **10**, 883–889.

Johnston, J. R., Mortimer, R. K. (1959) Use of snail digestive juice in isolation of yeast spore tetrads, *J. Bacteriol.* **78**, 292.

Kang, M. S., Cabib, E. (1986) Regulation of fungal cell wall growth: a guanine nucleotide-binding, proteinaceous component required for activity of (1 → 3)-β-D-glucan synthase, *Proc. Natl. Acad. Sci. USA* **83**, 5808–5812.

Kapteyn, J. C., Montijn, R. C., Vink, E., de la Cruz, J., Llobell, A., Douwes, J. F., Shimoi, H., Lipke, P. N., Klis, F. M. (1996) Retention of *Saccharomyces cerevisiae* cell wall proteins through a phophodiester-linked β-1,3-/β-1,6-glucan heteropolymer, *Glycobiology* **6**, 337–345.

Kapteyn, J. C., Ram, A. F. J., Groos, E. M., Kollár, R., Montijn, R., Van Den Ende, H., Llobell, A., Cabib, E., Klis, F. M. (1997) Altered extent of cross-linking of β1,6-glucosylated mannoproteins to chitin in *Saccharomyces cerevisiae* mutants with reduced cell wall β1,3-glucan content, *J. Bacteriol.* **179**, 6279–6284.

Kasahara, S., Yamada, H., Mio, T., Shiratori, Y., Miyamoto, C., Yabe, T., Nakajima, T., Ichishima, E., Furuichi, Y. (1994) Cloning of the *Saccharomyces cerevisiae* gene whose overexpression overcomes the effects of HM-1 killer toxin, which inhibits β-glucan synthesis, *J. Bacteriol.* **176**, 1488–1499.

Katohda, S., Abe, N., Matsui, M., Hayashibe, M. (1976) Polysaccharide composition of the cell wall of baker's yeast with special reference to cell walls obtained from large- and small-sized cells, *Plant Cell Physiol.* **17**, 909–919.

Kelly, R., Register, E., Hsu, M.-J., Kurtz, M., Nielsen, J. (1996) Isolation of a gene involved in 1,3-β-glucan synthesis in *Aspergillus nidulans* and purification of the corresponding protein, *J. Bacteriol.* **178**, 4381–4391.

Kessler, G., Nickerson, W. J. (1959) Glucomannan-protein complexes from cell walls of yeasts, *J. Biol. Chem.* **234**, 2281–2285.

Klebl, F., Tanner, W. (1989) Molecular cloning of a cell wall exo-β-1,3-glucanase from *Saccharomyces cerevisiae*, *J. Bacteriol.* **171**, 6259–6264.

Klis, F. M. (1994) Cell wall assembly in yeast, *Yeast* **10**, 851–869.

Ko, Y.-T., Frost, D. J., Ho, C.-T., Ludescher, R. D., Wasserman, B. P. (1994) Inhibition of yeast (1,3)-β-glucan synthase by phospholipase A_2 and its reaction products, *Biochim. Biophys. Acta* **1193**, 31–40.

Kollár, R., Petráková, E., Ashwell, G., Robbins, P. W., Cabib, E. (1995) Architecture of the yeast cell wall. The linkage between chitin and β(1 → 3)-glucan, *J. Biol. Chem.* **270**, 1170–1178.

Kollár, R., Reinhold, B. B., Petráková, E., Yeh, H. J. C., Ashwell, G., Drgonová, Kapteyn, J. C., Klis, F. M., Cabib, E. (1997) Architecture of the yeast cell wall. β(1 → 6)-Glucan interconnects mannoprotein, β(1 → 3)-glucan, and chitin, *J. Biol. Chem.* **272**, 17762–17775.

Kopecká, M., Phaff, H. J., Fleet, G. H. (1974) Demonstration of a fibrillar component in the cell wall of the yeast *Saccharomyces cerevisiae* and its chemical nature, *J. Cell Biol.* **62**, 66–76.

Larriba, G., Andaluz, E., Cueva, R., Basco, R. D. (1995) Molecular biology of yeast exoglucanases, *FEMS Microbiol. Lett.* **125**, 121–126.

Lee, D.-W., Ahn, G.-W., Kang, H.-G., Park, H.-M. (1999) Identification of a gene, *SOO1*, which complements osmo-sensitivity and defect in *in vitro* β1,3-glucan synthase activity in *Saccharomyces cerevisiae*, *Biochim. Biophys. Acta* **1450**, 145–154.

Li, H., Pagé, N., Bussey, H. (2002) Actin patch assembly proteins Las17p and Sla1p interact with *cis*-Golgi protein Kre6p and restrict cell wall growth to daughter cells (in preparation).

López-Romero, E., Ruiz-Herrera, J. (1977) Biosynthesis of β-glucans by cell-free extracts from *Saccharomyces cerevisiae*, *Biochim. Biophys. Acta* **500**, 372–384.

Lorenz, R. T., Parks, L. W. (1991) Physiological effects of fenpropimorph on wild-type *Saccharomyces cerevisiae* and fenpropimorph-resistant mutants, *Antimicrob. Agents Chemother.* **35**, 1532–1537.

Lu, C.-F., Montijn, R. C., Brown, J. L., Klis, F., Kurjan, J., Bussey, H., Lipke, P. N. (1995) Glycosyl phosphatidylinositol-dependent cross-linking of α-agglutinin and β1,6-glucan in the *Saccharomyces cerevisiae* cell wall, *J. Cell Biol.* **128**, 333–340.

Madden, K., Sheu, Y. J., Baetz, K., Andrews, B., Snyder, M. (1997) SBF cell cycle regulator as a target of the yeast PKC-MAP kinase pathway, *Science* **275**, 1781–1784.

Manners, D. J., Patterson, J. C. (1966) A re-examination of the molecular structure of yeast glucan, *Biochem. J.* **98**, 19c–20c.

Manners, D. J., Masson, A. J., Patterson, J. C. (1973a) The structure of a β-(1 → 3)-D-glucan from yeast cell walls, *Biochem. J.* **135**, 19–30.

Manners, D. J., Masson, A. J., Patterson, J. C. (1973b) The structure of a β-(1 → 6)-D-glucan from yeast cell walls, *Biochem. J.* **135**, 31–36.

Manning, B. D., Padmanabha, R., Snyder, M. (1997) The Rho-GEF Rom2p localizes to sites of polarized cell growth and participates in cytoskeletal functions in *Saccharomyces cerevisiae*, *Mol. Biol. Cell* **10**, 1829–1844.

Martin, H., Dagkessamanskaia, A., Satchanska, G., Dallies, N., François, J. (1999) *KNR4*, a suppressor of *Saccharomyces cerevisiae cwh* mutants, is involved in the transcriptional control of chitin synthase genes, *Microbiology* **145**, 249–258.

Martin-Yken, H., Martin, H., Dagkessamanskaia, A., Molina, M., Francois, J. (2001) KNR4 protein takes part in the cell integrity pathway and interacts with the map kinase Slt2p, *Yeast* **18**, S187 (Book of Abstracts of the Twentieth International Conference on Yeast Genetics and Molecular Biology, Prague, Czech Republic, August 26–31, 2001).

Matsui, Y., Toh-e, A. (1992) Yeast *RHO3* and *RHO4* ras superfamily genes are necessary for bud growth, and their defect is suppressed by a high dose of bud formation genes *CDC42* and *BEM1*, *Mol. Cell. Biol.* **12**, 5690–5699.

Mazur, P., Baginsky, W. (1996) *In vitro* activity of 1,3-β-D-glucan synthase requires the GTP-binding protein Rho1, *J. Biol. Chem.* **271**, 14604–14609.

Mazur, P., Morin, N., Baginsky, W., El-Sherbeini, M., Clemas, J. A., Nielsen, J. B., Foor, F. (1995) Differential expression and function of two homologous subunits of yeast 1,3-β-D-glucan synthase, *Mol. Cell. Biol.* **15**, 5671–5681.

McLellan, W. L., Lampen, J. O. (1968) Phosphomannanase (PR-factor), an enzyme required for the formation of yeast protoplasts, *J. Bacteriol.* **95**, 967–974.

Meaden, P., Hill, K., Wagner, J., Slipetz, D., Sommer, S. S., Bussey, H. (1990) The yeast *KRE5* gene encodes a probable endoplasmic reticulum protein required for (1 → 6)-β-D-glucan synthesis and normal cell growth, *Mol. Cell. Biol.* **10**, 3013–3019.

Merkenschlager, M., Schlossmann, K., Kurz, W. (1957) Ein mechanischer zellhomogenisator und seine anwendbarkeit auf biologische probleme, *Biochem. Z.* **329**, 332–340.

Mol, P. C., Wessels, J. G. H. (1987) Linkage between glucosaminoglycan and glucan determine alkali-insolubility of glucan in walls of *Saccharomyces cerevisiae*, *FEMS Microbiol. Lett.* **41**, 95–97.

Mol, P. C., Park, H.-M., Mullins, J. T., Cabib, E. (1994) A GTP-binding protein regulates the activity of (1 → 3)-β-glucan synthase, an enzyme directly involved in yeast cell wall morphogenesis, *J. Biol. Chem.* **269**, 31267–31274.

Montijn, R. C., Vink, E., Müller, W. H., Verkleij, A. J., Van Den Ende, H., Henrissat, B., Klis, F. M. (1999) Localization of synthesis of β1,6-glucan in *Saccharomyces cerevisiae*, *J. Bacteriol.* **181**, 7414–7420.

Mouassite, M., Camougrand, N., Schwob, E., Demaison, G., Laclau, M., Guerin, M. (2000) The 'SUN' family: yeast *SUN4/SCW3* is involved in cell septation, *Yeast* **16**, 905–919.

Mouyna, I., Fontaine, T., Vai, M., Monod, M., Fonzi, W. A., Diaquin, M., Popolo, L., Hartland, R. P., Latgé, J.-P. (2000) Glycosylphosphatidylinositol-anchored glucanosyltransferases play an active role in the biosynthesis of the fungal cell wall, *J. Biol. Chem.* **275**, 14882–14889.

Mundkur, B. (1960) Electron microscopical studies of frozen-dried yeast, *Exp. Cell. Res.* **20**, 28–42.

Muthukumar, G., Suhng, S. H., Magee, P. T., Jewell, R. D., Primerano, D. A. (1993) The *Saccharomyces cerevisiae SPR1* gene encodes a sporulation-specific exo-1,3-beta-glucanase which contributes to ascospore thermoresistance. *J. Bacteriol.* **175**, 386–394.

Necas, O. (1956) Regeneration of yeast cells from naked protoplasts, *Nature* **177**, 898–899.

Necas, O. (1961) Physical conditions as important factors for the regeneration of naked yeast protoplasts, *Nature* **192**, 580–581.

Nguyen, T. H., Fleet, G. H., Rogers, P. L. (1998) Composition of the cell walls of several yeast species, *Appl. Microbiol. Biotechnol.* **50**, 206–212.

Ni, L., Snyder, M. (2001) A genomic study of the bipolar bud site selection pattern in *Saccharomyces cerevisiae, Mol. Biol. Cell* **7**, 2147–2170.

Nonaka, H., Tanaka, K., Hirano, H., Fujiwara, T., Kohno, H., Umikawa, M., Mino A., Takai, Y. (1995) A downstream target of *RHO1* small GTP-binding protein is *PKC1*, a homolog of protein kinase C, which leads to activation of the MAP kinase cascade in *Saccharomyces cerevisiae, EMBO J.* **14**, 5931–5938.

Northcote, D. H., Horne, R. W. (1952) The chemical composition and structure of the yeast cell wall, *Biochem. J.* **51**, 232–236.

Oh, C.-S., Toke, D. A., Mandala, S., Martin, C. E. (1997) *ELO2* and *ELO3*, homologues of the *Saccharomyces cerevisiae ELO1* gene, function in fatty acid elongation and are required for sphingolipid formation, *J. Biol. Chem.* **272**, 17376–17384.

Ohya, Y., Caplin, B. E., Qadota, H., Tibbetts, M. F., Anraku, Y., Pringle, J. R., Marshall, M. S. (1996) Mutational analysis of the β-subunit of yeast geranylgeranyl transferase I, *Mol. Gen. Genet.* **252**, 1–10.

Onishi, J., Meinz, M., Thompson, J., Curotto, J., Dreikorn, S. et al. (2000) Discovery of novel antifungal (1,3)-β-D-glucan synthase inhibitors, *Antimicrob. Agents Chemother.* **44**, 739–746.

Orlean, P. (1997) Biogenesis of yeast wall and surface components, in: *The Molecular and Cellular Biology of the Yeast Saccharomyces – Cell Cycle and Cell Biology* (Pringle J. R., Broach J. R., Jones E. W, Eds.), Cold Spring Harbor, USA: Cold Spring Harbor Laboratory Press, 229–362.

Osmond, B. C., Specht, C. A., Robbins, P. W. (1999) Chitin synthase III: synthetic lethal mutants and "stress related" chitin synthesis that bypasses the *CSD3/CHS6* localization pathway, *Proc. Natl. Acad. Sci. USA* **96**, 11206–11210.

Osumi, M., Sato, M., Ishijima, S. A., Konomi, M., Takagi, T., Yaguchi, H. (1998) Dynamics of cell wall formation in fission yeast, *Schizosaccharomyces pombe, Fungal Gen. Biol.* **24**, 178–206.

Ozaki, K., Tanaka, K., Imamura, H., Hihara, T., Kameyama, T., Nonaka, H., Hirano, H., Matsuura, Y., Takai, Y. (1996) Rom1p and Rom2p are GDP/GTP exchange proteins (GEPs) for the Rho1p small GTP binding protein in *Saccharomyces cerevisiae, EMBO J.* **15**, 2196–2207.

Parent, S. A., Nielsen, J. B., Morin, N., Chrebet, G., Ramadan, N., Dahl, A. M., Hsu, M.-J., Bostian, K. A., Foor, F. (1993) Calcineurin-dependent growth of an FK506- and CsA-hypersensitive mutant of *Saccharomyces cerevisiae, J. Gen. Microbiol.* **139**, 2973–2984.

Parker, C. G., Fessler, L. I., Nelson, R. E., Fessler, J. H. (1995) *Drosophila* UDP-glucose:glycoprotein glucosyltransferase: sequence and characterization of an enzyme that distinguishes between denatured and native proteins, *EMBO J.* **14**, 1294–1303.

Parlati, F., Dominguez, M., Bergeron, J. J. M., Thomas, D. Y. (1995) *Saccharomyces cerevisiae CNE1* encodes an endoplasmic reticulum (ER) membrane protein with a sequence similarity to calnexin and calreticulin and functions as a constituent of the ER quality control apparatus, *J. Biol. Chem.* **270**, 244–253.

Parodi, A. J. (2000) Protein glucosylation and its role in protein folding, *Annu. Rev. Biochem.* **69**, 69–93.

Peat, S., Whelan, W. J., Edwards, T.E. (1958a) Polysaccharides of baker's yeast. Part II. Yeast glucan, *J. Chem. Soc.* 3862–3868.

Peat, S., Turvey, J. R., Evans, J. M. (1958b) Polysaccharides of baker's yeast. Part III. The presence of 1:6-linkages in yeast glucan, *J. Chem. Soc.* 3868–3870.

Peterson, J., Zheng, Y., Benderl, L., Myers, A., Cerione, R., Bender, A. (1994) Interactions between the bud emergence proteins Bem1p and Bem2p and Rho-type GTPases in yeast, *J. Cell Biol.* **127**, 1395–1406.

Phaff, H. J. (1971) Structure and biosynthesis of the yeast cell envelope, in: *The Yeasts – Physiology and Biochemistry of Yeasts* (Rose, A. H., Harrison, J. S., Eds.), London, New York: Academic Press, vol. 2, 135–210.

Philip, B., Levin, D. E. (2001) Wsc1 and Mid2 are cell surface sensors for cell wall integrity signaling that act through Rom2, a guanine nucleotide exchange factor for Rho1, *Mol. Cell. Biol.* **21**, 271–280.

Polizotto, R. S., Cyert, M. S. (2001) Calcineurin dependent nuclear import of the transcription factor Crz1p requires Nmd5p, *J. Cell Biol.* **154**, 951–960.

Popolo, L., Vai, M. (1999) The Gas1 glycoprotein, a putative wall polymer cross-linker, *Biochim. Biophys. Acta* **1426**, 385–400.

Qadota, H., Python, C. P., Inoue, S. B., Arisawa, M., Anraku, Y, Zheng, Y., Watanabe, T., Levin, D. E., Ohya, Y. (1996) Identification of yeast Rho1p GTPase as a regulatory subunit of 1,3-β-glucan synthase, *Science* **272**, 279–281.

Radding, J. A., Heidler, S. A., Turner, W. W. (1998) Photoaffinity analog of the semisynthetic echinocandin LY303366: identification of echinocandin targets in *Candida albicans*, *Antimicrob. Agents Chemother.* **42**, 1187–1194.

Ram, A. F. J., Wolters, A., ten Hoopen, R., Klis, F. M. (1994) A new approach for isolating cell wall mutants in *Saccharomyces cerevisiae* by screening for hypersensitivity to calcofluor white, *Yeast* **10**, 1019–1030.

Ram, A. F. J., Brekelmans, S. S. C., Oehlen, L. J. W. M., Klis, F. M. (1995) Identification of two cell cycle regulated genes affecting the β1,3-glucan content of cell walls in *Saccharomyces cerevisiae*, *FEBS Lett.* **358**, 165–170.

Ram, A. F. J., Kapteyn, J. C., Montijn, R. C., Caro, L. H. P., Douwes, J. E., Baginsky, W., Mazur, P., Van Den Ende, H., Klis, F. M. (1998) Loss of the plasma membrane-bound protein Gas1p in *Saccharomyces cerevisiae* results in the release of β1,3-glucan into the medium and induces a compensation mechanism to ensure cell wall integrity, *J. Bacteriol.* **180**, 1418–1424.

Robbins, E. A., Seeley, R. D. (1977) Cholesterol lowering effect of dietary yeast and yeast fractions, *J. Food Sci.* **42**, 694–698.

Rodríguez-Peña, J. M., Cid, V. J., Arroyo, J., Nombela, C. (2000) A novel family of cell wall-related proteins regulated differently during the yeast life cycle, *Mol. Cell. Biol.* **20**, 3245–3255.

Roelofsen, P. A. (1953) Yeast mannan, a cell wall constituent of baker's yeast, *Biochim. Biophys. Acta* **10**, 477–478.

Roemer, T., Bussey, H. (1991) Yeast β-glucan synthesis: *KRE6* encodes a predicted type II membrane protein required for glucan synthesis *in vivo* and for glucan synthase activity *in vitro*, *Proc. Natl. Acad. Sci. USA* **88**, 11295–11299.

Roemer, T., Bussey, H. (1995) Yeast Kre1p is a cell surface O-glycoprotein, *Mol. Gen. Genet.* **249**, 209–216.

Roemer, T., Delaney, S., Bussey, H. (1993) *SKN1* and *KRE6* define a pair of functional homologs encoding putative membrane proteins involved in β-glucan synthesis, *Mol. Cell. Biol.* **13**, 4039–4048.

Roemer, T., Paravicini, G., Payton, M. A., Bussey, H. (1994) Characterization of the yeast (1 → 6)-β-glucan biosynthetic components, Kre6p and Sknlp, and genetic interactions between the *PKC1* pathway and extracellular matrix assembly, *J. Cell Biol.* **127**, 567–579.

Romero, P. A., Dijkgraaf, G. J. P., Shahinian, S., Herscovics, A., Bussey, H. (1997) The yeast *CWH41* gene encodes glucosidase I. *Glycobiology* **7**, 997–1004.

Ross, G. D., Vetvicka, V., Yan, J., Xia, Y., Vetvickova, J. (1999) Therapeutic intervention with complement and β-glucan in cancer, *Immunopharmacology* **42**, 61–74.

Ruiz-Herrera, J. (1992) *Fungal cell wall: structure, synthesis and assembly*, Boca Raton: CRC Press, 59–88.

Sacher, M., Barrowman, J., Wang, W., Horecka, J., Zhang, Y., Pypaert, M., Ferro-Novick, S. (2001) TRAPP I implicated in the specificity of tethering in ER-to-Golgi transport, *Mol. Cell* **7**, 433–442.

Salkowski, E. (1894) Ueber die Kohlehydrate der Hefe, *Ber. Deutsch. Chem. Gess.* **27**, 497–502; 925–926 and 3325–3329.

Schmidt. A., Bickle, M., Beck, T., Hall, M. N. (1997) The yeast phosphatidylinositol kinase homolog TOR2 activates RHO1 and RHO2 via the exchange factor ROM2, *Cell* **88**, 531–542.

Shahinian, S., Bussey, H. (2000) β-1,6-Glucan synthesis in *Saccharomyces cerevisiae*, *Mol. Microbiol.* **35**, 477–489.

Shahinian, S., Dijkgraaf, G. J. P., Sdicu, A.-M., Thomas, D. Y., Jakob, C. A., Aebi, M., Bussey, H. (1998) Involvement of protein *N*-glycosyl chain glucosylation and processing in the biosynthesis of cell wall β-1,6-glucan of *Saccharomyces cerevisiae*, *Genetics* **149**, 843–856.

Shematek, E. M., Braatz, J. A., Cabib, E. (1980) Biosynthesis of the yeast cell wall. I. Preparation and properties of β-(1 → 3)glucan synthetase, *J. Biol. Chem.* **255**, 888–894.

Sietsma, J. H., Wessels, J. G. H. (1979) Evidence for covalent linkages between chitin and β-glucan in a fungal wall, *J. Gen. Microbiol.* **114**, 99–108.

Silve, S., Leplatois, P., Josse, A., Dupuy, P.-H., Lanau, C., Kaghad, M., Dhers, C., Picard, C., Rahier, A., Taton, M., Le Fur, G., Caput, D., Ferrara, P., Loison, G. (1996) The immunosuppressant SR 31747 blocks cell proliferation by inhibiting a steroid isomerase in *Saccharomyces cerevisiae*, *Mol. Cell. Biol.* **16**, 2719–2727.

Simpson, K. L., Wilson, A. W., Burton, E., Nakayama, T. O. M., Chichester, C.O. (1963) Modified French press for the disruption of microorganisms, *J. Bacteriol.* **86**, 1126–1127.

Spellman, P. T., Sherlock, G., Zhang, M. Q., Iyer, V. R., Anders, K., Eisen, M. B., Brown, P. O., Botstein, D., Futcher, B. (1998) Comprehensive identification of cell cycle-regulated genes of the yeast *Saccharomyces cerevisiae* by microarray hybridization, *Mol. Biol. Cell* **9**, 3273–3297.

Stasinopoulos, S. J., Fisher, P. R., Stone, B. A., Stanisich, V. A. (1999) Detection of two loci involved in (1 → 3)-β-glucan (curdlan) biosynthesis by *Agrobacterium* sp. ATCC31749, and comparative sequence analysis of the putative curdlan synthase gene, *Glycobiology* **9**, 31–41.

Stathopoulos, A. M., Cyert, M. S. (1997) Calcineurin acts through the *CRZ1/TCN1*-encoded transcription factor to regulate gene expression in yeast, *Genes Dev.* **11**, 3432–3444.

Stathopoulos-Gerontides, A., Guo, J. J., Cyert, M. S. (1999) Yeast calcineurin regulates nuclear localization of the Crz1p transcription factor through dephosphorylation, *Genes Dev.* **13**, 798–803.

Stock, S. D., Hama, H., Radding, J. A., Young, D. A., Takemoto, J. Y. (2000) Syringomycin E inhibition of *Saccharomyces cerevisiae*: requirement for biosynthesis of sphingolipids with very-long-chain fatty acids and mannose- and phosphoinositol-containing head groups, *Antimicrob. Agents Chemother.* **44**, 1174–1180.

Surarit, R., Gopal, P. K., Sheperd, M. G. (1988) Evidence for a glycosidic linkage between chitin and glucan in the cell wall of *Candida albicans*, *J. Gen. Microbiol.* **134**, 1723–1730.

Sutherland, I. W. (1998) Novel and established applications of microbial polysaccharides, *Trends Biotechnol.* **16**, 41–46.

Svoboda, A. (1966) Regeneration of yeast protoplasts in agar gels, *Exp. Cell. Res.* **44**, 640–642.

Szaniszlo, P. J., Kang, M. S., Cabib, E. (1985) Stimulation of β(1 → 3)glucan synthetase of various fungi by nucleoside triphosphates: generalized regulatory mechanism for cell wall biosynthesis, *J. Bacteriol.* **161**, 1188–1194.

Tanaka, H., Phaff, H. J. (1965) Enzymatic hydrolysis of yeast cell walls. I. Isolation of wall-decomposing organisms and separation and purification of lytic enzymes. *J. Bacteriol.* **89**, 1570–1580.

Terashima, H., Yabuki, N., Arisawa, M., Hamada, K., Kitada, K. (2000) Up-regulation of genes encoding glycosylphosphatidylinositol (GPI) attached proteins in response to cell wall damage caused by disruption of *FKS1* in *Saccharomyces cerevisiae*, *Mol. Gen. Genet.* **264**, 64–74.

Trombetta, E. S., Simons, J. F., Helenius, A. (1996) Endoplasmic reticulum glucosidase II is composed of a catalytic subunit, conserved from yeast to mammals, and a tightly bound noncatalytic HDEL-containing subunit, *J. Biol. Chem.* **271**, 27509–27516.

Uetz, P., Giot, L., Cagney, G., Mansfield, T. A., Judson, R. S., et al. (2000) A comprehensive analysis of protein–protein interactions in *Saccharomyces cerevisiae*, *Nature* **403**, 623–627.

Valdivieso, M. H., Ferrario, L., Vai, M., Duran, A., Popolo, L. (2000) Chitin synthesis in a *gas1* mutant of *Saccharomyces cerevisiae*, *J. Bacteriol.* **182**, 4752–4757.

Watanabe, D., Abe, M., Ohya, Y. (2001) Yeast Lrg1p acts as a specialized RhoGAP regulating 1,3-β-glucan synthesis, *Yeast* **18**, 943–951.

Watanabe, Y., Takaesu, G., Hagiwara, M., Irie, K., Matsumoto, K. (1997) Characterization of a serum response factor-like protein in *Saccharomyces cerevisiae*, Rlm1, which has transcriptional activity regulated by the Mpk1 (Slt2) mitogen-activated protein kinase pathway, *Mol. Cell. Biol.* **17**, 2615–2623.

Williams, D. L., Browder, I. W., Diluzio, deceased, Nicholas, R., Diluzio, N. M. representative of Estate. (1990) Soluble phosphorylated glucan: methods and composition for wound healing, US Patent 4, 975, 421.

Willment, J. A., Gordon, S., Brown, G. D. (2001) Characterization of the human β-glucan receptor and its alternatively spliced isoforms, *J. Biol. Chem.* **276**, 43818–43823.

Winzeler, E. A., Shoemaker, D. D., Astromoff, A., Liang, H., Anderson, K. et al. (1999) Functional characterization of the *S. cerevisiae* genome by gene deletion and parallel analysis, *Science* **285**, 901–906.

Yabe, T., Yamada-Okabe, T., Kasahara, S., Furuichi, Y., Nakajima, T., Ichishima, E., Arisawa, M., Yamada-Okabe, H. (1996) *HKR1* encodes a cell surface protein that regulates both cell wall β-glucan synthesis and budding pattern in the yeast *Saccharomyces cerevisiae*, *J. Bacteriol.* **178**, 477–483.

Yamochi, W., Tanaka, K., Nonaka, H., Maeda, A., Musha, T., Takai, Y. (1994) Growth site localization of *Rho1* small GTP-binding protein and its involvement in bud formation in *Saccharomyces cerevisiae*, *J. Cell Biol.* **125**, 1077–1093.

Zechmeister, L., Toth, G. (1934) Über die Polyose der Hefemembran I, *Biochem. Z.* **270**, 309–316.

Zhao, C., Jung, U. S., Garrett-Engele, P., Roe, T., Cyert, M. S., Levin, D. E. (1998) Temperature-induced expression of yeast *FKS2* is under the dual control of protein kinase C and calcineurin, *Mol. Cell. Biol.* **18**, 1013–1022.

17
Fungal Cell Wall Glycans

Dr. Shung-Chang Jong[1]
[1] American Type Culture Collection, 10801 University Boulevard, Manassas, VA 20110-2209, USA; Tel: +1-703-365-2742; Fax: +1-703-365-2730;
E-mail: sjong@atcc.org

Biopolymers for Medical and Pharmaceutical Applications. Edited by A. Steinbüchel and R. H. Marchessault

ISBN: 3-527-31154-8

AIDS	acquired immune deficiency syndrome
AZT	3′-azido-3′-deoxythymidine
BRM	biological response modifier
HIV	human immunodeficiency virus
IFN	interferon
IL	interleukin
KS-2	peptidomannan from *Lentinula edodes*
LDL	low-density-lipoprotein
NF	nuclear factor
NK cell	natural killer cell
PCV	polypeptide from *Trametes versicolor*
PSK	polysaccharide Kureha (trade name: Krestin) from *Trametes versicolor*
PSP	polysaccharopeptide from *Trametes versicolor*

1 Introduction

The cell wall is a vital (structural and functional) component of fungi and as important to the organism as any cytoplasmic organelle. The carbohydrate content of the fungal cell wall is basically the same in all genera: polysaccharides comprise about 80% of the dry weight. Long chains are arranged as microfibrils embedded in a matrix of glycoproteins, lipids, and nonfibrillar polysaccharides. Cell wall polysaccharides may be extracted, fractionally purified, and identified by a combination of techniques and instrumental analyses that include fractional solution, precipitation with ionic detergents or metallic ions, ultracentrifugation, electrophoresis, ultrafiltration through graded membranes, ion-exchange chromatography, gel filtration, and affinity chromatography. Methods of fractionation are based on the solubility in alkali or acid solutions. The alkali-soluble fraction will contain glycans, heteroglycans, and glycoproteins, while the alkali-insoluble fraction will have chitin and/or cellulose and the insoluble glucans that form the skeletal structure of the wall. Factors that affect isolation of various components and determination of their structure are the similarity and sequence of the monomers, the size of the ring involved, the positions of linkages, and anomeric configuration.

Carbohydrates include compounds with relatively small molecules, such as simple sugars (monosaccharides and disaccharides), as well as macromolecules, such as starch, glycogen, chitin, and cellulose (polysaccharides). Polysaccharides, which constitute a structurally and functionally diverse class of biological macromolecules, not only have practical applications (e.g. industrial water-soluble polymers), but they are potentially useful in the food, technology, and medical sectors (Franz, 1989). The term "glycan" is understood to mean a polysaccharide composed of monosaccharide residues joined by glycosidic linkages. It is derived from the word "glycose", meaning a simple sugar, plus "an" signifying a sugar polymer. Polysaccharides that produce one type of monosaccharide (D-glucose) on complete hydrolysis are called homoglycans (e.g. cellulose or starch); those with two or more different monosaccharides are heteroglycans (e.g. hyaluronic acid composed of *N*-acetylglucosamine and D-glucuronic acid). Polysaccharides that yield hexose monosaccharides upon hydrolysis are hexoglycans; those with pentose sugars are pentoglycans. Each class can be further divided according to the particular hexose or pentose involved. Polysaccharides composed of the hexose sugar D-glucose are called glucans; those made up of the pentose sugars D-xylose and L-arabinose are xylan and arabinan, respectively.

Glucans include cellulose made of β1–4-bound glucose, the β-glucans with variable proportions of β1–3 and β1–6 linkages, and α1–3 glucans. The most abundant and most thoroughly studied of the fungal glucans are the β-glucans, which range from water-soluble mucilaginous polymers to alkali-insoluble components. In general, high-molecular-weight β1–3 glucans with β1–6 branching are primarily responsible for the structural integrity of the cell wall. Chitin, chitosan, and cellulose composed of different sugar monomers have a β1–4 linkage and form straight fibers. Other types of linkages usually result in helical molecules. Chitin found at the hyphal apex of fungal mycelia is not only important for structure, but appears to be essential for growth and differentiation, as well as for pathogenesis of plants, animals, and humans. Cellulose is not considered part of the fungal glucan group, because it is found in the cell wall of a limited number of fungi and is usually accompanied by other glucose-containing polysaccharides. Those polysaccharides in the fungal cell wall that are covalently linked to polypeptides are called glycoproteins or peptidopolysaccharides, because they are composed of a polypeptide (protein) backbone with polysaccharide branches.

2
Historical Outline

In the last few decades fungi, particularly mushrooms, have increasingly been used as a source of dietary supplements, medical foods, and drugs (Jong and Birmingham, 1998). As dietary supplements, fungal extracts are concentrated and prepared for ingestion in a pill, capsule, tablet, or liquid form that contains one or more dietary ingredients (e.g. vitamins or amino acids) that increases the intake of those particular nutrients. As medical foods, fungal extracts are specially formulated and processed for the partial or exclusive feeding of a patient who may not be able to ingest, digest, absorb, or metabolize ordinary foodstuffs or to provide certain nutrients that cannot be provided by a normal diet alone. Drugs are intended for use in the diagnosis, cure, mitigation, treatment, or prevention of disease in humans or animals. Modern analytical techniques have revealed that many fungi contain biologically active polysaccharides. Some glycans extracted from fungal cell walls act as biological response modifiers (BRMs) *in vitro* and *in vivo* (Jong and Birmingham, 1990, 1992a,b, 1993a,b, 1994; Jong and Donovick, 1989; Jong and Yang, 1999; Jong et al., 1991; Ooi and Liu, 1999; Wasser and Weis, 1999a,b).

BRMs are vitally important for the maintenance of homeostasis in the interdependent body systems (e.g. the nervous system, the endocrine system, and the immune system) that helps the body adapt to environmental and psychological stress. In the immune system, BRMs activate a variety of specialized cells and components (e.g. natural killer cells, macrophages, lymphocytes, and cytokines) that identify and selectively destroy invading bacteria, fungi, yeast, viruses, parasites, and cancer cells. Cytokines, which are induced by a large panel of immunological stimuli and secreted by many cell types, are not only involved in regulating cellular survival, proliferation, differentiation, and death (apoptosis), but also play an important role in the control of cell migration, activation, and priming.

Most biological systems require cellular interactions for development and regulation. Since cytokine interactions (between cytokines and their receptors and between cytokine receptors and their signal transducers) control all physiological processes involving cellular interactions, each glycan possesses a certain degree of specificity in regulating and determining the immune response. Consequently, many biological and cellular functions of glycans are often modulated by the interactions with their specific receptors. For example, a β-glucan receptor on a monocyte does not recognize α-glucan, α-mannan, or α-galactan (Czop and Austen, 1985). When interacting with a β-glucan receptor, β-glucans exhibit biological activities that include the following: activation of macrophages, complement and/or the cytokine system (Abel and Czop, 1992), activation of a NF-κB-like nuclear transcription factor in purified human neutrophils (Wakshull et al., 1999), induction of hematopoiesis (Wakshull et al., 1999), and resistance to infections (Czop et al., 1989). Several different receptors for β-glucans have been identified on leukocytes, including 160- and 180-kDa proteins on human monocytes and U937 cells (Szabo et al., 1995), a glycosphingolipid β1–3 glucan receptor on human neutrophils (Wakshull et al., 1999), and a leukocyte complement receptor 3 protein on neutrophils, macrophages, and natural killer (NK) cells (Xia et al., 1999). In addition, a mannose receptor involved in the phagocytosis of pathogenic microorganisms has been found on macrophages (Astarie-Dequeker et al., 1999).

As BRMs, numerous fungal cell wall glycans and glycan–protein complexes exhibit significant antitumor, immunomodulating, antioxidant, free radical-scavenging, antiviral, antibacterial, antiparasitic, hepatoprotective, antifibrotic, antiinflammatory, antidiabetic, hypocholesterolemic, and antimetastatic activities (Jong et al., 1991; Mizuno, 1995a,b; Ooi and Liu, 1999; Wasser and Weis, 1999a,b). It has been shown that immune cells respond to fungal glycans by gene expression and the production of immunomodulatory cytokines that mediate the immunopotentiation of these glycans *in vivo* (Liu et al., 1996, 1999). Several glycan-based medical foods and antitumor drugs developed from mushrooms have become large nutraceutical and pharmaceutical market items in Japan, Russia, China, Korea, Hong Kong, Taiwan, Singapore, Canada, and the US.

3
Immunomodulatory and Antitumor Glycans

A number of glycans and glycan–protein complexes with immunomodulatory and antitumor activities *in vivo* and *in vitro* have been isolated from the fruiting body, sclerotium, mycelium, and liquid culture medium of filamentous fungi. Most of these active principles are associated with cell wall constituents and are found not only in glucans with different types of glycosidic linkages (e.g. β1–3 → 1–6 glucans and α1–3 glucans), but also in β-glucan–protein, α-manno-β-glucan, α-glucan–protein, hetero-β-glucans, heteroglycans, and heteroglycan–protein complexes (Ooi and Liu, 1999). The intensity of the activity is closely related to molecular weight, solubility in water, degree of branching, configuration, and binding modes of β1–6 branches to the principal β1–3 chain (Mizuno, 1995a). It appears that the greater the molecular weight and solubility of the glycan in water, the higher the immunomodulatory and antitumor activities (Jong et al., 1991; Sakagami et al., 1991).

The cell wall of mushrooms is a rich source of immunomodulatory and antitumor glucans. Marked antitumor activities have been found in the β1–3 glucans of *Grifola frondosa, Agaricus blazei, Lentinula edodes, Lyophyllum decastes, Ganoderma lucidum, G. applanatum, Volvariella volvacea, Pleurotus ostreatus, Flammulina velutipes, Hericium erinaceus,* and *Auricularia* spp.; the β1–3 glucan–protein complex from *Trametes versicolor, G. lucidum, G. applanatum, G. tsugae,* and *Flammulina velutipes*; the β1–6 glucan from *G. frondosa*; the β1–6 glucan–protein complex from *A. blazei*; heteroglucans from *G. lucidum, G. applanatum,* and *Tricholoma giganteum*; and heteroglucan–protein complexes from *A. blazei, L. edodes, G. lucidum, G. tsugae, Pleurotus sajor-caju, T. giganteum,* and *T. mongolicum.*

Four antitumor drugs incorporating fungal cell wall glycans (lentinan, PSK, PSP and Befungin) are commercially available. They are derived from the submerged cultured mycelia of *T. versicolor* (PSK, PSP), and the fruiting bodies of *L. edodes* (Lentinan) and *Inonotus obliquus* (Befungin) (Wasser and Weis, 1999a,b). These drugs are being used clinically either alone or in combination with chemotherapy and/or radiotherapy in the treatment of cancers of the head and neck, lung, stomach, colorectum, esophagus, breast, and cervix. In most cases, they are used after surgery to suppress metastasis and to relieve the side effects of chemotherapy and radiotherapy. These natural drugs are nontoxic and cause no adverse reactions or unwanted side effects, because they stimulate the immune function rather than attack cancer cells directly, as in chemotherapy.

3.1 β-Glucans and their Protein Complexes

Lentinan isolated from the fruiting body of *L. edodes* is a chemically pure β1–3 glucan with a backbone of β1–3 glucan and side chains of β1–3- and β1–6-linked D-glucose residues, together with a few internal β1–6 linkages. There are two β1–6 glucopyranoside branches for every five linear β1–3 glucopyranoside linkages. Lentinan contains only glucose and has a right-handed triple helix structure. The molecular formula is $(C_6H_{10}O_5)_n$; the mean molecular weight is about 1×10^6 (Jong and Birmingham, 1993a,b; Mizuno, 1995b); see Figure 1.

Lentinan is similar in chemical composition and biological activity to the extracellular polysaccharide schizophyllan, a water-soluble β1–3 glucan secreted by *Schizophyllum commune*. The antitumor activity of both lentinan and schizophyllan relates to their ordered triple helix conformation and is due mainly to host-mediated immune responses.

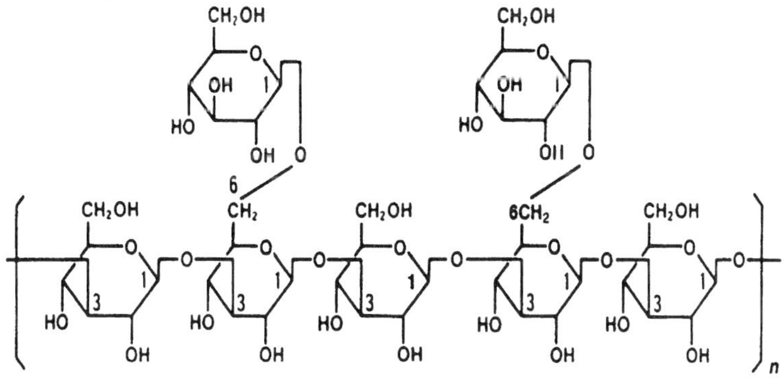

Fig. 1. Structure of lentinan isolated from the fruiting body of *L. edodes*.

β-D-glucan binds to the surface layer of lymphocytes or specific serum proteins that activate macrophages, T cells, and NK cells, and increase production of antibodies, interleukins (IL-1, IL-2), and interferon (IFN)-related effector cells (Mizuno, 1995b). Lentinan also suppresses cancer metastasis and recurrence in animal models (Sugano et al., 1989). Because lentinan has demonstrated marked antitumor activity, negligible side effects, and a BRM mode of action *in vitro* and *in vivo*, it has been clinically tested in cancer patients. It not only works against and, in some cases, even prevents the recurrence of gastrointestinal, colorectal, esophageal, lung, and breast cancers, acquired immune deficiency syndrome (AIDS), and other life-threatening conditions, with no dangerous side effects, but it also appears to boost the immune function in healthy people, helping to prevent infections and promote good health and general well-being (Chihara, 1993; Jong and Birmingham, 1993a,b).

Grifolan from *G. frondosa* is a β1–3 glucan similar to lentinan in primary structure (Adachi et al., 1990) and functions as a novel macrophage activator that increases cytokine production (Adachi et al., 1994). A branched β-glucan isolated from the cell wall of *Phytophthora parasitica* (A1) resembles schizophyllan with its β1–3 backbone and 1–6 branch points and β1–3 side chain linkage, and is a comparable or even more potent immunomodulating agent (Franz, 1989). Another highly branched β1–3→1–6 glucan (H11) with a molecular weight of 500 kDa with a remarkable antitumor effect has been isolated from mycelia of *Poria cocos* (Kanayama et al., 1983; Jong and Birmingham, 1994).

PSK or polysaccharide Kureha (trade name: Krestin) from the cultured mycelium of *T. versicolor* is composed of 62% water soluble, protein-bound polysaccharides with α1–4 glucan main component and β1–3 linkages plus 38% protein; see Figure 2.

PSK has shown the ability to promote cancer remission in human clinical trials and is being used as an oral antineoplastic agent for gastric, colorectal, and lung cancers. PSK has no substantial effect on immune responses of a normal host, but restores immune potential after the host has been depressed by tumor burden or anticancer chemotherapeutic agents (Kobayashi et al., 1993; Dong et al. 1996, 1997; Ooi and Liu, 1999). PSK is effective whether administered orally, intravenously, or intraperitoneally; local administration is more efficient than systemic (Mizutani and Yoshida, 1991). In addition, PSK suppresses pulmonary metastasis of methycholanthrene-induced sarcomas, human prostate cancer DU145M, and lymphatic metastasis of mouse leukemia P388 in spontaneous metastatic models, and inhibits metastasis of rat hepatoma (AH60C), mouse colon cancer (colon 26), and mouse leukemia (RL male 1) in artificial metastasis models (Kobayashi et al., 1995).

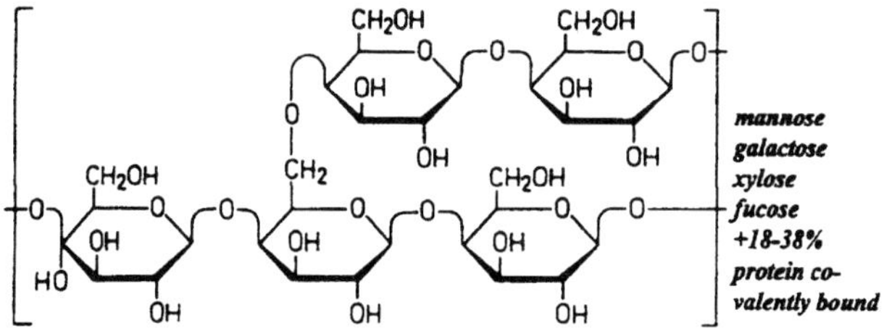

Fig. 2. Structure of PSK isolated from the cultured mycelium of *T. versicolor*.

The combined administration of PSK with cytokines, such as granulocyte colony-stimulating factor, granulocyte-macrophage colony-stimulating factor, and IL-3, improves recovery of myelosuppression following chemotherapy in mice (Kohgo, 1994).

PSP, similar to PSK in chemical composition, has been obtained from another strain of *T. versicolor* by a different method of extraction. PSK uses strain CM-101 isolated in Japan and is extracted by salting out with ammonium sulfate, whereas PSP uses strain Cov-1 isolated in China, and is prepared by water extraction and alcoholic precipitation. PSP is composed of 90% glucans and 10% peptides. In addition to glucose, PSP contains small amounts of five other monosaccharides, i.e. galactose, mannose, xylose, arabinose, and rhamnose. The main glucose chain is bound by $\alpha 1-4$ and $\beta 1-3$ glycosidic linkages. Both PSK and PSP are chemically homogenous with a molecular weight of approximately 100 kDa; their polypeptide moieties are rich in aspartic and glutamic acids. However, Japanese PSK lacks arabinose and rhamnose, and Chinese PSP does not contain fucose (Yang et al., 1992; Jong and Yang, 1999). PSP enhances immune functions in old tumor-bearing mice but not in young mice (Wu et al., 1998). It is effective in restoring immunosuppression (as indicated by depressed lymphocyte proliferation, NK cell function, the number of white blood cells, and the growth of spleen and thymus in rats, as well as increasing IgG, IL 2, and tumor necrosis factor-α production) induced by cytotoxic antitumor agents, such as 5-fluorouracil, cyclophosphamide, or bleomycin (Mori et al., 1987; Qian et al., 1997). PSK also enhances the anticancer activity of *cis*-diaminedichloroplatinum *in vitro* (Kobayashi et al., 1994). *in vitro* and animal and human studies document that PSP can enhance the immune function of normal animals and cells, antagonize immunosuppression caused by chemotherapeutic agents and tumor-induced immunosuppression in animals, and inhibit the growth of cancer cells in humans and animals (Jong and Yang, 1999). Today PSP is being used as an adjuvant with chemotherapy or radiotherapy in the clinical treatment of patients for lung, gastric, esophageal, and breast cancers. (Ng, 1998). Solubility in water and effective oral administration of both PSK and PSP are probably due to their protein-containing structure. PCV, a patented composition, contains *T. versicolor* polypeptides with molecular weights of 10 and 50 kDa (Yang and Chen, 1994). Cancer patients treated with purified PCV experience relatively low toxic side effects while benefiting therapeutically from antitumor effects superior to PSP and PSK. PCV induces tumor regression in some liver cancer patients (Chen et al., 1993).

As mentioned earlier, many biological functions of glucans are often achieved by interactions with their receptors. Complement receptor type 3 (CR3: Mac1– CD11/CD18) is a β-glucan receptor of macrophages (Xia et al., 1999) and neutrophil cells (Wakshull et al., 1999). A $\beta 1-3$ glucan obtained from *G. frondosa* has been shown to increase the expression of the β-glucan receptor CD11b on Kupffer cells (Adachi et al., 1998).

Several water-soluble and -insoluble (but soluble in salt or dilute alkali) polysaccharides with antitumor activity are obtained from *G. lucidum*, *G. applanatum*, and *G. tsugae*. They include a $\beta 1-3$ glucan, a homoglucan or its protein complex, and a heteroglucan or its protein complex (Mizuno et al., 1995b,c). Ganoderans, immunomodulatory β-glucans of *G. lucidum*, induce potent antitumor immunity in tumor-bearing mice (Han et al., 1995). Ganoderan B is a glucan–protein and Ganoderan C is a galactoglucan–protein with the same molecular weight of 400 kDa. Polysaccharides of both complexes are composed mainly of β1–6 and β1–3 glucan chains (Hikino et al., 1985). It has been reported that the antitumor effect of a polysaccharide extracted from fresh fruiting bodies of *G. lucidum* is mediated by cytokines released from activated macrophages and T lymphocytes (Wang et al., 1997). In addition to water-soluble β1–3 glucans, a water-soluble β1–6 glucan–protein obtained from the fruiting body of *Agaricus blazei* has been identified as the first β1–6 glucan with noticeable anticancer activity (Mizuno, 1995d). Polysaccharides from *A. blazei* significantly increase the percentage of CD3-, CD4-, and CD8-positive cells in mouse spleen cells after oral administration (Mizuno et al., 1998). EA 6, a β-glucan–protein isolated from the fruiting body of *Flammulina velutipes* contains 41.39% C, 6.92% H, 3.82% N, 70% saccharide, and 30% protein. It is composed of glucose, galactose, mannose, xylose, arabinose, and 16 amino acids (Ikekawa, 1995). EA 6 exhibits strong antitumor activity against sarcoma 180, Lewis lung cancer, and B-16 melanoma (Ikekawa, 1995).

Proflamin is an antitumor glycoprotein found in the cultured mycelium of *F. velutipes* (Ikekawa, 1995). It is effective against solid sarcoma 180, B-16 melanoma, adenocarcinoma 755, and Gardner lymphoma, and useful in combination therapy with other antitumor agents. Immunoactive polysaccharides, designated LEM and LAP, have been isolated from an extract of *L. edodes* mycelia grown on a solid medium of bagasse and rice bran (5:1 w/w). LEM is a light brown powder obtained from enzymes present in the mycelia by autolysis at 40–50°C for 60 h. LAP is a precipitate obtained from a water solution of LEM after addition of 4 volumes of ethanol. LEM represses carcinogenesis in the liver of rats by inhibiting the growth of ascites tumor cells, while LAP inhibits the growth of liver carcinoma cells and increases the survival rate of rats. Both LEM and LAP are glycoproteins consisting of glucose, galactose, xylose, arabinose, mannose, and fructose (Jong and Birmingham, 1993a,b).

3.2 Heteroglucans and their Protein Complexes

In addition to water-soluble β-D-glucans, several potent antitumor heteroglucans with heterosaccharide chains of xylose, mannose, galactose, and uronic acid, and their protein complexes have been extracted by salt and alkali from the cell walls of fungi (Mizuno et al., 1995b; Wasser and Weis, 1999a,b). Two heteroglucans from cultured mycelia of *T. versicolor* and *Tricholoma mongolicum* activate both lymphocytes and macrophages from BALK/c mice, and show no direct cytotoxic activity against fibroblasts, hepatoma cells, and choriocarcinoma cells (Wang et al., 1996a,b). KS-2, an α-mannan peptide, is obtained by hot water extraction from the cultured mycelium of *L. edodes*. Strong antitumor activity is achieved with both oral and intraperitoneal administration (Mizuno, 1995b) and is based on IFN-induction. KS-2 contains the amino acids serine, threonine, alanine, and proline, as well as residual amounts of the other amino acids in the peptide chain. ATOM, a glucomannan–protein complex from the cultured mycelium of *Agaricus blazei*, is highly effective against four kinds of established tumors, i.e. subcutaneously implanted Sarcoma 180 in mice, Ehrlich ascites carcinoma, Shionogi carcinoma 42, and Meth A fibrosarcinoma (Ito et al., 1997). A peptide–glucan preparation obtained from *A. blazei* is found to be immune enhancing (Ebina et al., 1998). Several antitumor polysaccharide components, such as glucuronoglucan, xyloglucan, mannoglucan, xylomannoglucan, and other active heteroglucans and their protein complexes have been extracted from *G. lucidum* and related *Ganoderma* species and purified for medicinal use (Saito et al., 1989; Jong and Birmingham, 1992b; Lei and Lin, 1992). Two strong antitumor heteroglucan–protein complexes from *G. tsugae* are identified as protein-containing glucogalactans associated with mannose and fucose (Zhang et al., 1994). Five heteroglycan–protein complexes with high antitumor activity have been isolated from the fruiting body of cultivated *Hericium erinaceus* using hot water at 100°C, followed by 1% ammonium oxalate (100°C), and finally 5% sodium hydroxide (30%). They are further fractionated and purified by ion-exchange chromatography, gel-filtration chromatography, and affinity chromatography. Infrared and nuclear magnetic resonance spectra identify these compounds as protein complexes of glucoxylan, xylan, heteroxyloglucan, and galactoxylogucan (Mizuno, 1995e). Two water-soluble heteroglycan–protein complexes with antitumor activity found in the fruiting body of *P. sajor-caju* are xyloglucan–protein and mannogalactan–protein (Mizuno and Zhuang, 1995b).

Eight high-molecular-weight polysaccharides extracted from cultured mycelia of *G. frondosa*, sequentially using hot water, 1% ammonium oxalate solution, and 5% sodium hydroxide solution, and then fractionated and purified, have high antitumor activity. They have been identified as heteroglycans or their protein complexes. The water-soluble fractions are heteromannan and its protein complexes, including a fucogalactomannan–protein complex, a mannogalactofucan, a galactoglucomannofucan–protein complex, and a glucogalactomannan. The water-insoluble fractions are heteroxylan and its protein complexes, including a mannofucoglucoxylan, a mannoglucofucoxylan–protein, a mannofucoglucoxylan–protein complex, and a glucomannofucoxylan–protein complex (Mizuno and Zhuang, 1995a).

3.3 Chemical Modification

Chemical modification has been used to enhance the antitumor activity of fungal β-glucans. β-glucans obtained from *Cordyceps ophioglossoides, Auricularia auricula,* and *Dendropolyporus umbellatus* have been made more water-soluble through several modifications. Ninety percent of the cell wall of *P. cocos* is composed of the polysaccharide pachyman, a linear β1–3 glucan possessing 9–10 branches of 1–6-linked β-D-glucopyransoyl groups and an internal β1–6 linkage. Pachyman has no antitumor effect, but after treatment with periodate or urea or after carboxymethylation, its derivatives (pachymaran and hydroxyethylpachymaran) exhibit strong antitumor activity (Kanayama et al., 1983; Jong and Birmingham, 1994). Apparently, severing the β1–6 linkages converts pachyman to a lentinan-like β1–3 linear glucan. It is interesting that such a marked difference in antitumor activity can result from such a small change in the fine structure of the β1–3 linked branching or some related structural difference arising from it. Similarly, linear α1–3 glucans from *Amanita muscaria* and *Agrocybe cylindracea* have little antitumor activity, but carboxymethylated linear α-1–3 glucans display potent antitumor activity (Yoshida et al., 1996).

Four water-soluble and five water-insoluble antitumor polysaccharides extracted from cultured mycelia of *G. frondosa* have been chemically modified by Smith degradation (polyaldehydration and polyalcoholization) and formic acid degradation (formylation and formolysis). Based on an antitumor assay against mouse solid tumor and an *in vivo* assay for mouse macrophage-mediated C3 complement-releasing activity, the modified polysaccharides have shown activity higher than found in the original compounds or exhibit new activity (Mizuno and Zhuang, 1995a,b).

4
Antiviral Glycans

Lentinan from *L. edodes* and some of its derivatives are effective against the AIDS virus by inhibiting viral replication and cell fusion (Jong and Birmingham, 1993a; Mizuno, 1995b). Lentinan reduces the toxicity of 3′-azido-3′-deoxythymidine (AZT). A combination of lentinan with AZT suppresses the surface expression of human immunodeficiency virus (HIV) antigens more strongly than does AZT alone. In addition, it can enhance the effect of AZT on replication of HIV in various human hematopoietic cell lines (Tochikura et al., 1987). Lentinan also exhibits antiviral activity against vesicular stomatitis virus, encephalitis virus, Abelson virus, and adenovirus type 12 in mice, stimulates nonspecific resistance against respiratory viral infections in mice, and confers complete protection against an LD_{75} challenge dose of virulent mouse influenza ASW15 (Wasser and Weis, 1999a,b). Lentinan sulfate suppresses replication of the HIV virus, inhibits the adsorption of the HIV virus into host cells, and inhibits the activity of reverse transcriptase when tested in MT-4 cells infected with HTLV-III. JLS-18, composed of 65–75% lignin, 15–30% polysaccharides, and 10–20% protein, is a water-soluble extract of the mycelium of *L. edodes* that inhibits herpes virus *in vitro* and *in vivo*. A compound designated Ac2P isolated from an aqueous extract of dried *L. edodes* mushrooms is a high molecular weight polysaccharide composed mainly of pentose sugars that inhibits virus replication. Tests in mice show that it is a selective inhibitor of orthomyxoviruses, such as influenza virus, but not paramyxo- or other viruses (Yamamura and Cochran, 1976). LEM-HT, extracted from cultured mycelia of *L. edodes*, can control the reduction of T lymphocytes due to infection with the AIDS virus by stimulating macrophages and IL-1 activity (Iizuka and Maeda, 1988). LEM-HT is 44% sugar, 24.6% protein, and 31.4% other components. Two peptidoglycans isolated from an extract of *L. edodes* mycelia enhance humoral and cell-medicated immunity to a wide range of diseases, e.g. hepatitis, influenza, herpes, cancer, AIDS, and mycotic infections (Sugano et al., 1985). A peptidomannan (KS-2) from cultured mycelia of *L. edodes* exerts antiviral activity by inducing interferon production and not by directly killing or inhibiting virus reproduction (Suzuki et al., 1979). When administrated orally or intraperitoneally to mice with influenza virus, KS-2 provides therapeutic as well as prophylactic protection.

PSK from *T. versicolor* has been reported to possess antiviral activities against ectromelia virus and cytomegalovirus infections (Tsukagoshi et al., 1984) and cell-free infection of HIV. While PSP does not exert any antibacterial activity at 128 mg l^{-1} (Ng et al., 1996), it has the potential to be a useful agent against viral infections, especially HIV-1 (Collins and Ng, 1997). The inhibition of the interaction between HIV-1 gp120 and immobilized CD4 receptor *in vitro* is very potent and demonstrates the potential of PSP to interfere with the binding of the HIV-1 virus to its cellular target *in vivo*. PSP also has a very potent inhibitory effect against HIV-1 reverse transcriptase *in vitro*. PSP and lentinan are the only natural products isolated so far that have specific reverse transcriptase inhibitory properties. Glycohydrolytic enzymes from the Golgi complex and associated with viral protein processing are also inhibited by PSP. The multivalent manner in which PSP acts *in vitro* to inhibit several different stages of the HIV life cycle makes it an ideal candidate for more detailed *in vivo* studies (Collins and Ng, 1997).

5 Antimicrobial Glycans

Lentinan and its derivatives have been shown to enhance host-resistance to various kinds of bacterial, fungal and parasitic infections (Wasser and Weis, 1999a,b). They increase resistance to *Schistosoma japonicum* and *S. mansoni*, exhibit activity against *Mycobacterium tuberculosis*, *Staphylococcus aureus*, *Micrococcus luteus*, *Pseudomonas aeruginosa*, *Klebsiella pneumoniae*, *Candida albicans*, and *Saccharomyces cerevisiae*, and increase host resistance to potentially lethal *Listeria monocytogenes* (Byram et al., 1979; Iguchi et al., 1985 Chen et al., 1987; Chihara, 1992). PSK exerts noticeable antimicrobial activity against *Escherichia coli*, *L. monocytogenes*, and *C. albicans* (Tsukagoshi et al., 1984; Sakagami et al., 1991).

6 Hepatoprotective Glycans

Lentinan has produced favorable results in the treatment of chronic persistent hepatitis and viral hepatitis B (Amagase, 1987). In combination with polysaccharides from *G. lucidum* and *T. versicolor*, lentinan has improved serum glutamate pyruvate transaminase and glutamate pyruvate transaminase levels in the livers of mice with toxic hepatitis (Zhang and Luan, 1986; Wasser and Weis, 1999a,b). A polysaccharide fraction from *L. edodes* exhibits protective action in the livers of animals, as well as the ability to improve liver function and enhance the production of antibodies to hepatitis B (Amagase, 1987; Mizuno, 1995a,b: Wasser and Weis, 1999a,b). A polysaccharide derivative from the alcoholic extract of *Dendropolyporus umbellatus* has been reported to possess hepatoprotective effects in mice (Lin and Wu, 1989). The carboxymethylpachymaran from *P. cocos* has produced an "immediate cure" of chronic viral hepatitis in human clinical studies (Jong and Birmingham, 1994). *Tremella* polysaccharide promotes the synthesis of protein and RNA in mouse liver cells and thus liver regeneration (Lin, 1993). PSP from *T. versicolor* is effective in protecting the liver from hepatotoxin-induced damage in the rat by decreasing the covalent binding of hepatoxin paracetamol (Yeung et al., 1994), and sulphation and glucuronidation of paracetamol (Yeung et al., 1995). Thus, PSP may be useful in the treatment of hepatitis and other symptoms associated with immunodeficiency. The crude polysaccharide extracts of *L. edodes*, *G. frondosa*, and *Tricholoma lobayense* exhibit a significant hepatoprotective effect in paracetamol-induced liver injury in the rat model (Ooi, 1996). A polysaccharide extracted from the fruiting body of *G. lucidum* has a good hepatoprotective ability (Kim et al., 1999; Lin et al., 1993).

7 Antifibrotic Glycans

Polysaccharides extracted from *G. lucidum* have been extensively investigated as a source of antifibrotic agents in a hepatic cirrhosis model. *Ganoderma* polysaccharides significantly reduced serum liver transaminases (AST and ALT) and also lowered the collagen content in the liver of rats subjected to bile duct ligation and scission-induced fibrosis (Park et al., 1997).

8 Antiinflammatory Glycans

A significant antiinflammatory function has been found in polysaccharides obtained from *Auricularia auricula-judae*, *Ganoderma japonicum*, and *Dictyophora indusiata* (Urai et al., 1983). Similarly, two polysaccharide fractions from *Trametes gibbosa* administered intravenously strongly inhibit vessel permeability in the rat pleural exudate model and effectively antagonize the inflammation mediator complex induced by carrageenan (Czarnecki and Grzybek, 1995). Antiinflammatory and antiallergy functions have also been discovered in a heterogalactan–protein complex isolated from *G. lucidum* (Kohda et al., 1985; Lin et al., 1993).

9 Antidiabetic and Hypoglycemic Glycans

High hypoglycemic activity in normal mice and in alloxan-induced hyperglycemic mice has been induced by Ganoderan B and C. These two glycan–protein complexes are obtained from the fruiting body of *G. lucidum* by adding ethanol to a hot water extract and separating the precipitated polymer substances by column chromatography (Hikino et al., 1985, 1989). Hypoglycemic activity is due to an increase in the plasma insulin level and acceleration of glucose metabolism. In addition, 3% ammonium-oxalate-soluble heteroglycans and a 5% NaOH-soluble peptidoglycan have been isolated from the water-soluble fractions of *G. lucidum* polysaccharides. Several fractions of these heteroglycans separated by various chromatographic methods have shown strong antitumor activity and hypoglycemic activity (Hikino et al., 1985; Mizuno et al., 1995a,b,c). However, no correlation has been found between the antitumor activity of these active peptidoglycans and their antihyperglycemic activity or the ratio of glycans and proteins in the complex. Lentinan inhibits the development of insulin-dependent (Type I) diabetes mellitus in non-obese diabetic mice (Mizuno, 1995b). Coreolan, a β-glucan–protein, obtained from submerged cultured mycelia of *T. versicolor* shows activity against experimental diabetes in animal and *in vitro* tests (Wasser and Weis, 1999a,b). An acidic polysaccharide isolated from the fruiting body of *Tremella aurantica* shows remarkable hypoglycemic activity in normal mice and two diabetic mouse models, streptozotocin-induced diabetes and genetic diabetes (Kiho et al., 1994). A high-molecular-weight glycoprotein prepared from hot water decoction and compounds of an ether–ethanol extract of *G. frondosa*, which is fed to genetically diabetic rats, has also lowered blood glucose levels in a noninsulin dependent diabetes mellitus model (Mizuno and Zhuang, 1995a).

10 Hypocholesterolemic Glycans

PSK induces a significant reduction in the level of low-density-lipoprotein (LDL) cholesterol in hyperlipidemia (stage I) patients (Tsukagoshi et al., 1984). Polysaccharides isolated from *Tremella fuciformis* have also been shown to effectively lower the LDL-cholesterol in rats (Cheung, 1996).

11 Patents

Patents contain the most complete and detailed information on the compounds with medicinal effects that have been extracted from fungal cell wall glycans (Table 1).

12 Outlook and Perspectives

Complex glycans not only have a greater potential for structural variability than either proteins or nucleic acids, but they also have a

Tab. 1 Patents for products or processes involving fungal cell wall glycans

US patent number	Assignee	Inventor(s)	Title	Date of publication
3759896		Komatsu, N., Kikumoto, S., Kimura, K., Sakai, S., Kamasuka, T., Momaki, Y., Takada, S., Yamamoto, T., Sugayama, J.	Process for manufacture of polysaccharides with antitumor action	September 18, 1973
4051314	Kureha Kagaku Kogyo Kabushiki Kaisha	Chtsuka, S., Uneo, S., Yoshikumi, C., Hiroshi, F., Chmura, Y., Wada, T., Fujii, T., Takahashi, E.	Polysaccharides and method for producing same	September 27, 1977
4271151		Hotta, T., Enomoto, S., Yoshikumi, C., Ohara, M., Uneo, S.	Protein-bound polysaccharides	June 2, 1981
4289688	Kureha Kagaku Kogyo Kabushiki Kaisha	Hotta, T., Enomoto, S., Yoshkumi, C., Ohara, M., Ueno, S.	Protein-bound polysaccharides	September 15, 1981
4409385	Takara Shuzo Co., Ltd.	Nakajima, K., Hirata, Y., Uchida, H., Kimizuka, Y., Taniguchi, T., Obayasi, A., Tanabe, O.	Polysaccharides having anticarcinogenic activity and method for producing same	October 11, 1983
4614733	Kureha Kagaku Kogyo Kabushiki Kaisha	Yoshikumi, C., Fujii, T., Fujii, M., Ohara, M., Kobayashi, A., Akatsu, T.	Polysaccharides pharmaceutical compositions and the use thereof	September 30, 1986
4820689	Kureha Kagaku Kogyo Kabushiki Kaisha	Ikuzawa, M., Oguchi, Y., Matsunaga, K., Toyoda, N., Furusho, T., Fujii, T., Yoshikumi, C.	Pharmaceutical composition containing a glycoprotein	April 11, 1989
5008243	Kureha Kagaku Kobushiki Kaisha	Ikuzawa, M., Oguchi, Y., Matsunaga, K., Toyoda, N., Furusho, T., Fujii, T., Yoshikumi, C.	Pharmaceutical composition containing a glycoprotein	Apr. 16, 1991
5374714		Yang, M. M. P., Chen, G.	Purified *Coriolus versicolor* polypeptide complex	Dec. 20, 1994
5585467	Il-Yang Pharm. Co., Ltd.	Lee, K. H., Chung, H., Lee, C. W., Chung, C. H.	Proteoglycan (G009) effective in enhancing antitumor immunity	December 17, 1996
5824648		Yang, M. M. P., Chen, G.	Rnase-CV (*Coriolus versicolor*)	October 20, 1998
6087335		Yang, M. M. P., Chen, G.	RNase-CV (*Coriolus versicolor*)	July 11, 2000
6090615		Nagaoka, H.	Method for extracting a Basidiomyces mycelium-containing culture medium using beta-1,3-glucanase	July 18, 2000

higher capacity for carrying biological information and participating in cellular interactions. Unlike the amino acids in proteins and the nucleotides in nucleic acids that bind to one another at a single site, the monosaccharide units in oligosaccharides and polysaccharides can link at several points to create a variety of linear or branched structures (Ooi and Liu, 1999). This potential combined with the wide range of biological activities controlled by the cytokine network provides the precise regulatory mechanisms necessary to maintain homeostasis in higher organisms. In addition, glycosylation of proteins and the cascade of signal transduction affect the activation, differentiation, and proliferation of lymphocytes (Benlagha et al., 1999). Glycosylated proteins are essential components of plasma membranes in eukaryotic cells (Kieda, 1998; Dunn and Dunn, 1999; Orntoft and Vestergaard, 1999) and protein kinase C is an essential element in the signal transduction pathways (Newton, 1995; Kurosaki, 1999). It is also known that the interactions of glycans with proteins or RNA play an important role in many biological and cellular processes (Hermann and Westhof, 1998; Nemoto and Irimura, 1999). Therefore, the wide range of physicochemical properties found in fungal cell wall glycans is the basis for different applications in the broad fields of pharmacy and medicine.

For instance, more than 50 mushroom species have yielded glycans that exhibit anticancer activity *in vitro* or in animal models. All are chemically β-D-glucan in nature or β-D-glucans linked to proteins, so called polysaccharide-peptides or proteoglycans. As a rule, the protein-linked glucans have greater immunopotentiating activity than the free glucans. Several glucans have shown clinically significant efficacy in human cancers. Because of their large molecular weight, lentinan and schizophyllan have little oral activity and must be administered intravenously or intramuscularly, respectively. However, PSK and PSP, the two proteoglycans from *Coriolous versicolor,* which can be administered orally, have been systematically investigated and exhibit the most promise as anticancer therapeutics.

Glucans and proteoglycans are prohomeostatic and compare favorably with classic biological response modifiers. They boost the immune function in healthy people, helping to prevent infections and promote good health and general well-being. At the same time they can enhance the immune system against cancer and protect against the immunosuppression that typically accompanies surgery and long-term chemotherapy. Their use alone or in combination with chemotherapy and radiation make them well suited for cancer management regimens. When administered as dietary supplements, they exhibit near perfect benefit–risk profiles (Kidd, 2000).

13 References

Abel, G., Czop, J. K. (1992) Stimulation of human monocyte beta-glucan receptors by glucan particles induces production of TNF-alpha and IL-1 beta, *Int. J. Immunopharmacol.* **14**, 1363–1373.

Adachi, Y., Ohno, N., Ohsawa, M., Oikawa, S., Yadomae, T. (1990) Change of biological activities of (1–3)-β-D-glucan from *Grifola frondosa* upon molecular weight reduction by heat treatment, *Chem. Pharmacol. Bull.* **38**, 477–481.

Adachi, Y., Okazaki, M., Ohno, N., Yadomae, T. (1994) Enhancement of cytokine production by macrophages stimulated with (1–3)-β-D-glucan, Grifolan (GRN), isolated from *Grifola frondosa*, *Biol. Pharmacol. Bull.* **17**, 1554–1560.

Adachi, Y., Ohno, N., Yadomae, T. (1998) Activation of murine kupffer cells by administration with gel-forming (1–3)-β-D-glucan from *Grifola frondosa*, *Biol. Pharmacol. Bull.* **21**, 278–283.

Amagase, H. (1987) Treatment of hepatitis B patients with *Lentinus edodes* mycelium, Proceedings of the XII International Congress of Gastroenterology, Lisbon, p. 197.

Astarie-Dequeker, C., N'Diaye, E. N., Le Cabec, V., Ritting, M. G., Prandi, J., and Maridonneau-Parini, I. (1999) The mannose receptor mediates uptake of pathogenic and nonpathogenic mycobacteria and bypasses bactericidal responses in human macrophages, *Infect. Immun.* **67**, 469–477.

Benlagha, L., Guglielmi, P., Cooper, M. D., Lassoued, K. (1999) Modifications of Igalpha and Igbeta expression as a function of B lineage differentiation, *J. Biol. Chem.* **174**, 19389–19396.

Byram, J. E., Sher, A., DiPietro, J., Von Lichtenberg, F. (1979) Potentiation of schistosome granuloma formation by lentinan – a T cell adjuvant, *Am. J. Pathol.* **94**, 201–222.

Chen, H. Y., Kaneda, S., Mikami, Y., Arai, T., Igarashi, K. (1987) Protective effects of various BRMs (biological response modifiers) against *Candida albicans* infection in mice, *Shinkin to Shinkinsho* **28**, 306–315.

Chen, C. N., Yang, M. M. P., Zhu, P. (1993) Research and application of the anticancerous mechanism of the purified PSP (PCV), in: *Proceedings of PSP International Symposium* (Yang, Q. Y., Kwok, C. Y., Eds.), Shanghai: Fudan University Press, 151–152.

Cheung, P. C. K. (1996) The hypocholesterolemic effect of two edible mushrooms: *Auricularia auricula* and *Tremella fuciformis* in hypercholesterolemic rats, *Nutr. Res.* **16**, 1721–1725.

Chihara, G. (1992) Immunopharmacology of lentinan, a polysaccharide isolated from *Lentinus edodes*: its application as a host defense potentiator, *Int. J. Oriental Med.* **17**, 55–77.

Chihara, G. (1993) Medical aspects of lentinan isolated from *Lentinus edodes* (Berk.) Sing, in: *Mushroom Biology and Mushroom Products* (Chang, S. T., Buswell, J. A., Chiu, S. W., Eds.), Hong Kong: The Chinese University of Hong Kong Press, 261–266.

Collins, R. A., Ng, T. B. (1997) Polysaccharopeptide from *Coriolus versicolor* has potential for use against human immunodeficiency virus type I infection, *Life Sci.* **60**, 383–387.

Czarnecki, R., Grzybek, J. (1995) Antiinflammatory and vasoprotective activities of polysaccharides isolated from fruiting bodies of higher fungi (P. I), Polysaccharides from *Trametes gibbosa* (Pers.:Fr) (Polyporaceae), *Phytother. Res.* **9**, 123–127.

Czop, J. K., Austen, K. F. (1985) A beta-glucan inhibitable-receptor on human monocytes: Its identity with the phagocytic receptor for particulate activators of the alternative complement pathway, *J. Immunol.* **134**, 2588–2593.

Czop, J. K., Valiante, N. M., Janusz, M. J. (1989) Phagocytosis of particulate activators of the human alternative complement pathway through monocyte beta-glucan receptors, *Prog. Clin. Biol. Res.* **297**, 287–296.

Dong, Y., Kwan, C. Y., Chen, Z. N., Yang, M. . M. (1996) Antitumor effects of a refined polysaccharide peptide fraction isolated from *Coriolus versicolor*: *in vitro* and *in vivo* studies, *Res. Commun. Mol. Pathol. Pharmacol.* **92**, 140–148.

Dong, Y., Yang, M. M. P., Kwan, C. Y. (1997) *In vitro* inhibition of proliferation of HL-60 cells by tetrandrine and *Coriolus versicolor* peptide derived from Chinese medicinal herbs, *Life Sci.* **60**, PL135–140.

Dunn, J. T., Dunn, A. D. (1999) The importance of thyroglobulin structure for thyroid hormone biosynthesis, *Biochimie* **81**, 505–509.

Ebina, T., Fujimiya, Y. (1998) Antitumor effect of a peptide–glucan preparation extracted from *Agaricus blazei* in a double-grafted tumor system in mice, *Biotherapy* **11**, 259–265.

Franz, G. (1989) Polysaccharides in pharmacy: current applications and future concepts, *Plant Med.* **55**, 493–497.

Han, M. D., Jeong, H., Lee, J. W., Back, S. J., Kim, S. U., Yoon, K. H. (1995) The composition and bioactivities of ganoderan by mycelial fraction of *Ganoderma lucidum* IY009, *Korean J. Mycol.* **23**, 285–297.

Hermann, T., Westhof, E. (1998) Saccharide–RNA recognition, *Biopolymers* **48**, 155–165.

Hikino, H., Konno, C., Mikrin, Y., Hayashi, T. (1985) Isolation and hypoglycemic activities of ganoderans A and B, glucans of *Ganoderma lucidum* fruit bodies, *Planta Med.* **4**, 339–340.

Hikino, H., Ishiyama, M., Suzuki, Y., Konno, C. (1989) Mechanisms of hypoglycemic activity of ganoderan B: A glycan of *Ganoderma lucidum* fruit bodies, *Planta Med.* **55**, 423–428.

Iguchi, Y. Kato, N., Yugari, Y., Kodama, K., Mitsuhashi, S. (1985) Enhancement of nonspecific resistance against bacterial infections in immuno-compromised mice by lentinan, an immunoactive polysaccharide, in: *Recent Advances in Chemotherapy: . Proceedings of the 14th International Congress on Chemotherapy* (Shigami, J., Ed.), Tokyo: University of Tokyo Press, 239–240.

Iizuka, C., Maeda, H. (1988) Anti-AIDS drug containing sugars and inorganic elements extracted from *Lentinus edodes*, European patent 292–601 (1988).

Ikekawa, T. (1995) Inokitake, *Flammulina velutipes*: host-mediated antitumor polysaccharides, *Food Rev. Int.* **11**, 203–206.

Ito, H., Shimura, K., Itoh, H., Kawade, M. (1997) Antitumor effects of new polysaccharide–protein complex (ATOM) prepared from *Agaricus blazei* (Iwade strain 101) "Himematsutake" and its mechanism in tumor-bearing mice, *Anticancer Res.* **17**, 277–284.

Jong, S. C., Birmingham, J. M. (1990) The medicinal value of the mushroom *Grifola*, *MIRCEN J. Appl. Microbiol. Biotechnol.* **6**, 227–235.

Jong, S. C., Birmingham, J. M. (1992a) Edible mushrooms in biotechnology, Proceedings of the Asian Mycological Symposium 1992, Seoul, pp. 18–35.

Jong, S. C., Birmingham, J. M. (1992b) Medicinal benefits of the mushroom *Ganoderma*, *Adv. Appl. Microbiol.* **37**, 101–134.

Jong, S. C., Birmingham, J. M. (1993a) Medicinal and therapeutic value of the shiitake mushroom, *Adv. Appl. Microbiol.* **39**, 153–184.

Jong, S. C., Birmingham, J. M. (1993b) Medicinal benefits of the shiitake mushroom, Proceedings of the National Shiitake Mushroom Symposium 1993, Alabama A & M University, pp. 120–134.

Jong, S. C., Birmingham, J. M. (1994) Medicinal and therapeutic value of the fungus *Poria cocos*, *SIM Ind. Microbiol. News.* **44**, 171–176.

Jong, S. C., Birmingham, J. M. (1998) Federal laws and regulations of health claims on mushrooms and their products, *Mushroom News* **46**, 25–31.

Jong, S. C., Donovick, R. (1989) Antitumor and antiviral substances from fungi, *Adv. Appl. Microbiol.* **34**, 183–262.

Jong, S. C., Yang, X. T. (1999) PSP – a powerful biological response modifier from the mushroom *Coriolus versicolor*, in: *Advanced Research in PSP 1999* (Yang, Q. Y., Ed.), Hong Kong: The Hong Kong Association for Health Care, 16–28.

Jong, S. C., Birmingham, J. M., Pai, S. H. (1991) Immunomodulatory substances of fungal origin, *EOS-J. Immunol. Immunopharmacol.* **11**, 115–122.

Kanayama, H., Adachi, N., Togami, M. (1983) A new antitumor polysaccharide from the mycelia of *Poria cocos* Wolf, *Chem. Pharmacol. Bull.* **31**, 1115–1118.

Kidd, P. M. (2000) The use of mushroom glucans and proteoglycans in cancer treatment, *Altern. Med. Rev.* **5**, 4–27.

Kieda, C. (1998) Role of lectin–glycoconjugate recognitions in cell–cell interactions leading to tissue invasion, *Adv. Exp. Med. Biol.* **435**, 75–82.

Kiho, T., Yoshida, I., Katsuragawa, M., Sakushima, M., Usui, S., Ukai, S. (1994) Polysaccharides in fungi, XXXIV, a polysaccharide from the fruiting bodies of *Amanita muscaria* and the antitumor activity of its carboxymethylated product, *Biol. Pharmacol. Bull.* **17**, 1460–1462.

Kim, D. H., Shim, S. B., Kim, N. J., Jang, I. S. (1999) Beta-glucuronidase-inhibitory activity and hepatoprotective effect of *Ganoderma lucidum*, *Biol. Pharmacol. Bull.* **22**, 162–164.

Kobayashi, H., Matsunaga, K., Fujii, M. (1993) PSK as a chemopreventive agent (review), *Cancer Epidemiol. Biomarkers Prevent.* **2**, 171–276.

Kobayashi, Y., Kariya, K., Saigenji, K., Nakamura, K. (1994) Enhancement of anti-cancer activity of cisdiaminedichloroplatinum by the protein-bound polysaccharide of *Coriolus versicolor* QUEL (PSK) *in vitro*, *Cancer Biother.* **9**, 351–358.

Kobayashi, H., Matsunaga, K., Oguchi, Y. (1995) Antimetastatic effects of PSK (Krestin), a protein-bound polysaccharide obtained from Basidiomyces: a review, *Cancer Epidemiol. Biomarkers Prevent.* **4**, 275–281.

Kohda, H., Tokumoto, W., Sakamoto, K., Fujii, M., Hirai, Y., Yamasaki, K., Komoda, Y., Nakamura, H., Ishihara, S., Uchida, M. (1985) The biologically active constituents of *Ganoderma lucidum* (Fr.) Karst. Histamine release-inhibitory triterpenes, *Chem. Pharmacol. Bull.* **33**, 1367–1374.

Kohgo, Y., Hirayama, Y., Sakamaki, S., Matsunaga, T., Ohi, S., Kuga, T., Kato, J., Niitsu, Y. (1994) Improved recovery of myelosuppression following chemotherapy in mice by combined administration of PSK and various cytokines, *Acta Haematol.* **92**, 130–135.

Kubo, K., Aoki, H., Noriba, H. (1994) Anti-diabetic activity present in the fruit body of *Grifola frondosa* (Maitake), *Biol. Pharmacol. Bull.* **17**, 1106–1110.

Kurosaki, T. (1999) Genetic analysis of B cell receptor signaling, *Annu. Rev. Immunol.* **17**, 555–592.

Lei, L. S., Lin, Z. B. (1992) Effect of *Ganoderma* polysaccharides on cell subpopulations and production of interleukin 2 in mixed lymphocyte response, *Acta Pharmacol. Sinica* **27**, 331–335.

Lin, J. M., Lin, C. C., Chiu, H. F., Yang, J. J., Lee, S. G. (1993) Evaluation of the anti-inflammatory and liver-protective effects of *Anoectochilus formosanus*, *Ganoderma lucidum* and *Gynostemma pentaphyllum* in rats, *Am. J. Chin. Med.* **21**, 59–69.

Lin, Z. B. (1993) Advances in the pharmacology of *Tremella* polysaccharides, in: *Mushroom Biology and Mushroom Products* (Chang, S. T., Buswall, J. A., Chiu, S. W., Eds.), Hong Kong: The Chinese University of Hong Kong Press, 293–299.

Lin, Y. F., Wu, L. (1989) Protective effect of *Polyporus umbellatus* on toxic hepatitis in mice, *Acta Pharmacol. Sinca* **9**, 345–348.

Liu, F., Fung, M. C., Ooi, V. E. C., Chang, S. T. (1996) Induction in the mouse of gene expression of immunomodulating cytokines by mushroom polysaccharide–protein complexes, *Life Sci.* **58**, 1795–1803.

Liu, F., Ooi, V. E. C., Fung, M. C. (1999) Analysis of immunomodulating cytokine mRNAs in the mouse induced by mushroom polysaccharides, *Life Sci.* **64**, 1005–1011.

Mizuno, T. (1995a) Bioactive biomolecules of mushrooms: food function and medicinal effect of mushroom fungi, *Food Rev. Int.* **11**, 7–21.

Mizuno, T. (1995b) Shiitake, *Lentinus edodes*: functional properties for medicinal and food purposes, *Food Rev. Int.* **11**, 111–128.

Mizuno, T. (1995c) Sarunokoshikake: polyporaecea fungi – Kofukisarunokoshikake, *Ganoderma applanatum* and Tsugasarunokoshikake, *Fomitopsis pinicola*, *Food Rev. Int.* **11**, 129–133.

Mizuno, T. (1995d) Kawariharatake, *Agaricus blazei* Murrill: medicinal and dietary effects, *Food Rev. Int.* **11**, 167–172.

Mizuno, T. (1995e) Yamabushitake, *Hericium erinaceum*: bioactive substances and medicinal utilization, *Food Rev. Int.* **11**, 173–178.

Mizuno, T., Zhuang, C. (1995a) Maitake, *Grifola frondosa*: pharmacological effects, *Food Rev. Int.* **11**, 135–150.

Mizuno, T. Zhuang, C. (1995b) Houbitake, *Pleurotus sajor-caju*: antitumor activity and utilization, *Food Rev. Int.* **11**, 185–188.

Mizuno, T., Sakai, K., Chihara, G. (1995a) Health foods and medicinal usages of mushrooms, *Food Rev. Int.* **11**, 69–82.

Mizuno, T., Sato, H., Nishitoba, T., Kawagishi, H. (1995b) Antitumor-active substances from mushrooms, *Food Rev. Int.* **11**, 13–62.

Mizuno, T., Wang, G., Zhang, J., Kawagishi, H., Nishitoba, T. (1995c) Reishi, *Ganoderma lucidum* and *Ganoderma tsugae*: bioactive substances and medicinal effects, *Food Rev. Int.* **11**, 151–166.

Mizuno, M., Morimoto, M., Minato, K., Tsuchida, H. (1998) Polysaccharides from *Agaricus blazei* stimulate lymphocyte T-cell subsets in mice, *Biosci. Biotechnol. Biochem.* **62**, 1607–1612.

Mizutani, Y., Yoshida, O. (1991) Activation by the protein-bound polysaccharide PSK (Krestin) of cytotoxic lymphocytes that act on fresh autologous tumor cells and T24 human urinary bladder transitional carcinoma cell line in patients with urinary bladder cancer, *J. Urol.* **145**, 1082–1087.

Mori, K., Toyomasu, T., Nanba, H., Kuroda, H. (1987) Antitumor activity of fruit bodies of edible mushrooms orally administered to mice. *Mushroom J. Tropics* **7**, 121–126.

Nemoto, Y., Irimura, T. (1999) Tumor metastases and adhesion molecules carbohydrates and lectins, *Gan To Kagaku Ryoho* **26**, 849–856.

Newton, A. C. (1995) Protein kinase C: structure, function, and regulation, *J. Biol. Chem.* **270**, 28495–28498.

Ng, T. B. (1998) A review of research on the protein-bound polysaccharide (polysaccharopeptide, PSP) from the mushroom *Coriolus versicolor* (Basidiomyces: Polyporaceae), *Gen. Pharmacol.* **30**, 1–4.

Ng, T. B., Ling, J. M. L., Wang, Z. T., Cai, J. N., Zy, G. J. (1996) Examination of coumarins, flavonoids and polysaccharopeptide for antibacterial activity, *Gen. Pharmacol.* **27**, 1237–1240.

Ooi, V. E. C. (1996) Hepatoprotective effect of some edible mushrooms, *Phytother. Res.* **10**, 536–538.

Ooi, V. E. C., Liu, F. (1999) A review of pharmacological activities of mushroom polysaccharides, *Int. J. Med. Mushrooms* **1**, 195–206.

Orntoft, T. F., Vestergarrd, E. M. (1999) Clinical aspects of altered glycosylation of glycoproteins in cancer, *Electrophoresis* **20**, 362–371.

Park, E. J., Ko, G., Kim, J., Sohn, D. H. (1997) Antifibrotic effects of a polysaccharide extracted from *Ganoderma lucidum*, glycyrrhizin, and pentoxifyline in rats with cirrhosis induced in biliary obstruction, *Biol. Pharmacol. Bull.* **37**, 3134–3136.

Qian, Z. M., Xy, M. F., Tang, P. L. (1997) Polysaccharide peptide (PSP) restores immunosuppression induced by cyclophosphamide in rats, *Am. J. Chin. Med.* **25**, 27–35.

Saito, K., Nishjima, M., Miyazaki, T. (1989) Studies on fungal polysaccharides. XXXV. Structural analysis of an acidic polysaccharide from *Ganoderma lucidum*, *Chem. Pharmacol. Bull.* **37**, 3134–3136.

Sakagami, H., Aoki, T., Simpson, A., Tanuma, S. (1991) Induction of immunopotentiation activity by a protein-bound polysaccharide, PSK [Review], *Anticancer Res.* **11**, 993–1000.

Sugano, N., Choji, Y., Takarada, T., Maeda, H. (1985) Immunopotentiation agents. European patent 154066.

Sugano, T. Yoshihama, T., Tsuchiya, T., Shio, T., Maeda, Y., Chihara, G. (1989) Prevention of tumour metastasis and recurrence of DBA/2, NC, CS-T fibrosarcoma, MH-134 hepatoma and other murine tumours using lentinan, *Int. J. Immunother.* **5**, 187–193.

Suzuki, F., Suzuki, C., Shimomura, E., Maeda, H., Fujii, T., Ishida, N. (1979) Antiviral and interferon-inducing activities of a new peptidomannan, KS-2, extracted from culture mycelia of *Lentinus edodes*, *J. Antibiot.* **32**, 1336–1345.

Szabo, T., Kadish, J. L., Czop, J. K. (1995) Biochemical properties of the ligand-binding 20-kDa subunit of the beta-glucan receptors on human mononuclear phagocytes, *J. Biol. Chem.* **270**, 2145–2151.

Tochikura, T. S., Naksshima, H., Kaneko, Y., Kobayashi, N., Yamamoto, N. (1987) Suppression of human immunodeficiency virus replication by 3′-azido-3′-deoxythymidine in various human hematopoietic cell lines *in vitro*: augmentation of effect by lentinan, *Jpn J. Cancer Res.* **78**, 583–589.

Tsukagoshi, S., Hashimota, Y., Fujii, G., Kobayashi, H., Nomoto, K., Orita, K. (1984) Krestin (PSK), *Cancer Treat. Rev.* **11**, 131–155.

Ukai, S., Kiho, T., Hara, C., Kuruma, Z., Tanaka, Y. (1983) Polysaccharides in fungi, XIV, Antiinflammatory effect of the polysaccharides from the fruit bodies of several fungi, *J. Pharmacol. Dynam.* **6**, 983–990.

Wakshull, E., Brunke-Reese, D., Lindermuth, J., Fisette, L., Nathans, R. S., Crowley, J. J., Tufts, J. C., Zimmerman, J., Mackin, W., Adams, D. S. (1999) PGG-glucan, a soluble beta-(1,3)-glucan, enhances the oxidative burst response, microbicidal activity, and activates an NF-kappa B-like factor in human PMN: evidence for a glycosphingolipid beta-(1,3)-glucan receptor, *Immunopharmacology* **41**, 89–107.

Wang, H. X., Ng, T. B., Ooi, V. E. C., Liu, W. K., Chang, S. T. (1996a) A polysaccharide-peptide complex from culture mycelia of the mushroom *Tricholoma mongolicum* with immunoenhancing and antitumor activities, *Biochem. Cell Biol.* **74**, 95–100.

Wang, H. X., Ng, T. B., Liu, W. K., Ooi, V. E. C., Chang, S. T. (1996b) Polysaccharide–peptide complexes from the cultured mycelia of the mushroom *Coriolus versicolor* and their culture medium activate mouse lymphocytes and macrophages, *Int. J. Biochem. Cell Biol.* **28**, 601–607.

Wang, S. Y., Hsu, M L., Hsu, H . C., Tzeng, C. H., Lee, S. S., Shiao, M . S., Ho, C. K. (1997) The antitumor effect of *Ganoderma lucidum* is mediated by cytokines released from activated macrophages and T. lymphocytes, *Int. J. Cancer* **70**, 699–705.

Wasser, S. P., Weis, A. L. (1999a) Therapeutic effects of substances occurring in higher Basidiomycetes mushrooms: a modern perspective, *Crit. Rev. Immunol.* **19**, 65–69.

Wasser, S. P., Weis (1999b) Medicinal properties of substances occurring in higher Basidiomycetes mushrooms: current perspectives [Review], *Int. J. Med. Mushrooms.* **1**, 31–62.

Wu, Z., Chen, C., Chen, Y. M. (1993) Clinical observation of esophageal cancer treatment by radiotherapy together with the oral administration of PSP, in: *Proceedings of PSP International Symposium* (Yang, Q. Y., Kwok, C. Y., Eds.), Shanghai: Fudan University Press, 265–268.

Wu, D., Han, S. N., Bronson, R. T., Smith, D. E., Meydani, S. N. (1998) Dietary supplementation with mushroom-derived protein-bound glucan does not enhance immune function in young and old mice, *J. Nutr.* **128**, 193–197.

Xia, Y., Vetvicka, V., Yan, J., Hankyrova, M., Mayadas, T., Ross, G. D. (1999) The beta-glucan-binding lectin site of mouse CR3 (CD11b/CD18) and its function in generating a primed state of the receptor that mediates cytotoxic activation in response to iC3b-opsonized target cells, *J. Immunol.* **162**, 2281–2290.

Yamamura, Y., Cochran, K. W. (1976) A selective inhibitor of myxoviruses from shiitake (*Lentinus edodes*), *Mushroom Sci.* **9**, 495–507.

Yang, Q. Y., Jong, S. C., Li, X. Y., Zhou, J. X., Chen, R. T., Xu, L. Z. (1992) Antitumor and immunomodulating activities of the polysaccharide–peptide (PSP) of *Coriolus versicolor, J. Immunol. Immunopharmacol.* **12**, 29–34.

Yeung, J. H. K., Chiu, L. C. M., Ooi, V. E. C. (1994) Effect of polysaccharide peptide (PSP) on glutathione and protection against paracetamol-induced hepatotoxicity in the rat, *Methods Find. Exp. Clin. Pharmacol.* **16**, 723–729.

Yeung, J. H. K., Chiu, L. C. M., Ooi, V. E. C. (1995) Effect of polysaccharide peptide (PSP) on glutathione and protection against paracetamol-induced hepatotoxicity in the rat, *Methods Find. Exp. Clin. Pharmacol.* **16**, 723–729.

Yoshida, I., Kiho, T., Usui, S., Sakushima, M., Ukai, S. (1996) Polysaccharides in fungi XXXVII, Immunomodulating activities of carboxymethylated derivatives of linear (1–3)-alpha-D-glucans extracted from the fruiting bodies of *Agrocybe cylindracea* and *Amanita muscaris, Biol. Pharmacol. Bull.* **19**, 114–121.

Zhang, X., Luan, H. (1986) Effects of some fungal polysaccharides on experimental hepatitis in mice, *J. Northeast Normal Univ.* **4**, 101–108.

Zhang, J., Wang, G., Li, H., Zhuang, C., Mizuno, T., Ito, H., Mayuzumi, H. G., Okamoto, H., Li, J. (1994) Antitumor active protein-containing glycans from the Chinese mushroom Songshan Lingzhi, *Ganoderma tsugae* mycelium, *Biosci. Biotechnol. Biochem.* **58**, 1202–1205.

18
Hyaluronan

Prof. Dr. Peter Prehm
Institut für Physiologische Chemie und Pathobiochemie, Waldeyerstr. 15,
D-48129 Münster, Germany; Tel.: +49-251-8355579; Fax: +49-251-8355596;
E-mail: prehm@uni-muenster.de

Biopolymers for Medical and Pharmaceutical Applications. Edited by A. Steinbüchel and R. H. Marchessault

ISBN: 3-527-31154-8

CHO	Chinese hamster ovary
PMA	phorbol-12-myristate-13-acetate
RHAMM	receptor for hyaluronan-mediated motility

1 Introduction

Although hyaluronan has a very simple structure, almost everything else concerning the molecule is unusual. Sometimes its role is mechanical and structural (as in the synovial fluid, the vitreous humor, or the umbilical cord), whereas sometimes it interacts in tiny concentrations in cells to trigger important responses. Hyaluronan has an unusual mechanism of biosynthesis and exceptional physical properties; consequently, research on this compound was cumbersome, with progress often impeded by failures – often because established procedures from other fields were not applicable and new techniques needed to be developed before any progress could be made.

During the decades of hyaluronan research, several books and reviews have marked such progress including Balazs (1970), Laurent (1989, 1998), Laurent and Fraser (1992), Goa and Benfield (1994), Lapcik et al. (1998), and Abatangelo and Weigel (2000), whilst reviews are published continually on a new web-site: http://www.glycoforum.gr.jp/science/hyaluronan.

2 Historical Outline

In 1934, Karl Meyer described a procedure for isolating a novel glycosaminoglycan from the vitreous humor of bovine eyes, and named it hyaluronic acid (from the Greek, hyalos = glassy, vitreous) (Meyer and Palmer, 1934). These authors showed that this substance contained a uronic acid and an amino sugar, but no sulfoesters. At physiological pH all carboxyl groups are dissociated, and hence the polysaccharide should be called hyaluronate. Today, this macromolecule is most frequently referred to as hyaluronan, in order to emphasize its polysaccharide nature. During the 1930s and 1940s, hyaluronan was isolated from many sources such as the vitreous body, synovial fluids, umbilical cord, skin, and rooster comb (Meyer, 1947) and also from streptococci (Kendall et al., 1937).

The physico-chemical characterization of hyaluranon was carried out during the 1950s and 1960s. The molecular weight is in the order of several millions, whilst in solution the chain behaves as an extended random coil, with a diameter of ~500 nm. At concentrations as low as 0.1%, the chains are entangled, and this results in extremely high, shear-dependent viscosity (Laurent, 1970). These properties enable hyaluronan to regulate water balance, osmotic pressure and flow resistance, to interact with proteins, and also to act as a sieve, as a lubricant, and to stabilize structures by virtue of electrostatic interactions (Comper and Laurent, 1978).

In 1972, Hardingham and Muir discovered that hyaluronan interacts with cartilage proteoglycans and serves as the central structural backbone of cartilage. This was the first example of a specific interaction between hyaluronan and a protein, and many more such interactions were discovered during the 1990s.

After 1980, the research spread in many directions, mainly because until that time it had been assumed that hyaluronan belonged to the proteoglycans, and that its biosynthesis proceeded in a similar manner. In fact, many studies were conducted to detect the protein moiety, but this assumption was disproved when a plausible mechanism of biosynthesis was proposed (Prehm, 1983a,b). It had also been assumed that the synthesis of hyaluronan occurred in the Golgi body – as was the case for all other secretory eucaryotic polysaccharides – until it was shown that hyaluronan was in fact synthesized at plasma membranes and the chains were extruded directly into the extracellular matrix (Prehm, 1984). The catabolic pathways of hyaluranon were also elucidated at about this time (Fraser et al., 1981).

Subsequently, it became possible to measure hyaluronan specifically in body fluids with high sensitivity (Tengblad et al., 1980), and also to visualize it histochemically. These advances opened the way to assess the role of hyaluronan in many pathological disturbances.

Balazs pioneered the application of hyaluronan for medical purposes, and produced highly viscous and noninflammatory preparations on a commercial scale both as an aid for ophthalmic surgery and as viscosupplementation for synovial fluids in patients with osteoarthritis (Balazs, 1982; Balazs and Denlinger, 1989).

Although the importance of hyaluronan in cellular behavior had been recognized for decades, it was not until 1986 that its requirement for detachment in mitotic cell division was proven (Brecht et al., 1986). Underhill and Toole (1979, 1982) reported that hyaluronan was an adhesive cell surface component that formed large coats around untransformed fibroblasts, and smaller coats around transformed cells.

Cell surface hyaluronan-binding proteins were discovered during the late 1980s (Turley et al., 1987; Aruffo et al., 1990), and studied intensively during the 1990s. Although hyaluranon was already known to be involved in both metastasis (Toole et al., 1979) and cell differentiation (Toole et al., 1977), it was investigations into the molecular biology of the receptors which led to a fundamental understanding of these processes. In particular, the receptors CD44 and RHAMM (Receptor for Hyaluronan-Mediated Motility) have attracted much enthusiasm, mainly because they are believed to be involved in cancer metastasis (Arch et al., 1992; Hall et al., 1995). However, a sobering shock reached the scientific community, when CD44-deficient mice were bred that had only marginal physiological impairments (Schmits et al., 1997). In addition, the receptor RHAMM became a matter of bitter scientific debate when it was found to be located mainly intracellularly (Hofmann et al., 1998a; Turley et al., 1998). Subsequently, a number of other intracellular hyaluronan-binding proteins have been found (Huang et al., 2000), though their function remains somewhat of a mystery. During the 1990s, hyaluronan synthases were cloned from different sources (Weigel et al., 1997), each capable of producing hyaluronan of different chain lengths and quantities (Itano et al., 1999).

The actions of hyaluronan as an adhesive component and also as a detachment factor appeared paradoxical. This paradox has recently been solved however, when it was realized that the cellular functions are mediated through cell-surface receptors that are susceptible to proteases (Dube et al., 2001). It appears that hyaluronan acts as an amplifier for active proteases, but as an attachment factor when proteases are inactive.

It has long been known that hyaluronan is very sensitive to breakdown by oxygen radicals (Wong et al., 1981), and it has become clear that it is the breakdown products which mediate important biological functions. Oligosaccharides of hyaluronan induce angiogenesis (West et al., 1985) and also activate lymphocytes (McKee et al., 1997; Termeer et al., 2000). Radical degradation generates reactive aldehydes (Uchiyama et al., 1990) which modify proteins into the main antigenic structures of rheumatoid arthritis (Prehm, 2000). This discovery finally terminated a long and oppressive period of ignorance in a medically important problem, and may eventually lead to a curative treatment of these diseases that currently are treated only symptomatically.

3 Chemical Structure

The complete structure of hyaluranon was elucidated by the group of Karl Meyer, who characterized the oligosaccharides obtained by the action of testicular hyaluronidases (Weissman and Meyer, 1954). Hyaluronan consists of basic disaccharide units of D-glucuronic acid and D-*N*-acetylglucosamine, these being linked together through alternating β-1,4 and β-1,3 glycosidic bonds (Figure 1).

The number of repeat disaccharides in a completed hyaluronan molecule can reach 10,000 or more, with a molecular mass of ~4×10^6 daltons (each disaccharide is ~400 daltons). In a physiological solution, the backbone of a hyaluronan molecule is stiffened by a combination of the chemical structure of the disaccharide, internal hydrogen bonds, and interactions with solvent. In addition, the preferred shape in water features hydrophobic patches on alternating sides of the flat, tape-like secondary structure. The two sides are identical, so that hyaluronan molecules are ambidextrous, enabling them to aggregate via specific interactions in water to form meshworks, even at low concentrations (Scott et al., 1991). In physiological solutions a hyaluronan molecule assumes an expanded random coil structure which occupies a very large domain. The actual mass of hyaluronan within this domain is very low, and ~0.1% molecules would overlap each other at a hyaluronan concentration of 1 mg mL^{-1}, or higher. This domain structure of hyaluronan has interesting and important consequences. Small molecules such as water, electrolytes and nutrients can diffuse freely through the solvent, within the domain; however, large molecules such as proteins will be partially excluded from the domain because of their hydrodynamic sizes in solution. This leads both to slower diffusion of macromolecules through the network and to their lower concentration in the network compared with the surrounding hyaluronan-free compartments. At pH 7, the carboxyl groups are predominantly ionized, and the hyaluronan molecule is a polyanion that has associated exchangeable cation counterions to maintain charge neutrality.

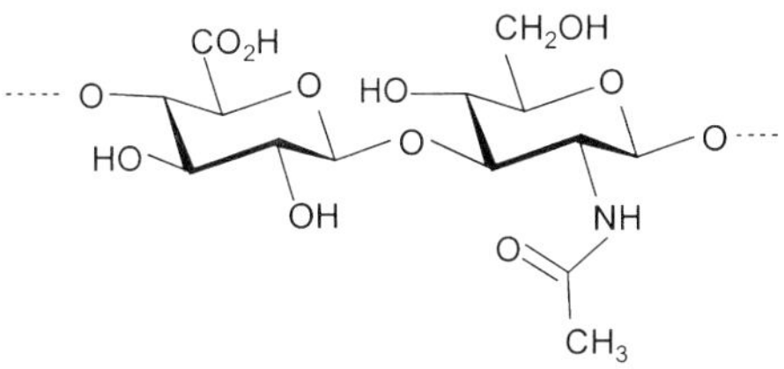

Fig. 1 Repeating unit of hyaluranon.

4 Occurrence of Hyaluronan

Hyaluronan is present in all vertebrates, and also in the capsule of some pathogenic bacteria such as *Streptococcus* sp. and *Pasteurella*. It is a component of extracellular matrices in most tissues, and in some tissues it is a major constituent. The concentration of hyaluronan is particularly high in rooster comb (7.5 mg mL^{-1}), in the synovial fluid (3–4 mg mL^{-1}), in umbilical cord (3 mg mL^{-1}), in the vitreous humor of the eye (0.2 mg mL^{-1}), and in skin (0.5 mg mL^{-1}). In other tissues that contain less hyaluronan, it forms an essential structural component of the matrix. In cartilage it forms the aggregation center for aggrecan, the large chondroitin sulfate proteoglycan, and re-

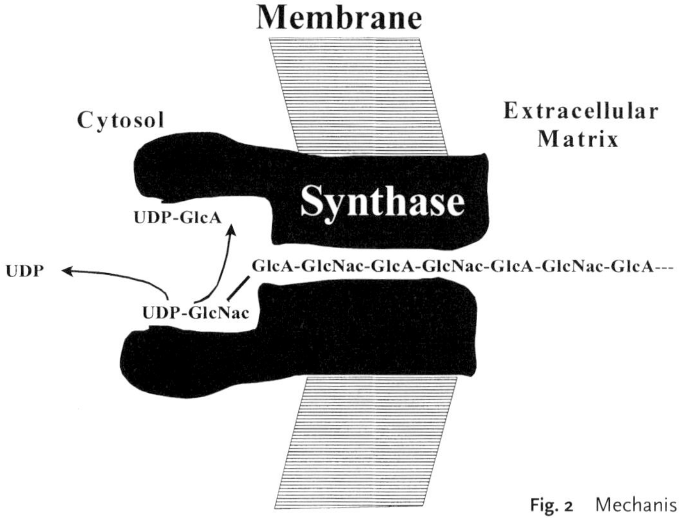

Fig. 2 Mechanism of hyaluranon synthesis.

tains this macromolecular assembly in the matrix by specific hyaluronan–protein interactions. It also forms a scaffold for binding of other matrix components around smooth muscle cells on the aorta, and on fibroblasts in the dermis of skin. The largest deposit of hyaluronan resides in the skin; in an adult human this totals ~8 g. Hyaluronan has also been detected intracellularly in proliferating cells (Evanko and Wight, 1999)

5
Mechanism of Hyaluronan Synthesis

The unusual mechanism of hyaluranon synthesis has impeded progress for a long time – a situation which has also occurred with other important polysaccharides such as cellulose and chitin. However, it now appears that two mechanisms have evolved independently – for mammalian cells and for streptococci on the one hand, and for *Pasteurella* on the other hand.

5.1
Chain Elongation

Hyaluronan synthesis in mammalian cells differs from that of other polysaccharides in many aspects. The molecule is elongated at the reducing end by alternate transfer of UDP-hyaluronan to the substrates UDP-GlcNac and UDP-GlcA, thereby liberating the UDP-moiety (Figure 2) (Prehm, 1983a,b).

Other glucosaminoglycans grow at the non-reducing end and hence require a protein backbone. Hyaluronan is synthesized at plasma membranes, the nascent chains being extruded directly into the extracellular matrix (Prehm, 1984). In contrast, other glucosaminoglycans are made in the Golgi body. Chain initiation does not require either a protein backbone (as for proteoglycans) or preformed oligosaccharides as starters; only the presence of the nucleotide sugar precursors is needed to initiate new chains. During elongation the chain is retained on the membrane-integrated synthase, this mechanism of synthesis being in operation for hyaluranon synthesis in both vertebrates and in Gram-positive streptococci. However, a different mechanism seems to exist for hyaluranon synthesis in Gram-negative *Pasteurella* in which the chains are elongated at the non-reducing end (DeAngelis, 1999).

5.2 Chain Size

One point for discussion is what determines the size of the synthesized hyaluronan, and this aspect of polymerization also applies to other macromolecular syntheses such as for proteins, DNA, or RNA. An answer was provided from experiments on isolated membranes from fibroblasts or streptococci, whereby the removal of nascent hyaluronan from the hyaluronan synthase enzyme stimulated its production. This was demonstrated in isolated streptococcal membranes (Nickel et al., 1998) and also in intact fibroblasts (Philipson et al., 1985). It thus appeared that high molecular-weight hyaluronan inhibited its own chain elongation, when it was retained on the synthase. This phenomenon may occur for solely thermodynamic reasons, because the decrease in entropy during the synthesis of a macromolecule must be compensated by free energy from cleavage of the nucleotide sugars and the subsequent formation of ordered structures. In fact, this explains why macromolecules such as proteins, RNA or DNA do not exceed a certain chain length (Peller, 1980).

5.3 Chain Export

The growing chain must be exported through a membrane pore, and consequently the proposal was made by Weigel that this pore is formed by the synthase itself, because the inactivation rate of the synthase by irradiation did not permit the participation of other proteins (Tlapak-Simmons et al., 1998). However, these authors did not show that transport of hyaluronan through the vesicle membrane was inactivated, and methods should be developed to confirm this finding.

5.4 Swelling

Hydration and swelling of nascent hyaluronan occurs at the site of synthesis on the cell surface. While swelling to enormous volumes (diameters up to 500 nm), one molecule of hyaluronan will displace many other cell-surface components by virtue of exclusion. Hence, it is conceivable that this swelling provides a mechanism whereby adhesive components are disrupted from the anchored cell.

5.5 Macromolecular Assembly

Macromolecular assembly with other matrix molecules such as proteoglycans also occurs at the cell surface. The compartmentation of hyaluronan and proteoglycan syntheses to the Golgi complex and the plasma membranes thus ensures that, during synthesis, very large aggregates are formed at the site of final deposition, and not intracellularly.

6 Hyaluronan Synthases

Hyaluronan is synthesized at the protoplast membrane of group A and group C streptococci (Markovitz and Dorfman, 1962), the enzymatic activity being solubilized by very mild detergents such as digitonin (Triscott and van de Rijn, 1986). Conventional purification procedures such as ion-exchange chromatography of detergent-solubilized membrane proteins yielded inhomogeneous protein mixtures that could not be separated into individual constituents without loss of enzymatic activity. Therefore, a new method, based on the phase separation of a detergent solution, was developed to allow purification of the synthase in its active form (Prehm et al., 1996). It was known that membrane proteins can be separated from soluble proteins by phase separation of a Triton X-114 extract. Phase separation can be induced in 1% Triton X-114 solutions by a temperature shift from 0 °C to 37 °C, with soluble proteins remaining in the aqueous phase and membrane proteins in the detergent phase. However, Triton X-114 was shown to inactivate the hyaluronan synthase. It was found that digitonin can undergo phase separation by the addition of polyethylene glycol 6000 at 0 °C, and that the synthase will remain in the aqueous phase, where it is associated with hyaluronan. Final purification of the hyaluronan synthase was achieved by ion-exchange chromatography and yielded an electrophoretically homogeneous protein of 42 kDa. This study proved that a single protein was sufficient to direct hyaluronan synthesis, and that the method may be generally applicable to other membrane proteins that are associated with polysaccharides, because it combines the advantages of the mild detergent digitonin with phase separation of all membrane proteins from polysaccharide-binding proteins.

Molecular cloning of the streptococcal hyaluronan synthase was reported independently by DeAngelis and van de Rijn (DeAngelis et al., 1993; Dougherty and van de Rijn, 1994). The gene was designated *HasA*. The *Streptococcus pyogenes* operon encodes two other proteins: HasB is a UDP-glucose-dehydrogenase, which is required to convert UDP-glucose to UDP-GlcA (Dougherty and van de Rijn, 1993), while HasC is a UDP-glucose-pyrophosphorylase, which is required to convert glucose-1-phosphate and UTP to UDP-glucose (Crater et al., 1995). Mammalian synthases were cloned simultaneously from a mutant mouse mammary carcinoma (Itano and Kimata, 1996) and *Xenopus laevis* (Meyer and Kreil, 1996). Now, three mammalian synthases are known and have been designated Has1, Has2, and Has3 (reviewed by Weigel et al., 1997). Because these proteins have 30% identity in terms of amino acid sequence with the streptococcal synthase, the genes may have a common ancestor. The synthase from *Pasteurella* has been cloned by DeAngelis et al. (1998), and is structurally unrelated to the other synthases.

7
Hyaluronan-binding Proteins and Receptors

Hyaluronan-binding proteins are constituents of the extracellular matrix, and stabilize its integrity. Hyaluronan receptors are involved in cellular signal transduction; one receptor family includes the binding proteins aggrecan, link protein, versican and neurocan and the receptors CD44, TSG6 (Lee et al., 1992), hyaluronectin (Delpech and Halavent, 1981), GHAP (Perides et al., 1990), and Lyve-1 (Banerji et al., 1999). The RHAMM receptor is an unrelated hyaluronan-binding protein, and the hyaluronan-binding sites contain a motif of a minimal site of interaction with hyaluronan. This is represented by B(X7)B, where B is any basic amino acid except histidine, and X is at least one basic amino acid and any other moiety except acidic residues. CD44 and RHAMM have attracted much attention, because they were believed to be involved in metastasis (Arch et al., 1992; Hall et al., 1995).

7.1
CD44

CD44 is a pleiomorphic extracellular matrix receptor that also binds to fibronectin and laminin, and also interacts with hyaluronan with relatively low affinity ($K_D = 10^{-8}$ M). CD44 exists as many isoforms, though some do not bind to hyaluronan. The hyaluronan-binding properties of CD44 have been implicated in promoting cell–cell aggregation and migration upon hyaluronan and collagen substrates. CD44 binds to oligosaccharides of up to 18 residues in a monovalent manner, and above 20 residues in a divalent manner, and this results in increased avidity (Lesley et al., 2000). When murine mammary carcinoma cells were transfected to produce high levels of soluble CD44, growth and metastasis *in vivo* was inhibited. In contrast, transfection with a CD44 mutant that does not bind CD44 has no effect (Peterson et al., 2000). The role of CD44 in hyaluronan synthesis shedding was analyzed in detail in metastatic and nonmetastatic melanoma cell lines that differed in degradation of CD44 and hyaluronan production (Lüke and Prehm, 1999). The nonmetastatic melanoma cell line IF6 did not significantly degrade CD44, while the metastatic cell line MV3 produced a soluble fragment. Increased hyaluronan synthesis and shedding correlated with proteolysis of CD44 on the melanoma cell lines. Intact cell surface CD44 retains hyaluronan to the vicinity of the synthase to inhibit shedding and the initiation of new chains.

The binding of hyaluronan to cells is also influenced by the level of CD44 expression (Takahashi et al., 1995; Miyake et al., 1998), by CD44 modifications such as expression of variants in different cell types (Van der Voort et al., 1995), by glycosylation (Skelton et al., 1998), by intracellular binding to ankyrin (Zhu and Bourguignon, 1998), or by phosphorylation (Puré et al., 1995). Treatment of cells with phorbol-12-myristate-13-acetate (PMA), calcium and forskolin stimulated phosphorylation of CD44, reduced the hyaluronan-binding activity (Liu et al., 1998), and stimulated hyaluronan synthesis (Klewes and Prehm, 1994). All these factors may be involved in the regulation of hyaluronan synthesis and exert profound effects on migration, growth and metastasis.

Reduced hyaluronan binding to CD44 might have two consequences:

- Loss of signal transduction from the extracellular hyaluronan to intracellular phosphorylation by the CD44-associated kinase.
- Overproduction of hyaluronan that swells to enormous volumes on the cell surface to displace cellular adhesion molecules.

Proteolysis of CD44 plays a key role in hyaluronan-mediated effects on cell growth, migration, metastasis, and adhesion (Dube et al., 2001) (Figure 3). Hyaluronan serves as an additional adhesive component that inhibits migration and proliferation, when it is retained on the cell surface by hyaluronan-binding proteins. In contrast, most transformed cells produce higher levels of surface proteases that also degrade CD44, resulting in the release of surface-bound hyaluronan and the stimulation of synthesis. In these cells, hyaluronan serves as an additional detachment factor, as the hyaluronan molecules which swell on the cell surface also displace undegraded adhesive components. Thus, hyaluronan acts as an amplifier of both active and inactive cell-surface proteases. It will therefore be difficult to identify those proteases which cleave CD44 at the surface of different cell lines. Recently, a chymotrypsin-like sheddase was thought responsible for shedding CD44 from a myoepithelial cell line (Lee et al., 2000).

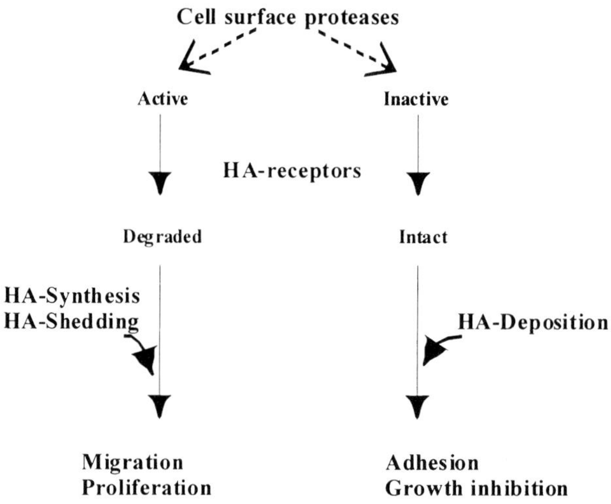

Fig. 3 Dual functions of hyaluronan.

The belief in a central role for CD44 in migration and metastasis met with severe disillusionment when CD44-deficient mice showed only marginal health disturbances (Schmits et al., 1997). This is an apparent contradiction to many studies which showed severe defects in the tissues of originally healthy animals when the function of CD44 was inhibited. A possible explanation might be that an unknown substitute of CD44 is expressed during development in CD44-defective mice (Ponta et al., 1998).

7.2 RHAMM

The hyaluronan-binding protein RHAMM was originally identified as a cell-surface protein that was also thought to be involved in hyaluronan-mediated migration and metastasis (Turley, 1989; Hall et al., 1995). Recently, a bitter scientific debate questioned its location and its function (Hofmann et al., 1998a; Turley et al., 1998), but it now appears that RHAMM is expressed either at the cell surface at low amounts under specific growth conditions, or is not expressed at all and is mainly localized intracellularly (Assmann et al., 1998; Hofmann et al., 1998b).

7.3 Other Hyaluronan-binding Proteins

Other intracellular hyaluronan-binding proteins have also been detected (Huang et al., 2000), but their functions remain elusive. Proliferating cells contain also detectable amounts of intracellular hyaluronan (Evanko and Wight, 1999) though again, the function of this is unknown.

8 Mechanisms of Hyaluronan Release from the Cell Surface

Hyaluronan could be released from cells by either enzymatic or radical degradation, or by dissociation as the intact macromolecule. Hyaluronidases were present in some transformed cells (Orkin et al., 1982), but their pH optima favor an intralysosomal function (Bernanke and Orkin, 1984); moreover, hyaluronidase-deficient cells also shed hyaluronan (Klein and von Figura, 1980). The mechanism of hyaluronan shedding from eucaryotic cells lines was analyzed by Prehm (1990). All cell lines shed identical sizes of hyaluronan as retained on the surface, but differed in the amount of hyaluronan synthesized and the ratio of released and retained hyaluronan. A method was developed which discriminated between intramolecular degradation and dissociation as the intact macromolecule. The cells were pulse-labeled to form hyaluronan chains with labeled and unlabeled segments, after which the sizes of labeled hyaluronan released into the media during the pulse extension period were determined by gel filtration. B6 cells released the same sizes as were retained on the cell surface, indicating that no intramolecular degradation occurred, and hyaluronan dissociated as the intact macromolecule. In contrast, SV3T3 cells released hyaluronan of varying molecular weight distribution during extension of the labeled segment. Shedding of smaller fragments could be prevented by radical scavengers such as superoxide dismutase and tocopherol. Therefore, SV3T3 cells released hyaluronan not only by dissociation, but also by radical degradation.

9 Regulation of Hyaluronan Synthesis

Almost any disturbance of cellular homeostasis results in a stimulation of hyaluronan production, though very few reports have investigated the inhibition of hyaluronan synthesis. Three levels of regulation have been recognized:

1) Expression of the synthase.
2) Stimulation and inhibition of the synthase by growth or differentiation factors acting on a specific target cell.
3) Disturbances in the integrity of the extracellular matrix, particularly the degradation of cell surface-bound hyaluronan.

9.1 Expression of the Synthase

The synthase has a high turnover rate and a half-life of 82 min, indicating a strict control of its activity by the transcription rate (Bansal and Mason, 1986). The three mammalian synthases are expressed in different tissues, thereby producing hyaluronan in different amounts and sizes (Itano et al., 1999).

9.2 Stimulation and Inhibition of the Synthase

Many growth or differentiation factors have been shown to stimulate hyaluronan synthesis (Tomida et al., 1975, 1977a,b; Lembach, 1976; Prehm, 1980; D'Arville and Mason, 1983; Hamerman and Wood, 1984; Hamerman et al., 1986; Pulkki, 1986; Tammi and Tammi, 1986; Wu and Wu, 1986; Heldin et al., 1989). During the early part of the 20th century, Kabat (1939) had already recognized enhanced hyaluronan production in tumors of sarcosis and leukosis. Viral transformation by Rous sarcoma virus (RSV) stimulates hyaluronan production in chondrocytes (Mikuni Takagaki and Toole, 1981), in chicken fibroblasts (Hamerman et al., 1965; Ishimoto et al., 1966), and in myoblasts (Yoshimura, 1985), whilst SV40 transformation induces a similar increase in 3T3 fibroblasts (Hopwood and Dorfman, 1977; Goldberg et al., 1984).

It is surprising that only two inhibitors of hyaluronan synthases have been discovered. Periodate-oxidized UDP-GlcA or UDP-GlcNac have been used as suicide inhibitors of the synthase (Prehm, 1985). These inhibitors do not surmount cell membranes, and must be imported into the cytoplasm by osmotic lysis of pinocytotic vesicles. Vesnarinone suppresses the synthase activity in intact myofibroblasts (Ueki et al., 2000). Hence, it may be worthwhile developing or isolating inhibitors of hyaluronan synthesis that might be used as curative drugs in pathological disturbances related to hyaluranon production, for example edema.

9.2.1 Signal Transduction at Membranes

The hyaluronan receptor CD44 is responsible for outside-in signaling (Perschl et al., 1995) and, intracellularly, this is associated with the tyrosine kinase $p56^{lck}$. Binding of hyaluronan stimulates intracellular phosphorylation (Taher et al., 1996). Similar phosphorylation was observed with bivalent antibodies to CD44, indicating that receptor cross-linking on the cell surface instigates the signal transduction cascade. Thus, the mechanism resembles growth factor-dependent signal transduction pathways. It is also possible that the angiogenic effects of hyaluronan oligosaccharides are transduced through CD44, though this mechanism does not apply to all cell types because CD44-deficient lymphocytes might also be activated by hyaluronan oligosaccharides (Termeer et al., 2000). Hence, other unknown receptors or signal transduction pathways must exist.

9.2.2 Intracellular Signal Transduction

Intracellular signal transduction pathways are dependent on protein synthesis and activation of protein kinase C (Heldin et al., 1992), and have been studied in detail in both B6 cells and 3T3 fibroblasts (Klewes and Prehm, 1994). Activation by fetal calf serum was inhibited by cycloheximide or by the protein kinase inhibitors H-7 or H-8, indicating that transcription as well as phosphorylation were required for activation. The activation by serum was markedly prolonged, when serum was added together with cholera toxin or theophylline. Without serum stimulation the hyaluronan synthase could also be activated by PMA, by dibutyryl-c-AMP, or forskolin. Increasing the intracellular Ca^{2+} concentration with a Ca-ionophore also led to an activation. In isolated plasma membranes the synthase activity could be decreased by phosphatase treatment, and enhanced by ATP in B6 cells and by ATP in the presence of PMA in 3T3 fibroblasts. In conclusion, hyaluronan synthase is induced by transcription and activated by phosphorylation by protein kinase C, c-AMP-dependent protein kinases or Ca^{2+}-dependent protein kinases.

9.3 Influence of Chain Length on Further Elongation

Another factor is the amount of cell surface hyaluronan which influenced its own production. Notably, hyaluronidase treatment of intact cells stimulated synthesis (Philipson et al., 1985). Interesting features on the regulation of hyaluronan synthesis by the growing hyaluronan chain itself have been obtained from studies with streptococci (Nickel et al., 1998). Group A and C streptococci differ in their capacity to retain hyaluronan as a coat on their cell surface. In group C streptococci, a 56-kDa hyaluronan receptor was closely associated with the synthase; this protein had an intrinsic kinase activity that performed autophosphorylation in response to extracellular ATP. Autophosphorylation of the 56-kDa protein led in turn to a reduction in hyaluronan binding and increased shedding of the hyaluronan capsule. Simultaneously, the synthase increased its activity to replace the lost hyaluronan chains. It thus appears that a large hyaluronan chain inhibited its own elongation, when it was retained in the vicinity of the synthase.

A similar mechanism for the regulation of hyaluronan synthesis operates on eucaryotic cells that express the CD44 receptor (Lüke and Prehm, 1999). Proteolytic degradation of CD44 in melanoma cells correlated with the metastatic potential and activation of the hyaluronan synthesis. This activation was the result of facilitated dissociation of growing hyaluronan chains from plasma membranes. These results add a new function to the CD44 receptor as a regulator of hyaluronan synthesis.

10 Turnover and Catabolism

Within the circulation of an adult human, a total of 34 mg of hyaluronan is turned over each day (Fraser and Laurent, 1989; Lebel et al., 1994). The major origins of hyaluronan are the joints, skin, eyes, and intestine. In skin and joints the half-life is about 12 h (Reed et al., 1990; Coleman et al., 2000), whilst in the anterior chamber of the eye it is 1–1.5 h (Laurent et al., 1993), and in the vitreous humor it is 70 days. The rapid turnover is somewhat surprising because hyaluronan has been regarded as a structural component of connective tissue.

Hyaluronan is drained through the lymph and reaches the circulating blood, where the serum concentration may reach ~31 ng L^{-1}. Hyaluranon is effectively endocytosed by the liver, the liver endothelial cells expressing a membrane receptor that endocytically clears hyaluronan from the circulation (Zhou et al., 1999). It is then transported into lysosomes that contain hyaluronidase, β-glucuronidase and β-*N*-acetyl-glucosaminidase (Roden et al., 1989).

11 Functions of Hyaluronan

Many different functions have been assigned to hyaluronan, but these can be grouped into cellular, physiological, and pathological functions. Most functions are determined either by the physical properties or by interactions with hyaluronan-binding proteins.

11.1 Cellular Functions

Fibroblasts are surrounded by a coat of hyaluronan that is lost upon transformation (Underhill and Toole, 1982) but which modifies cell–cell aggregation (Underhill, 1982) and cell–substratum adhesion (Erickson and Turley, 1983). Hyaluronan synthesis is coordinated with cell growth, with proliferating cells producing more hyaluronan than resting cells (Tomida et al., 1975; Mian, 1986). Synchronized fibroblasts show the highest hyaluronan production during mitosis, because hyaluronan synthesis is required for the detachment and mitosis of fibroblasts (Brecht et al., 1986).

The function of hyaluronan on cellular behavior appeared paradoxical, as hyaluronan synthesis was seen to be increased during cell migration (Toole, 1991), mitosis (Brecht et al., 1986), and tumor invasion (Toole et al., 1979). Overproduction of hyaluronan in the human tumor cell line HT1080 enhanced anchorage-independent growth and tumorigenicity (Kosaki et al., 1999). In contrast, hyaluronan promotes cell adhesion (Miyake et al., 1990) and cell–cell aggregation (Underhill, 1982), but inhibits cell proliferation (Goldberg and Toole, 1987; West and Kumar, 1989).

A similar paradoxical situation applies for hyaluronidases in tumor progression. In some tumors, hyaluronidase treatment blocks lymph node invasion by tumor cells for T-cell lymphoma (Zahalka et al., 1995), but overexpression of hyaluronidase correlates with disease progression in bladder, breast and prostate cancer (Lokeshwar et al., 1996, 1997; Bertrand et al., 1997).

A solution to this paradox was obtained from a comparison of the cellular behavior of hyaluronan-deficient Chinese hamster ovary (CHO) cells and CHO cells transfected with the hyaluronan synthase. Surprisingly, hyaluronan synthesis reduced initial cell adhesion, migration, growth and the density at contact inhibition. All these apparent contradictions were combined into a model for the cellular functions of hyaluronan (Dube et al., 2001). Thus, hyaluronan serves as an adhesive component when it is retained on the cell surface by intact CD44, and as a detachment factor when it is released. Migration- and growth- inhibitory effects of hyaluronan are mediated by the proteolytic cleavage of CD44. The cellular function of hyaluronan might also be an amplifier of cell-surface proteases: when proteases are inactive, hyaluranon amplifies the action of cellular adhesion factors by deposition and binding to receptors. However, when proteases are active, it amplifies cell detachment from the environment by activation of synthesis and shedding (see Figure 3).

11.2 Physiological Functions

A prominent physiological function of hyaluronan is the creation of hydrated pathways that allow the cells to penetrate cellular and fibrous barriers. Such hydrated pericellular matrices are required not only for cell rounding in mitosis, but also for cell migration during morphogenesis and wound healing.

11.2.1 Differentiation and Morphogenesis

Studies carried out on embryonic limb development have illustrated how hyaluronan modulates differentiation *in vivo* (Toole, 1991). Early limb mesodermal cells are surrounded and separated by an extensive, hyaluronan-enriched matrix. At this stage, pericellular hyaluronan appears to be tethered to the membrane-integrated synthase. This hydrated matrix facilitates the proliferation and migration of early mesenchymal precursors. Subsequently, the mesoderm condenses, i.e., the intercellular matrix decreases in volume at the sites of future cartilage and muscle differentiation. Mesodermal cells lose the ability to form hydrated matrices, the level of lysosomal hyaluronidases increases, and much of the pericellular hyaluronan is removed. The remaining hyaluronan is now retained at the cell surface by receptors. Similar events occur in the early mesodermal development of skin and teeth. Further differentiation of condensed limb mesoderm into cartilage is accompanied by recovery of matrix formation and by extensive production of proteoglycans that are tethered to the cell surface via hyaluronan and CD44.

$Has2^{-/-}$ mouse mutants contain virtually no hyaluranon at the E9.5 (embryo day 9.5) stage of development (Camenisch et al., 2000), and at this stage they exhibit multiple defects, including yolk sac, vasculature and heart abnormalities. The cardiac jelly and cardiac cushions fail to form valves and other structures.

11.2.2 Wound Healing

The following scenario has been proposed as a model for wound healing by (Weigel et al., 1986). After a wound has been sealed by the platelet plug and fibrin clot, platelet activators stimulate the inflammation and surrounding cells to synthesize hyaluronan into the matrix of the fibrin clot. The clot swells and becomes more porous in order to facilitate the migration of cells into the fibrin matrix. As more cells migrate into the wound, the hyaluronan–fibrin matrix is degraded and replaced by collagens and proteoglycans. The hyaluronan degradation products stimulate angiogenesis and the formation of new blood vessels.

Wound healing in fetal tissues occurs without scarring and is correlated with the prolonged presence of hyaluronan (Longaker et al., 1991). Hyaluronidases that are present in adult wounds were considered responsible for fibrotic healing (West et al., 1997).

11.2.3 Synovia

Hyaluronan is produced by the fibroblast-like type A cells in the upper synovial lining (Pitsillides et al., 1993), and not only serves as a lubricant but also enhances the resistance of the synovial linings to fluid outflow (Coleman et al., 1997). The concentration of hyaluronan in the synovial fluid ranges from 1.4 to 3.6 mg mL^{-1}. Its molecular mass is variable, and in normal synovial fluid has been estimated as 7.0×10^6 daltons, though this is decreased to $3-5 \times 10^6$ daltons in rheumatoid synovial fluids (Balazs et al., 1967; Dahl et al., 1985). The turnover rate depends on the size, indicating a partial reflection of hyaluronan by the lining (Coleman et al., 2000). Hyaluronan is drained through the lymphatic vessels and catabolized in lymph nodes (Fraser et al., 1996).

11.3 Pathological Functions

Aberrant hyaluronan synthesis will lead to disturbances of cell behavior and tissue integrity. For *Streptococcus* sp. and *Pasteurella*, this serves as a non-antigenic disguising capsule in the infected host.

11.3.1 Metastasis

Most malignant solid tumors contain elevated levels of hyaluronan; such enrichment of hyaluronan in tumors may be caused by increased production by the tumors themselves, or by induction in the surrounding stromal cells (Toole et al., 1979; Knudson et al., 1984). The mechanisms whereby hyaluronan–receptor interactions influence tumor cell behavior are not clearly understood, and this is currently a highly active area of investigation.

11.3.2 Edema

Inflammation of various organs is often accompanied with an accumulation of hyaluronan. This can cause edema due to the osmotic activity, and can in turn lead to dysfunction of the organs. For example, hyaluronan accumulation in the rheumatoid joint impedes flexibility of the joint (Engström-Laurent, 1997), whilst accumulation in rejected transplanted kidneys can cause edema and increased intracapsular pressure (Hällgren et al., 1990). Hyaluranon also accumulates in pulmonary edema (Nettelbladt et al., 1989) and during myocardial infarction (Waldenstrom et al., 1991).

11.3.3 Streptococci

Hyaluronan is produced by group A and C streptococci and deposited into a capsule which serves as a major virulence factor of pathogenic streptococci (Kass and Seastone, 1944; Wessels et al., 1991, 1994) because it protects the bacteria against phagocytosis (Whitnack et al., 1981) and oxygen damage (Cleary and Larkin, 1979). Another virulence factor is the hyaluronidase, which is a spreading factor and facilitates penetration of the bacteria through infected tissue (McClean, 1941; MacLennan, 1956). When streptococci enter the stationary phase, they lose the capsule (van de Rijn, 1983), but remain virulent. A constant rate of hyaluronan shedding from the capsule might be advantageous for bacteria, because they will eliminate host components such as attacking antibodies.

A 56-kDa protein was identified as the first example of a prokaryotic extracellular protein kinase. This is expressed on the surface of group C streptococci, and can bind and retain hyaluronan in the absence of ATP. The capsule is shed in the presence of extracellular ATP (Nickel et al., 1998). An equivalent hyaluronan-binding protein was not detected on group A streptococci that do not retain their capsule to the same extent.

Group A and C streptococci differ both in their hosts and their infection routes. Group A streptococci are virulent human pathogens and cause tonsilitis, scarlet fever and rheumatic fever, the infection route being via the throat. In contrast, group C streptococci are mainly animal pathogens, but can also infect humans; their infection route is mainly opportunistic through wounds in the skin.

Group C streptococci have clearly evolved a mechanism to retain their hyaluronan capsule that protects them against desiccation when they are localized on the skin surface. When they enter a wound, they encounter high concentrations of ATP from necrotic tissue and also need to defend themselves against attacking antibodies and macrophages. This they do by shedding the capsule, thereby preventing any attachment.

The cell surface-bound kinase on group C streptococci also elicits antibodies in the infected host. These antibodies show an immunological cross-reaction with cell surfaces proteins from fibroblasts, and are cytotoxic in the presence of complement (Prehm et al., 1995). This protein has recently been used as a vaccine to protect mice against fatal group C streptococci (Chanter et al., 1999).

12
Hyaluronan Degradation

As a very large molecule, hyaluronan is prone to mechanical degradation either by ultrasonic treatment or by thermal degradation at elevated temperatures. Physiologically, hyaluronan can be degraded by oxygen free radicals or by hyaluronidases.

12.1
Degradation by Free Radicals

Oxygen free radical degradation is mostly a side reaction of activated neutrophil granulocytes or monocytes in an inflammation. Radical degradation of hyaluronan results in a dramatic drop of the viscosity and function of the synovial fluid in the inflamed joints of patients with rheumatoid arthritis. This mechanism was hypothesized by Greenwald and Moak (1986). Although a hyaluronan structure modified by radical damage has been detected in the synovial fluid of the inflamed rheumatoid joint (Grootveld et al., 1991), direct biochemical proof for hyaluronan degradation was obtained from organ cultures of healthy and rheumatoid synovial tissue (Schenck et al., 1995). Healthy tissues and some arthritic tissues did not contain significant amounts of granulocytes and produced high molecular-weight hyaluronan. In contrast, arthritic tissue infiltrated with granulocytes released low molecular-weight hyaluronan. Hyaluronan degradation was accompanied by massive oxygen radical production. Radical scavengers protected hyaluronan from degradation in synovial tissue, in particular by the iron chelators DETAPAC or Desferal that block the formation of hydroxyl radicals. Hydroxyl radicals degrade hyaluronan with an efficiency of 100%, with 65% being cleaved within 15 min and the remaining radicals giving rise to thermally labile products that eventually lead to chain scission (Al Assaf et al., 2000). A variety of free aldehydes and oligosaccharides are produced in the rheumatoid synovial fluid (Chapman et al., 1989; Grootveld et al., 1991). The structures formed by oxygen radical damage of hyaluronan have been analyzed in detail by Uchiyama et al. (1990), the main degradation product being L-threotetradialdose.

12.2 Degradation by Hyaluronidases

Hyaluronidases are classified into three groups (Kreil, 1995).

1) Mammalian-type hyaluronidases (EC 3.2.1.35) are endo-β-*N*-acetylhexosaminidases and produce tetrasaccharides and hexasaccharides as the major end-products. They have both hydrolytic and transglycosidase activities, and can degrade hyaluronan and chondroitin sulfates.
2) Bacterial hyaluronidases (EC 4.2.99.1) are endo-β-*N*-acetylhexosaminidases and yield primarily disaccharides. They operate by a β-elimination reaction.
3) Hyaluronidases (EC 3.2.1.36) from leeches, other parasites, and crustaceans are endo-β-glucuronidases that generate tetrasaccharide and hexasaccharide end-products.

Hyaluronidase from testis has long been known as a spreading factor (Chain and Duthie, 1940). Other hyaluronidases from vertebrate tissues are more difficult to purify and analyze, because they occur in low concentrations and are often unstable (Csóka et al., 1997).

13 Production

Hyaluronan is produced either from rooster combs (Pharmacia AB, Uppsala, Sweden, have developed a special strain of roosters with highly luxuriant combs) or from streptococci. Commercially available hyaluranon is produced in molecular weights ranging from less than 10^6 daltons to as high as 8×10^6 daltons. There are many hyaluronan producers worldwide. For example, Genzyme Corp. (Framingham, MA, USA) merged with Biomatrix and operates a plant producing hyaluronan from mammalian sources in Canada. Anika (Woburn, MA, USA); Lifecore Biomedical (Chaska, MN, USA) are other suppliers. Pharmacia produces hyaluronan in Sweden, Fidia Advanced Biopolymers (Brindisi) in Italy, Bio-Technology General Corp. (Iselin, NJ, USA) in Israel, and a number of companies, including Kibun Food Chemifa Co. and Seikagaku Corp. (both Tokyo), in Japan.

Today there are still many products on the market that contain hyaluronan isolated from rooster combs, because this has set the standards for high molecular weight, purity and noninflammatory properties. Even if hyaluronan from streptococci meets these criteria, its market penetration is hampered by the reluctance of customers to change to cheaper medical or cosmetic products. In addition many of the streptococcal preparations are not characterized as thoroughly, or are less pure (Manna et al., 1999). The development of large-scale production from *Streptococcus zooepidemicus* cultures had to overcome many obstacles, including: growth in chemically defined media (Kjems and Lebech, 1976; van de Rijn and Kessler, 1980; Armstrong et al., 1997; Cooney et al., 1999); the production of high molecular-weight material (Kim et al., 1996); the elimination of toxic impurity such as streptolysin; and increasing the yield to about 7 g L^{-1} (Lowther and Rogers, 1956; Thonard et al., 1964; Gerlach and Kohler, 1970).

13.1 Patents

The patent literature concerning hyaluranon is extensive, the number of annual patent applications having increased from fewer than 10 before 1985 to about 200 in 2000. The major inventors in this field and their corporate affiliations are summarized in Table 1. Many different processing techniques and uses for hyaluronan have been developed and patented by Balazs and his coworkers. For example, the important Balazs patent (which was issued in 1979 but now has expired) on hyaluronan isolated from animal tissue that does not cause an inflammatory response when tested in the eye of the owl monkey (Balazs, 1979). This is marketed by Upjohn-Pharmacia as Healon®, a sterile, pyrogen-free, nonantigenic and noninflammatory, high molecular-weight fraction of hyaluronan.

13.2 Market

The world market of hyaluronan is difficult to estimate, because many companies produce it for medical and cosmetic applications. The current US market of $157 million for viscosupplementation in osteoarthritis is driven by two products: Synvisc® from Biomatrix (now Genzyme) and Hyalgan® from Fidia Pharmaceutical. In Europe, Fidia's Hyalgan® is the leading hyaluronan-based viscosupplement product. Details of the major pharmaceutical products are listed in Table 2.

Tab. 1 Some hyaluronan-related patents

Patent no.	*Patent holder*	*Inventors*	*Title*	*Date*
U.S. 4,141,973	Biomatrix	Balazs	Ultrapure hyaluronic acid and the use thereof	1979
U.S. 4,713,448	Biomatrix	Balazs	Chemically modified hyaluronic acid preparation and method of recovery thereof from animal tissues	1987
U.S. 4,957,744	Fidia	della Valle	Esters of hyaluronic acid	1990
U.S. 4,636,524	Biomatrix	Balazs	Cross-linked gels of hyaluronic acid and products containing such gels	1987
U.S. 4,937,270	Genzyme	Hamilton	Water-insoluble derivatives of hyaluronic acid	1990
U.S. 5,644,049	Murst Italian	Giusti	Biomaterial comprising hyaluronic acid and derivatives thereof in interpenetrating polymer networks	1997
U.S. 4,663,233	Biocaot	Beavers	Lens with hydrophilic coating	1987
U.S. 5,585,361	Burns	Genzyme	Method for the inhibition of platelet adherence and aggregation	1996

Tab. 2 Some commercially available pharmaceuticals containing hyaluronan

	Concentration [mg mL⁻¹]	*Size [mL]*	*Manufacturer*	*Application*	*Price (US $)*
AMO Vitrax Syringe	30	65	Allergen	Ophthalmology	138
Amvisc Plus Syringe	16	5	Chiron	Ophthalmology	145
Amvisc Plus Syringe	16	8	Chiron	Ophthalmology	190
Amvisc Syringe	12	5	Chiron	Ophthalmology	112
Healon GV Syringe	14	55	Kabi	Ophthalmology	101
Healon GV Syringe	14	85	Kabi	Ophthalmology	131
Healon Syringe	10	55	Kabi	Ophthalmology	94
Hyalgan SDV	10	2	Sanofi	Osteoarthritis	166
Hyalgan Syringe	10	2	Sanofi	Osteoarthritis	166
Provisc Syringe	10	4	Alcon	Ophthalmology	117
Provisc Syringe	10	55	Alcon	Ophthalmology	142
Provisc Syringe	10	85	Alcon	Ophthalmology	178
Synvisc Syringe	8	3×2.25	Wyeth	Osteoarthritis	705
Viscoat Syringe	40	5	Alcon	Ophthalmology	151

14 Medical Applications

Hyaluronan preparations or higher cross-linked products are increasingly used for many medical applications, such as ophthalmic viscosurgery, supplementation of the synovial fluid in patients with osteoarthritis (Balazs and Denlinger, 1989), as membranes for postsurgical separation of tissues (Burns et al., 1997), and as drug delivery systems (Vercruysse and Prestwich, 1998). Many cosmetics contain hyaluronan as an ingredient, because it is believed to keep skin young and fresh by preventing dryness as a result of its water-binding capacity, though scientific evidence on this subject is lacking.

14.1 Ophthalmics

Hyaluronan was first marketed in the early 1980s as Healon® in the ophthalmic field as a viscous gel which could be injected into the anterior chamber of the eye to protect tissues such as the corneal endothelium (Miller and Stegmann, 1983). A number of other hyaluronan products are now marketed by competing firms in the ophthalmic market.

14.2 Arthritis

Intra-articular administration of hyaluronan has been used in animals and man with reported clinical efficacy. In man, hyaluronan is being used to relieve pain and improve joint mobility in the treatment of osteoarthritis with intra-articular injections of Hyalgan® (Sanofi Pharmaceuticals), Orthovisc® (Anika Therapeutics), and SynVisc® (Biomatrix, now Genzyme). It has also been proposed for several degenerative joint diseases as an alternative to the traditional steroid therapy (Altman and Moskowitz, 1998; Wobig et al., 1999; Adams et al., 2000).

14.3 Wound Healing and Scarring

Hyaluronan products have been developed to foster the healing process. For burn and chronic ulcer patients, a line of modified hyaluronan products based on a HYAFF™

polymer is being marketed by Convatec in Europe, and is currently undergoing clinical trials in the US (Goa and Benfield, 1994).

14.4 Adhesion Prevention

Most surgical procedures are accompanied by undesired tissue damage caused by cutting, desiccation, lack of adequate blood supply and manipulative abrasion, and undesired connective tissue bridges (adhesions) are often formed on the damaged surfaces of organs. Hyaluronan preparations such as Seprafilm® from Genzyme can reduce such adhesions and improve the surgical outcome (Beck, 1997).

14.5 Drug Delivery

Hyaluronan is an ideal molecule for use as a carrier of drugs, particularly for local administration. As the polysaccharide is a ubiquitous component of tissues and fluids, it is immunologically inert. It can be metabolized in the lysosomes of certain cells, and the backbone of the molecule provides different chemical groups for drug attachment. Investigations are on-going for topical and intravenous drug delivery systems using modified hyaluronans (Vercruysse and Prestwich, 1998).

15 Effects of Hyaluronan Oligosaccharides

Hyaluronan oligosaccharides augment fibroblast proliferation, migration, stimulate the formation of new blood vessels, and also activate macrophages. These effects have been studied in detail in cell culture, where they stimulate proliferation of synovial fibroblasts *in vitro* (Goldberg and Toole, 1987), induce angiogenesis (Montesano et al., 1996), stimulate chemokine production on macrophages (McKee et al., 1996, 1997; Horton et al., 1998), and activate lymphocytes (Termeer et al., 2000).

16 Outlook and Perspectives

In times characterized by an enthusiasm for the achievements of molecular biology that are often announced with great fanfare for the welfare of mankind, it may be wise to issue a reminder that the understanding and management of some diseases require a horizon beyond the application of gene technology. Indeed, research on hyaluronan may be a typical illustration that pertinent yet silent progress contributes significantly to the solution of medical problems.

Nonetheless, many cellular and physiological functions of hyaluronan remain elusive. There is a notion amongst the scientific community that hyaluronan participates in the pathogenesis of metastasis and rheumatoid autoimmune diseases, and research into hyaluronan–cell interactions will clearly contribute to therapeutic interventions in these diseases. A second promising area is the development of hyaluronan-based biomaterials and hyaluronan-modified surfaces. Indeed, a number of important products have already reached the market, and the introduction of many hyaluronan-derived devices and drugs are eagerly anticipated during the next decade.

17
References

Abatangelo, G. and Weigel, P.H. (2000) *Redefining Hyaluronan*. Amsterdam: Wheley.

Adams, M. E., Lussier, A. J., Peyron, J. G. (2000) A risk-benefit assessment of injections of hyaluronan and its derivatives in the treatment of osteoarthritis of the knee. *Drug Safety* **23**, 115–130.

Al Assaf, S., Meadows, J., Phillips, G. O., Williams, P. A., Parsons, B. J. (2000) The effect of hydroxyl radicals on the rheological performance of hylan and hyaluronan. *Int. J. Biol. Macromol.* **27**, 337–348.

Altman, R. D., Moskowitz, R. (1998) Intraarticular sodium hyaluronate (Hyalgan®) in the treatment of patients with osteoarthritis of the knee: a randomized clinical trial. *J. Rheumatol.* **25**, 2203–2212.

Arch, R., Wirth, K., Hofmann, M., Ponta, H., Matzku, S., Herrlich, P., Zoller, M. (1992) Participation in normal immune responses of a metastasis-inducing splice variant of CD44. *Science* **257**, 682–685.

Armstrong, D. C., Cooney, M. J., Johns, M. R. (1997) Growth and amino acid requirements of hyaluronic-acid-producing *Streptococcus zooepidemicus. Appl. Microbiol. Biotechnol.* **47**, 309–312.

Aruffo, A., Stamenkovic, I., Melnick, M., Underhill, C. B., Seed, B. (1990) CD44 is the principal cell surface receptor for hyaluronate. *Cell* **61**, 1303–1313.

Assmann, V., Marshall, J. F., Fieber, C., Hofmann, M., Hart, I. R. (1998) The human hyaluronan receptor RHAMM is expressed as an intracellular protein in breast cancer cells. *J. Cell Sci.* **111**, 1685–1694.

Balazs, E. A. (1970) Chemistry and Molecular Biology of the Intercellular Matrix. London: Academic Press.

Balazs, E. A. (1979) Ultrapure hyaluronic acid and the use thereof. U.S. Patent 4,141,973.

Balazs, E. A. (1982) Use of hyaluronic acid in eye surgery. *Ann. Ther. Clin. Ophthalmol.* **33**, 95–110.

Balazs, E. A. (1987a) Chemically modified hyaluronic acid preparation and method of recovery thereof from animal tissues. U.S. Patent 4,713,448.

Balazs, E. A. (1987b) Cross-linked gels of hyaluronic acid and products containing such gels. U.S. Patent 4,636,524.

Balazs, E. A., Denlinger, J. L. (1989) Clinical uses of hyaluronan. *Ciba Found. Symp.* **143**, 265–275.

Balazs, E. A., Watson, D., Duff, I. F., Roseman, S. (1967) Hyaluronic acid in synovial fluid. I. Molecular parameters of hyaluronic acid in normal and arthritis human fluids. *Arthritis Rheum.* **10**, 357–376.

Banerji, S., Ni, J., Wang, S.X., Clasper, S., Su, J., Tammi, R., Jones, M., Jackson, D. G. (1999) LYVE-1, a new homologue of the CD44 glycoprotein, is a lymph-specific receptor for hyaluronan. *J. Cell Biol.* **144**, 789–801.

Bansal, M. K., Mason, R. M. (1986) Evidence for rapid metabolic turnover of hyaluronate synthetase in Swarm rat chondrosarcoma chondrocytes. *Biochem. J.* **236**, 515–519.

Beavers, E. M. (1987) Lens with hydrophilic coating. U.S. Patent 4,663,233.

Beck, D. E. (1997) The role of Seprafilm™ bioresorbable membrane in adhesion prevention. *Eur. J. Surg.*, **163** (Suppl. 577), 49–55.

Bernanke, D. H., Orkin, R. W. (1984) Hyaluronate binding and degradation by cultured embryonic chick cardiac cushion and myocardial cells. *Dev. Biol.*, **106**, 360–367.

Bertrand, P., Girard, N., Duval, C., D'Anjou, J., Chauzy, C., Ménard, J. F., Delpech, B. (1997)

Increased hyaluronidase levels in breast tumor metastases. *Int. J. Cancer* **73**, 327–331.
Brecht, M., Mayer, U., Schlosser, E., Prehm, P. (1986) Increased hyaluronate synthesis is required for fibroblast detachment and mitosis. *Biochem. J.* **239**, 445–450.
Burns, J. W., Valeri, C. R. (1996) Method for the inhibition of platelet adherence and aggregation. U.S. Patent 5,585,361.
Burns, J. W., Colt, M. J., Burgess, L. S., Skinner, K. C. (1997) Preclinical evaluation of Seprafilm™ bioresorbable membrane. *Eur. J. Surg.* **163** (Suppl. 577), 40–48.
Callegaro, L., Giusti, P. (1997) Biomaterial comprising hyaluronic acid and derivatives thereof in interpenetrating polymer networks. U.S. Patent 5,644,049.
Camenisch, T. D., Spicer, A. P., Brehm-Gibson, T., Biesterfeldt, J., Augustine, M. L., Calabro, A., Jr., Kubalak, S., Klewer, S. E., McDonald, J. A. (2000) Disruption of hyaluronan synthase-2 abrogates normal cardiac morphogenesis and hyaluronan-mediated transformation of epithelium to mesenchyme. *J. Clin. Invest.* **106**, 349–360.
Chain, E., Duthie, E. S. (1940) Identity of hyaluronidase and spreading factor. *Br. J. Exp. Pathol.* **21**, 324–338.
Chanter, N., Ward, C. L., Talbot, N. C., Flanagan, J. A., Binns, M., Houghton, S. B., Smith, K. C., Mumford, J. A. (1999) Recombinant hyaluronate associated protein as a protective immunogen against *Streptococcus equi* and *Streptococcus zooepidemicus* challenge in mice. *Microbiol. Pathog.* **27**, 133–143.
Chapman, M. L., Rubin, B. R., Gracy, R. W. (1989) Increased carbonyl content of proteins in synovial fluid from patients with rheumatoid arthritis. *J. Rheumatol.* **16**, 15–18.
Cleary, P. P., Larkin, A. (1979) Hyaluronic acid capsule: strategy for oxygen resistance in group A streptococci. *J. Bacteriol.* **140**, 1090–1097.
Coleman, P. J., Scott, D., Ray, J., Mason, R. M., Levick, J. R. (1997) Hyaluronan secretion into the synovial cavity of rabbit knees and comparison with albumin turnover. *J. Physiol.* **503** , 645–656.
Coleman, P. J., Scott, D., Mason, R. M., Levick, J. R. (2000) Role of hyaluronan chain length in buffering interstitial flow across synovium in rabbits. *J. Physiol.* **526**, 425–434.
Comper, W. D., Laurent, T. C. (1978) Physiological function of connective tissue polysaccharides. *Physiol. Rev.* **58**, 255–315.
Cooney, M. J., Goh, L. T., Lee, P. L., Johns, M. R. (1999) Structured model-based analysis and control of the hyaluronic acid fermentation by *Streptococcus zooepidemicus*: physiological implications of glucose and complex nitrogen-limited growth. *Biotechnol. Prog.* **15**, 898–910.
Crater, D. L., Dougherty, B. A., van de Rijn, I. (1995) Molecular characterization of *has*C from an operon required for hyaluronic acid synthesis in group A streptococci – Demonstration of UDP-glucose pyrophosphorylase activity. *J. Biol. Chem.* **270**, 28676–28680.
Csóka, T. B., Frost, G. I., Wong, T., Stern, R. (1997) Purification and microsequencing of hyaluronidase isozymes from human urine. *FEBS Lett.* **417**, 307–310.
D'Arville, C., Mason, R. M. (1983) Effects of serum and insulin on hyaluronate synthesis by cultures of chondrocytes from the Swarm rat chondrosarcoma. *Biochim. Biophys. Acta* **760**, 53–60.
Dahl, L. B., Dahl, I. M., Engstrom Laurent, A., Granath, K. (1985) Concentration and molecular weight of sodium hyaluronate in synovial fluid from patients with rheumatoid arthritis and other arthropathies. *Ann. Rheum. Dis.* **44**, 817–822.
DeAngelis, P. L. (1999) Molecular directionality of polysaccharide polymerization by the *Pasteurella multocida* hyaluronan synthase. *J. Biol. Chem.* **274**, 26557–26562.
DeAngelis, P. L., Papaconstantinou, J., Weigel, P. H. (1993) Molecular cloning, identification, and sequence of the hyaluronan synthase gene from group A *Streptococcus pyogenes*. *J. Biol. Chem.* **268**, 19181–19184.
DeAngelis, P. L., Jing, W., Drake, R. R., Achyuthan, A. M. (1998) Identification and molecular cloning of a unique hyaluronan synthase from *Pasteurella multocida*. *J. Biol. Chem.* **273**, 8454–8458.
della Valle, F. (2001) Esters of hyaluronic acid. U.S. Patent 4,957,744.
Delpech, B., Halavent, C. (1981) Characterization and purification from human brain of a hyaluronic acid-binding glycoprotein, hyaluronectin. *J. Neurochem.* **36**, 855–859.
Dougherty, B. A., van de Rijn, I. (1993) Molecular characterization of hasB from an operon required for hyaluronic acid synthesis in group A streptococci. Demonstration of UDP-glucose dehydrogenase activity. *J. Biol. Chem.* **268**, 7118–7124.
Dougherty, B. A., van de Rijn, I. (1994) Molecular characterization of hasA from an operon required for hyaluronic acid synthesis in group A streptococci. *J. Biol. Chem.* **269**, 169–175.
Dube, B., Luke, H. J., Aumailley, M., Prehm, P. (2001) Hyaluronan reduces migration and pro-

liferation in CHO cells. *Biochim. Biophys. Acta* **1538**, 283–289.

Engström-Laurent, A. (1997) Hyaluronan in joint disease. *J. Intern. Med.* **242**, 57–60.

Erickson, C. A., Turley, E. A. (1983) Substrata formed by combination of extracellular matrix components alter neural crest cell migration. *J. Cell Sci.* **61**, 299–323.

Evanko, S. P., Wight, T. N. (1999) Intracellular localization of hyaluronan in proliferating cells. *J. Histochem. Cytochem.* **47**, 1331–1341.

Fox, E. M., Walts, A. E., Acharya, R. A., Hamilton, R. (1990) Water insoluble derivatives of hyaluronic acid. U.S. Patent 4,937,270.

Fraser, J. R., Laurent, T. C. (1989) Turnover and metabolism of hyaluronan. *Ciba Found. Symp.* **143**, 41–53.

Fraser, J. R., Laurent, T. C., Pertoft, H., Baxter, E. (1981) Plasma clearance, tissue distribution and metabolism of hyaluronic acid injected intravenously in the rabbit. *Biochem. J.* **200**, 415–424.

Fraser, J. R. E., Cahill, R. N., Kimpton, W. G., Laurent, T. C. (1996) Lymphatic system, in: *Extracellular Matrix. 1. Tissue Function* (Comper, W. D., Ed.), Amsterdam: Harwood Academic Publications, 110–131.

Gerlach, D., Kohler, W. (1970) [Production and isolation of streptococcal hyaluronic acid]. *Zentralbl. Bakteriol. Orig.* **215**, 187–195.

Goa, K. L., Benfield, P. (1994) Hyaluronic acid. A review of its pharmacology and use as a surgical aid in ophthalmology, and its therapeutic potential in joint disease and wound healing. *Drugs* **47**, 536–566.

Goldberg, R. L., Toole, B. P. (1987) Hyaluronate inhibition of cell proliferation. *Arthritis Rheum.* **30**, 769–778.

Goldberg, R. L., Seidman, J. D., Chi Rosso, G., Toole, B. P. (1984) Endogenous hyaluronate-cell surface interactions in 3T3 and simian virus-transformed 3T3 cells. *J. Biol. Chem.* **259**, 9440–9446.

Greenwald, R. A., Moak, S. A. (1986) Degradation of hyaluronic acid by polymorphonuclear leukocytes. *Inflammation* **10**, 15–30.

Grootveld, M., Henderson, E. B., Farrell, A., Blake, D. R., Parkes, H. G. H.-P. (1991) Oxidative damage to hyaluronate and glucose in synovial fluid during exercise of the inflamed rheumatoid joint. Detection of abnormal low-molecular-mass metabolites by proton-n.m.r. spectroscopy. *Biochem. J.* **273**, 459–467.

Hall, C. L., Yang, B., Yang, X., Zhang, S., Turley, M., Samuel, S., Lange, L. A., Wang, C., Curpen, G. D., Savani, R. C., Greenberg, A. H., Turley, E. A. (1995) Overexpression of the hyaluronan receptor RHAMM is transforming and is also required for H-*ras* transformation. *Cell* **82**, 19–28.

Hamerman, D., Wood, D. D. (1984) Interleukin 1 enhances synovial cell hyaluronate synthesis. *Proc. Soc. Exp. Biol. Med.* **177**, 205–210.

Hamerman, D., Todaro, G. J., Green, H. (1965) The production of hyaluronate by spontaneously established cell lines and viral transformed lines of fibroblastic origin. *Biochim. Biophys. Acta* **101**, 343–351.

Hamerman, D., Sasse, J., Klagsbrun, M. (1986) A cartilage-derived growth factor enhances hyaluronate synthesis and diminishes sulfated glycosaminoglycan synthesis in chondrocytes. *J. Cell Physiol.* **127**, 317–322.

Hardingham, T. E., Muir, H. (1972) The specific interaction of hyaluronic acid with cartilage proteoglycans. *Biochim. Biophys. Acta* **279**, 401–405.

Hällgren, R., Gerdin, B., Tufveson, G. (1990) Hyaluronic acid accumulation and redistribution in rejecting rat kidney graft. Relationship to the transplantation edema. *J. Exp. Med.* **171**, 2063–2076.

Heldin, P., Laurent, T. C., Heldin, C. H. (1989) Effect of growth factors on hyaluronan synthesis in cultured human fibroblasts. *Biochem. J.* **258**, 919–922.

Heldin, P., Asplund, T., Ytterberg, D., Thelin, S., Laurent, T. C. (1992) Characterization of the molecular mechanism involved in the activation of hyaluronan synthetase by platelet-derived growth factor in human mesothelial cells. *Biochem. J.* **283**, 165–170.

Hofmann, M., Assmann, V., Fieber, C., Sleeman, J. P., Moll, J., Ponta, H., Hart, I. R., Herrlich, P. (1998a) Problems with RHAMM: a new link between surface adhesion and oncogenesis? *Cell* **95**, 591–592.

Hofmann, M., Fieber, C., Assmann, V., Göttlicher, M., Sleeman, J., Plug, R., Howells, N., Von Stein, O., Ponta, H., Herrlich, P. (1998b) Identification of IHABP, a 95kDa intracellular hyaluronate binding protein. *J. Cell Sci.* **111**, 1673–1684.

Hopwood, J. J., Dorfman, A. (1977) Glycosaminoglycan synthesis by cultured human skin fibroblasts after transformation with Simian virus 40. *J. Biol. Chem.* **252**, 4777–4785.

Horton, M. R., McKee, C. M., Bao, G., Liao, F., Farber, J. M., Hodge DuFour, J., Pure, E., Oliver, B. L., Wright, T. M., Noble, P. W. (1998) Hyaluronan fragments synergize with interferon-gamma to induce the C-X-C chemokines Mig and interferon-inducible protein-10 in mouse macrophages. *J. Biol. Chem.* **273**, 35088–35094.

Huang, L., Grammatikakis, N., Yoneda, M., Banerjee, S. D., Toole, B. P. (2000) Molecular characterization of a novel intracellular hyaluronan-binding protein. *J. Biol. Chem.* 275, 29829–29839.

Ishimoto, N., Temin, H. M., Strominger, J. L. (1966) Studies of carcinogenesis by avian sarcoma viruses. II. Virus- induced increase in hyaluronic acid synthetase in chicken fibroblasts. *J. Biol. Chem.* **241**, 2052–2057.

Itano, N., Kimata, K. (1996) Expression cloning and molecular characterization of HAS protein, a eukaryotic hyaluronan synthase. *J. Biol. Chem.* **271**, 9875–9878.

Itano, N., Sawai, T., Yoshida, M., Lenas, P., Yamada, Y., Imagawa, M., Shinomura, T., Hamaguchi, M., Yoshida, Y., Ohnuki, Y., Miyauchi, S., Spicer, A. P., McDonald, J. A., Kimata, K. (1999) Three isoforms of mammalian hyaluronan synthases have distinct enzymatic properties. *J. Biol. Chem.* **274**, 25085–25092.

Kabat, E. A. (1939) A polysaccharide in tumors due to a virus of leukosis and sarcoma in fowls. *J. Biol. Chem.* **130**, 143–147.

Kass, E. H., Seastone, C. V. (1944) The role of the mucoid polysaccharide hyaluronic acid in the virulence of group A hemolytic streptococci. *J. Exp. Med.* **70**, 319–330.

Kendall, F. E., Heidelberger, M., Dawson, M. H. (1937) A serologically inactive polysaccharide elaborated by mucoid strains of group A hemolytic streptococci. *J. Biol. Chem.* **118**, 61–69.

Kim, J. H., Yoo, S. J., Oh, D. K., Kweon, Y. G., Park, D. W., Lee, C. H., Gil, G. H. (1996) Selection of a *Streptococcus equi* mutant and optimization of culture conditions for the production of high molecular weight hyaluronic acid. *Enzyme Microb. Technol.* **19**, 440–445.

Kjems, E., Lebech, K. (1976) Isolation of hyaluronic acid from cultures of streptococci in a chemically defined medium. *Acta Pathol. Microbiol. Scand. B.* **84**, 162–164.

Klein, U., von Figura, K. (1980) Characterization of dermatan sulfate in mucopolysaccharidosis VI. Evidence for the absence of hyaluronidase-like enzymes in human skin fibroblasts. *Biochim. Biophys. Acta* **630**, 10–14.

Klewes, L., Prehm, P. (1994) Intracellular signal transduction for serum activation of the hyaluronan synthase in eukaryotic cell lines. *J. Cell Physiol.* **160**, 539–544.

Knudson, W., Biswas, C., Toole, B. P. (1984) Interactions between human tumor cells and fibroblasts stimulate hyaluronate synthesis. *Proc. Natl. Acad. Sci. USA* **81**, 6767–6771.

Kosaki, R., Watanabe, K., Yamaguchi, Y. (1999) Overproduction of hyaluronan by expression of the hyaluronan synthase Has2 enhances anchorage-independent growth and tumorigenicity. *Cancer Res.* **59**, 1141–1145.

Kreil, G. (1995) Hyaluronidases – A group of neglected enzymes. *Protein Sci.* **4**, 1666–1669.

Lapcik, L., De Smedt, S., Demeester, J., Chabrecek, P. (1998) Hyaluronan: preparation, structure, properties, and applications. *Chem. Rev.* **98**, 2663–2684.

Laurent, T. C. (1970) Structure of hyaluronic acid, in: *Chemistry and Molecular Biology of the Intercellular Matrix* (Balazs, E. A., Ed.), New York: Academic Press, 703–732.

Laurent, T. C. (1989) *The Biology of Hyaluronan.* Ciba Foundation Symposium **143**. Chichester: Wiley.

Laurent, T. C. (1998) *The Chemistry, Biology and Medical Applications of Hyaluronan and its Derivatives.* Wenner-Gren International Series **72**. London: Portland Press.

Laurent, T. C., Fraser, J. R. (1992) Hyaluronan. *FASEB J.* **6**, 2397–2404.

Laurent, T. C., Dahl, L. B., Lilja, K. (1993) Hyaluronan injected in the anterior chamber of the eye is catabolized in the liver. *Exp. Eye Res.* **57**, 435–440.

Lebel, L., Gabrielsson, J., Laurent, T. C., Gerdin, B. (1994) Kinetics of circulating hyaluronan in humans. *Eur. J. Clin. Invest.* **24**, 621–626.

Lee, M. C., Alpaugh, M. L., Nguyen, M., Deato, M., Dishakjian, L., Barsky, S. H. (2000) Myoepithelial-specific CD44 shedding is mediated by a putative chymotrypsin-like sheddase. *Biochem. Biophys. Res. Commun.* **279**, 116–123.

Lee, T. H., Wisniewski, H. G., Vilcek, J. (1992) A novel secretory tumor necrosis factor-inducible protein(TSG-6) is a member of the family of hyaluronate binding proteins, closely related to the adhesion receptor CD44. *J. Cell Biol.* **116**, 545–557.

Lembach, K. J. (1976) Enhanced synthesis and extracellular accumulation of hyaluronic acid during stimulation of quiescent human fibroblasts by mouse epidermal growth factor. *J. Cell Physiol.* **89**, 277–288.

Lesley, J., Hascall, V. C., Tammi, M., Hyman, R. (2000) Hyaluronan binding by cell surface CD44. *J. Biol. Chem.* **275**, 26967–26975.

Liu, D. C., Liu, T., Sy, M. S. (1998) Identification of two regions in the cytoplasmic domain of CD44 through which PMA, calcium, and forskolin differentially regulate the binding of CD44 to hyaluronic acid. *Cell. Immunol.* **190**, 132–140.

Lokeshwar, V. B., Lokeshwar, B. L., Pham, H. T., Block, N. L. (1996) Association of elevated levels of hyaluronidase, a matrix-degrading enzyme, with prostate cancer progression. *Cancer Res.* **56**, 651–657.

Lokeshwar, V. B., Öbek, C., Soloway, M. S., Block, N. L. (1997) Tumor-associated hyaluronic acid: a new sensitive and specific urine marker for bladder cancer. *Cancer Res.* **57**, 773–777.

Longaker, M. T., Chiu, E. S., Adzick, N. S., Stern, M., Harrison, M. R., Stern, R. (1991) Studies in fetal wound healing. V. A prolonged presence of hyaluronic acid characterizes fetal wound fluid. *Ann. Surg.* **213**, 292–296.

Lowther, D. A., Rogers, H. J. (1956) The role of glutamine in the biosynthesis of hyaluronate by streptococcal suspensions. *Biochem. J.* **62**, 304–314.

Lüke, H. J., Prehm, P. (1999) Synthesis and shedding of hyaluronan from plasma membranes of human fibroblasts and metastatic and non-metastatic melanoma cells. *Biochem. J.* **343**, 71–75.

MacLennan, A. P. (1956) The production of capsules, hyaluronic acid and hyaluronidase by group A and group C streptococci. *J. Gen. Microbiol.* **14**, 134–142.

Manna, F., Dentini, M., Desideri, P., De Pita, O., Mortilla, E., Maras, B. (1999) Comparative chemical evaluation of two commercially available derivatives of hyaluronic acid (hylaform from rooster combs and restylane from *Streptococcus*) used for soft tissue augmentation. *J. Eur. Acad. Dermatol. Venereol.* **13**, 183–192.

Markovitz, A., Dorfman, A. (1962) Synthesis of capsular polysaccharide hyaluronic acid by protoplast membrane preparations of group A streptococci. *J. Biol. Chem.* **237**, 273–279.

McClean, D. (1941) The capsulation of streptococci and its relation to diffusion factor (hyaluronidase). *J. Pathol. Bacteriol.* **53**, 13.

McKee, C. M., Penno, M. B., Cowman, M., Burdick, M. D., Strieter, R. M., Bao, C., Noble, P. W. (1996) Hyaluronan (HA) fragments induce chemokine gene expression in alveolar macrophages – The role of HA size and CD44. *J. Clin. Invest.* **98**, 2403–2413.

McKee, C. M., Lowenstein, C. J., Horton, M. R., Wu, J., Bao, C., Chin, B. Y., Choi, A. M. K., Noble, P. W. (1997) Hyaluronan fragments induce nitric-oxide synthase in murine macrophages through a nuclear factor kappaB-dependent mechanism. *J. Biol. Chem.* **272**, 8013–8018.

Meyer, K. (1947) The biological significance of hyaluronic acid and hyaluronidase. *Physiol. Rev.* **27**, 335–359X.

Meyer, K., Palmer, J. W. (1934) The polysaccharide of the vitreous humor. *J. Biol. Chem.* **107**, 629–634.

Meyer, M. F., Kreil, G. (1996) Cells expressing the DG42 gene from early *Xenopus* embryos synthesize hyaluronan. *Proc. Natl. Acad. Sci. USA* **93**, 4543–4547.

Mian, N. (1986) Analysis of cell-growth-phase-related variations in hyaluronate synthase activity of isolated plasma-membrane fractions of cultured human skin fibroblasts. *Biochem. J.* **237**, 333–342.

Mikuni Takagaki, Y., Toole, B. P. (1981) Hyaluronate-protein complex of Rous sarcoma virus-transformed chick embryo fibroblasts. *J. Biol. Chem.* **256**, 8463–8469.

Miller, D., Stegmann, R. (1983) Healon. A Guide to its use in Ophthalmic Surgery. New York: Wiley.

Miyake, H., Hara, I., Okamoto, I., Gohji, K., Yamanaka, K., Arakawa, S., Saya, H., Kamidono, S. (1998) Interaction between CD44 and hyaluronic acid regulates human prostate cancer development. *J. Urol.* **160**, 1562–1566.

Miyake, K., Underhill, C. B., Lesley, J., Kincade, P. W. (1990) Hyaluronate can function as a cell adhesion molecule and CD44 participates in hyaluronate recognition. *J. Exp. Med.* **172**, 69–75.

Montesano, R., Kumar, S., Orci, L., Pepper, M. S. (1996) Synergistic effect of hyaluronan oligosaccharides and vascular endothelial growth factor on angiogenesis in vitro. *Lab. Invest.* **75**, 249–262.

Nettelbladt, O., Tengblad, A., Hällgren, R. (1989) Lung accumulation of hyaluronan parallels pulmonary edema in experimental alveolitis. *Am. J. Physiol.* **257**, L379–L384.

Nickel, V., Prehm, S., Lansing, M., Mausolf, A., Podbielski, A., Deutscher, J., Prehm, P. (1998) An ectoprotein kinase of group C streptococci binds hyaluronan and regulates capsule formation. *J. Biol. Chem.* **273**, 23668–23673.

Orkin, R. W., Underhill, C. B., Toole, B. P. (1982) Hyaluronate degradation in 3T3 and simian virus-transformed3T3 cells. *J. Biol. Chem.* **257**, 5821–5826.

Peller, L. (1980) Thermodynamic considerations in the synthesis and assembly of biological macromolecules. *Macromolecules* **13**, 609–615.

Perides, G., Asher, R., Dahl, D., Bignami, A. (1990) Glial hyaluronate-binding protein (GHAP) in optic nerve and retina. *Brain Res.* **512**, 309–316.
Perschl, A., Lesley, J., English, N., Trowbridge, I., Hyman, R. (1995) Role of CD44 cytoplasmic domain in hyaluronan binding. *Eur. J. Immunol.* **25**, 495–501.
Peterson, R. M., Yu, Q., Stamenkovic, I., Toole, B. P. (2000) Perturbation of hyaluronan interactions by soluble CD44 inhibits growth of murine mammary carcinoma cells in ascites. *Am. J. Pathol.* **156**, 2159–2167.
Philipson, L. H., Westley, J., Schwartz, N. B. (1985) Effect of hyaluronidase treatment of intact cells on hyaluronate synthetase activity. *Biochemistry* **24**, 7899–7906.
Pitsillides, A. A., Wilkinson, L. S., Mehdizadeh, S., Bayliss, M. T., Edwards, J. C. (1993) Uridine diphosphoglucose dehydrogenase activity in normal and rheumatoid synovium: the description of a specialized synovial lining cell. *Int. J. Exp. Pathol.* **74**, 27–34.
Ponta, H., Wainwright, D., Herrlich, P. (1998) The CD44 protein family. *Int. J. Biochem. Cell Biol.*, **30**, 299–305.
Prehm, P. (1980) Induction of hyaluronic acid synthesis in teratocarcinoma stem cells by retinoic acid. *FEBS Lett.* **111**, 295–298.
Prehm, P. (1983a) Synthesis of hyaluronate in differentiated teratocarcinoma cells. Characterization of the synthase. *Biochem. J.* **211**, 181–189.
Prehm, P. (1983b) Synthesis of hyaluronate in differentiated teratocarcinoma cells. Mechanism of chain growth. *Biochem. J.* **211**, 191–198.
Prehm, P. (1984) Hyaluronate is synthesized at plasma membranes. *Biochem. J.* **220**, 597–600.
Prehm, P. (1985) Inhibition of hyaluronate synthesis. *Biochem. J.* **225**, 699–705.
Prehm, P. (1990) Release of hyaluronate from eukaryotic cells. *Biochem. J.* **267**, 185–189.
Prehm, P. (2000) Antigene von rheumatischen Autoimmunerkrankungen. Patent PCT/EP00/05279.
Prehm, S., Herrington, C., Nickel, V., Völker, W., Briko, N. I., Blinnikova, E. A., Schmiedel, A., Prehm, P. (1995) Antibodies against proteins of streptococcal hyaluronate synthase bind to human fibroblasts and are present in patients with rheumatic fever. *J. Anat.* **187**, 271–277.
Prehm, S., Nickel, V., Prehm, P. (1996) A mild purification method for polysaccharide binding membrane proteins: phase separation of digitonin extracts to isolate the hyaluronate synthase from *Streptococcus* sp. in active form. *Protein Expression and Purification* **7**, 343–346.
Pulkki, K. (1986) The effects of synovial fluid macrophages and interleukin-1 on hyaluronic acid synthesis by normal synovial fibroblasts. *Rheumatol. Int.* **6**, 121–125.
Puré, E., Camp, R. L., Peritt, D., Panettieri, R. A., Jr., Lazaar, A. L., Nayak, S. (1995) Defective phosphorylation and hyaluronate binding of CD44 with point mutations in the cytoplasmic domain. *J. Exp. Med.* **181**, 55–62.
Reed, R. K., Laurent, T. C., Taylor, A. E. (1990) Hyaluronan in prenodal lymph from skin: changes with lymph flow. *Am. J. Physiol.* **259**, H1097–H1100.
Roden, L., Campbell, P., Fraser, J. R., Laurent, T. C., Pertoft, H. T.-J. (1989) Enzymic pathways of hyaluronan catabolism. *Ciba Foundation Symposium* **143**, 60–76.
Schenck, P., Schneider, S., Miehlke, R., Prehm, P. (1995) Synthesis and degradation of hyaluronate by synovia from patients with rheumatoid arthritis. *J. Rheumatol.* **22**, 400–405.
Schmits, R., Filmus, J., Gerwin, N., Senaldi, G., Kiefer, F., Kundig, T., Wakeham, A., Shahinian, A., Catzavelos, C., Rak, J., Furlonger, C., Zakarian, A., Simard, J. J., Ohashi, P. S., Paige, C. J., Gutierrez, R. J., Mak, T. W. (1997) CD44 regulates hematopoietic progenitor distribution, granuloma formation, and tumorigenicity. *Blood* **90**, 2217–2233.
Scott, J. E., Cummings, C., Brass, A., Chen, Y. (1991) Secondary and tertiary structures of hyaluronan in aqueous solution, investigated by rotary shadowing-electron microscopy and computer simulation. Hyaluronan is a very efficient network-forming polymer. *Biochem. J.* **274**, 699–705.
Skelton, T. P., Zeng, C. X., Nocks, A., Stamenkovic, I. (1998) Glycosylation provides both stimulatory and inhibitory effects on cell surface and soluble CD44 binding to hyaluronan. *J. Cell Biol.* **140**, 431–446.
Taher, T. E. I., Smit, L., Griffioen, A. W., Schilder-Tol, E. J. M., Borst, J., Pals, S. T. (1996) Signaling through CD44 is mediated by tyrosine kinases – Association with $p56^{lck}$ in T lymphocytes. *J. Biol. Chem.* **271**, 2863–2867.
Takahashi, K., Stamenkovic, I., Cutler, M., Saya, H., Tanabe, K. K. (1995) CD44 hyaluronate binding influences growth kinetics and tumorigenicity of human colon carcinomas. *Oncogene* **11**, 2223–2232.
Tammi, R., Tammi, M. (1986) Influence of retinoic acid on the ultrastructure and hyaluronic acid synthesis of adult human epidermis in whole skin organ culture. *J. Cell Physiol.* **126**, 389–398.

Tengblad, A., Caputo, C. B., Raisz, L. G. (1980) Quantitative analysis of hyaluronate in nanogram amounts. *Biochem. J.* **185**, 101–105.

Termeer, C. C., Hennies, J., Voith, U., Ahrens, T., Weiss, M., Prehm, P., Simon, J. C. (2000) Oligosaccharides of hyaluronan are potent activators of dendritic cells. *J. Immunol.* **165**, 1863–1870.

Thonard, J. C., Migliore, S. A., Blustein, R. (1964) Isolation of hyaluronic acid from broth cultures of streptococci. *J. Biol. Chem.* **239**, 726–728.

Tlapak-Simmons, V. L., Kempner, E. S., Baggenstoss, B. A., Weigel, P. H. (1998) The active streptococcal hyaluronan synthases (HASs) contain a single HAS monomer and multiple cardiolipin molecules. *J. Biol. Chem.* **273**, 26100–26109.

Tomida, M., Koyama, H., Ono, T. (1975) Induction of hyaluronic acid synthetase activity in rat fibroblasts by medium change of confluent cultures. *J. Cell Physiol.* **86**, 121–130.

Tomida, M., Koyama, H., Ono, T. (1977a) A serum factor capable of stimulating hyaluronic acid synthesis in cultured rat fibroblasts. *J. Cell Physiol.* **91**, 323–328.

Tomida, M., Koyama, H., Ono, T. (1977b) Effects of adenosine 3′:5′-cyclic monophosphate and serum on synthesis of hyaluronic acid in confluent rat fibroblasts. *Biochem. J.* **162**, 539–543.

Toole, B. P. (1991) Proteoglycans and hyaluronan in morphogenesis and differentiation, in: *Cell Biology of the Extracellular Matrix* (Hay, E. D., Ed.), New York: Plenum Press, 305–341.

Toole, B. P., Okayama, M., Orkin, R. W., Yoshimura, M., Muto, M., Kaji, A. (1977) Developmental roles of hyaluronate and chondroitin sulfate proteoglycans. *Soc. Gen. Physiol. Ser.* **32**, 139–154.

Toole, B. P., Biswas, C., Gross, J. (1979) Hyaluronate and invasiveness of the rabbit V2 carcinoma. *Proc. Natl. Acad. Sci. USA* **76**, 6299–6303.

Triscott, M. X., van de Rijn, I. (1986) Solubilization of hyaluronic acid synthetic activity from streptococci and its activation with phospholipids. *J. Biol. Chem.* **261**, 6004–6009.

Turley, E. A. (1989) The role of a cell-associated hyaluronan-binding protein in fibroblast behaviour. *Ciba Foundation Symposium* **143**, 121–133.

Turley, E. A., Moore, D., Hayden, L. J. (1987) Characterization of hyaluronate binding proteins isolated from 3T3 and murine sarcoma virus transformed 3T3 cells. *Biochemistry* **26**, 2997–3005.

Turley, E. A., Pilarski, L., Nagy, J. I. (1998) Problems with RHAMM: a new link between surface adhesion and oncogenesis? Response. *Cell* **95**, 592–593.

Uchiyama, H., Dobashi, Y., Ohkouchi, K., Nagasawa, K. (1990) Chemical change involved in the oxidative reductive depolymerization of hyaluronic acid. *J. Biol. Chem.* **265**, 7753–7759.

Ueki, N., Taguchi, T., Takahashi, M., Adachi, M., Ohkawa, T., Amuro, Y., Hada, T., Higashino, K. (2000) Inhibition of hyaluronan synthesis by vesnarinone in cultured human myofibroblasts. *Biochim. Biophys. Acta* **1495**, 160–167.

Underhill, C. B. (1982) Interaction of hyaluronate with the surface of simian virus40- transformed 3T3 cells: aggregation and binding studies. *J. Cell Sci.* **56**, 177–189.

Underhill, C. B., Toole, B. P. (1979) Binding of hyaluronate to the surface of cultured cells. *J. Cell Biol.* **82**, 475–484.

Underhill, C. B., Toole, B. P. (1982) Transformation-dependent loss of the hyaluronate-containing coats of cultured cells. *J. Cell Physiol.* **110**, 123–128.

van de Rijn, I. (1983) Streptococcal hyaluronic acid: proposed mechanisms of degradation and loss of synthesis during stationary phase. *J. Bacteriol.* **156**, 1059–1065.

van de Rijn, I., Kessler, R. E. (1980) Growth characteristics of group A streptococci in a new chemically defined medium. *Infect. Immun.* **27**, 444–448.

Van der Voort, R., Manten-Horst, E., Smit, L., Ostermann, E., Van den Berg, F., Pals, S. T. (1995) Binding of cell-surface expressed CD44 to hyaluronate is dependent on splicing and cell type. *Biochem. Biophys. Res. Commun.* **214**, 137–144.

Vercruysse, K. P., Prestwich, G. D. (1998) Hyaluronate derivatives in drug delivery. *Crit. Rev. Therap. Drug Carrier Systems* **15**, 513–555.

Waldenstrom, A., Martinussen, H. J., Gerdin, B., Hällgren, R. (1991) Accumulation of hyaluronan and tissue edema in experimental myocardial infarction. *J. Clin. Invest.* **88**, 1622–1628.

Weigel, P. H., Fuller, G. M., LeBoeuf, R. D. (1986) A model for the role of hyaluronic acid and fibrin in the early events during the inflammatory response and wound healing. *J. Theor. Biol.* **119**, 219–234.

Weigel, P. H., Hascall, V. C., Tammi, M. (1997) Hyaluronan synthases. *J. Biol. Chem.* **272**, 13997–14000.

Weissman, B., Meyer, K. (1954) The structure of hyalobiuronic acid and of hyaluronic acid from umbilical cord. *J. Am. Chem. Soc.* **76**, 1753–1757.

Wessels, M. R., Moses, A. E., Goldberg, J. B., DiCesare, T. J. (1991) Hyaluronic acid capsule is a virulence factor for mucoid group A streptococci. *Proc. Natl. Acad. Sci. USA* **88**, 8317–8321.

Wessels, M. R., Goldberg, J. B., Moses, A. E., DiCesare, T. J. (1994) Effects on virulence of mutations in a locus essential for hyaluronic acid capsule expression in group A streptococci. *Infect. Immun.* **62**, 433–441.

West, D. C., Kumar, S. (1989) The effect of hyaluronate and its oligosaccharides on endothelial cell proliferation and monolayer integrity. *Exp. Cell Res.* **183**, 179–196.

West, D. C., Hampson, I. N., Arnold, F., Kumar, S. (1985) Angiogenesis induced by degradation products of hyaluronic acid. *Science* **228**, 1324–1326.

West, D. C., Shaw, D. M., Lorenz, P., Adzick, N. S., Longaker, M. T. (1997) Fibrotic healing of adult and late gestation fetal wounds correlates with increased hyaluronidase activity and removal of hyaluronan. *Int. J. Biochem. Cell Biol.* **29**, 201–210.

Whitnack, E., Bisno, A. L., Beachey, E. H. (1981) Hyaluronate capsule prevents attachment of group A streptococci to mouse peritoneal macrophages. *Infect. Immun.* **31**, 985–991.

Wobig, M., Bach, G., Beks, P., Dickhut, A., Runzheimer, J., Schwieger, G., Vetter, G., Balazs, E. A. (1999) The role of elastoviscosity in the efficacy of viscosupplementation for osteoarthritis of the knee: a comparison of hylan G-F 20 and a lower-molecular-weight hyaluronan. *Clin. Ther.* **21**, 1549–1562.

Wong, S. F., Halliwell, B., Richmond, R., Skowroneck, W. R. (1981) The role of superoxide and hydroxyl radicals in the degradation of hyaluronic acid induced by metal ions and by ascorbic acid. *J. Inorg. Biochem.* **14**, 127–134.

Wu, R., Wu, M. M. (1986) Effects of retinoids on human bronchial epithelial cells: differential regulation of hyaluronate synthesis and keratin protein synthesis. *J. Cell Physiol.* **127**, 73–82.

Yoshimura, M. (1985) Change of hyaluronic acid synthesis during differentiation of myogenic cells and its relation to transformation of myoblasts by Rous sarcoma virus. *Cell Differ.* **16**, 175–185.

Zahalka, M. A., Okon, E., Gosslar, U., Holzmann, B., Naor, D. (1995) Lymph node (but not spleen) invasion by murine lymphoma is both CD44- and hyaluronate-dependent. *J. Immunol.* **154**, 5345–5355.

Zhou, B., Oka, J. A., Singh, A., Weigel, P. H. (1999) Purification and subunit characterization of the rat liver endocytic hyaluronan receptor. *J. Biol. Chem.* **274**, 33831–33834.

Zhu, D., Bourguignon, L. Y. W. (1998) The ankyrin-binding domain of CD44s is involved in regulating hyaluronic acid-mediated functions and prostate tumor cell transformation. *Cell Motil. Cytoskeleton* **39**, 209–222.

19
Levan

Dr. Sang-Ki Rhee[1], Dr. Ki-Bang Song[2], Dr. Chul-Ho Kim[3], Dr. Buem-Seek Park[4], Ms. Eun-Kyung Jang[5], Dr. Ki-Hyo Jang[6]

[1] Biomolecular Engineering Laboratory, Korea Research Institute of Bioscience and Biotechnology (KRIBB), 52 Eoeun-dong, Yuseong, Daejeon 305-333, Korea; Tel.: +82-42-860-4450; Fax: +82-42-860-4594; E-mail: rheesk@mail.kribb.re.kr

[2] Biomolecular Engineering Laboratory, Korea Research Institute of Bioscience and Biotechnology (KRIBB), 52 Eoeun-dong, Yuseong, Daejeon 305-333, Korea; Tel.: +82-42-860-4457; Fax: +82-42-860-4594; E-mail: songkb@mail.kribb.re.kr

[3] Biomolecular Engineering Laboratory, Korea Research Institute of Bioscience and Biotechnology (KRIBB), RealBioTech Co., Ltd., #202 Bioventure Center, KRIBB, 52 Eoeun-dong, Yuseong, Daejeon 305-333, Korea; Tel.: +82-42-860-4452; Fax: +82-42-860-4594; E-mail: kim3641@mail.kribb.re.kr

[4] Biomolecular Engineering Laboratory, Korea Research Institute of Bioscience and Biotechnology (KRIBB), 52 Eoeun-dong, Yuseong, Daejeon 305-333, Korea; Tel.: +82-42-860-4454; Fax: +82-42-860-4594; E-mail: buemseekpk@mail.kribb.re.kr

[5] RealBioTech Co., Ltd., #202 Bioventure Center, KRIBB, 52 Eoeun-dong, Yuseong, Daejeon 305-333, Korea; Tel.: +82-42-863-4381; Fax: +82-42-863-4382; E-mail: levanis@realbio.com

[6] Department of Medical Nutrition, Graduate School of East-West Medical Science, Kyung Hee University, Suwon 449-701, Korea; Tel.: +82-2-961-0506; Fax: +82-2-961-9215; E-mail: kihyojang@hotmail.com

Biopolymers for Medical and Pharmaceutical Applications. Edited by A. Steinbüchel and R. H. Marchessault

ISBN: 3-527-31154-8

1-kestose	*O*-β-D-fructofuranosyl-(2 → 1)-β-D-fructofuranosyl-(2 → 1)-β-D-glucopyranoside
1-SST	sucrose:sucrose 1-fructosyltransferase
6-SFT	sucrose:fructan 6-fructosyltransferase
DFA	di-β-D-fructofuranose dianhydride
DP	degree of polymerization
EPS	exopolysaccharides
FFT	fructan:fructan fructosyltransferase
FOS	fructo-oligosaccharides
HPr	histidine-containing phosphocarrier protein
LBT	levanbiosyl transfer
LFT	levanfructosyl transfer
LFTase	levan fructotransferase
PEG	polyethylene glycol
PTS	phosphoenolpyruvate-dependent carbohydrate
RBT	RealBioTech Co., Ltd.
TLC	thin-layer chromatography

1 Introduction

Fructan, one of the most highly distributed biopolymers in nature, is a homopolysaccharide composed of D-fructofuranosyl residues joined by β-(2,6) and β-(2,1) linkages. Two types of fructan, distinguishable by the type of linkage present, are inulin and levan. The term levan is used to describe the microbial polyfructan which consists of D-fructofuranosyl residues linked predominantly by β-(2,6) linkage as a main chain, but with some β-(2,1) branching points. The other polyfructan, inulin, is mainly isolated from natural vegetable sources and serves as a reserve carbohydrate in the Compositate and Gramineae (Vandamme and Derycke, 1983), although inulins from the microbial origin have also been reported in *Streptococcus mutans* and *Streptococcus sanguis*, the human pathogens involved in dental caries (Birkhed et al., 1979). The fructose homopolymer, levan, is found in plants and especially in bioproducts of microorganisms. Plant levans, graminans, and phleins have shorter residues (varying from 10 to ~200 fructose residues) than microbial levans, of which molecular weights are up to several million daltons, with multiple branches. Microbial levans are produced extracellularly from sucrose- and raffinose-based substrates by levansucrase (sucrose 6-fructosyltransferase, EC 2.4.1.10) from a wide range of taxa such as bacteria, yeasts, and fungi (Han, 1990; Hendry and Wallace, 1993). Microbial levans are produced mainly by bacteria such as *Bacillus subtilis, Zymomonas mobilis, Bacillus polymyxa, Aerobacter levanicum, Erwinia amylovora, Rhanella aquatilis* and *Pseudomonas*. The production and utilization of levan in the industrial field have been strictly limited until very recently, and very few reports have been made on the production of levan using fermentation techniques (Elisashvili, 1984; Beker et al., 1990; Han, 1990; Keith et al., 1991; Ohtsuka et al., 1992; Uchiyama, 1993). Recently, great interest in this fructan has been renewed to discover novel applications for levan as a new industrial gum in the fields of cosmetics, foods (e.g., as dietary fiber), and pharmaceuticals. In this chapter, we describe the production and degradation of levan by use of enzymatic reactions, the genetic regulation and control of such reactions, and outline the properties of levan and the current status of its industrial applications.

2 Historical Outline

Historically, levan was generally considered to be an undesirable byproduct of sugar and juice processing because it increases the viscosity of the processing liquor (Fuchs, 1959; Avigad, 1965). Levan was first described by Lippmann in Germany in 1881, when the name "laevulan" was proposed. Greig-Smith (1901) later showed that a strain of *Bacillus*, when grown on sucrose, produced fructans, and the name "levan" was then introduced as being analogous to dextran. The term laevulan now denotes partially degraded levan fractions. However, early reports on levan were confusing because the microbial nomenclature was unsystematic and the materials were inadequately described.

The biosynthesis of levan was elucidated some years later. The mechanism was shown to involve two enzymes, sucrose fructosyltransferase and fructan fructosyltransferase, and was proposed by Edelman and Jefford in 1968. The enzyme kinetics of the transfructosylation reaction was revealed by Chambert and Gonzy-Treboul in 1976. The enzyme which is now generally recognized as levansucrase was named by Hestrin et al. in 1943, and is responsible for the synthesis of levan from sucrose. The most extensive studies of levansucrase were performed in *B. subtilis*, and focused on the localization of the enzyme as well as its properties, expression regulation, genetic organization, and kinetics (Suzuki and Chatterton, 1993). As levan began to receive more attention based on its potential applications, many levan-producing microorganisms were identified (Han, 1990; Hendry and Wallace, 1993). The mass production of levan from *Z. mobilis*, and the secretion of levansucrase were reported relatively recently in detail by Song and coworkers (1996) and Ananthalakshmy and Gunasekaran (1999).

Although extensive research and searches for industrial applications have been conducted with dextran (which is also known as a bacterial biopolymer), much less attention has been focused on levan, mainly because of the very poor yields obtained in its industrial production. Nonetheless, great interest has been expressed in the diverse aspects of this fructose homopolymer over the past decade, despite the applications of levan having remained relatively few in number because of the limited supplies. Levans are now available commercially in reagent grade from microbial sources (e.g., from Sigma Chemical Co., IGI Biotechnology), but these are used only for research purposes. Since the time when levan was first produced on a large scale by using levansucrase from genetically engineered *Escherichia coli* (Song et al., 1996), attention has been renewed on the potential industrial application of levan and its derivatives in the fields of agriculture, cosmetics, food ingredients, animal feed and pharmaceuticals (Clarke et al., 1997; Kim et al., 1998; Vijn and Smeekens, 1999; Rhee et al., 2000d), as well as being a good source of pure fructose and di-β-D-fructofuranose dianhydride (DFA) (Saito and Tomita, 2000).

3 Chemical Structures of Levan

Fructans are chemically versatile molecules, and consist of a single glucose unit attached to two or more fructose units. Three fructan trisaccharides are known, each being produced through a glycosidic linkage of fructose to one of the three primary hydroxyl groups of sucrose. Fructose linked to the primary carbon of the fructose moiety of sucrose forms 1-kestose (also called isokestose), while fructose linked to the sixth carbon of the fructose moiety of sucrose forms 6-kestose (also called kestose). Both of these trisaccharides have a terminal glucose and a terminal fructose. Linkage of a fructose moiety to the sixth carbon of glucose moiety of sucrose forms neokestose, with both end groups being fructose (Nelson and Spollen, 1987).

Chemically, levan consists of β-D-fructofuranosyl residues linked predominantly through β-(2,6) as 6-kestose of the basic trisaccharide, with extensive branching through β-(2,1) linkages (Figure 1). In contrast, inulin is composed of β-D-fructofuranose attached by β-(2,1) linkages. The first monomer of the chain is either a β-D-glucopyranosyl or a β-D-fructofuranosyl residue. Although they are similar fructose homopolymers, it is evident that levan is different from inulin-type fructan since microbial inulin contains 5–7% of β-(2,6)-linked branches (Wolff et al., 2000).

The molecular shape of levan, as visualized by electron microscopy (Newbrun et al., 1971), is spheroidal, indicating that the constituent chains are extended radially at the same synthetic rate. The molecular

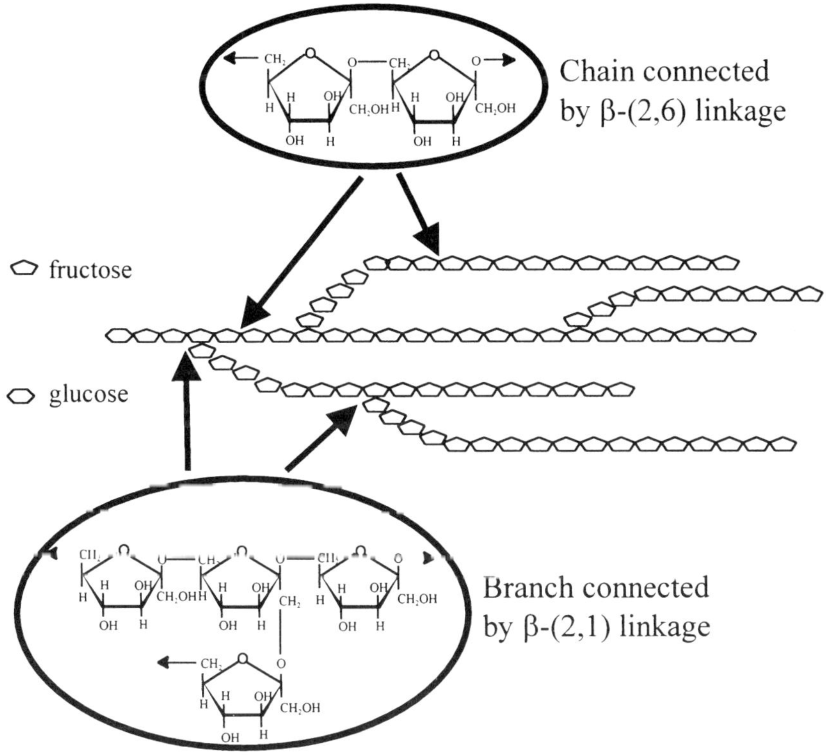

Fig. 1 Structure of levan. The main chain is connected by β-(2,6) linkages and the branch is connected to the main chain by a β-(2,1) linkage; the branch then continues with β-(2,6) linkages.

weight of bacterial levans is typically in the range of 2×10^6 to 10^8 (Keith et al., 1991), with the final molecular size being influenced by the synthesizing conditions such as ionic strength, temperature, and co-solutes. Although microbial levans have the similar structure, several types of (IX) levan are produced by different microorganisms, and this may be attributed to a varying degree of polymerization (DP) and branching of the repeating unit.

A cell-free enzyme system could be used to synthesize levans which have both β-(2,6) and β-(2,1) linked fructosyl units, and with similar structure to those of a whole-cell enzyme system. However, the structure of levans synthesized in the cell-free enzyme system is also known to differ in length compared with that synthesized by whole-cell systems (Han, 1990).

4 Occurrence

Various polysaccharides are produced as structural components in living organisms, and levan is one of the most diversely distributed components in plants, yeasts, fungi, and bacteria in particular. Levan is produced by grass (*Dactylis glomerata, Poa secunda* and *Agropyron cristatum*), wheat and barley (*Hordeum vulgare*), fungi (*Aspergillus sydawi* and *Aspergillus versicolor*) and yeasts (Han, 1990). Levans produced by microorganisms have been reported by Han (1990), and Hendry and Wallace (1993), and are listed in Table 1.

Previously, oral bacteria such as *Streptococcus, Rothis* and *Odontomyces* had received much attention due to their presence in human dental caries, together with soil microorganisms, especially *Bacillus.* Subsequently, focus was centered on the biological and functional aspects of levan rather than on its oral accumulation. The most extensive studies of levan were performed using *B. subtilis* (Suzuki and Chatterton, 1993). Furthermore, levans from *Bacillus polymyxa* (Aymerich, 1990) and *Pseudomonas* sp. (Hettwer et al., 1995, 1998) were identified as playing a role in the plant defense response. Levan from *B. subtilis* was shown to be tolerant against salt stress (Kunst and Rapoport, 1995), while that obtained from *Z. mobilis* exhibited antitumor activity (Calazans et al., 2000). The synthesis of levan using the genus *Lactobacillus* was also recently reported (Van Geel-Schutten et al., 1999).

Tab. 1 Levan-producing microorganisms

Microorganism	*Reference*
Acetobacter xylinum	Tajima et al. (1998)
Actinomyces naeslundii	Bergeron et al. (2000)
Bacillus circulans	Perez Oseguera et al. (1996)
Bacillus stearothermophilus	Li et al. (1997)
Gluconacetobacter (formerly *Acetobacter*) *diazotrophicus*	Arrieta et al. (1996)
Lactobacillus reuteri	Van Geel-Schutten et al. (1999)
Pseudomonas syringae pv. phaseolicola	Hettwer et al. (1995)
Pseudomonas syringae pv. glycinea	Hettwer et al. (1998)
Rahnella aquatilis	Ohtsuka et al. (1992); Song et al. (1998)
Serratia levanicum	Kojima et al. (1993)
Zymomonas mobilis	Song et al. (1993)

5 Physiological Functions of Levan

Bacterial polysaccharides are found either as a dense layer of more or less regularly arranged polymer structures attached to the bacterial cell walls (capsules) or as loosely associated exopolysaccharides (EPS) (Beveridge and Graham, 1991). Levans produced microbiologically have a number of interesting features. The levan which is synthesized extracellularly by bacteria may be visualized in a sucrose-containing medium, giving rise to a typical mucoid morphology. This type of mucoid feature provides a role in the symbiosis, phytopathogenesis, or participation in the defense mechanism against cold and dry conditions (Kunst and Rapoport, 1995). Extracellular levan produced by bacterial plant pathogens increases bacterial fitness and also acts as a detoxifying barrier against plant defence compounds (Hettwer et al., 1998). Among natural polysaccharides, glucans and fructans possess antitumor activity, and levan is included in this list. The antitumor activity of levan against sarcoma 180 depends on the molecular weight of the polysaccharides (Calazans et al., 2000), and this may indicate the polydiversity of levan.

Levans produced in plants are present as storage carbohydrates in the stem and leaf sheaths, and are degraded in a later stage of the growing season to provide plants with carbohydrates for grain filling (Pollock and Cairns, 1991). The biological role of polysaccharides in protection is less clearly understood, but a hypothesis has recently emerged. Levan penetrates into lipid membranes composed of monomolecular lipid layers, after which interactions occur which are orders of magnitude greater than the interaction between disaccharides and lipids. An extended layer of levan adheres to the lipids and partially protrudes into the aqueous phase. It is also possible that the membranes present are coated with levan; this coating of membranes imparts a reduction in accessibility of the membrane surface to proteins. In this way, in a biological system levan is able to protect membranes by interacting with the membrane lipid fraction (Vereyken et al., 2001).

6 Chemical Analysis and Detection

Several methods can be used in the qualitative and quantitative analysis of levan and the estimation of its concentration in solutions, with spectrophotometry and chromatography being the major techniques.

6.1 Spectrophotometry

Low concentrations of levan are measured by monitoring the optical density at 450–550 nm, as the presence of levan creates turbidity within the enzyme reaction mixture.

6.2 High-Performance Liquid Chromatography (HPLC)

The HPLC method is employed for both qualitative and quantitative determination of levan and other components (oligosaccharides, sucrose, fructose, and glucose). Details of the method are described here. The enzyme reaction mixture or levan solution is filtered using a 0.45 μm pore size membrane filter, and the filtrate is analyzed by HPLC equipped with a gel filtration column and refractive index detector (Shodex Ionpack KS-802, 300×8 mm; Showa Denko Co., Japan) (Jang et al., 2000). Deionized water is used as a mobile phase at a flow rate of

0.4 mL min^{-1}. The DP of levan is also determined by HPLC equipped with successive columns, GPC 4000–GPC 1000 (Polymer Laboratories, USA), and a refractive index detector (Jang et al., 2001). The analyses of sugar components and linkage type of levan are determined by using acid hydrolysis, methylation and nuclear magnetic resonance (NMR) shift experiments (Suzuki and Chatterton, 1993). In ^{13}C-NMR, signals of carbons of levan obtained from *Z. mobilis* and *Aerobacter levanicum* are identical, and show six main resonances at 104.9, 81.0, 76.9, 64.1, and 60.6 p.p.m. (Song and Rhee, 1994), these signals differing from those obtained with inulin.

6.3 Other Methods

Levan (nonmobiles) can be distinguished from oligosaccharides, sucrose, and other byproducts, either qualitatively or quantitatively, by the use of thin-layer chromatography (TLC). The sucrose-hydrolyzing activity of bacterial levansucrase is also used in the determination of levan concentration. The methods established are based on the fact that glucose is formed stoichiometrically in relation to the amount of fructose incorporated into levan (major product) and oligosaccharides (minor products). The amounts of glucose generated by the enzymatic reaction can be determined quantitatively by commercially available kits from the suppliers (Song et al., 1993).

7 Biosynthesis of Levan

The biosynthesis of levan requires the involvement of an extracellular enzyme levansucrase, which shows specificity for sucrose. Genetic characterization of the enzyme and the regulation of levan synthesis have been extensively studied, mostly using levansucrase genes from *B. subtilis* and *Z. mobilis*.

7.1 Enzymology of Levan Synthesis

Levansucrase (sucrose:2,6-β-D fructan:6-D-fructosyltransferase, sucrose 6-fructosyltransferase, EC 2.4.1.10.) was first named by Hestrin et al. (1943), and is responsible for the synthesis of levan from sucrose. Levansucrase exists as constituent intracellular and inducible extracellular forms in microorganisms (Han, 1990). The function of the levan-producing enzyme located intracellularly in some bacteria is not yet understood. The most abundant substrate for levansucrase in nature is sucrose, but raffinose also serves as a substrate.

Levansucrase is a type of transferase which catalyzes a fructosyl transfer from sucrose to various acceptor molecules. The enzyme catalyzes the following reactions:

1. Polymerization:

$$(\text{Sucrose})_n \rightarrow (\text{Glucose})_n + \text{Levan} + \text{Oligosaccharides}$$

2. Hydrolysis:

$$\text{Sucrose} + H_2O \rightarrow \text{Fructose} + \text{Glucose}$$

$$(\text{Levan})_n + H_2O \rightarrow (\text{Levan})_{n-1} + \text{Fructose}$$

3. Acceptor:

$$\text{Sucrose} + \text{Acceptor molecules} \rightarrow \text{Fructosyl-acceptor} + \text{Glucose}$$

4. Exchange:

$$\text{Sucrose} + [^{14}\text{C}]\text{Glucose} \rightarrow \text{Fructose-}[^{14}\text{C}]\text{Glucose} + \text{Glucose}$$

5. Disproportionation:

$$[\text{Levan}]_m + [\text{Levan}]_n \rightarrow [\text{Levan}]_{m-1} + [\text{Levan}]_{n+1}$$

The enzyme catalyzes hydrolysis and polymerization reactions concomitantly (Reaction 1), resulting in a fructose homopolymer (levan) and free glucose. This reaction occurs when sucrose exists as the sole fructosyl donor and acceptor, and involves three steps: initiation, propagation, and termination (Chambert et al., 1974). The chains of levan grow step-wise by repeated transfer of a hexosyl group from the donor to growing acceptor molecules. The enzyme primarily catalyzes a coupled reaction by a ping-pong mechanism, i.e., sucrose hydrolysis followed by transfructosylation involving a fructosyl-enzyme intermediate (Chambert et al., 1976).

When water acts as an acceptor, a free fructose is generated from both sucrose and levan (Reaction 2). This reaction occurs in all the levansucrase-catalyzed reactions mentioned above, but the rate is much slower when compared with a sugar acceptor. Reaction 3 occurs in the presence of an acceptor in the environment. The enzyme transfers the fructosyl residue of sucrose specifically to the C-1 hydroxyl group of aldose in the acceptor. Compounds containing hydroxyl groups, such as methanol, glycerol and oligosaccharides, can act as fructosyl acceptors.

The reaction mechanism yields a nonreducing sugar compound and a series of oligosaccharides, in which the sugar molecule with one more fructose moiety remains as a major reaction product. The reaction occurs predominantly in the presence of a high concentration of fructosyl donors, such as sucrose or raffinose. Reaction 4 might be considered analogous to Reactions 2 and 3, but differs in the regeneration of sucrose, which has a high-energy bond. The enzyme also catalyzes Reaction 5, a disproportionation reaction, in which the degree of polydispersity of levan or oligomers is modified. The above five reactions compete with one another, yielding a specific major product with some minor products but they are predominantly controlled by environmental factors.

At present, little is known of plant levans, and their biosynthesis is not fully understood (Heyer et al., 1999). One plant levan, known as "graminan", is synthesized by sucrose:fructan 6-fructosyltransferase (6-SFT) which catalyzes the formation and extension of β-(2,6)-linked fructans. The 6-SFT is closely related to vacuolar invertase and transfers the fructosyl residues from sucrose preferentially to 1-kestose or larger fructans (Sprenger et al., 1995). However, most fructan synthesis in plants occurs in two steps (Edelman and Jefford, 1968). Initially, sucrose:sucrose 1-fructosyltransferase (1-SST; EC 2.4.1.99.) catalyzes the formation of the trisaccharide 1-kestose and glucose from two molecules of sucrose. Later, fructan:fructan 1-fructosyltransferase (1-FFT; EC 2.4.1.100.) reversibly transfers fructosyl residues from one fructan with a DP of ≥ 3 to another DP of ≥ 2, producing a mixture of fructans with different chain lengths.

7.2 Genetic Basis of Levan Synthesis

As yet, levansucrase genes have been cloned and biochemically characterized in seven Gram-negative strains; namely, *Acetobacter diazotrophicus* (Arrieta et al., 1996), *Acetobacter xylinum* (Tajima et al., 2000), *Erwinia amylovora* (Geier and Geider, 1993), *P. syringae* pv. glycinea, *P. syringae* pv. phaseolicola (Hettwer et al., 1998), *Rahnella aquatilis* (Song et al., 1998) and *Z. mobilis* (Song et al., 1993). Several levansucrase genes have also been cloned in Gram-positive strains, such as *Bacillus* (Gay et al., 1983; Li et al., 1997) and *Streptococcus* species (Sato et al., 1984). All levansucrase genes share several conserved regions, which are thought to be important for the enzyme activity. Although conservation is observed, dissimilarity exists depending on the source of the enzyme. Levansucrase genes from a Gram-negative origin show relatively high similarity (>50%) when compared with the genes from Gram-positive bacterial enzymes. However, very little similarity (<30%) exists among the genes from two different sources (Song and Rhee, 1994). The deduced amino acid sequences are aligned in Figure 2.

Although the amino acid sequences of levansucrases do not show any considerable homology to those of sucrose-related enzymes, the third (-EWS/AGT/SP/A-) and the

```
                     I                                                  II
Psp-Lsc   22 YEPTVWSRAD ALKVNENDPT TT-Q-PLVSA DFPVM--SDT VF-IWDTMPL RELDGTVVSV NGWSVILTLT ADRHPNDPQY
Psg-Lsc    6 YAPTIWSRAD ALKVNENDPT TT-Q-PLVSP DFPVM--SDT VF-IWDTMPL RELDGTVVSV NGWSVIVTLT ADRHPDDPQY
Ra-LsrA    6 YTPTIWTRAD ALKVNENDPT TT-Q-PIVDA DFPVM--SDE VF-IWDTMPL RSLDGTVVSV DGWSVIFTLT AQRNNNNSEY
Ea-Lsc     6 YKPTLWTRAD ALKVHEDDPT TT-Q-PVIDI AFPVM--SEE VF-IWDTMPL RDFDGEIISV NGWCIIFTLT ADRNTDNPQF
Zm-LevU    8 AEPSLWTRAD AMKVHTDDPT AT-M-PTIDY DFPVM--TDK YW-VWDTWPL RDINGQVVSG QGWSVIFALV ADRTKY----
Bs-SacB   41 YGISHITRHD MLQIPEQQKN EKYQVPEFDS STIKNISSAK GLDVWDSWPL QNADGTVANY HGYHIVFALA GDPKNADDTS
Bst-SurB  41 YGISHITRHD MLQIPEQQKN EKYQVPEFDS STIKNISSAK GLDVWDSWPL QNADGTVANY HGYHIVFALA GDPKNADDTS

                                                          III
Psp-Lsc   97 LDANGRYDIK RDWEDRHGRA RMSYWYSRTG KDWIFGGRVM AEGVSPTTRE WAGTPILLND KGDIDLYYTC VTPGAAIAKV
Psg-Lsc   81 VGANGRYDIK RDWEDRHGRA RMCYWYSRTG KDWIFGGRVM AEGVSPTTRE WAGTPVLLND KGDIDLYYTC VTPGAAIAKV
Ra-LsrA   81 LDAEGNYDIT SDWNNRHGRA RICYWYSRTG KDWIFGGRVM AEGVSPTSRE WAGTPILLNE DGDIDLYYTF VTPGATIAKV
Ea-Lsc    81 QDENGNYDIT RDWEKRHGRA RICYWYSRTG KDWIFGGRVM AEGVAPTTRE WAGTPILLND RGDIDLYYTG VTPGATIAKV
Zm-LevU   79 ----G----- --WHNRNDGA RIGYFYSRGG SNWIFGGHLL KDGANPRSWE WSGCTIMAPG TANSVEVFFT SVNDTPSESV
Bs-SacB  121 IYMFYQKVGE TSIDSWKNAG RVFKDSDKFD AN-------- DSILKDQTQE WSGSATFTSD -GKIRLFYTD FSGKHYGKQT
Bst-SurB 121 IYMFYQKVGE TSIDSWKTPG RVFKDSDKFD AN-------- DSILKDQTQE WSGSATFTSD -GKIRLFYTD FSGKHYGKQT

                                                            IV
Psp-Lsc  177 ----RGRIVT SDQGVELKDF TQVKKLFEAD GTYYQTEAQ- --------NS SWNFRDPSPF IDPNDGKL-- ---------Y
Psg-Lsc  161 ----RGRIVT SDKGVELKDF TEVKTLFEAD GKYYQTEAQ- --------NS TWNFRDPSPF IDPNDGKL-- ---------Y
Ra-LsrA  161 ----RGKVLT SEEGVTLAGF NEVKSLFSAD GVYYQTESQ- --------NP YWNFRDPSPF IDPHDGKL-- ---------Y
Ea-Lsc   161 ----RGKIVT SDQSVSLEGF QQVTSLFSAD GTIYQTEEQ- --------NA FWNFRDPSPF IDRNDGKL-- ---------Y
Zm-LevU  148 PAQCKGYIYA DDKSVWFDGF DKVTDLFQAD GLYYADYAE- --------NN FWDFRDPHVF ITPKDGKL-- ---------Y
Bs-SacB  192 LTTAQVNVSA SDSSLNINGV EDYKSIFDGD GKTYQNVQQF IDEGNYSSGD NHTLRDPHYV EDKGHKYLVF EANTGTEDGY
Bst-SurB 192 LTTAQVNVSA SDSSLNINGV EDYKSIFDGD SKTYQNVQQF IDEGNYSSGD NHTLRDPHYV EDKGHKYLVF EANTGTEDGY

                                                                                V
Psp-Lsc  233 MVFEGNVAGE RGSHTVGAAE LGPVPPGHED VGGARFQVGC IGLAVAKDLS GEEWEILPPL VTAVGVNDQT ERPHYVFQDG
Psg-Lsc  212 MVFEGNVAGE RGTHTVGAAE LGPVPPGHEE TGGARFQVGC IGLAVAKDLS GDEWEILPPL VTAVGVNDQT ERPHYVFQDG
Ra-LsrA  217 MVFEGNVAGE RGSHVIFKQE MGTLPPGHRD VGNARYQAGC IGMAVAKDLS GDEWEILPPL VTAVGVNDQT ERPHEVFQDG
Ea-Lsc   217 MLFEGNVAGP RGSHEITQAE MGNVPPGYED VGGAKYQAGC VGLAVAKDLS GSEWQILPPL ITAVGVNDQT ERPHFVFQDG
Zm-LevU  208 ALFEGNVAME RGTVAVGEEE IGPVPPKTET PDGARYCAAA IGIAQALNEA RTEWKLLPPL VTAFGVNDQT ERPHVVFQNG
Bs-SacB  272 QGEESLFNKA YYGKSTSFFR QESQKLLQSD KKRTAELANG ALGMIELNDD YTLKKVMKPL IASNTVTDEI ERANVFKMNG
Bst-SurB 272 QGEESLFNKA YYGKSTSFFR QESQKLLQSD KNRTAELANG ALGMIELNDD YTLKKVMKPL IASNTVTDEI ERANVFKMNG

                 VI                              VII
Psp-Lsc  313 KYYLFTISHK -FTYAEGLTG PDGVY--GFV GEH-LFGPYR PMNASGLVLG NPPEQPFQTY SHCVMPNGLV TSFIDSVPTE
Psg-Lsc  297 KYYLFTISHK -FTYADGVTG PDGVY--GFV GEH-LFGPYR PMNASGLVLG NPPAQPFQTY SHCVMPNGLV TSFIDSVPTS
Ra-LsrA  297 KYYLFTISHK -ETYADGLTG PDGVY--GFL SDN-LTGPYS PMNGSGLVLG NPPSQPFQTY SHCVMPNGLV TSFIDNVPTS
Ea-Lsc   297 KYYLFTISHK -YTFADNLTG PDGVU--GFV SDK-LTGPYT PMNSSGLVLG NPSSQPFQTY SHYVMPNGLV TSFIDSVPWK
Zm-LevU  288 LTYLFTISHH S-TYADGLSG PDGVY--GFV SENGIFGPYE PLNGSGLVLG NPSSQPYQAY SHYVMINGLV TSFIDTIPSS
Bs-SacB  352 KWYLFTDSRG SKMTIDGITS NDI-YMLGYV S-NSLTGPYK PLNKTGLVLK MDLDPNDVTF TYSHFAVPQA KGNNVVITSY
Bst-SurB 352 KWYLSTDSRG SQMTIDGITS NDI-YMLGYV S-NSLTGPYK PLNKTGLVLK MDLDPNDVTF TYSHFAVPQA TGNNVVITSY

Psp-Lsc  389 GED-YRIGGT EAPTVRILLK GDRSFVQEEY DYGYIPAMKD VTLS 431  P.syringae pv.phaseolicola
Psg-Lsc  373 GED-YRIGGT EAPTVRILLE GDRSFVQEVY DYGYIPAMKN VVLS 415  P.syringae pv.glycinea
Ra-LsrA  373 DGN-YRIGGT EAPTVKIVLK GNRSFVERVF DYGYIPPMKN IILN 415  R.aquatilis
Ea-Lsc   373 GKD-YRIGGT EAPTVKILLK GDRSFIVDSF DYGYIPAMKD ITLK 415  E.amylovora
Zm-LevU  365 DPNVYRYGGT LAPTIKLELV GHRSFVTEVK GYGYIPPQIE WLAE 408  Z.mobilis
Bs-SacB  430 MTNRGFYADK QSTFAPSFLL NIKGKKTSVV KDSILEQGQL TVNK 473  B.subtilis
Bst-SurB 430 MTNRGFYADK QSTFAPSFLL NIQGKKTSVV KASILDQGQL TVNQ 473  B.stearothermophilus
```

Fig. 2 Multiple alignment of deduced amino acid sequences of bacterial levansucrases. Origins of levansucrase are indicated in ends of the sequences. Asterisks indicate identical- and similar-residues in all levansucrases. Regions considered as important for activity are boxed (I–VII). Amino acid residues that are different between Gram-negative and Gram-positive origin are indicated by dots.

fourth (-FRDP-) conserved regions are found in all fructosyl- and glucosyltransferases, sucrase, sucrose-phosphate hydrolase and even in fructan-hydrolyzing enzymes. The fact that the regions are preserved in all of the sucrose-related enzymes implies that they may be catalytically important regions for the hydrolysis of sucrose. The serine residue in the sixth region (-YLFTI/DS-) has been proposed as the putative residue of the catalytic site (Chambert and Petit-Glatron, 1991).

7.3 Regulation of Levan Synthesis

The genes encoding sucrose-hydrolyzing enzymes may not be linked to each other on the chromosome, but linked only with accessory genes coding for proteins belonging to the phosphoenolpyruvate-dependent carbohydrate phosphotransferase (PTS) system. The expression of these genes is regulated by many regulatory protein systems such as the *glk* operon of *Z. mobilis* and the pleiotropic system of *B. subtilis*, etc.

7.3.1 Regulation at the Protein Level

In microorganisms, two types of sucrose utilization were found: intra- and extracellular. Commonly, these sucrose-uptake utilization systems exist within the cell; sucrose is transported by the PTS system. The sucrase system of this type is well known in *B. subtilis* (Klier and Rapoport, 1988). In contract, some bacteria such as *Z. mobilis* that lack the PTS system first hydrolyze sucrose extracellularly to monomeric sugars, after which these sugars are transported inside the cell (Di Marco and Romano, 1985).

The co-contribution of both saccharolytic enzymes for the sucrose utilization of *Z. mobilis* has been well characterized. The glucose uptake and utilization system (*glk* operon), which is located very close to the *levU* operon and is also linked metabolically with the sucrose utilization system of *Z. mobilis*, is also regulated by the mechanism of tightly linked gene expression (Liu et al., 1992). In the intervening sequence of the *levU* and *glf* operon, two putative ORFs, encoding Lrp-like regulatory protein and aspartate racemase respectively, were found (Song et al., 1999).

At the molecular level, the genes encoding sucrose-hydrolyzing enzymes reported to date are not linked to each other on the chromosome, but are linked only with accessory genes coding for proteins belonging to the PTS system (Bruckner et al., 1993). The expression of these genes is modulated by regulatory mechanisms, such as anti-termination or repression, which is controlled by the complex regulatory network system including many regulatory proteins (Klier and Rapoport, 1988).

7.3.2 Regulation at Transcriptional and Translational Levels

The genes encoding the extracellular levansucrase and sucrase have been isolated and characterized. The nucleotide sequences of the DNA segment containing the genes encoding extracellular levansucrase and sucrase of *Z. mobilis* and *B. subtilis* were reported recently (Kyono et al., 1995). The two genes are located together in an operon on the chromosome, whereas almost all other genes coding for saccharolytic enzymes in other bacteria and yeasts are dispersed on the chromosomes (Carson and Botstein, 1983). The levansucrase gene of *B. subtilis* is activated in the presence of an inducer (sucrose or fructose), and is under a pleiotropic regulatory system controlling the expression of the sucrose operon (Lepesant et al., 1976; Shimotsu and Henner, 1986).

The pleiotropic system involving the *degS/degU*, *degQ* (formerly *sacU* and *sacQ*) and *degR* genes affects the expression of *sacB* (Débarbouillé et al., 1991). Levansucrase is encoded by the *sacB* gene and expressed from a constitutive promoter in the closely linked *sacR* locus. The *sacR* locus contains a palindromic structure acting as a transcription terminator. In the presence of sucrose, an anti-terminator, the *sacY* gene product that belongs to the *sacS* operon allows transcription of the *sacB* gene. The expression of this gene is also controlled by other regulatory genes such as two- component system *degS/degU* and also by *degQ* (Rapoport and Klier, 1990) (Figure 3).

8 Biodegradation of Levan

Although the biodegradation of levan involves several enzymes including levanase, levansucrase, and levan fructotransferase, the genetic characterization of these is limited to levanase.

8.1 Enzymology of Levan Degradation

Levan is degraded to D-fructose, levanbiose, sucrose, levan oligomers or low molecular-weight levan by the hydrolytic activity of levanase, levansucrase or levan fructotransferase from some plants and microorganisms. The mode and degree of hydrolysis depend on the enzyme sources and the reaction conditions.

8.1.1 Levanase

Many levan-forming microorganisms also produce hydrolytic enzymes–levanases–that degrade levan (Hestrin and Goldblum, 1953; Avigad, 1965). Certain strains of *Bacillus, Pseudomonas, Actinomyces, Aerobacter, Clostridium* and *Streptococcus* produce exocellular levanase (2,6-β-fructan 6-levanbiohydrolase, EC 3.2.1.64.) (Fuchs, 1959; Uchiyama, 1993). The enzyme hydrolyzes only levan, and the resulting product is usually levanbiose, indicating that a terminal fructosyl unit is removed. An exo-hydrolytic enzyme

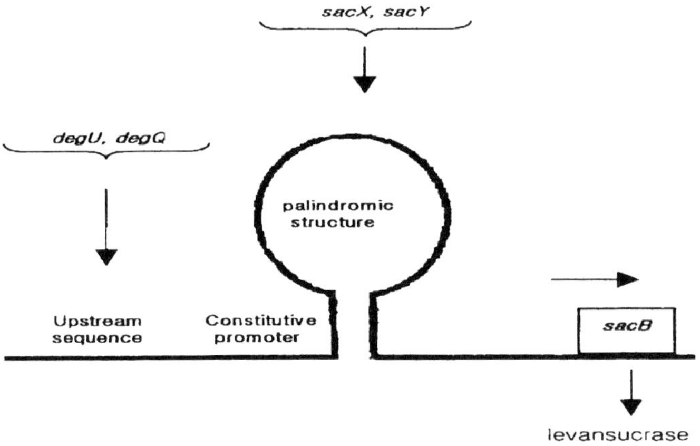

Fig. 3 Specific and pleiotropic control mechanisms affecting the *sacB* gene in *Bacillus subtilis*. *degU*: transcriptional regulator of degradation enzymes; *degQ*: pleiotropic regulatory gene; *sacX*: negative regulatory protein of *sacY*; *sacY*: positive levansucrase synthesis regulatory protein; *sacB*: gene encoding levansucrase. (Reprinted from Débarbouillé et al., 1991, p. 758, with permission from Elsevier Science.)

which has a 2,6-β-linkage-specific fructan-β-fructosidase activity (Marx et al., 1997) was reported from the grass *Lolium perenne*, and a 2,1-β-linkage-specific exohydrolase from Jerusalem artichoke. The other exo-levanase (fructan-β-fructosidase, EC 3.2.1.80; beta-D-fructofuranosidase, EC 3.2.1.26.) hydrolyzes levan to produce D-fructose. Endo-levanases (2,6-β-D-fructan fructanohydrolase, EC 3.2.1.65.) hydrolyze levan and levan oligomers consisting of more than three fructosyl units.

8.1.2 Levansucrase

Levan may be degraded not only by levanases, but also by levansucrase itself, which may catalyze the hydrolysis under certain conditions (Rapoport and Dedonder, 1963). The degree of levan hydrolysis depends on the enzyme sources and reaction conditions. For example, levansucrase from *R. aquatilis* showed a higher degradation activity than did that from *Z. mobilis* (Song et al., 1998), though for both enzymes a higher degradation activity of levan was seen as the reaction temperature was increased from 4 °C to 30 °C.

Although indirect evidence of the reversal of enzymatic synthesis of levan has been observed, little is known regarding the nature of such enzymatic degradation. Smith (1976) showed that beta-fructofuranosidase present in tall fescue degraded levan by removing one fructose residue at a time until a molecule of sucrose remained. Levansucrase of *B. subtilis* has a hydrolytic effect on small levans (Dedonder, 1966), the hydrolytic action stopping at branch points. Neither inulin, inulobiose, inulintriose, nor methyl D-fructofuranoside is hydrolyzed, despite these substrates being hydrolyzed by inulinase and yeast invertase. This hydrolytic activity may be responsible for the appearance of heterogeneous short-chain polysaccharides, rather than uniform high molecular-weight polymers, in the final product of many levan preparations.

8.1.3 Levan Fructotransferase

Microbial levan is an interesting starting material for the production of valuable oligosaccharides such as DFA IV (Yun, 1996; Saito and Tomita, 2000). DFA IV (di-D-fructose-2,6′:6,2′-dianhydride) is an oligosaccharide which is produced from levan by microbial enzymes, i.e., levan fructotransferase (LFTase) and a type of levanase (Tanaka et al., 1981, 1983; Saito et al., 1997). Currently, two LFTases have been isolated and cloned from *Arthrobacter nicotinovorans* GS-9 (Saito et al., 1997) and *A. ureafaciens* (Tanaka et al., 1981; Song et al., 2000). The enzymes have also been shown to degrade levan molecules from the nonreducing fructose end of the outer chains, and to catalyze intermolecular levanbiosyl and levanfructosyl transfer (LBT and LFT, respectively) reactions (Tanaka et al., 1983).

8.2 Genetic Basis of Levan Degradation

In *B. subtilis*, the expression of the levanase operon is inducible by fructose and is subjected to catabolite repression. A fructose-inducible promoter has been characterized 2.7 kb upstream from the gene *sacC*, which encodes levanase. The *sacC* gene is the distal gene of an operon containing five genes: *levD*, *levE*, *levF*, *levG*, and *sacC* (Martin et al., 1989; Débarbouillé et al., 1991) and is expressed under the regulated control of *sacR*, the inducible levansucrase leader region. The first four gene products are involved in a fructose-PTS system. In *Pseudomonas*, levanase is an exohydrolase of levan and produces levanbiose as a sole product; the limits of hydrolysis of levan from *Z. mobilis* and *Serratia* sp. were 65% and 80%, respectively (Jung et al., 1999).

8.3 Regulation of Levan Degradation

There are two levels on which the expression of the levanase operon in *B. subtilis* is controlled: (1) an induction by fructose, which involves a positive regulator, LevR, and the fructose phosphotransferase system encoded by this operon (lev-PTS); and (2) a global regulation of catabolite repression (Débarbouillé et al., 1991) (Figure 4).

The LevR protein is an activator for the expression of the levanase operon from *B. subtilis*. RNA polymerase containing the sigma 54-like factor sigma L recognizes the promoter of this operon. One domain of the LevR protein is homologous to activators of the NtrC family, and another resembles anti-terminator proteins of the BglG family (Débarbouillé et al., 1991). It has been proposed that the domain, which is similar to anti-terminators, is a target of phosphoenolpyruvate:sugar phosphotransferase system (PTS)-dependent regulation of LevR activity. The LevR protein is not only negatively regulated by the fructose-specific enzyme IIA/B of the phosphotransferase system encoded by the levanase operon (lev-PTS), but is also positively controlled by the histidine-containing phosphocarrier protein (HPr) of PTS (Martin et al., 1990; Stülke et al., 1995). This second type of control of LevR activity depends on phosphoenolpyruvate-dependent phosphorylation of HPr histidine 15, as demonstrated with point mutations in the ptsH gene encoding HPr. *In vitro* phosphorylation of partially purified LevR was obtained in the presence of phosphoenolpyruvate, enzyme I, and HPr. The dependence of truncated LevR polypeptides on stimulation by HPr indicates that the domain homologous to anti-terminators is the target of HPr-dependent regulation of LevR activity. This domain appears to be duplicated in the LevR protein. The first anti-terminator-like domain seems to be the target of enzyme I and HPr-dependent phosphorylation and the site of LevR activation, whereas the carboxy-terminal anti-terminator-like domain could be the target for negative regulation by lev-PTS (Débarbouillé et al., 1990).

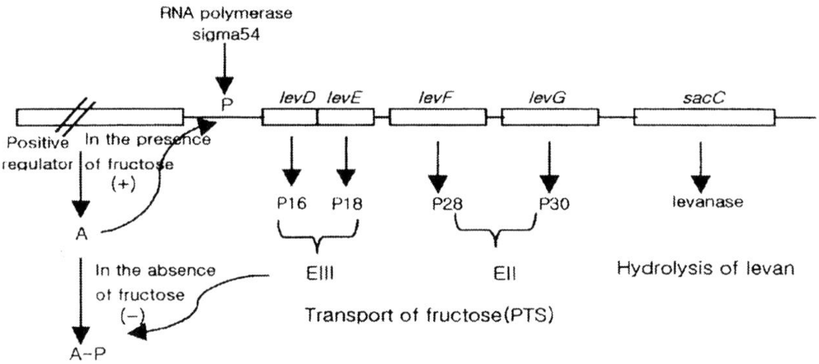

Fig. 4 Regulation model of levanase operon from *Bacillus subtilis*. P represents the fructose-inducible promoter. The *levD*, *levE*, *levF*, and *levG* gene products correspond to a fructose-specific phosphoenolpyruvate-dependent carbohydrate (PTS). *LevR* encodes a positive regulator. The activator may exist in two forms: (A-P), an inactive phosphorylated form; or (A), an active non-phosphorylated form. (Reprinted from Débarbouillé et al., 1991, p. 759, with permission from Elsevier Science.)

9 Biotechnological Production of Levan

Levan can be produced by either microbial fermentation or enzymatic synthesis, but the conversion yield of sucrose to levan is higher in the latter process than in the former.

9.1 Isolation and Screening for Levan-producing Strains

In the screening of levan-producing microbial strains, the levan formation activity can be determined using solid agar plates containing sucrose; the strains producing levan are then isolated by following the analytical procedures. The presence of levansucrase is positively selected by inducing mucoid morphology to the microorganisms. Subsequently, levan is collected from the agar plates by precipitation with alcohol (methanol, ethanol, or isopropanol). The identity of levan is then determined after acid hydrolysis of the polymers, followed by TLC analysis; fructose is identified as a single spot on the TLC plates.

9.2 Fermentative Production of Levan

The microbial production of levan requires fermentation and handling of highly viscous solutions. The conditions for producing levan by growing cultures of bacteria vary according to the microorganisms used, but yields of levan production are fairly low (Table 2); this is due to the utilization of sucrose as energy source, the formation of byproducts, the low level of levansucrase production, and the presence of levanase activity in bacteria. In addition, the recovery process of levan from the fermentation broth is often very difficult due to the high viscosity of levan. In theory, the yield of levan production by levansucrase is 50% (w/w) when sucrose is used as a substrate. Routinely, the yields of levan production based on the amount of sucrose consumed are no higher than 58% of the theoretical yield by fermentation (Table 2).

9.3 *In vitro* Biosynthesis of Levan

In the *in vitro* formation of levan by bacterial levansucrase, sucrose serves as fructosyl donor while the released glucose inhibits levan formation. The inhibitory action is influenced by competition with the glucose moiety of sucrose for the enzyme activity. The glucose moiety of sucrose can be replaced by D-xylose, L-arabinose, lactose, etc. Although the catalytic properties of levansucrase vary, the substrate specificity for acceptors is relatively broad where alco-

Tab. 2 Levan production by fermentation processes.

Strains	*Type of production*[a]	*Substrate conc. [%]*	*Levan Yield [%, w/w]*[b]	*Reference*
Bacillus spp.	BF	12	23.5	Elisashvili (1984)
Bacillus polymyxa	BF	15	26.6	Han (1990)
Erwinia herbicola	CF	5	19.2	Keith et al. (1991)
Gluconobacter oxydans	BF	6.2	23.3	Uchiyama (1993)
Rahnella aquatilis	BF	10	29	Ohtsuka et al. (1992)
Zymomonas mobilis	CF	12	23	Beker et al. (1990)

[a] BF, batch fermentation; CF, continuous fermentation. [b] Based on sucrose consumed.

hol, monosaccharides, disaccharides, sugar alcohols and levan are available, but not for levanbiose, levantriose, and levantetraose.

The optimal temperature range for the *in vitro* synthesis of levan is from 0 °C to 40 °C. Levansucrase prepared from *Z. mobilis* displayed an optimum levan formation activity at 0 °C (Song et al., 1996), from *B. subtilis* at > 10 °C (Elisashvili, 1984), from *Pseudomonas* at 18 °C (Hettwer et al., 1995), and from *Rahnella* at 40 °C (Ohtsuka et al., 1992). Most levansucrase activities are inactivated at temperatures higher than 45 °C, with the exception of *R. aquatilis* ATCC 33071, which shows the maximum velocity of levan formation at 50 °C within 3 h, after which the rate declines slightly. Interestingly, at lower temperatures, transfructosylation rather than hydrolysis of sucrose is preferentially catalyzed; however, at higher temperatures hydrolysis is preferentially catalyzed, and this thermolabile feature may have advantages for large-scale levan production. In particular, levan production by *Z. mobilis* levansucrase was most active at the lowest temperature (0 °C), so that it could provide a stable operating condition with minimized contamination opportunities (Song et al., 1996). Plant fructosyltransferases lose 50% of their activity at 5 °C compared with that obtained at the optimum temperature of 20–25 °C (Koops and Jonker, 1996). The *Z. mobilis* levansucrase is stable at pH 4–7, and no activity is observed below pH 3 and above pH 9, similar to the enzymes from *Bacillus*, *Pseudomonas*, and *Rahnella*, the optimum pH of which was 6.0. However, the enzyme from *B. licheniformis* NRRL B-18962 retains 50% of its maximal activity at 55 °C and pH 4.

When *Z. mobilis* levansucrase was immobilized onto hydroxyapatite, the enzymatic and biochemical properties were similar to those of native enzyme towards salt and detergent effects (Jang et al., 2000). However, immobilization of the enzyme on the surface of a matrix shifts the optimum pH to acidic conditions (pH 4.0). The cell-free system synthesized two types of levan which differ in molecular weight. Levans produced by the immobilized system consisted of a higher proportion of low molecular-weight levan to total levan generated than those obtained by the native enzyme. Toluene-permeabilized whole-cell systems produced levan similarly to immobilized systems (Jang et al., 2001).

9.4 Recovery and Purification of Levan

In the microbial production of levan, the yield is low and, as a consequence, costly processes are required to extract the levan from the fermentation broth. The separation process of levan with high purity from the reaction mixture containing sucrose, glucose, fructose, and fructo-oligosaccharides is both laborious and inefficient. Likewise, the separation of levan by using solvents requires huge amounts of ethanol, methanol, isopropanol, or acetone to be used (Rhee et al., 1998). Subsequently, this solvent is lost as waste or recovered by distillation. Recently, membrane processes have been developed to separate polysaccharides from the fermentation broth or enzymatic reaction mixture, without organic waste. However, the resulting solution contains a low concentration (< 5%) of levan, and this must be recovered using various types of drier.

9.5 Commercial Production of Levan

Currently, a Korean start-up company, RealBioTech Co., Ltd. (RBT), is the first and only company worldwide to produce levan on a commercial basis. RBT produces levan in large-scale quantities for supply to companies as a moisturizing ingredient in cosmetics, as a dietary fiber, as food and feed additives, and as a fertilizer (http://www.realbio.com). High-purity levan (>99%) is used in cosmetics and functional foods, while low-grade levan (<15%) containing glucose and oligosaccharides is used as a supplement for feeds and fertilizers.

9.6
Market Analysis and Cost of Levan Production

As levan became available commercially only recently, it is difficult to estimate the size of its current market. It is clear, however, that the expected world market is huge, since levan has a variety of functions and applications within the bioindustries. The production cost of levan depends mainly on the cost of the raw materials, including sucrose and levansucrase, for which the depreciation costs account for 30% and 18%, respectively (RealBioTech Co., unpublished data).

9.7
Levan Competitors

Many types of oligosaccharides and polysaccharides may represent potential competitors of levan in industrial applications. The strongest competitors in the food industry are the fructo-oligosaccharides (FOS), and especially inulin, which belongs to the same fructan category as levan. The use of inulin is limited as a dietary fiber by its insolubility in water at room temperature, whereas levan is a water-soluble and viscous fructan. Other potential competitors are dextran in the food and pharmaceutical industries, β-glucans in the feed industry, and hyaluronic acid in the cosmetic industry. In addition, levan could replace other potential competitors such as xanthan gum, pullulan, and mannan in the food industry. While most commercially available polysaccharides are produced either by microbial fermentation or direct extraction from natural sources, it is possible to produce levan from sucrose by a simple one-step enzyme reaction, which in turn makes it more competitive in terms of production costs.

10
Properties of Levan

While levan is highly soluble in water at room temperature, inulin is almost insoluble (<0.5%), this difference being most likely due to the presence of β-(2,6) linkages in levan. Despite their highly branched molecular structures, microbial levans have several common interesting features in soluble form (Kasapis et al., 1994), including an exceptionally low intrinsic viscosity for a polymer of high molecular weight, an unusual sensitivity of viscosity to increasing concentration between the beginning and end of the intermediate zone, and an extreme concentration-dependence of viscosity at the intermediate zone (Figure 5). These properties may be derived from intermolecular interaction by physical entanglement rather than from any form of specific noncovalent association.

The viscosity of a solution of bacterial levan, originating from *P. syringae* pv. phaseolicola, exhibited Newtonian characteristics up to a concentration of 20%. The concentration-dependence of the 'zero-shear' specific viscosity for the levan solution was unusually high, as would be expected from the molecule's branched structure (Kasapis et al., 1994). In Figure 5, there are three linear regions with changes of slope at levan concentrations of about 4% and 20%. The linear region, including the intermediate region, was also observed in xanthan gum (Milas et al., 1990). However, the concentration-dependence of viscosity may

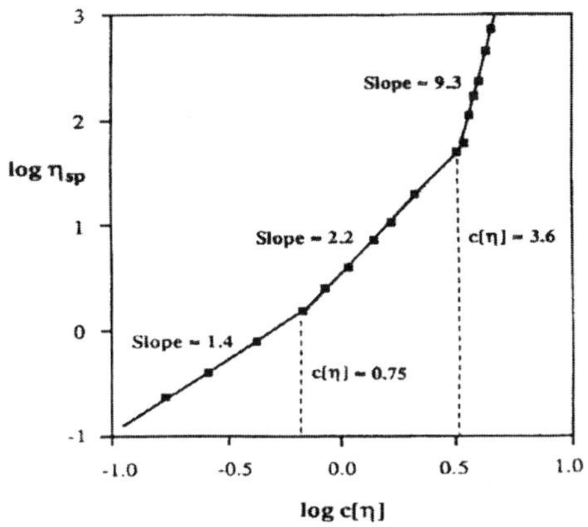

Fig. 5 Concentration dependence of 'zero-sheara-pos; specific viscosity for levan at 20 °C. The parameter η is used for solution viscosity and η_{sp} for specific viscosity which defines the fractional increase in viscosity due to the presence of the polymer ($\eta_{sp} = (\eta - \eta_s)/\eta_s = \eta_{rel} - 1$). η_{rel} indicates relative viscosity, which is the ratio of η to η_s (η_s for solvent viscosity). (Reprinted from Kasapis et al., 1994, p. 59, with permission from Elsevier Science.)

vary with DP, pH, temperature, and salt concentrations (Vina et al., 1998).

11 Applications of Levan

Levan has a wide range of industrial applications, for example in medicine, pharmacy, agriculture, and food. However, it is likely that the low production cost of levan will permit a much increased use of levan in the near future.

11.1 Medical Applications

Microbially produced levan has a direct effect on tumor cells that is related to a modification in the cell membrane, including changes in cell permeability (Leibovici and Stark, 1985; Calazans et al., 2000), as well as radioprotective and antibacterial activities (Vina et al., 1998). Levan derivatives have also been suggested as inhibitors of smooth muscle cell proliferation, as excipients in making tablets, and as agents to transit water into gels. Sulfated, phosphated and acetylated levans have also been suggested as anti-AIDS agents (Clarke et al., 1997).

11.2 Pharmaceutical Applications

Water-soluble polymers, including levan, can be used in a wide variety of applications in the pharmaceutical industry. Water-soluble polymers such as cellulose derivatives, pectin and carrageenan play key roles in the formulation of solid, liquid, semisolid and even controlled release dosage forms (Guo et al., 1998). The viscosity of levan varies with its DP and degree of branching, which relates to the number of side fructose chains attached to one fructose unit in the main fructose chain, and in this respect levan can be used in pharmaceutical formulations in various ways. Low molecular-weight, less branched levan usually provides a low viscosity, and can be used as a tablet binder in immediate-release dosage forms, while levans of medium- and high-viscosity grade are used in controlled-release matrix formulations. Levan has also been suggested as a possible substitute for blood expanders (Imam and Abd-Allah, 1974).

11.3 Agricultural Applications

Microbial levans were introduced in plants to promote their agronomic performance in temperature zones, as well as their natural storage capacities (Vijn and Smeekens, 1999). Transgenic tobacco plants expressing levansucrase genes from *B. subtilis* (Pilon-

Smits et al., 1995) or *Z. mobilis* (Park et al., 1999) showed an increased tolerance to drought and cold stresses. Transgenic plants accumulating fructan have been suggested as novel nutritional feed for ruminants (Biggs and Hancock, 1998, 2001). Recently, microbial levan produced enzymatically was developed as an animal feed (Rhee et al., 2000d) and also as a soil conditioner to improve the germination of various seeds (Imam and Abd-Allah, 1974).

11.4 Food Applications

A number of novel applications of levan have been suggested, particularly in food (Han, 1990; Suzuki and Chatterton, 1993). Levan may act as a prebiotic to change the intestinal microflora, thereby offering beneficial effects when present in the human diet. Levan and its partially hydrolyzed products are fermented by intestinal bacteria including bifidobacteria and *Lactobacillus* sp. (Müller and Seyfarth, 1997; Yamamoto et al., 1999; Marx et al., 2000). Levanheptaose was also suggested as a carbon source for selective intestinal microflora, including *Bifidobacterium adolescentis, Lactobacillus acidophilus,* and *Eubacterium limosum,* whereas *Clostridium perfringens, E. coli* and *Staphylococcus aureus* did not utilize levan (Kang et al., 2000). Cholesterol- and triacylglycerol-lowering effects of levan have also been reported (Yamamoto et al., 1999) and may be applied to develop levans as health foods or nutraceuticals.

Today, it also seems possible that levan might be used in the dairy industry, as *Lactobacillus reuteri* produces levan-type EPS. EPS-producing lactic acid bacteria, including the genera of *Streptococcus, Lactobacillus,* and *Lactococcus,* are used *in situ* to improve the texture of fermented dairy products such as yogurt and cheese. This group of food-grade bacteria produces a wide variety of structurally different polymers, including levan with potential use for new applications. A number of Japanese companies use microbial levans as additives in their milk products which contain *Lactobacillus* species. In addition, the replacement with levan of thickeners or stabilizers that are produced by nonfood-grade bacteria has recently emerged (Van Kranenburg et al., 1999).

11.5 Other Applications

One of the striking consequences of the densely branched structure of levan is its effectiveness in resisting interpenetration by other polymers, leading to macroscopic phase-separation (Kasapis et al., 1994; Chung et al., 1997). Dextran is often used to create two-phase liquid systems (e.g., polyethylene glycol (PEG)/dextran), and to purify biological materials of interest by selectively partitioning them into one phase (Albertsson and Tjerneld, 1994). Microbial levans also display phase-separation phenomena with pectin, locust bean gum, and PEG. Solutions of levan and locust bean gum showed a substantial reduction in viscosity, similar to the mixture levan/pectin. Levan/locust bean gum phases can be separated into discrete phases in mixed solutions in which the lower one consisted predominantly of the denser polymer, the levan phase (Kasapis et al., 1994). The PEG/levan two-phase system was prepared by combining PEG (60%, w/w) and levan (6.77%, w/w) (Chung et al., 1997). This aqueous two-phase system showed a good partitioning with six model proteins including horse heart cytochrome c, horse hemoglobin, horse heart myoglobin, hen egg albumin, bovine serum albumin, and hen egg lysozyme.

12 Patents

Many patents have been filed for the application of levan as functional food additives (Table 3). It was claimed that levan from *S. salivarius* can be used as a food additive with a hypocholesterolemic effect (Kazuoki, 1996). New applications of levan as food and feed additives (Rhee et al., 2000d) as well as a raw material for the production of difructose dianhydride IV (Rhee et al., 2000c) were also developed. Levan derivatives such as sulfated, phosphated or acetylated levans, were claimed to be anti-AIDS agents (Robert and Garegg, 1998), while a fructan *N*-alkylurethane which has excellent surface-active properties in combination with good biodegradability was patented as a surfactant for household use and industrial applications by means of replacing a hydroxyl group of fructose with an alkylaminocarbonyloxyl group (Stevens et al., 1999). A glycol/levan aqueous two-phase system which can substitute the glycerol/dextran system was developed for the partitioning of proteins (Rhee et al., 2000b). Besides levan, the fructosyl transferase activity of levansucrase has been used in the production of alkyl β-D-fructoside (Rhee et al., 2000a).

In spite of many studies on the production and application of levan, few of the levansucrase genes isolated from the microorganisms *Z. mobilis* (Rhee and Song, 1998), *R. aquatilis* (Rhee et al., 1999) and *A. diazotrophicus* (Juan et al., 1998) have been patented. A method for the production of levan using a recombinant levansucrase from *Z. mobilis* was claimed (Rhee et al., 1998), and a creative patent preparing transgenic plants harboring levansucrase gene was applied for by German researchers (Roeber et al., 1994). In the case of levansucrase, a process was claimed for the production of acid-stable

Tab. 3 Relevant patents for levan production and applications

Publication number	*Applicants*	*Inventors*	*Title of invention*	*Date*
US 4,769,254	IBI	Mays, T.D., Dally, E.L.	Microbial production of polyfructose.	September 6, 1988
US 4,927,757	IRFI	Hatcher, H.J. et al.	Production of substantially pure fructose.	May 22, 1990
WO 94/04692	IGF; Roeber, M.; Geier, G.; Geider, K.; Willmitzer, L.	Roeber, M. et al.	DNA sequences which lead to the formation of polyfructans (levans), plasmids containing these sequences as well as a process for preparing transgenic plants.	March 3, 1994
US 5,334,524	SOLVAY ENZYMES INC.	Robert, L.C.	Process for producing levansucrase using *Bacillus*.	August 2, 1994
US 5,527,784	–	Kazuoki, I.	Antihyperlipidemic and antiobesity agent comprising levan or hydrolysis products thereof obtained from *Streptococcus salivarius*.	June 17, 1996
US 5,547,863	USASA	Han, Y.W., Clarke, M.A.	Production of fructan (levan) polyfructose polymers using *Bacillus polymyxa*.	August 20, 1996

Tab. 3 (cont.)

Publication number	*Applicants*	*Inventors*	*Title of invention*	*Date*
WO 98/03184	Clarke, G; Margaret, A; SPRI -	Robert, E.J. et al.	Levan derivatives, their preparation, composition and applications including medical and food applications.	January 29, 1998
US 5,731,173		Juan, G.A.S. et al.	Fructosyltransferase enzyme, method for its production and DNA encoding the enzyme.	March 24, 1998
Korean patent 145946	KRIBB	Rhee, S. K. et al.	Method for production of levan using levansucrase.	May 6, 1998
Korean patent 176410	KRIBB	Rhee, S. K. et al.	A novel levansucrase.	November 13, 1998
Korean patent 0207960	KRIBB	Rhee, S. K. et al.	Base and amino acid sequence of levansucrase derived from *Rahnella aquatilis.*	April 14, 1999
EP 0964054 A1	TS N.V.	Stevens, C.V. et al.	Surface-active alkylurethanes of fructans.	December 15, 1999
Korean patent 0257118	KRIBB	Rhee, S. K. et al.	A process for preparation of alkyl β-D-fructoside using levansucrase.	Febraury 28, 2000
Korean patent 262769	KRIBB	Rhee, S. K. et al.	Novel polyethylene glycol/levan aqueous two-phase system and protein partitioning method using thereof.	May 6, 2000
WO 01/29185	KRIBB and RBT	Rhee, S. K. et al.	Enzymatic production of difructose dianhydride IV from sucrose and elevant enzymes and genes coding for them.	October 19, 2000
WO 01/49127	KRIBB and RBT	Rhee, S. K. et al.	Animal feed containing simple polysaccharides.	December 29, 2000

EP: European Patent; IBI: Igene Biotechnology Institute; IGF: INST GENBIOLOGISCHE FORSCHUNG (DE); IRFI: Idaho Research Foundation Institute; KRIBB: Korea Research Institute of Bioscience and Biotechnology; PCT: World Intellectual Property Organization; RBT: RealBioTech Co., Ltd.; SPRI: Sugar Processing Research Institute; TS N.V: Tiense Suikerraffinaderij N.V.; US: United States Patent; USASA: The United States of America as represented by the Secretary of the Agriculture; WO: World IPO(PCT).

levansucrase from *Bacillus* which is not induced by sucrose (Robert, 1994).

13 Current Problems and Limits

Although levan is a water-soluble and low-viscosity polysaccharide, its applications might be limited due to its turbidity and low fluidity. In order to expand the application areas of levan, the development of biosynthetic methods, including the involvement of novel enzymes, will be essential in order to control the DP (molecular weight) and the degree of branching. More important factors from a commercial aspect include process development for the large-scale production which is comparable with that used for other competitive polysaccharides, especially inulin from chicory. In order to produce levan more economically, a cost-effective purification process of levan from the reaction mixture containing sucrose, glucose, fructose and fructo-oligosaccharides must be developed. The membrane processes will likely serve as one of these solutions.

14
Outlook and Perspectives

Levan has a great potential as a functional biomaterial in the food, cosmetic, pharmaceutical and other industries. However, use of this biopolymer has yet not been practicable due to a paucity of information on its polymeric properties highlighting its industrial applicability, as well as a lack of feasible processes for large-scale production.

For technical applications, fructans with a high molecular mass and a low degree of branching would be desirable. However, microbial levans and their oligomers have been less well characterized with regard to carbohydrate structure analysis. In order to utilize the versatility of water-soluble levans to a maximum, a broader understanding of the behavior of levan is required. The characterization of polysaccharides has advanced considerably during the past two decades due to the introduction of powerful methods such as mass spectrometry, nuclear magnetic resonance, atomic force microscopy, scanning probe microscopy, small-angle X-ray scattering, small-angle neutron scattering, and molecular-mechanic-based carbohydrate modeling (Brant, 1999). As a consequence, the complete characterization of levan should be attained in the near future. Furthermore, the fundamental rheological properties of levan in solution, for example viscosity, thixotropy, dilatancy, elasticity, pseudoelasticity, and viscoelasticity will become increasingly important for new applications of this compound.

15 References

Albertsson, P. A., Tjerneld, F. (1994) Phase diagrams, *Methods Enzymol.* **228**, 3–13.

Ananthalakshmy, V. K., Gunasekaran, P. (1999) Overproduction of levan in *Zymomonas mobilis* by using cloned *sacB* gene, *Enzyme Microb. Technol.* **25**, 109–115.

Arrieta, J., Hernández, L., Coego, A., Suárez, V., Balmori, E., Menéndez, C., Petit-Glatron, M. F., Chambert, R., Selman-Housein, G. (1996) Molecular characterization of the levansucrase gene from the endophytic sugarcane bacterium *Acetobacter diazotrophicus* SRT4, *Microbiology* **142**, 1077–1085.

Avigad, G. (1965) In *Methods in Carbohydrate Chemistry*, New York, London: Academic Press, vol. V, 161–165.

Aymerich, S. (1990) What is the role of levansucrase in *Bacillus subtilis*? *Symbiosis* **9**, 179–184.

Beker, M. J., Shvinka, J. E., Pankova, L. M., Laivenienks, M. G., Mezhbarde, I. N. (1990) A simultaneous sucrose bioconversion into ethanol and levan by *Zymomonas mobilis*, *Appl. Biochem. Biotechnol.* **24/25**, 265–274.

Bergeron, L. J., Morou-Bermudez, E., Burne, R. A. (2000) Characterization of the fructosyltransferase gene of *Actinomyces naeslundii* WVU45, *J. Bacteriol.* **182**, 3649–3654.

Beveridge, T. J., Graham, L. L. (1991) Surface layers of bacteria, *Microbiol. Rev.* **55**, 684–705.

Biggs, D. R., Hancock, K. R. (1998) *In vitro* digestion of bacterial and plant fructans and effects on ammonia accumulation in cow and sheep rumen fluids, *J. Gen. Appl. Microbiol.* **44**, 167–171.

Biggs, D. R., Hancock, K. R. (2001) Fructan 2000, *Trends Plant Sci.* **6**, 8–9.

Birkhed, D., Rosell, K.-G, Granath, K. (1979) Structure of extracellular water-soluble polysaccharides synthesized from sucrose by oral strains of *Streptococcus mutans*, *Streptococcus salivarius*, *Streptococcus sanguis*, and *Actinomyces viscosus*, *Arch. Oral Biol.* **24**, 53–61.

Brant, D. A. (1999) Novel approaches to the analysis of polysaccharide structures, *Curr. Opin. Struct. Biol.* **9**, 556–562.

Bruckner, R., Wagner, E., Gotz, F. (1993) Cloning and characterization of the scrA gene encoding the sucrose-specific Enzyme II of the phosphotransferase system from *Staphylococcus xylosus*, *Mol. Gen. Genet.* **241**, 33–41.

Calazans, G. M. T., Lima, R. C., de França, F. P., Lopes, C. E. (2000) Molecular weight and antitumour activity of *Zymomonas mobilis* levans, *Int. J. Biol. Macromol.* **27**, 245–247.

Carson, M., Botstein, D. (1983) Organization of the SUC gene family in *Saccharomyces*, *Mol. Cell. Biol.* **3**, 351–359.

Chambert, R. G., Gonzy-Treboul, G. (1976) Levansucrase of *Bacillus subtilis*. Characterization of a stabilized fructosyl-enzyme complex and identification of an aspartyl residue as the binding site of the fructosyl group, *Eur. J. Biochem.* **71**, 493–508.

Chambert, R., Petit-Glatron, M. F. (1991) Polymerase and hydrolase activities of *Bacillus subtilis* levansucrase can be separately modulated by site-directed mutagenesis, *Biochem. J.* **279**, 35–41.

Chambert, R. G., Gonzy-Treboul, G., Dedonder, R. (1974) Kinetic studies of levansucrase of *Bacillus subtilis*, *Eur. J. Biochem.* **41**, 285

Chung, B. H., Kim, W. K., Song, K. B., Kim, C. H., Rhee, S. K. (1997) Novel polyethylene glycol/levan aqueous two-phase system for protein partitioning, *Biotech. Techn.* **11**, 327–329.

Clarke, M. A., Roberts, E. J., Garegg, P. J. (1997) New compounds from microbiological products of sucrose, *Carbohydr. Polym.* **34**, 425.

Débarbouillé, M., Arnaud, M., Fouet, A., Klier, A., Rapoport, G. (1990) The *sacT* genes regulating the *sacPA* operon in *Bacillus subtilis* shares strong homology with transcriptional antiterminators, *J. Bacteriol.* **172**, 3966–3973.

Débarbouillé, M., Martin. V, Arnaud, M., Klier, A., Rapoport, G. (1991) Positive and negative regulation controlling expression of the sac genes in *Bacillus subtilis, Res. Microbiol.* **142**, 757–764.

Dedonder, R. (1966) Levansucrase from *Bacillus subtilis, Methods Enzymol.* **8**, 500–505.

Di Marco, A. A., Romano, A. H. (1985) D-Glucose transport system of *Zymomonas mobilis, Appl. Environ. Microbiol.* **49**, 151–157.

Edelman, J., Jefford, T. G. (1968) The mechanism of fructosan metabolism in higher plants as exemplified in *Helianthus tuberosus, New Phytol.* **67**, 517–531.

Elisashvili, V. I. (1984) Levan synthesis by *Bacillus* sp., *Appl. Biochem. Microbiol.* **20**, 82–87.

Fuchs, A. (1959) *On the synthesis and breakdown of levan by bacteria*, Thesis, Uitgeverij Waltman, Delft.

Gay, P., Le Coq, D., Steinmetz, M., Ferrari, E., Hoch, J. A. (1983) Cloning structural gene *sacB*, which codes for exoenzyme levansucrase of *Bacillus subtilis*: expression of the gene in *Escherichia coli, J. Bacteriol.* **153**, 1424–1431.

Geier, G., Geider, K. (1993) Characterization and influence on virulence of levansucrase gene from the fireblight pathogen *Erwinia amylovora, Physiol. Mol. Plant Pathol.* **42**, 387–404.

Greig-Smith, R. (1901) The gum fermentation of sugar cane juice, *Proc. Linn. Soc. N.S.W.* **26**, 589.

Guo, J.-H, Skinner, G. W., Harcum, W. W., Barnum, P. E. (1998) Pharmaceutical applications of naturally occurring water-soluble polymers, *Pharmaceutical Science & Technology Today* **1**, 254–261.

Han, Y. W. (1989) Levan production by *Bacillus polymyxa, J. Ind. Microbiol.* **4**, 447–452.

Han, Y. W. (1990) Microbial levan, *Adv. Appl. Microbiol.* **35**, 171–194.

Han, Y. W., Clarke, M. A. (1996) Production of fructan(levan) polyfructose polymers using *Bacillus polymyxa*, U. S. Patent 5,547,863.

Hatcher, H. J., Gallian, J. J., Leeper, S. A. (1990) Production of substantially pure fructose, U. S. Patent 4,927,757.

Hendry, G. A. F., Wallace, R. K. (1993) The origin, distribution, and evolutionary significance of fructans, in: *Science and Technology of Fructans* (Suzuki, M., Chatterton, N. J., Eds.), Boca Raton: CRC Press, 119–139.

Hestrin, S., Goldblum, J. (1953) Levanpolyase, *Nature* **172**, 1047–1064.

Hestrin, S., Avineri-Shapiro, S., Aschner, M. (1943) The enzymatic production of levan, *Biochem. J.* **37**, 450.

Hettwer, U., Gross, M., Rudolph, K. (1995) Purification and characterization of an extracellular levansucrase from *Pseudomonas syringae* pv. phaseolicola, *J. Bacteriol.* **177**, 2834–2839.

Hettwer, U., Jaeckel, F. R., Boch, J., Meyer, M., Rudolph, K., Ullrich, M. S. (1998) Cloning, nucleotide sequence, and expression in *Escherichia coli* of levansucrase genes from the plant pathogens *Pseudomonas syringae* pv. glycinea and *P. syringae* pv. phaseolicola, *Appl. Environ. Microbiol.* **64**, 3180–3187.

Heyer, A. G., Lloyd, J. R., Kossmann, L. (1999) Production of polymeric carbohydrates, *Curr. Opin. Biotechnol.* **10**, 169–174.

Imam, G. M., Abd-Allah, N. M. (1974) Fructosan, a new soil conditioning polysaccharide isolated from the metabolites of *Bacillus polymyxa* AS-1 and its clinical applications, *Egypt. J. Bot.* **17**, 19–26.

Jang, K. H., Kim, J. S., Song, K. B., Kim, C. H., Chung, B. H., Rhee, S. K. (2000) Production of levan using recombinant levansucrase immobilized on hydroxyapatite, *Bioprocess Eng.* **23**, 89–93.

Jang, K. H, Song, K. B., Kim, C. H., Chung, B. H., Kang, S. A., Chun, U. H., Choue, R. W., Rhee, S. K. (2001) Comparison of characteristics of levan produced by different preparations of levansucrase from *Zymomonas mobilis, Biotechnol. Lett.* **23**, 339–344.

Juan, G. A. S., Lazaro, H. G., Alberto, C. G., Guillermo, S. H. S. (1998) Fructosyltransferase enzyme, method for its production and DNA encoding the enzyme, U. S. Patent 5,731,173.

Jung, K. E., Lee, S. O., Lee, J. D., Lee, T. H. (1999) Purification and characterization of a levanbiose-producing levanase from *Pseudomonas* sp. No. 43, *Biotechnol. Appl. Biochem.* **28**, 263–268.

Kang, S. K., Park, S. J., Lee, J. D., Lee, T. H. (2000) Physiological effects of levanoligosaccharide on growth of intestinal microflora, *J. Korean Soc. Food Sci. Nutr.* **29**, 35–40.

Kasapis, S., Morris, E. R., Gross, M., Rudolph, K. (1994) Solution properties of levan polysaccharide from *Pseudomonas syringae* pv. phaseolicola, and its possible primary role as a blocker of recognition during pathogenesis, *Carbohydr. Polym.* **23**, 55–64.

Kazuoki, I. (1996) Antihyperlipidemic and anti-obesity agent comprising levan or hydrolysis products thereof obtained from *Streptococcus salivarius*, U. S. Patent 5,527,784.

Keith, K., Wiley, B., Ball, D., Arcidiacono, S., Zorfass, D., Mayer, J., Kaplan, D. (1991) Continuous culture system for production of biopolymer levan using *Erwinia herbicola*, *Biotechnol. Bioeng.* **38**, 557–560.

Kim, M. G., Seo, J. W., Song, K.-B., Kim, C. H., Chung, B. H, Rhee, S. K. (1998) Levan and fructosyl derivatives formation by a recombinant levansucrase from *Rahnella aquatilis*, *Biotechnol. Lett.* **20**, 333–336.

Klier, A. F., Rapoport, G. (1988) Genetics and regulation of carbohydrate catabolism in *Bacillus*, *Annu. Rev. Microbiol.* **42**, 65–95.

Kojima, I., Saito, T., Iizuka, M., Minamiura, N., Ono, S. (1993) Characterization of levan produced by *Serratia* sp., *J. Ferment. Bioeng.* **75**, 9–12.

Koops, A. J., Jonker, H. H. (1996) Purification and characterization of the enzymes of fructan biosynthesis in tubers of *Helianthus tuberosus* Colombia. II. Purification of sucrose:sucrose 1-fructosyltransferase and reconstitution of fructan synthesis *in vitro* with purified sucrose:sucrose 1-fructosyltransferase and fructan:fructan 1-fructosyltransferase, *Plant Physiol.* **110**, 1167–1175.

Kunst, F., Rapoport, G. (1995) Salt stress is an environmental signal affecting degradative enzyme synthesis in *Bacillus subtilis*, *J. Bacteriol.* **177**, 2403–2407.

Kyono, K., Yanase, H., Tonomura, K., Kawasaki, H., Sakai, T. (1995) Cloning and characterization of *Zymomonas mobilis* genes encoding extracellular levansucrase and invertase, *Biosci. Biotechnol. Biochem.* **59**, 289–293.

Leibovici, J., Stark, Y. (1985) Increase in cell permeability to a cytotoxic agent by the polysaccharide levan, *Cell. Mol. Biol.* **31**, 337–341.

Lepesant, J. A., Kunst, F., Pascal, M., Steinmetz, M., Dedonder, R. (1976) Specific and pleiotropic regulatory mechanisms in the sucrose system of *Bacillus subtilis*, *Microbiology* **168**, 58–69.

Li, Y., Triccas, J. A., Ferenci, T. (1997) A novel levansucrase-levanase gene cluster in *Bacillus stearothermophilus* ATCC 12980, *Biochim. Biophys. Acta* **1353**, 203–208.

Liu, J., Barnell, W. O., Conway, T. (1992) The polycistronic mRNA of the *Zymomonas mobilis glf-zwf-edd-glk* operon is subject to complex transcript processing, *J. Bacteriol.* **174**, 2824–2833.

Martin, I., Débarbouillé, M., Klier, A., Rapoport, G. (1989) Induction and metabolic regulation of levanase synthesis in *Bacillus subtilis*, *J. Bacteriol.* **171**, 1885–1892.

Martin, I., Débarbouillé, M., Klier, A., Rapoport, G. (1990) Levanase operon of *Bacillus subtilis* includes a fructose-specific phosphotransferase system regulating the expression of the operon, *J. Mol. Biol.* **214**, 657–671.

Marx, S. P., Nosberger, J., Frehner, M. (1997) Hydrolysis of fructan in grasses: 2,6-β-linkage-specific fructan-β-fructosidase from stubble of *Lolium perenne*, *New Phytol.* **135**, 279–290.

Marx, S. P., Winkler, S., Hartmeier, W. (2000) Metabolization of β-(2,6)-linked fructose-oligosaccharides by different bifidobacteria, *FEMS Microbiol. Lett.* **182**, 163–169.

Mays, T. D., Dally, E. L. (1988) Microbial production of polyfructose, U.S. Patent 4,769,254.

Milas, M., Rinaudo, M., Knipper, M., Schuppise, J. L. (1990) Flow and viscoelastic properties of xanthan gum solutions, *Macromolecules* **23**, 2506–2511.

Müller, M., Seyfarth, W. (1997) Purification and substrate specificity of an extracellular fructanhydrolase from *Lactobacillus paracasei* ssp. *paracasei* P 4134, *New Phytol.* **136**, 89–96.

Nelson, C. J., Spollen, W. G. (1987) Fructans, *Physiol. Plant.* **71**, 512–516.

Newbrun, E., Lacy, R., Christie, T. M. (1971) The morphology and size of the extracellular polysaccharide from oral streptococci, *Arch. Oral Biol.* **16**, 863–872.

Ohtsuka, K., Hino, S., Fukushima, T., Ozawa, O., Kanematsu, T., Uchida, T. (1992) Characterization of levansucrase from *Rahnella aquatilis* JCM-1638, *Biosci. Biotechnol. Biochem.* **56**, 1373–1377.

Park, J. M., Kwon, S. Y., Song, K. B., Kwak, J. W., Lee, S. B., Nam, Y. W., Shin, J. S., Park, Y. I., Rhee, S. K., Paek, K. H. (1999) Transgenic tobacco plants expressing the bacterial levansucrase gene show enhanced tolerance to osmotic stress, *J. Microbiol. Biotechnol.* **9**, 213–218.

Perez Oseguera, M. A., Guereca, L., Lopez-Munguia, A. (1996) Properties of levansucrase from *Bacillus circulans*, *Appl. Microbiol. Biotechnol.* **45**, 465–471.

Pilon-Smits, E. A. H., Ebskamp, M. J. M., Paul, M. J., Jeuken, M. J. W., Weisbeek, P. J., Smeekens, J. C. M. (1995) Improved performance of transgenic fructan accumulating tobacco under drought stress, *Plant Physiol.* **107**, 125–130.

Pollock, C. J., Cairns, A. J. (1991) Fructan metabolism in grasses and cereals, *Annu. Rev. Plant Physiol. Plant Mol. Biol.* **42**, 77–101.

Rapoport, G., Dedonder, R. (1963) Le lévane-sucrase de *B. subtilis* III. Reaction d'hydrolyse, de transfert et d'échange avec des analogues du saccharose, *Bull. Soc. Chim. Biol.* (France) **45**, 515–535.

Rapoport, G., Klier, A. (1990) Gene expression using *Bacillus*, *Curr. Opin. Biotechnol.* **1**, 21–27.

Rhee, S. K., Song, K. B. (1998) A novel levansucrase, Korean Patent 176410.

Rhee, S. K., Song, K. B., Kim, C. H. (1998) Method for production of levan using levansucrase, Korean Patent 145946.

Rhee, S. K., Song, K. B., Seo, J. W., Kim, C. H., Chung, B. H. (1999) Base and amino acid sequence of levansucrase derived from *Rahnella aquatilis*, Korean Patent 0207960.

Rhee, S. K., Kim, C. H., Song, K. B., Kim, M. G., Seo, J. W., Chung, B. H. (2000a) A process for preparation of alkyl β-D-fructoside using levansucrase, Korean Patent 0257118.

Rhee, S. K, Chung, B. H., Kim, W. K., Song, K. B., Kim, C. H. (2000b) Novel polyethylene glycol/levan aqueous two-phase system and protein partitioning method using thereof, Korean Patent 262769.

Rhee, S. K., Song, K. B., Kim, C. H., Ryu, E. J., Lee, Y. B. (2000c) Enzymatic production of difructose dianhydride IV from sucrose and elevant enzymes and genes coding for them, PCT-KR00-01183.

Rhee, S. K., Song, K. B., Yoon, B. D., Kim, C. H. (2000d) Animal feed containing simple polysaccharides, PCT-KR00-01556.

Robert, E. J., Garegg, P. J. (1998) Levan derivatives, their preparation, composition and applications including medical and food applications, WO 98/03184.

Robert, L. C. (1994) Process for producing levan sucrase using *Bacillus*, U. S. Patent 5,334,524.

Roeber, M., Geier, G., Geider, K., Willmitzer, L. (1994) DNA sequences which lead to the formation of polyfructans (levans), plasmids containing these sequences as well as a process for preparing transgenic plants, WO-94/004692.

Saito, K., Tomita, F. (2000) Difructose anhydrides: their mass-production and physiological functions, *Biosci. Biotechnol. Biochem.* **64**, 1321–1327.

Saito, K., Goto, H., Yokoda, A., Tomita, F. (1997) Purification of levan fructotransferase from *Arthrobacter nicotinovorans* GS-9 and production of DFA IV from levan by the enzyme, *Biosci. Biotechnol. Biochem.* **61**, 1705–1709.

Sato, S., Koga, T., Inoue, M. (1984) Isolation and some properties of extracellular D-glucosyltransferases and D-fructosyltransferase from *Streptococcus mutans* serotypes c, e, and f, *Carbohydr. Res.* **134**, 293–304.

Shimotsu, H., Henner, D. J. (1986) Modulation of *Bacillus subtilis* levansucrase gene expression by sucrose and regulation of the steady-state mRNA level by *sacU* and *sacQ* genes, *J. Bacteriol.* **168**, 380–388.

Smith, A. E. (1976) Beta-fructofuranosidase and invertase activity in tall fescue culm bases, *J. Agric. Food Chem.* **24**, 476–478.

Song, K. B., Rhee, S. K. (1994) Enzymatic synthesis of levan by *Zymomonas mobilis* levansucrase overexpressed in *Escherichia coli*, *Biotechnol. Lett.* **16**, 1305–1310.

Song, K. B., Joo, H. K., Rhee, S. K. (1993) Nucleotide sequence of levansucrase gene (*levU*) of *Zymomonas mobilis* ZM1 (ATCC10988), *Biochim. Biophys. Acta* **1173**, 320–324.

Song, K. B., Belghith, H., Rhee, S. K. (1996) Production of levan, a fructose polymer, using an overexpressed recombinant levansucrase, *Ann. N. Y. Acad. Sci.* **799**, 601–607.

Song, K. B., Seo, J. W., Kim, M. K., Rhee, S. K. (1998) Levansucrase from *Rahnella aquatilis*: gene cloning, expression and levan formation, *Ann. N. Y. Acad. Sci.* **864**, 506–511.

Song, K. B., Seo, J. W., Rhee, S. K. (1999) Transcriptional analysis of *levU* operon encoding saccharolytic enzymes and two apparent genes involved in amino acid biosynthesis in *Zymomonas mobilis*, *Gene* **232**, 107–114.

Song, K. B., Bae, K. S., Lee, Y. B., Lee, K. Y., Rhee, S. K. (2000) Characteristics of levan fructotransferase from *Arthrobacter ureafaciens* K2032 and difructose anhydride IV formation from levan, *Enzyme Microb. Technol.* **27**, 212–218.

Sprenger, N., Bortlik, K., Brandt, A., Boller, T., Wiemken, A. (1995) Purification, cloning, and functional expression of sucrose:fructan 6-fructosyltransferase, a key enzyme of fructan synthesis in barley, *Proc. Natl. Acad. Sci. USA* **92**, 11652–11656.

Stevens, C. V., Karl, B., Isabelle, M.-A., Lucien, D. (1999) Surface-active alkylurethanes of fructans, EP 0964054 A1.

Stülke, J., Martin, V. I., Charrier, V., Klier, A., Deutscher, J., Rapoport, G. (1995) The HPr protein of the phosphotransferase system links induction and catabolite repression of the *Bacillus subtilis* levanase operon, *J. Bacteriol.* **177**, 6928–6936.

Suzuki, M., Chatterton, N. J. (1993) *Science and Technology of Fructans*. Boca Raton: CRC Press.

Tajima, K., Uenishi, N., Fujiwara, M., Erata, T., Munekata, M., Takai, M. (1998) The production of a new water-soluble polysaccharide by *Acetobacter xylinum* NCI 1005 and its structural analysis by NMR spectroscopy, *Carbohydr. Res.* **305**, 117–122.

Tajima, K., Tanio, T., Kobayashi, Y., Kohno, H., Fujiwara, M., Shiba, T., Erata, T., Munekata, M., Takai, M. (2000) Cloning and sequencing of the levansucrase gene from *Acetobacter xylinum* NCI 1005, *DNA Res.* **7**, 237–242.

Tanaka, K., Kawaguchi, H., Ohno, K., Shohji, K. (1981) Enzymatic formation of difructose IV from bacterial levan, *J. Biochem.* **90**, 1545–1548.

Tanaka, K., Karigane, T., Yamaguchi, F., Nishikawa, S., Yoshida, N. (1983) Action of levan fructotransferase of *Arthrobacter ureafaciens* on levanooligosaccharides, *J. Biochem.* **94**, 1569–1578.

Uchiyama, T. (1993) Metabolism in microorganisms. Part II. Biosynthesis and degradation of fructans by microbial enzymes other than levansucrase. in: *Science and Technology of Fructans* (Suzuki, M., Chatterton, N. J., Eds.), Boca Raton: CRC Press, 169–190.

Vandamme, E. J., Derycke, D. G. (1983) Microbial inulinase: fermentation process, properties, and applications, *Adv. Appl. Microbiol.* **29**, 139–176.

Van Geel-Schutten, G. H., Faber, E. J., Smit, E., Bonting, K., Smith, M. R., Ten Brink, B., Kamerling, J. P., Vliegenthart, J. F. G., Dijkhuizen, L. (1999) Biochemical and structural characterization of the glucan and fructan exopolysaccharides synthesized by the *Lactobacillus reuteri* wild-type strain and mutant strains, *Appl. Environ. Microbiol.* **65**, 3008–3014.

Van Kranenburg, R., Boels, I. C., Kleerebezem, M., de Vos, W. M. (1999) Genetics and engineering of microbial exopolysaccharides for food: approaches for the production of existing and novel polysaccharides, *Curr. Opin. Biotechnol.* **10**, 498–504.

Vereyken, I. J., Chupin, V., Demel, R. A., Smeekens, S. C. M., Kruijff, B. (2001) Fructans insert between the headgroups of phospholipids, *Biochim. Biophys. Acta* **1510**, 307–320.

Vijn, I., Smeekens, S. (1999) Fructan: more than reserve carbohydrate? *Plant Physiol.* **120**, 351–359.

Vina, I., Karsakevich, A., Gonta, S., Linde, R., Bekers, M. (1998) Influence of some physicochemical factors on the viscosity of aqueous levan solutions of *Zymomonas mobilis*, *Acta Biotechnol.* **18**, 167–174.

Wolff, D., Czapla, S., Heyer, A. G., Radosta, S., Mischnick, P., Springer, J. (2000) Globular shape of high molar mass inulin revealed by static light scattering and viscometry, *Polymer* **41**, 8009–8016.

Yamamoto, Y., Takahashi, Y., Kawano, M., Iizuka, M., Matsumoto, T., Saeki, S., Yamaguchi, H. (1999) *In vitro* digestibility and fermentability of levan and its hypocholesterolemic effects in rats, *J. Nutr. Biochem.* **10**, 13–18.

Yun, J. W. (1996) Fructooligosaccharides – occurrence, preparation, and application, *Enzyme Microb. Technol.* **19**, 107–117.

20 Proteoglycans (Glycosaminoglycans/ Mucopolysaccharides)

Dr. Takuo Nakano[1], Dr. Walter T. Dixon[2], Dr. Lech Ozimek[3]

[1] Department of Agricultural, Food and Nutritional Science, University of Alberta, Edmonton, Alberta T6G 2P5, Canada; Tel: +1-780-492-0381; Fax: +1-780-492-4265; E-mail: tnakano@ualberta.ca

[2] Department of Agricultural, Food and Nutritional Science, University of Alberta, Edmonton, Alberta T6G 2P5, Canada; Tel: +1-780-492-3233; Fax: +1-780-492-9130; E-mail: Walter.dixon@ualberta.ca

[3] Department of Agricultural, Food and Nutritional Science, University of Alberta, Edmonton, Alberta T6G 2P5, Canada; Tel: +1-780-492-2665; Fax: +1-780-492-8914; E-mail: Lech.ozimek@ualberta.ca

Biopolymers for Medical and Pharmaceutical Applications. Edited by A. Steinbüchel and R. H. Marchessault

ISBN: 3-527-31154-8

aFGF	acidic fibroblast growth factor
ADAMTS	a disintegrin and metalloproteinase with thrombospondin motifs
AT	antithrombin
bFGF	basic fibroblast growth factor
CPC	cetylpyridinium chloride
CTAB	cetyltrimethylammonium bromide
DEAE	diethylaminoethyl
ELISA	enzyme-linked immunosorbent assay
GAG	glycosaminoglycan
Gal	D-galactose
GalNAc	*N*-acetyl-D-galactosamine
GalNAc(4S)	*N*-acetyl-D-galactosamine 4-sulfate
GalNAc(6S)	*N*-acetyl-D-galactosamine 6-sulfate
GlcAT	glucuronosyl transferase
GlcNAc	*N*-acetyl-D-glucosamine
GlcNAc(6S)	*N*-acetyl-D-glucosamine 6-sulfate
$GlcNSO_3$	*N*-sulfated-D-glucosamine
$GlcNSO_3(3S)$	*N*-sulfated-D-glucosamine 3-sulfate
HC	heparin cofactor
IdoA	L-iduronic acid
IdoA(2S)	L-iduronic acid 2-sulfate
LDL	low-density lipoprotein
MMP	matrix metalloproteinase
UDP	uridinediphosphate
Xyl	D-xylose

1 Introduction

A proteoglycan is a heterogeneous biopolymer composed of a protein core to which one or more glycosaminoglycan (GAG) chains are covalently attached. GAGs are straight-chain acidic polysaccharides comprised of repeating disaccharide units of hexosamine and uronic acid (or galactose). The term mucopolysaccharide is from an old nomenclature used to describe materials that are hexosamine-containing polysaccharides of animal origin (Meyer, 1938). It was replaced by a widely accepted terminology, GAG, in the 1960s. Common GAGs found in mammals include hyaluronan (or hyaluronic acid), chondroitin sulfate, dermatan sulfate, keratan sulfate, heparan sulfate, and heparin. With the exception of hyaluronan, a nonsulfated GAG which exists as a free polymer, all GAGs are sulfated and involved in proteoglycan structures. Proteoglycans are synthesized by almost all types of cells, and are found in the extracellular matrix, at the cell surface, and within secretory granules and the cell nucleus. Their biological roles are highly diversified, ranging from tissue hydration to regulation of various cell functions. The structure and function of proteoglycans have been extensively reviewed by many researchers (Kjellén and Lindahl, 1991; Wight et al., 1991; Noonan and Hassel, 1993; Hardingham et al., 1994; Iozzo, 1998; Bernfield et al., 1999; Couchman and Woods, 2000; Gallagher and Lyon, 2000; Neame and Kay, 2000; Stevens et al., 2000; Vertel and Ratcliffe, 2000; Zimmermann, 2000). This chapter will discuss relatively diverse aspects of GAG and proteoglycan biology ranging from basic chemistry to industrial applications.

2 Historical Outline

Chondroitin sulfate was the first GAG found in mammalian cartilage (Fischer and Boedeker, 1861). Occurrence of three types of chondroitin sulfate, including chondroitin sulfate A, chondroitin sulfate B (or dermatan sulfate), and chondroitin sulfate C, was described later (Meyer and Rapport, 1951). GAGs other than chondroitin sulfate were discovered in the early half of the 20th century. In 1916, McLean extracted a material having anticoagulant activity from liver tissue. This material was later named "heparin" by Howell and Halt (1918). Meyer and Palmer (1934) isolated GAG containing *N*-acetylglucosamine and glucuronic acid from bovine vitreous humor, and called the substance "hyaluronic acid". Suzuki (1939) isolated a polysaccharide from bovine cornea and found that the compound contained equimolar amounts of galactose, glucosamine, acetyl groups, and sulfate. This polysaccharide was later called keratan sulfate by Meyer et al. (1953). A decade later, Jorpes and Gardell (1948) isolated a new GAG, heparan monosulfate (later called heparan sulfate), from bovine lung and liver.

Studies of cartilage GAG in the 1950s showed that chondroitin sulfate is complexed with protein (Shatton and Schubert, 1954), and that serine is involved in a covalent linkage between chondroitin sulfate and protein (Muir, 1958). The linkage structure was then reported to consist of a trisaccharide unit (galactosyl-galactosyl-xylose), in which the xylose residue is glycosidically linked to the hydroxyl group of serine (Rodén, 1968). Occurrence of the same trisaccharide between GAG and the protein core was demonstrated in dermatan sulfate (Fransson, 1968), heparan sulfate (Knecht et al., 1967), and heparin (Lindahl, 1966) proteoglycans.

Suzuki and coworkers (Saito et al., 1968; Yamagata et al., 1968) introduced bacterial chondroitinases as tools to identify disaccharide units involved in chondroitin sulfate and dermatan sulfate. Their studies revealed the occurrence of both chondroitin sulfate and dermatan sulfate disaccharides in the same GAG chain (chondroitin sulfate/dermatan sulfate copolymer). Linker and Hovingh (1972) isolated and characterized bacterial enzymes specific to heparan sulfate and heparin. These enzymes are also useful for structural studies of heparan sulfate and heparin.

In 1969, Sajdera and Hascall developed a technique to extract and isolate proteoglycan from cartilage using chaotropic solvents (e.g. 4 M guanidinium chloride) in combination with cesium chloride equilibrium density-gradient centrifugation. This facilitated the extraction and chemical characterization of intact proteoglycan from tissues. Hardingham and Muir (1974) then reported that cartilage proteoglycan (aggrecan), extracted and purified by the method of Sajdera and Hascall (1969), could bind hyaluronan. This interaction was shown to be stabilized by a link protein (Heinegård and Hascall, 1974). The structure of a proteoglycan aggregate was also visualized for the first time using electron microscopy (Rosenberg et al., 1975). The method of Sajdera and Hascall (1969) was improved by Oegema et al. (1975), who introduced the use of proteinase inhibitors during extraction and subsequent purification procedures to prevent the proteolytic degradation of proteoglycans. To date, 4 M guanidinium chloride and proteinase inhibitors have been used to extract proteoglycans from tissues and cells by many researchers.

During the 1980s, information concerning proteoglycan structure and function increased exponentially. Important findings include: (1) discovery of the pentasaccharide sequence of heparin involved in binding antithrombin III (AT III) (Thunberg et al., 1982), (2) demonstration of the interaction of decorin with collagen (Vogel et al., 1984; Scott et al., 1986), (3) development of monoclonal antibodies recognizing core proteins (Pringle et al., 1985; Sobue et al., 1988, 1989) or GAG chains (Caterson et al., 1983; Avnur and Geiger, 1984), (4) the discovery of new proteoglycans including biglycan (Rosenberg et al., 1985) and fibromodulin (Heinegård et al., 1986; Oldberg et al., 1989), and (5) the cloning of cDNAs encoding the core proteins of decorin (Krusius and Ruoslahti, 1986), biglycan (Fisher et al., 1989), and aggrecan (Doege et al., 1987).

Techniques utilizing molecular biology have significantly contributed to the advance of proteoglycan research in the past decade. The primary amino acid sequence of more than 30 distinct families of proteoglycan has been elucidated by cDNA cloning techniques. The cloning and sequencing of genes encoding a variety of enzymes involved in the biosynthesis of GAGs including chondroitin-6-sulfotransferase (Fukuta et al., 1995), heparan sulfate *N*-sulfotransferase (Hashimoto et al., 1992; Wei et al., 1993), and heparin *N*-sulfotransferase (Eriksson et al., 1994) were also determined. Studies of heparan sulfate proteoglycans from cell surface and basement membranes provided new knowledge on the biological functions of these biopolymers (Yanagishita and Hascall, 1992; Couchman and Woods, 1993, 2000; Rosenberg et al., 1997; Iozzo, 1998; Bernfield et al., 1999; Dunlevy and Hassell, 2000; Filmus and Song, 2000; Jaakkola et al., 2000).

3
Chemical Structures

The heterogeneity of proteoglycans largely depends on a number of different protein cores, different types of GAGs, and different numbers and lengths of individual GAG chains. This large structural diversity contributes to a wide variety of biological functions. In this section, we review the structure and composition of GAGs, oligosaccharides, core proteins, and linkage carbohydrates found between the GAG chain and the core protein.

3.1
GAGs

The chemical nature of the repeating disaccharide units of major GAGs, including hyaluronan, chondroitin-4-sulfate, chondroitin-6-sulfate, dermatan sulfate, keratan sulfate, heparan sulfate, and heparin, are shown in Figure 1. Hyaluronan is composed

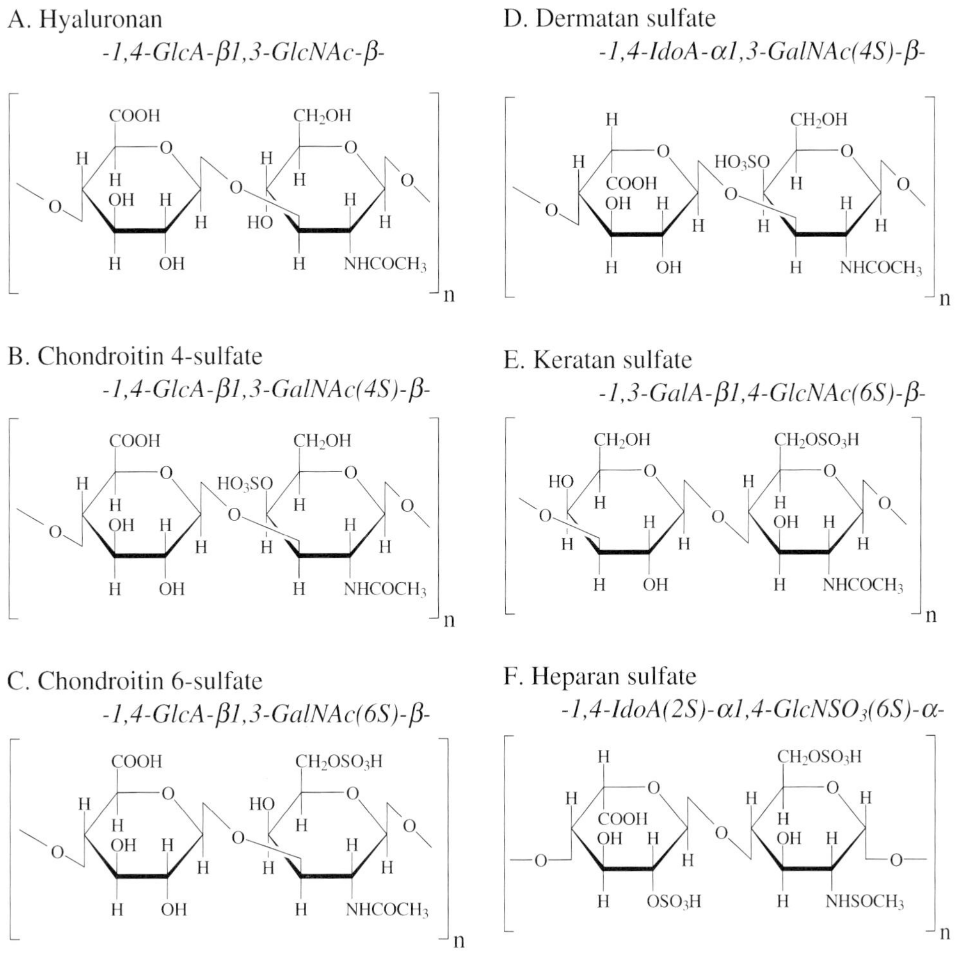

Fig. 1 Chemical structures of GAGs. The repeating disaccharide unit for each GAG is shown. See text for details.

of repeating a disaccharide units of *N*-acetylglucosamine linked through a β1–3 glycosidic bond to D-glucuronic acid (Rodén, 1980). The unit is polymerized via β1–4 linkages (Figure 1a). Hyaluronan has a much larger molecular size (10^2–10^4 kDa) than sulfated GAGs (5–50 kDa) (Muir and Hardingham, 1975; Toole, 2000).

Chondroitin sulfate is a galactosaminoglycan composed of a repeating disaccharide units of ester(*O*)-sulfated *N*-acetylgalactosamine linked through a β1–3 glycosidic bond to D glucuronic acid The linkage between the disaccharide units is β1–4. Chondroitin sulfate is usually sulfated on either the C4 (Figure 1b) or C6 (Figure 1c) of the *N*-acetylgalactosamine (Rodén, 1980). The degree of sulfation, however, may vary among tissues. For example, Meyer et al. (1953) reported the presence in bovine cornea of an undersulfated chain referred to as chondroitin with a molar ratio of sulfate to hexosamine as low as 0.12. Suzuki (1960) studied shark cartilage chondroitin sulfate and found oversulfated chondroitin sulfate with a molar ratio of sulfate to hexosamine of about 2. Studies of human costal cartilage suggest that the ratio of chondroitin-4-sulfate to chondroitin-6-sulfate decreases and the content of keratan sulfate increases with age (Mathews and Glagov, 1966; Iwata, 1969).

Dermatan sulfate (Figure 1d) is an isomer of chondroitin sulfate, in which a large proportion of the D-glucuronic acid is converted to L-iduronic acid by C5 epimerization catalyzed by a glucuronosylepimerase prior to sulfation (Malmström and Åberg, 1982). Pig epimysium dermatan sulfate is a copolymer composed of both dermatan sulfate and chondroitin sulfate disaccharides, in which 87 and 13% of total uronic acid are L-iduronic acid and D-glucuronic acid, respectively (Nakano et al., 1996b). Natural occurrence of a D-glucuronic acid-free dermatan sulfate chain (homopolymer) is unknown. Uronic acid composition varies among galactosaminoglycans. For example, chicken comb contains copolymers with low L-iduronic acid content (approximately 30%; Nakano and Sim, 1992), which may be called chondroitin sulfate/dermatan sulfate copolymers. Chain composition of galactosaminoglycan copolymers has been analyzed in various biological sources including bovine skin (Fransson and Rodén, 1967), periodontal ligament (Pearson and Gibson, 1982), and milk (Nakano and Ozimek, 1999).

Keratan sulfate has a repeating disaccharide unit of *N*-acetylglucosamine linked through a $\beta 1-4$ glycosidic bond to D-galactose, which is polymerized via $\beta 1-3$ linkages (Figure 1e). The disaccharide is also called *N*-acetyllactosamine. It is sulfated at the C6 position of the glucosamine, or the galactose, or both. Keratan sulfate also contains small amounts of D-galactosamine, D-mannose, L-fucose, and silaic acid (Stuhlsatz et al., 1989). Two forms of keratan sulfate, types I and II, are known. Type I is mainly found in cornea and type II in skeletal tissues. These forms are distinguished by the structure of the linkage to the core protein (see below).

Heparan sulfate consists of repeating disaccharide units of glucosamine and uronic acid. The latter is either L-iduronic acid linked $\alpha 1-4$ to glucosamine (Figure 1f) or D-glucuronic acid linked $\beta 1-4$ to glucosamine. The repeating disaccharides are joined by $\alpha 1-4$ and $\beta 1-4$ linkages. The glucosamine is either *N*-sulfated or *N*-acetylated. The C6 and C3 of glucosamine and C2 of L-iduronic acid or D-glucuronic acid can be *O*-sulfated (Gallagher et al., 1986). Heparin has a structure similar to that of heparan sulfate but contains more sulfate and iduronic acid.

3.2 Oligosaccharides

Most proteoglycans have been shown to contain oligosaccharides covalently attached to core proteins via either *O*-glycosidic or *N*-glycosidic linkages. Studies of cartilage proteoglycan (aggrecan) demonstrated that the *O*-linked oligosaccharides are mainly located in the chondroitin sulfate-rich domain, while the *N*-linked oligosaccharides are in the hyaluronan-binding region of the molecule (Hardingham et al., 1994). Decorin contains two to four *N*-linked oligosaccharides (Scott, 1993). Scott and Dodd (1990) found that the enzymatic removal of the oligosaccharides promotes the self-aggregation of bovine decorin *in vitro* and suggested a role of the oligosaccharides *in vivo* in preventing self-aggregation of decorin, which may promote the interaction of this proteoglycan with collagen.

3.3
Linkage of GAG to Core Protein

All sulfated GAGs are covalently attached to core proteins through carbohydrate linkage structures. Chondroitin sulfate, dermatan sulfate, heparan sulfate, and heparin have a common trisaccharide, galactosyl-galactosyl-xylose, in which the xylose residue is *O*-glycosidically linked to serine in the core protein. It has been demonstrated that varying proportions of the linkage regions are either phosphorylated or sulfated in some proteoglycans. For example, Oegema et al. (1984) reported the sequence of galactosyl-galactosyl-xylose-2-phosphate in a rat chondrosarcoma chondroitin sulfate proteoglycan and Fransson et al. (1985) in a bovine lung heparan sulfate proteoglycan. Sugahara et al. (1991) studied whale cartilage chondroitin sulfate proteoglycan and reported occurrence of galactosyl-galactose-4-sulfate attached to xylose. The biological significance for this phosphorylation or sulfation is not well understood.

The linkage to the core protein in keratan sulfate proteoglycan differs from that in other GAGs. In cornea, keratan sulfate is *N*-glycosidically linked to asparagine, while keratan sulfate in cartilage is *O*-glycosidically linked to serine or threonine. Further information on the linkage structures of these keratan sulfates is available elsewhere (Stuhlsatz et al., 1989).

3.4
Core Proteins

There are many genetically distinct proteoglycan core proteins. Table 1 lists the major proteoglycans described in the literature. These include five families such as hyaluronan-binding proteoglycans, small leucine-rich proteoglycans, and heparan sulfate/heparin proteoglycans found in the basement membrane, cell surface and secretory granules (Kjellén and Lindahl, 1991). Structures of some of the well-characterized core proteins are discussed below.

Aggrecan, a predominant large proteoglycan found in cartilaginous tissues, is the most extensively investigated proteoglycan (Hardingham et al., 1994; Vertel and Ratcliffe, 2000). It consists of an approximately 230 kDa core protein, to which GAG chains including approximately 100 chondroitin sulfate chains of molecular size 10–25 kDa, 30–60 keratan sulfate chains of 3–15 kDa, and N- and O-linked oligosaccharides are covalently attached. The core protein has three distinct globular domains, G1 and G2 at the N-terminus, and G3 at the C-terminus (Figure 2a). Most of the GAG chains are found between G2 and G3. All of the chondroitin sulfate and about half of the keratan sulfate chains are found in the chondroitin sulfate-rich region, accounting for more than half of the core protein. A second region, the keratan sulfate-rich region, is located between the chondroitin sulfate-rich region and the G2 globular domain of core protein. The G1 domain can interact noncovalently with hyaluronan to form large aggregates stabilized by around 40 kDa link proteins. The three globular domains of aggrecan contain discrete structural motifs found in other proteins.

Versican is another hyaluronan-binding large proteoglycan produced by fibroblasts. Its core protein has two globular domains at both the N- and C-termini (Figure 2b) which have significant homology with similar domains (G1 and G3, respectively) of aggrecan core proteins. Mammalian versican has four isoforms (splice variants), while in the chicken versican, six isoforms have been reported to occur (Zimmermann, 2000). Each isoform has a different structure in the GAG attachment domain. The molecular masses for the four isoforms of human

Tab. 1 Proteoglycans from mammalian tissues

	GAG	*Major source*
Hyaluronan-binding proteoglycans		
Aggrecan	CS/KS	cartilage
Versican	CS	fibroblasts
Neurocan	CS	brain
Brevican	CS	brain
Small leucine-rich proteoglycans		
Decorin	CS/DS	skin, bone
Biglycan	CS/DS	skin, bone
Fibromodulin	KS	cartilage
Lumican	KS	cornea
Epiphycan	CS/DS	
Osteoglycin	KS	
Basement membrane proteoglycans		
Perlecan	HS/CS/DS	basement membranes
Agrin	HS	neuromuscular junctions, renal basement membrane
Barmacan	CS	basement membranes
Cell surface proteoglycans		
Syndecan 1	HS/CS	epitherial cells, fibroblasts
Syndecan 2	HS	fibroblasts
Syndecan 3	HS/CS	Schwann cells
Syndecan 4	HS	fibroblasts
Glypican 1	HS	CNS (adult)
Glypican 2	HS	CNS (embryo)
Glypican 3	HS	CNS (embryo)
Glypican 4	HS	CNS (embryo)
Betaglycan	HS/CS	many types of cells
Thrombomodulin	CS/DS	vascular endothelium
Intracellular proteoglycans		
Serglycin	heparin, HS, CS/DS	hematopoietic cells

GAG, glycosaminoglycan; CS, chondroitin sulfate; KS, keratan sulfate; DS, dermatan sulfate; HS, heparan sulfate; CNS, central nervous system.

versican core protein are 370, 262, 180, and 72 kDa (Zimmermann, 2000). Neurocan and brevican (Table 1) are also known to have the capability to interact with hyaluronan. Since they share hyaluronan-binding properties, and structural domains at N- and C-terminal regions (e.g. the lectin domain), aggrecan versican, neurocan, and brevican have been proposed to belong to the same gene family called lectican (Ruoslahti, 1996) or hyalectan (Iozzo, 1998).

All small leucine-rich proteoglycans including decorin, biglycan, fibromodulin, and lumican have core proteins of about 40 kDa. Each protein has a central domain composed of seven to 11 leucine-rich repeats, both N- and C-terminal ends of the domain being attached by cysteine-rich domains (Figure 2c). In bovine tissues, decorin has a single galactosaminoglycan chain attached to the serine at residue 4 (Chopra et al., 1985), while biglycan has two galactosaminoglycan chains at serines 5 and 10 (Neame et al., 1989). Bovine fibromodulin has four keratan sulfate chains *N*-glycosidically linked to its leucine-rich domain (Plaas et al., 1990).

Perlecan is a large proteoglycan found in the basement membrane. It is composed of a protein core (400–467 kDa) to which approximately three or four heparan sulfate or chondroitin sulfate/dermatan sulfate chains are covalently attached (Noonan and Hassell, 1993; Iozzo, 1994). The core protein comprises five domains including a GAG attachment region and low-density lipoprotein (LDL) receptor-like structure (Figure 2d).

Syndecans are a family of cell surface proteoglycans with their core protein sizes ranging from approximately 20 to 40 kDa. They have an extracellular domain (ectodomain), a hydrophobic transmembrane domain, and a cytoplasmic domain (Figure 2e). The ectodomain contains several Ser–Gly sequences for GAG attachment sites. Heparan sulfate is the major GAG found in syndecans. All syndecan core proteins have an identical tetrapeptide sequence Gly–Phe–Tyr–Ala at their C-termini (Jaakkola et al., 2000).

Glypicans are another family of cell surface heparan sulfate proteoglycans. The protein core size ranges from 60–80 kDa. Each glypican is attached to the plasma membrane through a covalent link between the core protein and glycosylphosphatidylinositol anchor (Figure 2f). The core protein of each glypican contains an approximately 50 kDa domain with 14 conserved cysteines which may form disulfide bridges (Filmus and Song, 2000).

Serglycin is a proteoglycan found in the storage granules of mast cells. Its core protein has a Ser–Gly repeating sequence in the GAG attachment domain (Figure 2g). Serglycin is highly resistant to proteolytic degradation. There is no known endopeptidase that can cleave a protein composed of Ser–Gly repeats. Serglycin synthesized in connective tissue mast cells contains 10 or more heparin chains of around 100 kDa, which probably renders it more anionic than any other known biopolymer (Wight et al., 1991).

4
Biochemistry and Physiology

Proteoglycans and GAGs are essential for maintaining tissue integrity, for a number of cell functions, and for interactions between cells or cell and extracellular matrix. This section includes discussion on proteoglycans and GAGs in relation to biosynthesis and degradation, genes, methods of isolation and analysis, and functions.

4.1
Biosynthesis

The biosynthesis of a proteoglycan is initiated by the formation of its core protein in the rough endoplasmic reticulum. This is followed by the addition of a linkage region tetrasaccharide, glucuronosyl-galactosyl-galactosyl-xylose. The linkage is initiated by transfer of xylose from the nucleotide precursor, uridinediphosphate (UDP)-xylose, to the hydroxyl group of specific serine residue within a Ser–Gly sequence of the core protein. This is catalyzed by the enzyme xylosyl transferase. After the addition of the xylose, two galactose residues are incorporated to form a trisaccharide catalyzed by the two different galactosyltransferases I and II. Completion of the linkage region occurs by the addition of a glucuronic acid residue to the second galactose residue catalyzed by glucuronosyl transferase (GlcAT)-I. Tone et al. (1999) studied the specificity of a recombinant human GlcAT-I, and reported

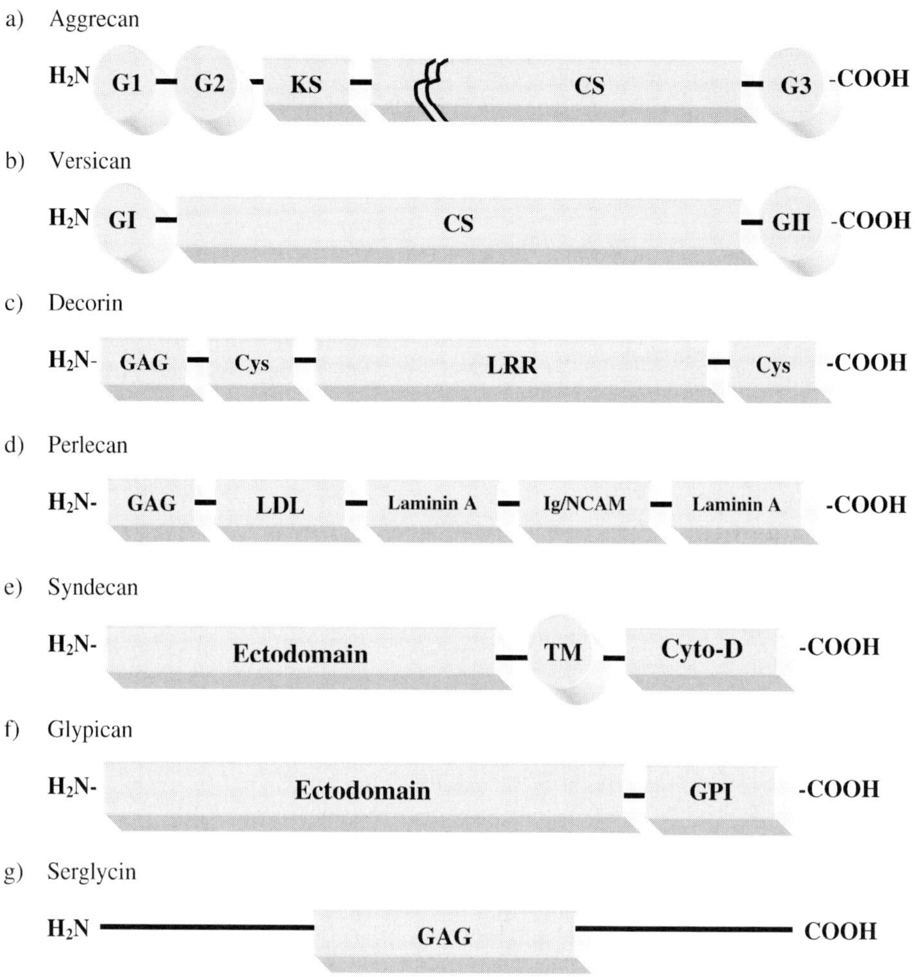

Fig. 2 Schematic representation of proteoglycans: G, globular domain; KS, keratan sulfate rich domain; CS, chondroitin sulfate rich domain; GAG, glycosaminoglycan binding domain; Cys, cysteine rich domain; LRR, leucine rich domain; LDL, domain homologous with LDL receptor; Laminin A, domain homologous with short arm of laminin A chain; Ig/NCAM, domain homologous with immunoglobulin superfamily and neural cell adhesion molecule; TM, transmembrane domain; Cyto-D, cytoplasmic domain; GPI, glycosylphosphatidylinositol anchor.

that the enzyme catalyzes transfer of glucuronic acid from UDP-glucuronic acid to the nonreducing end of Galβ1–3Galβ1–4Xyl (the linkage trisaccharide), but not to Galβ1–3Galβ1-*O*-benzyl or chondroitin (GalNAcβ1–4GlcA)$_n$ polymer. This suggests that GlcAT-I recognizes the whole trisaccharide including xylose and that this enzyme is distinct from GlcAT-II involved in the chain elongation of chondroitin sulfate.

The linkage tetrasaccharide is then extended by the sequential addition of *N*-acetylgalactosamine and glucuronic acid in the case of chondroitin sulfate and dermatan sulfate. Two enzymes, *N*-acetylgalactosaminyl transferase and GlcAT-II, are involved in the addition of the disaccharide unit. Heparan sulfate and heparin require addition of repeating units of *N*-acetylglucosamine and glucuronic acid catalyzed by *N*-acetylglucosaminyl transferase and GlcAT-II, respectively.

During chain elongation, sulfotransferases catalyze transfer of sulfate groups from 3′-phosphoadenosine 5′-phosphosulfate to C4 and C6 positions of the *N*-acetylgalactosamine residues in chondroitin chain. Dermatan sulfate is derived from chondroitin during biosynthesis through the activity of an enzyme, glucuronosyl epimerase (Malmström and Åberg, 1982), which inverts the configuration at the C5 of varying proportion of D-glucuronic acid residues to form L-iduronic acid. Biosynthesis of heparan sulfate and heparin requires further modifications including *N*-deacetylation followed by *N*-sulfation of the amino groups of the glucosamine, C5 epimerization of D-glucuronic acid to L-iduronic acid, and *O*-sulfation of the C2 position of L-iduronic acid, and the C6 and C3 positions of glucosamine (Rodén, 1980). Studies of enzymes from rat liver (Wei et al., 1993) and mouse mastocytoma (Pettersson et al., 1991; Orellana et al., 1994) suggest that a single protein catalyzes both *N*-deacetylation and *N*-sulfation during heparan sulfate/heparin synthesis. Investigations of heparan sulfate-*O*-sulfotransferases produced by Chinese hamster ovary cells demonstrated two genetically distinct enzymes catalyzing sulfation at the C2 of L-iduronic acid (Kobayashi et al., 1996, 1997) and the C6 of *N*-sulfoglucosamine (Habuchi et al., 1995, 1998). This is in contrast to the report of Wlad et al. (1994) who purified heparin *O*-sulfotransferase from mouse mastocytoma tissues, and reported that an enzyme of around 60 kDa catalyzes both the 2-*O*- and 6-*O*-sulfotransferase reactions. Due to the variable extent of epimerization and sulfation, the heparan sulfate or heparin chains produced display considerable microheterogeneity. When the GAG chain has more *O*- than *N*-sulfation, with the majority (more than 80%) of its glucosamines being *N*-sulfated, it is called heparin (Gallagher et al., 1986).

The biosynthetic pathway for oligosaccharides that link keratan sulfate to core protein resembles that for *N*- or *O*-linked oligosaccharide found in many glycoproteins (Hascall and Midura, 1989; Funderburgh, 2000). However, little is known about the mechanism by which a specific oligosaccharide is selected as the site of keratan sulfate formation. Extension of the keratan sulfate chain occurs by the action of glycosyltransferases to alternately add galactose and *N*-acetylglucosamine to the chain followed by sulfate addition by the action of specific sulfotransferases. Keratan sulfate galactose-6-sulfotransferase purified from bovine cornea (Ruter et al., 1984) and later cloned (Fukuta et al., 1997) catalyzes sulfation at the C6 of galactose in corneal keratan sulfate.

Hyaluronan is synthesized at the cell membrane by alternating addition of UDP-*N*-acetylglucosamine and UDP-glucuronic acid. This is in contrast to the other GAGs which are synthesized in the Golgi apparatus. The chain elongation is by the addition of the disaccharide units to the reducing end of the molecule. This process of chain growth is also different from that seen in the sulfated GAG chains which are elongated from the nonreducing end. Hyaluronan synthase, a membrane-spanning enzyme with large cytoplasmic loops, is involved in the chain elongation of hyaluronan (Itano and Kimata, 1998; Toole, 2000). Recently, three genes of hyaluronan synthase (Has1, Has2, and Has3) have been discovered in mice and humans and their chromosomal locations have been reported (Spicer et al., 1997).

4.2
Genes

The aggrecan gene was localized to human chromosome 15g26 (Korenberg, 1993; Just et al., 1993) and mouse chromosome 7 (Watanabe et al., 1994). The human gene contains 19 exons spanning 39.4 kb of the genome. Exon 1 is noncoding, while the remaining 18 encode the protein core of 2454 amino acids with a calculated mass of 254,379 Da. The gene for human versican was located to chromosome 5q12–14 (Iozzo et al., 1992) and that of mouse versican to chromosome 13 (Naso et al., 1995). Both the human and mouse genes extend over 90–100 kb and are divided into 15 exons. The mouse neurocan gene has been reported to map to chromosome 8 (Rauch et al., 1995). The different chromosomal location of these hyaluronan-binding proteoglycans indicates an early evolutionary divergence of this gene family (Iozzo, 1996).

Studies of small leucine-rich proteoglycans demonstrated location of human decorin, biglycan, fibromodulin, and lumican genes to chromosomes 12q23 (Danielson et al., 1993), Xq27 (Fisher et al., 1991), 1q32 (Sztrolovics et al., 1994), and 12q21.3–22 (Chakravarti et al., 1995), respectively. Both decorin and biglycan genes contain eight exons (spanning more than 38 kb), while fibromodulin and lumican contain three exons. Based on the homology of amino acid sequence, Iozzo (1998) calculated an evolutionary distance among the six leucine-rich proteoglycans and grouped them into three subfamilies: (1) decorin and biglycan, (2) fibromodulin and lumican, and (3) epiphycan and osteoglycin. The sequence homology between subfamily members ranged from 40 to 57%.

The human perlecan gene resides on chromosome 1p34-pter (Dodge et al., 1991). It is composed of 94 exons spanning over 120 kb of genomic DNA (Cohen et al., 1993). The agrin gene has been localized to human chromosome at 1p32-pter, and mouse chromosome 4 (Rupp et al., 1992).

The chromosomal location of cell surface proteoglycans has been reviewed (Bernfield et al., 1999; Filmes and Song, 2000; Jaakkola et al., 2000). The gene for syndecan 1 maps to human chromosome 2p23 and mouse chromosome 12, and that for syndecan 2 to human chromosome 8q23 and to mouse chromosome 15. Human glypican 1 is located on chromosome 2g35–37, and human glypicans 3 and 4 both on chromosome Xg 26.

The human gene of serglycine has been localized to chromosome 10q22.1 (Mattei et al., 1989). It consists of three exons spanning more than 16.7 kb (Humphries et al., 1992). Exon 1, present at the 5′ region, encodes the 25 amino acid signal peptide. The second exon codes the *N*-terminal domain of 48 amino acids, and the third exon codes for a 79 amino acid sequence, which includes the Ser–Gly repeat sequence and the C-terminal chain of 43 amino acids.

4.3
Degradation

Proteoglycan turnover is normally initiated in the extracellular matrix where proteinases cleave the core protein by the action of neutral proteinases with no detectable changes in GAG. Protein degradation has been studied more extensively in the cartilage proteoglycan, aggrecan, than in any other proteoglycan partially because aggrecan degradation is an important factor involved in the degradation of articular cartilage in arthritis. It has been reported that the interglobular domain between the *N*-terminal G1 and G2 domains of aggrecan (Figure 2a) has two sites highly susceptible to proteolysis (Roughley and Mort, 2000).

One site is between the Asn341 and Phe342, where matrix metalloproteinases (MMPs), including MMPs 1, 2, 3, 7, 8, 9 and 13, cleave aggrecan. The other site is between the Glu373 and Ala374, which is cleaved by aggrecanase. The aggrecanase is a member of a membrane-associated proteinase family called an ADAMTS (a disintegrin and metalloproteinase with thrombospondin motifs). Two members, ADAMTS-4 (aggrecanase-1; Tortorella et al., 1999) and ADAMTS-11 (aggrecanase-2; Abbaszade et al., 1999), have been cloned. Both share extensive homology and cleave aggrecan at the same site (Glu373–Ala374). Protein degradation in aggrecan appears to require both MMPs and aggrecanase as the major proteinases. Occurrence of the two G1 domains of human aggrecan resulting from the respective actions of MMPs and aggrecanase have been demonstrated in normal and arthritic articular cartilage (Lark et al., 1997) and intervertebral disk (Sztrolovics et al., 1997).

Relatively limited information is available on the degradation of proteoglycans other than aggrecan. However, evidence of protein core degradation has been demonstrated in hyaluronan-binding proteoglycans (versican, neurocan, and brevican), link protein and small leucine-rich proteoglycans (biglycan, decorin, and fibromodulin) (see Roughly and Mort, 2000, for a review). A cell surface proteoglycan, glypican, which is covalently linked to a glycosylphosphatidylinositol anchor, is cleaved by the action of a specific phospholipase (Filmus and Song, 2000). The GAG peptides liberated after proteoglycan cleavage are digested intracellulary with lysosomal proteinases, glycosidases, and sulfatases (Rodén, 1980; Kresse and Glössle, 1987).

4.4 Nonmammalian Enzymes

GAGs can be digested with nonmammalian enzymes, which have been used to characterize the chain composition of GAG. Chondroitin-ABC lyase (EC 4.2.2.4) produced by *Proteus vulgaris* cleaves the β1–4, *N*-acetylhexosamidic linkage to either D-glucuronic acid or L-iduronic acid (Yamagata et al., 1968). Chondroitinase-AC lyases (EC 4.2.2.5) from *Flavobacterium heparinum* and *Arthrobacter aurescens* cleave the β1–4, *N*-acetylhexosamidic linkage to D-glucuronic acid (Yamagata et al., 1968). The latter acts in a predominantly exolytic fashion, while the former acts in a randomly endolytic fashion (Ernst et al., 1995). Therefore, the *Arthrobacter* chondroitinase digests chondroitin sulfate/dermatan sulfate copolymer, with low iduronic acid content (30% total uronic acid), much less efficiently (about 1.6 times; Nakano et al., 2001) than does the *Flavobacterium* chondroitinase. Chondroitin-B lyase (EC 4.2.2) produced by *F. heparinum* splits the β1–4, *N*-acetylgalactosamidic linkage to L-iduronic acid (Yamagata et al., 1968). Chondro-4-sulfatase (EC 3.1.6.9) and chondro-6-sulfatase (EC 3.1.6.10) both produced by *P. vulgaris* catalyze the hydrolytic desulfation at C4 and C6, respectively, on the disaccharides of chondroitin sulfate or dermatan sulfate (Yamagata et al., 1968). These enzymes do not act on higher oligosaccharides from chondroitin sulfate or dermatan sulfate with the exception of tetrasaccharide sulfates derived from chondroitin-4-sulfate, which has been reported to be susceptible to chondro-4-sulfatase (Seno et al., 1974).

There are three distinct keratan sulfate-degrading enzymes. Endo-β-galactosidase (EC 3.2.1.103) produced by a variety of microorganisms including *Escherichia freundii* (Fukuda and Mastumura, 1975) and *Pseudomonas* sp. IFO-13309 (Nakazawa and

Suzuki, 1975) catalyzes hydrolytic cleavage of β1–4 galactosidic linkages of keratan sulfate with nonsulfated galactosyl residues. The second enzyme, keratanase I from *Pseudomonas* sp. also cleaves internal β1–4 galactosidic linkages in keratan sulfate but unlike endo-β-galactosidase, it requires at least one *N*-acetylglucosamine-6-sulfate residue adjacent to the galactose residue participating in the galactosidic linkage (Nakazawa et al., 1989). The third enzyme, keratanase II produced by a Gram-negative bacterium, cleaves β1–3 glycosidic linkages to galactose as well as those to galactose-6-sulfate in keratan sulfates, in which most, if not all, of the *N*-acetylglucosamine residues are sulfated (Nakazawa et al., 1989).

Heparinases I (EC 4.2.2.7) and III (EC 4.2.2.8) both produced by *F. heparinum* act on heparin and heparan sulfate, respectively, while heparinase II from the same bacteria degrades both GAGs (Linhardt et al., 1990; Ernst et al., 1995).

Hyaluronidase (EC 4.2.2.1) from *Streptomyces hyalurolyticus* cleaves *N*-acetylglucosamidic linkages in hyaluronan. This enzyme is specific to hyaluronan, and does not act on sulfated GAGs (Ohya and Kaneko, 1970).

4.5
Extraction of Proteoglycans

Animal tissues usually contain more than one type of proteoglycan. Various techniques to extract and purify individual proteoglycans have been reported. For extraction, tissues are cut into small pieces (e.g. around 1 mm^3 in size) at 4 °C or pulverized in liquid nitrogen. High-speed homogenization which disrupts proteoglycan molecules is not recommended. Buffers containing 4 M guanidinium chloride in combination with proteinase inhibitors are used in many laboratories to extract proteoglycans. Proteinase inhibitors that minimize protein degradation include EDTA (divalent cation chelator to inhibit metalloproteinases), *N*-ethylmaleimide (thiol proteinase inhibitor), phenylmethylsulfonyl fluoride (serine proteinase inhibitor), pepstatin (aspartic proteinase inhibitor), and leupeptin (inhibitor of both thiol proteinases and trypsin-like serine proteinases).

Successive extractions of tissues with solutions other than 4 M guanidinium chloride are also used to obtain fractions rich in a specific proteoglycan. For example, human articular cartilage proteoglycan has been sequentially extracted with phosphate-buffered saline, 7 M urea, and 4 M guanidinium chloride to give three proteoglycan fractions with different GAG composition (Vilim and Krajickova, 1991). By using 4 M guanidinium chloride with and without 0.5 M EDTA (a demineralizing agent), Goldberg et al. (1988) extracted proteoglycans related to mineralized and newly formed nonmineralized tissues, respectively. Heparan sulfate proteoglycans, associated with cell surface molecules through specific or nonspecific interactions, are extracted with NaCl, heparin, mannose-6-phosphate, or inositol hexaphosphate (Fedarko, 1994). Cell-membrane-bound and phosphatidylinositol-bound heparan sulfate proteoglycans are released by the actions of trypsin and phospholipase C, respectively (Yanagishita, 1992).

4.6
Extraction of GAGs

GAGs can be directly liberated from tissues by hydrolysis with exogenous enzymes (e.g., papain and pronase) or alkaline treatment (Rodén et al., 1972; Taniguchi, 1982). Proteolysis provides a single GAG chain, to which a small peptide consisting of several amino acids is covalently attached. On the other hand, alkali treatment results in cleavage of the linkage between xylose and a serine hydroxyl group to release the peptide-free GAG (β-elimination). GAGs are also liberated from tissues by activation of endogenous enzymes (autolysis). Studies of autolysis of bovine nasal cartilage by separately incubating tissues in 0.1M sodium acetate (Nakano et al., 1998) and acidified water (Nakano et al., 2000) at pH 4.5 and 37°C showed release of approximately 80 and 70% of total GAG, respectively.

4.7 Isolation of Proteoglycans

Cesium chloride density-gradient centrifugation has been used to isolate a large proteoglycan, such as aggrecan, which has a high buoyant density. By using either associative (0.5 M guanidinium chloride) or dissociative (4 M guanidinium chloride) buffer, aggrecan molecules can be isolated as aggregates (with hyaluronan and link protein) or as monomers. This method is useful for separating large proteoglycans with high buoyant density from small proteoglycans with low buoyant density (e.g. decorin). However, the method is not suitable for separation of small proteoglycans with a similar molecular weight.

Chromatography on diethylaminoethyl (DEAE) anion-exchange column is probably the most widely used method to separate proteoglycans. Both hyaluronan and proteoglycan are adsorbed on DEAE–Sephacel (Pharmacia) equilibrated in 7 M urea/ 0.15 M NaCl/0.05 M Tris–HCl/0.02% (w/v) sodium azide, pH 6.6. Hyaluronan is eluted with 0.2 M NaCl, while proteoglycan is eluted with 0.3–0.5 M NaCl (Nakano and Scott, 1989).

Proteoglycans with different molecular sizes can be separated from each other by gel chromatography. For example, using Sepharose CL-4B (Pharmacia), fibrocartilage proteoglycans, solubilized with 4 M guanidinium chloride, are separated into two fractions containing large and small proteoglycans (Nakano et al., 1997).

Hydrophobic interaction chromatography has been used to separate proteoglycans which exhibit variable hydrophobicity. Choi et al. (1989) and Westergren-Thorsson et al. (1991) separated the two small proteoglycans, biglycan and decorin, from each other by chromatography on octyl-Sepharose (Pharmacia). Scott et al. (1997) in a study of pig meniscus proteoglycan also used octyl-Sepharose to isolate biglycan, decorin, and fibromodulin. Other methods for the purification of proteoglycans include hydroxyapatite chromatography and ligand-affinity chromatography (see Fedarko, 1994, for a review).

During chromatography, eluates are monitored for protein by measuring absorbance at 280 nm and for GAG by either determining the content of its constituent sugar uronic acid (see below) or by 1,9-diethylmethylene blue dye binding assay (Farndale et al., 1982). Anions (e.g. Cl^- from NaCl or CN^- from KSCN) form metachromasia with 1,9-dimethylmethylene blue, and thus interfere with the quantitative assay of sulfated GAG. Safranin-O (Lammi and Tammi, 1988) and Alcian blue (Gold, 1979) are also used to monitor GAGs. Use of [^{35}S]sulfate or [^{3}H]glucosamine as metabolic precursors to radiolabel proteoglycan is another method to monitor the molecule (Beeley, 1985).

4.8 Characterization of Proteoglycan

The ability of proteoglycans to interact with hyaluronan can be determined by incubating the proteoglycan sample with hyaluronan followed by gel chromatography on Sepharose CL-2B (Handingham and Muir, 1974). The molecular size of the proteoglycan can be estimated by gel chromatography, sodium dodecylsulfate–polyacrylamide gel electrophoresis, ultracentrifugation methods, light scattering and matrix-assisted laser desorption/ionization techniques. The morphology of proteoglycan molecules can be examined by the rotary shadowing electron microscopy technique (Mörgelin et al., 1989). Immunochemical methods [e.g. enzyme-linked immunosorbent assay (ELISA) and immunoblotting] using specific polyclonal and monoclonal antibodies are useful in identifying proteoglycans and GAGs.

GAG chains can be liberated from the core protein by proteolytic digestion with papain or pronase or by the β-elimination reaction with alkali treatment (Rodén et al., 1972; Taniguchi, 1982). The released GAGs are recovered by one of the following methods: (1) precipitation with ethanol in the presence of sodium or potassium acetate, (2) precipitation with quaternary ammonium compound [cetylpyridinium chloride (CPC) or cetyltrimethylammonium bromide (CTAB)], and (3) adsorption on an anion exchanger. GAGs are then fractionated according to: solubility in different concentrations of ethanol; solubility of GAGs complexed with CPC or CTAB in different concentrations of NaCl; or elution of the GAGs with different concentrations of NaCl on an anion-exchange column (Rodén et al., 1972; Taniguchi, 1982).

Cellulose acetate electrophoresis is a common method to examine the type of GAGs. Electrophoresis of a GAG mixture in 0.1 M pyridine/1.2 M acetic acid, pH 3.5 separates hyaluronan, heparan sulfate, dermatan sulfate, chondroitin sulfate, and heparin from one another. Chondroitin sulfate and dermatan sulfate have the same mobility in 0.1 M potassium phosphate buffer, pH 7.0 (Nakano and Sim, 1989).

Characterization of GAGs using specific enzymes (see above) is essential in order to determine GAG composition and structure. After incubation of samples with an enzyme, the degradation of the GAG is determined by the turbidimetric method (DiFerrante, 1956) or by cellulose acetate electrophoresis. Copolymeric galactosaminoglycans composed of dermatan sulfate and chondroitin sulfate disaccharides can be detected by gel chromatography of GAG sample digested with chondroitinase-AC or chondroitinase-B lyase on Sephacryl S-300 (Nakano et al., 1996b).

Uronic acid content in GAG can be determined using the original (Dische, 1947) or modified (Kosakai and Yoshizawa, 1979) carbazole reaction. Uronic acid can also be analyzed by the orcinol (Davidson, 1966) or the meta-diphenyl reaction (Blumenkrantz and Asboe-Hansen, 1973). Filisetti-Cozzi and Carpita (1991) reported that interference with the assay caused by contaminating neutral sugars present with uronic acid can be eliminated by addition of sulfamate to the reaction mixture. The color yields for glucuronic acid and iduronic acid are the same among the carbazole method of Kosakai and Yoshizawa (1979), the orcinol method and the diphenyl method. However, the color yield is lower for iduronic acid than glucuronic acid in the carbazole reaction of Dische (1947). Therefore, the color yield ratio of the Dische method to Kosakai and Yoshizawa method or the diphenyl method is used to estimate the proportion of iduronic acid in total uronic acid in copolymeric galactosaminoglycans. The content of iduronic acid in the galactosaminoglycan may also be estimated by the periodate–Schiff reaction (DiFerrante et al., 1971) using hog skin dermatan sulfate as a standard (Nakano et al., 1996a). The content of total protein may be determined by measuring the absorbance at 280 nm. The method of Lowry et al., the bicinchoninic acid reaction and Bradford dye binding method are also used for protein determination (Bollag and Edelstein, 1991).

The total content of hexosamine in the GAG hydrolyzate is determined by the Elson–Morgan reaction (Boas, 1953) or the indole reaction (Dische and Borenfreund, 1950). The amino acids as well as glucosamine and galactosamine are commonly analyzed by using an amino acid analyzer (Pearson and Gibson, 1982). Total sulfate content in GAG hydrolyzate is determined by using reagents containing barium chloride (Dodgson and Price, 1962), sodium rhodisonate (Terho and Hartiala, 1971), or benzidine–HCl (Antonopoulos, 1962). The degree of sulfation of GAG is estimated using cellulose acetate electrophoresis in 0.1 M HCl, providing a pH at which only the sulfate group is ionized (Wessler, 1971). Chondrosulfatases are useful for identification of the site of attachment of *O*-sulfate groups in galactosaminoglycans (Saito et al., 1968). The method of analysis of *N*-sulfated groups in GAGs has been described elsewhere (Langunoff and Warren, 1962).

4.9 Microscopy

Localization of a proteoglycan in tissue sections is important for studying the composition and metabolism of the molecule *in situ*. In both light and electron microscopy, proteoglycans are histochemically stained by ionic interaction between anionic GAG and cationic dyes (e.g. Alcian blue). Staining reactions in different concentrations of salt (e.g. $MgCl_2$; Scott and Dorling, 1965) and treatments of sections with enzymes (e.g. chondroitinases) or chemicals (e.g. nitrous acid to degrade heparan sulfate/heparin) provide additional information (Kida, 1993). Immunohistochemical localization of proteoglycans with monoclonal (Nakano et al., 1996a) or polyclonal (Bianco et al., 1990) antibodies is more specific compared to histochemical localization of the molecule. *In situ* hybridization with complementary DNA or RNA is used to study gene expression for the protein core of proteoglycan (e.g. decorin and/or biglycan; Bianco et al., 1990; Takagi et al., 2000) or for enzymes (e.g. chondroitin-6-sulfotransferase; Habuchi and Habuchi, 1998) involved in GAG synthesis.

4.10 Functions

The biological functions of proteoglycans are either due to the polyanionic properties of GAGs or the structure of core proteins or both. Aggrecan, the major noncollagenous protein in the cartilage, plays an essential role in the mechanical function of this load-bearing tissue. Most aggrecan molecules are believed to be aggregated through noncovalent interaction with hyaluronan and link protein, and are entrapped in the collagen fibril network (Muir, 1977). The highly polyanionic GAG chains from each aggregate attract water and provide the tissue with swelling pressure necessary to maintain tissue resilience to resist compressive forces (Muir, 1977; Maroudas et al., 1992). Thus, the proteoglycan content is the highest in the zone subjected to maximum compressive load in articular (Kempson et al., 1970) and fibrocartilaginous (Nakano and Scott, 1996; Nakano et al., 1997) tissues. Proteoglycan is lost in degenerative articular cartilage (Mankin and Lippiello, 1970; Nakano and Aherne, 1993).

In the growth cartilage, the polyanionic aggrecan that can bind calcium may have a role in endochondral ossification, a process of long bone growth (Poole et al., 1989; Hunter, 1991). Aggrecan may also indirectly affect bone development since the absence of this proteoglycan due to a genetic defect in chicken (Strirpe et al., 1987) and mouse

(Kimata et al., 1981) causes stunted bone growth.

Shinomura et al. (1992), in a study of embryonic chick limb development, cultured chick limb bud mesenchymal cells with and without a chondroitin sulfate proteoglycan called PGM (chick versican). They observed that the differentiation of mesenchymal cells to chondrogenic cells was much faster in cells cultured with PGM (1 day) than those without PGM (5–6 days). These authors further reported that PGM has a role in regulating chondrogenesis in the chick limb bud.

The small leucine-rich proteoglycans, including decorin, fibromodulin, and lumican, that can bind to collagen fibrils (Vogel, 1994; Neame and Kay, 2000) may have roles in regulating collagen fibril formation. A study of decorin knockout mice showed that these animals have abnormally fragile skins due to aberrant organization of collagen fibrils (Danielson et al., 1997). Similarly, Chakravati et al. (1998) observed skin fragility in lumican knockout mice. These animals also showed corneal opacity, suggesting an important role of this proteoglycan in determining the transparency of the cornea. Decorin, biglycan, and fibromodulin can bind transforming growth factor-β, and may have a role in controlling the effect of this growth factor.

The physiological function of cell surface or basement membrane heparan sulfate proteoglycan depends on the specific binding properties of heparan sulfate. This GAG binds extracellular matrix proteins including collagens (types I, II, and V), fibronectin and tenascin, and plays a role in cell–cell and cell–matrix interactions. Heparan sulfate which interacts with various growth factors, including acidic and basic fibroblast growth factors (aFGF and bFGF) and hepatocyte growth factor, has significant effects on cell growth. For example, when cell surface heparan sulfate binds bFGF (inactive) it induces a conformational change of the growth factor, which becomes activated and is recognized by the signal transducing receptors in the plasma membrane (Gallagher and Turnbull, 1992). This suggests that heparan sulfate is essential for the bFGF activity. Heparan sulfate by binding FGFs may also protect the growth factors from proteolysis and thermal denaturation (Gallagher and Turnbull, 1992). Gallagher and Lyon (2000) suggest involvement of IdoA(2S)–$GlcSO_3$ repeats of heparan sulfate in binding to bFGF.

Heparan sulfate that interacts with apolipoproteins and lipoprotein lipase has been suggested to be involved in lipid metabolism (Wight, 1989; Olivecrona and Bengtsson-Olivecrona, 1989). Heparan sulfate on the surface of vascular endothelium regulates proteinase activity by binding AT III, an inhibitor of thrombin, which is a proteinase involved in blood coagulation. This binding is responsible for an accelerated formation of an enzyme–inhibitor complex to prevent thrombus formation. The high affinity binding of AT III to heparan sulfate (or heparin) requires a specific pentasaccharide sequence: GlcNAc(6S)–GlcA–$GlcNSO_3$(3,6-diS)–IdoA(2S)–$GlcNSO_3$(6S) (Thunberg et al., 1982). An anticoagulant activity is also found in dermatan sulfate proteoglycan, which activates another proteinase inhibitor, heparin cofactor II (HC II). The HC II rapidly inactivates thrombin after binding dermatan sulfate. The specific binding region of dermatan sulfate has been identified as a hexasaccharide sequence: $[\text{IdoA(2S)–GalNAc(4S)}]_3$ (Maimone and Tollefsen, 1990).

In the storage granules of connective tissue mast cells, highly anionic heparin chains of serglycin whose core protein is extremely resistant to proteolysis (see above) interact with cationic proteins (e.g. carboxypeptidases and histamines) to form macromolecular complexes. This may be important for concentrating and storing bioactive molecules without autolysis until they are released from the cell.

The polyanionic hyaluronan with high visco-elasticity occupies large solution domains under physiological conditions, and contributes to tissue hydration and lubrication. Hyaluronan is also involved in cell–cell and cell–matrix adhesion, as well as proliferation and migration by binding to cell surface receptors. In the cartilage, this GAG contributes to the load-bearing capacity of the tissue (see above). Extensive reviews of the function of hyaluronan are available elsewhere (Fraser et al., 1997; Toole, 2000).

5 Application and Production

GAGs have been reported to have a wide range applications in the pharmaceutical, cosmetic, and food industries. Sodium hyaluronate is used as an aid for eye surgery (Pharmacia, 1985), for treatment of osteoarthritis and joint injuries (Ghosh, 1994), and as moisturizer in cosmetics.

Chondroitin sulfate can be used for preservation of corneal tissues in eye banks (Keates and Rabin, 1988). This GAG can also be added to eye drops. It is noted that chondroitin sulfate helps protect the surface tissue of the cornea (Takaku and Kamoshita, 1999). Intraarticular or intramuscular administration of chondroitin sulfate with chondroprotective or antiarthritic potential has been used for treatment of degenerative joint diseases in animals (Dean et al., 1991) and humans (Adam, 1980). Oral administration of glucosamine, which may stimulate proteoglycan synthesis in articular cartilage, was suggested to be effective for prevention or treatment of osteoarthritis (Gottlieb, 1997; Kelly, 1998).

It has been reported that intramuscular administration of chondroitin sulfate suppresses LDL deposition in the arteries (Matsushima et al., 1987). This may be due to the interaction between the exogenous chondroitin sulfate and circulating LDL to produce chondroitin sulfate–LDL complex which cannot bind to arterial tissue GAG. This could lead to a reduced incidence of arthrosclerosis.

Chondroitin sulfate–iron complex has been reported as a potent antianemic agent, in which chondroitin sulfate contributes to an increased bioavailability of iron (Barone et al., 1988). In Japan, chondroitin sulfate can be added to mayonnaise and salad dressing to improve hydration and emulsification in these products (Shokuhin Kagaku Binran, 1978).

Heparin has been used for many years as an anticoagulant and antithrombotic drug in the management of cardiovascular diseases and an important drug for open-heart surgery.

Commercial GAGs including hyaluronan, chondroitin sulfate, dermatan sulfate, keratan sulfate, heparan sulfate and heparin are available as research chemicals from the following suppliers: Seikagaku Kogyo (Tokyo, Japan), Sigma Chemical Company (St. Louis, MO, USA), and ICN Biomedical Research Products (Costa Mesa, CA, USA).

Commercially available GAG products include: Healon® (sodium hyaluronate; Pharmacia, Uppsala, Sweden) and Opegan® (sodium hyaluronate; Seikagaku Kogyo, Tokyo, Japan) both used for eye surgery, and Artz® (sodium hyaluronate, Seikagaku Kogyo), Arteparon® (GAG polysulfate; Luitpold-Werk, Munich, Germany) and Rumalon® (GAG–peptide association complex; Robapharm, Basel, Switzerland) which are used for treatment of degenerative joint diseases. Glucosamine sulfate, a dietary supplement of glucosamine (see above), is available from many suppliers. Commercial heparin as an anticoagulant and antithrombotic drug, is available from a number of suppliers.

6 Outlook and Perspectives

Proteoglycans are highly diverse molecules found intracellularly, on cell membranes, and within the extracellular matrix. Many functions of proteoglycans are related to their location as well as the structures of GAG and core protein. Future research is needed in a variety of different areas to address a number of remaining challenges. Little is still known about the mechanisms of regulation of core protein synthesis and GAG addition in response to development, aging, diseases, and external factors. Utilization of mutant animals with targeted disruption of single proteoglycan-related genes will facilitate understanding of their functions. Research into the potential uses of proteoglycans and GAGs in the pharmaceutical, cosmetic, and functional food sectors will continue to be areas of rapid growth.

7 Patents

Table 2 lists some recent patents of interest related to GAGs and proteoglycans. Four products have been disclosed to stimulate

Tab. 2 Recent patents related to GAGs and proteoglycans

Patent number	*Patent holder*	*Inventor(s)*	*Title of patent*	*Date of publication*
US 5300490	Mochida Pharmaceutical Co., Ltd., Tokyo, Japan	Y. Kunihiro, R. Tanaka, M. Ichimura, A. Uemura, N. Ohzawa and E. Mochida	Anticoagulant substance obtained from urine	5 April 1994
WO 95/35092	Institute for Advanced Skin Research Inc., Yokohama, Japan	S. Tanaka, H. Doi and N. Yamamoto	Skin activator having glycosaminoglycan production accelerator activity	28 December 1995
US 5500409	Thomas Jefferson University	E. M. L. Tan and J. J. Uitto	Method for inhibiting collagen synthesis by keloid fibroblasts	19 March 1996
US 5541095	University of Massachusetts Medical Center	C. B. Hirschberg, A. Orellana, Y. Hashimoto, S. J. Swiedler, Z. Wei and M. Ishihara	Glycosaminoglycan specific sulfotransferases	30 July 1996
JP 8–239404	Kanebo Ltd., Tokyo, Japan	M. Kamio, S. Sakai, T. Sayo, S. Inoue and T. Matsui	Hyaluronic acid production accelerator	17 September 1996
US 5631241	Fidia S.p.A., Via ponte della Fabbrica, Italy	F. della Valle, A Romeo and S. Lorenzi	Pharmaceutical compositions containing hyaluronic acid fractions	20 May 1997

Tab. 2 (cont.)

Patent number	*Patent holder*	*Inventor(s)*	*Title of patent*	*Date of publication*
US 5712247	University of North Carolina	H. -F. Wu and F. C. Church	Use of lactoferrin to modulate and/or neutralize heparin activity	27 January 1998
JP 10–95735	Institute for Advanced Skin Research Inc., Yokohama, Japan	T. Souma and E. Sato	Hyaluronic acid productivity potentiator	14 April 1998
US 5808021	Glycomed Incorporated, Alameda, CA	K. R. Holme, W. Liang, D. L. Tyrell and P. N. Shaklee	Method controlling *O*-desulfation of heparin	15 Septembeer 1998
JP 10–295383	Institute for Advanced Skin Research Inc. Yokohama, Japan	E. Hara and E. Sato	Promoter for hyaluronic acid synthase gene	10 November 1998
US 5902785	Genetics institute, Inc. Cambridge, MA	G. Hattersley, N. W. Walfman, E. A. Morris and V. A. Rosen.	Cartilage induction by bone morphogenetic proteins	11 May 1999
US 5916557	The Trustees of Columbia University in the City of New York and International Technology Management Associates, Ltd., Harvard, MA	L. Berlowitz-Tarrant, A. Ratcliffe and S. Mizuno	Methods of repairing connective tissues	29 January 1999
JP 11–209261	Shiseido Co. Ltd., Tokyo, Japan	Y. Nakazawa, Y. Suzuki and T. Magara	Hyaluronic acid production promoter	3 August 1999
US 5935796	University of Melbourne	A. J. Fosang	Diagnostic methods and compositions relating to the proteoglycan proteins of cartilage breakdown	10 August 1999
US 6051701	Fidia Advanced Biopolymers SRL, Brindisi, Italy	G. Cialdi, R. Barbuci and A. Magnani	Sulfated hyalronic acid and esters	18 April 2000

biosynthesis of GAG in the human skin tissues. These include a specific lysophospholipid (Tanaka et al., 1995), a low-molecular-weight (5 kDa or below) fraction from bovine fetal or calf serum (Kamio et al., 1996), an extract from a plant belonging to *Labiatae* (Souma and Sato, 1998), and an extract from Cuachalalate (*Juliana adstringens*) (Nakazawa, 1999). Each material can be compounded with cosmetics or medical products, which may minimize wrinkling due to age-dependent decreases in hyaluronan content in the skin.

della Valle et al. (1997) have described methods of preparation of hyaluronan from hen crests. They demonstrated two fractions of hyaluronan, one with molecular weight between 50 and 100 kDa which may mitigate scar formation, and a second fraction having a molecular weight of 500–730 kDa, which is useful for eye surgery as well as the treatment of osteoarthritis. The inventors further describe the use of the hyaluronan fractions as vehicles of ophthalmic drugs.

Cialdi and Magnani (2000) sulfated hyaluronan using sulfur trioxide and pyridine. They reported that the sulfated hyaluronic acid esters have antithrombotic, anticoagulant, and antiviral activities, and that the sulfated polyanions may be used to prepare biomaterials.

Holme et al. (1999) developed a method of making 2-*O*,3-*O*-desulfated heparin. The product lacks anticoagulant activity but has significant anticancer activity *in vivo*.

Wu and Church (1998) developed a method to neutralize heparin activity using lactoferrin, a heparin-binding protein. The method may be used to correct the heparin-induced prolongation of blood coagulation in the patients undergoing cardiovascular surgery.

Osteoarthritis is a condition manifesting degenerative articular cartilage accompanied by loss of the proteoglycan, aggrecan from the cartilage. Fosang (1999) developed an ELISA technique to determine the content of aggrecan fragments in synovial fluids or sera from osteoarthritic patients. The inventors used a monoclonal antibody recognizing a sequence Phe–Phe–Gly–Val–Gly found in a peptide generated by cleavage of aggrecan between Asn341 and Phe342, the site susceptible to MMPs (see above).

Little progress has been made towards a cure for osteoarthritis. Recent patents disclose methods of cartilage repair by using bone morphogenetic proteins (Hattersley et al., 1999) and an extracellular matrix-altering enzyme activity (Berlowitz-Tarrant et al., 1999).

Acknowledgements

The authors wish to thank Mr. Eryck R. Silva-Hernandez for his expert clerical assistance in the preparation of this manuscript.

8 References

Abbaszade, I., Liu, R., Yang, F., Rosenfeld, S. A., Ross, O. H., Link J. R., Ellis, D. M., Tortorella, M. D., Pratta, M. A., Hollis, J. M., Wynn, R., Duke, J. L., George, H. J., Hillman, M. C., Jr., Murphy, K., Wiswall, B. H., Copeland, R. A., Decicco, C. P., Bruckner, R., Nagase, H., Itoh, Y., Newton, R. C., Magolda, R. L., Trzaskos, J. M., Hollis, G. F., Arner, E. C., Burn, T. C. (1999) Cloning and characterization of *ADAMTS11*, an aggrecanase from ADAMTS family, *J. Biol. Chem.* **274**, 23443–23450.

Adam, M., Krabcová, M., Musilová, J., Pešáková, V., Brettschneider, I., Deyl, Z. (1980) Contribution to the mode of action of glycosaminoglycan-polysulphate (GAGPS) upon human osteoarthroic cartilage, *Arzneim-Forsch. Drug Res.* **30**, 1730–1732.

Antonopoulos, C. A. (1962) A modification for the determination of sulphate in mucopolysaccharides by the benzidine method, *Acta Chem. Scand.* **16**, 1521–1522.

Avnur, Z., Geiger, B. (1984) Immunocytochemical localization of native chondroitin-sulfate in tissues and cultured cells using specific monoclonal antibody, *Cell* **38**, 811–822.

Barone, D., Orlando, L., Vigna, E., Baroni, S. Borghi, A. M. (1988) A new potent antianemic agent with a favourable pharmacokinetic profile, *Drugs Exp. Clin. Res.*, **14** (Suppl. 1), 1–14.

Beeley, J. G. (1985) Radioactive labeling techniques, in: *Laboratory Techniques in Biochemistry and Molecular Biology: Glycoprotein and Proteoglycan Techniques* (Burdon, R. H., van Knippenberg, P. H., Eds.), Amsterdam: Elsevier, 365–425.

Berlowitz-Tarrant, L., Ratcliffe, A., Mizuno, S. (1999) Methods of repairing connective tissues, US patent 5 916 557.

Bernfield, M., Götte, M., Park, P. W., Reizes, O., Fitzgerald, M. L., Lincecum, J., Zako, M. (1999) Functions of cell surface heparan sulfate proteoglycans, *Annu. Rev. Biochem.* **68**, 729–777.

Bianco, P., Fisher, L. W., Young, M. F., Termine, J. D., Robey, P. G. (1990) Expression and localization of the two small proteoglycans biglycan and decorin in developing human skeletal and non-skeletal tissues, *J. Histochem. Cytochem.* **38**, 1549–1563.

Blumenkrantz, N., Asboe-Hansen, G. (1973) New method for quantitative determination of uronic acids, *Anal. Biochem.* **54**, 484–489.

Boas, N. F. (1953) Methods for the determination of hexosamine in tissues, *J. Biol. Chem.* **204**, 553–563.

Bollag, D. M., Edelstein, S. J. (1991) *Protein Methods*. New York: Wiley-Liss.

Caterson, B., Christner, J. E., Baker, J. R. (1983) Identification of a monoclonal antibody that specifically recognizes corneal and skeletal keratan sulfate, *J. Biol. Chem.* **258**, 8848–8854.

Chakravati, S., Magnuson, T., Lass, J. S., Jepsen, K. J., LaMantia, C., Carrol, H. (1998) Lumican regulates collagen fibril assembly: skin fragility and corneal opacity in the absence of lumican, *J. Cell Biol.* **141**, 1277–1286.

Choi, H. U., Johnson, T. L., Pal, S., Tang, L.-H., Rosenberg, L., Neame, P. J. (1989) Characterization of the dermatan sulfate proteoglycans, DS-PG I and DS-PG II from bovine articular cartilage and skin isolated by octyle-Sepharose chromatography, *J. Biol. Chem.* **264**, 2876–2884.

Chopra, R. K., Pearson, C. H., Pringle, G. A., Fackre, D. S., Scott, P. G. (1985) Dermatan sulphate is located on serine-4 of bovine skin proteodermatan sulphate, *Biochem. J.* **232**, 277–279.

Cialdi, G., Barbuci, R., Magnani, A. (2000) Sulfated hyaluronic acid and esters thereof. US patent 6 027 741.

Couchman, J. R., Woods, A. (1993) Structure and biology of pericellular proteoglycans, in: *Cell Surface and Extracellular Glycoconjugates* (Roberts, D. D., Mecham, R. P., Eds.), San Diego, CA: Academic Press, 33–82.

Couchman, J. R., Woods, A. (2000) Signaling through the syndecan proteoglycans, in: *Proteoglycans, Structure, Biology, and Molecular Interactions* (Iozzo, R. V., Ed.), New York: Marcel Dekker, 147–159.

Danielson, K. G., Bairboult, H. Holmes, D. F., Graham, H., Kadler, K. E., Iozzo, R. V. (1997) Targeted disruption of decorin leads to abnormal collagen fibril morphology and skin fragility, *J. Cell Biol.* **136**, 729–743.

Davidson, E. A. (1966) Analysis of sugars found in mucopolysaccharides, *Methods Enzymol.* **8**, 52–60.

Dean, D. D., Mung, O. E., Rodriquez, I., Carreno, M. R., Morales, S., Agudez, A., Madan, M. E., Altman, R. D., Annefeld, M., Howell, D. S. (1991) Amelioration of lapine osteoarthritis by treatment with glycosaminoglycan–peptide association complex (Rumalon), *Arthritis Rheum.* **34**, 304–313.

della Valle, F., Rome, A., Lorenzi, S. (1997) Pharmaceutical compositions containing hyaluronic acid fractions, US patent 5 631 241.

DiFerrante, N. (1956) Turbidimetric measurement of acid mucopolysaccharides and hyaluronidase activity, *J. Biol. Chem.* **220**, 303–306.

DiFerrante, N., Donnelly, P. V., Berglund, R. K. (1971) Colorimetric measurement of dermatan sulphate, *Biochem. J.* **124**, 549–553.

Dische, Z. (1947) A new specific color reaction of hexuronic acids, *J. Biol. Chem.* **167**, 189–198.

Dische, Z. Borenfreund, E. (1950) A spectrophotometric method for the micro-determination of hexosamines, *J. Biol. Chem.* **184**, 517–522.

Dodgson, K. S., Price, R. G. (1962) A note on the determination of the ester sulphate content of sulphated polysaccharides, *Biochem J.* **84**, 106–110.

Doege, K., Sasaki, M., Horigan, E., Hassel, J. R., Yamada, Y. (1987) Complete primary structure of the rat cartilage proteoglycan core protein deduced from cDNA clones, *J. Biol. Chem.* **262**, 17757–17767.

Dunlevy, J. R., Hassell, J. R. (2000) Heparan sulfate proteoglycans in basement membranes: perlecan, agrin, and collagen XVIII, in: *Proteoglycans, Structure, Biology, and Molecular Interactions* (Iozzo, R. V., Ed.), New York: Marcel Dekker, 275–326.

Eriksson, I., Sandback, D., Ek, B., Lindahl, U., Kjellén, L. (1994) cDNA cloning and sequencing of mouse mastocytoma glucosaminyl *N*-deacetylase/*N*-sulfotransferase, an enzyme involved in the biosynthesis of heparin, *J. Biol. Chem.* **269**, 10438–10443.

Ernst, S., Langer, R., Cooney, C. L., Sasisekharan, R. (1995) Enzymatic degradation of glycosaminoglycans, *Crit. Rev. Biochem. Mol. Biol.* **30**, 387–444.

Farndale, R. W., Sayers, C. A., Barrett, A. J. (1982) A direct spectrophotometric assay for sulfated glycosaminoglycans in cartilage cultures, *Connect. Tissue Res.* **9**, 247–248.

Fedarko, N. S. (1994) Isolation and purification of proteoglycans, in: *Proteoglycans* (Jollès, P., Ed.), Basel: Birkhäuser Verlag, 9–35.

Filisetti-Cozzi, T. M. C. C., Carpita, M. C. (1991) Measurement of uronic acids without interference from neutral sugars, *Anal. Biochem.* **197**, 157–162.

Filmus, J., Song, H. H. (2000) Glypicans, in: *Proteoglycans, Structure, Biology, and Molecular Interactions* (Iozzo, R. V., Ed.), New York: Marcel Dekker, 161–176.

Fischer, G., Boedeker, C. (1861) Künstliche Bildung von Zucker aus Knorpel (Chondrogen), und über die Umsetzung des genossenen Knorpels in menschlichen Körper, *Ann. Chem. Pharm.* **117**, 111–118.

Fisher, L. W., Termine, J. D., Young, M. F. (1989) Deduced protein sequence of bone small proteoglycan I (biglycan) shows homology with proteoglycan II (decorin) and several nonconnective tissue proteins in a variety of species, *J. Biol. Chem.* **264**, 4571–4576.

Fosang, A. J. (1999) Diagnostic methods and compositions relating to the proteoglycan proteins of cartilage breakdown, US patent 5 935 796.

Fransson, L.-Å. (1968) Structure of dermatan sulfate. IV. Glycopeptide from the carbohydrate-protein linkage region of pig skin dermatan sulfate, *Biochim. Biophys. Acta* **156**, 311–316.

Fransson, L.-Å., Rodén, L. (1967) Structure of dermatan sulfate. I. Degradation by testicular hyaluronidase, *J. Biol. Chem.* **242**, 4161–4169.

Fransson, L.-Å., Silverberg, I., Carlstedt, I. (1985) Structure of the heparan-sulfate linkage region. Demonstration of the sequence galactosyl-galactosyl-xylose-2-phosphate, *J. Biol. Chem.* **260**, 14722–14726.

Fraser, J. R. E., Laurent, T. C., Laurent, U. B. G. (1997) Hyaluronan: its nature, distribution, functions and turnover, *J. Intern. Med.* **242**, 27–33.

Fukuda, M. N., Matsumura, G. (1975) Endo-β-galactosidase of *Escherichia freudii* hydrolysis of pig clonic mucin and milk oligosaccharides by endoglycosidic action, *Biochem. Biophys. Res. Commun.* **64**, 465–471.

Fukuta, M., Inazawa, J., Torii, T., Tsuzuki, K., Shimada, E., Habuchi, O. (1997) Molecular cloning and characterization of human keratan sulfate Gal-6-sulfotransferase, *J. Biol. Chem.* **272**, 32321–32328.

Fukuta, M., Uchimura, K., Nakashima, K., Kato, M., Kimata, K., Shinomura, T., Habuchi, O. (1995) Molecular cloning and expression of chick chondrocyte chondroitin 6-sulfotransferase, *J. Biol. Chem.* **270**, 18575–18580.

Funderburgh, J. L. (2000) Corneal proteoglycans, in: *Proteoglycans, Structure, Biology, and Molecular Interactions* (Iozzo, R. V., Ed.), New York: Marcel Dekker, 237–273.

Gallagher, J. T., Lyon, M. (2000) Heparan sulfate: Molecular structure and interactions with growth factors and morphogens, in: *Proteoglycans, Structure, Biology, and Molecular Interactions* (Iozzo, R. V., Ed.), New York: Marcel Dekker, 27–60.

Gallagher, J. T., Turnbull, J. E. (1992) Heparan sulphate in the binding and activation of basic fibroblast growth factor, *Glycobiology* **2**, 523–528.

Gallagher, J. T., Lyon, M., Steward, W. P. (1986) Structure and function of heparan sulphate proteoglycans, *Biochem. J.* **236**, 313–325.

Ghosh, P. (1994) The role of hyaluronic acid (hyaluronan) in health and disease. Interactions with cells, cartilage and components of synovial fluid, *Clin. Exp. Rheumatol.* **12**, 75–82.

Gold, E. W. (1979) A simple spectrophotometric method for estimating glycosaminoglycan concentration, *Anal. Biochem.* **99**, 183–188.

Goldberg, H. A., Domenicucci, C., Pringle, G. A., Sodek, J. (1988) Mineral-binding proteoglycans of fetal porcine calvarial bone, *J. Biol. Chem.* **263**, 12092–12101.

Gottlieb, M. S. (1997) Conservative management of spinal osteoarthritis with glucosamine sulfate and chiropractic treatment, *J. Manipulative Physiol. Ther.* **20**, 400–414.

Habuchi, H., Habuchi, O., Kimata, K. (1995) Purification and characterization of heparan sulfate 6-sulfotransferase from culture medium of Chinese hamster ovary cells, *J. Biol. Chem.* **270**, 4172–4179.

Habuchi, H., Kobayashi, M., Kimata, K. (1998) Molecular characterization and expression of heparan-sulfate 6-sulfotransferase. Complete cDNA cloning in human and partial cloning in Chinese hamster ovary cells, *J. Biol. Chem.* **273**, 9208–9213.

Habuchi, O., Habuchi, H. (1998) Sulfotransferases involved in the biosynthesis of glycosaminoglycans (in Japanese), *Tanpakushitsu Kakusan Koso* **43**, 2378–2386.

Hara, E., Sato, E. (1998) Promoter for hyaluronic acid synthase gene, Japanese patent 10–295383.

Hardingham, T., Muir, H. (1974) Hyaluronic acid in cartilage and proteoglycan aggregation, *Biochem. J.* **139**, 565–581.

Hardingham, T. E., Fosang, A. J., Dudhia, J. (1994) The structure, function and turnover of aggrecan, the large aggregating proteoglycan from cartilage, *Eur. J. Clin. Chem. Clin. Biochem.* **32**, 249–257.

Hascall, V. C., Midura, R. (1989) Keratan sulphate proteoglycans: Chemistry and biosynthesis of the linkage regions, in: *Keratan Sulphate, Chemistry, Biology, Chemical Pathology* (Greiling, H., Scott, J. E., Eds.), London: The Biochemical Society, 66–75.

Hashimoto, Y., Orellana, A., Gil, G., Hirschberg, C. B. (1992) Molecular cloning and expression of rat liver *N*-heparan sulfate sulfotransferase, *J. Biol. Chem.* **267**, 15744–15750.

Hattersley, G., Wolfman, N. M., Morris, E. A., Rosen, V. A. (1999) Cartilage induction by bone morphogenetic proteins, US patent 5 902 785.

Heinegård, D., Hascall, V. C. (1974) Aggregation of cartilage proteoglycans. III. Characteristics of the proteins isolated from trypsin digests of aggregates, *J. Biol. Chem.* **249**, 4250–4256.

Heinegård, D., Larsson, T., Sommarin, Y., Franzén, A., Paulsson, M., Hedbom, E. (1986) Two novel matrix proteins isolated from articular cartilage show wide distributions among connective tissues, *J. Biol. Chem.* **261**, 13866–13872.

Hirschberg, C. B., Orellana, A., Hashimoto, Y., Swiedler, S. J., Wei, Z., Ishihara, M. (1996) Glycosaminoglycan specific sulfotransferases, US patent 5 541 095.

Holme, K. R., Liang, W., Tyrrell, D. J., Shaklee, P. N. (1999) Method for controlling O desulfation of heparin, US patent 5 808 021.

Howell, W. H., Holt, E. (1918) Two new factors in blood coagulation-heparin and pro-antithrombin, *Am. J. Physiol.* **47**, 328–341.

Hunter, G. K. (1991) Role of proteoglycan in the provisional calcification of cartilage. A review and reinterpretation, *Clin. Orthop.* **262**, 256–280.

Iozzo, R. V. (1994) Perlecan: a gem of a proteoglycan, *Matrix Biol.* **14**, 203–208.

Iozzo, R. V. (1998) Matrix proteoglycans: from molecular design to cellular function, *Annu. Rev. Biochem.* **67**, 609–652.

Itano, N., Kimata, K. (1998) Hyaluronan synthase: new directions for hyaluronan research, *Trends Glycosci. Glycotechnol.* **10**, 23–28.

Iwata, H. (1969) The determination and the fine structures of chondroitin sulfate isomers of human cartilage and pathological tissues (in Japanese), *J. Jpn. Orthop. Ass.* **43**, 455–473.

Jaakkola, P., Kainulainen, V., Jalkanen, M. T. (2000) Syndecans, in: *Proteoglycans, Structure, Biology, and Molecular Interactions* (Iozzo, R. V., Ed.), New York: Marcel Dekker, 115–145.

Jorpes, E., Gardell, S. (1948) On heparin monosulphuric acid, *J. Biol. Chem.* **176**, 267–276.

Kamio, M. Sakai, S., Sayo, T., Inoue, S., Matsui, T. (1996) Hyaluronic acid production accelerator, Japanese patent 8–239404.

Keates, R. H., Rabin, B. (1988) Extending corneal storage with 2.5% chondroitin sulfate (K-Sol), *Ophthal. Surg.* **19**, 817–820.

Kelly, G. S. (1998) The role of glucosamine sulfate and chondroitin sulfates in the treatment of degenerative joint disease, *Alt. Med. Rev.* **3**, 27–39.

Kempson, G. E., Muir, H., Swanson, S. A. V., Freeman, M. A. R. (1970) Correlation between stiffness and the chemical constituents of cartilage on the human femoral head, *Biochim. Biophys. Acta* **215**, 70–77.

Kida, J. (1993) Electron microscopic cytochemical studies on sulfated glycosaminoglycans in rat mast cell granules, *Acta Histochem. Cytochem.* **26**, 135–146.

Kimata, K., Barrach, H.-J., Brown, K. S., Pennypacker, J. P. (1981) Absence of proteoglycan core protein in cartilage from the *cmd/cmd* (cartilage matrix deficiency) mouse, *J. Biol. Chem.* **256**, 6961–6968.

Kjellén, L., Lindhal, U. (1991) Proteoglycans: structures and interactions, *Annu. Rev. Biochem.* **60**, 443–475.

Knecht, J., Cifonelli, A., Dorfman, A. (1967) Structural studies on heparan sulfate of normal and Hurler tissues, *J. Biol. Chem.* **242**, 4652–4461.

Kobayashi, M., Habuchi, H., Habuchi, O., Saito, M., Kimata, K. (1996) Purification and characterization of heparan sulfate 2-sulfotransferase from cultured Chinese hamster ovary cells, *J. Biol. Chem.* **271**, 7645–7653.

Kobayashi, M., Habuchi, H., Yoneda, M., Habuchi, O., Kimata, K. (1997) Molecular cloning and expression of Chinese hamster ovary cell heparan sulfate 2-sulfotransferase, *J. Biol. Chem.* **272**, 13980–13985.

Kosakai, M., Yoshizawa, Z. (1979) A partial modification of the carbazole method of Bitter and Muir for quantitation of hexuronic acids, *Anal. Biochem.* **93**, 295–298.

Kresse, H., Glössle, J. (1987) Glycosaminoglycan degradation, *Adv. Enzymol. Relat. Areas Mol. Biol.* **60**, 217–311.

Krusius, T., Ruoslahti, E. (1986) Primary structure of an extracellular matrix proteoglycan core protein deduced from cloned cDNA, *Proc. Natl Acad. Sci. USA* **83**, 7683–7687.

Kunihiro, Y., Tanaka, R., Ichimura, M., Uemura, A., Ohzawa, N., Mochida, E. (1994) Anticoagulant substance obtained from urine, US patent 5 300 490.

Lammi, M., Tammi, M. (1988) Densitometric assay of nanogram quantities of proteoglycans precipitated on nitro-cellulose membrane with safranin-O, *Anal. Biochem.* **168**, 352–357.

Langunoff, D., Warren, G. (1962) Determination of 2-deoxy-2-sulfaminohexose content of mucopolysaccharides, *Arch. Biochem. Biophys.* **99**, 396–400.

Lark, M. W., Bayne, E. K., Flanagan, J., Harper, C. F., Hoerrner, L. A., Hutchinson, N. I., Singer, I. I., Donatelli, S. A., Weidner, J. R., Williams, H. R., Mumford, R. A., Lohmander, L. S. (1997) Aggrecan degradation in human cartilage. Evidence for both matrix metalloproteinase and aggrecanase activity in normal, osteoarthritic, and rheumatoid joints, *J. Clin. Invest.* **100**, 93–106.

Lindahl, U. (1966) Further characterization of the heparin–protein linkage region, *Biochim. Biophys. Acta* **130**, 368–382.

Linhardt, R. J. Turnbull, J. E., Wang, H. M., Loganathan, D., Gallagher, J. T. (1990) Examination of the substrate specificity of heparin and heparan sulfate lyases, *Biochemistry* **29**, 2611–2617.

Linker, A., Hovingh, P. (1972) Heparinase and heparitinase from Flavobacteria, *Methods Enzymol.* **28**, 902–911.

Maimone, M. M., Tollefsen, D. M. (1990) Structure of a dermatan sulfate hexasaccharide that binds to heparin cofactor II with high affinity, *J. Biol. Chem.* **265**, 18263–18271.

Malmström, A., Aberg, L. (1982) Biosynthesis of dermatan sulfate. Assay and properties of the uronosyl C-5 epimerase, *Biochem. J.* **201**, 489–493

Mankin, H. J., Lippiello, L. (1970) Biochemical and metabolic abnormalities in articular cartilage from osteo-arthritic human hips, *J. Bone Joint Surg.* **52A**, 424–434.

Maroudas, A., Schneiderman, R., Popper, O. (1992) The role of water, proteoglycan, and collagen in solute transport in cartilage, in: *Articular Cartilage and Osteoarthritis* (Kuettner, K., Ed.), New York: Raven Press, 355–371.

Mathews, M. B., Galgov, S. (1966) Acid mucopolysaccharide patterns in aging human cartilage, *J. Clin. Invest.* **45**, 1103–1111.

Matsushima, T., Nakashima, Y., Sugano, M., Tasaki, H., Kuroiwa, A., Koide, O. (1987) Suppression of atherogenesis in hypercholesterolemic rabbits by chondroitin-6-sulfate, *Artery* **14**, 316–337.

McLean, J. (1916) The thromboplastic action of cephalin, *Am. J. Physiol.* **41**, 250–257.

Meyer, K. (1938) The chemistry and biology of mucopolysaccharides and glycoproteins, *Cold Spring Harbor Symp. Quant. Biol.* **6**, 91–102.

Meyer, K., Palmer, J. W. (1934) The polysaccharide of vitreous humor, *J. Biol. Chem.* **107**, 629–634.

Meyer, K., Rapport, M. M. (1951) The mucopolysaccharides of the ground substance of connective tissue, *Science* **113**, 596–599.

Meyer, K., Linker, A., Davidson, E. A., Weissmann, B. (1953) The mucopolysaccharides of bovine cornea, *J. Biol. Chem.* **205**, 611–616.

Mörgelin, M., Paulsson, M., Malmström, A., Heinegård, D. (1989) Shared and distinct structural features of interstitial proteoglycans from different bovine tissues revealed by electron microscopy, *J. Biol. Chem.* **264**, 12080–12090.

Muir, H. (1958) The nature of the link between protein and carbohydrate of a chondroitin sulphate complex from hyaline cartilage, *Biochem. J.* **69**, 195–204.

Muir, H. (1977) Molecular approach to the understanding of osteoarthrosis, *Ann. Rheum. Dis.* **36**, 199–208.

Muir, H., Hardingham, T. E. (1975) Structure of proteoglycans, in: *MTP International Review of Science. Biochemistry of Carbohydrate.* Biochemistry Series Vol. 5 (Whelan, W. J., Ed.), London: Butterworths, 153–222.

Nakano, T., Aherne, F. X. (1993) Articular cartilage lesions in female breeding swine, *Can. J. Anim. Sci.* **73**, 1005–1008.

Nakano, T., Ozimek, L. (1999) Bovine milk glycosaminoglycans, *Milchwissewnschaft* **53**, 629–633.

Nakano, T., Scott, P. G. (1989) Proteoglycans of the articular disc of the bovine temporomandibular joint. I. High molecular weight chondroitin sulphate proteoglycan, *Matrix* **9**, 277–283.

Nakano, T., Scott, P. G. (1996) Changes in the chemical composition of the bovine temporomandibular joint disc with age, *Arch. Oral. Biol.* **41**, 845–853.

Nakano, T., Sim, J. S. (1989) Glycosaminoglycans from the rooster comb and wattle, *Poult. Sci.* **68**, 1303–1306.

Nakano, T., Sim, J. S. (1992) A quantitative chemical study of the comb and wattle galactosaminoglycans from Single Comb White Leghorn roosters, *Poult. Sci.* **71**, 1540–1547.

Nakano, T., Sim, J. S., Imai, S., Koga, T. (1996a) Lack of chondroitin sulphate epitope in the proliferating zone of the growth plate of chicken tibia, *Histochem. J.* **28**, 867–873.

Nakano, T., Sunwoo, H. H., Li, X., Price, M. A., Sim, J. S. (1996b) Study of sulfated glycosaminoglycans from porcine skeletal muscle epimysium including analysis of iduronosyl and glucuronosyl residues in galactosaminoglycan fractions, *J. Agric. Food Chem.* **44**, 1424–1434.

Nakano, T., Dodd, C. M., Scott, P. G. (1997) Glycosaminoglycans and proteoglycans from different zones of the porcine knee meniscus, *J. Orthop. Res.* **15**, 213–220.

Nakano, T., Nakano, K., Sim, J. S. (1998) Extraction of glycosaminoglycan peptide from bovine nasal cartilage with 0.1 M sodium acetate, *J. Agric. Food Chem.* **46**, 772–778.

Nakano, T., Ikawa, N., Ozimek, L. (2000) An economical method to extract chondroitin sulphate–peptide from bovine nasal cartilage, *Can. Agric. Eng.* **42**, 205–208.

Nakano, T., Ikawa, N., Ozimek, L. (2001) Extraction of glycosaminoglycans from chicken eggshell, *Poult. Sci.* **80**, 681–684.

Nakazawa, K., Suzuki, S. (1975) Purification of keratan sulfate-endogalactosidase and its action on keratan sulfates of different origin, *J. Biol. Chem.* **250**, 912–917.

Nakazawa, K., Ito, M., Yamagata, T., Suzuki, S. (1989) Substrate specificity of keratan sulphate-degrading enzymes (endo-β-galactosidase, keratanase and keratanase II) from microorganisms, in: *Keratan Sulphate, Chemistry, Biology, Chemical Pathology* (Greiling, H., Scott, J. E., Eds.), London: The Biochemical Society, pp. 99–110.

Nakazawa, Y., Suzuki, Y., Magara, T., Tajima, M. (1999) Hyaluronic acid production promoter. Japanese patent 11–209261.

Neame, P. J., Kay, C. J. (2000) Small leucine-rich proteoglycans, in: *Proteoglycans, Structure, Biology, and Molecular Interactions* (Iozzo, R. V., Ed.), New York: Marcel Dekker, pp. 201–235.

Neame, P. J., Choi, H. U., Rosenberg, L. C. (1989) The primary structure of the core protein of the small, leucine-rich proteoglycan (PG I) from bovine articular cartilage, *J. Biol. Chem.* **264**, 8653–8661.

Noonan, D. M., Hassel, J. R. (1993) Proteoglycans of the basement membrane, in: *Molecular and Cellular Aspects of Basement Membranes* (Rohrbach, D. H., Timple, R., Eds.), New York: Academic Press, 189–210.

Oegema, T. R., Jr., Hascall, V. C., Dziewiatkowski, D. D. (1975) Isolation and characterization of proteoglycans from the Swarm rat chondrosarcoma, *J. Biol. Chem.* **250**, 6151–6159.

Oegema, T. R., Jr., Kraft, E. L., Jourdian, G. W., VanValen, T. R. (1984) Phosphorylation of chondroitin sulfate in proteoglycans from the Swarm rat chondrosarcoma, *J. Biol. Chem.* **259**, 1720–1726.

Ohya, T., Kaneko, Y. (1970) Novel hyaluronidase from *Streptomyces, Biochim. Biophys. Acta* **198**, 607–609.

Oldberg, Å., Antonson, P., Lindblom, K., Heinergård, D. (1989) A collagen-binding 59-kd protein (fibromodulin) is structurally related to the small interstitial proteoglycans PG-S1 and PG-S2 (decorin), *EMBO J.* **8**, 2601–2604.

Olivecrona, T., Bengtsson-Olivecrona, G. (1989) Heparin and lipases, in: *Heparin. Chemical and Biological Properties, Clinical Applications* (Lane, D. A., Lindahl, U., Eds.), London: Edward Arnold, 335–361.

Orellana, A., Hirschberg, C. B., Wei, Z., Swiedler, S. J., Ishihara, M. (1994) Molecular cloning and expression of a glycosaminoglycan *N*-acetylglucosaminyl *N*-deacetylase/*N*-sulfotransferase from a heparin-producing cell line, *J. Biol. Chem.* **269**, 2270–2276.

Pearson, C. H., Gibson, G. J. (1982) Proteoglycans of bovine periodontal ligament and skin. Occurrence of different hybrid-sulphated galactosaminoglycans in distinct proteoglycans, *Biochem. J.* **201**, 27–37.

Pettersson, I., Kusche, M., Unger, E., Wlad, H., Nylund, L., Lindahl, U., Kjéllen, L. (1991) Biosynthesis of heparin. Purification of a 110-kDa mouse mastocytoma protein required for both glycosaminyl *N*-deacetylation and *N*-sulfation, *J. Biol. Chem.* **266**, 8044–8049.

Pharmacia (1985) *Healon® (Sodium Hyaluronate), Technical Information and Clinical Experience.* Baie d'Urfe, Quebec: Pharmacia Canada Inc.

Plaas, A. H. K., Neame, P. J., Nivens, C. M., Reiss, L. (1990) Identification of the keratan sulfate attachment sites on bovine fibromodulin, *J. Biol. Chem.* **265**, 20634–20640.

Poole, A. R. Matsui, Y., Hinex, A., Lee, E. R. (1989) Cartilage macromolecules and the calcification of cartilage matrix, *Anat. Rec.* **224**, 167–179.

Pringle, G. A., Dodd, C. M., Osborn, J. W., Pearson, C. H., Mossman, T. R. (1985) Production and characterization of monoclonal antibodies to bovine skin proteodermatan sulfate, *Collagen Relat. Res.* **5**, 23–39.

Rodén, L. (1968) The protein–carbohydrate linkages of acid mucopolysaccharides, in: *The Chemical Physiology of Mucopolysaccharides* (Quintarelli, G., Ed.), London: J. & A. Churchill, 17–32.

Rodén, L. (1980) Structure and metabolism of connective tissue proteoglycans, in: *The Biochemistry of Glycoproteins and Proteoglycans* (Lennarz, W. J., Ed.), New York: Plenum Press, 267–371.

Rodén, L., Baker, J. R., Cifonelli, A., Mathews, M. B. (1972) Isolation and characterization of connective tissue polysaccharides, *Methods Enzymol.* **53A**, 69–82.

Roughley, P. J., Mort, J. S. (2000) Catabolism of proteoglycans, in: *Proteoglycans, Structure, Biology, and Molecular Interactions* (Iozzo, R. V., Ed.), New York: Marcel Dekker, 93–113.

Rosenberg, L., Hellmann, W., Kleinschmidt, A. K. (1975) Electron microscopic studies of proteoglycan aggregates from bovine articular cartilage, *J. Biol. Chem.* **250**, 1877–1883.

Rosenberg, L. C., Choi, H. U., Tang, L. H., Johnson, T. L., Pal, S., Webber, C., Reiner, A., Poole, A. R. (1985) Isolation of dermatan sulfate proteoglycans from mature bovine articular cartilages, *J. Biol. Chem.* **260**, 6304–6313.

Rosenberg, R. D., Shworak, N. W., Liu, J., Schwartz, J. J., Zhang, L. (1997) Perspectives series: cell adhesion in vascular biology, *J. Clin. Invest.* **99**, 2062–2070.

Ruoslahti, E. (1996) Brain extracellular matrix, *Glycobiology* **6**, 489–492.

Rupp, F., Özçelik, T., Linial, M., Peterson, K., Franke, U., Scheller, R. (1992) Structure and chromosomal localization of the mammalian agrin gene, *J. Neurosci.* **12**, 3535–3544.

Ruter, E.-R., Kresse, H. (1984) Partial purification and characterization of 3′-phosphoadenylylsulfate: keratan sulfate sulfotransferases, *J. Biol. Chem.* **259**, 11771–11776.

Saito, H., Yamagata, T., Suzuki, S. (1968) Enzymatic methods for the determination of small quantities of isomeric chondroitin sulfates, *J. Biol. Chem.* **243**, 1534–1542.

Sajdera, S. W., Hascall, V. C. (1969) Protein–polysaccharide complex from bovine nasal cartilage, *J. Biol. Chem.* **244**, 77–87.

Scott, J. E., Dorling, J. D. (1965) Differential staining of acid glycosaminoglycan (mucopolysaccharides) by alcian blue in salt solutions, *Histochemie* **5**, 221–233.

Scott, P. G. (1993) Structures of the protein core of the small dermatan sulphate proteoglycans, in: *Dermatan Sulfate Proteoglycans, Chemistry, Biology, Chemical Pathology* (Scott, J. E., Ed.), London: Portland Press, 81–101.

Scott, P. G., Dodd, C. M. (1990) Removal of the three *N*-linked oligosaccharides promotes the self-aggregation of bovine skin proteodermatan sulphate, *Connect. Tissue Res.* **24**, 225–236.

Scott, P. G., Winterbottom, N., Dodd, C. M., Edward, E., Pearson, C. M. (1986) A role for disulphide bridges in the protein core in the interaction of proteodermatan sulphate and collagen, *Biochem. Biophys. Res. Commun.* **138**, 1348–1354.

Scott, P. G., Nakano, T., Dodd, C. M. (1997) Isolation and characterization of small proteoglycans from different zones of the porcine knee meniscus, *Biochim. Biophys. Acta* **1336**, 254–262.

Seno, N., Akiyama, F., Anno, K. (1974) Substrate specificity of chondrosulfatases from *Proteus vulgaris* from sulfated tetrasaccharides, *Biochim. Biophys. Acta* **362**, 290–298.

Shatton, J., Schubert, M. (1954) Isolation of mucoprotein from cartilage, *J. Biol. Chem.* **211**, 565–573.

Shinomura, T., Nishida, Y., Kimata, K. (1992) A large chondroitin sulfate proteoglycan (PG-M) and cartilage differentiation, in: *Articular Cartilage and Osteoarthritis* (Kuettner, K., Ed.), pp. 35–44. New York: Raven Press.

Shokuhin Kagaku Binran (1978) *Manual of Food Science* (in Japanese). Tokyo: Kyoritsu Shuppan.

Sobue, M., Nakashima, N., Fukatsu, T., Nagasaka, T., Fukata, S., Ohiwa, N., Nara, Y, Ogura, T., Katoh, T., Takeuchi, J. (1989) Production and immunohistochemical characterization of a monoclonal antibody raised to proteoglycan purified from a human yolk sac tumour, *Histochem. J.* **21**, 455–459.

Sobue, M., Nakashima, N., Fukatsu, T., Nagasaka, T., Katoh, T., Ogura, T., Takeuchi, J. (1988) Production and characterization of monoclonal antibody to dermatan sulfate proteoglycan, *J. Histochem. Cytochem.* **36**, 479–485.

Souma, T., Sato, E. (1998) Hyaluronic acid productivity potentiator. Japanese patent 10–95735.

Spicer, A. P., Salden, M. F., Olsenm, A. S., Brown, N., Wells, D. E., Doggett, N. A., Itano, N., Kimata, K., Inazawa, J., McDonald, J. A. (1997) Chromosomal localization of the human and mouse hyaluronan synthase genes, *Genomics* **41**, 493–497.

Stevens, R. L., Humphries, D. E., Wong, G. W. (2000) Serglycin proteoglycans: The family of proteoglycans stored in the secretory granules of certain effector cells of the immune system, in: *Proteoglycans, Structure, Biology, and Molecular Interactions* (Iozzo, R. V., Ed.), New York: Marcel Dekker, pp. 177–200.

Stripe, N. S., Argraves, W. S., Goetinck, P. F. (1987) Chondrocytes from the cartilage proteoglycan deficient mutant, nanomelia, synthesize greatly reduced levels of the proteoglycan core protein transcript, *Dev. Biol.* **124**, 77–81.

Stuhlsatz, H. W., Keller, R., Becker, G., Oeben, M., Lernarz, L., Fisher, D. C., Greiling, H. (1989) Structure of keratan sulphate proteoglycans: core proteins, linkage regions, carbohydrate chains, in: *Keratan Sulphate, Chemistry, Biology, Chemical Pathology* (Greiling, H., Scott, J. E., Eds.), London: The Biochemical Society, 1–11.

Sugahara, K., Masuda, M., Harada, T., Yamashina, I., de Waard, P., Vliegenthart, J. F. G. (1991) Structural studies on sulfated oligosaccharides derived from the carbohydrate–protein linkage region of chondroitin sulfate proteoglycans of whale cartilage, *Eur. J. Biochem.* **202**, 805–811.

Suzuki, S. (1960) Isolation of novel disaccharides from chondroitin sulfates, *J. Biol. Chem.* **235**, 3580–3588.

Suzuki, M. (1939) Biochemical studies on carbohydrates. L. Prosthetic group of corneamucoid, *J. Biochem. (Tokyo)* **30**, 185–191.

Sztrolovics, R., Alini, M., Roughley, P. J., Mort, J. S. (1997) Aggrecan degradation in human intervertebral disc and articular cartilage, *Biochem J.* **326**, 235–241.

Takagi, M., Kamiya, N., Urushizaki, T., Tada, Y., Tanaka, H. (2000) Gene expression and immunohistochemical localization of biglycan in association with mineralization in the matrix of epiphyseal cartilage, *Histochem. J.* **32**, 175–186.

Takaku, F., Kamoshita, S., Chief Eds. (1999) *Manual of Therapeutic Agents* (in Japanese). Tokyo: Igakushoin.

Tan, E. M. L., Uitto, J. J. (1996) Methods for inhibiting collagen synthesis by keloid fibroblasts, US patent 5 500 409.

Tanaka, S., Doi, H., Yamamoto, N. (1995) Skin activator having glycosaminoglycan production accelerator activity, WO 95/35092.

Taniguchi, N. (1982) Isolation and analysis of glycosaminoglycans, in: *Glycosaminoglycans and Proteoglycans in Physiological and Pathological Processes of Body Systems* (Varma, R. S., Varma, R., Eds), Basel: Karger, 20–40.

Thunberg, L., Bäckström, G., Lindahl, U. (1982) Further characterization of the antithrombin-binding sequence in heparin, *Carbohydr. Res.* **100**, 393–410.

Terho, T., Hartiala, K. (1971) Method of determination of the sulfate content of glycosaminoglycans, *Anal. Biochem.* **41**, 471–476.

Tone, Y., Kitagawa, H., Imiya, K., Oka, S., Kawasaki, T., Sugahara, K. (1999) Characterization of recombinant human glucuronosyltransferase I involved in the biosynthesis of the glycosaminoglycan–protein linkage region of proteoglycans, *FEBS Lett.* **459**, 415–420.

Toole, B. P. (2000) Hyaluronan, in: *Proteoglycans, Structure, Biology, and Molecular Interactions* (Iozzo, R. V., Ed.), New York: Marcel Dekker, 61–92.

Tortorella, M. D., Burn, T. C., Pratta, M. A., Abaszade, I., Hollis, J. M., Liu, R., Rosenfield, S. A., Copeland, R. A., Decicco, C. P., Wynn, R., Rockwell, A., Yang, F., Duke, J. L. Soloman, K., George, H., Bruckner, R., Nagase, H., Itoh, Y., Ellis, D. M., Ross, H., Wiswall, H., Murphy, K. (1999) Purification and cloning of aggrecanase-1: a member of the ADAMTS family of proteins, *Science* **284**, 1664–1666.

Vertel, B. M., Ratcliffe, A. (2000) Aggrecan, in: *Proteoglycans, Structure, Biology, and Molecular Interactions* (Iozzo, R. V., Ed.), New York: Marcel Dekker, 343–377.

Vilim, V., Krajickova, J. (1991) Proteoglycans of human articular cartilage. Identification of several populations of large and small proteoglycans and of hyaluronic acid-binding proteins in successive cartilage extracts, *Biochem. J.* **273**, 579–585.

Vogel, K. G. (1994) Glycosaminoglycans and proteoglycans, in: *Extracellular Assembly and Structure* (Yurchenco, P. D., Birk, D. E., Mecham, R. P., Eds.), San Diego, CA: Academic Press, 243–279.

Vogel, K. G., Paulsson, M., Heinegård, D. (1984) Specific inhibition of type I and type II collagen fibrillogenesis by the small proteoglycan of tendon, *Biochem. J.* **223**, 587–597.

Wei, Z., Swiedler, S. J., Ishihara, M., Orellana, A., Hirschberg, C. B. (1993) A single protein catalyzes both *N*-deacetylation and *N*-sulfation during the biosynthesis of heparan sulfate, *Proc. Natl Acad. Sci. USA* **90**, 3885–3888.

Wessler, E. (1971) Electrophoresis of acidic glycosaminoglycans in hydrochloric acid: micro method for sulfate determination, *Anal. Biochem.* **41**, 67–69.

Westergren-Thorsson, G., Antonson, P., Malmström, A., Heinegård, D., Oldberg, Å. (1991) The synthesis of a family of structurally related proteoglycans in fibroblasts is differently regulated by TGF-beta, *Matrix* **11**, 177–183.

Wight, T. N., Heinegård, D. K., Hascall, V. C. (1991) Proteoglycans, structure and functions, in: *Cell Biology of Extracellular Matrix*, 2nd edn. (Hay, E. D., Ed.) New York: Plenum Press, 45–78.

Wlad, H., Maccarana, M., Eriksson, I., Kjellén, L., Lindahl, U. (1994) Biosynthesis of heparin. Different molecular forms of *o*-sulfotransferases, *J. Biol. Chem.* **269**, 24538–24541.

Wu, H.-F., Church, F. C. (1998) Use of lactoferrin to modulate and/or neutralize heparin activity, US patent 5 712 247.

Yamagata, T., Saito, H., Habuchi, O., Suzuki, S. (1968) Purification and properties of bacterial chondroitinases and chondrosulfatases, *J. Biol. Chem.* **243**, 1523–1535.

Yanagishita, M. (1992) Glycosylphosphatidylinositol-anchored and core protein-intercalated heparan sulfate proteoglycans in rat ovarian granulosa cells have distinct secretory, endocytotic, and intracellular degradative pathways, *J. Biol. Chem.* **267**, 9505–9511.

Yanagishita, M., Hascall, V. C. (1992) Cell surface heparan sulfate proteoglycans, *J. Biol. Chem.* **267**, 9451–9454.

Zimmermann, D. R. (2000) Versican, in: *Proteoglycans, Structure, Biology, and Molecular Interactions* (Iozzo, R. V., Ed.), New York: Marcel Dekker, pp. 327–341.

21
Schizophyllan

PD Dr. Udo Rau
Technical University Braunschweig, Institute of Biochemistry and Biotechnology, Spielmannstr. 7, D-38106 Braunschweig, Germany; Tel.: +49-531/391-5740; Fax: +49-531/391-5763; E-mail: U.Rau@tu-bs.de

Biopolymers for Medical and Pharmaceutical Applications. Edited by A. Steinbüchel and R. H. Marchessault

ISBN: 3-527-31154-8

ADP	adenosine diphosphate
ATP	adenosine triphosphate
BDM	bio dry mass
cpm	counts per minute
DMSO	dimethylsulfoxide
EOR	enhanced oil recovery
GDP	guanosine diphosphate
Glc	glucose
GTP	guanosine triphosphate
L	liquid phase
Me	metal
MW	molecular weight
ORF	open reading frame
O_2	oxygen
P	phosphate
pO_2	oxygen partial pressure of the liquid phase
Pro-A	proteinic acceptor
Re	Reynold's number
r.p.m.	revolutions per minute
S	substrate
SDS-PAGE	sodium dodecyl sulfate-polyacrylamide gel electrophoresis
UDP	uridine diphosphate
UTP	uridine triphosphate
X	biomass
*	gas/liquid interface
C (g L^{-1})	concentration
k_La (h^{-1})	volume-related oxygen transfer coefficient
K_m	Michaelis-Menten constant
K_S (mol L^{-1})	substrate saturation constant
qO_2 (h^{-1})	specific oxygen uptake rate (oxygen uptake rate divided by actual biomass)
R	feedback rate
V (L)	volume
V (L h^{-1})	volume feed
$Y_{X/O2}$	yield coefficient (biomass formed/oxygen consumed)
δ (ppm)	chemical shift
[η] (cm^3 g^{-1})	intrinsic viscosity

1 Introduction

Many fungi are able to secrete exopolysaccharides, the molecular structure of which varies between homopolysaccharides, which are composed of a single monomer, and heteropolysaccharides, which contain chemically different monomers. However, in contrast to bacteria where for example xanthan is synthesized by *Xanthomonas campestris* in large amounts, fungi secrete only relatively small amounts of heteropolysaccharides (Seviour et al., 1992).

The fungal homopolysaccharides frequently contain D-glucose as the monomer which is connected glycosidically by either α- or β-linkages. Branched β-glucans are secreted (among other fungi) by *Sclerotium rolfsii* (Pilz et al., 1991), *Sclerotium glucanicum* (Rau et al., 1992b), *Monilinia fructigena* (Cordes, 1990), and *Botrytis cinerea* (Gawronski et al., 1996). These β-glucans possess a uniform, primary molecular structure that is identical with that of schizophyllan, which forms the subject of this review.

Acknowledgement

This review is dedicated to Prof. Dr. Wolf-Dieter Deckwer on the occasion of his 60th birthday.

2 Historical Outline

According to Essig (1922), the wood-rotting and filamentously growing basidiomycete *Schizophyllum commune* was first described by Dillenius in 1719, and was found to secrete a neutral homoglucan called schizophyllan.

Later, Wang and Miles (1964) performed a physiological characterization of the dikaryotic mycelia of *S. commune*, whilst during the early 1970s two Japanese articles not only described the formation and properties of schizophyllan (Kikumoto et al., 1970) but also elucidated its structure by means of enzymatic degradation studies (Kikumoto et al., 1971). These were the first and, for a long time also the last, publications that dealt with the production of schizophyllan. It was during the late 1970s that the group of Wessels (Sietsma and Wessels, 1977, 1979) first began to analyze the cell wall of *S. commune* and to investigate how the organism carried out β-glucan synthesis. During the 1980s, a number of Japanese authors investigated the physico-chemical characteristics of schizophyllan, notably the triple helical arrangement in aqueous solution (Norisuye et al., 1980; Kashiwagi et al., 1981), conformation analysis in gels at raised pH (Saito et al., 1979) and conformation transition by the addition of dimethylsulfoxide (Kitamura, 1989), as well as its controlled degradation by hydrodynamic shear (Kojima et al., 1984) and ultrasonic treatment (Tabata et al., 1981). The non-Newtonian flow behavior of colloid disperse aqueous schizophyllan solutions was investigated by Rau and Wagner (1987).

Studies on the production (Rau et al., 1989) and additional downstream processing (Cordes et al., 1989) of schizophyllan were continued by the group of Rau. These authors established an oxygen controlled batch process (Brandt et al., 1993; Rau and Brandt, 1994) as well as an oxygen-limited continuous process (including cell feedback) for the enhanced production of schizophyllan (Brandt, 1995), as well as creating downstream processing (Rau et al., 1992a; Rau, 1999). At the same time, Steiner and co-workers were investigating the cultivation of *S. commune* (Steiner et al., 1988, 1993), albeit with their emphasis directed towards the differing enzymatic activities of this fungus (Haltrich and Steiner, 1994; Steiner et al., 1987).

The oil crisis in 1973 provided the first initiative to use schizophyllan as a polymer additive in enhanced oil recovery (EOR) (Lindoerfer et al., 1988; Rau et al., 1992c), but this role declined as the price of oil subsequently fell. Although, depending on the availability and price of crude oil, the use of schizophyllan in EOR may well return. It are the immune-stimulating properties of schizophyllan and subsequent use in the pharmaceutical industry that have recently proved exciting (Tsuzuki et al., 1999; Kidd, 2000; Miura et al., 2000; Mueller et al., 2000; Ooi and Liu, 2000).

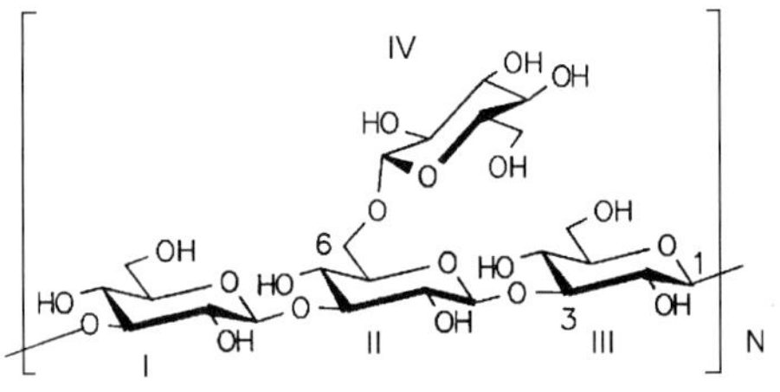

Fig. 1 Subunit of schizophyllan. N = 9000–18 000.

3 Chemical Structure

In contrast to xanthan, which is an anionic heteropolysaccharide, schizophyllan does not carry any charged groups and contains only β-D-glucose. This neutral homoglucan consists of a backbone chain of 1,3-β-D-glucopyranose units linked with single 1,6-bonded β-D-glucopyranoses at about every third glucose molecule in the basic chain (Figure 1). The molecular weight ranges from 6 to 12×10^6 g mol^{-1} (Rau et al., 1990).

The following data were obtained from the ^{13}C NMR-spectra (Figure 2) of schizophyllan: The signal at 102.7 and 102.8 ppm is due to C1 carbons of all sugars. The area of these signals was set to 100%. The signals in the 86 ppm region can be attributed to C3 carbons involved in the 1,3-β-glycosidic linkage. The chemical shifts of the free hydroxymethyl C6 carbons are found in the 60 ppm region (Rinaudo and Vincendon, 1982). The integral intensity of the signals from C1 and linked C3 gave a ratio of 1:0.74, and implies that a single glucose molecule is 1,6-β-linked

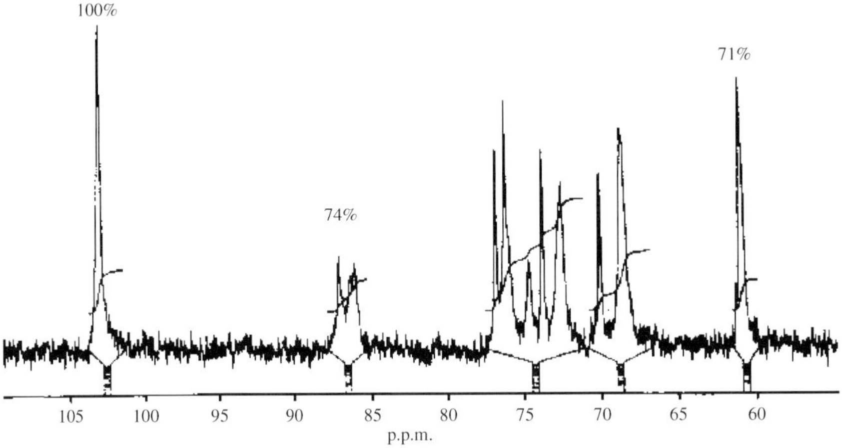

Fig. 2 ^{13}C-NMR spectra of schizophyllan in DMSO-d_6 at 80 °C. δ = 60.6 (C-6_{I}), 60.7 (C-6_{II}), 60.9 (C-6_{IV}), 68.2 (C-$4_{II,III}$), 68.4 (C-4_{I}), 70.0 (C-4_{IV}), 72.4 (C-$2_{II,III}$), 72.5 (C-2_{I}), 73.4 (C-2_{IV}), 74.5 (C-5_{III}), 75.9 (C-5_{II}), 76.1 (C-$5_{I,IV}$), 76.4 (C-3_{IV}), 85.8 (C-3_{III}), 86.1 (C-3_{II}), 86.4 (C-3_{I}), 102.7 (C-$1_{I\text{-}IV}$), 102.8 (C-$1_{I\text{-}IV}$). (Recorded by V. Wray, GBF, Braunschweig, Germany.)

to every third monomer of the main 1,3-β-chain.

Schizophyllan dissolves in water and dilute (≤ 0.01 M) sodium hydroxide solution to form highly viscous solutions. Norisuye et al. (1980) concluded from sedimentation equilibrium, light scattering and viscosity measurements that in aqueous

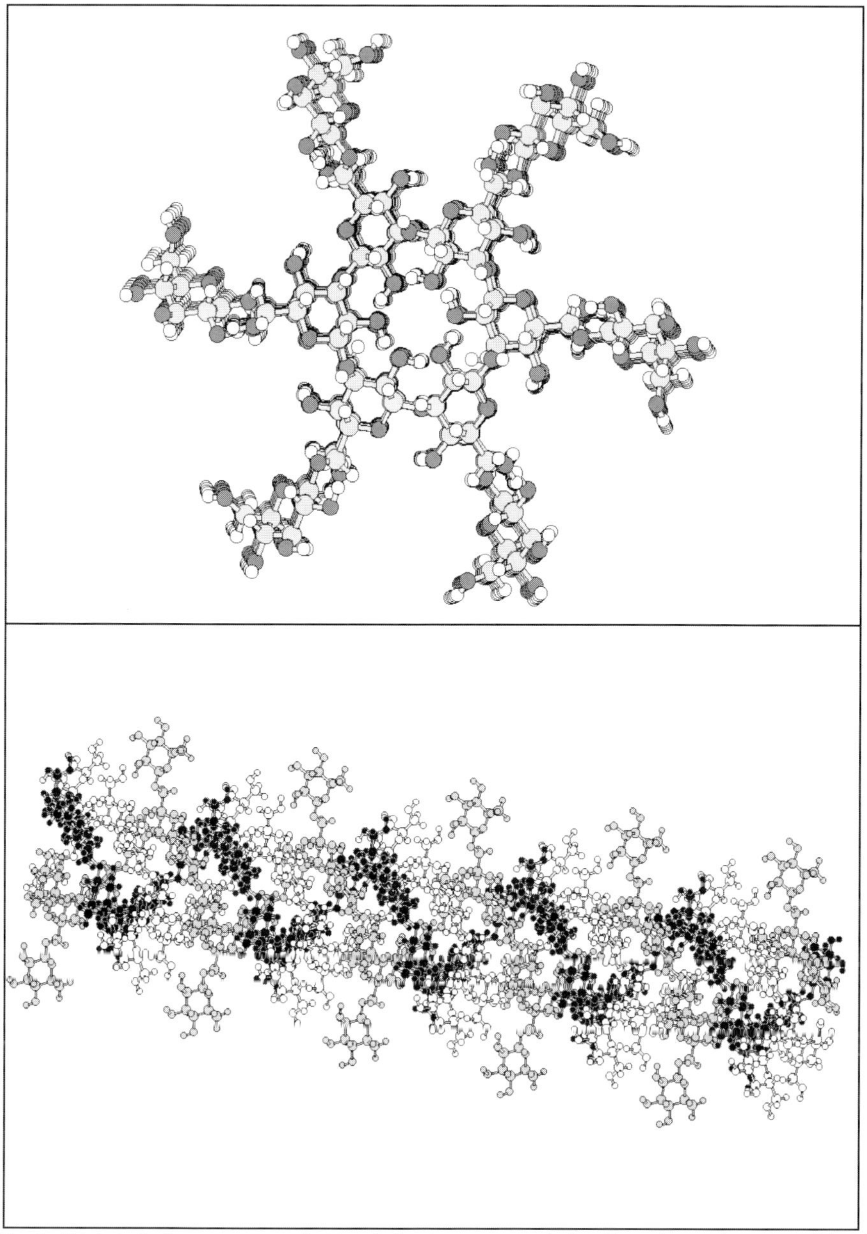

Fig. 3 Triple helical arrangement of schizophyllan in aqueous solution. Top: view to the center of the triplex. Bottom: side view.

solutions a trimer with triple helical structure is formed (Figure 3).

This structure (triple helix II) is stabilized by interchain hydrogen bonds at C-2 position with the β-1,6-glucose residues protruding outside the helix backbone. Using the data of Kashiwagi et al. (1981), it was calculated that there was a pitch of 0.3 ± 0.01 nm per glucose residue, a hydrodynamic diameter of 2.6 ± 0.4 nm, and a persistence length of 180 ± 30 nm for the triplex. Up to a molecular weight of 5×10^5 g mol^{-1}, the triple helix is almost perfectly rigid, underlined by a Mark-Houwink exponent of 1.49 (0.5 = random coil, 2 = rigid rod); however, the structure behaves like a semiflexible rod at increased molecular weights (Norisuye et al., 1980; Rau et al., 1992b).

At a midpoint temperature of 6 °C it is assumed that side-chain glucose residues slightly vary their positions, the result being a higher organized structure together with surrounding water molecules which form a helical chain at the outside of the triple helix. This ordered structure (triple helix I) surrounding the helical core increases the rigidity of the molecule and leads to gel formation (Asakawa et al., 1984; Van et al., 1984; Maeda et al., 1999; Sakurai and Shinkai, 2000).

High-sensitivity differential scanning calorimetry and optical rotatory dispersion (ORD) were used by Kitamura et al. (1996) to study the conformational transitions of schizophyllan in aqueous alkaline solution. They proposed a phase diagram for the conformation of schizophyllan as a function of temperature and pH. A more ordered structure (triple helix I) was attained at a temperature < 10 °C and pH < 11. This structure is stable up to 60 °C at pH 13. The triple helix II, attained at > 10 °C and pH < 11, converts to single coiled chains at > 135 °C or pH > 13 (Kitamura, 1989). The renatured samples at schizophyllan concentrations < 1 g L^{-1}, which were prepared by dissolution in 0.25 M KOH followed by neutralization with HCl, were observed as mixture of globular, linear and circular structures, and larger aggregates with less-defined morphology by electron microscopy. Subsequent annealing at 115–120 °C increases the proportion of circular species. This irreversibility of the triple helix reconversion is in sharp contrast to the investigations performed by Young and Jacobs (1998), who observed a slow (2–7 days) conversion of single helix to triple helix conformation. The renaturation took place between 39 and 84%, and was verified by fluorescence detection using aniline blue, which binds only to single-helix schizophyllan.

4 Chemical Analysis

Periodate cleaves oxidatively sugars with two vicinal hydroxyl groups. The β-1,3-backbone chain of schizophyllan does not carry adjacent hydroxyl groups and, therefore, will be not attacked. However, the side-chain glucose molecule is cleavable by 2 mol of periodate between C2–C3 and C3–C4 with simultaneous release of 1 mol of formic acid. Schizophyllan consumed between 0.52–0.55 mol periodate and released 0.22–0.26 mol formic acid per "anhydroglucose" unit, in accordance with the established structure (see Figure 1) of a 1,3-linked β-D-glucan with one β-D-glucopyranosyl group attached to position 6 of every third unit (Schulz and Rapp, 1991).

Partially oxidized schizophyllan is subsequently reduced by sodium borohydride (Smith degradation) and all nonacetalized aldehydes are transferred into hydroxy groups. Simultaneously, the acetals of the precleaved sugars are hydrolyzed and 0.23–0.25 mol glycerol is released per anhydroglucose unit. These procedures are described in detail elsewhere (Muenzer, 1989).

5 Occurrence and Physiological Function

The white rot basidiomycete *S. commune* belongs to the family *Aphyllophorales* (Donk, 1964) and forms "gilled" fruit-bodies (Figure 4). This fungus is a cosmopolitan organism that grows under extremely variable conditions throughout the temperature and tropic zones (Wessels, 1965). *S. commune* is frequently found on wood because it degrades lignin efficiently, but not cellulose, which is left as white fibers. Cooke (1961) reported that many native tribes used the fruit-bodies of *S. commune* as food or even as a type of chewing gum.

The water-insoluble portion of the hyphal wall consists of three types of polysaccharides: glucosoaminoglycans (chitin, chitosan and heteropolysaccharides containing both *N*-acetylglucosamine and glucosamine); the alkali-soluble *S*-glucan (α-1,3-glucan); and the alkali-insoluble or alkali-resistant *R*-glucan (β-1,3- and β-1,6-linked glucan), together comprising about 70% of the dry weight. The *R*-glucan is highly branched with β-1,3- and β-1,6-linkages; different structures may even be found in the same strain. The partial covalent linkage between the *R*-glucan and chitin is the reason for insolubility of the complex. Enzymatic or chemical degradation of the *R*-glucan chitin complex renders all of the β-glucan soluble in water or alkali (Sietsma and Wessels, 1979).

Schizophyllan is the water-soluble, extracellular glucan fraction excreted by the fungus that partially adheres to the walls as a jelly-like slime or mucilage (Figure 5). It is, therefore, a matter of opinion to consider schizophyllan a wall component, or not. The biological role of this gum is not yet well understood, though Wang and Miles (1964) assumed that it acts as a reserve carbon source because under glucose-limited conditions (Rau et al., 1990), β-glucan-degrading enzymes – the β-glucanases – are released (Rapp, 1992). The extracellular release of these enzymes underlies a glucose-induced repression. Wessels (1978) suggested that schizophyllan formation may be a result of unbalanced synthesis of *R*-glucan, or to a defect in the assembly of this component in the inner layer of the cell wall.

6 Biosynthesis

Selitrennikoff (1995) has provided an excellent review on the synthesis of β-glucans in

Fig. 4 Fruit-body of *Schizophyllum commune*. The characteristic lamellar structure is clearly recognizable.

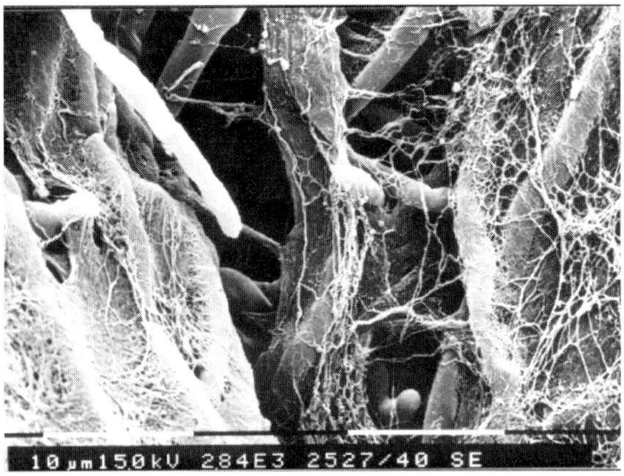

Fig. 5 Micrograph (×2800) of hyphae from *S. commune* covered with a mucilage of schizophyllan.

fungi. The 1,3-β-glucan synthase (E.C. 2.4.1.34; UDP-glucose: 1,3-β-D-glucan 3-β-D-glucosyltransferase) formed in the endoplasmic reticulum is transferred to the dictyosomes, which produce large apical vesicles (Ruiz-Herrera, 1991). The enzyme is transported by these vesicles to the hyphal tip, and is released at this location by fusing with the plasma membrane (Figure 6).

The β-glucan synthase is not active during transport, and only becomes active after plasma membrane–vesicle fusion. Enzyme activity does not require a divalent metal ion, and does not use a lipid-linked intermediate, as is known for prokaryotes (Sutherland, 1982). However, a proteinic acceptor molecule (Andaluz et al., 1986, 1988) has been identified. It is possible that this is the same as the GTP-binding protein which interacts with the enzyme and stimulates its activity (Mol et al., 1994). Interestingly, a primer is not required to induce enzyme activity.

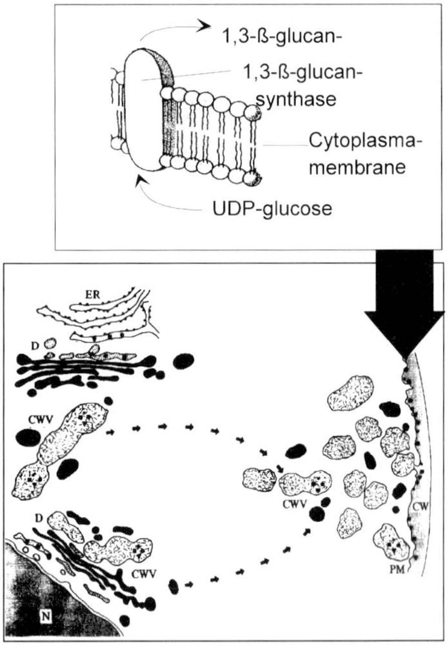

Fig. 6 Synthesis, transport and function of the β-glucan synthase. ER, endoplasmic reticulum; D, dictysome; N, nucleus; CWV, cell wall vesicle; PM, plasma membrane; CW, cell wall. (Adapted from Fèvre, 1979.)

The assay for 1,3-β-glucan synthase activity of *S. commune* was performed as follows (Kottuz and Rapp, 1990; Rau, 1997). Intensive mixing disrupted the micelles, and the enriched membrane fractions were isolated by high-speed centrifugation (34,000 r.p.m.) at 4 °C. The crude enzyme was dissolved in buffer (0.05 M Tris-HCl, pH 7.2, 1 mM EDTA, 7 mM cellobiose), and UDP-[U-^{14}C]-glucose was used as precursor. After incubation at 26 °C for 45 min, the unincorporated radioactive [^{14}C]-glucose was separated by filtration. The amount of radioactivity bound to the enzyme and retained on the filter was determined by scintillation counting.

The apparent K_m values ranged from 0.1 to 5 mM, depending on the source of the enzyme (Selitrennikoff, 1995). The crude enzyme is highly unstable, and has a half-life of only a few minutes at 25 °C. However, a substantial increase in stability, with only 20% loss of activity, is achieved during 19-day storage in buffer solution at low temperatures of –80 or –196 °C (Rau, 1997). Stimulation of the enzyme activity is also possible by the addition of nucleotide di- and triphosphates (Figure 7).

This behavior was also found for the glucan synthase of *Pyricularia oryzae* (Kominato et al., 1987), *Saccharomyces cerevisiae* (Shematec and Cabib, 1980), *Saprolegnia monoica* (Girard and Fèvre, 1984) and *Neurospora crassa* (Lourdes et al., 1995). Kang and Cabib (1986) showed that 1,3-β-glucan synthase is composed of multiple protein subunits, which can be separated by treatment of the membranes with salt and detergents. The soluble fraction contains at least one GTP-binding protein, while the membrane fraction retains the core catalytic center.

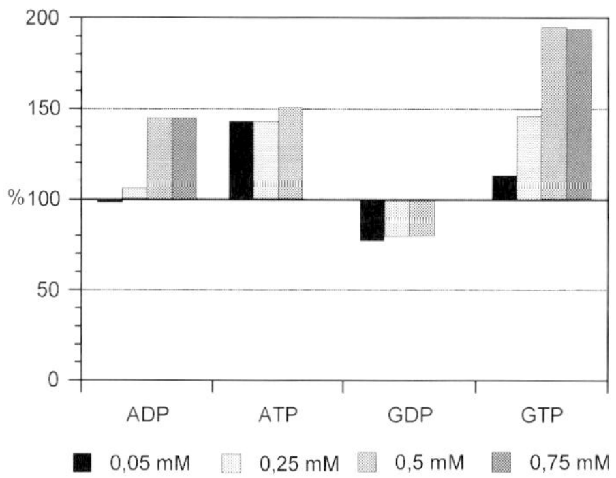

Fig. 7 Variation of 1,3-β-synthase activity by addition of nucleotide phosphates. Total (100%) activity is equal to a control assay without nucleotides (Rau, 1997).

These authors identified a 20-kDa GTP-binding protein that appeared to be the regulatory subunit, and a GTPase-activating protein that may regulate the GTP-binding protein (Mol et al., 1994).

β-Linked disaccharides such as cellobiose and laminaribiose also stimulate 1,3-β-glucan synthase activity (Wang and Bartnicki-Garcia, 1982). Quigley and Selitrennikoff (1984) found that disaccharides did not stimulate enzyme activity by acting as a primer, but rather interacted with the enzyme at a nonsubstrate site, thereby reducing the saturation constant for UDP-glucose.

The 1,3-β-glucan synthase operates as an integral membrane enzyme (Rau, 1997; Jabri et al., 1989) that uses UDP-Glc as substrate and vectorially synthesizes its β-1,3-glucan from the site of substrate hydrolysis (cytoplasmic facing) to the extracytoplasmic face of the plasma membrane (Jabri et al., 1989). Once external, the glucans are assembled into the cell wall by mechanisms that are not fully understood. The process that terminates the glucan polymerization process is also unknown. Chitin and 1,3-β-glucan become covalently cross-linked through lysine linkages (Sietsma and Wessels, 1979, 1981; Wessels and Sietsma, 1979, 1981), but the responsible enzyme(s) has (have) not been identified or assayed (Figure 8).

Different mechanisms have been proposed for the cell wall assembly, including: (1) a balance between synthesis and degradation of the polymers (Bartnicki-Garcia and Lippman, 1972); and (2) the steady-state model of Wessels (1988). These models were recently discussed by Bartnicki-Garcia (1999).

Sietsma et al. (1985) found, by using autoradiographic studies, that the alkali-insoluble β-glucan, which is linked with chitin and deposited at the apex in the extension zone, is mainly 1,3-β-linked and

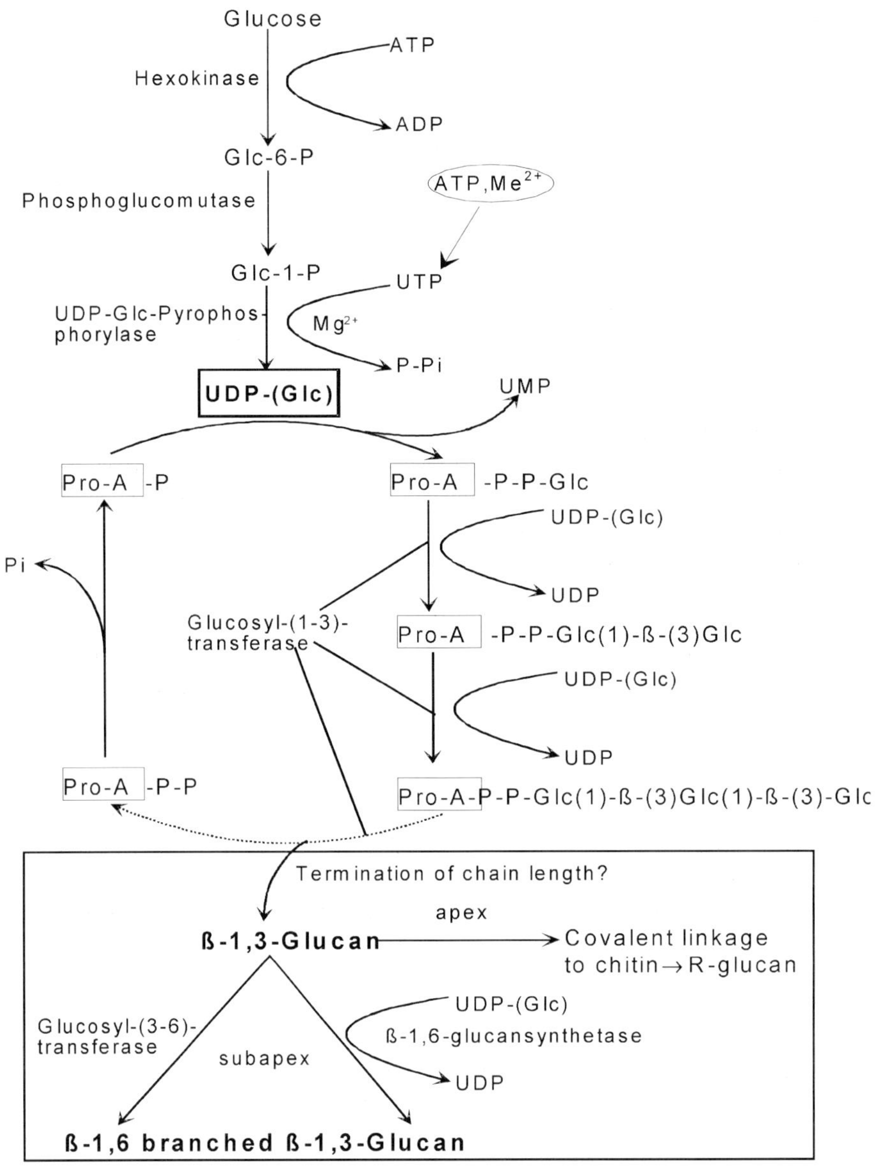

Fig. 8 Biosynthesis of schizophyllan. Pro-A, proteinic acceptor.

may be devoid of any 1,6-linkages at the extreme apex. This unbranched structure facilitates the formation of stable triple helices (Marchessault and Deslandes, 1980) from the chains that is required for cross-linking the chitin, and for subsequent microfibril formation. The mature wall contains very few pure 1,3-β-glucan chains. The 1,6-β-linked glucosyl residues occur in the extension zone, but form the majority in the subapically synthesized glucan. In the mature wall about half of the 1,3-β-glucan chains attached to chitin contain single glucose branches that are 1,6-β-linked to the main chain, while the other half also contains longer branches of 1,6-β-glucan chains. At least the presence of single glucose branches does not eliminate the tendency of the 1,3-β-glucan to form triple helices (Sato et al., 1981). These results do not permit conclusions to be made concerning the subapical synthesis of one or more glucans with variation in the 1,3-β- and 1,6-β-linkages that they contain. Thus, it seems probable that pure 1,3-β-glucans deposited in the cell wall are subsequently modified in subapical areas by attachment of 1,6-β-linked branches. Brown and Bussey (1993) proposed another concept, namely that the 1,6-β-glucans appear to be synthesized in the secretory pathway, transported to the plasma membrane, exocytosed, and then assembled into the cell wall. This is in sharp contrast to the procedure of 1,3-β-glucan synthesis mentioned earlier.

The characterization of 1,3-β-glucan synthase can be summarized as follows:

1) Enzyme activity is particulate, i.e., an integral membrane enzyme activity.
2) UDP-glucose is the preferred substrate.
3) The enzyme is not a zymogen, i.e., does not require proteolytic cleavage for *in vitro* activity.
4) The *in vitro* pH optimum is slightly basic (7.2).
5) The K_m value varies between 0.1 and 5 mM.
6) Enzyme activity can be stimulated by nucleoside phosphates and β-linked disaccharides.
7) A lipid-linked intermediate is not used.
8) The *in vitro* product is a linear 1,3-β-glucan with no detectable other linkages or branch points.

7
Molecular Genetics

S. commune has been used in many very different molecular genetics investigations, and it is not possible to cover all of these in this review. Hence, only a few examples will be described, including: the control of development (Marion et al., 1996; Shen et al., 1996; Lengeler and Kothe, 1999); the expression of heterologous genes (Schuren and Wessels, 1998; Scholtmeijer et al., 2001); the influence of the gene *FRT1* on homo- and dikaryotic fruiting (Horton et al., 1999); and the expression of the hydrophobin gene *SC3* (van Wetter et al., 2000; Wessels, 1999).

The 1,3-β-D-glucan synthase (EC 2.4.1.34) is a membrane enzyme which is activated by GTP and has been fractionated into soluble (GTP-binding) and membrane-bound (catalytic) components (Kang and Cabib, 1986; Mol et al., 1994). Genes encoding this enzyme have still not been isolated or sequenced in *S. commune*. However, this enzyme is essential for wall growth in yeasts and, therefore, related genetic investigations have been performed in other microorganisms, for example *S. cerevisiae* (Mazur and Baginski, 1996), *N. crassa* (Polizeli et al., 1995), and *Candida albicans* (Kondoh et al., 1997).

Two genes encode this multisubunit enzyme. *FKS1* (ORF: YLR342W) alias *GSC1*, *CND1*, *CWH53*, *ETG1* and *PBR1* encode a 215-kDa integral membrane protein (Fks1p) which mediates sensitivity to the echinocandin class of antifungal glucan synthase inhibitors and is a subunit of this enzyme. The residual glucan synthase activity present in *fks1* disruption mutants, the nonessential nature of the gene, and hybridization analysis of yeast chromosomal DNA pointed to the existence of a homologous gene encoding a functionally redundant product (Douglas et al., 1994). *FKS2* (ORF: YGR032W) alias *GSC2*, the homologue of *FKS1* encodes a 217-kDa also integral membrane protein (Fks2p) which at the amino acid level is 88% identical to Fks1p. The topological similarity of these proteins to many transporters suggests a possible role in transport of the growing glucan polymer across the membrane. Hydropathy profiles of Fks1p and Fks2p suggest that the genes encode integral membrane proteins which can be assumed to have approximately 16 transmembrane domains. The association of the *FKS* gene products with the catalytic activity suggests that these genes may encode catalytic subunits. The isolation of a neutralizing monoclonal antibody provides additional evidence that these subunits may be catalytic. However, the possibility of other glucan synthase subunits being present in the membrane fraction or the solubilized enzyme cannot be ruled out (Mazur et al., 1995). Disruption of each gene was not lethal, but disruption of both genes was lethal (Inoue et al., 1995).

An additional protein is necessary for activation of the 1,3-β-D-glucan synthase (Qadota et al., 1996). The gene *RHO1* (ORF: YPR165W) encodes a 22-kDa GTPase (Rho1p). The deduced amino acid sequence predicts that *C. albicans* Rho1p is 82.9% identical to *Saccharomyces* Rho1p and contains all the domains conserved among Rho-type GTPases from other organisms (Kondoh et al., 1997). *C. albicans* Rho1p was shown to interact directly with *C. albicans* Gsc1p in a ligand overlay assay and a cross-linking study. These results indicate that *C. albicans* Rho1p acts in the same manner as *S. cerevisiae* Rho1p to regulate as subunit the β-1,3-glucan synthesis (Arellano et al., 1996; Mazur and Baginsky, 1996).

8 Biodegradation

Schizophyllan is degradable by β-glucanases, which are secreted by *S. commune* when glucose is consumed as the carbon source. 1,3-β-glucanases are widely distributed among filamentous fungi (Fèvre, 1979; Prokop, 1990; Bielecki and Galas, 1991; Sutherland, 1999), and biodegradation of schizophyllan in soil, for example after use in enhanced oil recovery, is therefore possible.

Lo et al. (1990) described the kinetics and specificities of two closely related β-glucosidases secreted by *S. commune* that had similar molecular weights (102 and 96 kDa) and which were competitively inhibited by their glucose product. Both enzymes had binding sites for three glucose residues.

Chiu and Tzean (1995) investigated glucanolytic enzyme production by *S. commune* Fr. during mycoparasitism, and showed that the extracellular endo-1,3(4)-β-glucanase attacked 16 out of 50 fungi, representing oomycetes, zygomycetes and hyphomycetes, and which are either saprophytic, soilborne or foliar plant pathogens.

The production of extracellular β-1,3-glucanase activity by a monokaryotic *S. commune* strain was monitored by Prokop et al. (1994). The results showed that the β-glucanase activity consisted of an endo-β-1,3-glucanase activity, in addition to a negligible amount of β-1,6-glucanase and β-glucosidase activities. Unlike the β-1,3-glucanase production of the dikaryotic parent strain *S. commune* ATCC 38548, the β-1,3-glucanase formation of the monokaryon was not regulated by catabolite repression. The latter enzyme was purified from the culture filtrate by lyophilization, anion exchange chromatography on Mono Q, and gel filtration on Sephacryl S-100. It appeared homogeneous on SDS–PAGE, with a molecular mass of 35.5 kDa and an isoelectric point of

3.95. The enzyme was only active toward glucans containing β-1,3-linkages, including lichenan, a β-1,3/β-1,4-D-glucan. It attacked laminarin in an endo-like manner to form laminaribiose, laminaritriose, and high oligosaccharides. Whilst the extracellular β-glucanases from the dikaryotic *S. commune* ATCC 38548 degraded significant amounts of schizophyllan, the endo-β-1,3-glucanase from the monokaryon showed greatly reduced activity toward this high molecular mass β-1,3-/β-1,6-glucan, the K_m (using laminarin as substrate) being 0.28 mg mL^{-1}. The optimal pH was 5.5, and optimal temperature 50 °C; the enzyme was stable between pH 5.5 and 7.0.

9 Production

A wide range of carbon sources sustain both growth and schizophyllan formation. Simple mono- or disaccharides, such as glucose, xylose, mannose, sucrose, maltose and galactose, containing lactose, are utilized as well as polysaccharides such as starch or xylan (Steiner et al., 1993; Rau, 1997). In the following examples of cultivation only glucose was used as the carbon source, however.

9.1 Batch Cultivation

The production of schizophyllan is strongly coupled with growth and, as known for primary metabolites, secretion under nitrogen starvation reduces to zero when the stationary phase is reached. The medium for cultivation of *S. commune* is simple and requires only four components: glucose 30 g L^{-1}, yeast extract 3 g L^{-1}, KH_2PO_4 1 g L^{-1} and $MgSO_4.7H_2O$ 0.5 g L^{-1}.

If a filamentously growing fungus is used to produce highly viscous pseudoplastic polysaccharide suspensions, not only short mixing times and high mass transfer but also shear stress (which depends on the type of impeller used and stirrer speed influencing physiology of the fungus) must be considered. In other words, in proportion to the agitator used, the impeller must present a compromise between micro- and macromixing and schizophyllan release from the cell wall on the one hand, and low shear stress for the fungus and glucan on the other hand. Furthermore, the impeller should enable easy construction for a subsequent scale-up. Therefore, various types of impellers (Intermig™, Fundaspin™, helicon-ribbon, Rushton-turbine, fan) were tested in relation to their power requirement as well as yield, productivity and quality (expressed as specific shear viscosity) of the formed schizophyllan (Rau et al., 1992b; Gura and Rau, 1993). As a result of these investigations, the four-bladed fan impeller with a width ratio (impeller to vessel diameter) of 0.64 and a blade pitch of 45 ° gave the best product in relation to the highest specific shear viscosity (193 mPa s; shear rate 0.3 s^{-1}, 25 °C, 0.3 g L^{-1} aqueous schizophyllan solution) and highest yield at low formation of biomass (Figure 9).

The variation of intrinsic viscosity and molecular weight of schizophyllan during a cultivation run is shown in Figure 10.

The increase of these molecular data may be explained by premature release of mainly shorter schizophyllan molecules from the cell wall due to a low solution viscosity at the beginning of cultivation. The decrease at the end of cultivation is solely related to degradation by shear stress caused by the impeller and/or attack by released β-glucanases. Corresponding with intrinsic and shear viscosity, the flow behavior index shows a minimum of 0.18, i.e., a maximum of pseudoplasticity (data not shown).

As can be seen from Figure 9, most schizophyllan is produced during the oxy-

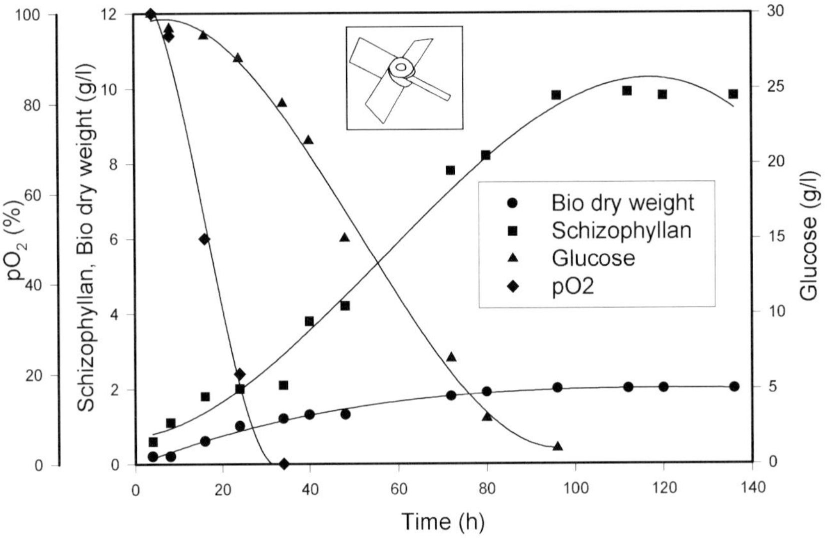

Fig. 9 A 30-L batch cultivation of *S. commune* equipped with three fan impellers at 100 r.p.m., 27 °C, an initial pH of 5.3 and an aeration rate of 150 L h^{-1}. pO_2, oxygen partial pressure of the liquid phase.

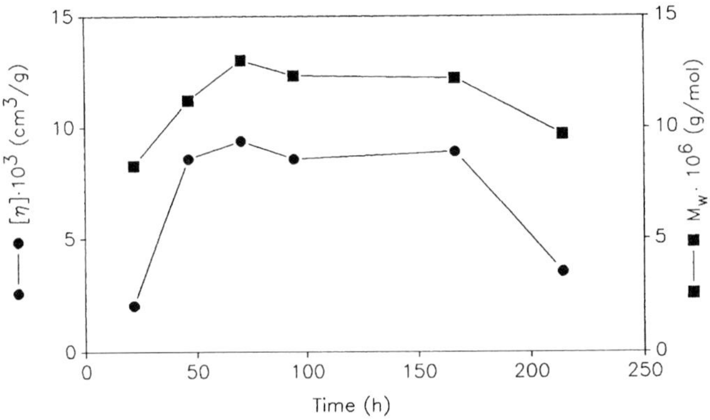

Fig. 10 A 50-L bioreactor cultivation of *S. commune*. Cell-free samples were used for the determination of molecular weight M_w (low angle light scattering) and intrinsic viscosity [η] (Rau et al., 1990).

gen-limited phase between 30 and 100 h. During this time interval, the partial pressure of oxygen in the liquid phase is almost zero and the additional formation of ethanol occurred (Figure 11). *S. commune* always produces ethanol as a result of a redundant fermentative pathway. Partly existing anaerobic domains inside the mycelia can never be fully avoided, as seen by the fact that ethanol is always formed in spite of a pO_2 $>5\%$ throughout the whole cultivation run. However, the pO_2 is not a representative quantity in this connection as it reflects only one local point of the liquid phase and not the situation inside the micelles. Otherwise, local excess of oxygen is primarily used for biomass formation. This characteristic indicates that the fungus requires an optimum oxygen supply for a maximum of schizophyllan secretion. This fact is proved by the determination of an optimum constant specific oxygen uptake rate which depends

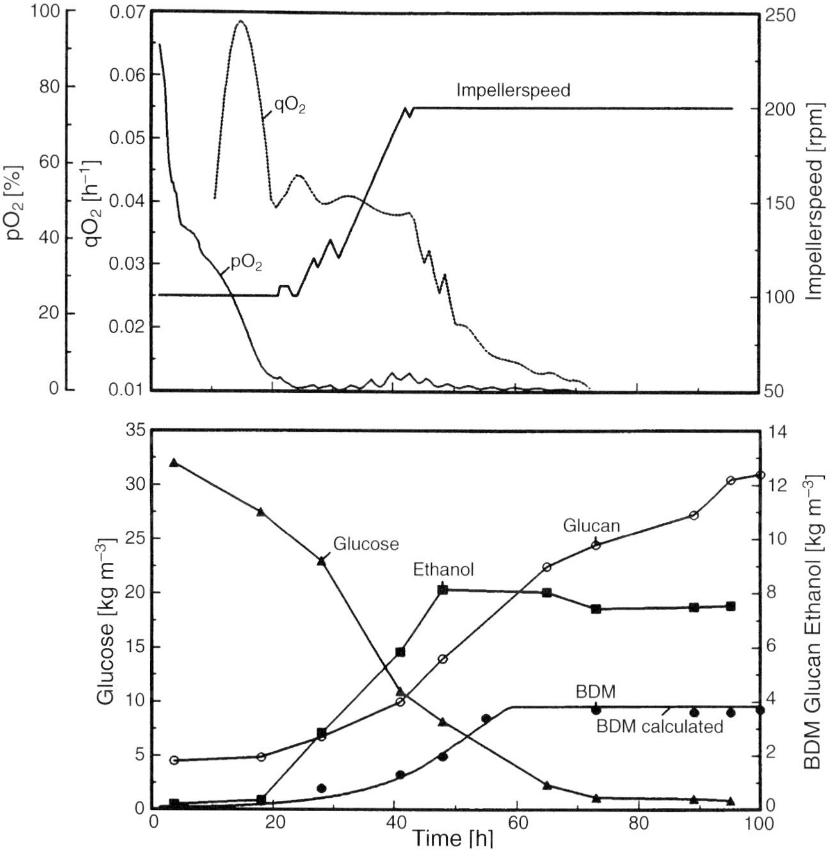

Fig. 11 A 30-L-batch cultivation of *S. commune* with controlled specific oxygen uptake rate (qO_2) by variation of impeller speed (fan impeller). For further cultivation conditions, see Figure 9.

on the process conditions (Rau et al., 1992d). Application of a constant specific oxygen uptake rate (oxygen consumed per time [g $L^{-1} h^{-1}$] divided by the actual biomass [g L^{-1}]) by controlling the stirrer speed resulted in an increased yield of schizophyllan (13 g L^{-1}) at equal biomass formed (see Figure 11). The on-line modeling of biomass for the calculation of the specific oxygen uptake rate was achieved by using the following equation (Rau and Brandt 1994):

$$\frac{dC_X}{dt} = k_L a \left(C^*_{O_2,L} - C_{O_2,L} \right) \cdot Y_{X/O_2} \cdot \frac{C_S}{K_S + C_S} \quad (1)$$

A further important result is the decreased amount of ethanol (from 12 to 7 g L^{-1}) that was channeled primarily into schizophyllan as a result of the improved oxygen supply (Rau and Brandt, 1994).

9.2 Continuous Cultivation

S. commune underlies a glucose-induced repression of the formation of β-glucanases (Rapp, 1992; Prokop et al., 1994), and therefore cultivations were terminated when glucose consumption was complete. Prolonged cultivation under carbon-limited conditions led to the release of β-glucan-degrading enzymes which caused a slight increase in glucose concentration, accompanied by a decrease in schizophyllan concentration as well as a sharp drop in the specific viscosity (mPa g^{-1}). Furthermore, Figure 11 shows that an increase of schizophyllan formation is possible by using a controlled oxygen supply. Independent of the process mode (batch or continuous), an optimum specific oxygen uptake rate exists for maximum yield and productivity of schizophyllan as well as minimum biomass formation because the biomass must be separated in further downstream processing. This characteristic was proven by variation of the oxygen supply during continuous cultivations in an oxygen-limited chemostat (Rau, 1997) (Table 1).

The addition of oxygen to the air feed enhanced growth and the associated schizophyllan formation until their maxima were reached. The specific oxygen uptake rate also increased with biomass, but specific schizophyllan productivity decreased. In spite of increased oxygen supply, all cultivations were oxygen-limited and therefore the various specific oxygen uptake rates reflected different states of oxygen limitation. Related to primary mycelial (homogeneous) growth of the fungus and using the same bioreactor set-up described in Figures 9 and 11, then the optimum specific oxygen uptake rate ranged between 0.04 and 0.06 h^{-1}. However, if another bioreactor or process configuration is used this specific value can change drastically and must be determined individually.

Compared with batch cultivation, the continuous mode revealed an increase in productivity (Rau et al., 1992d). The schematic set-up of the continuous process is shown in Figure 12.

In order to achieve a further increase of productivity combined with facilitated downstream processing, biomass feedback was used. A cross-flow filtration unit comprising a stainless steel membrane (200 μm) was employed for separation of biomass from the viscous culture suspension. For schizophyllan, a maximum productivity of 40 g L^{-1} per day was achieved at a feedback rate (permeate flow/medium input flow) of 0.92 and dilution rate of 0.2 h^{-1} (maximum specific growth rate 0.12 h^{-1}). Optimized process and filtration conditions resulted in a near cell-free and undiluted β-glucan solution at the outlet of the bioreactor (Figure 13).

10 Downstream Processing

An economic design for the downstream process of highly viscous and mycelia-con-

Tab. 1 Influence of oxygen supply on growth and formation of schizophyllan by addition of N_2 or O_2 to the air-feed during oxygen-limited continuous cultivations of *S. commune*. The 30-L bioreactor was equipped with three fan impellers (150 r.p.m.). All concentrations are related to stationary conditions. Dilution rate 0.04 h^{-1}, constant aeration rate 144 L h^{-1}. Note: all cultivations showed a $pO_2 \approx 0\%$.

$\%N_2$	40	10	–	–	–	–
%Air	60	90	100	90	70	30
$\%O_2$	–	–	–	10	30	70
Biomass [g L^{-1}]	0.65	0.9	1	1.2	1.7	2.4
Schizophyllan [g L^{-1}]	3.3	3.5	3.6	4.1	4.6	3.7
Productivity [g L^{-1} day^{-1}]	3.2	3.4	3.5	3.9	4.4	3.6
Specific productivity [g g^{-1} day^{-1}]	4.9	3.8	3.5	3.2	2.6	1.5
qO_2 [h^{-1}]	0.041	0.047	0.049	0.056	0.061	0.071

Specific producitivity: g schizophyllan formed per g biomass per day.qO_2: specific oxygen uptake rate.

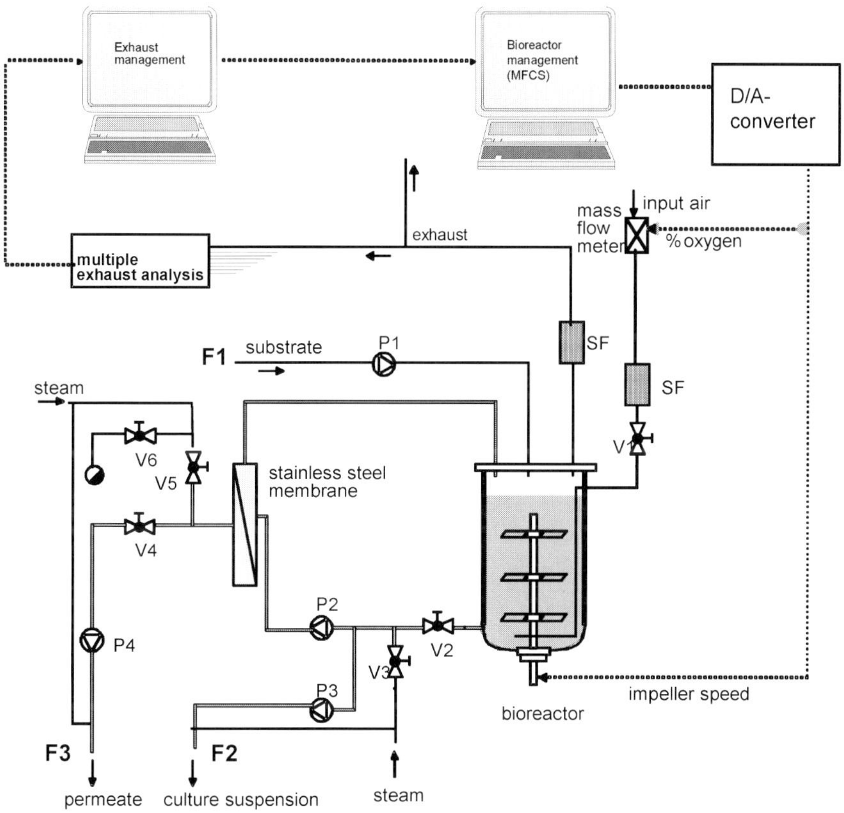

Fig. 12 Set-up of the continuous schizophyllan process including cell feedback. F1, medium input flow; F2, suspension output flow; F3, permeate; F1 = F2 + F3; feedback rate R = F3/F1; SF, sterile filter; V, valve; p, pump; MFCS, micro fermenter control system (B. Braun Biotech International; now Sartorius AG, Göttingen, Germany.)

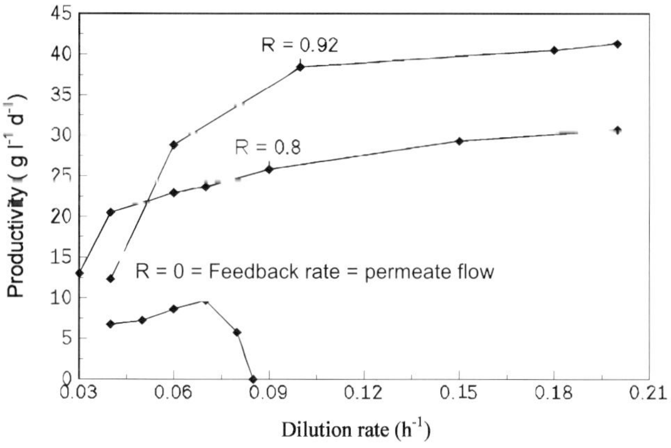

Fig. 13 Influence of feedback rate on the productivity of schizophyllan during 30-L oxygen-limited continuous cultivations with *S. commune*. Substrates (g L^{-1}) input feed: glucose 30; yeast extract 0.8; KH_2PO_4 0.2; $MgSO_4$ 0.1. For further cultivation conditions, see Figure 9.

taining culture suspensions is a challenge for the bioengineer. A three-stage, dead-end filtration (Jahn-Held et al., 1990) works in principle for cell separation, but this procedure is not recommended for scale-up. Therefore, centrifugation or high-speed cross-flow microfiltration were chosen which were also applicable for higher volumes. Again, microfiltration showed the best performance for purification and concentration of cell-free schizophyllan solutions.

10.1 Cell Separation

The suspensions harvested from batch cultivations contain cells that must be separated by either centrifugation or microfiltration. The best results with centrifugation are obtained when the diluted (≤ 1 g L^{-1} schizophyllan, ≤ 0.2 g L^{-1} biomass) and homogenized suspension is fed to a solid ejecting disc separator (e.g., CSA-1, 5700 r.p.m.; Westfalia, Oelde, Germany). The resulting supernatant contains only small amounts of hyphal fragments (concentration < 0.1 g L^{-1}), and this can easily be separated by dead-end filtration using glass-fiber filters.

A more effective alternative method for cell separation is cross-flow microfiltration. An undiluted suspension can be used if a sintered stainless steel membrane (10 µm; Krebsoege, Radevormwald, Germany) is used at high tangential feed velocity. Cell-free schizophyllan solutions without fragments, but with the same concentration as at the end of batch cultivation, are obtained as the permeate (Haarstrick et al., 1991; Haarstrick, 1992; Rau, 1997, 1999).

10.2 Purification and Concentration of Schizophyllan

The cell-free schizophyllan solution obtained either by high-speed microfiltration or by continuous cultivation with integrated biomass feedback must be purified (diafiltration mode) or eventually further concentrated (concentration mode) by using a cross-flow microfiltration technique (Figure 14).

Parallel investigations using different cross-flow systems (Haarstrick et al., 1991) led to the recommended use of low-shear PROSTAK™ (Millipore Corp., USA) flat membrane modules (0.1 µm). The best results related to a high permeation rate were achieved when the tangential feed velocity was at its individual maximum, avoiding a transmembrane pressure > 0.8 bar. Purification of the solution was attained by using the diafiltration mode when all schizophyllan molecules were fully rejected and low molecular-weight compounds (< 0.1 µm) such as proteins, glucose and salts, permeate the membrane. The permeate flow corresponds to the input solvent (water) flow, so that the volume of the retentate remains constant. The concentration mode was started by cutting the input solvent flow. During this process the negative influence of fouling at the membrane surface was increased, with the consequence of a continual decrease in permeation rate (Figure 15). This results in a highly viscous, colorless and transparent solution. Drying or lyophilization of the product solution must be avoided because only 50% (w/w) of the dried schizophyllan can be redissolved in water. Dimethylsulfoxide (DMSO) can be used to dissolve the dried schizophyllan totally, but this solvent degrades the triple helix to single coiled chains, with a drastic reduction in viscosity.

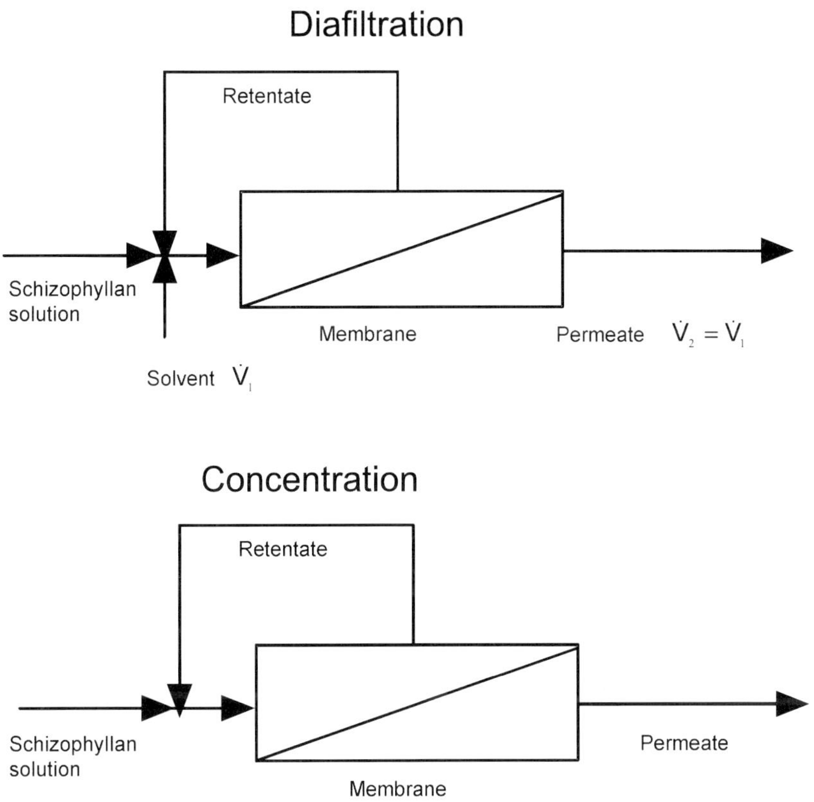

Fig. 14 Scheme of cross-flow microfiltration for the purification (diafiltration) and concentration of cell-free schizophyllan solutions. $\dot{V}$ = volume feed, L h^{-1}.

11 World Market

Schizophyllan is not the only trivial name for the exopolysaccharide of *S. commune*; elsewhere (primarily in Asiatic countries) it is also known as sizofilan and sizofiran. To the best of the author's knowledge, only two Japanese companies currently manufacture schizophyllan: Kaken Pharmaceutical Co., Ltd., Tokyo (Anonymous, 2001a); and Taito Co., Ltd., Kobe (Anonymous, 2001b).

Kaken offers schizophyllan/sizofiran under the brand name Sonifilan™ as an anti-malignant tumor agent. Each ampoule (2 mL) contains 20 mg sizofiran with a molecular weight of approx. 450,000 daltons that should be injected intramuscularly in one to two divided doses each week. Its use is indicated for the enhancement of the direct effect of radiotherapy in the treatment of uterine cervical carcinoma. In 1998, Sonifilan was licensed to Kwang Dong in Korea. Information concerning production facilities and prices are not currently available.

12 Properties

In aqueous solution, schizophyllan is arranged as a triple helix with protruding

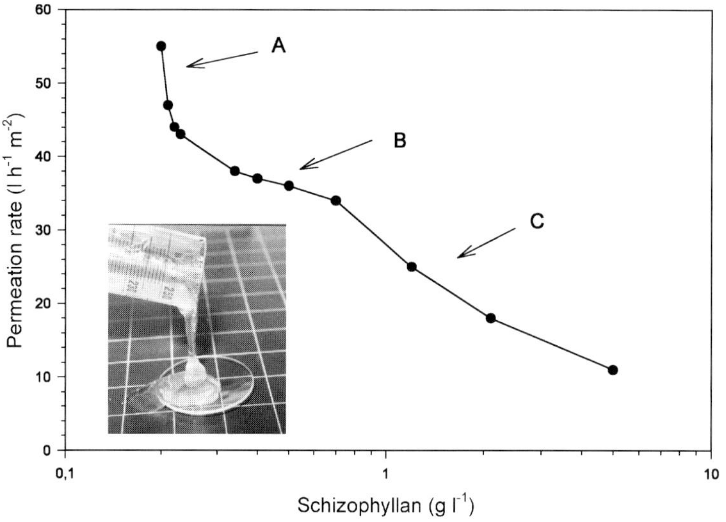

Fig. 15 Decrease of the permeation rate during the concentration run of a cell-free schizophyllan solution applying cross-flow microfiltration. Pore size 0.1 μm, transmembrane pressure 0.8 bar, tangential feed velocity 6.6 m s^{-1} (Rau, 1999). A = adsorption with increasing concentration polarization; B = steady-state of mass transport from and to membrane; C = gel-layer formation (fouling).

pendent β-1,6-linked D-glucose units originating from the outside of the triplex. In DMSO, at temperatures > 135 °C and at a pH > 12, the triple helix melts to single coiled strains, equivalent to reducing the average molecular weight by one-third (Norisuye et al., 1980). Thermal degradation is enhanced by the presence of oxygen (Zentz et al., 1992). Aqueous solutions show thixotropic, pseudoplastic (Figure 16) and viscoelastic behavior (Oertel and Kulicke, 1991). Native suspensions, additionally containing the producing fungus, reveal enhanced non-Newtonian characteristics due to the filamentous network of the internal woven hyphae.

The viscosity is decreased with increasing shear rate. Due to this flow behavior an individual viscosity is strongly connected to a single shear rate and characterizes the quality of schizophyllan because when used as a viscosifier, a high viscosity at low concentration is required. Furthermore, the shear viscosity depends on the concentration of the schizophyllan. For example, Figure 16 shows the flow behavior of a solution with 5 g L^{-1} schizophyllan. The comparison of different solutions > 5 g L^{-1} yielded viscosities > 10 Pa s at low shear rate (0.3 s^{-1}). The mean value of a 0.3 g L^{-1} solution (shear rate 0.3 s^{-1}) varies between 50 and 150 mPa s, depending on the quality of the schizophyllan.

Shear and intrinsic viscosity is slightly increased by the addition of NaCl or Mg^{2+} and Ca^{2+}ions. This is an unusual behavior, because synthetic polymers such as polyacrylamide show a slight decrease in intrinsic viscosity when salt concentration is increased, this being due to reduced solvent quality (coil contraction). An explanation for the increase in schizophyllan volume could be enhanced energetic interactions inside the triple helix in a poorer solvent. Additional intramolecular forces increase the stiffness of the triple helix and induce an expansion of the macromolecule (Rau et al.,

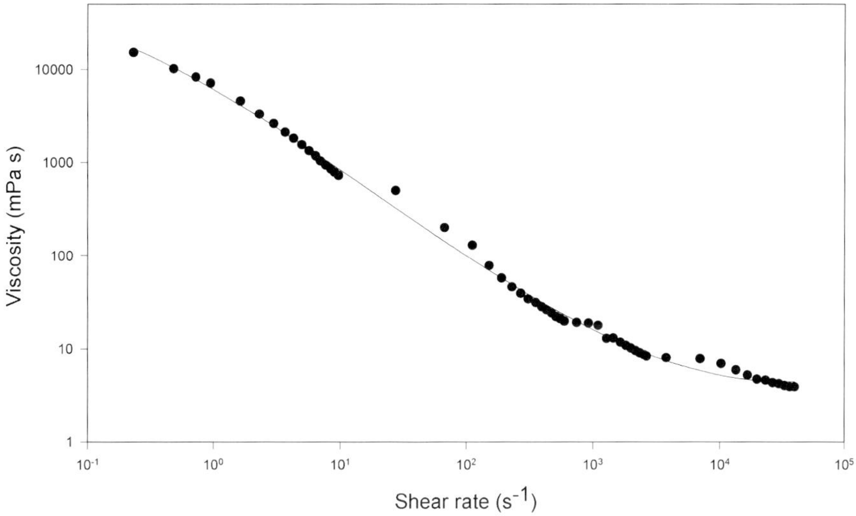

Fig. 16 Pseudoplastic flow behavior of an aqueous schizophyllan solution (5 g L^{-1}). Shear viscosity was measured by a rotary viscometer (Haake, Karlsruhe, Germany) at 25 °C and at different constant shear rates until a constant shear stress resulted (Rau, 1999).

1990). Although schizophyllan undergoes no gelation by itself in aqueous solution, the addition of sorbitol results in the formation of a transparent gel at lower temperatures. Based on the results of small-angle X-ray scattering, sorbitol is found to disentangle a part of triple-stranded helices and bridge the disentangled parts, which serve as a cross-linking domain (Maeda et al., 1999). Schizophyllan also forms gels in the presence of borate ions (Rau et al., 1992c; Grisel and Muller, 1997). The gelation occurs by both hydrogen bonding and chelation of borate ions through the hydroxyl groups of the biopolymer as far as they are in favorable position, e.g. in the side chain.

Sakurai and Shinkai (2000) found that a mixture of schizophyllan and poly(ethylene oxide) in aqueous solution underwent phase separation at around 3–4 °C, and this temperature was independent of both polymer concentration and the difference in poly(ethylene oxide) molecular weight (6000 and 70,000).

Oertel and Kulicke (1991) described the formation of aqueous, lyotropic phases of ultrasonically degraded (MW 335,000 g mol^{-1}) schizophyllan investigated by stationary shear flow, as well as with the aid of polarization microscopy. In oscillatory measurements, schizophyllan exhibited a maximum for the storage modulus. The ORD of liquid crystal solutions (MW 400,000 g mol^{-1}) was also investigated by Van et al. (1984). An abrupt change in ORD behavior occurred when an isotropic solution was cooled to a temperature close to the isotropic–biphasic boundary temperature, indicating the occurrence of a pretransition from isotropic to cholesteric phases. Yanaki (1982) claimed the cholesteric liquid crystal formation of schizophyllan. Furthermore, the rotational dynamics of schizophyllan was determined by transient electric birefringence (Fuglestad et al., 1996).

Schizophyllan possesses no thermoplastic characteristics; it has no specific melting point, but it decomposes at temperatures

> 180 °C. The moisture content of the dried product was ~10% (w/w), this being bound as water of crystallization (Schulz, 1992), while the density was 1429 kg m^{-3} (Creszenzi and Gamani, 1988).

Acetylation of the polymer yielding schizophyllan acetate can be carried out under strong acidic catalysis; the resultant product is only slightly soluble in water or DMSO (Albrecht and Rau, 1994).

Muenzberg et al. (1995) found that hydrolysis of aqueous schizophyllan solutions was possible by incubation in DURAN™ borosilicate glass 3.3 (DIN ISO 3585) at 121 °C and 1 bar. A slight decrease in pH, a rapid loss of viscosity, and a constant increase in reducing end-groups indicated that the degradation of schizophyllan proceeded regioselectively by cleaving only the main chain, although normally the protruding 1,6-linkages would be expected to be less stable against hydrolytic influences. Maintenance of the side chains was additionally verified by ^{13}C-NMR spectra. Stepwise ultrafiltration of degraded solutions yielded fractions with varying molar masses, with the mass ratio of fractions depending on the total incubation time. The regioselectivity of this degradation method was explained by a pore theory.

13 Applications and Patents

Claims exist for the production of schizophyllan using different strains of *S. commune*, and for its application as an additive for polymer flooding in the field of enhanced oil recovery (Lindoerfer et al., 1988, 1991). Due to the viscosity-related stability of schizophyllan against high shear rates, increased temperatures and high salinities, this material is very useful in this type of application (Rau et al., 1992c).

Aqueous schizophyllan solutions were shown effectively to reduce drag in pipe flow, the drag reduction depending on schizophyllan concentration and being minimal at 0.5 g L^{-1}. The transition from laminar to turbulent flow was delayed at a higher Reynold's (*Re*) value (Haarstrick and Rau, 1993).

Schulz et al. (1992) described the preparation of films with native schizophyllan and with a polyalcohol derived from schizophyllan by chemical treatment. The films could only be prepared by casting from aqueous solutions because the polymers were not thermoplastic. They possess a low permeability to oxygen (schizophyllan: < 2 mL m^{-2} day^{-1} bar^{-1}), but present a high permeability to water vapor and hence can be used to protect foods against oxygen-mediated spoilage. The tensile strength of the films was 45–58 N mm^{-2} for schizophyllan, and 12–18 N mm^{-2} for the polyalcohol.

Besides schizophyllan, *S. commune* also secretes a 24-kDa hydrophobin. Hydrophobins were initially discovered in the mid-1980s and represent a unique family of small, amphipathic proteins (about 100 amino acids) that play an important role in forming stable coatings on various surfaces. Enzymatic digestion of the hydrophobin eliminates the ability of the remaining schizophyllan to assemble as a stable entity on a hydrophobic surface. By using water contact angle measurements and atomic force microscopy, Martin et al. (1999) showed that schizophyllan and hydrophobin form a synergistic complex that allows facile surface modification of both hydrophilic and hydrophobic surfaces.

The tetrasaccharide subunit (see Figure 1) can be prepared by chemical degradation of schizophyllan, and is useful as oligomerization building block (Kunz et al., 1991). Schizophyllan, in combination with colloidal silica particles and water-soluble salts, can be used for the fine polishing of wafers (Sasaki, 1990) and as a membranous material in terms of a polysaccharide associate after first dissolving it in DMSO (random coil) with subsequent addition of water or methanol to return the polysaccharide to a state capable of assuming the helical structure (Yanaki, 1981). The production of a monoclonal antibody that reacts specifically with a β-1,3-glycosidic bond and could be used to determine small amounts of schizophyllan has been described (Hirata, 1992; Hirata et al., 1993). A polyclonal antibody was also used in a sandwich-type enzyme immunoassay (Adachi et al., 1999) for quantifying schizophyllan, as well as for estimating its ultrastructure (triple/single helix).

Due to its shear-thinning characteristics combined with a high specific viscosity, schizophyllan can be generally used as viscosifier, e.g., in cremes and lotions. Within the scope of special cosmetic applications, schizophyllan is also useful as a skin anti-aging, depigmenting and healing agent – generally termed 'skin care' (Kim et al., 1999). Schizophyllan is an active ingredient that can increase skin cell proliferation, collagen biosynthesis and also aid in recovery after sunburn. In addition, schizophyllan effectively reduces skin irritation and is therefore suitable for cosmetic and dermatological applications (Kim et al., 1999, 2000).

To date, a vast amount of data has been generated in relation to the use of schizophyllan as a pharmaceutical compound (for reviews, see Bohn and BeMiller, 1995; Kraus, 1990). Schizophyllan acts as biological response modifier and has been known to be a nonspecific stimulator of the immune system, generally as a result of macrophage activation. In 1991, Czop and Kay identified a macrophage cell surface receptor that is specific for a small oligosaccharide with β-1,3-D-glucose linkages. It is generally known that a glucan with β-1,3-linkages has greater macrophage stimulatory activity than any other linkage type. The receptor binding in a human monocyte-like cell line was confirmed specifically for schizophyllan by Mueller et al. (2000).

Schizophyllan has been used as an immunotherapeutic agent for cancer treatment in Japan since 1986. It is used in conjunction with chemotherapy or radiotherapy. The additional application of schizophyllan to radiation increased both macrophage and T-lymphocyte infiltration in local lung tumor cells in mice (Inomata et al., 1996). An overdose of schizophyllan must be avoided however as this reduces the antitumor activity (Miura et al., 2000). Clinical studies have shown that the administration of schizophyllan, along with antineoplastic drugs, prolongs the lives of patients with lung, gastric or uterine cancers (Furue, 1987; Yoshio et al., 1992; Kimura et al., 1994). Schizophyllan has no antigenicity or mitogenic effect on the T cell; its antitumor activity is exerted only through the activation of macrophages, which subsequently augment the T-cell cascade (Kidd, 2000; Ooi and Lui, 2000). However, an evaluation of schizophyllan in patients with advanced head and neck squamous cell carcinoma did not reveal any significant improvements in immunological parameters (Mantovani et al., 1997).

With regard to its immune-stimulating activities, schizophyllan was claimed as anti-(AIDS) virus agent (Shigero et al., 1989; Hagiwara and Kikuchi, 1992), as prevention against fish diseases (Yano, 1990), and as an immune effect enhancer for vaccines (Honma, 1994). An ultrasonicated schizophyllan, neoschizophyllan, with novel pharmacological activity was described by Kikumoto et al. (1978).

It has been shown that the molar mass, the degree of branching, conformation and chemical modification significantly affect the pharmaceutical activity of schizophyllan.

However, it is difficult to establish a uniform structure–functional activity relationship because reported results have differed widely. Structural features such as β-1,3-linkages in the main chain and β-1,6-branch points (branching degree 0.2–0.33) are needed because loss of the side-chain glucose reduced both antitumor action (Kishida, 1992; Suda et al., 1994) and water solubility.

Ooi and Lui (2000) stated that glucans with high molecular weight appear to be more effective than those with low molecular weight. They regarded glucans with a molecular mass of 800,000 g mol^{-1} to be maximally effective, though very small schizophyllan molecules also showed some efficacy. Investigations with native schizophyllan ($6-10 \times 10^6$ g mol^{-1}) showed that it is likely to be effective against attack by tobacco mosaic virus (Michiko, 1989; Stuebler and Buchenauer, 1996; Rau, 1997) in tobacco leaves (*Nicotiana tabacum*) as a regioselectively degraded (Muenzberg et al., 1995) and purified molar mass fraction (1000–5000 g mol^{-1}) of schizophyllan.

The conformation of schizophyllan also appears to influence the immunological activity, though this point is controversial. Native schizophyllan is arranged as triple helix, but molecules of molecular mass $<50{,}000$ g mol^{-1} (Kojima et al., 1984) or schizophyllan treated with NaOH show single-helix conformation. Kulicke et al. (1997) presented results, based on a Congo red assay, that helical structures are not essential – nor even advantageous – for immunological activity. Saito et al. (1991) also found that single-helix conformation stimulates antitumor activity. In contrast, Ooi and Lui (2000) stated that single-helix schizophyllan showed a reduced ability to inhibit tumor growth as compared with the native material. The truth will lie somewhere between these situations because, depending on the immunological assay used, either the triple- or single-helical conformer showed increased activities (Miura et al., 1995; Ohno et al., 1995; Tsuzuki et al., 1999).

Table 2 provides a summary of patents relating to the numerous applications of branched β-glucans.

14
Outlook and Perspectives

Using batch or continuous cultivations, schizophyllan can be produced from glucose in high yields (13 g L^{-1}) and high productivities (40 g L^{-1} per day). The downstream processing has been improved and can easily be carried out using high-speed, cross-flow filtration techniques without producing solvent waste that is harmful to the environment. In relation to these data, and depending on the scale of production, an economically realistic selling price for schizophyllan of ≤ 50 DM per kg should be possible.

In addition, depending on the political and economical situations, a prospective intensified use in enhanced oil recovery is conceivable. Should the application of schizophyllan as a bulk product prove unattractive however, its pharmaceutical properties might provide sufficient stimulus for its production, and hopefully industries outside of Japan will add schizophyllan to their product lists.

Tab. 2 Summary of patents related to schizophyllan, in order of date of publication

No. of patent	*Patent holder*	*Inventors*	*Title of patent*	*Date of publication*
US 4098661	Taito Co., Ltd. (JP)	Kikumoto et al.	Method of producing neoschizophyllan having novel pharmalogical activity	1978-07-04
JP 56127603	Taito Co., Ltd. (JP)	Yanaki	Preparation of polysaccharide associate	1981-10-06
JP 57147576	Taito Co., Ltd. (JP)	Yanaki	Novel liquid crystal of polysaccharide	1982-09-11
EP 0271907	Wintershall AG (GER)	Lindoerfer et al.	Homopolysaccharides with a high molecular weight, process for their extracellular preparation and their use, as well as the corresponding fungus strains	1988-06-22
JP 1272509	JAPAN TOBACCO INC.	Michiko	Production of controlling agent of plant virus by microorganism	1989-10-31
JP 1287031	Taito Co., Ltd. (JP)	Shigero et al.	Anti-AIDS virus agent	1989-11-23
US 4921615	Wintershall AG (GER)	Jahn-Held et al.	Separation of solid particles of various sizes from viscous liquids	1990-05-01
EP 0373501	Mitsubishi Monsanto Chem (JP)	Sasaki	Fine polishing composition for wafers	1990-06-20
JP 2218615	Taito Co., Ltd. (JP)	Yano	Preventive for fish desease comprising water-soluble glucan	1990-08-31
DE 4012 238	Wintershall AG (GER)	Lindoerfer et al.	Verfahren zur Erhöhung der volumenbezogenen Produktivität (g/ld) nichtionischer Biopolymere	1991-01-03
EP 0416396	Merck Patent GmbH (GER)	Kunz et al.	Tetrasaccharide and process for their preparation	1991-03-13
JP 4054124	Taito Co., Ltd. (JP)	Hagiwara et al.	Anti-virus agent	1992-01-02
JP 4346791	Taito Co., Ltd. (JP)	Hirata	Monoclonal antibody	1992-12-02
JP 6172217	Taito Co., Ltd. (JP)	Honma	Immune effect enhancer for vaccine	1994-06-21
JP 11313667	Pacific Co., Ltd. (KR)	Kim et al.	Production of beta-1,6-branched beta-1,3-glucan useful as skin anti-aging, depigmenting and healing agent	1999-10-07

15 References

Adachi, Y., Miura, N. N., Ohno, N., Tamura, H., Tanaka, S., Yadomae, T. (1999) Enzyme immunoassay system for estimating the ultrastructure of (1,6)-branched (1,3)-β-glucans, *Carbohydr. Polym.* **39**, 225–229.

Albrecht, A., Rau, U. (1994) Acetylation of β-1,6-branched β-1,3-glucan yielding schizophyllan-acetate, *Carbohydr. Polym.* **24**, 193–197.

Andaluz, E., Guillen, A., Larriba, G. (1986) Preliminary evidence for a glucan acceptor in the yeast *Candida albicans, Biochem. J.* **240**, 495–502.

Andaluz, E., Ridruejo, J. C., Ramirez, M., Ruiz-Herrera, J., Larriba, G. (1988) Initiation of glucan synthesis in yeast, *FEMS Microbiol. Lett.* **49**, 251–255.

Anonymous (2001a).

Anonymous (2001b).

Arellano, M., Duran, A., Perez, P. (1996) Rho1 GTPase activates 1–3-β-D-glucan synthase and is involved in *Schizosaccharomyces pombe* morphogenesis, *EMBO J.* **15**, 4584–4591.

Asakawa, T., Van, K., Teramoto, A. (1984) A thermal transition in a cholesteric liquid crystal of aqueous schizophyllan, *Mol. Cryst. Liq. Cryst.* **116**, 129–139.

Bartnicki-Garcia, S. (1999) Glucans, walls, and morphogenesis: on the contributions of J. G. H. Wessels to the golden decades of fungal physiology and beyond. *Fungal Genet. Biol.* **27**, 119–127.

Bartnicki-Garcia, S., Lippman, E. (1972) The bursting tendency of hyphal tips of fungi: presumptive evidence for a delicate balance between wall synthesis and wall lysis in apical growth, *J. Gen. Microbiol.* **73**, 487–500.

Bielecki, S., Galas, E. (1991) Microbial β-glucanases different from cellulases, *Crit. Rev. Biotechnol.* **10**, 275–304.

Bohn, J. A., BeMiller, J. A. (1995) 1,3-β-D-glucans as biological response modifiers: a review of structure-functional activity relationships, *Carbohydr. Polym.* **28**, 3–14.

Brandt, C. (1995). *O_2-geregelte β-Glucanproduktion mit Schizophyllum commune* ATCC 38548 *im Batch- und Chemostatbetrieb*, PhD Thesis, Technical University Braunschweig, GER.

Brandt, C., Schilling, B., Gura, E., Rau, U., Wagner, F. (1993) Definierte Sauerstoffversorgung im Batch und im O_2-limitierten Chemostaten mit Zellrückführung zur Produktion von β-1,3-Glucanen. *DECHEMA Biotechnology Conferences* **2**, 465–466.

Brown, J., Bussey, H. (1993) The yeast KRE9 gene encodes an O-glycoprotein involved in cell surface β-glucan assembly, *Mol. Cell. Biol.* **13**, 6346–6356.

Chiu, S. C., Tzean, S. S. (1995) Glucanolytic enzyme production by *Schizophyllum commune* Fr. during mycoparasitism, *Physiol. Mol. Plant Path.* **46**, 83–94.

Cooke, W. B. (1961). The genus *Schizophyllum, Mycologia* **53**, 575–599.

Cordes, K. (1990) Produktionsoptimierung und Charakterisierung der von *Monilinia fructigena* ATCC 24976 und ATCC 26106 gebildeten extrazellulären Glucane, PhD Thesis, Technical University of Braunschweig, GER.

Cordes, K., Rau, U., Wagner, F. (1989) Influence of processing parameters and downstream processing on viscometric data of aqueous glucan solutions, *DECHEMA Biotechnology Conference* **3**, 1067–1070.

Crescenzi, V., Gamini, A. (1988) On the solid state and solution conformations of a polycarboxylate derived from the polysaccharide scleroglucan, *Carbohydr. Polym.* **9**, 169–184.

Czop, J. K., Kay, J. (1991) Isolation and characterization of β-glucan receptors on human mononuclear phagocytes, *J. Exp. Med.***173**, 1511–1520.

Donk, M. A. (1964) A conspectus of the families of Aphyllophorales, *Persoonia* **3**, 199–324.

Douglas, C. M., Foor, F., Marrinan, J. A., Morin, N., Nielsen, J. B., Dahl, A. M., Mazur, P., Baginsky, W., Li, W., El-Sherbeini, M., Clemas, J. A., Mandala, S. M., Frommer, E. R., Kurtz, M. B. (1994) The *Saccharomyces cerevisiae FKS1* (*ETG1*) gene encodes an integral membrane protein which is a subunit of 1,3-β-D-glucan synthase, *Proc. Natl. Acad. Sci. USA* **91**, 12907–12911.

Essig, F. M. (1922) The morphology, development and economic aspects of *Schizophyllum commune* Fries, *Am. J. Bot.* **36**, 360–363.

Fèvre, M. (1979). Glucanases, glucan synthases and wall growth in *Saprolegnia monoica*, in: *Fungal Walls and Hyphal Growth* (Burnett, J. H., Trinci, A. P. J., Eds.), Cambridge: Cambridge University Press, 225–263.

Fuglestad, G. A., Mikkelsen, A., Elgsaeter, A., Stokke, B. T. (1996) Transient electric birefringence study of rod-like triple-helical polysaccharide schizophyllan, *Carbohydr. Polym.* **29**, 277–283.

Furue, H. (1987) Biological characteristics and clinical effect of schizophyllan, *Drugs Today* **23**, 335–346.

Gawronski, M., Conrad, H., Springer, T., Stahmann, K.-P. (1996) Conformational changes of the polysaccharide cinerean in different solvents from scattering methods, *Macromolecules* **24**, 7820–7825.

Girard, V., Fèvre, M. (1984) Distribution of (1-3)-β- and (1-4)-β-glucan synthases along the hyphae of *Saprolegnia monoica*, *J. Gen. Microbiol.* **130**, 1557–1562.

Grisel, M., Muller, G. (1997) The salt effect over the physical interactions occurring for schizophyllan in the presence of borate ions, *Macromol. Symp.* **114**, 127–132.

Gura, E., Rau U. (1993) Comparison of agitators for the production of branched β-1,3-D-glucans by *Schizophyllum commune*, *J. Biotechnol.* **27**, 193–201.

Haarstrick, A. (1992). Mechanische Trennverfahren zur Gewinnung zellfreier, hochviskoser Polysaccharidlösungen von *Schizophyllum commune ATCC 38548*, PhD Thesis, Technical University Braunschweig.

Haarstrick, A., Rau, U. (1993) Strömungscharakteristik pseudoplastischer Polysaccharidlösungen von *Schizophyllum commune* ATCC 38548, *Chem. Ing. Tech.* **65**, 556–559.

Haarstrick, A., Rau, U., Wagner, F. (1991) Cross-flow filtration as a method of separating fungal cells and purifying the polysaccharide produced, *Bioprocess Eng.* **6**, 179–186.

Hagiwara, K., Kikuchi, M. (1992) Anti-virus agent, JP 4054124.

Haltrich, D., Steiner, W. (1994) Formation of xylanase by *Schizophyllum commune*: effect of medium components, *Enzyme Microb. Technol.* **16**, 229–235.

Hirata, A. (1992) Monoclonal antibody, JP 4346791.

Hirata, A., Itoh, W., Tabata, K., Kojima, T., Itoyama, S., Sugawara, I. (1993) Preparation and characterization of murine anti-schizophyllan monoclonal antibody, *Biosci. Biotech. Biochem.* **57**, 125–126.

Honma, M. (1994) Immune effect enhancer for vaccine, JP 6172217.

Horton, J. S., Palmer, G. E., Smith, W. J. (1999) Regulation of dikaryon-expressed genes by *FRT1* in the basidiomycete *Schizophyllum commune*, *Fungal Genet. Biol.* **26**, 33–47.

Inomata, T., Goodman, G. B., Fryer, C. J., Chaplin, D. J., Palcic, B., Lam, G. K., Nishioka, A., Ogawa, Y. (1996) Immune reaction induced by X-rays and ions and its stimulation by schizophyllan (SPG), *Br. J. Cancer Suppl.* **27**, 122–125.

Inoue, S. B., Takewaki, N., Takasuka, T., Mio, T., Adachi, M., Fujii, Y., Miyamoto, C., Arisawa, M., Furuichi, Y., Watanabe, T. (1995) Characterization and gene cloning of 1,3-beta-D-glucan synthase from *Saccharomyces cerevisiae*, *Eur. J. Biochem.* **231**, 845–854.

Jabri, E., Quigley, D. R., Alders, M., Hrmova, M., Taft, C. S., Phelps, P., Selitrennikoff, C. P. (1989) (1-3)-β-Glucan synthesis of *Neurospora crassa*, *Curr. Microbiol.* **19**, 153–161.

Jahn-Held, W., Lindoerfer, W., Sewe, K.-U., Wagner, F., Ziebolz, B. (1990) Separation of solid particles of various sizes from viscous liquids, US Patent No. 4921615.

Kang, M., Cabib, R. (1986) Regulation of fungal cell wall growth: a guanine nucleotide-binding, proteinaceous component required for activity of 1,3-β-D-glucan synthase, *Proc. Natl. Acad. Sci. USA* **83**, 5808–5812.

Kashiwagi, Y., Norisuye, T., Fujita, H. (1981) Triple helix of *Schizophyllum commune* polysaccharide in dilute solution: light scattering and viscosity in dilute aqueous sodium hydroxide, *Macromolecules* **14**, 1220–1225.

Kidd, P. M. (2000) The use of mushroom glucans and proteoglycans in cancer treatment, *Altern. Med. Rev.* **5**, 4–27.

Kikumoto, S., Miyajima, T., Yoshizumi, S., Fujimoto, S., Kimura, K. (1970) Polysaccharide produced by *Schizophyllum commune*. I. Formation and some properties of an extracellular polysaccharide. *Nippon Nogei Kagaku Kaishi* **44**, 337–342.

Kikumoto, S., Miyajima, T., Kimura, K., Okubo, S., Komatsu, N. (1971) Polysaccharide produced by *Schizophyllum commune*. II. Chemical structure of an extracellular polysaccharide, *Nippon Nogei Kagaku Kaishi* **45**, 162–168.

Kikumoto, S., Yamamoto, O., Komatsu, N., Kobayashi, H., Kamasuka, T. (1978) Method of producing neoschizophyllan having novel pharmacological activity, US Patent No. 4098661.

Kim, J. S., Kim, M. S., Lee, D. C., Lee, S. G., So, S., Kim, Y. T., Park, B. H., Park, K. M. (1999) Production of beta-1,6-branched beta-1,3-glucan useful as skin anti-aging, depigmenting and healing agent, JP 11313667.

Kim, M.-S., Park, K. M., Chang, I.-S., Kang, H.-H., Sim, Y.-C. (2000) β-1,6-branched β-1,3-glucans in skin care, *Allured's Cosmetic & Toiletries Magazine* **115**, 79–86.

Kimura, Y., Tojima, H., Fukase, S., Takeda, K. (1994) Clinical evaluation of sizofilan as assistant immunotherapy, *Otolaryngology* (Stockholm) **511**, 192–195.

Kishida, E., Yoshiaki, S., Misaki, A. (1992) Effects of branch distribution and chemicals modifications of antitumor (1-3)-β-D-glucans, *Carbohydr. Polym.* **17**, 89–95.

Kitamura, S. (1989) A differential scanning calorimetric study of the conformational transitions of schizophyllan in mixtures of water and dimethylsulfoxide, *Biopolymers* **28**, 639–654.

Kitamura, S., Hirano, T., Takeo, K., Fukada, H., Takahashi, K., Falch, B. H., Stokke, B. T. (1996) Conformational transitions of schizophyllan in aqueous alkaline solutions, *Biopolymers* **39**, 407–416.

Kojima, T., Tabaka, K., Ikumoto, T., Yanaki, T. (1984) Depolymerization of schizophyllan by controlled hydrodynamic shear, *Agric. Biol. Chem.* **48**, 915–921.

Kominato, M., Kamimiy, S., Tanake, H. (1987) Preparation and properties of β-glucan synthase of *Pyricularia oryzae* P_2, *Agric. Biol. Chem.* **51**, 755–761.

Kondoh, O., Tachibana, Y., Ohya, Y., Arisawa, M., Watanabe, T. (1997) Cloning of the *RHO1* gene from *Candida albicans* and its regulation of beta-1,3-glucan synthesis, *J. Bacteriol.* **179**, 7734–7741.

Kottutz, E., Rapp, P. (1990) 1,3-β-glucan synthase in cell-free extracts from mycelium and protoplasts of *Sclerotium glucanicum*, *J. Gen. Microbiol.* **136**, 1517–1523.

Kraus, J. (1990) Biopolymere mit antitumoraler und immunmodulierender Wirkung, *Pharmazie in unserer Zeit* **19**, 157–164.

Kulicke, W. M., Lettau, A. I., Thielking, H. (1997) Correlation between immunological activity, molar mass, and molecular structure of different 1,3-β-D-glucans, *Carbohydr. Res.* **297**, 135–143.

Kunz, H., Klinkhammer, U., Kinzy, W., Neumann, S., Radunz, H.-E. (1991) Tetrasaccharide and process for their preparation, EP 0416396.

Lengeler, K. B., Kothe, E. (1999) Identification and characterization of *brt1*, a gene down-regulated during B-regulated development in *Schizophyllum commune*, *Curr. Genet.* **35**, 551–556.

Lindoerfer, W., Sewe, K.-U., Wagner, F., Münzer, S., Nachtwey, S., Rapp, P., Rau, U., Stephan D. (1988) Homopolysaccharides with a high molecular weight, process for their extracellular preparation and their use, as well as the corresponding fungus strains, EP 0 271 907.

Lindoerfer, W., Sewe, K.-D., Wagner, F., Münzer, S., Rau, U., Veuskens, J. (1991) Verfahren zur Erhöhung der volumenbezogenen Produktivität (g/ld) nichtionischer Biopolymere, DE- 40 12 238.

Lo, A. C., Barbier, J.-R., Willick, G. E. (1990). Kinetics and specificities of two closely related β-glucosidases secreted by *Schizophyllum commune*, *Eur. J. Biochem.* **192**, 175–181.

Lourdes, M., Polizeli, T. M., Noventa-Jordao, M. A., DaSilva, M. M., Jorge, J. A., Terenzi, H. F. (1995) 1,3-β-D-glucan synthase activity in mycelial and cell wall-less phenotypes of the fz, sg, os-1 ("slime") mutant strain of *Neurospora crassa*, *Exp. Mycol.* **19**, 35–47.

Maeda, H., Yuguchi, Y, Kitamura, S., Urakawa, H., Kajiwara, K., Richtering, W., Fuchs, T., Burchard, W. (1999) Structural aspects of gelation in schizophyllan/sorbitol aqueous solution, *Polymer J.* **31**, 530–534.

Mantovani, G., Bianchi, A., Curreli, L., Ghiani, M., Astara, G., Lampis, B., Santona, M. C., Dessi, D., Esu, S., Lai, P., Massa, E., Maccio, A., Proto, E. (1997) Clinical and immunological evaluation of schizophyllan (SPG) in combination with standard chemotherapy in patients with head and neck squamous cell carcinoma, *Int. J. Oncol.* **10**, 213–221.

Marchessault, R. H., Deslandes, Y. (1980). Texture and crystal structures of fungal polysaccharides, in: *Fungal Polysaccharides. A.C.S. Symposium series* (Sandford, P. A., Matsuda, K., Eds.), Washington, DC: American Chemical Society, 221–250, vol. 126.

Marion, A. L., Bartholomew, K. A., Wu, J., Yang, H., Novotny, C. P., Ullrich, R. C. (1996) The Aα mating type locus of *Schizophyllum commune*: structure and function of gene *X*, *Curr. Genet.* **29**, 143–149.

Martin, G. G., Cannon, G. C., McCormick, C. L. (1999) Adsorption of a fungal hydrophobin onto surfaces as mediated by the associated polysaccharide schizophyllan, *Biopolymers* **49**, 621–633.

Mazur, P., Baginsky, W. (1996) In vitro activity of 1,3-β-glucan synthase requires the GTP-binding protein *RHO1*, *J. Biol. Chem.* **271**, 14604–14609.

Mazur, P., Moring, N., Baginsky, W., El-Sherbeini, M., Clemas, J. A., Nielsen, J. B., Foor, F. (1995) Differential expression and function of two homologous subunits of yeast 1,3-β-D-glucan synthase, *Mol. Cell. Biol.* **15**, 5671–5681.

Michiko, A. (1989) Production of controlling agent of plant virus by microorganism, JP 1272509.

Miura, N. N., Ohno, N., Adachi, Y., Aketagawa, J., Tamura, H., Tanaka, S., Yadomae, T. (1995) Comparison of the blood clearance of triple- and single-helical schizophyllan in mice, *Biol. Pharm. Bull.* **18**, 185–189.

Miura, T., Miura, N. N., Ohno, N., Adachi, Y., Shimada, S., Yadomae, T. (2000) Failure in antitumor activity by overdose of an immunomodulating beta-glucan preparation, sonifilan, *Biol. Pharm. Bull.* **23**, 249–253.

Mol, P. C., Park, H. M., Mullins, J. T., Cabib, E. (1994) A GTP-binding protein regulates the activity of 1,3-beta-glucan synthase, an enzyme directly involved in yeast cell wall morphogenesis, *J. Biol. Chem.* **269**, 31267–31274.

Mueller, A., Raptis, J., Rice, P. J., Kalbfleisch, J. H., Stout, R. D., Ensley, H. E., Browder, W., Williams, D. L. (2000) The influence of glucan polymer structure and solution conformation on binding to (1 → 3)-beta-D-glucan receptors in a human monocyte-like cell line, *Glycobiology* **10**, 339–346.

Muenzberg, J., Rau, U., Wagner, F. (1995) Investigations to the regioselective hydrolysis of a branched β-1,3-glucan, *Carbohydr. Polym.* **27**, 271–276.

Muenzer, S. (1989) Produktion und Charakterisierung eines von *S. commune ATCC 38548* gebildeten extrazellulären β-1,3-Glucans, PhD Thesis, Technical University Braunschweig.

Norisuye, T., Yanaki, T., Fujita, H. (1980) Triple helix of a *Schizophyllum commune* polysaccharide in aqueous solution, *J. Polym. Sci. Polym. Phys.* **18**, 547–558.

Oertel, R., Kulicke, W.-M. (1991) Viscoelastic properties of liquid crystals of aqueous biopolymer solutions, *Rheol. Acta* **30**, 140–150.

Ohno, N., Miura, N. N., Chiba, N., Adachi, Y., Yadomae, T. (1995) Comparison of the immunopharmacological activities of triple and single-helical schizophyllan in mice, *Biol. Pharm. Bull.* **18**, 1242–1247.

Ooi, V. E., Liu, F. (2000) Immunomodulation and anti-cancer activity of polysaccharide-protein complexes, *Curr. Med. Chem.* **7**, 715–729

Pilz, F., Auling, G., Rau, U., Stephan, D., Wagner, F. (1991) A high affinity Zn^{2+} uptake system and oxygen supply control growth and biosyntheses of an extracellular, branched β-1,3-glucan in *Sclerotium rolfsii* ATCC 15205, *Exp. Mycol.* **15**, 181–192.

Polizeli, M. T. M., Noventa, J. M. A., Silva, M. M., Jorge, J. A., Terenzi, H. F. (1995) (1,3)-beta-D-glucan synthase activity in mycelial and cell wall-less phenotypes of the *fz, sg, os-1* ("Slime") mutant strain of *Neurospora crassa*, *Exp. Mycol.* **19**, 35–47.

Prokop, A. (1990) Protoplastenmutagenese und Protoplastenfusion von Schizophyllum commune: Einfluβ auf die Synthese von β-1,3-Glucanen sowie Reinigung und Charakterisierung einer Endo-β-1,3-Glucanase, PhD Thesis, Technical University Braunschweig.

Prokop, A. Rapp, P. Wagner, F. (1994) Production, purification, and characterization of an extracellular endo-beta-1,3-glucanase from a monokaryon of *Schizophyllum commune* ATCC 38548 defective in exo-beta-1,3-glucanase formation, *Can. J. Microbiol.* **40**, 18–23.

Qadota, H., Python, C. P., Inoue, S. B., Arisawa, M., Anraku, Y., Zheng, Y., Watanabe, T., Levin, D. E., Ohya, Y. (1996) Identification of yeast Rho1p GTPase as a regulatory subunit of 1,3-beta-glucan synthase, *Science* **272**, 279–281.

Quigley, D. R., Selitrennikoff, C. P. (1984) β(1-3)Glucan synthase activity of *Neurospora crassa*: stabilization and partial characterization, *Exp. Mycol.* **8**, 202–214.

Rapp, P. (1992) Formation, separation and characterization of three β-1,3-glucanases from *Sclerotium glucanicum*, *Biochim. Biophys. Acta* **1117**, 7–14.

Rau, U. (1997) Biosynthese, Produktion und Eigenschaften von extrazellulären Pilz-Glucanen. Aachen, Germany: Shaker Verlag.

Rau, U. (1999) Production of schizophyllan, in: *Methods in Biotechnology – Carbohydrate Biotechnology Protocols* (Bucke, C., Ed.), Totowa, NJ, USA: Humana Press, Inc., 43–57, vol. 10.

Rau, U., Brandt, C. (1994) Oxygen controlled batch cultivation of *Schizophyllum commune* for enhanced production of branched β-1,3-glucans, *Bioprocess Eng.* **11**, 161–165.

Rau, U., Wagner, F. (1987) Non-Newtonian flow behaviour of colloid-disperse glucan solutions, *Biotechnol. Lett.* **9**, 95–100.

Rau, U., Gura, E., Schliephaake, A., Wagner, F. (1989) Influence of processing on the formation of exopolysaccharides by filamentously growing fungi, *DECHEMA Biotechnology Conferences* **3**, 571–574.

Rau, U., Müller, R.-J., Cordes, K., Klein, J. (1990) Process and molecular data of branched 1,3-β-D-glucans in comparison with xanthan, *Bioprocess Eng.* **5**, 89–93.

Rau, U., Gura, E., Haarstrick, A. (1992a) Prozess-integrierte Aufarbeitung verzweigter β-1,3-Glucane (schizophyllan), *GIT* **12**, 1233–1238.

Rau, U., Gura, E., Olszewski, E., Wagner, F. (1992b) Enhanced glucan formation of filamentous fungi by effective mixing, oxygen limitation and fed-batch processing, *J. Ind. Microbiol.* **9**, 19–26.

Rau, U., Haarstrick, A., Wagner, F. (1992c) Eignung von Schizophyllanlösungen zum Polymerfluten von Lagerstätten mit hoher Temperatur und Salinität, *Chem.-Ing.-Tech.* **64**, 576–577.

Rau, U., Olszewski, E., Wagner, F. (1992d) Gesteigerte Produktion von verzweigten β-1,3-Glucanen mit *Schizophyllum commune* durch Sauerstofflimitierung, *GIT* **4**, 331–337.

Rinaudo, M., Vincendon, M. (1982) ^{13}C-NMR structural investigation of scleroglucan, *Carbohyd. Polym.* **2**, 135–144.

Ruiz-Herrera, J. (1991) Biosynthesis of β-glucans in fungi, *Antonie van Leeuwenhoek* **60**, 73–81.

Saito, H., Ohki, T., Sasaki, T. (1979) A 13C-Nuclear magnetic resonance study of polysaccharide gels, *Carbohydr. Res.* **74**, 227–240.

Saito, H., Yoshioka, Y., Uehara, N. (1991) Relationship between conformation and biological response for (1-3)-β-D-glucans in the activation of coagulation Factor G from limulus amebocyte lysate and host-mediated antitumor activity. Demonstration of single-helix conformation as a stimulant, *Carbohydr. Res.* **217**, 181–190.

Sakurai, K., Shinkai, S. (2000) Phase separation in the mixture of schizophyllan and poly(ethylene oxide) in aqueous solution driven by a specific interaction between the glucose side chain and poly(ethylene oxide), *Carbohydr. Res.* **324**, 136–140.

Sasaki, S. K. J. (1990) Fine polishing composition for wafers, EP 0373501.

Sato, T., Norisuye, T., Fukita, H. (1981) Melting behaviour of *Schizophyllum commune* polysaccharides in mixtures of water and dimethylsulfoxide, *Carbohydr. Res.* **95**, 195–204.

Scholtmeijer, K., Wösten, H. A. B., Springer, J., Wessels, J. G. H. (2001) Effect of introns and AT-rich sequences on expression of the bacterial hygromycin B resistance gene in the basidiomycete *Schizophyllum commune, Appl. Environ. Microbiol.* **67**, 481–483.

Schulz, D. (1992) Untersuchungen zur Folien- und Gelbildung von natürlichen und modifizierten β-1,3-Glucanen, PhD Thesis, Technical University Braunschweig, Germany.

Schulz, D., Rapp, P. (1991) Properties of the polyalcohol prepared from the β-D-glucan schizophyllan per periodate oxidation and borohydrate reduction, *Carbohydr. Res.* **222**, 223–231.

Schulz, D., Rau, U., Wagner, F. (1992) Characteristics of films prepared by native and modified branched β-1,3-D-glucans. *Carbohydr. Polym.* **18**, 295–299.

Schuren, F. H. J., Wessels, J. G. H. (1998) Expression of heterologous genes in *Schizophyllum commune* is often hampered by the formation of truncated transcripts, *Curr. Genet.* **33**, 151–156.

Selitrennikoff, C. P. (1995) Antifungal drugs: (1,3)β-glucan synthase inhibitors, in: *Molecular Biology Intelligence Unit* (Molsberry, D. M., Ed.), Austin, Texas, USA: L. R. G. Landes Company, 45–89.

Seviour, R. J., Stasinopoulos, S. J., Auer, D. P. F., Gibbs, P. A. (1992) Production of pullulan and other exopolysaccharides by filamentous fungi, *Crit. Rev. Biotechnol.* **12**, 279–298.

Shematec, E. M., Cabib, E. (1980) Biosynthesis of the yeast cell wall. II. Regulation of β-(1-3)glucan synthetase by ATP And GTP, *J. Biol. Chem.* **255**, 895–902.

Shen, G.-P., Park, D.-C., Ullrich, R. C., Novotny, C. P. (1996) Cloning and characterization of a *Schizophyllum* gene with *Aß6* mating-type activity, *Curr. Genet.* **29**, 136–142.

Shigero, M., Fisamu, S., Wataru, I. (1989) Anti-aids virus agent. JP 1287031.

Sietsma, J. H., Wessels, J. G. H. (1977) Chemical analysis of the hyphal wall of *Schizophyllum commune, Biochim. Biophys. Acta* **496**, 225–239.

Sietsma, J. H., Wessels, J. G. H. (1979) Evidence for covalent linkages between chitin and β-glucan in a fungal wall, *J. Gen. Microbiol.* **114**, 99–108.

Sietsma, J. H., Wessels, J. G. H. (1981) Solubility of (1-3)-β-D-(1-6)-β-D-glucan in fungal walls: importance of presumed linkage between glucan and chitin, *J. Gen. Microbiol.* **125**, 209–212.

Sietsma, J. H., Sonnenberg, A. M. S., Wessels, J. G. H. (1985) Localization by autoradiography of synthesis of (1-3)-β- and (1-6)-β linkages in a wall glucan during hyphal growth of *Schizophyllum commune*, *J. Gen. Microbiol.* **131**, 1331–1337.

Steiner, W., Lafferty, R. M., Gomes, I., Esterbauer, H. (1987) Studies on a wild strain of *Schizophyllum commune*: cellulase and xylanase production and formation of the extracellular polysaccharide schizophyllan, *Biotechnol. Bioeng.* **30**, 169–178.

Steiner, W., Divjak, H., Lafferty, R. M., Steiner, E., Esterbauer, H., Gomes, I. (1988) Production and properties of schizophyllan and scleroglucan, *DECHEMA Biotechnology Conferences* **1**, 149–154.

Steiner, W., Haltrich, D., Lafferty, R. M. (1993) Production, properties and practical applications of fungal polysaccharides, in: *Biosurfactants* (Kosaric, N., Ed.), New York, USA: Marcel Dekker, 175–204, vol. 48.

Stuebler, D., Buchenauer, H. (1996) Antiviral activity of the glucan lichenan (poly-β-(1,3,-1,4)D-anhydroglucose): 1. Biological activity in tobacco plants, *J. Phytopath.* **144**, 37–43.

Suda, M., Ohno, N. Adachi, Y. Yadomae, T. (1994) Preparation and properties of metabolically 3H- or 13C-labeled 1,3-β-D-glucan SSG from *Sclerotinia sclerotiorum* IFO 9395, *Carbohydr. Res.* **254**, 213–220.

Sutherland, I. W. (1982). Biosynthesis of microbial exopolysaccharides, in: *Advances in Microbial Physiology* (Rose, A. H., Gareth-Morris, J. G., Eds.), New York, USA: Academic Press, 79–150, vol. 23.

Sutherland, I. W. (1999) Polysaccharases for microbial exopolysaccharides. *Carbohydr. Polym.* **38**, 319–328.

Tabata, K., Ito, W., Kojima, T. (1981) Ultrasonic degradation of schizophyllan, an antitumor polysaccharide produced by *Schizophyllum commune* Fries, *Carbohydr. Res.* **89**, 121–135.

Tsuzuki, A., Tateishi, T., Ohno, N., Adachi, Y., Yadomae, T. (1999) Increase of hematopoietic responses by triple or single helical conformer of an antitumor 1,3-β-D-glucan preparation, Sonifilan, in cyclophosphamide-induced leukopenic mice, *Biosci. Biotechnol. Biochem.* **63**, 104–110.

Van, K., Asakawa, T., Teramota, A. (1984) Optical rotatory dispersion of liquid crystal solutions of a triple-helical polysaccharide schizophyllan, *Polymer J.* **16**, 61–69.

van Wetter, M.-A., Wösten, H. A. B., Sietsma, J. H., Wessels, J. G. H. (2000) Hydrophobin gene expression affects hyphal wall composition in *Schizophyllum commune*, *Fungal Genet. Biol.* **31**, 99–104.

Wang, C., Miles, P. G. (1964) The physiological characterization of dikaryotic mycelia of *Schizophyllum commune*, *Physiol. Plant.* **17**, 573–588.

Wang, M. C., Bartnicki-Garcia, S. (1982) Synthesis of noncellulose cell-wall β-glucan by cell-free extracts from zoospores and cysts of *Phytophtora palmivora*, *Exp. Mycol.* **6**, 125–135.

Wessels, J. G. H. (1965) Morphogenesis and biochemical processes in *S. commune*, *Wentia* **13**, 1–113.

Wessels, J. G. H. (1978) Incompatibility factors and the control of biochemical processes, in: *Genetics and Morphogenesis in Basidiomycetes* (Schwalb, M. N., Miles, P. G., Eds.), New York, USA: Academic Press, 81–104.

Wessels, J. G. H. (1988) A steady-state model for apical wall growth in fungi, *Acta Bot. Neerl.* **37**, 3–16.

Wessels, J. G. H. (1999) Fungi in their own right, *Fungal Genet. Biol.* **27**, 134–145.

Wessels, J. G. H., Sietsma, J. H. (1979) Wall structure and growth in *Schizophyllum commune*, in: *Fungal Walls and Hyphal Growth* (Burnett, J. H., Trinci, A. P. J., Eds.), Cambridge: Cambridge University Press, 27–48.

Wessels, J. G. H., Sietsma, J. H. (1981) Significance of linkages between chitin and β-glucan in fungal walls, *Microbiology*, **127**, 232–234.

Yanaki, T. (1981) Preparation of polysaccharide associate, JP 56127603.

Yanaki, T. (1982) Novel liquid crystal of polysaccharide, JP 57147576.

Yano, T. (1990) Preventive for fish disease comprising water-soluble glucan, JP 2218615.

Yoshio, S., Katsuhiko, H., Kazumasa, M. (1992) Augmenting the effect of sizofiran on the immunofunction of regional lymph nodes in cervical cancer, *Cancer* **69**, 1188–1194.

Young, S.-H., Jacobs, R. R. (1998) Sodium hydroxide-induced conformational change in schizophyllan detected by the fluorescence dye, aniline blue, *Carbohydr. Res.* **310**, 91–99.

Zentz, F., Verchere, J.-F., Muller, G. (1992) Thermal denaturation and degradation of schizophyllan, *Carbohydr. Polym.* **17**, 289–297.